AF327323

Merrill
Algebra Two
with Trigonometry

Foster • Rath • Winters

Merrill Publishing Co.
A Bell & Howell Information Company
Columbus, Ohio

Toronto • London • Sydney

ISBN 0-675-05616-0

Published by
Merrill Publishing Company
A Bell & Howell Information Company
Columbus, Ohio 43216

Printed in the United States of America

Authors

Alan G. Foster is chairman of the Mathematics Department at Addison Trail High School, Addison, Illinois. He has taught mathematics courses at every level of the high school curriculum. Mr. Foster obtained his B.S. from Illinois State University and his M.A. in mathematics from the University of Illinois. Mr. Foster is active in professional organizations at local, state, and national levels. He is past-president of the Illinois Council of Teachers of Mathematics. Mr. Foster is a recipient of the Illinois Council of Teachers of Mathematics T. E. Rine Award for excellence in the teaching of mathematics. He received the Illinois State Presidential Award for Excellence in Teaching of Mathematics in 1987. He is co-author of *Merrill Geometry*, *Merrill Algebra Essentials*, and *Merrill Algebra One*.

James N. Rath has 30 years of classroom experience in teaching mathematics at every level of the high school curriculum. Mr. Rath is a former head of the mathematics department at Darien High School, Darien, Connecticut. He earned his B.A. in philosophy from the Catholic University of America and his M.Ed. and M.A. in mathematics from Boston College. Mr. Rath is active in professional organizations at local, state, and national levels. He is a co-author of *Merrill Pre-Algebra*, *Merrill Algebra Essentials*, and *Merrill Algebra One*.

Leslie J. Winters is the Secondary Mathematics Specialist for the Los Angeles Unified School District. He has thirty years of classroom experience in teaching mathematics at every level from junior high school to college. He received his B.A. in mathematics from Pepperdine University, and advanced degrees from the University of Dayton, University of Southern California, and Boston College. He is a past president of the California Mathematics Council–Southern Section. He was a recipient of the 1983 Presidential Award for Excellence in Mathematics Teaching in the state of California. He is a co-author of *Merrill Algebra Essentials* and *Merrill Algebra One*.

Consultants

Anne Auman
Mathematics Education
  Program Specialist
Kansas State Department
  of Education
Topeka, Kansas

Bert K. Waits
Professor of Mathematics
The Ohio State University
Columbus, Ohio

Gregory D. Foley
Assistant Professor of
  Mathematics Education
The Ohio State University
Columbus, Ohio

Reviewers

Marsha Bird Bordas
Mathematics Teacher
Lafayette High School
Lexington, Kentucky

Abner A. Korsness
Teacher and Mathematics
    Department Chairperson
Rex Putnam High School
Milwaukee, Oregon

Bob Mora
Mathematics Coordinator, K-12
Carrollton-Farmers Branch
    Independent School District
Carrollton, Texas

Ella L. Greenberg
Teacher and Mathematics
    Department Chairperson
Highland High School
Bakersfield, California

Dennis K. Levin
Teacher and Mathematics
    Department Chairperson
Fargo South High School
Fargo, North Dakota

James J. Pender
Chairman Of Mathematics, K-12
Westwood Public Schools
Westwood, Massachusetts

Hector Hirigoyen
Assistant Principal, Curriculum
W. R. Thomas Junior High School
Miami, Florida

Janice A. Massey
Teacher
Northwestern High School
Rock Hill, South Carolina

Aldonia C. Winn
Teacher
J. S. Clark Senior High School
New Orleans, Louisiana

Classroom Learner Verification Teachers

Sanford Bearman, North Miami Beach High School, Miami, FL
Edwin S. Carr, Langley High School, Pittsburgh, PA
Michael Carter, Olympus Senior High School, Salt Lake City, UT
David Conrad, Bloomingdale High School, Valrico, FL
Michael Dailey, Beechcroft High School, Columbus, OH
David Ebersole, Lincoln Junior High School, Lancaster, PA
Gary Foland, St. Charles West High School, St. Charles, MO
Sarah Gibbs, Yates High School, Houston, TX
Debbie Jodziewicz, Central High School, Manchester, NH
Bob Johnson, Ford High School, Sterling Heights, MI
Becky Link, Briggs High School, Columbus, OH
Sherry Mahler, Coral Gables High School, Miami, FL
Jenni Marple, Monroe High School, Sepovada, CA
Barbara Martinelli, Kennedy High School, Grenada Hills, CA
Ron Mathison, Sious City East High School, Sioux City, IA
Nancy Mathras, Springfield Central High School, Springfield, MA
John A. Phillips, San Juan High School, Citrus Heights, CA
Frances Prendergast, Central High School, Philadelphia, PA
Susan Raudabaugh, Withrow High School, Cincinnati, OH
Jean Richmond, Escambia High School, Pensacola, FL
Robert Russell, West Roxbury High School, West Roxbury, MA
Jan Schultehenrich, Francis Howell High School, St. Charles, MO
Marie Sciabarassi, Coral Gables High School, Miami, FL
Betty Scott, South Oak Cliff High School, Dallas, TX
Ann Shealy, Richland Northeast High School, Columbia, SC
Cindy Slama, Lafayette High School, Ballwin, MO

This fourth edition of *Merrill Algebra Two with Trigonometry* is designed for use by students in second-year high school algebra courses. The text, which was developed in the classroom by experienced high school teachers, is based on the successful prior editions of *Merrill Algebra Two with Trigonometry*. The goals of the text are to develop proficiency with mathematical skills, to expand understanding of mathematical concepts, to improve logical thinking, and to promote success. To achieve these goals, the following strategies are used.

Build upon a Solid Foundation.   This program develops and utilizes the learning spiral. Students first review those concepts basic to the understanding of algebra. Students' understanding is thus solidified before the introduction of more difficult concepts.

Utilize Sound Pedagogy.   *Merrill Algebra Two with Trigonometry* covers, in logical sequence, all topics generally presented at this level. Concepts are introduced when they are needed. Each concept presented is then used within that lesson and in later lessons.

Gear Presentation for Learning.   An appropriate reading level has been maintained throughout the text. Furthermore, many photographs, illustrations, charts, graphs, and tables provide visual aids for the concepts and skills presented. Hence, students are able to read and learn with increased understanding.

Use Relevant Real-Life Applications.   Applications are provided not only for practice, but also to aid understanding of how concepts are used.

Give Ample Opportunity for Review.   A one-page cumulative review in each chapter provides students with a review of the skills and concepts presented up to that point. This research-proven method is most effective in increasing student retention of mathematical skills. Each chapter also includes two to four mini-reviews. Each of these provides students with a quick review of previously-presented concepts.

Use Relevant Technology.   Examples are provided to instruct students in using a calculator. The use of the calculator is related to the concepts taught within the lesson. A Using Computer feature in each chapter provides a computer program and exercises related to the objectives of the chapter. Graphing Calculator Applications appear periodically throughout the text. Each of these instructs students in the use of the graphing calculator as a problem-solving tool.

Teachers and students familiar with the earlier editions of this text will be pleased to see that the text has been updated in keeping with current trends. The clarity of explanations has been retained and the sequencing of topics improved in this new edition of *Merrill Algebra Two with Trigonometry*.

Table of Contents

1 Equations and Inequalities — 2

2 Linear Relations and Functions — 42

3 Systems of Equations and Inequalities___80

4 Matrices___120

9 Conics _______________________ 300

10 Polynomial Functions _______________________ 344

Symbols

a^n	the nth power of a		$!$	factorial
$\lvert a \rvert$	the absolute value of a		$f \circ g$	composition function f of g
$-a$	additive inverse of a or the opposite of a		$>$	is greater than
Cos^{-1}	Arccosine		$<$	is less than
$C(n, r)$	combinations of n elements taken r at a time		$\geq$	is greater than or equal to
$a + bi$	complex number		$\leq$	is less than or equal to
$^\circ$	degrees		$\log_b x$	the logarithm to the base b or x
det	determinant		$P(n, r)$	permutations of n things taken r at a time
e	base of natural logarithms		$\pm$	positive or negative
$\in$	is an element of		$\{\ \ \}$	set
$\varnothing$	empty set		$\sqrt{}$	the principal square root of
$=$	equals or is equal to		$\sqrt[n]{}$	the nth root of
$\neq$	does not equal		Σ	(sigma) summation symbol
$\approx$	approximately equal to			
$f(x)$	f of x or the value of f at x			
f^{-1}	inverse function of f			

Equations and Inequalities

Equations can be used in many areas of our life. For example, Julia wants to buy two $30 concert tickets. Her job as a waitress pays $3.75 an hour. In addition, she has already saved $15. How many hours must she work to earn enough money to buy the tickets? This problem can be solved using the equation $3.75h + 15 = 2(30)$.

1-1 Expressions and Formulas

The mathematical expression $4(x + 2y) - 5x^3$ contains the following kinds of symbols.

$$\begin{array}{rll}
\text{constants:} & 4 \quad 2 \quad 5 \quad 3 \\
\text{variables:} & x \quad y & \text{represent unknown quantities} \\
\text{operation symbols:} & + \quad - & \text{tell which operations are involved} \\
\text{grouping symbols:} & (\ \ ) & \text{tell order for doing operations}
\end{array}$$

The following numerical expressions all represent the number sixteen. Sixteen is called the **value** of each expression.

$$4^2 \qquad\qquad 9 + 7 \qquad\qquad (12 \div 2) + 10$$

In the expression 4^2, the **base** is 4 and the **exponent** is 2. An exponent indicates the number of times the base is used as a factor.

4^2 means $4 \cdot 4$. 10^1 means 10. x^5 means $x \cdot x \cdot x \cdot x \cdot x$.

What is the value of $4 + 3 \cdot 2$? You might evaluate the expression in one of the ways below. Which value is correct?

Multiply 3 and 2. Then add to 4.

$$\begin{aligned}
4 + 3 \cdot 2 &= 4 + 6 \\
&= 10
\end{aligned}$$

Add 4 and 3. Then multiply by 2.

$$\begin{aligned}
4 + 3 \cdot 2 &= 7 \cdot 2 \\
&= 14
\end{aligned}$$

A numerical expression should have a unique value. To find this value, you must follow the established order of operations.

1. **Evaluate all powers.**
2. **Do all multiplications and divisions from left to right.**
3. **Do all additions and subtractions from left to right.**

Order of Operations

Thus, the value of $4 + 3 \cdot 2$ is 10.

Example

1 **Find the value of $4 + 8^2 \div 4 \cdot 2$.**

$$\begin{aligned}
4 + 8^2 \div 4 \cdot 2 &= 4 + 64 \div 4 \cdot 2 \qquad &\textit{Evaluate all powers.} \\
&= 4 + 16 \cdot 2 \\
&= 4 + 32 \qquad &\textit{Do all multiplications and division left to right.} \\
&= 36 \qquad &\textit{Do all additions and subtractions left to right.}
\end{aligned}$$

The value is 36.

Grouping symbols are used to clarify or change the order of operations. When finding the value of an expression, start with the operations inside the innermost grouping symbols.

Two kinds of grouping symbols are parentheses ( ) and brackets [ ].

Example

2 **Find the value of $[(5 + 7)^2 \div 9] \cdot 6$.**

$$\begin{aligned}
[(5 + 7)^2 \div 9] \cdot 6 &= [(12)^2 \div 9] \cdot 6 & &\text{\textit{First add 5 and 7.}} \\
&= [144 \div 9] \cdot 6 & &\text{\textit{Then find } } 12^2. \\
&= [16] \cdot 6 & &\text{\textit{Now divide 144 by 9.}} \\
&= 96 & &\text{\textit{Multiply 16 by 6.}}
\end{aligned}$$

The value is 96.

Mathematical expressions that contain at least one variable are called **algebraic expressions**. You can evaluate an algebraic expression by replacing each variable with a value.

Each time a variable occurs, it should be replaced with the same value.

Example

3 **Evaluate $5x^2 + 3xy$ if $x = -4$ and $y = 6$.**

$$\begin{aligned}
5x^2 + 3xy &= 5(-4)^2 + 3(-4)(6) & &\text{\textit{Replace x by } }-4 \text{ \textit{and y by 6.}} \\
&= 5(16) + 3(-4)(6) & &\text{\textit{Find } }(-4)^2. \\
&= 80 + (-72) & &\text{\textit{Multiply left to right.}} \\
&= 8 & &\text{\textit{Add.}}
\end{aligned}$$

The value is 8.

A **formula** is a mathematical sentence about the relationships among certain quantities.

In a formula, if you know replacements for every variable except one, you can find a replacement for that variable.

The formula $A = \ell \cdot w$ relates the area of a rectangle to its length and width.

Example

4 **The relationship between Celsius temperature (C) and Fahrenheit temperature (F) is given by $C = \dfrac{5(F - 32)}{9}$. Find C if $F = 68$.**

$$\begin{aligned}
C &= \frac{5(F - 32)}{9} & &\text{\textit{Write the formula.}} \\
&= \frac{5(68 - 32)}{9} & &\text{\textit{Replace F by 68.}} \\
&= 20 & &\text{\textit{Subtract 32 from 68. Multiply 5 by 36. Divide 180 by 9.}}
\end{aligned}$$

The Celsius temperature is 20°.

5   **Simple interest is calculated by using the formula $I = prt$. In the formula, p represents the principal in dollars, r represents the annual interest rate, and t represents the time in years. Find the amount of interest for two years if the principal is \$6000 and the rate is 12%.**

$$\begin{aligned} I &= prt \\ &= 6000(0.12)(2) \qquad \textit{12\% = 0.12} \\ &= 1440 \end{aligned}$$

The interest on \$6000 at 12% for 2 years is \$1440.

6   **Using Calculators**

The volume of a sphere is calculated by using the formula $V = \frac{4}{3}\pi r^3$. In the formula, r represents the measure of the radius. Find the volume of a sphere with a radius of 7.6 centimeters. Use 3.14 to approximate π.

$$V = \frac{4}{3}\pi r^3$$

$$= \frac{4}{3}(3.14)(7.6)^3 \qquad\qquad \textit{Display is rounded to the}$$

nearest hundredth.

ENTER: 4 $\boxed{\div}$ 3 $\boxed{\times}$ 3.14 $\boxed{\times}$ 7.6 $\boxed{y^x}$ 3 $\boxed{=}$

DISPLAY: 4   3 1.33 3.14 4.19 7.6   3 1837.85

The volume of the sphere is approximately 1837.85 cubic centimeters.

Exploratory Exercises

Find the value of each expression.

1. $2 + 8 - 3$ **2.** $5 - 3 \cdot 2$ **3.** $5 - (4 + 3)^2$

4. $2(6 + 1)$ **5.** $2 \cdot 6 + 1$ **6.** $3^3 - 2^3$

7. $3(2^2 + 3)$ **8.** $2 + 5^2$ **9.** $2(3 + 8) - 1$

10. $2(8) - 8 \div 4$ **11.** $5^2 - (3 + 2)^2$ **12.** $3 + (3 - 3)^3 - 3$

Written Exercises

Find the value of each expression.

13. $(6 + 5) \cdot 4 - 3$ **14.** $(6 + 5)(4 - 3)$ **15.** $12 + 8 \div 4$

16. $12 + (8 \div 4)$ **17.** $12 \div 8 \cdot 4$ **18.** $12 \div (8 \cdot 4)$

19. $(5 + 3)^2 - 16 \div 4$ **20.** $5 + 3^2 - 16 \div 4$ **21.** $2 \cdot 9 \div 6 + 7$

22. $8 - 2 \cdot 3 - 3$ **23.** $12 + 18 \div 6 + 7$ **24.** $4 + 8 \cdot 4 \div 2 - 10$

25. $[19 - (8 - 1)] \div 3$ **26.** $-8 \div [20 \div (16 - 11)]$

27. $[(-8 + 3) \times 4 - 2] \div 6$ **28.** $3 + [8 \div (9 - 2 \times 4)]$

Evaluate if $a = 3$, $b = 7$, $c = -2$, $d = \frac{1}{2}$, and $e = 0.3$.

29. $6a^3 - 2b$ **30.** $3ab - 6bc$ **31.** $3ad + bc$

32. $c^2 - 5d$ **33.** $5c + be$ **34.** $8e - 3bc$

35. $(a + c)^3 + d^2$ **36.** $(a + b - d)^2$ **37.** $12a^2 + bc$

38. $4a - 12cd$ **39.** $\dfrac{6ac}{d}$ **40.** $\dfrac{3ab}{cd}$

41. $\dfrac{3ce}{d^2}$ **42.** $\dfrac{10e^2}{3d}$ **43.** $\dfrac{5a + 3c}{3b}$

44. $\dfrac{3ab^2 - c^3}{a + c}$ **45.** $(5a + 3d)^2 - e^2$ **46.** $(3b - 21d)^2$

Find the interest, I, given the following values for the principal, rate, and time.

47. $p = \$1000$, $r = 6\%$, and $t = 3$ years   **48.** $p = \$2500$, $r = 9\%$, and $t = 30$ months

49. $p = \$5000$, $r = 8\%$, and $t = 9$ months   **50.** $p = \$2000$, $r = 8\%$, and $t = 9$ months

51. $p = \$20{,}000$, $r = 14\frac{1}{2}\%$, and $t = 6$ years   **52.** $p = \$65{,}000$, $r = 16\%$, and $t = 63$ months

The formula for the total area of a right circular cylinder is $T = 2\pi rh + 2\pi r^2$. T represents the measure of the total area, r represents the measure of the radius, and h represents the measure of the height. Use a calculator to find the measure of the total area of each right circular cylinder given the following values. Use 3.14 to approximate π.

53. $r = 6$, and $h = 10.2$ **54.** $r = 4.5$ and $h = 7$ **55.** $r = 11.8$ and $h = 23$

56. $r = 26.3$ and $h = 26.3$ **57.** $r = 0.59$ and $h = 1.15$ **58.** $r = 12.7$ and $h = 32.6$

The formula for the area of a trapezoid is $A = \frac{h}{2}(b_1 + b_2)$. A represents the measure of the area, h represents the measure of the altitude, and b_1 and b_2 represent the measures of the bases. Find the measure of the area of each trapezoid given the following values.

59. $b_1 = 12$, $b_2 = 20$, and $h = 8$ **60.** $b_1 = 4$, $b_2 = 11$, and $h = 7$

61. $h = 8$, $b_1 = 8.6$, and $b_2 = 14.8$ **62.** $h = 12$, $b_2 = 9.7$, and $b_1 = 6.2$

63. $h = 9$, $b_2 = 7\frac{2}{3}$, and $b_1 = 4\frac{1}{6}$ **64.** $h = 6\frac{1}{2}$, $b_1 = 8\frac{1}{3}$, and $b_2 = 12$

Challenge Exercises

> **Example:** Use five 5's to write an expression that has a value of 10.
>
> $$\frac{55}{5} - \frac{5}{5} = 10 \qquad \text{or} \qquad 5 \cdot 5 - (5 + 5 + 5) = 10$$

65. Use nine 8's to write an expression that has a value of 14.

66. Use six 7's to write an expression that has a value of 110.

67. Use four 3's to write an expression for each integer from 1 to 10.

Understanding algebra depends largely on understanding the symbols used in algebra. The symbols of language are the letters of the alphabet. In English, there are just 26 letters, and most letters have one or two sounds. On the other hand, there are many symbols used in mathematics, and the meaning of these symbols frequently depends on how they are used. Notice how the symbol 4 is used in each expression.

Expression	*Translation*
4	four
40	forty
0.04	four hundredths
4^2	four squared
2^4	two to the fourth power
$\dfrac{1}{4}$	one-fourth
$(3 + 2)^4$	three plus two, quantity to the fourth power

In like manner, the symbol $-$ may be used in several ways.

Expression	*Translation*
$9 - 4$	nine minus four
-4	negative four
$\dfrac{3}{4}$	three-fourths or three divided by four

Exercises

Match each expression with the correct translation.

1. $3a$	**a.**	three times a squared plus one
2. a^3	**b.**	three times a
3. $3a^2 + 1$	**c.**	three times a plus one, quantity squared
4. $(3a + 1)^2$	**d.**	a divided by three
5. $(3a)^2 + 1$	**e.**	a to the third power
6. $\dfrac{a}{3}$	**f.**	three times a, quantity squared, plus one

Write each algebraic expression in words.

7. $3ad + bc$ **8.** $6a^3 - 2b$ **9.** $(a + b - c)^2$

10. $(a + c)^3 + d^2$ **11.** $\dfrac{6ac}{d}$ **12.** $\dfrac{5a + 3c}{3b}$

You should be familiar with the following sets of numbers and their graphs.

- The set of **natural numbers, N**

  $N = \{1, 2, 3, 4, 5, \ldots\}$

- The set of **whole numbers, W**

  $W = \{0, 1, 2, 3, 4, \ldots\}$

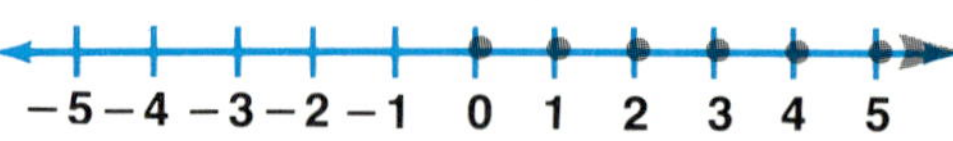

- The set of **integers, Z**

  $Z = \{\ldots, -2, -1, 0, 1, 2, \ldots\}$

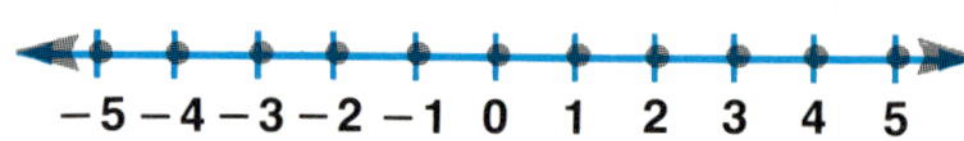

- The set of **rational numbers, Q**

  $Q = \left\{ \begin{array}{l} \text{all numbers that can be expressed} \\ \text{in the form } \dfrac{m}{n}, \text{ where } m \text{ and } n \\ \text{are integers and } n \text{ is } not \text{ zero} \end{array} \right\}$

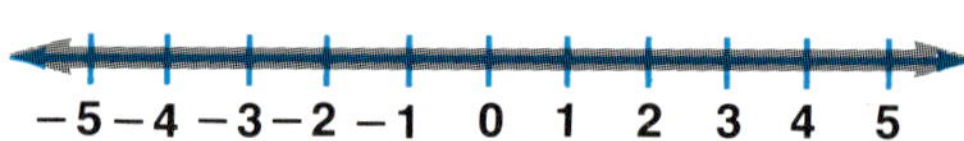

  Q = {all numbers that can be expressed as either terminating or repeating decimals.}

Examples of rational numbers are $-\dfrac{2}{3}$, 5.8, -7, and 0. Even if all the rational numbers could be graphed on the number line, there would still be "holes" in the number line.

Repeating decimals such as $8.\overline{34}$ and $1.2121\ldots$ are also examples of rational numbers.

- The set of **irrational numbers, I**

  $I = \{$all nonterminating, nonrepeating decimals$\}$

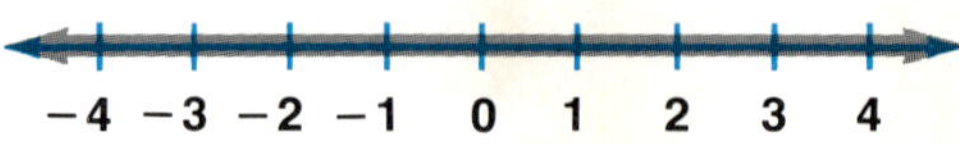

Examples of irrational numbers are $\sqrt{2}$, π, and $-\sqrt{7}$.
As with the rational numbers, the graph of the irrational numbers produces a number line with "holes."

Here are some examples of nonrepeating decimals: $1.010010001\ldots$ $0.123456789\ldots$

Irrational numbers combined with rational numbers form the set of **real numbers**. The graph of the real numbers is the entire number line.

- The set of **real numbers, R**

  $R = \{$all rationals and irrationals$\}$

The relationship between the real numbers and the number line can be summarized as follows.

- Each real number corresponds to exactly one point on the number line.
- Each point on the number line corresponds to exactly one real number.

1 Name the sets of numbers to which each number belongs.

a. $\sqrt{5}$

b. $\dfrac{3}{2}$

c. -7

irrationals (**I**)
reals (**R**)

rationals (**Q**)
reals (**R**)

integers (**Z**)
rationals (**Q**)
reals (**R**)

2 Evaluate each expression. Then name the sets of numbers to which each value belongs.

a. $\sqrt{9}$

$\sqrt{9} = 3$

natural numbers (**N**),
whole numbers (**W**), integers (**Z**),
rationals (**Q**), reals (**R**)

b. $6 \div 10$

$6 \div 10 = 0.6$ or $\dfrac{3}{5}$

rationals (**Q**)
reals (**R**)

Exploratory Exercises

Name the sets of numbers to which each number belongs.

1. -8

2. $\dfrac{4}{3}$

3. -2.6

4. π

5. $-\dfrac{7}{2}$

6. 10

7. $\sqrt{10}$

8. 5.11

9. 0

10. $-\dfrac{12}{3}$

11. -1.0

12. $\sqrt{16}$

13. $-\sqrt{49}$

14. $1.2222\ldots$

15. $3.13313331\ldots$

16. $-2.\overline{173}$

Written Exercises

Evaluate each expression. Then name the sets of numbers to which each value belongs.

17. $8 - 7$

18. $7 - 8$

19. $-54 \div 6$

20. $68 \div 100$

21. -2.4×10

22. $3.9 + 2.6$

23. $9.2 - 8.2$

24. $6 \div 2^2$

25. $\sqrt{25}$

26. $\sqrt{36 + 5}$

27. $\sqrt{100} - 100$

28. $0 - \sqrt{4}$

29. $\sqrt{2} - \sqrt{2}$

30. $\sqrt{16 + 9}$

31. $\sqrt{16} + \sqrt{9}$

32. $\sqrt{16} - \sqrt{9}$

33. $\sqrt{9} \div \sqrt{4}$

34. $\sqrt{9} \div 3$

35. $\sqrt{8(18)}$

36. $\sqrt{6}(\sqrt{9} - 3)$

37. $\sqrt{25 - 11} + \sqrt{25 + 11} - 6$

38. $\sqrt{20 + 16} - \sqrt{20 - 16} - 6$

Determine whether each statement is *true* or *false*. If *false*, give a counterexample.

39. Every whole number is an integer.
40. Every integer is a whole number.
41. Every rational number is an integer.
42. Every real number is irrational.
43. Every irrational number is a real number.
44. Every integer is a rational number.
45. Every real number is either a rational number or an irrational number.
46. Every rational number is either an integer or a whole number.

Find all irrational numbers of the form $\sqrt{n}$, where n is a natural number, such that $\sqrt{n}$ is between each pair of numbers.

47. 2 and 3
48. 6 and 7
49. 4.6 and 4.8
50. 10.1 and 10.2

mini-review

Find the value of each expression.

1. $(8 + 5) \cdot 3 - 7$
2. $4(8 - 3)^2$
3. $9 \cdot 6 \div 2 \cdot 7$
4. $5 \cdot 6 - 6 \div 3$
5. $17 - 4^2 + 6$
6. $-6(-3)^2 - 4^3$

Evaluate if $a = 4$, $b = -3$, and $c = \frac{1}{2}$.

7. $ab^2 - c$
8. $\dfrac{5a}{c}$
9. $\dfrac{12ac}{b}$
10. $3a + 4b$
11. $3c^2 - b$
12. $c(a + b)^2$
13. $3a - 4b$
14. $a^2 + b^2 + c^2$
15. $(a + b + c)^2$

Evaluate each expression. Then name the sets of numbers to which each value belongs.

16. $(8.7)^2$
17. $-\dfrac{51}{17}$
18. $-2.4\left(\dfrac{1}{3}\right)$
19. $\sqrt{25 - 16}$
20. $\sqrt{25} - \sqrt{16}$
21. $\sqrt{25 + 16}$
22. $-\sqrt{36 - 25}$
23. $36 - \sqrt{36}$
24. $1 + 7 \cdot 9 - (8 \div 2)^3$
25. $1 + (7 \cdot 9 - 8) \div 2^3$

Operations with real numbers have several important properties. One of the basic properties of addition and multiplication is **commutativity**. The order in which two real numbers are added or multiplied does not change their sum or their product.

$$8 + 3 = 3 + 8 \qquad 24 \cdot 9 = 9 \cdot 24$$

Another basic property of addition and multiplication is **associativity**. The way three or more real numbers are grouped, or associated, does not change their sum or their product.

$$(8 + 6) + 4 = 8 + (6 + 4) \qquad (11 \cdot 4) \cdot 3 = 11 \cdot (4 \cdot 3)$$

The sum of any real number and zero is identical to the original number. So, for real numbers the **additive identity** is zero.

$$9.2 + 0 = 9.2 \qquad 0 + \sqrt{5} = \sqrt{5}$$

The product of any real number and one is identical to the original number. So, for real numbers the **multiplicative identity** is one.

$$\frac{7}{9} \cdot 1 = \frac{7}{9} \qquad (1)(6.5) = 6.5$$

Each real number has a unique **additive inverse**. The sum of a number and its additive inverse is zero.

$$2 + (-2) = 0 \qquad -4.3 + 4.3 = 0$$

Each nonzero real number has a unique **multiplicative inverse**, or **reciprocal**. The product of a number and its reciprocal is one.

$$3\left(\frac{1}{3}\right) = 1 \qquad (-0.1)(-10) = 1$$

The following chart summarizes the properties for addition and multiplication.

For any real numbers a, b, and c		
	Addition	*Multiplication*
Commutative	$a + b = b + a$	$a \cdot b = b \cdot a$
Associative	$(a + b) + c = a + (b + c)$	$(a \cdot b) \cdot c = a \cdot (b \cdot c)$
Identity	$a + 0 = a = 0 + a$	$a \cdot 1 = a = 1 \cdot a$
Inverse	$a + (-a) = 0 = (-a) + a$	If a is *not* zero, then $a \cdot \frac{1}{a} = 1 = \frac{1}{a} \cdot a.$

Properties of Real Numbers

The **distributive property** relates the operations of addition and multiplication.

For all real numbers a, b, and c,
$$a(b + c) = ab + ac \text{ and } (b + c)a = ba + ca.$$

Distributive Property

$$5(8 + 3) = 5(8) + 5(3) \qquad (6 + (-2))(0.5) = 6(0.5) + (-2)(0.5)$$
$$5(11) = 40 + 15 \qquad\qquad (4)(0.5) = 3 + (-1)$$
$$55 = 55 \qquad\qquad 2 = 2$$

The distributive property can be used to simplify expressions containing like terms. **Like terms** are identical except for their numerical factors. In $5y + 3x + 7y + 2y^2$, for example, $5y$ and $7y$ are like terms. Notice that $5y$, $3x$, and $2y^2$ are unlike terms.

Examples

1 Simplify $5y + 3x + 7y + 2y^2$.

$$
\begin{aligned}
5y + 3x + 7y + 2y^2 &= 5y + (3x + 7y) + 2y^2 & \text{Associative Property for Addition} \\
&= 5y + (7y + 3x) + 2y^2 & \text{Commutative Property for Addition} \\
&= (5y + 7y) + 3x + 2y^2 & \text{Associative Property for Addition} \\
&= (5 + 7)y + 3x + 2y^2 & \text{Distributive Property} \\
&= 12y + 3x + 2y^2
\end{aligned}
$$

2 Simplify $\left(\frac{1}{2}a\right)(2b) + ab$.

$$
\begin{aligned}
\left(\tfrac{1}{2}a\right)(2b) + ab &= \tfrac{1}{2}(a \cdot 2)b + ab & \text{Associative Property for Multiplication} \\
&= \tfrac{1}{2}(2 \cdot a)b + ab & \text{Commutative Property for Multiplication} \\
&= \left(\tfrac{1}{2} \cdot 2\right)(ab) + ab & \text{Associative Property for Multiplication} \\
&= 1ab + 1ab & \text{Multiplicative Inverse and Multiplicative Identity} \\
&= (1 + 1)ab & \text{Distributive Property} \\
&= 2ab
\end{aligned}
$$

Exploratory Exercises

State the property shown in each equation.

1. $8 + (6 + 4) = (8 + 6) + 4$

2. $7(5) = 5(7)$

3. $a(3 - 2) = a \cdot 3 - a \cdot 2$

4. $8 + (1 + 6) = 8 + (6 + 1)$

5. $3 + (-3) = 0$

6. $2(3 + 6) = 2(3) + 2(6)$

7. $3 + 6 = 6 + 3$

8. $8(6 - 7) = (6 - 7)8$

9. $8\left(\frac{1}{8}\right) = 1$

10. $3(47) = 3(40) + 3(7)$

Written Exercises

State the property shown in each equation.

11. $3(9) = 9(3)$

12. $(5 + 6) + 3 = 5 + (6 + 3)$

13. $(4 + 11) \cdot 6 = 4(6) + 11(6)$

14. $(a + b) + [-(a + b)] = 0$

15. $11 + a = a + 11$

16. $(3 + 9) + 14 = 14 + (3 + 9)$

17. $3 + (a + b) = (a + b) + 3$

18. $7 \cdot 1 = 7$

19. $8(5 + 3) = 40 + 24$

20. $3a + 6 = 3(a + 2)$

21. $(11a + 3b) + 0 = 11a + 3b$

22. $8(4) = 4(8)$

23. $3\left(\frac{1}{3}\right) = 1$

24. $(4 + 9a)2b = 2b(4 + 9a)$

25. $a + b + 0 = a + b$

26. $\left(\frac{1}{m}\right)m = 1$

27. $11(3a + 2b) = 11(2b + 3a)$

28. $5a + (-5a) = 0$

29. $1 = ax^2 \cdot \frac{1}{ax^2}$

30. $0 + 7 = 7$

Simplify each expression.

31. $8 + 15 - 3$

32. $5(13 + 25)$

33. $7x + 8y + 9y - 5x$

34. $3a + 5b + 7a - 3b$

35. $3(5a + 6b) + 8(2a - b)$

36. $2(7c - 5d) - 3(d + 2c)$

37. $\frac{1}{4}(12 + 20a) + \frac{3}{4}(12 + 20a)$

38. $\frac{1}{2}(17 - 4x) - \frac{3}{4}(6 - 16x)$

39. $\frac{2}{3}\left(\frac{1}{2}a + 3b\right) + \frac{1}{2}\left(\frac{2}{3}a + b\right)$

40. $\frac{3}{4}(2x - 5y) + \frac{1}{2}\left(\frac{2}{3}x + 4y\right)$

41. $7(0.2m + 0.3n) + 5(0.6m - n)$

42. $9(0.6a - 0.2c) + 3(0.2a + 1.1c)$

43. If $a + b = a$, what is the value of b? What is b called?

44. If $ab = 1$, what is the value of a? What is a called?

45. If $ab = a$, what is the value of b? What is b called?

46. If $a + b = 0$, what is the value of a? What is a called?

Challenge Exercises

A set is *closed with respect to addition* if the sum of any two elements of the set is also in the set. A set is *closed with respect to multiplication* if the product of any two elements of the set is also in the set.

Determine whether each set is closed with respect to the given operation.

Examples:	**R**, addition	**I**, multiplication
	Yes, since the sum of any two real numbers is also a real number.	No, since the product of two irrational numbers is not always irrational.
		$\sqrt{5} \cdot \sqrt{5} = 5$ *This product is rational.*

47. R, multiplication

48. W, addition

49. I, addition

50. Q, multiplication

51. Z, addition

52. Z, multiplication

1-4 Solving Equations

In your study of arithmetic, you have assumed these properties of equality.

For any real number a, $\quad a = a$.	**Reflexive Property of Equality**
For all real numbers a and b, if $a = b$, then $b = a$.	**Symmetric Property of Equality**
For all real numbers a, b, and c, if $a = b$ and $b = c$, then $a = c$.	**Transitive Property of Equality**

Example

1 **Name the property of equality shown in each statement.**

a. $21.4 = 21.4$

a. reflexive property

b. If $36 \cdot 2 = 72$, then $72 = 36 \cdot 2$.

b. symmetric property

c. If $8 = 6 + 2$ and $6 + 2 = 5 + 3$, then $8 = 5 + 3$.

c. transitive property

Sentences with variables to be replaced, such as $5 = x + 4$, are called *open sentences*. Finding the replacements for the variable that make a sentence true is called *solving* the open sentence. Each of these is called a **solution** of the open sentence. The **solution set** of an open sentence is the set of *all* replacements for variables that make the sentence true.

An **equation** is a statement of equality between two mathematical expressions. Some equations may be solved by substitution. Recall that the **substitution property** allows you to replace an expression with an equivalent expression.

Example

2 **Solve $y = 8(0.3) + 1.2$.**

$$y = 8(0.3) + 1.2$$
$$= 2.4 + 1.2 \qquad \text{Substitute 2.4 for 8(0.3).}$$
$$= 3.6 \qquad \text{Substitute 3.6 for 2.4 + 1.2.}$$

The solution is 3.6

Sometimes an equation can be solved by adding, or subtracting, the same number to each side of the equation.

Example

3 **Solve $x + 28.3 = 56.0$.**

$$x + 28.3 = 56.0$$
$$x + 28.3 + (-28.3) = 56.0 + (-28.3) \qquad \textit{Add } -28.3 \textit{ to each side.}$$
$$x = 27.7 \qquad \textit{Substitution Property}$$

Check:
$$x + 28.3 = 56.0$$
$$27.7 + 28.3 \stackrel{?}{=} 56.0$$
$$56.0 = 56.0 \quad ✔$$

The solution is 27.7.

This equation could be solved by subtracting 28.3 from each side.

For any real numbers a, b, and c, if $a = b$, then $a + c = b + c$ and $a - c = b - c$.	*Addition and Subtraction Properties of Equality*

Sometimes an equation can be solved by multiplying, or dividing, each side by the same number.

Examples

4 **Solve $7x = 42$.**

$$7x = 42$$
$$\frac{1}{7} \cdot 7x = \frac{1}{7} \cdot 42 \qquad \textit{Multiply each side by } \frac{1}{7}.$$
$$x = 6 \qquad \textit{Substitution Property}$$

Check:
$$7x = 42$$
$$7(6) \stackrel{?}{=} 42$$
$$42 = 42 \quad ✔$$

The solution is 6.

This equation could be solved by dividing each side by 7.

5 **Solve $-\frac{2}{3}k = 14$.**

$$-\frac{2}{3}k = 14$$
$$-\frac{3}{2}\left(-\frac{2}{3}\right)k = \left(-\frac{3}{2}\right)(14) \qquad \textit{Multiply each side by } -\frac{3}{2}.$$
$$k = -21 \qquad \textit{Substitution Property}$$

Check:
$$-\frac{2}{3}k = 14$$
$$-\frac{2}{3}(-21) \stackrel{?}{=} 14$$
$$14 = 14 \quad ✔$$

The solution is -21.

For any real numbers a, b, and c, if $a = b$, then $a \cdot c = b \cdot c$ and, if c is not zero, $\frac{a}{c} = \frac{b}{c}$.	*Multiplication and Division Properties of Equality*

Some equations involve more than one operation. You may combine steps when solving equations.

Examples

6 **Solve $5x - 7 = 23$.**

$$5x - 7 = 23$$
$$5x - 7 + 7 = 23 + 7 \qquad \text{Addition Property}$$
$$5x = 30 \qquad \text{Substitution Property}$$
$$\frac{5x}{5} = \frac{30}{5} \qquad \text{Division Property}$$
$$x = 6 \qquad \text{Substitution Property}$$

Check:
$$5x - 7 = 23$$
$$5(6) - 7 \stackrel{?}{=} 23$$
$$30 - 7 \stackrel{?}{=} 23$$
$$23 = 23 \quad ✔$$

The solution is 6.

7 **Solve $0.75(8a + 20) - 2(a - 1) = 4$.**

$$0.75(8a + 20) - 2(a - 1) = 4$$
$$6a + 15 - 2a + 2 = 4 \qquad \text{Distributive and Substitution Properties}$$
$$4a + 17 = 4 \qquad \text{Commutative, Distributive and Substitution Properties}$$
$$4a = -13 \qquad \text{Subtraction and Substitution Properties}$$
$$a = -3.25 \qquad \text{Division and Substitution Properties}$$

The solution is -3.25. *Check this result.*

8 Using Calculators

Solve $68x + 373 = 802$.

$$68x + 373 = 802 \qquad \text{Rewrite the equation to isolate the variable, } x.$$
$$x = \frac{802 - 373}{68}$$

ENTER: (802 − 373) ÷ 68 =

DISPLAY: 802   373 429   68 6.309   *The result was rounded to 3 decimal places.*

The solution is approximately 6.309.

Exploratory Exercises

State the property shown in each statement.

1. $3 + (2 + 3) = 3 + (2 + 3)$
2. $3 + (2 + 3) = 3 + 5$
3. If $x + 3 = 7$, then $x = 4$.
4. If $5x = 40$, then $x = 8$.
5. If $7 = \sqrt{49}$, and $\sqrt{49} = 2 + 5$, then $7 = 2 + 5$.
6. If $8 + 1 = 9$ and $9 = 3 + 6$, then $8 + 1 = 3 + 6$.
7. If $3 = 2 + 1$, then $2 + 1 = 3$.
8. If $5 + 7 = 12$, then $12 = 7 + 5$.
9. If $2x = 6$, then $2x + 1 = 6 + 1$.
10. If $3a = 2y$ and $2y = 5n$, then $3a = 5n$.
11. If $7 = n$, then $n = 7$.
12. If $0.5x = 6$, then $x = 12$.

Written Exercises

State the property shown in each statement.

13. If $8 + 1 = x$, then $x = 8 + 1$.

14. If $6 + 9 = 5 + 10$ and $5 + 10 = 15$, then $6 + 9 = 15$.

15. If $7 + 4 = 7 + 3 + 1$ and $7 + 3 + 1 = 10 + 1$, then $7 + 4 = 10 + 1$.

16. $9 + (2 + 10) = 9 + 12$

17. $6 + a = 6 + a$

18. If $11 - 5 = 4 + 2$, then $4 + 2 = 11 - 5$.

19. $4 + 7 + 9 = 11 + 9$

20. $4 + 7 + 9 = 4 + 7 + 9$

21. If $5x + 7x = 4$, then $(5 + 7)x = 4$.

22. If $\frac{3}{4}x = \frac{2}{3}$, then $9x = 8$.

23. If $7x = 21$, then $x = 3$.

24. If $7b + 3 = 10$, then $7b = 7$.

25. $5x + 8x = 30 - 4$
   a. $(5 + 8)x = 30 - 4$
   b, $\quad\ 13x = 26$
   c. $\qquad\ x = 2$

26. $8a - 3 = 22$
   a. $8a = 22 + 3$
   b. $8a = 25$
   c. $a = \frac{25}{8}$

27. $6b - 7 = 3b + 2$
   a. $6b - 3b = 2 + 7$
   b. $(6 - 3)b = 9$
   c. $\quad\ 3b = 9$
   d. $\qquad\ b = 3$

28. $\frac{1}{2}(4d - 20) = 2d + d - 2$
   a. $2d - 10 = (2 + 1)d - 2$
   b. $2d - 10 = 3d - 2$
   c. $-10 + 2 = 3d - 2d$
   d. $\quad -8 = d$

Solve each equation.

29. $4x = 30$

30. $15 = -3y$

31. $a + 17 = 31$

32. $12 = 8 - r$

33. $\frac{3}{4}x = \frac{5}{7}$

34. $5q = \frac{2}{5}$

35. $\frac{2}{5}m = 1\frac{3}{4}$

36. $\frac{3}{7}x = 4\frac{1}{2}$

37. $3x + 8 = 29$

38. $4 - 8x = 36$

39. $2 + 12x = -142$

40. $3 - 2x = 18$

41. $\frac{3}{4}r + 1 = 10$

42. $1 + \frac{2}{3}y = 27$

43. $9x + 4 = 2\frac{1}{2}$

44. $4x + \frac{3}{5} = 1\frac{7}{10}$

45. $\frac{8e}{9} + \frac{1}{3} = \frac{3}{5}$

46. $\frac{3}{8} - \frac{1}{4}x = \frac{1}{16}$

47. $1.2x + 3.7 = 13.3$

48. $4.5 - 3.9m = 20.1$

49. $1.1x - 0.09 = 2.22$

50. $9 = 16d + 51$

51. $5t + 4 = 2t + 13$

52. $2y - 8 = 14 - 9y$

53. $3x + 5 = 9x + 2$

54. $3x - 4 = 7x - 11$

55. $\frac{3}{4}s - \frac{1}{2} = \frac{1}{4}s + 5$

56. $\frac{2}{3} - \frac{3}{5}x = \frac{2}{5}x + \frac{4}{3}$

57. $8 - x = 5x + 32$

58. $4 + 3p = p - 12$

59. $2(3d - 10) = 40$

60. $3 = -3(y + 5)$

61. $5(3x + 5) = 4x - 8$

62. $2(6 - 7k) = 2k - 4$

63. $8x - 3 = 5(2x + 1)$

64. $2x - 4(x + 2) = -2x - 8$

65. $285 - 38x = 2033$

66. $2467 - 897b = 10{,}091.5$

67. $610y + 950 = -5760$

68. $1476.4 + 348z = 6766$

69. $567.3n + 405.25 = 823.2n - 4328.9$

70. $847.6k - 3269.5 = 610.1k + 2406.75$

1-5 Problem Solving: Using Equations

In algebra, we often translate verbal expressions into algebraic expressions. Variables are used to represent numbers that are not known. Any letter may be used as a variable.

Verbal Expression	Algebraic Expression
a number increased by 5	$x + 5$
twice the cube of *a number*	$2n^3$
the square of *a number* increased by the cube of the *same number*	$x^2 + x^3$
three times the sum of *a number* and 7	$3(b + 7)$

Equations can be used to represent verbal mathematical sentences.

Verbal Sentence	Equation
Eight is equal to five plus three.	$8 = 5 + 3$
A number decreased by 7 is -3.	$y - 7 = -3$
A number divided by 3 is equal to $\frac{3}{4}$.	$\frac{x}{3} = \frac{3}{4}$

You can use equations to solve many verbal problems. Study the following problem and its solution.

Jeff Carter bought some 15¢ stamps and twice as many 25¢ stamps. He paid $1.95 for all the stamps. How many of each did he buy?

First, explore the problem carefully and choose a variable for one unspecified number. If possible, use the variable in writing expressions for other unspecified numbers in the problem.

The problem asks for how many stamps of each kind.

> Let n represent the number of 15¢ stamps.
> Since he bought twice as many 25¢ stamps,
> $2n$ represents the number of 25¢ stamps.

After exploring the problem you should plan the solution. Write an equation that describes the relationships in the problem.

> cost of 15¢ stamps + cost of 25¢ stamps = 195
> $15n$ + $25(2n)$ = 195

Next you should solve the equation and answer the problem.

$$15n + 25(2n) = 195$$
$$65n = 195$$
$$n = 3 \qquad \textit{Jeff bought three 15¢ stamps.}$$
$$2n = 6 \qquad \textit{He bought six 25¢ stamps.}$$

Since the cost of stamps is given in cents, express $1.95 as 195¢.

Finally, examine the solution. Check your solution in the words of the problem.

If n is 3, then Jeff bought three 15¢ stamps and $2n$, or six, 25¢ stamps.

$$3(15) + 6(25) \overset{?}{=} 195$$
$$45 + 150 \overset{?}{=} 195$$
$$195 = 195 \quad \text{✔} \quad \text{The solution is correct.}$$

The four steps for solving problems are summarized below.

> 1. **Explore the problem.**
> 2. **Plan the solution**
> 3. **Solve the problem.**
> 4. **Examine the solution.**

Problem-Solving Plan

Examples

1 **A number increased by 17 is 41. Find the number.**

Let n stand for the number.

A number increased by 17 is 41.
$$n \qquad + \qquad 17 = 41$$

$$n + 17 = 41$$
$$n = 24 \qquad \text{The number is 24.}$$

Does 24 increased by 17 equal 41?

$$24 + 17 \overset{?}{=} 41$$
$$41 = 41 \quad \text{✔}$$

2 **Henry Takahashi drove for 1.5 hours at 50 miles per hour. He drove 55 miles per hour for the rest of the trip. If Henry drove a total of 174 miles, how long did he drive at 55 miles per hour?**

Let t stand for the length of time Henry drove at 55 miles per hour.

total distance = distance at 50 mph + distance at 55 mph
$$174 \quad = \quad 50(1.5) \quad + \quad 55 \cdot t$$

$$174 = 50(1.5) + 55 \cdot t$$
$$174 = 75 + 55t$$
$$99 = 55t$$
$$1.8 = t \qquad \text{Henry drove 1.8 hours at 55 miles per hour.}$$

If Henry drives for 1.5 hours at 50 mph and for 1.8 hours at 55 mph, will he travel 174 miles?

$$1.5(50) + 1.8(55) \overset{?}{=} 174$$
$$75 + 99 \overset{?}{=} 174$$
$$174 = 174 \quad \text{✔}$$

Example

3 The sum of Laurie's and Peggy's ages is 34 years. In 5 years, twice Laurie's age increased by Peggy's age will be 67 years. How old is each now?

explore
Let n = Laurie's age now. Then, $34 - n$ = Peggy's age now.
$n + 5$ = Laurie's age in 5 years.
$(34 - n) + 5$ = Peggy's age in 5 years.

plan
$2(n + 5) + [(34 - n) + 5] = 67$

solve
$$2n + 10 + 34 - n + 5 = 67$$
$$n + 49 = 67$$
$$n = 18$$

examine
Laurie is 18 years old and Peggy is $34 - 18$ or 16 years old.
In 5 years, Laurie will be $18 + 5$ or 23. Peggy will be $16 + 5$ or 21. Will twice Laurie's age plus Peggy's age equal 67?
$$2(23) + 21 \stackrel{?}{=} 67$$
$$67 = 67 \quad \checkmark$$

Exploratory Exercises

State an algebraic expression for each of the following. Use any variable.

1. four times a number.
2. one third of a number
3. twice a number increased by 11
4. eight increased by the square of a number
5. twice the sum of a number and 7
6. five times a number decreased by 4
7. the sum of twice a number and 7
8. three decreased by twice a number
9. twelve decreased by the square of a number
10. the product of the square of a number and six
11. one-fifth the sum of 4 and a number
12. four times the sum of 8 and a number
13. eight times the sum of a number and its square
14. the sum of 8 and four times a number
15. the square of the sum of a number and 11
16. the sum of the square of a number and 11

Written Exercises

Solve each problem.

17. A number decreased by 89 is 29. Find the number.
18. The sum of twice a number and 3 is 49. Find the number.
19. Eighty-seven increased by three times a number is 165. Find the number.
20. Thirty-two decreased by twice a number is 18. Find the number.
21. Julia Blackford wants to pay about $6800 for a new car. The dealer says the car costs $8000. Julia's price is what percent of the dealer's price?
22. John Colston got $2000 trade-in on his car. The dealer is selling John's car for $1800. The dealer's selling price is what percent of the trade-in value?

23. Gunta's dad is 25 years older than Gunta. The sum of their ages is 61. How old is Gunta?

24. The sum of two consecutive odd integers is 124. What are the integers?

25. Mrs. Shieh was 24 years old when Ingrid was born. In three years the sum of their ages will be 68 years. How old is each now?

26. The sum of Paul and Beth Taulman's ages now is 67 years. Eight years ago Paul was twice as old as Beth. How old is each now?

27. Mary took five English tests. Her scores were 78%, 89%, 98%, 67%, and 90%. What must she score on the sixth test so that her average will be 85%?

28. Ben Jackson bought a microwave oven for $60 more than half its original price. He paid $274 for the oven. What was the original price?

29. Felipe Vasquez bought some 14¢ stamps and the same number of 22¢ stamps. He paid a total of $1.80 for all the stamps. How many of each kind did he buy?

30. Jenny Johnston bought some apples for 79¢ per pound and twice as many pounds of bananas for 39¢ per pound. If her total bill was $7.85, how many pounds of bananas did she buy?

31. Neva Fannin ordered concert tickets which cost $10.50 for adults and $7.50 for students. She ordered 4 more student tickets than adult tickets. Her total bill was $174. How many of each did she order?

32. The Drama Club sold 320 adult tickets and 153 student tickets for their last performance. Adult tickets were 75¢ more than student tickets. If the total receipts were $949.50, what was the price of each ticket?

33. Ron Dorris is on his way to San Diego, 300 miles away. He drives 45 miles per hour for 3 hours. He drives 55 miles per hour for the rest of the trip. How long does Ron drive at 55 miles per hour?

34. The width of a rectangle is 12 units less than its length. If you add 30 units to both the length and width, you double the perimeter. Find the length and width of the original rectangle.

35. Lita Vance is driving to Lake Tahoe, a distance of 662 miles. If she drives 55 miles per hour for 5 hours, at what speed must she travel to complete the total trip in 14 hours?

36. Two hours after a truck leaves Phoenix traveling at 45 miles per hour, a car leaves to overtake the truck. If it takes the car ten hours to catch the truck, what was the car's speed?

37. San Francisco and Los Angeles are 470 miles apart by train. An express train leaves Los Angeles at the same time a passenger train leaves San Francisco. The express train travels 10 miles per hour faster than the passenger train. The two trains pass each other in 2.5 hours. How fast is each train traveling?

Challenge Exercises

38. The perimeter of a square is 12 meters greater than the perimeter of another square. Its area is 45 square meters greater than the area of the smaller square. Find the perimeter of each square.

39. Two pieces of land have one side in common. One piece is in the shape of an equilateral triangle. The other piece is in the shape of a square. It takes 3125 feet of fence to go around the perimeter of the combined pieces of land. What is the length of each side of the square?

Find the value of each expression.

1. $15 + 9 \div 3$
2. $15 + (9 \div 3)$
3. $(15 + 9) \div 3$
4. $7 - 2^2 + 12 \div 4$
5. $(7 - 2)^2 + 12 \div 4$
6. $7 - (2^2 + 12) \div 4$

Evaluate if $a = -2$, $b = 3$, $c = 5$, $d = \frac{1}{3}$, and $e = 0.2$.

7. $5a^3 - 3b$
8. $3ac - bd$
9. $6be - a^2$
10. $5ad + 2bc$
11. $b^2 + c^2 + d^2$
12. $(b + c + d)^2$
13. $\dfrac{2ac}{d}$
14. $\dfrac{7ae}{bd^2}$
15. $\dfrac{3a^3 - 2b^3}{3c^2}$

Evaluate each expression. Then name the sets of numbers to which each value belongs. (Use N, W, Z, Q, I, and R.)

16. $2^3 - 3^2$
17. $4 + 2 \div 6$
18. $\sqrt{49 - 36}$
19. $\sqrt{49} - \sqrt{36}$
20. $\sqrt{25} \div \sqrt{4}$
21. $\sqrt{37 - 12} - \sqrt{37 + 12}$

State the property shown in each equation.

22. $(4 + 2) + 1 = 4 + (2 + 1)$
23. $xy + [-(xy)] = 0$
24. $5 \cdot (x + 2) = (x + 2) \cdot 5$
25. $5 \cdot (x + 2) = 5 \cdot (2 + x)$
26. $5 \cdot (x + 2) = 5x + 10$
27. $1 \cdot (5y^2) = 5y^2$

Simplify each expression.

28. $4x + 4y - x - 3y$
29. $3(4a - b) + 4a - c$
30. $\frac{1}{4}(16y + 12z) + \frac{1}{2}(4z - 6y)$
31. $\frac{1}{3}\left(\frac{3}{4}b - \frac{1}{2}c\right) - \frac{1}{2}\left(\frac{2}{3}b - \frac{1}{3}c\right)$
32. $5(1.3m - 0.7n) + 3(0.4m + n)$
33. $12(0.3x - 0.2y) - 6(1.2y - 3x)$

Solve each equation.

34. $21 = -7x$
35. $y - 12 = 11$
36. $\frac{2}{3}z = \frac{3}{4}$
37. $6 - 12t = 126$
38. $2a + 3 = a + 5$
39. $7b - 6 = 9 - 8b$
40. $\frac{2}{3}(5 - c) = -2$
41. $2(2x + 1) = 3x - 11$
42. $-3(x - 2) = 5(3 - x)$

Solve each problem.

43. Three times a number decreased by 7 is 14. Find the number.

44. Jim is 5 years older than Juan. In four years, the sum of their ages will be 35 years. How old is each now?

45. Ayako bought some grapes for 89¢ per pound and twice as many pounds of strawberries for $1.17 per pound. If her total bill was $6.46, how many pounds of strawberries did she buy?

46. Alan drove for 5 hours at 48 miles per hour. He then changed his speed for the remainder of the $8\frac{1}{2}$-hour trip. If the total trip covered 429 miles, what was the new speed?

1-6 Solving Absolute Value Equations

Certainly -5 and 5 are quite different, but they do have something in common. They are the same distance from 0 on the number line, but in opposite directions.

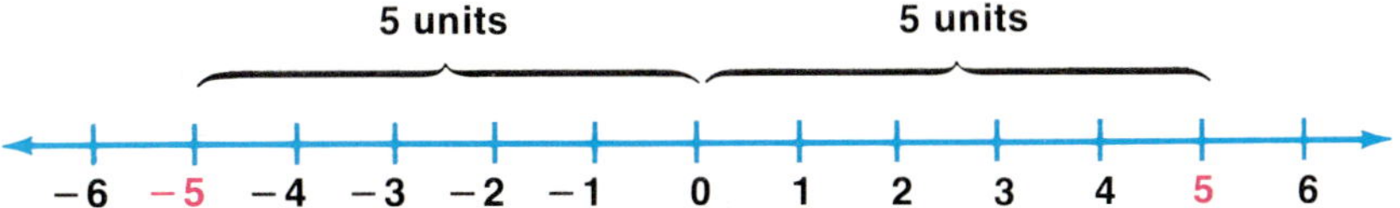

We say that -5 and 5 have the same **absolute value**. The absolute value of a number is the number of units it is from 0 on the number line.

The absolute value of -5 is 5.

$$|-5| = 5$$

The absolute value of 5 is 5.

$$|5| = 5$$

We can also define absolute value in the following way.

For any real number a:
If $a \geq 0$, then $|a| = a$.
If $a < 0$, then $|a| = -a$.

Definition of Absolute Value

Examples

1 **Find the absolute value of 3 and of -7.**

$$|3| = 3 \qquad |-7| = -(-7) \text{ or } 7$$

2 **Find the absolute value of $x - 5$.**

If x is 5 or greater, then $|x - 5| = x - 5$.
If x is less than 5, then $|x - 5| = -(x - 5)$.

3 **Evaluate $|3x - 2| + 7.2$ if $x = -3$.**

$$\begin{aligned}
|3x - 2| + 7.2 &= |3(-3) - 2| + 7.2 \\
&= |-9 - 2| + 7.2 \\
&= |-11| + 7.2 \\
&= 11 + 7.2 \\
&= 18.2
\end{aligned}$$

The value is 18.2.

Some equations contain absolute value expressions. The definition of absolute value is used in solving the equations.

4 **Solve $|x - 7| = 12$. Check each solution.**

$$|x - 7| = 12$$

If $x - 7$ is positive or zero $\quad x - 7 = 12 \quad$ or $\quad x - 7 = -12 \quad$ *If $x - 7$ is negative*
$$x = 19 \qquad\qquad x = -5$$

Check: $\quad |x - 7| = 12$
$$|19 - 7| \overset{?}{=} 12 \quad \text{or} \quad |-5 - 7| \overset{?}{=} 12$$
$$|12| \overset{?}{=} 12 \qquad\qquad |-12| \overset{?}{=} 12$$
$$12 = 12 \quad ✔ \qquad\qquad 12 = 12 \quad ✔$$

The solutions are 19 and -5. *The solution set is $\{19, -5\}$.*

5 **Solve $5|2x + 3| = 30$. Check each solution.**

$$5|2x + 3| = 30$$
$$|2x + 3| = 6$$

If $2x + 3$ is positive or zero $\quad 2x + 3 = 6 \quad$ or $\quad 2x + 3 = -6 \quad$ *If $2x + 3$ is negative*
$$2x = 3 \qquad\qquad 2x = -9$$
$$x = \frac{3}{2} \qquad\qquad x = -\frac{9}{2}$$

Check: $\quad 5|2x + 3| = 30$
$$5\left|2\left(\frac{3}{2}\right) + 3\right| \overset{?}{=} 30 \quad \text{or} \quad 5\left|2\left(-\frac{9}{2}\right) + 3\right| \overset{?}{=} 30$$
$$5|3 + 3| \overset{?}{=} 30 \qquad\qquad 5|-9 + 3| \overset{?}{=} 30$$
$$5|6| \overset{?}{=} 30 \qquad\qquad 5|-6| \overset{?}{=} 30$$
$$5(6) \overset{?}{=} 30 \qquad\qquad 5(6) \overset{?}{=} 30$$
$$30 = 30 \quad ✔ \qquad\qquad 30 = 30 \quad ✔$$

The solutions are $\frac{3}{2}$ and $-\frac{9}{2}$. *The solution set is $\left\{\frac{3}{2}, -\frac{9}{2}\right\}$.*

An absolute value equation may have *no solution*. For example, $|x| = -3$ is *never* true. Since the absolute value of a number is always positive or zero, there is *no* replacement for x that will make the sentence true. The equation $|x| = -3$ has no solution. The solution set has no members at all. This solution set is called the *empty set* and is symbolized by either $\{\ \}$ or $\emptyset$.

Another name for $\emptyset$ is the null set.

Example

6 **Solve $|3x + 7| + 4 = 0$.**

$$|3x + 7| + 4 = 0$$
$$|3x + 7| = -4$$

This sentence is *never* true, so the equation has *no solution*. The solution set is $\emptyset$.

It is important to check your answers when solving absolute value equations. Even if the correct procedure for solving the equation is used, the answers may not be actual solutions to the original equation. Study the following example.

Be especially careful when equations have variables on both sides.

Examples

7 Solve $|2x + 12| = 7x - 3$.

$$|2x + 12| = 7x - 3$$

$$2x + 12 = 7x - 3 \quad \text{or} \quad 2x + 12 = -(7x - 3)$$
$$15 = 5x \qquad\qquad 9x = -9$$
$$3 = x \qquad\qquad x = -1$$

Check: $\quad |2x + 12| = 7x - 3$

$$|2(3) + 12| \stackrel{?}{=} 7(3) - 3 \quad \text{or} \quad |2(-1) + 12| \stackrel{?}{=} 7(-1) - 3$$
$$18 = 18 \quad ✔ \qquad\qquad 10 \stackrel{?}{=} -10 \quad no$$

The only solution is 3.

8 **Using Calculators**

Evaluate $7|6y - 8|$ when $y = 1$.

First evaluate $6y - 8$.

ENTER: $6\ \boxed{\times}\ 1\ \boxed{-}\ 8\ \boxed{=}$

DISPLAY: $6 \qquad 1 \quad 6 \quad 8 \ \text{-}2$

Since an absolute value cannot be negative, change the sign of the number in the display. Then, multiply this number by 7.

ENTER: $\boxed{+/-}\ \boxed{\times}\ 7\ \boxed{=}$

DISPLAY: $\text{-}2 \quad 2 \qquad 7\ 14$

The value is 14.

Exploratory Exercises

Evaluate if $x = -5$.

1. $|x|$
2. $|4x|$
3. $|-2x|$
4. $|x + 6|$
5. $|7x - 1|$
6. $|-x|$
7. $|2x + 5|$
8. $|-2x + 5|$
9. $5 - |x|$
10. $5 - |-x|$
11. $|x| + x$
12. $|x - 7| - 8$
13. $7 - |3x + 10|$
14. $|x + 4| + |2x|$

For each equation, determine which numbers in $\{-2, -1, 0, 1, 2\}$ are solutions.

15. $|-x| = 2$
16. $|x - 2| = 1$
17. $|x| = x$
18. $|x| = -x$
19. $|x| = |x - 4|$
20. $-x = |x + 2|$

Solve each equation.

21. $|x + 11| = 42$
22. $|x + 6| = 19$
23. $|x - 5| = 11$
24. $|x - 3| = 17$

25. $3|x + 7| = 36$
26. $5|x + 4| = 45$
27. $8|x - 3| = 88$
28. $11|x - 9| = 121$

29. $\left|\frac{1}{2}x + 2\right| = 8$
30. $\left|x - \frac{7}{3}\right| = 6$
31. $\frac{1}{3}\left|6x + 5\right| = 7$
32. $\frac{1}{2}\left|3x - 4\right| = 3$

33. $|2x + 9| = 30$
34. $|2x - 37| = 15$
35. $|4x - 3| = -27$
36. $|2x + 7| = 0$

37. $3|5x + 2| = 51$
38. $8|4x - 3| = 64$
39. $7|3x + 5| = 25$
40. $4|6x - 1| = 29$

41. $-6|2x - 14| = -42$
42. $|2a + 7| = a - 4$
43. $|7 + 3a| = 11 - a$

44. $3|x + 6| = 9x - 6$
45. $|x - 4| = 3x$
46. $|3t - 5| = 2t$

47. $5|3x - 4| = x + 1$
48. $|x - 3| + 7 = 2$
49. $3|2x - 5| = -1$

Use a calculator to evaluate each expression.

50. $|7(-3) + 10|$
51. $|7(-3)| + 10$
52. $3|4x - 9|$ if $x = 2.8$

53. $3|4x - 9|$ if $x = 2.2$
54. $48|7k - 30|$ if $k = 14$
55. $4 - |5n - 8|$ if $n = 2$

Challenge Exercises

Solve each equation.

56. $|x + 3| = |5x - 9|$
57. $|x - 2| = |7x + 22|$
58. $|2x - 3| = |x + 7|$

59. $|3x + 2| = |5x - 12|$
60. $|3x - 1| = |5x - 2| - 4$
61. $|2x - 3| = |5x + 1| + 3$

mini-review

State the property shown in each equation or statement.

1. $x + 7 = x + 7$
2. $1(ab^2) = ab^2$
3. $1(ab^2) = (1a)b^2$
4. If $x = 7$, then $7 = x$
5. $5(4a + b) = 5(b + 4a)$
6. $5(4a + b) = 20a + 5b$

Solve each equation.

7. $11y = 132$
8. $3(2x - 5) = 3x - 8$
9. $3 + \frac{2}{3}x = 19$

10. $\frac{3}{5}t + 4 = \frac{t}{5} + 2$
11. $2|x - 3| = 32$
12. $-3|4 - 3x| = -27$

Solve each problem.

13. Twice a number increased by 17 is 51. Find the number.

14. Vanessa can type 62 words per minute. How long would it take her to type a manuscript of 1674 words?

15. Gladys is three times as old as Maria. In ten years, Gladys will be twice as old as Maria. How old is each one now?

1-7 Solving Inequalities

Akira and Mei are skaters. Suppose you compare their weights. You can make only one of the following statements.

Akira's weight *is less than* Mei's weight.	Akira's weight *is the same as* Mei's weight.	Akira's weight *is more than* Mei's weight.

Let j stand for Akira's weight and k stand for Mei's weight. Then you can use inequalities and an equation to compare their weights.

$$j < k \qquad j = k \qquad j > k$$

This is an illustration of the **trichotomy property**.

For any two real numbers a and b, exactly one of the following statements is true. $$a < b \qquad a = b \qquad a > b$$	*Trichotomy Property*

Adding the same number to each side of an inequality does *not* change the truth of the inequality.

For any real numbers, a, b, and c: 1. If $a > b$, then $a + c > b + c$ and $a - c > b - c$. 2. If $a < b$, then $a + c < b + c$ and $a - c < b - c$.	*Addition and Subtraction Properties for Inequalities*

These properties can be used to solve inequalities. Each solution set can be graphed on the number line.

Example

1 **Solve $9x + 7 < 8x - 2$. Graph the solution set.**

$$9x + 7 < 8x - 2$$
$$-8x + 9x + 7 < -8x + 8x - 2 \qquad \text{Add } -8x \text{ to each side.}$$
$$x + 7 < -2$$
$$x + 7 + (-7) < -2 + (-7) \qquad \text{Add } -7 \text{ to each side.}$$
$$x < -9$$

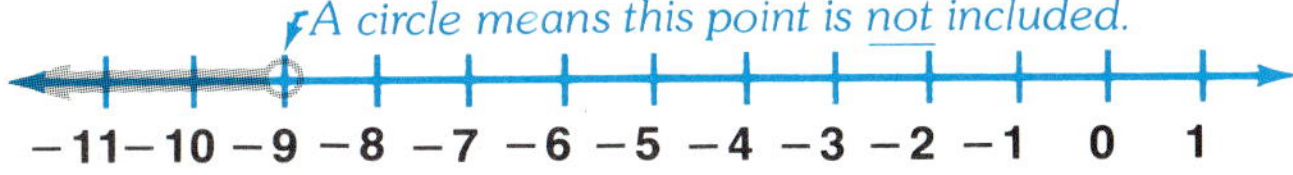

Any real number less than -9 is a solution.

To check, substitute -9 for x in $9x + 7 < 8x - 2$. The two sides should be equal. Then substitute a number less than -9. The inequality should be true.

You know that $18 > -11$ is a true inequality. If you multiply each side of this inequality by a positive number, the result is a true inequality.

$$18 > -11 \qquad \textit{A true inequality}$$
$$18(3) > -11(3) \qquad \textit{Multiply each side by 3.}$$
$$54 > -33 \qquad \textit{Another true inequality}$$

Suppose you multiply each side of a true inequality by a negative number. Try -2.

$$18 > -11 \qquad \textit{A true inequality}$$
$$18(-2) > -11(-2) \qquad \textit{Multiply each side by } -2.$$
$$-36 > 22 \qquad \textit{False!!!}$$

But, *reverse the inequality sign*, and the result is true.

$$-36 < 22 \qquad \textit{True.}$$

These examples suggest the following properties.

For any real numbers a, b, and c:

1. If c is positive and $a < b$, then $ac < bc$ and $\dfrac{a}{c} < \dfrac{b}{c}$.

2. If c is positive and $a > b$, then $ac > bc$ and $\dfrac{a}{c} > \dfrac{b}{c}$.

3. If c is negative and $a < b$, then $ac > bc$ and $\dfrac{a}{c} > \dfrac{b}{c}$.

4. If c is negative and $a > b$, then $ac < bc$ and $\dfrac{a}{c} < \dfrac{b}{c}$.

Multiplication and Division Properties for Inequalities

Examples 2 and 3 show how these properties are used to solve inequalities.

Example

2 **Solve $-0.5y < 6$. Graph the solution set.**

$$-0.5y < 6$$
$$(-2)(-0.5y) > (-2)(6) \qquad \textit{Reverse the inequality sign since each side is multiplied by } -2.$$
$$y > -12$$

Any real number greater than -12 is a solution.

The solution set can be written $\{y \mid y > -12\}$. It is read *the set of all numbers y such that y is greater than -12.*

The symbols $\neq$, $\leq$, and $\geq$ can also be used when comparing numbers. The symbol $\neq$ means *is not equal to.* The symbol $\leq$ means *is less than or equal to.* The symbol $\geq$ means *is greater than or equal to.*

3 Solve $-x \geq \dfrac{x - 11}{3}$. **Graph the solution set.**

$$-x \geq \frac{x - 11}{3}$$

$-3x \geq x - 11$ *Multiply each side by 3.*

$-4x \geq -11$ *Add $-x$ to each side.*

$x \leq \dfrac{11}{4}$ *Reverse the inequality sign.*

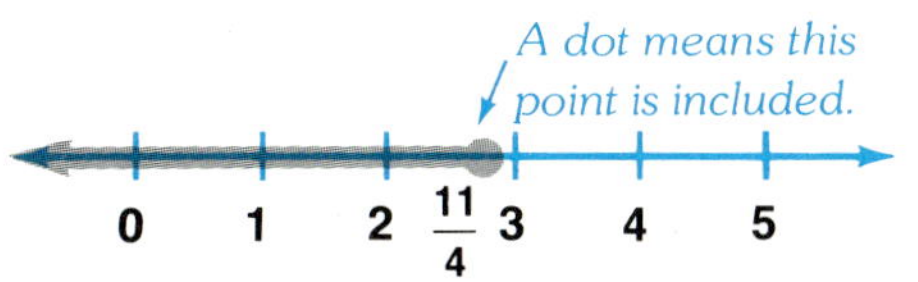

The solution set is $\left\{ x \mid x \leq \dfrac{11}{4} \right\}$.

Exploratory Exercises

Graph the solution set of each inequality.

1. $x > -3$
2. $a \leq 0$
3. $p \geq 4\frac{1}{2}$
4. $x < -7.5$
5. $\frac{5}{6}k \leq 10$
6. $-3n > 6$

Written Exercises

Solve each inequality. Graph the solution set.

7. $3x + 7 > 43$
8. $2t - 9 < 21$
9. $7n - 5 \geq 44$
10. $5x + 4 \geq 34$
11. $5r + 8 > 24$
12. $6s - 7 < 29$
13. $8 - 3x < 44$
14. $15 - 2t \geq 55$
15. $11 - 5y < -77$
16. $x - 5 < 0.1$
17. $29 - 7y < 24$
18. $5(x - 3) \geq 15$
19. $9(x + 2) < 72$
20. $-4(13 - 6t) < 26$
21. $3(3x + 2) > 7x - 2$

Solve each inequality.

22. $5(2x - 7) > 10$
23. $3(4x + 7) < 21$
24. $3(3w + 1) \geq 48$
25. $25 \leq -5(4 - p)$
26. $5(5z - 3) \leq 60$
27. $-42 > 7(2x + 3)$
28. $40 \leq -6(5r - 7)$
29. $7x - 5 > 3x + 4$
30. $3x + 1 < x + 5$
31. $1 - 2x \leq 5x - 2$
32. $3 - 2x \geq 0$
33. $2(r - 3) + 5 \geq 9$
34. $1 + 2(x + 4) \geq 1 + 3(x + 2)$
35. $4(4z + 5) - 5 > 3(4z - 1)$
36. $2(3m + 4) - 2 \leq 3(1 + 3m)$
37. $2(m - 5) - 3(2m - 5) < 5m + 1$
38. $3b - 2(b - 5) < 2(b + 4)$
39. $7 + 3y > 2(y + 3) - 2(-1 - y)$
40. $0.01x - 2.32 \geq 0$
41. $0.75x - 0.5 < 0$
42. $2.55x - 4.25 \leq 0$
43. $\dfrac{2x + 3}{5} \leq 0.03$
44. $\dfrac{2x + 3}{5} \geq -0.03$
45. $20\left(\dfrac{1}{5} - \dfrac{w}{4}\right) \geq -2w$
46. $\dfrac{3x - 3}{5} < \dfrac{4x - 2}{6}$
47. $\dfrac{x + 8}{16} - 1 > \dfrac{4 - x}{12}$
48. $\dfrac{2 - 3b}{4} \leq 2\left(\dfrac{2b}{3} - 4b\right)$

Challenge Exercises

Find the set of all numbers x satisfying the given conditions.

49. $x + 1 > 0$ and $x - 3 < 0$
50. $x - 1 < 0$ and $x + 2 > 0$
51. $3x - 2 \geq 0$ and $5x - 1 \leq 0$
52. $2x + 1 > 0$ and $x - 1 > 0$

1-8 Problem Solving: Using Inequalities

The symbols of inequality may be used to represent different types of verbal sentences. Study the verbal sentences for each inequality below.

$x < 14$

x is less than 14.
x is fewer than 14.

$x > 14$

x is greater than 14.
x is more than 14.

$x \leq 14$

x is less than or equal to 14.
x is at most 14.
x is no more than 14.
The greatest value of x is 14.
The maximum value of x is 14.

$x \geq 14$

x is greater than or equal to 14.
x is at least 14.
x is no less than 14.
The least value of x is 14.
The minimum value of x is 14.

Inequalities can be used to solve many verbal problems. The procedure used to solve problems with equations can also be used to solve problems with inequalities.

Example

1 **Corbie Cochran wants to invest \$10,000, some in bonds at 6% interest annually and the rest in stocks at 9% interest annually. If he wishes to earn at least \$720 in interest this year, what is the minimum he should invest in stocks?**

explore Let n = amount invested in stocks. *The phrase at least 720 means greater than or equal to 720.*
Then $10,000 - n$ = amount invested in bonds.

plan

$$\text{(rate)(amount)} + \text{(rate)(amount)} \geq \text{minimum desired}$$
$$(0.09)(n) + (0.06)(10,000 - n) \geq 720$$

solve

$$0.09n + 600 - 0.06n \geq 720$$
$$0.03n \geq 120$$
$$n \geq 4000$$

Corbie must invest at least \$4000 in stocks.

examine If Corbie invests \$4000 in stocks, then he will invest \$6000 in bonds. Find the amount he will earn from each investment. Is the total at least \$720?

$$6\% \text{ of } \$6000 = \$360$$
$$9\% \text{ of } \$4000 = \underline{\$360}$$
$$\$720 \quad \text{total}$$

Now check an amount greater than \$4000 for stocks. Be sure the total is greater than \$720.

2 **Kym Sutherland has 6 quarts of a 90% antifreeze solution. How much 40% antifreeze must she add to make a solution that is at most 70% antifreeze?**

explore Let x = number of quarts of 40% antifreeze to be added.
Write an expression for the amount of antifreeze in each solution.

$$90\% \text{ of } 6 \text{ qt} \quad \blacktriangleright \quad 0.90(6)$$

$$40\% \text{ of } x \text{ qt} \quad \blacktriangleright \quad 0.40(x)$$

$$70\% \text{ of } (6 + x) \text{ qt} \quad \blacktriangleright \quad 0.70(6 + x)$$

plan 90% of 6 qt + 40% of x qt is at most 70% of $(6 + x)$ qt
$$0.90(6) \quad + \quad 0.40(x) \quad \leq \quad 0.70(6 + x)$$

solve
$$0.90(6) + 0.40(x) \leq 0.70(6 + x)$$
$$5.4 + 0.4x \leq 4.2 + 0.7x$$
$$1.2 \leq 0.3x$$
$$4 \leq x$$

Therefore, Kym must add at least 4 quarts of 40% antifreeze solution.

examine Check whether adding 4 or more quarts of 40% solution to 6 quarts of 90% solution produces at most a 70% solution. Replace x by 4 in the expression below. Then try a number greater than 4.

$$\frac{0.90(6) + 0.40(x)}{6 + x}$$ This expression represents the strength of the antifreeze solution.

Try 4 $\dfrac{0.90(6) + 0.40(4)}{6 + 4} = 0.7$ or 70%   *Adding 4 quarts produces a 70% solution.*

Try 5 $\dfrac{0.90(6) + 0.40(5)}{6 + 5} \approx 0.67$ or 67%   *Adding 5 quarts produces a 67% solution.*

The answer is reasonable.

The table at the right is for state income tax. Corinne figured her tax to be $33.67. According to the table, this means her taxable income is at least $3925, but less than $3950.

Let c stand for Corinne's taxable income. Then two inequalities, $c \geq 3925$ and $c < 3950$, describe her taxable income. The sentence, $c \geq 3925$ and $c < 3950$, is called a **compound sentence**. A compound sentence containing *and* is true only if both parts of it are true. A compound sentence containing *or* is true if at least one part of it is true.

If Ohio taxable income (Line 6) is:		
At least	But less than	The tax is:
3,900	3,925	33.45
3,925	3,950	33.67
3,950	3,975	33.88
3,975	4,000	34.09
4,000	4,025	34.31
4,025	4,050	34.52
4,050	4,075	34.73
4,075	4,100	34.95
4,100	4,125	35.16
4,125	4,150	35.38
4,150	4,175	35.59
4,175	4,200	35.80
4,200	4,225	36.02
4,225	4,250	36.23

Another way of writing $c \geq 3925$ and $c < 3950$ is as follows.

$$3925 \leq c < 3950$$

The sentence is read, "c is greater than or equal to 3925 *and* less than 3950." (Inequality sentences containing **or** *cannot* be combined in this way.)

To solve a compound sentence, solve each part of the compound sentence.

Example

3 **Solve $3 < 2x - 1 < 9$.**

First, write the compound sentence using the word *and*.
Then solve each part of the sentence.

$$3 < 2x - 1 \quad \text{and} \quad 2x - 1 < 9$$
$$4 < 2x \qquad\quad \text{and} \qquad\quad 2x < 10$$
$$2 < x \qquad\quad \text{and} \qquad\quad x < 5 \qquad \text{The solution set is } \{x \mid 2 < x < 5\}.$$

Or both parts may be solved at the same time.

$$3 < 2x - 1 < 9$$
$4 < 2x < 10$ *Add 1 to each part of the compound sentence.*
$2 < x < 5$ *Divide each part by 2.*

Exploratory Exercises

State whether each compound sentence is *true* or *false*.

1. $9 > 8$ and $9 > 7$ **2.** $15 \geq 14$ and $5 < 4$ **3.** $-2 < 1$ and $1 > 2$

4. $5 < 10$ and $-5 > -10$ **5.** $-5 < -6$ or $3 > 0$ **6.** $-3 \leq -3$ or $8 > 11$

State a mathematical expression for each verbal expression.

7. The number of students, c, in the class is at least 30.

8. The number of students, m, in the math club is more than 50.

9. The number of students, s, that can ride the bus is no more than 48.

10. The maximum price, p, is \$250.

11. The minimum value, v, is 99¢.

12. The number of people, p, that can attend the show is no more than 400.

Rewrite each compound sentence using the word *and*.

13. $-3 < x < 2$ **14.** $-1 \leq b < 3$ **15.** $1 < 3y \leq 13$

16. $-12 < t \leq 4$ **17.** $5 \leq 3 - 2g < 1$ **18.** $-2 < 3(k + 2) < 6$

Rewrite each compound sentence without using the word *and*.

19. $-2 < x$ and $x < 10$ **20.** $0 \leq b$ and $b < 5$ **21.** $x \geq -2$ and $x \leq 10$

22. $y > 4$ and $y \leq 9$ **23.** $m \leq 5$ and $-5 \leq m$ **24.** $n > -15$ and $n < 7$

Solve each problem.

25. Diencha has $110.37 in her checking account. The bank does not charge for checks if $50 or more is in the account. What is the greatest amount for which she can write a check and not be charged?

26. The Municipal Parking Garage charges $1.50 for the first hour and $0.50 for each additional hour. For how many hours can you park your car if the most you pay is $4.50?

27. One number is twice another. Twice the lesser number increased by the greater number is at least 85. Find the *least* possible value for the lesser number.

28. Dave Sharp has $10 to buy stamps. He wants to buy twice as many 22¢ stamps as 14¢ stamps. What is the greatest number of 22¢ stamps he can buy?

29. Donna Pennycuff invested part of $8000 in stock that lost 2%. She invested the balance at 8% interest annually. If her total gain for the year was at least $400, what was the greatest amount of money that she could have invested in stock?

30. Apples cost 10¢ per pound more than grapefruit. Carleton Foushee buys 6 pounds of grapefruit and 8 pounds of apples and pays less than $10.50. What is the highest possible price per pound he could have paid for apples?

31. The Speedy Car Rental Company rents a car for $12.95 per day plus 15 cents per mile. Your company has limited you to a daily budget of $90. What is the maximum number of miles that you can drive each day?

32. A factory can produce a table in 30 minutes and a chair in 12 minutes. It plans to produce dining sets with 4 chairs for each table. What is the maximum number of sets that can be produced in an 8-hour shift?

33. You are enrolled in an algebra course where five tests will be given in the first quarter. You need 450 points to get an A. Your scores on the first four tests were 89, 87, 95, and 98. What is the minimum score that you can earn on the last test and still get an A for the quarter?

34. Harry Smith left an estate that was estimated to be worth at most $300,000. His will stated that one-fourth of the estate be given to his church and the remainder be divided equally among his four children. What are the maximum amounts to be paid to the church and to each child?

35. Susan's softball team has won 11 games and has lost 8 games. They have 11 more games to play. To win at least 50% of *all* games, how many more games must they win?

36. The Wildcats play 84 games this season. At midseason they had won 30 games. To win at least 60% of *all* games, how many of the remaining games must they win?

37. Roy has 8 gallons of 40% antifreeze solution. How much 100% antifreeze must he add to make a solution that is at least 60% antifreeze?

38. A 20 gallon salt solution is 30% salt. What is the greatest number of gallons of water that can be evaporated so the solution is still less than 45% salt?

Solve each compound sentence.

39. $2 < y + 4 < 11$

40. $1 < x - 2 < 7$

41. $x - 4 < 1$ or $x + 2 > 1$

42. $y + 6 > -1$ or $y - 2 < 4$

43. $4 < 2x - 2 < 10$

44. $-1 < 3m + 2 < 14$

45. $2x - 1 < -5$ or $3x + 2 \geq 5$

46. $5b < 9 + 2b$ or $9 - 2b > 11$

Applications in Energy BTU's

Energy is often measured in BTU's. BTU stands for British Thermal Unit. A BTU is the amount of heat required to raise the temperature of one pound of water 1° Fahrenheit.

Example 1: **How much energy (in BTU's) is required to heat 50 gallons of water from 40°F to 98°F? A gallon of water weighs about 8 pounds.**

energy required = weight $\times$ specific heat $\times$ temperature change
$$= 50(8 \text{ lb}) \times 1 \text{ BTU/lb°F} \times (98 - 40)°F$$
$$= 23{,}200 \text{ BTU}$$

The chart below shows the number of BTU's provided by various energy sources, *if* the heating system is 100% efficient.

> 1 pound of coal = 15,000 BTU
> 1 cubic foot of natural gas = 1000 BTU
> 1 gallon of gasoline = 150,000 BTU

Example 2: **How much natural gas is needed to produce 23,200 BTU of heat if the heating system is 60% efficient?**

Since the system is 60% efficient, each cubic foot of natural gas provides 60% of 1000 or 600 BTU. Divide 23,200 by 600.

$$23{,}200 \div 600 = 38\tfrac{2}{3} \qquad \text{About 39 ft}^3 \text{ of natural gas is needed.}$$

Exercises

Solve.

1. How much energy is required to heat 100 pounds of water from 40°F to 100°F?

2. How much energy is required to heat 25 gallons of water from 45°F to 103°F?

3. How much energy is required for 4 hot showers? Each shower uses 10 gallons of water heated from 50°F to 110°F.

4. A water heater holds 50 gallons of water. If it is 60% efficient, how much natural gas is needed to heat the water from 50°F to 130°F?

5. A coal stove is 20% efficient. How many pounds of coal are needed to heat 40 gallons of water from 50°F to 200°F?

6. A gasoline heater is 30% efficient. How many pounds of water can be heated from 40°F to 140°F with 4 gallons of gasoline?

1-9 Absolute Value Inequalities

The absolute value of a number represents its distance from zero on the number line. You can use this idea to help solve absolute value inequalities.

Examples

1 **Solve $|x| < 3$.**

$|x| < 3$ means the distance between x and 0 is less than 3 units. To make $|x| < 3$ true, you must substitute values for x that are less than 3 units from 0.

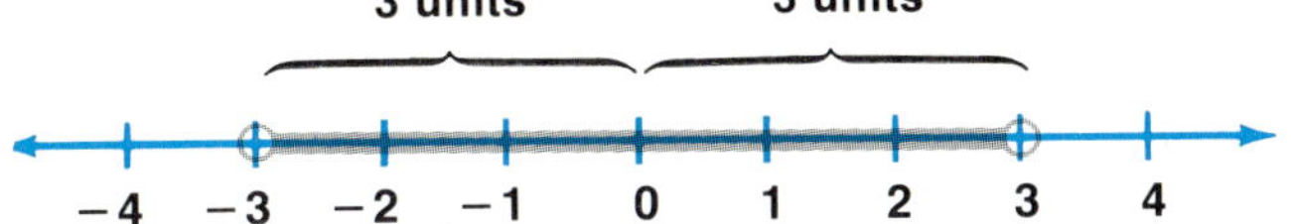

All the numbers between -3 and 3 are less than three units from zero. The solution set is $\{x | -3 < x < 3\}$.

2 **Solve $|x| \geq 2$.**

To make this true, you must substitute values for x that are 2 or more units from 0.

The solution set is $\{x | x \geq 2 \text{ or } x \leq -2\}$

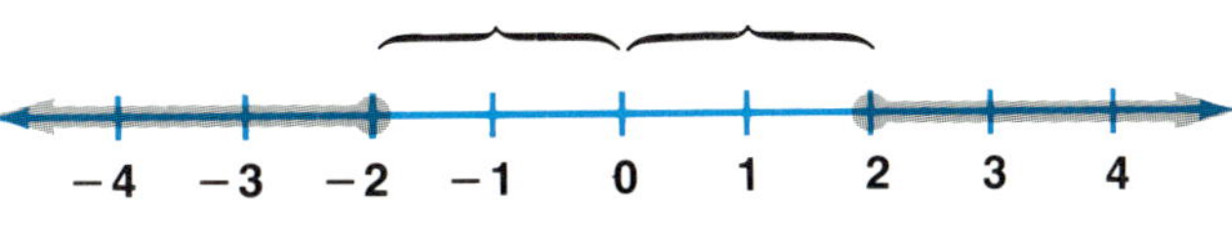

3 **Solve $|2x - 5| > 9$. Graph the solution set.**

The inequality $|2x - 5| > 9$ says that $2x - 5$ is more than 9 units from 0.

If $2x - 5 \geq 0$
$$2x - 5 > 9$$
$$2x > 14$$
$$x > 7$$

or

$$2x - 5 < -9$$
$$2x < -4$$
$$x < -2$$
If $2x - 5 < 0$

The solution set is $\{x | x < -2 \text{ or } x > 7\}$.

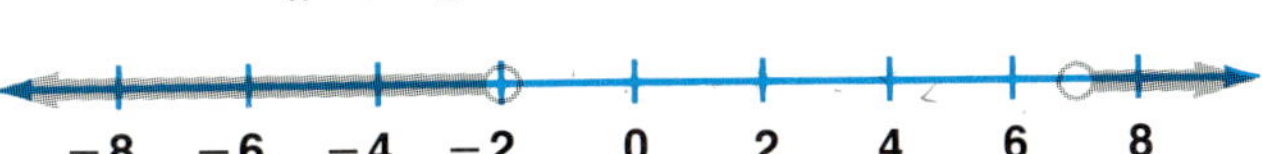

4 **Solve $|2x + 3| + 4 < 5$. Graph the solution set.**

$$|2x + 3| + 4 < 5$$
$$|2x + 3| < 1$$
Add -4 to each side to isolate the absolute value expression.

The inequality $|2x + 3| < 1$ says that $2x + 3$ is less than 1 unit from 0.

$$-1 < 2x + 3 < 1$$
$$-4 < 2x < -2$$
$$-2 < x < -1$$

The solution set is $\{x | -2 < x < -1\}$.

Some absolute value inequalities have *no solutions*. For example, $|4x - 9| < -7$ is *never* true. Since the absolute value of a number is always positive or zero, there is *no* replacement for x that will make the sentence true. The inequality $|4x - 9| < -7$ has no solution. Therefore, its solution set is $\emptyset$.

Some absolute value inequalities are *always* true. For example, $|10x + 3| > -5$ is *always* true. Since the absolute value of a number is always positive or zero, *any* replacement for x will make the sentence true. The solution set for $|10x + 3| > -5$ is the set of real numbers.

Exploratory Exercises

State an absolute value inequality for each of the following. Then graph each inequality.

1. all numbers between -3 and 3
2. all numbers less than 8 and greater than -8
3. all numbers greater than 6 or less than -6
4. all numbers less than or equal to 5, and greater than or equal to -5
5. $x > 3$ or $x < -3$
6. $x < 6$ and $x > -6$
7. $x \leq 4$ and $x \geq -4$
8. $x \geq 7$ or $x \leq -7$
9. $-6 < x < 6$
10. $-3 < x < 3$
11. $x > -2$ and $x < 2$
12. $x < 10$ and $x > -10$

Written Exercises

Solve each inequality. Graph each solution set.

13. $|x| < 9$
14. $|x| \geq 2$
15. $|x + 1| > 3$
16. $|x + 1| \leq 3$
17. $|x - 4| \leq 8$
18. $|3x| < 6$
19. $|7x| \geq 21$
20. $|x - 9| > 5$
21. $|x| > 7$
22. $|5x| < 35$
23. $|2x| \leq 26$
24. $|2x| \geq -64$
25. $|3x| < -15$
26. $|5x| < 15$
27. $|x + 3| > 17$
28. $|x - 4| \leq -12$
29. $|x - 12| < 42$
30. $|x + 9| \geq 17$
31. $|2x - 9| \leq 27$
32. $|3x + 11| > 1$
33. $|4x - 3| \geq 12$
34. $|3x + 7| \leq 26$
35. $|5x + 7| < 81$
36. $|3x + 11| > 42$
37. $|2x - 5| \leq 7$
38. $|6x + 25| + 14 < 6$
39. $6 + |3x| > 0$
40. $|4x| + 3 \leq 0$
41. $|x| \leq x$
42. $|x| > x$
43. $|x + 2| - x \geq 0$
44. $2 + |3 - 2x| > 0$

Challenge Exercises

Solve each inequality.

45. $|x + 1| + |x - 1| \leq 2$
46. $|x + 3| + |x - 3| > 8$
47. $|x + 2| + |x - 2| \leq 4$
48. $|x + 2| + |x - 2| > 16$

Solve each inequality. Graph the solution set.

1. $2x - 7 > -1$ **2.** $9 - 3x < 15$ **3.** $2x + 3 \leq 11$

Solve each inequality.

4. $2x + 13 < -7$ **5.** $15 - 4x < 3$ **6.** $4(3x + 5) \leq 48$

7. $3(x - 5) + 2 > 2(5x + 6)$ **8.** $8(x + 1) + 3x \geq 12 + x$

Solve each problem.

9. Becky has \$1 to buy stamps. She wants to buy three more 6¢ stamps than 4¢ stamps. What is the greatest number of 6¢ stamps she can buy?

10. One number is 8 more than twice another. If the sum of the numbers is at least 75, what is the *least* possible value of the lesser number?

Using Computers

Solving Inequalities

Trial-and-error can be used to find solutions to an inequality. Consider $|x - 3| \leq 4$.

Try -5.

$$|-5 - 3| \overset{?}{\leq} 4$$
$$|-8| \overset{?}{\leq} 4$$
$$8 \overset{?}{\leq} 4 \quad \text{no}$$

Try 0.

$$|0 - 3| \overset{?}{\leq} 4$$
$$|-3| \overset{?}{\leq} 4$$
$$3 \overset{?}{\leq} 4 \quad \text{yes}$$

Try 5.

$$|5 - 3| \overset{?}{\leq} 4$$
$$|2| \overset{?}{\leq} 4$$
$$2 \overset{?}{\leq} 4 \quad \text{yes}$$

So, two solutions are 0 and 5. There are many other solutions. Finding solutions by trial-and-error is time consuming unless you use a computer. A computer can test many values quickly and easily. The following BASIC program can be used to find all integer solutions between -20 and 20.

```
10   FOR X = -20 TO 20
20   IF ABS(X - 3) <= 4 THEN PRINT X
30   NEXT X
40   END
```

ABS means absolute value.
< = means $\leq$.
Lines 10 to 30 form a loop in the program.

Enter the program and type RUN.
The computer will print these solutions: $-1, 0, 1, 2, 3, 4, 5, 6, 7$

Exercises

For each inequality, write a new statement for line 20 of the program above. Then run the new program and list the integer solutions between -20 and 20.

1. $|x + 1| > 3$ **2.** $|2x - 3| \leq 5$ **3.** $|3x + 4| \geq 15$

4. $|8 - x| < 7$ **5.** $|x^2 + 5x - 3| > 3$ **6.** $|x^3 + 4x - 6| \leq 2$

value (3)
base (3)
exponent (3)
algebraic expression (4)
formula (4)
natural numbers, $\mathbf{N}$ (8)
whole numbers, $\mathbf{W}$ (8)
integers, $\mathbf{Z}$ (8)
rational numbers, $\mathbf{Q}$ (8)

irrational numbers, $\mathbf{I}$ (8)
real numbers, $\mathbf{R}$ (8)
commutativity (11)
associativity (11)
additive identity (11)
multiplicative identity (11)
additive inverse (11)
multiplicative inverse (11)
reciprocal (11)

Distributive Property (12)
like terms (12)
solution (14)
solution set (14)
equation (14)
Substitution Property (14)
absolute value (23)
Trichotomy Property (27)
compound sentence (31)

Chapter Summary

1. Order of Operations: 1. Evaluate all powers.   2. Do all multiplications and divisions from left to right.   3. Do all additions and subtractions from left to right.   (3)

2. In a formula, if you know replacements for every variable except one, you can find a replacement for that variable.   (4)

3. Each real number corresponds to exactly one point on the number line. Each point on the number line corresponds to exactly one real number.   (8)

4. Properties of Operations: (11–12)

	For any real numbers a, b, and c	
	Addition	*Multiplication*
Commutative	$a + b = b + a$	$a \cdot b = b \cdot a$
Associative	$(a + b) + c = a + (b + c)$	$(a \cdot b) \cdot c = a \cdot (b \cdot c)$
Identity	$a + 0 = a = 0 + a$	$a \cdot 1 = a = 1 \cdot a$
Inverse	$a + (-a) = 0 = -a + a$	If a is *not* zero, then $a \cdot \dfrac{1}{a} = 1 = \dfrac{1}{a} \cdot a.$
Distributive of multiplication over addition: $a(b + c) = ab + ac$ and $(b + c)a = ba + ca$		

5. Properties of Equality: (14)

	For any real numbers a, b, and c
Reflexive	$a = a$
Symmetric	If $a = b$, then $b = a.$
Transitive	If $a = b$ and $b = c$, then $a = c.$

6. Properties of Equality: (15)

<table>
<tr><td colspan="2" align="center">For any real numbers a, b, and c</td></tr>
<tr><td>Addition and Subtraction</td><td>If $a = b$, then $a + c = b + c$, and $a - c = b - c$.</td></tr>
<tr><td>Multiplication and Division</td><td>If $a = b$, then $a \cdot c = b \cdot c$, and if c is not zero, $\dfrac{a}{c} = \dfrac{b}{c}$.</td></tr>
</table>

7. Problem-Solving Plan: 1. Explore the problem.   2. Plan the solution. 3. Solve the problem.   4. Examine the solution.   (19)

8. The absolute value of a number is the number of units it is from zero on the number line. For any number a, if $a \geq 0$, then $|a| = a$. If $a < 0$, then $|a| = -a$.   (23)

9. Trichotomy Property: For any two numbers a and b, exactly one of the following statements is true. $a < b$, $a = b$, or $a > b$.   (27)

10. Properties of Inequalities: (27–28)

<table>
<tr><td colspan="2" align="center">For any real numbers a, b, and c</td></tr>
<tr><td>Addition and Subtraction</td><td>1. If $a > b$, then $a + c > b + c$ and $a - c > b - c$.
2. If $a < b$, then $a + c < b + c$ and $a - c < b - c$.</td></tr>
<tr><td>Multiplication and Division</td><td>1. If c is positive and $a < b$, then $ac < bc$ and $\dfrac{a}{c} < \dfrac{b}{c}$.
2. If c is positive and $a > b$, then $ac > bc$ and $\dfrac{a}{c} > \dfrac{b}{c}$.
3. If c is negative and $a < b$, then $ac > bc$ and $\dfrac{a}{c} > \dfrac{b}{c}$.
4. If c is negative and $a > b$, then $ac < bc$ and $\dfrac{a}{c} < \dfrac{b}{c}$.</td></tr>
</table>

Chapter Review

1–1   **Find the value of each expression.**

1. $(4 + 6)^2 - 24 \div 3$ **2.** $4 + 6^2 - 24 \div 3$ **3.** $(3 - 9)^2 \times 2 - 10 \div 6$

Evaluate if $a = -\dfrac{1}{2}$, $b = 4$, $c = 5$, and $d = -3$.

4. $\dfrac{3ab}{cd}$ **5.** $\dfrac{4a + 3c}{3b}$ **6.** $\dfrac{3ab^2 - d^3}{a}$

1–2   **Name the sets of numbers to which each number belongs. (Use N, W, Z, Q, I, and R.)**

7. -25.8 **8.** $\sqrt{13}$ **9.** $\dfrac{64}{8}$ **10.** -9.0

1–3 **State the property shown in each equation.**

11. $3 + (a + b) = (a + b) + 3$

12. $(3 + a) + b = 3 + (a + b)$

13. $(a + b) + 0 = (a + b)$

14. $(a + b) + [-(a + b)] = 0$

Simplify each expression.

15. $(9 + 48)7 + 3$

16. $15r + 18s + 16r - 8s$

17. $7p + 9q - 10p + 4q$

18. $0.2(14y + 7z) - 0.4(7z + 14y)$

1–4 **State the property shown in each equation or statement.**

19. $3r + (7 - 1)s = 3r + 6s$

20. $5x + 3y = 5x + 3y$

21. If $3a = b$, then $b = 3a$.

22. If $c = 7d$ and $7d = 1$, then $c = 1$.

Solve each equation.

23. $15x + 25 = 2(x - 4)$

24. $3(6 - 4x) = 4x + 2$

25. $2(3x - 1) = 3(x + 2)$

26. $9 = \dfrac{3y - 6}{2}$

27. $\dfrac{2}{7}a = \dfrac{16}{5}$

28. $\dfrac{3}{4}y + \dfrac{3}{4} = \dfrac{5}{2}$

1–5 **Solve each problem.**

29. The width of a rectangle is 4 meters more than one-third its length. The perimeter is 64 meters. Find the length and width.

30. Maureen's dad is 32 years older than Maureen. Six years ago, the sum of their ages was 52. How old is Maureen now?

31. Jeanne bought some $16 skirts and some $22 blouses. She bought 2 more blouses than skirts. Her total bill was $120. How many of each did she buy?

32. Two hours after a truck leaves Dallas, a car leaves to overtake the truck. The car travels 15 miles per hour faster than the truck. If it takes the car 6 hours to catch the truck, what was the car's speed?

1–6 **Solve each equation.**

33. $|2x - 37| = 15$

34. $|p - 3| + 7 = 2$

35. $8|2a - 3| = 64$

1–7 **Solve each inequality. Graph the solution set.**

36. $8(2x - 1) > 11x + 31$

37. $4(3x + 2) + 6 \geq 7x - 9$

38. $3 - 4x \leq 6x + 5$

1–8 **Solve each problem.**

39. Tania has 10 gallons of a 50% antifreeze solution. How much 100% antifreeze must she add to make a solution that is at least 80% antifreeze?

40. Roxie has $8.40 to spend on gasoline. Where Roxie lives gasoline costs between $1.40 and $1.60 per gallon. How many gallons of gasoline can Roxie buy?

Solve each compound sentence.

41. $-1 < 3(y - 2) \leq 9$

42. $2x - 5 < -5$ or $3x + 2 \geq 5$

1–9 **Solve each inequality. Graph the solution set.**

43. $|2x + 5| \leq 4$

44. $|3x + 7| \geq 26$

45. $7 + |9 - 2x| > 4$

State the property shown in each equation or statement.

1. $(5 \cdot r) \cdot s = 5 \cdot (r \cdot s)$

2. $(5 \cdot r) \cdot s = s \cdot (5 \cdot r)$

3. $\left(4 \cdot \frac{1}{4}\right) \cdot 3 = \left(4 \cdot \frac{1}{4}\right) \cdot 3$

4. $(6 - 2)a - 3b = 4a - 3b$

5. If $2(7) + 3 = 14 + 3$ and $14 + 3 = 17$, then $2(7) + 3 = 17$.

6. If $(a + b)c = ac + bc$, then $ac + bc = (a + b)c$.

Find the value of each expression.

7. $(2 + 3)^3 - 4 \div 2$

8. $[2 + 3^3 - 4] \div 2$

9. $(2^5 - 2^3) + 2^3$

10. $[5(19 - 4) \div 3] - 4^2$

Evaluate if $a = -9$, $b = \frac{2}{3}$, $c = 8$, and $d = -6$.

11. $\frac{a}{b^2} + c$

12. $\frac{db + 4c}{a}$

13. $2b(4a + d^2)$

14. $\frac{4a + 3c}{3b}$

Name the sets of numbers to which each number belongs. (Use N, W, Z, Q, I, and R.)

15. $\sqrt{13}$

16. 0.68

17. $\frac{-10}{2}$

18. $\sqrt{49}$

Solve each equation.

19. $2x - 7 - (x - 5) = 0$

20. $5y - 3 = -2y + 10$

21. $\frac{a}{4} + 3 = \frac{5}{2}$

22. $5t + 7 = 5t + 3$

23. $5r - (5 + 4r) = (3 + r) - 8$

24. $|5x + 10| - 3 = 0$

25. $|5x + 10| + 3 = 0$

26. $|4x - 5| + 4 = 7x + 8$

Solve each inequality. Graph the solution set.

27. $4 > m + 1$

28. $4p + 8 \geq 12$

29. $3(2 + 3r) + r < 2(9 - r)$

30. $-12 < 7x - 5 \leq 9$

31. $|5 + t| \leq 8$

32. $|9x - 4| + 8 > 4$

Solve each problem.

33. Mary scored 87, 89, 76, and 77 on her first four English tests. She needs 400 points to earn a grade of B for the course. What must be the minimum score on her next test in order for her to have a B average?

34. Gloria's softball team has won 13 games and lost 7 games. They have 12 more games to play. To win at least 50% of *all* games, how many more games must they win?

35. The formula $A = \frac{180(n - 2)}{n}$ relates the measure of one interior angle, A, of a regular polygon to the number of sides, n. If an interior angle of a regular polygon measures 150 degrees, find the number of sides.

Linear Relations and Functions

There is a relationship between the number of aluminum cans processed at a recycling plant and the amount of aluminum produced. Since the recycled cans vary in size and shape, the same number of cans may yield different amounts of aluminum. This relation is not considered a function.

2-1 The Coordinate Plane

Points in a plane can be located by using **ordered pairs** of real numbers.

Two perpendicular number lines separate the plane into four regions, called **quadrants**. The horizontal number line is usually called the **x-axis**. The vertical number line is usually called the **y-axis**. The **origin** is the intersection of these two axes. The plane determined by the perpendicular is called a **coordinate plane**.

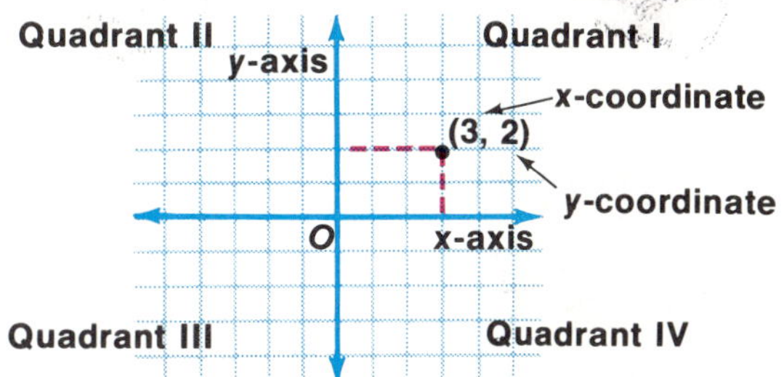

The two axes do not lie in any quadrant.

Each point in the coordinate plane corresponds to exactly one ordered pair of numbers. These numbers are called the **coordinates** of the point. The coordinates of a point are given in a particular order, (x, y). For example, $(-2, 3)$ and $(3, -2)$ do not represent the same point.

Each ordered pair of numbers corresponds to exactly one point in the coordinate plane. The point is the **graph** for the ordered pair.

When there are no labels on the axes, each unit is understood to represent an increment of one. Occasionally, you may need to label the axes using increments other than one.

The coordinates of the origin are (0, 0).

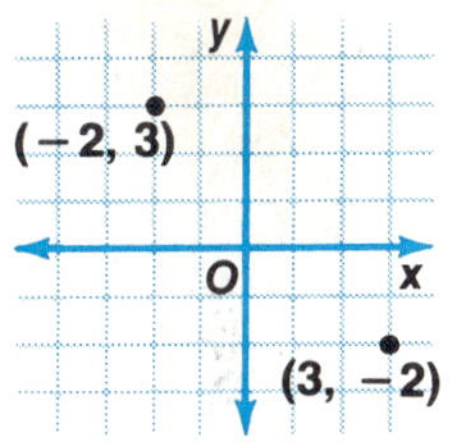

Example

1 **Graph the set of ordered pairs.**

$$\{(8, -6), (4, -2), (1, 1), (-3, 5), (-6, 8)\}$$

Label the axes by twos.
Then graph each ordered pair.

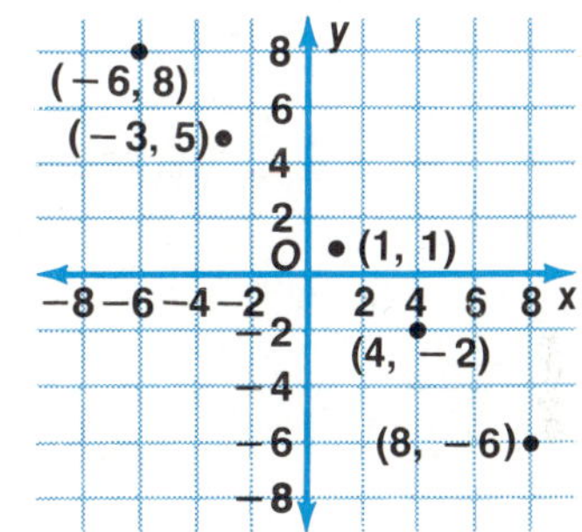

Exploratory Exercises

In which quadrant will (x, y) lie given the following conditions?

1. x is positive and y is positive.

2. x is positive and y is negative.

3. x is negative and y is negative.

4. x is negative and y is positive.

5. x is negative and y is 0.

6. x is 0 and y is positive.

Graph each ordered pair on the same coordinate plane.

7. $(7, 4)$ **8.** $(-7, 4)$ **9.** $(7, -4)$ **10.** $(-7, -4)$

11. $(2.5, 0)$ **12.** $(0, 2.5)$ **13.** $(1.5, 3)$ **14.** $(3, 1.5)$

Graph each set of ordered pairs on the same coordinate plane.

15. $\{(-8, 8), (-8, -8), (0, 4), (0, -4), (8, 0)\}$

16. $\{(-6, 12), (-3, 6), (0, 0), (6, -3), (12, -6)\}$

17. $\left\{(-1, 1), \left(-\frac{1}{4}, \frac{1}{2}\right), \left(\frac{1}{2}, 1\right), \left(1, -\frac{3}{4}\right)\right\}$

18. $\left\{\left(-2, \frac{1}{2}\right), \left(-1, 1\frac{1}{2}\right), \left(0, \frac{3}{4}\right), \left(1\frac{1}{2}, -2\right)\right\}$

19. $\{(3, 4), (9, 21), (15, 15), (21, 13), (25, 3)\}$

20. $\{(2, 12), (4, 3), (8, 11), (11, 3), (14, 12)\}$

21. $\{(0, -15), (5, -20), (10, -10), (12, 12)\}$

22. $\{(-40, 40), (20, 10), (10, -30), (-25, -25)\}$

23. $\{(0, -3), (3, 6), (6, 9), (9, 6), (12, -3)\}$

24. $\{(0.8, 0.6), (0.2, 0), (-0.2, -0.4), (-0.6, -0.6)\}$

25. $\{(-1.75, -2), (-1, -1.5), (0, -0.75), (1, 0), (2, 1.5)\}$

26. $\{(-9, 15), (-6, 0), (-3, -9), (0, -12), (3, -9), (6, 0), (9, 15)\}$

The coordinates of three vertices of a rectangle are given. Graph them. Then, find each fourth vertex.

27. $(3, 1), (3, -3), (-5, -3)$

28. $(2, 0), (0, 2), (-4, -2)$

29. $(1, 0), (3, 0), (3, 3)$

30. $(-1, 0), (1, 1), (0, 3)$

31. $(-3, 4), (5, 4), (5, -3)$

32. $(4, -1), (1, -4), (0, -3)$

mini-review

Find the value of each expression.

1. $(5 + 3)^2 - 36 \div 9$

2. $5 + 3^2 - 36 \div 9$

3. $[(3 + 7) - 4](4 + 2)$

4. $10 \div 6 - 4(8 - 4)^3$

Evaluate if $a = \frac{1}{2}$, $b = -3$, $c = -4$, and $d = 6$.

5. $\dfrac{3ab}{cd}$

6. $\dfrac{4a + 5c}{3d}$

7. $\dfrac{4ab^2 + b^2}{2cd}$

Solve each equation or inequality.

8. $5(x + 3) = 2x - 3$

9. $4(y - 1) = 6(y + 3) + 4$

10. $4|k - 3| = -8$

11. $|m - 5| - 3 = 4$

12. $-3(x - 2) + 1 \geq 4x$

13. $4t + 7 = 2(2t - 1)$

Solve each problem.

14. Five times the sum of a number and 17 is 155. Find the number.

15. Michelle's softball team has won 12 games and has lost 6 games. They have 12 more games to play. To win at least $66\frac{2}{3}\%$ of all games, how many of the remaining games must they win?

2-2 Relations and Functions

The amount of electricity produced in recent years in the U.S. can be shown using ordered pairs. The first number is the year, and the second number is the amount of electricity produced in billions of kilowatt hours.

Year	Kilowatt Hours (billions)
1980	2290
1981	2300
1982	2240
1983	2310
1984	2410
1985	2470
1986	2500

(1980, 2290), (1981, 2300), (1982, 2240), (1983, 2310)
(1984, 2410), (1985, 2470), (1986, 2500)

A set of ordered pairs is called a **relation**. The set of first coordinates, in this case years, is called the **domain** of the relation. The set of second coordinates, kilowatt hours (in billions), is called the **range** of the relation.

> **A relation is a set of ordered pairs. The domain is the set of all first coordinates of the ordered pairs. The range is the set of all second coordinates of the ordered pairs.**

Definition of Relation, Domain, and Range

A **mapping** illustrates how each element in the domain of a relation is paired with an element in the range. The following diagrams are mappings of the given relations.

{(3, 2), (2, 7), (5, 8)}

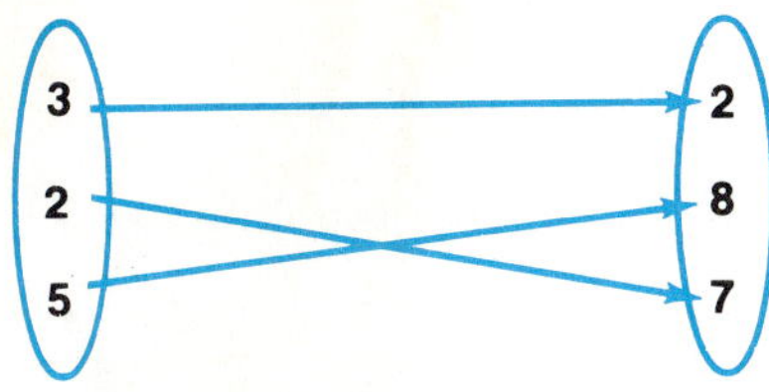

{(8, 4), (3, 9), (1, 2), (7, 4)}

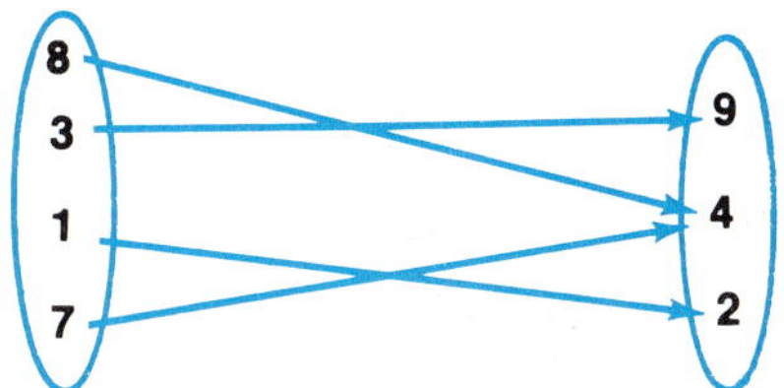

Example

1 State a relation shown by the graph. Then state the domain and range of the relation.

The relation is
{(−3, 7), (−2, 4), (−1, 1), (0, 2), (1, 5), (2, 8)}.

The domain is {−3, −2, −1, 0, 1, 2}.

The range is {7, 4, 1, 2, 5, 8}.

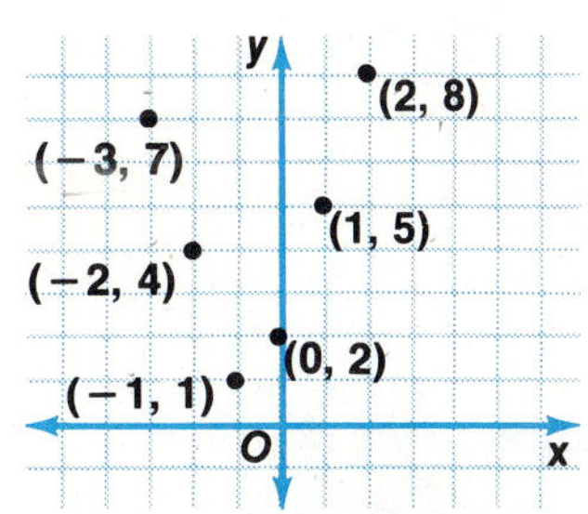

The graph below on the left shows how the amount of electricity consumed compares with the cost of electricity. The graph below on the right shows the amount of electricity used in a certain area at different hours in an average weekday.

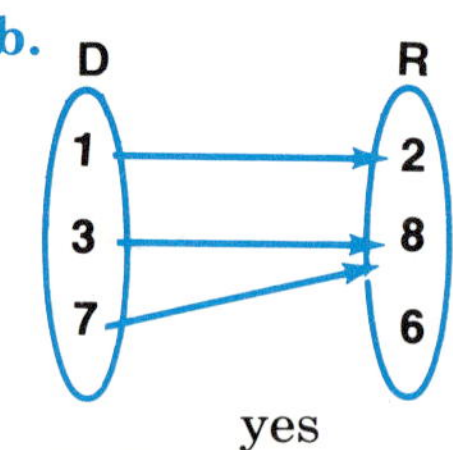

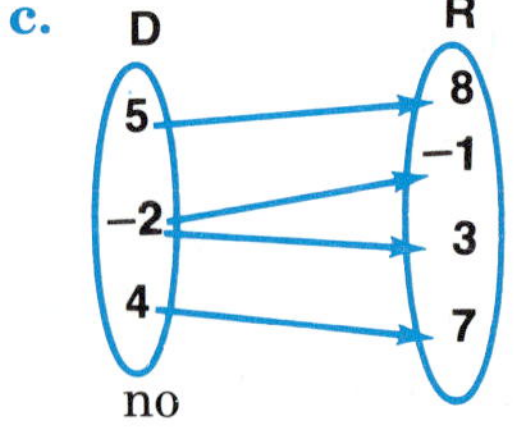

The first graph shows a special type of relation called a **function**.

> A function is a relation in which each element of the domain is paired with exactly one element of the range.

Definition of Function

Examples

2 Is {(2, 3), (3, −4), (4, 1), (1, 3)} a function?

This relation is a function since each element of the domain is paired with exactly one element of the range.

3 Is {(4, 4), (−2, 3), (4, 2), (3, 4), (1, 1)} a function?

This relation is *not* a function. The element 4 of the domain is paired with two different elements of the range, 4 and 2.

4 Which of the following mappings represent functions?

a.

D   R

1 → 2
2 → 3
4 → 4

yes

b.

D   R

1 → 2
3 → 8
7 → 6

yes

c.

D   R

5 → 8
−2 → −1
4 → 3
→ 7

no

−2 is paired with two elements of the range.

5 Is the set of ordered pairs that satisfy $y = 4x$ a function?

Suppose the value of x is 3. What is the corresponding value of y? Is there more than one value for y? When x is 3, y is 12. There is only one value for y. If you try other values of x, you will see that they are always paired with exactly one value of y. Thus, the equation $y = 4x$ represents a function.

The graph on the right represents the following relation.

$$\{(-2, 3), (-1, 1), (1, 2), (1, -1), (3, 1)\}$$

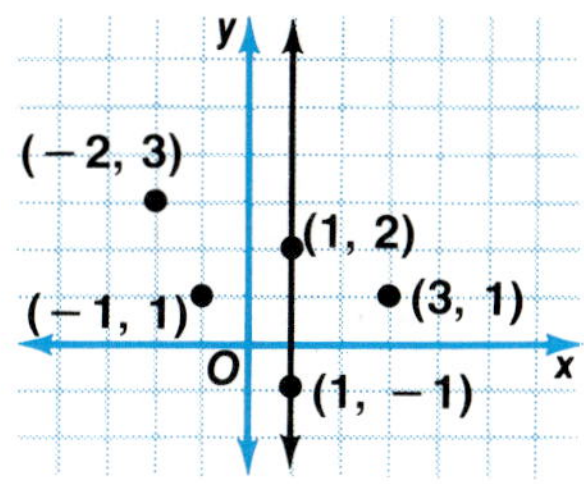

Suppose you drew a vertical line through each point on the graph. The vertical line through $(1, 2)$ would also pass through $(1, -1)$. This shows that the relation is *not* a function. There are two elements of the range, 2 and -1, that pair with one element of the domain, 1.

| If any vertical line drawn on the graph of a relation passes through no more than one point of that graph, then the relation is a function. | *Vertical Line Test for a Function* |

Example

6 **Use the vertical line test to determine if the relation graphed is a function.**

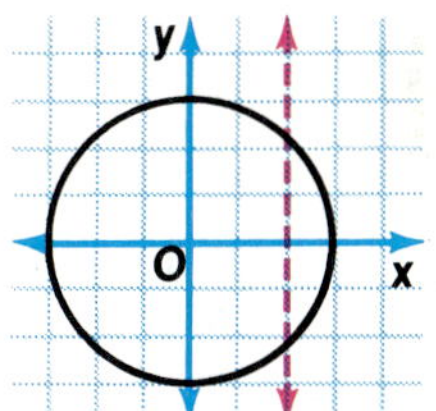

The vertical line whose equation is $x = 2$ intersects the graph at $(2, 2)$ and $(2, -2)$. Therefore, the relation is *not* a function.

Equations that represent functions often are written in a special way. The equation $y = 2x + 1$ can be written $f(x) = 2x + 1$. The symbol $f(x)$ is read "f of x." Likewise, the symbol $f(3)$ is read "f of 3." If 3 is an element of the domain of the function, then $f(3)$ is the corresponding element of the range. To indicate that the value of $f(3)$ is 7, you write $f(3) = 7$.

Letters other than f can be used to represent a function. For example, the equation $y = 4x + 3$ can be written $g(x) = 4x + 3$.

Examples

7 **Find $f(15)$ if $f(x) = 100x - 5x^2$.**

$$f(x) = 100x - 5x^2$$

$f(15) = 100(15) - 5(15)^2$ *Substitute 15 for x.*

$\quad\quad = 1500 - 5(225)$

$\quad\quad = 375$

Therefore, $f(15) = 375$.

8 **Using Calculators**

Find $f(1.6)$ if $f(x) = 3x^3 - 4x^2$.

ENTER: 3 [×] 1.6 [x^y] 3 [−]

DISPLAY: *3   1.6   3 12.288*

ENTER: 4 [×] 1.6 [x^2] [=]

DISPLAY: *4   1.6 2.56 2.048*

Therefore, $f(1.6) = 2.048$.

Example

9 Find $g(a + 2)$ if $g(x) = x^2 - 7$.

$$g(x) = x^2 - 7$$
$$g(a + 2) = (a + 2)^2 - 7 \qquad \textit{Substitute } a + 2 \textit{ for } x.$$
$$= a^2 + 4a + 4 - 7$$
$$= a^2 + 4a - 3$$

Therefore, $g(a + 2) = a^2 + 4a - 3$.

In this book, sometimes the equation for a function is given without a specified domain. In such cases, the domain is understood to be all real numbers for which the function is defined. The corresponding range values are also real numbers.

Exploratory Exercises

State the domain and range of each relation. Then state if the relation is a function.

1. $\{(4, 4), (1, 1), (3, 3)\}$

2. $\{(6, 4)\}$

3. $\{(4, 3), (8, -2), (-17, 4), (-17, 8)\}$

4. $\{(1, 5), (5, 1)\}$

5. $\{(-3, -3), (-2, -2), (2, 2), (4, 4)\}$

6. $\{(-3, 3), (-2, 2), (2, -2), (4, -4)\}$

7. $\{(5, -3), (-3, 5)\}$

8. $\{(-3, 3), (-2, 3), (2, 3), (4, 3)\}$

State the relation shown by each of the following mappings. Then state if the relation is a function.

9. 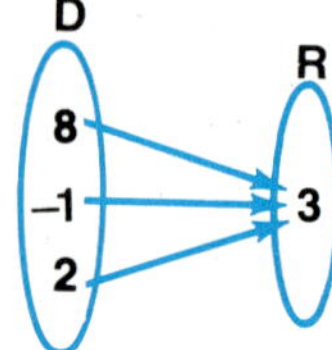

10. 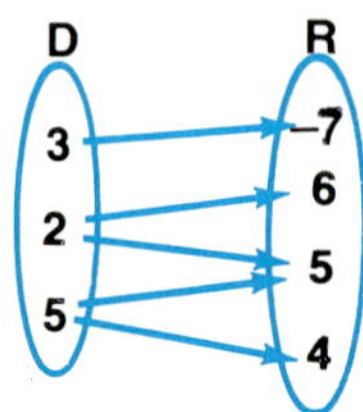

11. 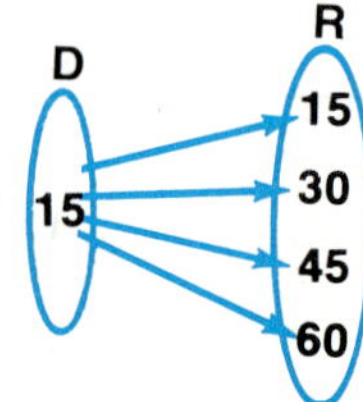

12. 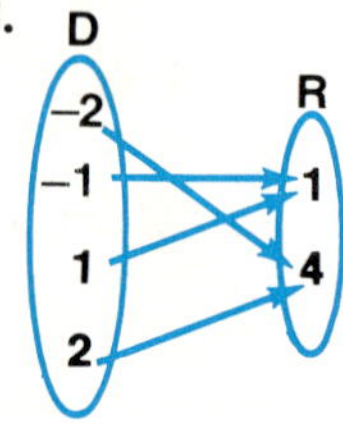

Written Exercises

State a relation shown by the graph. Then state the domain and range of the relation.

13. 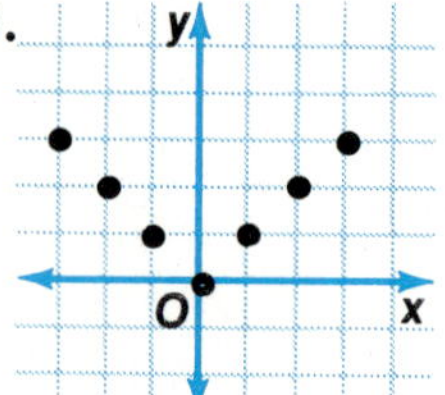

14. 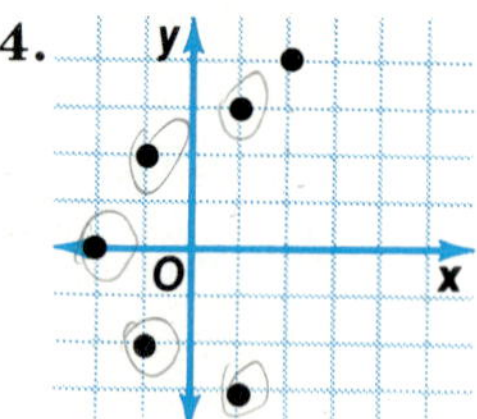

15. 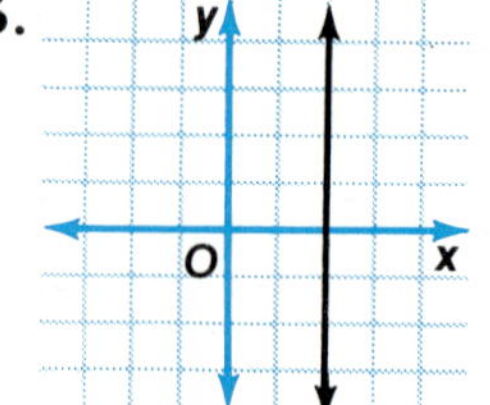

16. 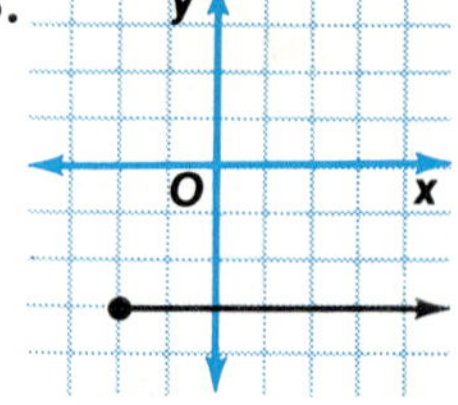

Use the vertical line test to determine if each relation is a function. Write *yes* or *no*.

17. 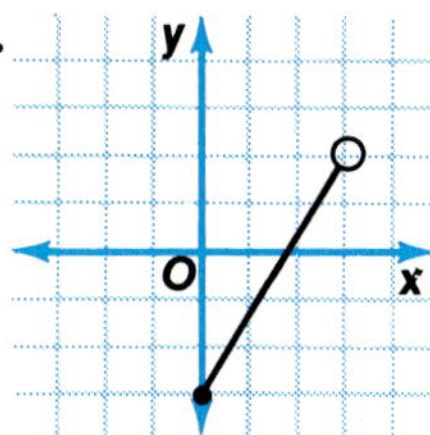

18. 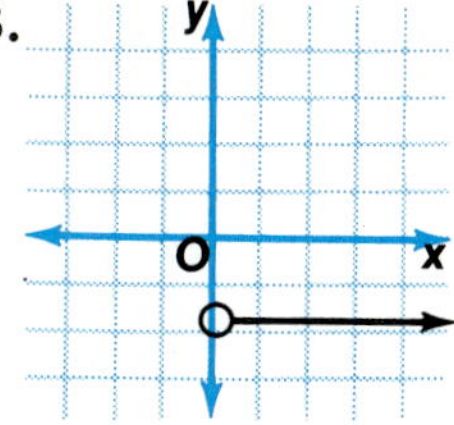

19. 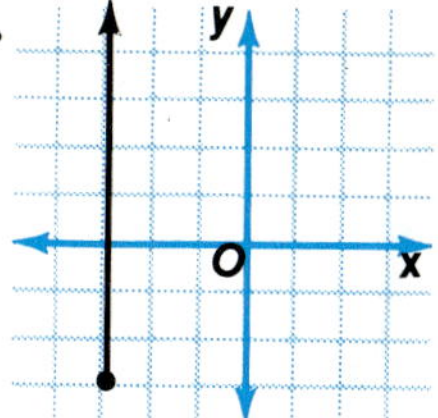

20. 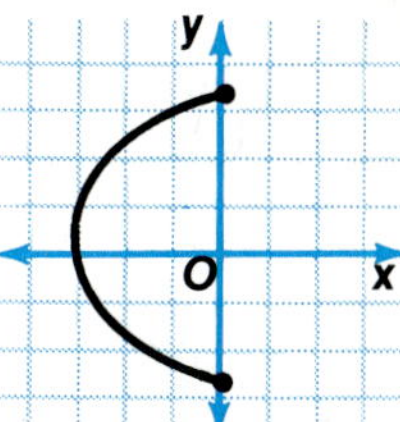

21. 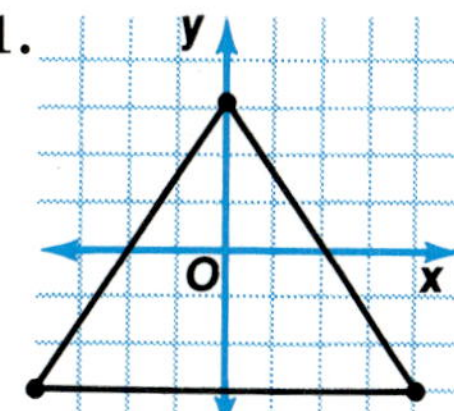

22. 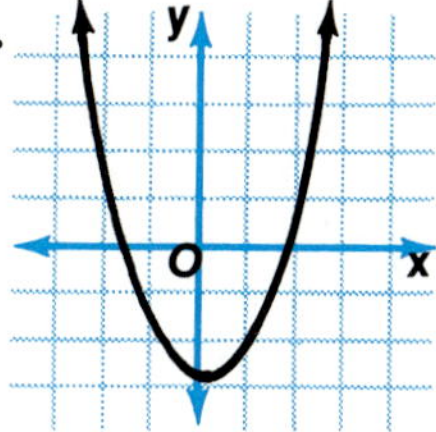

23. 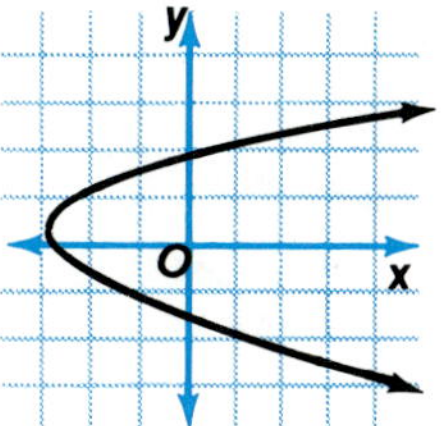

24. 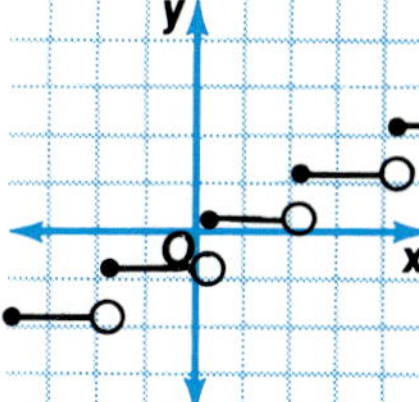

Given $f(x) = \dfrac{7}{x-2}$, find each value.

25. $f(12)$ **26.** $f(3)$ **27.** $f(-1)$ **28.** $f(5.5)$ **29.** $f(0)$ **30.** $f(1.3)$

31. $f\left(\dfrac{1}{2}\right)$ **32.** $f\left(\dfrac{2}{3}\right)$ **33.** $f(a)$ **34.** $f(u+2)$ **35.** $f(3a)$ **36.** $f(2)$

Given $g(x) = 4x^3 + 2x^2 + x - 7$, find each value.

37. $g(1)$ **38.** $g(-4)$ **39.** $g\left(-\dfrac{1}{2}\right)$ **40.** $g\left(\dfrac{1}{2}\right)$ **41.** $g(t)$ **42.** $g(2s)$

Given $h(x) = \dfrac{x^2 + 5x - 6}{x + 3}$, find each value.

43. $h(6)$ **44.** $h(-4)$ **45.** $h\left(\dfrac{1}{3}\right)$ **46.** $h\left(\dfrac{1}{2}\right)$ **47.** $h(a+1)$ **48.** $h(2m+3)$

Given $j(x) = x^4 - 3x^2 + 1$, use a calculator to find each value.

49. $j(3)$ **50.** $j(-7)$ **51.** $j(0.25)$ **52.** $j(3.75)$ **53.** $j\left(\dfrac{7}{9}\right)$ **54.** $j\left(-\dfrac{12}{5}\right)$

Challenge Exercises

State the domain of each relation.

> **Sample:** $f(x) = \dfrac{14}{x+4}$ The relation is undefined when the denominator is 0.
>
> $$x + 4 = 0 \quad \text{or} \quad x = -4$$
>
> Thus, the domain is the set of all real numbers except -4.

55. $f(x) = \dfrac{3}{x-5}$ **56.** $f(x) = \dfrac{8}{|2x-7|}$ **57.** $g(x) = \dfrac{3}{x^2}$ **58.** $g(x) = \dfrac{2x+3}{2x-1}$

59. $x = |y|$ **60.** $y = |x| - 1$ **61.** $x = |-y|$ **62.** $x = -|y+4|$

 Linear Functions

You can write the solutions to open sentences in two variables as sets of ordered pairs. These solutions can be graphed in the coordinate plane. The following graph represents the solutions to $3x - y = 1$.

An infinite number of ordered pairs will satisfy $3x - y = 1$. The graph of these ordered pairs is a straight line. An equation whose graph is a straight line is called a **linear equation**.

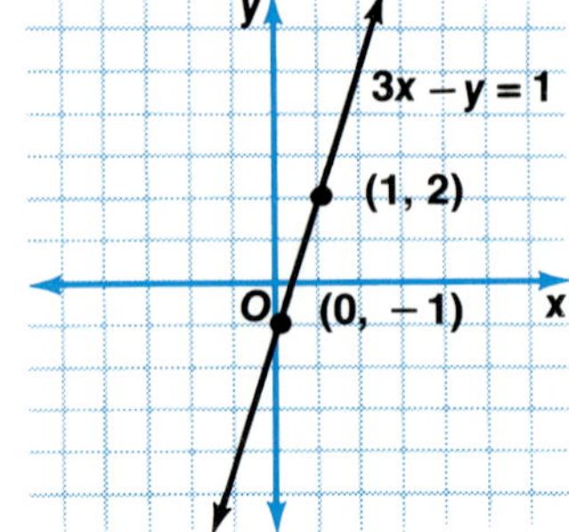

When variables other than x and y are used, assume that the letter coming first in the alphabet represents the domain or horizontal coordinate.

In a linear equation, each term is a constant, like 7, or a constant times a variable to the first power, like $3x$. Thus, $4x + 3y = 7$, $y = 8$, $5m - n = 1$, and $y = 7 + 2x$ are linear equations. But $3x + y^2 = y$ and $\frac{1}{x} + y = 4$ are *not* linear equations.   *Why?*

Any linear equation can be written in **standard form**.

> The standard form of a linear equation is
> $$Ax + By = C$$
> where **A**, **B**, and **C** are real numbers, and **A** and **B** are *not both* zero.

Standard Form of a Linear Equation

Usually A, B, and C are given as integers that have greatest common factor 1.

Example

1 **Write the equation $x = \frac{2}{3}y - 1$ in standard form.**

$$x = \frac{2}{3}y - 1$$
$$3x = 2y - 3 \qquad \text{Multiply each side by 3 to eliminate the fraction.}$$
$$3x - 2y = -3 \qquad \text{Add } -2y \text{ to each side.}$$

To graph a linear equation, it is helpful to make a table of ordered pairs that satisfy the equation. These ordered pairs can then be graphed and connected with a straight line. Since two points determine a line, you need only two points to graph a linear equation in two variables. In checking your work, it is helpful to use a third point.

2 **Graph $3y = -2x - 6$.**

First, solve the equation for y.

$$3y = -2x - 6$$
$$y = -\frac{2}{3}x - 2$$

Next, find three ordered pairs that satisfy the equation. Then, graph the ordered pairs and connect the points with a line.

x	$-\frac{2}{3}x - 2$	y	(x, y)
-3	$-\frac{2}{3}(-3) - 2$	0	$(-3, 0)$
0	$-\frac{2}{3}(0) - 2$	-2	$(0, -2)$
3	$-\frac{2}{3}(3) - 2$	-4	$(3, -4)$

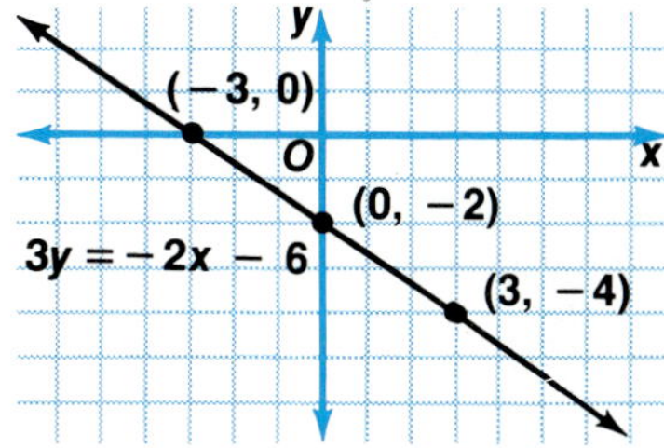

3 **Graph $2a = 3b - 4$.**

First, solve the equation for b.

$$2a = 3b - 4$$
$$-3b = -2a - 4$$
$$b = \frac{2}{3}a + \frac{4}{3}$$

Next, find three ordered pairs that satisfy the equation. Then graph the ordered pairs and connect the points with a line.

a	$\frac{2}{3}a + \frac{4}{3}$	b	(a, b)
-2	$\frac{2}{3}(-2) + \frac{4}{3}$	0	$(-2, 0)$
1	$\frac{2}{3}(1) + \frac{4}{3}$	2	$(1, 2)$
4	$\frac{2}{3}(4) + \frac{4}{3}$	4	$(4, 4)$

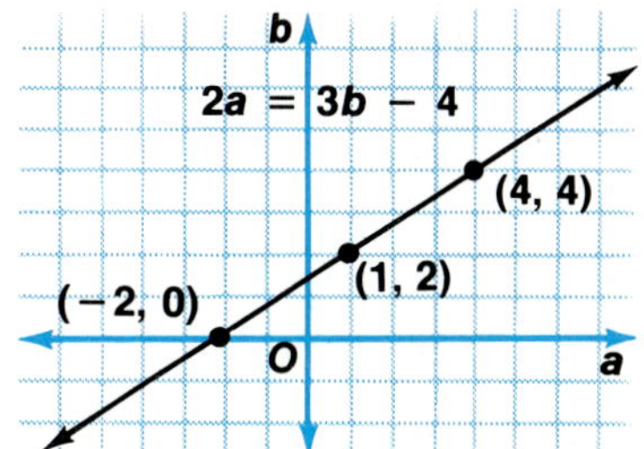

Not all linear equations represent functions. For example, consider the equation $x = 3$. Its graph is a vertical line. Any vertical line drawn on the graph of the equation passes through every point of that graph. The relation is *not* a function.

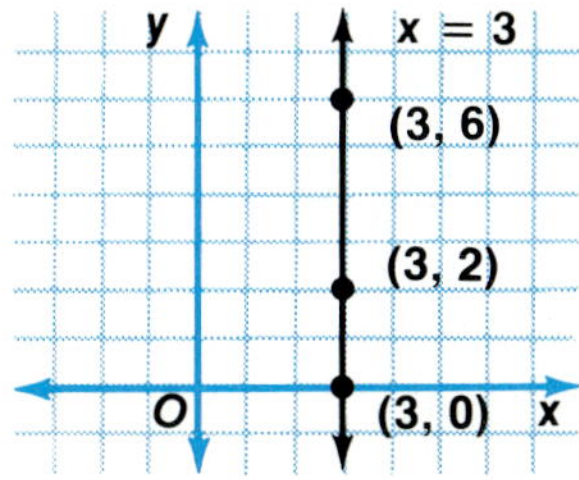

The equation $x = 3$ means that y can have any value as long as x is 3.

Any function whose ordered pairs satisfy a linear equation is called a
linear function.

> **A function is linear if it can be defined by $f(x) = mx + b$ where m and b are real numbers.**
>
> *Definition of Linear Function*

In the definition of a linear function, m or b may be zero.

If $m = 0$, then $f(x) = b$. The graph is a horizontal line. This function is called a **constant function**.

$f(x) = 0$ is called the zero function.

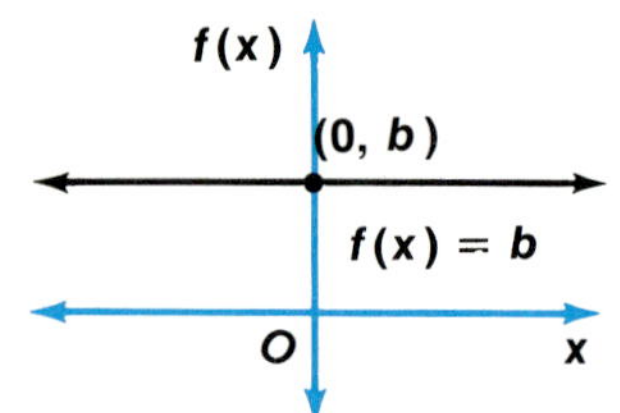

Examples

4 Which of the following are linear functions?

a. $f(x) = x^3 + 4$

This cannot be written in the form $f(x) = mx + b$. Thus, it is *not* a linear function.

b. $g(x) = 4 - x$

This can be written in the form $f(x) = mx + b$ where $m = -1$ and $b = 4$. Thus, it is a linear function.

Exploratory Exercises

State whether each equation is a linear equation.

1. $x^2 + y^2 = 7$
2. $x + y = 4$
3. $x - 2y = 5$
4. $x^2 = 9$
5. $a + 3b = 7$
6. $5m^2 = n^2$
7. $y = -4x$
8. $7 = 2y$

State whether each of the following is a linear function.

9. $f(x) = x^2 + 3$
10. $g(x) = 7$
11. $3[g(x)] = x$
12. $g(x) = x - 4$
13. $5[f(x)] = 5x^2 - 5$
14. $4[f(b)] = 2 - b$
15. $f(x) = 1.2 - 3.7x$
16. $g(x) = x(2 - x)$

Written Exercises

Write each equation in standard form.

17. $y = 2x - 6$
18. $y = -4x + 1$
19. $x = 5$
20. $y - 7 = 0$
21. $y = \frac{5}{8}x + 1$
22. $x = \frac{1}{3}y - 4$
23. $y = 3x$
24. $x = \frac{3}{5} + \frac{y}{4}$

Graph each equation.

25. $y = x$
26. $y = x + 1$
27. $y = 2x + 3$
28. $y = 5x - 4$
29. $b = 2a - 3$
30. $p = 5q + 1$
31. $x + y = 7$
32. $x - y = 4$
33. $4x + 3y = 12$
34. $2x - 5y = 10$
35. $2a + 3b = 6$
36. $5 = 5x$
37. $f(x) = 2x + 1$
38. $f(x) = 3x - 1$
39. $3s - 2t = 6$
40. $2r - 3s = 6$
41. $5x = 4$
42. $3y = 5$
43. $x + 1 = 2y$
44. $3x - 4 = 2y$
45. $\frac{1}{3}x + \frac{1}{2}y = 1$
46. $\frac{2}{3}x + \frac{1}{4}y = 2$
47. $\frac{x}{4} - \frac{y}{3} = 2$
48. $\frac{x}{3} + \frac{y}{2} = 3$

Applications in Recreation

Simulation games are games that resemble real life processes. The game called LIFE, created by John Horton Conway, is one example. It shows, in a simple way, the evolution of a society of living organisms as it ages with time.

You can play LIFE using a grid of squares and counters in two different colors, say black and red. The counters are placed on the grid, one to a square. Then you change the positions of the counters according to the following *genetic laws*, or rules for births, deaths, and survivals.

SURVIVALS Every counter with 2 or 3 neighboring counters survives for the next generation. *Neighboring counters have common sides or corners.*

DEATH Every counter with 4 or more neighbors dies from overpopulation. Every counter with only one neighbor or none dies from isolation.

BIRTHS Every empty cell with exactly 3 neighbors is a birth cell. A counter is placed on the cell for the next generation.

To help eliminate mistakes in play, the following procedure is suggested.

1. Start with a pattern of black counters.
2. Put a red counter where each birth cell will occur.
3. Put a second black counter on top of each death cell. Ignore the red counters when determining death cells.
4. Check the new pattern. Then, remove all dead counters, and replace the newborn red counters with black counters.

first generation   **second generation**

Notice that births and deaths occur simultaneously. Together, they are a move to the next generation.

When you play this game, you will find that many patterns either die, reach a stable repeating pattern, or become blinkers alternating between two patterns.

Computers are very helpful in simulations because they can display the changes, and a number of patterns can be observed in a short period of time.

Exercises

Find the next six generations for each of the following.

1. 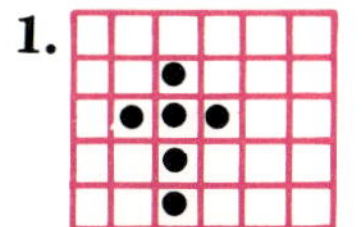 **2.** 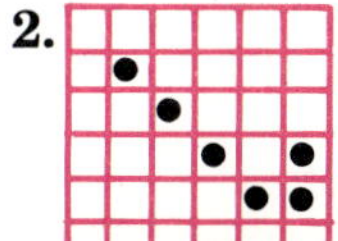 **3.** 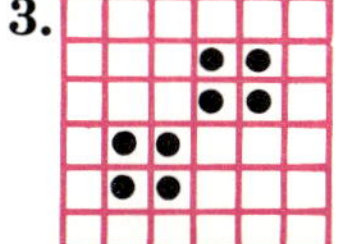 **4.** 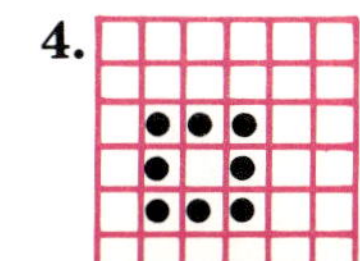 **5.** 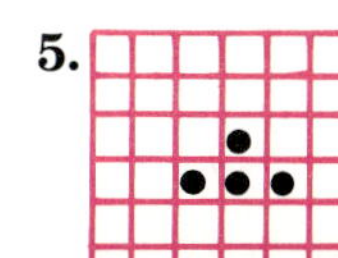

"

2-4 Slopes and Intercepts

A ramp installed to give people with a handicap access to a certain building has a base 12 meters long and an elevation of 2 meters. The steepness or **slope** of the ramp is found by using the following ratio.

$$\text{slope} = \frac{\text{change in vertical units}}{\text{change in horizontal units}}$$

Thus, the slope of the ramp is $\frac{2}{12}$ or $\frac{1}{6}$.

Slope is also defined for the graphs of linear functions. In the graph of $f(x) = 4x$ shown at the right, the y-coordinates increase 8 units for each 2 units increase in the corresponding x-coordinates. The slope of the line whose equation is $f(x) = 4x$ is $\frac{8}{2}$ or 4. The vertical change is the difference of the y-coordinates. The horizontal change is the difference of the corresponding x-coordinates.

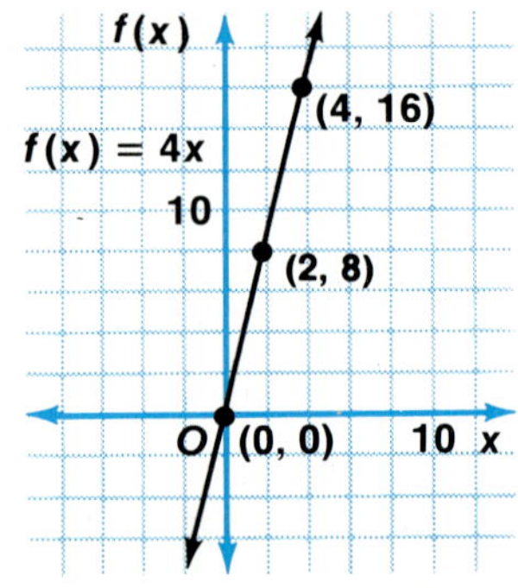

x	$4x$
0	0
2	8
4	16

$+2$ ... $+8$
$+2$ ... $+8$

> The slope, m, of a line passing through points (x_1, y_1) and (x_2, y_2) is given by
> $$m = \frac{y_2 - y_1}{x_2 - x_1}.$$
>
> *Definition of Slope*

Example

1 Determine the slope of the line that passes through $(1, -3)$ and $(0, -5)$.

$$m = \frac{y_2 - y_1}{x_2 - x_1}$$

Remember that m represents the slope of a line.

$$= \frac{-5 - (-3)}{0 - 1}$$

$$= \frac{-2}{-1}$$

$$= 2$$

The slope of the line is 2.

2 Determine the slope of each line.

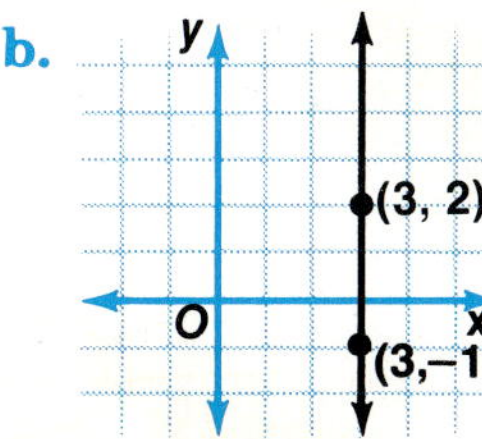

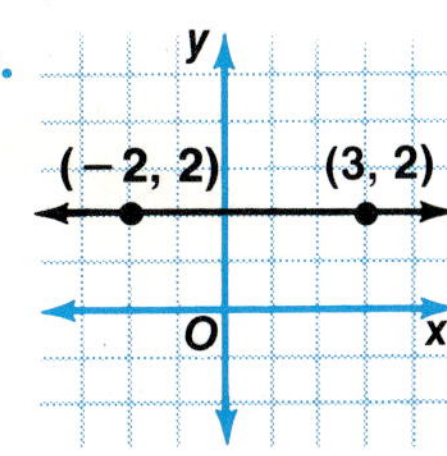

a. $m = \dfrac{4 - 0}{-6 - (-2)}$

$= \dfrac{4}{-4}$ or -1

The slope is -1.

b. $m = \dfrac{2 - (-1)}{3 - 3}$

$= \dfrac{3}{0}$

The slope is undefined.

c. $m = \dfrac{2 - 2}{-2 - 3}$

$= \dfrac{0}{-5}$ or 0

The slope is 0.

3 Graph the line that passes through $(-1, -2)$ and whose slope is $\dfrac{3}{4}$.

First graph the ordered pair $(-1, -2)$. Since the slope of the line is $\dfrac{3}{4}$, the vertical change is 3 and the horizontal change is 4. From $(-1, -2)$ move 3 units up and 4 units to the right. This point is $(3, 1)$.

Connect these two points with a straight line.

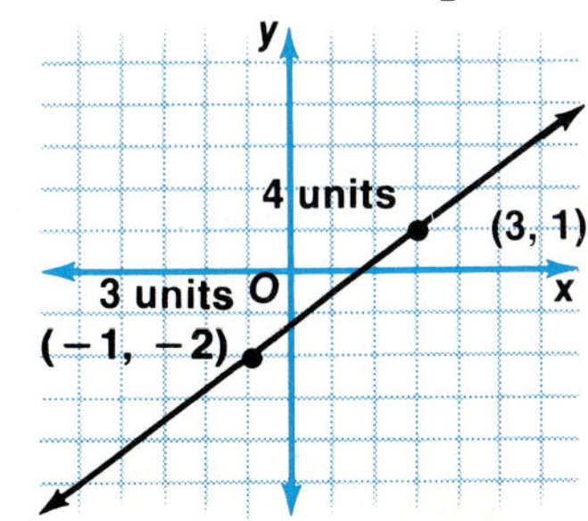

The previous examples suggest the following statements about the slope of a line.

If the line rises to the right, then the slope is positive.

If the line is horizontal, then the slope is zero.

If the line falls to the right, then the slope is negative.

If the line is vertical, then the slope is *undefined*.

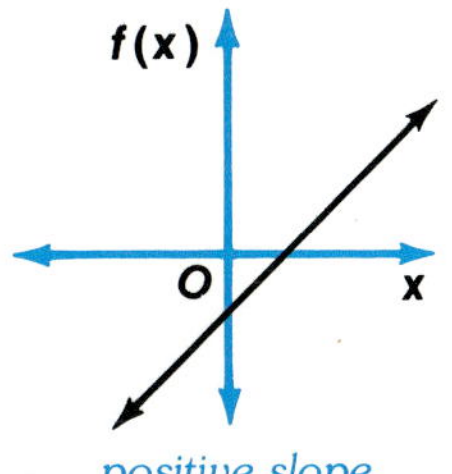
positive slope

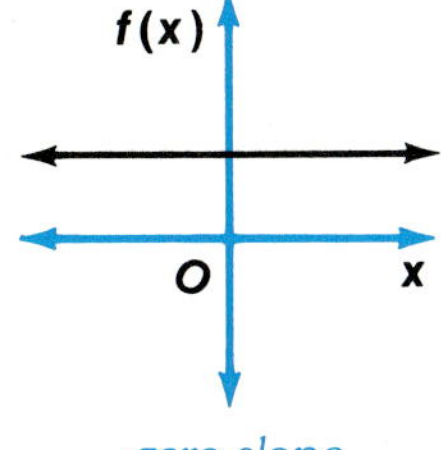
zero slope

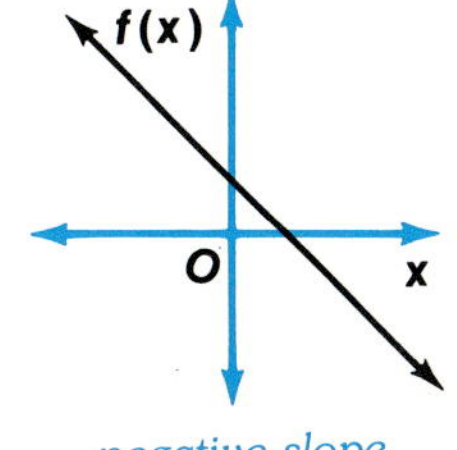
negative slope

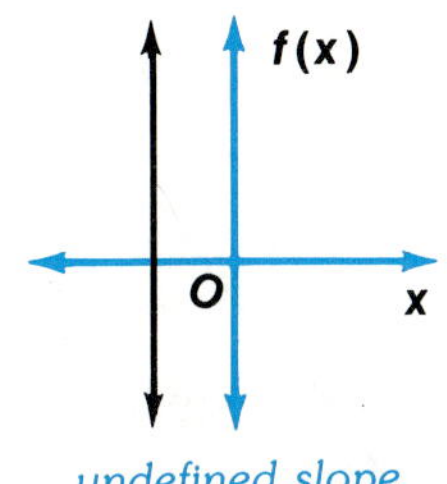
undefined slope

The graphs of $f(x) = 3x + 2$, $g(x) = 3x$, and $h(x) = 3x - 5$ are lines with the same slope. But, these lines do *not* pass through the same points.

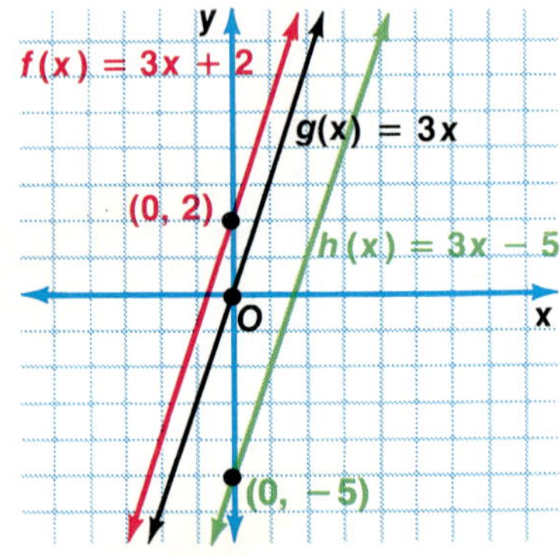

The slope of each line is 3. Note that 3 is also the coefficient of x in each equation. In general, the slope of a line is the coefficient of x when its equation is solved for y.

Consider the points where each line crosses the y-axis.

$$f(x) = 3x + 2 \quad \text{crosses at } (0, 2).$$
$$g(x) = 3x \qquad\quad \text{crosses at } (0, 0).$$
$$h(x) = 3x - 5 \quad \text{crosses at } (0, -5).$$

The x-coordinate of each point is 0.

The y-coordinates of these points are called the **y-intercepts** of the lines. The y-intercept is the value of y when x is zero.

$$f(x) = 3x + 2 \text{ has } y\text{-intercept } 2.$$
$$g(x) = 3x \text{ or } g(x) = 3x + 0 \text{ has } y\text{-intercept } 0.$$
$$h(x) = 3x - 5 \text{ or } h(x) = 3x + (-5) \text{ has } y\text{-intercept } -5.$$

Note that the y-intercept is the constant term of each equation when solved for y.

The **x-intercept** of a line is the value of x when y is zero. What are the x-intercepts of the lines described above?

$$f(x) = 3x + 2 \text{ has } x\text{-intercept } -\frac{2}{3}.$$

$$g(x) = 3x \text{ has } x\text{-intercept } 0.$$

$$h(x) = 3x - 5 \text{ has } x\text{-intercept } \frac{5}{3}.$$

Example

4 | **Find the y-intercept and x-intercept of the line whose equation is $5x + 3y = 9$.**

To find the y-intercept, let $x = 0$ and solve for y.

$$5(0) + 3y = 9$$
$$3y = 9$$
$$y = 3 \qquad \text{The } y\text{-intercept is 3.}$$

To find the x-intercept, let $y = 0$ and solve for x.

$$5x + 3(0) = 9$$
$$5x = 9$$
$$x = \frac{9}{5} \qquad \text{The } x\text{-intercept is } \frac{9}{5}.$$

Exploratory Exercises

State the slope, *y*-intercept, and *x*-intercept of each line.

1. 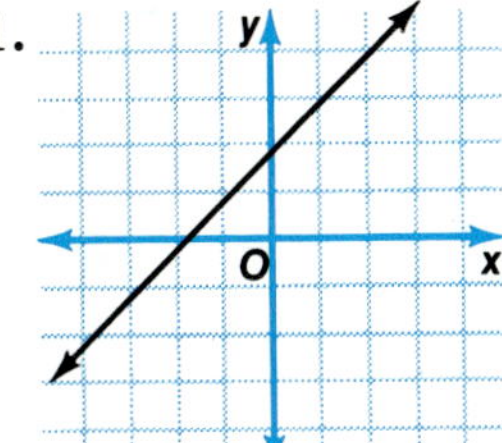

2. 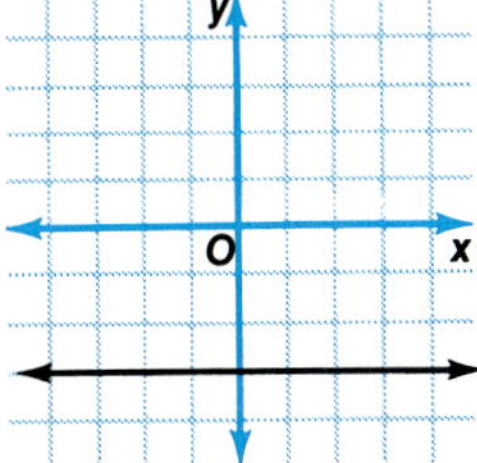

3. 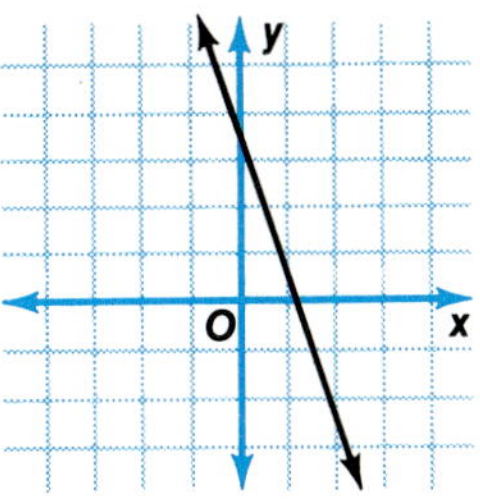

4. 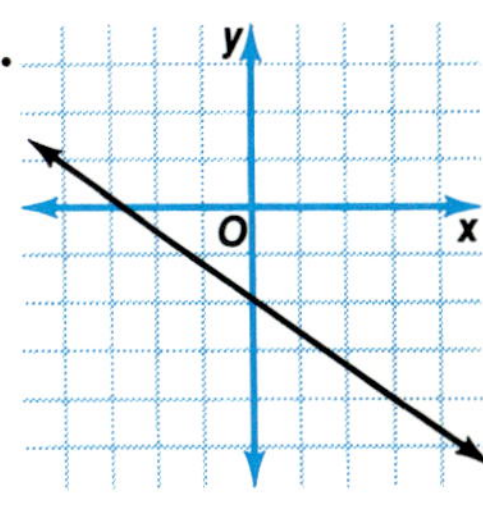

5. 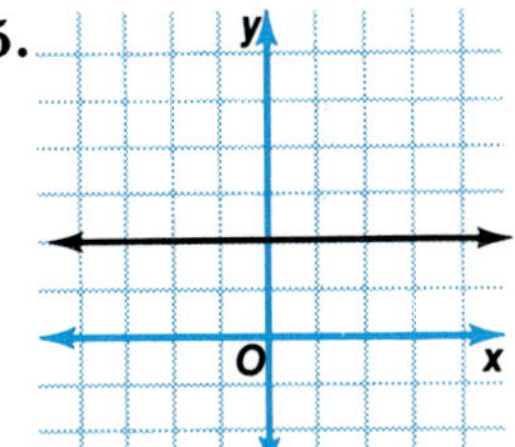

6. 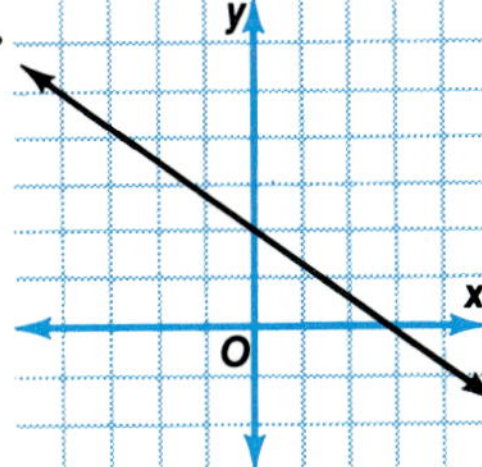

7. 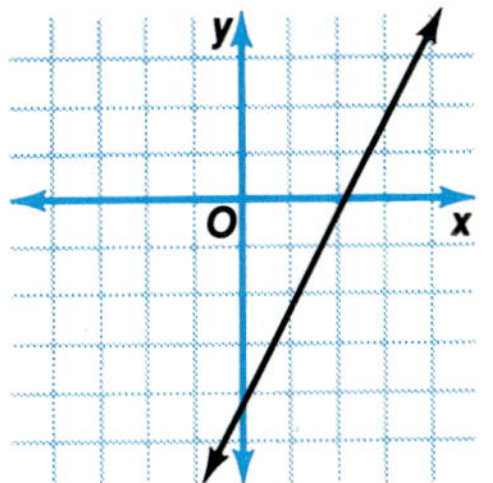

8. 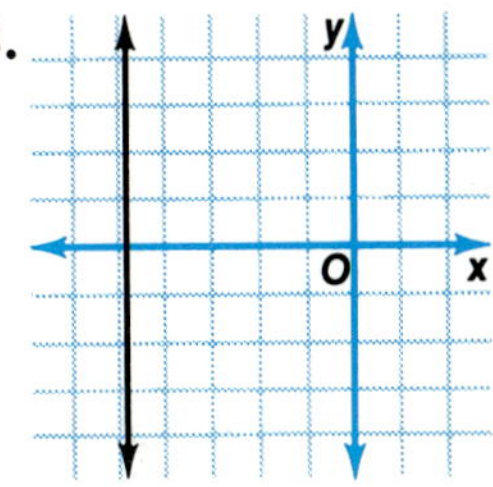

Written Exercises

Determine the slope of the line passing through each pair of points.

9. $(6, 1)$ and $(8, -4)$

10. $(5, 7)$ and $(4, -6)$

11. $(-3, 0)$ and $(8, 2)$

12. $(-6, -3)$ and $(4, 1)$

13. $(-8, -2)$ and $(-4, 8)$

14. $(6, 1)$ and $(6, 7)$

15. $(2.5, 3)$ and $(1, -9)$

16. $(0, 0)$ and $\left(\frac{3}{2}, \frac{1}{4}\right)$

17. $\left(1\frac{3}{4}, \frac{1}{3}\right)$ and $\left(2, \frac{1}{3}\right)$

18. $(1.8, 6)$ and $(-1, 3.2)$

19. $(a, 2)$ and $(a, -2)$

20. $(5, k)$ and $(k + 1, k)$

Find the *y*-intercept and *x*-intercept of each line whose equation is given below.

21. $y = 5x - 9$

22. $y = -3x - 5$

23. $y - 1 = 7x$

24. $y + 6 = 5x$

25. $3y = -2x - 15$

26. $3y = x + 4$

27. $y = -2$

28. $x = 4$

29. $f(x) = x - 2$

30. $g(x) = 4x - 1$

31. $x + 2y = 5$

32. $2x + 3y = 6$

33. $3x - 2y = 12$

34. $5x + 3y = 30$

35. $x + 2y = 7$

36. $3x + y = 6$

37. $2x - 3y = 12$

38. $5x - 2y = 20$

Graph the line that passes through the given point and has the given slope.

39. $(-3, 2)$, $m = -2$

40. $(0, 0)$, $m = 3$

41. $(3, 4)$, $m = -1$

42. $(-1, 1)$, $m = \frac{1}{4}$

43. $(2, -1)$, $m = 0$

44. $(-4, 1)$, $m = -\frac{5}{3}$

45. $(2, -3)$, $m = \frac{3}{5}$

46. $(-3, -1)$, undefined

47. $(2, 5)$, $m = -\frac{2}{3}$

48. $\left(4, 3\frac{1}{2}\right)$, $m = 0$

49. $(5, -2)$, undefined

50. $\left(\frac{1}{4}, 1\right)$, $m = 4$

Find the missing coordinate if the line passing through the two points has the given slope.

51. $(0, 0)$ and $(x, 7)$, $m = 7$

52. $(6, y)$ and $(2, -13)$, $m = 3$

53. $(-2, -7)$ and $(0, y)$, $m = -\frac{1}{2}$

54. $(x, -1)$ and $(3, 6)$, $m = 1$

55. $(5, 3)$ and $(3, y)$, $m = -3$

56. $\left(-17\frac{1}{2}, -35\right)$ and $(x, 0)$, $m = 2$

57. $\left(-10\frac{2}{3}, y\right)$ and $\left(-8\frac{1}{3}, 16\right)$, $m = -3$

58. $\left(4\frac{1}{2}, -3\right)$ and $\left(x, \frac{1}{2}\right)$, $m = -\frac{5}{7}$

mini-review

1. Evaluate $5x^2(11 - 2^3)$ if $x = 9$.

2. Is $\sqrt{16}$ a rational number?

3. Simplify $\frac{3}{4}x + \frac{1}{3}x$.

4. Solve $|3x - 1| \le 2$.

5. James took four science tests. His scores were 76, 90, 82, and 95. What must he score on a fifth test so that his average for all the tests will be at least 85?

State whether each equation is a linear equation.

6. $y = x + 2$

7. $y = x^3 - 2$

8. $4y = 2$

State whether each function is a linear function.

9. $g(x) = 5$

10. $r(x) = 232 - x$

11. $f(x) = 3 - x^2$

Write each equation in standard form.

12. $y = 2x + 3$

13. $y = -\frac{4}{5}x - \frac{2}{3}$

14. $4x = \frac{1}{2} - 3y$

Determine the slope of the line passing through each pair of points.

15. $(4, 7)$ and $(3, -5)$

16. $(8, -3)$ and $(13, -6)$

17. $\left(\frac{1}{2}, 7\right)$ and $\left(\frac{3}{4}, -5\right)$

18. $(-4, -5)$ and $(-6, -7)$

Excursions in Algebra — Age of Diophantus

The solution to this riddle is the age of the ancient Greek mathematician, Diophantus.

His youth lasted one-sixth of his life. He grew a beard after one-twelfth more. He married after one-seventh more. He had a son 5 years later. His son lived half as long as his father. Diophantus died 4 years after his son died.

2-5 Finding Linear Equations

In the figure at the right, $\overleftrightarrow{AB}$ passes through points $A(0, b)$ and $B(x, y)$. Notice that b is the y-intercept of $\overleftrightarrow{AB}$.

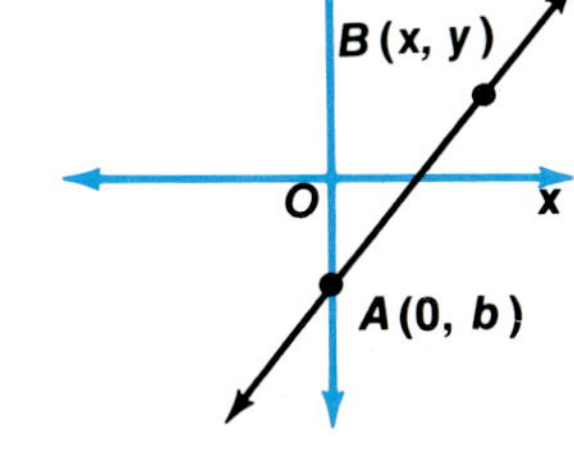

Suppose you want to find an equation for $\overleftrightarrow{AB}$. Let m represent the slope of the line.

$$m = \frac{y - b}{x - 0}$$ Use the definition of slope.
$\overleftrightarrow{AB}$ passes through (x, y)
$$m = \frac{y - b}{x}$$ and $(0, b)$, so that $x_2 = x$,
$x_1 = 0$, $y_2 = y$, and $y_1 = b$.
$$mx = y - b$$ Solve for y.
$$y = mx + b$$

This equation of a line, $y = mx + b$, is called the **slope-intercept form.**

> The slope-intercept form of the equation of a line is $y = mx + b$, where m is the slope and b is the y-intercept.

Slope-Intercept Form of a Linear Equation

Suppose you are given the slope and y-intercept of a line. You can find an equation of the line by substituting these values into the slope-intercept form of the equation. For example, if the slope of a line is $-\frac{2}{3}$ and its y-intercept is 14, an equation of the line is $y = -\frac{2}{3}x + 14$. The standard form of this equation is $2x + 3y = 42$.

The form $y = mx + b$ is often written as $f(x) = mx + b$.

Example

1 **Find the slope-intercept form and the standard form of the equation of the line that has a slope of $\frac{2}{3}$ and passes through (6, 7).**

First, substitute the slope and coordinates of the point into the slope-intercept form and solve for b.

$$y = mx + b$$
$$7 = \left(\frac{2}{3}\right)(6) + b$$ Substitute 7 for y, $\frac{2}{3}$ for m, and 6 for x.
$$7 = 4 + b$$
$$3 = b$$

Write the equation in slope-intercept form.

$$y = \frac{2}{3}x + 3$$ Substitute $\frac{2}{3}$ for m and 3 for b.

The standard form of this equation is $2x - 3y = -9$.

2 Find the slope-intercept form of the equation of a line passing through $(-2, 5)$ and $(3, 0)$.

First use the two given points to find the slope of the line.

$$m = \frac{5 - 0}{-2 - 3}$$ *Remember that m represents the slope of the line.*

$$= -1$$

Next, substitute the slope and the coordinates of one point in the slope-intercept form and solve the equation for b.

$$y = mx + b$$
$$5 = (-1)(-2) + b$$ *Substitute 5 for y, -1 for m, and -2 for x. Using the other*
$$3 = b$$ *point would give the same results.*

The slope-intercept form of the equation of the line is $y = -x + 3$.

3 Find the slope and y-intercept of a line with equation in standard form.

Recall that the standard form of the equation of a line is $Ax + By = C$. If $B \neq 0$, the slope and y-intercept can be found by changing the equation to slope-intercept form.

$$Ax + By = C$$
$$By = -Ax + C$$
$$y = -\frac{A}{B}x + \frac{C}{B}$$ The slope is $-\frac{A}{B}$ and the y-intercept is $\frac{C}{B}$, for $B \neq 0$.

Consider the line with equation $y = -2x + 4$. The standard form of this equation is $2x + y = 4$. The following calculations use the information from Example 3 to find the slope and y-intercept.

$$\text{slope} = -\frac{A}{B} \qquad\qquad y\text{-intercept} = \frac{C}{B}$$

$$= -\frac{2}{1} \text{ or } -2 \qquad\qquad = \frac{4}{1} \text{ or } 4$$

If $B = 0$ in the standard form of the equation of a line, the slope is undefined, the line has no y-intercept, and $C \neq 0$.

Notice that these values are the same as the values you would obtain from the slope-intercept form of the equation.

Exploratory Exercises

Find the slope-intercept form of the equation of the line for each slope, m, and each y-intercept, b.

1. $m = 5, b = -3$ **2.** $m = -1, b = 3$ **3.** $m = -1, b = 4$

4. $m = 8, b = 1$ **5.** $m = \frac{2}{3}, b = -7$ **6.** $m = \frac{1}{4}, b = 6$

7. $m = 2.5, b = 0$ **8.** $m = -4.1, b = -9$ **9.** $m = 0, b = 0$

State the slope and _y_-intercept of the graph of each equation.

10. $y = -2x + 5$
11. $y = -4x + 6$
12. $y = x - 8$
13. $y = -\frac{3}{4}x - 3$
14. $y = \frac{1}{3}x$
15. $\frac{1}{2}y = 7x - 1$
16. $-y = 0.2x + 6$
17. $-y = x$
18. $y = cx + t$

Find the slope-intercept form of each equation.

19. $2x + 5y = 10$
20. $3x - y = 6$
21. $2x + 3y = 4$
22. $4x + 8y = 11$
23. $2x - 2y = 4$
24. $2x = 11$

Written Exercises

State the slope and _y_-intercept of the graph of each equation.

25. $-2x + y = 15$
26. $0.4x - 0.4y = 0.4$
27. $5x - 3y = 0.6$
28. $3x - 4y = 1$
29. $8x = 1$
30. $x - \frac{1}{2}y = 2$
31. $2x - \frac{1}{3}y = -5$
32. $\frac{1}{5}x - \frac{1}{3}y = \frac{1}{7}$

Find the slope-intercept form of the equation of the line that satisfies the given conditions.

33. slope $= \frac{1}{2}$, passes through $(6, 4)$
34. slope $= \frac{3}{4}$, passes through $(8, -1)$

35. slope $= \frac{2}{3}$, passes through $(6, -2)$
36. slope $= -\frac{4}{5}$, passes through $(2, -3)$

37. slope $= 4$, passes through $(2, -3)$
38. slope $= 5$, passes through the origin
39. passes through $(6, 1)$ and $(8, -4)$
40. passes through $(-6, -3)$ and $(4, 1)$
41. passes through $(-3, 0)$ and $(8, 2)$
42. passes through $(6, 1)$ and $(6, 7)$
43. passes through $(-8, -2)$ and $(-4, 8)$
44. passes through $(4, 6)$ and $(0, 0)$
45. x-intercept $= -3$, y-intercept $= 6$
46. x-intercept $= 6$, y-intercept $= 4$
47. x-intercept $= \frac{3}{5}$, y-intercept $= 6$
48. x-intercept $= \frac{1}{3}$, y-intercept $= -\frac{1}{4}$
49. x-intercept $= 0$, y-intercept $= 2$
50. x-intercept $= -7$, y-intercept $= 0$

Find the standard form of the equation of the line that satisfies the given conditions.

51. slope $= 1$, passes through $(2, 3)$
52. slope $= 2$, passes through $(4, 6)$
53. slope $= -2$, passes through $(-1, 4)$
54. slope $= -3$, passes through $(2, 5)$
55. slope $= \frac{4}{3}$, passes through $(4, 3)$
56. slope $= -\frac{2}{3}$, passes through $(-3, 5)$

57. passes through $(2, 3)$ and $(1, 5)$
58. passes through $(2, 5)$ and $(3, 6)$
59. passes through $(1, -2)$ and $(3, 7)$
60. passes through $(1, 0)$ and $(2, -5)$
61. passes through $(-1, 5)$ and $(2, 3)$
62. passes through $(-1, -1)$ and $(-1, 8)$
63. x-intercept $= 4$, y-intercept $= 3$
64. x-intercept $= 9$, y-intercept $= 1$
65. x-intercept $= -6$, y-intercept $= 5$
66. x-intercept $= -1$, y-intercept $= 1$
67. x-intercept $= 0$, y-intercept $= -5$
68. x-intercept $= 0$, y-intercept $= 7$

Challenge Exercises

Find _k_ in each equation if the given ordered pair is a solution.

69. $5x + ky = 8$, $(3, -1)$
70. $4x - ky = 7$, $(4, 3)$
71. $2x - 3y = k$, $(-1, -4)$
72. $3x + 8y = k$, $(0, -0.5)$
73. $kx + 3y = 11$, $(7, 2)$
74. $kx - 5y = 21$, $(3, -2)$

Evaluate if $a = 4$, $b = -7$, $c = \frac{1}{3}$, and $d = -3$.

1. $\dfrac{a}{c^2}$

2. $\dfrac{ab + 4c}{d}$

3. $2a(4c + b^2) + |d|$

4. $b^3 - d^3$

Write *true* or *false*. If false, give a counter-example.

5. Every rational number is an integer.
6. Every irrational number is a real number.

Simplify.

7. $(9 - 29)7 - 12$

8. $\dfrac{2}{3}(15 - 9a) - (6a - 21)\left(\dfrac{1}{3}\right)$

9. $\dfrac{3}{4}\left(\dfrac{4}{3} - 12b\right) + \dfrac{2}{3}(6b - 30) - \dfrac{3}{8}(24 - 40b)$

Solve each equation or inequality.

10. $4 - 7a = 25$
11. $2(6 - 7x) = 2x - 4$
12. $|2r - 9| = 0$
13. $5x + 8 > 24$
14. $1 - 2y \le 5y - 2$
15. $|x - 4| \le -8$
16. $|7 + 3x| = 11 - x$
17. $y(y - 3) + 5 = y(y + 4) - 12y$

Graph each equation or inequality on the number line.

18. $x + 5 = 5(x + 5)$
19. $5(n - 3) \le 6(n - 3)$
20. $3(a - 3) - a = a + 1$
21. $5(y - 1) - (2y + 1) < 5(y + 2)$

Graph each set of ordered pairs.

22. $\{(2, 26), (1.5, 30), (1, 20), (0, 16)\}$
23. $\{(-2.2, 0), (-2.6, 1.8), (2, 3.2)\}$

Given $f(x) = 2x^3 + 4x^2 - x + 2$, find each value.

24. $f(3)$
25. $f(-4)$
26. $f\left(\dfrac{1}{2}\right)$
27. $f(0.2)$
28. $f(t)$
29. $f(2x)$

Graph each equation.

30. $y - 3x = 2$
31. $y = -5$
32. $y = \dfrac{1}{2}x$
33. $y = \dfrac{2}{3}x + 4$
34. $x = 4$
35. $2y + 3x = 12$

Find the x-intercept and y-intercept of the graph of each equation.

36. $3x + 5y = 10$
37. $\dfrac{1}{2}x + 4y = -4$
38. $y = 6$
39. $x = -12$

Find the slope-intercept form of the equation of the line that satisfies the given conditions.

40. x-intercept $= 3$, y-intercept $= 6$
41. x-intercept $= -4$, y-intercept $= -6$
42. slope $= -2$, passes through $(-3, 0)$
43. slope $= \dfrac{4}{5}$, passes through $(6, -2)$
44. passes through $(7, 3)$ and $(-3, 3)$
45. passes through $(6, 1)$ and $(-4, 8)$

Solve each problem.

46. Sam is driving a distance of 662 miles. If he drives 55 miles per hour for 5 hours, at what speed must he travel to complete the total trip in 14 hours?
47. The length of a garden is 3 yards more than twice the width. The perimeter is at most 84 yards. What is the maximum length of the garden?

2-6 Special Functions

Recall that a linear function can be written in the form $y = mx + b$ or $f(x) = mx + b$ where m and b are real numbers. Some linear functions have special names.

If $m = 0$, the function is called a **constant function**. Its graph is a horizontal line.

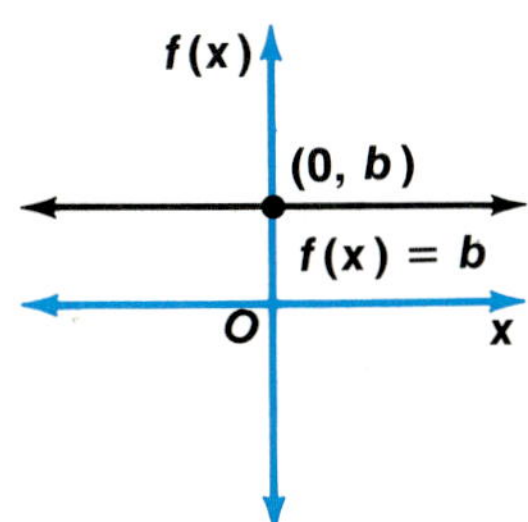

If $b = 0$ and $m = 1$, the function is called the **identity function**. Its graph passes through the origin and forms congruent angles with the axes.

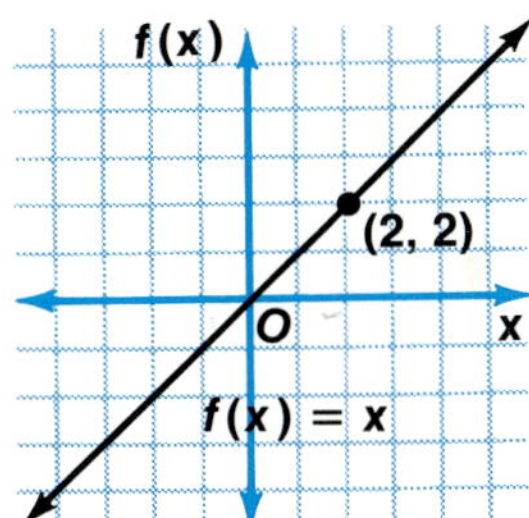

If $b = 0$ and $m \neq 0$, the function is called a **direct variation**. Its graph passes through the origin.

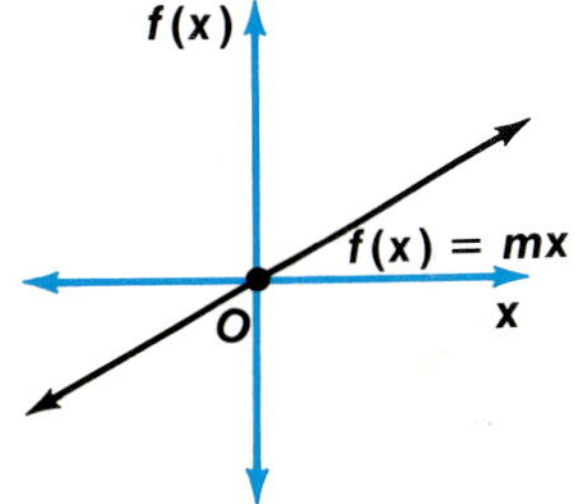

A linear function described by $y = b$ or $f(x) = b$ is called a constant function.

Definition of Constant Function

A linear function described by $y = x$ or $f(x) = x$ is called the identity function.

Definition of Identity Function

A linear function described by $y = mx$ or $f(x) = mx$ where $m \neq 0$ is called a direct variation. The constant m is called the constant of variation or constant of proportionality.

Definition of Direct Variation

Note that the identity function is a special case of direct variation.

A special language is often used with direct variation. The equation $y = mx$ means y varies directly with x, or y is *directly proportional* to x.

Several other functions are closely related to linear functions. Absolute value functions are one example.

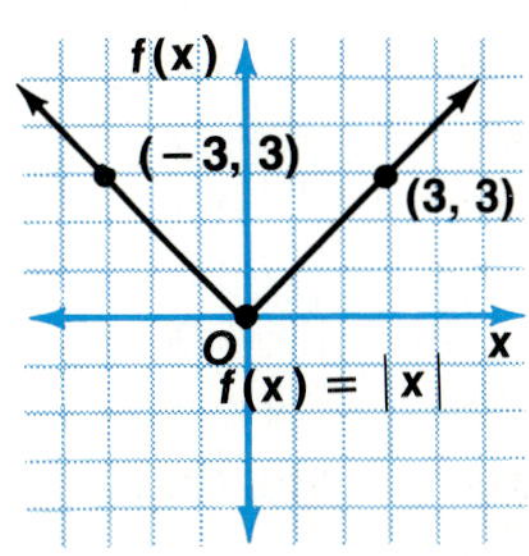

Consider $f(x) = |x|$ or $y = |x|$. When x is positive or zero, the function is like $y = x$. When x is negative, the function is like $y = -x$. The graph of $f(x) = |x|$ is shown on the right.

All of the following graphs represent absolute value functions.

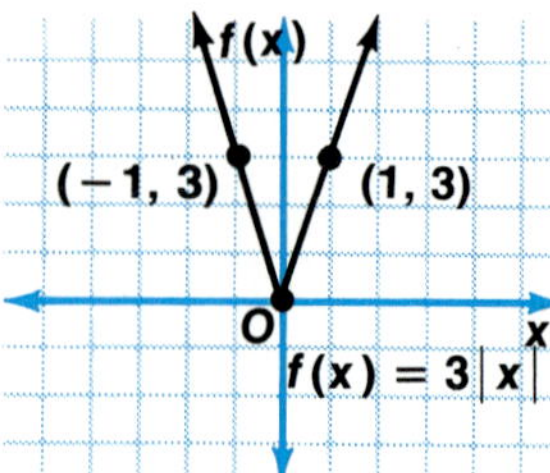

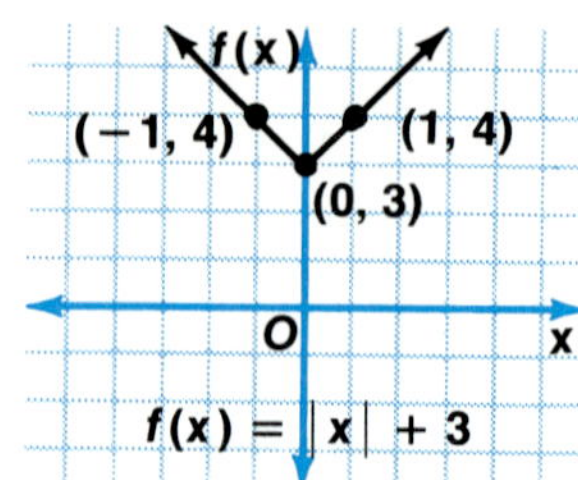

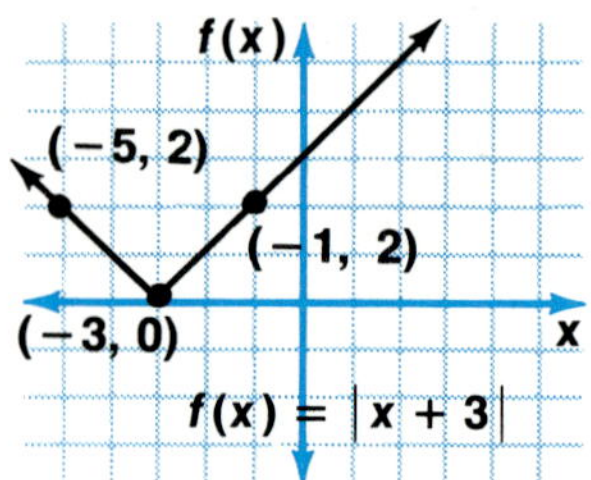

The domain is {x|x is real}.
The range is {y|y ≥ 0}.

The domain is {x|x is real}.
The range is {y|y ≥ 3}.

The domain is {x|x is real}.
The range is {y|y ≥ 0}.

Step functions like the ones graphed below are also related to linear functions.   *Remember that the open circle means that the point is not included in that part of the graph.*

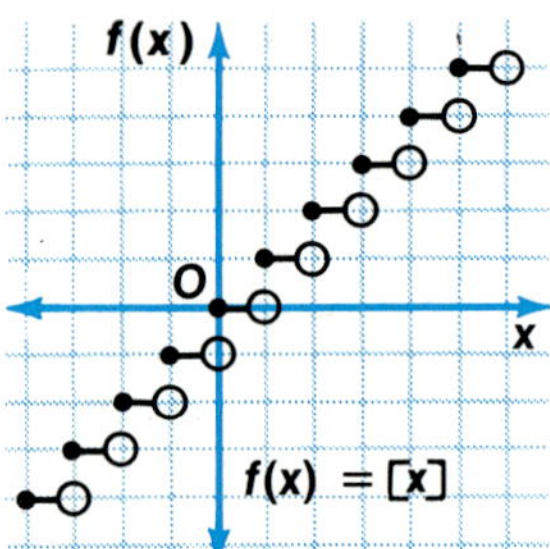

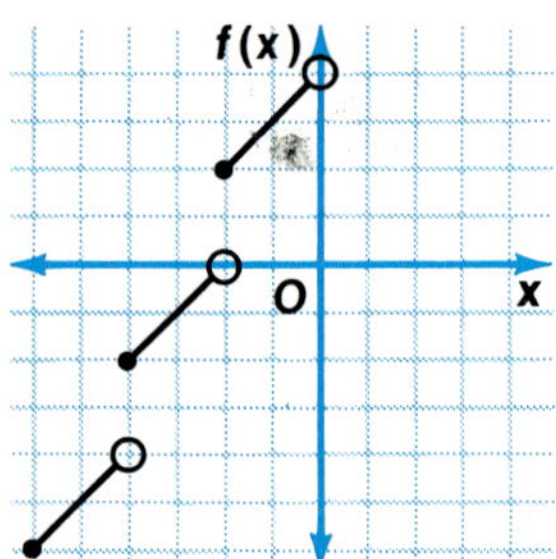

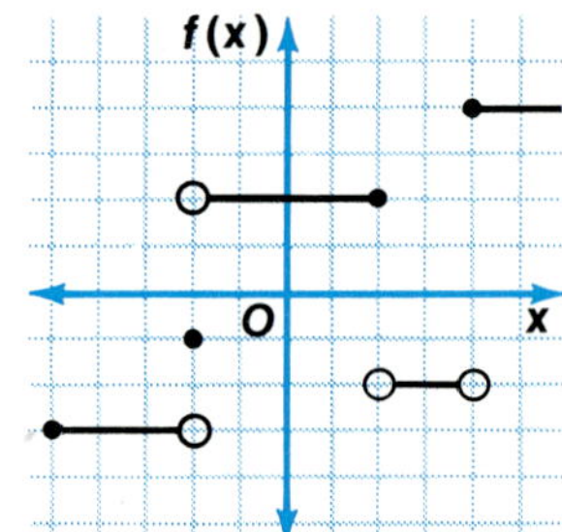

One type of step function is the **greatest integer function**. The *greatest integer of x* is written [x] and means the greatest integer *not* greater than x. For example, [6.2] is 6 and [−1.8] is −2. The greatest integer function is given by $f(x) = [x]$. Its graph is shown above on the left.

Example

1 **Graph $f(x) = [x] + 1$.**

x	[x]	[x] + 1
0	0	1
0.2	0	1
0.6	0	1
1.0	1	2
1.5	1	2
2.0	2	3
−1.0	−1	0
−1.5	−2	−1
−2.0	−2	−1

The domain is all real numbers.

The range is all integers.

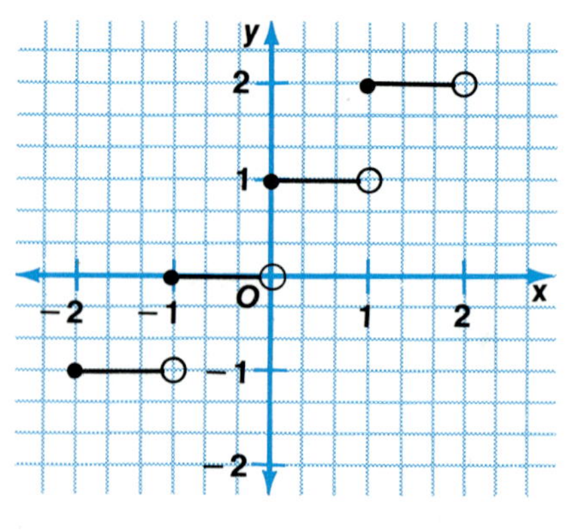

Exploratory Exercises

Identify each function as C for *constant*, as I for *identity*, as D for *direct variation*, as A for *absolute value*, or as G for *greatest integer function*.

1. $f(x) = |3x - 2|$

2. $f(x) = [-x]$

3. $g(x) = 3x$

4. $y = |2x|$

5. $g(x) = [2x]$

6. $g(x) = 19$

7. $f(x) = \left| x - \dfrac{2}{3} \right|$

8. $m(x) = \dfrac{2}{3}$

9. $f(x) = \dfrac{1}{2}x$

10. $g(x) = [2x + 1]$

11. $h(x) = -7$

12. $p(x) = x$

If $h(x) = [2x + 1]$, find each value.

13. $h(2)$

14. $h(-3)$

15. $h(1.4)$

16. $h(-3.6)$

17. $h\left(\dfrac{2}{3}\right)$

18. $h\left(-\dfrac{9}{7}\right)$

Written Exercises

Identify each function as C for *constant*, as I for *identity*, as D for *direct variation*, as A for *absolute value*, or as G for *greatest integer function*.

19. $g(x) = -x$

20. $k(x) = |x - 8|$

21. $f(x) = 0$

22. $h(x) = [x - 3]$

23. $f(x) = [3x - 5]$

24. $m(x) = |-3x|$

25. $f(x) = \left[x - \dfrac{1}{2} \right]$

26. $h(x) = -\dfrac{11}{4}x$

27. $f(x) = |2x + 1|$

Graph each function.

28. $f(x) = x$

29. $g(x) = 3x$

30. $h(x) = -2x$

31. $f(x) = \dfrac{1}{2}x$

32. $p(x) = |x|$

33. $r(x) = |2x|$

34. $g(x) = |-3x|$

35. $f(x) = |x + 3|$

36. $f(x) = \left| x - \dfrac{2}{3} \right|$

37. $f(x) = |2x + 1|$

38. $g(x) = [-x]$

39. $h(x) = -[x]$

40. $p(x) = [2x]$

41. $f(x) = 2[x]$

42. $r(x) = [x] - 3$

43. $p(x) = 3 - [x]$

44. $h(x) = |3x| - 2$

45. $g(x) = -|x| + 3$

46. $r(x) = 2|x| - 3$

47. $f(x) = |x - 3| + 2$

48. $f(x) = \left[x - \dfrac{1}{2} \right]$

49. $f(x) = -[x - 2]$

50. $m(x) = [3x - 5]$

51. $g(x) = [2x + 7]$

Explain how the graphs of each pair of equations differ.

52. $y = 2[x]$ and $y = [2x]$

53. $y = [x + 5]$ and $y = [x] + 5$

54. $y = |x - 3|$ and $y = |x| - 3$

55. $y = |3x|$ and $y = 3|x|$

56. $y = |2x + 5|$ and $y = |2x| + 5$

57. $y = [2x + 5]$ and $y = [2x] + 5$

58. $y = |ax|$ and $y = a|x|$

59. $y = |x + b|$ and $y = |x| + b$

60. $y = -3[x]$ and $y = [-3x]$

61. $y = -2|x|$ and $y = |-2x|$

Challenge Exercises

Graph each function.

62. $y = |[x]|$

63. $y = [|x|]$

64. $y = x - [x]$

65. $y = x + |x|$

Solve each equation or inequality.

1. $0.5y - 7.5 = -7.1$

2. $3(2x - 3) \geq 4x - 13$

3. $|2x - 3| = 11$

4. $|5x - 13| < 37$

5. Mr. Wiekart bought a dinette set for \$53 more than 50% of its original price. If he paid \$382, what was the original price of the dinette set?

If $f(x) = x^2 - 3x$, find each value.

6. $f(2)$

7. $f(-1)$

8. $f(0.3)$

9. $f\left(-\frac{1}{2}\right)$

10. Is $\{(2, 3), (3, -2), (4, 2), (3, 4)\}$ a function?

11. Find the slope of the line that passes through $(3, 5)$ and $(-4, 8)$.

Find the x-intercept and y-intercept of each line whose equation is given.

12. $x + 2y = 8$

13. $y = 7$

14. $3x = 5y + 9$

State the slope and y-intercept of the graph of each equation.

15. $x + 2y = 12$

16. $3y + 4x = 9$

17. $y = 3$

Find the slope-intercept form of the equation of the line that satisfies the given conditions.

18. slope $= \frac{1}{2}$, passes through $(2, 8)$

19. slope $= 4$, passes through $(-2, -5)$

20. passes through $(-3, 2)$ and $(5, 1)$

21. passes through $(3, 9)$ and $(-2, -1)$

Excursions in Algebra

History

René Descartes (1596–1650) was a French mathematician and philosopher. He is credited with the invention of a branch of mathematics called analytic geometry. Analytic geometry combines the following from algebra and geometry.

1. The coordinate plane
2. The correspondence of ordered pairs of numbers to points in the coordinate plane
3. Graphs of expressions like $f(x) = 2x + 1$

Descartes, in 1637, became the first mathematician to put the three ideas together.

Sometimes, ordered pairs are referred to as Cartesian coordinates. The word Cartesian is taken from the name Descartes.

An important strategy for solving problems is to look for a pattern. Sometimes a pattern may be obvious, but many times it is not. You may have to "play" with the facts and figures until you arrive at a pattern. Then you can solve the problem by generalizing and applying the pattern.

Example: **Write an equation showing the relationship between the variables in the chart at the right.**

x	1	2	3	4	5
y	17	21	25	29	33

Look for a pattern by finding differences between successive values of x and successive values of y.

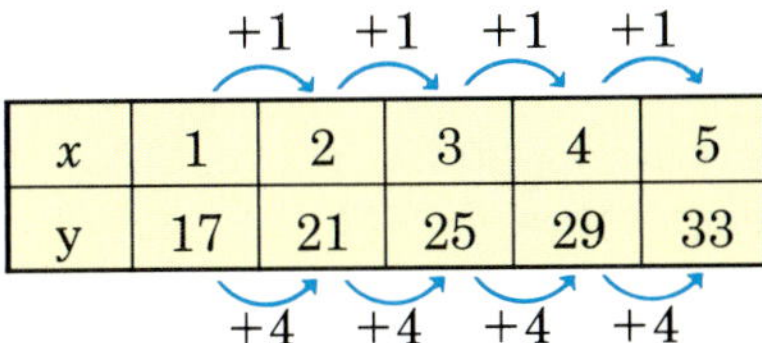

Notice that the changes in x are 1, while the changes in y are 4. This suggests that $y = 4x$ since the changes in y are 4 times the changes in x. Try $x = 1$ in $y = 4x$.

$$y = 4 \cdot 1 \text{ or } 4$$

But when x is 1, y is 17, not 4. To make y equal to 17 when x is equal to 1, you need to add 13. The equation becomes $y = 4x + 13$. This equation checks for each value of x.

Exercises

Write an equation showing the relationship between the variables in each chart. Then complete the chart.

1.

x	1	2	3			6
y	8	16		32	40	

2.

a	1	2	3	4	5	6
b	5	7	9			

3. The Worthington City Council has 12 members. After their last session, each member shook hands with each other member. How many handshakes were there in all?

4. At Dublin High School, there are 500 students and 500 lockers, numbered 1 through 500. Suppose the first student opens each locker. Then the second student closes every second locker. The third student changes the state of every third locker (the student closes the open lockers and opens the closed lockers). The fourth student changes the state of every fourth locker. This process continues until the five-hundredth student changes the state of the five-hundredth locker. After this process is completed, state which lockers are open.

2-7 Problem Solving: Linear Functions

Geothermal energy is generated wherever water comes into contact with heated underground rocks. This heat turns the water into steam which can be used to make electricity.

The underground temperature of rocks varies with their depth below the surface. The temperature at the surface is about 20°C. At a depth of 2 km, the temperature is about 90°C. This information can be used to find an equation for the relationship, assuming the relationship is linear.

explore

Read the problem carefully.
Let d represent the depth given in kilometers.
Let t represent the temperature in degrees Celsius.
Use ordered pairs of the form (d, t) to show the relationship between depth and temperature.

- *When the depth is 0, the temperature is 20°C.* ▸ (0, 20)
- *When the depth is 2 km, the temperature is 90°C.* ▸ (2, 90)

plan

Write a linear equation from the ordered pairs.

solve

A line through (0, 20) and (2, 90) has a slope of $\frac{90 - 20}{2 - 0}$ or 35. The line intersects the vertical axis at 20. So, the equation is $t = 35d + 20$.

examine

Check whether the solution makes sense. Find t if $d = 0$ and if $d = 2$.

When d is 0, t is $35(0) + 20$, or 20. ✔
When d is 2, t is $35(2) + 20$, or 90. ✔

So, the equation $t = 35d + 20$ is correct.

This equation gives a good approximation for depths less than 4 kilometers.

Examples

1 **Find the underground temperature 3600 meters below the surface.**

$t = 35d + 20$
$t = 35(3.6) + 20$ *3600 m = 3.6 km*
$t = 146$ The temperature is about 146°C.

2 **The employees at the Ace Repair Shop use the chart below to compute the customers' bills. Write an equation to describe the relationship between the number of labor hours and the charge. (Assume the relationship is linear.)**

Labor Hours	0.5	0.75	1	1.25	1.5	2
Charge	$29	$33.50	$38	$42.50	$47	$56

First draw a graph to show the relationship.

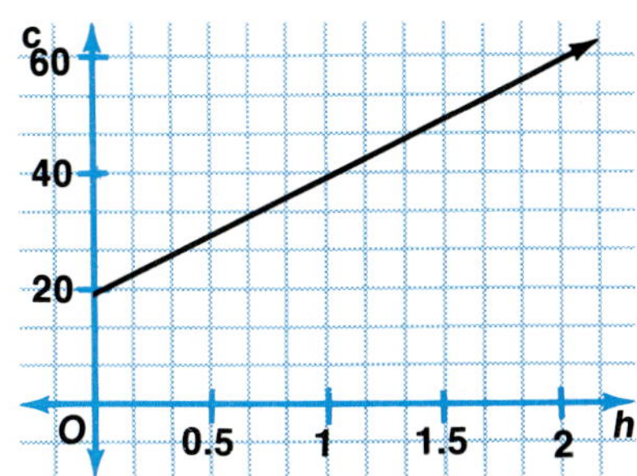

Using two points on the line, (2, 56) and (1, 38), find the slope of the line.

$$m = \frac{56 - 38}{2 - 1}$$
$$= \frac{18}{1} \text{ or } 18$$

Then, substitute the slope and the coordinates of one point in the slope-intercept form and solve for b.

$$y = mx + b$$
$$38 = (18)(1) + b$$
$$20 = b$$

Thus, the equation that describes the relationship is $c = 18h + 20$.

Exploratory Exercises

Solve each problem.

1. Inches of snowfall varies directly with inches of rainfall. Thirty-five inches of snow is about the same as 3.5 inches of rain. Write an equation to describe the relationship between inches of snowfall and inches of rainfall.

2. Twelve inches of snowfall is equivalent to how much rainfall?

3. Forty-eight inches of snowfall is equivalent to how much rainfall?

4. Eleven inches of rainfall is equivalent to how much snowfall?

5. Four and one-half inches of rainfall is equivalent to how much snowfall?

6. Frontier Auto Shop has a standard $12 shop charge for every job it takes. In addition, the mechanic working on the job charges $20 per hour. Write an equation to describe the relationship between time spent on a job and total charge for the job.

7. Cindy Baker is a mechanic at Frontier Auto Shop. She spent 4 hours on a job. How much will the customer be charged?

8. Joshua Levy is a mechanic at Frontier Auto Shop. He spent 2.5 hours on a job. How much will the customer be charged?

9. Frontier Auto Shop charges $47 for a job. How much time did the mechanic spend on the job?

10. Frontier Auto Shop charged $100 for a job. How much time did the mechanic spend on the job?

11. Light travels faster than sound. If you count the number of seconds between when you see lightning and when you hear it, you can estimate how far away it is. The distance d in kilometers between you and the lightning is estimated by $d = \frac{1}{3}s$ where s is the number of seconds. Complete the chart for estimating the distance.

Time (seconds)	2	3	4	5	10	15	20	25	30	40	50	60
Distance (kilometers)												

12. Crickets vary their number of chirps with the temperature. If you count the number of chirps in a minute, you can tell the temperature. The temperature t in degrees Celsius is estimated by $t = 0.2(n + 32)$ where n is the number of chirps in one minute. Complete the chart for estimating the temperature.

Chirps	50	60	70	80	90	100	110	120	130
Temperature (Celsius)									

Written Exercises

Write an equation to describe the relationship in each problem.

13. The present population of Whitehall is 47,000. The population increases by 550 each year. What will be the population in y years?

14. A telephone company charges $12 per month plus 10¢ for each local call. What is the monthly bill if c local calls are made?

15. A ranger calculates there are 6,000 deer in a preserve. She also estimates that 75 more deer die than are born each year. How many deer will be in the preserve x years from now?

16. The RD Rug Company produced 600 oriental rugs its first year of business. Each year after that, production increased by 100. What will be the annual production of oriental rugs after y years?

17. The ABC Car Rental Agency rents cars for $23 a day plus 18¢ a mile. How much does it cost to rent a car for a week and drive it t miles?

18. Emilio earns a flat salary of $150 per week for selling records. He receives an extra 30¢ for each record over 100 that he sells. How much will he make if he sells r records?

For each chart, write an equation to describe the relationship. Then, use the equation to complete each table. Assume each relationship is linear.

19. Cost of Bowling at Bingo's Bowling Lanes

Number of Games	1	2	3	4	5	6
Cost	$1.25	$2.50	$3.75			

20. Cost of Repair Work at Acme Auto Repair

Hours of Labor	1	2	3	4	6	8	12	15	20	25	30
Total Bill	$44	$56	$68	$80							

21. Cost of Long Distance Phone Call

Minutes	3	4	5	6	8	10	15	20	30	45	60
Cost	$1.38	$1.72	$2.06								

22. Number of Luncheon Specials Sold at Open Hearth Restaurant

Number of Customers	20	30	40	50	75	100	150	200
Number of Luncheon Specials	10	16	22	28				

2-8 Graphing Linear Inequalities in Two Variables

The graph of $y = -\frac{2}{3}x + \frac{5}{3}$ is a line that separates the coordinate plane into two regions.

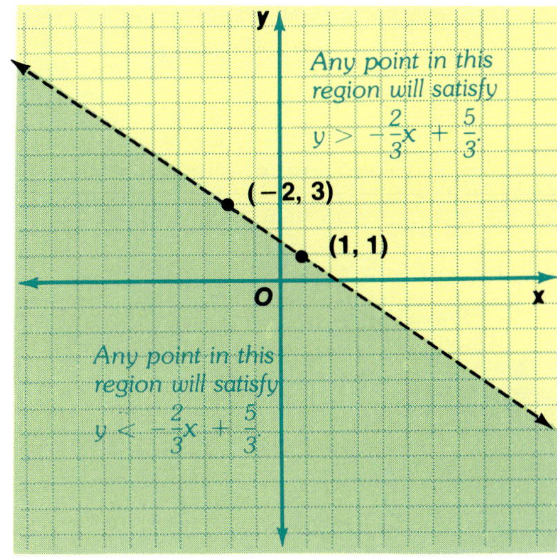

The graph of $y > -\frac{2}{3}x + \frac{5}{3}$

is the region *above* the line. In that region the value of y is greater than the value of $-\frac{2}{3}x + \frac{5}{3}$.

The graph of $y < -\frac{2}{3}x + \frac{5}{3}$

is the region *below* the line. In that region the value of y is less than the value of $-\frac{2}{3}x + \frac{5}{3}$.

The line described by $y = -\frac{2}{3}x + \frac{5}{3}$ is called the *boundary* of each region. If the boundary is part of a graph, it is drawn as a solid line. If the boundary is *not* part of a graph, it is drawn as a broken line.

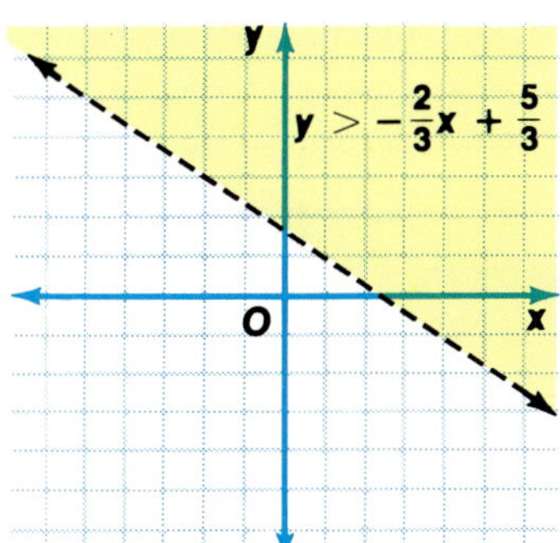

Note that $>$ tells you the boundary is <u>not</u> included.

Note that $\geq$ tells you the boundary is included.

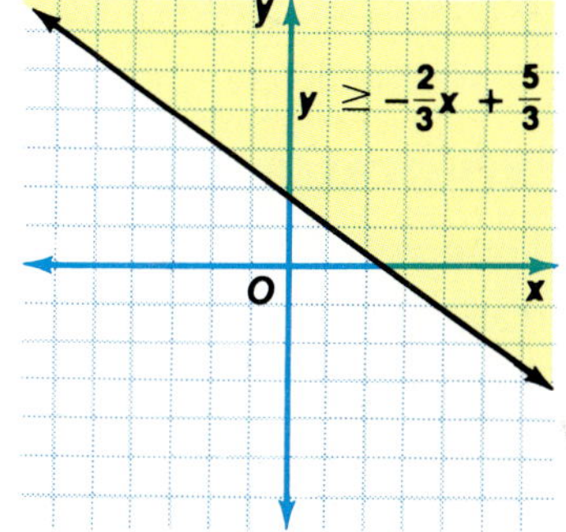

After graphing the boundary of an inequality, test a point *not* on the boundary. This will tell you whether the region that includes this point is part of the graph of the inequality. For the inequality $y > -\frac{2}{3}x + \frac{5}{3}$,

the ordered pair $(0, 0)$ does not satisfy the inequality $\left(0 > \frac{5}{3} \text{ is false}\right)$. So, the region to be shaded is the region that does not contain the origin.

The origin is often a convenient point to test.

1 **Graph $2y - 5x \le 1$.**

$$2y - 5x \le 1$$
$$2y \le 5x + 1$$
$$y \le \frac{5}{2}x + \frac{1}{2}$$

The slope is $\frac{5}{2}$.

The y-intercept is $\frac{1}{2}$.

Test $(3, 0)$:
$$2y - 5x \le 1$$
$$2(0) - 5(3) \overset{?}{\le} 1$$
$$-15 \le 1 \quad ✔$$

The shaded region should include $(3, 0)$.

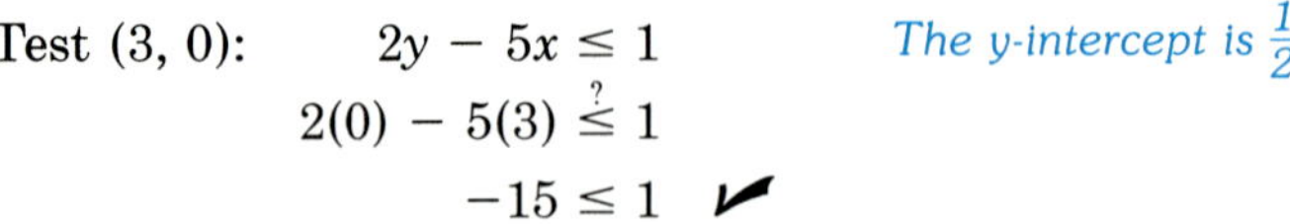

Note that the boundary is included.

2 **Graph $y > |x| - 1$.**

$$y > |x| - 1$$
$$|x| < y + 1$$

$$x < y + 1 \quad \text{and} \quad x > -y - 1$$
$$y > x - 1 \quad \text{and} \quad y > -x - 1$$

Test $(0, 0)$:
$$y > |x| - 1$$
$$0 \overset{?}{>} |0| - 1$$
$$0 > -1 \quad ✔$$

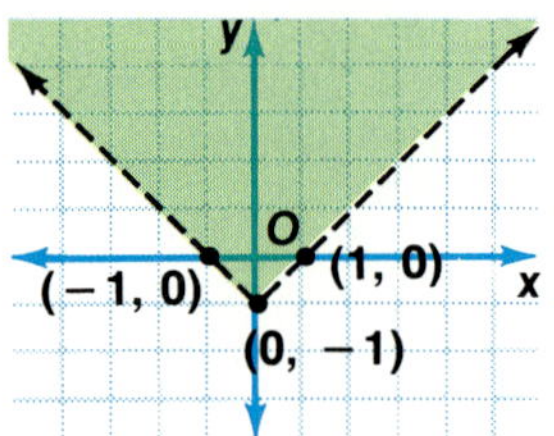

Note that the boundary is not included.

The shaded region should include $(0, 0)$.

Exploratory Exercises

Name which points, $(0, 0)$, $(2, -3)$, or $(-1, 2)$, satisfy each inequality.

1. $x + 2y < 7$ **2.** $3x - y \ge 2$ **3.** $3x + 2y \le 0$ **4.** $x + y \ge 7$

5. $4x + 2y \ge 7$ **6.** $5x - y > 2$ **7.** $y < 0$ **8.** $4x \ge -12$

Written Exercises

Draw the graph of each inequality.

9. $y < 3$ **10.** $x < 1$ **11.** $y > 5x - 3$ **12.** $y \ge x - 7$

13. $y \le -3x + 1$ **14.** $y < 2x + 3$ **15.** $2x - 5y \ge 4$ **16.** $2x + 5y \ge 3$

17. $y > \frac{1}{3}x + 7$ **18.** $y > \frac{1}{2}x - 3$ **19.** $x - 2y \le 2$ **20.** $-2x + 5 \le 3y$

21. $y \ge |x|$ **22.** $y \ge |x| - 3$ **23.** $|x| + y \ge 3$ **24.** $y + |x| < 2$

25. $y < |x| + 2$ **26.** $y < |x| - 2$ **27.** $y > |2x|$ **28.** $y \ge |3x|$

29. $|y| < x$ **30.** $|y| \ge 2x$ **31.** $|x| \le |y|$ **32.** $|x| > |y|$

33. Graph all points to the right of $x = 4$.

34. Graph all points in the second quadrant bounded by the lines $x = -2$, $x = -5$, and $y = 3$.

35. Graph all points in the first quadrant bounded by the two axes and the line $x + 2y = 4$.

36. Graph all points in the fourth quadrant bounded by the two axes and the lines $3x - y = 4$ and $x - y = 5$.

Challenge Exercises

Draw the graph of each equation or inequality.

37. $|x| + |y| = 1$ **38.** $|x| - |y| = 1$ **39.** $|x| + |y| \leq 1$ **40.** $|x| + |y| > 1$

mini-review

1. Solve $\left|\frac{1}{2}x + 3\right| = 0$.

2. Solve $6x + 1 > x + 21$ and graph the solution set.

3. Find $f(b - 2)$ if $f(x) = x^2 - 2x - 5$.

4. Find the x-intercept and y-intercept of the graph of $5y = x - 2$.

5. Find the slope and y-intercept of the graph of $3x - 4y = -10$.

Find the standard form of the equation of the line that satisfies the given conditions.

6. slope $= -\frac{3}{2}$, passes through $(4, -1)$

7. x-intercept $= -2$, y-intercept $= 5$

8. passes through $(-1, -6)$ and $(7, 4)$

Graph each function.

9. $f(x) = |2x| + 1$.   **10.** $g(x) = |x + 3|$ **11.** $h(x) = 2 - [x]$

12. Explain how the graphs of $y = [4x]$ and $y = 4[x]$ differ.

Write an equation to describe the relationship in each problem.

13. A forest ranger estimates that there are 500 deer in the national park. He also estimates that 50 more deer are born than die each year. How many deer will be in the park y years from now?

14. Joan earns $200 per week working in a record store. In addition, she earns 20¢ for each record over 50 that she sells. How much will she earn in one week if she sells r records?

15. The population of Georgetowne is 5250. The population decreases by 75 people each year. What will the population be in x years?

Slope and Y-Intercept

The computer program at the right can be used to find the slope and y-intercept of the line through any two points. The program uses the formula for slope.

$$m = \frac{y_2 - y_1}{x_2 - x_1} \qquad \textit{See line 40.}$$

To find a formula for y-intercept, substitute the coordinates of a specific point (x_1, y_1) into the slope-intercept form. Then solve for b.

$$y_1 = m_1 x + b$$
$$b = y_1 - mx_1 \qquad \textit{See line 60.}$$

```
10   PRINT "ENTER THE
     COORDINATES"
15   PRINT "OF TWO POINTS."
20   INPUT X1, Y1, X2, Y2
30   IF X1 = X2 THEN 90
40   LET M = (Y2 − Y1) /
     (X2 − X1)
44   PRINT
50   PRINT "SLOPE = ";M
60   LET B = Y1 − M * X1
70   PRINT "Y-INTERCEPT = ";B
80   GOTO 110
90   PRINT "UNDEFINED SLOPE."
100  PRINT "NO Y-INTERCEPT."
110  END
```

When you run the program, the computer will ask for the coordinates of two points and print a question mark. Suppose the points are $(3, 5)$ and $(7, -2)$. The coordinates should be entered as shown at the right. Study the output. Notice that the slope and y-intercept are expressed as decimals.

```
RUN
ENTER THE COORDINATES
OF TWO POINTS.
? 3,5,7,−2

SLOPE = −1.75
Y-INTERCEPT = 10.25
```

Exercises

Use the program above to find the slope and y-intercept of the line that passes through each pair of points. Then write the slope-intercept form of the equation of the line.

1. $(2, -7), (-2, -9)$ **2.** $(4, 9), (-1, 12)$ **3.** $(-2, 7), (-2, 2)$ **4.** $(-6, 3), (0, 3)$

Three or more points that lie on the same line are _collinear_. Decide if points A, B, and C are collinear. (Hint: Use the program to see if $\overleftrightarrow{AB}$ has the same slope as $\overleftrightarrow{BC}$.) Then write the equations for $\overleftrightarrow{AB}$, $\overleftrightarrow{BC}$, and $\overleftrightarrow{AC}$.

5. $A(1, 29), B(3, 61), C(7, 125)$ **6.** $A(0, 1), B(5, 2), C(6, 3)$

7. $A(-1, 4), B(-2, 0), C(-3, -2)$ **8.** $A(3, -7), B(6, -6), C(15, -3)$

9. $A(9, -3), B(0, 0), C(18, 6)$ **10.** $A(6, -5), B(6, -1), C(4, -5)$

Vocabulary

ordered pairs   (43)
quadrants   (43)
x-axis   (43)
y-axis   (43)
origin   (43)
coordinate plane   (43)
coordinate   (43)
graph   (43)
relation   (45)
domain   (45)
range   (45)
mapping   (45)
function   (46)

vertical line test   (47)
linear equation   (50)
standard form   (50)
linear function   (52)
slope   (54)
y-intercept   (56)
x-intercept   (56)
slope-intercept form   (59)
constant function   (63)
identity function   (63)
direct variation   (63)
greatest integer function   (64)

Chapter Summary

1. Each point in a coordinate plane corresponds to exactly one ordered pair of numbers. Each ordered pair of numbers corresponds to exactly one point in a coordinate plane.   (43)

2. A relation is a set of ordered pairs. The domain is the set of all first coordinates of the ordered pairs. The range is the set of all second coordinates of the ordered pairs.   (45)

3. A function is a relation in which each element of the domain is paired with exactly one element of the range.   (46)

4. Vertical Line Test for a Function: If any vertical line drawn on the graph of a relation passes through no more than one point of that graph, then the relation is a function.   (47)

5. An equation whose graph is a straight line is called a linear equation in two variables.   (50)

6. Standard Form of a Linear Equation: The standard form of a linear equation is $Ax + By = C$, where A, B, and C are real numbers, and A and B are *not both* zero.(50)

7. A linear function can be defined by $f(x) = mx + b$ where m and b are real numbers.   (52)

8. The slope, m, of a line passing through points (x_1, y_1) and (x_2, y_2) is given by

$$m = \frac{y_2 - y_1}{x_2 - x_1}.   (54)$$

9. The y-intercept is the value of a function when x is zero. The x-intercept is the value of a function when y is zero.   (56)

10. Slope-Intercept Form of a Linear Equation: The slope-intercept form of the equation of a line is $y = mx + b$. The slope is m and the y-intercept is b. (59)

11. A linear function described by $y = b$ or $f(x) = b$ is called a constant function. (63)

12. A linear function described by $y = x$ or $f(x) = x$ is called the identity function. (63)

13. A linear function described by $y = mx$ or $f(x) = mx$ where $m \neq 0$ is called a direct variation. The constant m is called the constant of variation or constant of proportionality. (63)

14. The greatest integer of x is written $[x]$ and means the greatest integer not greater than x. (64)

15. The graph of a linear inequality in two variables is the set of points in one of the two regions of the coordinate plane formed by the boundary line. Only the ordered pairs in one of the two regions satisfy the inequality. (71)

Chapter Review

2–1 **Graph each set of ordered pairs. Use appropriate labels on the axes.**

    1. $\left\{ (4, 2), (3, 1), (0, 5), \left(-5, \frac{1}{2}\right), \left(-2, -1\frac{1}{2}\right) \right\}$

    2. $\{(0, 40), (25, 10), (-30, 15), (-20, 30), (10, 20)\}$

    3. $\{(-9, -12), (4, -8), (3, -3), (6, 10)\}$

2–2 **State the domain and range of each relation. Then state if the relation is a function.**

    4. $\{(4.5, 1), (-4.5, 2), (4.5, 3), (-3.5, 4)\}$

    5. $\{(1, 4.5), (2, -4.5), (3, 4.5), (4, -3.5), (5, -3.5)\}$

    6. $\{(13, -8), (14, -9), (15, -9), (16, -8)\}$

Use the vertical line test to determine if each relation is a function.

7. 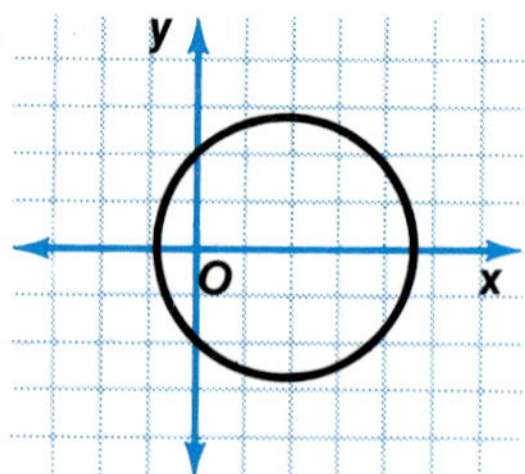   **8.** 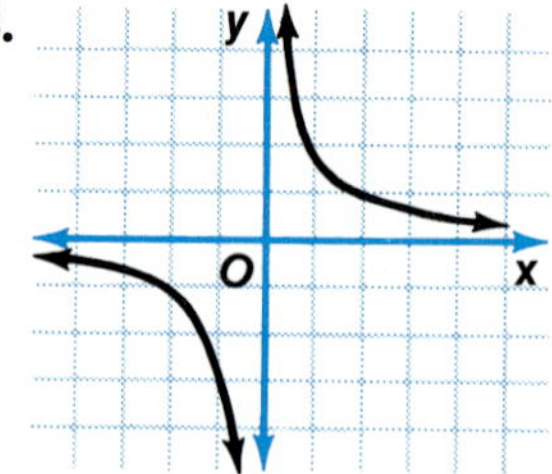   **9.** 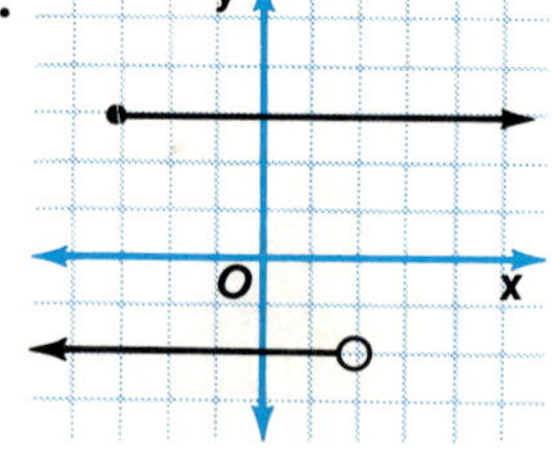

Given $f(x) = 2x^3 + 4x^2 + 4x + 1$, find each value.

    10. $f(0)$ **11.** $f(-3)$ **12.** $f(a)$

State whether each equation is a linear equation.

13. $x^2 + y^2 = 4$ **14.** $xy = 12$ **15.** $y = 5$ **16.** $x + y = 9$

Graph each equation.

17. $y = \frac{1}{3}x$ **18.** $y = 3x$ **19.** $y = 2x - 1$ **20.** $3y = 2x + 1$

2–4   **Find the slope, y-intercept and x-intercept of the graph of each of the following.**

21. $3x + 7y = 2$ **22.** $4x - 3y = 7$

23. $\frac{1}{3}x - \frac{2}{3}y = 5$ **24.** $\frac{1}{2}x + \frac{3}{4}y = 1$

Determine the slope of the line passing through each pair of points.

25. $(5, 1)$ and $(3, 7)$ **26.** $(-4, 2)$ and $(-4, 0)$ **27.** $(-3, -2)$ and $(0, 4)$

A line has a slope of $\frac{2}{3}$. For each pair of points on the line, find the missing coordinate.

28. $(0, 6)$ and $(-6, y)$ **29.** $(-9, 0)$ and $(x, 10)$

Graph the line that passes through the given point and has the given slope.

30. $(-3, 2)$, $m = 3$ **31.** $(3, 4)$, $m = \frac{1}{4}$ **32.** $(2, -1)$, $m = -\frac{5}{3}$

2–5   **Find the slope-intercept form of the equation of the line that satisfies the given conditions.**

33. slope $= 5$, y-intercept $= -7$ **34.** slope $= -3$, passes through $(1, -4)$

35. passes through $(-3, 0)$ and $(4, -6)$ **36.** x-intercept $= -2$, y-intercept $= 5$

Find the standard form of the equation of the line that satisfies the given conditions.

37. slope $= 4$, passes through $(5, -2)$ **38.** slope $= -\frac{3}{2}$, y-intercept $= 7$

39. x-intercept $= 2$, y-intercept $= -6$ **40.** passes through $(5, 3)$ and $(-2, 1)$

2–6   **Graph each function.**

41. $g(x) = |-x|$ **42.** $q(x) = |x - 3|$ **43.** $f(x) = |2x| - 1$

44. $p(x) = [-x]$ **45.** $f(x) = -[x]$ **46.** $f(x) = [3x]$

2–7   **47.** The Thomas Parcel Service charges \$2.50 per pound for the first three pounds and \$1.00 for each pound thereafter. Write an equation to find the cost of sending a package that weighs p pounds. Assume $p \geq 3$.

48. Use the equation in Exercise 47 to find the cost of sending packages that weigh 6 lb, 8 lb, 15 lb, and 42 lb.

2–8   **Graph each inequality.**

49. $3x + 4y < 9$ **50.** $2x - 5y > 4$

51. $y \geq |x| + 5$ **52.** $y \leq |x + 5|$

Graph each set of ordered pairs. Use appropriate labels on the axes.

1. $\{(-8, 1), (-4, 8), (3, 0), (8, -6), (0, -5)\}$

2. $\left\{\left(\frac{1}{5}, \frac{3}{5}\right), \left(\frac{2}{5}, \frac{4}{5}\right), \left(-\frac{3}{5}, 0\right), \left(0, \frac{2}{5}\right), \left(-\frac{1}{5}, -\frac{2}{5}\right)\right\}$

State the domain and range of each relation. Then state if the relation is a function.

3. $\{(1, 2), (7, 2), (9, -3), (2, 7)\}$ **4.** $\{(1.4, 1.4), (1.4, 1.5)\}$ **5.** $\{(3, -2), (5, 4), (7, -2)\}$

Use the vertical line test to determine if each relation is a function.

6. 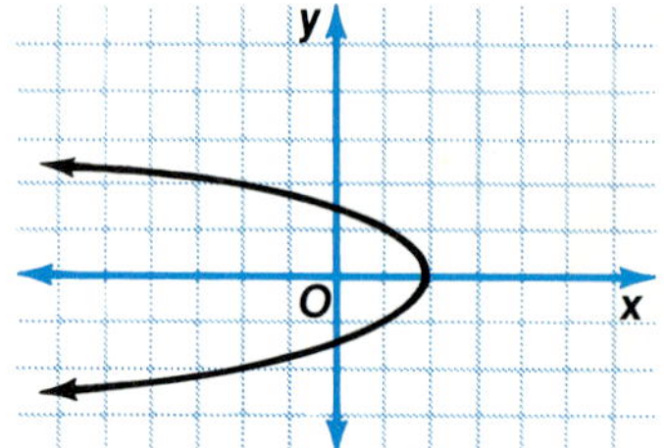   **7.** 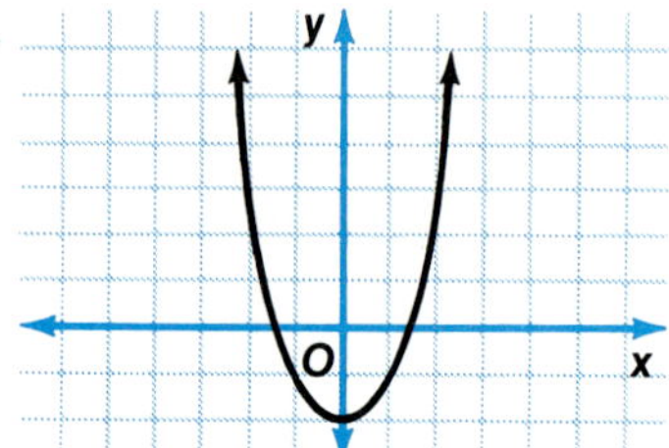   **8.** 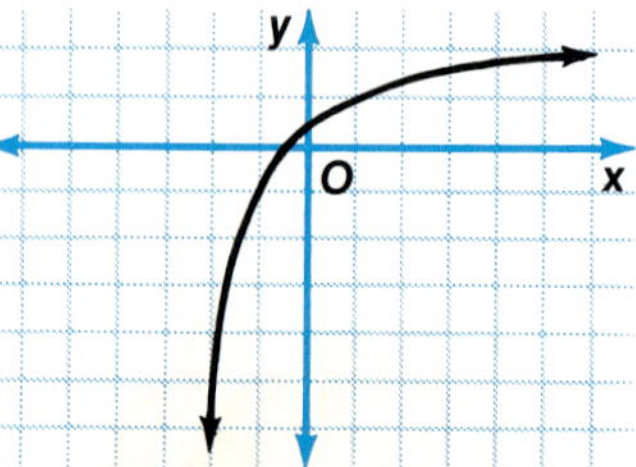

Given $f(x) = 2x^3 + 3x^2 - 5x - 4$, find each value.

9. $f(-3)$ **10.** $f\left(\frac{2}{3}\right)$ **11.** $f(2c)$

Graph each equation or inequality.

12. $y = \frac{12}{5}x$ **13.** $y = 7$ **14.** $y = -7x + 3$ **15.** $y = 3x - \frac{1}{2}$

16. $5x + 2y = 12$ **17.** $3y - 3x = 10$ **18.** $5y + 3x - 10 = 0$ **19.** $4x + 12y = -15$

20. $2x + 3y > 7$ **21.** $4x + 6y \leq 9$ **22.** $f(x) = |x + 2|$ **23.** $f(x) = -[x]$

24. $y < 3|x| - 1$ **25.** $y \geq |3x - 1|$ **26.** $y = |4x + 3| + 2$ **27.** $y = 2[x] - 7$

Determine the slope of the line passing through each pair of points.

28. $(7, 4)$ and $(-2, 5)$ **29.** $(0, 6)$ and $(1, -1)$

Find the slope-intercept form of the equation of the line that satisfies the given conditions.

30. slope $= 4$, passes through $(-3, 3)$ **31.** passes through $(0, 7)$ and $(5, 2)$

Find the standard form of the equation of the line that satisfies the given conditions.

32. slope $= \frac{1}{3}$, y-intercept $= 7$ **33.** x-intercept $= -6$, y-intercept $= -6$

A line has a slope $-\frac{2}{5}$. For each pair of points on the line, find the missing coordinate.

34. $(5, 1)$ and $(15, y)$ **35.** $(x, 2)$ and $(-5, 5)$

36. As a waiter, Ryan earns an average of \$28.00 a day in tips. If he also earns \$3.00 per hour, write an equation to determine the total amount he earns in an h hour day.

The test questions on this page deal with a variety of concepts from arithmetic and algebra.

Directions: Choose the one best answer. Write A, B, C, or D.

1. 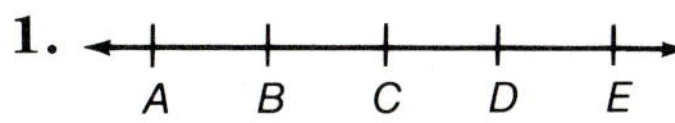
   Figure is not drawn to scale.

   Suppose numbers $\frac{5}{6}$, 1, $\frac{1}{3}$, $\frac{3}{2}$, and $\frac{6}{5}$ are arranged from least to greatest on the number line above, with each number corresponding to a letter. If the greatest number corresponds to E, which letter corresponds to $\frac{5}{6}$?

   (A) A **(B)** B **(C)** C **(D)** D

2. If $8x + 10y$ represents the perimeter of a rectangle, and $x + 3y$ represents its width, the length is
   (A) $3x + 2y$ **(B)** $7x + 7y$
   (C) $6x + 4y$ **(D)** $3.5x + 3.5y$

3. In the series 2, 6, 11, 17, 24, _____, _____, _____ the eighth term is
   (A) 41 **(B)** 45 **(C)** 51 **(D)** 62

4. If 9 less than the product of a number and -4 is greater than 7, which of the following could be that number?
   (A) -5 **(B)** -3 **(C)** 4 **(D)** 5

5. Which of the following is *not* equivalent to $-\frac{6}{8}$?

   (A) $-\frac{3}{4}$ **(B)** $\frac{3}{-4}$ **(C)** $-\frac{-12}{-16}$ **(D)** $-\frac{6}{-8}$

6. Which of the following is the least?

   (A) $\frac{1}{2}$ **(B)** $\frac{7}{13}$ **(C)** $\frac{4}{9}$ **(D)** $\frac{8}{15}$

7. If the radius of a circle is tripled, then the area is multiplied by
   (A) 3 **(B)** 6 **(C)** 9 **(D)** 27

8. A point on the graph of $x + 3y = 13$ is
   (A) $(4, 4)$ **(B)** $(-5, 6)$
   (C) $(-2, 3)$ **(D)** $(4, -3)$

9. Which represents an irrational number?

   (A) $-\frac{2}{3}$ **(B)** $\sqrt{4}$ **(C)** π **(D)** 0

10. For what values of y will $2y - 4$ be equal to $2y + 6$?
    (A) all negative values
    (B) 0
    (C) all positive values
    (D) no values

11. If a number is increased by 5 and the result is multiplied by 8, the product is 168. What is the original number?
    (A) 16 **(B)** 42 **(C)** 46 **(D)** 128

12. The average of 7, 5, 9, 3, and $2x$ is x. What is the value of x?
    (A) 2.4 **(B)** 4 **(C)** 6 **(D)** 8

13. The radius of a wheel is 4 cm. How many revolutions will it make if it is rolled a distance of 400π?
    (A) 25π **(B)** 25 **(C)** 50 **(D)** 100π

14. Which of the following is the difference of two consecutive prime numbers less than 40?
    (A) 1 **(B)** 5 **(C)** 8 **(D)** 11

15. How many integers between 199 and 301 are divisible by 4 or 10?
    (A) 26 **(B)** 31 **(C)** 35 **(D)** 37

16. A patient must be given medication every 7 hours starting at 7 A.M. Monday. On what day will the patient first receive medication at 6 P.M.?
    (A) Monday **(B)** Tuesday
    (C) Wednesday **(D)** Friday

Systems of Equations and Inequalities

One of the most practical applications of mathematics to business is the branch of mathematics called linear programming. A farmer can use it to decide which crops to plant and how many acres to use in order to minimize the costs and maximize the profit.

3-1 Parallel and Perpendicular Lines

The graphs of $y = -\frac{3}{4}x - 3$, $y = -\frac{3}{4}x$, and $y = -\frac{3}{4}x + 2$ are straight lines that have the same slope. They are **parallel lines**.

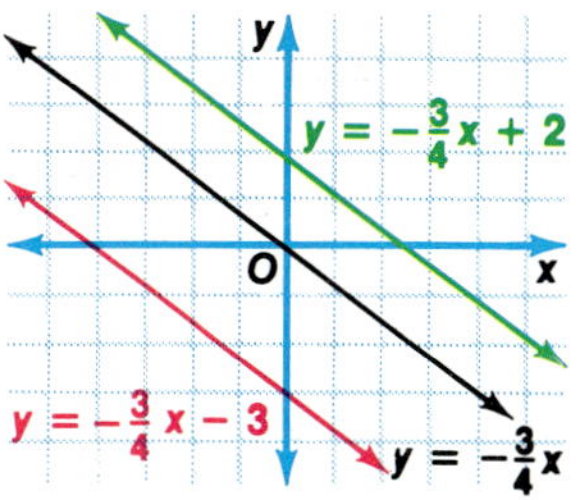

In a plane, lines with the same slope are parallel lines. Also, all vertical lines are parallel.

Definition of Parallel Lines

Examples

1 Find the slope of a line parallel to the line whose equation is $3y - 5x = 15$.

Parallel lines have the same slope. Find the slope of the line whose equation is $3y - 5x = 15$. To do so, write the equation in slope-intercept form. $y = mx + b$

$$3y - 5x = 15$$
$$3y = 5x + 15$$
$$y = \frac{5}{3}x + 5$$

The slope of any line parallel to the given line is $\frac{5}{3}$.

2 Find an equation of the line that passes through $(4, 6)$ and is parallel to the line whose equation is $y = \frac{2}{3}x + 5$.

The slope is $\frac{2}{3}$.   *Notice that $y = \frac{2}{3}x + 5$ is in slope-intercept form.*

Use $(4, 6)$ and the slope $\frac{2}{3}$ to find the y-intercept.

$$y = mx + b$$
$$6 = \left(\frac{2}{3}\right)(4) + b \qquad \textit{Substitution Property}$$
$$6 = \frac{8}{3} + b$$
$$\frac{10}{3} = b \qquad \textit{The y-intercept is } \frac{10}{3}.$$

An equation of the line is $y = \frac{2}{3}x + \frac{10}{3}$.

The graphs of $y = \frac{5}{3}x + 2$ and $y = -\frac{3}{5}x + 6$ are straight lines that are **perpendicular**. Notice how their slopes are related.

$$\left(\frac{5}{3}\right)\left(-\frac{3}{5}\right) = -1$$

The product of their slopes is -1.

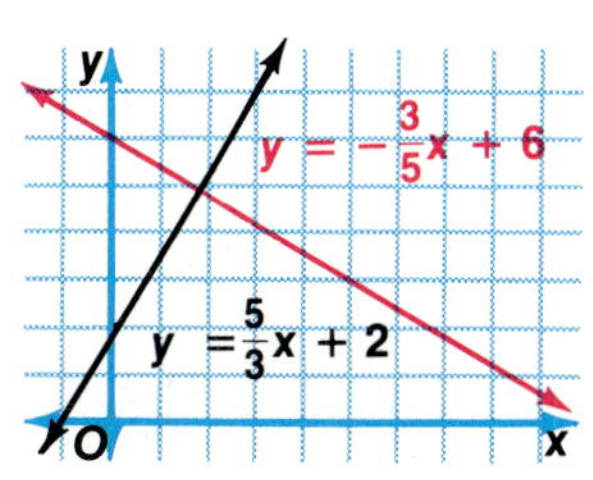

Here is a way to show that the slopes of any two nonvertical perpendicular lines in a plane have a product of -1. Consider a line that is neither vertical nor horizontal, with slope $\frac{r}{s}$. This is shown by the black line at the right. Now consider rotating the line 90°. Notice that the slope of the new line (shown in red) is $-\frac{s}{r}$. The product of the slopes of the two lines is $\left(\frac{r}{s}\right)\left(-\frac{s}{r}\right)$ or -1.

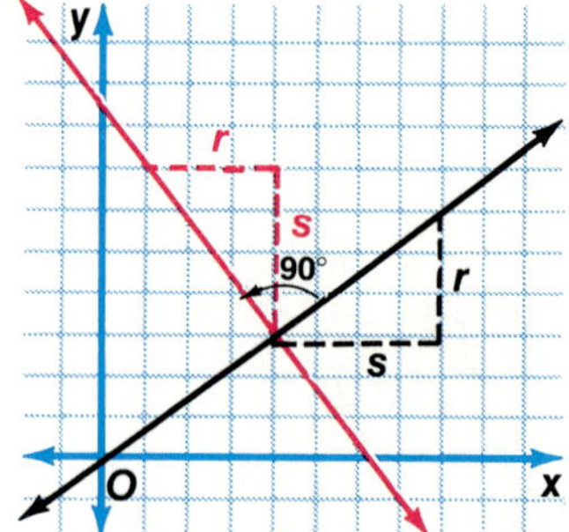

Examples

3 **Find the slope of a line perpendicular to the line whose equation is $y - 3x = 2$.**

$y - 3x = 2$
$y = 3x + 2$ *This equation is in slope-intercept form.*

The slope of the given line is 3.

$3 \cdot m = -1$ *Let m stand for the slope of the perpendicular line.*

$m = -\frac{1}{3}$ The slope of any line perpendicular to the given line is $-\frac{1}{3}$.

4 **Find an equation of the line that passes through (4, 6) and is perpendicular to the line whose equation is $y = \frac{2}{3}x + 5$.**

The slope of the given line is $\frac{2}{3}$.   *$y = \frac{2}{3}x - 5$ is in slope-intercept form.*

$\frac{2}{3} \cdot m = -1$ *Let m stand for the slope of the perpendicular line.*

$m = -\frac{3}{2}$ The slope is $-\frac{3}{2}$.

Use (4, 6) and the slope $-\frac{3}{2}$ to find the y-intercept.

$y = mx + b$

$6 = \left(-\frac{3}{2}\right)(4) + b$

$12 = b$ *The y-intercept is 12.*

An equation of the line is $y = -\frac{3}{2}x + 12$.   *This could be written $3x + 2y = 24$.*

Find the slope of a line that is parallel to each line whose equation is given.

1. $y = 4x + 2$
2. $y = 5 - 2x$
3. $2y = 3x - 8$
4. $6y - 5x = 0$
5. $\frac{1}{3}x - \frac{3}{8}y = 11$
6. $x = 4y + 7$

Find the slope of a line that is perpendicular to each line whose equation is given.

7. $y = 2x + 4$
8. $y = 2 - 5x$
9. $3y = 8x - 2$
10. $5y - 6x = 0$
11. $3x - 8y = 11$
12. $4x = 7y + 1$

State whether the graphs of the following equations are parallel, perpendicular, or neither.

13. $x + y = 5$
$x + y = -10$
14. $x + y = 5$
$x - y = 5$
15. $y = 2x$
$y = 2x - 4$
16. $2y + 3x = 5$
$3y - 2x = 5$
17. $3x - 8y = 11$
$3x - 6y = 10$
18. $2y + 3x = 5$
$3y + 3x = 5$
19. $\frac{1}{3}x + \frac{2}{3}y = \frac{3}{5}$
$2x + 4y = 7$
20. $\frac{1}{2}x + \frac{1}{3}y = 2$
$2x - 3y = 4$
21. $7x - 5y = 2$
$\frac{1}{7}x - \frac{1}{5}y = 2$

Find an equation of the line that passes through each given point and is parallel to the line with the given equation.

22. $(4, 2)$; $y = 2x - 4$
23. $(3, 1)$; $y = \frac{1}{3}x + 6$
24. $\left(\frac{1}{2}, \frac{1}{3}\right)$; $x + 2y = 5$
25. $(0, 0)$; $3x - y = 4$
26. $(-3, -1)$; $y + x = 6$
27. $(7, -1)$; $2y - 3x = 1$
28. $(4, -2)$; $\frac{x}{3} - \frac{y}{5} = 2$
29. $\left(\frac{1}{2}, \frac{1}{3}\right)$; $x + y = 4$
30. $(3, 7)$; $\frac{2}{3}x = \frac{1}{2}y$

Find an equation of the line that passes through each given point and is perpendicular to the line with the given equation.

31. $(-2, 0)$; $y = -3x + 7$
32. $(-3, 4)$; $3y = 2x + 3$
33. $(2, 5)$; $3x + 5y = 7$
34. $(3, 6)$; $4x - y = 2$
35. $(0, -4)$; $6x - 3y = 5$
36. $(5, -3)$; $4x + 7y = 11$
37. $(12, 6)$; $\frac{3}{4}x + \frac{1}{2}y = 2$
38. $(-10, 3)$; $\frac{2}{3}x + y = 6$
39. $(6, -5)$; $3x - \frac{1}{5}y = 3$

Find the value of a for which the graph of the first equation is perpendicular to the graph of the second equation.

40. $y = ax - 5$; $2y = 3x$
41. $y = ax + 2$; $3y - 4x = 7$
42. $y = \frac{a}{3}x - 6$; $4x + 2y = 6$
43. $3y + ax = 8$; $y = \frac{3}{4}x + 2$

44. Show that $(1, 3)$, $(4, 1)$, and $(5, 9)$ are vertices of a right triangle.

45. Show that $(-6, 5)$, $(-2, 7)$, $(5, 3)$, and $(1, 1)$ are vertices of a parallelogram.

46. Three vertices of a parallelogram are $(10, 3)$, $(-1, 2)$ and $(1, -1)$. Find the fourth vertex.

47. Two vertices of a square are $(3, 3)$ and $(11, 7)$. Find the other two vertices of the square.

3-2 Solving Systems of Equations Graphically

The cost of renting a car from ACE is \$18 per day plus 30¢ per mile driven. The cost of renting a similar car from QUALITY is \$20 per day plus 25¢ per mile driven. Caroline needs to rent a car for one day. Should she rent from ACE or QUALITY?

Let d = cost of renting a car for one day.
Let m = number of miles driven in one day.

You can write the following equations.

$d = 18 + 0.30m$ *Cost of renting car from ACE for one day.*
$d = 20 + 0.25m$ *Cost of renting car from QUALITY for one day.*

Graphing these two equations shows how the costs compare. The graphs show that the QUALITY car costs more if less than 40 miles are driven. The QUALITY and ACE cars cost the same if 40 miles are driven. The ACE car costs more if more than 40 miles are driven.

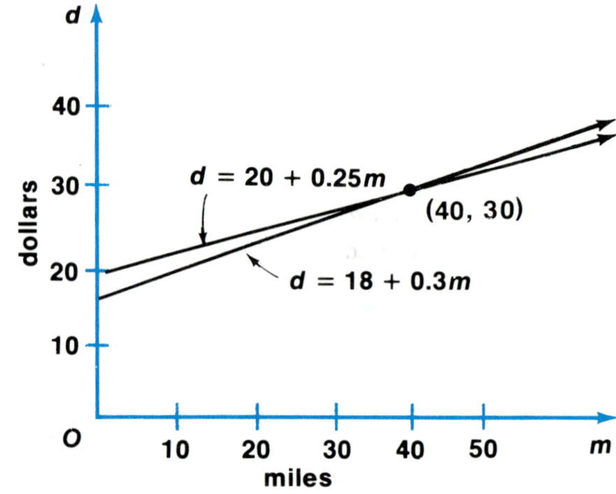

Each point on a line satisfies the equation of the line. Since (40, 30) is on both lines graphed at the right, it satisfies both equations.

Together the equations $d = 18 + 0.30m$ and $d = 20 + 0.25m$ are called a **system of equations**. The solution of the system is (40, 30).

Example

1   **Solve this system by graphing:** $3x - y = 1$
$$2x + y = 4$$

The slope-intercept form of $3x - y = 1$ is $y = 3x - 1$.

The slope-intercept form of $2x + y = 4$ is $y = -2x + 4$.

The two lines have different slopes. The graphs of the equations are intersecting lines.

The solution of the system is (1, 2).

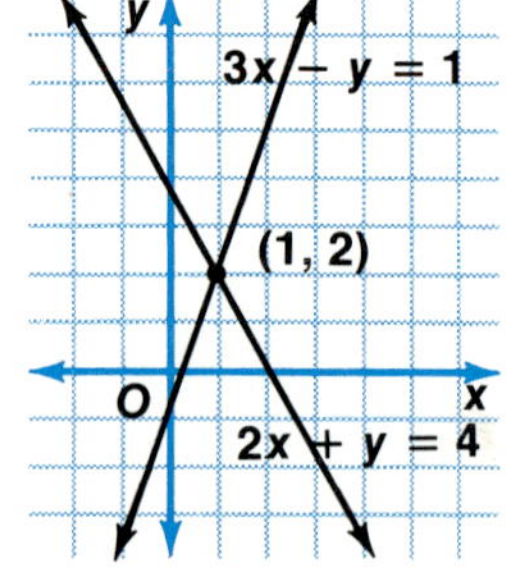

2   **Solve this system by graphing:** $2y + 3x = 6$
$$4y + 6x = 12$$

The slope-intercept form of $2y + 3x = 6$ is $y = -\frac{3}{2}x + 3$.

The slope-intercept form of $4y + 6x = 12$ is $y = -\frac{3}{2}x + 3$.

Both lines have the same slope and the same y-intercept. The graphs of the equations are the same line. Any ordered pair on the graph satisfies both equations. So, there is an *infinite* number of solutions to this system of equations.

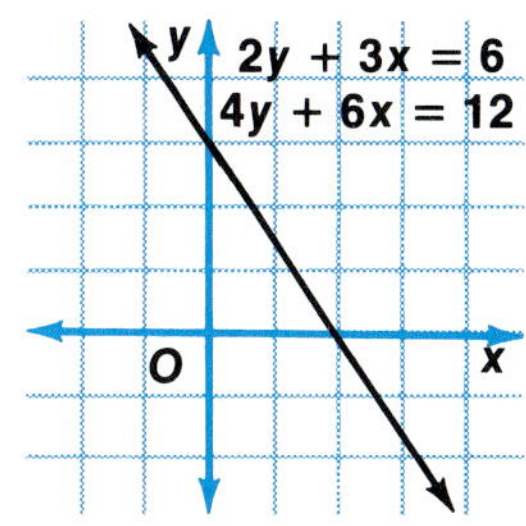

The solution set is $\{(x, y)\,|\,2y + 3x = 6\}$.

A **consistent** system of equations has at least one solution. For example, the systems of equations in Examples 1 and 2 are consistent. If there is exactly one solution, the system is **independent**. If there is an infinite number of solutions, the system is **dependent**. So, the system in Example 1 is *consistent and independent*. The system in Example 2 is *consistent and dependent*.

Example

3   **Solve this system by graphing:** $y = -3x - 2$
$$y = -3x + 3$$

Both lines have the same slope but different y-intercepts. The graphs of the equations are parallel lines. Since they do not intersect, there are *no solutions* to this system of equations. Such a system is said to be **inconsistent**.   *The solution set is the empty set, Ø.*

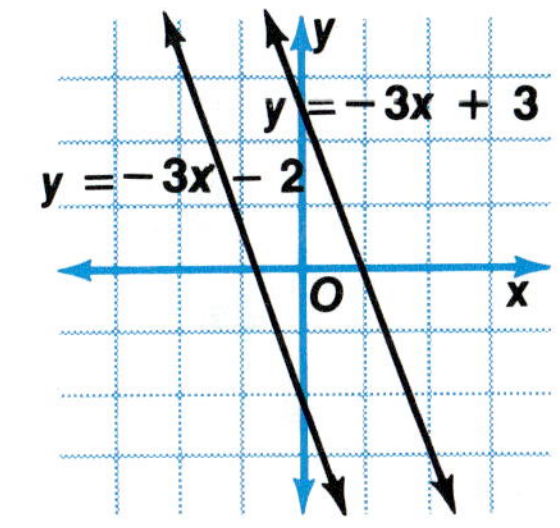

The following chart gives a summary of the possibilities for the graphs of two linear equations in two variables.

Graphs of Equations	Slopes of Lines	Name of System of Equations	Number of Solutions
lines intersect	different slopes	consistent and independent	one
lines coincide	same slope, same intercepts	consistent and dependent	infinite
lines parallel	same slope, different intercepts	inconsistent	none

Exploratory Exercises

State the ordered pair that is the intersection of each pair of lines.

1. a, b
2. a, e
3. a, d
4. b, c
5. c, d
6. e, d
7. b, f
8. f, d

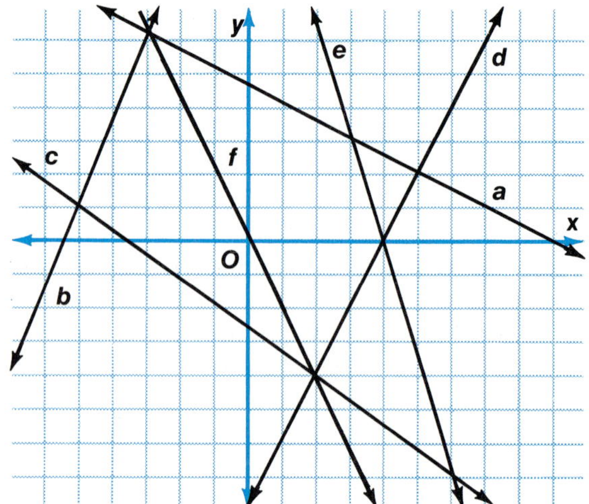

Written Exercises

Graph each system of equations and state its solution. Then state whether the system is *consistent and independent*, *consistent and dependent*, or *inconsistent*.

9. $x + y = 4$
$2x + 3y = 9$

10. $x + y = 6$
$x - y = 2$

11. $x + y = 6$
$3x + 3y = 3$

12. $x + 1 = y$
$2x - 2y = 8$

13. $x + 2y = 5$
$3x - 15 = -6y$

14. $2x + 4y = 8$
$x + 2y = 4$

15. $y = -3x$
$6y - x = -38$

16. $\frac{1}{2}x + \frac{1}{3}y = 2$
$x - y = -1$

17. $\frac{3}{4}x - y = 0$
$\frac{y}{3} + \frac{x}{2} = 6$

18. $x + y = 1$
$3x + 5y = 7$

19. $x + 5y = 10$
$x + 5y = 15$

20. $3x - 8y = 4$
$6x - 42 = 16y$

21. $x + y = -6$
$2x - y = 2$

22. $3x + 6 = 7y$
$x + 2y = 11$

23. $2x + 3y = 5$
$-6x - 9y = -15$

24. $\frac{2}{3}x = \frac{5}{3}y$
$2x - 5y = 0$

25. $y = \frac{x}{2}$
$2y = x + 4$

26. $9x + 8y = 8$
$\frac{3}{4}x + \frac{2}{3}y = 8$

27. $2x + 3y = -4$
$-3x + y = -5$

28. $-2x + 5y = -14$
$x - y = 1$

29. $9x - 5 = 7y$
$4\frac{1}{2}x - 3\frac{1}{2}y = 2\frac{1}{2}$

30. Write an equation of the line passing through $(5, -1)$ that is inconsistent with $5x - 3y = 2$.

31. Find a and b so that $ax + 5y = b$ and $6x + 10y = 16$ will be consistent and dependent equations.

Write a system of equations for each problem and solve by graphing.

32. George bought 7 quarts of cleaning fluid: x quarts at \$3.00 per quart and y quarts at \$2.00 per quart. Find x and y if the total cost was \$16.00.

33. Kenneth mixes r cups of cashews with s cups of peanuts. This yields 12 cups of a mixture having one part cashews to three parts peanuts. Find r and s.

The cost of driving a car is affected by the rate at which it uses gasoline. The graph at the right shows the cost of gasoline, based on the number of miles driven and the usage rate in miles per gallon (mpg). (The cost of gasoline was $1.50 per gallon.)

Suppose Car *A* gets an average of 20 miles per gallon and Car *B* gets 30 miles per gallon. Each car is driven 600 miles. From the graph, you can see that the gasoline cost for Car *A* is $45 and for Car *B* is $30.

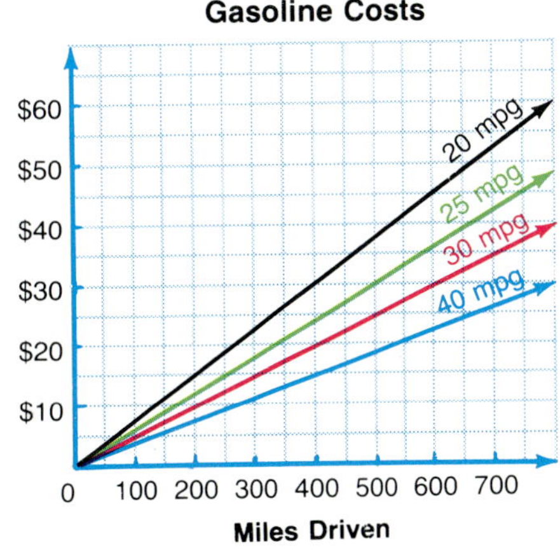

There are also fixed expenses for operating a car, such as loan repayment and insurance. The weekly fixed expenses are $40 for Car *A*, and $55 for Car *B*. This information can be used to draw a graph that shows the total weekly operating costs for the two cars.

Each line on the graph at the right has the same slope as the corresponding line on the graph above. However, the intercept of each line is the fixed cost per week.

Study the graph. The operating cost for Car *A* is less when a person drives less than 600 miles per week. When is the operating cost for Car *B* less than for Car *A*?

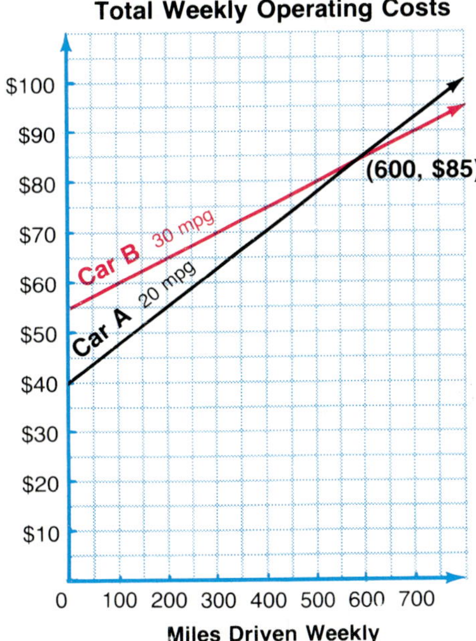

Exercises

1. Make a graph to compare the total costs of operating Cars *C* and *D*. Car *C* gets 40 mpg and has fixed costs of $50 per week. Car *D* gets 25 mpg and has fixed costs of $40 per week.

2. Make a graph to compare the total costs of operating Cars *Q* and *R*. Car *Q* gets 25 mpg and has fixed costs of $35 per week. Car *R* gets 20 mpg and has fixed costs of $45 per week.

3-3 Solving Systems of Equations Algebraically

Usually a system of equations is easier to solve by algebraic methods rather than by graphing. Two such methods are the **substitution method** and the **elimination method**.

Examples

1 **Use the substitution method to solve this system:** $3x - y = 1$
$$3x + 2y = 16$$

Solve the first equation for y.

$$3x - y = 1$$
$$y = 3x - 1$$

Find x by substituting $3x - 1$ for y in the second equation.

$$3x + 2y = 16$$
$$3x + 2(3x - 1) = 16 \qquad \text{Substitute } 3x - 1 \text{ for } y.$$
$$3x + 6x - 2 = 16 \qquad \text{The resulting equation now has only one variable, } x.$$
$$9x = 18$$
$$x = 2 \qquad \text{Solve for } x.$$

Now find y by substituting 2 for x in $3x - y = 1$.

$$3x - y = 1$$
$$3(2) - y = 1 \qquad \text{Substitute 2 for } x.$$
$$y = 5 \qquad \text{Solve for } y.$$

The solution is $(2, 5)$.

2 **Use the elimination method to solve this system:** $4x + 2y = -8$
$$x - 2y = -7$$

Add the second equation to the first equation to eliminate y.

$$4x + 2y = -8 \qquad \text{The } y\text{-coefficients, 2 and } -2\text{, are additive inverses.}$$
$$\underline{x - 2y = -7} \qquad \text{Add.}$$
$$5x = -15 \qquad \text{The variable } y \text{ is eliminated.}$$
$$x = -3 \qquad \text{Solve for } x.$$

Now find y by substituting -3 for x in $x - 2y = -7$.

$$x - 2y = -7$$
$$(-3) - 2y = -7 \qquad \text{Substitute } -3 \text{ for } x.$$
$$-2y = -4$$
$$y = 2 \qquad \text{Solve for } y.$$

The solution is $(-3, 2)$.

Example

3 Use the elimination method to solve this system: $2x + 3y = 2$
$$3x - 4y = -14$$

Adding the two equations does *not* eliminate either of the variables. However, if the first equation is multiplied by 4 and the second equation is multiplied by 3, the variable y can be eliminated by addition.

$2x + 3y = 2$ **Multiply by 4.** $8x + 12y = 8$

$3x - 4y = -14$ **Multiply by 3.** $9x - 12y = -42$

Now add to eliminate y.

$$\begin{aligned} 8x + 12y &= 8 \\ 9x - 12y &= -42 \qquad \text{Add.} \\ \hline 17x &= -34 \qquad \text{The variable } y \text{ is eliminated.} \\ x &= -2 \qquad \text{Solve for } x. \end{aligned}$$

An alternate method is to eliminate x first. Multiply the first equation by 3 and the second equation by −2.

$$\begin{aligned} 6x + 9y &= 6 \\ -6x + 8y &= 28 \\ \hline 17y &= 34 \\ y &= 2 \end{aligned}$$

Now find y by substituting -2 for x in $2x + 3y = 2$.

$$\begin{aligned} 2x + 3y &= 2 \\ 2(-2) + 3y &= 2 \qquad \text{Substitute } -2 \text{ for } x. \\ -4 + 3y &= 2 \\ 3y &= 6 \\ y &= 2 \qquad \text{Solve for } y. \end{aligned}$$

Then solve for x.

$$\begin{aligned} 2x + 3(2) &= 2 \\ 2x &= -4 \\ x &= -2 \end{aligned}$$

The solution is $(-2, 2)$.

Exploratory Exercises

For each system, state the multipliers you would use to eliminate each of the variables by addition.

1. $2x + 3y = 7$
$3x - 4y = 2$

2. $x - y = 1$
$3x - y = 3$

3. $6x - 4y = 20$
$4x + y = 6$

4. $x + 2y = 3$
$5x - 3y = 2$

5. $3x + 4y = 6$
$2x + 5y = 11$

6. $x + 8y = 12$
$-3x + 7y = -5$

7. $3x + 4y = 7$
$4x - 3y = 1$

8. $2x - 3y = 0$
$6x + 5y = 7$

9. $x + y = 6$
$-2x + y = -3$

Written Exercises

Solve each system of equations by the substitution method.

10. $y = 3x$
$x + 2y = -21$

11. $6x - 4y = -6$
$3x + y = 3$

12. $2x + 2y = 4$
$x - 2y = 0$

13. $2m + n = 1$
$m - n = 8$

14. $x - 2y = 5$
$3x - 5y = 8$

15. $3x + 4y = -7$
$2x + y = -3$

Solve each system of equations by the elimination method.

16. $4x + y = 9$
$3x - 2y = 4$

17. $x - y = 4$
$x + 2y = 1$

18. $2x + y = 0$
$5x + 3y = 1$

19. $3x + 2y = 9$
$2x - 3y = 19$

20. $4x - 3y = -4$
$3x - 2y = -4$

21. $8x + 3y = 4$
$4x - 9y = -5$

Solve each system of equations. (Use either method.)

22. $3x + 2y = 40$
$x - 7y = -2$

23. $2x + 3y = 8$
$x - y = 2$

24. $x + y = 6$
$x - y = 4.5$

25. $2x - y = 36$
$3x - \frac{1}{2}y = 26$

26. $3y - 2x = 4$
$\frac{1}{6}(3y - 4x) = 1$

27. $3x + \frac{1}{3}y = 10$
$2x - 5 = \frac{1}{3}y$

28. $5m + 2n = -8$
$4m + 3n = 2$

29. $2a + 2b = -3$
$5a + 3b = 6$

30. $3x - 5y = -13$
$4x + 3y = 2$

31. $-9x - 6y = -15$
$13x + 7y = 18\frac{1}{3}$

32. $3m + 7n = 5$
$2m = -7 - 3n$

33. $2y - 3x = 0$
$x - y + 2 = 0$

34. $\frac{2x + y}{3} = 15$
$\frac{3x - y}{5} = 1$

35. $\frac{1}{4}x + y = \frac{7}{2}$
$\frac{1}{2}x - \frac{1}{4}y = 1$

36. $\frac{x}{2} - \frac{2y}{3} = 2\frac{1}{3}$
$\frac{3x}{2} + 2y = -25$

37. $2x + 3y - 8 = 0$
$3x + 2y - 17 = 0$

38. $0.2a = 0.3b$
$0.4a - 0.2b = 0.2$

39. $\frac{1}{3}x + \frac{1}{3}y = 5$
$\frac{1}{6}x - \frac{1}{9}y = 0$

40. $\frac{1}{3}x + 5 = \frac{2}{3}y$
$\frac{1}{2}x + \frac{1}{3}y = \frac{1}{2}$

41. $\frac{a}{2} + \frac{b}{3} = 4$
$\frac{2a}{3} + \frac{3b}{2} = \frac{35}{3}$

42. $\frac{m}{3} + \frac{n}{5} = -\frac{1}{5}$
$\frac{2m}{3} - \frac{3n}{4} = -5$

43. $34x - 63y = -1063$
$14x + 43y = 2251$

44. $108x + 537y = -1395$
$-214x - 321y = 535$

45. $93a + 17b = -157.1$
$74a - 75b = -4392$

46. $4.3a + 6.4b = -2.85$
$5.2a - 6.5b = 42.12$

47. $3.2a + 3.5b = -6.365$
$-2.3a + 5.3b = 44.669$

48. $6.7a + 9.3b = -38.99$
$-7.2a - 3.1b = 18.46$

49. $6a + 7b = -10.15$
$9.2a - 6b = 69.944$

50. $4a + 8.3b = 48.799$
$5a + 3b = 15.2$

Solve each problem.

51. The sum of two numbers is 42. Their difference is 12. What are the two numbers?

52. The sum of Kari's age and her mother's age is 52. Kari's mother is 20 years older than Kari. How old is each?

53. The perimeter of a rectangle is 86 cm. Twice the width exceeds the length by 2 cm. Find the dimensions of the rectangle.

54. The hypotenuse of a right triangle measures 75 m. The length of one leg is four times one-third the other leg. What are the lengths of the legs?

Find the vertices of the triangles with sides contained in the lines whose equations are given.

55. $x - y - 7 = 0$
$3x - 11y + 11 = 0$
$x + y + 1 = 0$

56. $x + \dfrac{3}{2}y - 4 = 0$
$x + 4y + 6 = 0$
$x - \dfrac{1}{2}y - 8 = 0$

Find the vertices of the parallelograms with sides contained in the lines whose equations are given.

57. $2x + y + 12 = 0$
$2x - y + 8 = 0$
$2x - y - 4 = 0$
$2x + y - 12 = 0$

58. $x - y - 2 = 0$
$x + 2y + 3 = 0$
$x - y + 5 = 0$
$3x + 6y - 6 = 0$

mini-review

State the property shown.

1. $(9 \cdot x) \cdot y = y \cdot (9 \cdot x)$

2. $(5 - 3)f + 2g = 2f + 2g$

3. Evaluate $[(5 + 3^2) \div 7] + 8$.

Solve each equation.

4. $a - 2 = 3a - 8$

5. $|4x + 5| - 6 = 0$

Solve each inequality. Graph the solution set.

6. $5c - 1 \geq 8$

7. $12 < 2p + 4$

State the domain and range of each relation. Then state if the relation is a function.

8. $\{(0, 1), (2, 3), (3, 5)\}$

9. $\{(-2, 0), (-1, 1), (-2, 2), (0, 3)\}$

Determine the slope of the line that passes through each pair of points.

10. $(4, -1)$ and $(2, 1)$

11. $(0, 6)$ and $(5, 21)$

Find the standard form of the equation of the line that satisfies the given conditions.

12. slope $= \dfrac{2}{5}$, y-intercept $= 3$

13. x-intercept $= 8$, y-intercept $= \dfrac{1}{2}$

Find an equation of the line that passes through each given point and is parallel to the line with the given equation.

14. $(3, 2)$; $5x + 7y = 3$

15. $(-6, -1)$; $4x - y = -2$

Graph each system of equations. Then state whether the system is consistent and independent, consistent and dependent, or inconsistent.

16. $3x - 7y = -1$
$3x + 7y = 2$

17. $2x + 3y = 7$
$2x + 3y = -7$

3-4 Cramer's Rule

A **determinant** is a square array of numbers or variables enclosed between vertical lines. The following determinant has 2 rows and 2 columns and is called a *second order* determinant.

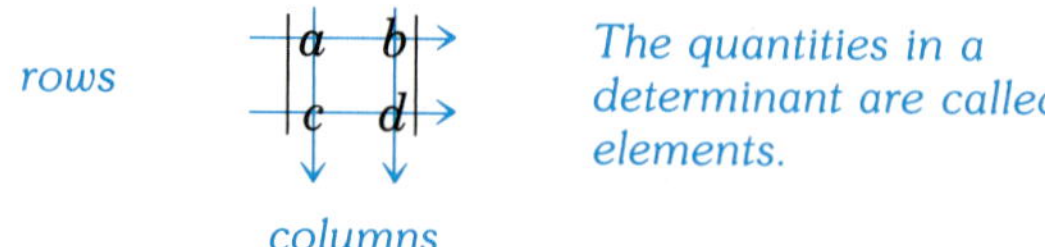

The value of a second order determinant is defined as follows.

$$\begin{vmatrix} a & b \\ c & d \end{vmatrix} = ad - bc$$

Value of a Second Order Determinant

Note that the value of a second order determinant is found using products along the diagonals. Subtract the value of bc from ad.

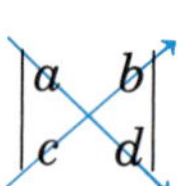

Examples

1 Find the value of $\begin{vmatrix} 2 & 1 \\ 3 & 4 \end{vmatrix}$.

$$\begin{vmatrix} 2 & 1 \\ 3 & 4 \end{vmatrix} = 2 \cdot 4 - 3 \cdot 1$$

$$= 5$$

2 Find the value of $\begin{vmatrix} -6 & 7 \\ 0 & -2 \end{vmatrix}$.

$$\begin{vmatrix} -6 & 7 \\ 0 & -2 \end{vmatrix} = -6(-2) - 0(7)$$

$$= 12$$

Determinants can be used to solve a system of linear equations. Consider solving the following system of two equations in two variables.

$$ax + by = c$$
$$dx + ey = f$$

a, b, c, d, e, and f represent constants, not variables.

Solve for y using the elimination method.

$$
\begin{array}{ll}
adx + bdy = cd & \text{Multiply the first equation by } d. \\
\underline{-adx - aey = -af} & \text{Multiply the second equation by } -a. \\
bdy - aey = cd - af & \text{Add.} \\
(bd - ae)y = cd - af & \text{Factor.}
\end{array}
$$

$$y = \frac{cd - af}{bd - ae} \text{ or } y = \frac{af - cd}{ae - bd}$$

Notice that $bd - ae$ cannot be zero.

Then solve for x in the same manner.

$$x = \frac{ce - bf}{ae - bd}$$

Thus, the solution to the system $\begin{aligned}ax + by &= c\\dx + ey &= f\end{aligned}$ is

$$\left(\frac{ce - bf}{ae - bd}, \frac{af - cd}{ae - bd}\right).$$

Notice that the denominator of each fraction is the same. It can be written as a determinant. Also, each numerator may be written as a determinant.

$$ae - bd = \begin{vmatrix} a & b \\ d & e \end{vmatrix} \qquad ce - bf = \begin{vmatrix} c & b \\ f & e \end{vmatrix} \qquad af - cd = \begin{vmatrix} a & c \\ d & f \end{vmatrix}$$

Therefore, the solution to a system of two linear equations in two variables can be found using determinants. This method is known as **Cramer's Rule**.

The solution to the system $\begin{cases}ax + by = c\\dx + ey = f\end{cases}$ is (x, y)

where $x = \dfrac{\begin{vmatrix} c & b \\ f & e \end{vmatrix}}{\begin{vmatrix} a & b \\ d & e \end{vmatrix}}$, $y = \dfrac{\begin{vmatrix} a & c \\ d & f \end{vmatrix}}{\begin{vmatrix} a & b \\ d & e \end{vmatrix}}$ and $\begin{vmatrix} a & b \\ d & e \end{vmatrix} \neq 0$.

Cramer's Rule

Example

3   **Use Cramer's Rule to solve this system:** $\begin{aligned}3x - 5y &= -7\\x + 2y &= 16\end{aligned}$

$$x = \frac{\begin{vmatrix} -7 & -5 \\ 16 & 2 \end{vmatrix}}{\begin{vmatrix} 3 & -5 \\ 1 & 2 \end{vmatrix}} \qquad\qquad y = \frac{\begin{vmatrix} 3 & -7 \\ 1 & 16 \end{vmatrix}}{\begin{vmatrix} 3 & -5 \\ 1 & 2 \end{vmatrix}}$$

$$= \frac{-7(2) - 16(-5)}{3(2) - 1(-5)} \qquad\qquad = \frac{3(16) - 1(-7)}{3(2) - 1(-5)}$$

$$= \frac{66}{11} \qquad\qquad\qquad\qquad = \frac{55}{11}$$

$$= 6 \qquad\qquad\qquad\qquad\quad = 5$$

The solution is $(6, 5)$.

Exploratory Exercises

Find the value of each determinant.

1. $\begin{vmatrix} 3 & 1 \\ 4 & 6 \end{vmatrix}$

2. $\begin{vmatrix} 7 & -3 \\ 0 & 1 \end{vmatrix}$

3. $\begin{vmatrix} 0 & 1 \\ 0 & 1 \end{vmatrix}$

4. $\begin{vmatrix} 11 & -2 \\ -3 & -5 \end{vmatrix}$

5. $\begin{vmatrix} 1 & 0 \\ 0 & 1 \end{vmatrix}$

6. $\begin{vmatrix} -8 & -7 \\ -4 & -6 \end{vmatrix}$

7. $\begin{vmatrix} 2 & 4 \\ -3 & 1 \end{vmatrix}$

8. $\begin{vmatrix} -5 & 3 \\ 1 & 2 \end{vmatrix}$

Write the determinants you would use to solve each system by Cramer's Rule.

9. $3x + 2y = 5$
$4x - y = 3$

10. $4x + 2y = 8$
$6x - 3y = 0$

11. $2a - 4b = 16$
$3a - 5b = 21$

12. $r - s = 0$
$2r + 5s = -3$

13. $3x + y = -8$
$4x - 2y = -14$

14. $x + y = 6$
$x - y = 2$

Written Exercises

Find the value of each determinant.

15. $\begin{vmatrix} 24 & 6 \\ -13 & -4 \end{vmatrix}$

16. $\begin{vmatrix} 18 & -5 \\ -9 & 11 \end{vmatrix}$

17. $\begin{vmatrix} -13 & -11 \\ 17 & -12 \end{vmatrix}$

18. $\begin{vmatrix} -6 & 7 \\ -9 & 10 \end{vmatrix}$

Solve each system of equations by using Cramer's Rule.

19. $5x + 4y = -1$
$2x - y = 10$

20. $x - 4y = 1$
$2x + 3y = 13$

21. $3m - n = 3$
$m + n = 5$

22. $2x - y = 7$
$x + 3y = 7$

23. $3x + 2y = 5$
$5x - 6y = 11$

24. $x + 11 = 8y$
$8(x - y) = 3$

25. $3x - 7y = 2$
$-6x + 13y = -4$

26. $x - y = 0$
$2x + 5y = -3$

27. $3a + 8 = -b$
$4a - 2b = -14$

28. $6a + 5b = -7$
$2a - 3b = 7$

29. $7y + 4x = 22$
$8x - 2y = -5$

30. $3x - 4y = 23$
$9x + 2y = -15$

31. $2x - 3y = 19$
$6x + 6y = -3$

32. Explain why Cramer's Rule will not work if a system of equations is dependent or inconsistent.

33. How do you determine if a system of equations is dependent by using Cramer's Rule?

34. How do you determine if a system of equations is inconsistent by using Cramer's Rule?

Excursions in Algebra

History

Gabriel Cramer was a Swiss mathematician of the 18th century. He was a professor of mathematics at the University of Geneva at the age of twenty. Although Cramer's Rule is named after him, he was not the first person to originate that result. Colin Maclaurin, a British mathematician, wrote the rule for solving systems of equations by determinants in his *Treatise of Algebra*. It was published in 1748, two years before Cramer published Cramer's Rule. Often in mathematics, the person who popularized a result had his or her name attached to it, although later it was learned that someone else had originally discovered the same result.

Find the value of each expression.

1. $8 + 6 \div 2 - 5$ **2.** $3(7 - 4)^2$

Evaluate if $a = -3$, $b = \frac{2}{3}$, and $c = -4$.

3. $b^2c - 2a$ **4.** $(a + b)^2$

State the property shown in each of the following.

5. $3(5 + 8) = 3(8 + 5)$

6. $3(5 + 8) = 3(5) + 3(8)$

7. If $8 - 2 = 6$ then $6 = 8 - 2$

8. $7(1) = 7$

Given $f(x) = x^2 - 6x + 31$, find each value.

9. $f(-7)$ **10.** $f(5)$ **11.** $f(2a + 3)$

12. Find the slope-intercept form of the equation $5x - 9y = 7$.

Solve each equation or inequality.

13. $|3x - 7| = 41$ **14.** $|2x + 3| = 6 - x$

15. $7x - 3 < 9x + 5$ **16.** $3|7x - 24| \le 48$

Solve each system of equations.

17. $x + 6y = 1$
$3x - 2y = 13$

18. $\frac{1}{3}x + \frac{1}{2}y = 7$
$2x - \frac{1}{5}y = 10$

Find the standard form of the equation of the line that satisfies the given conditions.

19. slope $= \frac{2}{3}$; passes through $(5, -2)$

20. passes through $(4, 7)$ and $(7, -2)$

21. x-intercept $= 6$, y-intercept $= -2$

22. Find the slope, x-intercept, and y-intercept of the graph of $6x - 5y = 12$.

Find an equation of the line that is parallel to the graph of the given equation and passes through the given point.

23. $6x - 7y = 4$; $(-3, 7)$

24. $y = \frac{3}{4}x - 2$; $(-3, 7)$

25. $y = -\frac{12}{7}x - \frac{1}{7}$; $(-2, 4)$

Find an equation of the line that is perpendicular to the graph of the given equation and passes through the given point.

26. $6x - 7y = 4$; $(1, -6)$

27. $y = \frac{3}{4}x - 2$; $(1, -6)$

28. $y = -\frac{2}{3}x - 1$; $(6, -1)$

Graph each equation or inequality.

29. $5x - 2y = 6$ **30.** $y = [2x] + 1$

31. $4x + 3y \ge 8$ **32.** $y \ge |x + 1|$

Find the value of each determinant.

33. $\begin{vmatrix} 8 & 7 \\ 3 & -1 \end{vmatrix}$ **34.** $\begin{vmatrix} -2 & 8 \\ 1 & -4 \end{vmatrix}$

35. Use Cramer's Rule to solve this system of equations.
$$-3x - 2y = -5$$
$$4x + y = 3$$

Solve each problem.

36. Luke is three years older than Theo. Ten years ago, he was twice as old as Theo. How old is each now?

37. Yasuko has 51 coins in nickels and dimes, and has \$4.15 in all. How many of each does she have?

38. Charlie has a 16 quart radiator in his car and knows that the antifreeze mixture is at least 40%. How much must he drain off and replace with 100% antifreeze to have at least 70%?

 Graphing Systems of Inequalities

Consider the following system of inequalities.

$$y \geq 2x - 1$$
$$y \leq -2x - 2$$

To solve this system, find the ordered pairs that satisfy *both* inequalities. One way is to graph each inequality and find the intersection of the two graphs.

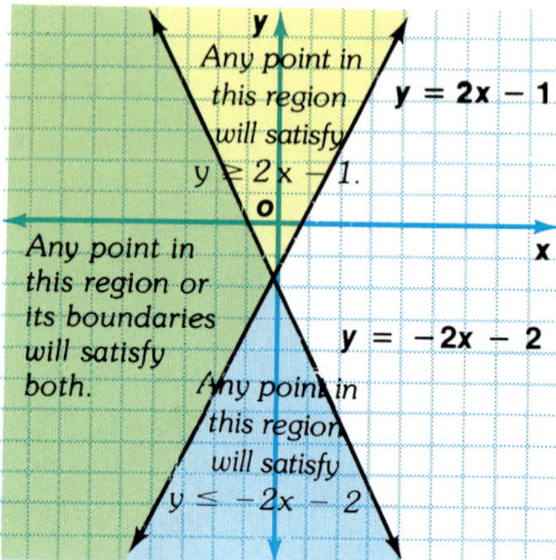

Some systems of inequalities have *no* solutions.

Example

1 Solve this system of inequalities by graphing: $\begin{aligned} y &> x + 1 \\ y &< x - 4 \end{aligned}$

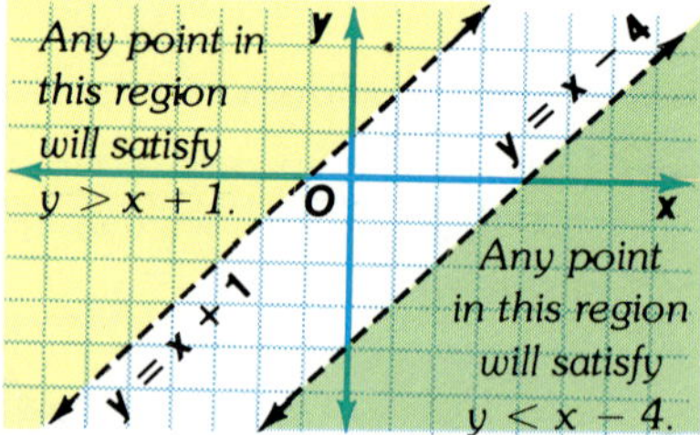

The broken lines indicate that the boundaries are not part of the graphs.

The graphs of the two inequalities have *no* points in common. So, *no* ordered pair will satisfy both inequalities. The solution set is the empty set, ∅.

Systems of more than two inequalities can also be graphed.

Example

2 **Solve this system of inequalities by graphing:** $y \leq 3$
$$x \geq -2$$
$$y > x$$

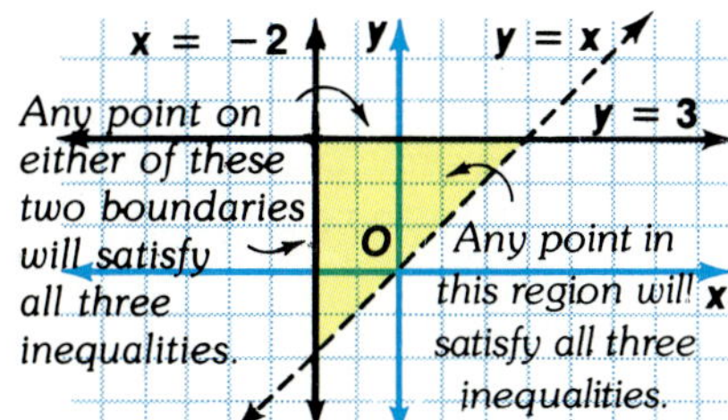

Exploratory Exercises

Does the point given in color satisfy the system of inequalities?

1. $y < \frac{1}{2}x + 2$ $(0, 0)$
$y > 3x - 2$

2. $y < x - 2$ $(1, 2)$
$y > -x$

3. $y > -2$ $(1, 1)$
$x \leq 1$

4. $y < 3$ $(4, 4)$
$x \geq -1$

Written Exercises

Solve each system of inequalities by graphing.

5. $y > 3$
$x \leq 1$

6. $y \leq 5$
$x \geq -1$

7. $y > 3$
$y + x > 2$

8. $y < -2$
$y - x > 1$

9. $y \geq 2x - 2$
$y \leq -x + 2$

10. $y \geq x - 3$
$y \geq -x + 1$

11. $y > x + 1$
$y < x - 3$

12. $y - x \leq 3$
$y \geq x + 2$

13. $y < -x - 3$
$x + 2 > y$

14. $x + y > 5$
$x - y \leq 3$

15. $x + y \geq 3$
$2x - 3y \leq 6$

16. $x + 2y \geq 7$
$3x - 4y < 12$

17. $y > x + 2$
$2y < x - 3$

18. $x + y > 7$
$x + y < 10$

19. $|x| > 5$
$x + y < 6$

20. $|x + 2| < 3$
$x + y \geq 1$

21. $x \geq 1$
$y < -1$
$y > x$

22. $y \leq 2$
$y \geq 2x$
$y \geq x + 1$

23. $x \geq -1$
$x \leq 1$
$y > 2$

24. $x \geq -2$
$2y \geq 3$
$x - y \leq -5$

25. $2y \leq x$
$3x + 5y \leq 10$
$x \leq 3$

26. $y < 2x + 1$
$y > 2x - 2$
$3x + y > 8$

27. $y > x + 3$
$y < x - 4$
$2y + 3x > 4$

28. $x + y < 9$
$x - y > 3$
$y - x > 4$

To solve many problems in mathematics, you can write an equation. Then you solve the equation and answer the problem. However, this procedure may not always be the best one to use.

There are other strategies that may be more suitable than writing an equation. One of these strategies is to solve a simpler problem. This strategy involves setting aside the original problem and solving a simpler or more familiar case of the problem. The same concepts and relationships that were used to solve the simpler problem can then be used to solve the original problem.

Example: **Find the sum of the numbers 1 through 1000.**

Obviously, this problem could be solved by actually adding all the numbers. But this process would be very tedious even if you used a calculator.

Consider a simpler problem. Find the sum of the numbers 1 through 10.

$$
\begin{aligned}
S &= 1 + 2 + 3 + \cdots + 10 \\
S &= 10 + 9 + 8 + \cdots + 1 \\
\hline
2S &= 11 + 11 + 11 + \cdots + 11 \\
2S &= 10 \cdot 11 \\
S &= 5 \cdot 11 \text{ or } 55
\end{aligned}
$$

Now extend this concept to the original problem.

$$
\begin{aligned}
S &= 1 + 2 + 3 + \cdots + 1000 \\
S &= 1000 + 999 + 998 + \cdots + 1 \\
\hline
2S &= 1001 + 1001 + 1001 + \cdots + 1001 \\
2S &= 1000 \cdot 1001 \\
S &= 500 \cdot 1001 \text{ or } 500{,}500
\end{aligned}
$$

This strategy may also involve breaking a complicated problem down into several easier problems. After these problems are solved, their solutions may be used to solve the original problem.

Exercises

Solve each problem.

1. A drain pipe is 750 cm long. A spider climbs up 100 cm during the day but falls back 80 cm during the night. If the spider begins at the bottom of the pipe, on what day will it get to the top?

2. Find the total number of squares in the checkerboard shown at the right.

3. Find the sum of the first n positive integers.

4. A total of 3001 digits were used to print the page numbers of the Northern College annual. How many pages are in the annual?

3-6 Linear Programming

Linear programming is a procedure for finding the maximum or the minimum value of a function in two variables, subject to given conditions on the variables, called **constraints**. The constraints are often expressed as linear inequalities.

Suppose you want to find the maximum or minimum value for the function $f(x, y) = 5x - 3y$. The values of x and y have the following constraints.

$$x \geq 0 \qquad 0 \leq y \leq 5 \qquad x + y \leq 7 \qquad 3y \geq x - 3$$

By graphing each inequality and finding the solution set for the system, you can determine the set of ordered pairs that satisfy all of the given constraints. The polygonal region shown at the right is the set of points which are common solutions to the above inequalities.

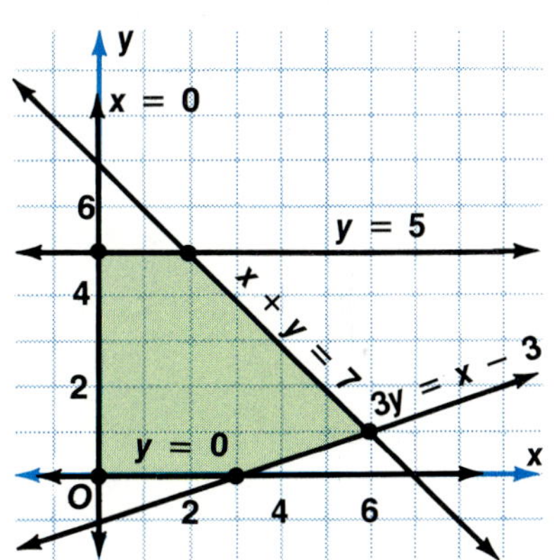

Every point in the shaded region has coordinates that satisfy each of the inequalities given above.

The region includes infinitely many points. It would be impossible to evaluate the function for all points within the region. Mathematicians have shown that the maximum and minimum values are always represented by vertices of the region. In this case there are five vertices. So, you need to find the value of $5x - 3y$ for $(0, 0)$, $(3, 0)$, $(6, 1)$, $(2, 5)$, and $(0, 5)$.

$$f(x, y) = 5x - 3y$$

$$f(6, 1) = 5(6) - 3(1) \qquad\qquad f(0, 0) = 5(0) - 3(0)$$
$$= 27 \qquad\qquad\qquad\qquad = 0$$

$$f(2, 5) = 5(2) - 3(5) \qquad\qquad f(3, 0) = 5(3) - 3(0)$$
$$= -5 \qquad\qquad\qquad\qquad = 15$$

$$f(0, 5) = 5(0) - 3(5)$$
$$= -15$$

Thus, the maximum value is 27 at $(6, 1)$ and the minimum value is -15 at $(0, 5)$.

1 Find the maximum and minimum values of $f(x, y) = x + 2y$ for the polygonal region determined by the following inequalities.

$$x \geq 0 \qquad y \geq 0 \qquad 2x - y \leq 4 \qquad x + 3y \leq 9$$

First, graph the inequalities and find the coordinates of the vertices of the resulting polygon.

The coordinates of the vertices are $(0, 0)$, $(2, 0)$, $(3, 2)$, and $(0, 3)$.

Then evaluate the function $f(x, y) = x + 2y$ at each vertex.

$$f(0, 0) = 0 + 2(0) = 0$$
$$f(2, 0) = 2 + 2(0) = 2$$
$$f(3, 2) = 3 + 2(2) = 7$$
$$f(0, 3) = 0 + 2(3) = 6$$

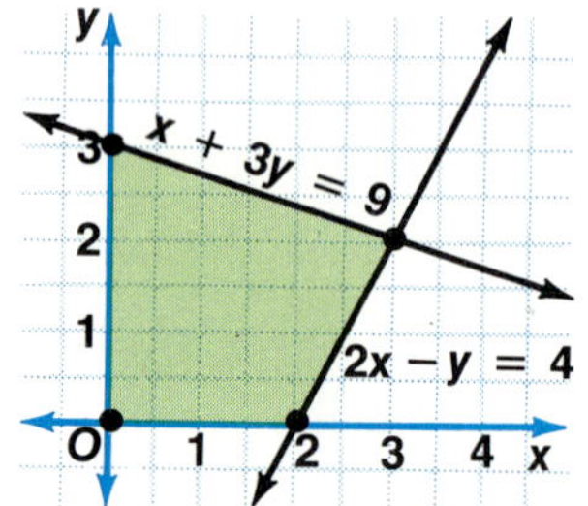

The maximum value of the function is 7 at $(3, 2)$ and the minimum value is 0 at $(0, 0)$.

2 Using Calculators

Find the maximum and minimum values of $f(x, y) = 3x + 4y$ for the polygonal region determined by these inequalities.

$$x \leq 6 \qquad y \leq 3 \qquad x - 3y \leq 9 \qquad 3x + y \leq 6$$

First, graph the inequalities and find the coordinates of the vertices of the resulting polygon.

The coordinates of the vertices are $(1, 3)$, $(6, 3)$, $(2.7, -2.1)$, and $(6, -1)$.

Then use a calculator to evaluate $f(x, y) = 3x + 4y$ at each vertex.

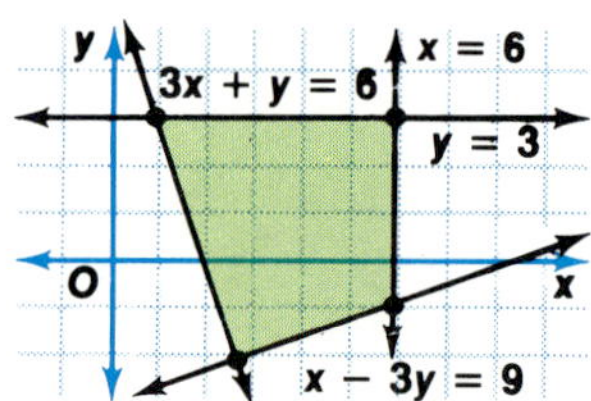

ENTER: 3 $\boxed{\times}$ 1 $\boxed{+}$ 4 $\boxed{\times}$ 3 $\boxed{=}$

DISPLAY: *3 3 1 3 4 4 3 15*

ENTER: 3 $\boxed{\times}$ 6 $\boxed{+}$ 4 $\boxed{\times}$ 3 $\boxed{=}$

DISPLAY: *3 3 6 18 4 4 3 30*

ENTER: 3 $\boxed{\times}$ 2.7 $\boxed{+}$ 4 $\boxed{\times}$ 2.1 $\boxed{+/-}$ $\boxed{=}$

DISPLAY: *3 3 2.7 8.1 4 4 2.1 −2.1 −0.3*

ENTER: 3 $\boxed{\times}$ 6 $\boxed{+}$ 4 $\boxed{\times}$ 1 $\boxed{+/-}$ $\boxed{=}$

DISPLAY: $3\ \ 3\ \ 6\ \ 18\ \ 4\ \ 4\ \ 1\ \ -1\ \ 14$

So, $f(1, 3) = 15$, $f(6, 3) = 30$, $f(2.7, -2.1) = -0.3$, and $f(6, -1) = 14$. The maximum value of the function is 30 at $(6, 3)$, and the minimum value is -0.3 at $(2.7, -2.1)$.

Exploratory Exercises

Given $f(x, y) = 3x + 2y$, find each value.

1. $f(3, 2)$
2. $f(4, 1)$
3. $f(-2, 1)$
4. $f(5, -2)$
5. $f(0, -5)$
6. $f(6, 0)$
7. $f(0.5, 1.2)$
8. $f(-1, 0.25)$

Given $f(x, y) = 5x - 2y$, use a calculator to find each value.

9. $f(4, 1)$
10. $f(0, 0)$
11. $f(-5, 4)$
12. $f(-2, -6)$
13. $f(5, -3)$
14. $f(3, 1.5)$
15. $f(0, -0.3)$
16. $f(-0.2, -1)$

Find the maximum and minimum values of each function defined for the polygonal region having vertices $(0, 0)$, $(5, 0)$, $(4, 6)$, and $(0, 6)$.

17. $f(x, y) = x + y$
18. $f(x, y) = y - x$
19. $f(x, y) = \frac{1}{2}x - y$
20. $f(x, y) = x + 3y$
21. $f(x, y) = -x - 3y$
22. $f(x, y) = 0.5x - 1.5y$

Written Exercises

Graph each system of inequalities. Name the vertices of the polygon formed.

23. $1 \le y \le 3$
$y \le 2x + 1$
$y \le -\frac{1}{2}x + 6$

24. $x \ge 0$
$y \ge 3$
$y \ge 2x + 1$
$y \le -\frac{1}{2}x + 6$

25. $y \ge 1$
$y \le 2x + 1$
$x \le 6$

26. $0 \le x \le 50$
$0 \le y \le 70$
$60 \le x + y \le 80$

Graph each system of inequalities. Name the vertices of the polygon formed. Find the maximum and minimum values of the given function.

27. $y \ge 2$
$1 \le x \le 5$
$y \le x + 3$
$f(x, y) = 3x - 2y$

28. $x + y \ge 2$
$4y \le x + 8$
$y \ge 2x - 5$
$f(x, y) = 4x + 3y$

29. $y \le x + 6$
$y + 2x \ge 6$
$2 \le x \le 6$
$f(x, y) = 3x + y$

30. $x + y \ge 2$
$4y \le x + 8$
$2y \ge 3x - 6$
$f(x, y) = 3y + x$

31. $y \le 7$
$y \le x + 4$
$y \ge -x + 6$
$x \le 5$
$f(x, y) = 2x - 3y$

32. $y \le x + 5$
$y \ge x$
$x \ge -3$
$y + 2x \le 5$
$f(x, y) = x - 2y$

Use a calculator to find the maximum and minimum values of each function for the polygonal region determined by the given inequalities.

33. $y \geq 0$
$0 \leq x \leq 5$
$-x + y \leq 2$
$x + y \leq 6$
$f(x, y) = 5x - 3y$

34. $x \geq 0$
$y \geq 0$
$x + 2y \leq 6$
$2y - x \leq 2$
$x + y \leq 5$
$f(x, y) = 3x - 5y$

35. $y \leq 1$
$5x \leq -2$
$y \geq -2$
$1.2x - y \geq -2.9$
$f(x, y) = 4x + 2y$

mini-review

Evaluate if $a = 4$, $b = -\frac{1}{2}$, $c = -3$, and $d = 5$.

1. $\dfrac{5ac}{bd}$

2. $\dfrac{4a + b}{cd}$

3. $\dfrac{5b^2a - d}{c}$

Name the sets of numbers to which each number belongs. (Use N, W, Z, Q, I, and R.)

4. $-\sqrt{36}$

5. $8.1\overline{24}$

6. 2π

Simplify.

7. $(63 - 12)2 - 30$

8. $6p + 0.8q - (2.3p - 2.2q)$

Solve each compound sentence.

9. $9 < 3x + 2 < 23$

10. $18 < 7y - 4 \leq 24$

Given $f(x) = 4x^3 + 3x^2 + 2x - 2$, find each value.

11. $f(0)$

12. $f(-2)$

13. $f(b)$

Find the slope, y-intercept, and x-intercept of the graph of each of the following.

14. $2x - 6y = 4$

15. $\frac{1}{2}x + \frac{1}{3}y = 6$

Solve each system of equations by the substitution method.

16. $4x + 7y = -1$
$2x + y = 7$

17. $x + 5y = 14$
$-2x + 6y = 4$

Solve each system of equations by the elimination method.

18. $3x + 10y = -16$
$2x + 5y = 7$

19. $-4x + 7y = 17$
$12x + 3y = 9$

Solve each system of equations by using Cramer's Rule.

20. $2x + 2y = 5$
$2x - 5y = 29$

21. $9a - b = 1$
$3a + 2b = 12$

22. Solve this system of inequalities by graphing: $3x - y \geq -3$
$3x + 2y \leq 6$

3-7 Problem Solving: Linear Programming

Many practical problems can be solved by linear programming. These problems are of such a nature that certain constraints exist for the variables, and some function of these variables must be maximized or minimized. Use the following method to solve linear programming problems.

1. **Define variables.**
2. **Write a system of inequalities.**
3. **Graph the system. Find vertices of the polygon formed.**
4. **Write an expression to be maximized or minimized.**
5. **Substitute values from vertices into the expression.**
6. **Select greatest or least result. Answer the problem.**

Linear Programming Procedure

Examples

1 The Blair Company makes two types of pianos: spinets and consoles. The equipment in the factory allows for making at most 450 spinets and 200 consoles in one month. The chart shows the cost of making each type of piano and the profit. During the month of June, the company can spend $360,000 to make these pianos. To make the greatest profit, how many of each type should be made in June?

Piano	Cost per Unit	Profit per Unit
Spinet	$600	$125
Console	$900	$200

Define variables.

Let s = the number of spinets made.
Let c = the number of consoles made.

Write inequalities.

$0 \le s \le 450$ — The number of spinets made is at most 450.

$0 \le c \le 200$ — The number of consoles made is at most 200.

$600s + 900c \le 360{,}000$ — The cost of spinets plus the cost of consoles does not exceed $360,000.

Graph the system.

Any point in the shaded region or its boundaries will satisfy the conditions of the problem. The vertices of the polygon are (0, 0), (0, 200), (300, 200), (450, 100), and (450, 0).

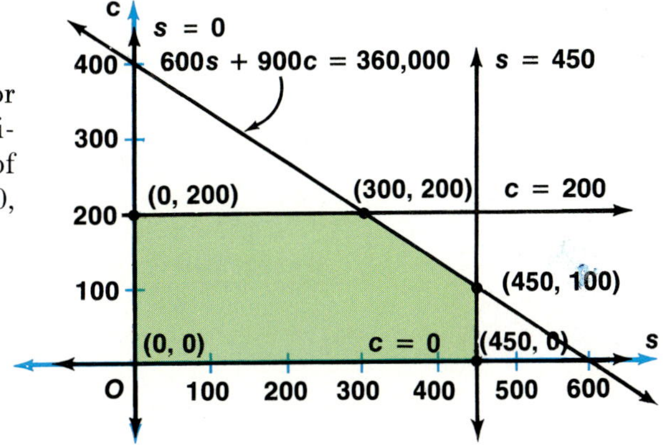

Recall that the maximum or minimum value is always at one of the vertices.

Substitute values from the vertices into the expression.

(s, c)	(0, 0)	(0, 200)	(300, 200)	(450, 100)	(450, 0)
125s + 200c	$0	$40,000	$77,500	$76,250	$56,250

Answer the problem.

The Blair Company will make the greatest profit by building 300 spinets and 200 consoles. This produces a profit of $77,500.

2 **Raw Materials A and B are used to make one of the Target Company's products. The product must contain no more than 9 units of A and at least 18 units of B. It must cost no more than $300. The chart at the right shows how much each unit of raw material costs and weighs. How much of each raw material should be used to maximize the weight?**

Material	Cost per Unit	Weight per Unit
A	$4	10 pounds
B	$12	20 pounds

Define variables.

Let a stand for amount of material A used.
Let b stand for amount of material B used.

Write inequalities.

$0 \le a \le 9$ The product contains no more than 9 units of A.

$b \ge 18$ The product contains at least 18 units of B.

$4a + 12b \le 300$ The cost can be no more than $300.

Graph the system.

The vertices are (0, 18), (0, 25), (9, 22), and (9, 18).

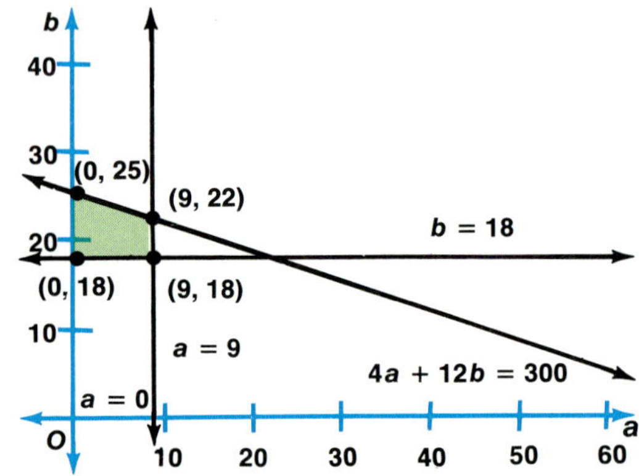

Write expression to be maximized.

Weight = Weight material A + Weight material B
$P(a, b) =$ 10a   +   20b

Substitute values into the expression.

(a, b)	(0, 18)	(0, 25)	(9, 22)	(9, 18)
10a + 20b	360	500	530	450

Answer the problem.

The Target Company should use 9 units of material A and 22 units of material B for each unit of product.

The available parking area of a parking lot is 600 square meters. A car requires 6 square meters and a bus requires 30 square meters of space. The attendant can handle no more than 60 vehicles. If a car is charged $2.50 and a bus $7.50, how many of each should be accepted to maximize income? Let c = the number of cars accepted. Let b = the number of buses accepted.

1. Write an expression to represent the total income.
2. Write an inequality to represent the total number of cars and buses.
3. Write an inequality to represent the amount of space required for the cars and buses.
4. Graph the system of inequalities. (Show only the first quadrant since b and c cannot be negative.)
5. Name the vertices of the polygon.
6. Evaluate the expression from Exercise 1 for each vertex.
7. To maximize income, how many cars and how many buses should be accepted?
8. What is the maximum income?

Written Exercises

Solve.

9. A painter has exactly 32 units of yellow dye and 54 units of green dye. He plans to mix as many gallons as possible of color A and color B. Each gallon of color A requires 4 units of yellow dye and 1 unit of green dye. Each gallon of color B requires 1 unit of yellow dye and 6 units of green dye.

   a. Let x be the number of gallons of color A and let y be the number of gallons of color B. Write the inequalities.

   b. Graph the system of inequalities and name the vertices of the polygon formed.

   c. Find the maximum number of gallons, $x + y$, possible.

10. A farmer has 20 days in which to plant corn and beans. The corn can be planted at a rate of 10 acres per day and the beans at a rate of 15 acres per day. The farm has 250 acres available.

   a. Let x be the number of acres of corn and let y be the number of acres of beans. Write the inequalities that represent the situation.

   b. Graph the system of inequalities and name the vertices of the polygon formed.

   c. If corn profits $30 per acre and beans profit $25 per acre, find the values of x and y that maximize the profit.

   d. If corn profits $29 per acre and beans profit $30 per acre, find the values of x and y that maximize the profit.

11. A dressmaking shop makes dresses and pantsuits. The equipment in the shop allows for making at most 30 dresses and 20 pantsuits in a week. It takes 10 worker-hours to make a dress and 20 worker-hours to make a pantsuit. There are 500 worker-hours available per week in the shop.

   a. If the profit on a dress and the profit on a pantsuit are the same, how many of each should be made to maximize the profit?

   b. If the profit on a pantsuit is three times the profit on a dress, how many of each should be made to maximize the profit?

12. Fashion Furniture makes two kinds of chairs, rockers and swivels. Two operations, A and B, are used. Operation A is limited to 20 hours a day. Operation B is limited to 15 hours per day. The following chart shows the amount of time each operation takes for one chair. It also shows the profit made on each chair.

Chair	Operation A	Operation B	Profit
Rocker	2 h	3 h	$12
Swivel	4 h	1 h	$10

How many chairs of each kind should Fashion Furniture make each day to maximize profit?

13. Recreation Unlimited produces footballs and basketballs. Producing a football requires 4 hours on machine A and 2 hours on machine B. Producing a basketball requires 6 hours on machine A, 6 hours on machine B, and 1 hour on machine C. Machine A is available 120 hours per week, machine B is available 72 hours per week, and machine C is available 10 hours per week. If the company profits $3 on each football and $2 on each basketball, how many of each should be produced to maximize the company's profit?

14. The table gives the amounts of ingredient A and ingredient B in two types of dog foods: X and Y.

Food Type	Amount of Ingredient A	Amount of Ingredient B
X	1 unit per pound	$\frac{1}{2}$ unit per pound
Y	$\frac{1}{3}$ unit per pound	1 unit per pound

The dogs in a kennel must get *at least* 40 pounds of food per day. The food may be a mixture of type X and type Y. The daily diet must contain at least 20 units of ingredient A and at least 30 units of ingredient B. The dogs must not get more than 100 pounds of food per day.

a. Type X costs $0.80 per pound and type Y costs $0.40 per pound. What is the least possible cost per day for feeding the dogs?

b. If the price on type X is raised to $1.00 per pound and the price on type Y remains the same, should the mixture be changed?

3-8 Equations of Planes

The equation $3x + 4y + z = 12$ is an open sentence in three variables, x, y, and z. You can write the solutions of open sentences in three variables as sets of **ordered triples**.

To draw the graph of a linear equation in three variables, it is necessary to use three dimensions. The graph of an equation of the form $Ax + By + Cz = D$ is a plane.

The graph of all ordered triples of real numbers is space.

If A, B, and C are real numbers and not all zero, then the graph of $Ax + By + Cz = D$ is a plane.	*Equation of a Plane*

To graph a plane it is necessary to separate space into eight regions, each called an **octant**. Think of three coordinate planes intersecting at right angles as shown below.

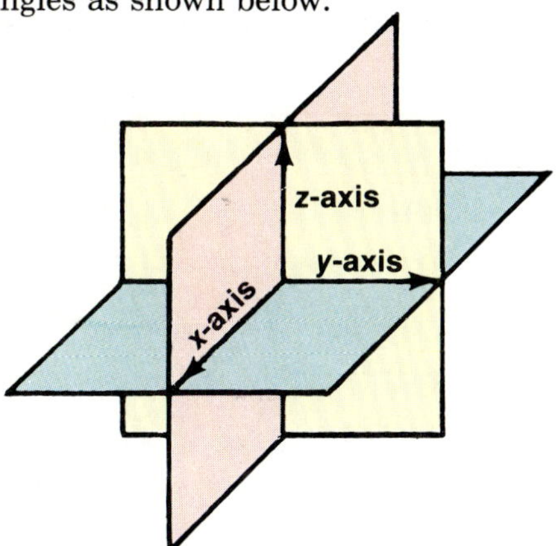

The octants are numbered as shown below. Any point lying in a coordinate plane is not in any octant. The signs of x, y, and z for each octant are shown as an ordered triple.

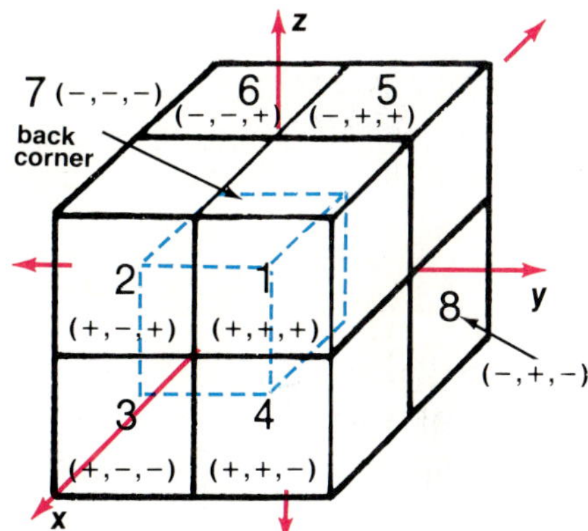

Example

1 **Locate the point (2, 4, 3).**

First, locate 2 on the positive x-axis, 4 on the positive y-axis, and 3 on the positive z-axis. Complete a "box" by drawing lines parallel to the axes through each intercept.

The point is in octant 1.

Octant 1 is also called the first octant.

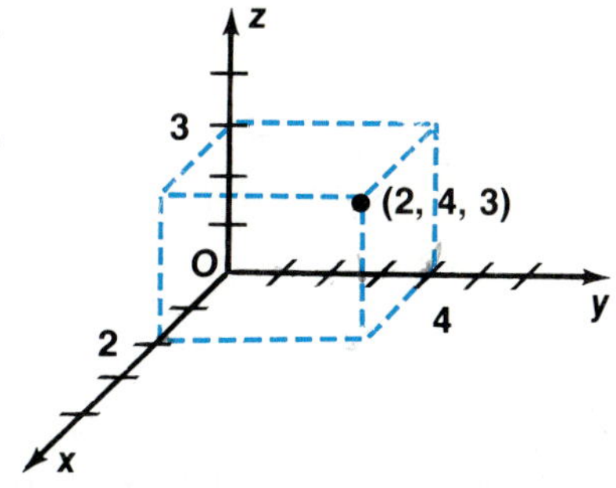

When graphing an ordered triple, it is not necessary to show the entire "box." Simply indicate the corner of the box. The desired point will always be the corner farthest from the origin.

Example

2 **Locate the point $(-2, -3, 1)$.**

First, locate -2 on the x-axis. Then draw a segment 3 units long in the negative direction, parallel to the y-axis. From that point, draw a segment 1 unit long in the positive direction, parallel to the z-axis.

The point is in octant 6.

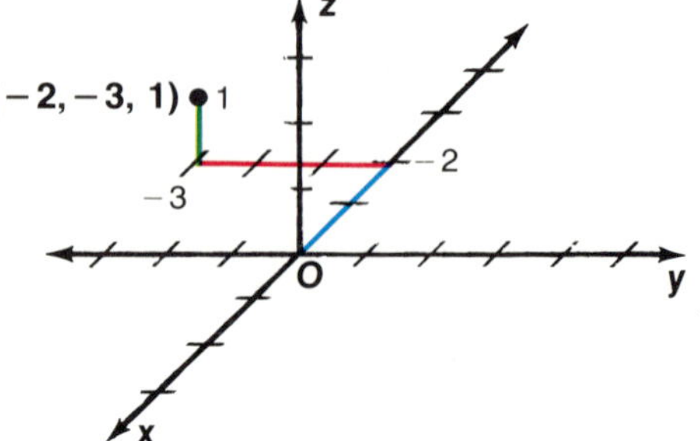

To graph a linear equation in three variables, first find the intercepts of the graph. Connect the intercepts on each axis. This forms the portion of a plane that lies in a single octant. Study the following example.

Example

3 **Graph $2x + 4y + 3z = 12$.**

To find the x-intercept, let $y = 0$ and $z = 0$.

$$2x = 12$$
$$x = 6$$

To find the y-intercept, let $x = 0$ and $z = 0$.

$$4y = 12$$
$$y = 3$$

To find the z-intercept, let $x = 0$ and $y = 0$.

$$3z = 12$$
$$z = 4$$

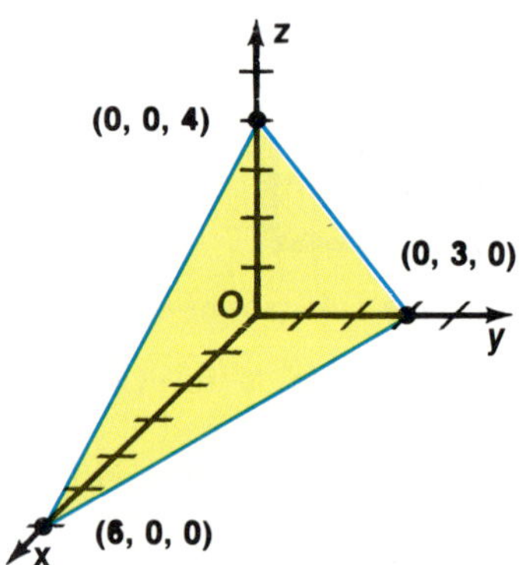

To indicate the plane, connect the intercepts.
Remember, a plane extends infinitely in two dimensions.

The intersection of a plane with one of the coordinate planes is a line called a **trace**. The xy-trace is the line formed by the intersection of a plane with the xy-plane. Since all points in the xy-trace have a z-coordinate of zero, the equation can be found by letting $z = 0$ in the equation of the plane. Equations for the other two traces can be found in a similar manner.

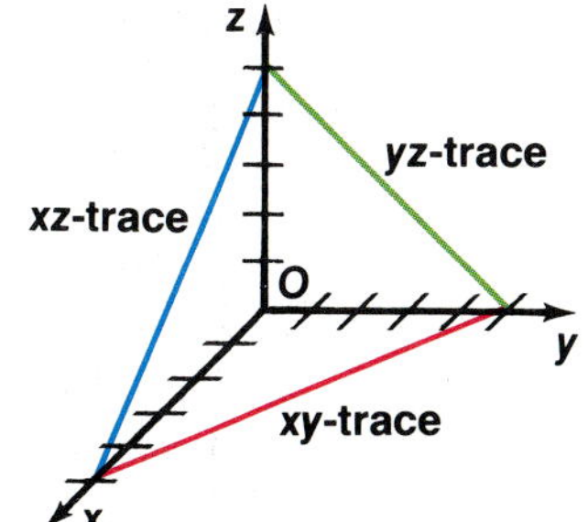

The portion in octant 1 is shown.

Example

4 **Graph $10x - 5y + 4z = 20$ and find the trace in each coordinate plane.**

The x-, y-, and z-intercepts are 2, -4, and 5.
To find the equation of the xy-trace, let $z = 0$.
$$10x - 5y = 20$$
$$2x - y = 4 \qquad \textit{Simplify.}$$

To find the equation of the xz-trace, let $y = 0$.
$$10x + 4z = 20$$
$$5x + 2z = 10 \qquad \textit{Simplify.}$$

To find the equation of the yz-trace, let $x = 0$.
$$-5y + 4z = 20$$

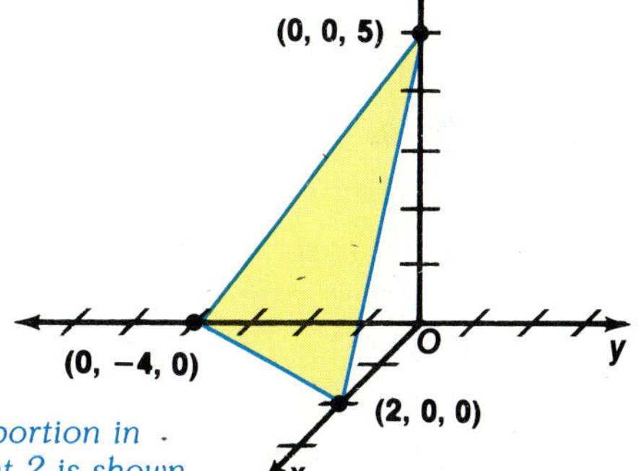

The portion in octant 2 is shown.

If you know the equations of two traces, you can write the equation of the plane containing the two traces.

Example

5 **The equations for two traces of a plane are $3x + 2y = -12$ and $3x - 4z = -12$. Find an equation of the plane.**

First, find the intercepts of the two traces.
The xy-trace is $3x + 2y = -12$. So, the x- and y-intercepts are -4 and -6.
The xz-trace is $3x - 4z = -12$. So, the x- and z-intercepts are -4 and 3.
Now, use the intercepts to write the equation of the plane.
The x-, y-, and z-intercepts of the plane are -4, -6, and 3. Since the least common multiple of -4, -6, and 3 is 12, use 12 as the constant in the equation. Divide 12 by each intercept to obtain the coefficients.

$$\frac{12}{-4} = -3 \qquad \frac{12}{-6} = -2 \qquad \frac{12}{3} = 4$$

The equation of the plane is $-3x - 2y + 4z = 12$.

In which octant does each point lie?

1. $(6, 4, 1)$
2. $(7, 5, -6)$
3. $(4, 0, 3)$
4. $(3, -7, 2)$
5. $(-3, 8, 2)$
6. $(0, -2, -4)$
7. $(-5, -6, 4)$
8. $(2, -5, -9)$
9. $(-3, -6, -1)$
10. $(-5, 8, 0)$
11. $(-5, 2, -3)$
12. $(-3, -2, 5)$

Find the x-, y-, and z-intercepts for each equation.

13. $10x - 2y + 5z = 10$
14. $2x + y - z = 12$
15. $x - y - z = 1$
16. $2x + y - 3z = 10$
17. $2x + 5y + z = 6$
18. $3x + 5y + 2z = 30$
19. $3x + 6y - 8z = 24$
20. $5x - 7y + 2z = 9$
21. $2x + 4y - z = 5$
22. $5x + y - 3z = 15$
23. $3y + 2z = 12$
24. $5x - 2y = 8$

Given the following conditions, name the octant in which the point (x, y, z) lies.

25. $x = 2, y = 3, z < 0$
26. $x < 0, y = 2, z > 0$
27. $x < 0, y > 0, z = -3$
28. $x = 3, y < 0, z > 0$
29. $x = -3, y > 0, z = 5$
30. $x < 0, y < 0, z < 0$
31. $x > 2, y < -1, z < -2$
32. $x > 3, y < -1, z = 4$
33. $x > 3, y > 0, z = 2$

Graph each equation. Find the x-, y-, and z-intercepts and the traces in the coordinate planes.

34. $2x - 3y + z = 12$
35. $4x - 2y - 3z = 24$
36. $2x + 8y - z = 8$
37. $-3x + 6y - 4z = 24$
38. $3x + 6y + 2z = 9$
39. $4x - y + 2z = 10$
40. $3x - y = 3$
41. $5y + 2z = 20$
42. $4x - 3y + 2z = 15$
43. $3x + 8y - 5z = 12$
44. $3z - 2x = 5$
45. $5x - 8y = -12$

Write an equation of the plane with the following x-, y-, and z-intercepts.

46. $4, -2, 3$
47. $2, -2, 5$
48. $-3, 1, 6$
49. $1, 3, 2$
50. $-1, -3, -4$
51. $7, 1, 4$
52. $\frac{1}{2}, 3, -2$
53. $-4, \frac{2}{3}, 1$
54. $4, \frac{1}{3}, \frac{1}{4}$
55. $\frac{3}{4}, -2, \frac{1}{3}$
56. $\frac{2}{3}, \frac{7}{3}, -\frac{1}{2}$
57. $-\frac{1}{2}, -\frac{4}{3}, -\frac{5}{8}$

Write an equation of the plane given two of its traces in the coordinate planes.

58. $3x - 5y = 8, 5y - 2z = -8$
59. $4x + 3y = 12, 3y - z = 12$
60. $2x - y = 6, 2y + 3z = -12$
61. $5x + 3z = 15, y + z = 5$
62. $5x - 4y = 20, 4y - 7z = -20$
63. $2x - 5z = 6, 3y - 5z = 6$
64. $x + 5y = 1, x + 3z = 1$
65. $x - 4y = 1, 8y + z = -2$
66. $2x + y = 4, x = 2$
67. $3y - 4z = 6, y = 2$
68. $x = 4, y = -3$
69. $x = -2, z = 5$

1. Find $f(6)$ if $f(x) = x^2 - 4x + 3$.

2. Find the slope, y-intercept, and x-intercept of the graph of $7x - 3y = 14$.

Find the value of each determinant.

3. $\begin{vmatrix} 4 & 5 \\ 0 & 3 \end{vmatrix}$

4. $\begin{vmatrix} 8 & 4 \\ -7 & 6 \end{vmatrix}$

5. $\begin{vmatrix} 11 & -13 \\ 6 & -11 \end{vmatrix}$

Solve each system of equations.

6. $3x - 5y = 4$
$4x - y = 11$

7. $7a + 9b = -2$
$-4a = 6b + 1$

8. $3m + 25n = 0$
$5m - 10n = 31$

Graph each system of inequalities. Name the vertices of the polygon formed. Find the maximum and minimum values of the given function.

9. $x + y \leq 5$
$x \geq 1$
$y \geq 2$
$f(x, y) = 3x + 2y$

10. $x \leq 7$
$y \geq 2$
$3x + 2y \leq 29$
$5x - 4y \geq -3$
$f(x, y) = 2x - 5y$

Solve each system of equations by using Cramer's Rule.

11. $x + y = 5$
$2x - 3y = 5$

12. $x - y = 7$
$2x - 3y = 4$

13. $2a - 3b = 23$
$3a - 5b = 8$

14. Write the equation of the plane with x-intercept 4, y-intercept 2, and z-intercept -5.

15. Write the equation of the plane with xz-trace $2x - 5z = 10$ and yz-trace $4y + 9z = -18$.

16. Sonia works at a clothing store. She earns \$8.00 an hour plus 40¢ for every item over 20 that she sells. She works 30 hours a week. How much money will she make if she sells c items?

Excursions in Algebra

Mathematics Contests

A number of mathematics contests are available to high school students. One of the best known contests is the Annual High School Mathematics Contest. Another contest is the U.S. Mathematical Olympiad, which is patterned after the International Mathematical Olympiad.

In one contest the following question appeared. The time limit for this question and another paired with it was 11 minutes. Eighty-six percent of the contestants answered the question correctly. Time yourself as you work the problem.

A club found that it could achieve a membership ratio of 2 adults for each minor either by inducting 24 adults or by expelling x minors. Find x.

 Solving Systems of Equations in Three Variables

The system below has three equations and three variables.

$$2x + 3y + z = 10$$
$$4x + 2y - z = 13$$
$$x + y + z = 5$$

Each equation is satisfied when x is 3, y is 1, and z is 1.

$$2(3) + 3(1) + (1) = 6 + 3 + 1 = 10$$
$$4(3) + 2(1) - (1) = 12 + 2 - 1 = 13$$
$$(3) + (1) + (1) = 3 + 1 + 1 = 5$$

Thus, the solution to this system is the ordered triple (3, 1, 1).

Systems of equations for three variables are solved using the same methods as those for equations in two variables.

Example

1 **Solve this system of equations.**

$$2x + y + z = 11$$
$$3y - z = -1$$
$$2z = 8$$

Solve the third equation, $2z = 8$.

$$2z = 8$$
$$z = 4$$

Substitute 4 for z in the second equation, $3y - z = -1$, to find y.

$$3y - (4) = -1$$
$$3y = 3$$
$$y = 1$$

Substitute 4 for z and 1 for y in the first equation, $2x + y + z = 11$, to find x.

$$2x + (1) + (4) = 11$$
$$2x = 6$$
$$x = 3$$

The solution is (3, 1, 4).

Check:

First equation	$2(3) + 1 + 4 = 11$
Second equation	$3(1) - 4 = -1$
Third equation	$2(4) = 8$

2 **Solve this system of equations.**

$$2x + 3y + 2z = 14$$
$$4x + 2y - z = 15$$
$$x + 2y + 3z = 10$$

Use elimination to make a system of two equations in two variables.

$2x + 3y + 2z = 14$
$4x + 2y - \ z = 15$ **Multiply by 2.** →

$2x + 3y + 2z = 14$
$\underline{8x + 4y - 2z = 30}$ *Add to*
$10x + 7y \qquad = 44$ *eliminate z.*

$4x + 2y - \ z = 15$
$x + 2y + 3z = 10$ **Multiply by 3.** →

$12x + 6y - 3z = 45$
$\underline{x + 2y + 3z = 10}$ *Add to*
$13x + 8y \qquad = 55$ *eliminate z.*

The result is two equations with the same two variables.

$$10x + 7y = 44$$
$$13x + 8y = 55$$

Use Cramer's Rule to solve this system for x and y.

$$x = \frac{\begin{vmatrix} 44 & 7 \\ 55 & 8 \end{vmatrix}}{\begin{vmatrix} 10 & 7 \\ 13 & 8 \end{vmatrix}} = \frac{44(8) - 7(55)}{10(8) - 7(13)}$$

$$= \frac{-33}{-11}$$

$$= 3$$

$$y = \frac{\begin{vmatrix} 10 & 44 \\ 13 & 55 \end{vmatrix}}{\begin{vmatrix} 10 & 7 \\ 13 & 8 \end{vmatrix}} = \frac{10(55) - 44(13)}{10(8) - 7(13)}$$

$$= \frac{-22}{-11}$$

$$= 2$$

Then substitute 3 for x and 2 for y in one of the original equations.

$$x + 2y + 3z = 10$$
$$3 + 2(2) + 3z = 10$$
$$3z = 3$$
$$z = 1$$

Substituting the values for x and y in any of the other original equations would give the same result.

The solution is $(3, 2, 1)$.

Exploratory Exercises

A possible solution for each system of equations is given in color. Check to see if it is the correct solution.

1. $x + y + z = 6$
$x - 3y + 2z = 1$
$2x - y + 2z = 0$ $(2, 2, 2)$

2. $2x + 3y - z = 0$
$x + 2y + z = 0$
$x - y + z = 0$ $(0, 0, 0)$

3. $x + y + z = 3$
$x - z = 1$
$y - z = -4$ $(3, -2, 2)$

4. $4x + y - 2z = 0$
$-2x + y + z = 0$
$x - 2y = 0$ $(3, 0, 6)$

5. $3x + 2y + z = 5$
$2x + y - z = 2$
$x + y + z = 0$ $(8, -11, 3)$

6. $x + y = -6$
$x + z = -2$
$y + z = 2$ $(-4, -2, 2)$

Written Exercises

Solve each system of equations.

7. $a - 2b + c = -9$
$2b + 3c = 16$
$4b = 8$

8. $x + y + z = -1$
$2x + y = 2$
$-3x = -9$

9. $2x + 4y - z = -3$
$y + z = 4$
$2y = -2$

10. $x + y - z = -1$
$x + y + z = 3$
$3x - 2y - z = -4$

11. $a + b + c = 0$
$2a + b - c = 2$
$2a + 2b + c = 5$

12. $x + y + z = 15$
$x + z = 12$
$y + z = 10$

13. $x - 2y + z = 3$
$2x + y - 2z = 31$
$-x + 2y + 3z = -23$

14. $x + y + z = 4$
$x - y + z = 0$
$x - y - z = -2$

15. $a + b - 2c = 4$
$2a + b + 2c = 0$
$a - 3b - 4c = -2$

16. $2x + 3y + z = 7$
$x + y + z = 4$
$3x + 4y - 2z = 6$

17. $x + y + z = -1$
$2x - y + z = 19$
$3x - 2y - 4z = 16$

18. $2x - y + 4z = 7$
$x - 3y + z = -2$
$3x - 2y + 2z = -2$

19. $x + 2y - 3z = 10$
$-4x + y - z = -10$
$3x - 7y + 2z = 5$

20. $x + 8y + 2z = -24$
$3x + y + 7z = -3$
$4x - 3y + 6z = 9$

21. $4r + 3s + 2t = 34$
$3r + 2s + 4t = 47$
$2r + 4s + 3t = 45$

22. $2a + b + c = 7$
$12a - 2b - 2c = 2$
$\dfrac{2a}{3} - b + \dfrac{c}{3} = -\dfrac{1}{3}$

23. $3r - 5s + 2t = 10$
$5r + 7s - 4t = -5$
$\dfrac{1}{2}r - \dfrac{1}{2}s + \dfrac{1}{4}t = \dfrac{3}{4}$

24. $2x + 3y + 4z = 3$
$5x - 9y + 6z = 1$
$\dfrac{1}{3}x - \dfrac{1}{2}y + \dfrac{2}{3}z = \dfrac{1}{6}$

Solve each problem.

25. The sum of three numbers is 6. The first number is twice the second, and the third number is three times the second. Find the numbers.

26. The sum of three numbers is 20. The first number is the sum of the second and the third, and the third number is three times the first. Find the numbers.

27. At Delicious Pizza, 5 slices of pizza, 2 salads, and 2 sodas cost $9.75. 3 slices, 2 salads, and 1 soda cost $7.15. 2 slices, 1 salad, and 1 soda cost $4.35. What are the prices of 1 slice of pizza, 1 salad, and 1 soda?

28. Dan and Jean want to buy a new living room set. A sofa, a love seat, and a tea table cost $1230. The sofa costs twice the price of the love seat. The tea table and the sofa cost $880. What are the prices of each piece of furniture?

29. The perimeter of a triangle is 18 feet. The longest side is twice as long as the shortest side. The length of the remaining side is the average of the lengths of the longest and shortest sides. Find the lengths of the 3 sides.

30. The perimeter of a triangle is 56 feet. The length of the longest side is three times the difference in the lengths of the other two sides. The longest side is also twice as long as the shortest side. Find the lengths of the 3 sides.

Challenge Exercises

Solve each system of equations.

31. $w + x + y + z = 2$
$2w - x - y + 2z = 7$
$2w + 3x + 2y - z = -2$
$3w - 2x - y - 3z = -2$

32. $4a + 6b + 2c + 4d = -12$
$-2a + 3b + c + d = -3$
$6a - 3b + c - d = 9$
$-10a - 6b - c - 2d = 1$

Parallel and Perpendicular Lines

The standard form of a linear equation is $Ax + By = C$. Suppose you are given the equation of a nonvertical, nonhorizontal line, and the coordinates of a point that is not on the line. The slope of a line parallel to the line that passes through the given point is $-\frac{A}{B}$. The slope of a perpendicular line that passes through the given point is $\frac{B}{A}$.

The following computer program can be used to find the equations of lines parallel and perpendicular to the given line, that pass through the given point.

```
10   INPUT "ENTER A, B, AND C OF A LIN
     EAR EQUATION IN STANDARD FORM:
     ";A, B, C
20   INPUT "ENTER THE COORDINATES OF A
     POINT THAT IS NOT ON THE LINE: ";
     P, Q
30   PRINT: PRINT "Y = "; - (A / B); "X + ";
     (Q + (A / B) * P);
35   PRINT " IS A LINE PARALLEL TO "
40   PRINT A; "X + "; B; "Y = "; C; " THAT
     PASSES THROUGH ("; P;", ";Q;")."
50   PRINT: PRINT "Y = "; (B / A);"X + ";
     (Q - (B / A) * P); PRINT " IS A LINE
     PERPENDICULAR TO "
60   PRINT A;"X + ";B;"Y = ";C;" THAT
     PASSES THROUGH ("; P;", ";Q;")."
70   END
```

Run the program for the linear equation $x + y = 5$ and the ordered pair $(4, 5)$.

```
RUN
ENTER A, B, AND C OF A LINEAR EQUATION
IN STANDARD FORM: 1, 1, 5
ENTER THE COORDINATES OF A POINT THAT
IS NOT ON THE LINE: 4, 5

Y = -1X + 9 IS A LINE PARALLEL TO
1X + 1Y = 5 THAT PASSES THROUGH
(4, 5).

Y = 1X + 1 IS A LINE PERPENDICULAR TO
1X + 1Y = 5 THAT PASSES THROUGH
(4, 5).
```

Exercises

Run the program for each linear equation and ordered pair. Write the equation of the parallel line and the equation of the perpendicular line.

1. $5x + y = -1$; $(3, 5)$ **2.** $-x + y = 6$; $(5, 2)$ **3.** $6x - 3y = 5$; $(1, 4)$

Lines that are parallel to the same line are parallel to each other. Lines that are perpendicular to the same line are also parallel to each other. A set of parallel lines is called a family of lines. Run the program several times for the linear equation $x + 4y = 8$ using the ordered pairs $(2, 3)$, $(4, 1)$, $(6, 2)$, and $(5, 5)$. Then complete the following exercises.

4. List equations for a family of lines that are parallel to $x + 4y = 8$.

5. List equations for a family of lines that are perpendicular to $x + 4y = 8$.

6. Do the four ordered pairs given above determine four distinct lines parallel to $x + 4y = 8$?

7. Do the four ordered pairs given above determine four distinct lines perpendicular to $x + 4y = 8$?

8. Write equations for four lines that are parallel to $x - 2y = 3$.

9. Write equations for four lines that are perpendicular to $x - 2y = 3$.

parallel lines   (81)
perpendicular lines   (82)
system of equations   (84)
consistent system   (85)
independent system   (85)
dependent system   (85)

inconsistent system   (85)
elimination method   (88)
substitution method   (88)
determinant   (92)
Cramer's Rule   (93)
linear programming   (99)

constraints   (99)
ordered triple   (107)
octant   (107)
trace   (108)

Chapter Summary

1. In a plane, lines with the same slope are parallel lines. Also, vertical lines are parallel.   (81)

2. In a plane, two nonvertical lines are perpendicular if and only if the product of their slopes is -1. Any vertical line is perpendicular to any horizontal line.   (82)

3. Possibilities for the graphs of two linear equations in two variables.   (85)

Graphs of Equations	Slopes of Lines	Name of System of Equations	Number of Solutions
lines intersect	different slopes	consistent and independent	one
lines coincide	same slope, same intercepts	consistent and dependent	infinite
lines parallel	same slope, different intercepts	inconsistent	none

4. The substitution method and the elimination method can be used to solve systems of equations.   (88)

5. A square array of numbers or variables enclosed between vertical lines is called a determinant.   (92)

6. The value of a second order determinant is defined as

$$\begin{vmatrix} a & b \\ c & d \end{vmatrix} = ad - bc. \quad (92)$$

7. Cramer's Rule: The solution to the system $\begin{cases} ax + by = c \\ dx + ey = f \end{cases}$ is (x, y)

$$\text{where } x = \frac{\begin{vmatrix} c & b \\ f & e \end{vmatrix}}{\begin{vmatrix} a & b \\ d & e \end{vmatrix}}, \ y = \frac{\begin{vmatrix} a & c \\ d & f \end{vmatrix}}{\begin{vmatrix} a & b \\ d & e \end{vmatrix}} \text{ and } \begin{vmatrix} a & b \\ d & e \end{vmatrix} \neq 0. \quad (93)$$

8. To solve a system of inequalities, find the ordered pairs that satisfy both inequalities.   (96)

9. Linear Programming Procedure:
   1. Define variables.
   2. Write a system of inequalities.
   3. Graph the system. Find vertices of the polygon formed.
   4. Write an expression to be maximized or minimized.
   5. Substitute values from vertices into the expression.
   6. Select the greatest or least result. Answer the problem.   (102)

10. If A, B, and C are real numbers and not all zero, then the graph of $Ax + By + Cz = D$ is a plane.   (107)

11. Suppose the equation of a plane is $Ax + By + Cz = D$. Then, the x-intercept is $\frac{D}{A}$, the y-intercept is $\frac{D}{B}$, and the z-intercept is $\frac{D}{C}$. Also, the xy-trace is $Ax + By = D$, the xz-trace is $Ax + Cz = D$, and the yz-trace is $By + Cz = D$. (108, 109)

Chapter Review

3–1 **Find an equation of the line that passes through each given point and is parallel to the line with the given equation.**

   1. $(4, 6); y = 3x - 2$
   2. $(1, 1); y = -3x - 1$
   3. $(-1, -1); y = 6x - 8$
   4. $(7, 7); 2x + 3y = 6$

Find an equation of the line that passes through each given point and is perpendicular to the line with the given equation.

   5. $(3, 5); y = 2x - 5$
   6. $(1, 4); y = 3x - 1$
   7. $(-1, -1); 2y + 3x = 10$
   8. $(0, 10); y = 4x$

3–2 **Graph each system of equations and state its solution. Then, state whether the system is *consistent and independent*, *consistent and dependent*, or *inconsistent*.**

   9. $x + y = -8$
        $2x - y = 2$
   10. $x + y = 6$
         $x - y = 2$
   11. $x + y = 11$
         $3x - 3y = 3$
   12. $x + y = 6$
         $2x + 2y = 12$

3–3 **Solve each system of equations algebraically.**

   13. $x + y = 6$
         $x - y = 4\frac{1}{2}$
   14. $3x - 5y = -13$
         $4x + 3y = 2$
   15. $2x + 3y = 8$
         $x - y = 2$
   16. $\frac{1}{3}x + \frac{1}{3}y = 5$
         $\frac{1}{6}x - \frac{1}{9}y = 0$

3–4   Find the value of each determinant.

17. $\begin{vmatrix} 2 & 3 \\ 4 & 5 \end{vmatrix}$

18. $\begin{vmatrix} 6 & -1 \\ 3 & -2 \end{vmatrix}$

Solve each system of equations using Cramer's Rule.

19. $7x - 8y = 11$
$9x - 2y = 3$

20. $x + 2y = 5$
$x - y = 6$

3–5   Solve each system of inequalities by graphing.

21. $x + y < 2$
$x + 2y > -3$

22. $y < -2$
$y - x > 1$

23. $y \geq x - 3$
$y \geq -x + 1$

24. $x + y < 4$
$y \geq -3x + 1$

3–6   **Graph each system of inequalities. Name the vertices of the polygon formed. Find the maximum and minimum values of the given function.**

25. $x \geq 0$
$y \geq 0$
$x + y \leq 3$
$3x + y \leq 6$
$f(x, y) = 2x + 4y$

26. $0 \leq x \leq 5$
$0 \leq y \leq 6$
$x + y \leq 9$
$f(x, y) = 2x + 3y$

3–7   Solve.

27. A farmer has 90 acres on which he may raise peanuts and corn. He has accepted orders requiring at least 10 acres of peanuts and 5 acres of corn. He also must follow a regulation that the acreage for corn must be at least twice the acreage for peanuts. If the profit is $100 per acre of corn and $200 per acre of peanuts, how many acres of each will give him the greatest profit?

3–8   **State the octant in which each point lies.**

28. $(8, -2, 3)$ **29.** $(4, 2, -1)$ **30.** $(-3, -6, -1)$ **31.** $(-2, 1, -3)$

Graph each equation. Find the x-, y-, and z-intercepts and the traces in the coordinate plane.

32. $2x - 3y + 4z = 12$

33. $x + y - 2z = 4$

34. $5x - 2y + 3z = 6$

35. $3x + 4y - 7z = 11$

3–9   Solve each system of equations.

36. $x + y + z = -1$
$2x + 4y + z = 1$
$3x - y - z = -15$

37. $x + y + 2z = 3$
$2x - 2y - 3z = 2$
$3x - y - 2z = 1$

1. Find an equation of the line that passes through $(2, 4)$ and is parallel to the line whose equation is $y = 3x - 5$.

Graph each system of equations. Then state the solution to the system.

2. $x + y = 7$
$x - y = 1$

3. $2x + 3y = 5$
$-3x + 6y = 12$

Solve each system of equations algebraically.

4. $3x + 8y = -6$
$4x - 2y = 11$

5. $x + 3y + z = 5$
$2x + y - 5z = 4$
$3x - y - 4z = -11$

Find the value of each determinant.

6. $\begin{vmatrix} 3 & 1 \\ -2 & 4 \end{vmatrix}$

7. $\begin{vmatrix} 9 & 0 \\ 5 & -6 \end{vmatrix}$

Solve each system of equations using Cramer's Rule.

8. $-2x + 3y = 5$
$x + 4y = 14$

9. $4x + y = 5$
$-8x + 3y = 10$

Solve each system of inequalities by graphing.

10. $y > 2x + 1$
$y \leq -3x + 4$

11. $x + y \leq 6$
$x - y \geq 4$

Graph each system of inequalities. Name the vertices of the polygon formed. Find the maximum and minimum values of the given function.

12. $0 \leq x \leq 6$
$0 \leq y \leq 2$
$x + y \leq 4$
$f(x, y) = 2x + y$

13. $0 \leq x \leq 4$
$y + x \leq 6$
$y \geq 0$
$f(x, y) = 3x - 2y$

State the octant in which each point lies.

14. $(-5, -4, 3)$

15. $(8, 2, -1)$

16. $(6, -4, 2)$

17. Graph $4x - 2y + 3z = 6$.

18. Find the x-, y-, and z-intercepts for $5x - 7y + 8z = 20$.

19. Find the traces in the coordinate planes for $3x - 5y + 6z = 10$.

20. A builder has 60 lots on which she can build houses with one house on each lot. She builds two types of houses, colonial and ranch. Sales experience has taught her that she should plan to build at least 3 times as many ranch-style houses as colonial. If she makes a profit of \$5000 on each colonial and \$4500 on each ranch, how many of each kind should she build to maximize profit?

CHAPTER 4

Matrices

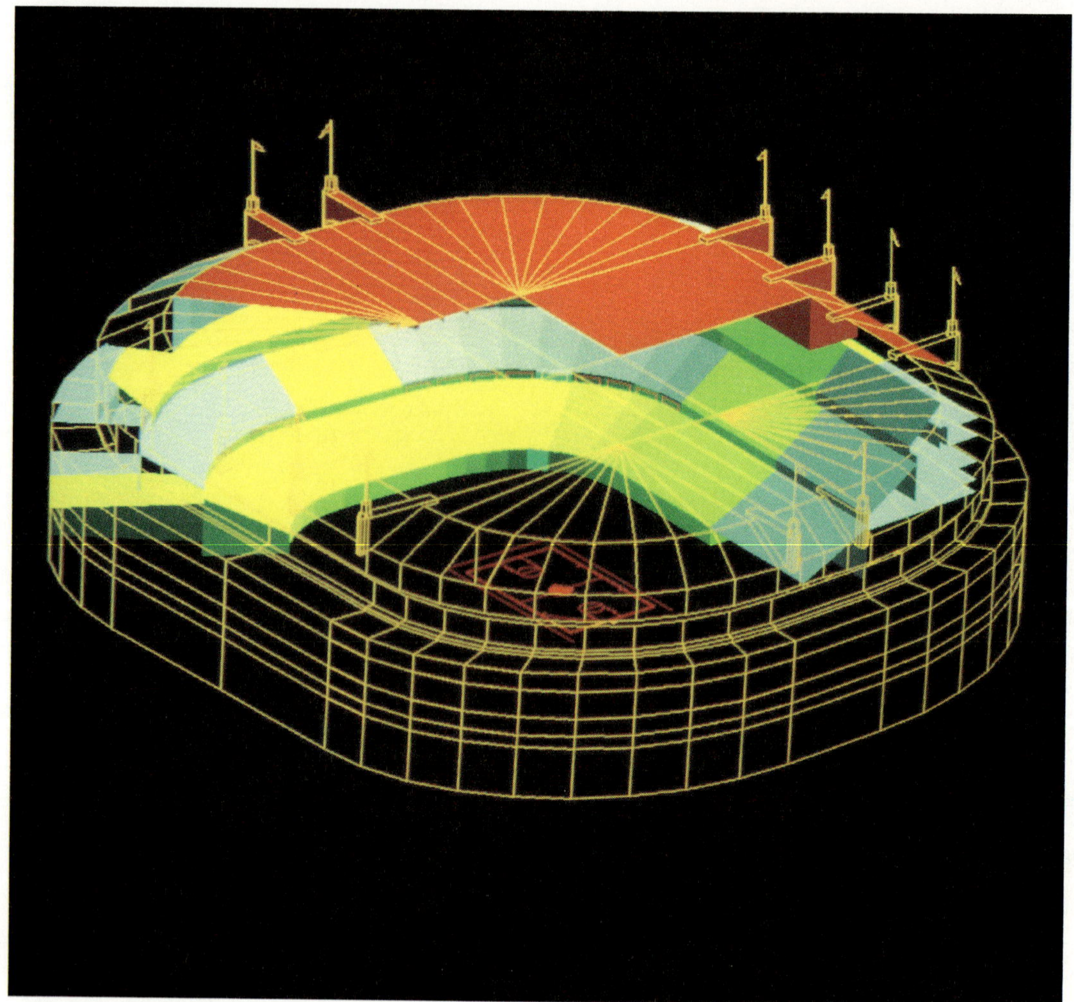

A computer can be programmed to make drawings such as the one shown above. The program involves many variables and many systems of equations. Arrays of numbers called matrices can represent systems of equations and are also used with three-dimensional computer graphics. By multiplying matrices, the size or perspective of a drawing can be changed.

4-1 Matrices and Determinants

A **matrix** is a rectangular array of terms called **elements**. The elements of a matrix are arranged in rows and columns and are usually enclosed by brackets or large parentheses. Matrices are often named by capital letters such as A, B, and M.

The matrix shown at the right has two rows and four columns. It is therefore called a 2×4 matrix (read "2 by 4"). The **dimension** of this matrix is 2×4.

$$\text{rows} \quad \begin{array}{c} \text{columns} \\ \begin{bmatrix} 1 & 9 & 5 & 7 \\ 6 & 4 & 12 & 2 \end{bmatrix} \end{array}$$

Example

1 **State the dimension of matrix A if $A = \begin{bmatrix} 1 & 3 \\ -5 & -18 \\ 11 & 7 \\ 0 & -2 \end{bmatrix}$.**

Since matrix A has 4 rows and 2 columns, the dimension of matrix A is 4×2.

When a capital letter is used to name a matrix, the dimension of the matrix may be written as a subscript. Thus, $A_{4 \times 2}$ can be used to represent matrix A from Example 1.

Certain matrices have special names. A matrix that has only one row is a *row matrix*, and a matrix that has only one column is a *column matrix*. A matrix that has the same number of rows as columns is a **square matrix**.

An important property of square matrices is that they all have determinants. The determinant of a 2×2 matrix can be evaluated by using the formula you learned in Chapter 3 for finding the value of a second-order determinant.

Example

2 **Evaluate the determinant of $\begin{bmatrix} 3 & -2 \\ 17 & 11 \end{bmatrix}$.**

The determinant of $\begin{bmatrix} 3 & -2 \\ 17 & 11 \end{bmatrix}$ is $\begin{vmatrix} 3 & -2 \\ 17 & 11 \end{vmatrix}$.

$$\begin{vmatrix} 3 & -2 \\ 17 & 11 \end{vmatrix} = 3(11) - (-2)(17)$$

$$= 33 - (-34) \text{ or } 67$$

A method called **expansion by minors** may be used to evaluate the determinant of any 3×3 matrix.

In a determinant, the **minor** of an element is the determinant formed when the row and column containing the element are deleted. For the determinant shown at the right, the minor of 1 and the minor of 5 are found as follows.

$$\begin{vmatrix} 1 & 3 & 7 \\ 4 & 8 & 2 \\ 9 & 5 & 6 \end{vmatrix}$$

The minor of 1 is $\begin{vmatrix} 8 & 2 \\ 5 & 6 \end{vmatrix}$.

The minor of 5 is $\begin{vmatrix} 1 & 7 \\ 4 & 2 \end{vmatrix}$.

The expansion of a third-order determinant given below uses the first row of the determinant. When the first row is used, each element in the row is multiplied by its minor and the signs of the term alternate, with the first term being positive.

Determinants of 3 × 3 matrices are called third-order determinants.

$$\begin{vmatrix} a & b & c \\ d & e & f \\ g & h & i \end{vmatrix} = a \begin{vmatrix} e & f \\ h & i \end{vmatrix} - b \begin{vmatrix} d & f \\ g & i \end{vmatrix} + c \begin{vmatrix} d & e \\ g & h \end{vmatrix}$$

Expansion of a Third Order Determinant

Example

3   **Evaluate the determinant of** $\begin{bmatrix} 2 & 3 & 4 \\ 6 & 5 & 7 \\ 1 & 2 & 8 \end{bmatrix}$ **using expansion by minors.**

$$\begin{vmatrix} 2 & 3 & 4 \\ 6 & 5 & 7 \\ 1 & 2 & 8 \end{vmatrix} = 2 \begin{vmatrix} 5 & 7 \\ 2 & 8 \end{vmatrix} - 3 \begin{vmatrix} 6 & 7 \\ 1 & 8 \end{vmatrix} + 4 \begin{vmatrix} 6 & 5 \\ 1 & 2 \end{vmatrix}$$

$$= 2(40 - 14) - 3(48 - 7) + 4(12 - 5)$$
$$= 52 - 123 + 28 \text{ or } -43$$

The determinant of $\begin{bmatrix} 2 & 3 & 4 \\ 6 & 5 & 7 \\ 1 & 2 & 8 \end{bmatrix}$ *is* $\begin{vmatrix} 2 & 3 & 4 \\ 6 & 5 & 7 \\ 1 & 2 & 8 \end{vmatrix}$.

Another method for evaluating third-order determinants is to use diagonals. To find the determinant of the matrix shown at the right, begin by writing the first two columns on the right side of the determinant. Then find the product of the elements in each diagonal. Add the products of the diagonals that extend from upper left to lower right. Then subtract the products of the other diagonals. The result is the value of the determinant.

$$\begin{bmatrix} a & b & c \\ d & e & f \\ g & h & i \end{bmatrix}$$

$$\begin{vmatrix} a & b & c \\ d & e & f \\ g & h & i \end{vmatrix} \begin{matrix} a & b \\ d & e \\ g & h \end{matrix}$$

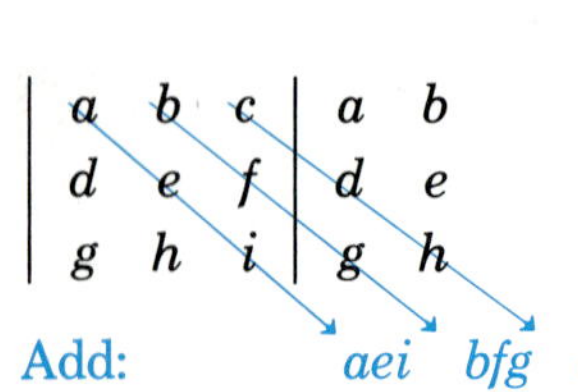

$$= aei + bfg + cdh - gec - hfa - idb$$

4 Evaluate the determinants of $\begin{bmatrix} 2 & 3 & 4 \\ 6 & 5 & 7 \\ 1 & 2 & 8 \end{bmatrix}$ and $\begin{bmatrix} -1 & 4 & 0 \\ 3 & -2 & -5 \\ -3 & 1 & 2 \end{bmatrix}$ using diagonals.

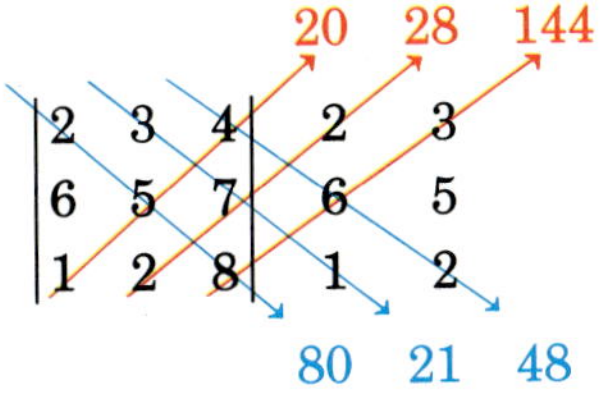

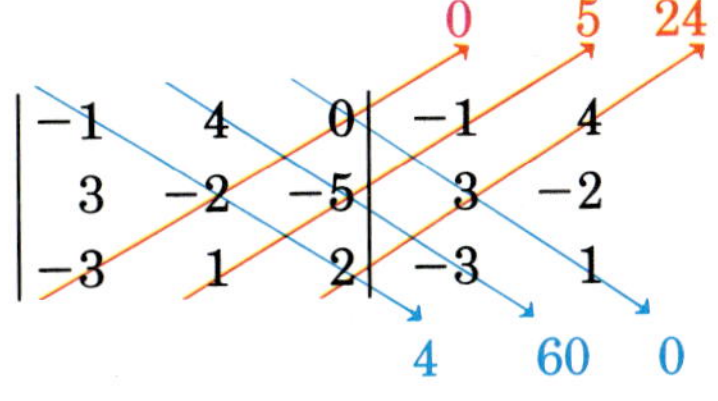

Find the products of the diagonals.

$= 80 + 21 + 48 - 20 - 28 - 144$
$= -43$

$= 4 + 60 + 0 - 0 - 5 - 24$
$= 35$

Exploratory Exercises

State the dimension of each matrix.

1. $\begin{bmatrix} 7 & -2 & -4 \\ -8 & 9 & 10 \\ 1 & 17 & 10 \\ 21 & -3 & 6 \end{bmatrix}$ **2.** $\begin{bmatrix} -11 \\ 6 \\ 45 \end{bmatrix}$ **3.** $\begin{bmatrix} 1 & 7 \\ 10 & 30 \end{bmatrix}$ **4.** $[-7 \quad 5 \quad 26]$

Find the value of each determinant.

5. $\begin{vmatrix} 5 & 7 \\ -3 & 2 \end{vmatrix}$ **6.** $\begin{vmatrix} 6 & 0 \\ 9 & -1 \end{vmatrix}$ **7.** $\begin{vmatrix} 2 & 0 & 2 \\ 0 & 3 & -3 \\ -3 & -2 & 0 \end{vmatrix}$ **8.** $\begin{vmatrix} 2 & 3 & 4 \\ 3 & 2 & -1 \\ 4 & 3 & 7 \end{vmatrix}$

Written Exercises

State the dimension of each matrix. Then evaluate its determinant (if one exists).

9. $\begin{bmatrix} 1 & -2 \\ 2 & -1 \end{bmatrix}$ **10.** $\begin{bmatrix} 5 & 7 & 9 \\ -1 & 3 & 3 \end{bmatrix}$ **11.** $\begin{bmatrix} 26 \\ -4 \end{bmatrix}$ **12.** $\begin{bmatrix} 3 & 4 \\ -6 & -2 \end{bmatrix}$

13. $\begin{bmatrix} 1 & 3 & -2 \\ 2 & -1 & 1 \\ -2 & 2 & 3 \end{bmatrix}$ **14.** $\begin{bmatrix} -4 & 1 & 3 \\ 2 & 0 & 1 \\ 4 & -5 & 0 \end{bmatrix}$ **15.** $\begin{bmatrix} -1 & 1 & 2 \\ 2 & 1 & 0 \\ 3 & 6 & -2 \end{bmatrix}$ **16.** $\begin{bmatrix} 2 & -1 & 3 \\ 3 & 2 & 1 \\ 1 & 3 & -2 \end{bmatrix}$

17. $\begin{bmatrix} 1 & -1 & 1 \\ 4 & 3 & 1 \\ 0 & 5 & 2 \end{bmatrix}$ **18.** $\begin{bmatrix} 3 & -1 & 2 \\ 0 & 4 & 1 \\ 5 & -2 & -3 \end{bmatrix}$ **19.** $\begin{bmatrix} -6 & 7 \\ -2 & -8 \\ 4 & 10 \end{bmatrix}$ **20.** $\begin{bmatrix} 1 & 2 & -3 \\ 3 & -5 & -1 \\ 4 & 4 & 1 \end{bmatrix}$

21. $\begin{bmatrix} 3 & 2 & 5 \\ -1 & 1 & 1 \\ 4 & 3 & 3 \end{bmatrix}$ **22.** $\begin{bmatrix} 5 & -1 & 2 \\ 2 & -3 & 5 \\ 3 & 2 & -3 \end{bmatrix}$ **23.** $\begin{bmatrix} 40 & 20 & -25 \\ 10 & 15 & 55 \\ -5 & -10 & -30 \end{bmatrix}$ **24.** $\begin{bmatrix} 44 & 41 & 46 \\ 32 & -59 & 36 \\ 72 & -61 & 84 \end{bmatrix}$

 Operations with Matrices

The properties and operations of real numbers may be used to define properties and operations of matrices. For example, the definition for equality of real numbers is used to define equality of matrices.

Two matrices are equal if and only if they have the same dimension and their corresponding elements are equal.	*Definition of Equal Matrices*

Example

1 **Find the values of x and y for which the following equation is true.**

$$\begin{bmatrix} 3x \\ y \end{bmatrix} = \begin{bmatrix} 31 + 4y \\ 6 - 2x \end{bmatrix}$$

Since matrices are equal if and only if the corresponding elements are equal, the following equations may be written.

$$3x = 31 + 4y$$
$$y = 6 - 2x$$

Substitute $6 - 2x$ for y in the first equation and solve for x.

$$3x = 31 + 4y$$
$$3x = 31 + 4(6 - 2x) \qquad \text{Substitute } 6 - 2x \text{ for } y.$$
$$3x = 31 + 24 - 8x$$
$$11x = 55$$
$$x = 5$$

Now substitute 5 for x in $y = 6 - 2x$.

$$y = 6 - 2x$$
$$y = 6 - 2(5) \qquad \text{Substitute 5 for } x.$$
$$y = -4$$

Therefore the matrices are equal if $x = 5$ and $y = -4$.

Multiplication of a matrix by a constant is performed according to the following rule.

The product of an $m \times n$ matrix, A, and a constant, k, is an $m \times n$ matrix kA. Each element of kA is the product of k and the corresponding element of A.	*Multiplying a Matrix by a Constant*

2 Find $5\begin{bmatrix} -1 & 3 & 5 \\ 2 & 8 & -4 \end{bmatrix}$.

Multiply each element of the matrix by 5.

$$5\begin{bmatrix} -1 & 3 & 5 \\ 2 & 8 & -4 \end{bmatrix} = \begin{bmatrix} 5(-1) & 5(3) & 5(5) \\ 5(2) & 5(8) & 5(-4) \end{bmatrix} = \begin{bmatrix} -5 & 15 & 25 \\ 10 & 40 & -20 \end{bmatrix}$$

Addition of matrices is defined for matrices with the same dimensions.

Addition of matrices with different dimensions is not defined.

> If A and B are two $m \times n$ matrices, then $A + B$ is an $m \times n$ matrix where each element is the sum of the corresponding elements of A and B.

Adding Matrices

Examples

3 Find $\begin{bmatrix} 5 & 4 \\ 3 & -1 \\ 2 & 7 \end{bmatrix} + \begin{bmatrix} -8 & 3 \\ -2 & 5 \\ 2 & 3 \end{bmatrix}$.

Add the corresponding elements.

$$\begin{bmatrix} 5 & 4 \\ 3 & -1 \\ 2 & 7 \end{bmatrix} + \begin{bmatrix} -8 & 3 \\ -2 & 5 \\ 2 & 3 \end{bmatrix} = \begin{bmatrix} 5 + (-8) & 4 + 3 \\ 3 + (-2) & -1 + 5 \\ 2 + 2 & 7 + 3 \end{bmatrix} = \begin{bmatrix} -3 & 7 \\ 1 & 4 \\ 4 & 10 \end{bmatrix}$$

4 Find $3\begin{bmatrix} 4 & -7 \\ 3 & 8 \end{bmatrix} - 5\begin{bmatrix} -3 & 6 \\ 5 & -2 \end{bmatrix}$.

$$3\begin{bmatrix} 4 & -7 \\ 3 & 8 \end{bmatrix} - 5\begin{bmatrix} -3 & 6 \\ 5 & -2 \end{bmatrix} = \begin{bmatrix} 12 & -21 \\ 9 & 24 \end{bmatrix} - \begin{bmatrix} -15 & 30 \\ 25 & -10 \end{bmatrix} = \begin{bmatrix} 27 & -51 \\ -16 & 34 \end{bmatrix}$$

Under special circumstances, matrices can be multiplied. The number of columns in the first matrix must be the same as the number of rows in the second matrix.

$$A_{m \times n} \qquad B_{n \times r}$$

> The product of an $m \times n$ matrix, A, and an $n \times r$ matrix, B, is the $m \times r$ matrix AB. The element in the ith row and the jth column of AB is the sum of the products of the corresponding elements in the ith row of A and the jth column of B.

Multiplying Matrices

5 If $A = \begin{bmatrix} 2 & -1 \\ 3 & 4 \end{bmatrix}$ and $B = \begin{bmatrix} 3 & -9 & 2 \\ 5 & 7 & -6 \end{bmatrix}$, find AB.   *Matrix BA is not defined.*

$$AB = \begin{bmatrix} 2(3) + (-1)(5) & 2(-9) + (-1)(7) & (2)(2) + (-1)(-6) \\ 3(3) + 4(5) & 3(-9) + 4(7) & (3)(2) + 4(-6) \end{bmatrix}$$

$$= \begin{bmatrix} 6 - 5 & -18 - 7 & 4 + 6 \\ 9 + 20 & -27 + 28 & 6 - 24 \end{bmatrix}$$

$$= \begin{bmatrix} 1 & -25 & 10 \\ 29 & 1 & -18 \end{bmatrix}$$

6 If $A = \begin{bmatrix} 2 & 1 & -3 \\ 5 & -4 & 1 \end{bmatrix}$ and $B = \begin{bmatrix} 2 & 5 & 2 \\ 3 & -6 & -3 \end{bmatrix}$, find AB.

The dimension of A is 2×3 and the dimension of B is 2×3. Since the number of columns of A, 3, is not the same as the number of rows of B, 2, multiplication of A and B is not defined.

7 Find the dimension of matrix M, if $A_{3 \times 2} \cdot B_{2 \times 4} = M$.

According to the definition of the product of two matrices, the dimension of M would be 3×4.

Exploratory Exercises

Find each product.

1. $3 \begin{bmatrix} 4 & 1 \\ -2 & 3 \end{bmatrix}$

2. $5 \begin{bmatrix} -3 & 4 & 1 \\ 2 & 7 & 0 \end{bmatrix}$

3. $\frac{1}{2} [5 \quad -4 \quad \sqrt{2}]$

4. $-4 \begin{bmatrix} 2 & 5 \\ -3 & 2 \\ 6 & -4 \end{bmatrix}$

5. $\frac{2}{3} \begin{bmatrix} 9 \\ -12 \\ 3 \end{bmatrix}$

6. $\frac{4}{5} \begin{bmatrix} 10 & -3 \\ 20 & 5 \end{bmatrix}$

Perform the indicated operations.

7. $\begin{bmatrix} 2 & -1 \\ 3 & 5 \end{bmatrix} + \begin{bmatrix} 4 & -1 \\ -3 & 2 \end{bmatrix}$

8. $\begin{bmatrix} 4 & 1 \\ 3 & 8 \\ -2 & 9 \end{bmatrix} + \begin{bmatrix} 3 & -7 \\ -4 & 2 \\ 0 & -5 \end{bmatrix}$

9. $[4 \quad 0 \quad 2 \quad -7] - [2 \quad 5 \quad -4 \quad 6]$

10. $\begin{bmatrix} 4 & 1 \\ -9 & 3 \end{bmatrix} - \begin{bmatrix} 8 & -4 \\ 2 & 3 \end{bmatrix}$

11. $\begin{bmatrix} 2 \\ -3 \\ 5 \end{bmatrix} + \begin{bmatrix} 4 \\ 7 \\ -6 \end{bmatrix} - \begin{bmatrix} -6 \\ 2 \\ -2 \end{bmatrix}$

12. $\begin{bmatrix} 2 & -4 \\ 6 & 5 \end{bmatrix} - \begin{bmatrix} 11 & 4 \\ -3 & 7 \end{bmatrix} - \begin{bmatrix} -1 & -7 \\ 10 & -3 \end{bmatrix}$

Find the dimension of each matrix M.

13. $A_{3\times 2} \cdot B_{2\times 3} = M$

14. $A_{4\times 1} \cdot B_{1\times 3} = M$

15. $A_{3\times 4} \cdot M = B_{3\times 5}$

16. $A_{3\times 1} \cdot M = B_{3\times 4}$

17. $M \cdot A_{2\times 3} = B_{3\times 3}$

18. $M \cdot A_{4\times 2} = B_{4\times 2}$

Written Exercises

Find the dimension of each matrix M.

19. $A_{5\times 1} \cdot B_{1\times 7} = M$

20. $A_{5\times 2} \cdot M = B_{5\times 4}$

21. $M \cdot A_{4\times 1} = B_{4\times 1}$

22. $A_{4\times 3} \cdot B_{3\times 1} = M$

23. $A_{3\times 2} \cdot M = B_{3\times 3}$

24. $M \cdot A_{3\times 5} = B_{5\times 5}$

Use matrices A, B, C, and D to evaluate each expression. Write "not defined" if the sum or product does not exist.

$$A = \begin{bmatrix} 3 & -1 \\ 2 & 4 \end{bmatrix} \quad B = \begin{bmatrix} 4 & 0 & -3 \\ 7 & -5 & 9 \end{bmatrix} \quad C = \begin{bmatrix} -6 & 4 \\ -2 & 8 \\ 3 & 0 \end{bmatrix} \quad D = \begin{bmatrix} -1 & 0 \\ 3 & 7 \end{bmatrix}$$

25. $2A$

26. $3D$

27. $\frac{1}{3}C$

28. $\frac{1}{2}B$

29. $A + D$

30. $D - A$

31. $A + B$

32. $2A + 3D$

33. $4D - 3A$

34. $5C - 3B$

35. AB

36. BA

37. AA

38. DD

39. AD

40. DA

41. BC

42. CB

43. AC

44. CD

45. $AB + B$

46. $CB + B$

47. $AD + CB$

48. $AD + BC$

Find the values of x, y, and z for which the following equations are true.

49. $y\begin{bmatrix} 3 & -4 \\ 2 & x \end{bmatrix} = \begin{bmatrix} 15 & -20 \\ z & 5 \end{bmatrix}$

50. $\begin{bmatrix} 2x \\ y+1 \end{bmatrix} = \begin{bmatrix} 8 \\ 3 \end{bmatrix}$

51. $4\begin{bmatrix} x & y-1 \\ 3 & z \end{bmatrix} = \begin{bmatrix} 20 & 8 \\ 6z & x+y \end{bmatrix}$

52. $3\begin{bmatrix} x+y & z \\ y & 0 \end{bmatrix} = \begin{bmatrix} 15 & 3 \\ z-x & 0 \end{bmatrix}$

53. $3\begin{bmatrix} x \\ 3x \end{bmatrix} + 2\begin{bmatrix} 3y \\ 5y-1 \end{bmatrix} = \begin{bmatrix} -6 \\ 4 \end{bmatrix}$

54. $4\begin{bmatrix} x \\ 2y \end{bmatrix} - 3\begin{bmatrix} 2y \\ 3x \end{bmatrix} = \begin{bmatrix} 26 \\ -53 \end{bmatrix}$

55. $\begin{bmatrix} x \\ 7z \\ 2y \end{bmatrix} - \begin{bmatrix} 4z \\ -3y \\ 3x \end{bmatrix} + \begin{bmatrix} -2y \\ 2x \\ -5z \end{bmatrix} = \begin{bmatrix} -4 \\ 11 \\ 18 \end{bmatrix}$

56. $3\begin{bmatrix} z-y \\ 3x+z \\ x-y \end{bmatrix} - 2\begin{bmatrix} 2z-x \\ x+y \\ y-2z \end{bmatrix} = \begin{bmatrix} -5 \\ 11 \\ 32 \end{bmatrix}$

Challenge Exercises

57. Show that the sum of two matrices is commutative.

58. Show that the product of two matrices is *not* commutative.

The elements of an $m \times n$ matrix can be represented using double subscript notation. An $m \times n$ matrix has m rows and n columns.

$$\begin{bmatrix} e_{11} & e_{12} & e_{13} \cdots e_{1n} \\ e_{21} & e_{22} & e_{23} \cdots e_{2n} \\ \cdot & \cdot & \cdot & \cdot \\ \cdot & \cdot & \cdot & \cdot \\ \cdot & \cdot & \cdot & \cdot \\ e_{m1} & e_{m2} & e_{m3} \cdots e_{mn} \end{bmatrix}$$

e_{ij} represents the element in row i and column j.

For example, the element in the third row and fourth column is e_{34}. A matrix can be multiplied by another matrix if the first matrix has the same number of columns as the second matrix has rows. Consider the two matrices, A and B, shown below. The product, AB, will have the same number of rows as A and the same number of columns as B.

$$A_{2\times3} \qquad\qquad B_{3\times4} \qquad\qquad AB_{2\times4}$$

$$\begin{bmatrix} -3 & 9 & 4 \\ 8 & 5 & -7 \end{bmatrix} \times \begin{bmatrix} 5 & 8 & 1 & 7 \\ 9 & 2 & -6 & -4 \\ -1 & 3 & 0 & 6 \end{bmatrix} = \begin{bmatrix} e_{11} & e_{12} & e_{13} & e_{14} \\ e_{21} & e_{22} & e_{23} & e_{24} \end{bmatrix}$$

Element e_{ij} can be obtained by multiplying row i of matrix A times column j of matrix B. The product of a row and a column is a real number, as shown in the following example.

Example: **Find e_{23} of matrix AB. Use matrices A and B shown above.**

Multiply the second row of A by the third column of B.

$$\begin{bmatrix} 8 & 5 & -7 \end{bmatrix} \cdot \begin{bmatrix} 1 \\ -6 \\ 0 \end{bmatrix} = 8(1) + 5(-6) + (-7)(0)$$

Find the sum of the products of the corresponding elements.

$$= 8 - 30 + 0$$
$$= -22$$

Exercises

Multiply a row of A by a column of B to obtain the indicated element of matrix AB.

1. e_{11} **2.** e_{21} **3.** e_{12} **4.** e_{22} **5.** e_{14} **6.** e_{24} **7.** e_{13}

8. Show the complete product matrix, AB.

4-3 Identities and Inverses

Since $\begin{bmatrix} a & b \\ c & d \end{bmatrix} \cdot \begin{bmatrix} 1 & 0 \\ 0 & 1 \end{bmatrix} = \begin{bmatrix} a & b \\ c & d \end{bmatrix}$ and $\begin{bmatrix} 1 & 0 \\ 0 & 1 \end{bmatrix} \cdot \begin{bmatrix} a & b \\ c & d \end{bmatrix} = \begin{bmatrix} a & b \\ c & d \end{bmatrix}$, then $\begin{bmatrix} 1 & 0 \\ 0 & 1 \end{bmatrix}$ is the **identity matrix** for multiplication of 2×2 matrices. This concept can be extended to the general case.

The identity matrix, *I*, for multiplication is a square matrix with a 1 for every element of the principal diagonal and a 0 in all other positions. The principal diagonal extends from upper left to lower right.

Identity Matrix for Multiplication

Recall that any nonzero real number a has a multiplicative inverse, $\frac{1}{a}$. That is, $a \cdot \frac{1}{a} = \frac{1}{a} \cdot a = 1$.

Similarly, any matrix A has an **inverse matrix** A^{-1} such that $A \cdot A^{-1} = A^{-1} \cdot A = I$.

Example

1 If $A = \begin{bmatrix} 5 & 3 \\ 2 & 1 \end{bmatrix}$, find A^{-1} and check.

Let $A^{-1} = \begin{bmatrix} x & y \\ z & w \end{bmatrix}$, then $\begin{bmatrix} 5 & 3 \\ 2 & 1 \end{bmatrix} \cdot \begin{bmatrix} x & y \\ z & w \end{bmatrix} = \begin{bmatrix} 1 & 0 \\ 0 & 1 \end{bmatrix}$.

Expanding yields the following equations.

(1) $5x + 3z = 1$ (2) $5y + 3w = 0$
(3) $2x + z = 0$ (4) $2y + w = 1$

Use equations (1) and (3) to find x and z.

$$5x + 3z = 1$$
$$2x + z = 0$$

$$5x + 3z = 1$$
$$-6x - 3z = 0$$
$$\text{Add.} \quad -x \phantom{{}+3z} = 1$$
$$x = -1$$

$$2(-1) + z = 0$$
$$z = 2$$

Use equations (2) and (4) to find y and w.

$$5y + 3w = 0$$
$$2y + w = 1$$

$$5y + 3w = 0$$
$$-6y - 3w = -3$$
$$\text{Add.} \quad -y \phantom{{}+3w} = -3$$
$$y = 3$$

$$2(3) + w = 1$$
$$w = -5$$

Check: $\begin{bmatrix} 5 & 3 \\ 2 & 1 \end{bmatrix} \cdot \begin{bmatrix} -1 & 3 \\ 2 & -5 \end{bmatrix} = \begin{bmatrix} -5 + 6 & 15 - 15 \\ -2 + 2 & 6 - 5 \end{bmatrix}$

$$= \begin{bmatrix} 1 & 0 \\ 0 & 1 \end{bmatrix}$$

Thus, $A^{-1} = \begin{bmatrix} -1 & 3 \\ 2 & -5 \end{bmatrix}$.

2 **Find the inverse of** $\begin{bmatrix} a & b \\ c & d \end{bmatrix}$.

Let $\begin{bmatrix} x & y \\ z & w \end{bmatrix}$ be the inverse of $\begin{bmatrix} a & b \\ c & d \end{bmatrix}$.

$$\begin{bmatrix} a & b \\ c & d \end{bmatrix}\begin{bmatrix} x & y \\ z & w \end{bmatrix} = \begin{bmatrix} 1 & 0 \\ 0 & 1 \end{bmatrix}$$

$\quad$ (1) $ax + bz = 1$ $\qquad$ (2) $ay + bw = 0$

$\quad$ (3) $cx + dz = 0$ $\qquad$ (4) $cy + dw = 1$

Use equations (1) and (3) to find x and z.

$$\begin{array}{l} ax + bz = 1 \\ cx + dz = 0 \end{array} \quad\Rightarrow\quad \begin{array}{l} adx + bdz = d \\ bcx + bdz = 0 \end{array}$$

Subtract. $\quad (ad - bc)x = d$

$$x = \frac{d}{ad - bc}$$

$$\begin{array}{l} ax + bz = 1 \\ cx + dz = 0 \end{array} \quad\Rightarrow\quad \begin{array}{l} acx + bcz = c \\ acx + adz = 0 \end{array}$$

Subtract. $\quad (bc - ad)z = c$

$$z = \frac{c}{bc - ad}$$

$$= \frac{-c}{ad - bc}$$

Use equations (2) and (4) to find y and w.

$$\begin{array}{l} ay + bw = 0 \\ cy + dw = 1 \end{array} \quad\Rightarrow\quad \begin{array}{l} ady + bdw = 0 \\ bcy + bdw = b \end{array}$$

Subtract. $\quad (ad - bc)y = -b$

$$y = \frac{-b}{ad - bc}$$

$$\begin{array}{l} ay + bw = 0 \\ cy + dw = 1 \end{array} \quad\Rightarrow\quad \begin{array}{l} acy + bcw = 0 \\ acy + adw = a \end{array}$$

Subtract. $\quad (bc - ad)w = -a$

$$w = \frac{-a}{bc - ad}$$

$$= \frac{a}{ad - bc}$$

Thus, the inverse of $\begin{bmatrix} a & b \\ c & d \end{bmatrix}$ is $\begin{bmatrix} \dfrac{d}{ad - bc} & \dfrac{-b}{ad - bc} \\ \dfrac{-c}{ad - bc} & \dfrac{a}{ad - bc} \end{bmatrix}$ or $\dfrac{1}{ad - bc}\begin{bmatrix} d & -b \\ -c & a \end{bmatrix}$.

Notice that $ad - bc$ is the value of the determinant of the matrix. Since $\dfrac{1}{ad - bc}$ is not defined when $ad - bc = 0$, the value of the determinant of the matrix must not be zero for the matrix to have an inverse.

Any matrix M, $\begin{bmatrix} a & b \\ c & d \end{bmatrix}$, **will have an inverse** M^{-1} **if and only if** $\begin{vmatrix} a & b \\ c & d \end{vmatrix} \neq 0$. **Then** $M^{-1} = \dfrac{1}{ad - bc}\begin{bmatrix} d & -b \\ -c & a \end{bmatrix}$.

Inverse of a 2 × 2 Matrix

The **transpose** of any matrix is the matrix formed by interchanging the rows and the columns. If the dimension of a matrix is $m \times n$, then the dimension of its transpose is $n \times m$. The transpose of matrix A is written A^T.

The transpose of a square matrix is also a square matrix.

Example

4   **If** $A = \begin{bmatrix} 5 & 0 & -3 \\ -2 & 1 & 7 \end{bmatrix}$ **and** $M = \begin{bmatrix} 2 & -1 \\ 3 & 5 \end{bmatrix}$, **find** A^T **and** M^T.

$$A^T = \begin{bmatrix} 5 & -2 \\ 0 & 1 \\ -3 & 7 \end{bmatrix} \qquad\qquad M^T = \begin{bmatrix} 2 & 3 \\ -1 & 5 \end{bmatrix}$$

Recall that a minor of an element of a determinant is the determinant obtained by removing the row and column containing the element. Each element of a square matrix also has a minor that is found in the same manner. A positive or negative sign is associated with the minor of an element, depending on the position of the element in the matrix. Study the diagram at the right. Consider the element in row r and column c, written e_{rc}.

$$\begin{bmatrix} + & - & + \\ - & + & - \\ + & - & + \end{bmatrix}$$

> The sign of the minor of e_{rc} is positive if $r + c$ is even.
> The sign of the minor of e_{rc} is negative if $r + c$ is odd.

The **signed minor** of an element is the minor multiplied by 1 if its sign is positive or -1 if its sign is negative. For the matrix at the right,

$$\begin{bmatrix} 5 & 1 & 3 \\ 2 & 3 & -7 \\ 4 & 9 & 3 \end{bmatrix}$$

the minor of the element 9 is $\begin{vmatrix} 5 & 3 \\ 2 & -7 \end{vmatrix}$. Since 9 occupies a negative position in this matrix, its signed minor is $-\begin{vmatrix} 5 & 3 \\ 2 & -7 \end{vmatrix}$.

Example

5 Find the signed minor of each element in the second row of $\begin{bmatrix} 5 & -2 & 6 \\ -3 & 4 & 5 \\ 8 & 2 & 0 \end{bmatrix}$.

Then evaluate each signed minor.

The signed minor of e_{21} is $-\begin{vmatrix} -2 & 6 \\ 2 & 0 \end{vmatrix}$. $\qquad -\begin{vmatrix} -2 & 6 \\ 2 & 0 \end{vmatrix} = -(0 - 12) = 12$

The signed minor of e_{22} is $+\begin{vmatrix} 5 & 6 \\ 8 & 0 \end{vmatrix}$. $\qquad +\begin{vmatrix} 5 & 6 \\ 8 & 0 \end{vmatrix} = 0 - 48 = -48$

The signed minor of e_{23} is $-\begin{vmatrix} 5 & -2 \\ 8 & 2 \end{vmatrix}$. $\qquad -\begin{vmatrix} 5 & -2 \\ 8 & 2 \end{vmatrix} = -(10 + 16) = -26$

Several steps are needed to find the inverse of a 3×3 matrix. An inverse exists if the determinant of the matrix does *not* equal zero.

Finding the Inverse of a 3×3 Matrix

To find the inverse of a 3×3 matrix with a nonzero determinant, D:

1. Replace each element by the value of its signed minor.
2. Find the transpose for the new matrix.
3. Multiply this transpose by the reciprocal of the determinant, D.

Example

6 Find the inverse of $\begin{bmatrix} 2 & 3 & -1 \\ 2 & 4 & 5 \\ 0 & -3 & 7 \end{bmatrix}$.

To be sure an inverse exists, evaluate the determinant, D, of the matrix.

$$\begin{vmatrix} 2 & 3 & -1 \\ 2 & 4 & 5 \\ 0 & -3 & 7 \end{vmatrix} = 2\begin{vmatrix} 4 & 5 \\ -3 & 7 \end{vmatrix} - 3\begin{vmatrix} 2 & 5 \\ 0 & 7 \end{vmatrix} + (-1)\begin{vmatrix} 2 & 4 \\ 0 & -3 \end{vmatrix}$$

$$= 2(43) - 3(14) + (-1)(-6)$$
$$= 50 \qquad \textit{Since the value is not zero, the matrix has an inverse.}$$

Now, find and evaluate the signed minor of each element of the matrix.

$$e_{11}: +\begin{vmatrix} 4 & 5 \\ -3 & 7 \end{vmatrix} = 43 \qquad e_{12}: -\begin{vmatrix} 2 & 5 \\ 0 & 7 \end{vmatrix} = -14 \qquad e_{13}: +\begin{vmatrix} 2 & 4 \\ 0 & -3 \end{vmatrix} = -6$$

$$e_{21}: -\begin{vmatrix} 3 & -1 \\ -3 & 7 \end{vmatrix} = -18 \qquad e_{22}: +\begin{vmatrix} 2 & -1 \\ 0 & 7 \end{vmatrix} = 14 \qquad e_{23}: -\begin{vmatrix} 2 & 3 \\ 0 & -3 \end{vmatrix} = 6$$

$$e_{31}: +\begin{vmatrix} 3 & -1 \\ 4 & 5 \end{vmatrix} = 19 \qquad e_{32}: -\begin{vmatrix} 2 & -1 \\ 2 & 5 \end{vmatrix} = -12 \qquad e_{33}: +\begin{vmatrix} 2 & 3 \\ 2 & 4 \end{vmatrix} = 2$$

Replace each element by the value of its signed minor.	Find the transpose for the new matrix.	Multiply by the reciprocal of the determinant, D.
$\begin{bmatrix} 43 & -14 & -6 \\ -18 & 14 & 6 \\ 19 & -12 & 2 \end{bmatrix}$	$\begin{bmatrix} 43 & -18 & 19 \\ -14 & 14 & -12 \\ -6 & 6 & 2 \end{bmatrix}$	$\dfrac{1}{50}\begin{bmatrix} 43 & -18 & 19 \\ -14 & 14 & -12 \\ -6 & 6 & 2 \end{bmatrix}$

Check that the product of this matrix and the original matrix is the identity matrix.

Exploratory Exercises

What is the sign of the minor for each indicated element in a 4 × 4 matrix?

1. e_{31} **2.** e_{43} **3.** e_{22} **4.** e_{11} **5.** e_{33} **6.** e_{14}

7. Write a 2 × 2 identity matrix.

8. Write a 3 × 3 identity matrix.

9. Can a matrix have a transpose and not an inverse?

10. Can a matrix have an inverse and not a transpose?

11. Can a non-square matrix have an inverse? Why or why not?

Find the transpose and inverse for each matrix.

12. $\begin{bmatrix} 24 & 4 \\ -3 & 7 \end{bmatrix}$ **13.** $\begin{bmatrix} 0 & 1 \\ 1 & 0 \end{bmatrix}$ **14.** $\begin{bmatrix} 4 & -3 \\ 3 & 8 \end{bmatrix}$ **15.** $\begin{bmatrix} 2 & -5 \\ 6 & 1 \end{bmatrix}$

Find the signed minor for each indicated element of the matrix at the right.

$$\begin{bmatrix} 1 & 4 & -2 \\ 3 & -2 & 6 \\ 8 & 0 & -6 \end{bmatrix}$$

16. e_{21} **17.** e_{11} **18.** e_{33}

19. e_{12} **20.** e_{13} **21.** e_{22}

Written Exercises

Find the signed minor for each element in the second column of each matrix.

22. $\begin{bmatrix} 1 & -3 & 2 \\ 5 & -6 & -4 \\ 9 & 3 & 7 \end{bmatrix}$ **23.** $\begin{bmatrix} 2 & -3 & 1 \\ 4 & 0 & -2 \\ 4 & 6 & -3 \end{bmatrix}$ **24.** $\begin{bmatrix} 1 & -6 & 5 & 0 \\ 3 & 3 & 7 & -3 \\ -4 & 1 & -2 & 5 \\ 2 & 3 & 4 & -2 \end{bmatrix}$

25-28. Evaluate the signed minor of each element in the second column of the matrix in Exercise 24.

Find the transpose and the inverse (if one exists) for each matrix.

29. $\begin{bmatrix} 4 & 3 \\ -2 & 8 \end{bmatrix}$

30. $\begin{bmatrix} 3 & 2 \end{bmatrix}$

31. $\begin{bmatrix} 2 & 1 & 5 \\ -3 & 4 & 7 \end{bmatrix}$

32. $\begin{bmatrix} -2 & 5 \\ 3 & 1 \end{bmatrix}$

33. $\begin{bmatrix} 9 & 3 \\ 1 & 4 \end{bmatrix}$

34. $\begin{bmatrix} 2 & -4 \\ -1 & 2 \end{bmatrix}$

35. $\begin{bmatrix} 12 \\ -7 \\ 9 \end{bmatrix}$

36. $\begin{bmatrix} 1 & 0 & 0 \\ 0 & 1 & 0 \\ 0 & 0 & 1 \end{bmatrix}$

37. $\begin{bmatrix} 3 & 1 & 2 \\ -2 & 0 & 4 \\ 3 & 5 & 2 \end{bmatrix}$

38. $\begin{bmatrix} -2 & 3 \\ 1 & 10 \\ 0 & -6 \end{bmatrix}$

39. $\begin{bmatrix} 1 & 0 & 2 \\ 0 & 4 & 2 \\ 3 & 5 & 0 \end{bmatrix}$

40. $\begin{bmatrix} 1 & -4 & 3 \\ 2 & 3 & -7 \\ 3 & -1 & -4 \end{bmatrix}$

41. $\begin{bmatrix} 4 & -2 & 5 \\ 0 & 6 & 1 \\ 5 & -2 & 3 \end{bmatrix}$

42. $\begin{bmatrix} 7 & 4 & 2 \\ 0 & 2 & 3 \\ 1 & 5 & -2 \end{bmatrix}$

mini-review

Solve each equation or inequality.

1. $|9 - 4y| = 7$

2. $-3 < 2y - 9 \le 6$

3. $|5t - 16| \ge 14$

4. Find the standard form of the equation of the line that passes through $(-3, 7)$ and $(6, -1)$.

5. Find the slope, x-intercept, and y-intercept of the graph of $4x + 7y - 21 = 0$.

Graph each equation or inequality.

6. $6x + 2y = 11$

7. $f(x) = |x + 2|$

8. $4x - 2y \ge 2$

Solve each system of equations.

9. $x + 3y = 11$
$2x - 5y = -33$

10. $2p - 5q = -2$
$4p + 8q = 5$

11. $2a + 8b + c = 10$
$a - 2b + 2c = -22$
$-3a + 5b + 3c = 9$

12. Find an equation of the line that passes through $(9, -4)$ and is perpendicular to the graph of $3x + 5y = 1$.

13. Write the equation for a plane with xz-trace $x + z = -5$ and yz-trace $5y - 3z = 15$.

14. Solve this system of inequalities by graphing: $2x + y \ge 5$
$3y - x \le -6$

15. Evaluate $\begin{vmatrix} 4 & -2 \\ 3 & 7 \end{vmatrix}$.

16. Find $\begin{bmatrix} 5 & -2 \\ 3 & 8 \end{bmatrix} + 2 \begin{bmatrix} 6 & 1 \\ 0 & -4 \end{bmatrix}$.

17. Evaluate $\begin{vmatrix} 2 & -3 & -5 \\ -4 & 3 & 7 \\ 1 & 0 & -3 \end{vmatrix}$.

18. Find $\begin{bmatrix} 4 & 1 \\ -2 & 3 \\ 3 & 1 \end{bmatrix} \cdot \begin{bmatrix} 6 \\ -4 \end{bmatrix}$.

4-4 Using Cramer's Rule to Solve Systems of Equations

In Chapter 3, you learned that a system of two linear equations in two variables does not always have a solution that is a unique ordered pair. Similarly, a system of three linear equations in three variables does not always have a solution that is a unique ordered triple.

The graph of each equation in a system of three linear equations in three variables is a plane. Depending on the constants involved, one of the following possibilities occurs.

The three planes intersect at one point. So, the system has a unique solution.

The three planes intersect at a line. There are an infinite number of solutions to the system.

Each diagram below shows three planes that have no points in common. These systems of equations have no solutions.

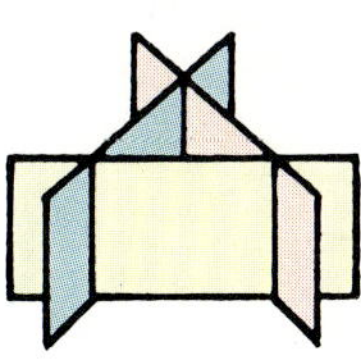  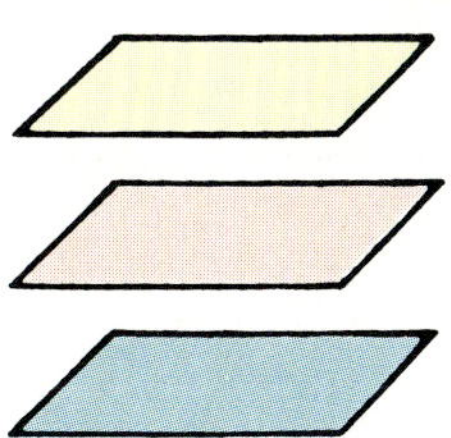

In Chapter 3, you also learned to solve systems of two linear equations in two variables using Cramer's Rule. To solve the system $\begin{cases} ax + by = c \\ dx + ey = f \end{cases}$ using Cramer's Rule, the determinant $\begin{vmatrix} a & b \\ d & e \end{vmatrix}$ must not equal zero. If this determinant does equal zero, then the system does not have a unique solution. This test can be extended to systems of three linear equations in three variables.

The system of equations $\begin{cases} a_1x + b_1y + c_1z = d_1 \\ a_2x + b_2y + c_2z = d_2 \\ a_3x + b_3y + c_3z = d_3 \end{cases}$ has a unique solution if and only if $\begin{vmatrix} a_1 & b_1 & c_1 \\ a_2 & b_2 & c_2 \\ a_3 & b_3 & c_3 \end{vmatrix} \neq 0.$	*Test for Unique Solutions*

Cramer's Rule can also be extended to solve any system of three linear equations in three variables that has a unique solution. The determinant in each denominator contains the coefficients of the variables. The determinant in each numerator is the same determinant except for the column of the coefficients of the variable to be found. This column is replaced by the column of constant terms from the system of equations.

The determinant in each denominator is the same one used in the Test for Unique Solutions.

Example

1 **Determine whether this system of equations has a unique solution. If so, solve the system using Cramer's Rule.**

$$a + 2b - c = -7$$
$$2a + 3b + 2c = -3$$
$$a - 2b - 2c = 3$$

Each determinant in this example is evaluated by using diagonals.

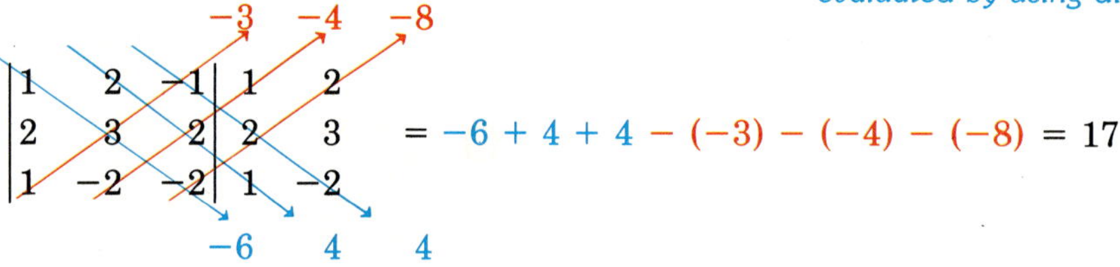

$$= -6 + 4 + 4 - (-3) - (-4) - (-8) = 17$$

Since the value of the determinant for this system is not zero, the system has a unique solution.

$$a = \frac{\begin{vmatrix} -7 & 2 & -1 \\ -3 & 3 & 2 \\ 3 & -2 & -2 \end{vmatrix}}{\begin{vmatrix} 1 & 2 & -1 \\ 2 & 3 & 2 \\ 1 & -2 & -2 \end{vmatrix}} = \frac{42 + 12 + (-6) - (-9) - 28 - 12}{17} = \frac{17}{17} \text{ or } 1$$

$$b = \frac{\begin{vmatrix} 1 & -7 & -1 \\ 2 & -3 & 2 \\ 1 & 3 & -2 \end{vmatrix}}{\begin{vmatrix} 1 & 2 & -1 \\ 2 & 3 & 2 \\ 1 & -2 & -2 \end{vmatrix}} = \frac{6 + (-14) + (-6) - 3 - 6 - 28}{17} = \frac{-51}{17} \text{ or } -3$$

$$c = \frac{\begin{vmatrix} 1 & 2 & -7 \\ 2 & 3 & -3 \\ 1 & -2 & 3 \end{vmatrix}}{\begin{vmatrix} 1 & 2 & -1 \\ 2 & 3 & 2 \\ 1 & -2 & -2 \end{vmatrix}} = \frac{9 + (-6) + 28 - (-21) - 6 - 12}{17} = \frac{34}{17} \text{ or } 2$$

The solution is $(1, -3, 2)$.   *Check this solution.*

Exploratory Exercises

Determine whether each system of equations has a unique solution.

1. $2x + 2z = 5$
$3y - 3z = -2$
$-3x - 2y = 11$

2. $2a + 4b - c = -6$
$a - 2b + 3c = 2$
$a + 2b - 4c = -10$

3. $-r + 7s + 2t = 6$
$2r + s = -11$
$3r - 6s - 2t = -25$

4. $2x - y + 3z = 5$
$3x + 2y - 5z = 4$
$x - 4y + 11z = 3$

5. $2a + 3b + 4c = 1$
$3a + 2b - c = -3$
$4a + 3b + 7c = 6$

6. $m - 2n + 3p = -4$
$2m - n + 4p = -1$
$2m + 3n + 5p = 1$

Name the determinants you would use to solve each system by Cramer's Rule.

7. $a - 2b + 3c = -4$
$2a - b + 4c = -1$
$2a + 3b + 5c = 1$

8. $2x + 4y - z = -6$
$x - 2y + 3z = 2$
$x + 2y - 4z = -10$

Written Exercises

Determine whether each system of equations has a unique solution. If so, solve the system using Cramer's Rule.

9. $2x - y + z = -2$
$x + 2y + 6z = 3$
$3x - y + 2z = -1$

10. $4a + b + 3c = 1$
$2a + c = 3$
$4a - 6b = 8$

11. $a + 2b - 3c = -13$
$2a - b + 3c = 23$
$3a + b - 3c = -8$

12. $3x - y + 2z = 11$
$6x - 3y + z = -1$
$-3x - 2y + 2z = 11$

13. $2x - y + 3z = 4$
$3x + 2y + z = 7$
$x + 3y - 2z = 3$

14. $x + 9y - 2z = 2$
$-x - 3y + 4z = 1$
$2x + 3y - 6z = -5$

15. $x + 4y + 3z = 10$
$2x - 2y + z = 15$
$x + 2y - 3z = -1$

16. $5x - y + 2z = 5$
$2x - 3y + 5z = 1$
$3x + 2y - 3z = 4$

17. $3x + 2y - 2z = 5$
$6x - 5y + 2z = 7$
$3x - 4y + 8z = 0$

18. $x + 5y = -1$
$2x - 5y + 3z = 11$
$x + 10y - 6z = -7$

19. $2x + 3y + 4z = 4$
$2x - 8z = -1$
$4x - 6y + 4z = -1$

20. $3x + 4y + z = 10$
$6x - 2y - z = 6$
$3x + 6y - 2z = 2$

Write a system of equations for each problem and solve using Cramer's Rule.

21. Floyd has 16 coins in pennies, nickels, and dimes. The number of dimes is equal to the sum of the number of pennies and number of nickels. If the total value of the coins is $1.08, how many of each kind does he have?

22. Meiko bought 97 cans of soft drink. The number of root beers exceeded the number of colas by 15. The total number of colas and orange drinks was 23 less than twice the number of root beers. How many cans of each did she buy?

23. At Burger Heaven, 2 cheeseburgers and 3 orders of fries cost $3.65. A cheeseburger and 2 milkshakes cost $2.47. A cheeseburger, 2 orders of fries, and a milkshake cost $3.01. What is the cost of each item?

24. At Tapeland, 60-minute blank tapes cost $3.19, 90-minute blank tapes cost $3.89, and 120-minute blank tapes cost $4.59. Carl bought 10 blank tapes at Tapeland for $40.30. If he bought twice as many 90-minute tapes as 120-minute tapes, how many of each tape did he buy?

Solve each equation or inequality.

1. $\frac{3}{4}(x - 5) = \frac{4}{5}(x + 4)$

2. $|3y - 7| = 5$

3. $5a + 1 \le 8a - 3$

4. $5 < 2x - 9 < 11$

5. $|9 - 3t| > 5$

6. If $f(x) = 2x^2 + 5x - 3$, find $f\left(\frac{2}{3}\right)$.

7. Find the slope, x-intercept, and y-intercept of the line whose equation is $9x - 2y = 4$.

Find the standard form of the equation of the line that satisfies the given conditions.

8. x-intercept $= 6$, y-intercept $= -5$

9. passes through $(4, -1)$ and $(-3, 2)$

Graph each equation or inequality.

10. $5x - 3y = 6$

11. $f(x) = |x - 2|$

12. $y < 2x + 6$

13. $y \ge |3x|$

14. Find an equation of the line that passes through $(-1, 2)$ and is parallel to the graph of $2x - 7y = 11$.

15. Find the value of a for which the graph of $y = ax - 3$ is perpendicular to the graph of $-6x + 11y = 4$.

16. Solve this system of inequalities by graphing:

$$y > x$$
$$y < x - 3$$

17. Graph this system of inequalities and name the vertices of the polygon formed. Then find the maximum and minimum values of $f(x, y) = 4x - 3y$ for the region.

$$2y \le x + 7 \qquad y \ge x - 4$$
$$x \le 5 \qquad y \ge -3x$$

Solve each system of equations.

18. $2x - 3y = -9$
    $x + 7y = -13$

19. $x + 5y + 2z = 10$
    $3x - 3y + 2z = 2$
    $2x + 4y - z = -15$

20. Graph $15x + 10y + 6z = 30$.

21. Evaluate $\begin{vmatrix} 2 & -3 & 1 \\ 3 & 5 & 2 \\ 1 & 0 & -3 \end{vmatrix}$.

22. Find $\begin{bmatrix} -3 & 14 & 12 \\ -2 & -1 & 7 \end{bmatrix} + \begin{bmatrix} 1 & -5 & 10 \\ 22 & 13 & -8 \end{bmatrix}$.

23. Find $\begin{bmatrix} -2 & 1 \\ 3 & -6 \\ 4 & 5 \end{bmatrix} \cdot \begin{bmatrix} 1 & 2 & -3 & 7 \\ -3 & 2 & 9 & -1 \end{bmatrix}$.

24. Find the transpose of $\begin{bmatrix} -7 & 5 & -11 \\ 2 & -4 & 9 \end{bmatrix}$.

Find the inverse of each matrix.

25. $\begin{bmatrix} 4 & -2 \\ -3 & 6 \end{bmatrix}$

26. $\begin{bmatrix} 2 & -3 & 1 \\ 3 & 5 & 2 \\ 1 & 0 & -3 \end{bmatrix}$

Solve each problem.

27. Leon bought a 10-speed bicycle on sale for 75% of its original price. The sale price was $41 less than the original price. Find the original price and the sale price.

28. A certain telephone call costs $1.38 for the first three minutes and $0.34 for each minute thereafter. What is the cost of a 17 minute phone call?

29. Sherri bought 24 cans of soda at the store. She bought r cans at $0.19 per can and t cans at $0.29 per can. Find r and t if she spent $5.46 on soda.

30. One angle of a triangle is twice another. The third angle exceeds four times the smaller by 12 degrees. Find the measurement of each angle.

A system of equations may be represented by a matrix called an **augmented matrix**. Each row in the matrix corresponds to the coefficients of an equation. Each column corresponds to the coefficients of a given variable or the constant term.

System of Equations

$$3x + 4y - 2z = 5$$
$$2x + y - z = 1$$
$$-x - y - 2z = -9$$

Augmented Matrix

$$\begin{bmatrix} 3 & 4 & -2 & 5 \\ 2 & 1 & -1 & 1 \\ -1 & -1 & -2 & -9 \end{bmatrix}$$

The system of equations can be solved by using the augmented matrix rather than the equations themselves. This matrix can be modified by transforming rows since each row represents an equation. Each change of the matrix represents a corresponding change of the system. Any operation that results in an equivalent system of equations is permitted for the matrix.

An equivalent system of equations is a system that has the same solution as the original system.

In general, you may use any of the following row operations on an augmented matrix. The resulting matrix yields the same solution as the original matrix.

1. **Interchange any two rows.**
2. **Replace any row with a nonzero multiple of that row.**
3. **Replace any row with the sum of that row and a multiple of another row.**

Row Operations on Matrices

Example

1 Use an augmented matrix to solve this system: $x + y + 2z = 9$
$$2x + y - z = 1$$
$$3x + 4y - 2z = 5$$

Write the augmented matrix for the system.

$$\begin{bmatrix} 1 & 1 & 2 & 9 \\ 2 & 1 & -1 & 1 \\ 3 & 4 & -2 & 5 \end{bmatrix}$$

Multiply row 1 by -2 and add to row 2.

$$\begin{bmatrix} 1 & 1 & 2 & 9 \\ 0 & -1 & -5 & -17 \\ 3 & 4 & -2 & 5 \end{bmatrix}$$

Multiply row 1 by -3 and add to row 3.

$$\begin{bmatrix} 1 & 1 & 2 & 9 \\ 0 & -1 & -5 & -17 \\ 0 & 1 & -8 & -22 \end{bmatrix}$$

Notice that the number of rows and number of columns of the matrix remain the same.

Add row 2 to row 3.

$$\begin{bmatrix} 1 & 1 & 2 & 9 \\ 0 & -1 & -5 & -17 \\ 0 & 0 & -13 & -39 \end{bmatrix}$$

The last augmented matrix represents the following system of equations.

$$x + y + 2z = 9$$
$$-y - 5z = -17$$
$$-13z = -39$$

This system has the same solution as the original system but can be more easily solved.

$$-13z = -39 \qquad \text{\textit{Solve for z.}}$$
$$z = 3$$

$$-y - 5(3) = -17 \qquad \text{\textit{To find y, substitute 3 for z}}$$
$$-y - 15 = -17 \qquad \text{\textit{in }} -y - 5z = -17.$$
$$y = 2$$

$$x + (2) + 2(3) = 9$$
$$x + 8 = 9 \qquad \text{\textit{To find x, substitute 3 for z and}}$$
$$x = 1 \qquad \text{\textit{2 for y in x + y + 2z = 9.}}$$

The solution is (1, 2, 3). *Check this solution.*

The goal of the augmented matrix method is to produce a system of equations that can be more easily solved than the original system. One way of accomplishing this is to obtain all zeros in either of the two "triangles" shown at the right. In Example 1, all zeros were obtained in the "bottom triangle."

$$\begin{bmatrix} a & b & c & d \\ e & f & g & h \\ i & j & k & l \end{bmatrix}$$

The simplest system of equations is produced by obtaining all zeros in both "triangles" and ones in place of a, f, and k.

Example

2 **Use an augmented matrix to solve this system:** $x + 2y + z = 0$
$$2x + 5y + 4z = -1$$
$$x - y - 9z = -5$$

Write the augmented matrix.

$$\begin{bmatrix} 1 & 2 & 1 & 0 \\ 2 & 5 & 4 & -1 \\ 1 & -1 & -9 & -5 \end{bmatrix}$$

The first element in row 1 is already 1.

Multiply row 1 by -1 and add to row 3.

$$\begin{bmatrix} 1 & 2 & 1 & 0 \\ 2 & 5 & 4 & -1 \\ 0 & -3 & -10 & -5 \end{bmatrix}$$

The first element in row 3 is now 0.

Multiply row 1 by -2 and add to row 2.

$$\begin{bmatrix} 1 & 2 & 1 & 0 \\ 0 & 1 & 2 & -1 \\ 0 & -3 & -10 & -5 \end{bmatrix}$$

The first element in row 2 is now 0, and the second element in row 2 is now 1.

Multiply row 2 by -2 and add to row 1.

$$\begin{bmatrix} 1 & 0 & -3 & 2 \\ 0 & 1 & 2 & -1 \\ 0 & -3 & -10 & -5 \end{bmatrix}$$

The second element in row 1 is now 0.

Multiply row 2 by 3 and add to row 3.

$$\begin{bmatrix} 1 & 0 & -3 & 2 \\ 0 & 1 & 2 & -1 \\ 0 & 0 & -4 & -8 \end{bmatrix}$$

The second element in row 3 is now 0.

Multiply row 3 by $-\frac{1}{4}$.

$$\begin{bmatrix} 1 & 0 & -3 & 2 \\ 0 & 1 & 2 & -1 \\ 0 & 0 & 1 & 2 \end{bmatrix}$$

The third element in row 3 is now 1.

Multiply row 3 by 3 and add to row 1.

$$\begin{bmatrix} 1 & 0 & 0 & 8 \\ 0 & 1 & 2 & -1 \\ 0 & 0 & 1 & 2 \end{bmatrix}$$

The third element in row 1 is now 0.

Multiply row 3 by -2 and add to row 2.

$$\begin{bmatrix} 1 & 0 & 0 & 8 \\ 0 & 1 & 0 & -5 \\ 0 & 0 & 1 & 2 \end{bmatrix}$$

This matrix has zeros in both "triangles." So, the solution is readily apparent.

The last augmented matrix represents $x = 8$, $y = -5$, and $z = 2$. The solution is $(8, -5, 2)$.

Exploratory Exercises

State the row operations you would use to obtain zero in the second column of row one.

1. $\begin{bmatrix} -3 & 2 & 1 \\ 4 & 2 & 6 \end{bmatrix}$

2. $\begin{bmatrix} 2 & 4 & 3 \\ -2 & -3 & 1 \end{bmatrix}$

3. $\begin{bmatrix} -6 & -2 & -3 \\ 4 & 3 & 11 \end{bmatrix}$

4. $\begin{bmatrix} 2 & -1 & 3 \\ 4 & 2 & -6 \end{bmatrix}$

5. $\begin{bmatrix} -9 & -1 & 6 \\ 0 & -4 & -8 \end{bmatrix}$

6. $\begin{bmatrix} 1 & -6 & 0 \\ -2 & 4 & -3 \end{bmatrix}$

Written Exercises

Solve each system of equations using the augmented matrix method.

7. $6x + y = 9$
$3x + 2y = 0$

8. $-2x - 3y = -11$
$3x + y = -1$

9. $3x + 2y = 5$
$4x - 3y = 1$

10. $2a - 3b = 19$
$3a + 2b = 9$

11. $8x - 2y = 3$
$4x + 12y = -5$

12. $4m + 6n = 7$
$8m - 3n = 4$

13. $x + y + z = -2$
$2x - 3y + z = -11$
$-x + 2y - z = 8$

14. $x + y + z = 6$
$2x - 3y + 4z = 3$
$4x - 8y + 4z = 12$

15. $x + 2y + z = 24$
$2x - 3y + z = -1$
$x - 2y + 2z = 7$

16. $x - 2y - 4z = -3$
$2x + 3y + 7z = 13$
$3x - 2y + 5z = -15$

17. $a + b + c = 0$
$3a - 2b + 5c = 1$
$2a + b + 2c = -1$

18. $2x + y + z = 0$
$3x - 2y - 3z = -21$
$4x + 5y + 3z = -2$

19. $4x + 3y + z = -10$
$x - 12y + 2z = -5$
$x + 18y + z = 4$

20. $2x + 6y + 8z = 5$
$-2x + 9y - 12z = -1$
$4x + 6y - 4z = 3$

21. $8m - 3n - 4p = 6$
$4m + 9n - 2p = -4$
$6m + 12n + 5p = -1$

22. $4x + 2y + 3z = 6$
$2x + 7y - 3z = 0$
$-3x - 9y + 2z = -13$

23. $2x - 9y + 5z = -5$
$-5x - y + 2z = -20$
$3x - 7y + 2z = 5$

24. $4a - 3b + c = -1$
$5a + 9b + 3c = 2$
$2a - 6b - 3c = 0$

4-6 Using Inverse Matrices to Solve Systems of Equations

Consider the following matrices:

$$A = \begin{bmatrix} 5 & 3 \\ 7 & 5 \end{bmatrix} \qquad X = \begin{bmatrix} x \\ y \end{bmatrix} \qquad C = \begin{bmatrix} -5 \\ -11 \end{bmatrix}$$

If $AX = C$, then $\begin{bmatrix} 5 & 3 \\ 7 & 5 \end{bmatrix} \cdot \begin{bmatrix} x \\ y \end{bmatrix} = \begin{bmatrix} -5 \\ -11 \end{bmatrix}$.

By multiplying the matrices, you can obtain a system of equations.

$$\begin{bmatrix} 5x + 3y \\ 7x + 5y \end{bmatrix} = \begin{bmatrix} -5 \\ -11 \end{bmatrix} \quad\blacktriangleright\quad \begin{aligned} 5x + 3y &= -5 \\ 7x + 5y &= -11 \end{aligned}$$

The equation $\begin{bmatrix} 5 & 3 \\ 7 & 5 \end{bmatrix} \cdot \begin{bmatrix} x \\ y \end{bmatrix} = \begin{bmatrix} -5 \\ -11 \end{bmatrix}$ is called a **matrix equation.**

$\begin{bmatrix} 5 & 3 \\ 7 & 5 \end{bmatrix}$ *is the coefficient matrix of the system.*

This matrix equation represents the system $\begin{cases} 5x + 3y = -5 \\ 7x + 5y = -11 \end{cases}$.

Example

1 **Write the following systems of equations as matrix equations.**

a. $5x + 2y = 8$
$\quad\ 3x - y = 7$

The matrix equation is

$$\begin{bmatrix} 5 & 2 \\ 3 & -1 \end{bmatrix} \cdot \begin{bmatrix} x \\ y \end{bmatrix} = \begin{bmatrix} 8 \\ 7 \end{bmatrix}.$$

b. $2x + y - z = 9$
$\quad\ x - 3y + 2z = 16$
$\quad\ 3x + 2y - z = 5$

The matrix equation is

$$\begin{bmatrix} 2 & 1 & -1 \\ 1 & -3 & 2 \\ 3 & 2 & -1 \end{bmatrix} \cdot \begin{bmatrix} x \\ y \\ z \end{bmatrix} = \begin{bmatrix} 9 \\ 16 \\ 5 \end{bmatrix}.$$

The matrix equation $\begin{bmatrix} 1 & 0 \\ 0 & 1 \end{bmatrix} \cdot \begin{bmatrix} x \\ y \end{bmatrix} = \begin{bmatrix} a \\ b \end{bmatrix}$ represents the linear system

$\begin{cases} x = a \\ y = b \end{cases}$. This relationship suggests a method for solving matrix equations. If the coefficient matrix can be changed into the appropriate identity matrix, then the solution is readily apparent. This change can be accomplished by multiplying each side of the matrix equation by the inverse of the coefficient matrix. Study the following examples.

Examples

2 Solve $\begin{bmatrix} 5 & 3 \\ 7 & 5 \end{bmatrix} \cdot \begin{bmatrix} x \\ y \end{bmatrix} = \begin{bmatrix} -5 \\ -11 \end{bmatrix}$.

The inverse of the coefficient matrix $\begin{bmatrix} 5 & 3 \\ 7 & 5 \end{bmatrix}$ is $\frac{1}{4}\begin{bmatrix} 5 & -3 \\ -7 & 5 \end{bmatrix}$.

Multiply each side of the equation by this inverse.

$$\frac{1}{4}\begin{bmatrix} 5 & -3 \\ -7 & 5 \end{bmatrix} \cdot \begin{bmatrix} 5 & 3 \\ 7 & 5 \end{bmatrix} \cdot \begin{bmatrix} x \\ y \end{bmatrix} = \frac{1}{4}\begin{bmatrix} 5 & -3 \\ -7 & 5 \end{bmatrix} \cdot \begin{bmatrix} -5 \\ -11 \end{bmatrix}$$

$$\begin{bmatrix} 1 & 0 \\ 0 & 1 \end{bmatrix} \cdot \begin{bmatrix} x \\ y \end{bmatrix} = \frac{1}{4}\begin{bmatrix} 8 \\ -20 \end{bmatrix} \qquad \frac{1}{4}\begin{bmatrix} 5 & -3 \\ -7 & 5 \end{bmatrix} \cdot \begin{bmatrix} 5 & 3 \\ 7 & 5 \end{bmatrix} = \frac{1}{4}\begin{bmatrix} 4 & 0 \\ 0 & 4 \end{bmatrix}$$

$$\begin{bmatrix} x \\ y \end{bmatrix} = \begin{bmatrix} 2 \\ -5 \end{bmatrix} \qquad\qquad\qquad = \begin{bmatrix} 1 & 0 \\ 0 & 1 \end{bmatrix}$$

Thus, the solution is $(2, -5)$.

3 Use a matrix equation to solve this system: $3x - 2y + z = 0$
$$2x + 3y = 12$$
$$y + 4z = -18$$

The matrix equation for this system is $\begin{bmatrix} 3 & -2 & 1 \\ 2 & 3 & 0 \\ 0 & 1 & 4 \end{bmatrix} \cdot \begin{bmatrix} x \\ y \\ z \end{bmatrix} = \begin{bmatrix} 0 \\ 12 \\ -18 \end{bmatrix}$.

To find the inverse of $\begin{bmatrix} 3 & -2 & 1 \\ 2 & 3 & 0 \\ 0 & 1 & 4 \end{bmatrix}$, first evaluate its determinant, D.

$$\begin{vmatrix} 3 & -2 & 1 \\ 2 & 3 & 0 \\ 0 & 1 & 4 \end{vmatrix} = 36 + 0 + 2 - 0 - 0 - (-16) = 54$$

Now, find and evaluate the signed minor of each element in the matrix.

$$e_{11}: +\begin{vmatrix} 3 & 0 \\ 1 & 4 \end{vmatrix} = 12 \qquad e_{12}: -\begin{vmatrix} 2 & 0 \\ 0 & 4 \end{vmatrix} = -8 \qquad e_{13}: +\begin{vmatrix} 2 & 3 \\ 0 & 1 \end{vmatrix} = 2$$

$$e_{21}: -\begin{vmatrix} -2 & 1 \\ 1 & 4 \end{vmatrix} = 9 \qquad e_{22}: +\begin{vmatrix} 3 & 1 \\ 0 & 4 \end{vmatrix} = 12 \qquad e_{23}: -\begin{vmatrix} 3 & -2 \\ 0 & 1 \end{vmatrix} = -3$$

$$e_{31}: +\begin{vmatrix} -2 & 1 \\ 3 & 0 \end{vmatrix} = -3 \qquad e_{32}: -\begin{vmatrix} 3 & 1 \\ 2 & 0 \end{vmatrix} = 2 \qquad e_{33}: +\begin{vmatrix} 3 & -2 \\ 2 & 3 \end{vmatrix} = 13$$

The matrix with each element replaced by the value of its signed

minor is $\begin{bmatrix} 12 & -8 & 2 \\ 9 & 12 & -3 \\ -3 & 2 & 13 \end{bmatrix}$.

The inverse of the coefficient matrix is the transpose of this matrix multiplied by the reciprocal of the value of the determinant, D.

Thus, the inverse of $\begin{bmatrix} 3 & -2 & 1 \\ 2 & 3 & 0 \\ 0 & 1 & 4 \end{bmatrix}$ is $\dfrac{1}{54}\begin{bmatrix} 12 & 9 & -3 \\ -8 & 12 & 2 \\ 2 & -3 & 13 \end{bmatrix}$.

Multiply each side of the equation by this inverse.

$$\frac{1}{54}\begin{bmatrix} 12 & 9 & -3 \\ -8 & 12 & 2 \\ 2 & -3 & 13 \end{bmatrix}\begin{bmatrix} 3 & -2 & 1 \\ 2 & 3 & 0 \\ 0 & 1 & 4 \end{bmatrix}\begin{bmatrix} x \\ y \\ z \end{bmatrix} = \frac{1}{54}\begin{bmatrix} 12 & 9 & -3 \\ -8 & 12 & 2 \\ 2 & -3 & 13 \end{bmatrix}\begin{bmatrix} 0 \\ 12 \\ -18 \end{bmatrix}$$

$$\begin{bmatrix} 1 & 0 & 0 \\ 0 & 1 & 0 \\ 0 & 0 & 1 \end{bmatrix}\begin{bmatrix} x \\ y \\ z \end{bmatrix} = \frac{1}{54}\begin{bmatrix} 162 \\ 108 \\ -270 \end{bmatrix}$$

$$\begin{bmatrix} x \\ y \\ z \end{bmatrix} = \begin{bmatrix} 3 \\ 2 \\ -5 \end{bmatrix}$$

Remember that the product of a matrix and its inverse is the identity matrix.

The solution is $(3, 2, -5)$.

Exploratory Exercises

State the system of linear equations represented by each matrix equation.

1. $\begin{bmatrix} 3 & 1 \\ 4 & -2 \end{bmatrix}\begin{bmatrix} x \\ y \end{bmatrix} = \begin{bmatrix} 13 \\ 24 \end{bmatrix}$

2. $\begin{bmatrix} 5 & 4 \\ 3 & -5 \end{bmatrix}\begin{bmatrix} x \\ y \end{bmatrix} = \begin{bmatrix} -3 \\ -24 \end{bmatrix}$

3. $\begin{bmatrix} 2 & -11 \\ 1 & 2 \end{bmatrix}\begin{bmatrix} x \\ y \end{bmatrix} = \begin{bmatrix} 3 \\ 9 \end{bmatrix}$

4. $\begin{bmatrix} 2 & 3 \\ 0 & 1 \end{bmatrix}\begin{bmatrix} x \\ y \end{bmatrix} = \begin{bmatrix} -16 \\ 0 \end{bmatrix}$

5. $\begin{bmatrix} 1 & -2 & 0 \\ 3 & 1 & 2 \\ 4 & -3 & 3 \end{bmatrix}\begin{bmatrix} x \\ y \\ z \end{bmatrix} = \begin{bmatrix} -8 \\ 9 \\ 1 \end{bmatrix}$

6. $\begin{bmatrix} 2 & 0 & 5 \\ 1 & 8 & 2 \\ 3 & -5 & 7 \end{bmatrix}\begin{bmatrix} x \\ y \\ z \end{bmatrix} = \begin{bmatrix} 1 \\ 2 \\ 3 \end{bmatrix}$

7. $\begin{bmatrix} 1 & 1 & 2 \\ 3 & -6 & 4 \\ 4 & -5 & -2 \end{bmatrix}\begin{bmatrix} x \\ y \\ z \end{bmatrix} = \begin{bmatrix} 5 \\ -17 \\ 4 \end{bmatrix}$

8. $\begin{bmatrix} 5 & 3 & -2 \\ 2 & 3 & 2 \\ 1 & -2 & -6 \end{bmatrix}\begin{bmatrix} x \\ y \\ z \end{bmatrix} = \begin{bmatrix} 0 \\ -5 \\ 1 \end{bmatrix}$

State a matrix equation that represents each system of linear equations.

9. $2x - y = 11$
$3x + 2y = -1$

10. $3x - y = 5$
$2x + 3y = 29$

11. $2x + 5y = 1$
$3x + 4y = 12$

12. $x + 7y = 1$
$2x + 5y = -7$

13. $6a + 9b = 6$
$4a + 6b = 8$

14. $4x - 10y = 12$
$6x - 15y = 18$

15. $x + y - z = 2$
$2x + 3y - 4z = -1$
$x - 6y + z = 1$

16. $2x + 5y + 6z = -2$
$x - 3y + 2z = 1$
$5x + 2y + 8z = 9$

17. $2a - 3b + 3c = 8$
$3a + b - c = -3$
$a - 7b + 7c = 19$

18–25. Solve each matrix equation in Exercises 1–8.

26–34. Use matrix equations to solve each system of equations in Exercises 9–17.

Solve each matrix equation.

35. $\begin{bmatrix} 4 & 8 \\ 2 & -3 \end{bmatrix} \begin{bmatrix} x \\ y \end{bmatrix} = \begin{bmatrix} 7 \\ 0 \end{bmatrix}$

36. $\begin{bmatrix} 5 & 1 \\ 9 & 3 \end{bmatrix} \begin{bmatrix} x \\ y \end{bmatrix} = \begin{bmatrix} 1 \\ 1 \end{bmatrix}$

37. $\begin{bmatrix} 3 & 1 & 1 \\ -6 & 5 & 3 \\ 9 & -2 & -1 \end{bmatrix} \begin{bmatrix} x \\ y \\ z \end{bmatrix} = \begin{bmatrix} -1 \\ -2 \\ 2 \end{bmatrix}$

38. $\begin{bmatrix} 1 & 4 & 2 \\ 2 & -4 & 6 \\ -1 & 8 & -4 \end{bmatrix} \begin{bmatrix} x \\ y \\ z \end{bmatrix} = \begin{bmatrix} 3 \\ 1 \\ 2 \end{bmatrix}$

39. $\begin{bmatrix} 2 & 4 & -1 \\ 6 & 2 & 3 \\ 7 & 2 & -2 \end{bmatrix} \begin{bmatrix} x \\ y \\ z \end{bmatrix} = \begin{bmatrix} -2 \\ 3 \\ 0 \end{bmatrix}$

40. $\begin{bmatrix} 1 & 2 & 2 \\ 2 & -1 & 1 \\ 3 & -2 & 3 \end{bmatrix} \begin{bmatrix} x \\ y \\ z \end{bmatrix} = \begin{bmatrix} 5 \\ 1 \\ 0 \end{bmatrix}$

Solve each system of equations using matrix equations.

41. $4x + 3y = 5$
$8x - 9y = 0$

42. $6a + 2b = 11$
$3a - 8b = 1$

43. $4x - 2y + 3z = -5$
$2x + y + 5z = 2$
$8x - 3y + z = -1$

44. $2m + 2n + 3p = 3$
$5m - 4n + 6p = -3$
$m - 2n = 0$

45. $3a + 6b + 2c = -5$
$6a - 6b + 4c = 5$
$9a - 12b = -5$

46. $x - y + 3z = 4$
$5x + 3y - 6z = 10$
$3x - y + 3z = 9$

mini-review

1. Solve $3x - 5 < 1$ or $2x + 7 > 13$.

2. Solve $|8 - 5a| \leq 7$.

3. Find x if the line that passes through $(15, x)$ and $(7, -3)$ has slope -4.

4. Graph $f(x) = |3x - 5|$.

5. Graph $y \leq |2x + 3|$.

6. Solve this system by graphing:
$5x + 3y = 9$
$x - 6y = 15$

7. Solve this system using Cramer's Rule: $3x - 5y = -7$
$2x + 7y = 47$

8. Graph this system of inequalities and name the vertices of the polygon formed. Then find the maximum and minimum values of $f(x, y) = x - 5y$ for the region.
$$x \geq 0 \qquad y \geq 3 \qquad 2x - y \leq -5 \qquad 3x + 2y \leq 24$$

9. Write the equation of a plane with xy-trace $2x + 3y = 6$ and yz-trace $5y - 2z = 10$.

10. Find the transpose of $\begin{bmatrix} -1 & 7 \\ 4 & -6 \\ 5 & -9 \end{bmatrix}$.

11. Find the inverse of $\begin{bmatrix} 3 & 4 & -2 \\ 2 & 1 & -1 \\ -1 & -1 & -2 \end{bmatrix}$.

12. Use Cramer's Rule to solve this system: $5m + 3n - 2p = 0$
$2m + 3n + 5p = 7$
$m - 2n - 6p = 1$

13. Use an augmented matrix to solve this system: $3x + 5y + 6z = 3$
$x - 3y + 2z = 1$
$5x + 2y + 8z = 9$

For a fund raising project the members of the classes of Mrs. Clef and Mr. Steele sold cookies as shown in the chart below.

	Chocolate Chips	Peanut Creams	Mint Wafers
Mrs. Clef's Class	32 boxes	20 boxes	30 boxes
Mr. Steele's Class	28 boxes	22 boxes	34 boxes

The profit per box is shown below.

Chocolate Chips $0.40
Peanut Creams   0.20
Mint Wafers   0.50

The profit for each class is found as follows.

                                                               Profit

Mrs. Clef's Class $32(0.40) + 20(0.20) + 30(0.50) = 31.80$ ▶ $31.80

Mr. Steele's Class: $28(0.40) + 22(0.20) + 34(0.50) = 32.60$ ▶ $32.60

Another way to find the profit for each class is to use multiplication of matrices.

Sales per Class × Profit per Box = Profit per Class ($)

$$\begin{bmatrix} 32 & 20 & 30 \\ 28 & 22 & 34 \end{bmatrix} \times \begin{bmatrix} 0.40 \\ 0.20 \\ 0.50 \end{bmatrix} = \begin{bmatrix} 31.80 \\ 32.60 \end{bmatrix}$$

Exercises

For each exercise, copy the two matrices above (Sales per Class and Profit per Box), changing the appropriate elements. Then multiply to find the profit for each class.

1. The profit on each box of Peanut Creams is 45¢ instead of 20¢.

2. The profit on each type of box is 15¢ more.

3. Boxes of Peanut Creams Sold:
     Mrs. Clef's Class   42
     Mr. Steele's Class   35

4. Boxes of Chocolate Chips Sold:
     Mrs. Clef's Class   25
     Mr. Steele's Class   30

4-7 Problem Solving: Using Three Variables

A system of three linear equations in three variables can be used to
solve many types of problems.

Example

1 **The sum of the digits of a three-digit number is 10. When the digits are reversed, the new number formed is 99 less than the original number. The hundreds digit equals the sum of the tens digit and the units digit. Find the number.**

explore Let h represent the hundreds digit, t represent the tens digit, and u represent the units digit.
Then, $100h + 10t + u$ represents the number.

plan Write a system of equations.

$$h + t + u = 10$$
$$100u + 10t + h = (100h + 10t + u) - 99$$
$$h = t + u$$

solve First, simplify the second equation.

$$100u + 10t + h = (100h + 10t + u) - 99$$
$$99u - 99h = -99$$
$$h - u = 1$$

Now solve the system of equations using the substitution method. Substitute $t + u$ for h in the first equation and simplify.

$$h + t + u = 10$$
$$(t + u) + t + u = 10$$
$$2(t + u) = 10$$
$$t + u = 5$$
$$h = 5$$

Now substitute 5 for h in $h - u = 1$ and solve for u.

$$5 - u = 1$$
$$u = 4$$

Finally, substitute 5 for h and 4 for u in $h = t + u$ to find t.

$$5 = t + 4$$
$$t = 1$$

Since $h = 5$, $t = 1$, and $u = 4$, the number is 514.

examine The hundreds digit, 5, equals the sum of the tens digit, 1, and the units digit, 4.
When the digits of 514 are reversed, the number 415 is formed. The number 415 is 99 less than the original number, 514.

2 Last year, Kathie Faught invested $48,000, some in stocks, some in bonds, and the remainder in a term account. She earned 4% on the stocks, 7% on the bonds, and 6% on the term account. For the year, she earned a total of $2860. She earned three times as much from the term account as she did from the stocks. How much did she invest in each?

explore Let s represent the amount in stocks, b represent the amount in bonds, and t represent the amount in the term account.

plan Write a system of equations.

$$s + b + t = 48000$$
$$0.04s + 0.07b + 0.06t = 2860$$
$$0.06t = 3(0.04s) \quad \text{or} \quad 0.12s - 0.06t = 0$$

solve Solve the system of equations. Use Cramer's Rule.

Each determinant will be evaluated using diagonals.

$$\begin{vmatrix} 1 & 1 & 1 \\ 0.04 & 0.07 & 0.06 \\ 0.12 & 0 & -0.06 \end{vmatrix} = -0.0042 + 0.0072 - (0.0084 - 0.0024)$$

$$= -0.003$$

Since the value of the determinant is not zero, the system has a unique solution.

$$s = \frac{\begin{vmatrix} 48000 & 1 & 1 \\ 2860 & 0.07 & 0.06 \\ 0 & 0 & -0.06 \end{vmatrix}}{-0.003}$$

$$= \frac{-201.6 - (-171.6)}{-0.003}$$

$$= 10000$$

$$t = \frac{\begin{vmatrix} 1 & 1 & 48000 \\ 0.04 & 0.07 & 2860 \\ 0.12 & 0 & 0 \end{vmatrix}}{-0.003}$$

$$= \frac{343.2 - 403.2}{-0.003}$$

$$= 20000$$

$$b = \frac{\begin{vmatrix} 1 & 48000 & 1 \\ 0.04 & 2860 & 0.06 \\ 0.12 & 0 & -0.06 \end{vmatrix}}{-0.003}$$

$$= \frac{-171.6 + 345.6 - (343.2 - 115.2)}{-0.003}$$

$$= 18000$$

Kathie invested $10,000 in stocks, $18,000 in bonds, and $20,000 in a term account.

examine Find the earnings from stocks, bonds, and the term account.

Stocks: 4% of $10,000 = $400
Bonds: 7% of $18,000 = $1260
Term: 6% of $20,000 = $1200

Three times the amount earned in stocks is 3($400) or $1200.

$$\overline{\hspace{2em}\$2860}$$ ✔ The total is correct.

Exploratory Exercises

State an equation using three variables to illustrate each sentence.

1. The sum of Rose's and Bill's ages exceeds twice Maggie's age by 14 years.
2. The sum of the angles of a triangle is 180°.
3. Annie's weight increased by twice Betty's weight is 80 pounds greater than Colleen's weight.
4. The perimeter of a triangle is 42 inches.
5. The sum of two angles of a triangle exceeds the third by 62°.

Written Exercises

Write three equations in three variables for each problem. Then solve.

6. The sum of the digits of a three-digit number is 20. The tens digit exceeds twice the units digit by 1. When the digits are reversed, the new number formed is 297 less than the original number. Find the number.

7. The sum of the digits of a three-digit number is 17. The tens digit is three times the hundreds digit. The sum of the hundreds and the tens digits is one less than the units digit. Find the number.

8. In a three-digit number, twice the hundreds digit exceeds the sum of the tens and units digit by 3. When the digits are reversed, the new number formed is 198 greater than the original number. The sum of the digits of the new number is 12. Find the original number.

9. The Ice Cream Shoppe sells the following sizes of ice cream cones: single, 89¢; double, $1.19; triple, $1.39. One day, Kyong Mae sold 52 ice cream cones. She sold two more than twice as many doubles as triples. If she sold $58.98 in cones, how many of each size did she sell?

10. At the student-faculty basketball game, Mr. Foster, Mr. Waits, and Mrs. Leshnock scored a total of 63 points for the faculty team. Mr. Foster scored twice as many points as Mrs. Leshnock. Mr. Waits scored three points more than the sum of the points scored by Mrs. Leshnock and Mr. Foster. How many points were scored by each person?

11. Jim Hunt sells three models of toasters. On Wednesday, he sold 8 of model A, 3 of model B, and 6 of model C. On Thursday, he sold 6 of A, 4 of B, and 10 of C. On Friday, he sold 10 of A, 7 of B, and 6 of C. His total sales for these three days were $400, $488, and $568, respectively. Find the cost of each model of toaster.

12. If Ann buys 6 apples, 5 bananas, and 2 pears, she will pay $4.15. If she buys 3 apples, 7 bananas, and 4 pears, she will pay $4.10. If apples are 11¢ less than twice as expensive as pears, what is the cost of each item?

13. At Frank's Fried Chicken, a salad, 2 rolls, and 2 pieces of chicken cost $3.65. A roll and 3 pieces of chicken cost $3.20. If a salad is three times as expensive as a roll, what is the cost of each item?

14. The sum of the ages of Mark, Laurie, and Peggy is 79 years. The sum of Mark's and Peggy's ages exceeds twice Laurie's age by one year. Five years ago, Mark was the same age as Peggy is now. Find their ages now.

15. The sum of the ages of Arturo, Benny, and Carlos is 41 years. Twice Arturo's age exceeds the sum of Benny's and Carlos's age by one year. Five years ago, Benny was 2 years more than twice as old as Carlos. Find their ages now.

16. The perimeter of a triangle is 45 cm. The two shorter sides differ by 2 cm. The longest side is 7 cm less than the sum of the other two sides. Find the length of each side.

17. The measure of one angle of a triangle is twice the measure of another. The measure of the third angle exceeds four times the measure of the smaller by 12 degrees. Find the measure of each angle.

18. The largest angle of a triangle is 15 degrees greater than the smallest. The sum of the two larger angles exceeds twice the smaller by 24°. Find the measurement of each angle.

19. The perimeter of a triangle is 83 in. The longest side is three times the length of the shortest side and 17 in. more than one-half the sum of the other two sides. Find the length of each side.

> → / ↓ Using Computers ← * ↑ <

Cramer's Rule

The computer program below can be used to solve a system of three equations in three variables.

The DATA line in the program shows the coefficients of the system of equations shown at the right.

$$x + 2y - z = -7$$
$$2x + 3y + 2z = -3$$
$$x - 2y - 2z = 3$$

```
5    DIM A(3, 4), B(3, 4)
10   FOR I = 1 TO 3
20   FOR J = 1 TO 4
30   READ A(I,J)
40   NEXT J
50   NEXT I
55   DATA 1,2,-1,-7,2,3,2,-3,1,-2,-2,3
60   FOR K = 0 TO 3
70   FOR J = 1 TO 3
80   FOR I = 1 TO 3
90   IF J = K THEN 120
100    LET B(I,J) = A(I,J)
110    GOTO 130
120    LET B(I,J) = A(I,4)
130    NEXT I
140    NEXT J
150    LET M1 = B(2,2) * B(3,3) - B(3,2) * B(2,3)
160    LET M2 = B(2,1) * B(3,3) - B(3,1) * B(2,3)
170    LET M3 = B(2,1) * B(3,2) - B(3,1) * B(2,2)
180    LET D(K) = B(1,1) * M1 - B(1,2) * M2 + B(1,3) * M3
190    NEXT K
200    IF D(0) = 0 THEN 250
210    PRINT "X = ";D(1);"/";D(0); " OR ";D(1) / D(0)
220    PRINT "Y = ";D(2);"/";D(0); " OR ";D(2) / D(0)
230    PRINT "Z = ";D(3);"/";D(0); " OR ";D(3) / D(0)
240    GOTO 260
250    PRINT "NO UNIQUE SOLUTION"
260    END
```

In lines 10–50, the computer interprets the data as an augmented matrix.

In the program, A(I,J) is the element in row
I and column J of the augmented matrix
shown at the right.

$$\begin{bmatrix} 1 & 2 & -1 & -7 \\ 2 & 3 & 2 & -3 \\ 1 & -2 & -2 & 3 \end{bmatrix}$$

Lines 60–190 of the program form a loop for evaluating four different 3×3 determinants. For each determinant, three columns of the augmented matrix are used. The four values are stored as D(0), D(1), D(2), and D(3).

Line 200 checks whether the system has a unique solution by looking at the value of the determinant D(0). If D(0) is 0, then the computer will print "NO UNIQUE SOLUTION." Otherwise, the computer will read lines 210–230 and print the solution of the system.

The output for the program is shown at the right.

```
]RUN
X = 17/17 OR 1
Y = −51/17 OR −3
Z = 34/17 OR 2
```

Exercises

1–12. Use the program to solve each system of equations in Exercises 9–20 on page 137.

Vocabulary

matrix (121)

elements (121)

dimension (121)

square matrix (121)

expansion by minors (121)

minor (122)

identity matrix (129)

inverse matrix (129)

transpose (131)

signed minor (131)

augmented matrix (139)

matrix equation (142)

Chapter Summary

1. All square matrices have determinants. (121)

2. Expansion of a Third Order Determinant: (122)

$$\begin{vmatrix} a & b & c \\ d & e & f \\ g & h & i \end{vmatrix} = a\begin{vmatrix} e & f \\ h & i \end{vmatrix} - b\begin{vmatrix} d & f \\ g & i \end{vmatrix} + c\begin{vmatrix} d & e \\ g & h \end{vmatrix}$$

3. Evaluating a third order determinant using diagonals: (122)

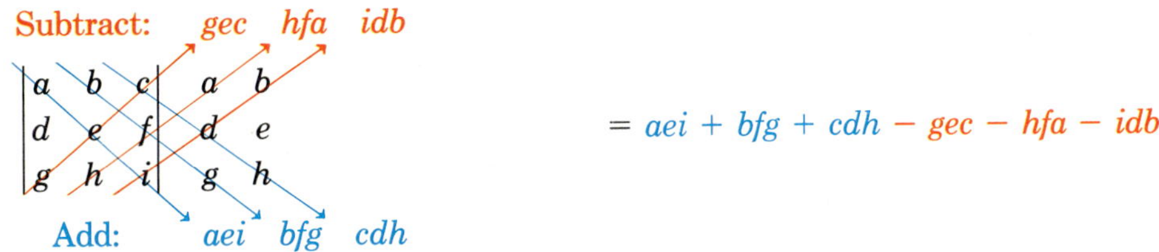

4. Two matrices are equal if and only if they have the same dimension and their corresponding elements are equal. (124)

5. The product of an $m \times n$ matrix, A, and a constant, k, is an $m \times n$ matrix kA. Each element of kA is the product of k and the corresponding element of A.   (124)

6. If A and B are two $m \times n$ matrices, then $A + B$ is an $m \times n$ matrix where each element is the sum of the corresponding elements of A and B.   (125)

7. The product of an $m \times n$ matrix, A, and an $n \times r$ matrix, B, is the $m \times r$ matrix AB. The element in the ith row and j^{th} column of AB is the sum of the products of the corresponding elements in the ith row of A and the jth column of B.   (125)

8. The identity matrix, I, for multiplication is a square matrix with a 1 for every element of the principal diagonal and 0 in all other positions. The principal diagonal extends from upper left to lower right.   (129)

9. Any matrix M, $\begin{bmatrix} a & b \\ c & d \end{bmatrix}$, will have an inverse M^{-1} if and only if

$$\begin{vmatrix} a & b \\ c & d \end{vmatrix} \neq 0. \text{ Then } M^{-1} = \frac{1}{ad - bc} \begin{bmatrix} d & -b \\ -c & a \end{bmatrix}.   \quad (130)$$

10. To find the inverse of a 3×3 matrix with a nonzero determinant, D: (132)
   1. Replace each element by the value of its signed minor.
   2. Find the transpose for the new matrix.
   3. Multiply this transpose by the reciprocal of the determinant, D.

11. The system of equations $\begin{cases} a_1x + b_1y + c_1z = d_1 \\ a_2x + b_2y + c_2z = d_2 \\ a_3x + b_3y + c_3z = d_3 \end{cases}$ has a unique solution if and

only if $\begin{vmatrix} a_1 & b_1 & c_1 \\ a_2 & b_2 & c_2 \\ a_3 & b_3 & c_3 \end{vmatrix} \neq 0.$ (135)

12. Cramer's Rule can be extended to solve any system of three linear equations in three variables that has a unique solution. The determinant in each denominator contains the coefficients of the variables. The determinant in each numerator is the same determinant except for the column of coefficients of the variable to be found. This column is replaced by the column of constant terms from the system of equations.   (136)

13. Row Operations on Matrices: (139)
   1. Interchange any two rows.
   2. Replace any row with a nonzero multiple of that row.
   3. Replace any row with the sum of that row and a multiple of another row.

14. The goal of the augmented matrix method is to produce a system of equations that can be more easily solved than the original system.   (140)

15. The equation $\begin{bmatrix} a & b \\ d & e \end{bmatrix} \cdot \begin{bmatrix} x \\ y \end{bmatrix} = \begin{bmatrix} c \\ f \end{bmatrix}$ represents the system $\begin{cases} ax + by = c \\ dx + ey = f \end{cases}$. Inverse matrices are used to solve matrix equations.   (142)

Chapter Review

4–1 **State the dimension and evaluate the determinant (if one exists) for each matrix.**

1. $\begin{bmatrix} 5 & -11 \\ 17 & 4 \end{bmatrix}$
2. $\begin{bmatrix} 7 & 1 & 6 \\ 3 & -1 & 4 \\ -2 & 3 & 0 \end{bmatrix}$
3. $\begin{bmatrix} 5 & 8 & -3 \\ -2 & -1 & 4 \\ 3 & 2 & -4 \end{bmatrix}$
4. $\begin{bmatrix} 2 & 7 & -1 \\ 4 & 5 & 10 \end{bmatrix}$

4–2 **Perform the indicated operations.**

5. $3\begin{bmatrix} 4 & 1 & -2 & -1 \\ 2 & -3 & 4 & 0 \end{bmatrix}$
6. $\begin{bmatrix} 4 & 2 \\ -2 & -3 \\ -1 & 5 \end{bmatrix} - \begin{bmatrix} 7 & -5 \\ 0 & 4 \\ 9 & -1 \end{bmatrix}$
7. $\begin{bmatrix} 5 & 0 & 7 \\ 4 & 1 & 3 \end{bmatrix} \cdot \begin{bmatrix} 3 & -2 \\ -1 & 4 \\ -2 & 0 \end{bmatrix}$

Find the values of x, y, and z for which the following equations are true.

8. $\begin{bmatrix} 2x & 5y - 3 \\ 8y & 3z \end{bmatrix} = \begin{bmatrix} 16 & x - 6 \\ x & x + y \end{bmatrix}$
9. $4\begin{bmatrix} x \\ 2y \end{bmatrix} - 5\begin{bmatrix} y \\ 3x \end{bmatrix} = \begin{bmatrix} 8 \\ 13 \end{bmatrix}$

4–3 **Find the transpose and the inverse (if one exists) for each matrix.**

10. $\begin{bmatrix} 3 & -2 \\ 4 & 1 \end{bmatrix}$
11. $\begin{bmatrix} 4 & -3 & 5 \\ 7 & 0 & -2 \end{bmatrix}$
12. $\begin{bmatrix} 1 & -3 & 5 \\ 2 & 4 & -7 \\ 7 & -1 & 1 \end{bmatrix}$
13. $\begin{bmatrix} 1 & 3 & 0 \\ 1 & 4 & -2 \\ 2 & 1 & 2 \end{bmatrix}$

4–4 **Determine whether each system of equations has a unique solution. If so, solve the system using Cramer's Rule.**

14. $x + y - z = -5$
$2x - 2y - 3z = -2$
$2x + y + 2z = 9$

15. $x + y + 2z = 3$
$2x - 2y - 3z = 2$
$3x - y - 2z = 1$

16. $4x - 2y + z = 4$
$3x + y - 2z = 1$
$x - 3y + 3z = 3$

4–5 **Solve each system of equations using the augmented matrix method.**

17. $x + y = 1$
$2x - 3y = 17$

18. $x + y - 2z = -3$
$2x - 3y + 4z = 12$
$5x - 2y - 3z = 6$

19. $2x + 3y + 2z = -5$
$3x - y + 6z = -6$
$x - 4y + 2z = -2$

4–6 **Solve each system of equations using matrix equations.**

20. $2x - 9y = 5$
$x + 2y = 9$

21. $x + y + z = -1$
$2x + 4y + z = 1$
$3x - y - z = -15$

22. $2x + 3y - z = -2$
$3x + y + z = 9$
$2x - 4y - 3z = 8$

4–7 **Write three equations in three variables for each problem. Then solve.**

23. The sum of the digits of a three-digit number is 18. The units digit is one less than twice the tens digit. When the digits are reversed, the new number formed is the same as the original. Find the number.

24. At Chester's Pizza, a drink, a salad, and 2 slices of pizza cost $4.12. A drink, 2 salads, and 3 slices of pizza cost $6.30. A salad, 2 drinks, and 4 slices of pizza cost $7.25. Find the cost of each item.

State the dimension and evaluate the determinant (if one exists) for each matrix.

1. $\begin{bmatrix} 2 & -3 & 1 \\ 3 & -1 & 2 \\ 1 & 2 & -3 \end{bmatrix}$
  2. $\begin{bmatrix} 7 & -10 & 1 & 4 \\ 6 & 8 & 5 & -1 \end{bmatrix}$
  3. $\begin{bmatrix} 1 & 3 & 1 \\ 2 & 1 & -5 \\ 3 & -1 & -4 \end{bmatrix}$

Perform the indicated operations.

4. $\begin{bmatrix} 2 & -4 & 1 \\ 3 & 8 & -2 \end{bmatrix} - 2\begin{bmatrix} 1 & 2 & -4 \\ -2 & 3 & 7 \end{bmatrix}$
  5. $\begin{bmatrix} 1 & 2 \\ -4 & 3 \\ 5 & 2 \end{bmatrix}\begin{bmatrix} 5 \\ 4 \end{bmatrix}$

6. Find the value of x and y for which $\begin{bmatrix} x \\ 2y \end{bmatrix} - 3\begin{bmatrix} y+1 \\ 2x \end{bmatrix} = \begin{bmatrix} 5 \\ -16 \end{bmatrix}$ is true.

Find the inverse (if one exists) for each matrix.

7. $\begin{bmatrix} 5 & -2 \\ 6 & 3 \end{bmatrix}$
  8. $\begin{bmatrix} 1 & 0 & 2 \\ 1 & 2 & -3 \\ 2 & 0 & 3 \end{bmatrix}$
  9. $\begin{bmatrix} 1 & 2 & -3 \\ 3 & -1 & 2 \\ 9 & 4 & -5 \end{bmatrix}$

Determine whether each system of equations has a unique solution. If so, solve the system using Cramer's Rule.

10. $r + s + t = 7$
$3r - 7s + 2t = 11$
$-9r + 21s + 3t = -3$

11. $2x + 4y - 3z = -8$
$x - 2y + 6z = -1$
$x + 4y - 9z = -8$

12. $x + 2y - 5z = 1$
$2x - y + 2z = -4$
$8x + y - 4z = 10$

Solve each system of equations using the augmented matrix method.

13. $6x - y = -15$
$5x + 2y = -4$

14. $2x - 3y + z = 7$
$3x - y + 2z = 1$
$x + 2y - 3z = -14$

15. $x + 3y - z = 0$
$3x + 5y + 2z = 9$
$2x - y - 8z = 1$

Solve each system of equations using matrix equations.

16. $x + 8y = -3$
$2x - 6y = -17$

17. $x + 3y + z = 5$
$2x + y - 5z = 4$
$3x - y - 4z = -11$

18. $3x + y + 2z = 4$
$4x + 2y - z = -4$
$x - 3y - 4z = 2$

Write three equations in three variables for each problem. Then solve.

19. The sum of the digits of a three digit number is 12. The units digit is twice the hundreds digit. Three times the sum of the units and the hundreds digit equals the tens digit. Find the number.

20. Sue's Ice Cream Parlor sells the following sizes of ice cream cones: single, 85¢; double, $1.15; triple, $1.40. One day, Sal sold 71 ice cream cones. He sold one more than twice as many singles as doubles. If he sold $71.85 in cones, how many of each size did he sell?

The test questions on this page deal with ratios, proportions, and percents.

Directions: Choose the one best answer. Write A, B, C, or D.

1. 40% of 10 inches is how many sixths of 2 feet?

   (A) $\frac{1}{3}$ **(B)** 1   **(C)** 2   **(D)** 4

2. For nonzero numbers a, b, c, and d, $\frac{a}{b} = \frac{c}{d}$. Which of the following must be true?

   (A) $\frac{a}{b} = \frac{b}{c}$ **(B)** $\frac{a+b}{b} = \frac{c+b}{d}$

   (C) $\frac{d}{b} = \frac{c}{a}$ **(D)** $\frac{b}{c+d} = \frac{d}{a+b}$

3. Find the percent of increase if your salary increases from $250 a week to $300 a week.

   (A) $16\frac{2}{3}\%$ **(B)** 20%

   (C) 22%   **(D)** 25%

4. If your grade was 90 and is now 75, find the percent of decrease.

   (A) $16\frac{2}{3}\%$ **(B)** 18%

   (C) 20%   **(D)** 22%

5. 9 is 6% of what number?

   (A) 110   **(B)** 120   **(C)** 130   **(D)** 150

6. If $\frac{x}{y} = \frac{5}{6}$, then $18x =$

   (A) $\frac{5}{3}y$ **(B)** $90y$ **(C)** $15y$ **(D)** $\frac{5y}{6}$

7. The price of an item was reduced by 20% then later reduced by 5%. The two reductions were equivalent to the single reduction of

   (A) 15%   **(B)** 24%   **(C)** 25%   **(D)** 75%

8. Ten gallons of gas were added to a tank that had been $\frac{1}{4}$ full. If it is now $\frac{7}{8}$ full, how many gallons does the tank hold?

   (A) 16   **(B)** 18   **(C)** 20   **(D)** 24

9. The ratio of Jean's weight to Jim's weight is 3:4. If Jean gains 30 pounds and Jim does not gain any, the ratio will be 7:8. How much does Jim weigh?

   (A) 60   **(B)** 170   **(C)** 180   **(D)** 240

10. Last year Joe attended one-half the number of sporting events that Jan did. George attended one-third the number that Jan did. If George attended 8 sporting events, how many did Joe attend?

    (A) 1   **(B)** 4   **(C)** 8   **(D)** 12

11. If $\frac{1}{9} = \frac{x}{.45}$, what is the value of x?

    (A) .05   **(B)** .5   **(C)** 5   **(D)** 6

12. In a class of 54 students, 12 are honor students. What part of the class are *not* honor students?

    (A) $\frac{21}{33}$ **(B)** $\frac{7}{9}$ **(C)** $\frac{2}{7}$ **(D)** $\frac{2}{9}$

13. 22% of 440 is 4.4% of

    (A) 96.8   **(B)** 425.92
    (C) 220   **(D)** 2200

14. 75% of $10a$ is b. What percent of $2b$ is a?

    (A) $13\frac{1}{3}$ **(B)** $6\frac{2}{3}$ **(C)** $7\frac{1}{2}$ **(D)** 15

15. John spent $\frac{1}{4}$ of his money on a book and then spent $\frac{2}{5}$ of the remaining money for lunch. What fractional part of the original amount is left?

    (A) $\frac{3}{20}$ **(B)** $\frac{3}{8}$ **(C)** $\frac{9}{20}$ **(D)** $\frac{4}{9}$

CHAPTER 5

Polynomials

In the field of genetics, scientists study the dominant and recessive traits of a population, such as eye color and hair color. They can predict the frequency of occurrence for a specific trait in a population by expanding certain binomial expressions and then evaluating the polynomial. In this chapter, you will learn about addition, subtraction, multiplication, and division of polynomials.

5-1 Monomials

A **monomial** is an expression that is a number, a variable, or the product of a number and one or more variables. Expressions like -3, y, m^7, $-4x^2$, and $\frac{3}{5}ab^3$ are monomials. Expressions like $2x + 1$, $\frac{3}{x}$, and $\sqrt{x}$ are *not* monomials.

A monomial that contains no variable is called a **constant**. The numerical factor of a monomial is called the **coefficient**. For example, the coefficient of $-4x^2$ is -4. The **degree of a monomial** is the sum of the exponents of its variables. The degree of a nonzero constant is 0. The constant 0 has no degree.

Monomial	Coefficient	Variables	Exponent(s)	Degree
y	1	y	1	1
$-4x^2$	-4	x	2	2
m^7	1	m	7	7
$\frac{3}{5}ab^3$	$\frac{3}{5}$	a and b	1 and 3	4

Recall that $y = 1 \cdot y^1$.

If two monomials are the same or differ only by their coefficients, they are called **like terms**. For example, $6x^3y$ and $17x^3y$ are *like terms*, and $3a^2b$ and $4ab^2$ are *unlike terms*.

Examples

1 **Simplify $6x^3y - 17x^3y$.**

$$6x^3y - 17x^3y = (6 - 17)x^3y \qquad \text{\textit{Distributive Property}}$$
$$= -11x^3y$$

2 **Simplify $3n^4p^5 - 7n^4p^5 + 8n^4p^5$.**

$$3n^4p^5 - 7n^4p^5 + 8n^4p^5 = (3 - 7 + 8)n^4p^5 \qquad \text{\textit{Distributive Property}}$$
$$= 4n^4p^5$$

The expression x^3 means $x \cdot x \cdot x$. You can use the meaning of exponents to discover how to multiply powers.

Example

3 **Simplify $(s^2t^3)(s^4t^5)$.**

$$(s^2t^3)(s^4t^5) = (s \cdot s \cdot t \cdot t \cdot t)(s \cdot s \cdot s \cdot s \cdot t \cdot t \cdot t \cdot t \cdot t)$$
$$= s \cdot s \cdot s \cdot s \cdot s \cdot s \cdot t \cdot t \cdot t \cdot t \cdot t \cdot t \cdot t \cdot t$$
$$= s^6t^8 \qquad \text{\textit{Notice that } } 2 + 4 = 6 \text{ and } 3 + 5 = 8. \text{ Also, } s^2 \cdot s^4 = s^6 \text{ and } t^3 \cdot t^5 = t^8.$$

Example 3 suggests the following property.

> **For any real number a, and positive integers m and n,**
> $$a^m \cdot a^n = a^{m+n}.$$

Multiplying Powers

Examples

4 **Simplify $(4^2)^3$.**

$$(4^2)^3 = 4^2 \cdot 4^2 \cdot 4^2$$
$$= 4^{2+2+2} \quad \text{\textit{Multiplying Powers}}$$
$$= 4^6 \qquad\quad \text{\textit{Property}}$$

5 **Simplify $(h^4)^5$.**

$$(h^4)^5 = h^4 \cdot h^4 \cdot h^4 \cdot h^4 \cdot h^4$$
$$= h^{4+4+4+4+4}$$
$$= h^{20}$$

Examples 4 and 5 suggest the following property.

> **For any real number a, and positive integers m and n,**
> $$(a^m)^n = a^{mn}.$$

Raising a Power to a Power

Examples

6 **Simplify $(ab)^4$.**

$$(ab)^4 = a \cdot b \cdot a \cdot b \cdot a \cdot b \cdot a \cdot b$$
$$= a \cdot a \cdot a \cdot a \cdot b \cdot b \cdot b \cdot b$$
$$= a^4 b^4$$

7 **Simplify $\left(\frac{1}{2}x^2\right)^3$.**

$$\left(\frac{1}{2}x^2\right)^3 = \frac{1}{2} \cdot x^2 \cdot \frac{1}{2} \cdot x^2 \cdot \frac{1}{2} \cdot x^2$$
$$= \left(\frac{1}{2}\right)^3 (x^2)^3 \text{ or } \frac{1}{8}x^6$$

Examples 6 and 7 suggest the following property.

> **For any real numbers a, b, and positive integer m,**
> $$(ab)^m = a^m b^m.$$

Finding a Power of a Product

You can simplify many kinds of expressions using the properties of exponents along with the commutative and associative properties.

Examples

8 **Simplify $(4x^2y)(-3x^3y^4)$.**

$$(4x^2y)(-3x^3y^4) = 4 \cdot (-3) \cdot x^2 \cdot x^3 \cdot y \cdot y^4$$
$$= -12 \cdot x^{2+3} \cdot y^{1+4}$$
$$= -12x^5y^5$$

9 **Simplify $(-2r^2s^3)^3$.**

$$(-2r^2s^3)^3 = (-2)^3(r^2)^3(s^3)^3$$
$$= -8 \cdot r^{(2\cdot3)} \cdot s^{(3\cdot3)}$$
$$= -8r^6s^9$$

Exploratory Exercises

State whether each expression is a monomial. If it is, name its coefficient and degree.

1. $7x$ **2.** y^2 **3.** $3xy + y$ **4.** $-5ab$ **5.** $\dfrac{11xy}{7}$ **6.** $\sqrt{cd}$

7. $\dfrac{3}{x}$ **8.** -8 **9.** $5x^3$ **10.** $11m$ **11.** $5x^3y^2z^4$ **12.** $4xy$

13. 17 **14.** 0 **15.** $-24p^4q$ **16.** $-b$ **17.** $\dfrac{3st}{q}$ **18.** $\dfrac{12a^2b}{6}$

Written Exercises

Simplify.

19. $4m + 7m + (-3m)$ **20.** $2x^3 + 3x^3 + (-6x^3)$ **21.** $4d^3 - d^3 + 2d^3$

22. $4ab^2 - 3ab^2$ **23.** $27x^2 - 3y^2 + 12x^2$ **24.** $3x^2y + 4 - 3x^2y$

25. $y^5 \cdot y^7$ **26.** $n^4 \cdot n^3 \cdot n^2$ **27.** $2^3 \cdot 2^4$ **28.** $t^{13} \cdot t^{15} \cdot t^{18}$

29. $8^6 \cdot 8^4 \cdot (8^2)^2$ **30.** $(m^3)^2$ **31.** $(y^5)^2$ **32.** $(2a)^3$

33. $(3a)^4$ **34.** $(rs^3)(-5r^2s^3)$ **35.** $(5m^2k^2)(4mk^3)$ **36.** $(x^2y^2)^2x^3y^3$

37. $\left(-\dfrac{3}{4}x^2y^3\right)^2\left(\dfrac{8}{9}xy^4\right)$ **38.** $(4rs^2t)^2\left(-\dfrac{1}{2}r^2t\right)^3(3st^3)^4$ **39.** $\left(\dfrac{3}{5}c^2f\right)\left(\dfrac{4}{3}cd\right)^2$

40. $(3x^2y)^2(5xy^2z)^4$ **41.** $(-2ab^2)^3(6a^2)^4$ **42.** $(-4a)(a^2)(-a^3) + 3a^2a^4$

43. $2(rk)^2(5rt^2) - k(2rk)(2rt)^2$ **44.** $(2xy^2)^3 + (2xy^2)^2(6xy^2)$ **45.** $(3a)(a^2b)^3 + (2a)^2(-a^5b^3)$

46. $(5a)(6a^2b)(3ab^3) + (4a^2)(3b^3)(2a^2b)$ **47.** $(5mn^2)(m^3n)(-3p^2) + (8np)(3mp)(m^3n^2)$

mini-review

Write the equation for the line with the given conditions.

1. passes through $(6, 4)$ and perpendicular to $3x - 5y = 7$.

2. passes through the points $(8, -1)$ and $(6, 4)$.

Solve each system of equations.

3. $9x + 6y = 7$
$6x - 15y = -8$

4. $x + 4y + 5z = 0$
$3x - 9y - z = 6$
$2x - 7y - 3z = 9$

Find the slope-intercept form of the equation of the line that satisfies the given conditions.

5. slope $= \dfrac{1}{3}$, passes through $(5, 3)$ **6.** x-intercept $= 5$, y-intercept $= -3$

7. Solve $\begin{bmatrix} 2 & -3 & 5 \\ -3 & 4 & 3 \\ 7 & 4 & 2 \end{bmatrix} \cdot \begin{bmatrix} x \\ y \\ z \end{bmatrix} = \begin{bmatrix} -17 \\ 14 \\ -15 \end{bmatrix}$.
     8. Evaluate $\begin{vmatrix} 5 & 4 & -3 \\ -2 & -1 & 8 \\ 3 & 2 & -4 \end{vmatrix}$.

9. Given $f(x) = 2x^3 + 3x^2 - 4$, find $f(3t)$.

Find the y-intercept and x-intercept of each line whose equation is given below.

10. $4y = -3x - 12$ **11.** $y = 7$ **12.** $g(x) = 3x + 2$

5-2 Dividing Monomials

If you add exponents when you multiply powers, then it seems reasonable to subtract exponents when you divide powers.

$$\frac{x^5}{x^3} = \frac{x \cdot x \cdot x \cdot x \cdot x}{x \cdot x \cdot x} \qquad \leftarrow 5 \text{ factors}$$
$$\leftarrow 3 \text{ factors}$$
$$= x \cdot x \qquad \leftarrow 2 \text{ or } (5 - 3) \text{ factors}$$
$$= x^{5-3} \text{ or } x^2 \qquad \frac{x \cdot x \cdot x}{x \cdot x \cdot x} = \frac{x^3}{x^3} = 1$$

This and other similar examples suggest the following property.

> **For any real number a, except $a = 0$, and integers m and n,**
> $$\frac{a^m}{a^n} = a^{m-n}$$

Dividing Powers

Why is 0 not an acceptable value for a?

Examples

1 Simplify $\dfrac{p^9}{p^4}$.

$$\frac{p^9}{p^4} = p^{9-4} \qquad \text{To divide powers,}$$
$$\text{subtract the exponents.}$$
$$= p^5$$

2 Simplify $\dfrac{(2xy)^5}{(x^2y)^2}$.

$$\frac{(2xy)^5}{(x^2y)^2} = \frac{(2)^5(x)^5(y)^5}{(x^2)^2(y)^2}$$
$$= \frac{32x^5y^5}{x^4y^2}$$
$$= 32x^{(5-4)}y^{(5-2)}$$
$$= 32x^1y^3 \text{ or } 32xy^3$$

Study the two ways to simplify $\dfrac{x^4}{x^4}$ shown below.

$$\frac{x^4}{x^4} = \frac{x \cdot x \cdot x \cdot x}{x \cdot x \cdot x \cdot x} \qquad \qquad \frac{x^4}{x^4} = x^{4-4} \qquad \textit{Dividing Powers Property}$$
$$= 1 \qquad \qquad \qquad \qquad = x^0$$

Since $\dfrac{x^4}{x^4}$ cannot have two different values, you can conclude that $x^0 = 1$, where x is not equal to zero. In general, any nonzero number raised to the zero power is equal to 1.

> **For any real number a, except $a = 0$,**
> $$a^0 = 1.$$

Zero Exponent

Now examine why 0^0 is not defined. Logically 0^0 could stand for 0^{m-m} or $\dfrac{0^m}{0^m}$. The expression $\dfrac{0^m}{0^m}$ would imply division by zero since $0^m = 0$. Division by zero is not defined. Thus, 0^0 is not defined.

Study the two ways to simplify $\frac{x^3}{x^7}$ shown below. Assume x is a nonzero real number.

$$\frac{x^3}{x^7} = \frac{x \cdot x \cdot x}{x \cdot x \cdot x \cdot x \cdot x \cdot x \cdot x} \qquad\qquad \frac{x^3}{x^7} = x^{3-7} \qquad \textit{Dividing Powers Property}$$

$$= \frac{1}{x \cdot x \cdot x \cdot x} \qquad\qquad\qquad\quad = x^{-4}$$

$$= \frac{1}{x^4}$$

Since $\frac{x^3}{x^7}$ cannot have two different values, you can conclude that $x^{-4} = \frac{1}{x^4}$. By taking the reciprocal of each expression, you can conclude that $\frac{1}{x^{-4}} = x^4$. These and other similar examples suggest the following properties.

<table>
<tr><td>

For any real number a, except $a = 0$, and any integer n,

$$a^{-n} = \frac{1}{a^n} \text{ and } \frac{1}{a^{-n}} = a^n.$$

</td><td>

Negative Exponents

</td></tr>
</table>

To simplify a quotient of monomials, write an equivalent form that uses only positive exponents and no parentheses. Also, each base should appear only once and all fractions should be in simplest form. The rules for positive integral exponents can be extended to include negative exponents.

Examples

3 Simplify $\dfrac{xy^{-4}}{x^6y^4z^{-2}}$.

$$\frac{xy^{-4}}{x^6y^4z^{-2}} = \frac{x}{x^6} \cdot \frac{y^{-4}}{y^4} \cdot \frac{1}{z^{-2}}$$

$$= x^{1-6} \cdot y^{-4-4} \cdot z^2$$

$$= x^{-5} \cdot y^{-8} \cdot z^2$$

$$= \frac{z^2}{x^5 y^8}$$

4 Simplify $\dfrac{(2ab)^2(a^3c)^2}{10b^2c^3}$.

$$\frac{(2ab)^2(a^3c)^2}{10b^2c^3} = \frac{(4a^2b^2)(a^6c^2)}{10b^2c^3} \quad \textit{Multiplying Powers Property}$$

$$= \frac{4}{10}\left(\frac{a^8}{1}\right)\left(\frac{b^2}{b^2}\right)\left(\frac{c^2}{c^3}\right)$$

$$= \frac{2}{5}a^8b^0c^{-1} \quad \textit{Dividing Powers Property}$$

$$= \frac{2a^8}{5c}$$

The properties of exponents may be used to simplify many different types of expressions.

5 Simplify $\dfrac{5^{2k}}{5^{2k-3}}$.

$$\dfrac{5^{2k}}{5^{2k-3}} = 5^{2k-(2k-3)} \quad \textit{Dividing Powers Property}$$
$$= 5^3$$
$$= 125$$

6 Simplify $-49r^3s^5(7rs^2)^{-1}$.

$$-49r^3s^5(7rs^2)^{-1} = -49r^3s^5\left(\dfrac{1}{7rs^2}\right)$$
$$= \dfrac{-49r^3s^5}{7rs^2} \quad r \neq 0, s \neq 0$$
$$= -7r^2s^3$$

You can use the properties of exponents to simplify and evaluate powers of rational numbers.

7 Evaluate $\left(\dfrac{3}{5}\right)^4$.

$$\left(\dfrac{3}{5}\right)^4 = \dfrac{3}{5} \cdot \dfrac{3}{5} \cdot \dfrac{3}{5} \cdot \dfrac{3}{5}$$
$$= \dfrac{3^4}{5^4}$$
$$= \dfrac{81}{625}$$

8 Evaluate $\left(\dfrac{2}{3}\right)^{-5}$.

$$\left(\dfrac{2}{3}\right)^{-5} = \left[\left(\dfrac{2}{3}\right)^{-1}\right]^5 \quad \textit{Use the power property.}$$
$$\left(\dfrac{2}{3}\right)^{-1} = \dfrac{1}{\frac{2}{3}} = \dfrac{3}{2}$$
$$= \left(\dfrac{3}{2}\right)^5$$
$$= \dfrac{3}{2} \cdot \dfrac{3}{2} \cdot \dfrac{3}{2} \cdot \dfrac{3}{2} \cdot \dfrac{3}{2}$$
$$= \dfrac{3^5}{2^5}$$
$$= \dfrac{243}{32}$$

These and other similar examples suggest the following properties.

> **For any nonzero real numbers a and b, and integer n,**
> $$\left(\dfrac{a}{b}\right)^n = \dfrac{a^n}{b^n} \text{ and } \left(\dfrac{a}{b}\right)^{-n} = \left(\dfrac{b}{a}\right)^n \text{ or } \dfrac{b^n}{a^n}.$$

Powers of Quotients

Sometimes several properties may be used to simplify an expression.

9 Simplify $\left(\dfrac{1}{k}\right)^{-3}$.

$$\left(\dfrac{1}{k}\right)^{-3} = \left(\dfrac{k}{1}\right)^3$$
$$= k^3$$

10 Simplify $\left(\dfrac{3y^2}{2x}\right)^{-2}$.

$$\left(\dfrac{3y^2}{2x}\right)^{-2} = \dfrac{(2x)^2}{(3y^2)^2} \quad \textit{Powers of Quotients Property}$$
$$= \dfrac{4x^2}{9y^4} \quad \textit{Finding a Power of a Product Property}$$

Simplify.

1. $\dfrac{r^4}{r}$　　2. $\dfrac{n^5}{n^5}$　　3. $\dfrac{x^6}{x^8}$　　4. $\dfrac{5y^{10}}{y^{13}}$　　5. $\dfrac{3m^6}{m^6}$　　6. 6^{-2}

7. $\dfrac{1}{m^{-2}}$　　8. $\dfrac{6^{-2}}{6^{-4}}$　　9. $\dfrac{3^{-3}}{3^{-2}}$　　10. $\left(\dfrac{1}{2}\right)^{-2}$　　11. $\left(\dfrac{1}{5}\right)^3$　　12. $\left(\dfrac{1}{10}\right)^6$

13. $\left(\dfrac{1}{10}\right)^{-4}$　　14. $\left(\dfrac{2}{3}\right)^0$　　15. $\left(\dfrac{3}{b}\right)^6$　　16. $\left(\dfrac{7}{b}\right)^{-5}$　　17. $\left(\dfrac{k}{4}\right)^{-3}$　　18. $\left(\dfrac{1}{k}\right)^0$

Written Exercises

Simplify.

19. $m^{-8}m^3$　　20. $r^{-2}r^4$　　21. $\dfrac{12n^8}{4n^3}$　　22. $\dfrac{-24s^8}{2s^5}$

23. $\dfrac{6mn^2}{3m}$　　24. $\dfrac{an^6}{n^5}$　　25. $\dfrac{xy^7}{x^4}$　　26. $\dfrac{48a^8}{12a^{11}}$

27. $\dfrac{15b^9}{3b^{12}}$　　28. $\dfrac{4x^3}{28x^5}$　　29. $\dfrac{12b^4}{60b^6}$　　30. $\dfrac{-20y^5}{40y^2}$

31. $\dfrac{2x^{-3}}{6(x^2)^2}$　　32. $\dfrac{8(m^{-2})^2}{4m^{-2}}$　　33. $\dfrac{16b^6c^5}{4b^4c^2}$　　34. $\dfrac{1}{m^0 + n^0}$

35. $\dfrac{4}{x^0 + y^0}$　　36. $\dfrac{-15r^5s^2}{5r^5s^{-4}}$　　37. $\dfrac{-27w^3t^7}{-3w^3t^{12}}$　　38. $\dfrac{-2a^3b^6}{24a^2b^2}$

39. $\dfrac{(3c^2)^2(-d^5)}{-45c^7d^3}$　　40. $\dfrac{-66p^3(mp)^{10}}{33(mp)^2}$　　41. $\dfrac{20n^5m^9}{20nm^7}$　　42. $\dfrac{16b^6c^5}{(2b^2c)^2}$

43. $\dfrac{3^{xy+5}}{3^{xy}}$　　44. $\dfrac{r^{2a}}{r^{2a-3}}$　　45. $\dfrac{x^{3a}}{x^{3a-2}}$　　46. $\dfrac{5^{2x}}{5^{2x+2}}$

47. $(x^3y^2)^{-1}$　　48. $(m^4n^5)^{-2}$　　49. $5^{-3}b^3x^4y^{-1}$　　50. $8a^2b^4(-2b)^{-1}$

51. $3^3r^3s^3(3r)^{-2}$　　52. $36x^3y^5(12x^2y^2)^{-1}$　　53. $(6z)^{-4}x^3y^0$　　54. $(2)^{-7}b^{-6}c^0$

55. $\left(\dfrac{x}{k^{-1}}\right)^{-1}$　　56. $\left(\dfrac{2}{d^3f}\right)^5$　　57. $\left(\dfrac{1}{x^2y^3}\right)^3$　　58. $\left(\dfrac{3}{2x^{-2}}\right)^{-1}$

59. $\left(\dfrac{x}{y^{-1}z^2}\right)^{-1}$　　60. $\left(\dfrac{-3y^4}{2y^2}\right)^{-2}$　　61. $\left(\dfrac{1}{5}\right)^{-2} + \left(\dfrac{1}{4}\right)^{-1}$　　62. $\left(\dfrac{1}{2}\right)^{-2} + \left(\dfrac{1}{3}\right)^2$

63. $\dfrac{-15r^5s^8(r^3s^2)}{45r^4s}$　　64. $\dfrac{-3w^6t^7}{(-27w^3t^2)(wt)^2}$　　65. $\dfrac{(-2r^3)^2(r^{-2})^{-1}}{(r^2)^{-3}}$　　66. $\dfrac{(4x^3y)(4^2x^{-1}y)}{4^3xy^2}$

Challenge Exercises

Simplify.

67. $m^{-3}(m^2 + m^4 - m^{-1})$　　　68. $a^3(a^{-2} + a^{-5} + a)$　　　69. $y^{-4}(y^{-2} - y^{-3} - y^0)$

5-3 Scientific Notation

An important application of exponents is **scientific notation**. Recall that very large or very small numbers are often written in scientific notation.

Study these examples.

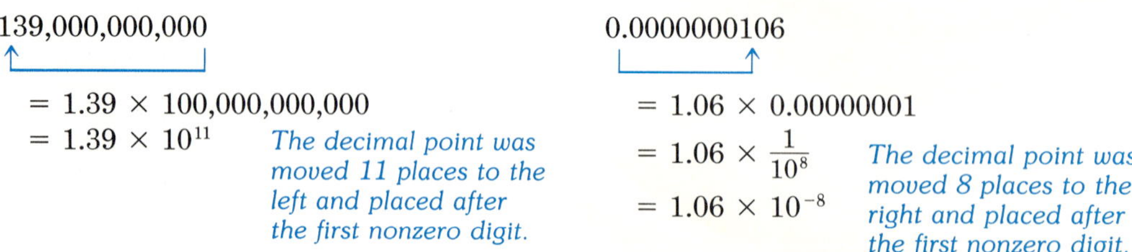

$$139{,}000{,}000{,}000$$
$$= 1.39 \times 100{,}000{,}000{,}000$$
$$= 1.39 \times 10^{11}$$

The decimal point was moved 11 places to the left and placed after the first nonzero digit.

$$0.0000000106$$
$$= 1.06 \times 0.00000001$$
$$= 1.06 \times \frac{1}{10^8}$$
$$= 1.06 \times 10^{-8}$$

The decimal point was moved 8 places to the right and placed after the first nonzero digit.

A number is expressed in scientific notation when it is in the form $a \times 10^n$, where $1 \le a < 10$ and n is an integer.	*Definition of Scientific Notation*

Examples

1 **Multiply 0.543 by 617,000 using scientific notation.**

$$0.543 \times 617{,}000 = (5.43 \times 10^{-1})(6.17 \times 10^5)$$

Express 0.543 and 617,000 in scientific notation.

$$= 5.43 \times 6.17 \times 10^{-1} \times 10^5$$
$$= 33.5031 \times 10^4$$
$$= 335{,}031$$

2 **Using Calculators**

Divide 8.6×10^{-4} by 3.1×10^{-2}.

The exponential shift key may vary depending upon the calculator you are using.

ENTER: 8.6   [EE]   [+/−]   4   [÷] 8.6×10^{-4} is expressed as 8.6 − 04.

DISPLAY: *8.6   8.600   8.6−00   8.6−04*

ENTER: 3.1   [EE]   [+/−]   2   [=] 3.1×10^{-2} is expressed as 3.1 − 02.

DISPLAY: *3.1   3.100   3.1−00   3.1−02   0.02774*

The quotient is rounded to 5 decimal places.

The answer is about 0.02774 or 2.774×10^{-2}.

Express each of the following in scientific notation.

1. 0.0021 **2.** 0.0692 **3.** 810.4 **4.** 482.09

5. 9,000,000,000 **6.** 786,500,000 **7.** 0.000000721 **8.** 0.0000528

Express each of the following in decimal notation.

9. 6×10^3 **10.** 9.8×10^4 **11.** 5.7×10^{-4} **12.** 5.4×10^{-3}

13. 3.21×10^6 **14.** 7.2×10^{-5} **15.** 4.27×10^{-2} **16.** 6.7×10^6

Written Exercises

Evaluate. Express each answer both in scientific and in decimal notation.

17. $(7.2 \times 10^5)(8.1 \times 10^3)$ **18.** $(9.5 \times 10^3)^2$ **19.** $\dfrac{8 \times 10^{-1}}{16 \times 10^{-2}}$ **20.** $\dfrac{15 \times 10^4}{6 \times 10^{-2}}$

21. $(4.5 \times 10^3)(7.5 \times 10^2)^{-1}$ **22.** $(9 \times 10^3)^{-1}(3.5 \times 10^{-2})$ **23.** $(6.9 \times 10^3)(1.4 \times 10^3)^{-1}$

24. $(4,300)(0.02)$ **25.** $(34,000)(0.0056)$ **26.** $\dfrac{0.000000036}{0.00011}$

27. $\dfrac{5,600,000,000}{60,000}$ **28.** $\dfrac{(93,000,000)(0.0005)}{0.0015}$ **29.** $\dfrac{(84,000,000)(0.00004)}{0.0016}$

Solve each problem.

30. Light from a laser will travel about 300,000 kilometers per second. How many kilometers can this light travel in a day?

31. In astronomy, a light year is the distance light travels in one year. If light travels about 186,000 miles per second, how many miles are in a light year?

32. A heart beats about once every second. If a person is 78 years old, about how many beats has the person's heart made?

33. The mass of a proton is 1.672×10^{-24} g. If the mass of Earth's moon is 7.35×10^{22} kg, how many times larger is its mass than that of a proton?

34. Wavelengths of light are measured in Angstrom units. An Angstrom unit is 10^{-8} cm. The wavelength of cadmium's green line is 5085.8 Angstrom units. How many wavelengths of cadmium's green line are there in one meter?

35. Venus has an average distance of 1.08×10^8 kilometers from the Sun. Saturn has an average distance of 1.428×10^9 kilometers from the Sun. About how much closer to the Sun is Venus?

36. Metal expands and contracts with changes in temperature. The change in the length of steel per degree Celsius is given by the constant 11×10^{-6}. A steel bridge 200 meters long varies in temperature 70° Celsius. What is the change in the length of the bridge in centimeters?

37. Newton's law of gravitation can be used to compute the mass of Earth in grams. His formula applied is:

$$908 = \frac{6.67 \times 10^{-8} \times 1 \times M}{(6.37 \times 10^8)^2}$$

If M represents mass, what is the mass of Earth in grams?

5-4 Polynomials

A **polynomial** is either a monomial or the sum of monomials. For example, $8x + y$, $3x^2 + 2x + 4$, and $5 - 3y + 5xy^2$ are polynomials. The expression $x^2 + \frac{2}{x}$ is *not* a polynomial since $\frac{2}{x}$ is not a monomial.

Each monomial in a polynomial is called a **term** of the polynomial. A polynomial with two unlike terms is called a *binomial*. A polynomial with three unlike terms is called a *trinomial*.

The **degree of a polynomial** is the degree of the monomial of greatest degree.

Example

1 **Find the degree of $5x^3 + 3x^2y^2 + 4xy^2 + 3x - 2$.**

$5x^3$ has degree 3.
$3x^2y^2$ has degree 4.
$4xy^2$ has degree 3.
$3x$ has degree 1.
-2 has degree 0.

The terms of polynomials are usually arranged so that the powers of one of the variables are in ascending or descending order.

The degree of the polynomial is 4.

You can simplify polynomials with like terms by using the Distributive Property or by adding or subtracting the coefficients of like terms. Both methods are shown in the examples below.

Examples

2 **Simplify $6x^2y + 3xy^4 + 7y + 5xy^4 - 9x^2y + 8y$.**

$$6x^2y + 3xy^4 + 7y + 5xy^4 - 9x^2y + 8y = (6x^2y - 9x^2y) + (3xy^4 + 5xy^4) + (7y + 8y)$$
$$= (6 - 9)x^2y + (3 + 5)xy^4 + (7 + 8)y$$
$$= -3x^2y + 8xy^4 + 15y$$

3 **Simplify $(7m^2k - 8mk^2 + 19k) + (18mk^2 - 3k)$.**

$$(7m^2k - 8mk^2 + 19k) + (18mk^2 - 3k) = 7m^2k - 8mk^2 + 18mk^2 + 19k - 3k$$
$$= 7m^2k + 10mk^2 + 16k$$

4 **Simplify $(2x^2 - 3xy + 5y^2) - (4x^2 - 3xy - 2y^2)$.**

$$(2x^2 - 3xy + 5y^2) - (4x^2 - 3xy - 2y^2) = 2x^2 - 4x^2 - 3xy + 3xy + 5y^2 + 2y^2$$
$$= -2x^2 + 0xy + 7y^2$$
$$= -2x^2 + 7y^2$$

You can use the Distributive Property to multiply polynomials.

Examples

5 **Find $3x(4xy^3 - 7x^2y - 3y)$.**

$$3x(4xy^3 - 7x^2y - 3y) = 3x \cdot 4xy^3 - 3x \cdot 7x^2y - 3x \cdot 3y$$
$$= 12x^2y^3 - 21x^3y - 9xy$$

6 **Find $(2a - 3b)(3a + 4ab + b)$.**

$$(2a - 3b)(3a + 4ab + b)$$
$$= 2a(3a + 4ab + b) - 3b(3a + 4ab + b)$$
$$= 2a(3a) + 2a(4ab) + 2a(b) - 3b(3a) - 3b(4ab) - 3b(b)$$
$$= 6a^2 + 8a^2b + 2ab - 9ab - 12ab^2 - 3b^2$$
$$= 6a^2 + 8a^2b - 7ab - 12ab^2 - 3b^2$$

7 **Find $(x + 4)(x + 11)$.**

$$(x + 4)(x + 11) = (x + 4) \cdot x + (x + 4) \cdot 11$$
$$= (x \cdot x) + (4 \cdot x) + (x \cdot 11) + (4 \cdot 11)$$
$$= x^2 + 4x + 11x + 44$$
$$= x^2 + 15x + 44$$

The following process can also be used to multiply binomials.

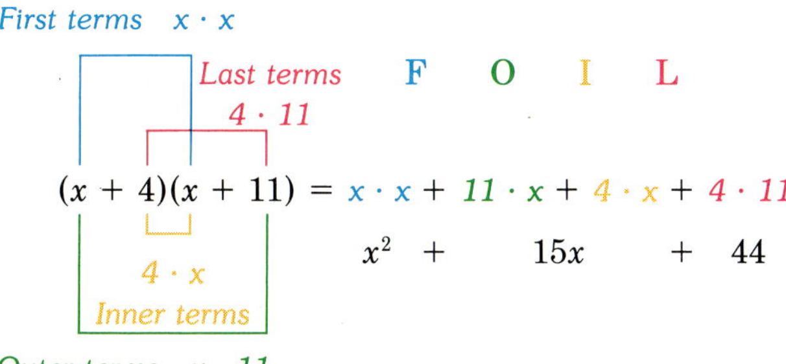

This process is called the **FOIL** method.

The product of two binomials is the sum of the products of

    F the *first* terms,

    O the *outer* terms,

    I the *inner* terms, and

    L the *last* terms.

FOIL Method for Multiplying Binomials

8 Find $(2x + 3)(x - 5)$.

$$
\begin{array}{ccccccccc}
& & F & & O & & I & & L \\
(2x + 3)(x - 5) &=& 2x \cdot x &+& 2x \cdot (-5) &+& 3 \cdot x &+& 3 \cdot (-5) \\
&=& 2x^2 &-& 10x &+& 3x &-& 15 \\
&=& 2x^2 &-& & & 7x & & - \quad 15
\end{array}
$$

9 Find $(5x - 3)(2x - 7)$.

$$
\begin{array}{ccccccccc}
& & F & & O & & I & & L \\
(5x - 3)(2x - 7) &=& 5x \cdot 2x &+& 5x \cdot (-7) &+& (-3) \cdot 2x &+& (-3) \cdot (-7) \\
&=& 10x^2 &-& 35x &-& 6x &+& 21 \\
&=& 10x^2 &-& & 41x & & &+ \quad 21
\end{array}
$$

Exploratory Exercises

Find the degree of each polynomial.

1. $3x^2 + 27xy$

2. $x^2 + 2x + 3$

3. $a^8 + a^7b + a^6b^2 - a^2b^6 - ab^7 - b^8$

4. $3x^4y^2 - 5x^2y + 3$

5. $3r^5 - 3r^4 - 7r - 5$

6. $z^5 + 5z^4 + 9z^3 + 9z^2$

7. $m^3 + 2mn^2 + 4n^3$

8. $5xy - 2x^2 - 3y^2$

9. $13xy^7 + 36x^3y^5 - 2x^4y^5 - xy$

10. $16x^4yz + 12x^2y^3z - 24x^3y^2z - 18xy^4z$

Use the FOIL method to verify each product.

11. $(a + b)^2 = a^2 + 2ab + b^2$ **12.** $(a - b)^2 = a^2 - 2ab + b^2$ **13.** $(a + b)(a - b) = a^2 - b^2$

Written Exercises

Simplify.

14. $(7x - 2y) + (9x + 4y)$

15. $(-12y - 6y^2) + (-7y + 6y^2)$

16. $(-3x + 5y - 10z) + (-6x + 8z - 6y)$

17. $(7t + 4m + a) + (-3a - 7t + m)$

18. $(7m^2 + 9m + 3) - (3m^2 + 8m + 2)$

19. $(3a^2 - 5d + 17) - (-a^2 + 5d - 3)$

20. $(2x^2 + x + 5) + (3x^2 - x - 4)$

21. $(3y^2 + 5y - 7) + (2y^2 - 7y + 10)$

22. $-4(a + 2b) + 7(a + b)$

23. $4(a^2 - b^2) + 3(a^2 + b^2)$

24. $(8r^2 + 5r + 14) - (7r^2 + 6r + 8)$

25. $(4k^2 + 10k - 14) - (3k^2 + 7k - 5)$

26. $(4x + 11) - (3x^2 + 7x - 3)$

27. $(10n^2 - 3nt + 4t^2) - (3n^2 + 5nt)$

28. $(4x^4 + 3x^3 + x - 7) + (3x^4 - 5x^3 + 7x^2 - 3x + 8)$

29. $(8b^5 + 4b^3 + 7b^2 - 15) + (3b^5 + 7b^3 - b + 14)$

30. $(2m^7 + 3m^5 + 2m^3 - 18) - (m^7 + 3m^4 + m^3 + 2m^2)$

31. $(x^3 - 3x^2y + 4xy^2 + y^3) - (7x^3 + x^2y - 9xy^2 + y^3)$

32. $-7a^2(a^3 - ab)$

33. $4f(gf^2 - bh)$

34. $3a^3b^2(-2ab^2 + 4a^2b - 7a)$

35. $-5mn^2(-3m^2n + 6m^3n - 3m^4n^4)$

36. $4ax^2(-9a^3x^2 + 8a^2x^3 + 6a^4x^4)$

37. $17b^3d^2(-4b^2d^2 - 11b^3d^3 - 5bd^4)$

38. $x^{-4}(x^2 + x - 3)$

39. $r^{-3}(r^5 - 2r^3 + r^{-1})$

40. $4a^{-1}b^2(a^2b^{-1} + 3a^3b^{-2} + 4^{-2}ab^{-1})$

41. $y^2x^{-3}(yx^4 + y^{-1}x^3 + y^{-2}x^2)$

42. $(x + 7)(x + 2)$

43. $(m - 7)(m + 5)$

44. $(m^2 + 5)(m^2 - 4)$

45. $(y^2 + y)(y^2 + 5)$

46. $(3y - 8)(2y + 7)$

47. $(2x + 7)(3x + 5)$

48. $(2r^2 + 3)(r^2 - 5)$

49. $(2x + 3y)(3x - 5y)$

50. $(6a - 5)(7a - 9)$

Use the FOIL method to find each product.

51. $(m + 4)^2$

52. $(x - 8)^2$

53. $(y - 2)^2$

54. $(k + 6)^2$

55. $(y - 5)(y + 5)$

56. $(2p + q^3)^2$

57. $(x - 3y)^2$

58. $(a + 6b)(a - 6b)$

59. $(4m - 3n)^2$

60. $(5r - 2)^2$

61. $(1 + 4r)^2$

62. $(6a + 2)^2$

63. $(4a - 2b)(4a + 2b)$

64. $(5x + 12)(5x - 12)$

65. $(x^3 - y)(x^3 + y)$

Find each product.

66. $(a - 1)(a^2 - 2a - 1)$

67. $(2x - 3)(x^2 - 3x - 8)$

68. $(x - y)(x^2 + xy + y^3)$

69. $(a + b)(a^2 - ab + b^2)$

70. $(m - 4)(3m^2 + 5m - 4)$

71. $(2t - 5)(t^2 + 7t + 8)$

72. $r(r - 2)(r - 3)$

73. $p(p + 5)(p - 1)$

74. $(b + 1)(b - 2)(b + 3)$

75. $(2x - 3)(x + 1)(3x - 2)$

76. $(2a + 1)(a - 2)^2$

77. $(a - 2b)^2(2a + 3b)$

78. $(a - b)(a^2 + ab + b^2)$

79. $(2k + 3)(k^2 - 7k + 21)$

80. $(x^2 + y)(x^2 + xy + y^2)$

81. $(z^2 + r)(z + zr + r^2)$

82. $(a - 2)(a^2 + 6a + 9)$

83. $(y + 4)(y^2 - 7y + 12)$

Challenge Exercises

Find each product.

84. $(x^2 + 2x - 3)(5x^2 + 3x - 7)$

85. $(3r^2 + 2d + 1)(5r^2 - 2r - 6)$

86. $(a^2 - ab + b^2)(a^2 + ab + b^2)$

87. $(4m^2 - m + 8)(m^3 + 2m^2 + 3m + 4)$

Excursions in Algebra   Special Products

Certain special products can help you multiply numbers mentally.
You know that $(x - y)(x + y) = x^2 - y^2$. Suppose you want to find $36 \cdot 44$. Notice that $36 = 40 - 4$ and $44 = 40 + 4$.

$$36 \cdot 44 = (40 - 4)(40 + 4)$$
$$= 40^2 - 4^2$$
$$= 1600 - 16 \text{ or } 1584$$

Try this method to calculate $27 \cdot 33$, $19 \cdot 21$, and $53 \cdot 67$ mentally.
You also know that $(x + 1)^2 = x^2 + 2x + 1$ and $(x - 1)^2 = x^2 - 2x + 1$.
Suppose you want to find $(101)^2$ and $(99)^2$. Notice that $101 = 100 + 1$ and $99 = 100 - 1$.

$$(101)^2 = (100)^2 + 2(100) + 1 \qquad (99)^2 = (100)^2 - 2(100) + 1$$
$$= 10000 + 200 + 1 \qquad\qquad = 10000 - 200 + 1$$
$$= 10{,}201 \qquad\qquad\qquad = 9801$$

Try these methods to calculate 31^2, 13^2, 201^2, 39^2, and 399^2 mentally.

Find the slope, y-intercept, and x-intercept of each line whose equation is given below.

1. $4x - 5y = 15$

2. $-3x + 7y = 12$

Graph each equation or inequality.

3. $3x - 2y = 4$

4. $g(x) = |x| - 2$

5. $f(x) \geq |x + 2|$

Given $f(x) = 12x^2 - 13x + 5$, find each value.

6. $f(-3)$

7. $f(1) - f(-1)$

8. $f(3)$

9. $f(6) - f(3)$

Given the following conditions, write the equation of the line for each.

10. passes through $(6, 2)$ and parallel to the line $3x - 2y = 7$.

11. passes through $(8, 5)$ and perpendicular to the line $-3x + 4y = 12$.

12. passes through $(4, 7)$ and $(6, 3)$.

Solve each system using Cramer's Rule.

13. $13x - 8y = -2$
$45x + 75y = 41$

14. $4x + y = 6$
$-2x + 3y = 10$

Find the value of each determinant.

15. $\begin{vmatrix} 4 & -2 \\ 3 & 2 \end{vmatrix}$

16. $\begin{vmatrix} 3 & 1 \\ 4 & 7 \end{vmatrix}$

17. $\begin{vmatrix} 1 & 2 & -2 \\ 2 & -1 & 1 \\ 3 & -3 & 4 \end{vmatrix}$

18. $\begin{vmatrix} 6 & 2 & -4 \\ 9 & 3 & 6 \\ -11 & 5 & 1 \end{vmatrix}$

19. Graph $6x - 3y + 2z = 12$ and find the trace in each coordinate plane.

Evaluate the signed minor for each indicated element of this determinant.

$\begin{vmatrix} 1 & -2 & 3 \\ 4 & -3 & 1 \\ 6 & 5 & 4 \end{vmatrix}$

20. e_{11}

21. e_{23}

22. e_{33}

Solve each system of equations.

23. $5x + 3y = 5$
$2x - 5y = 33$

24. $4x + 3y = 13$
$-5x - 7y = -39$

25. $2x - 3y + 7z = -23$
$3x - 4y + 2z = 7$
$9x + 5y - z = 22$

26. Graph each system of inequalities. Name the vertices of the polygon formed.

$y \geq x - 4$
$y \geq -3x$

$y \leq -\frac{1}{2}x + \frac{7}{2}$
$x \geq 0$

Simplify.

27. $\left(\dfrac{4}{x}\right)^{-2}$

28. $\left(\dfrac{2}{3}\right)^{-4}$

Evaluate. Express each answer in scientific notation.

29. $\dfrac{5 \times 3^{-1}}{15 \times 3^{-3}}$

30. $(2,800)(0.02)$

31. Find the product of $(3x - 7)(4x^2 + 5x + 6)$.

32. Find $\begin{bmatrix} 4 & 1 & 3 \\ 2 & -5 & 4 \end{bmatrix} \cdot \begin{bmatrix} 3 & -2 \\ -1 & 4 \\ -2 & 0 \end{bmatrix}$.

33. Solve $\begin{bmatrix} 1 & 0 & 2 \\ 2 & -1 & 1 \\ 3 & 2 & 0 \end{bmatrix} \cdot \begin{bmatrix} x \\ y \\ z \end{bmatrix} = \begin{bmatrix} 12 \\ 10 \\ 4 \end{bmatrix}$.

Find the inverse of each matrix.

34. $\begin{bmatrix} 3 & -4 \\ 2 & -6 \end{bmatrix}$

35. $\begin{bmatrix} 2 & -4 & 6 \\ 1 & 3 & -5 \\ 3 & 2 & 4 \end{bmatrix}$

Solve the problem.

36. The perimeter of a triangle is 51 cm. The two shorter sides differ by 3. The longest side is 5 cm less than the sum of the other two. Find the length of each side.

5-5 Factoring

The factored form of a polynomial is the set of prime polynomials whose product is equivalent to that polynomial. To write $10x^2 + 6x$ in factored form, first find the greatest common factor (GCF) of $10x^2$ and $6x$.

GCF is the greatest factor that a set of terms has in common.

$$10x^2 = 2 \cdot 5 \cdot x \cdot x$$
$$\updownarrow \qquad \updownarrow$$
$$6x = 2 \cdot 3 \cdot x$$

The greatest common factor (GCF) of $10x^2$ and $6x$ is $2x$.

Now use the Distributive Property to factor the expression.

$$10x^2 + 6x = (2x \cdot 5x) + (2x \cdot 3)$$
$$= 2x(5x + 3)$$

$10x^2 + 6x$ written in factored form is $2x(5x + 3)$.

Examples

1 **Factor $12xy^2 - 8x^2y$.**

$12xy^2 - 8x^2y = (2 \cdot 2 \cdot 3 \cdot x \cdot y \cdot y) - (2 \cdot 2 \cdot 2 \cdot x \cdot x \cdot y)$
$\qquad = (4xy \cdot 3y) - (4xy \cdot 2x)$ *4xy is the GCF.*
$\qquad = 4xy(3y - 2x)$

2 **Factor $2n^2y + 3ny^2 - 5n^2y^2 + 7ny^3$.**

$2n^2y + 3ny^2 - 5n^2y^2 + 7ny^3$
$\qquad = (2 \cdot n \cdot n \cdot y) + (3 \cdot n \cdot y \cdot y) - (5 \cdot n \cdot n \cdot y \cdot y) + (7 \cdot n \cdot y \cdot y \cdot y)$
$\qquad = (ny \cdot 2n) + (ny \cdot 3y) - (ny \cdot 5ny) + (ny \cdot 7y^2)$ *ny is the GCF.*
$\qquad = ny(2n + 3y - 5ny + 7y^2)$

Rearranging and grouping terms is helpful when factoring polynomials that have more than three terms.

Example

3 **Factor $a^2 - 2ab + a - 2b$.**

$a^2 - 2ab + a - 2b = a(a - 2b) + 1(a - 2b)$ *Group the terms in pairs and factor the GCF.*
$\qquad = (a + 1)(a - 2b)$ *(a − 2b) is the common binomial factor.*

These factors could be found another way.

$a^2 - 2ab + a - 2b = a^2 + a - 2ab - 2b$ *Commutative Property*
$\qquad = a(a + 1) - 2b(a + 1)$
$\qquad = (a - 2b)(a + 1)$ *(a + 1) is the common binomial factor.*

You have found products like $(x - 7)(x + 7)$.

$$(x - 7)(x + 7) = x \cdot x + 7 \cdot x - 7 \cdot x - 7 \cdot 7$$
$$= x^2 \qquad\qquad\qquad - 7^2$$
$$= x^2 - 49$$

To factor $x^2 - 49$, reverse the steps.

$$x^2 - 49 = x^2 - 7^2$$
$$= x^2 + 7x - 7x - 7^2$$
$$= x(x + 7) - 7(x + 7)$$
$$= (x - 7)(x + 7)$$

Factoring $x^2 - 49$ shows the following pattern.

For any numbers a and b, $a^2 - b^2 = (a - b)(a + b)$.	**Factoring Difference of Squares**

Examples

4   **Factor $16a^2 - 4$.**

$$16a^2 - 4 = 4(4a^2 - 1) \qquad \textit{4 is the GCF.} \qquad \textit{The GCF should always}$$
$$= 4[(2a)^2 - (1)^2] \qquad\qquad\qquad\quad \textit{be factored first.}$$
$$= 4(2a - 1)(2a + 1)$$

5   **Factor $4x^2 - \dfrac{1}{9}$.**

$$4x^2 - \frac{1}{9} = (2x)^2 - \left(\frac{1}{3}\right)^2$$
$$= \left(2x - \frac{1}{3}\right)\left(2x + \frac{1}{3}\right)$$

6   **Factor $x^3 + 2x^2 - x - 2$.**

To factor this expression, first group the terms in pairs.

$$x^3 + 2x^2 - x - 2 = x^2(x + 2) - 1(x + 2)$$
$$= (x^2 - 1)(x + 2) \qquad \textit{(x + 2) is the common binomial factor.}$$
$$= (x - 1)(x + 1)(x + 2) \qquad \textit{Factor } x^2 - 1.$$

These factors could be found another way.

$$x^3 + 2x^2 - x - 2 = x^3 - x + 2x^2 - 2 \qquad \textit{Commutative Property}$$
$$= x(x^2 - 1) + 2(x^2 - 1) \qquad \textit{Group and factor.}$$
$$= (x^2 - 1)(x + 2) \qquad \textit{(}x^2 - 1\textit{) is the common binomial factor.}$$
$$= (x - 1)(x + 1)(x + 2) \qquad \textit{Factor } x^2 - 1.$$

You know how to factor the difference of two squares. Can you factor the difference of two cubes? Consider $a^3 - b^3$.

$$a^3 - b^3 = a^3 - a^2b + a^2b - b^3$$
$$= a^2(a - b) + b(a^2 - b^2)$$
$$= a^2(a - b) + b(a - b)(a + b)$$
$$= (a - b)[a^2 + b(a + b)]$$
$$= (a - b)(a^2 + ab + b^2)$$

Notice that $-a^2b + a^2b = 0$.
Distributive Property
Factor.
Distributive Property
Multiply.

A similar method can be used to show how to factor the sum of two cubes. That is, $a^3 + b^3 = (a + b)(a^2 - ab + b^2)$.

> **For any numbers a and b,**
>
> $$a^3 + b^3 = (a + b)(a^2 - ab + b^2), \text{ and}$$
>
> $$a^3 - b^3 = (a - b)(a^2 + ab + b^2).$$

Factoring Sum or Difference of Cubes

Examples

7 **Factor $m^3 + 27$.**   *This is the sum of cubes. The GCF is 1.*

$$m^3 + 27 = m^3 + 3^3$$
$$= (m + 3)(m^2 - m \cdot 3 + 3^2)$$
$$= (m + 3)(m^2 - 3m + 9)$$

8 **Factor $27y^3 - 8x^3$.**   *This is the difference of cubes. The GCF is 1.*

$$27y^3 - 8x^3 = (3y)^3 - (2x)^3$$
$$= (3y - 2x)[(3y)^2 + 3y \cdot 2x + (2x)^2]$$
$$= (3y - 2x)(9y^2 + 6xy + 4x^2)$$

Exploratory Exercises

Factor.

1. $6a + 6b$

2. $8m - 2n$

3. $ab + ac$

4. $y^3 - y^2$

5. $r^2 - 9$

6. $x^2 - 49$

7. $y(3y - 2) + 4k(3y - 2)$

8. $3m(m - 7) + k(m - 7)$

9. $9r(t + v) - w(t + v)$

10. $100 - m^2$

11. $y^2 - 81z^2$

12. $2x^2 + 6y + 8b$

13. $x^2 + xy + 3x$

14. $3a^2 + 6a + 9y$

15. $r^4 + r^3s + r^2s^2$

16. $25a^2 - b^2$

17. $49s^2 - 100$

18. $5x^2y - 10xy^2$

19. $2x^2(x - 3) + (x - 3)$

20. $5a(b + c) + (b + c)$

21. $b(3b - 2y) - (3b - 2y)$

22. $x^3 + 8$

23. $b^3 - 27$

24. $r^3 - 1$

Factor.

25. $b^2 - 144$

26. $y^3 - 1$

27. $1 + r^3$

28. $3y^2 + 12yk - 2y - 8k$

29. $3ax - 6bx - 4a + 8b$

30. $ay - ab - cy + cb$

31. $3m^2 - 21m + mk - 7k$

32. $a^2 + ab - 2a - 2b$

33. $r^2 - 4r - rp + 4p$

34. $m^2 - 121$

35. $8 - x^3$

36. $8a^3 + 1$

37. $2y^3 - 98y$

38. $a^3b^3 - 27$

39. $8 + x^3$

40. $2x^3 - 6x^2 + x - 3$

41. $3b^2 - 2by - 3b + 2y$

42. $tw - 3w - w^2 + 3t$

43. $8b^3 - 27x^3$

44. $9p^2 - 4q^2$

45. $ab - a^4b$

46. $16s^2 - 81r^2$

47. $r^3s^3 - 8s^3$

48. $m^6 - 27$

49. $k^3 + 4k^2 - 9k - 36$

50. $a^2x - b^2x + a^2y - b^2y$

51. $x^2 - y^2 + 4y - 4x$

52. $r^4 - s^4$

53. $64y^3 - 1$

54. $y^6 + 125$

55. $16y^4 - k^4$

56. $1 - 8m^6$

57. $16x^4 - 196y^4$

58. $16m^3 - 2$

59. $3r - 81r^4$

60. $7p^3 + 56s^3$

61. $(a + b)^2 - m^2$

62. $(x - y)^2 - z^2$

63. $(c + d)^3 + f^3$

mini-review

Solve each system of equations.

1. $33x - 22y = -4$
$2x - 3y = -1$

2. $8x - y + 2z = 7$
$-5x + y - z = -2$
$4x - 2y + 4z = 2$

Find the slope, y-intercept, and x-intercept of each line whose equation is given below.

3. $4x + y = 8$

4. $5x - 3y = 7$

5. $\frac{1}{4}x - \frac{1}{2}y = 6$

6. Find the value of $\begin{vmatrix} 3 & 1 & -2 \\ 4 & 2 & -1 \\ 1 & -3 & -4 \end{vmatrix}$.

7. If $A = \begin{bmatrix} 6 & -5 & 3 \\ 2 & 1 & 4 \end{bmatrix}$ and $B = \begin{bmatrix} 6 & 4 \\ -2 & 8 \\ 3 & 0 \end{bmatrix}$, find $A \cdot B$ and $B \cdot A$.

8. Write an equation of the line that passes through $(4, 2)$ and is inconsistent with $4x + 8y = 10$.

9. Graph this system of inequalities. Name the vertices of the polygon formed.
$y \geq 2; \; y \leq 3x + 4; \; y \leq -x + 6$

Given $g(x) = 10x^2 - 3x + 2$, find each value.

10. $g(5)$

11. $g(-4)$

12. $g(-2) + g(2)$

Simplify.

13. $\left(\dfrac{2a}{b}\right)^{-2}$

14. $\left(\dfrac{1}{2}rs\right)^2 (4r^3s^3)^2$

15. $\dfrac{20xy^3}{(4x^2y^2)^3}$

Some problems may provide more information than is actually necessary to solve the problem. Other problems may not provide enough information to solve the problem. And yet other problems may be misleading because they lead to unwarranted assumptions. Consider the following example.

Example: Suppose two trains, each traveling uniformly, start towards each other at the same time along a straight track. One train is traveling at 60 mph and the other at 90 mph. At the start, with the trains 300 miles apart, a bee travels from one train to the other at the rate of 200 mph. When the bee reaches the second train, it immediately returns to the first train. This continues until the trains meet. How many miles did the bee travel?

A first reaction may be to try to track the path of the bee. Instead, simply calculate the bee's distance by multiplying its rate of speed by the time it traveled.

To find the time it traveled, calculate the time it took the trains to meet.

Let t = the number of hours the trains traveled before meeting.

$$60t + 90t = 300$$
$$150t = 300$$
$$t = 2$$

The bee was traveling 200 mph. In two hours, the bee traveled a total distance of 400 miles.

Exercises

Solve each problem.

1. Samuel went into a hardware store and asked the clerk how much 1 cost. The clerk said, "25¢." Then he asked how much 10 cost, and the clerk said, "50¢." "Good," he replied, "I'll take 1025," and then paid the clerk $1.00. What did he buy?

2. Kandy and Mark are both chess players. They have completed 7 games. Each has won the same number of games and there were no ties. How did this happen?

3. How much will it cost to cut a log into eight equal pieces, if cutting it into four equal pieces costs 90¢?

4. Find the digit that each letter represents so that the following equation is true.
$$(HE)^2 = SHE$$

5-6 More Factoring

When factoring trinomials, first look for the greatest common monomial factor.

$$18x^3 - 48x^2y + 32xy^2 = 2x \cdot 9x^2 - 2x \cdot 24xy + 2x \cdot 16y^2$$
$$= 2x(9x^2 - 24xy + 16y^2)$$

The trinomial $9x^2 - 24xy + 16y^2$ can be factored further. Use one of the following patterns.

For any numbers a and b, $a^2 + 2ab + b^2 = (a + b)^2$, **and** $a^2 - 2ab + b^2 = (a - b)^2$	*Factoring Perfect Square Trinomials*

Example

1 **Factor $9x^2 - 24xy + 16y^2$.**

$$9x^2 - 24xy + 16y^2 = (3x)^2 - 2(12xy) + (4y)^2$$
$$= (3x)^2 - 2(3x)(4y) + (4y)^2$$
$$= (3x - 4y)^2$$

Use the pattern $a^2 - 2ab + b^2 = (a - b)^2$.

The factored form of the original trinomial, $18x^3 - 48x^2y + 32xy^2$, is $2x(3x - 4y)^2$.

Many trinomials like $x^2 - 5x + 6$ are *not* perfect squares. They can be factored by reversing the FOIL method.

$$(x + r)(x + s) = x \cdot x + x \cdot s + r \cdot x + r \cdot s$$
$$= x^2 + (r + s)x + rs$$

For $x^2 - 5x + 6$, the middle coefficient -5 corresponds to $r + s$. The 6 corresponds to rs. You must find two numbers, r and s, whose sum is -5 and whose product is 6.

Factors of 6	Sum of Factors
1, 6	7
$-1, -6$	-7
2, 3	5
$-2, -3$	-5

The two numbers are -2 and -3.   *Check this using the FOIL method.*

$$x^2 - 5x + 6 = (x - 2)(x - 3)$$

2 Factor $x^2 - 3x - 18$.

Factors of -18	Sum of Factors
-1, 18	17
1, -18	-17
-2, 9	7
2, -9	-7
-3, 6	3
3, -6	-3

The two numbers are 3 and -6. The product of 3 and -6 is -18 and the sum of 3 and -6 is -3.

$$x^2 - 3x - 18 = (x + 3)(x - 6)$$

Consider $2x^2 + 7x + 6$. The coefficient of x^2 is *not* 1. The factors of $2x^2 + 7x + 6$ can be found by reversing the following pattern.

$$(ax + b)(cx + d) = (ax + b)cx + (ax + b)d$$
$$= acx^2 + bcx + adx + bd$$

Notice that the product of the *coefficient* of the x^2 term and the *constant* term is $abcd$. The product of the two coefficients of the x term, bc and ad, is also $abcd$.

For $2x^2 + 7x + 6$, the product of the x^2 *coefficient* and the *constant* is 12.

$$2x^2 + 7x + 6$$

The two coefficients of x must have a sum of 7 and a product of 12. The only possibility is 3 and 4 because $3 + 4 = 7$ and $3 \cdot 4 = 12$.

$$2x^2 + (4x + 3x) + 6$$

Consider the factors of 12 that have a sum of 7.

$$
\begin{aligned}
2x^2 + 7x + 6 &= 2x^2 + (4x + 3x) + 6 \\
&= (2x^2 + 4x) + (3x + 6) \qquad \text{\textit{Associative Property}} \\
&= 2x(x + 2) + 3(x + 2) \qquad \text{\textit{Distributive Property (x + 2 is a common binomial factor.)}} \\
&= (2x + 3)(x + 2) \qquad \text{\textit{Distributive Property}}
\end{aligned}
$$

3   **Factor $6m^2 + 19m + 10$.**

The product of the m^2 coefficient, 6, and the constant, 10, is 60. Thus, the coefficients of m must have a sum of 19 and a product of 60. The only possibility is 4 and 15, because $4 + 15 = 19$ and $4 \cdot 15 = 60$.

$$
\begin{aligned}
6m^2 + 19m + 10 &= 6m^2 + (4m + 15m) + 10 \\
&= (6m^2 + 4m) + (15m + 10) \\
&= 2m(3m + 2) + 5(3m + 2) \\
&= (2m + 5)(3m + 2)
\end{aligned}
$$

4   **Factor $8z^2 - 27z - 20$.**

$$
\begin{aligned}
8z^2 - 27z - 20 &= 8z^2 + (-32z + 5z) - 20 \\
&= (8z^2 - 32z) + (5z - 20) \\
&= 8z(z - 4) + 5(z - 4) \\
&= (8z + 5)(z - 4)
\end{aligned}
$$

Notice that $-32 + 5 = -27$ and $(-32)(5) = -160$.

The next example shows how to group, find a perfect square trinomial, and factor the difference of squares.

5   **Factor $a^2 + 4ab - 9x^2 + 4b^2$.**

$$
\begin{aligned}
a^2 + 4ab - 9x^2 + 4b^2 &= (a^2 + 4ab + 4b^2) - 9x^2 \\
&= [(a)^2 + 2(a \cdot 2b) + (2b)^2] - (3x)^2 \\
&= (a + 2b)^2 - (3x)^2 \\
&= [(a + 2b) - 3x][(a + 2b) + 3x] \\
&= (a + 2b - 3x)(a + 2b + 3x)
\end{aligned}
$$

Group the terms of the trinomial square.

Factor the difference of squares.

The following checklist can be used to help you factor a given polynomial.

1. Check for the greatest common factor.
2. Check for special products.
   a. If there are *two terms*, look for difference of squares, sum of cubes, or difference of cubes.
   b. If there are *three terms*, look for a perfect square trinomial.
3. Try other factoring methods.
   a. If there are *three terms*, try the trinomial pattern.
   b. If there are *four or more terms*, try grouping.

6   Factor $m^3 - 3m^2a + 3ma^2 - a^3$.

$$
\begin{aligned}
m^3 - 3m^2a + 3ma^2 - a^3 &= (m^3 - a^3) - (3m^2a - 3ma^2) \\
&= (m^2 + ma + a^2)(m - a) - 3ma(m - a) \qquad \text{\textit{3ma is the GCF.}} \\
&= (m^2 + ma + a^2 - 3ma)(m - a) \\
&= (m^2 - 2ma + a^2)(m - a) \\
&= (m - a)^2(m - a) \\
&= (m - a)^3
\end{aligned}
$$

7   Factor $6a^2 + 27a - 15$.

$$
\begin{aligned}
6a^2 + 27a - 15 &= 3(2a^2 + 9a - 5) \qquad \text{\textit{3 is the GCF.}} \\
&= 3(2a - 1)(a + 5)
\end{aligned}
$$

8   Factor $r^3 - r^2 - 30r$.

$$
\begin{aligned}
r^3 - r^2 - 30r &= r(r^2 - r - 30) \qquad \text{\textit{r is the GCF.}} \\
&= r(r - 6)(r + 5)
\end{aligned}
$$

9   Factor $5y^6 - 5y^2$.

$$
\begin{aligned}
5y^6 - 5y^2 &= 5y^2(y^4 - 1) \qquad \text{\textit{$5y^2$ is the GCF.}} \\
&= 5y^2(y^2 + 1)(y^2 - 1) \qquad \text{\textit{Factor the difference of squares.}} \\
&= 5y^2(y^2 + 1)(y + 1)(y - 1)
\end{aligned}
$$

10   Factor $x^2z^2 - 8x^2z + 16x^2$.

$$
\begin{aligned}
x^2z^2 - 8x^2z + 16x^2 &= x^2[z^2 - 8z + 16] \qquad \text{\textit{x^2 is the GCF.}} \\
&= x^2[(z)^2 - 2(4z) + (4)^2] \qquad \text{\textit{Use the pattern}} \\
& \quad \text{\textit{$a^2 - 2ab + b^2 = (a - b)^2$.}} \\
&= x^2(z - 4)^2
\end{aligned}
$$

Exploratory Exercises

Factor.

1. $20d^2 + 10d$ **2.** $-15x^2 - 5x$

3. $ab(c - d) + 5d(c - d)$ **4.** $a(a + b) - 2(a + b)$

5. $a^2 + 5a + 6$ **6.** $s^2 - 6s + 8$

7. $y^2 + 6y + 9$ **8.** $r^2 + 16r + 64$

9. $3x(a - 2b) - 4(a - 2b)$ **10.** $a(y - b) - c(y - b)$

11. $b^2 + 7b + 6$ **12.** $8m^2 + 4am + 16my$

13. $7pm + 2p^2 - 14px$ **14.** $5(c^2 + d) + 6(c^2 + d)$

15. $k^2 - 8k + 16$ **16.** $p^2 - 5p + 4$

Written Exercises

Factor.

17. $f^2 - 18f + 81$	**18.** $r^2 - 6r + 9$
19. $a^2 + 12a + 35$	**20.** $d^2 + 4d - 21$
21. $k^2 + 12k + 36$	**22.** $p^2 + 14p + 49$
23. $3y^2 + 5y + 2$	**24.** $4x^2 + 11x + 6$
25. $4x^2 - 9$	**26.** $9y^2 - 64$
27. $4z^2 - 20z + 21$	**28.** $3t^2 + 13t + 12$
29. $a^2 + 4ab + 4b^2$	**30.** $m^2 - 6mk + 9k^2$
31. $3d^2 - 48$	**32.** $a^2b^2 - 25a^2$
33. $p^2 - 4bp + 4b^2$	**34.** $9a^2 - 12ab + 4b^2$
35. $4r^2 - 20rs + 25s^2$	**36.** $4k^2 + 26k + 30$
37. $x^3 + 2x^2 - 35x$	**38.** $2a^3 - 7a^2 - 15a$
39. $6d^2 + 33d - 63$	**40.** $4h^2 + 8h - 96$
41. $x^4 - 13x^2 + 36$	**42.** $y^4 - 14y^2 + 45$
43. $2y^3 - 8y^2 - 42y$	**44.** $3m^3 + 21m^2 + 36m$
45. $18d^2 - 19d - 12$	**46.** $9g^2 - 12g + 4$
47. $2r^3 - 16s^3$	**48.** $3m^3 + 24p^3$
49. $(x + y)^2 - \dfrac{1}{4}$	**50.** $(2a + b)^2 - \dfrac{1}{16}$
51. $4a^2 + 4ab - y^2 + b^2$	**52.** $a^2 - b^2 + 8b - 16$
53. $m^2 - k^2 + 6k - 9$	**54.** $4a^2 - 6b - 9b^2 - 1$
55. $a + b + 3a^2 - 3b^2$	**56.** $a^2 - 4a + 4 - 25x^2$
57. $2ab + 2am - b - m$	**58.** $y^3 + y^2 - y - 1$
59. $a^2 - a + \dfrac{1}{4} - y^2$	**60.** $\dfrac{1}{16} - 9x^2 + 12xy - 4y^2$
61. $2ab + 2am + b^2 - m^2$	**62.** $3pq + 3ps + q^2 - s^2$
63. $4ax + 14ay - 10bx - 35by$	**64.** $r^2 - rt - rt^2 + t^3$
65. $x^3 + y^3 - x^2y - xy^2$	**66.** $t^3 + 125 + 5t^2 + 25t$
67. $(r - p)^3 + 4rp(r - p)$	**68.** $a(a + 1)(a + 2) - 3a(a + 1)$
69. $10x^2 - 14xy - 15x + 21y$	**70.** $8ax - 6x - 12a + 9$

Challenge Exercises

Factor.

71. $a^{2n} - 64$	**72.** $x^{3n} - y^{3n}$
73. $a^4 - 12a^3b + 24a^2b^2 - 8ab^3$	**74.** $m^3n + m^2n - mn^3 - mn^2$
75. $a^4 - 16a^2 + 3a^3 - 48a$	**76.** $4x^2 + 12xy + 9y^2 - z^2$
77. $x^3y - 3x^2y - 6xy + 8y$	**78.** $m^2 - 8m + 16 - 4a^2 + 28ab - 49b^2$

Applications in Biology Genetics

Every reproductive cell contains genes in pairs. When genes split and recombine, the combinations may be dominant, hybrid, or recessive. The following diagram shows the genes splitting to form the three genotypes.

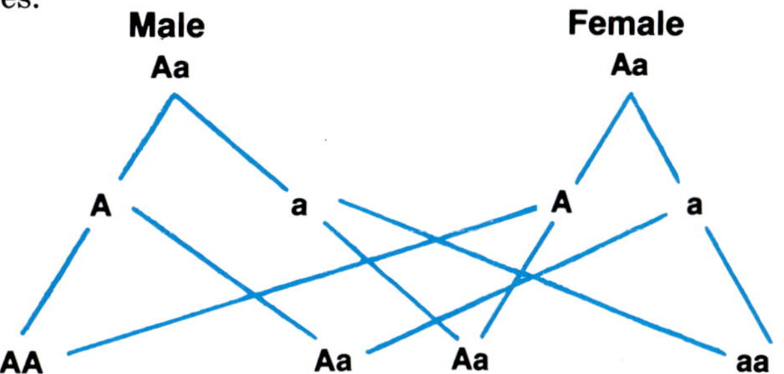

The presence of a dominant gene means that the trait is visible. This is called the phenotype.

A - dominant gene in a pair

a - recessive gene in a pair

AA - pure dominant genotype   **Aa - hybrid genotype**   **aa - pure recessive genotype**

The Hardy-Weinberg Law is used to study dominant and recessive traits in a population. The frequency of the members of a pair of genes for a specific trait in a population is described by the expansion of a binomial expression, $(p\,\mathbf{A} + q\,\mathbf{a})^2$.

$$(p\,\mathbf{A} + q\,\mathbf{a})(p\,\mathbf{A} + q\,\mathbf{a}) = p^2\,\mathbf{AA} + 2pq\,\mathbf{Aa} + q^2\,\mathbf{aa}$$

p is proportion of dominant "A" gene in population.

q is proportion of recessive "a" gene in population.

$p + q = 1$ The sum of the proportions must equal the whole population.

Example: In the original population on a remote island, the recessive blue-eyed gene had a frequency of 1:4 while the dominant brown-eyed gene had a frequency of 3:4. After several generations, what would be the expected proportion of genotypes in the population?

Brown-eyed, pure dominant

$$p^2\,\mathbf{AA} = \left(\tfrac{3}{4}\right)^2 \mathbf{AA}$$
$$= \tfrac{9}{16}\,\mathbf{AA}$$
9 out of 16

Brown-eyed, hybrid

$$2pq\,\mathbf{Aa} = 2\left(\tfrac{3}{4}\right)\left(\tfrac{1}{4}\right) \mathbf{Aa}$$
$$= \tfrac{3}{8}\,\mathbf{Aa}$$
6 out of 16

Blue-eyed, pure recessive

$$q^2\,\mathbf{aa} = \left(\tfrac{1}{4}\right)^2 \mathbf{aa}$$
$$= \tfrac{1}{16}\,\mathbf{aa}$$
1 out of 16

Exercises

1. Assume that a certain recessive trait occurs in $\frac{1}{25}$ of the population. That is, $q^2 = \frac{1}{25}$. What proportion of the population is pure dominant genotype and what proportion carries the hybrid genotype?

2. Assume that a hybrid genotype occurs in $\frac{12}{25}$ of the population. That is, $2pq = \frac{12}{25}$. Since $p + q = 1$, or $p = 1 - q$, substitute for p and determine the pure dominant and pure recessive proportions of the population.

5-7 Dividing Polynomials

You can use the properties of exponents to divide a monomial by a monomial. You also use these properties to divide a polynomial by a monomial.

Examples

1 **Divide $16x^4$ by $8x^3$.**

$$16x^4 \div 8x^3 = \frac{16x^4}{8x^3} \qquad \textit{The denominator cannot have a value of 0. } (x \neq 0)$$

$$= \frac{16}{8} \cdot x^{4-3}$$

$$= 2x$$

2 **Simplify $(2x^2 + x - 15)(x - 3)^{-1}$.** $\qquad x \neq 3$

$$(2x^2 + x - 15)(x - 3)^{-1} = \frac{2x^2 + x - 15}{x - 3}$$

$$= \frac{(2x + 5)(x - 3)}{x - 3} \qquad \textit{Factor the numerator.}$$

$$= 2x + 5 \qquad \textit{Simplify } \frac{x - 3}{x - 3} = 1.$$

3 **Simplify $\dfrac{36m^4y^4 - 18m^3y}{6m^2y}$.** $\qquad m \neq 0, y \neq 0$

$$\frac{36m^4y^4 - 18m^3y}{6m^2y} = \frac{36m^4y^4}{6m^2y} - \frac{18m^3y}{6m^2y}$$

$$= \frac{36}{6} \cdot m^{4-2}y^{4-1} - \frac{18}{6} \cdot m^{3-2}y^{1-1}$$

$$= 6m^2y^3 - 3m \qquad y^{1-1} = y^0 \textit{ or } 1$$

$$= 3m(2my^3 - 1)$$

Sometimes factoring cannot be used to divide a polynomial by a polynomial. When this happens, a process similar to long division is used. The example at the right reviews long division with numbers.

Divide 883 by 21.

$$\begin{array}{r} 42 \\ 21\overline{)883} \\ \underline{84} \qquad \textit{Subtract } 4 \cdot 21. \\ 43 \\ \underline{42} \qquad \textit{Subtract } 2 \cdot 21. \\ 1 \qquad \textit{Stop when remainder is less than 21.} \end{array}$$

When applying this division process to polynomials, remember that you must have like terms to add or subtract.

$$\frac{883}{21} = 42\frac{1}{21}$$

4 **Divide $8y^2 + 8y + 3$ by $2y + 1$.**

$$
\begin{array}{r}
4y + 2 \\
2y + 1 \overline{)\, 8y^2 + 8y + 3} \\
\underline{8y^2 + 4y} \\
4y + 3 \\
\underline{4y + 2} \\
1
\end{array}
$$

Subtract $4y(2y + 1)$.

Subtract $2(2y + 1)$.

Stop when degree of remainder is less than degree of $2y + 1$.

$$\frac{8y^2 + 8y + 3}{2y + 1} = 4y + 2 + \frac{1}{2y + 1} \qquad y \neq -\frac{1}{2}$$

5 **Simplify $(x^3 + 5x^2 + 5x + 16)(x^2 + 3)^{-1}$.**

$$(x^3 + 5x^2 + 5x + 16)(x^2 + 3)^{-1} = (x^3 + 5x^2 + 5x + 16) \div (x^2 + 3)$$

$$
\begin{array}{r}
x + 5 \\
x^2 + 3 \overline{)\, x^3 + 5x^2 + 5x + 16} \\
\underline{x^3 \qquad\quad + 3x} \\
5x^2 + 2x + 16 \\
\underline{5x^2 \qquad\; + 15} \\
2x + 1
\end{array}
$$

Align like terms.

$$(x^3 + 5x^2 + 5x + 16)(x^2 + 3)^{-1} = x + 5 + \frac{2x + 1}{x^2 + 3}$$

If the remainder upon division is zero, then the divisor is a factor of the polynomial.

6 **Show that $4 - h$ is a factor of $h^3 - 6h^2 + 13h - 20$.**

$$
\begin{array}{r}
-h^2 + 2h - 5 \\
-h + 4 \overline{)\, h^3 - 6h^2 + 13h - 20} \\
\underline{h^3 - 4h^2} \\
-2h^2 + 13h \\
\underline{-2h^2 + 8h} \\
5h - 20 \\
\underline{5h - 20} \\
0
\end{array}
$$

Rewrite $4 - h$ as $-h + 4$.

Because the remainder is 0, $-h + 4$ or $4 - h$ is a factor of $h^3 - 6h^2 + 13h - 20$.

Exploratory Exercises

Use division to simplify each expression.

1. $\dfrac{6xy^2 - 3xy + 2x^2y}{xy}$

2. $\dfrac{a^3b^2 - a^2b + 2a}{-ab}$

3. $\dfrac{6r^2s^2 + 3rs^2 - 9r^2s}{3rs}$

4. $\dfrac{3(x - 7)^6}{(x - 7)^{10}}$

5. $\dfrac{2(x + 3)^4}{10(x + 3)^2}$

6. $\dfrac{2(y^2 - 5)^3}{8(y^2 - 5)^5}$

7. $\dfrac{c^2 - c - 30}{c - 6}$

8. $\dfrac{h^2 - 11h + 28}{h - 4}$

9. $\dfrac{r^2 + 8r + 16}{r + 4}$

10. $(2a^2b - 4ab^2)(a - 2b)^{-1}$

11. $(c^2 - c^3)(c^2 - 1)^{-1}$

12. $(a^3 - b^3)(a - b)^{-2}$

Written Exercises

Simplify.

13. $\dfrac{6p^4q^2 + 4p^2q + 5pq^3}{pq}$

14. $\dfrac{12mz^3 + 9m^2z^2 - 15m^2z}{-3mz}$

15. $\dfrac{28k^3py - 42kp^2y^2 + 56kp^3y^2}{14kpy}$

16. $\dfrac{18k^3lm^2 + 27k^2lm + 45k^2l^2m^2}{9klm}$

17. $\dfrac{15r^2s + 23rs^2 + 6s^3}{3rs}$

18. $\dfrac{4a^2b^3c^4 + 13ab - 12a^4b^2c}{2abc}$

19. $(a^2 - 5a - 84)(a + 7)^{-1}$

20. $(x^2 + 20x + 91)(x + 7)^{-1}$

21. $(a^2 - 5ab + 6b^2) \div (a - 3b)$

22. $(12x^2 - 4xy - y^2) \div (2x - y)$

23. $(x^2 - 12x - 45) \div (x + 3)$

24. $(a^2 + 7a - 60) \div (a + 12)$

25. $(6y^2 + 7y - 3) \div (2y + 3)$

26. $(15b^2 + 14b - 8) \div (5b - 2)$

27. $(8a^2 + 34a + 19) \div (2a + 7)$

28. $(20b^2 - 17b - 61) \div (5b + 7)$

29. $(28y^2 + 23y - 12) \div (7y - 3)$

30. $(80x^2 + 6x - 4) \div (10x - 3)$

31. $(a^2 + 4a - 16) \div (6 - a)$

32. $(y^2 + y - 8) \div (3 - y)$

33. $(8x^2 - 4x + 11)(x + 5)^{-1}$

34. $(126k^2 + 113k - 26)(14k - 3)^{-1}$

35. $(56m^2 - 113m + 59) \div (8m - 7)$

36. $(20r^2 + 7r - 10) \div (5r - 2)$

37. $(6y^3 + 11y^2 - 4y - 4)(3y - 2)^{-1}$

38. $(8x^3 - 22x^2 - 5x + 12)(4x + 3)^{-1}$

39. $(6a^3 + 5a^2 + 9) \div (2a + 3)$

40. $(8b^2 - 4b + 1) \div (2b - 1)$

41. $(m^3 - 1) \div (m - 1)$

42. $(a^3 - 8) \div (a - 2)$

43. $(y^3 - 9y^2 + 27y - 28) \div (y - 3)$

44. $(x^3 + 6x^2 + 12x + 12) \div (x + 2)$

45. $(6a^3 - 5a^2 - 12a - 4) \div (3a + 2)$

46. $(2p^3 + 7p^2 - 29p + 29) \div (2p - 3)$

47. $(m^3 - 7m + 3m^2 - 21)(m^2 - 7)^{-1}$

48. $(48p^3 - 15 + 6p^2 - 40p)(6p^2 - 5)^{-1}$

49. $(x^3 + 4x - 4) \div (x + 2)$

50. $(2t^3 - 2t - 3) \div (t - 1)$

51. $(x^4 + 4) \div (x^2 - 2x + 2)$

52. $(y^4 + 4y^3 + 10y^2 + 12y + 9) \div (y^2 + 2y + 3)$

53. $(a^4 - 3a^2 + 1) \div (a^2 + a - 1)$

54. $(x^4 - 4x^2 + 12x - 9) \div (x^2 + 2x - 3)$

55. Is $3y - 2$ a factor of $6y^3 - y^2 - 5y + 2$? Write *yes* or *no*.

56. Is $4x + 5$ a factor of $4x^3 + x^2 + 10$? Write *yes* or *no*.

57. One factor of $a^3 - 2a^2 - a + 2$ is $a - 2$. Find the other factors.

58. One factor of $2m^3 - 11m^2 + 18m - 9$ is $m - 3$. Find the other factors.

59. Find the remainder when dividing $x^2 + 3x + 5$ by $x - 2$. If $f(x) = x^2 + 3x + 5$, find $f(2)$. Compare the answers.

60. Find the remainder when dividing $x^3 - 5x^2 + 6x - 4$ by $x - 3$. If $f(x) = x^3 - 5x^2 + 6x - 4$, find $f(3)$. Compare the answers.

Challenge Exercises

Find the values for k so that each remainder is zero.

61. $(x^2 + 8x + k) \div (x - 4)$

62. $(x^2 + kx + 6) \div (x - 1)$

63. $(b^3 + 6b^2 + 12b + k) \div (b + 3)$

64. $(a^3 + 7a^2 + ka - 8) \div (a + 4)$

Simplify.

65. $\left(\dfrac{y^2 + 2y - 15}{y^2 + 3y - 10}\right)\left(\dfrac{y^2 - 9}{y^2 - 9y + 14}\right)^{-1}$

66. $\left(\dfrac{x^3 - y^3}{x^2 + xy + y^2}\right)\left(\dfrac{(x - y)^2}{4x + 4y}\right)^{-1}$

mini-review

Solve each system of equations.

1. $2x + 9y = -5$
$3x + 14y = -7$

2. $3x - 4y - 5z = -1$
$-2x + 4y + 7z = 15$
$x - y + 3z = 27$

3. Graph $4x + 3y + 6z = 12$ and find the traces in the coordinate plane.

4. Solve using Cramer's Rule.
$7x + 2y = 11$
$3x + 8y = -16$

Find the standard form of the equation of the line that satisfies the given conditions.

5. slope $= \dfrac{1}{4}$, y-intercept $= 5$

6. x-intercept $= -4$, y-intercept $= 3$

7. Find $\begin{vmatrix} 2 & 4 & -3 \\ 1 & -2 & 6 \\ 1 & 4 & -9 \end{vmatrix}$.

8. Find the inverse of $\begin{bmatrix} 3 & 6 \\ -2 & 5 \end{bmatrix}$.

Simplify.

9. $\left(\dfrac{a^{-2}}{bx^{-3}}\right)^{-4}$

10. $\left(\dfrac{3y^2}{2x}\right)^{-2}$

11. $\dfrac{-6a^3bc^3}{48a^2bc^4}$

12. $r(r - 8)(r + 10)$

13. $-4c^3d^2(c^4d - 8c^3d^2 + 7d^4)$

14. $(3mn)(3m^3n)^2 + (3m^3)(-m^4n^3)$

5-8 Synthetic Division

There is another method for dividing a polynomial by a binomial of the form $x - r$, called **synthetic division**. To divide the polynomial $3x^3 - 4x^2 - 3x - 2$ by $x - 3$, follow the steps below.

Step 1   Write the terms of the polynomial in descending order. Then write the coefficients as shown.

$$3 \quad -4 \quad -3 \quad -2$$

Step 2   Write the constant, r, of the divisor $x - r$ to the left. For the divisor $x - 3$, r is 3.

$$3 \,\rfloor\; 3 \quad -4 \quad -3 \quad -2$$

Step 3   Bring down the first coefficient.

$$\begin{array}{r|rrrr} 3 & 3 & -4 & -3 & -2 \\ \hline & 3 & & & \end{array}$$

Step 4   Multiply the first coefficient by r. Then write the product under the second coefficient.

$$\begin{array}{r|rrrr} 3 & 3 & -4 & -3 & -2 \\ & & 9 & & \\ \hline & 3 & & & \end{array}$$

Step 5   Add. $-4 + 9 = 5$

$$\begin{array}{r|rrrr} 3 & 3 & -4 & -3 & -2 \\ & & 9 & & \\ \hline & 3 & 5 & & \end{array}$$

Step 6   Multiply the sum, 5, by r. Write the product under the next coefficient.

$$\begin{array}{r|rrrr} 3 & 3 & -4 & -3 & -2 \\ & & 9 & 15 & \\ \hline & 3 & 5 & & \end{array}$$

Step 7   Add. $-3 + 15 = 12$

$$\begin{array}{r|rrrr} 3 & 3 & -4 & -3 & -2 \\ & & 9 & 15 & \\ \hline & 3 & 5 & 12 & \end{array}$$

Step 8   Again multiply the sum, 12, by r. Write the product under the next coefficient.

$$\begin{array}{r|rrrr} 3 & 3 & -4 & -3 & -2 \\ & & 9 & 15 & 36 \\ \hline & 3 & 5 & 12 & \end{array}$$

Step 9   Add. $-2 + 36 = 34$
The remainder is 34.

$$\begin{array}{r|rrrr} 3 & 3 & -4 & -3 & -2 \\ & & 9 & 15 & 36 \\ \hline & 3 & 5 & 12 & 34 \end{array}$$

Step 10   Write the result. $3x^2 + 5x + 12 + \dfrac{34}{x - 3}$

Compare this process to the long division process.

$$\begin{array}{r|rrrr} 3 & 3 & -4 & -3 & -2 \\ & & 9 & 15 & 36 \\ \hline & 3 & 5 & 12 & 34 \end{array}$$

$$\require{enclose}\begin{array}{r} 3x^2 + 5x + 12 \\ x - 3 \enclose{longdiv}{3x^3 - 4x^2 - 3x - 2} \\ \underline{3x^3 - 9x^2 } \\ 5x^2 - 3x \\ \underline{5x^2 - 15x } \\ 12x - 2 \\ \underline{12x - 36} \\ 34 \end{array}$$

1 Find $(2y^3 - 3y^2 - 8y + 4) \div (y + 2)$.

$$
\begin{array}{r|rrrr}
-2 & 2 & -3 & -8 & 4 \\
   &   & -4 & 14 & -12 \\
\hline
   & 2 & -7 & 6\ | & -8
\end{array}
$$

y + 2 is the same as y − (−2).

The result is $2y^2 - 7y + 6 - \dfrac{8}{y + 2}$.

Check: $(y + 2)(2y^2 - 7y + 6) - 8$ *Divisor × quotient + remainder*
$\quad = 2y^3 - 7y^2 + 6y + 4y^2 - 14y + 12 - 8$ *= dividend*
$\quad = 2y^3 - 3y^2 - 8y + 4$ ✔

2 Find $(2a^3 + a^2 + 12) \div (a + 2)$.

$$
\begin{array}{r|rrrr}
-2 & 2 & 1 & 0 & 12 \\
   &   & -4 & 6 & -12 \\
\hline
   & 2 & -3 & 6\ | & 0
\end{array}
$$

The coefficient of a is zero in $2a^3 + a^2 + 12$. Zero coefficients must be included when you do synthetic division.

The result is $2a^2 - 3a + 6$.

Check: $(a + 2)(2a^2 - 3a + 6)$
$\quad = 2a^3 - 3a^2 + 6a + 4a^2 - 6a + 12$
$\quad = 2a^3 + a^2 + 12$ ✔

3 Find $(2t^5 - 13t^3 - 70t^2 - 23t - 4) \div (t - 4)$.

$$
\begin{array}{r|rrrrrr}
4 & 2 & 0 & -13 & -70 & -23 & -4 \\
  &   & 8 & 32 & 76 & 24 & 4 \\
\hline
  & 2 & 8 & 19 & 6 & 1\ | & 0
\end{array}
$$

A remainder of zero indicates that $t - 4$ is a factor of the polynomial.

The result is $2t^4 + 8t^3 + 19t^2 + 6t + 1$.

Check: $(t - 4)(2t^4 + 8t^3 + 19t^2 + 6t + 1)$
$\quad = 2t^5 + 8t^4 + 19t^3 + 6t^2 + t - 8t^4 - 32t^3 - 76t^2 - 24t - 4$
$\quad = 2t^5 - 13t^3 - 70t^2 - 23t - 4$ ✔

Some divisors, such as $2x - 1$, have leading coefficients other than one. For example, suppose you want to divide $4x^3 + x - 1$ by $2x - 1$. You can display the intended division in the following manner, factoring the leading coefficient of the divisor from the divisor and dividend.

$$
\frac{4x^3 + x - 1}{2x - 1} = \frac{2\left(2x^3 + \frac{1}{2}x - \frac{1}{2}\right)}{2\left(x - \frac{1}{2}\right)} = \frac{2x^3 + \frac{1}{2}x - \frac{1}{2}}{x - \frac{1}{2}}
$$

Factor 2 from both the divisor and the dividend. Then simplify the expression.

Now, use synthetic division to divide $2x^3 + \frac{1}{2}x - \frac{1}{2}$ by $x - \frac{1}{2}$.

4 **Divide $4x^3 + x - 1$ by $2x - 1$.**

$$\frac{4x^3 + x - 1}{2x - 1} = \frac{2\left(2x^3 + \frac{1}{2}x - \frac{1}{2}\right)}{2\left(x - \frac{1}{2}\right)} \text{ or } \frac{2x^3 + \frac{1}{2}x - \frac{1}{2}}{x - \frac{1}{2}}$$

$$\frac{1}{2} \,\bigg|\, \begin{array}{cccc} 2 & 0 & \frac{1}{2} & -\frac{1}{2} \\[4pt] & 1 & \frac{1}{2} & \frac{1}{2} \\ \hline 2 & 1 & 1 & 0 \end{array}$$

The result is $2x^2 + x + 1$.   *Check the solution.*
Does $(2x - 1)(2x^2 + x + 1) = 4x^3 + x - 1$?

5 **Divide $3a^4 - 2a^3 + 5a^2 - 4a - 2$ by $3a + 1$.**

$$\frac{3a^4 - 2a^3 + 5a^2 - 4a - 2}{3a + 1} = \frac{3\left(a^4 - \frac{2}{3}a^3 + \frac{5}{3}a^2 - \frac{4}{3}a - \frac{2}{3}\right)}{3\left(a + \frac{1}{3}\right)}$$

$$= \frac{a^4 - \frac{2}{3}a^3 + \frac{5}{3}a^2 - \frac{4}{3}a - \frac{2}{3}}{a + \frac{1}{3}}$$

$$-\frac{1}{3} \,\bigg|\, \begin{array}{ccccc} 1 & -\frac{2}{3} & \frac{5}{3} & -\frac{4}{3} & -\frac{2}{3} \\[4pt] & -\frac{1}{3} & \frac{1}{3} & -\frac{2}{3} & \frac{2}{3} \\ \hline 1 & -1 & 2 & -2 & 0 \end{array}$$

The result is $a^3 - a^2 + 2a - 2$.   *Check the solution. Does*
$(3a + 1)(a^3 - a^2 + 2a - 2) = 3a^4 - 2a^3 + 5a^2 - 4a - 2$?

6 **Using Calculators**

Divide $x^3 + 13x^2 - 12x - 8$ by $x + 2$.

ENTER: 2 [+/−] [×] 1 [=] [+] 13 [=]

DISPLAY: $2 \quad -2 \qquad 1 \quad -2 \qquad 13 \quad 11$

$$-2 \,\bigg|\, \begin{array}{cc} 1 & 13 \\ & -2 \\ \hline 1 & 11 \end{array}$$

ENTER: [×] 2 [+/−] [=] [+] 12 [+/−] [=]

DISPLAY: $2 \quad -2 \quad -22 \qquad 12 \quad -12 \quad -34$

$$-2 \,\bigg|\, \begin{array}{ccc} 1 & 13 & -12 \\ & -2 & -22 \\ \hline 1 & 11 & -34 \end{array}$$

ENTER: [×] 2 [+/−] [=] [+] 8 [+/−] [=]

DISPLAY: $2 \quad -2 \quad 68 \qquad 8 \quad -8 \quad 60$

$$-2 \,\bigg|\, \begin{array}{cccc} 1 & 13 & -12 & -8 \\ & -2 & -22 & 68 \\ \hline 1 & 11 & -34 & 60 \end{array}$$

The result is $x^2 + 11x - 34 + \dfrac{60}{x + 2}$.

Exploratory Exercises

Use synthetic division to determine which of the following binomials are factors of $2a^2 - 7a - 4$.

1. $a + 4$ **2.** $a - 4$ **3.** $2a + 1$ **4.** $2a - 1$

Use synthetic division to determine which of the following binomials are factors of $2x^3 - 5x^2 - x + 6$.

5. $x + 1$ **6.** $x - 1$ **7.** $x + 2$ **8.** $2x - 3$

Written Exercises

Divide using synthetic division.

9. $(2x^3 - 3x^2 + 3x - 4) \div (x - 2)$

10. $(3y^3 + 2y^2 - 32y + 2) \div (y - 3)$

11. $(2a^3 + a^2 - 2a + 3) \div (a + 1)$

12. $(3m^3 - 2m^2 + 2m - 1) \div (m - 1)$

13. $(x^4 - 2x^3 + x^2 - 3x + 2) \div (x - 2)$

14. $(3y^4 - 6y^3 - 2y^2 + y - 6) \div (y + 1)$

15. $(6k^3 - 19k^2 + k + 6) \div (k - 3)$

16. $(z^4 - 3z^3 - z^2 - 11z - 4) \div (z - 4)$

17. $(2b^3 - 11b^2 + 12b + 9) \div (b - 3)$

18. $(x^3 + 2x^2 - 5x - 6) \div (x - 2)$

19. $(y^4 - 16y^3 + 86y^2 - 176y + 105) \div (y - 5)$

20. $(a^4 - 5a^3 - 13a^2 + 53a + 60) \div (a + 1)$

21. $(2x^4 - 5x^3 - 10x + 8) \div (x - 3)$

22. $(2a^4 - 5a^3 + 2a - 3) \div (a - 1)$

23. $(y^4 + 6y^3 - 7y^2 + 7y - 1) \div (y + 3)$

24. $(h^5 - 6h^3 + 4h^2 - 3) \div (h - 2)$

25. $(4x^4 - 5x^2 + 2x + 3) \div (2x - 1)$

26. $(2b^3 - 3b^2 - 8b + 4) \div (2b + 1)$

27. $(6x^3 - 28x^2 + 19x + 3) \div (3x - 2)$

28. $(4y^4 - 5y^2 - 8y - 10) \div (2y - 3)$

29. $(x^5 + 32) \div (x + 2)$

30. $(x^5 - 3x^2 - 20) \div (x - 2)$

31. Use synthetic division when dividing $3y^3 - 5y - 2$ by $y - 2$. Then, for $f(y) = 3y^3 - 5y - 2$, find $f(2)$. Compare answers.

32. Use synthetic division when dividing $2h^4 - 3h^2 + 1$ by $h - 1$. Then, for $f(h) = 2h^4 - 3h^2 + 1$, find $f(1)$ and $f(-1)$. Compare answers.

Divide using synthetic division and a calculator.

33. $(4x^3 + 6.7x^2 - 7.3x + 1.45) \div (x - 0.2)$

34. $(5x^5 + 7.7x^4 - x^2 + 7.59) \div (x - 0.5)$

35. $(x^3 - 2.8) \div (x + 0.4)$

36. $(2x^4 + 1.4) \div (2x - 1)$

Challenge Exercises

Use synthetic division to find the value for k so that each remainder is zero.

37. $(x^3 - 3x^2 - 11x + k) \div (x + 3)$

38. $(b^4 - 10b^3 + 18b^2 + kb + 40) \div (b - 8)$

Synthetic Division

The following is a simple program in BASIC to compute the coefficients of the
quotient and the remainder with synthetic division.

```
10   INPUT "DEGREE OF POLYNOMIAL: ";N
20   INPUT "CONSTANT R: ";R
30   PRINT "ENTER COEFFICIENTS:"
40   FOR X = 1 TO N + 1
50   INPUT A(X)
60   NEXT X
70   LET B(1) = A(1)
80   PRINT "COEFFICIENTS OF QUOTIENT
        ARE:"
90   PRINT B(1);"   ";
100   FOR X = 1 TO N − 1
110   B(X + 1) = A(X + 1) + R * B(X)
120   PRINT B(X + 1);"   ";
130   NEXT X
134   PRINT
140   PRINT "REMAINDER: ";A(N + 1) +
        R * B(N)
150   END
```

Input loop (lines 40–60)

Computation and Print loop (lines 100–130)

*R is the constant in the divisor $X - R$.
For example, in exercise 1
the divisor is $x - 3$, so $R = 3$.*

Exercises

Enter the program on a computer and use it to execute the following divisions.

1. $(x^4 + 2x^3 - 7x^2 + 2x - 8) \div (x - 3)$ (Hint: Let $R = 3$.)
2. $(x^4 + 8x^3 + 22x^2 + 24x + 9) \div (x + 1)$
3. $(2x^5 + 3x^4 - 6x^3 + 6x^2 - 8x + 3) \div (x + 1)$
4. $(x^5 - 3x^2 - 20) \div (x - 2)$
5. $(x^5 - 15x^3 - 10x^2 + 60x + 72) \div (x + 3)$
6. $(2y^4 - 2.5y^2 - 4y + 1.5) \div (y - 1.5)$
7. $(x^4 - 6x^2 + 8) \div (x - 2)$

This program will work up to the 10th degree. For a degree greater than 10, the statement 5 DIM A(n),
B(n) is required. The value of n must be at least one greater than the degree of the equation. Add the
DIM statement to the program and execute the following divisions.

8. $(x^{12} - 16x^8 - 256x^4 + 4096) \div (x - 2)$
9. $(x^{12} - 10) \div (x + 2)$

Vocabulary

monomial   (157)
constant   (157)
coefficient   (157)
degree of monomial   (157)
like terms   (157)

negative integer
   exponents   (161)
scientific notation   (164)
polynomial   (166)
term   (166)

binomial, trinomial   (166)
degree of a polynomial
(166)
FOIL   (167)
synthetic division   (186)

Chapter Summary

1. **Properties of Exponents:** For any nonzero numbers a and b, and integers m and n, the following hold.

    1. $a^m \cdot a^n = a^{m+n}$ (158)
    2. $(a^m)^n = a^{mn}$ (158)
    3. $(ab)^m = a^m b^m$ (158)
    4. $\dfrac{a^m}{a^n} = a^{m-n}$ (160)
    5. $a^0 = 1$ (160)
    6. $a^{-n} = \dfrac{1}{a^n}$ (161)
    7. $\dfrac{1}{a^{-n}} = a^n$ (167)
    8. $\left(\dfrac{a}{b}\right)^n = \dfrac{a^n}{b^n}$ (162)
    9. $\left(\dfrac{a}{b}\right)^{-n} = \left(\dfrac{b}{a}\right)^n = \dfrac{b^n}{a^n}$ (162)

2. The degree of a monomial is the sum of the exponents of its variables. The degree of a polynomial is the degree of the monomial of greatest degree. (157, 166)

3. A number is expressed in scientific notation when it is in the form $a \times 10^n$, where $1 \le a < 10$ and n is an integer.   (164)

4. The product of two binomials is the sum of the product of the following: (167)

F	O	I	L
the first terms	the outer terms	the inner terms	the last terms

5. **Factoring Difference of Squares:** For any numbers a and b, $a^2 - b^2 = (a - b)(a + b)$.   (172)

6. **Factoring Sum or Difference of Cubes:** For any numbers a and b, $a^3 + b^3 = (a + b)(a^2 - ab + b^2)$, and $a^3 - b^3 = (a - b)(a^2 + ab + b^2)$.   (173)

7. **Factoring Perfect Square Trinomials:** For any numbers a and b, $a^2 + 2ab + b^2 = (a + b)^2$, and $a^2 - 2ab + b^2 = (a - b)^2$.   (176)

8. Checklist for factoring polynomials.

    1. Check for the greatest common monomial factor.
    2. Check for special products.
        a. If there are *two terms*, look for difference of squares, sum of cubes, difference of cubes.
        b. If there are *three terms*, look for a perfect square trinomial.
    3. Try other factoring methods.
        a. If there are *three terms*, try the trinomial pattern.
        b. If there are *four or more terms*, try grouping.   (178)

5–1 **Simplify.**

    1. $y^9 \cdot y^2$ **2.** $(xy^4)(-5x^2y^3)$ **3.** $(x^3)^2$ **4.** $(4a^2)^3$

    5. $(5a)(6a^2b)(3ab^3) + (4a^2)(3b^3)(2a^2b)$ **6.** $(3a)(a^2b)^3 + (2a)^2(-a^5b^3)$

5–2 **Simplify.**

    7. $\dfrac{a^6}{a^2}$ **8.** $\dfrac{14a^4b^3}{(7ab)^{-2}}$ **9.** $\dfrac{m^3n^2}{2m^3n}$ **10.** $3x^0$

    11. $(3x)^0$ **12.** $\dfrac{-3x^3yz^4}{12x^2y^{-1}}$ **13.** $\left(\dfrac{1}{m}\right)^{-6}$ **14.** $\left(\dfrac{2}{3x^2}\right)^{-1}$

5–3 **15.** Change to scientific notation: $3{,}176{,}000{,}000$

    16. Change to decimal notation: 1.592×10^{-4}

    17. Express the answer in scientific notation: $\dfrac{72{,}000{,}000 \times 0.005}{0.0015}$

5–4 **Simplify.**

    18. $(4b^3 + 7b^2 - 3b + 5) + (-3b^3 + 8b^2 - 7)$

    19. $(p^4 + 5p^2 - 3p + 7) - (p^3 + p^2 - 3p + 5)$

    20. $(4a - 5)(a + 7)$ **21.** $(3m - 7)^2$ **22.** $(y + 7)(y^2 - 3y + 5)$

    23. $(2x - 5)(x^2 + 8x - 7)$ **24.** $(m + 1)(2m + 7)(m + 3)$ **25.** $(2z - 5)(2z + 5)(z - 6)$

5–5 **Factor.**

    26. $y^2 - 25$ **27.** $3a^2s - 6as^2 + 3s^3$ **28.** $m^3 + 8$

    29. $x^4 - y^4$ **30.** $8m^3 - 27$ **31.** $p^3q^3 - 27q^3$

    32. $x^2 - 2xy + x - 2y$ **33.** $3a^2 + 12ab - 2a - 8b$

5–6 **Factor.**

    34. $x^2 - 7x + 10$ **35.** $2x^2 + 7xy + 3y^2$ **36.** $9p^2 - 30pt + 25t^2$

    37. $r^3 + 6r^2s + 8rs^2$ **38.** $5b^2 - 19ab - 4a^2$ **39.** $6x^2 + 11xy + 4y^2$

    40. $-b^2 + 8b + a^2 - 16$ **41.** $4a^2 + 4ab - a^2 + b^2$

5–7 **Find each solution using long division. Show your work.**

    42. $(8y^3 - 22y^2 - 5y + 15) \div (4y + 3)$

    43. $(2r^3 + 11r^2 - 9r - 18) \div (2r + 3)$

    44. Show that $x + 1$ is *not* a factor of $2x^3 + x^2 - 11x - 30$.

    45. Show that $x - 3$ is a factor of $2x^3 - 11x^2 + 12x + 9$.

5–8 **Find each solution using synthetic division. Show your work.**

    46. $(2m^3 - 3m^2 - 8m + 1) \div (m - 4)$

    47. $(2r^3 + r^2 - 13r + 6) \div (2r - 1)$

    48. Show that $2x + 1$ is a factor of $2x^3 - 11x^2 + 12x + 9$.

Simplify.

1. $(m^2)^5$

2. $\dfrac{16b^7}{2b^5}$

3. $\left(\dfrac{m}{4}\right)^{-2}$

4. $\dfrac{x^3}{x^{-4}}$

5. $\dfrac{7m}{m^{-3}} + \dfrac{3m^3}{m^{-1}}$

6. $\left(\dfrac{2y^{-2}}{2}\right)^{-1}\left(\dfrac{m^2n}{y}\right)$

7. $(4a^2b^2)(5ab^3)$

8. $(5x)(6x^2y^3) + (2xy)^3 - x^2y^0$

9. $3x^2y + 4 - 3x^2y$

10. $(3y^4 + 5y^3 + y - 7) + (4y^4 - 3y^3 + 7y^2 - 3y + 8)$

11. $(4b^3 + 7b^2 - 3b + 5) - (2b^3 - 5b^2 + 2b + 7)$

12. $(2y + 7)(y - 3)$

13. $(3m + 5)^2$

14. $(y - 4)(3y^2 - 5y + 4)$

15. $(x + 3)(2x - 5)(3x + 4)$

Factor.

16. $a^2 - 121$

17. $k^2 + 3k - 40$

18. $a^2 + 6am + 9m^2$

19. $y^3 + 125$

20. $3y^2 + 15y + 18$

21. $64a^3 - 1$

22. $8a^3 - 12a^2b + 6ab^2 - b^3$

23. $6y^2 + 17y + 10$

24. $2m^3 - 6m^2 + m - 3$

25. $8m^2 - 14m - 15$

26. $r^2 + 4rs + 4s^2 - 9y^2$

27. $3y^2 - 19y + 28$

28. Change to scientific notation: 5,906,000,000

29. Change to decimal notation: 1.672×10^{-3}

30. Express the answer in scientific notation: $\dfrac{84{,}000{,}000 \times 0.0013}{0.021}$

31. Use long division to find $(6y^3 - 5y^2 - 12y - 17) \div (3y + 2)$. Show your work.

32. Use synthetic division to find $(5m^3 - 3m^2 + 2m - 5) \div (m + 2)$. Show your work.

33. Show that $a - 3$ is a factor of $a^4 - a^3 + a^2 - 25a + 12$.

34. One factor of $2y^3 + y^2 - 13y + 6$ is $y + 3$. Find the other factors.

35. One factor of $4b^3 + 6b^2 + 40b - 22$ is $2b - 1$. Find the other factors.

Roots

How fast is an airplane traveling at takeoff? Its velocity, v, in meters per second is found using the formula $v^2 = 2ad$ where a is the acceleration of the airplane in meters per second per second and d is the distance traveled on the runway in meters. Solving this equation for v involves finding roots. That is, $v = \sqrt{2ad}$. In this chapter you will learn more about roots.

6-1 Roots

Squaring a number means using that number as a factor two times.
Cubing a number means using that number as a factor three times.

$$6^2 = 6 \cdot 6 \text{ or } 36 \qquad \textit{6 is used as a factor two times.}$$
$$6^3 = 6 \cdot 6 \cdot 6 \text{ or } 216 \qquad \textit{6 is used as a factor three times.}$$

Raising a number to the *n*th power means using that number as a factor *n* times.

$$5^4 = 5 \cdot 5 \cdot 5 \cdot 5 \text{ or } 625 \qquad \textit{5 is used as a factor four times. } n = 4$$
$$2^8 = 2 \cdot 2 \cdot 2 \cdot 2 \cdot 2 \cdot 2 \cdot 2 \cdot 2 \qquad \textit{2 is used as a factor eight times. } n = 8$$
$$6^n = \underbrace{6 \cdot 6 \cdot 6 \cdot \ldots \cdot 6}_{n \text{ factors}} \qquad \textit{6 is used as a factor n times.}$$

The inverse of raising a number to the *n*th power is finding the **nth root** of that number. For example, the inverse of squaring a number is finding a **square root** of that number.

To find a square root of 36, you must find two equal factors whose product is 36.

$$x^2 = 36 \qquad x \cdot x = 36$$

Since 6 times 6 is 36, one square root of 36 is 6. Since -6 times -6 is 36, another square root of 36 is -6.

Any number that is the square of an integer is called a perfect square.

> **For any real numbers a and b,**
> **if $a^2 = b$, then a is a square root of b.**

Definition of Square Root

To find the cube root of 125, you must find three equal factors whose product is 125.

$$x^3 = 125 \qquad x \cdot x \cdot x = 125$$

Since 5 times 5 times 5 is 125, the cube root of 125 is 5.

> **For any real numbers a and b, and any positive integer**
> **n, if $a^n = b$, then a is an nth root of b.**

Definition of nth Root

The symbol $\sqrt[n]{}$ indicates an *n*th root.

$$\text{index} \longrightarrow \qquad \longleftarrow \text{radical sign}$$
$$\sqrt[n]{256} \longleftarrow \text{radicand}$$

When *no* index appears, the radical sign $\sqrt{}$ indicates a nonnegative square root.

Some numbers have more than one real root. For example, 36 has two real square roots, 6 and -6.

The symbol $\sqrt[n]{b}$ indicates the *principal* nth root of b. The principal nth root of b is a nonnegative number *unless* n is odd and b is negative. In this case the principal root is negative.

$\sqrt{36} = 6$ $\sqrt{36}$ indicates the principal square root of 36.

$-\sqrt{36} = -6$ $-\sqrt{36}$ indicates the negative of the principal square root of 36.

$\pm\sqrt{36} = \pm 6$ $\pm\sqrt{36}$ indicates both square roots of 36.   *$\pm$ means positive or negative.*

$\sqrt[3]{-27} = -3$ $\sqrt[3]{-27}$ indicates the principal cube root of -27.

$-\sqrt[4]{16} = -2$ $-\sqrt[4]{16}$ indicates the negative of the principal fourth root of 16.

The following chart gives a summary of the real nth roots of a number b.

The Real nth Roots of b

The real nth roots can be written as $\sqrt[n]{b}$ or $-\sqrt[n]{b}$.

	$b > 0$	$b < 0$	$b = 0$
n even	one positive root one negative root	no real roots	one real root, 0
n odd	one positive root no negative roots	no positive roots one negative root	one real root, 0

Examples

1   **Find $\pm\sqrt{64b^2}$.**

$$\begin{aligned} \pm\sqrt{64b^2} &= \pm\sqrt{(8b)^2} \\ &= \pm 8b \end{aligned}$$

The square roots of $64b^2$ are $\pm 8b$.

2   **Find $-\sqrt{(x + 3)^4}$.**

$$\begin{aligned} -\sqrt{(x + 3)^4} &= -\sqrt{[(x + 3)^2]^2} \\ &= -(x + 3)^2 \end{aligned}$$

The negative of the principal square root of $(x + 3)^4$ is $-(x + 3)^2$.

3   **Find $\sqrt[3]{27x^6}$.**

$$\begin{aligned} \sqrt[3]{27x^6} &= \sqrt[3]{(3x^2)^3} \\ &= 3x^2 \end{aligned}$$

The principal cube root of $27x^6$ is $3x^2$.

4   **Find $\sqrt[6]{c^6}$.**

Since $(c)^6 = c^6$, c is a sixth root of c^6.
Thus, $\sqrt[6]{c^6} = |c|$.

6 is an even number. So the principal sixth root of c^6 is a positive number.

These and other similar examples suggest the following property.

For any real number a, and any integer n, $n > 1$,
1. If n is even, then $\sqrt[n]{a^n} = |a|$, and
2. If n is odd, then $\sqrt[n]{a^n} = a$.

Property of nth Roots

Expressions such as $\sqrt{64}$ and $\sqrt[3]{-\frac{1}{8}}$ name rational numbers.

$$\sqrt{64} = 8 \qquad\qquad \sqrt[3]{-\frac{1}{8}} = -\frac{1}{2}$$

Real numbers that cannot be written as terminating or repeating decimals are **irrational numbers**. Numbers such as $\sqrt{2}$ and $\sqrt{3}$ are irrational. To compute with irrational numbers, decimal approximations are often used. A calculator or a table of roots such as the one on page 672 can be used to find decimal approximations.

Examples

5 Using Calculators

Find a decimal approximation for $\sqrt[3]{682}$.

Use the root key.   or   Use the inverse and power keys.

ENTER: 682 $\boxed{\sqrt[x]{y}}$ 3 $\boxed{=}$ 682 $\boxed{\text{INV}}$ $\boxed{y^x}$ 3 $\boxed{=}$

DISPLAY: *682*   *3 8.80227214*   *682*   *3 8.80227214*

To check, enter the following:

     8.80227214 $\boxed{y^x}$ 3 $\boxed{=}$

The result should be about 682.

6 Use the table of roots on page 672 to find a decimal approximation for $\sqrt[3]{28}$.

n	$\sqrt[3]{n}$	$\sqrt[3]{10n}$	$\sqrt[3]{100n}$
1.0	1.000	2.154	4.642
1.1	1.032	2.224	4.791
1.2	1.063	2.289	4.932
1.3	1.091	2.351	5.066
2.5	1.357	2.924	6.300
2.6	1.375	2.962	6.383
2.7	1.392	3.000	6.463
2.8	1.409	3.037	6.542
2.9	1.426	3.072	6.619

Let $n = 2.8$.
Then $10n = 10(2.8)$ or 28.

So $\sqrt[3]{28} = \sqrt[3]{10(2.8)}$
$\phantom{So \sqrt[3]{28}} = 3.037$

Check: $(3.037)^3 = 28.01137165$

Exploratory Exercises

Simplify.

1. $\sqrt{121}$
2. $-\sqrt{144}$
3. $\sqrt[3]{8}$
4. $\sqrt[4]{16}$
5. $\sqrt[3]{y^3}$
6. $-\sqrt[4]{y^4}$
7. $\sqrt[4]{y^8}$
8. $\sqrt[3]{-64}$
9. $\sqrt[5]{32n^5}$
10. $\sqrt{16a^2b^4}$
11. $\sqrt{(x-2)^2}$
12. $\sqrt{x^2 + 6x + 9}$

Use a calculator or the table of roots on page 672 to find each value.

13. $\sqrt{47}$
14. $\sqrt[3]{18}$
15. $\sqrt[3]{30}$
16. $-\sqrt{89}$
17. $-\sqrt{67}$
18. $-\sqrt[3]{56}$

Written Exercises

Simplify.

19. $-\sqrt{81}$
20. $\sqrt{169}$
21. $\sqrt{225}$
22. $-\sqrt[3]{27}$
23. $\sqrt[4]{81}$
24. $\sqrt[3]{64}$
25. $\sqrt[5]{-1}$
26. $\sqrt[3]{-1000}$
27. $\sqrt{0.49}$
28. $\sqrt[3]{0.125}$
29. $\sqrt{121n^2}$
30. $\sqrt{144x^6}$
31. $\sqrt{(3s)^4}$
32. $\sqrt{(5b)^4}$
33. $\sqrt{576}$
34. $\sqrt{676}$
35. $\sqrt{64a^2b^4}$
36. $-\sqrt{121b^2c^6}$
37. $\sqrt[3]{-8b^3m^3}$
38. $\sqrt[3]{-27r^3s^3}$
39. $\sqrt[3]{64a^6b^3}$
40. $\sqrt{(x+y)^2}$
41. $\sqrt{(3p+q)^2}$
42. $\sqrt[3]{(2m+n)^3}$
43. $\sqrt[3]{(z+a)^3}$
44. $\sqrt[4]{(r+s)^4}$
45. $\sqrt[5]{(2m-3)^5}$
46. $\sqrt{x^2 + 10x + 25}$
47. $\sqrt{x^2 + 6x + 9}$
48. $\sqrt{4r^2 + 12r + 9}$
49. $\sqrt{9x^2 + 6x + 1}$
50. $\sqrt{x^2 - 6xy + 9y^2}$
51. $\sqrt{4x^2 + 12xy + 9y^2}$

Use a calculator or the table of roots on page 672 to find each value.

52. $-\sqrt{99}$
53. $\sqrt{83}$
54. $\sqrt{9.5}$
55. $-\sqrt[3]{-41}$
56. $\sqrt[3]{23}$
57. $\sqrt[3]{8.1}$
58. $-\sqrt[3]{120}$
59. $\sqrt[3]{-60}$
60. $\sqrt[3]{300}$

Use a calculator to find each value to three places. Check your approximations by using the power key.

61. $\sqrt[4]{70}$
62. $\sqrt[5]{413}$
63. $\sqrt[4]{21}$
64. $\sqrt[7]{16,384}$
65. $\sqrt[8]{6581}$
66. $\sqrt[6]{8549}$

Challenge Exercises

67. Does $\sqrt[4]{(-x)^4} = x$ no matter what value x represents? Explain.
68. Does $\sqrt[5]{(-x)^5} = x$ no matter what value x represents? Explain.
69. Under what circumstances is $\sqrt[n]{(-x)^n} = x$?

6-2 Multiplying and Dividing Radicals

The following examples show an important property of radicals.

$$\sqrt{4} \cdot \sqrt{9} = 2 \cdot 3 \text{ or } 6$$

$$\sqrt{4 \cdot 9} = \sqrt{36} \text{ or } 6$$

$$\sqrt[3]{-8} \cdot \sqrt[3]{27} = -2 \cdot 3 \text{ or } -6$$

$$\sqrt[3]{-8 \cdot 27} = \sqrt[3]{-216} \text{ or } -6$$

> **For any real numbers a and b, and any integer n, $n > 1$,**
> 1. **If n is even, then $\sqrt[n]{ab} = \sqrt[n]{a} \cdot \sqrt[n]{b}$ as long as a and b are both nonnegative, and**
> 2. **If n is odd, then $\sqrt[n]{ab} = \sqrt[n]{a} \cdot \sqrt[n]{b}$.**

Product Property of Radicals

To simplify a square root, first write the prime factorization of the radicand. Then use the Product Property of Radicals to isolate perfect squares. Finally, simplify each radical.

Examples

1 **Simplify $\sqrt{63}$.**

$$\begin{aligned} \sqrt{63} &= \sqrt{3^2 \cdot 7} \\ &= \sqrt{3^2} \cdot \sqrt{7} \\ &= 3\sqrt{7} \end{aligned}$$

The prime factorization of 63 is $3^2 \cdot 7$.
Product Property of Radicals

2 **Simplify $\sqrt{45x^3y^2}$.**

$$\begin{aligned} \sqrt{45x^3y^2} &= \sqrt{3^2 \cdot 5 \cdot x^2 \cdot x \cdot y^2} \\ &= \sqrt{3^2} \cdot \sqrt{5} \cdot \sqrt{x^2} \cdot \sqrt{x} \cdot \sqrt{y^2} \\ &= 3x|y|\sqrt{5x} \end{aligned}$$

If $x < 0$, then $\sqrt{x^3}$ has no real roots.
Therefore, you must assume that $x \geq 0$.
Thus, it is not necessary to write $\sqrt{x^3} = |x|\sqrt{x}$.

To simplify nth roots, find the factors that are nth powers and use the Product Property.

Examples

3 **Simplify $\sqrt[4]{2m} \cdot \sqrt[4]{5m^3}$.**

$$\begin{aligned} \sqrt[4]{2m} \cdot \sqrt[4]{5m^3} &= \sqrt[4]{2m \cdot 5m^3} \\ &= \sqrt[4]{10m^4} \\ &= \sqrt[4]{10} \cdot \sqrt[4]{m^4} \\ &= m\sqrt[4]{10} \end{aligned}$$

4 **Simplify $\sqrt[3]{54x^3y^5}$.**

$$\begin{aligned} \sqrt[3]{54x^3y^5} &= \sqrt[3]{3^3 \cdot 2 \cdot x^3 \cdot y^3 \cdot y^2} \\ &= \sqrt[3]{3^3} \cdot \sqrt[3]{2} \cdot \sqrt[3]{x^3} \cdot \sqrt[3]{y^3} \cdot \sqrt[3]{y^2} \\ &= 3 \cdot \sqrt[3]{2} \cdot x \cdot y \cdot \sqrt[3]{y^2} \\ &= 3xy\sqrt[3]{2y^2} \end{aligned}$$

When multiplying rational numbers and radicals, multiply each separately and then simplify.

$$2\sqrt{2} \cdot 4\sqrt{6} = 2 \cdot 4 \cdot \sqrt{2} \cdot \sqrt{6}$$
$$= 8\sqrt{12}$$
$$= 8\sqrt{2^2} \cdot \sqrt{3}$$
$$= 8 \cdot 2 \cdot \sqrt{3} \text{ or } 16\sqrt{3}$$

You can use the Distributive Property to help simplify radicals.

Example

5   **Simplify $\sqrt{6}(\sqrt{3} + 2\sqrt{15})$.**

$$\sqrt{6}(\sqrt{3} + 2\sqrt{15}) = \sqrt{6} \cdot \sqrt{3} + \sqrt{6} \cdot 2\sqrt{15}$$
$$= \sqrt{18} + 2\sqrt{90}$$
$$= \sqrt{3^2 \cdot 2} + 2\sqrt{3^2 \cdot 10}$$
$$= 3\sqrt{2} + 2 \cdot 3\sqrt{10}$$
$$= 3\sqrt{2} + 6\sqrt{10}$$

The following examples show another important property of radicals.

$$\frac{\sqrt{100}}{\sqrt{4}} = \frac{10}{2} \text{ or } 5 \qquad\qquad \frac{\sqrt[3]{216}}{\sqrt[3]{-27}} = \frac{6}{-3} \text{ or } -2$$

$$\sqrt{\frac{100}{4}} = \sqrt{25} \text{ or } 5 \qquad\qquad \sqrt[3]{\frac{216}{-27}} = \sqrt[3]{-8} \text{ or } -2$$

For any real numbers a and b, $b \neq 0$, and any integer n, $n > 1$,

$$\sqrt[n]{\frac{a}{b}} = \frac{\sqrt[n]{a}}{\sqrt[n]{b}}$$

if all roots are defined.

Quotient Property of Radicals

Examples

6   **Simplify $\sqrt[3]{\dfrac{3}{8}}$.**

$$\sqrt[3]{\frac{3}{8}} = \frac{\sqrt[3]{3}}{\sqrt[3]{8}} \qquad \textit{Quotient Property of Radicals}$$
$$= \frac{\sqrt[3]{3}}{2}$$

7   **Simplify $\dfrac{6\sqrt{15}}{2\sqrt{3}}$.**

$$\frac{6\sqrt{15}}{2\sqrt{3}} = \frac{6}{2}\sqrt{\frac{15}{3}} \qquad \textit{Quotient Property of Radicals}$$
$$= 3\sqrt{5}$$

Fractions are usually written without radicals in the denominator. Similarly, radicands are not left in fraction form. The process of eliminating radicals from the denominator or fractions from the radicand is called **rationalizing the denominator**.

Examples

8 Simplify $\dfrac{3}{2\sqrt{5}}$.

$$\frac{3}{2\sqrt{5}} = \frac{3}{2\sqrt{5}} \cdot \frac{\sqrt{5}}{\sqrt{5}} \qquad \textit{Why is } \frac{\sqrt{5}}{\sqrt{5}} \textit{ used?}$$

$$= \frac{3\sqrt{5}}{2\sqrt{5 \cdot 5}}$$

$$= \frac{3\sqrt{5}}{10}$$

9 Simplify $\sqrt{\dfrac{5}{a}}$.

$$\sqrt{\frac{5}{a}} = \frac{\sqrt{5}}{\sqrt{a}} \cdot \frac{\sqrt{a}}{\sqrt{a}} \qquad \textit{Why is } \frac{\sqrt{a}}{\sqrt{a}} \textit{ used?}$$

$$= \frac{\sqrt{5 \cdot a}}{\sqrt{a \cdot a}}$$

$$= \frac{\sqrt{5a}}{a}$$

In general, a radical expression is simplified when the following conditions are met.

1. The index, n, is as small as possible.
2. The radicand contains no factor (other than one) which is the nth power of an integer or polynomial.
3. The radicand contains no fractions.
4. No radicals appear in the denominator.

Conditions for Simplified Radicals

Examples

10 Simplify $\dfrac{7}{\sqrt[4]{3}}$.

$$\frac{7}{\sqrt[4]{3}} = \frac{7}{\sqrt[4]{3}} \cdot \frac{\sqrt[4]{3^3}}{\sqrt[4]{3^3}} \qquad \textit{Why is } \frac{\sqrt[4]{3^3}}{\sqrt[4]{3^3}} \textit{ used?}$$

$$= \frac{7\sqrt[4]{3^3}}{\sqrt[4]{3 \cdot 3^3}}$$

$$= \frac{7\sqrt[4]{27}}{3}$$

11 Simplify $\sqrt[3]{\dfrac{5}{3b}}$.

$$\sqrt[3]{\frac{5}{3b}} = \frac{\sqrt[3]{5}}{\sqrt[3]{3b}} \cdot \frac{\sqrt[3]{3^2 b^2}}{\sqrt[3]{3^2 b^2}} \qquad \textit{Why is } \frac{\sqrt[3]{3^2 b^2}}{\sqrt[3]{3^2 b^2}} \textit{ used?}$$

$$= \frac{\sqrt[3]{5 \cdot 3^2 b^2}}{\sqrt[3]{3^3 b^3}}$$

$$= \frac{\sqrt[3]{45b^2}}{3b}$$

Simplify.

1. $\sqrt{8}$
2. $\sqrt{32}$
3. $\sqrt{50x^2}$
4. $\sqrt{98y^4}$

5. $\sqrt[3]{16}$
6. $\sqrt[3]{54}$
7. $\sqrt[4]{48}$
8. $\sqrt[5]{32}$

9. $\sqrt{b^3}$
10. $\sqrt{y^5}$
11. $\sqrt[4]{a^5}$
12. $\sqrt[3]{m^4}$

13. $\sqrt[5]{r^7}$
14. $\sqrt[4]{k^6}$
15. $\sqrt{3} \cdot \sqrt{15}$
16. $\sqrt{6} \cdot \sqrt{3}$

17. $\sqrt[4]{3} \cdot \sqrt[4]{54}$
18. $\sqrt[3]{9} \cdot \sqrt[3]{6}$
19. $\sqrt{5}(\sqrt{5} - \sqrt{3})$
20. $\sqrt{5}(\sqrt{7} + \sqrt{5})$

21. $\dfrac{\sqrt{6}}{\sqrt{3}}$
22. $\dfrac{\sqrt{10}}{\sqrt{2}}$
23. $\dfrac{\sqrt[3]{18y}}{\sqrt[3]{6}}$
24. $\dfrac{\sqrt[4]{35x^5}}{\sqrt[4]{7}}$

25. $\sqrt{\dfrac{5}{4}}$
26. $\sqrt{\dfrac{7}{9}}$
27. $\sqrt[3]{\dfrac{5}{8}}$
28. $\sqrt[3]{\dfrac{4}{27}}$

State the fraction that each of the following expressions should be multiplied by to rationalize the denominator.

29. $\dfrac{2}{\sqrt{3}}$
30. $\dfrac{4}{\sqrt{2}}$
31. $\dfrac{1}{\sqrt{a}}$
32. $\dfrac{3}{\sqrt{b}}$

33. $\dfrac{3}{\sqrt[3]{4}}$
34. $\dfrac{4}{\sqrt[3]{2}}$
35. $\dfrac{7}{\sqrt[3]{9}}$
36. $\dfrac{4}{\sqrt[3]{16}}$

Simplify.

37. $5\sqrt{54}$
38. $4\sqrt{50}$
39. $\sqrt[3]{24}$

40. $\sqrt[3]{56}$
41. $\sqrt{162}$
42. $\sqrt{450}$

43. $\sqrt[3]{-192}$
44. $3\sqrt{242}$
45. $6\sqrt{216}$

46. $\sqrt[4]{112}$
47. $(4\sqrt{18})(2\sqrt{14})$
48. $(-3\sqrt{24})(5\sqrt{20})$

49. $\sqrt[3]{121} \cdot \sqrt[3]{88}$
50. $(7\sqrt[3]{16})(5\sqrt[3]{20})$
51. $\sqrt{3}(\sqrt{6} - 2)$

52. $\sqrt{7}(3 + \sqrt{7})$
53. $\sqrt{7}(\sqrt{14} + \sqrt{21})$
54. $-\sqrt{2}(\sqrt{3} + \sqrt{2})$

55. $\sqrt[3]{2}(3\sqrt[3]{4} + 2\sqrt[3]{32})$
56. $\sqrt[3]{9}(4\sqrt[3]{9} + 2\sqrt[3]{6})$
57. $\sqrt[3]{8a^4b^7}$

58. $\sqrt{8m^2b^3}$
59. $\sqrt[4]{81m^4p^5}$
60. $\sqrt{8x^2y} \cdot \sqrt{2xy}$

61. $\sqrt{3x^2z^3} \cdot \sqrt{15x^2z}$
62. $\sqrt[3]{3ab^5} \cdot \sqrt[3]{24a^2b^2}$
63. $\sqrt[4]{5m^3b^5} \cdot \sqrt[4]{125m^2b^3}$

64. $\sqrt[4]{3b^6r^7} \cdot \sqrt[4]{81b^2r^2}$
65. $\sqrt{125m^2n} \cdot \sqrt{32m^4n^6}$
66. $\sqrt[4]{32a^5b^3} \cdot \sqrt[4]{162a^3b^2}$

67. $\sqrt{r}(\sqrt{r} + r\sqrt{s})$
68. $\sqrt{b}(b + a\sqrt{b})$
69. $\sqrt{m}(\sqrt{p} + \sqrt{mq})$

70. $\dfrac{\sqrt{10}}{\sqrt{2}}$
71. $\dfrac{\sqrt{12}}{\sqrt{3}}$
72. $\dfrac{\sqrt{14}}{\sqrt{2}}$
73. $\dfrac{\sqrt{21}}{\sqrt{7}}$
74. $\dfrac{\sqrt[3]{81}}{\sqrt[3]{9}}$

75. $\dfrac{\sqrt[3]{54}}{\sqrt[3]{6}}$
76. $\sqrt{\dfrac{5}{4}}$
77. $\sqrt{\dfrac{7}{16}}$
78. $\sqrt{\dfrac{8}{9}}$
79. $\sqrt{\dfrac{21}{12}}$

80. $\sqrt[3]{\dfrac{5}{8}}$ 81. $\sqrt[3]{\dfrac{2}{27}}$ 82. $\sqrt[3]{\dfrac{54}{125}}$ 83. $\sqrt[3]{\dfrac{16}{27}}$ 84. $\sqrt[4]{\dfrac{5}{16}}$

85. $\sqrt[4]{\dfrac{7}{81}}$ 86. $\sqrt{\dfrac{1}{3}}$ 87. $\sqrt{\dfrac{1}{5}}$ 88. $\sqrt{\dfrac{2}{m}}$ 89. $\sqrt{\dfrac{3}{r}}$

90. $\sqrt{\dfrac{5}{12a}}$ 91. $\sqrt{\dfrac{5}{32b}}$ 92. $\sqrt[3]{\dfrac{5}{9p^2}}$ 93. $\sqrt[3]{\dfrac{9}{4m^2}}$ 94. $\sqrt[4]{\dfrac{2}{3}}$

Solve each problem. Round all answers to the nearest hundredth.

95. Find the time, T, in seconds for a complete swing (back and forth) of a pendulum whose length is 6 feet.

 Let $T = 2\pi\sqrt{\dfrac{L}{32}}$, where $\pi \approx 3.14$.

96. Find the time, T, in seconds for a complete swing of a pendulum whose length is 98 centimeters.

 Let $T = 2\pi\sqrt{\dfrac{L}{980}}$, where $\pi \approx 3.14$.

97. Find the radius, r, of a sphere whose surface area, S, is 616 square inches.

 Let $r = \dfrac{1}{2}\sqrt{\dfrac{S}{\pi}}$, where $\pi \approx \dfrac{22}{7}$.

98. Find the time, t, in seconds required for a freely falling body to fall a distance s of 150 feet. Let $t = \dfrac{1}{4}\sqrt{s}$.

mini-review

Solve each system of equations algebraically.

1. $x - y = -6$
 $3x + 3y = 2$

2. $x + y = 8$
 $x - y = 10$

3. $-\dfrac{1}{2}x + 5y = 15$
 $2x - 6y = -12$

Find the value of each determinant.

4. $\begin{vmatrix} -2 & -3 \\ 4 & 5 \end{vmatrix}$

5. $\begin{vmatrix} -5 & 2 \\ 3 & 4 \end{vmatrix}$

6. $\begin{vmatrix} 8 & 12 \\ 9 & -13 \end{vmatrix}$

State the dimension and evaluate the determinant for each matrix.

7. $\begin{bmatrix} 2 & -18 \\ 11 & 3 \end{bmatrix}$

8. $\begin{bmatrix} 5 & 3 & 4 \\ 2 & -3 & 6 \\ -2 & 1 & 0 \end{bmatrix}$

Perform the indicated operations.

9. $4\begin{bmatrix} 2 & -5 & -3 & 4 \\ 1 & 4 & -3 & 0 \end{bmatrix}$

10. $\begin{bmatrix} 1 & 0 & 3 \\ -1 & 6 & -7 \end{bmatrix}\begin{bmatrix} 1 & -9 \\ 2 & -8 \\ 3 & 7 \end{bmatrix}$

Simplify.

11. $y^4 \cdot y^5$

12. $(ab^2)(-3a^4b^5)$

13. $(-7x^{-2})^4$

6-3 Computing with Radicals

Two radical expressions are called **like radical expressions** if the indexes are alike and the radicands are alike.

$7\sqrt[3]{2}$ and $6\sqrt[3]{2}$ are like expressions.	*Both the indexes and radicands are alike.*
$\sqrt[4]{9}$ and $\sqrt[5]{9}$ are *not* like expressions.	*The indexes are not alike.*
$5\sqrt{3x}$ and $-5\sqrt{3y}$ are *not* like expressions.	*The radicands are not alike.*
$\sqrt[3]{2a}$ and $\sqrt[4]{2b}$ are *not* like expressions.	*Neither the indexes nor the radicands are alike.*

Radicals are added or subtracted the same way monomials are added or subtracted.

Combine like terms.

$$3x + 2x + 4y = (3 + 2)x + 4y$$
$$= 5x + 4y$$

Combine like radicals.

$$3\sqrt{6} + 2\sqrt{6} + 4\sqrt{7} = (3 + 2)\sqrt{6} + 4\sqrt{7}$$
$$= 5\sqrt{6} + 4\sqrt{7}$$

Examples

1 Simplify $3 + 4\sqrt{7} + 5 + 6\sqrt{7}$.

$$3 + 4\sqrt{7} + 5 + 6\sqrt{7}$$
$$= (3 + 5) + (4 + 6)\sqrt{7}$$
$$= 8 + 10\sqrt{7}$$

2 Simplify $3 + 4\sqrt{a} + 6\sqrt{a}$.

$$3 + 4\sqrt{a} + 6\sqrt{a}$$
$$= 3 + (4 + 6)\sqrt{a}$$
$$= 3 + 10\sqrt{a}$$

3 Simplify $5\sqrt{27} + 2\sqrt{3} - 7\sqrt{48}$.

$$5\sqrt{27} + 2\sqrt{3} - 7\sqrt{48} = 5\sqrt{3^2 \cdot 3} + 2\sqrt{3} - 7\sqrt{4^2 \cdot 3}$$
$$= 5\sqrt{3^2}\sqrt{3} + 2\sqrt{3} - 7\sqrt{4^2}\sqrt{3}$$
$$= 5 \cdot 3\sqrt{3} + 2\sqrt{3} - 7 \cdot 4\sqrt{3}$$
$$= 15\sqrt{3} + 2\sqrt{3} - 28\sqrt{3}$$
$$= -11\sqrt{3}$$

Simplify each radical. Then add or subtract.

4 Simplify $\sqrt[3]{40a} + \sqrt[3]{135a}$.

$$\sqrt[3]{40a} + \sqrt[3]{135a} = \sqrt[3]{2^3 \cdot 5a} + \sqrt[3]{3^3 \cdot 5a}$$
$$= \sqrt[3]{2^3} \cdot \sqrt[3]{5a} + \sqrt[3]{3^3} \cdot \sqrt[3]{5a}$$
$$= 2\sqrt[3]{5a} + 3\sqrt[3]{5a}$$
$$= 5\sqrt[3]{5a}$$

Expressions such as $(5 + 3\sqrt{2})(3 + \sqrt{2})$ can be simplified using the FOIL Method.

$$(5 + 3\sqrt{2})(3 + \sqrt{2}) = \overset{F}{5 \cdot 3} + \overset{O}{5 \cdot \sqrt{2}} + \overset{I}{3\sqrt{2} \cdot 3} + \overset{L}{3\sqrt{2} \cdot \sqrt{2}}$$
$$= 15 + 5\sqrt{2} + 9\sqrt{2} + 6$$
$$= 21 + 14\sqrt{2}$$

5 Simplify $(6 + \sqrt{2})(\sqrt{10} + \sqrt{5})$.

$$\begin{aligned}
& \qquad\qquad\qquad\quad \text{F} \qquad\quad \text{O} \qquad\quad\;\; \text{I} \qquad\qquad \text{L} \\
(6 + \sqrt{2})(\sqrt{10} + \sqrt{5}) &= 6\sqrt{10} + 6\sqrt{5} + \sqrt{2} \cdot \sqrt{10} + \sqrt{2} \cdot \sqrt{5} \\
&= 6\sqrt{10} + 6\sqrt{5} + \quad \sqrt{20} \quad + \quad \sqrt{10} \\
&= 6\sqrt{10} + 6\sqrt{5} + \quad 2\sqrt{5} \quad + \quad \sqrt{10} \\
&= 7\sqrt{10} + 8\sqrt{5}
\end{aligned}$$

Binomials that are of the form $a\sqrt{b} + c\sqrt{d}$ and $a\sqrt{b} - c\sqrt{d}$ are called **conjugates** of each other. Examples 6 and 7 show that the product of conjugates is a rational number.

6 Simplify $(7 + \sqrt{2})(7 - \sqrt{2})$.

$$\begin{aligned}
(7 + \sqrt{2})(7 - \sqrt{2}) &= 7 \cdot 7 - 7\sqrt{2} + 7\sqrt{2} - \sqrt{2} \cdot \sqrt{2} \qquad \textit{FOIL Method} \\
&= 49 - \sqrt{2^2} \\
&= 49 - 2 \text{ or } 47
\end{aligned}$$

7 Simplify $(2a + 7\sqrt{b})(2a - 7\sqrt{b})$.

$$\begin{aligned}
(2a + 7\sqrt{b})(2a - 7\sqrt{b}) &= 4a^2 - 14a\sqrt{b} + 14a\sqrt{b} - 49(\sqrt{b})^2 \qquad \textit{FOIL Method} \\
&= 4a^2 - 49b
\end{aligned}$$

Conjugates can be used to rationalize the denominator.

8 Simplify $\dfrac{1 - \sqrt{3}}{5 + 2\sqrt{3}}$.

$$\begin{aligned}
\frac{1 - \sqrt{3}}{5 + 2\sqrt{3}} &= \frac{1 - \sqrt{3}}{5 + 2\sqrt{3}} \cdot \frac{5 - 2\sqrt{3}}{5 - 2\sqrt{3}} \\
&= \frac{5 - 2\sqrt{3} - 5\sqrt{3} + 2\sqrt{3^2}}{25 - (2\sqrt{3})^2} \\
&= \frac{5 - 2\sqrt{3} - 5\sqrt{3} + 6}{25 - 4(3)} \\
&= \frac{11 - 7\sqrt{3}}{13}
\end{aligned}$$

Exploratory Exercises

Name the conjugate of each expression.

1. $1 + \sqrt{3}$ **2.** $4 - \sqrt{5}$ **3.** $1 - \sqrt{2}$
4. $5 + 3\sqrt{3}$ **5.** $5 + 2\sqrt{5}$ **6.** $2\sqrt{2} - 3$
7. $2\sqrt{7} - 5$ **8.** $\sqrt{2} - 5\sqrt{3}$ **9.** $\sqrt{7} + \sqrt{2}$

Simplify.

10. $3\sqrt{7} - 4\sqrt{7}$ **11.** $8\sqrt[3]{6} + 3\sqrt[3]{6}$ **12.** $3\sqrt[4]{5} - 10\sqrt[4]{5}$ **13.** $7\sqrt{y} + 4\sqrt{y}$
14. $\sqrt[5]{3} + 4\sqrt[5]{3}$ **15.** $7\sqrt[3]{2} - 3\sqrt[3]{2}$ **16.** $2\sqrt{2} + \sqrt{8}$
17. $5\sqrt[3]{x} + 4\sqrt[3]{x} - 6\sqrt[3]{x}$ **18.** $8\sqrt{5} + \sqrt{75}$ **19.** $\sqrt[3]{40} - 2\sqrt[3]{5}$
20. $(3 + \sqrt{5})(4 + \sqrt{5})$ **21.** $(5 + \sqrt{3})(3 - \sqrt{3})$ **22.** $(3a + \sqrt{5b})(3a - \sqrt{5b})$
23. $(6 + \sqrt{2})(6 - \sqrt{2})$ **24.** $(4 + \sqrt{3})^2$ **25.** $(m + \sqrt{y})^2$

Written Exercises

Simplify.

26. $5\sqrt{2} + 3\sqrt{2} - 8$ **27.** $-3\sqrt{5} + 5\sqrt{2} + 4\sqrt{20} - 3\sqrt{50}$
28. $8\sqrt{3} - 3\sqrt{75}$ **29.** $3\sqrt{7} - 5\sqrt{28}$
30. $5\sqrt{20} + \sqrt{24} - \sqrt{180} + 7\sqrt{54}$ **31.** $7\sqrt[3]{5b} + 4\sqrt[3]{5b}$
32. $\sqrt[3]{48} - \sqrt[3]{6}$ **33.** $8\sqrt[3]{2a} + 3\sqrt[3]{2a} - 8\sqrt[3]{2a}$
34. $\sqrt[3]{54} - \sqrt[3]{128}$ **35.** $7\sqrt[3]{2} + 6\sqrt[3]{150}$ **36.** $5\sqrt[3]{135} - 2\sqrt[3]{81}$
37. $7\sqrt[3]{24} + \sqrt[3]{24}$ **38.** $\sqrt[3]{16} - \sqrt[3]{32}$ **39.** $\sqrt{98} - \sqrt{72} + \sqrt{32}$
40. $\sqrt{108} - \sqrt{48} + (\sqrt{3})^3$ **41.** $7\sqrt[4]{2} + 8\sqrt[4]{2}$ **42.** $\sqrt[4]{5} + 6\sqrt[4]{5} - 2\sqrt[4]{5}$
43. $\sqrt[4]{x^2} + \sqrt[4]{x^6}$ **44.** $-\sqrt{2x^2y^4} + \sqrt{8x^2y^4}$ **45.** $\sqrt[4]{y^4z^6} + \sqrt[4]{16y^4z^6}$
46. $\sqrt[3]{27m^5n^6} + \sqrt[3]{8m^8n^3}$ **47.** $\sqrt[4]{z^4} + \sqrt[3]{z^6} + \sqrt{z^8}$ **48.** $\sqrt{100m^3n} - \sqrt{64mn^3}$
49. $(5 + \sqrt{2})(3 + \sqrt{2})$ **50.** $(4 + \sqrt{3})(3 + \sqrt{6})$ **51.** $(5 + \sqrt{6})(5 - \sqrt{2})$
52. $(8 - \sqrt{3})(6 + \sqrt{3})$ **53.** $(7 + \sqrt{11p})(7 - \sqrt{11p})$ **54.** $(5 - 3\sqrt{5})(3 + \sqrt{5})$
55. $(\sqrt{3} + \sqrt{5})(\sqrt{12} - \sqrt{5})$ **56.** $(4 + \sqrt{5})^2$ **57.** $(1 - \sqrt{3})^2$
58. $(4\sqrt{5} - 3\sqrt{2})(2\sqrt{5} + 2\sqrt{2})$ **59.** $(\sqrt{3a} + \sqrt{2b})(\sqrt{15a} - \sqrt{3b})$
60. $(3 - \sqrt[3]{4})(\sqrt[3]{2} + \sqrt[3]{16})$ **61.** $(4 - \sqrt[3]{9})(\sqrt[3]{3} + \sqrt[3]{81})$
62. $(y + \sqrt[3]{4})(y^2 - y\sqrt[3]{4} + \sqrt[3]{16})$ **63.** $(x - \sqrt[3]{3})(x^2 + x\sqrt[3]{3} + \sqrt[3]{9})$
64. $(m + \sqrt[3]{a})(m^2 - m\sqrt[3]{a} + \sqrt[3]{a^2})$ **65.** $(2 + \sqrt[3]{k})(4 - 2\sqrt[3]{k} + \sqrt[3]{k^2})$

66. $\dfrac{1}{3 + \sqrt{5}}$ **67.** $\dfrac{3}{5 - \sqrt{2}}$ **68.** $\dfrac{2}{3 - \sqrt{5}}$ **69.** $\dfrac{7}{4 - \sqrt{3}}$

70. $\dfrac{1 + \sqrt{2}}{3 - \sqrt{2}}$ **71.** $\dfrac{2 + \sqrt{6}}{2 - \sqrt{6}}$ **72.** $\dfrac{2 - \sqrt{3}}{5 + 3\sqrt{3}}$ **73.** $\dfrac{3 + 4\sqrt{5}}{5 + 2\sqrt{5}}$

74. $\dfrac{\sqrt{x} + 1}{\sqrt{x} - 1}$ **75.** $\dfrac{m + \sqrt{a}}{2\sqrt{a} - p}$ **76.** $\dfrac{1 - 3\sqrt{k}}{5 + 2\sqrt{k}}$ **77.** $\dfrac{\sqrt{3} + n\sqrt{6}}{4 - \sqrt{n}}$

78. $\sqrt{\dfrac{2}{5}} + \sqrt{40} + \sqrt{10}$ **79.** $\sqrt[3]{\dfrac{1}{4}} + \sqrt[3]{54} - \sqrt[3]{16}$ **80.** $\sqrt[3]{\dfrac{2}{3}} + \sqrt[3]{144} - \sqrt[3]{243}$

Factor over the real numbers.

81. $x^2 - 5$ **82.** $a^2 + 4a\sqrt{5} + 20$ **83.** $b^2 - 10b\sqrt{2} + 50$ **84.** $r^3 - 2r$

In mathematics, many words have specific definitions. However, when these words are used in everyday language, they frequently have a different meaning. Study each pair of sentences. How does the meaning of the word in boldface differ?

A. Plants receive nourishment and water from their **roots**.
B. The square **roots** of 36 are 6 and -6.

A. The United States is a major world **power**.
B. Raising a number to the nth **power** means using that number as a factor n times.

A. I am **positive** I left my homework in my locker.
B. For any numbers a and b, and any **positive** integer n, if $a^n = b$, then a is an nth root of b.

A. The **principal** will speak at the school assembly.
B. The symbol $\sqrt[n]{b}$ indicates the **principal** nth root of b.

A. The soup tastes a little **odd**.
B. For any number a and any integer n greater than one, if n is **odd**, then $\sqrt[n]{a^n} = a$.

Read the following property and the paragraph below. Which words are mathematical words? Which words are ordinary words? Which mathematical words have another meaning in everyday language?

> **For any nonnegative real numbers a and b and any integer n, $n > 1$, $\sqrt[n]{ab} = \sqrt[n]{a} \cdot \sqrt[n]{b}$.**

Product Property of Radicals

Simplifying a square root means finding the square root of the greatest perfect square factor of the radicand. You use the Product Property of Radicals to simplify square roots.

Exercises

Write two sentences for each word. First, use the word in everyday language. Then use the word in a mathematical context.

1. index	**2.** negative	**3.** even	**4.** rational
5. irrational	**6.** like	**7.** rationalize	**8.** coordinates
9. real	**10.** degree	**11.** absolute	**12.** identity

6-4 Rational Exponents

The properties of exponents can also be extended to include rational number exponents. Study the examples below.

$$5^{\frac{1}{2}} \cdot 5^{\frac{1}{2}} = 5^{\frac{1}{2}+\frac{1}{2}} \quad \text{or} \quad 5^{\frac{1}{2}} \cdot 5^{\frac{1}{2}} = (5^{\frac{1}{2}})^2$$
$$= 5^1 \qquad\qquad\qquad = 5^1$$
$$= 5 \qquad\qquad\qquad = 5$$

But, it is also true that $\sqrt{5} \cdot \sqrt{5} = 5$. Therefore, the values of $5^{\frac{1}{2}}$ and $\sqrt{5}$ must be the same.

$$4^{\frac{1}{3}} \cdot 4^{\frac{1}{3}} \cdot 4^{\frac{1}{3}} = 4^{\frac{1}{3}+\frac{1}{3}+\frac{1}{3}} \quad \text{or} \quad 4^{\frac{1}{3}} \cdot 4^{\frac{1}{3}} \cdot 4^{\frac{1}{3}} = (4^{\frac{1}{3}})^3$$
$$= 4^1 \qquad\qquad\qquad = 4^1$$
$$= 4 \qquad\qquad\qquad = 4$$

But, it is also true that $\sqrt[3]{4} \cdot \sqrt[3]{4} \cdot \sqrt[3]{4} = 4$. Therefore, the values of $4^{\frac{1}{3}}$ and $\sqrt[3]{4}$ must be the same.

These examples and other similar examples suggest the following definition.

> For any real number b and for any integer n, $n > 1$,
> $$b^{\frac{1}{n}} = \sqrt[n]{b}$$
> except when $b < 0$ and n is even.

Definition of $b^{\frac{1}{n}}$

From the definition you can conclude that $7^{\frac{1}{2}} = \sqrt{7}$ and $(-8)^{\frac{1}{3}} = \sqrt[3]{-8}$ or -2. Since $-16 < 0$, the expression $(-16)^{\frac{1}{4}}$ is not defined.

Examples

1 Evaluate $27^{\frac{1}{3}}$.

$$27^{\frac{1}{3}} = \sqrt[3]{27} \quad \text{or} \quad 27^{\frac{1}{3}} = (3^3)^{\frac{1}{3}}$$
$$= \sqrt[3]{3^3} \qquad\qquad = 3^1$$
$$= 3 \qquad\qquad\quad = 3$$

2 Evaluate $256^{-\frac{1}{4}}$.

$$256^{-\frac{1}{4}} = \frac{1}{256^{\frac{1}{4}}} \quad \text{or} \quad 256^{-\frac{1}{4}} = (4^4)^{-\frac{1}{4}}$$
$$= \frac{1}{\sqrt[4]{256}} \qquad\qquad = 4^{-1}$$
$$= \frac{1}{\sqrt[4]{4^4}} \qquad\qquad = \frac{1}{4}$$
$$= \frac{1}{4}$$

3 Using Calculators

Evaluate $1000^{\frac{1}{3}}$.

ENTER: 1000 $\boxed{y^x}$ $\boxed{(}$ 1 $\boxed{\div}$ 3 $\boxed{)}$ $\boxed{=}$

DISPLAY: *1000*   *1*   *3 0.333333333 10*

The following results show how the properties of exponents can be extended even further.

$$5^{\frac{2}{3}} = (5^{\frac{1}{2}})^3 \quad \text{or} \quad (\sqrt{5})^3 \qquad\qquad 5^{\frac{3}{2}} = (5^3)^{\frac{1}{2}} \text{ or } \sqrt{5^3}$$

Therefore, the values of $(\sqrt{5})^3$ and $\sqrt{5^3}$ must be the same.

$$7^{\frac{2}{3}} = (7^{\frac{1}{3}})^2 \quad \text{or} \quad (\sqrt[3]{7})^2 \qquad\qquad 7^{\frac{2}{3}} = (7^2)^{\frac{1}{3}} \text{ or } \sqrt[3]{7^2}$$

Therefore, the values of $(\sqrt[3]{7})^2$ and $\sqrt[3]{7^2}$ must be the same. These and other similar examples suggest the following definition.

> **For any nonzero real number b, and any integers m and n, with $n > 1$,**
> $$b^{\frac{m}{n}} = \sqrt[n]{b^m} = (\sqrt[n]{b})^m$$
> **except when $b < 0$ and n is even.**

Definition of Rational Exponents

Examples

4 **Evaluate $27^{\frac{2}{3}}$.**

$$27^{\frac{2}{3}} = (\sqrt[3]{27})^2 \quad \text{or} \quad 27^{\frac{2}{3}} = (3^3)^{\frac{2}{3}}$$
$$= (3)^2 \qquad\qquad\qquad = 3^2$$
$$= 9 \qquad\qquad\qquad\quad = 9$$

5 **Evaluate $8^{\frac{1}{3}} \cdot 8^{\frac{4}{3}}$.**

$$8^{\frac{1}{3}} \cdot 8^{\frac{4}{3}} = 8^{\frac{5}{3}}$$
$$8^{\frac{5}{3}} = (\sqrt[3]{8})^5 \quad \text{or} \quad 8^{\frac{5}{3}} = (2^3)^{\frac{5}{3}}$$
$$= (2)^5 \qquad\qquad\qquad = 2^5$$
$$= 32 \qquad\qquad\qquad = 32$$

6 **Using Calculators**

Evaluate $6561^{\frac{3}{4}}$.

ENTER: 6561 $\boxed{y^x}$ $\boxed{(}$ 3 $\boxed{\div}$ 4 $\boxed{)}$ $\boxed{=}$

DISPLAY: *6561 6561   3 3 4 0.75 729*

For a radical to be in simplest radical form, its index must be as small as possible.

Example

7 **Express $\sqrt[6]{25}$ in simplest radical form.**

$$\sqrt[6]{25} = 25^{\frac{1}{6}} \quad \textit{The index is 6. Is it possible to find a smaller index?}$$

$$25^{\frac{1}{6}} = (25^{\frac{1}{2}})^{\frac{1}{3}} \qquad \text{or} \qquad 25^{\frac{1}{6}} = (5^2)^{\frac{1}{6}}$$
$$= (\sqrt{25})^{\frac{1}{3}} \qquad\qquad\qquad\quad = 5^{\frac{1}{3}}$$
$$= 5^{\frac{1}{3}} \qquad\qquad\qquad\qquad\; = \sqrt[3]{5}$$
$$= \sqrt[3]{5}$$

Examples

8 Express $3^{\frac{1}{2}}x^{\frac{2}{3}}y^{\frac{1}{6}}$ in simplest radical form.

$$3^{\frac{1}{2}}x^{\frac{2}{3}}y^{\frac{1}{6}} = 3^{\frac{3}{6}}x^{\frac{4}{6}}y^{\frac{1}{6}}$$
$$= (3^3x^4y^1)^{\frac{1}{6}}$$
$$= \sqrt[6]{27x^4y}$$

Rewrite all exponents using the least common denominator.

9 Express $\sqrt[5]{(32x)^2}$ using rational exponents.

$$\sqrt[5]{(32x)^2} = (32x)^{\frac{2}{5}}$$
$$= (2^5)^{\frac{2}{5}} \cdot x^{\frac{2}{5}}$$
$$= 2^2x^{\frac{2}{5}} \quad \text{or} \quad 4x^{\frac{2}{5}}$$

Exploratory Exercises

Evaluate.

1. $4^{\frac{3}{2}}$ **2.** $9^{\frac{3}{2}}$ **3.** $8^{-\frac{1}{3}}$ **4.** $16^{-\frac{3}{4}}$

5. $(16^{\frac{1}{2}})^{-\frac{1}{2}}$ **6.** $27^{-\frac{2}{3}}$ **7.** $64^{\frac{5}{6}}$ **8.** $64^{-\frac{1}{3}}$

9. $\sqrt[3]{8^2}$ **10.** $343^{\frac{2}{3}}$ **11.** $\sqrt[4]{81}$ **12.** $\sqrt[3]{216}$

13. $(4^{\frac{2}{3}})^3$ **14.** $9^{\frac{1}{3}} \cdot 9^{\frac{5}{3}}$ **15.** $36^{\frac{3}{4}} \div 36^{\frac{1}{4}}$ **16.** $16^{-\frac{3}{2}}$

Express in simplest radical form.

17. $\sqrt[4]{36}$ **18.** $\sqrt[6]{49}$ **19.** $\sqrt[6]{81}$ **20.** $\sqrt[4]{25}$

Written Exercises

Express using rational exponents.

21. $\sqrt{21}$ **22.** $\sqrt[3]{30}$ **23.** $\sqrt[6]{32}$ **24.** $\sqrt[4]{x}$

25. $\sqrt[3]{y}$ **26.** $\sqrt{25x^3y^4}$ **27.** $\sqrt[3]{8m^3r^6}$ **28.** $\sqrt[4]{8x^3y^5}$

29. $\sqrt[4]{27}$ **30.** $\sqrt[3]{16a^5b^7}$ **31.** $\sqrt[3]{n^2}$ **32.** $\sqrt[6]{b^3}$

Express in simplest radical form.

33. $36^{\frac{1}{4}}$ **34.** $5^{\frac{1}{2}}$ **35.** $6^{\frac{1}{3}}$ **36.** $x^{\frac{3}{4}}$ **37.** $a^{\frac{3}{2}}b^{\frac{5}{2}}$

38. $4^{\frac{1}{3}}x^{\frac{2}{3}}y^{\frac{4}{3}}$ **39.** $2^{\frac{5}{3}}x^{\frac{7}{3}}$ **40.** $(2x)^{\frac{1}{2}}x^{\frac{1}{2}}$ **41.** $5^{\frac{1}{3}}p^{\frac{2}{3}}q^{\frac{1}{3}}$ **42.** $(3m)^{\frac{2}{5}}n^{\frac{3}{5}}$

43. $r^{\frac{5}{2}}q^{\frac{3}{4}}$ **44.** $w^{\frac{4}{7}}y^{\frac{3}{7}}$ **45.** $x^{\frac{1}{3}}y^{\frac{1}{2}}$ **46.** $a^{\frac{5}{6}}b^{\frac{3}{2}}x^{\frac{7}{3}}$ **47.** $5^2b^{\frac{1}{2}}c^{\frac{1}{4}}$

48. $x^{\frac{3}{4}}y^{\frac{1}{3}}z^{\frac{5}{6}}$ **49.** $\sqrt[4]{9}$ **50.** $\sqrt[4]{49}$ **51.** $\sqrt[6]{8}$ **52.** $\sqrt[8]{16}$

Evaluate each expression.

53. $121^{\frac{1}{2}}$ **54.** $\left(\frac{1}{32}\right)^{\frac{1}{5}}$ **55.** $\left(\frac{343}{64}\right)^{\frac{1}{3}}$ **56.** $\left(\frac{216}{729}\right)^{\frac{2}{3}}$

57. $\sqrt[3]{12^3}$ **58.** $\sqrt[4]{256}$ **59.** $(6^{\frac{2}{3}})^3$ **60.** $(9^{\frac{3}{4}})^{\frac{2}{3}}$

61. $(0.125)^{\frac{2}{3}}$ **62.** $(0.008)^{\frac{1}{3}}$ **63.** $(0.027)^{\frac{1}{3}}$ **64.** $(0.0016)^{\frac{1}{4}}$

Challenge Exercises

Write each expression in simplest radical form.

65. $\sqrt[3]{2^5} \cdot \sqrt[4]{2}$ **66.** $\sqrt[3]{2^2} \cdot \sqrt[6]{2^7}$ **67.** $\sqrt{3} \cdot \sqrt[3]{3^2}$ **68.** $\sqrt[3]{5^2} \cdot \sqrt{5}$

69. $\sqrt[5]{16} \cdot \sqrt[5]{2}$ **70.** $\sqrt[3]{32} \cdot \sqrt[3]{2}$ **71.** $\sqrt[3]{\sqrt{27}}$ **72.** $\sqrt{\sqrt[3]{36}}$

State the domain and range of each relation. Then state if the relation is a function.

1. $\{(2, -1), (5, 3), (-1, 2), (3, 1)\}$
2. $\{(a, d), (b, z), (c, x), (a, z)\}$

Graph each system of equations and state its solution. Then state whether the system is consistent and independent, consistent and dependent, or inconsistent.

3. $5x + 4y = 6$
$-3x + 5y = -11$

4. $4x + 3y = 9$
$12x + 9y = 27$

5. Determine if this system of equations has a unique solution. If so, solve the system using Cramer's Rule.
$$3x + 2y - 5z = -11$$
$$-3x - 4y + 3z = 3$$
$$7x + 6y - 4z = -14$$

Solve each system of equations by using Cramer's Rule.

6. $39x + 65y = 41$
$-26x + 39y = 17$

7. $45x + 60y = 32$
$120x - 420y = 37$

8. In which octant does $(4, -2, 2)$ lie?

Solve each system of inequalities by graphing. For exercise 10, name the vertices of the polygonal region.

9. $2x + 3y < 6$
$y > -x + 4$

10. $x \geq 0$
$y \geq 1$
$x + y \leq 4$

Find the inverse of each matrix.

11. $\begin{bmatrix} 4 & 6 \\ -3 & 2 \end{bmatrix}$

12. $\begin{bmatrix} 1 & -6 & 8 \\ 3 & 5 & 2 \\ 2 & 4 & 1 \end{bmatrix}$

Solve the following matrix equation.

13. $\begin{bmatrix} 5 & 4 \\ 3 & -2 \end{bmatrix} \cdot \begin{bmatrix} x \\ y \end{bmatrix} = \begin{bmatrix} 7 \\ 13 \end{bmatrix}$

Simplify.

14. $\left(\dfrac{2}{3}c^2f\right)\left(\dfrac{3}{8}cd\right)$

15. $\dfrac{-4p^{-3}y^4}{(2y^{-2})^{-2}}$

16. $\sqrt[3]{\dfrac{4}{5x^2}}$

17. $(r + s)^3$

18. $(4x + 11) - (3x^2 + 7x - 3)$

19. $\sqrt[3]{27x^6y^3}$

20. $\sqrt[3]{135r^7s^5}$

21. $(\sqrt{3})(\sqrt{8})$

22. $\dfrac{1 + \sqrt{3}}{2 - \sqrt{3}}$

Factor.

23. $9a^3b^2 + 12ab^3$

24. $z^2 + 10z + 16$

25. $2r + 2s + rt + st$

26. $m^3 - 125$

27. $6a^2 + 4ay + 9ab + 6by$

Simplify using long division.

28. $(10m^3 - 21m^2 + 18m - 5) \div (2m + 3)$

29. $(6x^3 - 16x^2 + 5x - 4) \div (3x - 2)$

Simplify using synthetic division.

30. $(8m^3 + 50m^2 + 47m - 15) \div (m + 5)$

31. Show that $(x - 5)$ is a factor of $6x^3 - 41x^2 + 48x + 35$.

Solve each problem.

32. The length of a rectangle is 48 meters more than twice its width. Its perimeter is 570 meters. Find the dimensions.

33. Thirty-four decreased by five times a number is -6. Find the number.

34. The Marinacci family has invested $35,000. Part of the money is invested at 8% and part at 10%. If the total income from these investments is $3,100 how much is invested at each rate?

 Simplifying Expressions with Rational Exponents

A rational expression that contains a fractional exponent in the denominator must also be rationalized.

Examples

1 Simplify $\dfrac{1}{3^{\frac{1}{2}}}$.

$$\dfrac{1}{3^{\frac{1}{2}}} = \dfrac{1}{3^{\frac{1}{2}}} \cdot 1$$

$$= \dfrac{1}{3^{\frac{1}{2}}} \cdot \dfrac{3^{\frac{1}{2}}}{3^{\frac{1}{2}}} \qquad \textit{Why is } \dfrac{3^{\frac{1}{2}}}{3^{\frac{1}{2}}} \textit{ chosen?}$$

$$= \dfrac{3^{\frac{1}{2}}}{3}$$

2 Simplify $\dfrac{1}{4^{\frac{1}{3}}}$.

$$\dfrac{1}{4^{\frac{1}{3}}} = \dfrac{1}{4^{\frac{1}{3}}} \cdot 1$$

$$= \dfrac{1}{4^{\frac{1}{3}}} \cdot \dfrac{4^{\frac{2}{3}}}{4^{\frac{2}{3}}} \qquad \textit{Why is } \dfrac{4^{\frac{2}{3}}}{4^{\frac{2}{3}}} \textit{ chosen?}$$

$$= \dfrac{4^{\frac{2}{3}}}{4}$$

It is important to choose a multiplier carefully. Study the following ways to simplify $\dfrac{1}{5^{\frac{3}{2}}}$.

$$\dfrac{1}{5^{\frac{3}{2}}} = \dfrac{1}{5^{\frac{3}{2}}}\left(\dfrac{5^{\frac{3}{2}}}{5^{\frac{3}{2}}}\right)$$

$$= \dfrac{5^{\frac{3}{2}}}{5^3}$$

$$= \dfrac{5 \cdot 5^{\frac{1}{2}}}{5 \cdot 5^2} \text{ or } \dfrac{5^{\frac{1}{2}}}{5^2}$$

$$\dfrac{1}{5^{\frac{3}{2}}} = \dfrac{1}{5^{\frac{3}{2}}}\left(\dfrac{5^{\frac{1}{2}}}{5^{\frac{1}{2}}}\right)$$

$$= \dfrac{5^{\frac{1}{2}}}{5^2}$$

Notice that there are fewer steps in simplifying $\dfrac{1}{5^{\frac{3}{2}}}$ when 1 is written in the form $\dfrac{5^{\frac{1}{2}}}{5^{\frac{1}{2}}}$.

An expression is simplified when it meets these conditions.

1. **It has no negative exponents.**
2. **It has no fractional exponents in the denominator.**
3. **It is not a complex fraction.**
4. **The index of any remaining radical is as small as possible.**

Conditions for Simplified Expressions

When you simplify an expression, be sure your answer meets all of the above conditions. In some problems, the simplest form may not always be the most convenient to use. Sometimes the content of a problem determines the most appropriate form for the answer.

3 Simplify $r^{-\frac{1}{9}}$.

$$r^{-\frac{1}{9}} = \frac{1}{r^{\frac{1}{9}}}$$
$$= \frac{1}{r^{\frac{1}{9}}}\left(\frac{r^{\frac{8}{9}}}{r^{\frac{8}{9}}}\right)$$
$$= \frac{r^{\frac{8}{9}}}{r}$$

Why choose to multiply by $\dfrac{r^{\frac{8}{9}}}{r^{\frac{8}{9}}}$?

4 Simplify $\dfrac{a^{\frac{1}{2}} - b^{\frac{1}{2}}}{a^{\frac{1}{2}} + b^{\frac{1}{2}}}$.

$$\frac{a^{\frac{1}{2}} - b^{\frac{1}{2}}}{a^{\frac{1}{2}} + b^{\frac{1}{2}}} = \frac{a^{\frac{1}{2}} - b^{\frac{1}{2}}}{a^{\frac{1}{2}} + b^{\frac{1}{2}}} \cdot \frac{a^{\frac{1}{2}} - b^{\frac{1}{2}}}{a^{\frac{1}{2}} - b^{\frac{1}{2}}}$$
$$= \frac{a - 2a^{\frac{1}{2}}b^{\frac{1}{2}} + b}{a - b}$$

Exploratory Exercises

State a factor that can be used to rationalize each expression.

1. $\dfrac{6}{3^{\frac{1}{2}}}$ 2. $\dfrac{10}{5^{\frac{2}{3}}}$ 3. $\dfrac{16}{4^{\frac{3}{2}}}$ 4. $\dfrac{1}{x^{\frac{1}{3}}}$ 5. $\dfrac{1}{y^{\frac{2}{3}}}$ 6. $a^{-\frac{1}{5}}$

7. $p^{-\frac{3}{2}}$ 8. $\dfrac{1}{x^{\frac{1}{2}} + 1}$ 9. $\dfrac{m + p}{m^{\frac{1}{2}} + p}$ 10. $\dfrac{r}{r^{\frac{1}{2}} - s^{\frac{1}{2}}}$ 11. $\dfrac{2}{t^{\frac{3}{2}} + s^{\frac{1}{2}}}$ 12. $\dfrac{1}{b^{\frac{3}{2}} + b^{\frac{1}{2}}}$

Written Exercises

13–24. Simplify each expression in Exploratory Exercises 1–12.

Simplify.

25. $\dfrac{1}{y^{\frac{2}{5}}}$ 26. $\dfrac{3}{r^{\frac{4}{5}}}$ 27. $b^{-\frac{1}{4}}$ 28. $m^{-\frac{5}{6}}$

29. $\dfrac{15}{5^{\frac{2}{3}}}$ 30. $\dfrac{24}{6^{\frac{2}{3}}}$ 31. $\dfrac{rm^{\frac{1}{2}}}{b^{\frac{3}{2}}}$ 32. $\dfrac{pq}{\sqrt[3]{a}}$

33. $\dfrac{b^{\frac{3}{2}} + 3b^{-\frac{1}{2}}}{b^{\frac{1}{2}}}$ 34. $\dfrac{a^{\frac{5}{3}}m + 3a^{-\frac{1}{3}}}{a^{\frac{2}{3}}}$ 35. $\dfrac{3x + 4x^2}{x^{-\frac{2}{3}}}$ 36. $\dfrac{3m}{b^{-\frac{3}{2}} \cdot \sqrt[3]{a}}$

37. $(r^{-\frac{1}{6}})^{-\frac{2}{3}}$ 38. $(y^{\frac{1}{3}})^{-\frac{3}{4}}$ 39. $\dfrac{r^{\frac{3}{2}}}{r^{\frac{1}{2}} + 2}$ 40. $\dfrac{x^{\frac{1}{2}} + y^{\frac{1}{2}}}{x^{\frac{1}{2}} - y^{\frac{1}{2}}}$

41. $\dfrac{m^{\frac{1}{3}} - p^{\frac{1}{3}}}{m^{\frac{1}{2}} + p^{\frac{3}{2}}}$ 42. $\dfrac{x^{\frac{1}{2}} + 1}{x^{\frac{1}{2}} - 1}$ 43. $\dfrac{rs}{r^{\frac{1}{2}} + r^{\frac{3}{2}}}$ 44. $\dfrac{x^{\frac{1}{3}}}{x^{\frac{2}{3}} - x^{-\frac{1}{3}}}$

45. $\dfrac{b^{\frac{1}{2}}}{b^{\frac{3}{2}} - b^{\frac{1}{2}}}$ 46. $\dfrac{a^{-\frac{2}{3}}b^{\frac{1}{2}}}{b^{-\frac{3}{2}} \cdot \sqrt[3]{a}}$ 47. $\left(\dfrac{x^{-2}y^{-6}}{9}\right)^{-\frac{1}{2}}$ 48. $\left(\dfrac{z^{-\frac{2}{3}}}{5^{-1}z^{\frac{1}{3}}}\right)^{-2}$

49. $(\sqrt[6]{5}x^{\frac{7}{4}}y^{-\frac{2}{3}})^{12}$ 50. $\dfrac{8^{\frac{1}{6}} - 9^{\frac{1}{4}}}{\sqrt{3} + \sqrt{2}}$ 51. $\dfrac{9x^{-\frac{4}{3}} - 4y^{-2}}{3x^{-\frac{2}{3}} + 2y^{-1}}$ 52. $\dfrac{a^{\frac{5}{3}} - a^{\frac{1}{3}}b^{\frac{4}{3}}}{a^{\frac{2}{3}} + b^{\frac{2}{3}}}$

53. Evaluate $-\dfrac{4}{9}x^9\left(\dfrac{3}{x^2} - \dfrac{1}{\sqrt[3]{2}}\right)$ when $x = \sqrt[6]{2}$.

54. Evaluate $\dfrac{3^0 y + 4y^{-1}}{y^{-\frac{2}{3}}}$ when $y = 8$.

Challenge Exercises

Simplify.

55. $\dfrac{1}{x^{\frac{1}{3}} - y^{\frac{1}{3}}}$ 56. $\dfrac{a}{a^{\frac{1}{3}} - b^{\frac{2}{3}}}$ 57. $\dfrac{1}{a^{\frac{2}{3}} - b^{\frac{2}{3}}}$ 58. $\dfrac{z^{\frac{1}{3}}}{z^{\frac{2}{3}} - z^{\frac{1}{3}}}$

6-6 Equations with Radicals

The properties of radicals can be used to solve equations.

1 Solve $x + 2 = x\sqrt{3}$.

$$x + 2 = x\sqrt{3}$$
$$x - x\sqrt{3} = -2$$
$$x(1 - \sqrt{3}) = -2 \qquad \textit{Distributive Property}$$
$$x = \frac{-2}{1 - \sqrt{3}} \qquad \textit{The conjugate of } 1 - \sqrt{3} \textit{ is } 1 + \sqrt{3}.$$
$$x = \frac{-2}{1 - \sqrt{3}} \cdot \frac{1 + \sqrt{3}}{1 + \sqrt{3}}$$
$$x = \frac{-2(1 + \sqrt{3})}{1^2 - (\sqrt{3})^2}$$
$$x = \frac{-2(1 + \sqrt{3})}{-2} \text{ or } x = 1 + \sqrt{3}$$

Check:
$$x + 2 = x\sqrt{3}$$
$$(1 + \sqrt{3}) + 2 \stackrel{?}{=} (1 + \sqrt{3})\sqrt{3}$$
$$\sqrt{3} + 3 = \sqrt{3} + 3$$

The solution is $1 + \sqrt{3}$.

Variables may appear in the radicand of a radical. Equations containing such radicals are called **radical equations**.

Squaring each side of an equation may produce results that do *not* satisfy the original equation.

You must check all possible solutions in the original equation.

$$x = 2 \qquad \text{This equation has } \textit{one} \text{ solution, 2.}$$
$$(x)^2 = (2)^2 \qquad \text{Square each side.}$$
$$x^2 = 4 \qquad \text{This equation has } \textit{two} \text{ solutions, 2 and } -2.$$

2 Solve $7 + \sqrt{a - 3} = 1$.

$$7 + \sqrt{a - 3} = 1$$
$$\sqrt{a - 3} = -6 \qquad \textit{Isolate the radical.}$$
$$(\sqrt{a - 3})^2 = (-6)^2 \qquad \textit{Square each side.}$$
$$a - 3 = 36$$
$$a = 39$$

Check: $7 + \sqrt{a - 3} = 1$
$$7 + \sqrt{39 - 3} \stackrel{?}{=} 1$$
$$7 + \sqrt{36} \stackrel{?}{=} 1$$
$$13 \neq 1$$

The answer does *not* check.
The equation has *no* solutions.

3 Solve $3 - \sqrt{x - 2} = 0$.

$$3 - \sqrt{x - 2} = 0$$
$$3 = \sqrt{x - 2}$$
$$3^2 = (\sqrt{x - 2})^2 \qquad \textit{Square each side.}$$
$$9 = x - 2$$
$$11 = x$$

Check: $3 - \sqrt{x - 2} = 0$
$$3 - \sqrt{11 - 2} \stackrel{?}{=} 0$$
$$3 - \sqrt{9} \stackrel{?}{=} 0$$
$$3 - 3 \stackrel{?}{=} 0$$
$$0 = 0$$

The solution is 11.

4 **Solve $\sqrt{2y - 3} - \sqrt{2y + 3} = -1$.**

$$\sqrt{2y - 3} - \sqrt{2y + 3} = -1$$
$$\sqrt{2y - 3} = \sqrt{2y + 3} - 1 \qquad \textit{Isolate one radical.}$$
$$2y - 3 = 2y + 3 - 2\sqrt{2y + 3} + 1 \qquad \textit{Square each side.}$$
$$-7 = -2\sqrt{2y + 3} \qquad \textit{Note middle term from expansion.}$$
$$\frac{7}{2} = \sqrt{2y + 3} \qquad \textit{Isolate the remaining radical.}$$
$$\frac{49}{4} = 2y + 3 \qquad \textit{Square each side.}$$
$$\frac{37}{4} = 2y$$
$$y = \frac{37}{8}$$

The solution is $\frac{37}{8}$ or $4\frac{5}{8}$.

Check: $\quad \sqrt{2y - 3} - \sqrt{2y + 3} = -1$

$$\sqrt{2\left(\frac{37}{8}\right) - 3} - \sqrt{2\left(\frac{37}{8}\right) + 3} \stackrel{?}{=} -1$$
$$\sqrt{\frac{25}{4}} - \sqrt{\frac{49}{4}} \stackrel{?}{=} -1$$
$$\frac{5}{2} - \frac{7}{2} \stackrel{?}{=} -1$$
$$-1 = -1$$

5 **Solve $\sqrt[3]{3y - 1} - 2 = 0$.**

$$\sqrt[3]{3y - 1} - 2 = 0$$
$$\sqrt[3]{3y - 1} = 2$$
$$(\sqrt[3]{3y - 1})^3 = 2^3 \qquad \textit{Cube each side.}$$
$$3y - 1 = 8$$
$$3y = 9$$
$$y = 3$$

The solution is 3.

Check: $\quad \sqrt[3]{3y - 1} - 2 = 0$

$$\sqrt[3]{3(3) - 1} - 2 \stackrel{?}{=} 0$$
$$\sqrt[3]{8} - 2 \stackrel{?}{=} 0$$
$$2 - 2 \stackrel{?}{=} 0$$
$$0 = 0$$

6 **Solve $r = \sqrt[3]{\dfrac{3w}{4\pi d}}$ for d.**

$$r = \sqrt[3]{\frac{3w}{4\pi d}}$$
$$r^3 = \frac{3w}{4\pi d} \qquad \textit{Cube each side.}$$
$$r^3 \cdot d = \frac{3w}{4\pi d} \cdot d \qquad \textit{Multiply each side by d.}$$
$$\frac{r^3 d}{r^3} = \frac{3w}{4\pi r^3} \qquad \textit{Divide each side by } r^3.$$
$$d = \frac{3w}{4\pi r^3}.$$

Solve each equation.

1. $\sqrt{x} = 2$
2. $\sqrt{y} = 3$
3. $\sqrt{m} - 8 = 0$
4. $\sqrt{t} - 4 = 0$
5. $\sqrt{2x + 7} = 3$
6. $\sqrt{3x + 7} = 7$
7. $\sqrt[3]{x - 2} = 3$
8. $\sqrt[4]{2x + 7} = 2$
9. $x\sqrt{3} - x = 7$
10. $x\sqrt{5} + x = 3$
11. $x\sqrt{3} + 4 = 7 + \sqrt{3}$
12. $2z\sqrt{7} + 3 = 5 + 6\sqrt{7}$

Written Exercises

Solve each equation.

13. $1 + x\sqrt{2} = 0$
14. $3x\sqrt{3} + 2 = 0$
15. $7 + 6x\sqrt{5} = 0$
16. $6 + 2x\sqrt{3} = 0$
17. $2 + 5n\sqrt{10} = 0$
18. $x\sqrt{2} + 3x = 4$
19. $x - x\sqrt{5} = 2$
20. $3x + 5 = x\sqrt{3}$
21. $2x + 7 = -x\sqrt{2}$
22. $2x - x\sqrt{11} = 13$
23. $13 - 3x = x\sqrt{5}$
24. $\sqrt{y - 5} - 7 = 0$
25. $\sqrt{x - 4} - 3 = 0$
26. $\sqrt[3]{y + 1} = 2$
27. $\sqrt[3]{m - 1} = 3$
28. $\sqrt[4]{2a} = 3$
29. $\sqrt[4]{3p} - 2 = 0$
30. $\sqrt{2x + 3} - 7 = 0$
31. $\sqrt{3y - 5} - 3 = 1$
32. $\sqrt{5y + 1} + 6 = 10$
33. $\sqrt{1 + 2r} - 4 = -1$
34. $\sqrt{3x + 1} - 2 = 6$
35. $\sqrt{4a + 8} + 5 = 7$
36. $\sqrt[3]{m + 5} + 6 = 4$
37. $\sqrt[4]{2x + 3} + 5 = 4$
38. $\sqrt{x + 5} = \sqrt{2x - 3}$
39. $\sqrt{x - 4} = \sqrt{2x - 3}$
40. $\sqrt{x - 5} - \sqrt{x} = 1$
41. $\sqrt{m + 12} - \sqrt{m} = 2$
42. $\sqrt{b + 4} = \sqrt{b + 20} - 2$
43. $\sqrt{y + 6} - \sqrt{y} = \sqrt{2}$
44. $\sqrt{x - 1} + \sqrt{x + 3} = 5$
45. $\sqrt{x + 1} + \sqrt{x - 3} - 5 = 0$
46. $\sqrt{y^2 + 5y} + y + 10 = 0$
47. $\sqrt{4x + 1} - \sqrt{4x - 2} = 3$
48. $\sqrt{5x^2 + 7x - 2} - x\sqrt{5} + 4 = 0$
49. $\sqrt{3x^2 + 11x - 5} = x\sqrt{3} + 1$
50. $\sqrt{x + 8} + 3 = \sqrt{x + 35}$
51. $\sqrt{y + 12} + 1 = \sqrt{y + 21}$
52. $\sqrt{x - 4} - 6 = \sqrt{x + 20}$

Solve each equation for the variable indicated.

53. $y = \sqrt{r^2 + s^2}$ for r
54. $t = \sqrt{\dfrac{2s}{g}}$ for s
55. $r = \sqrt[3]{\dfrac{2mM}{c}}$ for c
56. $T = \dfrac{1}{2}\sqrt{\dfrac{u}{g}}$ for g
57. $v = \dfrac{1}{2}\sqrt{1 + \dfrac{T}{\ell}}$ for ℓ
58. $m^2 = \sqrt[3]{\dfrac{rp}{g^2}}$ for p

Solve.

59. Find the radius of a circle whose area is 616 cm^2. Let $A = \pi r^2$. Use $\frac{22}{7}$ as an approximation for π.

60. The area of a square is 54 m^2. Use the formula $A = s^2$ to approximate the length of each side of the square.

mini-review

Find an equation of the line that passes through each given point and is parallel to the line with the given equation.

1. $(2, 1)$; $y = x - 2$

2. $(6, 6)$; $3x + y = -6$

Solve each system of equations using Cramer's Rule.

3. $6x - 7y = 10$
$8x - 2y = 4$

4. $x + 3y = -5$
$x - 2y = 8$

Find the transpose and the inverse (if one exists) for each matrix.

5. $\begin{bmatrix} -2 & 2 & 6 \\ 4 & 3 & 7 \end{bmatrix}$

6. $\begin{bmatrix} 0 & 12 \\ -1 & 8 \\ 5 & -4 \end{bmatrix}$

7. $\begin{bmatrix} -6 & 5 & 4 \\ 1 & 0 & 8 \\ 2 & 3 & -4 \end{bmatrix}$

Solve each system of equations using the augmented matrix method.

8. $x + 3y - z = -5$
$3x + 2y + 2z = 15$
$-4x - 5y - 3z = 8$

9. $2x - 2y + z = 12$
$x + y - z = -6$
$4x + y + 2z = 9$

Simplify.

10. $\dfrac{c^2}{c^6}$

11. $\dfrac{8d^3 f^5}{(2df)^{-2}}$

12. $\dfrac{(5x^2)^0}{5x^2}$

Factor.

13. $12x^2 - 17x - 5$

14. $6x^2 + 13x - 5$

15. $4a^2 + 4ab - 9 + b^2$

16. Simplify $(14x^3 + 81x^2 + 40x - 75) \div (2x + 3)$ using long division.

Express in simplest radical form.

17. $\sqrt[3]{3^4} \cdot \sqrt[5]{3^2}$

18. $a^{\frac{1}{6}} b^{\frac{2}{3}} c^4$

19. $\sqrt[4]{100}$

Solve each problem. Round all answers to the nearest hundredth.

20. Find the length of the hypotenuse of a right triangle whose legs, a and b, measure 13 and 12 meters. Let $c^2 = a^2 + b^2$, where c is the hypotenuse.

21. Find the time, t, in seconds required for a freely falling body to fall a distance of 210 feet. Let $t = \frac{1}{4}\sqrt{s}$.

6-7 Pure Imaginary Numbers

Some equations have irrational solutions. For example, the solutions to $x^2 - 5 = 0$ are $\sqrt{5}$ and $-\sqrt{5}$.

$\sqrt{5}$ and π are examples of irrational numbers.

The equation $x^2 = -1$ has *no* solution among the real numbers. This is because the square of a real number is nonnegative.

The number i is defined to be a solution to $x^2 = -1$ and is *not* a real number. It is called the **imaginary unit**.

Using i as you would any constant, you can define square roots of negative numbers.

$$i^2 = -1 \quad \text{so} \quad \sqrt{-1} = i$$
$$(2i)^2 = 2^2 i^2 \text{ or } -4 \quad \text{so} \quad \sqrt{-4} = \sqrt{4} \cdot \sqrt{-1} \text{ or } 2i$$
$$(i\sqrt{3})^2 = i^2(\sqrt{3})^2 \text{ or } -3 \quad \text{so} \quad \sqrt{-3} = \sqrt{3} \cdot \sqrt{-1} \text{ or } i\sqrt{3}$$

To avoid confusion between $\sqrt{3}i$ and $\sqrt{3i}$, express $\sqrt{3}i$ as $i\sqrt{3}$.

> For any positive real number b,
> $$\sqrt{-(b^2)} = \sqrt{b^2} \cdot \sqrt{-1} \text{ or } bi$$
> where i is the imaginary unit, and bi is called a pure imaginary number.

Definition of Pure Imaginary Number

Pure imaginary numbers are simplified by rewriting them as the product of i and a real number. For example, $3i$, $-i\sqrt{3}$, and $5i\sqrt{10}$ are in simplest form.

Examples

1 Simplify $\sqrt{-16}$.
$$\sqrt{-16} = \sqrt{16} \cdot \sqrt{-1}$$
$$= 4 \cdot i$$
$$= 4i$$

2 Simplify $\sqrt{-24}$.
$$\sqrt{-24} = \sqrt{24} \cdot \sqrt{-1}$$
$$= (\sqrt{4} \cdot \sqrt{6})i$$
$$= 2i\sqrt{6}$$

Pure imaginary numbers can be multiplied by using the Commutative and Associative Properties for Multiplication.

Examples

3 Simplify $3i \cdot 5i$.
$$3i \cdot 5i = (3 \cdot 5)(i \cdot i)$$
$$= 15i^2$$
$$= 15(-1) \quad i^2 = -1$$
$$= -15$$

4 Simplify $\sqrt{-3} \cdot \sqrt{-12}$.

Change $\sqrt{-3}$ and $\sqrt{-12}$ to imaginary form before multiplying.

$$\sqrt{-3} \cdot \sqrt{-12} = i\sqrt{3} \cdot i\sqrt{12}$$
$$= i^2\sqrt{36}$$
$$= -1 \cdot 6 \text{ or } -6$$

Simplifying powers of i reveals an interesting pattern.

$$i^1 = i \qquad\qquad i^5 = i^4 \cdot i = 1 \cdot i = i$$
$$i^2 = -1 \qquad\qquad i^6 = i^4 \cdot i^2 = 1 \cdot (-1) = -1$$
$$i^3 = i^2 \cdot i = -1 \cdot i = -i \qquad i^7 = i^4 \cdot i^3 = 1 \cdot (-i) = -i$$
$$i^4 = i^2 \cdot i^2 = -1 \cdot (-1) = 1 \qquad i^8 = i^4 \cdot i^4 = 1 \cdot 1 = 1$$

The values i, -1, $-i$, and 1 repeat in cycles of four.

Examples

5 **Simplify i^{15}.**

$$i^{15} = i^4 \cdot i^4 \cdot i^4 \cdot i^3 \quad \text{or} \quad i^{15} = i^{14} \cdot i^1$$
$$= 1 \cdot 1 \cdot 1 \cdot (-i) \qquad\qquad = (i^2)^7 \cdot i$$
$$= -i \qquad\qquad\qquad = (-1)^7 \cdot i$$
$$\qquad\qquad\qquad\qquad = -i$$

6 **Solve $x^2 + 5 = 0$**

$$x^2 + 5 = 0$$
$$x^2 = -5$$
$$x = \pm\sqrt{-5} \qquad \sqrt{-5} = \sqrt{-1} \cdot \sqrt{5}$$
$$= \pm i\sqrt{5}$$

The solutions, both pure imaginary numbers, are $i\sqrt{5}$ and $-i\sqrt{5}$.

Exploratory Exercises

Simplify.

1. $\sqrt{-36}$
2. $\sqrt{-64}$
3. $4\sqrt{-2}$
4. $6\sqrt{-4}$
5. $\sqrt{-3} \cdot \sqrt{-3}$
6. $\sqrt{-2} \cdot \sqrt{-2}$
7. $\sqrt{-5} \cdot \sqrt{5}$
8. $\sqrt{-7} \cdot \sqrt{7}$
9. $3 \cdot 2i$
10. $5 \cdot 7i$
11. i^6
12. i^{91}

Written Exercises

Simplify.

13. $\sqrt{-81}$
14. $\sqrt{-121}$
15. $\sqrt{-50}$
16. $\sqrt{-98}$
17. $\sqrt{\dfrac{-4}{9}}$
18. $\sqrt{\dfrac{-9}{16}}$
19. $\sqrt{\dfrac{-1}{3}}$
20. $\sqrt{\dfrac{-1}{2}}$
21. i^5
22. i^{10}
23. i^{11}
24. i^{43}
25. i^{71}
26. i^{112}
27. i^{82}
28. i^{243}
29. $\sqrt{-8} \cdot \sqrt{-2}$
30. $\sqrt{-15} \cdot \sqrt{-5}$
31. $\sqrt{-14} \cdot \sqrt{-7}$
32. $\sqrt{-3} \cdot \sqrt{-18}$
33. $(\sqrt{-3})^2$
34. $(\sqrt{-12})^2$
35. $(\sqrt{-3})^3$
36. $(\sqrt{-4})^3$
37. $(-2\sqrt{-8})(3\sqrt{-2})$
38. $(4\sqrt{-12})(-2\sqrt{-3})$
39. $(6\sqrt{-24})(-3\sqrt{6})$
40. $(2\sqrt{15})(-3\sqrt{-15})$
41. $(2i)(3i)^2$
42. $5i(-2i)^2$

Solve each equation.

43. $x^2 + 16 = 0$
44. $a^2 + 49 = 0$
45. $z^2 + 169 = 0$
46. $c^2 + 144 = 0$
47. $n^2 + 3 = 0$
48. $t^2 + 12 = 0$
49. $2y^2 + 8 = 0$
50. $3b^2 + 18 = 0$
51. $5x^2 + 125 = 0$
52. $3z^2 + 24 = 0$
53. $4m^2 + 5 = 0$
54. $9k^2 + 32 = 0$

6-8 Complex Numbers

Numbers such as $5i$ and $27 - 2i$ are imaginary numbers since they contain the imaginary unit, i. The real numbers together with the imaginary numbers form the set of **complex numbers**.

> **A complex number is any number that can be written in the form $a + bi$ where a and b are real numbers and i is the imaginary unit. a is called the real part, and bi is called the imaginary part.**

Definition of Complex Number

Any real number is also a complex number. For example, $\sqrt{2}$ can be written as $\sqrt{2} + 0i$. Its imaginary part is $0i$ or 0. A complex number is a real number *only if its imaginary part is 0.*

Any two complex numbers denoted by $a + bi$ and $c + di$, are equal if and only if their real parts are equal and their imaginary parts are equal. That is,

$$a + bi = c + di \text{ if and only if } a = c \text{ and } b = d.$$

If $b \neq 0$, the complex number $a + bi$ is also called an imaginary number.

Example

1 **Find values for x and y such that $2x + 3yi = 6 + 2i$.**

$$2x + 3yi = 6 + 2i$$
$$2x = 6 \quad \text{and} \quad 3y = 2$$
$$x = 3 \qquad y = \frac{2}{3}$$

Check: $2(3) + 3\left(\frac{2}{3}\right)i \stackrel{?}{=} 6 + 2i$

$6 + 2i = 6 + 2i$

To add or subtract complex numbers, combine their real parts and combine their imaginary parts. To multiply complex numbers, use the FOIL method.

Examples

2 **Simplify $(3 + 6i) + (7 - 2i)$.**

$(3 + 6i) + (7 - 2i)$
$\quad = (3 + 7) + (6i - 2i)$
$\quad = 10 + 4i$

3 **Simplify $(6 - 5i) - (3 - 2i)$.**

$(6 - 5i) - (3 - 2i)$
$\quad = (6 - 3) + (-5i - (-2i))$
$\quad = 3 - 3i$

4 **Simplify $(6 - 7i)(4 + 3i)$.**

$(6 - 7i)(4 + 3i) = 6 \cdot 4 + 6 \cdot 3i + (-7i) \cdot 4 + (-7i) \cdot 3i$ *FOIL Method*
$\qquad = 24 + 18i - 28i - 21i^2$
$\qquad = (24 + 21) + (18i - 28i)$ $-21i^2 = 21$
$\qquad = 45 - 10i$

This chart summarizes addition, subtraction, and multiplication of complex numbers.

For any complex numbers $a + bi$ and $c + di$:
$(a + bi) + (c + di) = (a + c) + (b + d)i$
$(a + bi) - (c + di) = (a - c) + (b - d)i$
$(a + bi)(c + di) = (ac - bd) + (ad + bc)i$

Exploratory Exercises

Simplify.

1. $(6 + 3i) + (2 + 8i)$
2. $(4 - i) + (3 + 3i)$
3. $(5 + 2i) - (2 + 2i)$
4. $(7 - 6i) - (5 - 6i)$
5. $(7 + 3i) + (3 - 3i)$
6. $(2 + 4i) + (2 - 4i)$
7. $4(5 + 3i)$
8. $-6(2 - 3i)$
9. $(1 + 3i)(2 + 4i)$
10. $(2 - 3i)(1 - 4i)$
11. $(3 + 2i)(4 - i)$
12. $(6 + i)(6 - i)$

Find values of x and y for which each equation is true.

13. $x - yi = 5 + 6i$
14. $x + yi = 2 - 3i$
15. $x - yi = 7 - 2i$
16. $x - yi = 4 + 5i$
17. $x + 2yi = 3$
18. $2x + yi = 5i$

Written Exercises

Simplify.

19. $(3 + 2i) + (4 + 5i)$
20. $(2 + 6i) + (4 + 3i)$
21. $(9 + 6i) - (3 + 2i)$
22. $(11 - \sqrt{-3}) - (-4 + \sqrt{-5})$
23. $(5 + \sqrt{-7}) + (-3 + \sqrt{-2})$
24. $(8 - 7i) + (-5 - i)$
25. $(3 - 11i) - (-5 + 4i)$
26. $(-6 - 2i) - (-8 - 3i)$
27. $(4 + 2i\sqrt{3}) + (1 - 5i\sqrt{3})$
28. $(8 - 3i\sqrt{5}) + (-3 + 2i\sqrt{5})$
29. $2(-3 + 2i) + 3(-5 - 2i)$
30. $-6(2 - i) + 3(4 - 5i)$
31. $(2 - 3i)(5 + i)$
32. $(5 + 3i)(6 - i)$
33. $(6 - 2i)^2$
34. $(2 + i\sqrt{3})^2$
35. $(7 - i\sqrt{2})(5 + i\sqrt{2})$
36. $(4 - 3i)(7 - 2i)$
37. $(3 + 2i)^2$
38. $(3 + 4i)^2$
39. $(\sqrt{2} + i)(\sqrt{2} - i)$
40. $(3 + 2i)(3 - 2i)$
41. $(2 - \sqrt{-3})(2 + \sqrt{-3})$
42. $(3 + \sqrt{-2})(3 - \sqrt{-2})$
43. $(2 + i)(3 - 4i)(1 + 2i)$
44. $(6 - i)(5 + 2i)(3 + 3i)$
45. $(7 - 5i)(2 - 3i)(7 + 5i)$
46. $(9 + 2i)(5 + i)(9 - 2i)$
47. $(4 + 3i)(2 - 7i)(3 + i)$
48. $(7 - i)(4 + 2i)(5 + 2i)$

Find values of x and y for which each sentence is true.

49. $2x + 5yi = 4 + 15i$
50. $3x + 2yi = 18 + 7i$
51. $(x - y) + (x + y)i = 2 - 4i$
52. $(2x + y) + (x - y)i = 7 - i$
53. $(x + 2y) + (2x - y)i = 5 + 5i$
54. $(x + 4y) + (2x - 3y)i = 13 + 7i$

55. Write an expression for the additive inverse of the complex number $a + bi$.

56. Show that 0 is the additive identity for complex numbers.

57. Show that 1 is the multiplicative identity for complex numbers.

In which octant does each point lie?

1. $(-4, 3, 2)$ **2.** $(5, 6, -2)$ **3.** $(-1, -6, -8)$

4. Find the x-, y-, and z-intercepts for $4x + 6y - 4z = 30$.

Use matrix equations to solve each system of equations.

5. $x - 5y = -20$
    $x + 2y = 6$

6. $x + y + z = 8$
    $2x - 3y - 3z = -15$
    $-x + 15y + 5z = 12$

7. One factor of $2a^3 + 2a^2 - 8a - 8$ is $a - 2$. Find the other factors.

8. Show that $m - 3$ is a factor of $6m^3 - 13m^2 - 14m - 3$.

Solve each equation.

9. $\sqrt{x + 7} - 5 = 6$ **10.** $\sqrt{3m - 2} = 10$ **11.** $7y^2 + 9 = 0$

Solve the equation for the variable indicated.

12. $q = \sqrt{\dfrac{Fr^2}{k}}$ for k **13.** $r = \sqrt[3]{\dfrac{3V}{4\pi}}$ for V **14.** $v = \sqrt{\dfrac{2kE}{m}}$ for E

Excursions in Algebra   Kepler's Laws

Johannes Kepler, a German astronomer and mathematician, stated the three laws of planetary motion. These laws describe the shape of the planets' orbits around the sun, the velocities, and the distances of the planets with respect to the sun.

Kepler's third law states that the ratio of the square of a planet's period of revolution about the sun to the cube of its mean distance from the sun is constant for all planets. This law can be represented by the following formula.

$$\frac{P^2}{D^3} = K$$

P is the period of revolution, D is the mean distance, and K is a constant.

Exercises

1. Determine the value of K in Kepler's third law using Earth's period of revolution about the sun, approximately 365.25 days, and Earth's mean distance from the sun, approximately 93 million miles.

Use Kepler's law and the value of K from Exercise 1 to find the mean distances of each planet from the sun, given its approximate period of revolution about the sun.

2. Mercury; 88 days   **3.** Jupiter; 4332 days   **4.** Venus; 225 days

5. Mars; 687 days   **6.** Pluto; 90,582 days

In a simplified electrical circuit there are three basic things to be considered: flow of electric current, I, resistance to that flow, Z, called impedance, and electromotive force, E, called voltage. All three are related in the formula $E = I \cdot Z$. This basic formula can be expressed in several ways.

Current is measured in amps. Impedance is measured in ohms. Voltage is measured in volts.

$$E = I \cdot Z \qquad \frac{E}{Z} = I \qquad \frac{E}{I} = Z$$

The current, impedance, and voltage are often expressed as complex numbers. Electrical engineers use j instead of i to represent the imaginary unit. For electrical engineers, $j = \sqrt{-1}$ and $j^2 = -1$.

Example: **Compute the voltage, E, when $I = (35 - j40)$ amps and $Z = (10 + j2)$ ohms.**

$$\begin{aligned}
E &= I \cdot Z \\
&= (35 - j40)(10 + j2) \\
&= 350 + j70 - j400 - j^2 80 \\
&= 350 - j330 + 80 \\
&= 430 - j330
\end{aligned}$$

The voltage is $(430 - j330)$ volts.

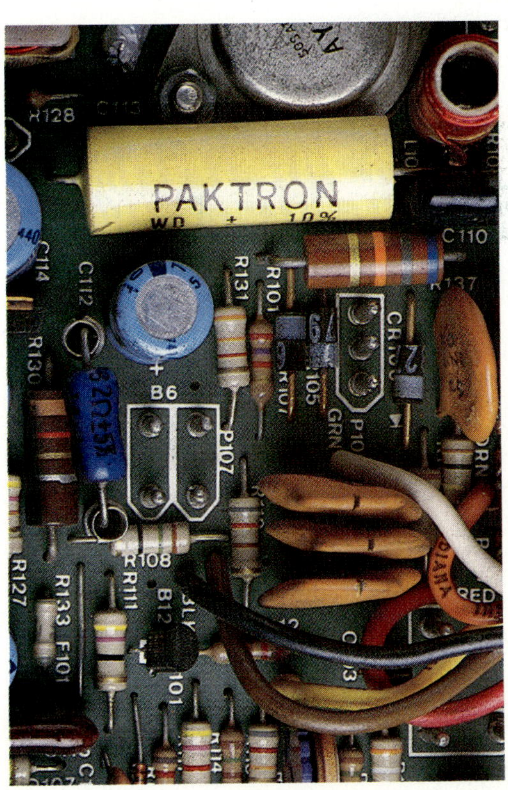

Exercises

Find E given the following values.

1. $I = (4 + j3)$ amps, $Z = (16 - j28)$ ohms
2. $I = (6 - j8)$ amps, $Z = (14 + j8)$ ohms
3. $I = (20 + j12)$ amps, $Z = (15 + j3)$ ohms

Find I given the following values.

4. $E = (70 + j226)$ volts, $Z = (6 + j8)$ ohms
5. $E = (85 + j110)$ volts, $Z = (3 - j4)$ ohms
6. $E = (60 + j112)$ volts, $Z = (10 - j6)$ ohms

Find Z given the following values.

7. $E = (-50 + j100)$ volts, $I = (-6 - j2)$ amps
8. $E = (100 + j10)$ volts, $I = (-8 + j3)$ amps
9. $E = (-70 + j240)$ volts, $I = (-5 + j4)$ amps

6-9 More about Complex Numbers

Complex numbers of the form $a + bi$ and $a - bi$ are called
conjugates of each other. Notice that the product of complex conjugates is always a real number.

Examples

1 Find $(3 + 5i)(3 - 5i)$.

$$(3 + 5i)(3 - 5i)$$
$$= 9 - 25i^2$$
$$= 9 - (-25)$$
$$= 34$$

2 Find $(a + bi)(a - bi)$.

$$(a + bi)(a - bi)$$
$$= a^2 - b^2i^2$$
$$= a^2 - (-b^2)$$
$$= a^2 + b^2$$

Sometimes rational expressions contain complex numbers. Since i
represents a radical, rational expressions are usually written without
imaginary numbers in the denominator. As with radicals, the denominator should be rationalized.

Examples

3 Simplify $\dfrac{3 + 7i}{2i}$.

$$\frac{3 + 7i}{2i} = \frac{3 + 7i}{2i} \cdot \frac{i}{i}$$
$$= \frac{3i + 7i^2}{2i^2}$$
$$= \frac{-7 + 3i}{-2}$$
$$= \frac{7 - 3i}{2}$$

4 Simplify $\dfrac{4 + 3i}{1 - 2i}$. *Conjugates are used to rationalize the denominator.*

$$\frac{4 + 3i}{1 - 2i} = \frac{4 + 3i}{1 - 2i} \cdot \frac{1 + 2i}{1 + 2i}$$
$$= \frac{4 + 8i + 3i - 6i^2}{1 + 2i - 2i - 4i^2}$$
$$= \frac{-2 + 11i}{5}$$

5 Find the multiplicative inverse of $2 - 3i$.

The multiplicative inverse of $2 - 3i$ is $\dfrac{1}{2 - 3i}$.

Now simplify.

$$\frac{1}{2 - 3i} = \frac{1}{2 - 3i} \cdot \frac{2 + 3i}{2 + 3i}$$
$$= \frac{2 + 3i}{4 + 6i - 6i - 9i^2}$$
$$= \frac{2 + 3i}{13}$$

Check: $2 - 3i \cdot \dfrac{2 + 3i}{13} = \dfrac{4 + 9}{13}$

$$= 1$$

The product of a number and its multiplicative inverse is 1.

The inverse is $\dfrac{2 + 3i}{13}$.

Exploratory Exercises

Find the conjugate of each complex number.

1. $2 + i$ **2.** $1 + 3i$ **3.** $5 - 4i$ **4.** $3 - 2i$ **5.** $4i$ **6.** $7i$

7. $-5i$ **8.** $-3i$ **9.** 6 **10.** 8 **11.** $5 - 6i$ **12.** $12 + i$

Show that each of the following are multiplicative inverses of one another.

13. $3 + 2i;\ \dfrac{3 - 2i}{13}$ **14.** $5 - 4i;\ \dfrac{5 + 4i}{41}$ **15.** $6 + 8i;\ \dfrac{3 - 4i}{50}$

Written Exercises

Find the product of each complex number and its conjugate.

16. $3 - 7i$ **17.** $6 + 5i$ **18.** $2 + 9i$ **19.** $17 - i$

20. $2 - 3i$ **21.** $7 - 7i$ **22.** $-2i$ **23.** $-10i$

Simplify.

24. $\dfrac{3 - 2i}{1 - i}$ **25.** $\dfrac{4 + 5i}{1 + i}$ **26.** $\dfrac{1 + i}{3 + 2i}$ **27.** $\dfrac{1 - i}{4 - 5i}$ **28.** $\dfrac{3 + 5i}{2i}$

29. $\dfrac{4 - 7i}{-3i}$ **30.** $\dfrac{5 - 6i}{-3i}$ **31.** $\dfrac{2 + i}{5i}$ **32.** $\dfrac{3}{4 - i}$ **33.** $\dfrac{2}{6 + 5i}$

34. $\dfrac{4}{\sqrt{3} + 2i}$ **35.** $\dfrac{7}{\sqrt{2} - 3i}$ **36.** $\dfrac{2 + i\sqrt{3}}{2 - i\sqrt{3}}$ **37.** $\dfrac{1 + i\sqrt{2}}{1 - i\sqrt{2}}$ **38.** $\dfrac{3 - i\sqrt{5}}{3 + i\sqrt{5}}$

39. $\dfrac{2 - i\sqrt{7}}{2 + i\sqrt{7}}$ **40.** $\dfrac{(2 + 3i)^2}{(3 + i)^2}$ **41.** $\dfrac{(3 + 3i)^2}{(1 + i)^2}$ **42.** $\dfrac{1 - i}{(1 + i)^2}$ **43.** $\dfrac{(4 + 3i)^2}{(3 - i)^2}$

Find the multiplicative inverse of each complex number.

44. $3 + i$ **45.** $2 - 5i$ **46.** $7 - 3i$ **47.** $3 + 7i$

48. $\dfrac{4i}{3 + i}$ **49.** $\dfrac{2i}{5 - i}$ **50.** $\dfrac{-i}{2 - 3i}$ **51.** $\dfrac{-3i}{3 + 4i}$

52. Write an expression for the multiplicative inverse of $a + bi$.

Challenge Exercises

53. Show that $-\dfrac{1}{2} + \dfrac{1}{2}i\sqrt{3}$ is a cube root of 1. (Hint: Find the cube of the number.)

54. Show that $-\dfrac{1}{2} - \dfrac{1}{2}i\sqrt{3}$ is another cube root of 1.

55. Show that $1 + i\sqrt{3}$ is a cube root of -8.

56. Show that $1 - i\sqrt{3}$ is another cube root of -8.

Excursions in Algebra   Mathematics Contests

The following problem appeared in a high school contest of the Mathematical Association of America.

The formula $\sqrt{x^3} = \dfrac{8 \cdot 10^8}{N}$ gives, for a certain group, the number, N, of individuals whose income exceeds x dollars. What is the lowest income, in dollars, of the wealthiest 800 individuals?

A **field** is a mathematical system consisting of a set of numbers, S, together with the operations of addition and multiplication. The following properties must hold for all elements a, b, and c in S.

Field Properties	Addition	Multiplication
Closure	$a + b$ is in S.	ab is in S.
Commutative	$a + b = b + a$	$a \cdot b = b \cdot a$
Associative	$(a + b) + c = a + (b + c)$	$(a \cdot b) \cdot c = a \cdot (b \cdot c)$
Identity	0 is in S. $a + 0 = a = 0 + a$	1 is in S. $a \cdot 1 = a = 1 \cdot a$
Inverse	$-a$ is in S. $a + {-a} = 0 = -a + a$	$\frac{1}{a}$ is in S. $a \cdot \frac{1}{a} = 1 = \frac{1}{a} \cdot a$ if $a \neq 0$
Distributive of Multiplication over Addition	$a(b + c) = ab + ac$ and $(b + c)a = ba + ca$	

These properties hold for all real numbers. So, the real number system is a field.

To determine if a number system is a field, decide whether the Closure Property holds for addition and multiplication. Then check to see if the commutative, associative, and distributive properties hold. (These three properties are always true for subsets of the real numbers.) Also, check whether identity elements and inverses exist. If you can find a counterexample, the system is *not* a field.

Example: **Is the whole number system a field?**

The additive inverse of 5 is -5. Since -5 is not a whole number, the whole number system is not a field.

Exercises

Show that each property holds for the set of complex numbers.

1. Commutative Property of Addition
2. Commutative Property of Multiplication
3. Associative Property of Addition
4. Associative Property of Multiplication
5. Distributive Property of Multiplication over Addition

6. Show that the set of complex numbers is closed under addition.
7. Show that the set of complex numbers is closed under multiplication.

Decide whether each number system is a field. If the system is not a field, give a counterexample for each field property that is not satisfied.

8. complex numbers
9. natural numbers
10. integers
11. positive real numbers
12. irrational numbers
13. rational numbers
14. imaginary numbers
15. negative rational numbers

Iteration

Computers and calculators do not store tables. They use *iteration* to find roots. An iteration method involves a repeated series of operations. Each repetition produces a more precise result. One iteration method for finding square roots is the divide-and-average method. The following program can be used to find $\sqrt{N}$, given N and an approximate value, A.

```
10   PRINT "WHAT IS THE RADICAND?"
20   INPUT N
30   PRINT "CHOOSE A FIRST
        APPROXIMATION:"
40   INPUT A
50   PRINT "SUCCESSIVE APPROXIMATIONS
        ARE: "
60   LET L = (N / A + A) / 2
70   PRINT L
80   IF ABS (L − A) < 0.00001 THEN
        110
90   LET A = L
100  GOTO 60
110  END
```

Enter positive numbers for N and A.

The computer will find the average of $\frac{N}{A}$ and A, and print it. If L is close to A, the computer will go to line 110. If not, the computer will replace A by L and return to line 60.

This method works even if the first approximation is poor. Suppose the program is used to find $\sqrt{875}$. The first column below shows the output when 30 is used as the first approximation. The second column shows the output when a poor choice, 9, is used. Compare the results.

```
]RUN                              ]RUN
WHAT IS THE RADICAND?             WHAT IS THE RADICAND?
?875                             ?875
CHOOSE A FIRST APPROXIMATION.     CHOOSE A FIRST APPROXIMATION.
?30                              ?9
SUCCESSIVE APPROXIMATIONS ARE:    SUCCESSIVE APPROXIMATIONS ARE:
29.5833334                       53.1111111
29.5803991                       34.7930033
29.5803989                       29.9708689
                                 29.5829425
                                 29.580399
                                 29.5803989
```

Exercises

Use the program above to approximate each square root.

1. $\sqrt{68}$ **2.** $\sqrt{947}$ **3.** $\sqrt{2050}$ **4.** $\sqrt{413}$ **5.** $\sqrt{1794}$ **6.** $\sqrt{4608}$

7. How can you change the program so that only the final approximation will be printed?

square root   (195)
*n*th root   (195)
irrational numbers   (197)
rationalizing the
  denominator   (201)

like radical
  expressions   (204)
conjugates   (205)
rational exponents
  (208, 209)

radical equations   (214)
imaginary unit   (218)
pure imaginary number
  (218)
complex numbers   (220)

Chapter Summary

1. For any real numbers a and b, if $a^2 = b$, then a is a square root of b.   (195)
2. For any real numbers a and b, and any positive integer n, if $a^n = b$, then a is an *n*th root of b.   (195)
3. The Real *n*th Roots of b (196)

	$b > 0$	$b < 0$	$b = 0$
n even	one positive root one negative root	no real roots	one real root, 0
n odd	one positive root no negative roots	no positive roots one negative root	one real root, 0

4. Property of *n*th Roots: For any real number a, and any integer n, $n > 1$,
   1. If n is even, then $\sqrt[n]{a^n} = |a|$, and
   2. If n is odd, then $\sqrt[n]{a^n} = a$.   (196)
5. Product Property of Radicals: For any real numbers a and b, and any integer n, $n > 1$,
   1. If n is even, then $\sqrt[n]{ab} = \sqrt[n]{a} \cdot \sqrt[n]{b}$ as long as a and b are both nonnegative, and
   2. If n is odd, then $\sqrt[n]{ab} = \sqrt[n]{a} \cdot \sqrt[n]{b}$.   (199)
6. Quotient Property of Radicals: For any real numbers a and b, $b \neq 0$, and any integer n, $n > 1$, $\sqrt[n]{\dfrac{a}{b}} = \dfrac{\sqrt[n]{a}}{\sqrt[n]{b}}$ if all roots are defined.   (200)
7. Conditions for Simplified Radicals:
   1. The index n is as small as possible.
   2. The radicand contains no factor (other than one) that is the *n*th power of an integer or polynomial.
   3. The radicand contains no fractions.
   4. No radicals appear in the denominator.   (201)
8. For any real number b and for any integer n, $n > 1$, $b^{\frac{1}{n}} = \sqrt[n]{b}$ except when $b < 0$ and n is even.   (208)
9. For any nonzero real number b, and any integers m and n, with $n > 1$, $b^{\frac{m}{n}} = \sqrt[n]{b^m} = (\sqrt[n]{b})^m$ except when $b < 0$ and n is even.   (209)

10. Conditions for Simplified Expressions:
   1. It has no negative exponents.
   2. It has no fractional exponents in the denominator.
   3. It is not a complex fraction.
   4. The index of any remaining radical is as small as possible.   (212)
11. These steps are usually used to solve a radical equation.
   1. Isolate the radical.   4. Solve for the variable.
   2. Raise both sides to a power.   5. Check each solution.   (214)
   3. Combine terms.
12. For any positive real number b, $\sqrt{-(b^2)} = \sqrt{b^2} \cdot \sqrt{-1}$ or bi where i is the imaginary unit, and bi is called a pure imaginary number.   (218)
13. A complex number is any number that can be written in the form $a + bi$ where a and b are real numbers and i is the imaginary unit. a is called the real part, and bi is called the imaginary part.   (220)

Chapter Review

6–1 **Find the principal roots of each expression.**

1. $\sqrt{49a^2}$ 2. $\sqrt[3]{-27}$ 3. $\sqrt[3]{8x^3y^6}$ 4. $\sqrt[4]{16}$

Use a calculator or the table of roots on page 672 to find each value.

5. $\sqrt[3]{39}$ 6. $\sqrt[3]{290}$ 7. $(9.8)^3$ 8. $(1.5)^3$

6–2 **Simplify.**

9. $\sqrt[3]{48a^2}$ 10. $\sqrt[4]{32m^5}$ 11. $-3\sqrt{18} \cdot 5\sqrt{15}$
12. $2\sqrt[3]{54} \cdot 4\sqrt[3]{16}$ 13. $\sqrt{5}(\sqrt{10} + 2)$ 14. $\sqrt{3}(\sqrt{6} + \sqrt{12})$
15. $\sqrt[3]{3}(\sqrt[3]{16} + \sqrt[3]{9})$ 16. $\sqrt{3m^3n^4} \cdot \sqrt{24m^6n}$ 17. $\sqrt[3]{2xy^4} \cdot \sqrt[3]{36x^2y^2}$

18. $\sqrt{\dfrac{10}{2}}$ 19. $\dfrac{\sqrt[3]{80}}{\sqrt[3]{2}}$ 20. $\dfrac{4}{\sqrt[3]{16}}$ 21. $\dfrac{\sqrt{3} + \sqrt{2}}{4 + \sqrt{3}}$

6–3 **Simplify.**

22. $3\sqrt{12} - 4\sqrt{75} + 4$ 23. $(6 + \sqrt{2})(10 + \sqrt{5})$ 24. $(3 + \sqrt{5})(3 - \sqrt{5})$

6–4 **Express in simplest radical form or using rational exponents.**

25. $r^2 s^{\frac{1}{3}} y^{\frac{1}{2}}$ 26. $\sqrt[6]{27}$ 27. $(3x)^{\frac{1}{2}} x^{\frac{1}{4}}$
28. $\sqrt[4]{x^4y^3}$ 29. $\sqrt[3]{27m^3n^2}$ 30. $\sqrt[5]{32w^{10}r^5}$

Evaluate.

31. $6^{\frac{1}{2}} \cdot 6^{\frac{3}{2}}$ 32. $125^{\frac{1}{3}}$ 33. $(0.008)^{\frac{2}{3}}$ 34. $(25^{\frac{3}{4}})^{\frac{2}{3}}$

6–5 **Simplify.**

35. $\dfrac{1}{z^{\frac{3}{5}}}$ 36. $\dfrac{w^{-\frac{2}{3}} r^{\frac{1}{2}}}{wr^2}$ 37. $\dfrac{z^{\frac{1}{3}}}{z^{\frac{2}{3}} - z^{-\frac{1}{3}}}$ 38. $\dfrac{x^{\frac{1}{2}} - y^{\frac{1}{2}}}{x^{\frac{1}{2}} + y^{\frac{1}{2}}}$

6–6   Solve each equation.

39. $\sqrt{2x + 7} = 3$ **40.** $\sqrt[3]{y + 2} = 4$ **41.** $\sqrt{r + 12} - \sqrt{r} = 2$

6–7   Simplify.

42. $\sqrt{-24}$ **43.** i^7 **44.** $\sqrt{-3} \cdot \sqrt{-3}$ **45.** $(2\sqrt{5})(-3\sqrt{-5})$

6–8   Find the sum, difference, and product for each pair of complex numbers.

46. $7 + 2i, \, 5 - 3i$ **47.** $3 + 8i, \, 3 - 8i$

6–9   Simplify.

48. $\dfrac{4 + 3i}{1 - 2i}$ **49.** $\dfrac{7 - 3i}{2i}$ **50.** $\dfrac{(4 - 3i)^2}{(3 + i)^2}$

Chapter Test

Find the principal root of each expression.

1. $\sqrt{81x^2}$ **2.** $\sqrt[3]{-27y^3}$ **3.** $\sqrt{x^2 - 8xy + 16y^2}$

Simplify.

4. $\sqrt[3]{108m^4}$

5. $7\sqrt[3]{16} \cdot 5\sqrt[3]{20}$

6. $\sqrt{125x^2y} \cdot 3\sqrt{81x^2y^2}$

7. $2\sqrt{27} + 2\sqrt{3} - 7\sqrt{48}$

8. $(4 - \sqrt{3})(4 + \sqrt{3})$

9. $(\sqrt{3} - \sqrt{5})(\sqrt{15} + 3)$

10. $\dfrac{3}{\sqrt[3]{4}}$

11. $\dfrac{3}{\sqrt[4]{8}}$

12. $\dfrac{2 + \sqrt{2}}{1 - \sqrt{2}}$

13. $\sqrt[3]{2}(\sqrt[3]{24} - \sqrt[3]{4})$

14. $\sqrt[3]{16x^3} + 5\sqrt[3]{54x^3} - 3\sqrt[3]{128x^3}$

15. $\sqrt{\dfrac{5}{32}} + \sqrt{90} + \dfrac{\sqrt{30}}{\sqrt{3}} - \sqrt{\dfrac{2}{5}}$

16. $\dfrac{1}{y^{\frac{2}{3}}}$ **17.** $\dfrac{15}{5^{\frac{2}{3}}}$ **18.** $\dfrac{r + s}{r^{\frac{1}{2}} + s^{\frac{1}{2}}}$ **19.** $\dfrac{x^{\frac{5}{3}}y + 3x^{-\frac{1}{3}}}{x^{\frac{2}{3}}}$

Evaluate.

20. $27^{-\frac{4}{3}}$ **21.** $(\sqrt[3]{27})^2$ **22.** $\dfrac{16}{4^{\frac{3}{2}}}$ **23.** $\left(\dfrac{1}{64}\right)^{-\frac{2}{3}}$

Express using rational exponents.

24. $\sqrt[3]{n^2}$ **25.** $\sqrt{a^6}$ **26.** $\sqrt[3]{8m^2r^7}$ **27.** $\sqrt[4]{x^4y^6b^2}$

Solve each equation.

28. $\sqrt{2x - 5} = 9$ **29.** $\sqrt[3]{a + 3} - 1 = 2$

Simplify.

30. $\sqrt{-50}$ **31.** $3i + \sqrt{-4}$ **32.** $(7 - 3i) + (4 + 5i)$

33. $\sqrt{-3} \cdot \sqrt{-24}$ **34.** $\dfrac{4 - 5i}{3 + 7i}$ **35.** $(4 - 6i) - (4 + 6i)$

36. $\dfrac{4 - 7i}{2i}$ **37.** $(6 + i)(7 - 3i)$ **38.** $\dfrac{(2 + 3i)^2}{(1 - i)^2}$

The questions on this page involve comparing two quantities, one in Column A and one in Column B. In certain questions, information related to one or both quantities is centered above them. All variables used stand for real numbers.

Directions:

Write A if the quantity in Column A is greater.

Write B if the quantity in Column B is greater.

Write C if the quantities are equal.

Write D if there is not enough information to determine the relationship.

Examples

Column A	Column B
I. $\dfrac{x}{y}$	$\dfrac{y}{x}$

The answer is D because it is not known whether x and y are positive or negative. The values of $\dfrac{x}{y}$ and $\dfrac{y}{x}$ depend on the choices for x and y.

Column A	Column B
II.	$a > 0$
$4a - 3a$	$4 \cdot 3a$

The answer is B because $12a$ is greater than a for all positive values of a.

	Column A	Column B
1.	$\dfrac{8a + 12}{2}$	$4a + 6$
2.	$(-9)^{72}$	$(-9)^{83}$
3.	$-3 < n < 0$	
	$-3(n + n)$	$(n)(n)(n)$
4.	$n + 11 > 12$	
	$3n + 8$	$16 - 2n$
5.	$a < 0 < b$	
	$\dfrac{a}{2}$	b^2
6.	a if $3a < 23$	b if $23 < 3b$
7.	$x^2 > y^2$	
	$(y + 1)^2$	$(x + 1)^2$
8.	$\dfrac{1}{2}\%$ of 400	0.5×400
9.	$\dfrac{.05}{x} = \dfrac{.2}{.06}$	
	x	0.2

	Column A	Column B
10.	$6 > b > -5$	
	$\dfrac{b}{5}$	$\dfrac{5}{b}$
11.	30% of r is 6	
	The percent that 60 is of r.	The percent that r is of 6.
12.	$.8 \times 6d$	$\dfrac{3}{5}$ of $8d$
13.	$23 - 4(2)0$	$6 + 10(3 - 2)$
14.	$\dfrac{x}{2} = z^2$	
	x	z
15.	area of a circle with diameter 10	area of a right triangle with hypotenuse 10
16.	$\dfrac{1}{3} < a < \dfrac{2}{3}$	
	$\dfrac{2}{3} < b < 1$	
	$a + b$	$a^2 + b^2$

CHAPTER 7

Quadratic Equations

Quadratic equations can be used in many situations. For example, Joel Bellman wants to build a swimming pool with a sidewalk of uniform width surrounding it. Joel wants the dimensions of the pool and sidewalk to be 15 meters by 20 meters. The pool must have an area of 155 square meters. To find out how wide the sidewalk must be, Joel uses a quadratic equation.

7-1 Quadratic Equations

Any equation that can be written in the form $ax^2 + bx + c = 0$, $a \neq 0$, is a **quadratic equation**. Equations such as $x^2 - 8 = 0$ and $x^2 - 2x - 15 = 0$ are quadratic equations. You have learned how to solve some types of quadratic equations.

Example

1 Solve $x^2 - 8 = 0$.

$$x^2 - 8 = 0 \qquad \textit{In this quadratic equation, } b = 0.$$
$$x^2 = 8$$
$$x = \pm\sqrt{8} \text{ or } \pm 2\sqrt{2} \qquad \text{The solutions are } 2\sqrt{2} \text{ and } -2\sqrt{2}.$$

Quadratic equations can be solved by factoring. The equation from Example 1, $x^2 - 8 = 0$, can be factored as the difference of two squares.

$$(x - \sqrt{8})(x + \sqrt{8}) = 0 \qquad \textit{The solutions are the same as in Example 1.}$$

The factoring method depends on the Zero Product Property.

For any real numbers a and b, if $ab = 0$, then $a = 0$ or $b = 0$.	*Zero Product Property*

Examples

2 Solve $x^2 - 2x - 15 = 0$.

$$x^2 - 2x - 15 = 0$$
$$(x - 5)(x + 3) = 0 \qquad \textit{Factor.}$$
$$x - 5 = 0 \quad \text{or} \quad x + 3 = 0 \qquad \textit{Zero Product}$$
$$x = 5 \quad \text{or} \qquad x = -3 \qquad \textit{Property}$$

The solutions are 5 and -3.

Check: $x^2 - 2x - 15 = 0$
$$(5)^2 - 2(5) - 15 \overset{?}{=} 0$$
$$0 = 0 \quad ✔$$
$$(-3)^2 - 2(-3) - 15 \overset{?}{=} 0$$
$$0 = 0 \quad ✔$$

3 Solve $9x^2 - 24x = -16$.

$$9x^2 - 24x = -16$$
$$9x^2 - 24x + 16 = 0$$
$$(3x - 4)(3x - 4) = 0$$
$$3x - 4 = 0 \quad \text{or} \quad 3x - 4 = 0$$
$$x = \frac{4}{3} \qquad\qquad x = \frac{4}{3}$$

The solution is $\frac{4}{3}$.

Check: $9x^2 - 24x = -16$
$$9\left(\frac{4}{3}\right)^2 - 24\left(\frac{4}{3}\right) \overset{?}{=} -16$$
$$16 - 32 \overset{?}{=} -16$$
$$-16 = -16 \quad ✔$$

Identify which of the following are quadratic equations.

1. $4x^2 + 7x - 3 = 0$
2. $5y^4 - 7y^2 = 0$
3. $2a^2 + 7a = 0$
4. $4x^2 - 2x + 11 = 0$
5. $3z^3 + 4 = 0$
6. $\frac{1}{2}y^2 + \frac{3}{4} = 0$
7. $\frac{2}{3}c^3 - 3c^2 + 2c = 0$
8. $z^2 + 7z - 3 = z^3$
9. $3d^2 - 9 = d$

Solve each equation.

10. $(x - 4)(x + 5) = 0$
11. $(y - 3)(y + 7) = 0$
12. $(a + 6)(a + 2) = 0$
13. $(m - 8)(m - 1) = 0$
14. $z(z - 1)^2 = 0$
15. $s(s + 4) = 0$
16. $(3y + 7)(y + 5) = 0$
17. $(2b + 5)(b - 1) = 0$
18. $(2x + 3)(3x - 1) = 0$

Solve each equation.

19. $x^2 + 6x + 8 = 0$
20. $z^2 + 4z + 3 = 0$
21. $a^2 - 9a + 20 = 0$
22. $m^2 - 8m + 12 = 0$
23. $b^2 + 3b - 10 = 0$
24. $y^2 - 4y - 21 = 0$
25. $x^2 + 4x + 4 = 0$
26. $c^2 - 12c + 36 = 0$
27. $n^2 + 3n = 0$
28. $z^2 - 5z = 0$
29. $d^2 - 3d = 4$
30. $m^2 + 6m = 27$
31. $z^2 + z = 30$
32. $y^2 - y = 12$
33. $2x^2 + 5x + 3 = 0$
34. $2a^2 + 9a + 4 = 0$
35. $2z^2 - 3z - 9 = 0$
36. $3b^2 + 13b - 10 = 0$
37. $3c^2 = 5c$
38. $10y^2 = y$
39. $6d^2 + 13d + 6 = 0$
40. $12m^2 + 25m + 12 = 0$
41. $4s^2 - 11s - 3 = 0$
42. $4a^2 - 17a + 4 = 0$
43. $z^2 + 3z - 40 = 0$
44. $6m^2 + 7m - 3 = 0$
45. $3r^2 - 14r + 8 = 0$
46. $2y^2 + 11y - 21 = 0$
47. $12c^2 - 17c - 5 = 0$
48. $3t^2 + 4t - 15 = 0$
49. $12p^2 + 8p = 15$
50. $18n^2 - 3n = 1$
51. $4b^2 - 13b = 12$
52. $18m^2 - 3m = 15$
53. $4x^2 + 9 = 12x$
54. $9y^2 + 16 = -24y$
55. $121 = 16b^2$
56. $4t^2 = 25$

Solve each equation.

57. $n^3 = 9n$
58. $a^3 = 81a$
59. $35z^3 + 16z^2 = 12z$
60. $18r^3 + 16r = 34r^2$
61. $25d^3 + 9d = 30d^2$
62. $16t^3 = 40t^2 - 25t$
63. $\frac{1}{4}x^2 - \frac{1}{5}x + 1 = 0$
64. $0.25y^2 = 0.6\,y - 0.36$

Problem Solving

Guess and Check

Some problems may be solved by using a guess-and-check strategy. This procedure involves first making an educated guess for the solution to a problem. The next step is to check your guess in terms of the problem. Even though it may not be correct, it may give you an indication of how to improve your next guess. This procedure is repeated until you arrive at the correct answer.

Example: **Find the least prime number greater than 720.**

One way to solve this problem is to check each integer beginning with 721 for prime divisors until you find a prime number. However, if known information is used, time and effort in this guess-and-check method can be saved. Any even integer greater than 720 is not prime. Since 720 is divisible by 3, every third integer greater than 720 will not be prime. Any integer greater than 720 whose last digit is 5 or 0 is not prime. Thus, many possibilities are eliminated.

Try 721.   *Using a calculator is helpful.*

$$721 \div 7 = 103 \quad \textit{not prime}$$

Try 727.

$$727 \div 7 \approx 103.9 \qquad 727 \div 11 \approx 66.1 \qquad 727 \div 13 \approx 55.9$$
$$727 \div 17 \approx 42.8 \qquad 727 \div 19 \approx 38.3 \qquad 727 \div 23 \approx 31.6$$
$$727 \div 29 \approx 25.1 \qquad \textit{Why can you stop after this division?}$$

Thus, 727 is the least prime number greater than 720.

Exercises

1. Find the least prime number greater than 840.

2. Rich and Peg raise cows and chickens. They counted all the heads and got 12. They counted all the feet and got 40. How many cows and chickens do they have?

3. Supply a digit for each letter so the multiplication problem is correct. Each letter represents a different digit.

$$\begin{array}{r} \text{ABCDE} \\ \times 4 \\ \hline \text{EDCBA} \end{array}$$

4. Write an eight-digit number using the digits 1, 2, 3, and 4 each twice so that the 1's are separated by 1 digit, the 2's are separated by 2 digits, the 3's are separated by 3 digits, and the 4's are separated by 4 digits.

5. Copy the diagram below. In each block write a digit such that the digit in the first block indicates the total number of zeros in the entire ten-digit number, the digit in the block marked "1" indicates the total number of 1's in the number, and so on to the last block whose digit indicates the total number of 9's in the number.

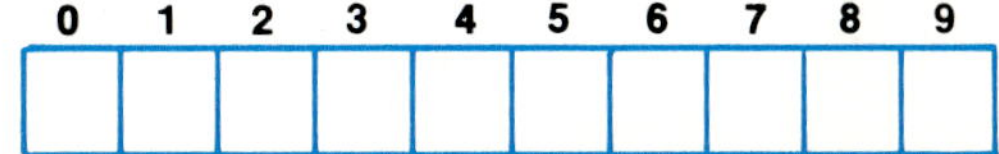

7-2 Completing the Square

An equation like $(x - 4)^2 = 3$ can be solved by taking the square root of each side.

$$
\begin{aligned}
(x - 4)^2 &= 3 \\
\sqrt{(x - 4)^2} &= \sqrt{3} \\
|x - 4| &= \sqrt{3} \\
x - 4 &= \pm\sqrt{3} \\
x &= 4 \pm \sqrt{3}
\end{aligned}
$$

These two steps may be omitted.

The solutions are $4 + \sqrt{3}$ and $4 - \sqrt{3}$.

The equation $x^2 - 6x + 9 = 2$ can be solved in a similar way.

$$
\begin{aligned}
x^2 - 6x + 9 &= 2 \\
(x - 3)^2 &= 2 \\
x - 3 &= \pm\sqrt{2} \\
x &= 3 \pm \sqrt{2}
\end{aligned}
$$

Factor.

The solutions are $3 + \sqrt{2}$ and $3 - \sqrt{2}$.

To solve a quadratic equation by taking the square root, you must have a perfect square equal to a constant. A method called **completing the square** is used to obtain this situation.

Consider the following perfect squares.

$$(x + 7)^2 = x^2 + 14x + 49 \qquad\qquad (x + b)^2 = x^2 + 2bx + b^2$$

$$\left(\frac{14}{2}\right)^2 \rightarrow (7)^2 \qquad\qquad\qquad \left(\frac{2b}{2}\right)^2 \rightarrow (b)^2$$

Notice that the coefficient of the linear term is divided by 2, then squared.

The pattern shown above can be used to complete the square when two terms are known. Complete the square for $x^2 - 8x$.

$$x^2 - 8x + \underline{\;\;?\;\;}$$

$$\left(\frac{8}{2}\right)^2 \rightarrow (4)^2 \text{ or } 16 \qquad \text{The answer is } x^2 - 8x + 16.$$

$x^2 - 8x + 16$ is a perfect square trinomial.

Example

1 Solve $x^2 - 8x + 11 = 0$ by completing the square.

$$
\begin{aligned}
x^2 - 8x + 11 &= 0 \\
x^2 - 8x &= -11 \\
x^2 - 8x + 16 &= -11 + 16 \\
(x - 4)^2 &= 5 \\
x - 4 &= \pm\sqrt{5} \\
x &= 4 \pm \sqrt{5}
\end{aligned}
$$

The left side of the equation is not a perfect square.

Subtract 11 from each side.

Add $\left(\frac{-8}{2}\right)^2$ or 16 to each side.

Factor. Then solve.

Take the square root of each side.

Add 4 to each side.

The solutions are $4 + \sqrt{5}$ and $4 - \sqrt{5}$.

When the coefficient of the second degree term is not 1, another step is required. In the following example, it is necessary to write an equivalent equation that has 1 as the coefficient of the second degree term.

Examples

2 **Solve $2m^2 - 8m + 3 = 0$.**

$$2m^2 - 8m + 3 = 0$$

$$m^2 - 4m + \frac{3}{2} = 0 \qquad \text{\textit{Divide each side by 2.}}$$

$$m^2 - 4m = -\frac{3}{2}$$

$$m^2 - 4m + 4 = -\frac{3}{2} + 4 \qquad \text{\textit{Add } } \left(\frac{-4}{2}\right)^2 \text{ \textit{or 4 to each side.}}$$

$$(m - 2)^2 = \frac{5}{2} \qquad \text{\textit{Factor.}}$$

$$m - 2 = \pm\sqrt{\frac{5}{2}} \qquad \text{\textit{Take the square root of each side.}}$$

$$m - 2 = \pm\frac{\sqrt{10}}{2} \qquad \text{\textit{Rationalize the denominator.}}$$

$$m = 2 \pm \frac{\sqrt{10}}{2} \qquad \text{\textit{Solve for m.}}$$

The solutions are $2 + \dfrac{\sqrt{10}}{2}$ and $2 - \dfrac{\sqrt{10}}{2}$.

3 **Solve $3x^2 - 11x - 4 = 0$.**

$$3x^2 - 11x - 4 = 0$$

$$x^2 - \frac{11}{3}x - \frac{4}{3} = 0 \qquad \text{\textit{Divide each side by 3.}}$$

$$x^2 - \frac{11}{3}x = \frac{4}{3}$$

$$x^2 - \frac{11}{3}x + \frac{121}{36} = \frac{4}{3} + \frac{121}{36} \qquad \text{\textit{Add } } \left(-\frac{11}{3} \div 2\right)^2 \text{ \textit{or }} \frac{121}{36} \text{ \textit{to each side.}}$$

$$\left(x - \frac{11}{6}\right)^2 = \frac{169}{36} \qquad \text{\textit{Factor.}}$$

$$x - \frac{11}{6} = \pm\frac{13}{6} \qquad \text{\textit{Take the square root of each side.}}$$

$$x = \frac{11}{6} \pm \frac{13}{6} \qquad \text{\textit{Add } } \frac{11}{6} \text{ \textit{to each side.}}$$

$$x = 4 \quad \text{or} \quad x = -\frac{1}{3}$$

The solutions are 4 and $-\dfrac{1}{3}$.

Exploratory Exercises

State whether or not each trinomial is a perfect square.

1. $x^2 + 4x + 4$
2. $a^2 + 4a + 28$
3. $b^2 - 6b - 9$
4. $x^2 - x + \frac{1}{4}$
5. $m^2 - 10m + 25$
6. $a^2 - 3a + \frac{9}{2}$

Find the value of c that makes each trinomial a perfect square.

7. $y^2 - 6y + c$
8. $x^2 + 2x + c$
9. $m^2 - 20m + c$
10. $t^2 + 40t + c$
11. $n^2 + 12n + c$
12. $x^2 + 18x + c$
13. $y^2 + 3y + c$
14. $r^2 - 9r + c$
15. $s^2 + 11s + c$
16. $a^2 - 100a + c$
17. $n^2 - n + c$
18. $x^2 + 15x + c$

Written Exercises

Find the value of c that makes each trinomial a perfect square.

19. $x^2 + 6x + c$
20. $x^2 - 10x + c$
21. $y^2 + \frac{1}{2}y + c$
22. $a^2 + 9a + c$
23. $r^2 + r + c$
24. $x^2 - 16x + c$
25. $y^2 - 3y + c$
26. $m^2 - \frac{2}{3}m + c$
27. $b^2 + 7b + c$
28. $a^2 - \frac{4}{5}a + c$
29. $r^2 + 50r + c$
30. $n^2 - 30n + c$

Solve each equation by completing the square.

31. $y^2 - 2y = 24$
32. $t^2 + 4t = 96$
33. $z^2 + 3z = 88$
34. $x^2 - 3x = 10$
35. $x^2 - 8x + 15 = 0$
36. $r^2 - 6r + 8 = 0$
37. $c^2 + 8c - 20 = 0$
38. $b^2 + 2b - 48 = 0$
39. $x^2 - 7x + 12 = 0$
40. $s^2 - 10s + 21 = 0$
41. $x^2 + 8x - 84 = 0$
42. $x^2 + 3x - 40 = 0$
43. $m^2 + 3m - 180 = 0$
44. $y^2 + 12y + 4 = 0$
45. $n^2 - 8n + 14 = 0$
46. $r^2 + 5r - 8 = 0$
47. $x^2 - 7x + 5 = 0$
48. $t^2 + 3t - 8 = 0$
49. $a^2 - 5a - 10 = 0$
50. $b^2 - \frac{3}{4}b + \frac{1}{8} = 0$
51. $y^2 + \frac{7}{3}y + \frac{2}{3} = 0$
52. $4x^2 + 19x - 5 = 0$
53. $6m^2 + 7m - 3 = 0$
54. $3c^2 - 14c + 8 = 0$
55. $2y^2 + 11y - 21 = 0$
56. $12r^2 - 17r - 5 = 0$
57. $3t^2 + 4t - 15 = 0$
58. $6s^2 + 2s + 3 = 0$
59. $3z^2 - 12z + 4 = 0$
60. $2x^2 - 3x + 4 = 0$

Challenge Exercises

Solve each equation by completing the square.

61. $x^2 + ax + b = 0$
62. $ax^2 + bx + c = 0$
63. $ax^2 + b = 0$

7-3 The Quadratic Formula

Completing the square can be used to develop a general formula for solving quadratic equations.

$$ax^2 + bx + c = 0 \quad (a \neq 0)$$ *Start with the general form of a quadratic equation.*

$$x^2 + \frac{b}{a}x + \frac{c}{a} = 0$$ *Divide by a so that the coefficient of x^2 is 1.*

$$x^2 + \frac{b}{a}x = -\frac{c}{a}$$ *Subtract $\frac{c}{a}$ from each side.*

$$x^2 + \frac{b}{a}x + \left(\frac{b}{2a}\right)^2 = -\frac{c}{a} + \left(\frac{b}{2a}\right)^2$$ *Complete the square by adding $\left(\frac{b}{a} \div 2\right)^2$ or $\left(\frac{b}{2a}\right)^2$ to each side.*

$$\left(x + \frac{b}{2a}\right)^2 = -\frac{c}{a} + \frac{b^2}{4a^2}$$ *Factor the left side.*

$$\left(x + \frac{b}{2a}\right)^2 = \frac{b^2 - 4ac}{4a^2}$$ *Add fractions on the right side.*

$$\left|x + \frac{b}{2a}\right| = \sqrt{\frac{b^2 - 4ac}{4a^2}}$$ *Take the square root of each side.*

$$x + \frac{b}{2a} = \pm\frac{\sqrt{b^2 - 4ac}}{2a}$$ *Simplify.*

$$x = \frac{-b \pm \sqrt{b^2 - 4ac}}{2a}$$ *Solve for x.*

The result is called the **quadratic formula**. It can be used to solve *any* quadratic equation.

The solutions of a quadratic equation of the form $ax^2 + bx + c = 0$ with $a \neq 0$ are given by this formula.

$$x = \frac{-b \pm \sqrt{b^2 - 4ac}}{2a}$$

Quadratic Formula

Example

1 **Solve $x^2 - 3x - 28 = 0$ using the quadratic formula.**

$$x = \frac{-b \pm \sqrt{b^2 - 4ac}}{2a}$$

$$= \frac{-(-3) \pm \sqrt{(-3)^2 - 4(1)(-28)}}{2(1)}$$ *Substitute the following values into the formula.*
$a = 1, b = -3, c = -28$

$$= \frac{3 \pm \sqrt{121}}{2}$$

$$= \frac{3 \pm 11}{2}$$ The solutions are $\frac{3 + 11}{2}$ and $\frac{3 - 11}{2}$ or 7 and -4.

The quadratic formula yields *both* solutions to a quadratic equation, even if those solutions are imaginary.

2   **Solve $2n^2 + 3n - 7 = 0$.**

$$n = \frac{-b \pm \sqrt{b^2 - 4ac}}{2a}$$

$$= \frac{-3 \pm \sqrt{3^2 - 4(2)(-7)}}{2(2)} \qquad a = 2, b = 3, c = -7$$

$$= \frac{-3 \pm \sqrt{65}}{4}$$

The solutions are $\dfrac{-3 + \sqrt{65}}{4}$ and $\dfrac{-3 - \sqrt{65}}{4}$.

3   **Solve $3x^2 - 5x + 9 = 0$.**

$$x = \frac{-b \pm \sqrt{b^2 - 4ac}}{2a}$$

$$= \frac{-(-5) \pm \sqrt{(-5)^2 - 4(3)(9)}}{2(3)} \qquad a = 3, b = -5, c = 9$$

$$= \frac{5 \pm \sqrt{-83}}{6}$$

The solutions are $\dfrac{5 + i\sqrt{83}}{6}$ and $\dfrac{5 - i\sqrt{83}}{6}$.   *Notice that the imaginary solutions always appear as conjugate pairs.*

4   **Using Calculators**

Solve $4x^2 - 70x + 145 = 0$. Round each solution to the nearest hundredth.

$$x = \frac{-b \pm \sqrt{b^2 - 4ac}}{2a} \qquad a = 4, b = -70, c = 145$$

First evaluate $\sqrt{b^2 - 4ac}$ and store the result in memory.

ENTER: 70 [+/-] [x²] [−] 4 [×] 4 [×] 145 [=] [√x] [STO]

DISPLAY: 50.7937004

Then find the solutions.

ENTER: 70 [+] [RCL] [=] [÷] [(] 2 [×] 4 [)] [=]

DISPLAY: 15.0992126

ENTER: 70 [−] [RCL] [=] [÷] [(] 2 [×] 4 [)] [=]

DISPLAY: 2.4007875

The solutions are approximately 15.10 and 2.40.

Some cubic equations can be solved using the quadratic formula. First, a binomial factor must be found.

Example

5 Solve $x^3 - 27 = 0$.

$$x^3 - 27 = 0 \qquad \text{\textit{The left side is the difference of cubes.}}$$
$$(x - 3)(x^2 + 3x + 9) = 0 \qquad \text{\textit{Factor.}}$$
$$x - 3 = 0 \quad \text{or} \quad x^2 + 3x + 9 = 0 \qquad \text{\textit{Zero Product Property}}$$

$$x = 3 \qquad\qquad x = \frac{-3 \pm \sqrt{(3)^2 - 4(1)(9)}}{2(1)}$$

In the equation $x^2 + 3x + 9 = 0$, $a = 1$, $b = 3$, $c = 9$.

$$= \frac{-3 \pm \sqrt{-27}}{2}$$

$$= \frac{-3 \pm 3i\sqrt{3}}{2}$$

The solutions are 3, $\dfrac{-3 + 3i\sqrt{3}}{2}$, and $\dfrac{-3 - 3i\sqrt{3}}{2}$.

Exploratory Exercises

State the values of a, b, and c for each quadratic equation.

1. $5x^2 - 3x + 7 = 0$ **2.** $2y^2 + y - 3 = 0$ **3.** $z^2 + 2z - 1 = 0$

4. $3r^2 - 4r = -1$ **5.** $3z^2 = 2z - 7$ **6.** $x^2 = 1 - x$

Written Exercises

Solve each equation.

7. $x^2 - x - 30 = 0$ **8.** $x^2 + 10x + 16 = 0$ **9.** $y^2 + 2y - 15 = 0$

10. $r^2 + 13r + 42 = 0$ **11.** $t^2 - 10t + 24 = 0$ **12.** $s^2 + 5s - 24 = 0$

13. $x^2 - 5x + 4 = 0$ **14.** $5x^2 - x - 4 = 0$ **15.** $3x^2 - 7x - 20 = 0$

16. $4x^2 - 11x - 3 = 0$ **17.** $6m^2 - m - 15 = 0$ **18.** $24x^2 - 14x - 5 = 0$

19. $14r^2 + 33r - 5 = 0$ **20.** $6y^2 + 19y + 15 = 0$ **21.** $20a^2 + 3a - 2 = 0$

22. $2x^2 + 3x + 3 = 0$ **23.** $6y^2 + 8y + 5 = 0$ **24.** $5m^2 + 7m + 3 = 0$

25. $2x^2 - 5x + 4 = 0$ **26.** $x^2 - 9x + 21 = 0$ **27.** $2z^2 + 2z + 3 = 0$

28. $6t^2 = 2t - 1$ **29.** $7y^2 = y + 2$ **30.** $8a^2 = -2a$

31. $24t = 7t^2$ **32.** $8r^2 + 6r + 1 = 0$ **33.** $3x^2 - 6x + 8 = 0$

34. $x^3 - 8 = 0$ **35.** $x^3 + 8 = 0$ **36.** $x^3 + 64 = 0$

37. $x^3 - 64 = 0$ **38.** $a^3 = 125$ **39.** $a^3 = 1$

Use a calculator to solve each equation. Round each solution to the nearest hundredth.

40. $2j^2 - 8j - 5 = 0$ **41.** $z^2 + 9z + 2 = 0$ **42.** $2x^2 + 5x - 9 = 0$

43. $11m^2 - 12m - 10 = 0$ **44.** $4a^2 + 3a - 2 = 0$ **45.** $t^2 - 16t + 4 = 0$

Solve each problem.

46. Find the length and width of a rectangle if its perimeter is 44 inches and its area is 117 square inches.

47. A rectangular plot of grass is 30 meters by 50 meters. The grass is cut in a uniform strip on all four sides. How wide is the strip when one-third of the grass has been cut?

---

mini-review

Simplify.

1. $3\sqrt{18} + 8\sqrt{8}$

2. $\sqrt{3}(5\sqrt{6} + \sqrt{15})$

3. $\sqrt{2p^2q^3} \cdot \sqrt{12p^5q}$

Factor.

4. $y^2 + 6y - 27$

5. $r^2 - 4s^2$

6. $12t^2 + 14t - 6$

Perform the given operation.

7. $3\begin{bmatrix} -4 & 0 & 1 \\ 7 & -2 & 5 \\ 1 & 1 & 4 \end{bmatrix} + \begin{bmatrix} 8 & 0 & 6 \\ -5 & 2 & -1 \\ 4 & -4 & 7 \end{bmatrix}$

8. $\begin{bmatrix} 2 & -1 \\ 5 & 3 \end{bmatrix} \cdot \begin{bmatrix} -2 \\ -3 \end{bmatrix}$

Find the sum, difference, and product for each pair of complex numbers.

9. $6 + i,\ 5 - 2i$

10. $4 + 3i,\ 4 - 3i$

Find each solution using long division. Show your work.

11. $(10s^3 - 3s^2 - 31s - 6) \div (2s + 3)$

12. $(6b^3 - 17b^2 + 8) \div (3b - 4)$

State the dimension of each matrix. Then evaluate its determinant (if one exists).

13. $\begin{bmatrix} 5 & 6 & 9 & 4 \\ -1 & -3 & 2 & 2 \end{bmatrix}$

14. $\begin{bmatrix} 3 & 2 & -2 \\ 1 & -3 & 5 \\ 4 & 0 & -1 \end{bmatrix}$

15. $\begin{bmatrix} 7 & -5 \\ 6 & 3 \end{bmatrix}$

Excursions in Algebra · Handshakes

At the conclusion of a committee meeting, a total of 28 handshakes were exchanged. Assuming each person was equally polite toward all the others, how many people were present? (Hint: Assume n persons were at the meeting. With how many persons did each person shake hands? Use a quadratic equation.)

7-4 The Discriminant

In the quadratic formula the expression under the radical sign, $b^2 - 4ac$, is called the **discriminant**. For example, the discriminant of the equation $2x^2 + 6x - 7 = 0$ is $6^2 - 4(2)(-7)$, or 92.

The discriminant can give information about the solutions, or **roots**, of a quadratic equation.

A root of an equation is a number that satisfies the equation.

Equation	Value of the Discriminant	Roots
$4x^2 + 20x + 25 = 0$	0	$-\dfrac{5}{2}$
$a^2 + a - 12 = 0$	49	$-4, 3$
$x^2 + 5x - 3 = 0$	37	$\dfrac{-5 + \sqrt{37}}{2}, \dfrac{-5 - \sqrt{37}}{2}$
$3y^2 + 4y + 5 = 0$	-44	$\dfrac{-2 + i\sqrt{11}}{3}, \dfrac{-2 - i\sqrt{11}}{3}$

The chart below summarizes the information the discriminant of a quadratic equation gives about its solutions. The coefficients of the variables in the equation must be real numbers.

Discriminant	Nature of the Roots
zero	one real root
positive	two real roots
negative	two imaginary roots

A quadratic equation with *integral* coefficients has rational roots if and only if its discriminant is a perfect square or 0.

Examples

1 **Find the value of the discriminant of $2x^2 + x - 3 = 0$. Then describe the nature of its roots.**

$a = 2, b = 1, c = -3$

$$b^2 - 4ac = (1)^2 - 4(2)(-3)$$
$$= 1 + 24$$
$$= 25$$

The value of the discriminant is a positive perfect square. So, $2x^2 + x - 3 = 0$ has two real roots and they are rational.

2 **Find the value of the discriminant of $x^2 + 8 = 0$. Then describe the nature of its roots.**

$a = 1, b = 0, c = 8$

$$b^2 - 4ac = (0)^2 - 4(1)(8)$$
$$= 0 - 32$$
$$= -32$$

The value of the discriminant is negative. So, $x^2 + 8 = 0$ has two imaginary roots.

Exploratory Exercises

Find the value of the discriminant for each quadratic equation.

1. $x^2 + 5x - 2 = 0$
2. $y^2 + 6y + 9 = 0$
3. $2x^2 - 5x + 3 = 0$
4. $a^2 = 16$
5. $t^2 - 8t + 16 = 0$
6. $5x^2 + 16x + 3 = 0$
7. $2y^2 + y - 10 = 0$
8. $6a^2 + 2a + 1 = 0$
9. $12a^2 - 7a + 1 = 0$
10. $-3x^2 + x - 2 = 0$
11. $x^2 + 4 = 0$
12. $3a^2 - a + 3 = 0$

Written Exercises

Find the value of the discriminant for each quadratic equation. Describe the nature of the roots. If the roots are real, tell whether they are rational or irrational. Then solve each equation.

13. $x^2 - 2x - 35 = 0$
14. $a^2 + 12a + 32 = 0$
15. $y^2 - 4y + 4 = 0$
16. $x^2 - 10x + 25 = 0$
17. $x^2 - 4x + 1 = 0$
18. $m^2 - 6m + 4 = 0$
19. $4x^2 + 8x + 3 = 0$
20. $4y^2 + 16y + 15 = 0$
21. $3x^2 + 11x + 4 = 0$
22. $z^2 + 4z + 2 = 0$
23. $m^2 - 2m + 5 = 0$
24. $y^2 - 6y + 13 = 0$
25. $a^2 + 9a - 2 = 0$
26. $c^2 - 12c + 42 = 0$
27. $a^2 = 6a$
28. $3m^2 = 108m$
29. $4x^2 - 8x + 13 = 0$
30. $x^2 + 4x + 53 = 0$
31. $3n^2 - 19n = -6$
32. $2a^2 - 13a = 7$
33. $x^2 - x + 1 = 0$
34. $n^2 + 4n + 29 = 0$
35. $a^2 + a - 5 = 0$
36. $3b^2 + 7b + 3 = 0$

Challenge Exercises

Find a value of k so that each given equation will have (a) one real root, (b) two real roots, and (c) two imaginary roots.

37. $x^2 + 3x + k = 0$
38. $kx^2 + 3x - 2 = 0$
39. $2x^2 - 5x - k = 0$

Excursions in Algebra

Perfect Numbers

Hrotsvitha (932–1002) was a nun who lived in a Benedictine Abbey in Saxony. She was one of the first persons to write about **perfect numbers**. A perfect number is one that is equal to the sum of its *aliquot* parts. That is, it is equal to the sum of all its factors including 1, but *not* including itself. Consider this example.

$$6 = 1 + 2 + 3$$

The factors of 6, other than 6, are 1, 2, and 3.

Hrotsvitha wrote about three perfect numbers other than 6, namely 28, 496, and 8128.

Exercises

Show that each number is a perfect number.

1. 28
2. 496
3. 8128

Simplify.

1. $(x^3)^2$
2. $\dfrac{a^7}{a^3}$
3. $4a^0$

4. $cd(c^2 + 4cd + 5d^2)$
5. $(4a + 7)(3a - 9)$
6. $(3a - 2)(2a^2 + a - 5)$
7. $(3a)(5a^2b) + (6ab)(10a^2)$
8. $(5t - 9)^2$
9. $(a + 4)(a - 3)(2a - 6)$

Factor.

10. $6a^2 + a - 35$
11. $5z^2 + 4z - 12$
12. $ab + 7a + 4b + 28$
13. $x^2 - x + 3xy - 3y$

14. Use synthetic division to show that $x + 3$ is a factor of $2x^3 + 15x^2 + 22x - 15$.

15. Use long division to show that $a - 5$ is a factor of $a^3 - 5a^2 - 6a + 30$.

Simplify.

16. $\sqrt{600}$
17. $\sqrt[3]{48a^3}$
18. $3\sqrt{18} + 5\sqrt{27}$
19. $\sqrt{3mn^4} \cdot \sqrt{25m^6}$
20. $(6 + \sqrt{3})(7 - \sqrt{2})$
21. $(5 + \sqrt{2})(5 - \sqrt{2})$
22. $\dfrac{(3 + \sqrt{5})}{(1 + \sqrt{2})}$
23. $\dfrac{4}{\sqrt[3]{8}}$
24. $\sqrt{-48}$
25. $(2\sqrt{5})(-5\sqrt{-5})$
26. $\dfrac{(5 + 2i)}{(1 - 2i)}$
27. $\dfrac{(6 - 3i)}{2i}$

Solve each equation.

28. $\sqrt{3w - 2} = 8$
29. $\sqrt[3]{2x + 1} = 3$
30. $\sqrt{z + 12} + \sqrt{z} = 2$
31. $\sqrt[3]{3y - 1} = 2$

32. Solve this system of equations:
$$6x - 2y - 3z = -10$$
$$-6x + y + 9z = 3$$
$$8x - 3y = -16$$

33. Solve this system of inequalities by graphing: $|x + 3| < 4$
$$x + y \geq 2$$

Solve each equation by completing the square.

34. $x^2 + 14x - 12 = 0$
35. $4y^2 - 8y - 7 = 0$

Solve each equation.

36. $x^2 - 20x = -75$
37. $d^2 - 5d - 24 = 0$
38. $2t^2 - 9 = 0$
39. $3c^2 + 5c = -2$
40. $3a^2 - 11a + 10 = 0$
41. $2x^2 - 5x + 4 = 0$

42. Sam bought 6 deluxe, 5 glazed and 2 cake doughnuts for $4.00. If he had bought 4 deluxe, 2 glazed, and 7 cake doughnuts, the cost would have been $3.40. A deluxe doughnut costs 5 cents less than twice the cost of a cake doughnut. Find the cost of each type of doughnut.

7-5 Sum and Product of Roots

Sometimes an equation must be found to fit certain conditions. For example, suppose you know that the roots of a quadratic equation are 5 and -7. Find a quadratic equation with these roots.

Let x stand for a root of the equation. $x = 5$ or $x = -7$

If $x = 5$, then $x - 5 = 0$. If $x = -7$, then $x + 7 = 0$.

$$(x - 5)(x + 7) = 0 \qquad \textit{Why?}$$
$$x^2 + 2x - 35 = 0$$

The quadratic equation $x^2 + 2x - 35 = 0$ has roots 5 and -7. Solve that equation as a check.

Consider the sum and product of 5 and -7.

$$\text{sum} = 5 + (-7) = -2 \qquad \text{product} = 5 \cdot (-7) = -35$$

How are the sum and product related to $x^2 + 2x - 35 = 0$?

In general, the roots of $ax^2 + bx + c = 0$ are $\dfrac{-b + \sqrt{b^2 - 4ac}}{2a}$ and $\dfrac{-b - \sqrt{b^2 - 4ac}}{2a}$. If these roots are called s_1 and s_2, then the sum and product of these roots can be found as follows.

$$s_1 = \frac{-b + \sqrt{b^2 - 4ac}}{2a}$$

$$s_2 = \frac{-b - \sqrt{b^2 - 4ac}}{2a}$$

$$s_1 + s_2 = \frac{-b + \sqrt{b^2 - 4ac}}{2a} + \frac{-b - \sqrt{b^2 - 4ac}}{2a}$$

$$= \frac{-2b}{2a} \qquad\qquad +\sqrt{b^2 - 4ac} - \sqrt{b^2 - 4ac} = 0$$

$$= -\frac{b}{a}$$

$$s_1 s_2 = \frac{-b + \sqrt{b^2 - 4ac}}{2a} \cdot \frac{-b - \sqrt{b^2 - 4ac}}{2a}$$

$$= \frac{b^2 - (b^2 - 4ac)}{4a^2} \qquad \textit{Use Distributive Property.}$$

$$= \frac{c}{a} \qquad\qquad \textit{Simplify.}$$

These conclusions can now be summarized.

If the roots of $ax^2 + bx + c = 0$ with $a \neq 0$ are s_1 and s_2, then

$$s_1 + s_2 = -\frac{b}{a} \text{ and } s_1 s_2 = \frac{c}{a}.$$

Sum and Product of Roots

1 Find the sum and product of the roots of $3x^2 - 16x - 12 = 0$. Then solve the equation.

$$s_1 + s_2 = -\frac{b}{a} \qquad\qquad s_1 s_2 = \frac{c}{a}$$

$$= -\frac{-16}{3} \text{ or } \frac{16}{3} \qquad\qquad = \frac{-12}{3} \text{ or } -4$$

$$x = \frac{-b \pm \sqrt{b^2 - 4ac}}{2a}$$

$$= \frac{-(-16) \pm \sqrt{(-16)^2 - 4(3)(-12)}}{2(3)}$$

In the equation $3x^2 - 16x - 12 = 0$, $a = 3$, $b = -16$, and $c = -12$.

$$= \frac{16 \pm \sqrt{400}}{6} \text{ or } \frac{16 \pm 20}{6}$$

The roots are $\dfrac{16 + 20}{6}$ and $\dfrac{16 - 20}{6}$ or 6 and $-\dfrac{2}{3}$.

2 Find a quadratic equation that has roots $-\dfrac{5}{4}$ and $\dfrac{16}{5}$.

$$s_1 + s_2 = -\frac{5}{4} + \frac{16}{5} \qquad\qquad s_1 s_2 = \left(-\frac{5}{4}\right)\left(\frac{16}{5}\right)$$

$$= \frac{39}{20} \qquad -\frac{b}{a} = \frac{39}{20} \qquad\qquad = \frac{-80}{20} \qquad \frac{c}{a} = \frac{-80}{20}$$

Therefore, $a = 20$, $b = -39$, and $c = -80$.
The equation is $20x^2 - 39x - 80 = 0$.

3 Find a quadratic equation that has roots $5 + 2i$ and $5 - 2i$.

$$s_1 + s_2 = (5 + 2i) + (5 - 2i) \qquad\qquad s_1 s_2 = (5 + 2i)(5 - 2i)$$

$$= 25 + 4$$

$$= 10 \qquad -\frac{b}{a} = 10 \text{ or } \frac{10}{1} \qquad\qquad = 29 \qquad \frac{c}{a} = 29 \text{ or } \frac{29}{1}$$

Therefore, $a = 1$, $b = -10$, and $c = 29$.
The equation is $x^2 - 10x + 29 = 0$.

4 Find k such that -3 is a root of $x^2 + kx - 24 = 0$.

Let $s_1 = -3$. Solve $s_1 s_2 = \dfrac{c}{a}$ for s_2. Then solve for k.

$$-3s_2 = \frac{-24}{1}$$

$$s_2 = 8$$

$$s_1 + s_2 = -\frac{b}{a}$$

$$-3 + 8 = -\frac{k}{1}$$

$$-5 = k \qquad \text{The value of } k \text{ is } -5.$$

State the sum and the product of the roots of each quadratic equation.

1. $x^2 + 7x - 4 = 0$
2. $x^2 + 8x + 7 = 0$
3. $x^2 - 3x + 5 = 0$
4. $2x^2 + 8x - 3 = 0$
5. $3x^2 + 7x - 9 = 0$
6. $2x^2 + 7 = 0$
7. $5x^2 - 3x = 0$
8. $4x^2 + 3x - 12 = 0$
9. $5x^2 = 3$
10. $2x^2 + 9x = 0$
11. $3x^2 - 2x + 11 = 0$
12. $7x^2 = 0$
13. $2x^2 - \frac{1}{2}x + \frac{2}{3} = 0$
14. $x^2 + 4x - \frac{5}{3} = 0$
15. $3x^2 - \frac{x}{5} - \frac{4}{5} = 0$

Written Exercises

Find the sum and the product of the roots of each quadratic equation. Then solve each equation.

16. $x^2 + 6x - 7 = 0$
17. $y^2 + 5y + 6 = 0$
18. $2z^2 - 5z - 3 = 0$
19. $6t^2 + 28t - 10 = 0$
20. $x^2 - 3x + 1 = 0$
21. $2c^2 - 5c + 1 = 0$
22. $4a^2 + 21a = 18$
23. $3b^2 - 8b = 35$
24. $2x^2 - 6x + 5 = 0$
25. $y^2 + 9y + 25 = 0$
26. $9n^2 - 1 = 0$
27. $s^2 - 16 = 0$
28. $2x^2 - 7x = 15$
29. $8m^2 + 6m = -1$
30. $15c^2 - 2c - 8 = 0$
31. $4k^2 + 27k - 7 = 0$
32. $a^2 + 25a + 156 = 0$
33. $x^2 + 4x - 77 = 0$
34. $3z^2 - 7z + 3 = 0$
35. $7s^2 + 5s - 1 = 0$
36. $12x^2 + 19x + 4 = 0$

Find a quadratic equation having the given roots.

37. $8, -2$
38. $5, -2$
39. $6, 4$
40. $-2, 3$
41. $6, -6$
42. $-9, -4$
43. $3, \frac{1}{2}$
44. $5, \frac{2}{3}$
45. $-\frac{3}{4}, 12$
46. $\frac{3}{4}, -4$
47. $-\frac{1}{2}, \frac{1}{2}$
48. $-\frac{2}{5}, \frac{2}{5}$
49. $\frac{5}{8}, \frac{1}{4}$
50. $\frac{1}{3}, \frac{1}{2}$
51. $\sqrt{3}, 2\sqrt{3}$
52. $\sqrt{2}, -5\sqrt{2}$
53. $2 + \sqrt{3}, 2 - \sqrt{3}$
54. $5 - \sqrt{2}, 5 + \sqrt{2}$
55. $3i, -3i$
56. $-6i, 6i$
57. $3 + 7i, 3 - 7i$
58. $5 + i\sqrt{3}, 5 - i\sqrt{3}$
59. $\frac{1 + \sqrt{7}}{2}, \frac{1 - \sqrt{7}}{2}$
60. $\frac{5 - 3i}{4}, \frac{5 + 3i}{4}$

Find k such that the number given is a root of the equation given.

61. $3; x^2 + kx - 21 = 0$
62. $1; x^2 + kx - 5 = 0$
63. $3; x^2 + 6x - k = 0$
64. $-\frac{3}{2}; 2x^2 + kx - 12 = 0$
65. $3; 6x^2 + kx - 5 = 0$
66. $-5; x^2 + 12x + k = 0$

Excursions in Algebra

Contest Problem

The following problem appeared in a high school contest of the Mathematical Association of America.

Suppose s_1 and s_2 are roots of the equation $ax^2 + bx + c = 0$. Find the value of $\frac{s_1^2 + s_2^2}{s_1^2 s_2^2}$ in terms of a, b, and c. (Hint: $s_1^2 + s_2^2 = (s_1 + s_2)^2 - 2s_1 s_2$)

Gravity affects falling objects and objects propelled upwards, like rockets and baseballs. On Earth, the acceleration due to gravity, g, is about 32 ft/s^2. The formula below can be used to compute an object's distance from the ground, s, after t seconds. The initial velocity, v_0, is given in feet per second.

$$s = v_0 t - \frac{1}{2}gt^2 \qquad \text{Replace } g \text{ by 32.} \qquad s = v_0 t - 16t^2$$

Example 1: **An object is propelled upwards from the ground with an initial velocity of 192 feet per second. How long will it take to return to ground level?**

$$s = v_0 t - 16t^2$$
$$0 = 192t - 16t^2$$

Replace s by 0 since the distance is 0 when the object returns to ground level. Replace v_0 by 192.

$$0 = t(192 - 16t)$$

Factor and use the Zero Product Property.

$$t = 0 \quad \text{or} \quad 192 - 16t = 0$$
$$-16t = -192$$
$$t = 12$$

The object will return to the ground in 12 seconds.

Example 2: **A rocket is fired upwards from a platform 288 feet above ground with an initial velocity of 960 feet per second. Find the number of seconds it will take the rocket to return to ground level.**

$$s = v_0 t - 16t^2$$
$$-288 = 960t - 16t^2$$

Ground level is 288 feet below the platform, so replace s by -288. Replace v_0 by 960.

$$16t^2 - 960t - 288 = 0$$

Rewrite the equation so the coefficient of t^2 is positive.

$$t^2 - 60t - 18 = 0$$

This is not easily factorable, so use the quadratic formula.

$$t \approx -0.3 \quad \text{or} \quad t \approx 60.3$$

Reject the negative root since negative time values represent the time before the rocket is fired. The rocket will return to the ground in about 60 seconds.

Exercises

Given the height of the platform from which each rocket is launched and the initial velocity below, find the number of seconds it will take the rocket to return to ground level.

1. height: 96 ft
velocity: 80 ft/s

2. height: 480 ft
velocity: 112 ft/s

3. height: 288 ft
velocity: 112 ft/s

7-6 Problem Solving: Quadratic Equations

Often a drawing will help you write an equation to solve a problem. Consider the following problem.

> The Pinetown Recreation Bureau planned to build an ice-skating rink with dimensions 30 meters by 60 meters. Their budget has been cut, so they must reduce the area of the rink to 1000 square meters. A strip will be removed from one end, and a strip of the same width will be removed from one side. Find the width of the strips.

explore   The problem asks for the width of the strips. Let w stand for the width of the strips.

Make a drawing.

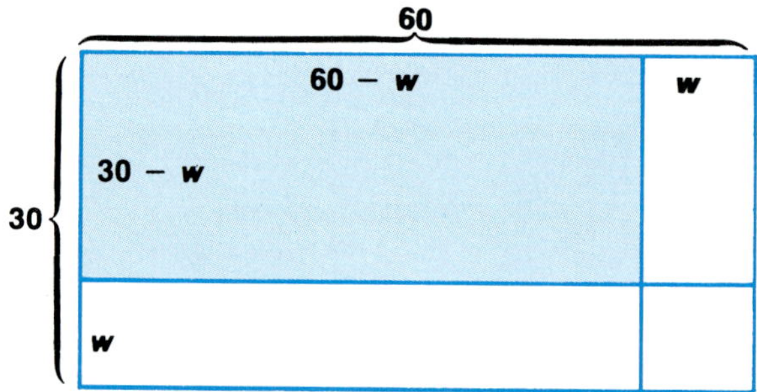

plan   The length of the blue rectangle is $60 - w$. The width of the blue rectangle is $30 - w$.

$$\text{length} \times \text{width} = \text{area}$$
$$(60 - w)(30 - w) = 1000$$

solve
$$(60 - w)(30 - w) = 1000$$
$$1800 - 60w - 30w + w^2 = 1000$$
$$800 - 90w + w^2 = 0$$
$$(10 - w)(80 - w) = 0$$

$$10 - w = 0 \quad \text{or} \quad 80 - w = 0$$
$$10 = w \quad \text{or} \quad 80 = w$$

Check all solutions to the problem. In this case, 80 is not a reasonable solution, so the strip will be 10 meters wide.

examine   If a 10-meter strip is removed from 30 meters, the width will be 20 meters. Likewise, the 60-meter length minus the 10 meters will leave a length of 50 meters.

$$20 \text{ m} \times 50 \text{ m} = 1000 \text{ m}^2$$

Written Exercises

Solve each problem.

1. Find two consecutive integers whose product is 702.

2. Find two consecutive even integers whose product is 288.

3. Find two consecutive odd integers whose product is 1443.

4. Find two consecutive integers whose product is 552.

5. If the product of two consecutive odd integers is decreased by one-third the lesser integer, the result is 250. Find the integers.

6. If the product of two consecutive integers is decreased by 20 times the greater integer, the result is 442. Find the integers.

7. The sum of the squares of two consecutive integers is 265. Find the integers.

8. The difference of the squares of two consecutive integers is 21. Find the integers.

9. A local rectangular park is 30 meters long by 20 meters wide. Plans are being made to double the area by adding a strip at one end and another of the same width on one side. Find the width of the strips.

10. A rectangular picture is 12 by 16 inches. If a frame of uniform width contains an area of 165 square inches, what is the width of the frame?

11. The length of a rectangular garden is 6 feet more than its width. A walkway 3 feet wide surrounds the outside of the garden. The total area of the walkway is 288 square feet. Find the dimensions of the garden.

12. Jackie is building a playhouse. She wants each rectangular window to have an area of 315 square inches. She also wants each window to be 6 inches higher than it is wide. What are the dimensions of each window?

13. Three times the square of a number equals 21 times the number. Find the number.

14. The difference of the squares of two consecutive odd integers is 48. Find the integers.

15. If a number is increased by its square, the result is 72. Find the number.

16. The square of a number exceeds 11 times the number by 312. Find the number.

17. If a number is decreased by its square, the result is $\frac{2}{9}$. Find the number.

18. The square of a number decreased by the square of one-half the number is 108. Find the number.

19. The sum of a number and its reciprocal is $\frac{10}{3}$. Find the number.

20. A number increased by 4 times its reciprocal is $\frac{20}{3}$. Find the number.

21. The Hillside Garden Club wants to double the area of its rectangular display of roses. If it is now 6 meters by 4 meters, by what equal amount must each dimension be increased?

22. A rectangular garden 25 feet by 50 feet is increased on all sides by the same amount. Its area increases 400 square feet. By how much is each dimension increased?

23. If a number is decreased by its reciprocal, the result is $\frac{11}{30}$. Find the number.

24. The difference between a number and its reciprocal is $\frac{16}{15}$. Find the number.

25. Jim Finley is a professional photographer. He has a photo 8 centimeters long and 6 centimeters wide. A customer wants a print of the photo. The print is to have half the area of the original. Jim plans to reduce the length and width of the photo by the same amount. What are the dimensions of the print?

26. Gary and Jan's family room has a rug that is 9 feet by 12 feet. A strip of floor of equal width is uncovered along all edges of the rug. If the area of the uncovered floor is 270 square feet, how wide is the strip?

27. The square of a number increased by 21 is equal to 7 times the number increased by 21. Find the number.

28. Thirty decreased by one-half of a number is one-sixth of the square of the number. Find the number.

29. A rectangular flower bed in Lake Park is 20 by 28 meters. A walk of uniform width surrounds the flower bed. If the area of the flower bed and walk is 1008 square meters, what is the width of the walk?

30. Three boys are to mow a rectangular lawn with dimensions 100 by 120 feet. Bill is going to mow one-third of the lawn by mowing a strip of uniform width around the outer edge of the lawn. What is the width of the strip?

Challenge Exercise

31. In the diagram at the right, the rectangle created by the dotted line and shaded red is similar to the rectangle with dimensions 1 by x. The value of x represents the "golden ratio." Write a proportion and solve for x to find the golden ratio. Round to the nearest thousandth.

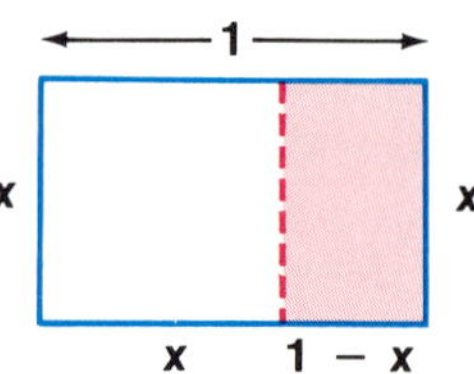

mini-review

Simplify.

1. $\dfrac{1}{a^{\frac{3}{4}}}$

2. $\dfrac{cd}{\sqrt[3]{e}}$

3. $\dfrac{f^{\frac{4}{3}} + 2f^{-\frac{2}{3}}}{f^{\frac{1}{3}}}$

4. $\dfrac{gh}{g^{\frac{1}{2}} + 5}$

5. $\dfrac{5 - 2i}{3i}$

6. $\dfrac{7 + 7i}{1 - 4i}$

7. Evaluate $\dfrac{84,000,000 \times 0.05}{0.00015}$. Express the result in scientific notation.

8. Use long division to show that $j + 8$ is a factor of $j^3 + 2j^2 - 55j - 56$.

9. Use synthetic division to show that $k - 3$ is not a factor of $k^3 + k^2 - 22k + 21$.

10. At Freda's Fancy Fruits, 2 cantaloupes, 2 nectarines, and 4 tangerines cost $3.54. A cantaloupe, 5 nectarines, and 6 tangerines cost $4.21. If the cost of a nectarine is 20 cents less than twice the cost of a tangerine, find the cost of each item.

7-7 Quadratic Form

The equation $x^4 - 20x^2 + 64 = 0$ is *not* a quadratic equation. But it looks very much like a quadratic equation. It is in **quadratic form** because it can be written as $(x^2)^2 - 20(x^2) + 64 = 0$.

> **For any numbers a, b, and c, except $a = 0$, an equation that may be written as $a[f(x)]^2 + b[f(x)] + c = 0$, where $f(x)$ is some expression in x, is in quadratic form.**
>
> *Definition of Quadratic Form*

An equation in quadratic form can be solved by the same methods used for solving quadratic equations.

In an equation such as $x^2 + 2x + 1 = 0$, $f(x)$ is just x.

Example

1   **Solve $x^4 - 13x^2 + 36 = 0$.**

$$x^4 - 13x^2 + 36 = 0$$
$$(x^2)^2 - 13(x^2) + 36 = 0 \qquad \text{\textit{The equation is in quadratic form. } } f(x) = x^2$$
$$(x^2 - 9)(x^2 - 4) = 0$$
$$(x + 3)(x - 3)(x + 2)(x - 2) = 0$$

$$x + 3 = 0 \quad \text{or} \quad x - 3 = 0 \quad \text{or} \quad x + 2 = 0 \quad \text{or} \quad x - 2 = 0$$
$$x = -3 \quad \text{or} \quad x = 3 \quad \text{or} \quad x = -2 \quad \text{or} \quad x = 2$$

The solutions or roots are -3, 3, -2, and 2.

Recall that $(a^m)^n$ equals a^{mn} for any positive number a and any rational numbers n and m. This property of exponents is often used when solving equations.

Example

2   **Solve $t^{\frac{2}{3}} = 16$.**

$$t^{\frac{2}{3}} = 16$$
$$(t^{\frac{2}{3}})^{\frac{3}{2}} = \pm(16)^{\frac{3}{2}} \qquad \text{\textit{Raise each side to the power } } \tfrac{3}{2}. \text{ \textit{Why is the } } \pm \text{ \textit{necessary?}}$$
$$t = \pm(16^{\frac{1}{2}})^3$$
$$t = \pm(4)^3$$
$$t = \pm 64$$

Check:

$$t^{\frac{2}{3}} = 16 \qquad\qquad t^{\frac{2}{3}} = 16$$
$$64^{\frac{2}{3}} \stackrel{?}{=} 16 \qquad\qquad (-64)^{\frac{2}{3}} \stackrel{?}{=} 16$$
$$(\sqrt[3]{64})^2 \stackrel{?}{=} 16 \qquad\qquad (\sqrt[3]{-64})^2 \stackrel{?}{=} 16$$
$$4^2 \stackrel{?}{=} 16 \qquad\qquad (-4)^2 \stackrel{?}{=} 16$$
$$16 = 16 \ \checkmark \qquad\qquad 16 = 16 \ \checkmark$$

The solutions or roots are 64 and -64.

3 **Solve $y^{-2} - 64 = 0$.**

$$y^{-2} - 64 = 0$$
$$(y^{-1})^2 - 64 = 0$$
$$(y^{-1} + 8)(y^{-1} - 8) = 0$$
$$y^{-1} = -8 \quad \text{or} \quad y^{-1} = 8$$
$$y = -\frac{1}{8} \qquad y = \frac{1}{8}$$

The equation may also be solved as follows.

$$y^{-2} = 64$$
$$(y^{-2})^{-\frac{1}{2}} = 64^{-\frac{1}{2}}$$
$$y = \pm\frac{1}{8}$$

The solutions are $-\frac{1}{8}$ and $\frac{1}{8}$.

4 **Solve $x^{\frac{1}{2}} - 6x^{\frac{1}{4}} + 8 = 0$.**

$$x^{\frac{1}{2}} - 6x^{\frac{1}{4}} + 8 = 0$$
$$(x^{\frac{1}{4}})^2 - 6(x^{\frac{1}{4}}) + 8 = 0 \qquad \textit{f(x) is } x^{\frac{1}{4}}.$$
$$(x^{\frac{1}{4}} - 2)(x^{\frac{1}{4}} - 4) = 0 \qquad \textit{Factor to solve for f(x).}$$
$$x^{\frac{1}{4}} - 2 = 0 \quad \text{or} \quad x^{\frac{1}{4}} - 4 = 0$$
$$x^{\frac{1}{4}} = 2 \qquad\qquad x^{\frac{1}{4}} = 4$$
$$(x^{\frac{1}{4}})^4 = (2)^4 \qquad (x^{\frac{1}{4}})^4 = 4^4$$
$$x = 2^4 \text{ or } 16 \qquad x = 4^4 \text{ or } 256$$

Check:
$$x^{\frac{1}{2}} - 6x^{\frac{1}{4}} + 8 = 0$$
$$16^{\frac{1}{2}} - 6(16^{\frac{1}{4}}) + 8 \stackrel{?}{=} 0$$
$$4 - 6(2) + 8 \stackrel{?}{=} 0$$
$$0 = 0 \quad ✔$$

$$256^{\frac{1}{2}} - 6(256^{\frac{1}{4}}) + 8 \stackrel{?}{=} 0$$
$$16 - 6(4) + 8 \stackrel{?}{=} 0$$
$$0 = 0 \quad ✔$$

The solutions are 16 and 256.

The quadratic formula can be used to solve equations that are expressed in quadratic form.

Remember to check each solution carefully.

5 **Solve $x - 7\sqrt{x} - 8 = 0$.**

$$x - 7\sqrt{x} - 8 = 0$$
$$(\sqrt{x})^2 - 7(\sqrt{x}) - 8 = 0 \qquad \textit{f(x) is } \sqrt{x}.$$
$$\sqrt{x} = \frac{-b \pm \sqrt{b^2 - 4ac}}{2a}$$
$$= \frac{-(-7) \pm \sqrt{(-7)^2 - 4(1)(-8)}}{2(1)} \qquad \begin{array}{l} a = 1, \\ b = -7, \\ c = -8 \end{array}$$
$$= \frac{7 \pm \sqrt{81}}{2} \text{ or } \frac{7 \pm 9}{2}$$
$$\sqrt{x} = 8 \quad \text{or} \quad \sqrt{x} = -1$$
$$x = 64 \quad \text{or} \quad x = 1$$

Check: $x - 7\sqrt{x} - 8 = 0$
$$64 - 7\sqrt{64} - 8 \stackrel{?}{=} 0$$
$$64 - 7 \cdot 8 - 8 \stackrel{?}{=} 0$$
$$0 = 0 \quad ✔$$

$$1 - 7\sqrt{1} - 8 \stackrel{?}{=} 0$$
$$1 - 7 - 8 \stackrel{?}{=} 0$$
$$-14 \neq 0$$

The only solution is 64.

State whether each equation is in quadratic form.

1. $x^4 + 5x^2 + 3 = 0$
2. $4y^4 - 3y^2 + 2 = 0$
3. $6x^4 + 7x - 8 = 0$
4. $6x^4 + 8x^2 = 0$
5. $6x + 5\sqrt{x} - 2 = 0$
6. $2p + 5\sqrt{p} = 9$

Solve each equation.

7. $r^{\frac{1}{3}} = 2$
8. $x^{\frac{1}{3}} = 3$
9. $y^{\frac{1}{2}} = 5$
10. $x^{-\frac{1}{2}} = 4$
11. $z^{-2} = 25$
12. $r^{-3} = 27$
13. $y^{\frac{3}{2}} - 8 = 0$
14. $z^{-\frac{1}{3}} - 2 = 0$
15. $p^{-2} = 169$

Written Exercises

Express each equation in quadratic form.

16. $x^{\frac{4}{3}} - 7x^{\frac{2}{3}} + 12 = 0$
17. $x^{-6} - 8x^{-3} + 16 = 0$
18. $x - 10x^{\frac{1}{2}} + 25 = 0$
19. $x^{\frac{1}{2}} + 7x^{\frac{1}{4}} + 12 = 0$
20. $x^{\frac{1}{2}} - 8x^{\frac{1}{4}} + 15 = 0$
21. $y^{\frac{1}{2}} - 10y^{\frac{1}{4}} + 16 = 0$
22. $r^{\frac{2}{3}} - 5r^{\frac{1}{3}} + 6 = 0$
23. $s^{\frac{2}{3}} - 9s^{\frac{1}{3}} + 20 = 0$
24. $a^{-\frac{2}{3}} - 11a^{-\frac{1}{3}} + 28 = 0$
25. $k^{-\frac{4}{3}} - 10k^{-\frac{2}{3}} + 21 = 0$

Solve each equation.

26. $x^4 - 5x^2 + 4 = 0$
27. $y^4 - 3y^2 + 2 = 0$
28. $z^4 - 25z^2 + 144 = 0$
29. $m^4 - 40m^2 + 144 = 0$
30. $s^4 - 25 = 0$
31. $x^4 - 16 = 0$
32. $y^4 - 9 = 0$
33. $a^4 - 36 = 0$
34. $x^4 - 25x^2 = 0$
35. $z^4 - 9z^2 = 0$
36. $b^4 + 9b^2 + 18 = 0$
37. $c^4 - 2c^2 - 8 = 0$
38. $x^4 - 6x^2 + 8 = 0$
39. $y^4 - 11y^2 + 24 = 0$
40. $m - 9\sqrt{m} + 8 = 0$
41. $s - 13\sqrt{s} + 36 = 0$
42. $x - 2\sqrt{x} + 1 = 0$
43. $z - 16\sqrt{z} + 64 = 0$
44. $a^6 - 64a^3 = 0$
45. $m^6 - 64 = 0$
46. $z^6 - 7z^3 - 8 = 0$
47. $a^6 + 26a^3 - 27 = 0$
48. $y^6 - 10y^3 + 16 = 0$
49. $m^6 - 2m^3 + 1 = 0$
50. $x^{\frac{1}{2}} - 10x^{\frac{1}{4}} + 16 = 0$
51. $x^{\frac{2}{3}} - 8x^{\frac{1}{3}} + 15 = 0$
52. $r^{\frac{2}{3}} - 12r^{\frac{1}{3}} + 20 = 0$
53. $b^{\frac{2}{3}} - 7b^{\frac{1}{3}} + 10 = 0$
54. $m - 11m^{\frac{1}{2}} + 30 = 0$
55. $s - 5s^{\frac{1}{2}} + 6 = 0$
56. $x^{\frac{4}{3}} - 8x^{\frac{2}{3}} + 16 = 0$
57. $y^{\frac{4}{3}} - 13y^{\frac{2}{3}} + 36 = 0$
58. $a^{-\frac{2}{3}} - 10a^{-\frac{1}{3}} + 21 = 0$
59. $y^{-\frac{2}{5}} - 4y^{-\frac{1}{5}} + 4 = 0$
60. $y^{-1} - 5y^{-\frac{1}{2}} + 6 = 0$
61. $y^3 - 16y^{\frac{3}{2}} + 64 = 0$

Challenge Exercises

Solve each equation.

62. $3g^{\frac{2}{3}} - 10g^{\frac{1}{3}} + 8 = 0$
63. $2a^{\frac{1}{2}} - 13a^{\frac{1}{4}} + 20 = 0$
64. $6x - 19x^{\frac{1}{2}} + 15 = 0$
65. $3m + m^{\frac{1}{2}} - 2 = 0$
66. $2r^{\frac{1}{2}} + r^{\frac{1}{4}} - 15 = 0$
67. $2y^{\frac{2}{3}} - 5y^{\frac{1}{3}} - 12 = 0$
68. $|x + 1|^2 = 4$
69. $|y - 2|^2 - 9 = 0$
70. $|m + 3|^2 - 2|m + 3| = -1$
71. $|a - 4|^2 - 7|a - 4| = -6$

Discriminant and Roots

The computer program at the right can be used to find the roots of a quadratic equation in decimal form. The program also states whether the roots are real or imaginary and prints the discriminant.

A quadratic equation in the form $ax^2 + bx + c = 0$ provides the values for A, B, and C.

Example: $x^2 - 14x + 49 = 0$

$$A = 1, B = -14, C = 49$$

Enter and run this program on your computer using these values. You should get the following output.

```
ONE ROOT: 7
THE DISCRIMINANT IS 0
```

```
10    INPUT "ENTER A,B,C, OF
         QUADRATIC FORMULA: ";A,B,C
20    LET D = B * B − 4 * A * C
30    IF D < 0 THEN 110
40    IF D > 0 THEN 70
50    PRINT "ONE ROOT: "; −B / (2 * A)
60    GOTO 150
70    LET X = (−B + SQR (D)) / (2 * A)
80    LET Y = (−B − SQR (D)) / (2 * A)
90    PRINT "TWO REAL ROOTS: ";X;" " ;Y
100   GOTO 150
110   PRINT "TWO COMPLEX ROOTS: ";
120   PRINT −B / (2 * A);"+";SQR
         (−D) / (2 * A);"I"
130   PRINT −B / (2 * A);"−";SQR
         (−D) / (2 * A);"I"
150   PRINT "THE DISCRIMINANT IS ";D
160   END
```

Exercises

Use the computer program above to find the values of the discriminant and roots of each quadratic equation. Describe the nature of the roots. If the roots are real, decide whether or not they are rational.

1. $10x^2 + 33x - 7 = 0$

2. $6n^2 + 8n + 1 = 0$

3. $3y^2 + 3y = -2$

4. $x^2 = 6x - 13$

Find the value of the discriminant and roots of each equation. What do you notice about the roots and discriminants of each group of equations?

5. $m^2 - 4m + 6 = 0$
$m^2 + 4m + 6 = 0$

6. $x^2 - 3x - 28 = 0$
$2x^2 - 6x - 56 = 0$
$5x^2 - 15x - 140 = 0$

7. $x^2 - 25 = 0$
$2x^2 - 10 = 0$
$16x^2 - 12 = 0$

8. $x^2 + 4x + 4 = 0$
$x^2 - 10x + 25 = 0$
$4x^2 + 32x + 64 = 0$

quadratic equation   (233)
zero product property   (233)
completing the square   (236)
quadratic formula   (239)

discriminant   (243)
roots   (243)
quadratic form   (253)

Chapter Summary

1. Zero Product Property: For any numbers a and b, if $ab = 0$, then $a = 0$ or $b = 0$.   (233)

2. Completing the square can be used to solve quadratic equations.   (236)

3. Quadratic Formula: The roots of a quadratic equation of the form $ax^2 + bx + c = 0$ with $a \neq 0$ are given by the following formula.

$$x = \frac{-b \pm \sqrt{b^2 - 4ac}}{2a} \quad (239)$$

4. The discriminant, $b^2 - 4ac$, gives information about the solutions or roots of a quadratic equation whose coefficients are real numbers.   (243)

Discriminant	Nature of the Roots
zero	one real root
positive	two real roots
negative	two imaginary roots

5. Sum and Product of Roots: If the roots of $ax^2 + bx + c = 0$, with $a \neq 0$, are s_1 and s_2, then

$$s_1 + s_2 = -\frac{b}{a} \text{ and } s_1 s_2 = \frac{c}{a}. \quad (246)$$

6. For any numbers a, b, and c, except $a = 0$, an equation that may be written as $a[f(x)]^2 + b[f(x)] + c = 0$, where $f(x)$ is some expression in x, is in quadratic form.   (253)

7. An equation in quadratic form can be solved by the same methods used for solving quadratic equations.   (253)

7–1   **Solve each equation.**

1. $(2x + 3)(3x - 1) = 0$
2. $(x + 7)(4x - 5) = 0$
3. $2x^2 + 5x + 3 = 0$
4. $2x^2 + 9x + 4 = 0$
5. $15a^2 + 13a = 6$
6. $8b^2 + 10b = 3$

7–2   **Find the value of c that makes each trinomial a perfect square.**

7. $x^2 + 14x + c$
8. $a^2 - 7a + c$

Solve each equation by completing the square.

9. $x^2 - 20x + 75 = 0$
10. $x^2 - 5x - 24 = 0$
11. $2t^2 + t - 21 = 0$
12. $r^2 + 4r = 96$

7–3   **Solve each equation using the quadratic formula.**

13. $3x^2 - 11x + 10 = 0$
14. $2x^2 - 8x = 0$
15. $2p^2 - 9 = 0$
16. $2q^2 - 5q + 4 = 0$

7–4   **Find the value of the discriminant for each quadratic equation. Describe the nature of the roots. If the roots are real, tell whether they are rational or irrational. Then solve each equation.**

17. $4x^2 - 40x + 25 = 0$
18. $2y^2 + 6y + 5 = 0$
19. $n^2 = 8n - 16$
20. $7b^2 = 4b$

7–5   **Find the sum and the product of the roots for each quadratic equation. Then solve each equation.**

21. $x^2 - 12x - 45 = 0$
22. $2m^2 - 10m + 9 = 0$
23. $3s^2 - 11 = 0$
24. $2x^2 = 3 - 5x$

Find a quadratic equation having the given roots.

25. $4, -6$
26. $\dfrac{3}{4}, \dfrac{1}{3}$
27. $5 - 3i, 5 + 3i$
28. $2 - \sqrt{3}, 2 + \sqrt{3}$

7–6   **Solve each problem.**

29. The square of a number decreased by twenty times that number is 384. Find the number.

30. Find five consecutive integers such that the sum of the squares of the smallest and largest is 208.

31. A rectangular lawn has dimensions 24 feet by 32 feet. A sidewalk will be constructed along the inside edges of all four sides. The remaining lawn will have an area of 425 square feet. How wide will the walk be?

7–7   **Solve each equation.**

32. $x^4 - 8x^2 + 16 = 0$
33. $x^4 - 12x^2 + 27 = 0$
34. $p - 4\sqrt{p} - 45 = 0$
35. $r + 9\sqrt{r} = -8$
36. $x^{\frac{1}{2}} - 15x^{\frac{1}{4}} + 36 = 0$
37. $x^{\frac{2}{3}} - 9x^{\frac{1}{3}} + 20 = 0$

Chapter Test

Solve each equation.

1. $a^2 + 8a - 33 = 0$

2. $6y^2 - y - 15 = 0$

3. $x^2 - 6x + 8 = 0$

4. $y^2 + 7y - 18 = 0$

5. $3b^2 + b - 14 = 0$

6. $12p^2 - 5p = 3$

7. $5x^2 - 125 = 0$

8. $4x^2 = 324$

9. $c^2 + 6c - 216 = 0$

10. $3x^2 + 4x + 2 = 0$

11. $x^4 - 9x^2 + 20 = 0$

12. $r - 9\sqrt{r} + 8 = 0$

13. $2d + 3\sqrt{d} = 9$

14. $x^4 - 11x^2 - 80 = 0$

15. $x^{\frac{1}{2}} - 15x^{\frac{1}{4}} + 50 = 0$

16. $s - 11s^{\frac{1}{2}} + 30 = 0$

17. $3f + f^{\frac{1}{2}} - 2 = 0$

18. $2m^{\frac{2}{3}} - 5m^{\frac{1}{3}} - 12 = 0$

Find the value of c that makes each trinomial a perfect square.

19. $n^2 + 6n + c$

20. $x^2 - 5x + c$

Find the value of the discriminant for each quadratic equation. Describe the nature of the roots. If the roots are real, tell whether they are rational or irrational.

21. $6x^2 + 7x - 5 = 0$

22. $2y^2 - 9y + 11 = 0$

23. $9a^2 - 30a + 25 = 0$

24. $7m^2 = 4m + 1$

Find the sum and the product of the roots for each quadratic equation.

25. $x^2 - 15x + 56 = 0$

26. $2z^2 - 3z - 12 = 0$

27. $n + 7 = 4n^2$

28. $2x^2 = 3 - 5x$

Find a quadratic equation having the given roots.

29. $0, -3$

30. $8, -3$

31. $\frac{4}{3}, \frac{2}{3}$

32. $5 + 2i, 5 - 2i$

Solve each problem.

33. The sum of the squares of two consecutive odd integers is 1154. Find the integers.

34. The Dolphin Pool Company will build a pool for Sally Wadman having 600 square feet of surface. Ms. Wadman's pool, along with a deck of uniform width, will have dimensions 30 feet by 40 feet. What will be the width of the deck around the pool?

CHAPTER 8

Quadratic Relations and Functions

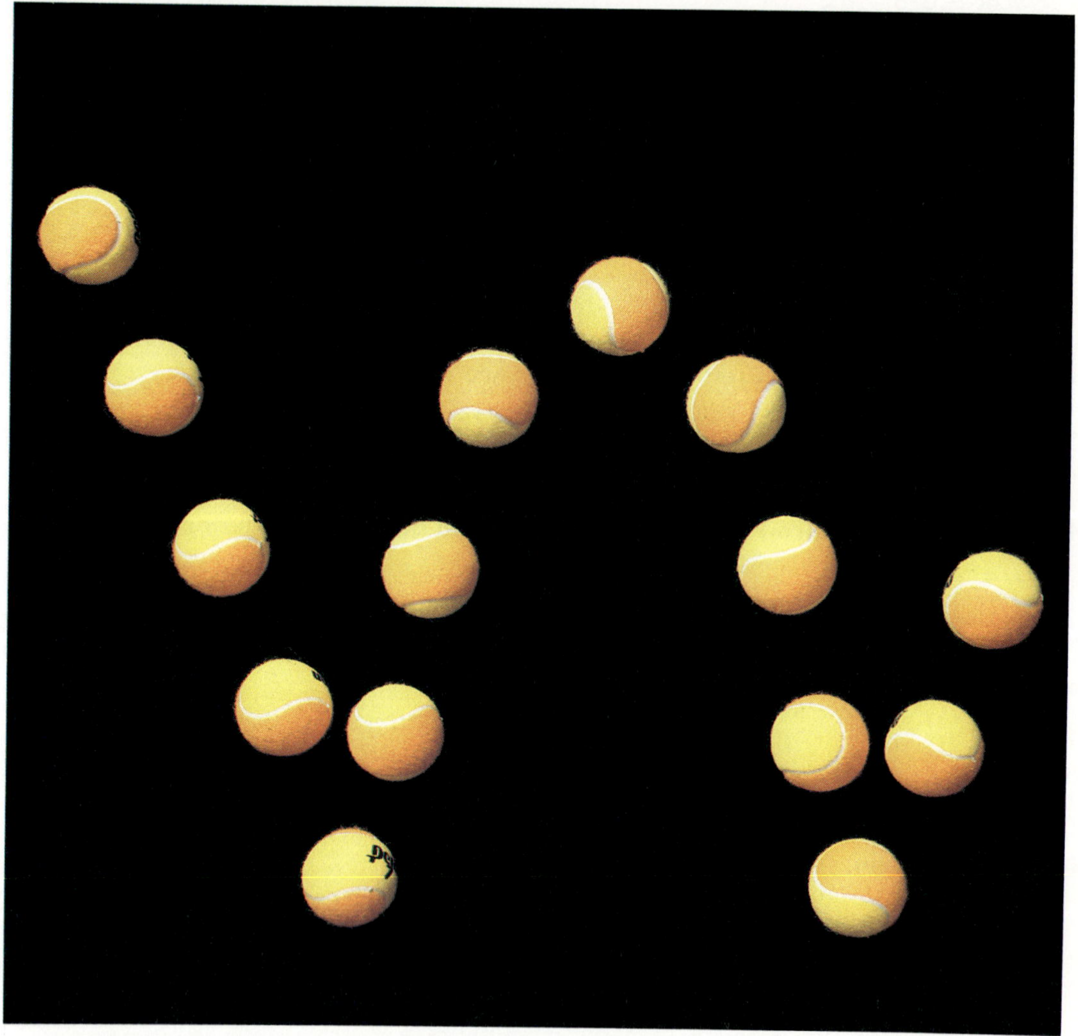

In physics, the path an object follows when thrown or dropped is called a trajectory . A bouncing object, like the tennis ball in the photo, will have a trajectory in the shape of a parabola.

8-1 Quadratic Functions

A model rocket is launched with an initial velocity of 50 meters per second. The height of the rocket is a function of the time after blast off. The height, $h(t)$, of the rocket t seconds after blast off is given by the equation $h(t) = 50t - 5t^2$. This equation is an example of a **quadratic function**.

A quadratic function is a function described by an equation of the form $f(x) = ax^2 + bx + c$ where $a \neq 0$.	*Definition of Quadratic Function*

In a quadratic function, ax^2 is called the *quadratic term*, bx is called the *linear term*, and c is called the *constant term*.

$f(x) = bx + c$ describes a linear function. $f(x) = c$ describes a constant function.

Examples

1 Write $f(x) = (x - 3)^2 + 5$ in quadratic form. Identify the quadratic term, the linear term, and the constant term.

$$f(x) = (x - 3)^2 + 5$$
$$= x^2 - 6x + 9 + 5$$
$$= x^2 - 6x + 14$$

The quadratic term is x^2.
The linear term is $-6x$.
The constant term is 14.

2 A theater has seats for 500 people. It is filled to capacity for each show and tickets cost \$3.00 per show. The owner wants to increase ticket prices. She estimates that for each \$0.20 increase in price, 25 fewer people will attend. Define a variable and write a quadratic function to describe the owner's income after she increases her prices.

Let p = number of \$0.20 price increases.
Then $3.00 + 0.20p$ = ticket price.
And $500 - 25p$ = number tickets sold.

Income = (number of tickets sold) × (ticket price)
$$I(p) = (500 - 25p) \times (3.00 + 0.20p)$$
$$= 1500 + 100p - 75p - 5p^2$$
$$= 1500 + 25p - 5p^2$$

Let $I(p)$ represent income to show that income is a function of price.

Exploratory Exercises

State whether each equation describes a quadratic function.

1. $f(x) = x^2 + 3x + 5$

2. $f(x) = -3x^2 - 8x - 7$

3. $f(x) = 2x - 6$

4. $f(x) = (x - 4)^2$

5. $g(x) = -3(x - 4)^2 - 6$

6. $p(x) = x + 1$

7. $m(x) = 3x^2$

8. $r(s) = s^2 + 2s$

9. $f(x) = \dfrac{1}{x^2} + \dfrac{1}{x} + 1$

10. $g(x) = -\dfrac{1}{3}x + \dfrac{4}{5}$

For each function identify the quadratic term, the linear term, and the constant term.

11. $f(x) = x^2 + 3x - \dfrac{1}{4}$

12. $f(x) = 4x^2 - 8x - 2$

13. $m(x) = x^2 - 3x - \dfrac{1}{4}$

14. $g(p) = \dfrac{1}{3}p + 4$

15. $g(a) = 3a^2 - 2$

16. $n(x) = -4x^2 - 8x - 9$

17. $j(z) = z^2 + 3z$

18. $q(d) = -4d^2 - 2d$

19. $h(x) = (x + 3)^2$

20. $h(x) = (2x - 5)^2$

21. $f(r) = (r - 2)^2 + 5$

22. $g(t) = (3t + 1)^2 - 8$

Written Exercises

Express each function in quadratic form.

23. $f(x) = (x - 2)^2$

24. $f(x) = (x + 4)^2$

25. $f(x) = (3x + 2)^2$

26. $f(x) = (2x - 5)^2$

27. $f(x) = 2(4x + 1)^2$

28. $f(x) = -4(2x - 4)^2$

29. $f(x) = 3(x - 4)^2 - 6$

30. $f(x) = 4(x + 1)^2 + 10$

31. $f(x) = 5(3x - 2)^2 + 4$

32. $f(x) = -3(2x + 2)^2 + 6$

33. $f(x) = \dfrac{1}{5}(10x - 5)^2 + 8$

34. $f(x) = \dfrac{1}{6}(6x + 12)^2 + 5$

35. Write a quadratic function to describe the area of a circle in terms of its radius.

36. Write a quadratic function to describe the area of an isosceles right triangle in terms of its legs.

Define a variable and write a quadratic function to describe each situation.

37. the product of two numbers whose sum is 40

38. the product of two numbers whose sum is 36

39. the product of two numbers whose difference is 64

40. the product of two numbers whose difference is 25

41. the area of a rectangle whose perimeter is 20 centimeters

42. the area of a rectangle whose perimeter is 64 millimeters

43. the sum of the squares of two numbers whose sum is 10

44. the sum of the squares of two numbers whose difference is 18

45. Ms. Morrison has 120 meters of fence to make a rectangular pen for her ducks. She will use the side of a shed for one side of the pen. Write a quadratic function to describe the area of the pen.

46. Bill Taylor's rectangular garden has a fence along one side. He wishes to fence in the other three sides with 200 feet of fencing. Write a quadratic function to describe the area of the garden.

47. A taxi service transports 300 passengers a day between two airports. The charge is $8.00. The owner estimates that for each $1 increase in fare, he will lose 20 passengers. Write a quadratic function to describe the owner's income after he increases his prices.

48. Last year 200 people came to see the fall play at Jones High School. The cost per ticket was $2.00. This year the drama teacher estimates that for each $0.25 increase in price, 10 fewer people will come to the play. Write a quadratic function to describe the income after the price is increased.

mini-review

Simplify.

1. $\left(\frac{4}{3}xy^2\right)^3\left(\frac{9}{8}x^4y^2\right)^2$

2. $\frac{36a^3b^{-2}}{(3a^{-2}b)^3}$

3. $\frac{4^{2x-3}}{4^{2x}}$

4. $(3y - 5)(y + 8)$

5. $\sqrt[4]{16b^8r^4}$

6. $\frac{x^{\frac{2}{3}}}{x^{\frac{1}{3}} - x^{-\frac{2}{3}}}$

7. $\sqrt{-24} \cdot \sqrt{-8}$

8. $\sqrt{\frac{3}{5}} + \sqrt{60} - \sqrt{15}$

9. $\frac{1 - 2i}{2 - i}$

Factor.

10. $2a^2 - 6ab + a - 3b$

11. $x^3 - 4x^2 - 4x + 16$

12. $3z^2 + 5z - 2$

13. $x^2 + 8x + 16 - 4y^2$

14. Evaluate $(216)^{\frac{2}{3}}$.

15. Express $\sqrt[3]{27r^2s^4}$ using exponents.

16. Express $4^{\frac{2}{3}}x^{\frac{5}{3}}y^{\frac{5}{6}}$ in simplest radical form.

17. Find the value of c that makes $n^2 + 10n + c$ a perfect square trinomial.

Solve each equation.

18. $3x^2 = 7x$

19. $\sqrt{2y + 7} + 3 = 8$

20. $9b^2 + 25 = 30b$

21. $m^2 + 12m + 16 = 0$

22. $8x^3 + 125 = 0$

23. $a^4 - 9a^2 + 20 = 0$

Solve each problem.

24. If the product of two consecutive even integers is increased by 8 times the lesser integer, the result is 200. Find the integers.

25. Inez and Julio's living room has a rug that is 12 feet by 9 feet. A strip of equal width is uncovered along all edges of the rug. If the area of the room is 418 square feet, how wide is the strip?

Graphing Calculator Application: Graphing Quadratic Functions

The graphing calculator is a powerful tool for studying graphs of equations and for solving systems of equations graphically. The graphics function of this calculator makes it possible to produce a wide variety of graphs quickly and easily. These graphs are drawn using a series of dots, or pixels, on the calculator's *graphics screen*.

On most graphing calculators, the GRAPH key is pressed first and then the function to be graphed is entered. Finally, a key such as EXE or DRAW is pressed to produce a graph of the function on the graphics screen.

The size of the graphics screen varies from calculator to calculator. The graphics screens shown are for a 95 × 63 dot screen.

Each graphing calculator will automatically graph certain *built-in functions*, such as the quadratic function, $y = x^2$. Most calculators have approximately 20 built-in functions. Functions that are not built-in are called *user-generated functions*.

Check the calculator's manual to determine all of its built-in functions.

The calculator will produce the same graph each time it graphs a built-in function. This graph can be produced by pressing a single key on the calculator that corresponds to the function.

Example

1   **Graph $y = x^2$.**

     ENTER: GRAPH x^2 EXE

When a built-in function is graphed, any graph previously on the graphics screen is cleared.

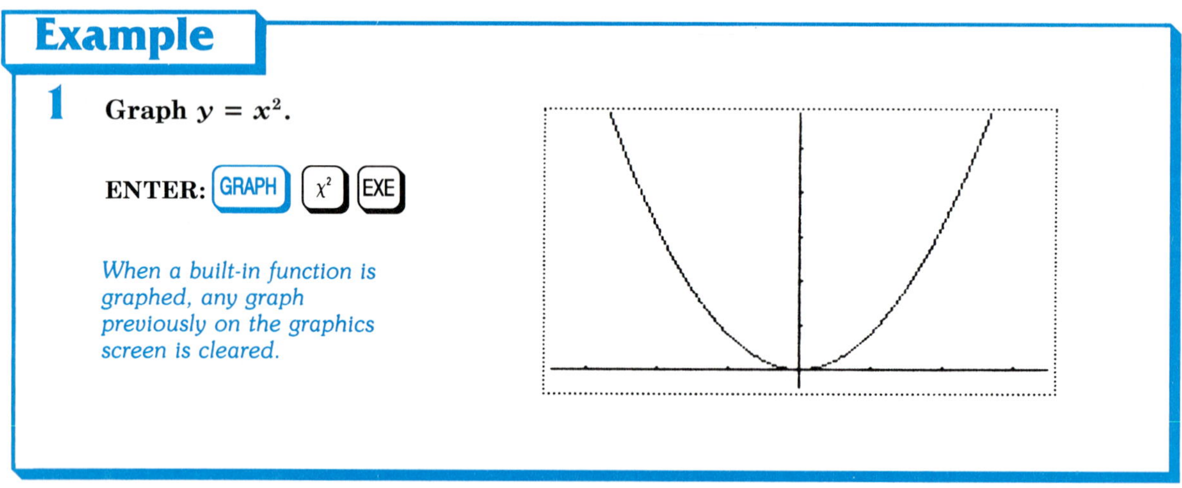

When the calculator produces the graph of a built-in function, the x-axis scale, y-axis scale, and position of the x-axis and y-axis on the graphics screen are automatically set. To graph user-generated functions, the x-axis scale and y-axis scale may need to be manually set to desired values. These scales can be set on the calculator's *range screen*.

The range screen is displayed after pressing the [RANGE] key. This screen lists the values of the following six range parameters.

Xmin: minimum value of x-axis
Xmax: maximum value of x-axis
Xscl: x-axis scale
Ymin: minimum value of y-axis
Ymax: maximum value of y-axis
Yscl: y-axis scale

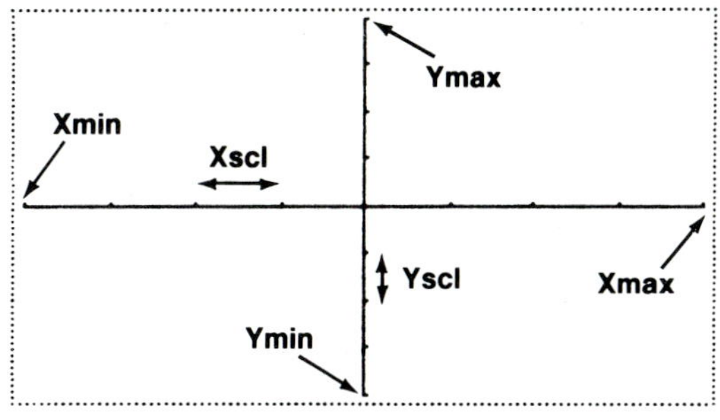

The values of the parameters Xmin, Xmax, Ymin, and Ymax determine the position of the x-axis and the y-axis on the graphics screen.

In some cases, the x-axis or y-axis will not appear on the graphics screen.

When the calculator produces the graph of a built-in function, each range parameter value for that function is automatically entered on the range screen. These values are chosen because they produce an appropriate portion of the graph of the function on the graphics screen.

User-generated functions differ from built-in functions because the same graphical representation is not given each time a user-generated function is graphed. When the calculator produces the graph of a user-generated function, the range parameter values are not automatically set. Instead, the function is graphed using the last set of values entered on the range screen. This may produce only a small portion, or none, of the graph of the function, as shown in the following example.

The range parameter values for $y = x^2$ on a 95 × 63 dot screen are as follows:
Xmin: -7 Ymin: -2
Xmax: 7   Ymax: 29
Xscl: 2   Yscl: 5
These values may be different for a calculator with a different-sized graphics screen.

Example

2 **Graph $y = x^2 - 16x - 2$.**

If the range parameter values have not been changed since Example 1, the function will be graphed using the values for the built-in function $y = x^2$.

ENTER: [GRAPH]  [X] [x^2] [−]

16 [X] [−] 2 [EXE]

On many calculators, alphabetic characters cannot be entered by pressing a single key. Check the calculator's manual to determine the correct procedure.

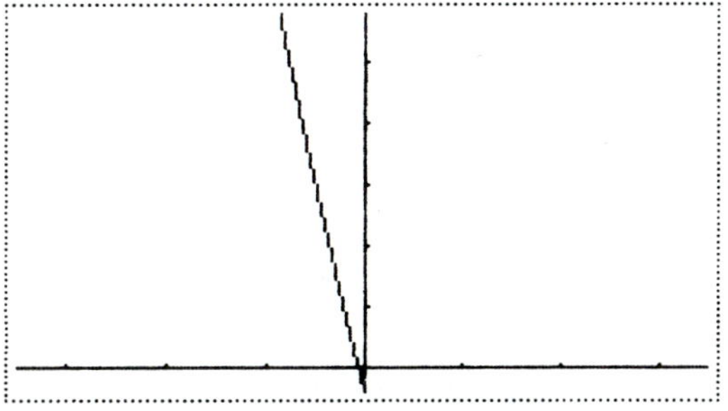

To produce an appropriate portion of the graph of a user-generated function, it is usually necessary to manually set the range parameter values. On most calculators, these values are entered on the range screen in the following order.

$$\text{Xmin} \to \text{Xmax} \to \text{Xscl} \to \text{Ymin} \to \text{Ymax} \to \text{Yscl}$$

After pressing the [RANGE] key to get the range screen display, enter the value for Xmin and press the [EXE] or [=] key. The remaining parameter values are entered in the same manner.

Example

3 Set the range parameters so that Xmin is -10, Xmax is 20, Xscl is 3, Ymin is -70, Ymax is 30, and Yscl is 10. Then graph $y = x^2 - 16x - 2$.

ENTER: [RANGE] [(-)] 10 [EXE] 20 [EXE]

3 [EXE] [(-)] 70 [EXE]

30 [EXE] 10 [EXE]

[GRAPH] [X] [x^2] [−]

16 [X] [−] 2 [EXE]

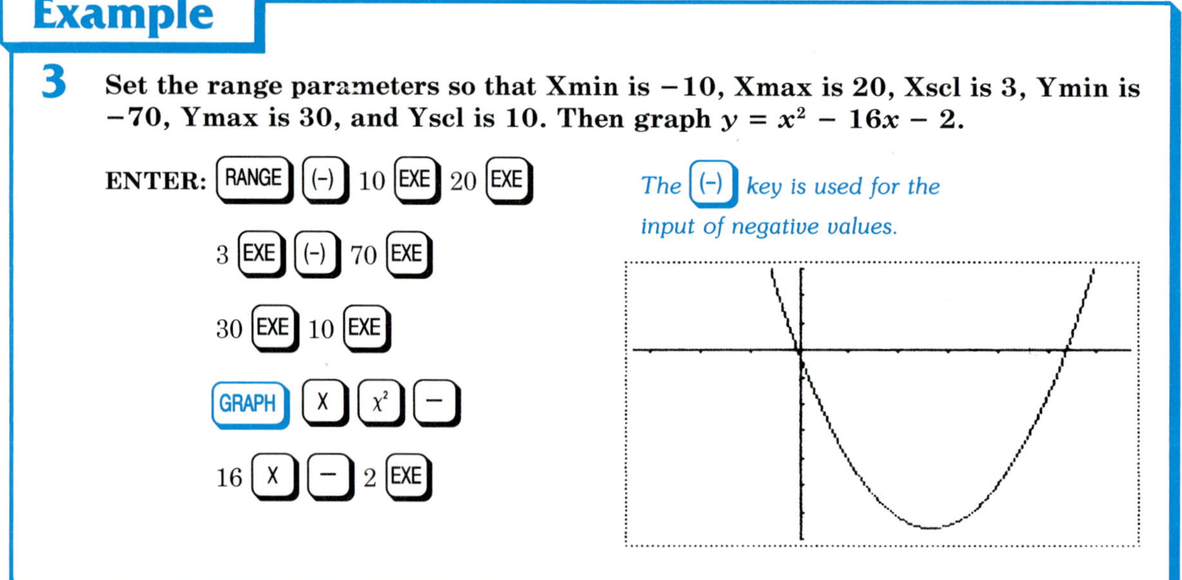

When a built-in function is graphed, the range parameter values automatically change to the values for that function. However, if the [X] key is pressed when entering a built-in function, then the range parameter values will not be changed. For example, the range parameters will not change to the values for $y = x^2$ if the [X] key is pressed before the [x^2] key is pressed.

Once range parameter values have been set, they will remain the same until the values are either manually changed or automatically set by graphing a built-in function. Thus, multiple functions can be graphed using the same set of range parameters. If the graphics screen is not cleared after a graph is drawn, then all of the functions will appear on the same set of axes.

4 **Graph $y = x^2$, $y = x^2 - 8$, $y = (x - 3)^2$, and $y = (x - 3)^2 - 8$ on the same set of axes.**

Before graphing these functions, try different sets of range parameter values to see which values will produce a complete graph of each function. One possible set of values is given below. Set the range parameters to these values and then graph each function.

> Xmin: -6.5 Xmax: 13.5 Xscl: 2 Ymin: -10 Ymax: 30 Yscl: 5

ENTER: GRAPH X x^2 EXE   GRAPH X x^2 − 8 EXE

GRAPH (X −

3) x^2 EXE

GRAPH (X − 3

) x^2 − 8 EXE

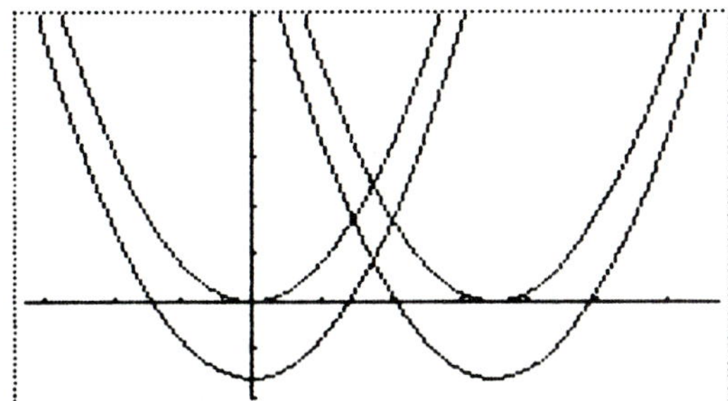

5 **Graph $y = x^2$, $y = -x^2$, $y = 3x^2$, $y = -3x^2$, $y = 0.2x^2$, and $y = -0.2x^2$ on the same set of axes.**

One possible set of range parameter values for these functions is given below. Set the range parameters to these values and then graph each function.

> Xmin: -10 Xmax: 10 Xscl: 2 Ymin: -30 Ymax: 30 Yscl: 5

ENTER: GRAPH X x^2 EXE   GRAPH (−) X x^2 EXE

GRAPH 3 X x^2 EXE   GRAPH (−) 3 X x^2 EXE

GRAPH 0.2 X x^2 EXE

GRAPH (−) 0.2 X x^2

EXE

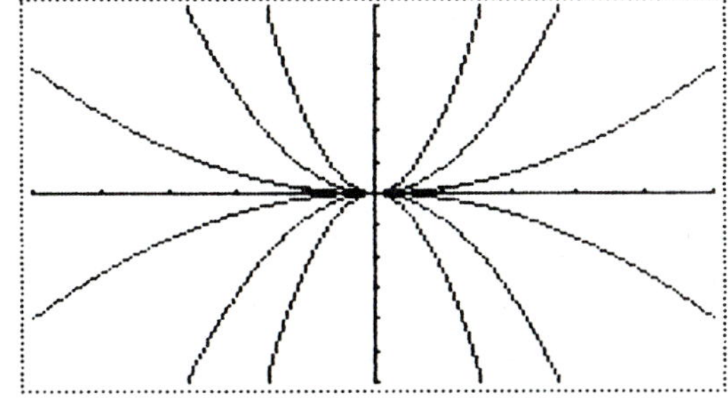

Exploratory Exercises

Graph $y = x^2 - 6x + 7$ for each set of range parameter values. Then sketch the graph shown on the graphics screen, indicating the x-axis scale and the y-axis scale.

1. Xmin: -1, Xmax: 9, Xscl: 1, Ymin: 0, Ymax: 35, Yscl: 5
2. Xmin: -10, Xmax: 16, Xscl: 2, Ymin: -5, Ymax: 160, Yscl: 10
3. Xmin: -100, Xmax: 100, Xscl: 20, Ymin: -100, Ymax: 2000, Yscl: 400
4. Xmin: -25, Xmax: 25, Xscl: 5, Ymin: -50, Ymax: 1200, Yscl: 200

Written Exercises

Graph each pair of equations on the same set of axes on a graphing calculator. Then compare the graph of the first equation to the graph of the second equation.

5. $y = (x - 5)^2$, $y = x^2$
6. $y = (x + 8)^2$, $y = x^2$
7. $y = x^2 + 3$, $y = x^2$
8. $y = x^2 - 6$, $y = x^2$
9. $y = (x - 2)^2 + 11$, $y = x^2 + 11$
10. $y = (x + 7)^2 - 18$, $y = x^2 - 18$
11. $y = (x + 1)^2 + 8$, $y = (x + 1)^2$
12. $y = (x - 4)^2 - 13$, $y = (x - 4)^2$
13. $y = (x - 10)^2 + 9$, $y = x^2$
14. $y = (x + 6)^2 - 5$, $y = x^2$
15. $y = (x + 9)^2 + 4$, $y = x^2$
16. $y = (x - 11)^2 - 7$, $y = x^2$
17. $y = 2x^2$, $y = x^2$
18. $y = 0.25x^2$, $y = x^2$
19. $y = -x^2$, $y = x^2$
20. $y = -\frac{5}{3}x^2$, $y = x^2$
21. $y = 5x^2 + 1$, $y = x^2 + 1$
22. $y = -0.55x^2 - 10$, $y = x^2 - 10$
23. $y = -\frac{1}{3}x^2 + 14$, $y = x^2$
24. $y = \frac{7}{4}x^2 - 2$, $y = x^2$
25. $y = -7(x - 13)^2$, $y = (x - 13)^2$
26. $y = 0.375(x + 2)^2$, $y = (x + 2)^2$
27. $y = \frac{6}{7}(x + 5)^2$, $y = x^2$
28. $y = \frac{19}{6}(x - 8)^2$, $y = x^2$
29. $y = 3(x - 9)^2 + 5$, $y = (x - 9)^2 + 5$
30. $y = -0.1(x + 1)^2 + 3$, $y = (x + 1)^2 + 3$
31. $y = -\frac{3}{4}(x + 12)^2 - 12$, $y = x^2$
32. $y = \frac{5}{2}(x - 3)^2 - 9$, $y = x^2$

Use the results of Exercises 5–32 to answer each question. Assume that h, k, and a are real numbers and $a \neq 0$.

33. How does the graph of $y = (x - h)^2$ compare to the graph of $y = x^2$?
34. How does the graph of $y = x^2 + k$ compare to the graph of $y = x^2$?
35. How does the graph of $y = (x - h)^2 + k$ compare to the graph of $y = (x - h)^2$?
36. How does the graph of $y = (x - h)^2 + k$ compare to the graph of $y = x^2 + k$?
37. How does the graph of $y = (x - h)^2 + k$ compare to the graph of $y = x^2$?
38. How does the graph of $y = ax^2$ compare to the graph of $y = x^2$?
39. How does the graph of $y = a(x - h)^2$ compare to the graph of $y = (x - h)^2$?
40. How does the graph of $y = ax^2 + k$ compare to the graph of $y = x^2 + k$?
41. How does the graph of $y = a(x - h)^2 + k$ compare to the graph of $y = (x - h)^2 + k$?
42. How does the graph of $y = a(x - h)^2 + k$ compare to the graph of $y = x^2$?

Communication in everyday life depends on a knowledge of English grammar. In the same manner, communication in algebra depends on a knowledge of mathematical grammar.

Arrangement of symbols is an important part of grammar and music. Examples in music, English, and mathematics are given below. In each case, the statement on the left communicates a thought. The statement on the right is nonsense.

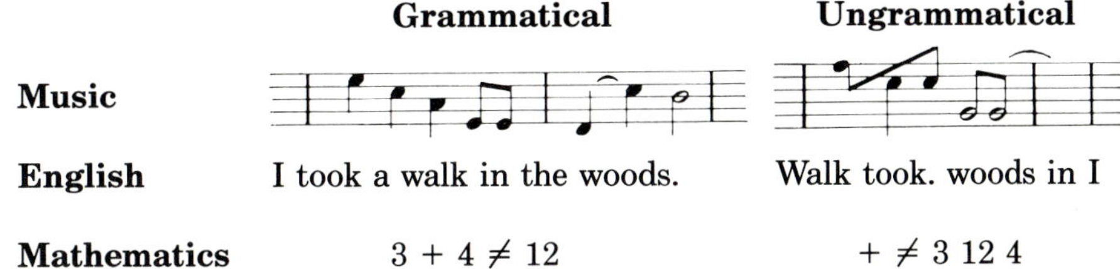

	Grammatical	**Ungrammatical**
Music		
English	I took a walk in the woods.	Walk took. woods in I
Mathematics	$3 + 4 \neq 12$	$+ \neq 3\ 12\ 4$

Punctuation also plays an important role in grammar. Study the two sentences.

My mother, my sister Carol, Andrew, and I went to the theater.

My mother, my sister, Carol, Andrew, and I went to the theater.

In the first sentence, four people, namely my mother, my sister whose name is Carol, Andrew, and I, went to the theater. In the second sentence, five people, namely my mother, my sister, Carol, Andrew, and I, went to the theater.

Now study the following mathematical sentences.

$$g(a) = 3a^2 - 2 \qquad G(a) = (3a)^2 - 2$$

In the first sentence, g of a is equal to three times a squared minus two. In the second sentence, G of a is equal to three times a, quantity squared, minus two. In other words, in the first sentence, square a and then multiply by three. In the second sentence, multiply three times a and then square the product.

Exercises

Explain how the equations in each pair differ.

1. $f(x) = (x - 2)^2$ $F(x) = x^2 - 2$

2. $f(x) = 3(x + 4)^2$ $F(x) = (3x + 4)^2$

3. $f(x) = x^2 + 3x + 5$ $F(x) = x^2 + 3(x + 5)$

4. $f(x) = (x - 4)^2$ $F(x) = (4 - x)^2$

5. $h(x) = (2x - 5)^2$ $H(x) = (2x)^2 - 5^2$

6. $m(x) = (-2x)^2 + 3$ $M(x) = -(2x)^2 + 3$

7. $g(x) = 4x^2 - 2$ $G(x) = 4(x^2 - 2)$

8. $n(x) = (x + 2)^2$ $N(x) = x^2 + 4x$

8-2 Parabolas

Consider the following situation.

A rocket is launched with an initial velocity of 50 meters per second. The height, $h(t)$, of the rocket t seconds after blast-off is given by the quadratic equation $h(t) = 50t - 5t^2$.

A graph of this equation can be drawn by making a table of values, plotting the points $(t, h(t))$, and connecting them with a smooth curve. The resulting graph is called a **parabola**.

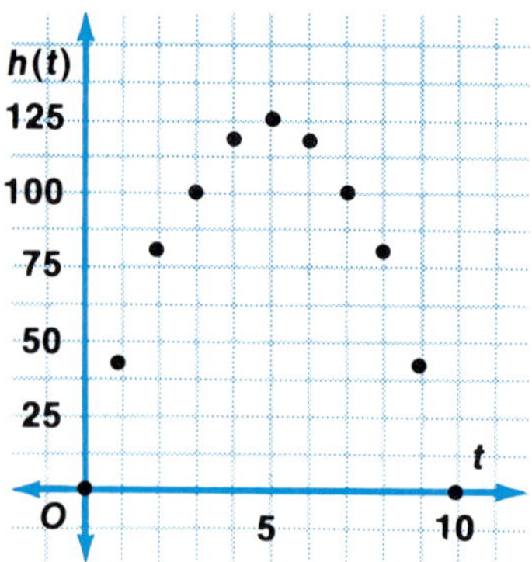

t	$50t - 5t^2$	$h(t)$
0	$50(0) - 5(0)^2$	0
1	$50(1) - 5(1)^2$	45
2	$50(2) - 5(2)^2$	80
3	$50(3) - 5(3)^2$	105
4	$50(4) - 5(4)^2$	120
5	$50(5) - 5(5)^2$	125
6	$50(6) - 5(6)^2$	120
7	$50(7) - 5(7)^2$	105
8	$50(8) - 5(8)^2$	80
9	$50(9) - 5(9)^2$	45
10	$50(10) - 5(10)^2$	0

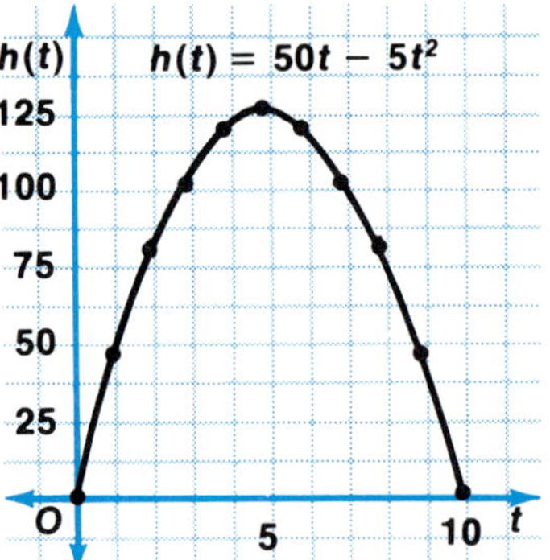

Notice that 10 seconds after blast-off, the height of the rocket is zero. This means that it has returned to earth. It appears to have reached its maximum height of 125 meters at 5 seconds after blast-off. Test some other values of t between 4 and 6 to check this conclusion.

Test values of t like 4.9 and 5.1 to determine if the maximum height is reached at 5 seconds.

The graph of any quadratic function is a parabola. Parabolas have certain common characteristics. They all have an **axis of symmetry** and a **vertex**.

The axis of symmetry of a parabola is the line about which the parabola is symmetric. If the coordinate plane could be folded along the axis of symmetry, then the portions of the parabola on each side of the axis would match.

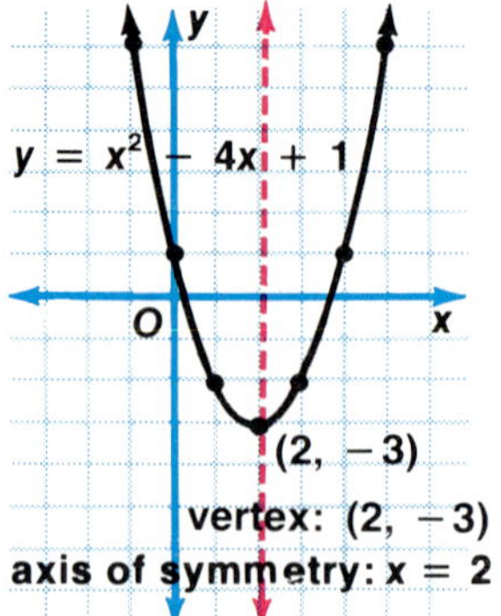

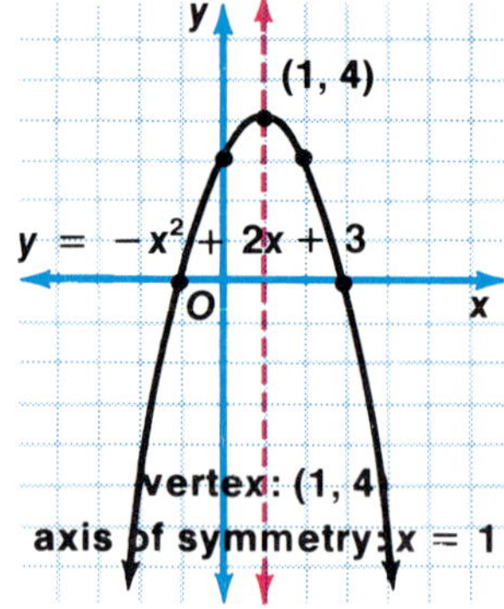

The vertex of a parabola is the point of intersection of the parabola and its axis of symmetry.

The graphs of $y = x^2$, $y = (x - 3)^2$, $y = x^2 + 4$, and $y = (x - 3)^2 + 4$ are shown at the right on the same set of axes. Study these graphs.

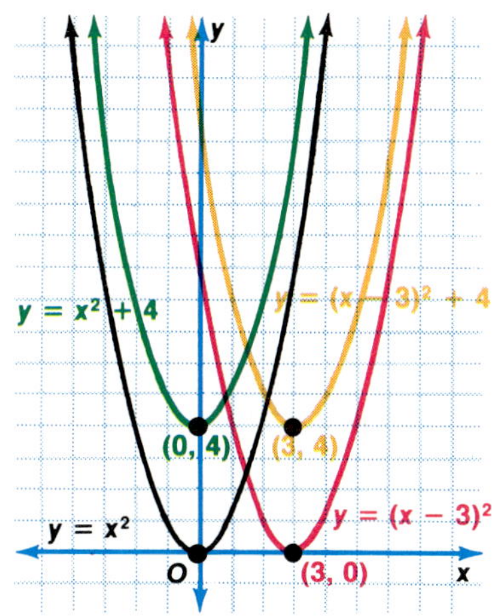

Equation	Vertex	Axis of Symmetry
$y = x^2$	$(0, 0)$	$x = 0$
$y = (x - 3)^2$	$(3, 0)$	$x = 3$
$y = x^2 + 4$	$(0, 4)$	$x = 0$
$y = (x - 3)^2 + 4$	$(3, 4)$	$x = 3$

These four equations can be expressed in the following general form.

$$y = (x - h)^2 + k$$

Once a quadratic equation is written in this form, the values of h and k are readily apparent.

Notice that each graph has the same shape. The only difference is their horizontal or vertical position.

$y = x^2$	$y = (x - 0)^2 + 0$	$h = 0, k = 0$
$y = (x - 3)^2$	$y = (x - 3)^2 + 0$	$h = 3, k = 0$
$y = x^2 + 4$	$y = (x - 0)^2 + 4$	$h = 0, k = 4$
$y = (x - 3)^2 + 4$	$y = (x - 3)^2 + 4$	$h = 3, k = 4$

Now, compare the values of h and k for these equations with the coordinates of the vertex of each graph shown above. Also, compare the value of h with each axis of symmetry. Finally, compare the shapes of the graphs. What do you notice?

In general, as values of h and k change, the graph of $y = (x - h)^2 + k$ is translated $|h|$ units to the left or right and $|k|$ units up or down.

$y = (x - h)^2 + k$	
Vertex	(h, k)
Axis of Symmetry	$x = h$

The graphs of $y = (x - h)^2 + k$ all have the same shape.

Example

1 **Name the vertex and axis of symmetry for the graph of $y = (x + 5)^2 - 3$.**

First, rewrite the equation in the form $y = (x - h)^2 + k$.

$$y = (x - (-5))^2 + (-3)$$

From this equation, $h = -5$ and $k = -3$.

Therefore, the vertex is $(-5, -3)$ and the axis of symmetry is $x = -5$.

2 **Name the vertex and axis of symmetry for the graph of $f(x) = (x + 2)^2 + 1$. Then draw the graph.**

This function can be written in the form $f(x) = (x - (-2))^2 + 1$.

From this equation, $h = -2$ and $k = 1$.

Therefore, the vertex is $(-2, 1)$ and
the axis of symmetry is $x = -2$.

It is helpful to find several points on
the graph other than the vertex.

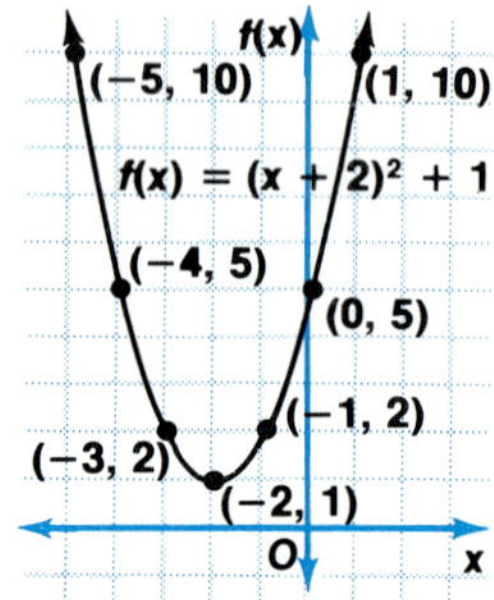

$$f(-3) = (-3 + 2)^2 + 1 \quad \text{or} \quad 2$$
$$f(-1) = (-1 + 2)^2 + 1 \quad \text{or} \quad 2$$
$$f(-4) = (-4 + 2)^2 + 1 \quad \text{or} \quad 5$$
$$f(0) = (0 + 2)^2 + 1 \quad \text{or} \quad 5$$
$$f(-5) = (-5 + 2)^2 + 1 \quad \text{or} \quad 10$$
$$f(1) = (1 + 2)^2 + 1 \quad \text{or} \quad 10$$

Notice that whenever two points have x-coordinates that are the same distance from the
axis of symmetry, $x = -2$, the points have the same y-coordinate.

3 **Graph $f(x) = x^2 + 8x + 15$.**

First, write the equation in the form $f(x) = (x - h)^2 + k$.
To do this, you must complete the square.

$$f(x) = x^2 + 8x + 15$$
$$= (x^2 + 8x + \underline{16}) + 15 - \underline{16} \qquad \textit{By adding and subtracting } \left(\tfrac{8}{2}\right)^2 \textit{ or 16,}$$
$$= (x + 4)^2 - 1 \qquad\qquad\qquad \textit{the function is not changed.}$$
$$= (x - (-4))^2 + (-1)$$

From this equation, $h = -4$ and $k = -1$.
Therefore, the vertex is $(-4, -1)$ and
the axis of symmetry is $x = -4$.

It is helpful to find several points on
the graph other than the vertex.

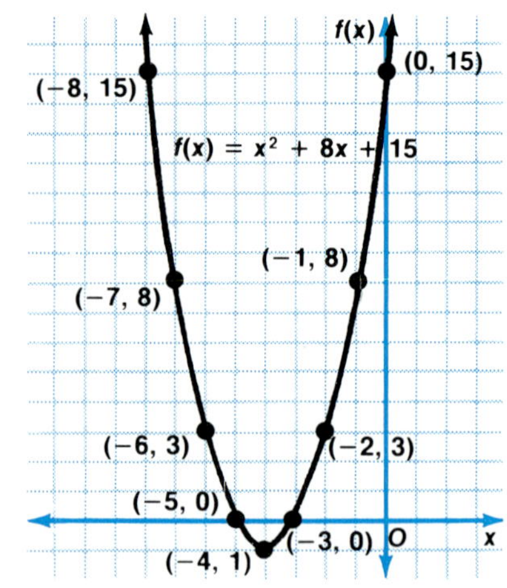

$$f(-3) = (-3 + 4)^2 - 1 \quad \text{or} \quad 0$$
$$f(-5) = (-5 + 4)^2 - 1 \quad \text{or} \quad 0$$
$$f(-2) = (-2 + 4)^2 - 1 \quad \text{or} \quad 3$$
$$f(-6) = (-6 + 4)^2 - 1 \quad \text{or} \quad 3$$
$$f(-1) = (-1 + 4)^2 - 1 \quad \text{or} \quad 8$$
$$f(-7) = (-7 + 4)^2 - 1 \quad \text{or} \quad 8$$
$$f(0) = (0 + 4)^2 - 1 \quad \text{or} \quad 15$$
$$f(-8) = (-8 + 4)^2 - 1 \quad \text{or} \quad 15$$

Exploratory Exercises

Name the vertex and axis of symmetry for the graph of each equation.

1. $f(x) = x^2$

2. $f(x) = (x + 2)^2$

3. $g(x) = (x - 8)^2$

4. $y = x^2 - 6$

5. $f(x) = x^2 + 11$

6. $y = (x - 5)^2 - 3$

7. $y = (x + 1.5)^2 - 3.2$

8. $f(x) = (x + 2)^2 + \dfrac{3}{2}$

9. $y = \left(x - \dfrac{1}{5}\right)^2 + 1$

Write the equation of each parabola if its equation has the form $y = (x - h)^2 + k$.

10. 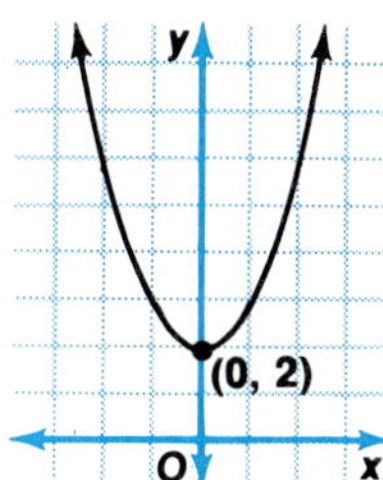

11. 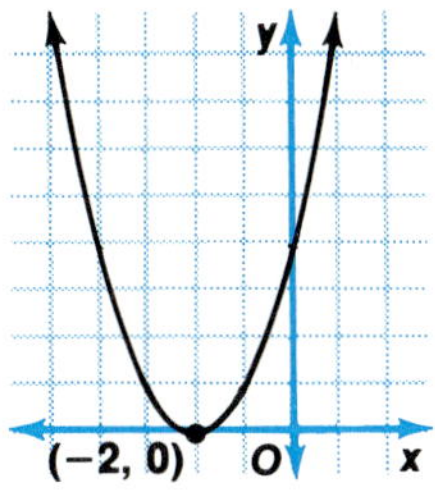

12. 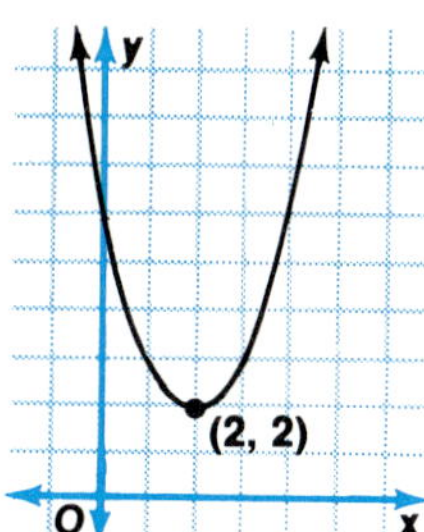

13. 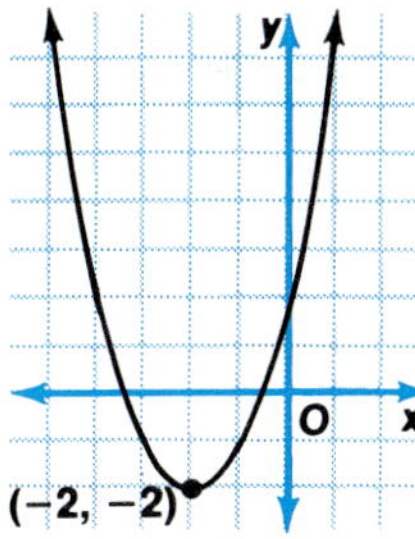

Written Exercises

Write each equation in the form $f(x) = (x - h)^2 + k$. Then name the vertex and axis of symmetry for the graph of each function.

14. $f(x) = x^2 - 2x + 1$

15. $f(x) = x^2 + 6x + 9$

16. $f(x) = x^2 + 6$

17. $f(x) = x^2 - 7$

18. $f(x) = x^2 + 8x$

19. $f(x) = x^2 - 12x$

20. $f(x) = x^2 + 16x + 67$

21. $f(x) = x^2 - 3x + 3$

22. $f(x) = x^2 - 5x - 4$

Graph each equation.

23. $y = (x - 3)^2$

24. $y = (x + 7)^2$

25. $f(x) = x^2 - 4$

26. $f(x) = x^2 + 3$

27. $f(x) = (x - 3)^2 + 5$

28. $y = (x + 2)^2 - 3$

29. $y = (x + 4)^2 + 1$

30. $y = (x - 1)^2 - 4$

31. $f(x) = x^2 + 6x + 2$

32. $f(x) = x^2 - 2x + 7$

33. $f(x) = x^2 + 10x + 27$

34. $y = x^2 - 4x - 2$

35. $y = x^2 - 5x$

36. $y = x^2 + 3x$

37. $f(x) = x^2 - x - 3$

38. $f(x) = x^2 + 7x + 15$

Challenge Exercises

Complete the following table for parabolas with equations of the form $y = (x - h)^2 + k$.

	axis of symmetry	contains the point	vertex	equation of the parabola
39.	$x = 0$	$(2, -1)$		
40.	$x = -2$	$(-5, 9)$		
41.	$x = -3$	$(1, 18)$		
42.	$x = 1$	$(-1, -2)$		

8-3 More Parabolas

In the previous lesson, you studied parabolas whose equations were of the form $y = x^2 + bx + c$. These equations can also be expressed in the form $y = (x - h)^2 + k$. Similarly, equations of the form $y = ax^2 + bx + c$, where $a \neq 0$, can be expressed in the following general form.

$$y = a(x - h)^2 + k$$

Since all quadratic functions can be described by an equation of the form $f(x) = ax^2 + bx + c$, where $a \neq 0$, all quadratic functions can be expressed in the general form $f(x) = a(x - h)^2 + k$.

The graph of $f(x) = a(x - h)^2 + k$ has the same vertex, (h, k), and axis of symmetry, $x = h$, as the graph of $f(x) = (x - h)^2 + k$. The value of the constant a affects the direction of opening and the shape of the parabola. To see this, study the graphs of the following quadratic functions. Note that these functions differ only in their values of a.

Recall that the graphs of $y = (x - h)^2 + k$ all have the same shape.

$$f(x) = (x - 1)^2 + 7 \qquad a = 1$$
$$f(x) = 2(x - 1)^2 + 7 \qquad a = 2$$
$$f(x) = \tfrac{1}{4}(x - 1)^2 + 7 \qquad a = \tfrac{1}{4}$$
$$f(x) = -(x - 1)^2 + 7 \qquad a = -1$$
$$f(x) = -2(x - 1)^2 + 7 \qquad a = -2$$
$$f(x) = -\tfrac{1}{4}(x - 1)^2 + 7 \qquad a = -\tfrac{1}{4}$$

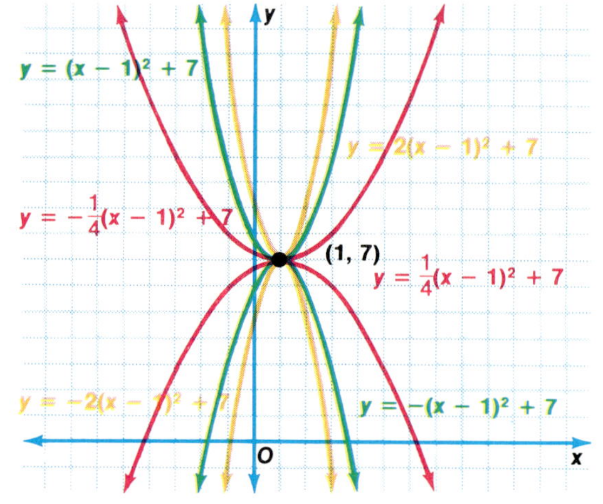

Each graph has the same vertex, $(1, 7)$, and the same axis of symmetry, $x = 1$.

Consider each value of a.

For what values of a do the graphs open upward? $a > 0$

For what values of a do the graphs open downward? $a < 0$

Finally, notice that as the value of $|a|$ increases, the graphs become narrower.

These and other examples suggest the following conclusions about the graph of $f(x) = a(x - h)^2 + k$.

$y = a(x - h)^2 + k$	a is positive	a is negative
Vertex	(h, k)	(h, k)
Axis of symmetry	$x = h$	$x = h$
Direction of opening	upward	downward
As the value of $|a|$ increases, the graphs of $y = a(x - h)^2 + k$ narrow.		

Examples

1 **Name the vertex, axis of symmetry, and direction of opening for the graph of $f(x) = -2(x - 5)^2$.**

This function can be written in the form $f(x) = -2(x - 5)^2 + 0$.

From this equation, $a = -2$, $h = 5$, and $k = 0$.

Therefore, the vertex is $(5, 0)$, the axis of symmetry is $x = 5$, and, since a is negative, the graph opens downward.

2 **Name the vertex, axis of symmetry, and direction of opening for the graph of $y = \frac{1}{3}x^2 + 4x + 7$.**

First, write the equation in the form $y = a(x - h)^2 + k$.
To do this, you must complete the square.

$$y = \frac{1}{3}x^2 + 4x + 7$$

$$= \frac{1}{3}(x^2 + 12x) + 7$$

$$= \frac{1}{3}(x^2 + 12x + \underline{36}) + 7 - \frac{1}{3}\underline{(36)}$$

$$= \frac{1}{3}(x + 6)^2 - 5$$

$$= \frac{1}{3}(x - (-6))^2 + (-5)$$

By adding and subtracting $\frac{1}{3}\left(\frac{12}{2}\right)^2$ or $\frac{1}{3}(36)$, the equation is not changed.

From this equation, $a = \frac{1}{3}$, $h = -6$, and $k = -5$.

Therefore, the vertex is $(-6, -5)$, the axis of symmetry is $x = -6$, and, since a is positive, the graph opens upward.

3 **Write the equation of the parabola shown below.**

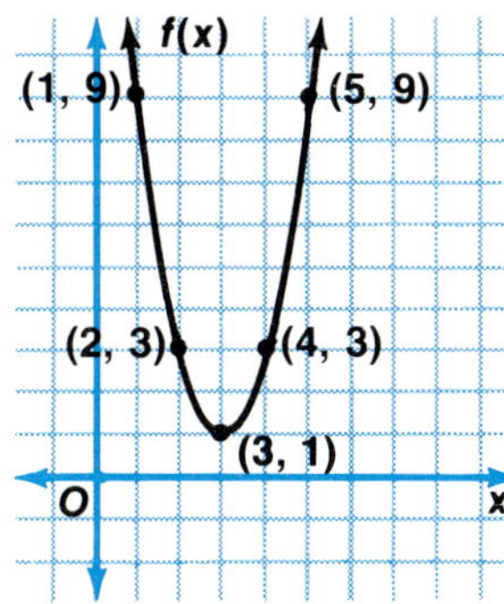

The vertex of the parabola is $(3, 1)$.

Therefore, $h = 3$ and $k = 1$.

Substitute the values of h and k and the coordinates of one point on the graph, other than the vertex, in the general form $y = a(x - h)^2 + k$. Then solve for a. *Use the ordered pair (2, 3).*

$$y = a(x - h)^2 + k$$
$$3 = a(2 - 3)^2 + 1$$
$$2 = a(-1)^2$$
$$a = 2$$

Substitute 3 for h, 1 for k, 2 for x, and 3 for y.

The equation of the parabola is $y = 2(x - 3)^2 + 1$ or $y = 2x^2 - 12x + 19$.

4 **Graph $f(x) = \frac{1}{2}(x - 2)^2 - 5$.**

This function can be written in the form $f(x) = \frac{1}{2}(x - 2)^2 + (-5)$.

From this equation, $a = \frac{1}{2}$, $h = 2$,

and $k = -5$.
Therefore, the vertex is $(2, -5)$ and
the axis of symmetry is $x = 2$.

Since $a = \frac{1}{2}$, the graph opens upward

and is wider than the graph of
$f(x) = (x - 2)^2 - 5$.

$$f(4) = \frac{1}{2}(4 - 2)^2 - 5 \quad \text{or} \quad -3$$

$$f(0) = \frac{1}{2}(0 - 2)^2 - 5 \quad \text{or} \quad -3$$

$$f(6) = \frac{1}{2}(6 - 2)^2 - 5 \quad \text{or} \quad 3$$

$$f(-2) = \frac{1}{2}(-2 - 2)^2 - 5 \quad \text{or} \quad 3$$

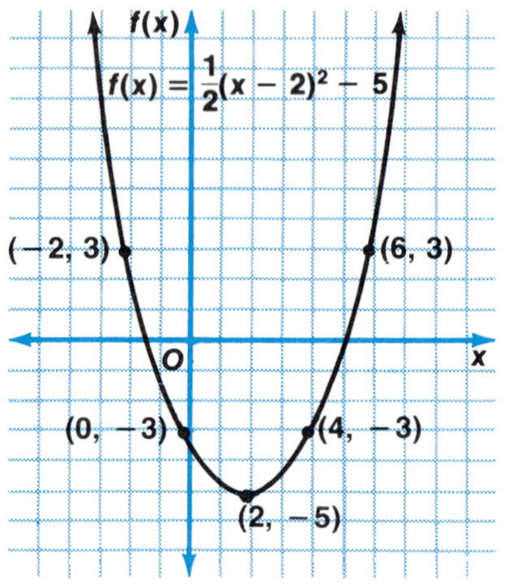

5 **Graph $f(x) = -3x^2 - 18x - 14$.**

First, write the equation in the form $f(x) = a(x - h)^2 + k$.
To do this, you must complete the square.

$$
\begin{aligned}
f(x) &= -3x^2 - 18x - 14 \\
&= -3(x^2 + 6x) - 14 \\
&= -3(x^2 + 6x + 9) - 14 - (-3)(9) \\
&= -3(x + 3)^2 + 13 \\
&= -3(x - (-3))^2 + 13
\end{aligned}
$$

From this equation, $a = -3$, $h = -3$,
and $k = 13$.

Therefore, the vertex is $(-3, 13)$ and
the axis of symmetry is $x = -3$.

Since $a = -3$, the graph opens downward
and is narrower than the graph of
$f(x) = (x + 3)^2 + 13$.

$$
\begin{aligned}
f(-2) &= -3(-2 + 3)^2 + 13 \quad \text{or} \quad 10 \\
f(-4) &= -3(-4 + 3)^2 + 13 \quad \text{or} \quad 10 \\
f(-1) &= -3(-1 + 3)^2 + 13 \quad \text{or} \quad 1 \\
f(-5) &= -3(-5 + 3)^2 + 13 \quad \text{or} \quad 1
\end{aligned}
$$

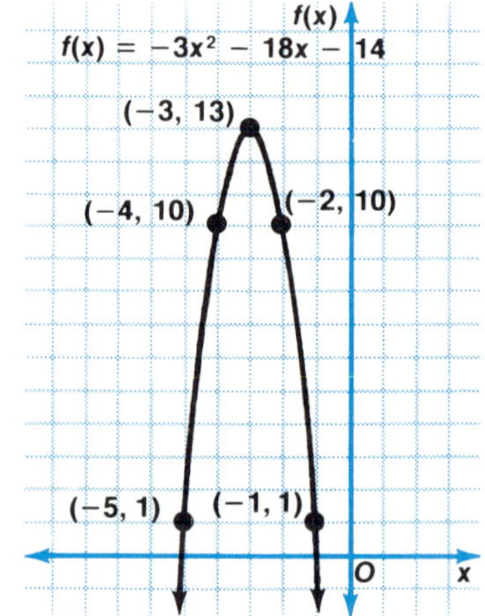

Exploratory Exercises

Name the vertex, axis of symmetry and direction of opening for the graph of each equation.

1. $y = 2x^2$

2. $f(x) = 4(x - 8)^2$

3. $f(x) = -1(x + 2)^2$

4. $f(x) = -3x^2 + 6$

5. $g(x) = 5x^2 - 6$

6. $y = 5(x + 3)^2 - 1$

7. $y = -2(x - 2)^2 - 2$

8. $f(x) = -\frac{1}{3}(x + 2)^2 - \frac{4}{3}$

9. $g(x) = 3\left(x - \frac{1}{2}\right)^2 + \frac{1}{4}$

10. Write the equation of a parabola with position 3 units to the right of the parabola with equation $f(x) = x^2$.

11. Write the equation of a parabola with position 4 units above the parabola with equation $f(x) = -2x^2$.

12. Write the equation of a parabola with position 6 units to the left of and 2 units above the parabola with equation $f(x) = -3x^2$.

13. Write the equation of a parabola with position 1 unit to the right of and 8 units below the parabola with equation $f(x) = 5x^2$.

Written Exercises

Write each equation in the form $f(x) = a(x - h)^2 + k$. Then name the vertex, axis of symmetry, and direction of opening for the graph of each equation.

14. $f(x) = x^2 + 2x - 2$

15. $f(x) = -x^2 - 2x + 2$

16. $f(x) = -3x^2 + 12x$

17. $f(x) = 4x^2 + 24x$

18. $f(x) = -2x^2 - 20x - 50$

19. $f(x) = 3x^2 - 18x + 11$

20. $f(x) = \frac{1}{3}x^2 - 4x + 15$

21. $f(x) = -\frac{1}{2}x^2 + 5x - \frac{27}{2}$

22. $f(x) = \frac{3}{4}x^2 + 9x + 10$

Write the equation of each parabola shown.

23. 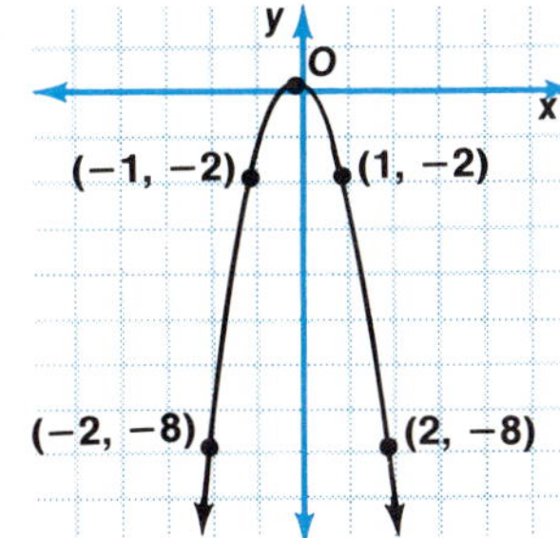

24. 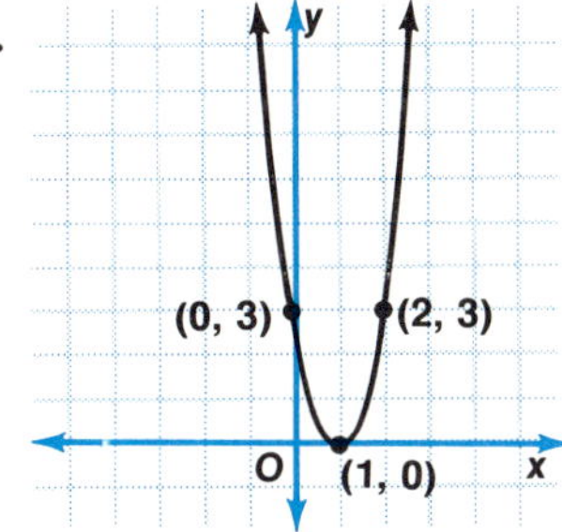

25. 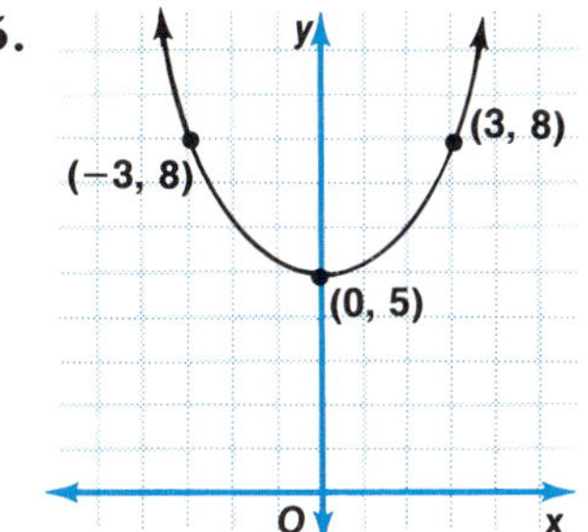

26. 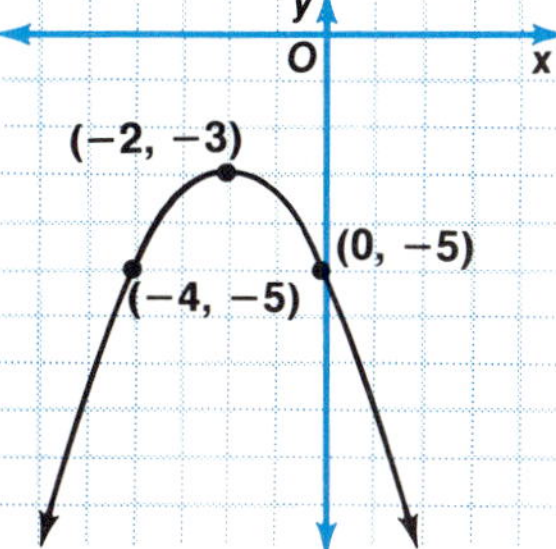

27. 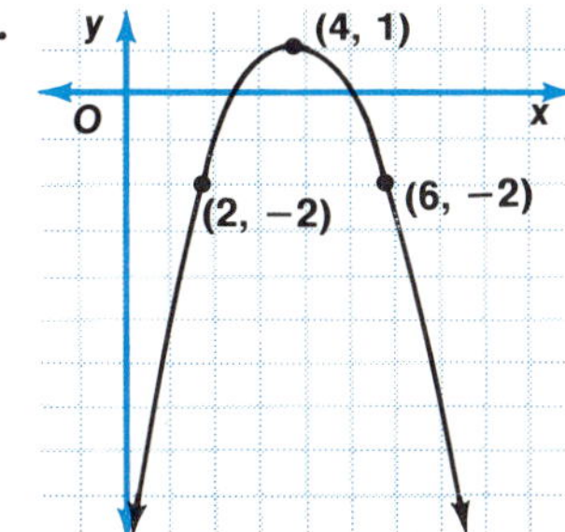

28. 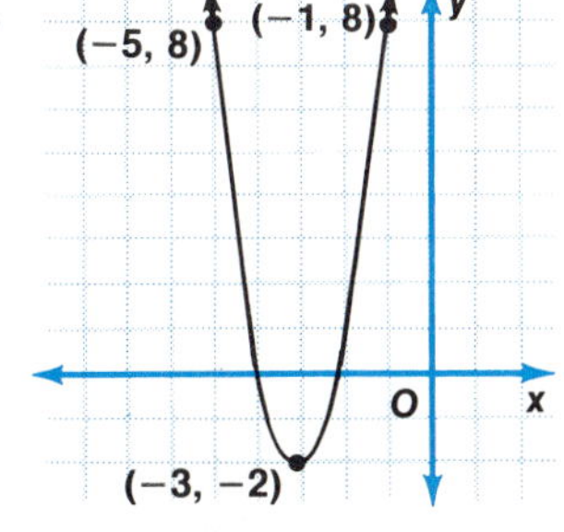

Graph each equation.

29. $f(x) = 3x^2 + 18x + 27$

30. $f(x) = 2x^2 + 3$

31. $f(x) = 2(x + 3)^2 - 5$

32. $f(x) = 3(x - 1)^2 + 2$

33. $f(x) = \frac{1}{2}(x + 3)^2 - 5$

34. $f(x) = \frac{1}{3}(x - 1)^2 + 2$

35. $f(x) = x^2 + 6x + 2$

36. $f(x) = x^2 - 2x + 7$

37. $f(x) = -2x^2 + 16x - 31$

38. $f(x) = -x^2 - 4x - 10$

39. $f(x) = 2x^2 + 8x + 10$

40. $f(x) = -5x^2 - 40x - 80$

41. $f(x) = -9x^2 - 18x - 6$

42. $f(x) = -4x^2 - 16x + 2$

43. $f(x) = -0.25x^2 - 2.5x - 0.25$

44. $f(x) = -\frac{2}{3}x^2 + 4x - 9$

Challenge Exercises

45. Write an equation of the parabola that passes through $(9, -3)$, $(6, 3)$ and $(4, 27)$.

46. Given $f(x) = ax^2 + bx + c$ with $a \neq 0$, complete the square to rewrite the equation in the form $f(x) = a(x - h)^2 + k$. State an expression for h and k in terms of a, b, and c.

mini-review

Simplify.

1. $\left(\dfrac{a^{-1}b^2}{c^{-2}}\right)^{-2}$

2. $54a^2b^3(6a^2b)^{-2}$

3. $\sqrt[3]{\dfrac{8}{49m}}$

4. $\dfrac{b^{-\frac{3}{2}} - 3b^{\frac{1}{2}}}{b^{-\frac{1}{2}}}$

5. $\dfrac{1}{3 - 7i}$

6. $\dfrac{1 - \sqrt{3}}{1 + \sqrt{3}}$

7. $(2 + 5\sqrt{3})(1 - 2\sqrt{3})$

8. $(5 - 2i)(2 - 5i)$

9. $\sqrt[3]{18x^3y^4} \cdot \sqrt[3]{48x^2y}$

Factor.

10. $x^3 + 3x^2 - 4x - 12$

11. $27a^3 - 64$

12. $27y^2 - 27y + 6$

13. $2r^4 - 13r^3 + 15r^2$

14. Use synthetic division to find $(3x^3 + 2x^2 - 7x + 8) \div (x - 3)$.

15. Find a quadratic equation that has roots $1 - 2i$ and $1 + 2i$.

16. Find k such that -3 is a root of $2x^2 + kx - 15 = 0$.

Solve each equation.

17. $20x^2 = 41x - 20$

18. $9z^2 + 4 = 12z$

19. $t^2 - 10t = -26$

20. $3x - 6\sqrt{x} - 45 = 0$

21. $2d^2 - 11d - 4 = 0$

Graph each equation. Then name the vertex and axis of symmetry for each graph.

22. $f(x) = (x - 6)^2 - 4$

23. $y = x^2 + 10x + 22$

24. Write a quadratic function to describe the product of two numbers whose sum is 125.

8-4 Problem Solving: Using Parabolas

A ball is thrown upward at an initial speed of 80 feet per second. The height it will reach after t seconds is given by the following function.

$$h(t) = 80t - 16t^2 \qquad \text{\textit{This function represents height in feet.}}$$

At what time does the ball reach its maximum height?

Compare the time-lapse photograph to the graph of $h(t) = 80t - 16t^2$. They both give a visual description of the height of the ball at a certain time.

The graph is a parabola. Its highest point is at the vertex (2.5, 100). This shows that the ball reaches a maximum height of 100 feet in 2.5 seconds.

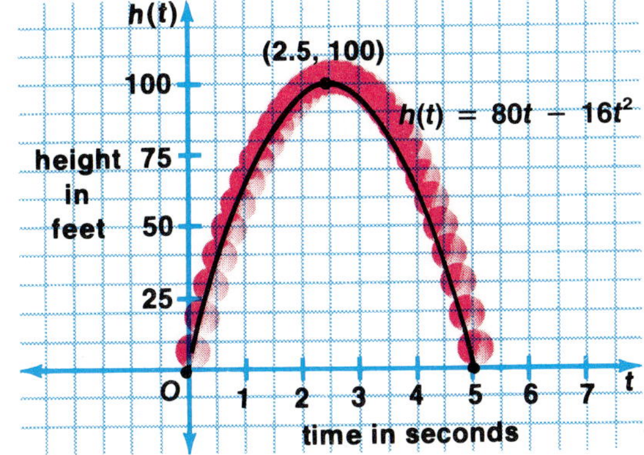

Many problem situations involve finding a **maximum** or **minimum value**. If such a problem can be described by a quadratic function, it often can be solved by finding the vertex of the graph of the function.

Examples

1 Find two numbers whose difference is 64 and whose product is a minimum.

explore
Let x represent the lesser number.
Then, $64 + x$ represents the greater number.

plan
$$\begin{aligned} \text{product} &= (x) \cdot (64 + x) \\ &= 64x + x^2 \end{aligned}$$

solve
The equation describing the product is a quadratic function. Find the vertex by writing the equation in the form $y = a(x - h)^2 + k$. Then the vertex is (h, k). Let y represent the product.

$$\begin{aligned} y &= 64x + x^2 \\ &= x^2 + 64x + \left(\frac{64}{2}\right)^2 - \left(\frac{64}{2}\right)^2 \qquad \textit{Complete the square.} \\ &= (x + 32)^2 - 1024 \end{aligned}$$

The vertex is $(-32, -1024)$.
Thus, the minimum product, -1024, occurs when x is -32.
When $x = -32$, $64 + x = 64 + (-32)$ or 32.

The two numbers are -32 and 32.

Check some other pairs of numbers whose difference is 64 to see that their product is not less than -1024.

$$(-33)(31) = -1023 \qquad (-30)(34) = -1020 \qquad (-11)(53) = -583$$

2 A theater has seats for 500 people. It is filled to capacity for each show and tickets cost \$3.00 per show. The owner wants to increase ticket prices. She estimates that for each \$0.20 increase in price, 25 fewer people will attend. What ticket price will maximize her income?

Let p represent the number of \$0.20 price increases. Then $3.00 + 0.20p$ represents the ticket price, and $500 - 25p$ represents the number of tickets sold.

Income = (number of ticket sold) × (ticket price)

$$\begin{aligned} \text{Income} &= (500 - 25p) \times (3.00 + 0.20p) \\ &= 1500 + 100p - 75p - 5p^2 \\ &= 1500 + 25p - 5p^2 \end{aligned}$$

The equation describing the income is a quadratic function. Find the vertex by writing the equation in the form $y = a(x - h)^2 + k$. Let y represent the income.

$$\begin{aligned} y &= 1500 + 25p - 5p^2 \\ &= -5(p^2 - 5p) + 1500 \\ &= -5\left[p^2 - 5p + \left(-\tfrac{5}{2}\right)^2\right] + 1500 + 5\left(-\tfrac{5}{2}\right)^2 \qquad \textit{Complete the square.} \\ &= -5\left(p - \tfrac{5}{2}\right)^2 + \frac{6125}{4} \end{aligned}$$

The vertex is $\left(\dfrac{5}{2}, \dfrac{6125}{4}\right)$ or $(2.5, 1531.25)$.

The maximum income, \$1531.25, occurs when the owner makes 2.5 prices increases. Since each increase is to be \$0.20, the total increase is $2.5(0.20)$ or \$0.50. Therefore, a ticket price of \$3.00 + \$0.50 or \$3.50 will maximize the owners income.

Try other values for p in the income equation to see whether her income could be greater than \$1531.25.

Exploratory Exercises

A newsletter has a circulation of 50,000 and sells for 40¢ a copy. Due to increased labor and production costs the publisher will raise the price of the newsletter. A publisher's survey indicates that for each 10¢ increase in price, the circulation decreases by 5000.

1. Let x stand for the number of 10¢ price increases. Write an algebraic expression to describe the increased price per copy.

2. Write an algebraic expression to describe the reduced circulation after the price increase.

3. The publisher's income on the newsletter is the product of the price per copy and the circulation. It is also a function of the number of price increases. Write an equation to describe this function.

4. Draw a graph relating the publisher's income to the number of price increases. Use number of price increases for the x-axis and income in dollars for the y-axis.

5. What price per copy will maximize the publisher's income on the newsletter?

A manufacturer can sell x items per month at a price of $(300 - 2x)$ dollars per item. It costs the manufacturer $(20x + 1000)$ dollars to produce x items.

6. One item sells for $(300 - 2x)$ dollars. Write an algebraic expression to describe the price of x items.

7. The manufacturer's profit is the selling price minus the cost of producing the items. Write an algebraic expression to describe the manufacturer's profit on x items in one month.

8. The manufacturer's profit is a function of the number of items sold. Write an equation to describe this function.

9. Draw a graph relating the manufacturer's profit to the number of items produced. Use the number of items for the x-axis and profit in dollars for the y-axis.

10. How many items should the manufacturer produce in one month to maximize profit?

Written Exercises

Solve each problem.

11. Find two numbers whose difference is 40 and whose product is a minimum.

12. Find two numbers whose sum is 36 and whose product is a maximum.

13. Find two numbers whose sum is 37 and whose product is a maximum.

14. Find two numbers whose difference is 25 and whose product is a minimum.

15. Find two numbers whose difference is -20 and whose product is a minimum.

16. Find two numbers whose sum is -16 and whose product is a maximum.

17. Find the dimensions and maximum area of a rectangle, if its perimeter is 40 centimeters.

18. Find the dimensions and maximum area of a rectangle, if its perimeter is 24 inches.

19. George Polo has 120 meters of fence to make a rectangular pen for rabbits. If a shed is used as one side of the pen, what would be the length and width for maximum area?

20. Sara Meyer has 150 feet of fence to put around a rectangular garden. If a 10 foot opening is left on one side for a gate, what would be the length and width for maximum area?

21. A taxi service operates between two airports, transporting 300 passengers a day. The charge is $8.00. The owner estimates that 20 passengers will be lost for each $1 increase in the fare. What charge would be most profitable for the service?

22. An airline transports 800 people a week between two cities. A round-trip ticket costs $300. The company wants to increase the price. They estimate that for each $5 increase 10 passengers will be lost. What ticket price will maximize their income?

23. An object is fired upward from the top of a tower at a velocity of 80 feet per second. The tower is 200 feet high. The height of the object above the ground t seconds after firing is given by the formula $h(t) = -16t^2 + 80t + 200$. What is the maximum height reached by the object? How long after firing does it reach maximum height?

24. A ball is thrown upward into the air with an initial velocity of 64 feet per second. The formula $h(t) = 64t - 16t^2$ gives its height above the ground after t seconds. What is its height after 1.5 seconds? What is its maximum height? How many seconds will pass before it returns to the ground?

25. Ken Graham has 1200 meters of fence to put around two rectangular yards. If the yards are to be separated by part of the fence, what would be the length and width for maximum area?

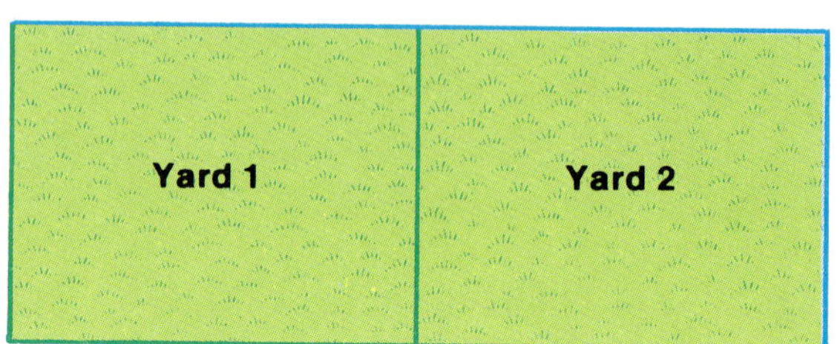

26. A square, which is 5 centimeters by 5 centimeters, is cut from each corner of a rectangular piece of cardboard and the sides are folded up to make a box. If the bottom of the box must have a perimeter of 50 centimeters, what would be the length, width, and height for maximum volume?

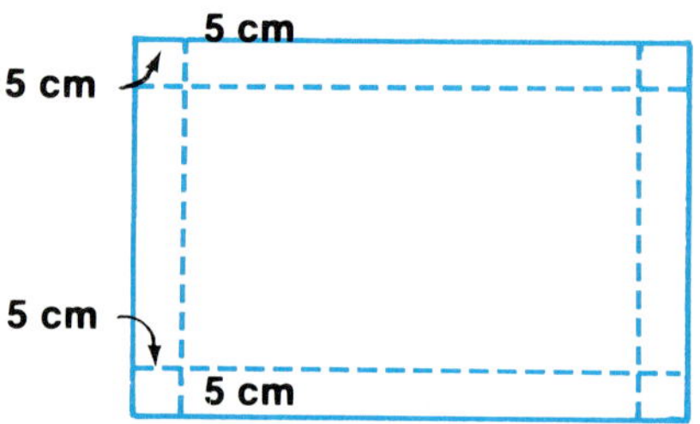

27. A rectangle is inscribed in an isosceles triangle. The base of the triangle is 8 cm and the height is 10 cm. Find the length of the inscribed rectangle with maximum area.

28. A wire 36 cm long is cut into 2 pieces and each piece is bent to form a square. How long should each piece be to minimize the sum of the areas of the two squares?

29. A wire 32 in. long is cut into 2 pieces. One piece is bent to form a square. The other piece is bent to form a rectangle which is 2 in. longer than it is wide. How long should each piece be to minimize the sum of the areas of the square and the rectangle?

30. A helicopter service transports passengers to an island during vacation season. The round trip fare is $20 and they transport 500 passengers per day. The owner estimates that for each $1 increase in fare, they will lose 20 passengers. What should they charge for maximum income?

31. Last year 200 people came to see the fall play at Jones High School. The cost per ticket was $2.00. This year the drama teacher wants to increase the ticket price. He estimates that for each $0.25 increase, 10 fewer people will come to the play. What ticket price will maximize the income?

32. The steamship, *The Golden Conifer*, is rented to take 100 campers to Pine Mountain. The fare is $5 per camper. The steamship company has agreed to reduce the fare by two cents per passenger for every camper over 100. How many passengers will produce a maximum rental for the company?

Solve each equation or inequality.

1. $|5x - 7| = 22$
2. $3(3a - 10) \le 2(4 - 5a)$
3. $|4r - 11| > 23$
4. $\sqrt{m + 5} - \sqrt{m + 1} = 2$
5. $d^2 - 11d + 24 = 0$
6. $3x^2 + 5 = -4x$
7. $z^{\frac{1}{2}} - 2z^{\frac{1}{4}} - 15 = 0$

8. State if the relation is a function.
$$\{(2, 6), (1, 5), (-2, 5), (-1, 5)\}$$

9. Find the standard form of the equation of the line passing through $(3, -4)$ and $(-2, 7)$.

Graph each equation or inequality.

10. $y = |3x| - 1$
11. $5x - 2y \le 3$

12. Solve this system: $5x + 3y = -7$
$3x - 2y = 11$

13. Write the equation for the plane with x-intercept $\frac{1}{4}$, y-intercept $\frac{1}{3}$, and z-intercept 4.

Solve each system of equations using Cramer's Rule.

14. $3x + 4y = -25$
$2x - 3y = 6$

15. $3x - 2y - z = 8$
$x + 2y = -1$
$x - y + 4z = 25$

16. Evaluate $\begin{vmatrix} -3 & -2 & 6 \\ 2 & 4 & -1 \\ 1 & 5 & -2 \end{vmatrix}$.

17. Simplify $\begin{bmatrix} 2 & -5 \\ 3 & 1 \end{bmatrix} \cdot \begin{bmatrix} 1 & -4 & -6 \\ -2 & 10 & 7 \end{bmatrix}$.

18. Find the inverse of $\begin{bmatrix} 1 & -2 & -2 \\ 3 & 4 & 3 \\ 2 & -1 & -3 \end{bmatrix}$.

19. Divide $(2z^4 - 7z^3 - 2z^2 + 3z + 1)$ by $(2z + 1)$ using synthetic division.

Factor.

20. $4x^2 - 11xy - 3y^2$
21. $r^3 - 3r^2s + 3rs^2 - s^3$

Simplify each expression.

22. $(-2ab^2)^5(6a^3)^2$
23. $(3 - 2\sqrt{5})^2$
24. $\dfrac{x^{\frac{1}{2}} - 1}{x^{\frac{1}{2}} + 1}$
25. $\left(\dfrac{r^2 s^{-1} t^{-2}}{r^{-1} s t^{-3}}\right)^{-2}$
26. $\sqrt{\dfrac{50}{27}}$
27. $\dfrac{2 + i\sqrt{3}}{2 - i\sqrt{3}}$

28. Find the value of the discriminant for $3y^2 - 2y + 5 = 0$. Then describe the nature of the roots.

29. Find a quadratic equation that has roots -5 and $\frac{2}{3}$.

Graph each equation. Then name the vertex, axis of symmetry and direction of opening for each graph.

30. $y = -(x - 3)^2 + 5$
31. $f(x) = 2x^2 + 8x + 1$

Solve each problem.

32. Rae has 6 gallons of 25% antifreeze solution. How much 100% antifreeze must she add to make a solution that is at least 60% antifreeze?

33. A certain positive number decreased by 3 is multiplied by the same number increased by 3. The product is 27. What is the number?

34. Find the two numbers whose sum is 81 and whose product is a maximum.

35. ABC Manufacturing produces bicycles and mopeds. The plant can produce no more than 24 bicycles and 16 mopeds per day. It takes 3 work-hours to make a bicycle and 6 to make a moped. There are 150 work-hours per day. If profits are equal on both items, how many of each should be made to maximize profit?

Graphing Calculator Application: Finding the Vertex of a Parabola

Graphing calculators have a tracing function that is used to determine the x- and y-coordinates of points on a graph displayed on the graphics screen. When the tracing function is activated, a pointer (blinking dot) will appear at either the extreme left point or extreme right point of the graph. The x-coordinate of this point will also appear at the bottom of the graphics screen. The pointer can be moved by using the arrow, or cursor, keys, ⬅ and ➡.

The x-coordinate value changes as the pointer moves, giving the x-coordinate of the point at the current position of the pointer. Pressing the [$X{\leftrightarrow}Y$] key will cause the y-coordinate of the point to be displayed instead of the x-coordinate.

Pressing the [$X{\leftrightarrow}Y$] key switches the display between the y-coordinate value and the x-coordinate value.

The tracing function can only be activated after a function is drawn on the graphics screen. Usually, a special key, such as a [TRACE] key, or one of the arrow keys is used to make the pointer appear on the graph. Check the calculator's manual to determine the correct procedure for that calculator.

The tracing function can determine the coordinates of points on any graph displayed on the graphics screen. However, it does have limitations. The tracing function will not be able to determine the coordinates of every point on a graph. Also, it usually determines only the approximate coordinates for a point of intersection of two graphs.

Examples

1   **Set the range parameter values so that Xmin is -8, Xmax is 8, Xscl is 2, Ymin is 0, Ymax is 35, and Yscl is 5. Graph $y = x^2 + 4$. Then activate the tracing function and move the pointer along this graph.**

ENTER: 

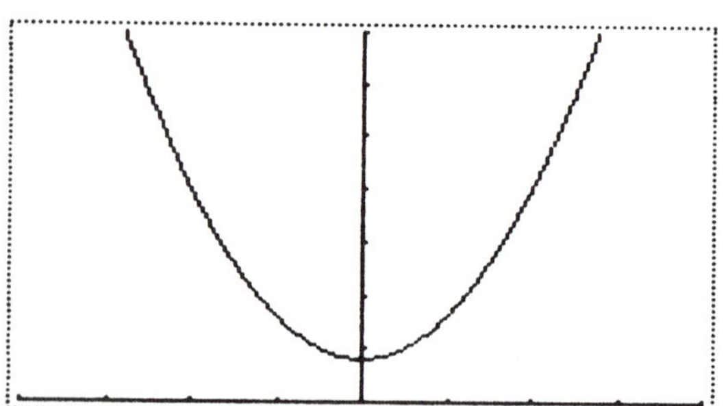

The range parameter values for this function were chosen to help illustrate the limitation of the tracing function.

Activate the tracing function so that the pointer appears at the extreme left point on the graph. Then press the ▷ key.

Notice that the pointer *did not* move to the next point on the graph but to a point in the next column of dots on the graphics screen. The tracing function moves the pointer across the screen one *column* of dots at a time. Thus, if a graph has more than one point in a column, the pointer will only "land" on one of those points. Any other points in that column cannot be "reached" by the pointer, and the x- and y-coordinates of those points cannot be determined using the tracing function.

2 **Use the tracing function to determine the coordinates of the vertex for the graph of $y = (x - 2)^2 - 3$.**

This equation is in the form $y = a(x - h)^2 + k$. $y = 1(x - 2)^2 + (-3)$

Thus, $h = 2$, $k = -3$, and the vertex of the graph is $(2, -3)$.

The range parameter values must be chosen so that the vertex is on the screen and easy to "land" on. One possible set of values for this equation is given below. Set the range parameters to these values and graph the equation.

   Xmin: −1   Xmax: 7   Xscl: 1   Ymin: −4   Ymax: 2   Yscl: 1

ENTER: GRAPH (X − 2) x^2 − 3 EXE

Activate the tracing function. Use the ▷ key to move the pointer to the vertex. The vertex is the "middle" point in the line of dots at the bottom of the graph of the parabola. The x-coordinate value displayed is 1.978723404.

Now press the X↔Y key to display the y-coordinate value. This value is −2.999547306.

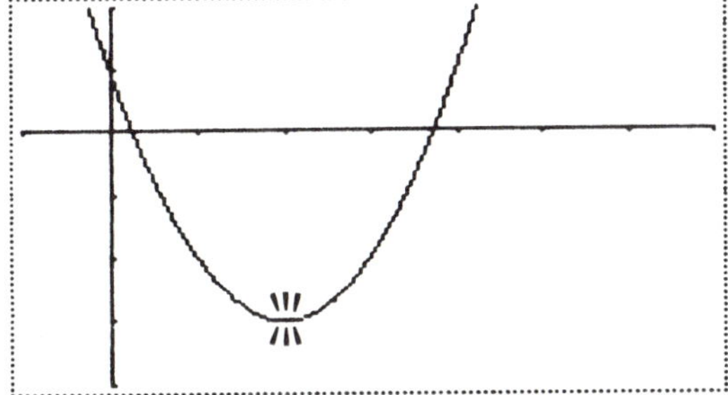

The vertex of the parabola, as determined by the calculator, is (1.978723404, −2.999547306), an approximation of the actual vertex, (2, −3).

In spite of its limitations, the tracing function provides a quick method for approximating the coordinates of the vertex of a parabola. This is especially helpful when finding the exact coordinates is a more difficult or time-consuming task.

Remember that the vertex is the point of intersection of the parabola and its axis of symmetry.

3 **Find the coordinates of the vertex for the graph of $y = 3x^2 + 13x - 8$.**

The exact coordinates of the vertex can be determined by writing the equation in the form $y = a(x - h)^2 + k$. However, approximate coordinates can be found more quickly by drawing the graph and using the tracing function.

One possible set of range parameter values for this function is given below. Set the range parameters to these values and graph the function.

Xmin: -6 Xmax: 2 Xscl: 1   Ymin: -30 Ymax: 30 Yscl: 5

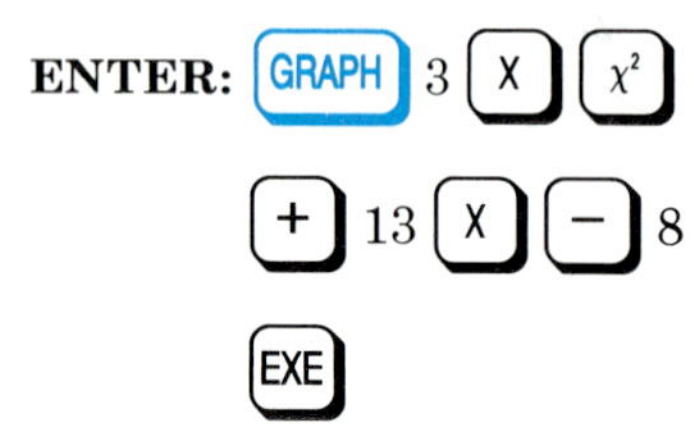

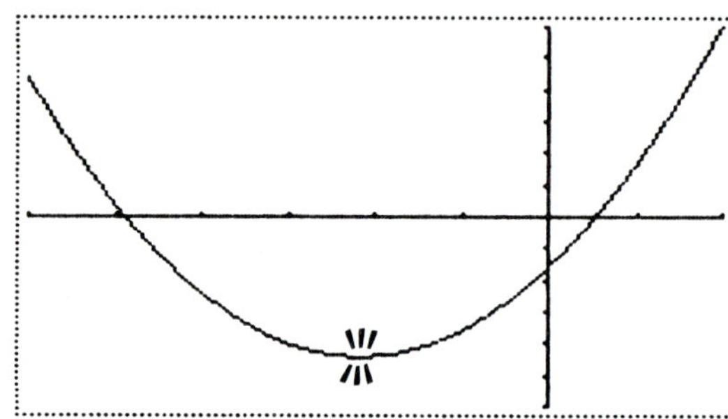

Activate the tracing function and move the pointer to the vertex. Record the x-coordinate value and then press the $\boxed{X \leftrightarrow Y}$ key to display the y-coordinate value.

The vertex is approximately $(-2.2, -22.1)$.

The exact coordinates are $\left(-\dfrac{13}{6}, -\dfrac{265}{12}\right)$.

4 **Mr. Pearson has 1275 yards of fence to make a rectangular corral for horses. If an existing fence is used as one side of the corral, what would be the length and width of the corral with maximum area?**

Let x represent the width, in yards, of the corral. Then $1275 - 2x$ represents the length of the corral.

Area = (length) × (width)
$$y = (1275 - 2x)x$$
$$= 1275x - 2x^2$$

Let y represent the area.

Many graphing calculators only allow the variable x to be used when entering functions for graphing. Thus, it is best to use x as the variable in your equations.

The equation describing the area is a quadratic function. Thus, the problem can be solved by finding the vertex of its graph using the tracing function.

First, choose a set of range parameters values that will give you a rough approximation for the position of the vertex. One possible set of values is given below. Set the range parameters to these values. Then graph $y = 1275x - 2x^2$.

Xmin: 0 Xmax: 600 Xscl: 50   Ymin: 0 Ymax: 250,000 Yscl: 25,000

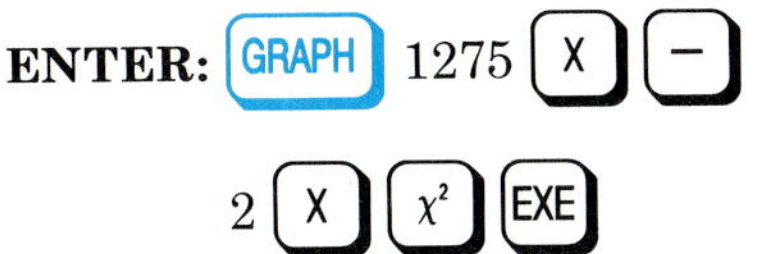

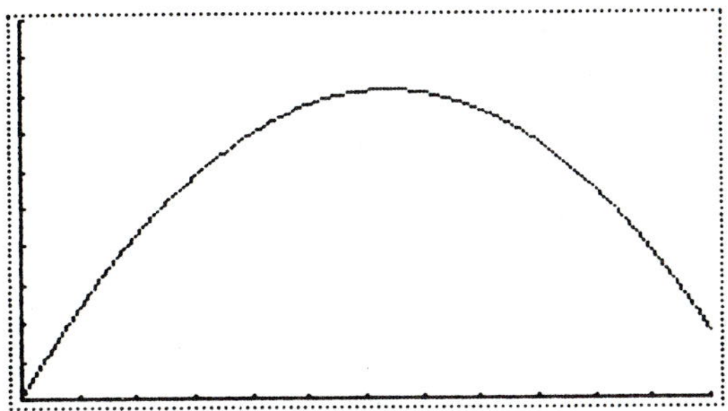

From the graph, you know that the x-coordinate of the vertex is between 250 and 400. Also, the y-coordinate is between 175,000 and 225,000. Use this information to choose new range parameter values for approximating the coordinates of the vertex. One possible set of values is given below.

Xmin: 250	Ymin: 175,000
Xmax: 400	Ymax: 225,000
Xscl: 15	Yscl: 5,000

Set the range parameters to these values. To graph $1275x - 2x^2$ using these new values, press the EXE (or DRAW) key.

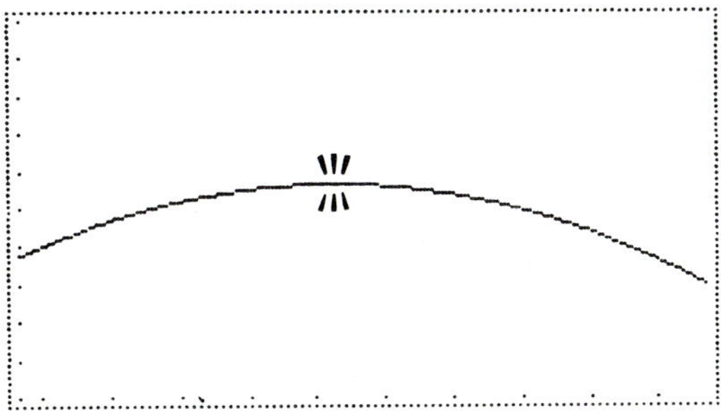

Activate the tracing function and move the pointer to the vertex. Record the x-coordinate value and then press the X↔Y key to display the y-coordinate value.

The exact coordinates of the vertex are (318.75, 203,203.125).

   The vertex is approximately (319, 203,203).

The maximum area, about 203,203 square yards, occurs when the width is about 319 yards and the length is about $1275 - 2(319)$ or 637 yards.

Exploratory Exercises

Graph $y = x^2 - 48x + 27$ for each set of range parameter values. Then use the tracing function to determine the coordinates of the vertex of the graph.

1. Xmin: −80, Xmax: 80, Xscl: 20, Ymin: −1000, Ymax: 500, Yscl: 200

2. Xmin: 0, Xmax: 60, Xscl: 10, Ymin: −800, Ymax: 50, Yscl: 100

3. Xmin: 0, Xmax: 50, Xscl: 10, Ymin: −700, Ymax: 0, Yscl: 100

4. Xmin: 0, Xmax: 48, Xscl: 10, Ymin: −650, Ymax: 0, Yscl: 100

Graph each equation on a graphing calculator. Then use the tracing function to approximate the coordinates of the vertex of the graph.

5. $y = (x + 8)^2 - 5$

6. $y = 2(x - 10)^2 + 14$

7. $y = x^2 - 14x + 70$

8. $y = 3x^2 + 36x + 133$

9. $y = \frac{1}{3}x^2 - 8x + 39$

10. $y = \frac{7}{5}x^2 + 42x + 297$

Written Exercises

Graph each equation on a graphing calculator. Then use the tracing function to approximate the coordinates of the vertex of the graph.

11. $y = x^2 - 87x - 23$

12. $y = x^2 + 103x - 12$

13. $y = x^2 + 236x + 102$

14. $y = 5x^2 - 302x - 321$

15. $y = 12x^2 + 813x + 100,101$

16. $y = 0.5x^2 - 37x + 777$

17. $y = \frac{2}{3}x^2 - \frac{3}{4}x + 6$

18. $y = \frac{4}{11}x^2 + 32x - 29$

19. $y = 25x^2 + 3720x + 10,255$

20. $y = 100x^2 - 7530x + 187,334)$

Solve each problem. Use the tracing function on a graphing calculator.

21. Mr. Ortiz has 287 meters of fence to make a rectangular pen for chickens. If a barn is used as one side of the pen, what would be the length and width for the maximum area?

22. Mrs. Harrison has 195 feet of fence to put around a rectangular garden. If a 12-foot opening is left on one side for a gate, what would be the length and width for the maximum area?

23. A taxi service operates between two airports, transporting 415 passengers a day. The charge is $10.50. The owner estimates that 23 passengers will be lost for each $1 increase in the fare. What charge would be the most profitable for the service?

24. An airline transports 1275 people a week between two cities. A round trip ticket costs $428. The company wants to increase the price. They estimate that for each $25 increase 48 passengers will be lost. What ticket price will maximize their income?

25. Find the dimensions and maximum area of a rectangle if its perimeter is 1575 centimeters.

26. Find the dimensions and maximum area of a rectangle if its perimeter is 10,760 inches.

27. James Mallory has 7638 meters of fence to put around two rectangular yards. If the yards are to be separated by part of the fence, what would be the length and width for maximum area?

28. A wire 125 cm long is cut into 2 pieces and each piece is bent to form a square. How long should each piece be to minimize the sum of the areas of the two squares?

29. A wire 97.5 in. long is cut into 2 pieces. One piece is bent to form a square. The other piece is bent to form a rectangle that is 2.5 in. longer than it is wide. How long should each piece be to minimize the sum of the areas of the square and the rectangle?

30. A square, which is 15 cm by 15 cm, is cut from each corner of a rectangular piece of cardboard and the sides are folded up to make a box. If the bottom of the box must have a perimeter of 725 cm, what would be the length, width, and height for maximum volume?

Applications in Archeology Babylonian Quadratics

Archaeologists have found artifacts that show how ancient civilizations solved mathematical problems. For example, the tablet in the photograph below contains one quadratic equation and its solution written sometime between 1900 and 1600 B.C. The solution is completely without algebraic notation. (Algebraic symbols were not in general usage until the middle of the 17th century A.D.)

This tablet explains the solution to a problem about a rectangle that has area 60 square units. The English translation, using our base ten number system, is given below.

Line	Front Side	Line	Reverse Side
1	The length exceeded the width by 7.	1–2	Subtract 3.5 from the one;
2	What are the length and width?	3	add it to the other.
3–5	As for you—halve 7, by which the length exceeded the width, and the result is 3.5.	4	One is 12, and the other is 5.
6–7	Multiply together 3.5 with 3.5, and the result is 12.25.	5	12 is the length, 5 the width.
8	To 12.25 which resulted for you,		
9	add 60, the product, and the result is 72.25.		
10	What is the square root of 72.25? 8.5.		
11	Lay down 8.5 and 8.5 its equal.		

The standard form of this type of quadratic equation is $x^2 + ax = b$, $b > 0$. In this example, if y is the length and x is the width, then $y = x + 7$ and $xy = 60$, or $x(x + 7) = 60$. The solution is given below.

$$x = \sqrt{\left(\frac{a}{2}\right)^2 + b} - \frac{a}{2} \qquad\qquad y = \sqrt{\left(\frac{a}{2}\right)^2 + b} + \frac{a}{2}$$

$$= \sqrt{\left(\frac{7}{2}\right)^2 + 60} - \frac{7}{2} \qquad\qquad = \sqrt{\left(\frac{7}{2}\right)^2 + 60} + \frac{7}{2}$$

$$= 5 \qquad\qquad\qquad\qquad = 12$$

Exercises

1. The length of a rectangle is 16 cm longer than the width. The area is 260 cm^2. Find the length and width of the rectangle using the method shown above.

2. The length of a rectangle is $\frac{2}{3}$ ft longer than the width. The area is $\frac{5}{3}$ ft^2. Find the length and width of the rectangle using the method shown above.

 Graphing Quadratic Inequalities

The graph of $y = x^2 - 6x + 5$ is a parabola that separates the coordinate plane into two regions. The graph of $y > x^2 - 6x + 5$ is the region *inside* the parabola. The graph of $y < x^2 - 6x + 5$ is the region *outside* the parabola.

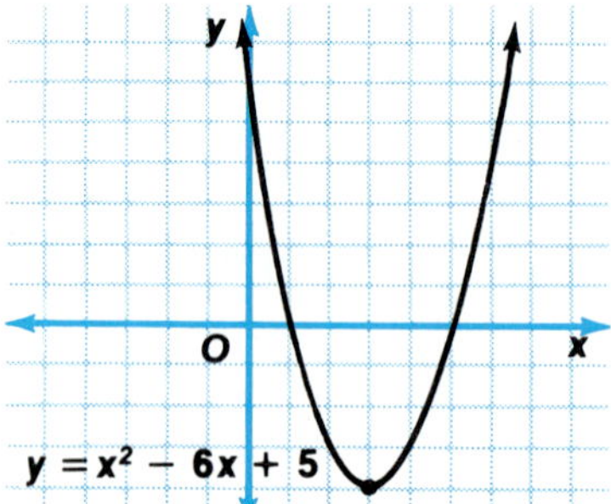

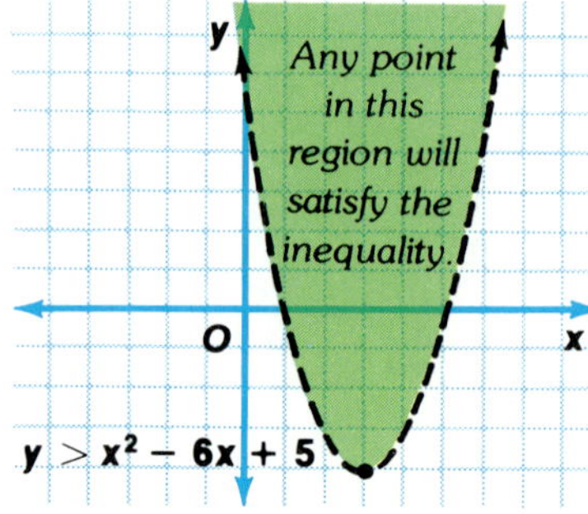

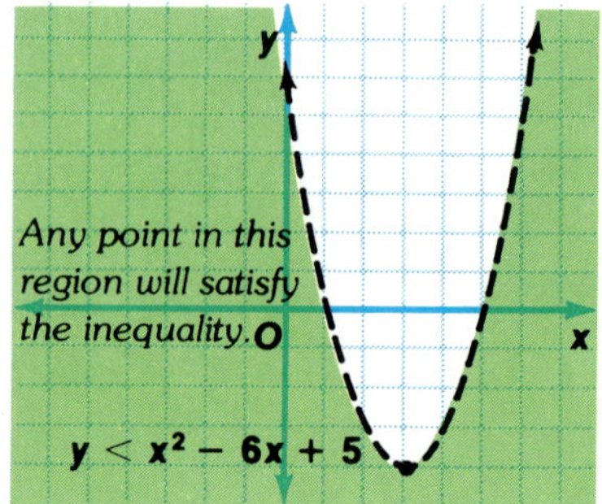

The parabola described by $y = x^2 - 6x + 5$ is called the **boundary** of each region. If it is broken, it is not part of the graph. If it is solid, it is part of the graph.

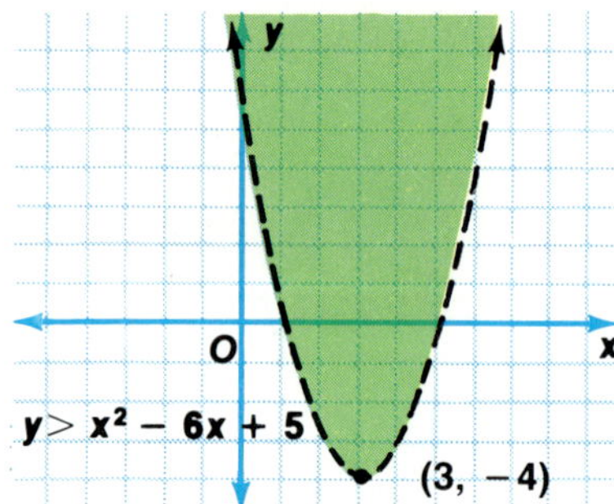

Note that $>$ tells you the boundary is not <u>included</u>.

Note that $\geq$ tells you the boundary is included.

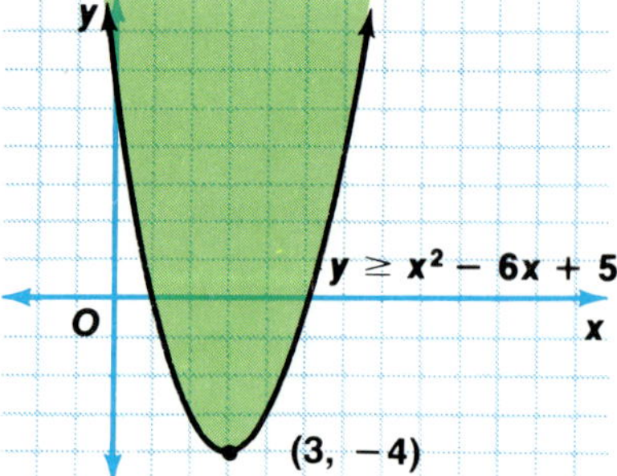

Examples

1 **Graph $y \leq -2x^2 + 8x - 4$.**

$$y \leq -2x^2 + 8x - 4$$
$$y \leq -2(x^2 - 4x) - 4$$
$$y \leq -2\left[x^2 - 4x + \left(\frac{4}{2}\right)^2\right] - 4 - (-2)\left(\frac{4}{2}\right)^2$$
$$y \leq -2(x - 2)^2 + 4$$

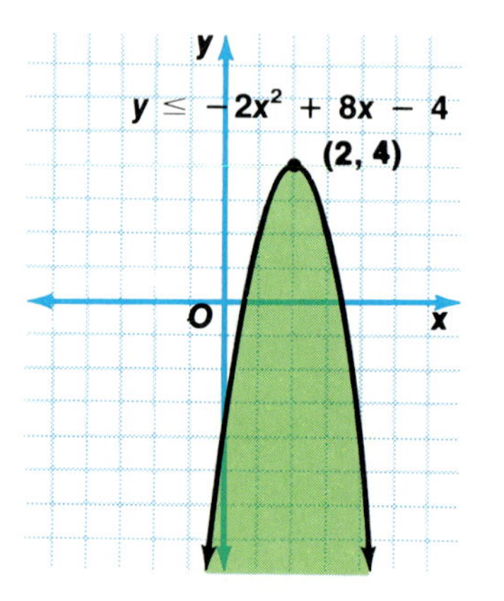

The boundary is a parabola with vertex $(2, 4)$, axis of symmetry $x = 2$, which opens downward.

Test points not on the boundary to determine if the region inside the parabola or the region outside the parabola belongs to the graph.

region inside the parabola

Test (2, 0): $y \leq -2x^2 + 8x - 4$
$0 \overset{?}{\leq} -2(2)^2 + 8(2) - 4$
$0 \leq 4$ ✔
belongs

region outside the parabola

Test (−2, 0): $y \leq -2x^2 + 8x - 4$
$0 \overset{?}{\leq} -2(-2)^2 + 8(-2) - 4$
$0 \not\leq -28$
does not belong

2 **Graph $y > x^2 - 7x + 10$.**

$y > x^2 - 7x + 10$

$y > \left[x^2 - 7x + \left(\frac{7}{2}\right)^2\right] + 10 - \left(\frac{7}{2}\right)^2$

$y > \left(x - \frac{7}{2}\right)^2 - \frac{9}{4}$

The boundary is a parabola with vertex $\left(\frac{7}{2}, -\frac{9}{4}\right)$,

axis of symmetry $x = \frac{7}{2}$, which opens upward.

Test points not on the boundary to determine if
the region inside the parabola or the region
outside the parabola belongs to the graph.

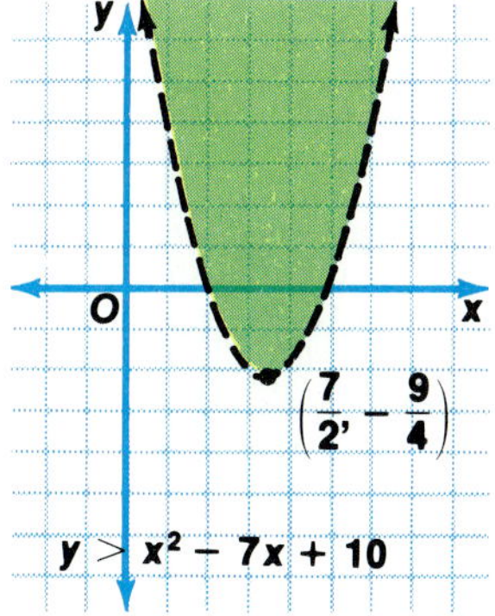

region inside the parabola

Test (4, 3): $y > x^2 - 7x + 10$
$3 \overset{?}{>} (4)^2 - 7(4) + 10$
$3 > -2$ ✔
belongs

region outside the parabola

Test (0, 0): $y > x^2 - 7x + 10$
$0 \overset{?}{>} (0)^2 - 7(0) + 10$
$0 \not> 10$
does not belong

Exploratory Exercises

A possible solution for each inequality is given in color. Determine if it is a correct solution.

1. $y > (x - 2)^2$ $(0, 5)$ **2.** $y \leq 2(x + 3)^2$ $(-2, 3)$ **3.** $y \geq x^2 - 16$ $(-3, 0)$

4. $y < -3x^2 + 5$ $(2, -7)$ **5.** $y \leq -x^2 + 5x$ $(4, -4)$ **6.** $y \geq 5x^2 - 6x$ $(1, 1)$

7. $y > x^2 + 8x + 17$ $(-1, 9)$ **8.** $y < -2x^2 - 8x + 11$ $(-5, 1)$

9. $y < -0.5x^2 + 9x - 2$ $(16, 15)$ **10.** $y \geq 4x^2 - 8x - 13$ $(3, -1)$

Written Exercises

Graph each inequality.

11. $y > x^2 + 2x + 1$ **12.** $y \leq x^2 + 6x + 9$ **13.** $y \geq x^2 + 8x + 16$

14. $y > x^2 - 10x + 25$ **15.** $y \leq x^2 - 16$ **16.** $y \geq x^2 - 49$

17. $y \leq x^2 - 13x + 36$ **18.** $y > x^2 + x - 30$ **19.** $y \geq x^2 + 3x - 18$

20. $y < x^2 - x - 20$ **21.** $y > x^2 + 3x - 4$ **22.** $y < x^2 + 4x + 3$

23. $y > 4x^2 - 8x + 3$ **24.** $y \leq 2x^2 + x - 3$ **25.** $y \geq -x^2 + 6x + 8$

26. $y \leq -x^2 - 7x + 10$ **27.** $y < -3x^2 + 5x + 2$ **28.** $y > -4x^2 - 3x - 6$

8-6 Solving Quadratic Inequalities

Consider the quadratic inequality $0 \geq x^2 + 5x - 6$. This inequality can be solved graphically.

The graph of $y \geq x^2 + 5x - 6$ is used to solve $0 \geq x^2 + 5x - 6$.

$$y \geq x^2 + 5x - 6$$
$$y \geq \left[\left(x^2 + 5x + \left(\frac{5}{2}\right)^2\right)\right] - 6 - \left(\frac{5}{2}\right)^2$$
$$y \geq \left(x + \frac{5}{2}\right)^2 - \frac{49}{4}$$

Each point on the x-axis has a y-coordinate of 0. Thus, those points on the x-axis that also satisfy $y \geq x^2 + 5x - 6$ are solutions to the inequality $0 \geq x^2 + 5x - 6$.

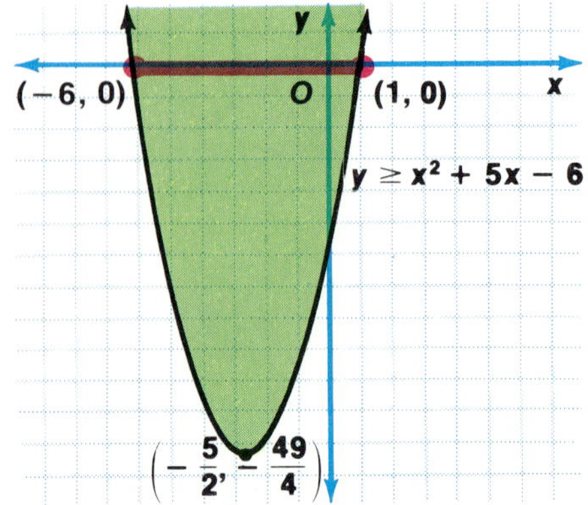

The points on the x-axis that satisfy the inequality are shown in red.

The inequality can also be solved algebraically.

$$x^2 + 5x - 6 \leq 0$$
$$(x - 1)(x + 6) \leq 0 \qquad \textit{Factor.}$$

Recall that the product of two factors is negative only if one factor is positive and one factor is negative.

$$x - 1 \leq 0 \text{ and } x + 6 \geq 0 \quad \text{or} \quad x - 1 \geq 0 \text{ and } x + 6 \leq 0$$
$$x \leq 1 \text{ and } \quad x \geq -6 \quad \text{or} \quad x \geq 1 \text{ and } \quad x \leq -6$$
$$-6 \leq x \leq 1 \qquad\qquad\qquad\qquad \textbf{never true}$$

The graphs of $x \geq 1$ and $x \leq -6$ do not intersect. Therefore, the statement $x \geq 1$ and $x \leq -6$ is never true.

Solving $0 \geq x^2 + 5x - 6$ either graphically or algebraically produces the same solution, $\{x | -6 \leq x \leq 1\}$.

Example

1 **Solve $x^2 + 8x > -15$ graphically.**

$$x^2 + 8x > -15$$
$$x^2 + 8x + 15 > 0$$
$$0 < x^2 + 8x + 15$$

Graph $y < x^2 + 8x + 15$.

$$y < x^2 + 8x + 15$$
$$y < (x^2 + 8x + 16) + 15 - 16$$
$$y < (x - 4)^2 - 1$$

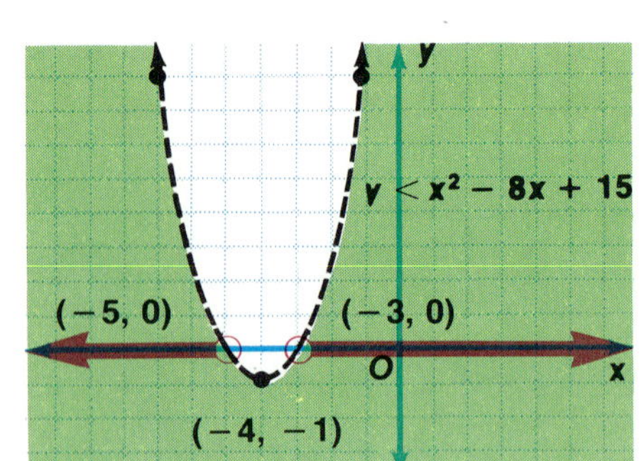

The points on the x-axis that also satisfy $y < x^2 + 8x + 15$ are solutions of $0 < x^2 + 8x + 15$.

Therefore, the solution set is $\{x | x < -5 \text{ or } x > -3\}$.

2 **Solve $x^2 > 2x + 8$ algebraically.**

$x^2 - 2x - 8 > 0$ *The product of two factors is positive only if both*
$(x - 4)(x + 2) > 0$ *factors are positive or both factors are negative.*

$x - 4 > 0$ and $x + 2 > 0$ or $x - 4 < 0$ and $x + 2 < 0$ *Solve each compound*
$x > 4$ and $x > -2$ or $x < 4$ and $x < -2$ *sentence shown in color. In each case, the values of x must satisfy each*

$x > 4$ or $x < -2$ *inequality.*

The solution set is $\{x \mid x > 4 \text{ or } x < -2\}$.

3 **Solve $(x - 5)(x + 3) > 0$.**

An alternate method for solving quadratic inequalities uses three test points.

First, solve the equation $(x - 5)(x + 3) = 0$.

$(x - 5)(x + 3) = 0$ *The solutions to this equation are often called critical points.*

$x - 5 = 0$ or $x + 3 = 0$
$x = 5$ or $x = -3$

The points 5 and -3 separate a number line into three parts, $x < -3$, $-3 < x < 5$, and $x > 5$, as shown below.

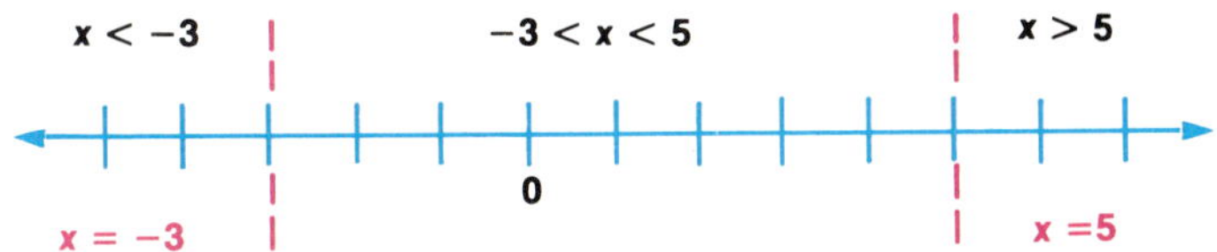

Choose a value of x from each part and substitute into $(x - 5)(x + 3) > 0$.

For $x < -3$, try -4.

$(x - 5)(x + 3) > 0$
$(-4 - 5)(-4 + 3) \overset{?}{>} 0$
$9 > 0$ ✔

For $-3 < x < 5$, try 0.

$(x - 5)(x + 3) > 0$
$(0 - 5)(0 + 3) \overset{?}{>} 0$
$-15 \not> 0$

For $x > 5$, try 7.

$(x - 5)(x + 3) > 0$
$(7 - 5)(7 + 3) \overset{?}{>} 0$
$20 > 0$ ✔

The solution set is
$\{x \mid x < -3 \text{ or } x > 5\}$.

Exploratory Exercises

Indicate if the factors must be positive or negative in each quadratic inequality.

1. $(x - 8)(x + 2) < 0$

2. $(x + 4)(x - 3) > 0$

3. $(x + 6)(x + 2) > 0$

4. $(x - 8)(x - 7) < 0$

5. $(x + 2)(x - 3) \geq 0$

6. $(x + 7)(x - 11) \leq 0$

7. $x^2 + 10x + 25 \leq 0$

8. $x^2 - 11x - 26 > 0$

9. $x^2 + 3x - 18 \geq 0$

Written Exercises

Solve each inequality.

10. $(x + 3)(x + 7) > 0$

11. $(y - 12)(y - 5) \leq 0$

12. $(a - 1.5)(a + 2.5) \geq 0$

13. $x^2 + x - 6 > 0$

14. $y^2 + 4y - 21 < 0$

15. $p^2 + 2p \geq 24$

16. $m^2 - 4m \leq 5$

17. $2b^2 - b < 6$

18. $6r^2 + 5r > 4$

19. $x^2 - 4x \leq 0$

20. $z^2 \geq 2z$

21. $t^2 \leq 36$

22. $b^2 \geq 3b + 28$

23. $r^2 + 12r \leq -27$

24. $a^2 - 10a + 25 \geq 0$

25. $2x^2 - 25 > 0$

26. $d^2 + 36 < -12d$

27. $9n^2 - 6n + 1 \leq 0$

28. $5c - 2c^2 > -3$

29. $-5y - 3y^2 < -2$

30. $b^2 + 8b \geq -16$

31. $m^2 \leq 3$

32. $4t^2 - 9 < -4t$

33. $9s^2 - 2 > -6s$

Challenge Exercises

Solve each inequality.

34. $(x - 3)(x + 4)(x - 1) > 0$

35. $(x + 2)(x - 3)(x + 6) < 0$

36. $(x - 8)(x + 4)(x + 2) \leq 0$

37. $(x + 5)(x + 6)(x + 7) \geq 0$

38. $(x + 2)(x + 3)(x - 1)(x - 2) \geq 0$

39. $(x - 6)(x + 5)(x - 4)(x + 1) > 0$

mini-review

Factor.

1. $x^3 + 3x^2y - 4xy^2 - 12y^3$

2. $(r + t)^3 - t^2(r + t)$

3. Use long division to find $(2x^4 + 8x^3 + 7x - 5) \div (x + 5)$.

4. Is $3y - 4$ a factor of $3y^3 + 14y^2 - 48y + 32$? Write *yes* or *no*.

5. One factor of $4a^3 + 16a^2 - 9a - 36$ is $2a + 3$. Find the other factors.

Simplify.

6. $\sqrt{x^2 - 8xy + 16y^2}$

7. $\sqrt{108a^2b^3c^4}$

8. $\left(4^{\frac{1}{3}}r^{-\frac{5}{6}}s^{-\frac{3}{4}}\right)^{-2}$

9. $(5 - 2i)(2 + i)(1 - 5i)$

10. $(\sqrt{-8})^2(\sqrt{-6})^3$

Solve each equation.

11. $\sqrt{3y - 4} = \sqrt{3y + 4} - 2$

12. $a + 9\sqrt{a} = -20$

13. $5m^2 + m = 7$

14. $6x^2 - 11x - 72 = 0$

Find the value of the discriminant for each quadratic equation. Describe the nature of the roots. If the roots are real, tell whether they are rational or irrational.

15. $2x^2 + 7x + 7 = 0$

16. $3z^2 - 11z + 9 = 0$

Graph each equation or inequality.

17. $y = -(x + 2)^2 + 3$

18. $f(x) = 3x^2 - 6x + 2$

19. $y < 2(x + 3)^2 - 1$

20. $y \leq -x^2 + 8x - 11$

Graphing Quadratic Functions

The computer program below can be used to find the vertex, axis of symmetry, and direction of opening for the graph of any quadratic function. The program also lists six points on the graph, three points on each side of the axis of symmetry.

```
10    INPUT "ENTER THE COEFFICIENTS OF THE EQUATION: ";A, B, C
20    LET H = −B / (2 ∗ A)
30    LET K = A ∗ H ∗ H + B ∗ H + C
40    PRINT "THE VERTEX IS ('';H; '', '';K; '')."
50    PRINT " THE AXIS OF SYMMETRY IS X = '';H; ''."
60    IF A > 0, THEN 80
70    PRINT "THE GRAPH OPENS DOWNWARD.": GOTO 90
80    PRINT " THE GRAPH OPENS UPWARD."
90    PRINT "HERE ARE THE COORDINATES OF SIX ADDITIONAL POINTS ON THE GRAPH:"
100   FOR N = H − 3 TO H + 3
110   IF H = INT(H) THEN 130
120   IF N = H − 3 THEN 170
130   LET X = INT(N)
140   IF X = H THEN 170
150   LET Y = A ∗ X ∗ X + B ∗ X + C
160   PRINT "(''; X; '','''; Y; ''),"
170   NEXT N
180   END
```

The quadratic function must be in the form $f(x) = ax^2 + bx + c$ where $a \neq 0$, in order to provide the values of A, B, and C.

Enter and run the program for the function $f(x) = 2x^2 - 5x + 3$. You should get the following results.

```
]RUN
ENTER THE COEFFICIENTS OF THE EQUATION: 2, −5, 3
THE VERTEX IS (1.25, −0.125).
THE AXIS OF SYMMETRY IS X = 1.25.
THE GRAPH OPENS UPWARD.
HERE ARE THE COORDINATES OF SIX ADDITIONAL POINTS ON THE GRAPH
(−1, 10)        (0, 3)          (1, 0)          (2, 1)
(3, 6)          (4, 15)
```

Exercises

Use the program to find the vertex, axis of symmetry, direction of opening, and six additional points on the graph of each function. Then graph the function.

1. $f(x) = x^2 - 28x + 186$

2. $f(x) = -x^2 - 16x + 36$

3. $f(x) = -5x^2 + 22x - 3$

4. $f(x) = 10x^2 - 37x - 7$

5. Name a point of intersection for the graphs of $f(x) = -5x^2 + 22x - 3$ and $f(x) = 10x^2 - 37x - 7$. Use the results from Exercises 3 and 4.

6. How can you change the program so that twelve additional points are found?

quadratic function (261)   axis of symmetry (270)   minimum value (279)
parabola (270)   maximum value (279)   boundary (290)
vertex (270)

Chapter Summary

1. A quadratic function is a function described by an equation of the form $f(x) = ax^2 + bx + c$ where $a \neq 0$. In this function, ax^2 is called the quadratic term, bx is called the linear term, and c is called the constant term. (261)

2. Graphing calculators can produce the graphs of a wide variety of functions and relations, including quadratic functions. (264–268)

3. The graph of a quadratic function is called a parabola. (270)

4. If the coordinate plane could be folded along the axis of symmetry of a parabola, then the portions of the parabola on each side of the axis would match. (270)

5. In general, as the values of h and k change, the graph of $y = a(x - h)^2 + k$ is translated $|h|$ units to the left or right and $|k|$ units up or down. The conclusion about the graph of $y = a(x - h)^2 + k$ can be summarized as follows: (274)

$y = a(x - h)^2 + k$	a is positive	a is negative
Vertex	(h, k)	(h, k)
Axis of symmetry	$x = h$	$x = h$
Direction of opening	upward	downward

As the value of $|a|$ increases the graphs of $y = a(x - h)^2 + k$ narrow.

6. If a maximum or minimum problem can be described by a quadratic function, it can often be solved by finding the vertex of the graph of the function. (279)

7. The tracing function on a graphing calculator can be used to determine the approximate coordinates of the vertex of a parabola. (284–288)

8. The graph of a quadratic inequality will include either the region inside the boundary or outside the boundary. The boundary itself may or may not be included. (290)

9. Quadratic inequalities can be solved either graphically or algebraically. (292)

8–1 **Express each function in quadratic form.**

1. $f(x) = 3(x + 2)^2 - 7$

2. $f(x) = -3(x - 7)^2 + 6$

3. $f(x) = -2(x - 3)^2 + 9$

4. $f(x) = 0.5(x + 6)^2 - 21$

Define a variable and write a quadratic function to describe each situation.

5. the product of two numbers whose difference is 110

6. the area of a rectangle whose perimeter is 39 centimeters

8–2 **Graph each equation. Then name the vertex and axis of symmetry for each graph.**

7. $y = (x - 3)^2$

8. $y = (x - 2)^2 - 3$

9. $f(x) = (x + 3)^2 - 7$

10. $f(x) = x^2 + 8x + 21$

11. $y = x^2 - 5x + 6$

12. $f(x) = x^2 + 9x + 25$

8–3 **Graph each equation. Then name the vertex, axis of symmetry, and direction of opening for each graph.**

13. $y = -2x^2 - 4$

14. $f(x) = 2(x + 1)^2 - 2$

15. $y = \frac{1}{2}(x + 2)^2 - 1$

16. $y = -\frac{4}{3}x^2 + 8x + 9$

17. $f(x) = 3x^2 - 6x + 10$

18. $f(x) = -5x^2 - 20x + 2$

8–4 **Solve each problem.**

19. Find two numbers whose difference is 5 and whose product is a minimum.

20. Find the dimensions and maximum area of a rectangle, if its perimeter is 30 centimeters.

21. A taxi service transports 320 passengers a day between two airports. The charge is $11.50. The owner estimates that 30 passengers will be lost for each $1.50 increase in the fare. What price will maximize the owner's income?

22. Franco has 200 meters of fence to make a rectangular yard for his dogs. If an existing fence is used as one side of the yard, what would be the length and width for maximum area?

8–5 **Graph each inequality.**

23. $y > x^2 + 3x - 4$

24. $y < x^2 + 4x + 3$

25. $y > -2x^2 + 9$

26. $y \geq 3x^2 - 15x + 22$

8–6 **Solve each inequality.**

27. $(x - 4)(x + 2) < 0$

28. $x^2 + 8x - 9 > 0$

29. $y^2 \geq 12y$

30. $6 - 5a - 4a^2 \leq 0$

31. $m^2 - 20m \geq -100$

32. $r^2 + 16r + 64 < 0$

Express each function in quadratic form.

1. $f(x) = (x + 2)^2 + 8$

2. $f(x) = 2(x - 3)^2 + 5$

3. $f(x) = \frac{1}{2}(x + 4)^2 - 7$

Write the equation of each parabola shown.

4. 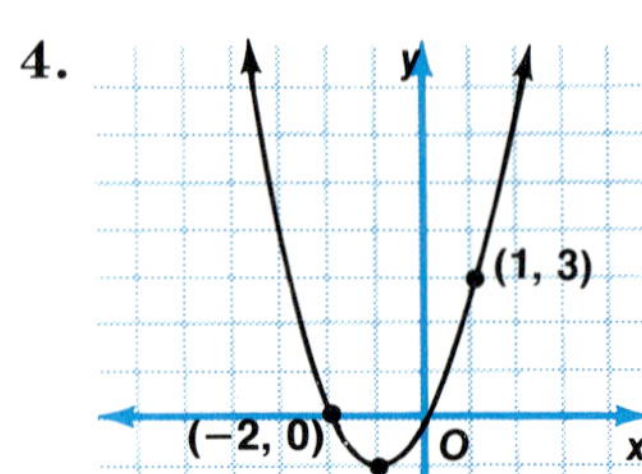

5. 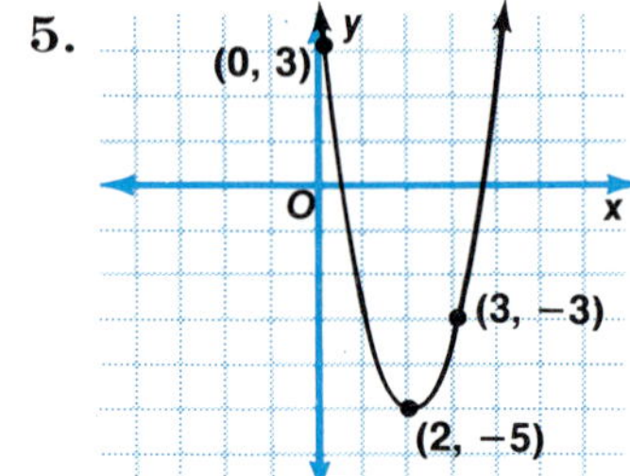

6. 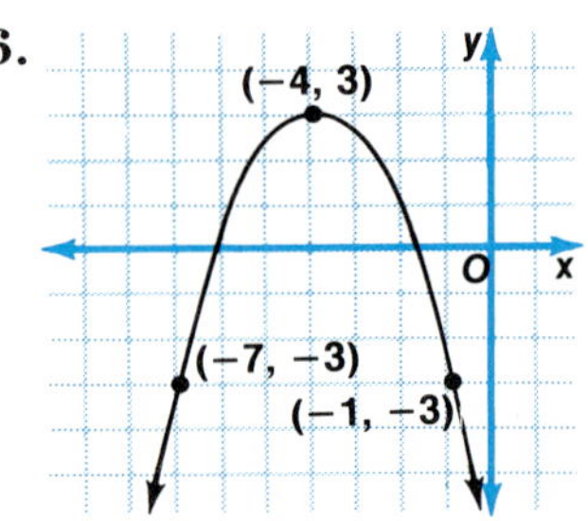

Graph each equation. Then name the vertex, axis of symmetry, and direction of opening for each graph.

7. $y = (x + 3)^2$

8. $y = (x + 2)^2 + 1$

9. $f(x) = -(x - 2)^2$

10. $f(x) = -2(x + 4)^2 + 6$

11. $y = \frac{1}{2}(x + 6)^2 - 3$

12. $y = x^2 + 3x + 6$

13. $y = x^2 - 8x + 9$

14. $f(x) = 2x^2 + 8x + 9$

15. $f(x) = -x^2 - 10x + 10$

16. $f(x) = -\frac{2}{3}x^2 + 4x - 3$

Solve each problem.

17. Find two numbers whose difference is -18 and whose product is a minimum.

18. Find the dimensions and maximum area of a rectangle, if its perimeter is 60 inches.

19. A rocket is shot upward with an initial velocity of 40 feet per second. Its height above the ground after t seconds is given by $h(t) = 40t - 16t^2$. What is its maximum height? When will it return to earth?

20. Harry plans to build a rectangular pen with 80 feet of fence. He will use the side of a barn as one side of the pen. Find the length and width of the pen with maximum area.

Graph each inequality.

21. $y \leq x^2 + 6x - 7$

22. $y < -x^2 + 4x - 4$

Solve each inequality.

23. $(x + 5)(x + 3) < 0$

24. $2x^2 + 3x - 2 > 0$

25. $y^2 - 10y \geq -25$

The test questions on this page deal with expressions and equations. The information at the right may help you with some of the questions.

Directions: Choose the one best answer. Write A, B, C, or D.

1. If $10 - 4y = 18 + 2y$, then what is the value of $3y$?

(A) -4 (B) -3 (C) $-\frac{4}{3}$ (D) $\frac{4}{3}$

2. If it takes 6 hours for 4 people to paint a room, how many hours will it take 5 people, working at the same rate, to paint a room that is the same size?

(A) $3\frac{1}{3}$ (B) $4\frac{4}{5}$ (C) $6\frac{1}{2}$ (D) $7\frac{1}{2}$

3. In a college lecture hall, the number of seats in each row is 25 less than the number of rows. If there are 350 seats in all, find the number of rows.

(A) 10   (B) 14   (C) 25   (D) 35

4. A grocery store sold 500 candy bars for $105.00. A small candy bar sells for 10¢ and a large bar sells for 35¢. How many of the small bars were sold?

(A) 55   (B) 77   (C) 220   (D) 280

5. If $x > 1$, which of the following increases as x increases?

I. $x - \frac{1}{x}$

II. $\frac{1}{x^2 - x}$

III. $4x^3 - 2x^2$

(A) I only   (B) II only
(C) I and III only   (D) I, II and III

6. Evaluate $3x^3 - 2x^2 + x - 1$, if $x = -1$.

(A) -7 (B) -3 (C) -2 (D) -1

---

1. Many problems can be solved without much calculating if the basic mathematical concepts are understood. Always look carefully at what is asked, and think of possible shortcuts for solving the problem.

2. Check your solutions by substituting values for the variables.

7. A bottle of type B perfume costs two dollars more than 2 bottles of type C perfume. If the total cost for one bottle of each type is $32, how much more does type B cost than type C?

(A) 10   (B) 12   (C) 22   (D) 32

8. Eight pencils and five pens cost $5.41, while nine pencils and three pens cost $3.75. What is the cost of a pencil?

(A) 12¢   (B) 15¢   (C) 38¢   (D) 89¢

9. John sold 3 less than twice the number of pizzas that Sam sold. If Sam sold x pizzas, how many more did John sell?

(A) $3x - 3$ (B) $2x - 3$
(C) $x + 3$ (D) $x - 3$

10. What is the value of $(x - y)^3$ if $y = x + 3$?

(A) -27 (B) 0   (C) 6   (D) 9

11. What is the value of xy in the equation $21xy + 77 = 32xy$?

(A) $-\frac{1}{7}$ (B) $\frac{1}{7}$ (C) -7 (D) 7

12. If $a^3 = 7$, then what is the value of $4a^6$?

(A) 28   (B) 56   (C) 196   (D) 1372

13. Which is the equation of a line that passes through $(2, -5)$ and is parallel to the graph of $y - 3x = 2$?

(A) $y = -3x - 11$ (B) $y = 3x - 11$
(C) $y = 3x - 5$ (D) $y = x - 7$

Conics

The planets travel about the sun following elliptical orbits. The ellipse is one of four curves formed by the intersection of a plane and a surface of a cone. The other three curves are the parabola, circle, and hyperbola. These curves are called conic sections.

9-1 Distance

The distance between two points on a number line can be found using absolute value.

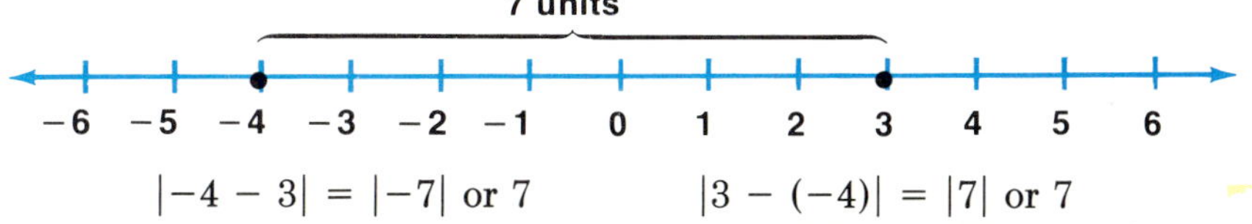

$$|-4 - 3| = |-7| \text{ or } 7 \qquad |3 - (-4)| = |7| \text{ or } 7$$

On a number line, the distance between two points whose coordinates are a and b is $\lvert a - b \rvert$ or $\lvert b - a \rvert$.	*Distance Between Points on a Number Line*

Consider two points in a plane with coordinates $(-2, -6)$ and $(3, -6)$. These points lie on a horizontal line. You can use absolute value to find the distance between the points.

$$|-2 - 3| = |-5| \text{ or } 5 \qquad$$ *Find the absolute value of the difference between the x-coordinates.*

The points with coordinates $(-2, 3)$ and $(-2, -6)$ lie on a vertical line. The distance between these two points is 9 units.

$$|3 - (-6)| = |9| \text{ or } 9 \qquad$$ *Find the absolute value of the difference between the y-coordinates.*

The Pythagorean Theorem may be used to find the distance between any two points in the coordinate plane.

Example

1 **Find the distance between two points with coordinates $(-2, 3)$ and $(2, -4)$**

Draw vertical and horizontal lines from each point to form a right triangle.

The square of the hypotenuse of a right triangle equals the sum of the squares of the other two sides.

$$d^2 = |3 - (-4)|^2 + |-2 - 2|^2$$
$$d^2 = \quad 7^2 \quad + \quad 4^2$$
$$d^2 = 65$$
$$d = \sqrt{65}$$
$$d \approx 8.062 \qquad$$ *Distance is positive.*

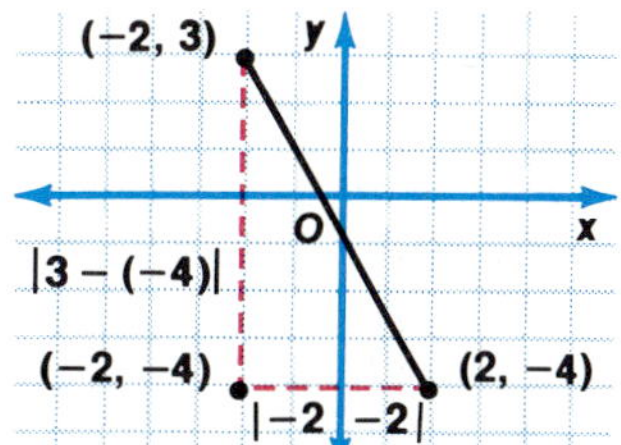

Suppose (x_1, y_1) and (x_2, y_2) name two points in the plane. Form a right triangle by drawing a vertical line through (x_1, y_1) and drawing a horizontal line through (x_2, y_2). These lines intersect at the point (x_1, y_2).   *Why?*

$$d^2 = |x_2 - x_1|^2 + |y_2 - y_1|^2$$
$$d^2 = (x_2 - x_1)^2 + (y_2 - y_1)^2$$
$$d = \sqrt{(x_2 - x_1)^2 + (y_2 - y_1)^2}$$

Why can $(x_2 - x_1)^2$ be substituted for $|x_2 - x_1|^2$?

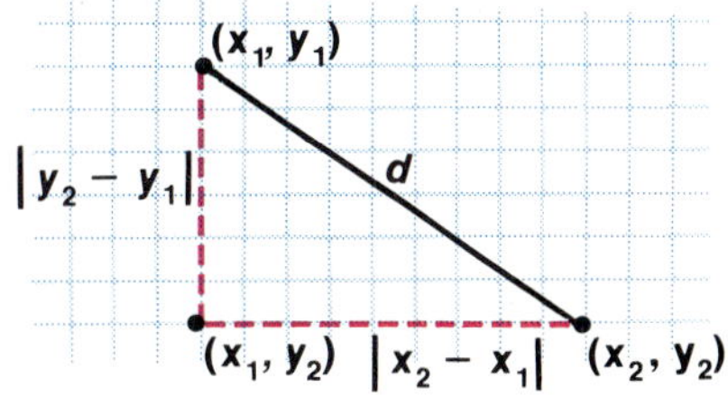

Examples

2 Use the distance formula to find the distance between $(-1, 6)$ and $(5, -4)$.

$$
\begin{aligned}
d &= \sqrt{(x_2 - x_1)^2 + (y_2 - y_1)^2} \\
&= \sqrt{(5 - (-1))^2 + (-4 - 6)^2} \\
&= \sqrt{(6)^2 + (-10)^2} \\
&= \sqrt{36 + 100} \\
&= \sqrt{136} \\
&= 2\sqrt{34}
\end{aligned}
$$

The distance is $2\sqrt{34}$ or about 11.66 units.

3 The point $(3, 1)$ lies on the line segment having endpoints at $(6, -1)$ and $(0, 3)$. Show that $(3, 1)$ is the midpoint of the line segment.

distance between $(3, 1)$ and $(6, -1)$

$$
\begin{aligned}
d &= \sqrt{(6 - 3)^2 + (-1 - 1)^2} \\
&= \sqrt{(3)^2 + (-2)^2} \\
&= \sqrt{9 + 4} \\
&= \sqrt{13} \qquad \sqrt{13} \approx 3.6
\end{aligned}
$$

distance between $(3, 1)$ and $(0, 3)$

$$
\begin{aligned}
d &= \sqrt{(0 - 3)^2 + (3 - 1)^2} \\
&= \sqrt{(-3)^2 + (2)^2} \\
&= \sqrt{9 + 4} \\
&= \sqrt{13}
\end{aligned}
$$

Since the distances are equal, $(3, 1)$ is the midpoint.

You can find the coordinates of the midpoint of a line segment as follows.

Example

4 Find the midpoint of a line segment having endpoints at $(1, -5)$ and $(-4, -7)$.

The coordinates of the midpoint are $\left(\dfrac{1 + (-4)}{2}, \dfrac{-5 + (-7)}{2}\right)$ or $\left(-\dfrac{3}{2}, -6\right)$.

Exploratory Exercises

The coordinates of two points on the number line are given. Find the distance between each pair of points.

1. 3, 5   **2.** $-4, -8$ **3.** $-3, 6$ **4.** $-6, 9$

5. $-11, 0$ **6.** $-16, 0$ **7.** $-32, -16$ **8.** $-19, 14$

9. $16.2, -14.9$ **10.** $7.5, -7.5$ **11.** $14\frac{2}{5}, -8\frac{3}{10}$ **12.** $3\frac{1}{2}, -6\frac{1}{3}$

Written Exercises

Use the distance formula to find the distance between each pair of points.

13. $(3, 6), (7, -8)$ **14.** $(4, 2), (-3, -6)$ **15.** $(-3, 1), (4, -2)$

16. $(-8, -7), (-2, -1)$ **17.** $(6, 7), (8, 0)$ **18.** $(9, 3), (-6, -8)$

19. $\left(\frac{1}{3}, \frac{1}{5}\right), (2, -4)$ **20.** $\left(1, \frac{1}{2}\right), \left(\frac{1}{3}, -2\right)$ **21.** $(0.2, 0.6), (0.3, 0.4)$

22. $(-0.2, 0.4), (-0.5, -0.6)$ **23.** $(-2.4, 0.6), (1.7, 0.8)$ **24.** $(3,3), (\sqrt{3}, \sqrt{3})$

25. $(3, \sqrt{3}), (4, \sqrt{3})$ **26.** $(-2\sqrt{7}, 10), (4\sqrt{7}, 8)$ **27.** $(2\sqrt{3}, 4\sqrt{3}), (2\sqrt{3}, -\sqrt{3})$

Find the value of c such that each pair of points is 5 units apart.

28. $(3, 5), (c, 2)$ **29.** $(-4, c), (-7, 7)$ **30.** $(c, 1.9), (1.2, 5.9)$ **31.** $(13, 10.1), (9, c)$

Find the midpoint of each line segment whose endpoints are given below.

32. $(6, 7), (8, 0)$ **33.** $(9, 3), (-6, -8)$ **34.** $\left(\frac{1}{3}, \frac{1}{5}\right), (2, -4)$

35. $\left(1, \frac{1}{2}\right), \left(\frac{1}{3}, -2\right)$ **36.** $(-2.4, 0.6), (1.7, 0.8)$ **37.** $(3, 3), (\sqrt{2}, -\sqrt{2})$

38. Find the perimeter of a quadrilateral with vertices at $(6, 3)$, $(4, 5)$, $(-4, 6)$, and $(-5, -8)$.

39. Find the lengths of the diagonals of a parallelogram with vertices at $(6, 8)$, $(-14, 8)$, $(8, -2)$, and $(-12, -2)$.

40. Triangle ABC has vertices at $A(-2, 8)$, $B(3, 5)$ and $C(7, -4)$. Find the coordinates of the midpoint of each side.

41. Show that triangle ABC with vertices $A(-3, 0)$, $B(-1, 4)$, and $C(1, -2)$ is an isosceles triangle.

42. Parallelogram $ABCD$ has vertices at $A(-5, 1)$, $B(0, 2)$, $C(-3, 6)$, and $D(-8, 5)$. Show that the diagonals of $ABCD$ bisect each other.

43. Right triangle DEF has vertices at $D(0, 1)$, $E(4, 1)$, and $F(0, 7)$. Show that the midpoint of the hypotenuse is the same distance from each vertex.

44. Show that $\left(\frac{x_1 + x_2}{2}, \frac{y_1 + y_2}{2}\right)$ is the midpoint of a line segment having endpoints with coordinates at (x_1, y_1) and (x_2, y_2).

45. Show that the points at $(-1, 3)$, $(3, 6)$, $(6, 2)$, and $(2, -1)$ are vertices of a square.

Challenge Exercises

46. Find the coordinates of a point one-fourth of the distance from $A(3, -2)$ to $B(11, 2)$.

47. Find the coordinates of a point three-fourths of the distance from $X(5, 8)$ to $Y(-7, 16)$.

9-2 Parabolas

The shape of the reflectors in automobile headlights is based on the **parabola**. The diagram at the right shows a cross section of a reflector. The light source is placed at a special point so the light is reflected in parallel rays. In this way, a straight beam of light is formed.

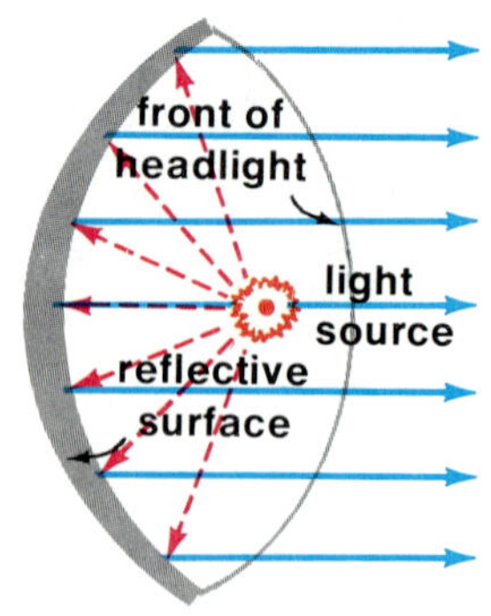

The point where the light source is placed is called the **focus** of the parabola. Parabolas can be defined in terms of the focus.

> **A parabola is the set of all points in a plane that are the same distance from a given point called the *focus* and a given line called the *directrix*.**

Definition of Parabola

The parabola at the right has focus at $(3, 4)$ and directrix $y = -2$. You can use the distance formula and the definition of a parabola to find the equation of this parabola. Let (x, y) be a point on the parabola. This point must be the same distance from the focus, $(3, 4)$, as it is from the directrix, $y = -2$.

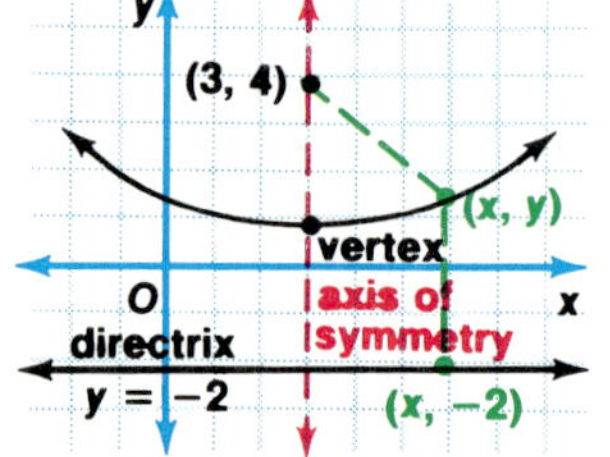

distance between (x, y) and (3, 4) = distance between (x, y) and (x, −2)

$$\sqrt{(x - 3)^2 + (y - 4)^2} = \sqrt{(x - x)^2 + (y - (-2))^2}$$
$$(x - 3)^2 + (y - 4)^2 = (x - x)^2 + (y + 2)^2 \quad \textit{Square each side.}$$
$$(x - 3)^2 + y^2 - 8y + 16 = y^2 + 4y + 4$$
$$(x - 3)^2 + 12 = 12y \quad \textit{Simplify.}$$
$$\frac{1}{12}(x - 3)^2 + 1 = y \quad \textit{Divide by 12.}$$

The equation of a parabola with focus at $(3, 4)$ and directrix $y = -2$ is $y = \frac{1}{12}(x - 3)^2 + 1$. The parabola is symmetric with respect to the line $x = 3$. This line is called the axis of symmetry of the parabola. The point of intersection of the axis of symmetry and the parabola is $(3, 1)$. This point is called the *vertex*.

Consider the line segment through the focus of a parabola perpendicular to its axis of symmetry with endpoints on the parabola. This segment is called the **latus rectum**. In the figure at the right, the latus rectum is $\overline{AB}$. The length of the latus rectum of the parabola given by the equation $y = a(x - h)^2 + k$ is $\left|\frac{1}{a}\right|$. Points A and B are $\left(\frac{1}{2}\right) \cdot \left(\frac{1}{a}\right)$ units each from the focus.

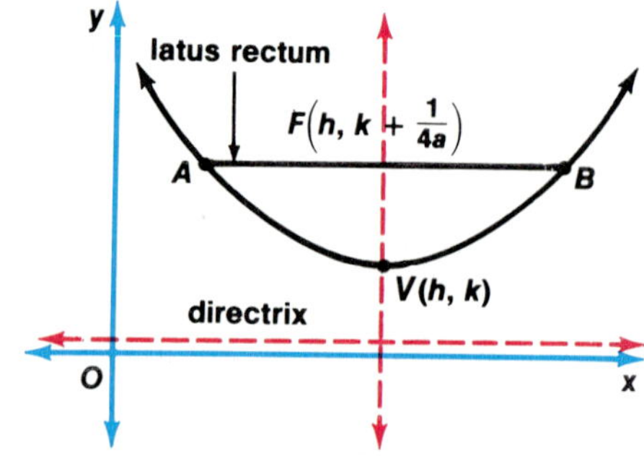

The equation for a parabola can be written in the form $y = a(x - h)^2 + k$ or in the form $x = a(y - k)^2 + h$. Each form provides valuable information about the graph.

Information about Parabolas		
form of equation	$y = a(x - h)^2 + k$	$x = a(y - k)^2 + h$
axis of symmetry	$x = h$	$y = k$
vertex	(h, k)	(h, k)
focus	$\left(h, k + \dfrac{1}{4a}\right)$	$\left(h + \dfrac{1}{4a}, k\right)$
directrix	$y = k - \dfrac{1}{4a}$	$x = h - \dfrac{1}{4a}$
direction of opening	upward if $a > 0$, downward if $a < 0$	right if $a > 0$, left if $a < 0$
length of latus rectum	$\left|\dfrac{1}{a}\right|$ units	$\left|\dfrac{1}{a}\right|$ units

Examples

1 **Graph $y = \frac{1}{4}(x - 2)^2 - 3$.** *The graph will be a parabola.*

vertex: $(2, -3)$

axis of symmetry: $x = 2$

$\dfrac{1}{4a} = \dfrac{1}{4\left(\frac{1}{4}\right)}$ *or 1*

focus: $(2, -3 + 1)$ or $(2, -2)$

directrix: $y = -3 - 1$ or $y = -4$

direction of opening: upward since $a > 0$

Length of latus rectum: $\left|\dfrac{1}{\frac{1}{4}}\right|$ or 4 units

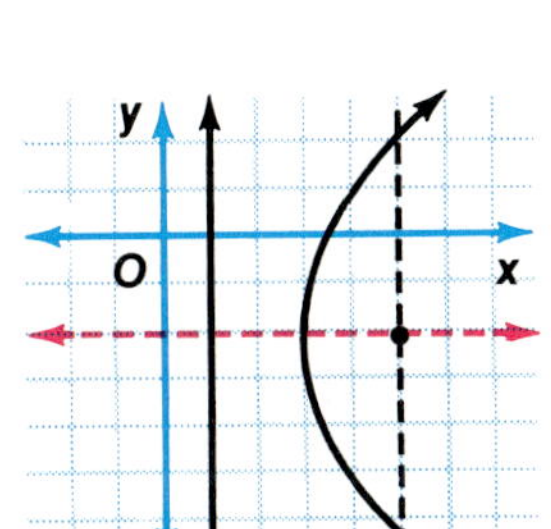

2 **Graph $8x = y^2 + 4y + 28$.**

First write the equation in the form $x = a(y - k)^2 + h$.

$$8x = y^2 + 4y + 28$$
$$8x = y^2 + 4y + \left(\frac{4}{2}\right)^2 - \left(\frac{4}{2}\right)^2 + 28 \quad \textit{Complete the square.}$$
$$8x = (y + 2)^2 + 24$$
$$x = \frac{1}{8}(y + 2)^2 + 3$$

vertex: $(3, -2)$

axis of symmetry: $y = -2$

$\dfrac{1}{4a} = \dfrac{1}{4\left(\frac{1}{8}\right)}$ *or 2*

focus: $(3 + 2, -2)$ or $(5, -2)$

directrix: $x = 3 - 2$ or 1

direction of opening: right since $a > 0$

length of latus rectum: $\left|\dfrac{1}{\frac{1}{8}}\right|$ or 8 units

Exploratory Exercises

Find the value of c that makes each trinomial a perfect square.

1. $x^2 + 4x + c$

2. $x^2 + 6x + c$

3. $y^2 - 8y + c$

4. $p^2 - 10p + c$

5. $r^2 + 3r + c$

6. $m^2 - 3m + c$

7. $x^2 - 7x + c$

8. $t^2 + 15t + c$

Change each equation to the form $y = a(x - h)^2 + k$.

9. $x^2 = 10y$

10. $x^2 = -2y$

11. $y = x^2 - 6x + 33$

12. $y = x^2 + 4x + 1$

13. $y = 3x^2 - 24x + 50$

14. $y = \frac{1}{2}x^2 - 3x + \frac{19}{2}$

Change each equation to the form $x = a(y - k)^2 + h$.

15. $6x = y^2$

16. $y^2 = -12x$

17. $x = y^2 + 8y + 20$

18. $x = y^2 - 14y + 25$

19. $x = \frac{1}{4}y^2 - \frac{1}{2}y - 3$

20. $x = 5y^2 - 25y + 60$

Written Exercises

Name the vertex, axis of symmetry, focus, directrix, and direction of opening of the parabola whose equation is given. Then find the length of the latus rectum and draw the graph.

21. $x^2 = 6y$

22. $y^2 = -8x$

23. $(x + 2)^2 = y - 3$

24. $(x - 4)^2 = 4(y + 2)$

25. $(x - 8)^2 = \frac{1}{2}(y + 1)$

26. $(x + 3)^2 = \frac{1}{4}(y - 2)$

27. $x^2 = (y - 1)$

28. $(x + 2)^2 = 6y$

29. $(y + 3)^2 = 4(x - 2)$

30. $(y - 8)^2 = -4(x - 4)$

31. $y = x^2 - 6x + 33$

32. $x = y^2 + 8y + 20$

33. $x = y^2 - 14y + 25$

34. $y = \frac{1}{2}x^2 - 3x + \frac{19}{2}$

35. $x = \frac{1}{4}y^2 - \frac{1}{2}y - 3$

36. $y = x^2 + 4x + 1$

37. $y = 3x^2 - 24x + 50$

38. $x = 5y^2 - 25y + 60$

The focus and directrix of a parabola are given. Write an equation for each parabola. Then draw the graph.

39. $(2, 4),\ y = 6$

40. $(3, 5),\ y = 1$

41. $(8, 0),\ y = 4$

42. $(0, 3),\ y = -1$

43. $(5, 5),\ y = -3$

44. $(6, 2),\ x = 4$

45. $(3, -1),\ x = -2$

46. $(4, -3),\ y = 6$

47. $(0, 4),\ x = 1$

Write the equation of each parabola described below. Then draw the graph.

48. vertex $(0, 0)$, focus $(0, -4)$

49. vertex $(5, -1)$, focus $(3, -1)$

50. vertex $(4, 3)$, axis $y = 3$, length of latus rectum 4, $a > 0$

51. vertex $(-7, 4)$, axis $x = -7$, length of latus rectum 6, $a < 0$

52. focus $(2, 4)$, directrix $y = 2$.

53. focus $(3, 6)$, directrix $x = 5$.

54. focus $(-1, 2)$, directrix $y = -1$.

55. focus $(2, -5)$, directrix $x = -4$.

Challenge Exercise

56. If the equation of a parabola is $y = a(x - h)^2 + k$, show that the length of the latus rectum is $\left|\frac{1}{a}\right|$.

The parabola is a very practical curve. Recall that if a light source is placed at the focus of a parabolic reflector, the light emitted is reflected off the parabola and straight ahead.

Conversely, with sunlight directed at a reflector, temperatures at the focal point can reach over one thousand degrees Fahrenheit, hot enough to melt steel.

Sound can be reflected in a similar way. If a microphone is placed at the focus of a parabolic reflector, it can pick up very faint noises. You may have seen such a reflector being used at a televised football game. It will pick up the signal calls all the way from the sidelines.

This receiver gathers the sound waves collected by the parabolic reflector. The focus concentrates the waves and produces a stronger signal.

Radar antennas also use parabolic reflectors. Radio signals are focused on a receiver placed at the focus.

The focus for a parabola with equation $y = ax^2 + bx + c$ has the following coordinates.

$$\left(-\frac{b}{2a}, \frac{4ac - b^2 + 1}{4a} \right)$$

Exercises

Find the coordinates of the focus for the parabola with the given equation.

1. $y = x^2 + 6x + 9$ **2.** $y = 2x^2 + 4x + 7$ **3.** $y = \frac{1}{2}x^2 - 2x - \frac{5}{2}$

4. $y = \frac{1}{4}x^2 - 2x + 3$ **5.** $y = x^2 + 2x + 4$ **6.** $8y = x^2$

 Circles

A **circle** is the set of points in a plane each of which is the same distance from a given point in the plane. The given distance is the **radius** of the circle, and the given point is the **center** of the circle.

The circle at the right has center at $(2, -3)$ and radius of 4 units. You can use the distance formula and the definition of a circle to find the equation of this circle.

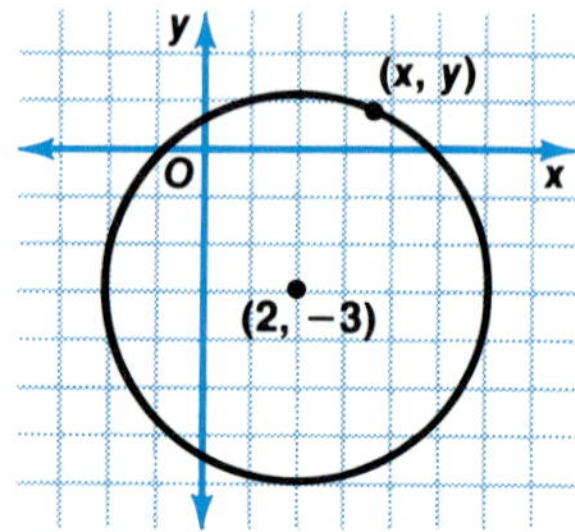

Let (x, y) name a point on the circle. The distance between (x, y) and the center of the circle, $(2, -3)$, must be 4 units.

distance between (x, y) and (2, −3) = 4

$$\sqrt{(x - 2)^2 + (y - (-3))^2} = 4 \quad \text{Apply the distance formula.}$$
$$(x - 2)^2 + (y - (-3))^2 = 4^2 \quad \text{Square each side.}$$
$$(x - 2)^2 + (y + 3)^2 = 16 \quad \text{Simplify.}$$

The equation of a circle with center at $(2, -3)$ and radius of 4 units is $(x - 2)^2 + (y + 3)^2 = 16$.

> **The equation of a circle with center at (h, k) and radius of r units is $(x - h)^2 + (y - k)^2 = r^2$.**
>
> *Equation of Circle with Center at (h, k)*

Example

1 **Graph $(x - 8)^2 + (y - 10)^2 = 100$.**

Center: $(8, 10)$
Radius: 10 units

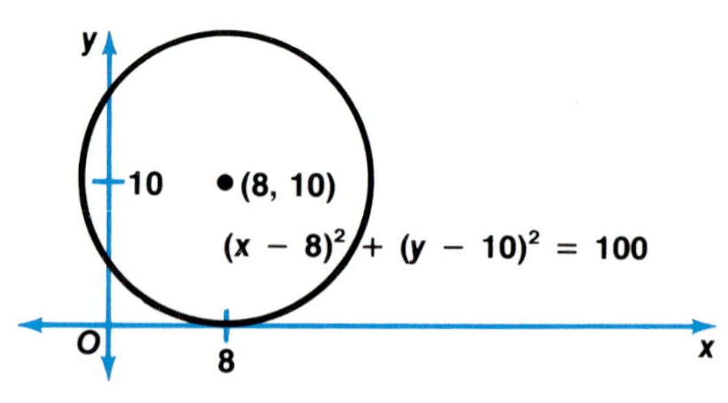

The graph of $x^2 + y^2 - 4x + 10y = 9$ is a circle. To write the equation in the form $(x - h)^2 + (y - k)^2 = r^2$, complete the square for each variable.

$$x^2 + y^2 - 4x + 10y = 9$$
$$x^2 - 4x + \blacksquare + y^2 + 10y + \square = 9 + \blacksquare + \square$$
$$x^2 - 4x + 4 + y^2 + 10y + 25 = 9 + 4 + 25 \qquad \textit{Complete the square.}$$
$$(x - 2)^2 + (y + 5)^2 = 38$$

The circle has center at $(2, -5)$ and radius of $\sqrt{38}$ units. $\qquad \sqrt{38} \approx 6.2$

Example

2   **Find the center and radius of the circle whose equation is $x^2 + 3 + y^2 + 9y - 10x = 2$. Then draw the graph.**

$$x^2 + 3 + y^2 + 9y - 10x = 2$$
$$x^2 - 10x + \blacksquare + y^2 + 9y + \square = 2 - 3 + \blacksquare + \square$$
$$x^2 - 10x + 25 + y^2 + 9y + \left(\frac{9}{2}\right)^2 = -1 + 25 + \left(\frac{9}{2}\right)^2$$
$$(x - 5)^2 + \left(y + \frac{9}{2}\right)^2 = \frac{177}{4}$$

The circle has center at $\left(5, -\frac{9}{2}\right)$ and radius of $\frac{\sqrt{177}}{2}$ or about 6.7 units.

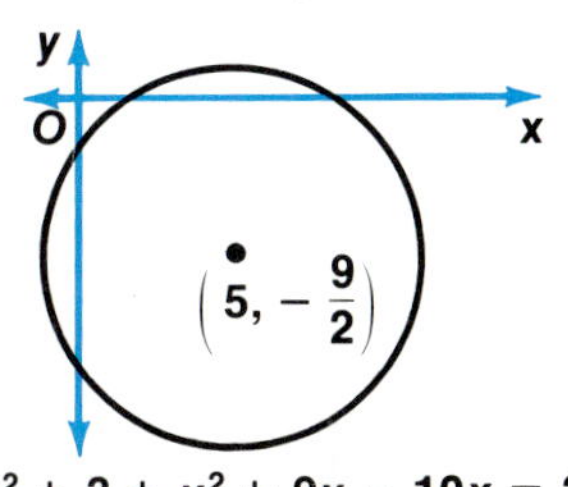

Exploratory Exercises

State whether the graph for each equation is a circle or a parabola.

1. $x^2 + y^2 + 7x - 5 = 0$ **2.** $x^2 + 4x + 4 = 9y + 27$ **3.** $y^2 = 6x - 4$

4. $y = x^2 + 8x + y^2$ **5.** $x^2 + y + y^2 = 12 - 3x$ **6.** $x^2 = 5y$

State the center and radius of each circle whose equation is given.

7. $x^2 + y^2 = 16$ **8.** $x^2 + (y - 2)^2 = 25$

9. $(x - 2)^2 + y^2 = 9$ **10.** $x^2 + y^2 = 40$

11. $(x - 10)^2 + (y + 10)^2 = 100$ **12.** $(x + 2)^2 + (y - 3)^2 = 81$

13. $(x + 4)^2 + \left(y - \frac{1}{2}\right)^2 = 6$ **14.** $(x - 4)^2 + y^2 = \frac{16}{25}$

15. $(x + 5)^2 + (y - 2)^2 = \frac{3}{4}$ **16.** $x^2 + (y + 5)^2 = \frac{81}{64}$

Written Exercises

Find the center and radius of each circle whose equation is given. Then draw the graph.

17. $(x - 2)^2 + y^2 = 9$ **18.** $(x + 4)^2 + y^2 = 49$

19. $x^2 + (y - 8)^2 = 64$ **20.** $x^2 + (y + 2)^2 = 4$

21. $x^2 + y^2 = 64$ **22.** $x^2 + y^2 = 121$

23. $(x - 2)^2 + (y - 5)^2 = 16$ **24.** $(x + 2)^2 + (y - 1)^2 = 81$

25. $(x + 8)^2 + (y - 3)^2 = 25$ **26.** $(x - 3)^2 + (y + 2)^2 = 169$

27. $(x + 1)^2 + (y + 9)^2 = 36$

28. $(x - 5)^2 + (y - 7)^2 = 49$

29. $x^2 + y^2 - 12x - 16y + 84 = 0$

30. $x^2 + y^2 - 18x - 18y + 53 = 0$

31. $x^2 + y^2 + 8x - 6y = 0$

32. $x^2 + y^2 + 14x + 6y = 23$

33. $x^2 + y^2 - 4x = 9$

34. $x^2 + y^2 - 6y = 16$

35. $3x^2 + 3y^2 + 6y + 9x = 2$

36. $x^2 + y^2 + 9x - 8y = -4$

37. $y^2 + 3 + x^2 + 9x - 10y = 6.75$

38. $4x^2 + 4y^2 + 36y = -5$

39. $x^2 + 2x + y^2 + 4y = 9$

40. $x^2 + y^2 + 4x = 8$

41. $x^2 + 2x + y^2 = 10$

42. $x^2 + y^2 + 14x + 6y = -50$

Write an equation for each circle whose center and radius are given.

43. $(6, 2)$, 5 units

44. $(6, 0)$, 6 in.

45. $(0, 3)$, 2 km

46. $(-3, -5)$, 5 cm

47. $(-6, 2)$, $\frac{1}{4}$ mi

48. $(-1, -3)$, $\frac{2}{3}$ yd

Write the equation of each circle described below.

49. The circle has center at $(1, 5)$ and passes through the origin.

50. The circle has center at $(4, -2)$ and passes through $(9, -3)$.

51. The endpoints of a diameter of the circle are at $(5, 2)$ and $(-1, 2)$.

52. The endpoints of a diameter of the circle are at $(4, -3)$ and $(8, 5)$.

53. The circle has center at $(-3, 8)$ and is tangent to the x-axis.

54. The circle has center at $(4, -3)$ and is tangent to the y-axis.

mini-review

Simplify.

1. $\sqrt[3]{54b^5}$

2. $\sqrt{21m^4n^3} \cdot \sqrt{63}$

3. $(5 + \sqrt[3]{3})(8 + \sqrt[3]{5})$

Write each expression in simplest radical form.

4. $\sqrt[6]{125}$

5. $a^{\frac{1}{2}}b^{\frac{1}{5}}$

6. $x^2(rt)^{\frac{1}{3}}r^{\frac{1}{6}}s^{\frac{1}{2}}$

Find the sum, difference, and product for each pair of complex numbers.

7. $-4 + 6i$, $2 - 5i$

8. $1 + 12i$, $3 + 4i$

Solve each equation using the quadratic formula.

9. $5x^2 + 12x - 10 = 0$

10. $4p^2 - 7p + 2 = 0$

Find the value of the discriminant for each quadratic equation. Describe the nature of the roots. If the roots are real, state whether they are rational or irrational.

11. $3t^2 + 4t - 15 = 0$

12. $-4x^2 - 8x - 9 = 0$

Graph each equation. Then name the vertex, axis of symmetry, and direction of opening.

13. $f(x) = -5x^2 + 30x - 45$

14. $y = (x - 2)^2$

15. Solve $(x + 3)(x - 2) < 0$.

9-4 Ellipses

A circle can be considered as a special case of a more general curve called an **ellipse**. A circle is defined in terms of a given point and a given distance. An ellipse is defined in terms of *two* given points and *two* distances. The illustration below shows one way to draw an ellipse.

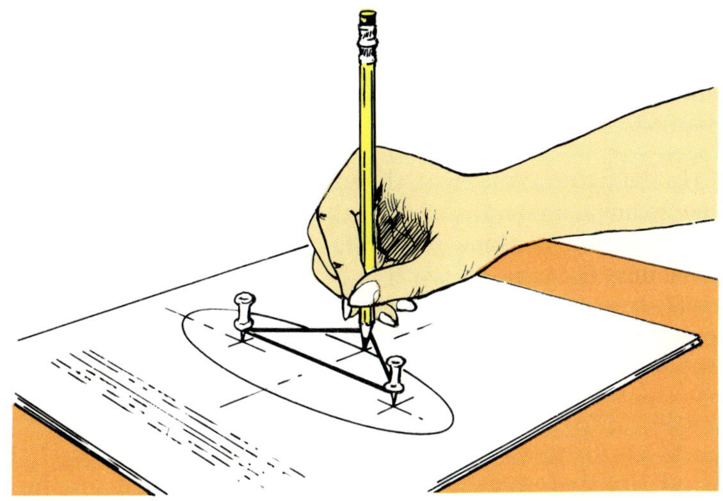

Use a piece of string about 25 cm long. Tie a knot in the string to make a loop. Place two thumbtacks in a piece of paper, about 10 cm apart. Loop the string around the tacks. Place the pencil in the loop. Keep the string tight and draw around the tacks.

The points where the tacks are placed are called the **foci** (plural of focus) of the ellipse. Ellipses can be defined in terms of their foci. *Foci is pronounced fō-sī.*

> **An ellipse is the set of all points in a plane such that the sum of the distances from two given points in the plane, called the *foci*, is constant.**

Definition of Ellipse

The ellipse at the right has foci at $(-3, 0)$ and $(3, 0)$. The sum of the distances to (x, y) from the two foci is 8 units. You can use the distance formula and the definition of an ellipse to find the equation of this ellipse.

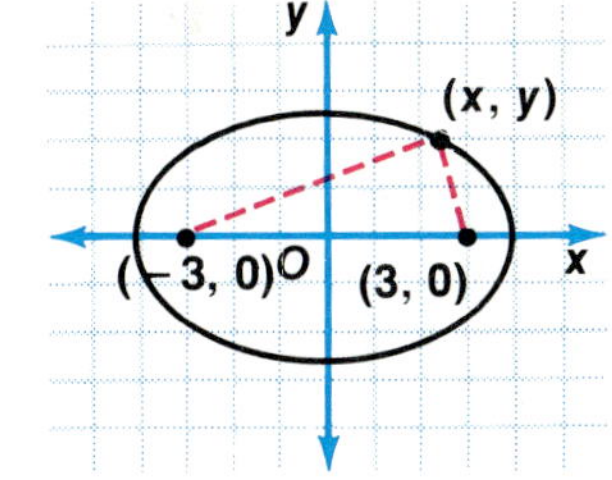

Let (x, y) name a point on the ellipse. The distance between (x, y) and $(-3, 0)$ *plus* the distance between (x, y) and $(3, 0)$ is 8 units.

$$\underset{\substack{\text{distance between}\\(x,\ y)\text{ and }(-3,\ 0)}}{} + \underset{\substack{\text{distance between}\\(x,\ y)\text{ and }(3,\ 0)}}{} = 8$$

$$\sqrt{(x - (-3))^2 + (y - 0)^2} + \sqrt{(x - 3)^2 + (y - 0)^2} = 8 \qquad \textit{Apply the distance formula.}$$

$$\sqrt{(x + 3)^2 + y^2} = 8 - \sqrt{(x - 3)^2 + y^2}$$

$$(x + 3)^2 + y^2 = 64 - 16\sqrt{(x - 3)^2 + y^2} + (x - 3)^2 + y^2 \qquad \textit{Square each side.}$$

$$3x - 16 = -4\sqrt{(x - 3)^2 + y^2} \qquad \textit{Simplify.}$$

$$9x^2 - 96x + 256 = 16[(x - 3)^2 + y^2] \qquad \textit{Square each side.}$$

$$112 = 7x^2 + 16y^2 \qquad \textit{Simplify.}$$

$$1 = \frac{x^2}{16} + \frac{y^2}{7} \qquad \textit{Divide by 112.}$$

The equation of an ellipse with foci at $(-3, 0)$ and $(3, 0)$, and with 8 units as the sum of the distances from the two foci is $\frac{x^2}{16} + \frac{y^2}{7} = 1$.

An ellipse has two axes of symmetry. The ellipse intersects the axes to define two segments whose endpoints lie on the ellipse. The longer segment is called the **major axis** and the shorter segment is called the **minor axis**. The intersection of the two axes is the **center** of the ellipse. The foci of an ellipse are always on the major axis.

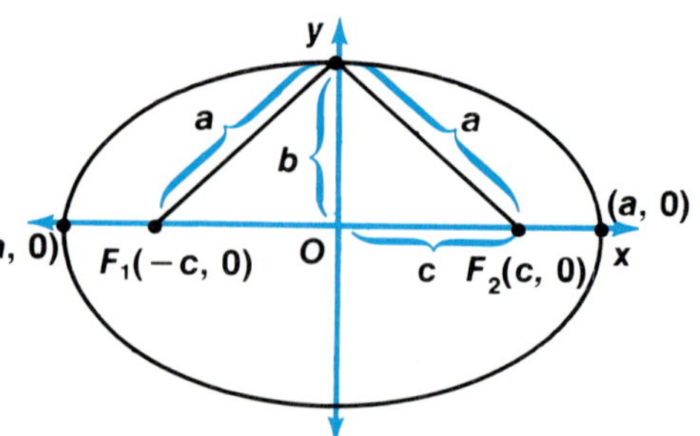

In the figure at the right, the sum of the distances from the foci to any point on the ellipse is $2a$ units. The distance from the center to a focus is c units. Using the Pythagorean Theorem, it can be shown that $b^2 = a^2 - c^2$.

The method used to find the equation of the ellipse on the previous page can be used to find the standard equation of an ellipse.

The length of the major axis is 2a. The length of the minor axis is 2b. Notice $a > b$.

If an ellipse has foci at $(-c, 0)$ and $(c, 0)$ and if the sum of the distances from the foci to any point on the ellipse is $2a$ units, then the standard equation of the ellipse is

$$\frac{x^2}{a^2} + \frac{y^2}{b^2} = 1, \text{ where } b^2 = a^2 - c^2.$$

If an ellipse has foci at $(0, -c)$ and $(0, c)$ and if the sum of the distances from the foci to any point on the ellipse is $2a$ units, then the standard equation of the ellipse is

$$\frac{x^2}{b^2} + \frac{y^2}{a^2} = 1, \text{ where } b^2 = a^2 - c^2.$$

Standard Equation of Ellipse with Center at the Origin

For ellipses, $a^2 > b^2$.

Example

1   Write the equation of the ellipse shown below.

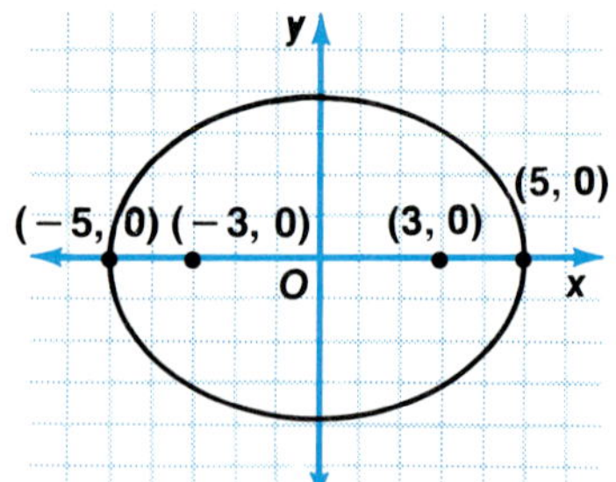

First, find the length of the major axis. The distance between $(-5, 0)$ and $(5, 0)$ is 10 units.

$$2a = 10$$
$$a = 5 \qquad a^2 = 25$$

Since the foci are at $(-3, 0)$ and $(3, 0)$, $c = 3$.

$$b^2 = a^2 - c^2$$
$$b^2 = 5^2 - 3^2 \text{ or } 16$$

The equation is $\frac{x^2}{25} + \frac{y^2}{16} = 1$.

In the equation of an ellipse, $a^2 > b^2$. This makes it easy to decide in an equation whether the foci are on the x-axis or y-axis.

Example

2 **Find the foci and the lengths of the major axis and minor axis of the ellipse whose equation is $9x^2 + y^2 = 36$. Then draw the graph.**

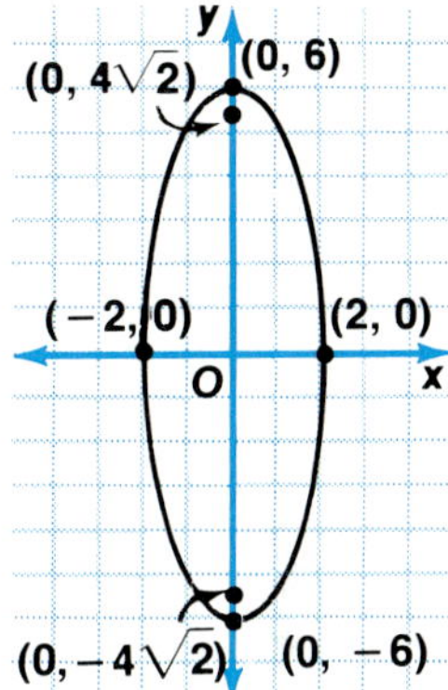

$9x^2 + y^2 = 36$ *Write the equation in standard form.*

$\dfrac{x^2}{4} + \dfrac{y^2}{36} = 1$ *Divide each side by 36.*

Since $36 > 4$, a^2 must be 36. So, $a = 6$, $b = 2$, and the foci are on the y-axis.

$b^2 = a^2 - c^2$
$4 = 36 - c^2$ $c^2 = 32$
$c = \sqrt{32}$ or $4\sqrt{2}$ $4\sqrt{2} \approx 5.7$

The foci are at $(0, 4\sqrt{2})$ and $(0, -4\sqrt{2})$.
The length of the major axis is $2a$ or 12 units.
The length of the minor axis is $2b$ or 4 units.

An equation of the form $\dfrac{x^2}{a^2} + \dfrac{y^2}{b^2} = 1$, or $\dfrac{x^2}{b^2} + \dfrac{y^2}{a^2} = 1$, represents an ellipse centered at the origin. Suppose an ellipse has the same shape but its center is at (h, k). The standard equations below can be obtained by replacing x with $x - h$ and y with $y - k$.

The equation of an ellipse whose center is at (h, k) and with a horizontal major axis is $\dfrac{(x - h)^2}{a^2} + \dfrac{(y - k)^2}{b^2} = 1$.

The equation of an ellipse whose center is at (h, k) and with a vertical major axis is $\dfrac{(x - h)^2}{b^2} + \dfrac{(y - k)^2}{a^2} = 1$.

Standard Equation of Ellipse with Center at (h, k)

For ellipses, $a^2 > b^2$.

Example

3 **Graph $\dfrac{(x + 2)^2}{36} + \dfrac{(y - 3)^2}{9} = 1$.**

The graph has the same shape as the graph of $\dfrac{x^2}{36} + \dfrac{y^2}{9} = 1$, but has center at $(-2, 3)$ rather than at the origin.

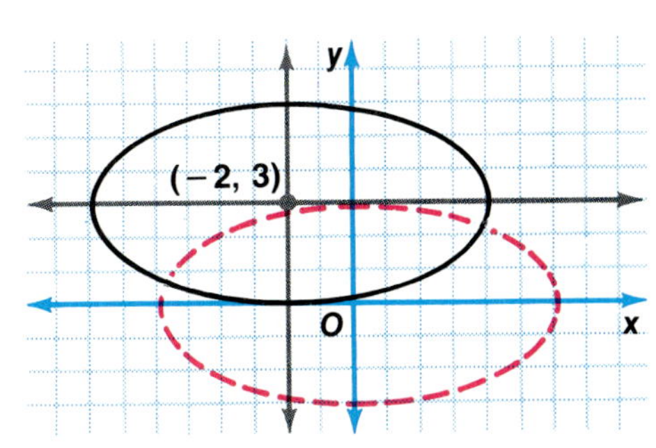

The graph of $x^2 + 2y^2 - 2x + 8y - 11 = 0$ is an ellipse. To write the equation in the form $\dfrac{(x - h)^2}{a^2} + \dfrac{(y - k)^2}{b^2} = 1$, complete the square for each variable.

$$x^2 + 2y^2 - 2x + 8y - 11 = 0$$
$$x^2 - 2x + \blacksquare + 2(y^2 + 4y + \square) = 11 + \blacksquare + 2\square \qquad \textit{Complete the square.}$$
$$x^2 - 2x + 1 + 2(y^2 + 4y + 4) = 11 + 1 + 2(4)$$
$$(x - 1)^2 + 2(y + 2)^2 = 20$$
$$\dfrac{(x - 1)^2}{20} + \dfrac{(y + 2)^2}{10} = 1$$

The ellipse has center at $(1, -2)$ and a horizontal major axis. Since $20 > 10$, $a^2 = 20$ and $b^2 = 10$. Thus, the major axis is $2\sqrt{20}$ or $4\sqrt{5}$ units long and the minor axis is $2\sqrt{10}$ units long.

$$4\sqrt{5} \approx 8.9$$
$$2\sqrt{10} \approx 6.5$$

Example

4 **Find the center, foci, and the lengths of the major axis and minor axis of the ellipse whose equation is $4x^2 + y^2 + 24x - 10y + 45 = 0$. Then draw the graph.**

$$4x^2 + y^2 + 24x - 10y + 45 = 0$$
$$4(x^2 + 6x + \blacksquare) + (y^2 - 10y + \square) = -45 + 4\blacksquare + \square$$
$$4(x^2 + 6x + 9) + (y^2 - 10y + 25) = -45 + 4(9) + 25$$
$$4(x + 3)^2 + (y - 5)^2 = 16$$
$$\dfrac{(x + 3)^2}{4} + \dfrac{(y - 5)^2}{16} = 1$$

The ellipse has center at $(-3, 5)$.

Since $16 > 4$, a^2 must be 16. So, $a = 4$, $b = 2$, and the line containing the foci is parallel to the y-axis.

$$b^2 = a^2 - c^2$$
$$2^2 = 4^2 - c^2$$
$$c^2 = 12$$
$$c = \sqrt{12} \text{ or } 2\sqrt{3} \qquad 2\sqrt{3} \approx 3.5$$

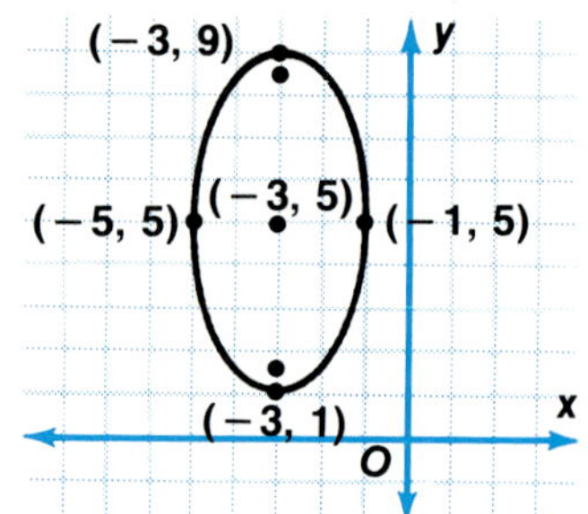

The foci are $2\sqrt{3}$ units above and below the center. The foci are at $(-3, 5 + 2\sqrt{3})$ and $(-3, 5 - 2\sqrt{3})$ or about $(-3, 8.5)$ and $(-3, 1.5)$. The length of the major axis is $2a$ or 8 units. The length of the minor axis is $2b$ or 4 units.

Name the center of each ellipse with the given equation and state whether the major axis is horizontal or vertical.

1. $\dfrac{x^2}{9} + \dfrac{y^2}{4} = 1$

2. $\dfrac{x^2}{16} + \dfrac{y^2}{1} = 1$

3. $\dfrac{x^2}{9} + \dfrac{y^2}{25} = 1$

4. $\dfrac{x^2}{10} + \dfrac{y^2}{36} = 1$

5. $\dfrac{x^2}{81} + \dfrac{(y-5)^2}{49} = 1$

6. $\dfrac{(x+3)^2}{25} + \dfrac{y^2}{9} = 1$

7. $\dfrac{(x-2)^2}{36} + \dfrac{(y+5)^2}{16} = 1$

8. $\dfrac{(x-4)^2}{16} + \dfrac{(y-4)^2}{121} = 1$

9. $\dfrac{(x+2)^2}{81} + \dfrac{(y+3)^2}{144} = 1$

Name the foci of each ellipse whose equation is given.

10. $\dfrac{x^2}{9} + \dfrac{y^2}{4} = 1$

11. $\dfrac{x^2}{16} + \dfrac{y^2}{1} = 1$

12. $\dfrac{x^2}{9} + \dfrac{y^2}{25} = 1$

13. $\dfrac{x^2}{10} + \dfrac{y^2}{36} = 1$

14. $\dfrac{x^2}{81} + \dfrac{(y-5)^2}{49} = 1$

15. $\dfrac{(x+3)^2}{25} + \dfrac{y^2}{9} = 1$

Write the equation of each ellipse in standard form.

16. $7x^2 + 14y^2 = 28$

17. $5x^2 + 15y^2 = 225$

18. $12x^2 + 6y^2 = 168$

19. $3x^2 + 6y^2 - 12x + 36y = 81$

20. $8x^2 + 4y^2 - 16x - 20y = 7$

21. $5x^2 + 15y^2 + 10x + 30y = 100$

22. $10x^2 + 6y^2 + 20x + 54y = 122$

Write the equation of each ellipse.

23. 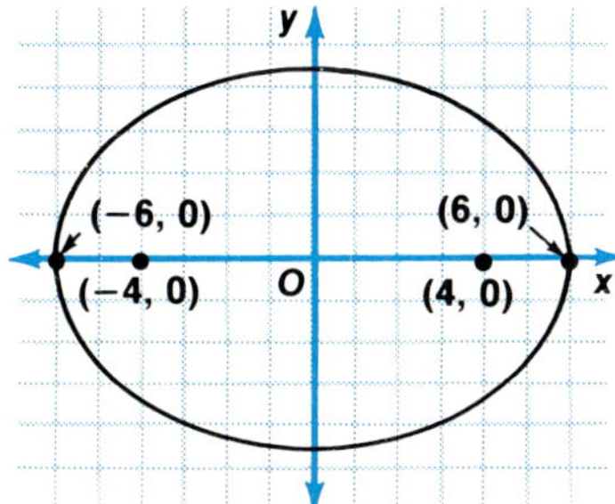

24. 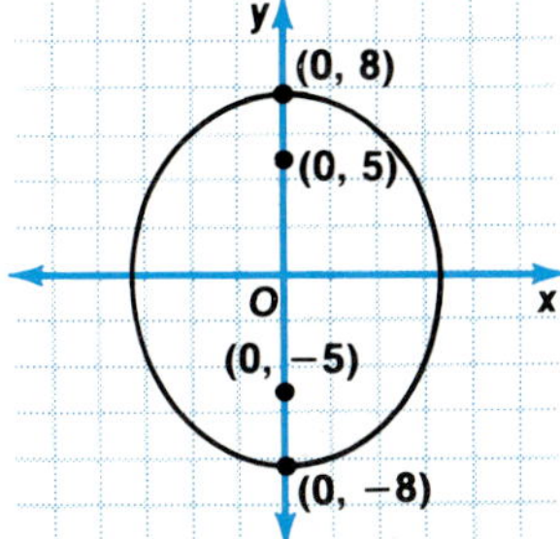

25. 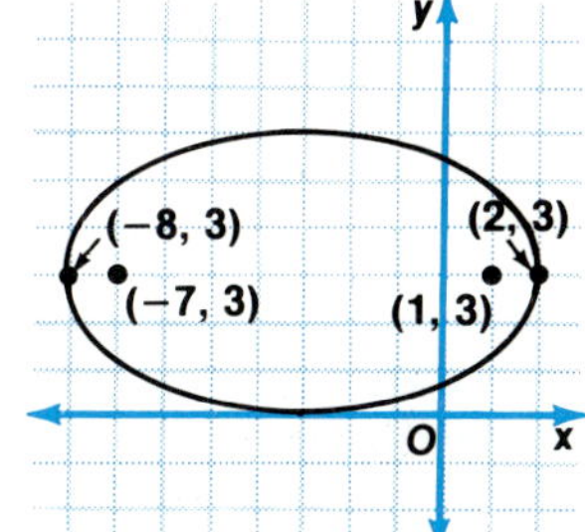

26. 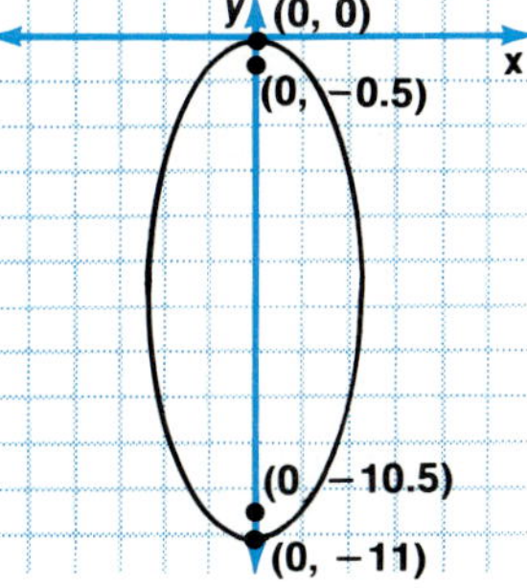

27. 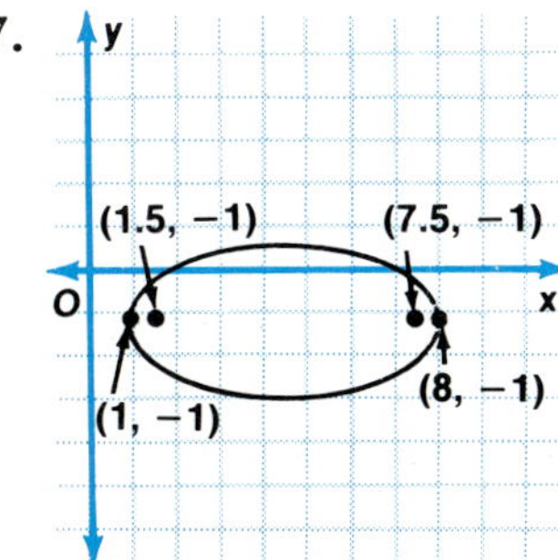

28. 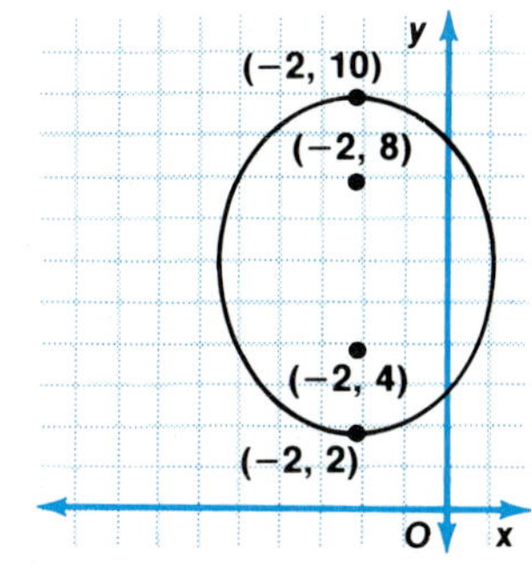

Find the center, foci, and lengths of the major axis and minor axis of each ellipse whose equation is given. Then draw the graph.

29. $\dfrac{x^2}{4} + \dfrac{y^2}{25} = 1$ **30.** $\dfrac{x^2}{36} + \dfrac{y^2}{16} = 1$ **31.** $\dfrac{x^2}{25} + \dfrac{y^2}{9} = 1$ **32.** $\dfrac{x^2}{10} + \dfrac{y^2}{5} = 1$

33. $9x^2 + 16y^2 = 144$ **34.** $3x^2 + 9y^2 = 27$ **35.** $4x^2 + 9y^2 = 36$ **36.** $4x^2 + y^2 = 4$

37. $36x^2 + 81y^2 = 2916$ **38.** $x^2 + 16y^2 = 16$ **39.** $27x^2 + 9y^2 = 81$

40. $\dfrac{(x + 3)^2}{36} + \dfrac{(y - 4)^2}{9} = 1$ **41.** $\dfrac{(x + 2)^2}{20} + \dfrac{(y + 3)^2}{40} = 1$ **42.** $\dfrac{(x - 8)^2}{4} + \dfrac{(y + 8)^2}{1} = 1$

43. $\dfrac{(x + 2)^2}{5} + \dfrac{(y - 3)^2}{2} = 1$ **44.** $\dfrac{(x - 4)^2}{121} + \dfrac{(y + 5)^2}{64} = 1$ **45.** $\dfrac{(x - 2)^2}{16} + \dfrac{(y - 3)^2}{9} = 1$

46. $9x^2 + 4y^2 - 18x + 16y = 11$ **47.** $3x^2 + 7y^2 - 12x - 28y = -19$

48. $9x^2 + 16y^2 - 18x + 64y = 71$ **49.** $16x^2 + 25y^2 + 32x - 150y = 159$

Write the equation of each ellipse described below.

50. The foci are at (0, 8) and (0, −8). The endpoints of the major axis are at (0, 10) and (0, −10).

51. The foci are at (12, 0) and (−12, 0). The endpoints of the minor axis are at (0, 5) and (0, −5).

52. The center is at (5, 4). The major axis is 16 units long and parallel to the x-axis. The minor axis is 9 units long.

53. The center is at (−2, 3). The major axis is 12 units long and parallel to the y-axis. The minor axis is 8 units long.

54. The endpoints of the major axis are at (2, 12) and (2, −4). The endpoints of the minor axis are at (4, 4) and (0, 4).

55. The endpoints of the major axis are at (−9, 2) and (5, 2). The endpoints of the minor axis are at (−2, 5) and (−2, −1).

56. The foci are at (5, 4) and (−3, 4). The major axis is 10 units long.

57. The foci are at (3, 8) and (3, −6). The major axis is 18 units long.

58. The foci are at (−3, 0) and (3, 0). The minor axis is 4 units long.

59. The foci are at (−1, 4) and (−1, −6). The minor axis is 6 units long.

Excursions in Algebra   The Capitol

If rays of light or sound are emitted from one focus of an elliptical reflector, these rays are concentrated at the other focus.

The elliptical chamber of the United States Capitol has this property. A person standing at one focus and whispering is easily heard by a person standing at the other focus.

9-5 Hyperbolas

When the light from a table lamp hits the wall, a curve called a **hyperbola** is formed.

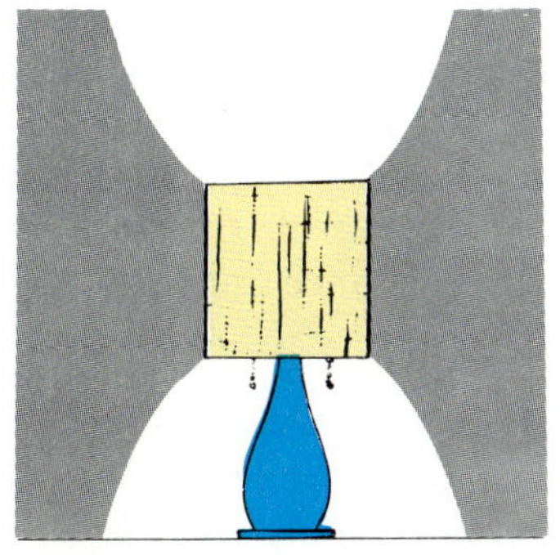

A hyperbola is the set of all points in a plane such that the absolute value of the difference of the distances from any point on the hyperbola to two given points in the plane, called the *foci*, is constant.

Definition of Hyperbola

The hyperbola at the right has foci at $(-5, 0)$ and $(5, 0)$. The absolute value of the difference of the distances from the two foci is 8. You can use the distance formula and the definition of a hyperbola to find the equation of this hyperbola.

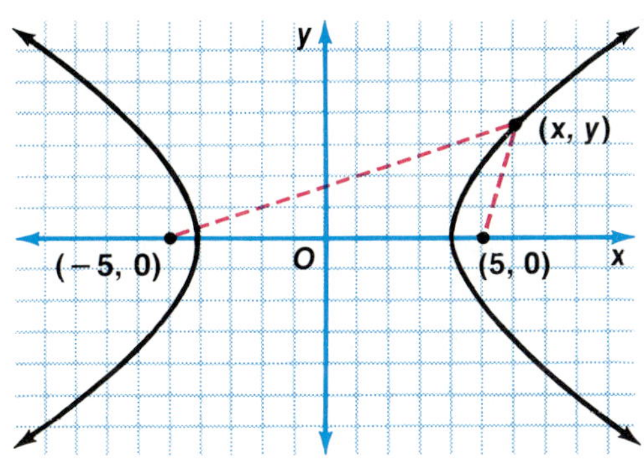

Let (x, y) name a point on the hyperbola. The distance between (x, y) and $(-5, 0)$ *minus* the distance between (x, y) and $(5, 0)$ is ± 8.

$$\underbrace{\text{distance between}}_{(x,\,y) \text{ and } (-5,\,0)} - \underbrace{\text{distance between}}_{(x,\,y) \text{ and } (5,\,0)} = \pm 8$$

$\sqrt{(x - (-5))^2 + (y - 0)^2} - \sqrt{(x - 5)^2 + (y - 0)^2} = \pm 8$ *Apply the distance formula.*

$\sqrt{(x + 5)^2 + y^2} = \pm 8 + \sqrt{(x - 5)^2 + y^2}$ *Rearrange terms.*

$(x + 5)^2 + y^2 = 64 \pm 16\sqrt{(x - 5)^2 + y^2} + (x - 5)^2 + y^2$ *Square each side.*

$5x - 16 = \pm 4\sqrt{(x - 5)^2 + y^2}$ *Simplify.*

$25x^2 - 160x + 256 = 16[(x - 5)^2 + y^2]$ *Square each side.*

$9x^2 - 16y^2 = 144$ *Simplify.*

$\dfrac{x^2}{16} - \dfrac{y^2}{9} = 1$ *Divide each side by 144.*

The equation of a hyperbola with foci at $(-5, 0)$ and $(5, 0)$, and with 8 as the absolute value of the difference between the distances from the two foci is $\dfrac{x^2}{16} - \dfrac{y^2}{9} = 1$.

The **center** of a hyperbola is the midpoint of the segment connecting the foci. The point on each branch of the hyperbola nearest the center is called a **vertex**. The **asymptotes** of a hyperbola are lines that the branches of the curve approach as the curve recedes from the center. Like the ellipse, the distance from the center to a vertex is a units and the distance from the center to a focus is c units.

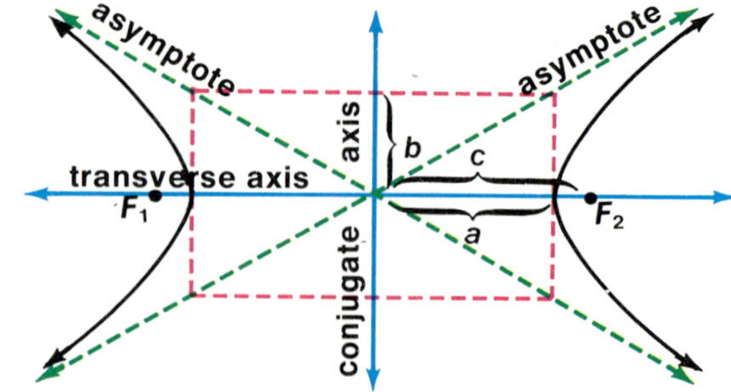

A hyperbola has two axes of symmetry, as shown above. The line segment length $2a$ that has its endpoints at the vertices is called the **transverse axis**. The segment of length $2b$ perpendicular to the transverse axis at its center is called the **conjugate axis**. For a hyperbola, the lengths a, b, and c are related by the formula $a^2 + b^2 = c^2$.

The asymptotes pass through the center of the hyperbola and form the diagonals of the rectangle whose sides are 2a and 2b units.

The method used to find the equation of the hyperbola on the previous page can also be used to find the standard equation of a hyperbola.

> If a hyperbola has foci at $(-c, 0)$ and $(c, 0)$ and if the absolute value of the difference of the distances from any point on the hyperbola to the two foci is $2a$ units, then the standard equation of the hyperbola is $\dfrac{x^2}{a^2} - \dfrac{y^2}{b^2} = 1$, where $c^2 = a^2 + b^2$.
>
> If a hyperbola has foci at $(0, -c)$ and $(0, c)$ and if the absolute value of the difference of the distances from any point on the hyperbola to the two foci is $2a$ units, then the standard equation of the hyperbola is $\dfrac{y^2}{a^2} - \dfrac{x^2}{b^2} = 1$, where $c^2 = a^2 + b^2$.

Standard Equation of Hyperbola with Center at the Origin

Example

1 **Write the equation of the hyperbola shown below.**

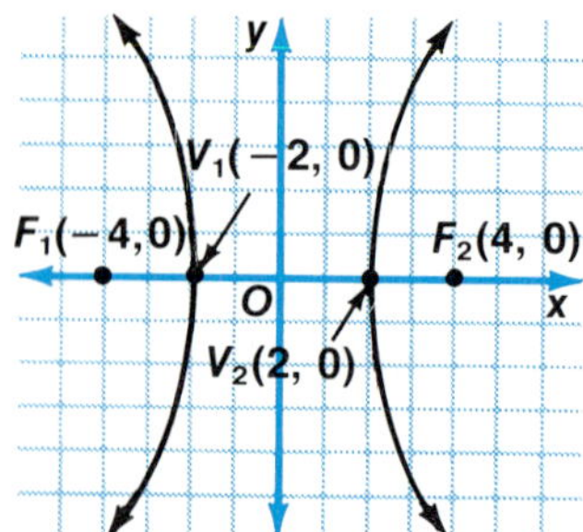

Since the distance from the center to a vertex is a units, and the distance from the center to a focus is c units, $a = 2$ and $c = 4$.

Now find b^2.

$$c^2 = a^2 + b^2$$
$$4^2 = 2^2 + b^2$$
$$b^2 = 12$$

The equation is $\dfrac{x^2}{4} - \dfrac{y^2}{12} = 1$.

Before graphing a hyperbola, it is helpful to sketch the asymptotes. The equations of the asymptotes are shown in the chart below.

The slopes of the asymptotes are shown in color.

Equation of the Hyperbola	$\dfrac{x^2}{a^2} - \dfrac{y^2}{b^2} = 1$	$\dfrac{y^2}{a^2} - \dfrac{x^2}{b^2} = 1$
Equation of the Asymptotes	$y = \pm\dfrac{b}{a}x$	$y = \pm\dfrac{a}{b}x$
Transverse Axis	Horizontal	Vertical

Example

2 **Find the vertices, foci, and the equations of the asymptotes of the hyperbola whose equation is $\dfrac{y^2}{16} - \dfrac{x^2}{9} = 1$. Then draw the graph.**

The hyperbola has center at $(0, 0)$ and a vertical transverse axis. Since $a = 4$ and $b = 3$, the equations of the asymptotes are $y = \dfrac{4}{3}x$ and $y = -\dfrac{4}{3}x$. Sketch a rectangle $2a$ by $2b$ and draw the diagonals with the correct slopes. Since $a = 4$, the distance from the center to a vertex is 4 units. Thus the vertices are at $(0, 4)$ and $(0, -4)$.

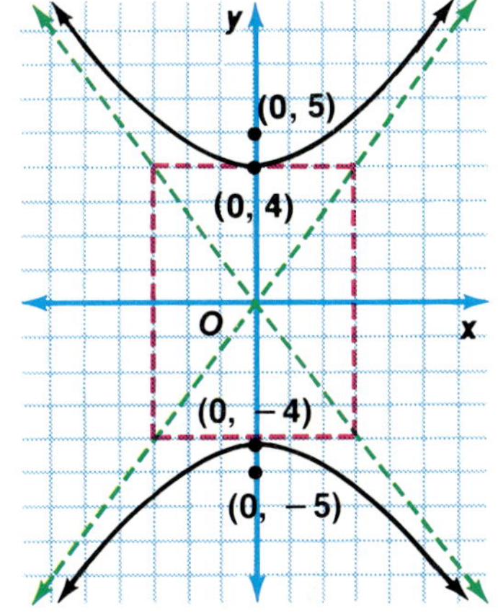

To locate the foci, find the value of c.

$$c^2 = a^2 + b^2$$
$$c^2 = 4^2 + 3^2$$
$$c = 5$$

So, the foci are at $(0, 5)$ and $(0, -5)$.

An equation of the form $\dfrac{x^2}{a^2} - \dfrac{y^2}{b^2} = 1$ or $\dfrac{y^2}{a^2} - \dfrac{x^2}{b^2} = 1$ represents a hyperbola centered at the origin. Suppose a hyperbola has the same shape but its center is at (h, k). The standard equation below can be obtained by replacing x with $x - h$ and y with $y - k$.

The equation of a hyperbola whose center is at (h, k) and with a horizontal transverse axis is
$$\frac{(x - h)^2}{a^2} - \frac{(y - k)^2}{b^2} = 1.$$

The equation of a hyperbola whose center is at (h, k) and with a vertical transverse axis is
$$\frac{(y - k)^2}{a^2} - \frac{(x - h)^2}{b^2} = 1.$$

Standard Equation of Hyperbola with Center at (h, k)

For any hyperbola with a vertical transverse axis, the slopes of the asymptotes are $\pm\frac{a}{b}$. For any hyperbola with a horizontal transverse axis, the slopes of the asymptotes are $\pm\frac{b}{a}$.

Example

3   Graph $\dfrac{(y-1)^2}{16} - \dfrac{(x+2)^2}{9} = 1$.

The graph has the same shape as the graph of $\dfrac{y^2}{16} - \dfrac{x^2}{9} = 1$, but has center at $(-2, 1)$ rather than at the origin.

The slopes of the asymptotes are $\pm\dfrac{4}{3}$.

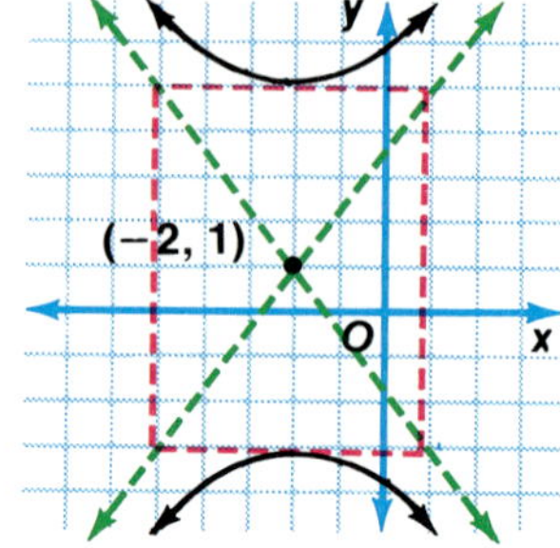

The graph of $x^2 - 4y^2 + 6x + 16y - 11 = 0$ is a hyperbola. To write the equation in the form $\dfrac{(x-h)^2}{a^2} - \dfrac{(y-k)^2}{b^2} = 1$, complete the square for each variable.

$$x^2 - 4y^2 + 6x + 16y - 11 = 0$$
$$x^2 + 6x + \blacksquare - 4(y^2 - 4y + \square) = 11 + \blacksquare - 4\square$$
$$x^2 + 6x + 9 - 4(y^2 - 4y + 4) = 11 + 9 + (-4)(4) \quad \textit{Complete the square.}$$
$$(x + 3)^2 - 4(y - 2)^2 = 4$$
$$\frac{(x+3)^2}{4} - \frac{(y-2)^2}{1} = 1$$

The hyperbola has center at $(-3, 2)$ and a horizontal transverse axis. Since $a^2 = 4$ and $b^2 = 1$, the transverse axis is 4 units long, and the conjugate axis is 2 units long.

Example

4   **Find the vertices, foci and slopes of the asymptotes of the hyperbola whose equation is $9x^2 - 4y^2 - 54x - 40y - 55 = 0$. Then draw the graph.**

$$9x^2 - 4y^2 - 54x - 40y - 55 = 0$$
$$9(x^2 - 6x + \blacksquare) - 4(y^2 + 10y + \square) = 55 + 9\blacksquare - 4\square$$
$$9(x^2 - 6x + 9) - 4(y^2 + 10y + 25) = 55 + 9(9) + (-4)(25)$$
$$9(x - 3)^2 - 4(y + 5)^2 = 36$$
$$\frac{(x-3)^2}{4} - \frac{(y+5)^2}{9} = 1$$

The hyperbola has center at $(3, -5)$ and a horizontal transverse axis. Since $a = 2$ and $b = 3$, the slopes of the asymptotes are $\frac{3}{2}$ and $-\frac{3}{2}$.

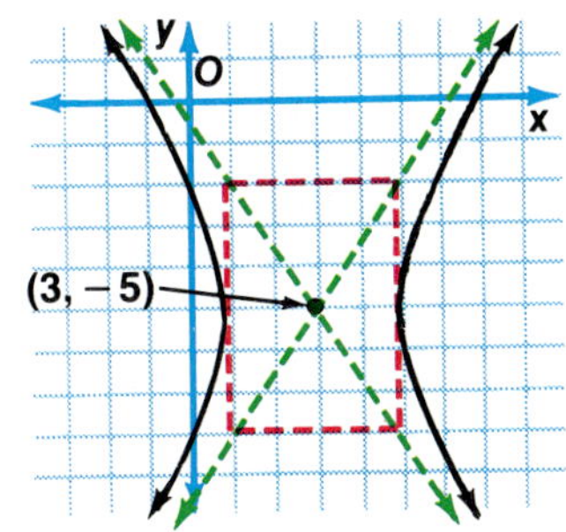

Since $a = 2$, the distance from the center to a vertex is 2 units. Thus, the vertices are at $(5, -5)$ and $(1, -5)$.

To locate the foci, find the value of c.

$$c^2 = a^2 + b^2$$
$$c^2 = 2^2 + 3^2$$
$$c = \sqrt{13} \qquad \sqrt{13} \approx 3.6$$

The foci are $\sqrt{13}$ units to the left and right of the center. Thus, the foci are at $(3 + \sqrt{13}, -5)$ and $(3 - \sqrt{13}, -5)$ or about $(6.6, -5)$ and $(-0.6, -5)$.

Exploratory Exercises

State whether the graph of each equation is an ellipse or a hyperbola.

1. $\dfrac{x^2}{9} + \dfrac{y^2}{4} = 1$
2. $\dfrac{x^2}{16} - \dfrac{y^2}{4} = 1$
3. $\dfrac{y^2}{18} - \dfrac{x^2}{20} = 1$
4. $\dfrac{x^2}{5} + \dfrac{y^2}{5} = 1$

5. $\dfrac{x^2}{36} - \dfrac{y^2}{1} = 1$
6. $\dfrac{x^2}{10} + \dfrac{y^2}{36} = 1$
7. $\dfrac{y^2}{25} + \dfrac{x^2}{9} = 1$
8. $\dfrac{x^2}{81} - \dfrac{y^2}{36} = 1$

Write the equation of each hyperbola in standard form.

9. $34x^2 - 6y^2 = 204$
10. $2x^2 - y^2 = 10$
11. $12x^2 - 8y^2 = 144$
12. $45y^2 - 5x^2 = 225$
13. $4(x - 2)^2 - (y - 1)^2 = 168$
14. $12x^2 - 3y^2 - 60x - 45y - 60 = 0$
15. $y^2 - 5x^2 + 10y + 10x = 24$
16. $y^2 - 20x^2 - 4x - 6y = 36$

Written Exercises

Write the equation of each hyperbola.

17. 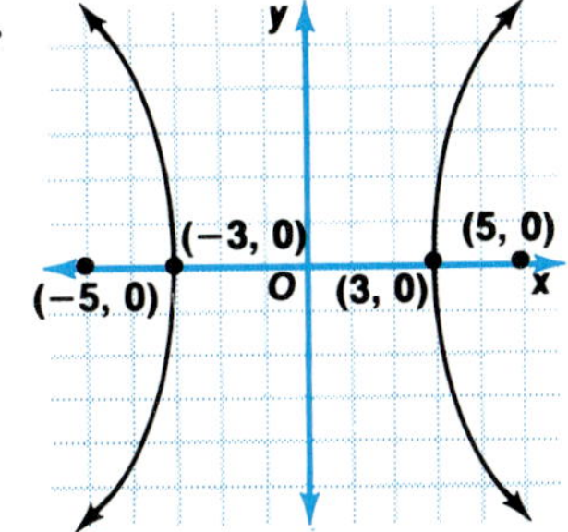

18. 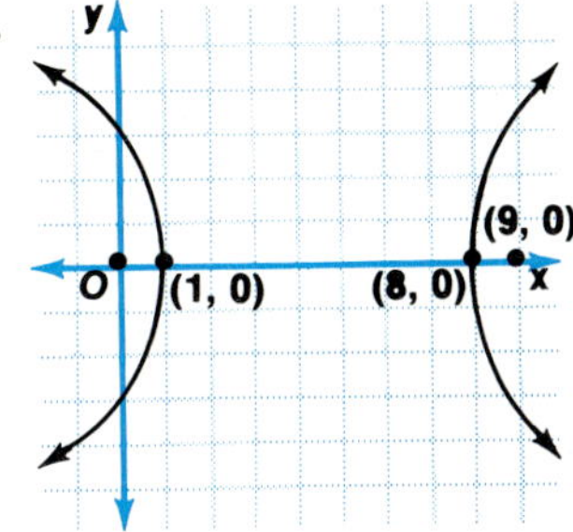

19. 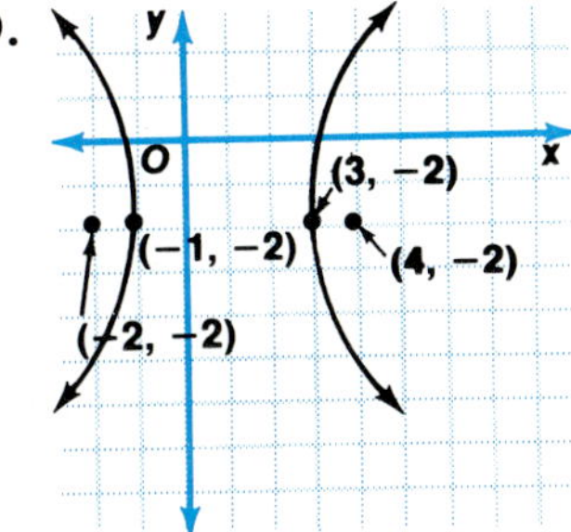

20. 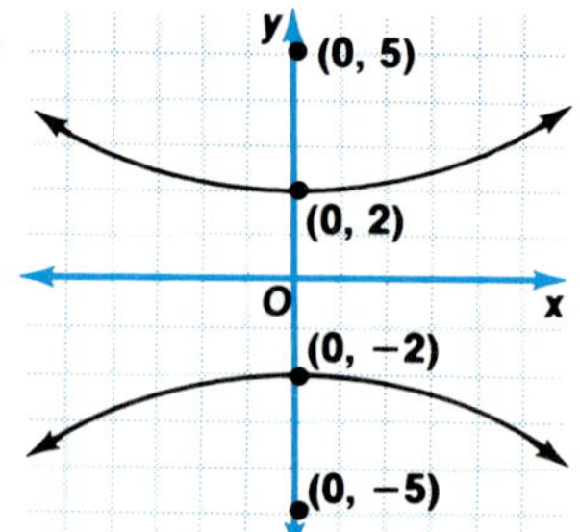

21. 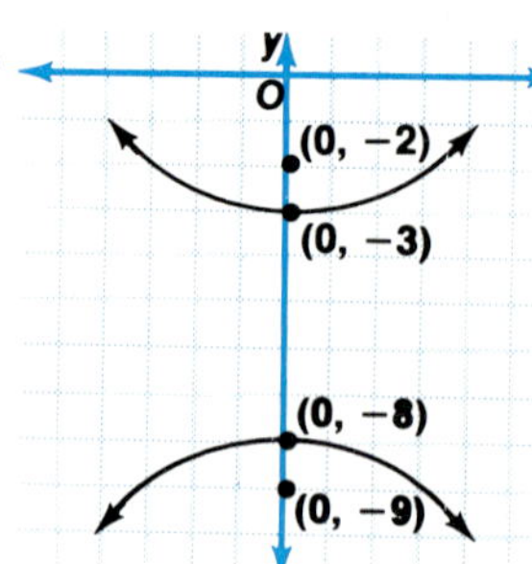

22. 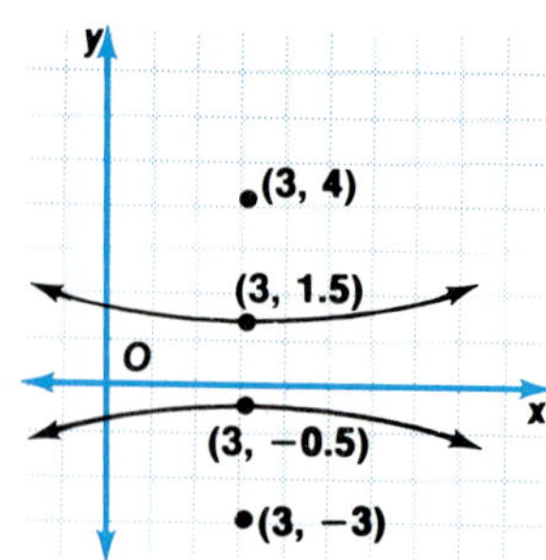

Find the vertices, foci and slopes of the asymptotes for each hyperbola whose equation is given. Then draw the graph.

23. $\dfrac{x^2}{9} - \dfrac{y^2}{25} = 1$ **24.** $\dfrac{x^2}{16} - \dfrac{y^2}{4} = 1$ **25.** $\dfrac{x^2}{36} - \dfrac{y^2}{1} = 1$ **26.** $\dfrac{y^2}{6} - \dfrac{x^2}{2} = 1$

27. $\dfrac{y^2}{81} - \dfrac{x^2}{25} = 1$ **28.** $\dfrac{y^2}{18} - \dfrac{x^2}{20} = 1$ **29.** $\dfrac{x^2}{4} - \dfrac{y^2}{9} = 1$ **30.** $\dfrac{y^2}{16} - \dfrac{x^2}{25} = 1$

31. $\dfrac{x^2}{81} - \dfrac{y^2}{36} = 1$ **32.** $\dfrac{y^2}{100} - \dfrac{x^2}{144} = 1$ **33.** $\dfrac{x^2}{9} - \dfrac{y^2}{16} = 1$ **34.** $\dfrac{x^2}{9} - \dfrac{y^2}{4} = 1$

35. $25x^2 - 4y^2 = 100$ **36.** $x^2 - y^2 = 4$ **37.** $x^2 - 2y^2 = 2$ **38.** $y^2 - 4x^2 = 4$

39. $y^2 = 36 + 4x^2$ **40.** $36y^2 - 81x^2 = 2916$ **41.** $\dfrac{(x + 6)^2}{36} - \dfrac{(y + 3)^2}{9} = 1$

42. $\dfrac{(y - 3)^2}{25} - \dfrac{(x - 2)^2}{16} = 1$ **43.** $\dfrac{(y - 4)^2}{16} - \dfrac{(x + 2)^2}{9} = 1$ **44.** $\dfrac{(x + 1)^2}{4} - \dfrac{(y - 4)^2}{9} = 1$

45. $5(x - 4)^2 - 4(y + 2)^2 = 100$ **46.** $x^2 - 4y^2 + 6x + 16y - 11 = 0$

47. $y^2 - 3x^2 + 6y + 6x = 18$ **48.** $y^2 - 4x^2 - 2y - 16x = -1$

Write the equation of each hyperbola described below.

49. center $(0, 0)$, $a = 1$, $b = 4$, horizontal transverse axis

50. center $(5, 4)$, $a = 2$, $b = 6$, vertical transverse axis

51. center $(-2, 2)$, $a = 6$, $c = 10$, vertical transverse axis

52. equations of the asymptotes $3x - 2y = 0$ and $3x + 2y = 0$, horizontal transverse axis, passes through $(2, 0)$

An equation of the form $xy = c$ when $c \neq 0$ represents a hyperbola having the x-axis and y-axis as asymptotes.

53. Find the domain for $xy = c$ when $c \neq 0$.   **54.** Find the range for $xy = c$ when $c \neq 0$.

55. Sketch the graph of $xy = 2$.   **56.** Sketch the graph of $xy = -3$.

Challenge Exercises

57. Find the equation of a hyperbola whose asymptotes are given by the equations $5x + 3y = -1$ and $5x - 3y = 11$, and that passes through $(4, -2)$.

58. Find the equation of a hyperbola whose asymptotes are the same as in Exercise 57, and that passes through $(1, 3)$.

Simplify.

1. $7\sqrt{27} + 5\sqrt{12}$

2. $\dfrac{7}{3 + \sqrt{2}}$

3. $(7\sqrt{-6})(5\sqrt{-8})$

4. $(2 + 4i)(5 + 4i)$

Solve.

5. $\sqrt{5a + 4} - 5 = 7$

6. $2m^2 + 15m + 3 = 0$

7. $x^2 - 3x < 10$

8. $|5x - 3| > 22$

For each equation, identify the quadratic term, the linear term, and the constant term.

9. $y = x^2 + 4x + 7$

10. $y = -2x^2 + 6$

Write each equation in quadratic form.

11. $f(y) = 5(y + 3)^2 - 9$

12. $f(a) = -2(a - 6)^2 + 10$

Find the value of c that makes each trinomial a perfect square.

13. $x^2 + 62x + c$

14. $x^2 - \dfrac{3}{5}x + c$

Find the value of the discriminant for each quadratic equation. Describe the nature of the roots.

15. $9m^2 = 3m$

16. $x^2 = 8x - 16$

Draw the graph of each equation. Then name the vertex, axis of symmetry, and direction of opening.

17. $f(x) = \dfrac{1}{2}(x + 1)^2 - 2$

18. $f(x) = 6(x - 3)^2$

Use the distance formula to find the distance between each pair of points.

19. $(5, -3), (8, 2)$

20. $(-4, -2), (3, 5)$

Find the midpoint of each line segment having the following endpoints.

21. $(7, 6), (8, 7)$

22. $(-3.5, 0.4), (2.2, 0.6)$

Find each of the following.

23. the axis of symmetry of the parabola $y = x^2 + 8x - 3$

24. the center and radius of the circle $x^2 + y^2 - 6x + 7 = 0$

25. the center and foci of the ellipse $\dfrac{x^2}{81} + \dfrac{y^2}{25} = 1$

Write the equation of each of the following. Then draw the graph.

26. a parabola whose vertex is at $(3, -1)$ and focus is at $(3, 2)$

27. a circle whose center is at $(-1, 4)$ and that passes through $(8, -1)$

28. an ellipse whose foci are at $(3, 2)$ and $(-5, 2)$ with a major axis 12 units long

Solve each problem.

29. The sum of the squares of two consecutive odd integers is 802. Find the integers.

30. Find two numbers whose sum is 42 and whose product is a maximum.

31. Write a quadratic equation having integral coefficients whose roots are $\dfrac{2}{3}$ and $\dfrac{1}{2}$.

32. Find the radius, r, of a sphere whose surface area S is 440 square inches. Let $r = \dfrac{1}{2}\sqrt{\dfrac{S}{\pi}}$, where $\pi \approx \dfrac{22}{7}$.

9-6 Conic Sections

Parabolas, circles, ellipses, and hyperbolas can be formed by slicing a hollow double cone in different directions. The curves, therefore, are called **conic sections**.

circle

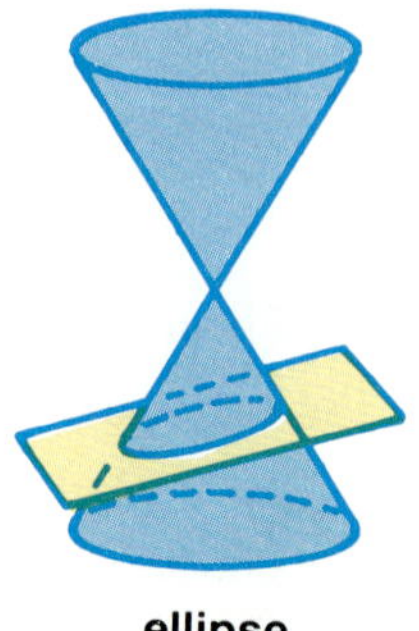

ellipse

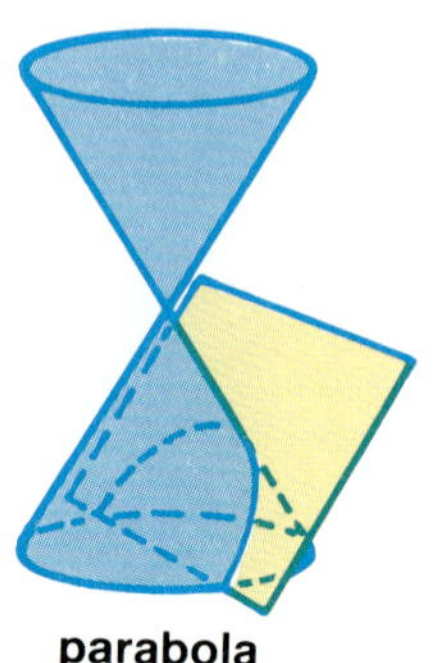

parabola

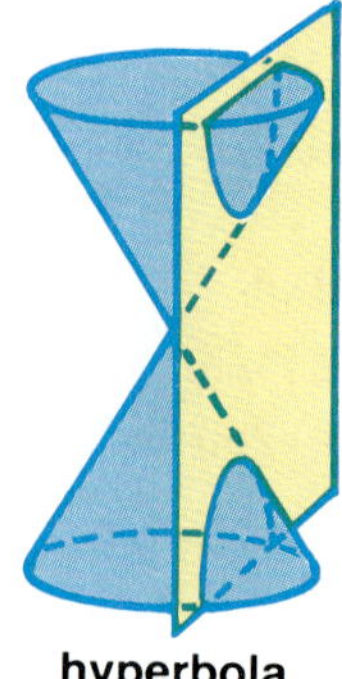

hyperbola

The conic sections are related in another way, too. Each conic section can be described by a quadratic equation in two variables.

> **The equation of a conic section can be written in the form $Ax^2 + Bxy + Cy^2 + Dx + Ey + F = 0$ where A, B, and C are not all zero.**

Equation of a Conic Section

To identify the conic section a quadratic equation in two variables represents, rewrite the equation in the forms you have learned. The following table summarizes these forms.

Most of the conic sections you have studied have equations with $B = 0$.

Rewriting the equations into these forms often involves completing the square.

Conic Section	Standard Form of Equation
parabola	$y = a(x - h)^2 + k$ or $x = a(y - k)^2 + h$
circle	$(x - h)^2 + (y - k)^2 = r^2$
ellipse	$\dfrac{(x - h)^2}{a^2} + \dfrac{(y - k)^2}{b^2} = 1$ or $\dfrac{(x - h)^2}{b^2} + \dfrac{(y - k)^2}{a^2} = 1$
hyperbola	$\dfrac{(x - h)^2}{a^2} - \dfrac{(y - k)^2}{b^2} = 1$ or $\dfrac{(y - k)^2}{a^2} - \dfrac{(x - h)^2}{b^2} = 1$ or $xy = c$, when $c \neq 0$

1 Is the graph of $x^2 + y^2 = x + 2$ a parabola, a circle, an ellipse, or a hyperbola?

$$x^2 + y^2 = x + 2$$
$$x^2 - x + \blacksquare + y^2 = 2 + \blacksquare$$
$$x^2 - x + \left(-\frac{1}{2}\right)^2 + y^2 = 2 + \left(-\frac{1}{2}\right)^2 \qquad \textit{Complete the square.}$$
$$\left(x - \frac{1}{2}\right)^2 + y^2 = \frac{9}{4}$$
$$\left(x - \frac{1}{2}\right)^2 + (y - 0)^2 = \left(\frac{3}{2}\right)^2 \qquad \textit{h is } \tfrac{1}{2}, \textit{ k is 0, r is } \tfrac{3}{2}.$$

The graph is a circle. The graph is also an ellipse with a and b both equal to $\frac{3}{2}$.

2 Is the graph of $6x^2 - 24x - 5y^2 - 10y - 11 = 0$ a parabola, a circle, an ellipse, or a hyperbola?

$$6x^2 - 24x - 5y^2 - 10y - 11 = 0$$
$$6(x^2 - 4x + \blacksquare) - 5(y^2 + 2y + \square) = 11 + 6\blacksquare - 5\square$$
$$6(x^2 - 4x + 4) - 5(y^2 + 2y + 1) = 11 + 6(4) - 5(1) \qquad \textit{Complete the square.}$$
$$6(x - 2)^2 - 5(y + 1)^2 = 30$$
$$\frac{(x - 2)^2}{5} - \frac{(y + 1)^2}{6} = 1 \qquad \textit{h is 2, k is } -1, \textit{ a is } \sqrt{5}, \textit{ b is } \sqrt{6}$$

The graph is a hyperbola.

Consider an equation of the form $Ax^2 + Bxy + Cy^2 + Dx + Ey + F = 0$ when $B = 0$. By comparing A and C, you can determine the type of conic section the equation represents.

If $A = C$, the equation represents a circle.
If A and C have the same sign and $A \neq C$, the equation represents an ellipse.
If A and C have opposite signs, the equation represents a hyperbola.
If $A = 0$ or $C = 0$, but not both, the equation represents a parabola.

Exploratory Exercises

State whether the graph of each equation is a parabola, a circle, an ellipse, or a hyperbola.

1. $x^2 + y^2 = 9$

2. $x^2 + y^2 = 4$

3. $y = (x - 3)^2 + 25$

4. $\dfrac{y^2}{8} - \dfrac{x^2}{10} = 1$

5. $\dfrac{x^2}{4} + \dfrac{y^2}{6} = 1$

6. $x = \left(y + \dfrac{1}{2}\right)^2$

7. $\dfrac{(x + 3)^2}{1} - \dfrac{(y - 4)^2}{9} = 1$

8. $\dfrac{(y - 7)^2}{3} + \dfrac{(x + 2)^2}{2} = 1$

Written Exercises

Write the standard form of each equation. Graph each equation and state whether the graph is a parabola, a circle, an ellipse, or a hyperbola.

9. $x^2 = 8y$

10. $4x^2 + 2y^2 = 8$

11. $3x^2 + 3y^2 = 81$

12. $9x^2 - 4y^2 = 4$

13. $3x^2 + 4y^2 + 8y = 8$

14. $13x^2 - 49 = -13y^2$

15. $y^2 - 2x^2 - 16 = 0$

16. $y = x^2 + 3x + 1$

17. $x^2 - 8y + y^2 = -11$

18. $\dfrac{(y-5)^2}{4} - (x+1)^2 = 4$

19. $x^2 + y = x + 2$

20. $(y-4)^2 = 9(x-4)$

21. $9x^2 + 25y^2 - 54x - 50y = 119$

22. $x^2 - 4y^2 + 10x - 16y = -5$

23. $3y^2 + 24y - x^2 - 2x = -41$

24. $x^2 + y^2 + 6y - 8x = -24$

Challenge Exercises

The graph of an equation of the form $Ax^2 + Bxy + Cy^2 + Dx + Ey + F = 0$ is either a conic section or a *degenerate case*. The degenerate cases for the conic sections are stated below. Graph each equation and identify the result.

	Conic	*Degenerate Case*
	ellipse or circle	isolated point
	hyperbola	two intersecting lines
	parabola	two parallel lines or one line

25. $9x^2 - y^2 = 0$

26. $3x^2 - 6x + 4y^2 + 32y + 67 = 0$

27. $y^2 - y = 0$

---

mini-review

Use the table of roots or a calculator to find each value.

1. $\sqrt[3]{43}$

2. $\sqrt{410}$

3. $(5.3)^3$

Find the sum and the product of the roots for each quadratic equation. Then solve the equation.

4. $x^2 - 10x - 100 = 0$

5. $2m^2 - 12m + 16 = 0$

Find a quadratic equation having the given roots.

6. $5, -8$

7. $2 - 2i, 2 + 2i$

8. $1 + \sqrt{2}, 1 - \sqrt{2}$

Determine the value of a so that each point is on the graph of $y = ax^2$.

9. $(5, 2)$

10. $(-4, -4)$

11. $(3, 1)$

Solve.

12. An object is fired upward from the top of a tower at a velocity of 90 feet per second. The tower is 120 feet high. The height of the object above the ground t seconds after firing is given by the formula $h(t) = -12t^2 + 90t + 120$. What is the maximum height reached by the object? How long after firing does it reach maximum height?

Graphing Calculator Application: Conic Sections

Parabolas, circles, ellipses, and hyperbolas can all be drawn on the graphics screen of a graphing calculator.

Examples

1 **Graph $5x = (y - 3)^2$.** *This graph is a parabola.*

This equation cannot be entered into the graphing calculator for graphing. Only equations that represent functions can be entered. Thus, you must first solve $5x = (y - 3)^2$ for y.

$$5x = (y - 3)^2$$
$$\pm\sqrt{5x} = y - 3$$
$$y = 3 \pm \sqrt{5x}$$

The equation $y = 3 \pm \sqrt{5x}$ represents two equations, $y = 3 + \sqrt{5x}$ and $y = 3 - \sqrt{5x}$. Choose a set of range parameter values that will produce a complete graph of each equation. One possible set of values is given below. Set the range parameters to these values and graph each equation.

Xmin: -2 Xmax: 30   Xscl: 3   Ymin: -15 Ymax: 15   Yscl: 3

ENTER: GRAPH 3 + √
(5 X) EXE
GRAPH 3 − √
(5 X) EXE

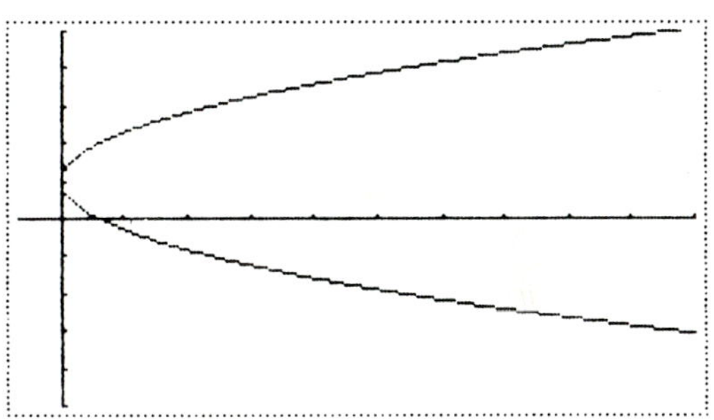

2 **Graph $x^2 + 2y^2 + 2x + 8y - 15 = 0$** *This graph is an ellipse.*

$$x^2 + 2y^2 + 2x + 8y - 15 = 0$$
$$x^2 + 2x + 2(y^2 + 4y) = 15$$
$$x^2 + 2x + 1 + 2(y^2 + 4y + 4) = 15 + 1 + 2(4) \qquad \text{Complete the square.}$$
$$(x + 1)^2 + 2(y + 2)^2 = 24$$
$$(y + 2)^2 = 12 - 0.5(x + 1)^2 \qquad \text{Solve for y.}$$
$$y + 2 = \pm\sqrt{12 - 0.5(x + 1)^2}$$
$$y = -2 \pm \sqrt{12 - 0.5(x + 1)^2}$$

One possible set of range parameter values for graphing both equations $y = -2 + \sqrt{12 - 0.5(x + 1)^2}$ and $y = -2 - \sqrt{12 - 0.5(x + 1)^2}$ is given below. Set the range parameters to these values and graph each equation.

Xmin: -6.9 Xmax: 4.9   Xscl: 2   Ymin: -7 Ymax: 2   Yscl: 1

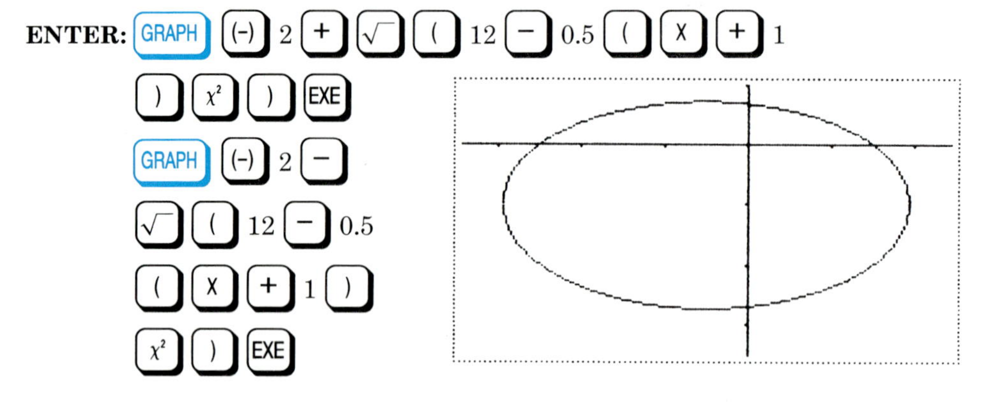

Exploratory Exercises

Graph $x^2 + y^2 = 64$ for each set of range parameter values. Then sketch the graph shown on the graphics screen, indicating the x-axis scale and y-axis scale.

1. Xmin: -8, Xmax: 8, Xscl: 2, Ymin: -8, Ymax: 8, Yscl: 2
2. Xmin: -15, Xmax: 15, Xscl: 3, Ymin: -15, Ymax: 15, Yscl: 3
3. Xmin: -8, Xmax: 8, Xscl: 2, Ymin: -15, Ymax: 15, Yscl: 3

4. The graph of $x^2 + y^2 = 64$ is a circle. Explain why the graphs in Exercises 1–3 appear as ellipses and not circles on the graphics screen.

5. Find a set of range parameter values that will make the graph of $x^2 + y^2 = 64$ appear as a circle on the graphics screen.

Written Exercises

Solve each equation for y. Then graph each equation on a graphing calculator. Sketch the graph shown on the graphics screen, indicating the x-axis scale and y-axis scale. State whether the graph is a parabola, circle, ellipse, or hyperbola.

(Hint: You may need to try many different sets of range parameter values before finding values that produce complete graphs of the equations and make circles appear as circles on the graphics screen.)

6. $x^2 = 10y$

7. $x^2 + y^2 = 100$

8. $\dfrac{x^2}{16} + \dfrac{y^2}{36} = 1$

9. $\dfrac{x^2}{49} - \dfrac{y^2}{25} = 1$

10. $(y + 5)^2 = 4x$

11. $3x^2 + 9y^2 = 81$

12. $4y^2 - 10x^2 = 100$

13. $12x^2 - y^2 = 24$

14. $(x + 3)^2 = \dfrac{1}{3}(y - 14)$

15. $\dfrac{(y + 6)^2}{16} + \dfrac{(x - 3)^2}{40} = 1$

16. $\dfrac{(y - 1)^2}{21} - \dfrac{(x - 2)^2}{10} = 1$

17. $(2y + 5)^2 = 4(x + 8)$

18. $(x - 8)^2 + 12y^2 = 36$

19. $10x^2 - 8(y + 11)^2 = 40$

20. $(x + 10)^2 + (y + 20)^2 = 75$

21. $24(y - 7)^2 + 60(x + 15)^2 = 480$

22. $x^2 + y^2 + 12x + 32y = 384$

23. $y^2 + 12x + 18y = 63$

24. $25x^2 + 16y^2 - 150x - 48y = 539$

25. $108y^2 - 72x^2 + 108y + 360x = 819$

9-7 Graphing Quadratic Systems

Consider the following system of equations.

$$y = x^2 - 4$$
$$y = -2x - 1$$

To solve this system you must find the ordered pairs that satisfy *both* equations. One way is to graph each equation and find the intersection of the two graphs.

The graph of $y = x^2 - 4$ is a parabola. The graph of $y = -2x - 1$ is a line. The parabola and the line intersect at two points, $(-3, 5)$ and $(1, -3)$. The solutions of the system are $(-3, 5)$ and $(1, -3)$.

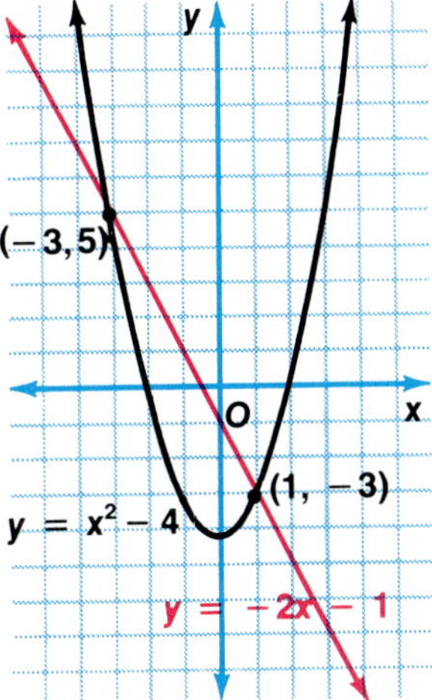

If the graphs of a system of equations are a conic section and a straight line, the system will have zero, one, or two solutions.

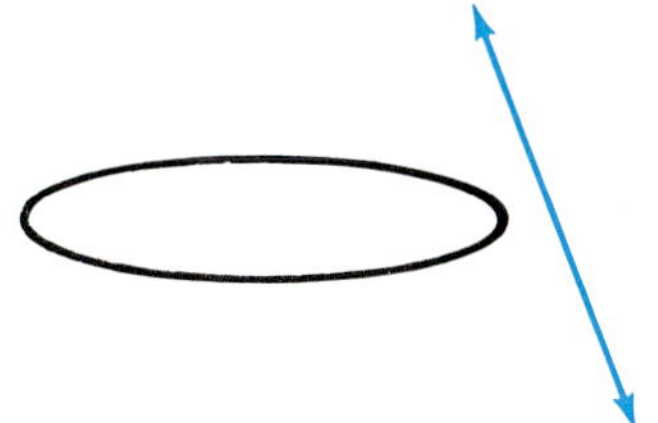

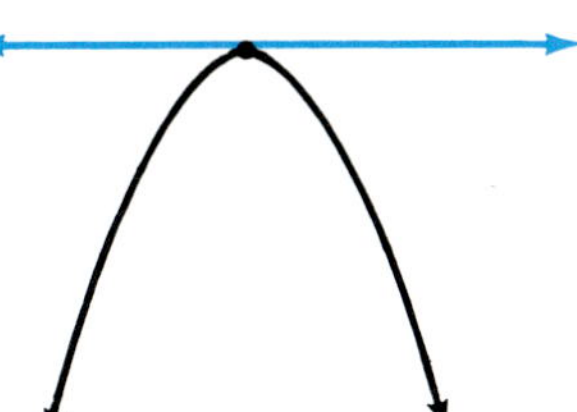

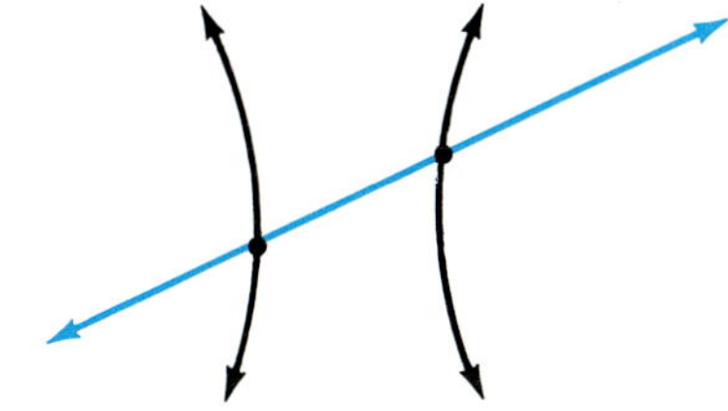

no solutions   **one solution**   **two solutions**

Example

1 **Graph the following system of equations and then find its solutions.**

$$x^2 + y^2 = 25$$
$$y - x = 1$$

The graph of $x^2 + y^2 = 25$ is a circle centered at the origin with a radius of 5 units.

The graph of $y - x = 1$ is a line with slope 1 and y-intercept 1.

The solutions of the system are $(3, 4)$ and $(-4, -3)$.

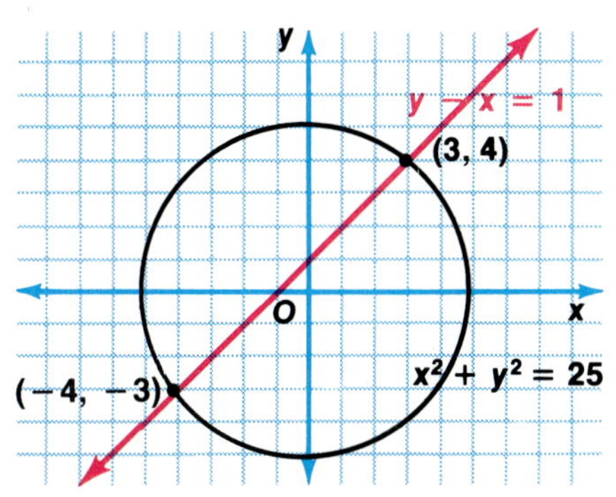

If the graphs of a system of equations are two conic sections, the system will have zero, one, two, three or four solutions.

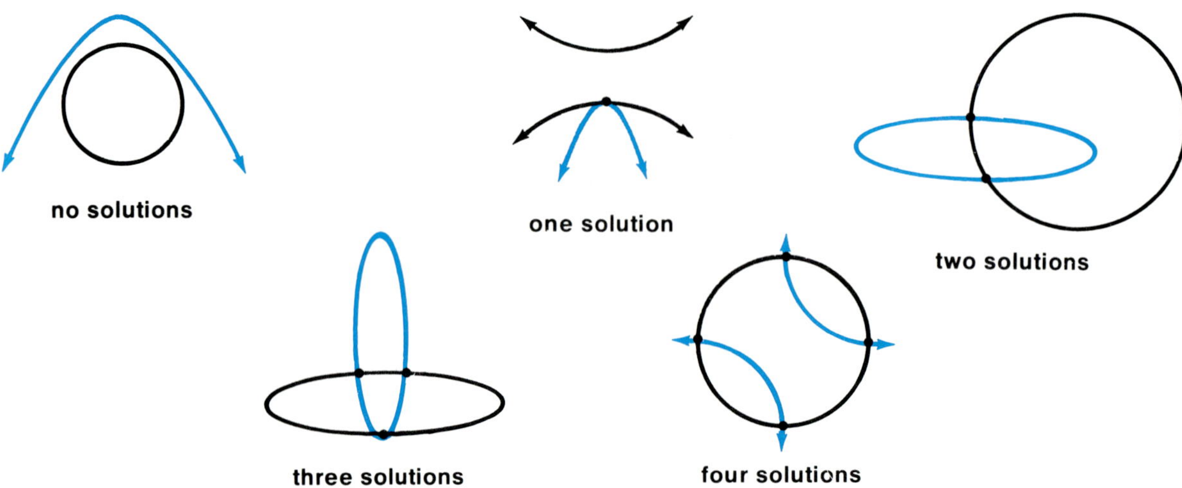

Examples

2 Graph the following system of equations and find its solutions.

$$x^2 + y^2 = 25$$
$$4y + x^2 = 25$$

The graph of $x^2 + y^2 = 25$ is a circle centered at $(0, 0)$ with a radius of 5 units. The graph of $4y + x^2 = 25$ is a parabola that opens downward, has vertex at $\left(0, 6\frac{1}{4}\right)$, and x-intercepts 5 and -5. The solutions of the system are $(5, 0)$, $(-5, 0)$, $(3, 4)$, and $(-3, 4)$.

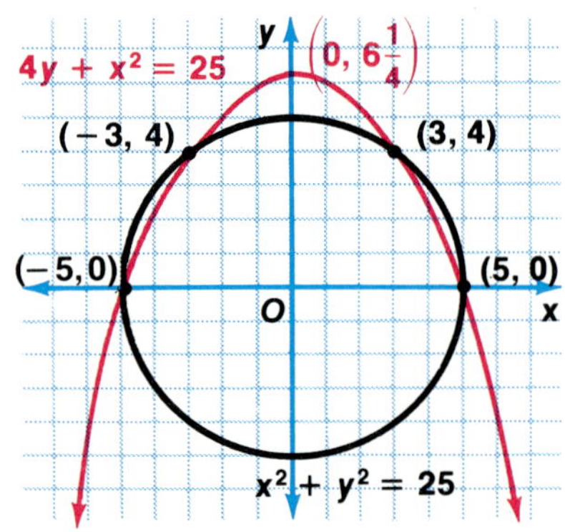

3 Graph the following system of equations and find its solutions.

$$y^2 - x^2 = 16$$
$$x^2 - y^2 = 16$$

The graph of $y^2 - x^2 = 16$ is a hyperbola with y-intercepts 4 and -4. The graph of $x^2 - y^2 = 16$ is a hyperbola with x-intercepts 4 and -4. Both hyperbolas have the same asymptotes, $y = \pm x$. The two hyperbolas do not intersect. Thus, the system has no solutions.

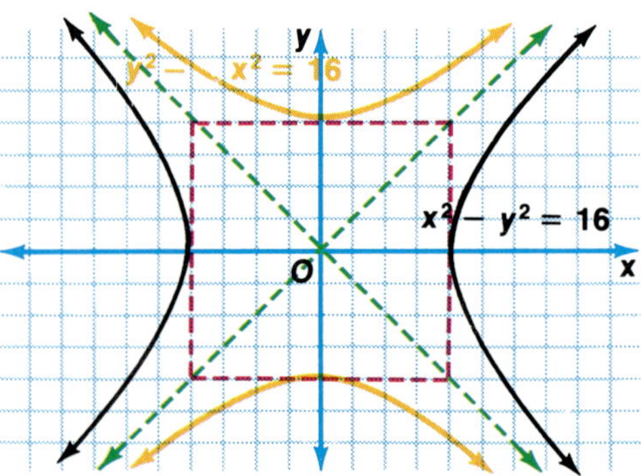

Graph each system of equations. Then find the solutions of each system.

1. $x^2 + y^2 = 16$
$y = 2$

2. $y = x^2$
$y - 2 = x$

3. $x^2 - y = 4$
$y = 3x$

4. $x = y$
$\dfrac{x^2}{20} + \dfrac{y^2}{5} = 1$

5. $\dfrac{x^2}{9} - \dfrac{y^2}{9} = 1$
$\dfrac{2}{3}y = \dfrac{1}{3}x - 1$

6. $x^2 + y^2 = 25$
$x + y = -7$

7. $(y - 1)^2 = x + 4$
$y + x = -1$

8. $\dfrac{x^2}{16} + \dfrac{y^2}{4} = 1$
$3y + 5x = 6$

9. $y^2 - x^2 = 9$
$y = 6$

10. $x^2 + y^2 = 100$
$x - y = 2$

11. $x^2 + y^2 = 9$
$x + y = 7$

12. $\dfrac{x^2}{4} + \dfrac{y^2}{1} = 1$
$x - y = 6$

13. $\dfrac{x^2}{16} - \dfrac{y^2}{4} = 1$
$y = 3x - 3$

14. $x^2 + 4y^2 = 25$
$2y = 1 - x$

15. $\dfrac{(x - 2)^2}{16} + \dfrac{y^2}{16} = 1$
$y - x = 2$

16. $y = -x^2$
$y = -x - 2$

17. $(x - 1)^2 + 4(y - 1)^2 = 20$
$x = y$

18. $\dfrac{x^2}{36} - \dfrac{y^2}{4} = 1$
$y = x$

19. $\dfrac{(x - 3)^2}{25} + \dfrac{(y - 4)^2}{9} = 1$
$5y + 3x = 44$

20. $(x - 3)^2 + (y + 6)^2 = 36$
$y + 3 = x$

21. $5x^2 + y^2 = 30$
$y^2 - 16 = 9x^2$

22. $x^2 + y^2 = 5$
$2x^2 + y = 0$

23. $2y^2 = 10 - x^2$
$3x^2 - 9 = y^2$

24. $4x^2 + 9y^2 = 36$
$4x^2 - 9y^2 = 36$

25. $x^2 + 4y^2 = 4$
$(x - 2)^2 + (y - 2)^2 = 1$

26. $x^2 + 4y^2 = 36$
$y = -x^2 + 3$

27. $x^2 - y^2 = 25$
$x^2 - y^2 = 7$

28. $x^2 + y^2 = 16$
$x^2 + y^2 = 9$

29. $x^2 + y^2 = 64$
$x^2 + 64y^2 = 64$

30. $x^2 - y^2 = 16$
$y^2 - x^2 = 16$

31. $x^2 + y^2 = 25$
$x^2 - y^2 = 7$

32. $2x^2 + y^2 = 17$
$x^2 + 2y^2 = 16$

33. $16x^2 = 4y^2 + 64$
$49x^2 = 4(y^2 - 49)$

Challenge Exercises

Solve each system of equations by graphing.

34. $x - y + 10 = 0$
$y^2 = 16(x + 10)$

35. $(x + 2)^2 + 16(y + 4)^2 = 16$
$y = -x$

The first step in solving some problems is to organize the given information in one of several ways. Sometimes making a list to organize the information is helpful. Another way to classify information is to make a table or chart.

Example: Rachel is beginning to jog. The first week she jogs 1 block each day. The second week she jogs 3 blocks each day. The third week she jogs 5 blocks each day, and so on. That is, on each successive week she jogs 2 more blocks per day than each day of the previous week. How many blocks will she jog each day of the twelfth week? How many blocks will she have jogged altogether after the twelfth week?

Make a table of the information given. Notice the pattern that develops.

Week	Blocks Jogged Each Day	Total Blocks Jogged
1	1	$7 \times 1 = 7$ or $7 \cdot 1$
2	3	$7 + (7 \times 3) = 28$ or $7 \cdot 4$
3	5	$28 + (7 \times 5) = 63$ or $7 \cdot 9$
4	7	$63 + (7 \times 7) = 112$ or $7 \cdot 16$
.	.	.
.	.	.
n	$2n - 1$	$7n^2$

Thus, the twelfth week she will jog $2(12) - 1$ or 23 blocks each day. After the twelfth week she will have jogged a total of $7 \cdot 12^2$ or 1008 blocks.

Exercises

1. A large sheet of paper is 0.15 mm thick. Suppose it is torn in half and the two pieces are placed together and torn in half again. Suppose this process continues for a total of 20 tears. How many pieces are in the stack of paper? How high is the stack of paper?

2. Classic Cleaners must clean 105 apartments in 5 days. Each day the owner puts 2 more employees on the job. However, each day after the first, each employee cleans 1 less apartment than each did on the previous day. What is the greatest number of apartments cleaned on any one day?

3. Lou Ann has a box of oranges to be divided among her seven piano students, who are all of different ages. She gives each student the number of oranges by which his age can be divided into the total number of oranges. All the oranges are to be distributed. Tara, who is the middle child in terms of age, receives 18 oranges. How many years older is Tara than the youngest student?

4. Paul questions Bill about the ages of his three children. Bill replies, "The product of their ages is 36 and the sum of their ages is the same as today's date." But Paul says he still needs more information. Then Bill tells him that the oldest child has blonde hair. Paul says that he now can determine their ages. How old are the three children?

Graphing Calculator Application: Graphing and Solving Quadratic Systems

The tracing function on a graphing calculator can be used to approximate the solutions to a quadratic system of equations.

Examples

1 **Use the tracing function to find approximate solutions to the following system of equations.**

$$y = 3x - 5$$
$$y = (x - 2)^2 - 3$$

Both equations represent functions.

Choose a set of range parameter values that will produce a complete graph of each equation. One possible set of values is given below. Set the range parameters to these values and graph each equation.

Xmin: -8 Xmax: 8   Xscl: 2   Ymin: -8 Ymax: 20   Yscl: 4

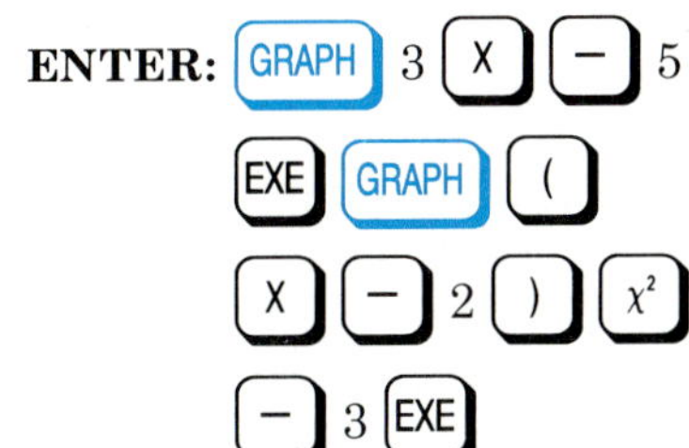

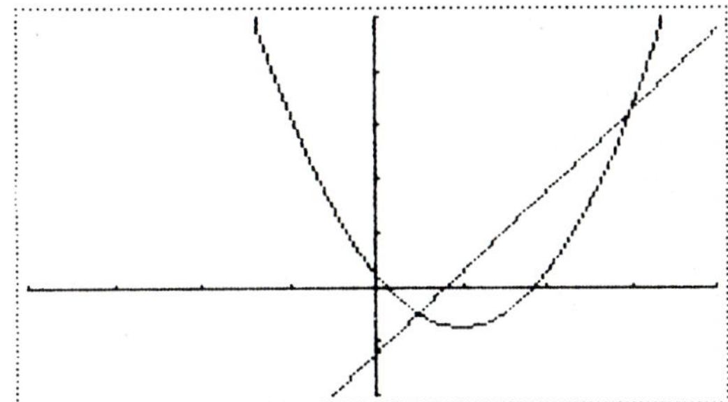

These graphs have two points of intersection. Activate the tracing function and move the pointer to each point of intersection, using the $X{\leftrightarrow}Y$ key to switch the display between the x- and y-coordinate values for the points.

Approximate solutions are $(1.0, -2.0)$ and $(6.0, 12.7)$. In this case there are exact solutions. Can you determine them?   *$(1, -2)$ and $(6, 13)$ are the exact solutions.*

2 **Use the tracing function to find approximate solutions to the following system of equations.**

$$9x^2 + 12y^2 = 216$$
$$16(y - 1)^2 - 25x^2 = 100$$

Neither one of the equations represents a function.

First solve each equation for y.

$$9x^2 + 12y^2 = 216 \qquad\qquad 16(y - 1)^2 - 25x^2 = 100$$
$$12y^2 = 216 - 9x^2 \qquad\qquad 16(y - 1)^2 = 100 + 25x^2$$
$$y^2 = 18 - 0.75x^2 \qquad\qquad (y - 1)^2 = 6.25 + 1.5625x^2$$
$$y = \pm\sqrt{18 - 0.75x^2} \qquad\qquad y - 1 = \pm\sqrt{6.25 + 1.5625x^2}$$
$$y = 1 \pm \sqrt{6.25 + 1.5625x^2}$$

The tracing function cannot always locate all points of intersection when more than two equations are graphed at the same time. In order to find all the solutions to this system, graph $y = \sqrt{18 - 0.75x^2}$ and $y = 1 + \sqrt{6.25 + 1.5625x^2}$ and find any points of intersection. Then graph $y = -\sqrt{18 - 0.75x^2}$ and $y = 1 - \sqrt{6.25 + 1.5625x^2}$ and find any additional points of intersection. One possible set of range parameter values for graphing these equations is given below. Set the range parameters to these values and graph the first two equations.

Xmin: -10 Xmax: 10   Xscl: 2   Ymin: -8 Ymax: 8   Yscl: 2

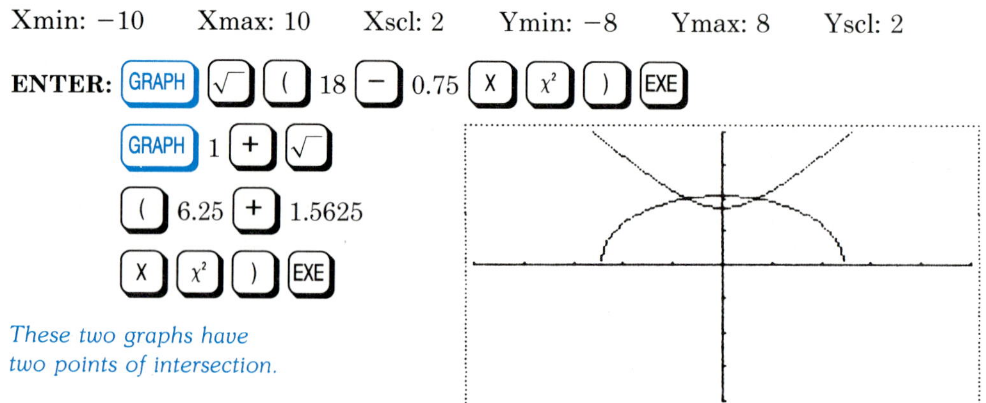

ENTER: GRAPH $\sqrt{}$ (18 $-$ 0.75 X x^2) EXE

GRAPH 1 $+$ $\sqrt{}$

(6.25 $+$ 1.5625

X x^2) EXE

These two graphs have two points of intersection.

Activate the tracing function and move the pointer to each point of intersection. Record the coordinates of these points. Then graph the other two equations.

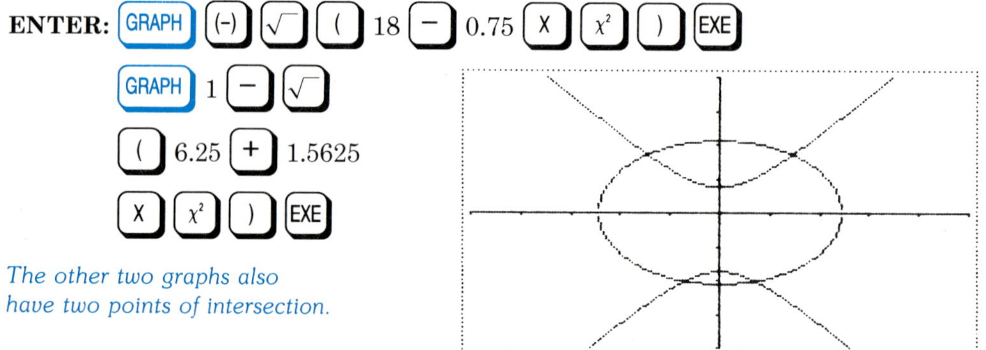

ENTER: GRAPH (-) $\sqrt{}$ (18 $-$ 0.75 X x^2) EXE

GRAPH 1 $-$ $\sqrt{}$

(6.25 $+$ 1.5625

X x^2) EXE

The other two graphs also have two points of intersection.

Again activate the tracing function and move the pointer to each point of intersection. Record the coordinates of these points.

The approximate solutions, as determined by the calculator, are $(-1.5, 4.1)$, $(1.5, 4.1)$, $(-3.0, -3.5)$, and $(3.0, -3.5)$.

Written Exercises

1–35. Graph each system of equations in Exercises 1–35 on page 331 on a graphing calculator. Then use the tracing function to find approximate solutions to each system.

9-8 Solving Quadratic Systems

You can use graphs to help find the solutions of a quadratic system. Often you must use algebra to find the exact solutions.

Examples

1 **Find the solutions of the following system of equations.**

$$x^2 + y^2 = 25$$
$$y - x = 1$$

The graphs of the equations are a circle and a straight line. The graphs show that there are two solutions to the system. Also, the values of x are between -5 and 5. And the values of y are between -5 and 5.

Use the substitution method to find the exact solutions. First rewrite $y - x = 1$ as $y = x + 1$.

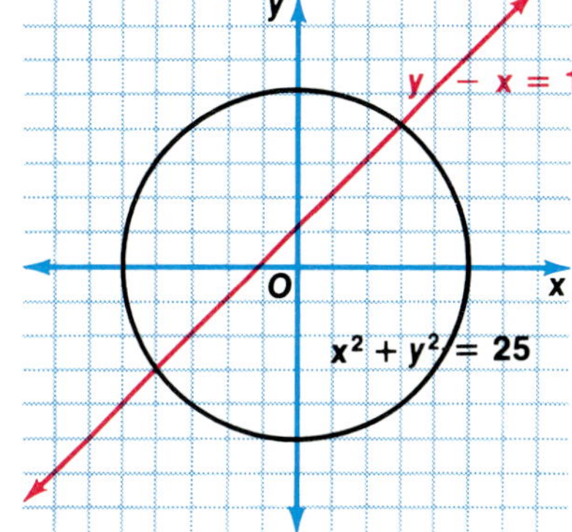

$$x^2 + y^2 = 25$$
$$x^2 + (x + 1)^2 = 25 \qquad \text{Substitute } x + 1 \text{ for } y.$$
$$x^2 + x^2 + 2x + 1 = 25$$
$$2x^2 + 2x - 24 = 0$$
$$2(x + 4)(x - 3) = 0$$

$$
\begin{array}{ccc}
x + 4 = 0 & \text{or} & x - 3 = 0 \\
x = -4 & & x = 3 \\
y = x + 1 & & y = x + 1 \\
= -4 + 1 & & = 3 + 1 \\
= -3 & & = 4
\end{array}
$$

The solutions are $(-4, -3)$ and $(3, 4)$.

2 **Find the solutions of the following system of equations.**

$$x^2 + 2y^2 = 10$$
$$3x^2 - y^2 = 9$$

The graphs of the equations are an ellipse and a hyperbola. The graphs show that there are four solutions to the system.

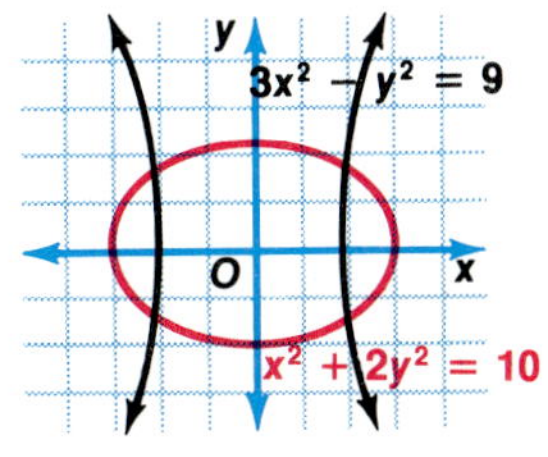

Use the elimination method to find the exact solutions.

$$x^2 + 2y^2 = 10$$
$$3x^2 - y^2 = 9$$

Multiply by 2.

$$x^2 + 2y^2 = 10$$
$$6x^2 - 2y^2 = 18$$

Add the two equations to eliminate y^2.

$$7x^2 = 28$$
$$x^2 = 4$$
$$x = \pm 2$$

$$x^2 + 2y^2 = 10$$
$$2^2 + 2y^2 = 10$$
$$2y^2 = 6$$
$$y^2 = 3$$
$$y = \pm\sqrt{3}$$

$$x^2 + 2y^2 = 10$$
$$(-2)^2 + 2y^2 = 10$$
$$2y^2 = 6$$
$$y^2 = 3$$
$$y = \pm\sqrt{3}$$

The solutions are $(2, \sqrt{3})$, $(-2, \sqrt{3})$, $(2, -\sqrt{3})$, and $(-2, -\sqrt{3})$.

To graph the solutions of a system of inequalities, graph each inequality and find the intersection of the two graphs.

$$x^2 + y^2 \geq 16$$
$$x + y = 2$$

The graph os $x^2 + y^2 \geq 16$ consists of all points on or outside the circle $x^2 + y^2 = 16$. This region is shaded green. The graph of $x + y = 2$ is a straight line with slope -1 and y-intercept 2.

The intersection of the shaded region and the straight line is the graph of the solutions. The graph of the solutions is indicated by the thick rays.

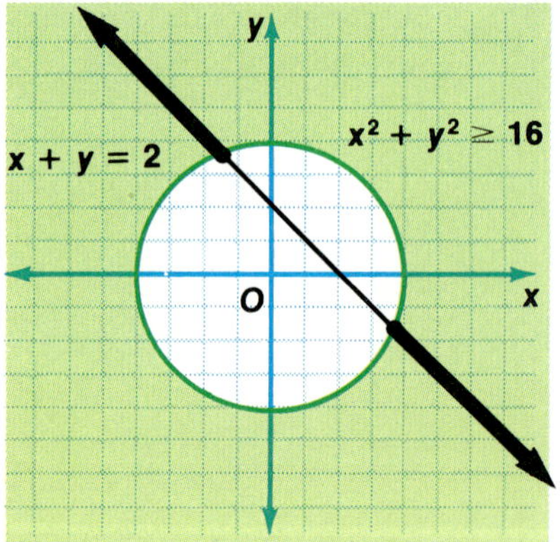

Example

3 **Solve this system of inequalities by graphing.**

$$x^2 + y^2 \leq 25$$
$$4y + x^2 \leq 25$$

The graph of $x^2 + y^2 \leq 25$ consists of all points on or within the circle $x^2 + y^2 = 25$. This region is shaded red.

The graph of $4y + x^2 \leq 25$ consists of all points on or within the parabola $4y + x^2 = 25$. This region is shaded gray.

The intersection of these two graphs represents the solutions for the system of inequalities.

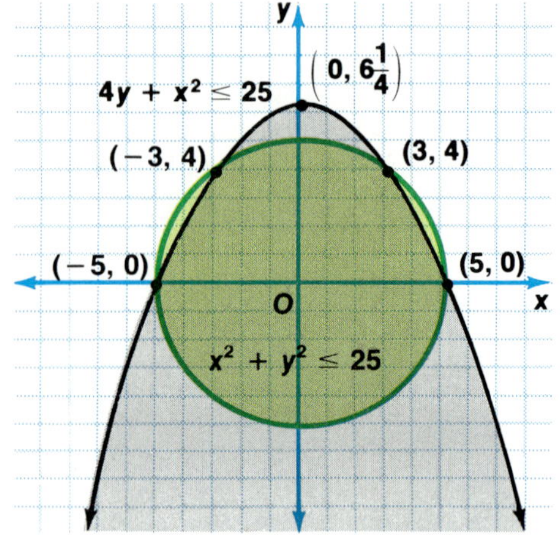

Written Exercises

Find the solutions of each system of equations.

1. $x^2 + y^2 = 16$
$x = 2$

2. $y = x^2$
$y - 2 = x$

3. $x = y$
$\dfrac{x^2}{20} + \dfrac{y^2}{5} = 1$

4. $x^2 - y = 4$
$y = 3x$

5. $x^2 - y^2 = 9$
$8y = 4x - 12$

6. $x^2 + y^2 = 25$
$x + y = -7$

7. $(y - 1)^2 = x + 4$
$y + x = -1$

8. $\dfrac{x^2}{16} + \dfrac{y^2}{4} = 1$
$2y + 5x = 4$

9. $y^2 - x^2 = 9$
$y = 6$

10. $x^2 + y^2 = 100$
$x - y = 2$

11. $x^2 + y^2 = 9$
$x + y = 7$

12. $x^2 + 4y^2 = 4$
$x - y = 6$

13. $x^2 - 4y^2 = 16$
$y = 3x - 3$

14. $x^2 + 4y^2 = 25$
$2y = 1 - x$

15. $(x - 2)^2 + y^2 = 16$
$y - x = 2$

16. $y = -x^2$
$y = -x - 2$

17. $x^2 - 4y = 0$
$y - 2x = -3$

18. $x^2 - 9y^2 = 36$
$y = x$

19. $\dfrac{(x - 3)^2}{25} + \dfrac{(y - 4)^2}{9} = 1$
$5y + 3x = 44$

20. $(x - 3)^2 + (y + 6)^2 = 36$
$y + 3 = x$

21. $5x^2 + y^2 = 30$
$y^2 - 16 = 9x^2$

22. $x^2 + y^2 = 5$
$2x^2 + y = 0$

23. $2y^2 = 10 - x^2$
$3x^2 - 9 = y^2$

24. $4x^2 + 9y^2 = 36$
$4x^2 - 9y^2 = 36$

25. $x^2 + y^2 = 16$
$x^2 + y^2 = 9$

26. $x^2 + y^2 = 64$
$x^2 + 64y^2 = 64$

27. $x^2 - y^2 = 25$
$x^2 - y^2 = 7$

Solve each system of inequalities by graphing.

28. $x^2 + y^2 < 9$
$y < -x^2$

29. $\dfrac{x^2}{9} - \dfrac{y^2}{4} < 1$
$x^2 + y^2 < 25$

30. $x^2 + y^2 \geq 4$
$x^2 + y^2 \leq 36$

31. $\dfrac{x^2}{16} - \dfrac{y^2}{1} \geq 1$
$x^2 + y^2 \geq 49$

32. $\dfrac{x^2}{25} - \dfrac{y^2}{16} \geq 1$
$x - y \geq 2$

33. $y \geq x^2 - 4$
$(y - 3)^2 \geq x + 2$

34. $x^2 + y^2 > 16$
$81x^2 + 9y^2 < 729$

35. $x^2 - 4y^2 < 16$
$x > y^2$

36. $x + 3 = y$
$x^2 + y^2 < 25$

37. $9x^2 + 4y^2 \leq 36$
$4x^2 + 9y^2 \geq 36$

38. $x + 2y > 1$
$x^2 + y^2 < 25$

39. $9x^2 - 4y^2 \geq 36$
$x + y = 4$

40. $4x^2 + 9y^2 \leq 36$
$x = 2$

41. $4x^2 + (y - 3)^2 \leq 16$
$2y = x + 1$

42. $y = -x$
$\dfrac{(x + 2)^2}{16} + \dfrac{(y + 3)^2}{1} \geq 1$

43. $5x^2 + 5y^2 \geq 25$
$2x - 3y = 5$

44. $x^2 + y^2 \geq 16$
$x = 4$

45. $x^2 + y^2 \leq 36$
$y = 6$

Challenge Exercises

Find the solutions of each system of equations or inequalities.

46. $x^2 + 4y^2 = 4$
$(x - 1)^2 + y^2 = 1$

47. $x = -y^2 + 2$
$2y - 2\sqrt{2} = x(\sqrt{2} + 2)$

48. $y \geq x^2 - 4$
$(y - 3)^2 \geq x + 2$

Using Computers

Conic Sections

Given an equation of the form $Ax^2 + Cy^2 + Dx + Ey + F = 0$, the following computer program can be used to find what type of conic section the equation represents or if it is a degenerate case.

```
5    READ A,C,D,E,F
10   IF A < > 0 THEN 30
20   IF D = 0 THEN 70
25   GOTO 80
30   IF C < > 0 THEN 50
40   IF E = 0 THEN 70
45   GOTO 80
50   LET F1 = D ^ 2 / (4 * A) + E ^ 2 /
        (4 * C)
60   IF F1 < > F THEN 80
70   PRINT "THIS IS A DEGENERATE CASE.":
        GOTO 120
80   IF A = C THEN PRINT "IT IS A CIRCLE.":
        GOTO 120
90   IF A * C > 0 THEN PRINT "IT IS AN EL-
        LIPSE.": GOTO 120
100  IF A * C < 0 THEN PRINT "IT IS A HYPER-
        BOLA.": GOTO 120
105  PRINT "IT IS A PARABOLA."
110  DATA 1,0,0,8,0
120  END
```

The equation used for this program is $x^2 = 8y$.

$A = 1, C = 0, D = 0,$
$E = 8, F = 0$

```
]RUN
IT IS A PARABOLA.
```

Exercises

Use the computer program at the left to find whether each of the following equations represents a conic section or a degenerate case.

1. $12x^2 + 36x + 16y^2 + 32y - 5 = 0$

2. $25x^2 - 4y^2 = 100$

3. $x^2 + 12x + y^2 - 8y = -44$

4. $4x^2 + 24x + y^2 - 10y + 45 = 0$

5. $(y + 3)^2 = -12(x - 2)$

6. $16x^2 + 8x + 16y^2 - 32y + 17 = 0$

parabola (304)
focus of the parabola (304)
directrix (304)
latus rectum (304)
circle (308)
radius (308)
center of the circle (308)
ellipse (311)
foci of the ellipse (311)
center of the ellipse (312)

major axis (312)
minor axis (312)
hyperbola (317)
foci of the hyperbola (317)
center of the hyperbola (318)
vertex of the hyperbola (318)
asymptotes of the hyperbola (318)
transverse axis (318)
conjugate axis (318)
conic sections (324)

Chapter Summary

1. **Distance Between Points on a Number Line:** On a number line, the distance between two points whose coordinates are a and b is $|a - b|$ or $|b - a|$. (301)

2. **Distance Formula for Two Points in a Plane:** The distance between two points with coordinates (x_1, y_1) and (x_2, y_2) is given by
$$d = \sqrt{(x_2 - x_1)^2 + (y_2 - y_1)^2}.$$ (302)

3. **Midpoint of a Line Segment:** If a line segment has endpoints at (x_1, y_1) and (x_2, y_2), then the midpoint of the line segment has coordinates $\left(\dfrac{x_1 + x_2}{2}, \dfrac{y_1 + y_2}{2}\right)$. (302)

4. A parabola is the set of all points in a plane that are the same distance from a given point called the *focus* and a given line called the *directrix*. (304)

5.

<table>
<tr><td colspan="3" align="center">Information about Parabolas</td></tr>
<tr><td>form of equation</td><td>$y = a(x - h)^2 + k$</td><td>$x = a(y - k)^2 + h$</td></tr>
<tr><td>axis of symmetry</td><td>$x = h$</td><td>$y = k$</td></tr>
<tr><td>vertex</td><td>(h, k)</td><td>(h, k)</td></tr>
<tr><td>focus</td><td>$\left(h,\ k + \dfrac{1}{4a}\right)$</td><td>$\left(h + \dfrac{1}{4a},\ k\right)$</td></tr>
<tr><td>directrix</td><td>$y = k - \dfrac{1}{4a}$</td><td>$x = h - \dfrac{1}{4a}$</td></tr>
<tr><td>direction of opening</td><td>upward if $a > 0$,
downward if $a < 0$</td><td>right if $a > 0$,
left if $a < 0$</td></tr>
<tr><td>length of latus rectum</td><td>$\left|\dfrac{1}{a}\right|$ units</td><td>$\left|\dfrac{1}{a}\right|$ units</td></tr>
</table>

(305)

6. A circle is the set of all points in a plane each of which is the same distance from a given point in the plane. (308)

7. The equation of a circle with center at (h, k) and radius of r units is $(x - h)^2 + (y - k)^2 = r^2$. (308)

8. An ellipse is the set of all points in a plane such that the sum of the distances from two given points in the plane, called the *foci*, is constant. (311)

9. If an ellipse with center at the origin has foci at $(-c, 0)$ and $(c, 0)$ and if the sum of the distances from the foci to any point on the ellipse is $2a$ units, then the standard equation of the ellipse is $\dfrac{x^2}{a^2} + \dfrac{y^2}{b^2} = 1$, where $b^2 = a^2 - c^2$.

If an ellipse with center at the origin has foci at $(0, -c)$ and $(0, c)$ and if the sum of the distances from the foci to any point on the ellipse is $2a$ units, then the standard equation of the ellipse is $\dfrac{x^2}{b^2} + \dfrac{y^2}{a^2} = 1$, where $b^2 = a^2 - c^2$. (312)

10. The standard equation of an ellipse whose center is at (h, k) and with a horizontal major axis is $\dfrac{(x - h)^2}{a^2} + \dfrac{(y - k)^2}{b^2} = 1$. The standard equation of an ellipse whose center is at (h, k) and with a vertical major axis is $\dfrac{(x - h)^2}{b^2} + \dfrac{(y - k)^2}{a^2} = 1$. (313)

11. A hyperbola is a set of all points in a plane such that the absolute value of the difference of the distances from any point on the hyperbola to two given points in the plane, called the *foci*, is constant. (317)

12. If a hyperbola with center at the origin has foci at $(-c, 0)$ and $(c, 0)$ and if the absolute value of the difference of the distances from any point on the hyperbola to the two foci is $2a$ units, then the standard equation of the hyperbola is $\dfrac{x^2}{a^2} - \dfrac{y^2}{b^2} = 1$, where $c^2 = a^2 + b^2$.

If a hyperbola with center at the origin has foci at $(0, -c)$ and $(0, c)$ and if the absolute value of the difference of the distances from any point on the hyperbola to the two foci is $2a$ units, then the standard equation of the hyperbola is $\dfrac{y^2}{a^2} - \dfrac{x^2}{b^2} = 1$, where $c^2 = a^2 + b^2$.   (318)

13. The standard equation of a hyperbola whose center is at (h, k) and with a horizontal transverse axis is $\dfrac{(x - h)^2}{a^2} - \dfrac{(y - k)^2}{b^2} = 1.$

The standard equation of a hyperbola whose center is at (h, k) and with a vertical transverse axis is $\dfrac{(y - k)^2}{a^2} - \dfrac{(x - h)^2}{b^2} = 1.$ (319)

14. Equation of a Conic Section: The equation of a conic section can be written in the form $Ax^2 + Bxy + Cy^2 + Dx + Ey + F = 0$ where A, B, and C are not all zero.   (324)

15. If the graphs of a system of equations are a conic section and a straight line, the system will have zero, one, or two solutions.   (329)

16. If the graphs of a system of equations are two conic sections, the system will have zero, one, two, three, or four solutions.   (330)

17. Parabolas, circles, ellipses, and hyperbolas can be drawn on the graphics screen of a graphing calculator.   (327–328)

18. The tracing function on a graphing calculator can be used to approximate the solutions of a quadratic system of equations.   (333–334)

Chapter Review

9–1 **Use the distance formula to find the distance between each pair of points.**

1. $(3, 6)$, $(7, -8)$

2. $(-8, -7)$, $(-2, -1)$

3. $(-2.4, 0.6)$, $(1.7, 0.8)$

4. $(2\sqrt{3}, 4\sqrt{3})$, $(2\sqrt{3}, -\sqrt{3})$

Find the midpoint of each line segment with endpoints having the following coordinates.

5. $(5, 2)$, $(-3, 1)$

6. $(17, -8)$, $(-13, 1)$

7. $(0.2, 0.6)$, $(0.3, 0.4)$

8. $(2, 2)$, $(\sqrt{2}, \sqrt{2})$

9–2 Name the vertex, axis of symmetry, focus, directrix, and direction of opening of the parabola whose equation is given. Then find the length of the latus rectum and draw the graph.

9. $4y = x^2$ **10.** $y^2 = -8x$ **11.** $(y - 8)^2 = -4(x - 4)$

9–3 Find the center and radius of each circle whose equation is given. Then draw the graph.

12. $x^2 + y^2 = 25$ **13.** $(x - 3)^2 + (y + 7)^2 = 81$ **14.** $x^2 + y^2 - 8x + 10y = 1$

9–4 Find the center, foci, and lengths of the major axis and minor axis for each ellipse whose equation is given. Then draw the graph.

15. $\dfrac{x^2}{8} + \dfrac{y^2}{16} = 1$ **16.** $\dfrac{(x - 3)^2}{25} + \dfrac{(y + 1)^2}{4} = 1$ **17.** $9x^2 + 16y^2 = 144$

9–5 Find the vertices, foci, and slopes of the asymptotes for each hyperbola whose equation is given. Then draw the graph.

18. $\dfrac{x^2}{16} - \dfrac{y^2}{81} = 1$ **19.** $25(y + 6)^2 - 20(x - 1)^2 = 500$ **20.** $49x^2 - 16y^2 = 784$

9–6 State whether the graph of each equation is a parabola, a circle, an ellipse, or a hyperbola.

21. $(x - 3)^2 = 4y - 4$ **22.** $3x^2 - 16 = -3y^2$

23. $4x^2 + 5y^2 = 20$ **24.** $3y^2 - 7x^2 = 21$

9–7 Graph each system of equations. Then state the solutions of each system.

25. $x + y = 1$ **26.** $x + y = 4$
$\quad\;\; x^2 + y^2 = 9$ $\qquad\;\; y = x^2$

9–8 Find the solutions of each system of equations.

27. $(x - 2)^2 + y^2 = 16$ **28.** $x^2 - y^2 = 16$
$\quad\;\; y - x = 2$ $\qquad\;\; y^2 - x^2 = 16$

Solve each system of inequalities by graphing.

29. $x^2 + y^2 < 25$ **30.** $y \geq x^2 + 4$
$\quad\;\; x + y > 5$ $\qquad\;\; x^2 + y^2 < 49$

Use the distance formula to find the distance between each pair of points.

1. $(2, 6)$, $(-7, -2)$

2. $(-3, 5)$, $(-11, -16)$

Find the midpoint of each line segment with endpoints having the following coordinates.

3. $(6, -4)$, $(-8, 3)$

4. $(-5.5, -7.8)$, $(1.3, -9.6)$

State whether the graph of each equation is a parabola, a circle, an ellipse, or a hyperbola. Then draw the graph.

5. $2y = x^2$

6. $x^2 + y^2 + 4x = 6$

7. $9x^2 + 49y^2 = 441$

8. $x^2 - 4y^2 = 4$

9. $(x - 3)^2 = 4(y + 1)$

10. $x^2 + 4x + y^2 - 8y = 2$

11. $6x^2 + 6y^2 = 6$

12. $\dfrac{(x - 3)^2}{81} + \dfrac{(y + 4)^2}{16} = 1$

13. $13y^2 - 2x^2 = 5$

14. $y - x^2 = x + 3$

15. $x^2 + 5y^2 - 16 = 0$

16. $16x^2 - 4y^2 = 64$

Find the solutions of each system of equations.

17. $x^2 + y^2 = 25$
$x + y = -1$

18. $x^2 + y^2 = 16$
$\dfrac{x^2}{16} - \dfrac{y^2}{9} = 1$

Solve each system of inequalities by graphing.

19. $x^2 + y^2 < 49$
$y < -x^2 + 2$

20. $x + y = 5$
$x^2 + y^2 \geq 49$

Find the equations for each conic section described below.

21. the parabola with vertex at $(6, -1)$ and focus at $(3, -1)$

22. the circle that has a diameter with endpoints at $(-2, -3)$ and $(4, 5)$

23. the ellipse with center at $(3, 1)$, major axis 12 units long, parallel to the y-axis, and minor axis $8\sqrt{2}$ units long

24. the hyperbola with center at $(2, -4)$, horizontal transverse axis 6 units long, and conjugate axis 10 units long

CHAPTER 10

Polynomial Functions

Suppose a certain packing crate must have a volume of 3 cubic meters. The width of the box is half the height, and the length is 0.5 meters more than the width. You can write a polynomial equation and apply the theorems in this chapter to find the dimensions of the packing crate.

10-1 Polynomial Functions

The following expressions are examples of **polynomials in one variable**.

$$\frac{1}{3}x + 2$$

$$4a^2 + 8a - 7$$

$$19x^3 + 7x^2 + x - 75$$

$$6y^5 + 62y^3 - 12y + 13$$

> A polynomial in one variable, x, is an expression of the form
>
> $$a_0x^n + a_1x^{n-1} + \cdots + a_{n-2}x^2 + a_{n-1}x + a_n.$$
>
> The coefficients a_0, a_1, a_2, $\ldots$, a_n represent real numbers, a_0 is not zero, and n represents a nonnegative integer.

Definition of Polynomial in One Variable

Example

1 **Determine if each expression is a polynomial in one variable.**

a. $7x^4 - 9x^2 + 2x + 1$

This is a polynomial in one variable, x.

b. $6x^2y^3 - 8xy + 7z + 2$

This is not a polynomial in one variable. There are three variables, x, y, and z.

c. $x + \frac{1}{x} - 2$

This is *not* a polynomial in one variable. The term $\frac{1}{x}$ cannot be written in the form x^n, where n is a nonnegative integer.

d. $y^2 + 2y + 3$

This is a polynomial in one variable, y.

The degree of a polynomial in one variable is the greatest exponent of its variable.

$$5 \qquad \text{has degree 0.}$$
$$3x + 2 \qquad \text{has degree 1.}$$
$$4x^2 + 8x + 7 \qquad \text{has degree 2.}$$
$$6x^5 + 62x^3 - 12x + 13 \qquad \text{has degree 5.}$$
$$a_0x^n + a_1x^{n-1} + \ldots + a_{n-1}x + a_n \qquad \text{has degree } n.$$

Example

5 Is $y + 5$ a factor of $3y^5 + y^3 - 2y^2 - 6y + 550$?

$$
\begin{array}{r|rrrrrr}
-5 & 3 & 0 & 1 & -2 & -6 & 550 \\
   &   & -15 & 75 & -380 & 1910 & -9520 \\
\hline
   & 3 & -15 & 76 & -382 & 1904 & -8970
\end{array}
$$

Use synthetic division.

Since the remainder is not 0, $y + 5$ is *not* a factor.

When a polynomial is divided by one of its factors, the quotient is called a **depressed polynomial**. The depressed polynomial can sometimes be factored to find the remaining factors. Study how this technique is used in Example 6.

Example

6 **Show that $x + 1$ is a factor of $x^3 - x^2 - 10x - 8$. Then find the remaining factors.**

$$
\begin{array}{r|rrrr}
-1 & 1 & -1 & -10 & -8 \\
   &   & -1 & 2 & 8 \\
\hline
   & 1 & -2 & -8 & 0
\end{array}
$$

The remainder is 0, so $x + 1$ is a factor of $x^3 - x^2 - 10x - 8$.

Thus, $x^3 - x^2 - 10x - 8 = (x^2 - 2x - 8)(x + 1)$.

The quotient, $x^2 - 2x - 8$, is a depressed polynomial. Can it be factored?

$$x^2 - 2x - 8 = (x - 4)(x + 2)$$

Thus, $x^3 - x^2 - 10x - 8 = (x - 4)(x + 2)(x + 1)$.

Exploratory Exercises

Divide using synthetic division. Write your answer in the form
dividend = quotient · divisor + remainder.

1. $(x^2 - 3x + 1) \div (x - 2)$

2. $(x^3 - 4x^2 + 2x - 6) \div (x - 4)$

3. $(x^3 - 8x^2 + 2x - 1) \div (x + 1)$

4. $(x^2 + 8x - 1) \div (x + 3)$

5. $(x^5 + x^4 + 2x - 1) \div (x - 2)$

6. $(2x^4 - x^2 + 1) \div (x + 1)$

7. $(x^5 + 32) \div (x + 2)$

8. $(x^5 - 3x^2 - 20) \div (x - 2)$

Written Exercises

Divide. Write your answer in the form *dividend = quotient · divisor + remainder.*

9. $(2x^3 + 8x^2 - 3x - 1) \div (x - 2)$

10. $(x^3 - 64) \div (x - 4)$

11. $(x^4 - 16) \div (x - 2)$

12. $(x^3 + 27) \div (x + 3)$

13. $(4x^4 + 3x^3 - 2x^2 + x + 1) \div (x - 1)$

14. $(6x^3 + 9x^2 - 6x + 2) \div (x + 2)$

15. $(3x^3 + 2x^2 - 4x - 1) \div \left(x + \dfrac{1}{2}\right)$

16. $(x^4 - 2x^3 + 4x^2 + 6x - 8) \div \left(x - \dfrac{1}{2}\right)$

Use synthetic substitution to find $f(3)$ and $f(-2)$ for each function f.

17. $f(x) = x^3 + 2x^2 - 3x + 1$

18. $f(x) = 2x^2 - 8x + 6$

19. $f(x) = x^3 - 8x^2 - 2x + 5$

20. $f(x) = x^3 + 8x + 1$

21. $f(x) = 3x^4 + 8x^2 - 1$

22. $f(x) = x^4 + x^3 + x^2 + x + 1$

23. $f(x) = x^5 + 8x^3 + 2$

24. $f(x) = 2x^3 + 2x^2 - 2x - 2$

Given a polynomial and one of its factors, find the remaining factors of the polynomial. (Some factors may not be binomials.)

25. $x^3 + x^2 - 4x - 4;\ x + 1$

26. $x^3 - 6x^2 + 11x - 6;\ x - 2$

27. $x^3 + 2x^2 - x - 2;\ x - 1$

28. $2x^3 + 17x^2 + 23x - 42;\ x + 6$

29. $x^3 - 3x + 2;\ x - 1$

30. $x^3 - x^2 - 5x - 3;\ x - 3$

31. $x^4 + 2x^3 - 8x - 16;\ x + 2$

32. $8x^4 + 32x^3 + x + 4;\ x + 4$

33. $x^5 + x^4 - x - 1;\ x + 1$

34. $16x^5 - 32x^4 - 81x + 162;\ x - 2$

Find values for k so that each remainder is 3.

35. $(x^2 + 8x + k) \div (x - 2)$

36. $(x^2 + kx + 3) \div (x - 1)$

37. $(x^3 + 8x^2 + kx + 4) \div (x + 2)$

38. $(x^3 + 4x^2 + kx + 1) \div (x + 1)$

39. $(x^2 + 3x - 37) \div (x - k)$

40. $(x^2 + 5x + 7) \div (x + k)$

mini-review

Solve each equation or inequality.

1. $x^2 + 12x + 32 = 0$

2. $2a^2 - 9a = 17$

3. $5z^2 + 11 = 14z$

4. $y^4 - 11y^2 = 42$

5. $x^2 < -5x + 14$

6. $2b^2 + 7b \geq 15$

7. Find a quadratic equation that has roots $-3 + 2i$ and $-3 - 2i$.

8. If June has 140 meters of fence to make a rectangular corral, find the dimensions of the corral that has the maximum area.

Graph each equation.

9. $y = (x - 2)^2 - 5$

10. $y = -3x^2 - 18x - 28$

11. $\dfrac{(x + 1)^2}{16} + \dfrac{(y - 2)^2}{9} = 1$

12. $\dfrac{(y + 2)^2}{4} - \dfrac{(x - 3)^2}{16} = 1$

Write the equation of each conic section described.

13. the parabola with vertex at $(0, 0)$ and focus at $(4, 0)$

14. the circle that has center at $(1, -1)$ and passes through $(-2, 3)$

15. the hyperbola with center at $(2, -3)$, vertical transverse axis 6 units long, and conjugate axis 8 units long

Solve each system of equations.

16. $x^2 + y^2 = 169$
$x - y = -7$

17. $(y - 2)^2 = x$
$x + 3y = 10$

18. $5x^2 + 9y^2 = 81$
$7x^2 - 3y^2 = 51$

Applications in Life Insurance Premiums

Life insurance premiums are determined by measuring the risks and calculating the amount of money needed to pay for the coverage of these risks.

The table at the right shows that the death rate for persons 18 years old is 1.69 per 1000. Suppose a company insures one thousand 18-year-olds for $5000 each. The company can expect to pay benefits of 1.69 × $5000 or $8450. On this basis only, each insured person would be expected to pay the company a premium of $8.45 for one year.

Age	Deaths per 1000
10	1.21
18	1.69
42	4.17

Because the premiums are invested in a variety of ways, the premium may be less. Suppose the premiums earn 10% annual interest, after profit. For each dollar the company needs at the end of a year, how much must be collected in premiums? Use the formula at the right.

$$A = P(1 + r)^t$$
$$\$1 = P(1.10)^1$$
$$\$0.9091 = P$$

Replace A by $1. The rate is 10% and the time is 1 year.

Interest Formula
$$A = P(1 + r)^t$$
P = original amount
A = amount of money at the end of specified time
t = time in years
r = yearly interest rate

The company must collect $0.9091 per dollar. Since $8450 is needed, the premiums must be 0.9091 × $8450, or $7682. Under these conditions, each of the 18-year-olds would pay $7.68.

Exercises

Use the above information to figure the yearly premium for each of the following.

1. a $10,000 policy with a 10% annual interest return for an 18-year-old
2. a $5,000 policy with a 10% annual interest return for a 42-year-old
3. a $5,000 policy with a 15% annual interest return for an 18-year-old
4. a $10,000 policy with a 15% annual interest return for an 18-year-old
5. a $10,000 policy with an 8% annual interest return for a 10-year-old
6. a $40,000 policy with a 15% annual interest return for a 42-year-old

10-3 Roots and Zeros

Some polynomial equations, such as $4x^2 - 1 = 0$, have no integral roots. Some, such as $x^2 - 2 = 0$, have no rational roots. Some, such as $x^2 + 1 = 0$, have no real roots. All polynomial equations have *at least one* root in the set of complex numbers. This is stated in the Fundamental Theorem of Algebra.

> **Every polynomial equation with degree greater than zero has at least one root in the set of complex numbers.**
>
> *The Fundamental Theorem of Algebra*

Another interesting theorem comes from The Fundamental Theorem of Algebra. It states that a polynomial equation has n complex roots if its polynomial has degree n.

For example, $x^3 + 5x^2 + 4x + 20 = 0$ has three roots in the set of complex numbers, $2i$, $-2i$, and -5. Notice that the equation $x^3 + 7x^2 + 15x + 9 = 0$ has three roots since -3 is a double root.

Karl Friedrich Gauss (1777–1855) is credited with the first proof of the Fundamental Theorem of Algebra.

$$x^3 + 5x^2 + 4x + 20 = 0$$
$$(x - 2i)(x + 2i)(x + 5) = 0$$

$x - 2i = 0$ or $x + 2i = 0$ or $x + 5 = 0$
$\quad x = 2i \qquad\qquad x = -2i \qquad\qquad x = -5$

$$x^3 + 7x^2 + 15x + 9 = 0$$
$$(x + 1)(x + 3)^2 = 0$$

$x + 1 = 0$ or $(x + 3)^2 = 0$
$\quad x = -1 \qquad\qquad x = -3$

Consider the polynomial function $f(x) = x^3 + 5x^2 + 4x + 20$. If $f(a) = 0$, then a is called a **zero of the function**. We know that $2i$, $-2i$, and -5 are roots of the equation $x^3 + 5x^2 + 4x + 20 = 0$, so $f(2i) = 0$, $f(-2i) = 0$, and $f(-5) = 0$. Thus, $2i$, $-2i$, and -5 are zeros of the function f. In general, the zeros of a function defined by $y = f(x)$ are roots of the equation $f(x) = 0$.

Example

1   **Find the zeros of $f(x) = (2x + 5)(x^2 - 6x + 13)$.**

Solve $(2x + 5)(x^2 - 6x + 13) = 0$.

$2x + 5 = 0$ or $x^2 - 6x + 13 = 0$ *Why?*
$\quad 2x = -5$
$\qquad x = -\dfrac{5}{2}$

$$x = \frac{-(-6) \pm \sqrt{(-6)^2 - 4(1)(13)}}{2(1)}$$

Use the quadratic formula.

$$= \frac{6 \pm \sqrt{-16}}{2}$$

$$= 3 \pm 2i$$

The zeros of the function are $-\dfrac{5}{2}$, $3 + 2i$, and $3 - 2i$.   *How can you check this?*

You may have noticed that imaginary roots of quadratic functions come in pairs. If an imaginary number is a zero of a polynomial function, then its conjugate is also a zero.

Suppose a and b are real numbers with $b \neq 0$. Then, if $a + bi$ is a zero of a polynomial function, $a - bi$ is also a zero of the function.	*Complex Conjugates Theorem*

Example

2   **Find all zeros of $f(x) = x^3 - 7x^2 + 17x - 15$ if $2 - i$ is one zero of $f(x)$.**

Since $2 - i$ is a zero, $2 + i$ also is a zero. Thus, both $x - (2 - i)$ and $x - (2 + i)$ are factors of the polynomial.

$$f(x) = [x - (2 - i)][x - (2 + i)][\quad ? \quad]$$
$$= (x^2 - 4x + 5)(\quad ? \quad)$$

Use division to find the other factor.

$$f(x) = (x^2 - 4x + 5)(x - 3)$$

Since $x - 3$ is a factor, 3 is a zero.

$$
\begin{array}{r}
x - 3 \\
x^2 - 4x + 5 \overline{\smash{)}\, x^3 - 7x^2 + 17x - 15} \\
\underline{x^3 - 4x^2 + 5x} \\
-3x^2 + 12x - 15 \\
\underline{-3x^2 + 12x - 15} \\
0
\end{array}
$$

The zeros are $2 - i$, $2 + i$, and 3.   *The polynomial has degree 3, so f(x) has 3 zeros.*

Suppose you wish to find the zeros of $f(x) = x^3 - 4x^2 - 7x + 10$. Synthetic division can be used to find factors of the polynomial. A combined and shortened form of synthetic division is shown below for binomial divisors of $x^3 - 4x^2 - 7x + 10$. Values of r in $x - r$ are shown in the first column of the table below. Beside each value is the last line of the synthetic division procedure. The last numeral in each row is the remainder. This procedure can be used to determine factors of a polynomial.

r	1	-4	-7	10
1	1	-3	-10	0
2	1	-2	-11	-12
3	1	-1	-10	-20
4	1	0	-7	-18
5	1	1	-2	0
-1	1	-5	-2	12
-2	1	-6	5	0

$x - 1$ is a factor.

$x - 5$ is a factor.

$x - (-2)$, or $x + 2$, is a factor.

Thus, the zeros of $f(x) = x^3 - 4x^2 - 7x + 10$ are 1, 5, and -2.

More discoveries about the zeros of polynomial functions were made by the French mathematician, René Descartes (1596–1650).

Descartes' Rule of Signs

Zero coefficients are ignored.

Example

3 **State the number of positive and negative real zeros for $p(x) = 2x^4 - x^3 + 5x^2 + 3x - 9$.**

Count the number of changes in sign for the coefficients of $p(x)$. Then find $p(-x)$ and count the number of changes in sign for its coefficients.

$p(x)$
$= 2x^4 - x^3 + 5x^2 + 3x - 9$

$+2 \quad -1 \quad +5 \quad +3 \quad -9$

yes   yes   no   yes

Since $p(x)$ has 3 sign changes, there are 3 or 1 positive real zeros.

$p(-x)$
$= 2(-x)^4 - (-x)^3 + 5(-x)^2 + 3(-x) - 9$
$= 2x^4 + x^3 + 5x^2 - 3x - 9$

$+2 \quad +1 \quad +5 \quad -3 \quad -9$

no   no   yes   no

Since $p(-x)$ has 1 sign change, there is 1 negative real zero.

The polynomial in Example 3 has degree 4. Thus, it has 4 zeros. The possibilities for the nature of these zeros can be given in a table.

Number of Positive Real Zeros	Number of Negative Real Zeros	Number of Imaginary Zeros	
3	1	0	*3 + 1 + 0 = 4*
1	1	2	*1 + 1 + 2 = 4*

Example

4 **Write the simplest polynomial function with integral coefficients whose zeros are 4 and $2 - 3i$.**

If $2 - 3i$ is a zero, then $2 + 3i$ is also a zero.   *Why?*

$$f(x) = [x - (2 - 3i)][x - (2 + 3i)](x - 4)$$
$$= (x^2 - 4x + 13)(x - 4)$$
$$= x^3 - 8x^2 + 29x - 52$$

Exploratory Exercises

State the number of positive real zeros and negative real zeros for each function.

1. $f(x) = x^4 - 2x^3 + x^2 - 1$
2. $f(x) = 3x^5 + 7x^2 - 8x + 1$
3. $f(x) = 4x^4 - 3x^3 + 2x^2 - x + 1$
4. $f(x) = x^7 - x^3 + 2x - 1$
5. $f(x) = x^4 - x^3 + x^2 + x + 1$
6. $f(x) = -x^4 - x^2 - x - 1$
7. $f(x) = x^6 - 2x^5 + 3x^4 - 8x^3 + 7x^2 - 1$
8. $f(x) = x^3 + x^2 + x + 1$
9. $f(x) = x^4 + x^3 - 7x - 1$
10. $f(x) = x^{10} - 1$

Find all zeros for each function.

11. $f(x) = (x - 3)(x + 5)(2x + 5)$
12. $f(x) = (x - 8)(7x - 5)(x + 3)$
13. $f(x) = (x - 3)^2(x + 2)(2x - 1)(3x - 2)$
14. $f(x) = (x + 5)^3(4x - 1)^2(5x + 3)$
15. $f(x) = 2(3x - 2)(x - 5)(x + 1)$
16. $f(x) = (x + 1)(2x + 3)(x - 5)$

Written Exercises

For each function, state the number of positive real zeros, negative real zeros, and imaginary zeros.

17. $f(x) = -x^3 + x^2 - x + 1$
18. $f(x) = x^3 + x^2 - 12$
19. $f(x) = 3x^4 + 2x^3 - 3x^2 - 4x + 1$
20. $f(x) = x^4 + x^3 + 2x^2 - 3x - 1$
21. $f(x) = 4x^5 - x^2 + 1$
22. $f(x) = x^5 - x^3 - x + 1$
23. $f(x) = 3x^5 - 8x^2 + x$
24. $f(x) = 3x^6 - x^4 + x^3 - x^2$
25. $f(x) = -7x^3 - 6x + 1$
26. $f(x) = x^3 + 1$
27. $f(x) = x^{10} - x^8 + x^6 - x^4 + x^2 - 1$
28. $f(x) = x^{14} + x^{10} - x^9 + x - 1$

Given a function and one of its zeros, find all the zeros of the function.

29. $f(x) = x^3 - 10x^2 + 34x - 40; \; 3 + i$
30. $f(x) = x^3 - 3x^2 + 9x + 13; \; 2 - 3i$
31. $g(x) = x^3 + 2x^2 - 3x + 20; \; 1 + 2i$
32. $g(x) = 2x^3 - 17x^2 + 90x - 41; \; 4 + 5i$
33. $f(x) = x^3 + 6x^2 + 21x + 26; \; -2$
34. $f(x) = x^3 - 6x^2 + 10x - 8; \; 4$
35. $g(x) = 2x^3 - x^2 + 28x + 51; \; -\dfrac{3}{2}$
36. $g(x) = 4x^4 + 17x^2 + 4; \; \dfrac{i}{2}$
37. $h(x) = x^4 - 6x^3 + 12x^2 + 6x - 13; \; 3 - 2i$

Write the simplest polynomial function with integral coefficients that has the given zeros.

38. $3, 2i$
39. $2, 1 - i$
40. $-2, 2 + 3i$
41. $-1, 1, 2 - i$
42. $-3, -1, 3 - 4i$
43. $-2 - i, 1 + 3i$
44. $4, i, -1 + i$
45. $-2i, 3i, 1 - i$

46. Explain why $4x^3 + 2x^2 + 1 = 0$ must have two complex roots.

Challenge Exercises

Find all zeros for each function. Use synthetic division.

47. $f(x) = x^3 + 4x^2 - x - 4$
48. $g(x) = x^3 - 3x^2 + 2x$
49. $h(x) = x^3 - 4x^2 - 39x + 126$
50. $f(x) = x^4 - 13x^2 + 36$
51. $g(x) = x^4 + 4x^3 - 7x^2 - 22x + 24$
52. $h(x) = x^3 - 7x^2 - 17x - 9$

The function $f(x) = x^3 - 7x + 6$ has real zeros 1, 2, and -3. An upper bound for the zeros of a function is a number for which no real zero *greater* than that number exists. For example 2, 3, and 10.9 are upper bounds for the zeros of f. A lower bound for the zeros of a function is a number for which no real zero *less* than that number exists. Some lower bounds for the zeros of f are -3, -17, and -125.4.

Upper and lower bounds may be found by using synthetic division.

Suppose c is a real number and $P(x)$ is divided by $x - c$ using synthetic division.
  a. If the resulting quotient and remainder have no change in sign, then c is an upper bound of the zeros of $P(x)$.
  b. If the resulting quotient and remainder have alternating signs, then c is a lower bound of the zeros of $P(x)$.

Test for Upper and Lower Bounds

Notice that there can be more than one upper bound or lower bound. Therefore, it is helpful to find the least integral upper bound and greatest integral lower bound of the zeros of a function. For $f(x) = x^3 - 7x + 6$, the least integral upper bound is 2, and greatest integral lower bound is -3.

Example: **Find the least positive integral upper bound and the greatest integral lower bound of the zeros of $f(x) = x^4 - 3x^3 - 2x^2 + 3x - 5$ using the Test for Upper and Lower Bounds.**

To find the upper bound, divide by increasing values of c until the quotient polynomial and the remainder have no change in sign.

c	1	-3	-2	3	-5
1	1	-2	-4	-1	-6
2	1	-1	-4	-5	-15
3	1	0	-2	-3	-14
4	1	1	2	11	39

The least positive integral upper bound is 4.

To find the lower bound, divide by decreasing values of c until the quotient polynomial and the remainder have alternating signs.

c	1	-3	-2	3	-5
-1	1	-4	2	1	-6
-2	1	-5	8	-13	21

The greatest negative integral lower bound is -2.

Any value less than -2 is also a lower bound. Any value greater than 4 is also an upper bound.

Exercises

Find the least positive integral upper bound and greatest negative integral lower bound of the zeros of each function using the Test for Upper and Lower Bounds.

1. $f(x) = x^3 + 3x^2 - 5x - 10$

2. $f(x) = x^4 - 8x + 2$

3. $f(x) = 3x^3 - 2x^2 + 5x - 1$

4. $f(x) = x^5 + 5x^4 - 3x^3 + 20x^2 - 15$

5. $f(x) = x^3 - 4x + 6$

6. $f(x) = 2x^3 - 4x^2 - 3$

10-4 The Rational Zero Theorem

The Rational Zero Theorem can be used to identify possible zeros of polynomial equations that have integral coefficients.

> **Let $f(x) = a_0x^n + a_1x^{n-1} + \cdots + a_{n-1}x + a_n$ represent a polynomial function with integral coefficients. If $\frac{p}{q}$ is a rational number in simplest form and is a zero of $y = f(x)$, then p is a factor of a_n, and q is a factor of a_0.**

Rational Zero Theorem

The Rational Zero Theorem can be proved as follows. Assume $\frac{p}{q}$ is a zero and p and q have no common factors.

$$a_0\frac{p^n}{q^n} + a_1\frac{p^{n-1}}{q^{n-1}} + \cdots + a_{n-1}\frac{p}{q} + a_n = 0 \qquad \text{Substitute } \frac{p}{q} \text{ for } x.$$

$$a_0p^n + a_1p^{n-1}q + \cdots + a_{n-1}pq^{n-1} + a_nq^n = 0 \qquad \text{Multiply each side by } a.$$

$$a_0p^n + a_1p^{n-1}q + \cdots + a_{n-1}pq^{n-1} = -a_nq^n \qquad \text{Subtract } a_nq^n \text{ from each side.}$$

$$p(a_0p^{n-1} + a_1p^{n-2}q + \cdots + a_{n-1}q^{n-1}) = -a_nq^n \qquad \text{Factor } p \text{ from the terms on the left side.}$$

Since p is a factor of the left side, it is also a factor of the right side. But p and q have no common factors, so p is a factor of a_n rather than q^n. Using a similar approach but factoring q from the last n terms, you can prove that q is a factor of a_0. Therefore, p is a factor of a_n and q is a factor of a_0.

Example

1 **Find all rational zeros of $f(x) = 2x^3 - 3x^2 + 4x + 3$.**

According to the Rational Zero Theorem, if $\frac{p}{q}$ is a zero of the function, then p is a factor of 3 and q is a factor of 2. Thus, p is ±1 or ±3, and q is ±1 or ±2. The possible rational zeros are ±1, ±3, $\pm\frac{1}{2}$, and $\pm\frac{3}{2}$.

You can test each possible zero using substitution or synthetic division.

$\frac{p}{q}$	2	-3	4	3
1	2	-1	3	6
3	2	3	13	42
$\frac{1}{2}$	2	-2	3	$\frac{9}{2}$
$\frac{3}{2}$	2	0	4	9

$\frac{p}{q}$	2	-3	4	3
-1	2	-5	9	-6
-3	2	-9	31	-90
$-\frac{1}{2}$	2	-4	6	0
$-\frac{3}{2}$	2	-6	13	$-\frac{33}{2}$

$-\frac{1}{2}$ is a zero.

The only rational zero of $f(x) = 2x^3 - 3x^2 + 4x + 3$ is $-\frac{1}{2}$.

When a zero is found, it may be unnecessary to continue testing for other possible zeros. Study the following examples.

Examples

2 **Find all rational zeros of $g(x) = 6x^3 + 11x^2 - 3x - 2$.**

The possible rational zeros are
$\pm 1, \pm 2, \pm \frac{1}{2}, \pm \frac{1}{3}, \pm \frac{2}{3}$, and $\pm \frac{1}{6}$.

p is ± 1 or ± 2.
q is $\pm 1, \pm 2, \pm 3,$ or ± 6.

$\frac{p}{q}$	6	11	-3	-2
1	6	17	14	12
-1	6	5	-8	6
$\frac{1}{2}$	6	14	4	0

$\frac{1}{2}$ is a zero.

The depressed polynomial is $6x^2 + 14x + 4$.
Now solve $6x^2 + 14x + 4 = 0$.

The quotient of the synthetic division provides the coefficients of the depressed polynomial.

$$6x^2 + 14x + 4 = 0$$
$$2(3x + 1)(x + 2) = 0$$
$$3x + 1 = 0 \quad \text{or} \quad x + 2 = 0$$
$$x = -\frac{1}{3} \quad \text{or} \quad x = -2$$

The rational zeros are $\frac{1}{2}, -\frac{1}{3}$, and -2.

Are these also all the zeros of the function?

3 **Find all zeros of $p(x) = 4x^4 - 7x^2 + 2x + 5$.**

The possible rational zeros are
$\pm 1, \pm 5, \pm \frac{1}{2}, \pm \frac{5}{2}, \pm \frac{1}{4}$, and $\pm \frac{5}{4}$.

$\frac{p}{q}$	4	0	-7	2	5
1	4	4	-3	-1	4
-1	4	-4	-3	5	0

-1 is a zero.

The depressed polynomial is $4x^3 - 4x^2 - 3x + 5$.

The possible rational zeros are $\pm 1, \pm 5, \pm \frac{1}{2}, \pm \frac{5}{2}, \pm \frac{1}{4}$, and $\pm \frac{5}{4}$.

$\frac{p}{q}$	4	-4	-3	5
-1	4	-8	5	0

-1 is a zero.

You do not need to test 1 again. Since 1 is not a zero of p(x), it cannot be a root of the depressed polynomial.

The depressed polynomial is $4x^2 - 8x + 5$. Now solve $4x^2 - 8x + 5 = 0$.

$$x = \frac{-(-8) \pm \sqrt{(-8)^2 - 4(4)(5)}}{2(4)}$$

$$= \frac{8 \pm \sqrt{-16}}{8} \text{ or } 1 \pm \frac{i}{2}$$

$\frac{8 \pm \sqrt{-16}}{8} = \frac{8 \pm 4i}{8}$ or $1 \pm \frac{i}{2}$

The zeros are $-1, -1, 1 + \frac{i}{2}$, and $1 - \frac{i}{2}$.

Exploratory Exercises

State all *possible* rational zeros for each function.

1. $f(x) = x^4 + x^2 - 2$

2. $g(x) = x^3 + 2x^2 - 3x + 5$

3. $h(x) = x^2 - 8x + 6$

4. $p(x) = x^3 + 5x^2 - 3$

5. $f(x) = x^3 - 2x^2 + 3x - 8$

6. $g(x) = x^3 - 4x + 10$

7. $p(x) = x^3 + 8x^2 - 3x + 1$

8. $h(x) = x^3 - 2x^2 - 5x - 9$

9. $g(x) = x^3 - 8x^2 - 11x + 20$

10. $f(x) = x^4 + 2x + 15$

11. $h(x) = 6x^4 + 35x^3 - x^2 - 7x - 1$

12. $p(x) = 6x^3 + 4x^2 - 14x - 2$

13. $g(x) = 3x^4 - 5x^2 + 4$

14. $f(x) = 2x^3 + x^2 + 5x - 3$

Written Exercises

Find all rational zeros for each function.

15. $f(x) = x^3 - x^2 - 34x - 56$

16. $g(x) = x^3 + x^2 - 80x - 300$

17. $h(x) = 2x^3 - 11x^2 + 12x + 9$

18. $f(x) = x^3 - 3x - 2$

19. $g(x) = x^4 - 3x^3 + x^2 - 3x$

20. $h(x) = x^4 - 3x^3 - 53x^2 - 9x$

21. $p(x) = x^4 + 10x^3 + 33x^2 + 38x + 8$

22. $f(x) = x^4 + x^3 - 9x^2 - 17x - 8$

23. $g(x) = x^4 + x^2 - 2$

24. $h(x) = x^4 - 6x^3 - 3x^2 - 24x - 28$

25. $f(x) = x^3 - 2x^2 - 13x - 10$

26. $g(x) = x^3 + 4x^2 - 3x - 18$

27. $h(x) = x^3 - x^2 - 40x + 12$

28. $p(x) = x^4 - 13x^2 + 36$

29. $f(x) = 12x^4 + 4x^3 - 3x^2 - x$

30. $g(x) = 48x^4 - 52x^3 + 13x - 3$

31. $h(x) = x^5 - 6x^3 + 8x$

32. $f(x) = 2x^5 - x^4 - 2x + 1$

Find all zeros of each function.

33. $f(x) = 6x^3 + 5x^2 - 9x + 2$

34. $g(x) = 8x^3 - 36x^2 + 22x + 21$

35. $h(x) = 5x^4 - 29x^3 + 55x^2 - 28x$

36. $p(x) = 6x^4 + 22x^3 + 11x^2 - 38x - 40$

37. $g(x) = 24x^4 - 94x^3 + 61x^2 + 21x - 18$

38. $f(x) = 4x^4 - 35x^3 + 78x^2 + 28x - 165$

39. $p(x) = 9x^5 - 94x^3 + 27x^2 + 40x - 12$

40. $h(x) = 4x^6 - x^4 - 50x^2 + 72$

Solve each problem.

41. A box is to have a volume of 72 m³. The width is 2 m longer than the height, and the length is 7 m longer than the height. Find the dimensions of the box.

42. A box is to have a volume of 144 cm³. The width is 3 cm longer than the height, and the length is 2 cm longer than the width. Find the dimensions of the box.

Excursions in Algebra

Contest Problem

The following problem appeared in a high school mathematics contest sponsored by the Mathematical Association of America.

Suppose $f(x)$ is a polynomial function such that for all real values of x, $f(x^2 + 1) = x^4 + 3x^2 + 3$. Find $f(x^2 + 2)$.

Simplify.

1. $5\sqrt[3]{16}$

2. $\sqrt[3]{81a^4b^2}$

3. $\sqrt[4]{4x^2y^2z}\,\sqrt[4]{8xy^2z^2}$

4. $(7 + \sqrt{3})(8 - \sqrt{6})$

5. $\sqrt{\dfrac{32}{81}}$

6. $\dfrac{1}{3 - \sqrt{3}}$

7. $\sqrt{-49}$

8. $\dfrac{4 + 3i}{5 - 2i}$

9. Find the value of the discriminant for $a^2 + a - 5 = 0$. Describe the nature of the roots.

Solve each equation.

10. $4 - \sqrt{3x - 6} = 1$

11. $y^2 - 12y + 36 = 0$

12. $2a^2 - 5a - 8 = 0$

13. $3m^2 + 7m = 6$

14. $n^2 + 10n + 29 = 0$

15. $4x - 4\sqrt{x} - 3 = 0$

16. Find the value of c that makes $x^2 + 6x + c$ a perfect square.

Find a quadratic equation having the given roots.

17. $4, 7$

18. $-\dfrac{2}{3}, \dfrac{4}{5}$

19. $-2 + 3i, -2 - 3i$

20. Identify the quadratic term, the linear term, and the constant term of the equation $q = -4x^2 - 2x$.

Solve each inequality.

21. $(x - 3)(x + 2) > 0$

22. $x^2 + 11x - 42 \leq 0$

23. Graph $y = 2(x - 3)^2 - 4$. Then state the vertex, axis of symmetry, and direction of opening.

Graph each equation or inequality.

24. $\dfrac{x^2}{9} + \dfrac{y^2}{25} = 1$

25. $4x^2 - y^2 = 4$

26. $x^2 + y^2 - 4x + 2y - 4 = 0$

27. $y \geq -2x^2 + 16x - 30$

28. $x = y^2 + 4y - 3$

Find the solutions of each system of equations.

29. $y + 6 = (x - 2)^2$
$x - y = 2$

30. $x^2 + 8y^2 = 25$
$4x^2 - y^2 = 1$

31. $8x^2 + 25y^2 = 200$
$8x = 5y^2 - 40$

32. Given $p(x) = x^2 - 2x + 6$, find the values of $p(-2)$ and $p(x - 1)$.

33. State the number of positive real zeros, negative real zeros, and imaginary zeros for $f(x) = x^4 + 3x^3 - 2x - 4$.

34. Find all zeros of $f(x) = x^3 + 4x + 80$ if $2 + 4i$ is one zero of $f(x)$.

35. Find all rational zeros of $g(x) = 3x^3 + x^2 - 26x - 12$.

36. Find all zeros of $p(x) = 2x^4 - 3x^3 - 35x^2 - 18x + 18$.

Solve each problem.

37. The length of the top of Ron's bookcase is 30 cm more than the width. The area of the top is 1800 cm^2. What are its dimensions?

38. Janessa has 120 feet of fence to put around a rectangular garden. If an 8 foot opening is left on one side for a gate, what would be the length and width for maximum area?

10-5 Approximating Zeros

The graph of a continuous curve drawn from a lower point to a higher point must *cross* every horizontal line in between. Thus, if the graph of a polynomial function is in part below the x-axis and in part above the x-axis, it must also *cross* the x-axis. You can use the **Location Principle** to approximate real zeros of a polynomial function.

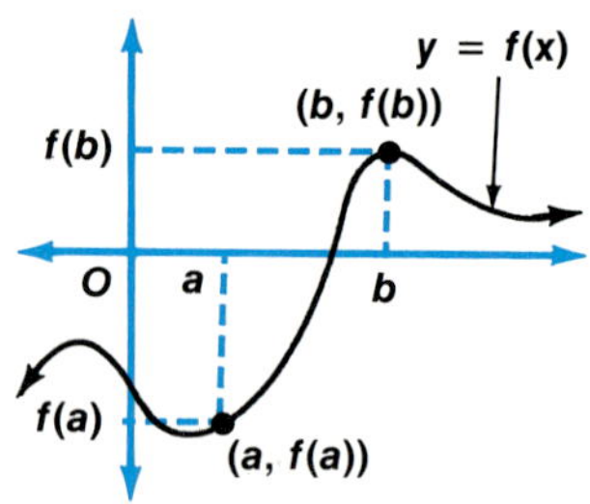

> Suppose $y = f(x)$ represents a polynomial function and a and b are two numbers such that $f(a) < 0$ and $f(b) > 0$. Then the function has at least one real zero between a and b.

The Location Principle

Example

1 **Approximate to the nearest tenth the real zeros of $f(x) = x^4 + x^2 - 6$.**

By Descartes' Rule of Signs, there is one positive real zero and one negative real zero. The other zeros are imaginary. By the Rational Zero Theorem, the possible rational zeros are $\pm 1, \pm 2, \pm 3,$ and ± 6. So, evaluate $f(x)$ for integral values of x from -6 to 6 or until both real zeros are located.

x	1	0	1	0	-6	$f(x)$
-6	1	-6	37	-222	1326	1326
-5	1	-5	26	-130	644	644
-4	1	-4	17	-68	266	266
-3	1	-3	10	-30	84	84
-2	1	-2	5	-10	14	14
-1	1	-1	2	-2	-4	-4
0	1	0	1	0	-6	-6
1	1	1	2	2	-4	-4
2	1	2	5	10	14	14

By the Location Principle, the negative real zero is between -1 and -2 and the positive real zero is between 1 and 2.

Locate the negative zero.

x	$f(x)$
-1.5	1.3125
-1.4	-0.1984

$f(-1.45) \approx 0.5$

The zero is "closer" to -1.4.

Why aren't values less than -1.5 or greater than 1.5 tested?

Locate the positive zero.

x	$f(x)$
1.0	-4
1.1	-3.3259
1.2	-2.4864
1.3	-1.4539
1.4	-0.1984
1.5	1.3125

The zero is "closer" to 1.4.

$f(1.45) \approx 0.5$

The real zeros are approximately -1.4 and 1.4.

2 Using Calculators

Approximate to the nearest tenth the real zeros of
$$f(x) = x^4 - x^3 - 4x^2 + 8x - 4.$$

By Descartes' Rule of Signs, there are three or one positive real zeros and one negative real zero. By the Rational Zero Theorem, the possible rational zeros are ± 1, ± 2, and ± 4. So, evaluate $f(x)$ from -4 to 4 or until the real zeros are located.

The procedure for filling out the synthetic substitution chart can be performed quickly on a calculator. The process for finding $f(-4)$ is shown below. The other values are computed in a similar manner.

ENTER: 4 [+/-] [STO] [×] 1 [+] 1 [+/-] [=]   *The display shows the second coefficient of the quotient.*

DISPLAY: 4 −4 −4   1 −4 1 −1 −5

ENTER: [×] [RCL] [+] 4 [+/-] [=]   *The display shows the third coefficient of the quotient.*

DISPLAY: −4 20 4 −4 16

ENTER: [×] [RCL] [+] 8 [=]   *The display shows the fourth coefficient of the quotient.*

DISPLAY: −4 −64 8 −56

ENTER: [×] [RCL] [+] 4 [+/-] [=]

DISPLAY: −4 224 4 −4 220 $f(-4) = 220$

x	1	−1	−4	8	−4	$f(x)$
−4	1	−5	16	−56	220	220
−3	1	−4	8	−16	44	44
−2	1	−3	2	4	−12	−12
−1	1	−2	−2	10	−14	−14
0	1	−1	−4	8	−4	−4
1	1	0	−4	4	0	0
2	1	1	−2	4	4	4
3	1	2	2	14	38	38
4	1	3	8	40	156	156

By the Location Principle there is a zero between −3 and −2.

1 is a zero.

The other two zeros are imaginary.

Locate the negative zero.

x	$f(x)$
−2.5	5.6875
−2.4	0.7616
−2.3	−3.4089

$f(-2.35) \approx -1.414$
The zero is "closer" to −2.4

The real zeros are 1 and approximately −2.4.

Written Exercises

Approximate to the nearest tenth the real zeros of each function. Use a calculator as necessary.

1. $f(x) = x^3 - 2x^2 + 6$
2. $g(x) = x^4 - 4x^2 + 3$
3. $h(x) = 2x^5 + 3x - 2$
4. $p(x) = x^4 - x^2 + 6$
5. $g(x) = x^3 + 2x^2 - 3x - 5$
6. $f(x) = x^3 - 5$
7. $p(x) = x^5 - 6$
8. $h(x) = 3x^3 - 16x^2 + 12x + 6$
9. $r(x) = x^3 - x^2 + 1$
10. $p(x) = x^4 - 4x^2 + 6$
11. $f(x) = 3x^2 - 8x + 1$
12. $g(x) = -7x^3 - 6x + 1$
13. $h(x) = x^5 - x^3 - x + 1$
14. $p(x) = 3x^4 - x^2 + x - 1$
15. $r(x) = x^3 + 1$
16. $f(x) = x^4 - x^2 - 6$
17. $f(x) = x^3 - 4x + 4$
18. $g(x) = x^3 - 3$
19. $h(x) = x^4 - 9x^3 + 25x^2 - 24x + 6$
20. $p(x) = x^5 + 4x^4 - x^3 - 9x^2 + 3$

mini-review

Solve each equation or inequality.

1. $x^2 - 13x = 48$
2. $2c^2 + 11 = 9c$
3. $3y^2 = 8y - 3$
4. $x^2 + 10x \leq 24$
5. $4z^2 + 25 \geq 20z$
6. $m^2 < 5m$

Graph each equation. Then name the vertex, axis of symmetry, and direction of opening for each graph.

7. $y = -(x + 5)^2 + 1$
8. $y = 2x^2 - 16x + 29$

State whether the graph of each equation is a parabola, a circle, an ellipse, or a hyperbola.

9. $x^2 + y^2 + 4x = 12$
10. $2x^2 - 3y^2 + 6x + 12y = 4$

Find the solutions of each system of equations.

11. $x^2 - 6y^2 = 25$
    $x - 6y = 5$
12. $4x^2 + 9y^2 = 81$
    $4x^2 + y^2 = 49$

For each function, state the number of positive real zeros, negative real zeros, and imaginary zeros.

13. $f(x) = x^3 - 3x^2 - 7x + 11$
14. $g(x) = 2x^4 + 2x^2 + x + 1$

15. Graph the solutions for the system $x^2 + 4y^2 \leq 16$ and $y > x^2 - 4$.

16. Find all zeros of $g(x) = 4x^3 - 29x^2 + 76x + 75$ if $4 - 3i$ is one zero of $g(x)$.

17. Find all rational zeros of $f(x) = 12x^3 + 16x^2 - 31x + 10$.

18. Find all zeros of $p(x) = 4x^4 - 20x^3 + 19x^2 + 45x - 63$.

A graphing calculator can generate graphs of polynomial functions on its graphics screen. It is often necessary to try many different sets of range parameter values before finding values that will produce a complete graph of a polynomial function.

Example

1 **Graph $y = 3x^3 + 2x^2 - 8x + 7$.**

First graph the equation using a "large" range along the x-axis and y-axis in order to get a general idea of the shape of the curve.

One possible set of range parameter values is given below. Set the range parameters to these values and graph the equation.

Xmin: -10 Xmax: 10 Xscl: 2
Ymin: -20 Ymax: 20 Yscl: 2

ENTER: 

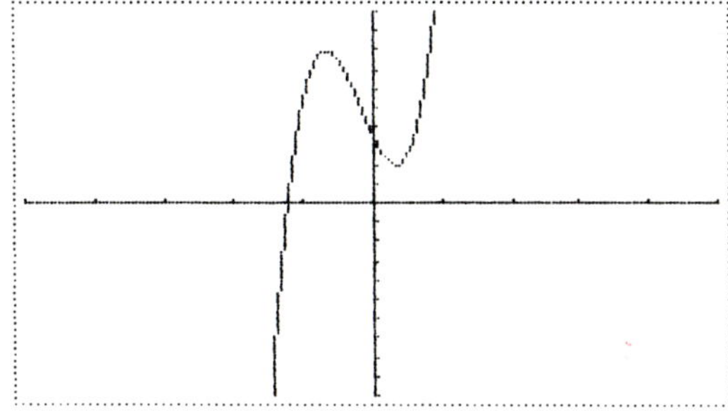

Now use the information from this graph to choose another set of range parameter values that will produce a complete graph.
Two possible sets of range parameter values are given below.

Xmin: -4 Xmax: 4   Xscl: 1
Ymin: -10 Ymax: 30 Yscl: 2

Xmin: -5 Xmax: 5   Xscl: 1
Ymin: -75 Ymax: 75 Yscl: 10

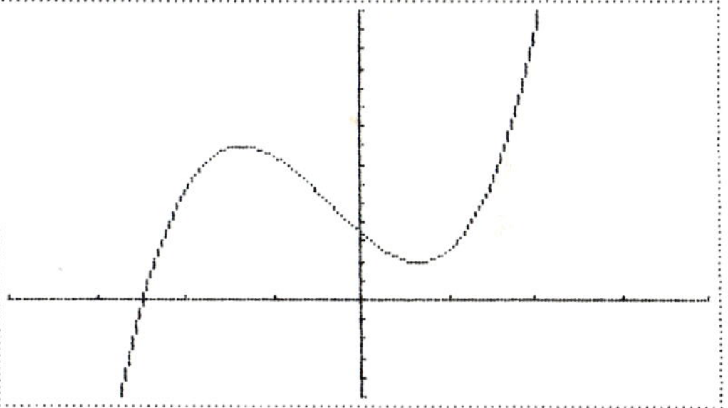

Graph the equation using each set of values. To do this, change the range parameters to each set of values and graph the equation by pressing the EXE (or =) key.

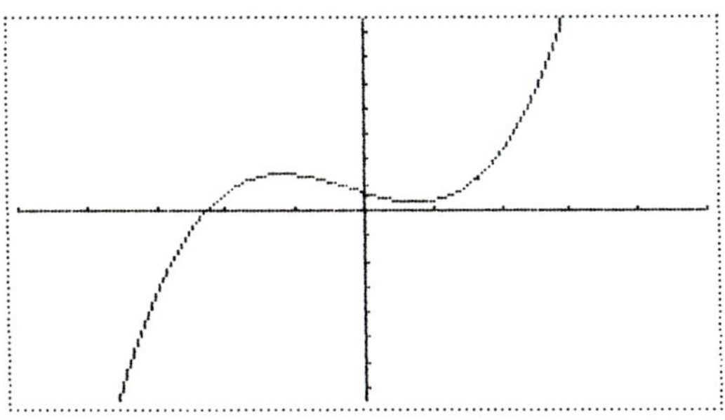

The graphing calculator can also be used to approximate the real zeros of a polynomial function to any desired accuracy. One method for approximating zeros is to graph the function and then use the tracing function to determine the point(s) of intersection of the graph and the x-axis. A more accurate method for approximating zeros is called "zooming-in."

To "zoom-in" on a real zero of a function, first graph the function and determine the approximate location of the zero. Then graph the function again using a smaller range along the x-axis that contains the zero. Continue this process until the zero is "squeezed" between two values on the x-axis that are as close together as desired.

Example

2 **Approximate to the nearest hundredth the real zeros of**
$f(x) = 3x^3 - 3x^2 + 2x - 1.$

Choose a set of range parameter values that will produce a graph that allows you to make a rough approximation for the location of any real zeros of the function. One possible set of values is given below. Set the range parameters to these values and graph $y = 3x^3 - 3x^2 + 2x - 1.$

Xmin: -3 Xmax: 3   Xscl: 1   Ymin: -20 Ymax: 20   Yscl: 4

ENTER: GRAPH 3 X x^y 3

$-$ 3 X x^2 $+$ 2

X $-$ 1 EXE

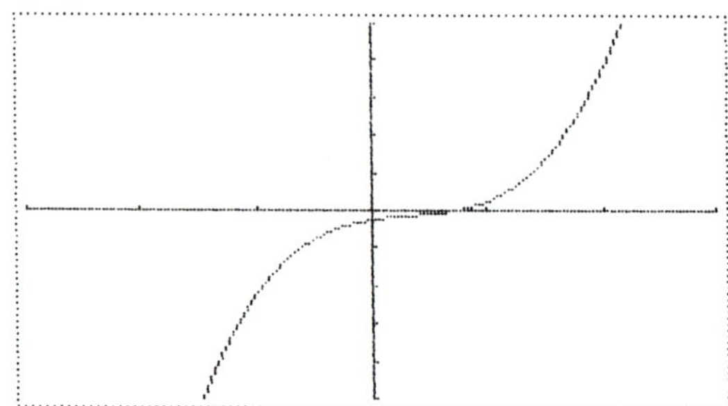

Based on this graph, the function f has a real zero between 0 and 1. To begin "zooming-in" on this zero, graph the function again using a smaller x-axis range that contains the zero. One possible set of range parameter values for the graph is given below.

Xmin: 0   Xmax: 1   Xscl: 0.1
Ymin: -1 Ymax: 1   Yscl: 0.2

Set the range parameters to these values and graph the function by pressing the EXE (or $=$) key.

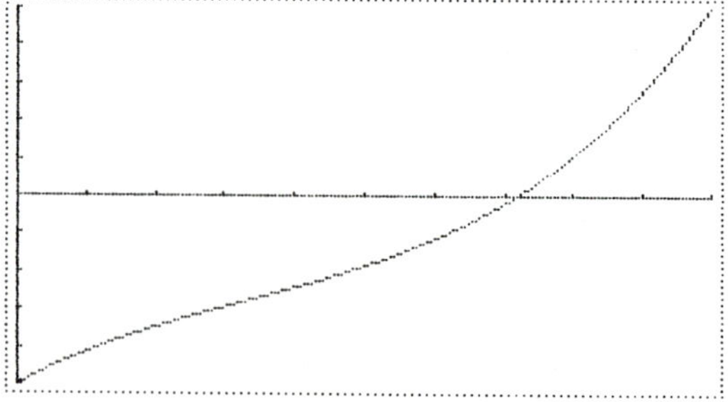

Based on this graph, the zero is between 0.7 and 0.8. Now graph the equation using a smaller x-axis range, divided into hundredths, that contains the zero. One possible set of range parameter values for the graph is given below.

Xmin: 0.7   Ymin: -0.1

Xmax: 0.8   Ymax: 0.1

Xscl: 0.01   Yscl: 0.02

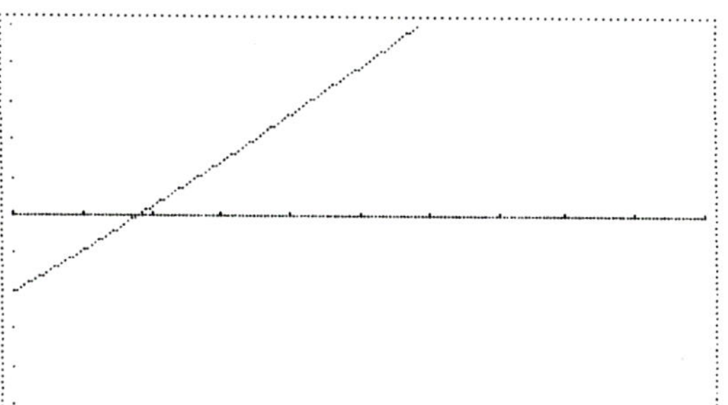

Set the range parameters to these values and graph the function by pressing the $\boxed{\text{EXE}}$ (or $\boxed{=}$)key.

Based on this graph, the zero is between 0.71 and 0.72, and it appears to be closer to 0.72. To check this, graph the equation again, changing only Xmin, Xmax, and Xscl so that Xmin is 0.71, Xmax is 0.72, and Xscl is 0.005.

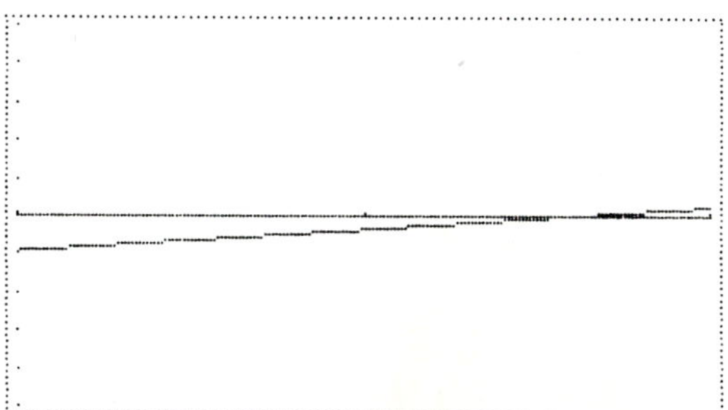

After changing these values, press the $\boxed{\text{EXE}}$ (or $\boxed{=}$) key.

Based on this graph, the zero is approximately 0.72.

Written Exercises

Graph each function on a graphing calculator using two different sets of range parameter values. Sketch each graph shown on the graphics screen, indicating the x-axis scale and the y-axis scale.

1. $f(x) = x^3 - 5$

2. $g(x) = x^3 - 3$

3. $h(x) = x^3 - 2x^2 + 6$

4. $p(x) = x^3 - 4x + 4$

5. $r(x) = x^3 - x^2 + 1$

6. $f(x) = -7x^3 - 6x + 1$

7. $g(x) = x^5 - 6$

8. $h(x) = 2x^5 + 3x - 2$

9. $p(x) = 3x^4 - x^2 + x - 1$

10. $r(x) = x^4 - x^2 - 6$

11. $f(x) = x^4 + 3x^3 - 4x^2 - 7$

12. $g(x) = x^4 - 4x^3 - 4x^2 + 24x - 6$

13. $h(x) = x^3 + 2x^2 - 3x - 5$

14. $p(x) = 3x^3 - 16x^2 + 12x + 6$

15. $r(x) = x^5 - x^3 - x + 1$

16. $f(x) = x^5 - 3x^4 + x^3 + 5x^2 - 6x - 1$

17. $g(x) = x^4 - 4x^2 + 3$

18. $h(x) = x^4 - 9x^3 + 25x^2 - 24x + 6$

19. $p(x) = x^5 + 4x^4 - x^3 - 9x^2 + 3$

20. $r(x) = x^5 + 2x^4 - 10x^3 - 20x^2 + 9x + 15$

21–40. Using a graphing calculator, approximate to the nearest hundredth the real zeros of each function in Exercises 1–20.

10-6 Graphing Polynomials

The simplest polynomial graphs are those with equations of the form $f(x) = x^n$ where n is a positive integer.

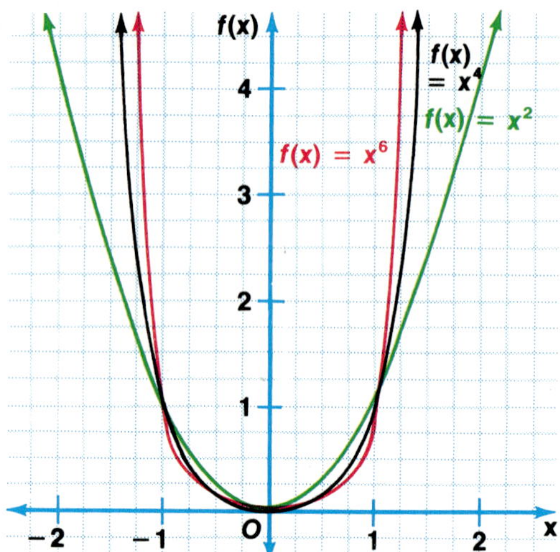

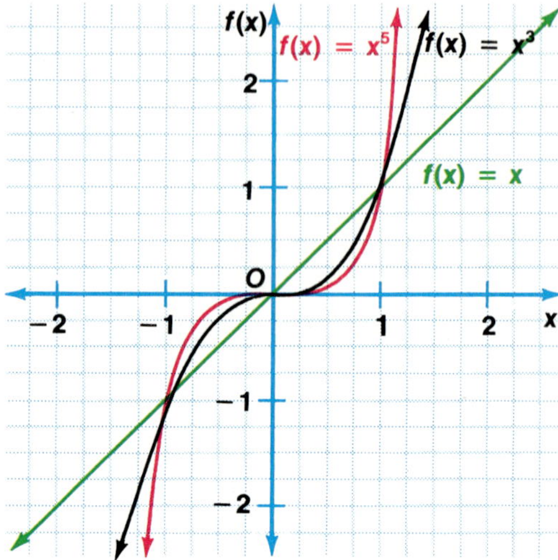

Notice that each graph has only one x-intercept, the origin. To sketch each graph, first find ordered pairs that satisfy each equation. Then graph the ordered pairs and connect them with a smooth continuous curve.

If a polynomial $f(x)$ can be factored into linear factors, much information about the graph of $y = f(x)$ is available. For example, consider $y = (x - 2)(x - 3)(x + 5)$.

1. The function has three zeros, 2, 3, and -5.
2. The function has negative values when x is less than -5 and when x is between 2 and 3.
3. The function has positive values when x is between -5 and 2 and when x is greater than 3.

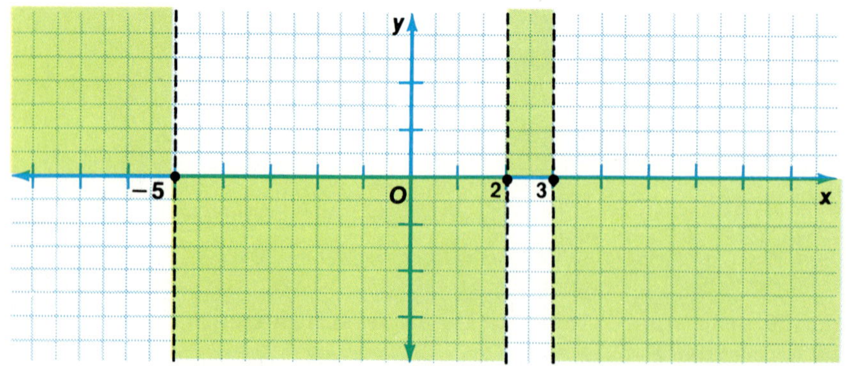

The graph crosses the x-axis at -5, 2 and 3. The shaded regions contain no points of the graph.

To sketch the graph, evaluate $f(x)$ for several integral values of x. Graph the ordered pairs formed from these values and connect them with a smooth, continuous curve. Then find and graph other ordered pairs that seem important to the graph.

You may want to use a calculator to find the y-coordinates.

x	y
-6	-72
-5	0
-4	42
-3	60
-2.5	61.875
-2	60
-1	48
0	30
1	12
2	0
2.5	-1.875
3	0
4	18
5	60

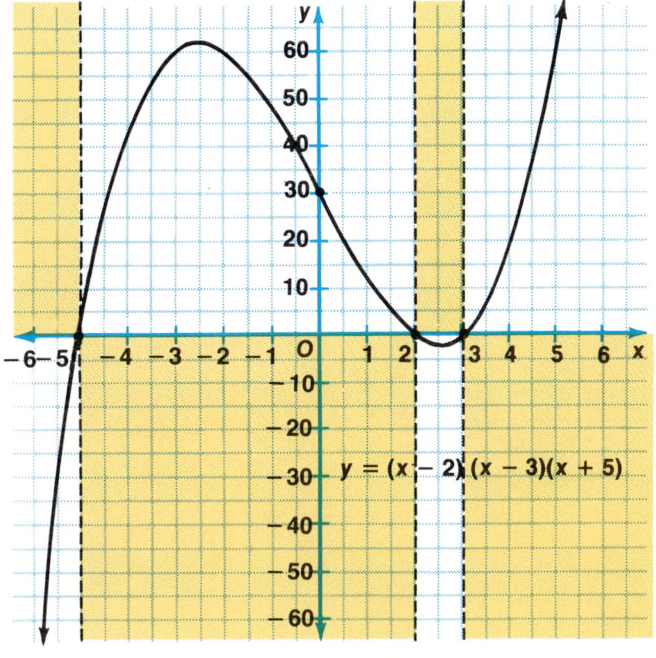

Notice that the vertical scale is different from the horizontal scale.
The vertical scale has been "condensed" so the graph will fit into the space provided.

Example

1 **Graph $f(x) = -x^3 + 2x^2 + 4x - 8$.**

First, factor the polynomial.

$$f(x) = -(x - 2)^2(x + 2)$$

1. The function has two zeros, 2 and -2.
2. The function has positive values when x is less than -2.
3. The function has negative or zero values when x is greater than -2.

x	-4	-3	-2	-1	0	1	2	3	4
$f(x)$	72	25	0	-9	-8	-3	0	-5	-24

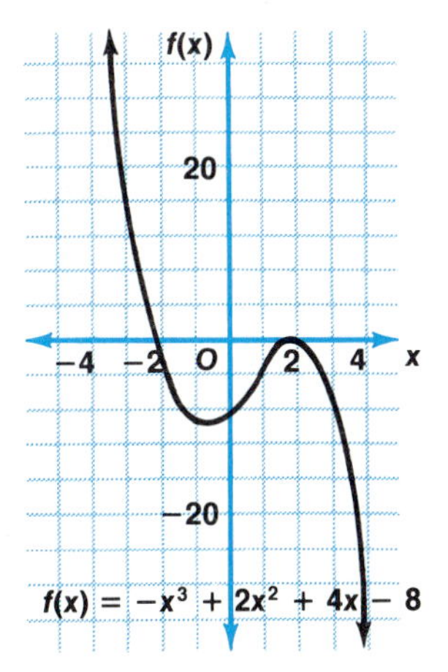

In general, the graph of a cubic polynomial function has a *sideways* S shape. Point A on each graph is called a **relative maximum** since there are no nearby points that have a y-coordinate greater than the y-coordinate of point A. Likewise, each point B is called a **relative minimum**. You can use this information to help graph functions that have imaginary zeros.

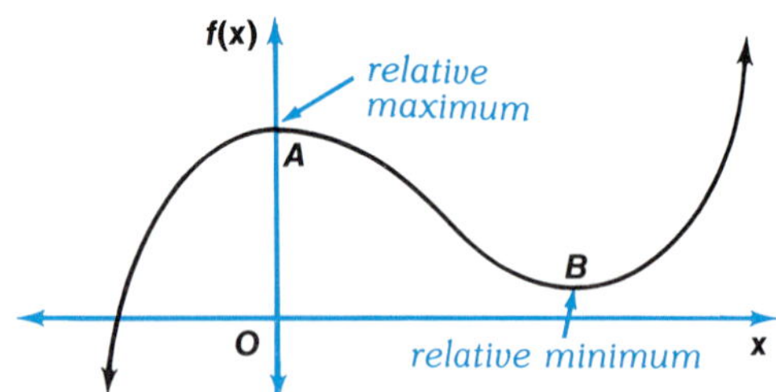

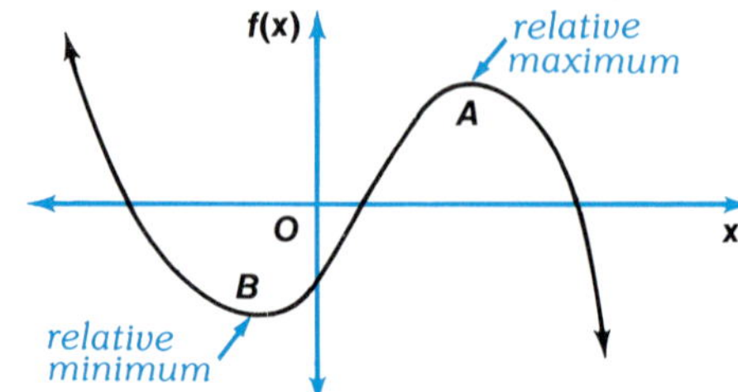

$f(x) = ax^3 + bx^2 + cx + d$, a positive

$f(x) = ax^3 + bx^2 + cx + d$, a negative

Example

2 **Graph $f(x) = x^3 - 6x - 9$.**

$f(x) = x^3 - 6x - 9$
$f(x) = (x - 3)(x^2 + 3x + 3)$

The function has one real zero, 3.

The graph of this cubic polynomial function has a *sideways* S shape. Thus, it has one relative maximum and one relative minimum.

The values of $f(-1.5)$ and $f(1.5)$ were computed to approximate the maximum and minimum more closely.

x	$f(x)$
-3	-18
-2	-5
-1.5	-3.375
-1	-4
0	-9
1	-14
1.5	-14.625
2	-13
3	0
4	31

-2, -1.5, -1 indicates a relative maximum

1, 1.5, 2 indicates a relative minimum

$\longleftarrow$ 3 is a zero.

$f(x) = x^3 - 6x - 9$

Written Exercises

Graph each function. Use a calculator as necessary to help find points on the graph.

1. $f(x) = x^3$
2. $f(x) = x^6$
3. $f(x) = 4x^6$
4. $f(x) = 3x^5$
5. $f(x) = (x - 1)(x - 2)(x + 2)$
6. $f(x) = (x + 4)(x - 1)(x + 1)$
7. $f(x) = (x - 2)^2(x + 3)$
8. $f(x) = (x - 3)^2(x + 1)$
9. $f(x) = x^3 - x$
10. $f(x) = -x^3 - x$
11. $f(x) = x^3 - x^2 - 8x + 12$
12. $f(x) = x^4 - 81$
13. $f(x) = x^4 - 10x^2 + 9$
14. $f(x) = x^3 + 5$
15. $f(x) = 15x^3 - 16x^2 - x + 2$
16. $f(x) = x^3 - 3x - 4$
17. $f(x) = -x^3 - 13x - 12$
18. $f(x) = -x^3 - 4x^2 - 8x - 8$

Study the mathematical terms used in the statements below.

$x^3 - 7x + 6$ is a **polynomial expression**, or a **polynomial**. Its factors are $(x - 2)$, $(x - 1)$, and $(x + 3)$.

When a polynomial is divided by one of its factors, the remainder is zero.

$x^3 - 7x + 6 = 0$ is a polynomial equation.

The **factored form** of the equation is
$$(x - 2)(x - 1)(x + 3) = 0.$$
Its **roots**, or **solutions**, are 2, 1, and -3.

When a root of an equation is substituted for the variable, the two sides are equal.

$f(x) = x^3 - 7x + 6$ is a **function**, or **polynomial function**. The **zeros** of the function are 2, 1, and -3.

When a zero of a function is substituted for the variable, the result is zero.

The function $f(x) = x^3 - 7x + 6$ may be expressed as $y = x^3 - 7x + 6$. When a value is substituted for x, exactly one value is obtained for y. Each ordered pair (x, y) is a **solution** to the equation $y = x^3 - 7x + 6$. For example, when x is -1, the value of y is 12. So, the ordered pair $(-1, 12)$ is a solution. Some other solutions are $(2, 0)$, $(1, 0)$, and $(-3, 0)$.

We do not use the term root for solutions that are ordered pairs.

Exercises

Choose the correct term in parentheses to complete each statement.

1. $x - 2$ is a (*factor*, *root*) of the polynomial $x^2 - 4x + 4$.

2. 2 is a (*root*, *zero*) of the function $f(x) = x^2 - 4x + 4$.

3. $(2, 0)$ is a (*solution*, *root*) of the equation $y = x^2 - 4x + 4$.

4. 2 is a (*root*, *zero*) of $x^2 - 4x + 4 = 0$.

5. $y - 6$ is a (*factor*, *solution*) of $y^2 - 5y - 6$.

6. The zeros of the (*function*, *expression*) $f(y) = y^2 - 5y - 6$ are 6 and -1.

7. The roots of the (*polynomial*, *equation*) $y^2 - 5y - 6 = 0$ are 6 and -1.

8. 7 is a (*factor*, *zero*) of $f(k) = k - 7$.

9. 7 is a (*solution*, *factor*) of $k - 7 = 0$.

10. $(3, -4)$ is a (*solution*, *root*) of $y = k - 7$.

11. $f(k) = k - 7$ is a(n) (*function*, *expression*).

12. The factors of the (*equation*, *expression*) $x^2 - 1$ are $x - 1$ and $x + 1$.

13. The roots of the (*function*, *equation*) $x^2 - 1 = 0$ are 1 and -1.

14. The equation $y = x^2 - 1$ has two (*roots*, *zeros*).

10-7 Composition of Functions

Scientists often use the Kelvin temperature scale. Kelvin temperature readings and Celsius temperature readings are related in the following way.

$$K = C + 273$$

Celsius readings and Fahrenheit readings are related in the following way.

$$C = \frac{5}{9}(F - 32)$$

Using these equations, you can write a new equation that shows how Kelvin readings and Fahrenheit readings are related.

$$
\begin{aligned}
K &= C + 273 \\
&= \frac{5}{9}(F - 32) + 273 \qquad \text{\textit{Substitute } } \tfrac{5}{9}(F - 32) \text{ \textit{for } } C. \\
&= \frac{5}{9}F + \frac{2297}{9}
\end{aligned}
$$

The above example illustrates **composition of functions**.

<table>
<tr>
<td>

Suppose f and g are functions such that the range of g is a subset of the domain of f. Then the composite function, $f \circ g$, can be described by the equation

$$[f \circ g](x) = f[g(x)].$$

</td>
<td>

Composition of Functions

</td>
</tr>
</table>

The symbol $f \circ g$ is read "f composition g", "the composite of f and g", or "f of g".

Examples

1 **If $f(x) = x + 273$ and $g(x) = \frac{5}{9}(x - 32)$, find $[f \circ g](x)$.**

$$
\begin{aligned}
[f \circ g](x) &= f[g(x)] \\
&= f\left[\frac{5}{9}(x - 32)\right] \qquad \text{\textit{Substitute } } \tfrac{5}{9}(x - 32) \text{ \textit{for } } g(x). \\
&= \left[\frac{5}{9}(x - 32)\right] + 273 \qquad \text{\textit{Evaluate } } f \text{ \textit{when x is } } \tfrac{5}{9}(x - 32). \\
&= \frac{5}{9}x + \frac{2297}{9} \qquad \text{\textit{Simplify.}}
\end{aligned}
$$

2 **If $f(x) = x^2 + 3$ and $h(x) = 2x - 1$, find $[f \circ h](2)$.**

$$
\begin{aligned}
[f \circ h](2) &= f[h(2)] \\
&= f[2(2) - 1] \qquad \text{\textit{Substitute } } 2(2) - 1 \text{ \textit{for } } h(2). \\
&= f[3] \qquad \text{\textit{Simplify.}} \\
&= 12 \qquad \text{\textit{Evaluate } } f \text{ \textit{when x is 3.}}
\end{aligned}
$$

3 If $f(x) = x + 7$ and $g(x) = 3 + 2x$, find $f[g(x)]$ and $g[f(x)]$.

$$f[g(x)] = f[3 + 2x] \qquad\qquad g[f(x)] = g[x + 7]$$
$$= (3 + 2x) + 7 \qquad\qquad\qquad = 3 + 2(x + 7)$$
$$= 2x + 10 \qquad\qquad\qquad\qquad = 2x + 17$$

Mappings can be used to show composition of functions. Suppose $f = \{(1, 2), (2, 3), (3, 4)\}$ and $g = \{(2, 3), (3, 1), (4, 2)\}$.

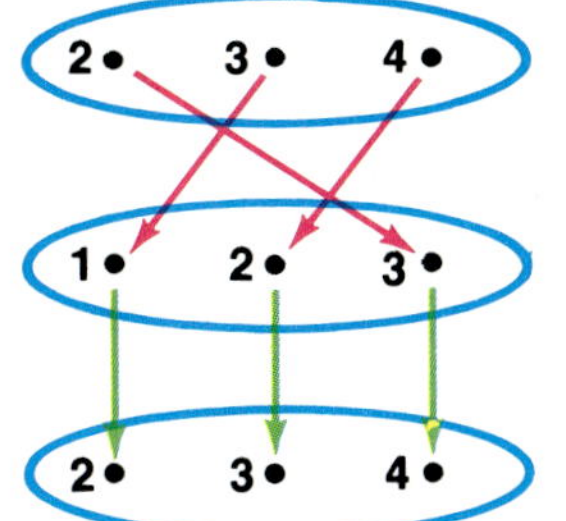

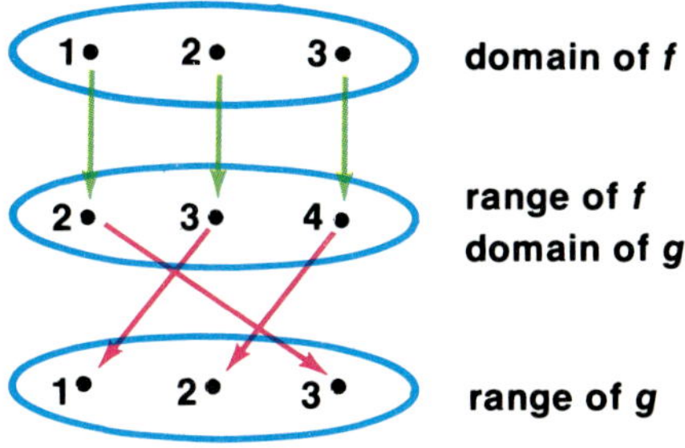

$$f \circ g = \{(2, 4), (3, 2), (4, 3)\} \qquad\qquad g \circ f = \{(1, 3), (2, 1), (3, 2)\}$$

4 If $f = \{(1, 5), (2, 5), (4, 3)\}$ and $g = \{(3, 1), (5, 4)\}$, find $f \circ g$ and $g \circ f$.

$$f \circ g = \{(3, 5), (5, 3)\} \qquad\qquad g \circ f = \{(1, 4), (2, 4), (4, 1)\}$$

Given two functions h and k, the composite functions $h \circ k$ or $k \circ h$ may not even exist. For example, let $h = \{(2, 4), (4, 6), (6, 8), (8, 10)\}$ and $k = \{(4, 5), (6, 5), (8, 12), (10, 12)\}$. The range of k is *not* a subset of the domain of h. So, $h \circ k$ *does not exist*. The range of h is a subset of the domain of k. So, $k \circ h$ *does exist*.

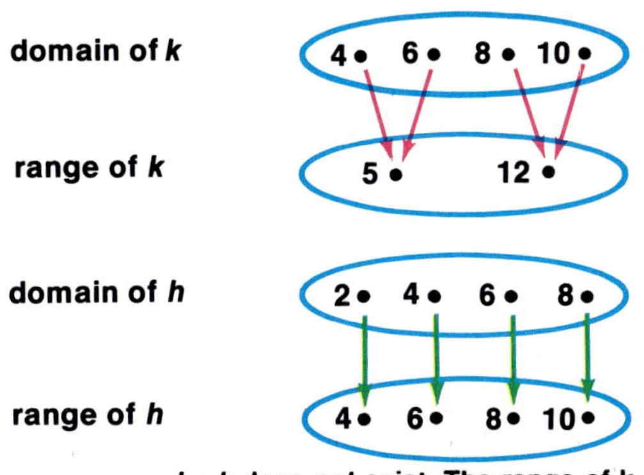

h ∘ k does *not* exist. The range of k is not a subset of the domain of h.

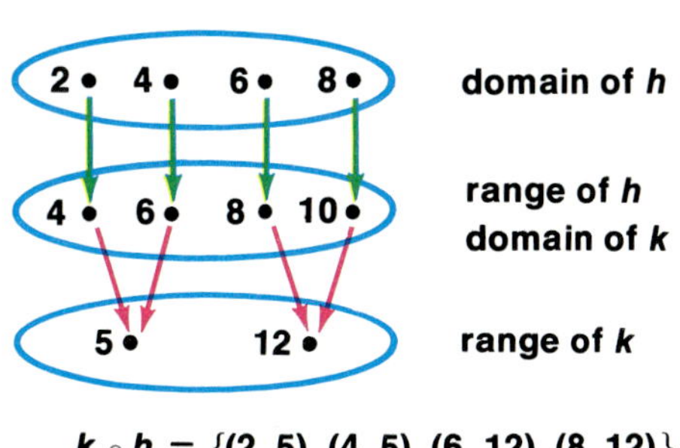

$$k \circ h = \{(2, 5), (4, 5), (6, 12), (8, 12)\}$$

For each function f, find $f(2)$, $f(-2)$, and $f(0)$.

1. $f(x) = x - 3$
2. $f(x) = x^2 + 4$
3. $f(x) = x^2 - 2x + 1$
4. $f(x) = |x + 2|$
5. $f(x) = |x| + 2$
6. $f(x) = x^3 - x^2 + x - 1$
7. $f(x) = x^4 - 6$
8. $f(x) = x^2 - 8x + 4$
9. $f(x) = x^5 - x^2 + 2$
10. $f(x) = x^2 - x^3$
11. $f = \{(0, 3), (2, 4), (-2, 9)\}$
12. $f = \{(0, 2), (-2, 2), (5, 2)\}$

Written Exercises

For each pair of functions, f and g, find $[f \circ g](3)$ and $[g \circ f](3)$.

13. $f(x) = x + 2$
    $g(x) = x - 1$
14. $f(x) = x^2 + 8$
    $g(x) = x - 3$
15. $f(x) = x^3 - 1$
    $g(x) = x + 1$
16. $f(x) = x$
    $g(x) = x$
17. $f(x) = 2x^2 + 1$
    $g(x) = x^2 - 1$
18. $f(x) = x^2$
    $g(x) = x^3$

For each pair of functions, f and g, find $f[g(x)]$ and $g[f(x)]$.

19. $f(x) = 2x + 1$
    $g(x) = x - 3$
20. $f(x) = 3x - 4$
    $g(x) = 2x + 5$
21. $f(x) = x^2 + 3$
    $g(x) = 2x - 1$
22. $f(x) = x + 2$
    $g(x) = 3x^2 - 1$
23. $f(x) = -x^2 - 8$
    $g(x) = x^2 - 1$
24. $f(x) = x + 2$
    $g(x) = x - 2$

For each pair of functions, f and g, find $f[g(-1)]$ and $g[f(-1)]$.

25. $f(x) = x - 1$
    $g(x) = x + 1$
26. $f(x) = x^2 + 2x + 1$
    $g(x) = -2x^2 - 1$
27. $f(x) = 3x^2 + 2$
    $g(x) = x - 3$
28. $f(x) = 2x^2 + 4x^3 + 1$
    $g(x) = x^2 + 1$
29. $f(x) = x - 8$
    $g(x) = |x|$
30. $f(x) = |x + 1|$
    $g(x) = |x + 1|$

If $f(x) = x^2$, $g(x) = 3x$, and $h(x) = x - 1$, find each value.

31. $[f \circ g](1)$
32. $[g \circ f](1)$
33. $[h \circ f](3)$
34. $[f \circ h](3)$
35. $g[f(-2)]$
36. $f[h(-3)]$
37. $g[h(-2)]$
38. $h[g(-2)]$
39. $f\left[h\left(-\frac{1}{2}\right)\right]$
40. $g\left[f\left(-\frac{1}{2}\right)\right]$
41. $f[h(\sqrt{2} + 3)]$
42. $f[g(1 + \sqrt{2})]$
43. $f[g(x)]$
44. $g[h(x)]$
45. $[f \circ (g \circ h)](x)$

Express $g \circ f$ and $f \circ g$, if they exist, as sets of ordered pairs.

46. $f = \{(2, 1), (3, 4), (6, -2)\}$
    $g = \{(1, 5), (4, -7), (-2, -3)\}$
47. $f = \{(3, 8), (4, 0), (6, 3), (7, -1)\}$
    $g = \{(8, 6), (0, 4), (3, 6), (-1, -8)\}$
48. $f = \{(0, 2), (1, 1), (3, 3), (4, -1)\}$
    $g = \{(0, 0), (1, 1), (2, 4), (3, 9)\}$
49. $f = \{(1, 0), (-1, 7), (5, 0), (9, 1)\}$
    $g = \{(6, 5), (0, -8), (7, -2), (1, -3)\}$

10-8 Inverse Functions and Relations

You have learned that two numbers are additive inverses if their sum is 0, the additive identity. Two numbers are multiplicative inverses if their product is 1, the multiplicative identity. The function $I(x) = x$ is called the **identity function**, since for any function f, $[f \circ I](x) = f(x)$ and $[I \circ f](x) = f(x)$. Two functions are **inverse functions** if both their compositions are the identity function.

Suppose $f(x) = 3x - 2$ and $g(x) = \dfrac{x + 2}{3}$. You can determine if they are inverse functions by finding both their compositions.

$$[f \circ g](x) = f[g(x)] \qquad\qquad [g \circ f](x) = g[f(x)]$$

$$= f\left[\frac{x + 2}{3}\right] \qquad\qquad = g[3x - 2]$$

$$= 3\left(\frac{x + 2}{3}\right) - 2 \qquad\qquad = \frac{(3x - 2) + 2}{3}$$

$$= x \quad [f \circ g](x) = I(x) \qquad\qquad = x \quad [g \circ f](x) = I(x)$$

Since both their compositions are the identity function, they are inverse functions.

Two functions, f and g, are inverse functions if and only if both their compositions are the identity function. That is, $$[f \circ g](x) = x \quad \text{and} \quad [g \circ f](x) = x.$$	*Definition of Inverse Functions*

A special notation often is used to show that two functions f and g are inverse functions.

$$g = f^{-1} \text{ and } f = g^{-1}$$

The notation f^{-1} is read "f inverse" or "the inverse of f." The -1 is not an exponent.

The ordered pairs of a function and its inverse are related in a special way. Consider the function $f(x) = 3x - 2$ and its inverse $f^{-1}(x) = \dfrac{x + 2}{3}$.

$$f(5) = 3 \cdot 5 - 2 \qquad\qquad f^{-1}(13) = \frac{13 + 2}{3}$$

$$= 13 \qquad\qquad\qquad = 5$$

The ordered pair $(5, 13)$ belongs to f.

The ordered pair $(13, 5)$ belongs to f^{-1}.

Suppose f and f^{-1} are inverse functions. Then $f(a) = b$ if and only if $f^{-1}(b) = a$.	*Property of Inverse Functions*

Thus, the inverse of a function can be found by reversing the order of each pair in the given function.

You may recall that any *relation* is a set of ordered pairs. The order of each pair in a relation can be reversed to create a new relation. These two relations are **inverse relations**.

Remember that a function is a special type of relation.

> Two relations are inverse relations if and only if whenever one relation contains the element (a, b), the other relation contains the element (b, a).

Definition of Inverse Relation

Example

1 **Find the inverse of each relation and determine if the inverse is a function.**

a. $\{(-1, -3), (3, 2), (-1, 5)\}$

The inverse is
$\{(-3, -1), (2, 3), (5, -1)\}$.
The inverse is a function.

b. $\{(1, 2), (2, 7), (3, 12), (4, 2)\}$

The inverse is
$\{(2, 1), (7, 2), (12, 3), (2, 4)\}$.
The inverse is not a function.

Suppose you wish to find the inverse of a function, f. By interchanging the variables in the equation for f, the order of each pair in f is interchanged. Thus, you can obtain the ordered pairs in the inverse and the equation for the inverse.

Example

2 **Find the inverse of $f(x) = 3x - 5$. Then show that f and its inverse are inverse functions.**

Rewrite $f(x)$ as $y = 3x - 5$. Then interchange the variables and solve for y.

$x = 3y - 5$ *Interchange variables x and y in $y = 3x - 5$.*

$3y = x + 5$ *Solve for y.*

$y = \dfrac{x + 5}{3}$ *Notice that the equation defines a function.*

Thus, the inverse of f is the function $f^{-1}(x) = \dfrac{x + 5}{3}$.

Now show that the compositions of f and f^{-1} are identity functions.

$[f \circ f^{-1}](x) = f[f^{-1}(x)]$

$\qquad = f\left[\dfrac{x + 5}{3}\right]$

$\qquad = 3\left(\dfrac{x + 5}{3}\right) - 5$

$\qquad = x \qquad [f \circ f^{-1}](x) = I(x)$

$[f^{-1} \circ f](x) = f^{-1}[f(x)]$

$\qquad = f^{-1}[3x - 5]$

$\qquad = \dfrac{(3x - 5) + 5}{3}$

$\qquad = x \qquad [f^{-1} \circ f](x) = I(x)$

Not all functions have inverses that are functions.

Example

3   **Find the inverse of $f(x) = x^2 + 3$.**

$$f(x) = x^2 + 3$$
$$y = x^2 + 3 \qquad \text{\textit{Rewrite the function f(x) using y.}}$$
$$\textit{inverse:} \quad x = y^2 + 3 \qquad \text{\textit{Interchange variables x and y.}}$$
$$y^2 = x - 3 \qquad \text{\textit{Solve for y.}}$$
$$y = \pm\sqrt{x - 3}$$

Thus, the inverse of f is the *equation* $y = \pm\sqrt{x - 3}$.

This equation *does not* define a function. For a given value of x, there is more than one value of y. For example, if x is 7, then y is 2 or -2.

The following example shows how the graphs of a function and its inverse are related.

Example

4   **Find the inverse of $f(x) = 2x + 1$. Then graph $f(x)$ and its inverse on the same coordinate plane.**

$$f(x) = 2x + 1$$
$$y = 2x + 1 \qquad \text{\textit{Rewrite the function f(x) using y.}}$$
$$\textit{inverse:} \quad x = 2y + 1 \qquad \text{\textit{Interchange the variables.}}$$
$$2y = x - 1 \qquad \text{\textit{Solve for y.}}$$
$$y = \frac{x - 1}{2} \qquad \text{\textit{This equation defines a function.}}$$

Thus, the inverse of f is the function $f^{-1}(x) = \frac{x - 1}{2}$.

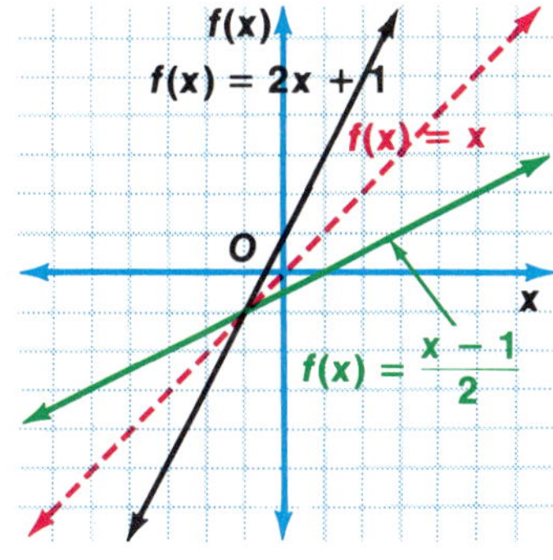

Now graph both $f(x)$ and $f^{-1}(x)$. Suppose the plane containing the graphs could be folded along the line $f(x) = x$. Then the graphs would coincide.

In general, the graphs of a relation and its inverse are mirror images, or reflections, of each other with respect to the graph of the identity function, $I(x) = x$.

Exploratory Exercises

Find the inverse of each relation and determine whether the inverse is a function.

1. $\{(3, 1), (2, 4), (1, 5)\}$
2. $\{(3, 2), (4, 2)\}$
3. $\{(3, 8), (4, -2), (5, -3)\}$
4. $\{(-1, -2), (-3, -2), (-1, -4), (0, 6)\}$
5. $\{(-3, 1), (2, 4), (2, 8)\}$
6. $\{(4, -2), (3, 7), (5, 7), (3, 8)\}$
7. $\{(-5, 3), (-2, 7), (0, 3), (6, 11)\}$
8. $\{(1, 3), (1, 1), (1, -1), (1, -3)\}$

Written Exercises

Find the inverse of each function.

9. $y = 2x$
10. $f(x) = x + 2$
11. $g(x) = -6x - 5$
12. $y = 3$
13. $f(x) = 0$
14. $y = 3x - 7$
15. $g(x) = \frac{1}{2}x + 4$
16. $h(x) = \frac{2}{5}x - 3$
17. $f(x) = -3x - 1$
18. $f(x) = x^2$
19. $y = x^2 - 4$
20. $g(x) = (x - 4)^2$

Graph each function and its inverse.

21. $f = \{(2, 1), (3, -4), (0, 1)\}$
22. $f = \left\{\left(\frac{1}{2}, 3\right), (0, -2), (7, 6)\right\}$
23. $y = x$
24. $y = -2x - 1$
25. $y = 3x$
26. $f(x) = x + 2$
27. $f(x) = \frac{3x + 1}{2}$
28. $y = \frac{x - 3}{5}$
29. $y = x^2 + 1$
30. $f(x) = (x - 4)^2$
31. $g(x) = (x + 2)^2 - 3$

Sketch the graph of the inverse of each relation. Then determine if the inverse is a function.

32. 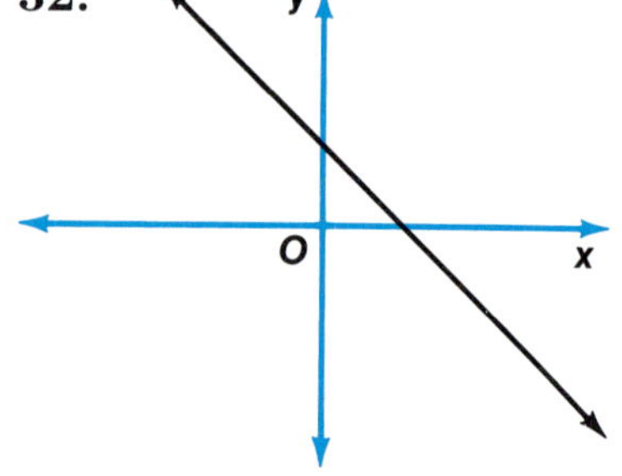

33. 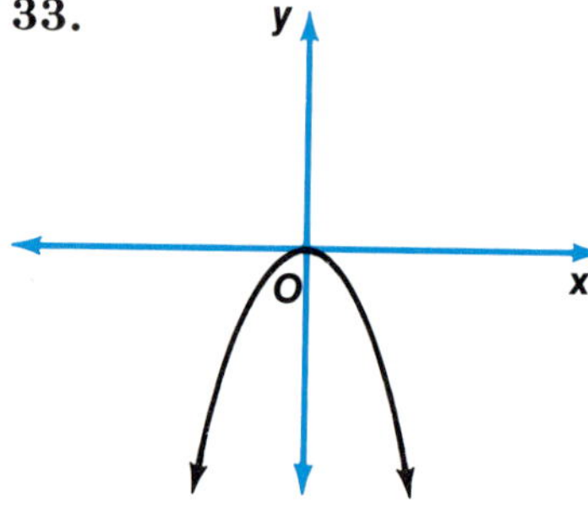

34. 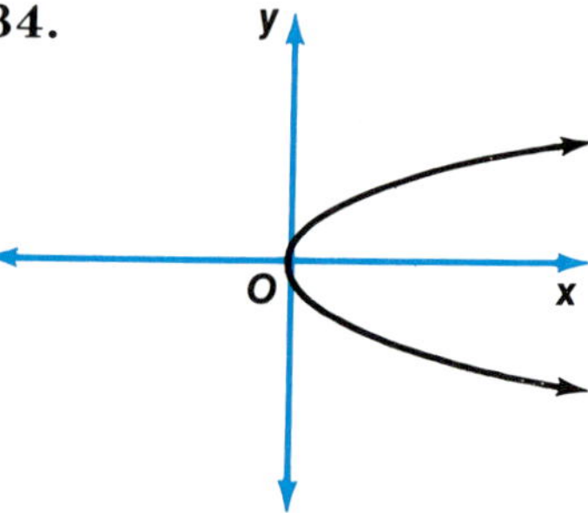

35. 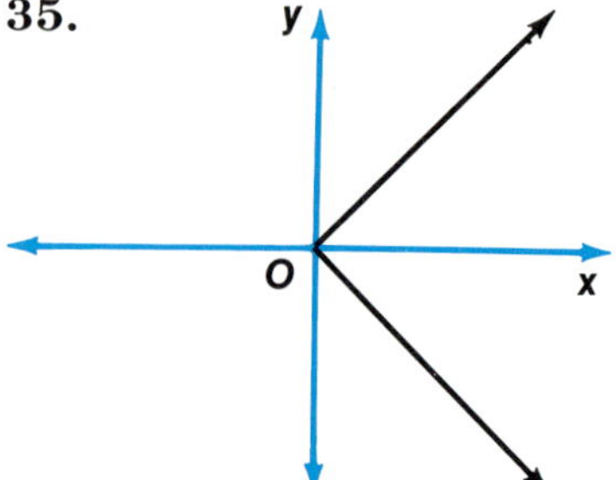

36. 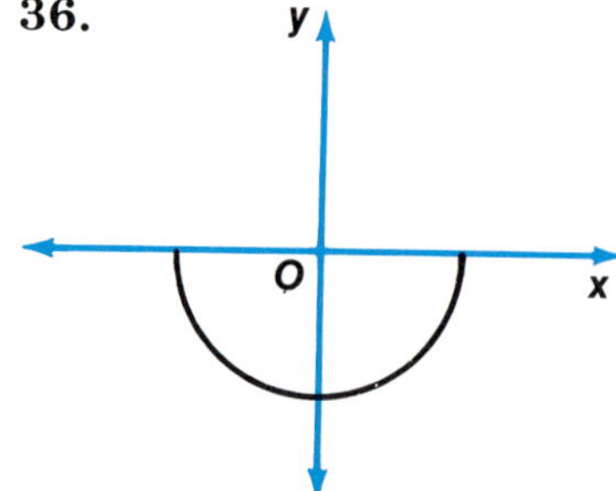

37. 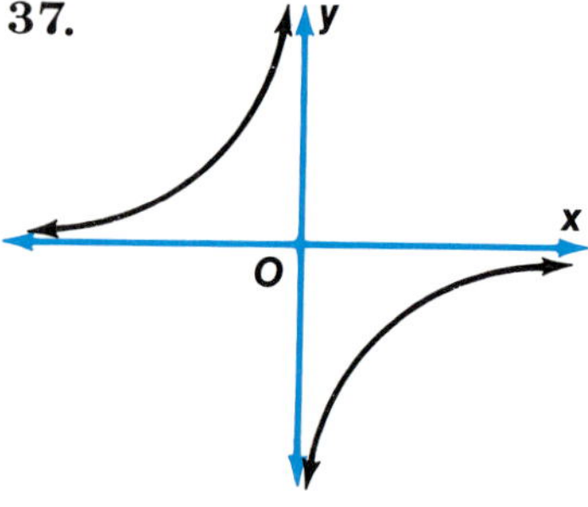

Determine whether each pair of functions are inverse functions.

38. $f(x) = x + 1$
$\quad\;\; f(x) = x - 1$

39. $y = x + 4$
$\quad\;\; y = x - 4$

40. $y = -2x + 3$
$\quad\;\; y = 2x - 3$

41. $f(x) = x$
$\quad\;\; f(x) = -x$

42. $y = 2x + 1$
$\quad\;\; y = \dfrac{x - 1}{2}$

43. $f(x) = 4x - 5$
$\quad\;\; f(x) = \dfrac{x + 5}{4}$

44. $f(x) = x - \dfrac{1}{2}$
$\quad\;\; f(x) = 2x + 1$

45. $y = \dfrac{x}{3} - \dfrac{2}{3}$
$\quad\;\; y = 3x - 2$

46. $f(x) = \dfrac{x}{5} + \dfrac{7}{5}$
$\quad\;\; f(x) = -5x + 7$

Solve each problem. In each case, assume that f and f^{-1} are functions.

47. If $f(3) = 4$, find $f^{-1}(4)$.

48. If $f(0.5) = 6$, find $f^{-1}(6)$.

49. If $f(a) = b$, find $f^{-1}(b)$.

50. If $f(r) = s$, find $f^{-1}(s)$.

51. If $f(a + 1) = 2$, find $f^{-1}(2)$.

52. If $f(m + n) = p$, find $f^{-1}(p)$.

mini-review

Solve each equation or inequality.

1. $6x^2 + x - 12 = 0$

2. $b^2 - 7b = -17$

3. $3y^2 = 6 - 5y$

4. $x^2 \le 4x + 60$

5. $z^2 + 12z < -36$

6. $n^2 - 11n > 0$

Find the value of the discriminant for each quadratic equation. Describe the nature of the roots. If the roots are real, tell whether they are rational or irrational.

7. $x^2 + 4x + 13 = 0$

8. $y^2 - 6y = 4$

State whether the graph of each equation is a parabola, a circle, an ellipse, or a hyperbola.

9. $3x - y^2 = 2x + 1$

10. $3x^2 + 3y^2 - 6x + 12y = 32$

11. $10x^2 + 40 = 4y^2 - 8y$

12. $x^2 + 7y^2 - 8x = 11$

13. Find the solutions for the system $5x^2 - 4y^2 = 121$ and $3x + 4y = 11$.

14. If $f(x) = 2x^2 + 3x - 7$, find $f(-2)$ and $f(x + 3)$.

15. Find all zeros of $g(x) = x^4 + x^3 - 6x^2 + 16x + 48$.

Approximate to the nearest tenth the real zeros of each function.

16. $2x^3 - x^2 - x - 6$

17. $h(x) = x^4 - 26x^2 - 75x - 56$

Graph each equation.

18. $y = 3(x - 2)^2 - 1$

19. $y = -x^2 - 2x + 3$

20. $9x^2 + 4y^2 = 36$

21. $25(y - 2)^2 - 9(x + 1)^2 = 225$

22. $f(x) = (x - 1)(x + 2)(x - 4)$

23. $p(x) = x^3 + 2x^2 - 2x - 12$

Finding the Real Zeros of a Polynomial Function

The computer program below can be used to approximate to the nearest hundredth the real zeros of a polynomial function.

This program approximates the real zeros of $f(x) = x^4 - 9x^2 + 20$. You determine the range of x values to be tested.

Enter and run this program using the range -4 to 4.

```
RUN
F(X) = X ^ 4 - 9 * X ^ 2 + 20
ENTER THE LOWEST VALUE OF X TO
BE TESTED: -4
ENTER THE HIGHEST VALUE OF X TO
BE TESTED: 4

-2.24 IS A ZERO OF F(X).
-2 IS A ZERO OF F(X).
2 IS A ZERO OF F(X).
2.24 IS A ZERO OF F(X).
```

The zeros of any function can be approximated using this program by changing statements 10 and 230.

Exercises

Make the necessary changes to the computer program to approximate to the nearest hundredth the real zeros of each function.

1. $f(x) = x^4 + 2x^3 - 7x^2 - 8x + 12$
2. $g(x) = x^4 + 5x^3 - 37x^2 - 5x + 36$
3. $h(x) = 8x^3 - 36x^2 + 22x + 21$
4. $p(x) = x^4 + x^3 + 5x^2 + 2x + 12$
5. $r(x) = 4x^4 - 35x^3 + 78x^2 + 28x - 165$
6. $f(x) = x^3 + 2x^2 - 3x - 7$
7. $g(x) = 3x^4 - 7x^3 + 2x^2 - 3x + 6$
8. $h(x) = x^5 - x^3 - x + 1$
9. $p(x) = x^5 + 4x^4 - x^3 - 9x^2 + 3$
10. $r(x) = x^4 - 4x^3 - 4x^2 + 24x - 6$
11. $f(x) = x^5 - 3x^4 + x^3 + 5x^2 - 6x - 1$
12. $g(x) = x^6 - 4x^5 + x^3 + 32x^2 - 72$

```
10   PRINT "F(X) = X ^ 4 - 9 * X ^ 2 + 20"
20   INPUT "ENTER LOWEST VALUE OF X TO
        BE TESTED: "; X1
30   INPUT "ENTER HIGHEST VALUE OF X TO
        BE TESTED: "; X2
40   LET Y = X1
45   PRINT
50   LET I = 1
60   LET X = Y
65   GOSUB 230
70   IF F = 0 THEN 210
80   IF Y = X2 THEN 260
90   LET FA = F
100  LET X = INT((Y + I) / I + .0005) * I
105  GOSUB 230
110  IF F <> 0 THEN 140
120  IF I < 1 THEN 140
130  LET X = X - .001
135  GOSUB 230
140  IF F * FA < 0 THEN 160
150  LET Y = INT ((Y + I) / I + .0005) * I
155  GOTO 60
160  LET I = I / 10
170  IF I > .001 THEN 100
180  LET X = (Y + X) / 2
190  GOSUB 230
200  IF F * FA < 0 THEN 210
205  LET Y = 2 * X - Y
210  LET C = 1
215  PRINT Y; " IS A ZERO OF F(X)."
220  LET Y = X + .0049
225  GOTO 50
230  LET F = X ^ 4 - 9 * X ^ 2 + 20
240  LET F = INT (10000 * F + .5) / 10000
250  RETURN
260  IF C = 1 THEN 280
270  PRINT "NO ZEROS FOUND BETWEEN ";
        X1; " AND "; X2; "."
280  END
```

Chapter Summary

1. A polynomial in one variable, x, is an expression of the form $a_0x^n + a_1x^{n-1} + \cdots + a_{n-2}x^2 + a_{n-1}x + a_n$. The coefficients $a_0, a_1, a_2, \ldots, a_n$ represent real numbers, a_0 is not zero, and n represents a nonnegative integer.   (345)

2. A polynomial function can be described by an equation of the form $p(x) = a_0x^n + a_1x^{n-1} + \ldots + a_{n-1}x + a_0$. The coefficients $a_0, a_1, \ldots, a_{n-1}$, and a_n represent real numbers, a_0 is not zero, and n represents a nonnegative integer.   (346)

3. The Remainder Theorem: If a polynomial $f(x)$ is divided by $x - a$, the remainder is the constant, $f(a)$, and $f(x) = q(x) \cdot (x - a) + f(a)$ where $q(x)$ is a polynomial with degree one less than the degree of $f(x)$.   (348)

4. The Factor Theorem: The binomial $x - a$ is a factor of the polynomial $f(x)$ if and only if $f(a) = 0$.   (349)

5. The Fundamental Theorem of Algebra: Every polynomial equation with degree greater than zero has at least one root in the set of complex numbers.   (353)

6. Complex Conjugates Theorem: Suppose a and b are real numbers with $b \neq 0$. Then, if $a + bi$ is a zero of a polynomial function, $a - bi$ is also a zero of the function.   (354)

7. Descartes' Rule of Signs: Suppose $p(x)$ is a polynomial whose terms are arranged in descending powers of the variable. The number of positive real zeros of $y = p(x)$ is the same as the number of changes in sign of the coefficients of the terms, or is less than this by an even number. The number of negative real zeros of $y = p(x)$ is the same as the number of changes in sign of the coefficients of the terms of $p(-x)$, or is less than this by an even number.   (355)

8. The Rational Zero Theorem: Let $f(x) = a_0x^n + a_1x^{n-1} + \cdots + a_{n-1}x + a_n$ represent a polynomial function with integral coefficients. If $\frac{p}{q}$ is a rational number in simplest form and a zero of $y = f(x)$, then p is a factor of a_n and q is a factor of a_0.   (358)

9. The Location Principle: Suppose $y = f(x)$ represents a polynomial function and a and b are two numbers such that $f(a) < 0$ and $f(b) > 0$. Then the function has at least one real zero between a and b.   (362)

10. Polynomial functions can be drawn on the graphics screen of a graphing calculator. The graphing calculator can also be used to approximate the zeros of a polynomial function.   (365–367)

11. Suppose f and g are functions such that the range of g is a subset of the domain of f. Then the composite function, $f \circ g$, can be described by the equation $[f \circ g](x) = f[g(x)]$.   (372)

12. Two functions, f and g, are inverse functions if and only if both their compositions are the identity function. That is,
$$[f \circ g](x) = x \text{ and } [g \circ f](x) = x. \quad (375)$$

13. Property of Inverse Functions: Suppose f and f^{-1} are inverse functions. Then $f(a) = b$ if and only if $f^{-1}(b) = a$.   (375)

14. Two relations are inverse relations if and only if whenever one relation contains (a, b), the other relation contains (b, a).   (376)

Chapter Review

10–1 **Find $p(-2)$ for each function p.**

1. $p(x) = 2x^3 + x^2 - 1$

2. $p(x) = 2x^4 - 3x^3 + 8$

Find $f(m + 2)$ for each function f.

3. $f(x) = x^2 + 3x - 7$

4. $f(x) = -4x^3 - 5$

Find $g(x + h) - g(x)$ for each function g.

5. $g(x) = 3x - 5$

6. $g(x) = x^2 - 2x + 4$

10–2 **Divide. Write your answer in the form $dividend = quotient \cdot divisor + remainder$.**

7. $(x^2 + 5x + 6) \div (x + 1)$

8. $(x^3 + 8x + 1) \div (x - 2)$

9. $(x^5 + 8x^3 + 2) \div (x + 2)$

10. $(x^4 + 3x^2 - 1) \div (x - 4)$

Given a polynomial and one of its factors, find the remaining factors of the polynomial.

11. $x^3 + 2x^2 - x - 2; x + 2$

12. $x^3 + 5x^2 + 8x + 4; x + 1$

13. $2x^3 - 13x^2 + 17x + 12; x - 3$

14. $x^4 - 6x^3 + 22x + 15; x + 1$

10–3 For each function, state the number of positive real zeros, negative real zeros, and imaginary zeros.

15. $f(x) = 2x^4 - x^3 + 5x^2 + 3x - 9$

16. $f(x) = -x^4 - x^2 - x - 1$

17. $f(x) = 7x^3 + 6x - 1$

18. $f(x) = x^4 + x^3 - 7x + 1$

Given a function and one of its zeros, find all the zeros of the function.

19. $f(x) = x^3 - 7x^2 + 17x - 15; \ 2 + i$

20. $g(x) = 2x^3 + x^2 + 46x - 78; \ -1 + 5i$

21. $h(x) = 3x^3 - 2x^2 - 5x - 6; \ 2$

22. $p(x) = x^4 - 5x^3 + 55x - 91; \ 3 - 2i$

10–4 Find all rational zeros of each function.

23. $f(x) = 2x^3 - 13x^2 + 17x + 12$

24. $g(x) = x^4 + 5x^3 + 15x^2 + 19x + 8$

Find all zeros of each function.

25. $f(x) = x^3 - 3x^2 - 10x + 24$

26. $g(x) = x^3 - 5x^2 + x + 7$

27. $h(x) = 6x^3 - 41x^2 + 58x - 15$

28. $p(x) = 2x^3 - 5x^2 - 28x + 15$

29. $r(x) = 12x^4 - 19x^3 + 9x - 2$

30. $t(x) = 4x^4 + 11x^3 + 10x^2 - 69x - 54$

10–5 Approximate to the nearest tenth the real zeros of each function.

31. $f(x) = x^3 - x^2 + 1$

32. $g(x) = 3x^4 - x^2 + x - 1$

33. $h(x) = x^4 - 2x^3 + 6x + 1$

34. $p(x) = 4x^3 + x^2 - 11x + 3$

10–6 Graph each function.

35. $f(x) = (x - 3)^2(x + 2)$

36. $f(x) = x^3 - 3x - 4$

37. $f(x) = 4x - x^3$

38. $f(x) = x^4 - 8x^2 + 16$

10–7 For each pair of functions, f and g, find $[f \circ g](3)$, $[g \circ f](4)$, and $[f \circ g](x)$.

39. $f(x) = 2x - 1$
$g(x) = 3x + 4$

40. $f(x) = |x - 2|$
$g(x) = 5x - 7$

41. $f(x) = x^2 + 2$
$g(x) = x - 3$

42. $f(x) = x^2 + 2x + 1$
$g(x) = 2x^2 - 1$

Express $f \circ g$ and $g \circ f$, if they exist, as sets of ordered pairs.

43. $f = \{(2, 1), (3, 2), (-1, 6)\}$
$g = \{(2, 2), (6, -1), (1, 3)\}$

44. $f = \{(-1, 9), (0, -3), (1, 1)\}$
$g = \{(-3, 1), (1, 0), (9, -2)\}$

10–8 Graph each function and its inverse.

45. $f(x) = 1 - 4x$

46. $f(x) = \dfrac{3x + 2}{2}$

47. $f(x) = (x - 6)^2$

Determine whether each pair of functions are inverse functions.

48. $f(x) = 3x - 4$
$g(x) = \dfrac{x - 4}{3}$

49. $f(x) = 4x - 5$
$g(x) = \dfrac{x + 5}{4}$

50. $f(x) = -2x - 3$
$g(x) = \dfrac{-x - 3}{2}$

Find $p(3)$ for each function p.

1. $p(x) = x^4 - x^3 + x - 1$

2. $p(x) = x^3 - 27$

Find $f(m - 1)$ for each function f.

3. $f(x) = 3x^2 - 4x + 5$

4. $f(x) = x^3 - x + 7$

Divide. Write your answer in the form *dividend = quotient · divisor + remainder*.

5. $(x^2 + 4x + 4) \div (x + 2)$

6. $(x^3 + 8x + 1) \div (x + 2)$

7. $(x^5 + 8x^3 + 2) \div (x - 2)$

8. $(x^4 + x^3 + x^2 + x + 1) \div (x + 1)$

9. If $x + 1$ is a factor of $x^3 - x^2 - 5x - 3$, find the remaining factors.

For each function, state the number of positive real zeros, negative real zeros, and imaginary zeros.

10. $f(x) = x^3 - x^2 - 14x + 24$

11. $g(x) = x^4 + x^3 - 9x^2 - 17x - 8$

12. Find all zeros of $f(x) = x^3 - 10x^2 + 34x - 40$ if $3 + i$ is one zero of $f(x)$.

Find all rational zeros of each function.

13. $f(x) = x^3 - 3x - 52$

14. $g(x) = 6x^3 + 4x^2 - 14x + 4$

Find all zeros of each function.

15. $f(x) = x^3 - 3x^2 - 53x - 9$

16. $g(x) = 4x^4 + x^2 + 11x - 6$

17. Approximate to the nearest tenth the real zeros of $f(x) = x^3 + 3x^2 - 2x + 1$.

Graph each function.

18. $f(x) = x^3 + 6x^2 + 6x - 4$

19. $f(x) = x^3 + 2x^2 - 8x - 21$

If $f(x) = 2x$ and $g(x) = x^2 - 1$, find each value.

20. $[f \circ g](-2)$

21. $g[f(3)]$

22. $[g \circ f](x)$

Find the inverse of each function.

23. $f(x) = 3x - 4$

24. $f(x) = x^2 + 2$

25. Show that $f(x) = 2x - 3$ and $g(x) = \dfrac{x + 3}{2}$ are inverse functions.

26. Graph $f(x) = 3x - 2$ and its inverse.

27. Graph $f(x) = -x^3 + 2x^2 + 4x - 10$. Then approximate to the nearest tenth the real zero of $f(x)$.

The test questions on this page deal with co-ordinates and geometry. The figures shown may not be drawn to scale.

Directions: Choose the one best answer. Write A, B, C, or D.

1. What is the length of a diagonal of a square, if the area is $25x^2$?

   (A) $5x$ **(B)** $5x\sqrt{2}$ **(C)** $10x$ **(D)** $5x^2$

2. What percent of rectangle $ACDF$ is shaded?

   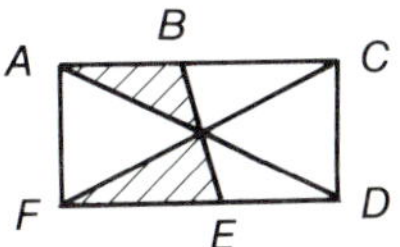

   (A) 25   **(B)** 30   **(C)** $33\frac{1}{3}$ **(D)** 50

3. In $1\frac{1}{2}$ hours, the minute hand of a clock rotates through an angle of how many degrees?

   (A) 60   **(B)** 180   **(C)** 540   **(D)** 720

4. The three straight lines intersect at the same point.

   $a + b =$

   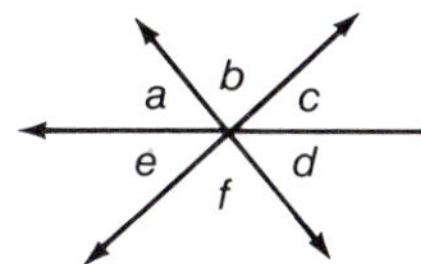

   (A) $b + f$ **(B)** $e + d$
   (C) $f + d$ **(D)** $c + e$

5. Line p is parallel to line q.

   $c - a =$

   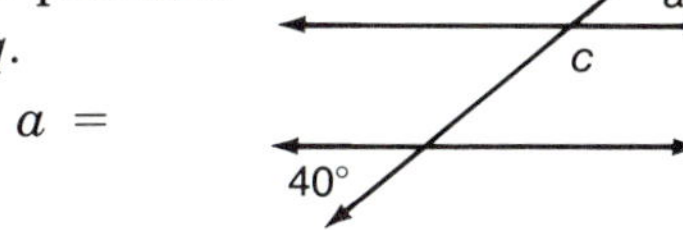

   (A) 0   **(B)** 100   **(C)** 120   **(D)** 150

6. 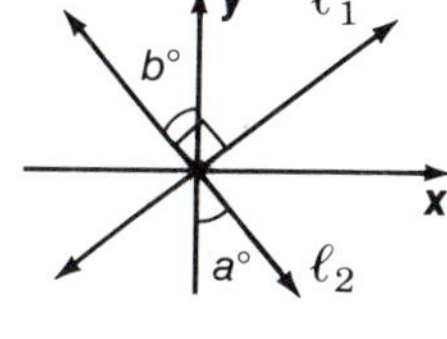

   On line segment PR in the figure above, $PQ = 8$ and $QR = 6$. What is the length of the segment joining the midpoints of segments PQ and QR?

   (A) 3   **(B)** 4   **(C)** 7   **(D)** 14

7. 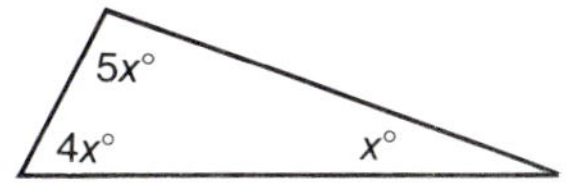

   In the triangle above, $4x =$

   (A) 18   **(B)** 32   **(C)** 40   **(D)** 72

8. 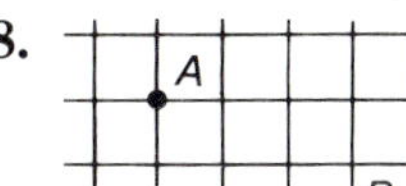

   Above is a section of a graph without the x and y axes. If A has coordinates $(4, 6)$, state the coordinates of B.

   (A) $(4, 7)$ **(B)** $(4, 1)$
   (C) $(9, 2)$ **(D)** $(7, 4)$

9. 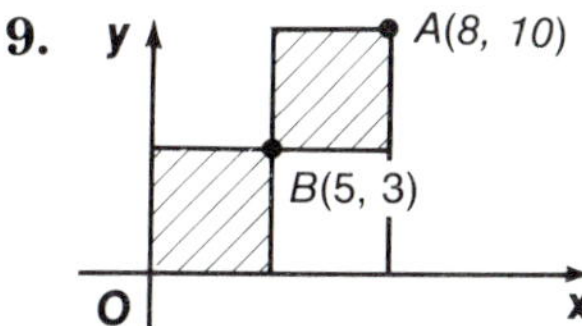

   Point A in the figure above has coordinates $(8, 10)$ and point B has coordinates $(5, 3)$. What is the total area of the shaded rectangles?

   (A) 36   **(B)** 40   **(C)** 70   **(D)** 80

10. If $\ell_1 \perp \ell_2$, what is the value of $a + b$?

    (A) 45   **(B)** 90
    (C) 110
    (D) cannot be determined

11.
    $\overline{QR} \perp \overline{RS}$
    $\overline{QR} \perp \overline{PQ}$
    $PQ = 10$
    $QR = 7$
    $RS = 14$

    What is the shortest distance from P to S?

    (A) 24   **(B)** 25   **(C)** 28   **(D)** 31

CHAPTER 11

Rational Polynomial Expressions

Jaime can paint the outside of his garage in 20 hours. His wife, Rosa, can paint the outside of the garage in 18 hours. Suppose they work together. The rational equation $\frac{x}{20} + \frac{x}{18} = 1$ can be used to determine how many hours it would take them to complete the job.

11-1 Multiplying and Dividing Rational Expressions

Rational numbers and **rational algebraic expressions** are similar. A rational number can be expressed as the quotient of two integers.

$$\frac{2}{3} \qquad \frac{415}{100} \qquad \frac{-6}{11}$$

All polynomials, including constants, are rational expressions. But not all rational expressions are polynomials.

A rational algebraic expression can be expressed as the quotient of two polynomials. As with fractions, the denominator cannot be 0.

$$\frac{6}{k} \qquad \frac{2x}{x-5} \qquad \frac{p^2-25}{p+6}$$

0 cannot be used as a replacement for k. 5 cannot be used as a replacement for x. What number <u>cannot</u> replace p?

To simplify a rational algebraic expression, divide both numerator and denominator by their greatest common factor (GCF).

The process is similar to simplifying fractions.

Examples

1 **Simplify** $\dfrac{2x(x-5)}{(x-5)(x^2-1)}$.

$$\frac{2x(x-5)}{(x-5)(x^2-1)} = \frac{2x}{x^2-1}$$

The GCF is $(x-5)$. The expression is undefined when x is 5, 1, or -1. Why?

2 **Simplify** $\dfrac{x^2y-x^2}{x^3-x^3y}$.

$$\frac{x^2y-x^2}{x^3-x^3y} = \frac{x^2(y-1)}{x^3(1-y)}$$

For what values is the expression undefined?

$$= \frac{-1 \cdot x^2(1-y)}{x \cdot x^2(1-y)}$$

$y - 1 = -1(1-y)$

$$= -\frac{1}{x}$$

The GCF is $x^2(1-y)$.

Both rational numbers and rational expressions are multiplied the same way. Multiply numerators and multiply denominators.

$$\frac{2}{3} \cdot \frac{4}{5} = \frac{2 \cdot 4}{3 \cdot 5} \text{ or } \frac{8}{15}$$

$$\frac{(x+1)}{(x-4)} \cdot \frac{(2x+3)}{(3x-1)} = \frac{(x+1)(2x+3)}{(x-4)(3x-1)} \text{ or } \frac{2x^2+5x+3}{3x^2-13x+4}$$

This method can be generalized as follows.

For all rational expressions $\dfrac{a}{b}$ and $\dfrac{c}{d}$, $b \neq 0$ and $d \neq 0$,

$$\frac{a}{b} \cdot \frac{c}{d} = \frac{ac}{bd}.$$

Multiplying Rational Expressions

3 Find $\dfrac{4a}{5b} \cdot \dfrac{15b}{16a}$. Write the answer in simplest form.

$$\dfrac{4a}{5b} \cdot \dfrac{15b}{16a} = \dfrac{60ab}{80ab} \qquad \textit{For what values is the expression undefined?}$$

$$= \dfrac{3}{4} \qquad \textit{The GCF is 20ab.}$$

4 Find $\dfrac{x^2 - 9}{x^2 + x - 12} \cdot \dfrac{x + 2}{x + 3}$. Write the answer in simplest form.

$$\dfrac{x^2 - 9}{x^2 + x - 12} \cdot \dfrac{x + 2}{x + 3} = \dfrac{(x^2 - 9)(x + 2)}{(x^2 + x - 12)(x + 3)}$$

$$= \dfrac{(x - 3)(x + 3)(x + 2)}{(x - 3)(x + 4)(x + 3)} \qquad \textit{Factor. For what values is the expression undefined?}$$

$$= \dfrac{x + 2}{x + 4} \qquad \textit{The GCF is } (x - 3)(x + 3).$$

Recall that dividing by a fraction is the same as multiplying by its multiplicative inverse.

$$\dfrac{3}{8} \div \dfrac{5}{3} = \dfrac{3}{8} \cdot \dfrac{3}{5} \qquad \textit{The multiplicative inverse of } \tfrac{5}{3} \textit{ is } \tfrac{3}{5}.$$

$$= \dfrac{9}{40}$$

Similarly, dividing by a rational algebraic expression is the same as multiplying by its multiplicative inverse.

Recall that a fraction names a rational number.

For all rational expressions $\dfrac{a}{b}$ and $\dfrac{c}{d}$, $b \neq 0$, $c \neq 0$, and $d \neq 0$,

$$\dfrac{a}{b} \div \dfrac{c}{d} = \dfrac{a}{b} \cdot \dfrac{d}{c}.$$

Dividing Rational Expressions

5 Find $\dfrac{4x^2y}{15a^3b^3} \div \dfrac{2xy^2}{5ab^3}$. Write the answer in simplest form.

$$\dfrac{4x^2y}{15a^3b^3} \div \dfrac{2xy^2}{5ab^3} = \dfrac{4x^2y}{15a^3b^3} \cdot \dfrac{5ab^3}{2xy^2} \qquad \textit{The inverse of } \dfrac{2xy^2}{5ab^3} \textit{ is } \dfrac{5ab^3}{2xy^2}.$$

$$= \dfrac{20x^2yab^3}{30xy^2a^3b^3}$$

$$= \dfrac{2x}{3ya^2} \qquad \textit{The GCF is } 10xyab^3.$$

Example

6 Find $\dfrac{x^2}{x^2 - 25y^2} \div \dfrac{x}{x + 5y}$.

$$\dfrac{x^2}{x^2 - 25y^2} \div \dfrac{x}{x + 5y} = \dfrac{x^2}{x^2 - 25y^2} \cdot \dfrac{x + 5y}{x} \qquad \text{The inverse of } \dfrac{x}{x + 5y} \text{ is } \dfrac{x + 5y}{x}$$

$$= \dfrac{x^2(x + 5y)}{x(x^2 - 25y^2)} \qquad \text{Factor } x^2 - 25y^2.$$

$$= \dfrac{x^2(x + 5y)}{x(x + 5y)(x - 5y)} \qquad \text{The GCF is } x(x + 5y).$$

$$= \dfrac{x}{x - 5y}$$

A complex rational expression, also called a **complex fraction**, is an expression whose numerator or denominator, or both, contain rational expressions.

$$\dfrac{\dfrac{x^2 - 4}{2}}{\dfrac{2 - x}{5}} \qquad\qquad \dfrac{\dfrac{1}{x} + 3}{\dfrac{2}{x} + 5} \qquad\qquad \dfrac{8}{\dfrac{x}{3 - y}}$$

To simplify a complex fraction, treat it as a division problem.

Example

7 Simplify $\dfrac{\dfrac{x^2 - 4}{2}}{\dfrac{2 - x}{5}}$.

$$\dfrac{\dfrac{x^2 - 4}{2}}{\dfrac{2 - x}{5}} = \dfrac{x^2 - 4}{2} \div \dfrac{2 - x}{5}$$

$$= \dfrac{x^2 - 4}{2} \cdot \dfrac{5}{2 - x} \qquad \text{The inverse of } \dfrac{2 - x}{5} \text{ is } \dfrac{5}{2 - x}.$$

$$= \dfrac{5(x^2 - 4)}{2(2 - x)}$$

$$= \dfrac{5(x + 2)(x - 2)}{2(2 - x)} \qquad \text{Factor.}$$

$$= \dfrac{5(x + 2)(x - 2)}{-2(x - 2)} \qquad 2 - x = -1(x - 2)$$

$$= -\dfrac{5(x + 2)}{2} \qquad \text{The GCF is } (x - 2).$$

Exploratory Exercises

For each expression, find the greatest common factor (GCF) of the numerator and denominator. Then simplify each expression.

1. $\dfrac{42y}{18xy}$
2. $\dfrac{42y^3x}{18y^7}$
3. $\dfrac{38a^2}{42ab}$
4. $\dfrac{-3x^2y^5}{18x^5y^2}$
5. $\dfrac{(-2x^2y)^3}{4x^5y}$

6. $\dfrac{a^3b^2}{(-ab)^3}$
7. $\dfrac{(2xy)^4}{(x^2y)^2}$
8. $\dfrac{(-3t^2u)^3}{(6tu^2)^2}$
9. $\dfrac{m+5}{2m+10}$
10. $\dfrac{4x}{x^2-x}$

State the multiplicative inverse of each expression.

11. $\dfrac{7x}{9y}$
12. $\dfrac{18x}{7}$
13. $\dfrac{x+y}{2}$
14. $\dfrac{x^2-8}{x-1}$

15. $\dfrac{(x+4)^2}{(x-3)^2}$
16. $x+a$
17. $\dfrac{1}{x-y}$
18. $-\dfrac{2x}{a^2-b^2}$

Write each quotient as a product.

19. $\dfrac{a^2}{b^2} \div \dfrac{b^2}{a^2}$
20. $\dfrac{p^3}{2q} \div \dfrac{-p^2}{4q}$
21. $\dfrac{3h}{h+1} \div (h-2)$
22. $\dfrac{y^2}{x+2} \div \dfrac{y}{x+2}$

Written Exercises

Find the multiplicative inverse of each expression. Write each answer in simplest form.

23. $\dfrac{14}{a^2}$
24. $\dfrac{a+b}{a^2-b^2}$
25. $\dfrac{1-x^2}{x+1}$
26. $\dfrac{8x^2-x^3}{16-2x}$

27. $\dfrac{a^2+2ab+b^2}{a^2-b^2}$
28. $\dfrac{y^2+8y-20}{y^2-4}$
29. $\dfrac{2x^2+8x}{x^2+x-12}$
30. $\dfrac{6y^3-9y^2}{2y^2+5y-12}$

31. $\dfrac{x^2-x-20}{x^2+7x+12}$
32. $\dfrac{y^2+4y+4}{3y^2+5y-2}$
33. $\dfrac{a^2+2a+1}{2a^2+3a+1}$
34. $\dfrac{2b^2-9b+9}{b^2-6b+9}$

Multiply or divide. Write each answer in simplest form.

35. $\dfrac{3ab}{4ac} \cdot \dfrac{6a^2}{3b^2}$
36. $\dfrac{-4ab}{21c} \cdot \dfrac{14c^2}{18a^2}$
37. $-\dfrac{3}{5a} \div \left(-\dfrac{9}{15ab}\right)$
38. $\dfrac{1}{x} \div \left(-\dfrac{3}{x^2}\right)$

39. $\dfrac{3d^3c}{a^4} \div \left(-\dfrac{6dc}{a^5}\right)$
40. $\dfrac{(ab)^3}{d^3} \div \dfrac{a^2b^4}{(cd)^4}$
41. $\dfrac{(cd)^3}{a} \cdot \dfrac{ax^2}{xc^2d}$
42. $\dfrac{(3a)^3}{18b^3} \cdot \dfrac{12a^4b^5}{(3a)^2}$

43. $\left(\dfrac{x^2}{y}\right)^2 \cdot \dfrac{5}{3x}$
44. $\left(\dfrac{3a^3}{b^2}\right)^3 \cdot \dfrac{4b^2}{3a^7}$
45. $\dfrac{5}{m-3} \div \dfrac{10}{m-3}$
46. $\dfrac{x+y}{a} \div \dfrac{x+y}{a^2}$

47. $\dfrac{4x+40}{3x} \cdot \dfrac{9x}{3x+30}$
48. $\dfrac{4y^2-9}{4y^2} \cdot \dfrac{8y}{2y-3}$
49. $\dfrac{3x-21}{x^2-49} \div \dfrac{3x}{x^2+7x}$

50. $\dfrac{2x+2}{x^2+5x+6} \div \dfrac{3x+3}{x^2+2x-3}$
51. $\dfrac{x^2-y^2}{y^2} \cdot \dfrac{y^3}{y-x}$
52. $-\dfrac{x^2-y^2}{x+y} \cdot \dfrac{1}{x-y}$

53. $\dfrac{a^2+2a-15}{a-3} \div \dfrac{a^2-4}{2}$
54. $\dfrac{a^2-b^2}{14} \cdot \dfrac{35}{a+b}$
55. $\dfrac{y^2-y}{w^2-y^2} \div \dfrac{y^2-2y+1}{1-y}$

56. $\dfrac{a^2-b^2}{2a} \div \dfrac{a-b}{6a}$
57. $\dfrac{x^2-y^2}{x+y} \cdot \dfrac{11}{x-y}$
58. $\dfrac{(y-2)^2}{(x-4)^2} \cdot \dfrac{x-4}{y-2}$

59. $\dfrac{x^2 + 3x - 10}{x^2 + 8x + 15} \cdot \dfrac{x^2 + 5x + 6}{x^2 + 4x + 4}$

60. $\dfrac{x^3 + y^3}{x^2 - y^2} \cdot \dfrac{3a}{9ab} \cdot \dfrac{6xb}{7x}$

61. $\dfrac{a^3 - b^3}{a + b} \cdot \dfrac{a^2 - b^2}{a^2 + ab + b^2}$

62. $\dfrac{y^2 - y - 12}{y + 12} \cdot \dfrac{y^2 - 4y - 12}{y - 4}$

63. $\dfrac{w^2 - 11w + 24}{w^2 - 18w + 80} \cdot \dfrac{w^2 - 15w + 50}{w^2 - 9w + 20}$

64. $\dfrac{2x^2 + x - 15}{4x^2 + 2x - 30} \cdot \dfrac{6x^2 - 8x + 2}{3x^2 + 8x - 3}$

Simplify each expression.

65. $\dfrac{\dfrac{x^2 - y^2}{2}}{\dfrac{x - y}{4}}$

66. $\dfrac{\dfrac{w^2 + 2w + 1}{w + 1}}{3}$

67. $\dfrac{\dfrac{5a^2 - 20}{2a + 2}}{\dfrac{10a - 20}{4a}}$

68. $\dfrac{\dfrac{x^2 - 1}{x^2 - 3x - 10}}{\dfrac{x^2 - 12x + 35}{x^2 + 3x + 2}}$

69. $\dfrac{\dfrac{2y}{y^2 - 4}}{\dfrac{3}{y^2 - 4y + 4}}$

70. $\dfrac{\dfrac{c^2 + 2c - 3}{3c + 3}}{\dfrac{c^2 + 5c + 6}{2c + 2}}$

71. $\dfrac{\dfrac{p^2 + 7p}{3p}}{\dfrac{49 - p^2}{3p - 21}}$

72. $\dfrac{\dfrac{9 - 4t^2}{t^2 + 6t + 9}}{\dfrac{8t - 12}{2t^2 + 5t - 3}}$

73. $\dfrac{\dfrac{3 + 10t^2 - 17t}{5t^2 + 4t - 1}}{\dfrac{4t^2 - 9}{3 + 5t + 2t^2}}$

74. $\dfrac{\dfrac{m^2 + 15m + 54}{m + 6}}{\dfrac{m + 9}{3}}$

75. $\dfrac{\dfrac{x}{x^2 - 9}}{\dfrac{3x - 3}{x^2 - x - 6}}$

76. $\dfrac{\dfrac{3m}{2m^2 + 7m - 15}}{\dfrac{6}{2m^2 - 3m}}$

Challenge Exercises

Simplify each expression.

77. $\dfrac{(a^2 - 5a + 6)^{-1}}{(a - 2)^{-2}} \div \dfrac{(a - 3)^{-1}}{(a - 2)^{-2}}$

78. $\dfrac{(x^2 + 7x + 10)^{-1}}{(x + 2)(x - 5)^{-2}} \div \dfrac{(x + 2)^{-2}}{(x - 5)^{-2}}$

mini-review

State whether the graph of each equation is a parabola, a circle, an ellipse, or a hyperbola.

1. $4y^2 - x^2 - 24y + 6x = 11$

2. $3x^2 + 5y^2 - 30x + 20y = -80$

3. $x = \frac{1}{3}y^2 + 2y + 7$

4. $x^2 + y^2 - 14x + 2y = -25$

If $f(x) = 3x^2$, $g(x) = x^2 - 1$, $h(x) = 4x + 2$, find each value.

5. $[f \circ g](2)$

6. $[g \circ h]\left(\dfrac{1}{2}\right)$

7. $[f \circ (h \circ g)](x)$

Name the vertex, axis of symmetry, and the direction of opening of the graph of each equation.

8. $y = -8(x - 3)^2 - 1$

9. $f(x) = x^2 + 22x + 121$

10. $y = -2x^2 - 12x - 23$

Find all zeros of each function.

11. $f(x) = x^3 - 7x^2 - 5x + 30$

12. $g(x) = x^3 + 5x^2 + 8x + 6$

The position of symbols is very important in algebra. Study the list of expressions below.

$$2x \qquad 2 - x \qquad x - 2 \qquad \frac{x}{2} \qquad \frac{2}{x} \qquad x^2 \qquad \frac{x^2}{x + 2}$$

Each expression contains the symbols 2 and x. However, each expression says something different. The same expressions are listed below. The arrows indicate the order in which the symbols usually are read, and the words on the right are English translations of the expressions.

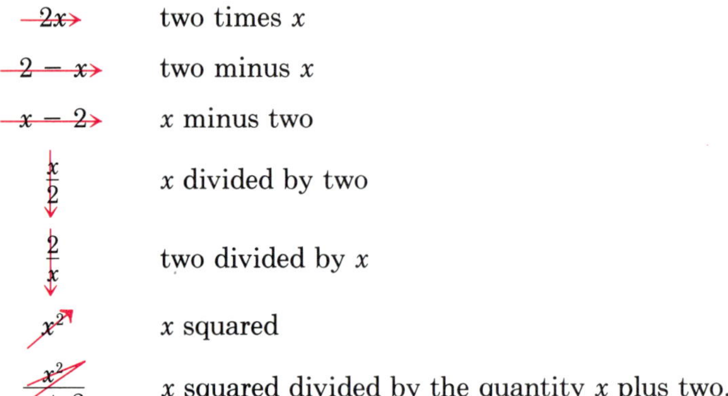

$2x$	two times x
$2 - x$	two minus x
$x - 2$	x minus two
$\dfrac{x}{2}$	x divided by two
$\dfrac{2}{x}$	two divided by x
x^2	x squared
$\dfrac{x^2}{x + 2}$	x squared divided by the quantity x plus two.

Exercises

As quickly as possible, look at each list of expressions. Which expressions in the list are the same as the first expression?

1. $\dfrac{x^3}{3}$ **a.** $\dfrac{x}{3}$ **b.** $\dfrac{x^3}{3}$ **c.** $\dfrac{x}{3^3}$ **d.** $\dfrac{3x}{3}$

2. $\dfrac{(cd)^3}{a}$ **a.** $\dfrac{(cd)^3}{a}$ **b.** $\dfrac{cd^3}{a}$ **c.** $\dfrac{3(cd)}{a}$ **d.** $\dfrac{(cd)^3}{a}$

3. $\dfrac{1}{x - y}$ **a.** $\dfrac{1}{y - x}$ **b.** $\dfrac{1}{x + y}$ **c.** $\dfrac{x}{1 - y}$ **d.** $\dfrac{1}{x - y}$

4. $\dfrac{x^2 - 4}{x - 1}$ **a.** $\dfrac{x - 4}{x - 1}$ **b.** $\dfrac{x^2 - 4}{x - 1}$ **c.** $\dfrac{x^2 - 4}{x^2 - 1}$ **d.** $\dfrac{4 - x^2}{x^2 - 1}$

5. $\dfrac{8x^2 - x^3}{16 - 2x}$ **a.** $\dfrac{8x^2 - x^3}{2x - 16}$ **b.** $\dfrac{8x^3 - x^2}{16 - 2x}$ **c.** $\dfrac{8x^2 - x^3}{16 - 2x}$ **d.** $\dfrac{8x^3 - x^2}{2x - 16}$

6–10. In problems 1–5, which expressions are different from the first expression? How are they different?

11. Turn to page 390. Copy each expression in Exploratory Exercises 1–10. Draw an arrow to indicate the order you would read the symbols in each expression.

12. Write an English translation for each expression in problem 11.

11-2 Adding and Subtracting Rational Expressions

You add rational numbers with common denomonators, such as $\frac{7}{13}$ and $\frac{4}{13}$, in the following way.

$$\frac{7}{13} + \frac{4}{13} = \frac{7 + 4}{13} \quad \text{or} \quad \frac{11}{13}$$

If two rational expressions have common denominators, add them in a similar fashion.

$$\frac{7x}{13y^2} + \frac{4x}{13y^2} = \frac{7x + 4x}{13y^2} \quad \text{or} \quad \frac{11x}{13y^2}$$

The sum of two rational expressions with a common denominator is the sum of the numerators over the common denominator.

To add two rational numbers with different denominators, first find two equivalent rational numbers with common denominators. Then add the equivalent fractions.

$$\frac{3}{8} + \frac{2}{3} = \frac{3 \cdot 3}{8 \cdot 3} + \frac{2 \cdot 8}{3 \cdot 8}$$ *The least common denominator is 24.*

$$= \frac{9}{24} + \frac{16}{24}$$

$$= \frac{9 + 16}{24}$$

$$= \frac{25}{24}$$

On way to find the least common denominator (LCD) of two or more rational numbers is to first factor each denominator into its prime factors. For example, suppose you wish to add $\frac{5}{36}$ and $\frac{7}{24}$.

The prime factors of 36 and 24 are shown below.

$$36 = 2^2 \cdot 3^2$$
$$24 = 2^3 \cdot 3$$

The least common denominator of 36 and 24 contains each prime factor the greatest number of times that it appears.

$$\text{LCD} = 2^3 \cdot 3^2 \text{ or } 72$$

The greatest number of times 3 appears is twice. The greatest number of times 2 appears is three times.

Now, rename the fractions using the LCD. Then add.

$$\frac{5}{36} + \frac{7}{24} = \frac{10}{72} + \frac{21}{72} \qquad \frac{5}{36} = \frac{5 \cdot 2}{36 \cdot 2} = \frac{10}{72} \qquad \frac{7}{24} = \frac{7 \cdot 3}{24 \cdot 3} = \frac{21}{72}$$

$$= \frac{31}{72}$$

To add rational expressions with different denominators, follow the same procedure. First find two equivalent expressions with a common denominator. Then add the equivalent expressions.

1 Find $\dfrac{7x}{15y^2} + \dfrac{y}{18xy}$.

First find the LCD of $15y^2$ and $18xy$.

$$15y^2 = 3 \cdot 5 \cdot y \cdot y \qquad 18xy = 2 \cdot 3 \cdot 3 \cdot x \cdot y$$

The LCD is $2 \cdot 3^2 \cdot 5 \cdot x \cdot y^2$ or $90xy^2$.

$$\dfrac{7x}{15y^2} + \dfrac{y}{18xy} = \dfrac{7x \cdot 6x}{15y^2 \cdot 6x} + \dfrac{y \cdot 5y}{18xy \cdot 5y} \qquad \textit{Rewrite each term using the LCD.}$$

$$= \dfrac{42x^2}{90xy^2} + \dfrac{5y^2}{90xy^2}$$

$$= \dfrac{42x^2 + 5y^2}{90xy^2}$$

2 Find $\dfrac{1}{x^2 - 2x - 15} + \dfrac{3x + 1}{2x - 10}$.

$$\dfrac{1}{x^2 - 2x - 15} + \dfrac{3x + 1}{2x - 10} = \dfrac{1}{(x - 5)(x + 3)} + \dfrac{3x + 1}{2(x - 5)}$$

$$= \dfrac{2}{2(x - 5)(x + 3)} + \dfrac{(3x + 1)(x + 3)}{2(x - 5)(x + 3)} \qquad \textit{The LCD is } 2(x - 5)(x + 3).$$

$$= \dfrac{2 + (3x + 1)(x + 3)}{2(x - 5)(x + 3)}$$

$$= \dfrac{2 + 3x^2 + 9x + x + 3}{2(x - 5)(x + 3)}$$

$$= \dfrac{3x^2 + 10x + 5}{2(x - 5)(x + 3)} \qquad \textit{Simplify the numerator.}$$

Use a similar method to subtract rational expressions.

3 Find $\dfrac{x + 4}{2x - 8} - \dfrac{x + 12}{4x - 16}$.

$$\dfrac{x + 4}{2x - 8} - \dfrac{x + 12}{4x - 16} = \dfrac{x + 4}{2(x - 4)} - \dfrac{x + 12}{4(x - 4)} \qquad \textit{Factor each denominator.}$$

$$= \dfrac{2(x + 4)}{4(x - 4)} - \dfrac{x + 12}{4(x - 4)} \qquad \textit{The LCD is } 4(x - 4).$$

$$= \dfrac{(2x + 8) - (x + 12)}{4(x - 4)} \qquad \textit{Remember that } -(x + 12) = -x - 12.$$

$$= \dfrac{x - 4}{4(x - 4)} \quad \text{or} \quad \dfrac{1}{4} \qquad \textit{Simplify.}$$

Complex fractions may involve sums or differences.

4 Simplify $\dfrac{\dfrac{1}{x} - \dfrac{1}{y}}{1 + \dfrac{1}{x}}$.

$$\frac{\dfrac{1}{x} - \dfrac{1}{y}}{1 + \dfrac{1}{x}} = \frac{\dfrac{y}{xy} - \dfrac{x}{xy}}{\dfrac{x}{x} + \dfrac{1}{x}}$$

In the numerator, the LCD is xy.

In the denominator, the LCD is x.

$$= \frac{\dfrac{y - x}{xy}}{\dfrac{x + 1}{x}}$$

Simplify the numerator and denominator.

$$= \frac{y - x}{xy} \div \frac{x + 1}{x}$$

Write the complex fraction as a division problem.

$$= \frac{y - x}{xy} \cdot \frac{x}{x + 1}$$

The GCF is x.

$$= \frac{y - x}{y(x + 1)} \quad \text{or} \quad \frac{y - x}{xy + y}$$

Exploratory Exercises

Find the least common denominator (LCD) for each pair of denominators.

1. 54, 28
2. 78, 39
3. 80, 125
4. 12, 27
5. $7a^2$, $14ab$
6. $36x^2y$, $20xyz$
7. $x(x - 2)$, $x^2 - 4$
8. $(x + 2)(x + 1)$, $x^2 - 1$
9. $x^2 + 2x + 1$, $x^2 - 9$
10. $3x + 15$, $x^2 + 3x - 15$
11. $x^2 - 8x$, $y^2 - 8y$
12. $96x^2$, $16(x + 9)$

Written Exercises

Add. Write each answer in simplest form.

13. $\dfrac{5}{6a} + \dfrac{7}{4a}$

14. $\dfrac{7}{ab} + \dfrac{9}{b}$

15. $\dfrac{5}{a} + 7$

16. $3t - 7 + \dfrac{3t + 1}{t - 5}$

17. $\dfrac{3x}{x - y} + \dfrac{4x}{y - x}$

18. $\dfrac{3a + 2}{a + b} + \dfrac{4}{2a + 2b}$

19. $-\dfrac{18}{9xy} + \dfrac{7}{2x} - \dfrac{2}{3x^2}$

20. $\dfrac{1}{8b} + \dfrac{5 - 3b}{2b^2} + \dfrac{3b - 2}{16b^3}$

21. $\dfrac{x}{x^2 - 9} + \dfrac{1}{2x + 6}$

22. $y - 1 + \dfrac{1}{y - 1}$

23. $\dfrac{3}{a - 2} + \dfrac{2}{a - 3}$

24. $\dfrac{6}{x^2 + 4x + 4} + \dfrac{5}{x + 2}$

25. $\dfrac{3}{x^2 - 25} + \dfrac{6}{x - 5}$

26. $\dfrac{x - 1}{x^2 - 1} + \dfrac{3}{5x + 5}$

27. $\dfrac{5}{x^2 - 3x - 28} + \dfrac{7}{2x - 14}$

28. $\dfrac{a}{a + 4} + \dfrac{2}{a^2 + 8a + 16}$

29. $\dfrac{2}{y^2 - 4y - 5} + \dfrac{5}{y^2 - 2y - 15}$

30. $\dfrac{2a}{3a - 15} + \dfrac{-16a + 20}{3a^2 - 12a - 15}$

31. $\dfrac{m^2 + n^2}{m^2 - n^2} + \dfrac{m}{n - m} + \dfrac{n}{m + n}$

32. $\dfrac{x}{x - y} + \dfrac{y}{y^2 - x^2} + \dfrac{2x}{x + y}$

33. $\dfrac{x + 1}{x - 1} + \dfrac{x + 2}{x - 2} + \dfrac{x}{x^2 - 3x + 2}$

Subtract. Write each answer in simplest form.

34. $\dfrac{3}{4a} - \dfrac{2}{5a} - \dfrac{1}{2a}$

35. $\dfrac{11}{9} - \dfrac{7}{2a} - \dfrac{6}{5a}$

36. $\dfrac{9}{y-2} - \dfrac{2}{1-y}$

37. $\dfrac{7}{y-8} - \dfrac{6}{8-y}$

38. $\dfrac{y}{y-9} - \dfrac{-9}{9-y}$

39. $\dfrac{8}{2y-16} - \dfrac{y}{8-y}$

40. $\dfrac{-4y}{y^2-4} - \dfrac{y}{y+2}$

41. $\dfrac{x}{x+3} - \dfrac{6x}{x^2-9}$

42. $3m+1 - \dfrac{2m}{3m+1}$

43. $\dfrac{y-2}{y-4} - \dfrac{y-8}{y-4}$

44. $\dfrac{6-y}{y-2} - \dfrac{3+y}{2-y}$

45. $\dfrac{x}{x^2+2x+1} - \dfrac{x+2}{x+1} - \dfrac{3x}{x+1}$

46. $\dfrac{x^2-3x+1}{x^2-4} - \dfrac{x^2+2x+4}{2-x} - \dfrac{x-4}{x-2}$

47. $\dfrac{m+3}{m^2-6m+9} - \dfrac{8m-24}{9-m^2}$

48. $\dfrac{3b-1}{b^2-49} - \dfrac{3b+2}{14+5b-b^2}$

Simplify.

49. $\dfrac{\dfrac{x}{2} - \dfrac{x}{5}}{\dfrac{x}{3} - \dfrac{x}{6}}$

50. $\dfrac{\dfrac{2x}{ab} - \dfrac{5x}{4}}{\dfrac{3x}{a} - \dfrac{6x}{5}}$

51. $\dfrac{\dfrac{x+y}{x}}{\dfrac{1}{x} + \dfrac{1}{y}}$

52. $\dfrac{\dfrac{x}{y} - \dfrac{y}{x}}{\dfrac{1}{x} + \dfrac{1}{y}}$

53. $\dfrac{\dfrac{2}{x} + \dfrac{9}{x^2} - \dfrac{5}{x^3}}{2x - \dfrac{1}{2x}}$

54. $\dfrac{\dfrac{1}{a} - \dfrac{1}{a^2} - \dfrac{20}{a^3}}{\dfrac{1}{a} + \dfrac{8}{a^2} - \dfrac{16}{a^3}}$

55. $\dfrac{3 + \dfrac{5}{a+2}}{3 - \dfrac{10}{a+7}}$

56. $\dfrac{\dfrac{2x}{2x+1} - 1}{1 + \dfrac{2x}{1-2x}}$

57. $\dfrac{\dfrac{m+4}{m} - \dfrac{3}{m+5}}{\dfrac{m-1}{m^2+5m} + \dfrac{3}{m+5}}$

58. $\dfrac{\dfrac{5x}{x^2-16}}{\dfrac{10}{x-4} + \dfrac{10}{x+4}}$

59. $\dfrac{\dfrac{x+4}{x-6} + \dfrac{x+1}{x+2}}{\dfrac{4}{x^2-4x-12}}$

60. $\dfrac{\dfrac{5}{y+7} + \dfrac{2}{y}}{\dfrac{2}{y} - \dfrac{10}{y(y+7)}}$

61. $\dfrac{y+2 + \dfrac{3}{y-5}}{\dfrac{4}{y-5} + 2}$

62. $\dfrac{(x+y)\left(\dfrac{1}{x} - \dfrac{1}{y}\right)}{(x-y)\left(\dfrac{1}{x} + \dfrac{1}{y}\right)}$

63. $\dfrac{n+1 - \dfrac{2}{n}}{n+4 + \dfrac{4}{n}}$

64. $\dfrac{m - \dfrac{1}{m}}{1 + \dfrac{4}{m} - \dfrac{5}{m^2}}$

65. $\dfrac{\dfrac{1}{x+5} + \dfrac{1}{x-3}}{\dfrac{2x^2-3x-5}{x^2+2x-15}}$

Challenge Exercises

Find each sum. Write each answer in simplest form.

66. $\dfrac{4b-1}{b^2-7b} + \dfrac{2}{b^2-49}$

67. $\dfrac{2x}{x^2+7x+10} + \dfrac{x-1}{x^2-25}$

68. $\dfrac{y}{y^2+4y-21} + \dfrac{4}{y^2+7y} + \dfrac{2y}{y^2-9}$

69. $\dfrac{3x}{4x^2-1} + \dfrac{5}{2x^2-x} + \dfrac{2x+1}{2x^2+5x+2}$

11-3 Solving Rational Equations

An equation that contains one or more rational expressions is called a **rational equation**. One way to begin to solve a rational equation is to multiply each side of the equation by the least common denominator (LCD).

Examples

1 Solve $\frac{x}{9} + \frac{x}{7} = 1$.

$$\frac{x}{9} + \frac{x}{7} = 1$$

$$63\left(\frac{x}{9} + \frac{x}{7}\right) = 63(1)$$

The LCD is $3 \cdot 3 \cdot 7$ or 63.

$$63 \cdot \frac{x}{9} + 63 \cdot \frac{x}{7} = 63$$

$$7x + 9x = 63$$
$$16x = 63$$
$$x = \frac{63}{16}$$

The solution is $\frac{63}{16}$.

Check:
$$\frac{x}{9} + \frac{x}{7} = 1$$

$$\frac{\frac{63}{16}}{9} + \frac{\frac{63}{16}}{7} \overset{?}{=} 1$$

$$\frac{7}{16} + \frac{9}{16} \overset{?}{=} 1$$

$$1 = 1 \quad ✔$$

2 Solve $\frac{3}{4} - \frac{3}{y + 2} = \frac{9}{28}$.

$$\frac{3}{4} - \frac{3}{y + 2} = \frac{9}{28}$$

$$28(y + 2)\left(\frac{3}{4} - \frac{3}{y + 2}\right) = 28(y + 2)\left(\frac{9}{28}\right)$$

The LCD is $2 \cdot 2 \cdot 7 \cdot (y + 2)$ or $28(y + 2)$.

$$28(y + 2)\left(\frac{3}{4}\right) - 28(y + 2)\left(\frac{3}{y + 2}\right) = 9(y + 2)$$

$$21y + 42 - 84 = 9y + 18$$
$$12y = 60$$
$$y = 5$$

Check:
$$\frac{3}{4} - \frac{3}{y + 2} = \frac{9}{28}$$

$$\frac{3}{4} - \frac{3}{5 + 2} \overset{?}{=} \frac{9}{28}$$

Substitute 5 for y.

$$\frac{3}{4} - \frac{3}{7} \overset{?}{=} \frac{9}{28}$$

$$\frac{21}{28} - \frac{12}{28} \overset{?}{=} \frac{9}{28}$$

$$\frac{9}{28} = \frac{9}{28} \quad ✔$$

The solution is 5.

A rational expression is not defined when the denominator is zero. Therefore, any values assigned to the variables that result in a denominator of zero must be excluded. After both sides of the equation are multiplied by the LCD, some of these values may appear as solutions of the new equation.

Examples

3   **Solve** $\dfrac{7}{x-3} = \dfrac{x+4}{x-3}$.

If x is 3, the denominator, $x - 3$, is 0. Since division by 0 is not defined, 3 must be an excluded value.

$$\frac{7}{x-3} = \frac{x+4}{x-3}$$

$$(x-3)\left(\frac{7}{x-3}\right) = (x-3)\left(\frac{x+4}{x-3}\right) \qquad \textit{The LCD is } x-3.$$

$$7 = x + 4$$

$$x = 3$$

Since 3 is an excluded value, the equation has no solution.

4   **Solve** $r + \dfrac{r^2 - 5}{r^2 - 1} = \dfrac{r^2 + r + 2}{r + 1}$.

The excluded values of this equation are -1 and 1.   *Why?*

$$r + \frac{r^2 - 5}{r^2 - 1} = \frac{r^2 + r + 2}{r + 1}$$

$$(r^2 - 1)\left(r + \frac{r^2 - 5}{r^2 - 1}\right) = (r^2 - 1)\left(\frac{r^2 + r + 2}{r + 1}\right) \qquad \textit{The LCD is } (r^2 - 1).$$

$$(r^2 - 1)r + (r^2 - 1)\left(\frac{r^2 - 5}{r^2 - 1}\right) = (r - 1)(r + 1)\left(\frac{r^2 + r + 2}{r + 1}\right)$$

$$r^3 - r + r^2 - 5 = (r - 1)(r^2 + r + 2)$$

$$r^3 + r^2 - r - 5 = r^3 + r - 2$$

$$r^2 - 2r - 3 = 0$$

$$(r - 3)(r + 1) = 0$$

$$r - 3 = 0 \quad \text{or} \quad r + 1 = 0$$

$$r = 3 \qquad\qquad r = -1$$

Check: $r + \dfrac{r^2 - 5}{r^2 - 1} = \dfrac{r^2 + r + 2}{r + 1}$ *Since -1 is an excluded value, it cannot be a solution to the equation.*

$$3 + \frac{3^2 - 5}{3^2 - 1} \overset{?}{=} \frac{3^2 + 3 + 2}{3 + 1}$$

$$3 + \frac{4}{8} \overset{?}{=} \frac{14}{4}$$

$$\frac{7}{2} = \frac{7}{2} \quad ✔$$

The solution is 3.

Exploratory Exercises

Find the least common denominator (LCD) for the expressions in each equation. State the excluded values.

1. $\dfrac{1}{x} + \dfrac{1}{2} = \dfrac{2}{x}$

2. $\dfrac{1}{5} = \dfrac{2}{10y}$

3. $\dfrac{1}{4} = \dfrac{s - 3}{8s}$

4. $\dfrac{6}{x} = \dfrac{9}{x^2}$

5. $\dfrac{9}{x + 5} = \dfrac{6}{x - 3}$

6. $\dfrac{11}{2y} - \dfrac{2}{3y} = \dfrac{1}{6}$

7. $\dfrac{3m}{2 + m} - \dfrac{5}{7} = 4$

8. $\dfrac{1 - b}{1 + b} = \dfrac{2b}{2b + 3}$

Written Exercises

State the excluded values. Then solve and check.

9. $\dfrac{2}{x} + \dfrac{1}{4} = \dfrac{11}{12}$

10. $\dfrac{x^2}{6} - \dfrac{x}{3} = \dfrac{1}{2}$

11. $r^2 + \dfrac{17r}{6} = \dfrac{1}{2}$

12. $\dfrac{2y}{3} - \dfrac{y + 3}{6} = 2$

13. $\dfrac{y + 1}{3} + \dfrac{y - 1}{3} = \dfrac{4}{3}$

14. $\dfrac{2y + 1}{5} - \dfrac{2 + 7y}{15} = \dfrac{2}{3}$

15. $\dfrac{2y - 5}{6} - \dfrac{y - 5}{4} = \dfrac{3}{4}$

16. $\dfrac{4t - 3}{5} - \dfrac{4 - 2t}{3} = 1$

17. $\dfrac{5 + 7p}{8} - \dfrac{3(5 + p)}{10} = 2$

18. $\dfrac{2q - 1}{3} - \dfrac{4q + 5}{8} = -\dfrac{19}{24}$

19. $8 - \dfrac{2 - 5x}{4} = \dfrac{4x + 9}{3}$

20. $\dfrac{3x + 1}{4} - \dfrac{x + 5}{5} = -2$

21. $x + 5 = \dfrac{6}{x}$

22. $a + 1 = \dfrac{6}{a}$

23. $\dfrac{1}{y^2 - 1} = \dfrac{2}{y^2 + y - 2}$

24. $\dfrac{x}{x^2 - 3} = \dfrac{5}{x + 4}$

25. $x + \dfrac{12}{x} - 8 = 0$

26. $\dfrac{a}{a - 1} + a = \dfrac{4a - 3}{a - 1}$

27. $\dfrac{5}{6} - \dfrac{2m}{2m + 3} = \dfrac{19}{6}$

28. $\dfrac{5}{2x} - \dfrac{3}{10} = \dfrac{1}{x}$

29. $\dfrac{1}{9} + \dfrac{1}{2a} = \dfrac{1}{a^2}$

30. $\dfrac{1}{x - 1} + \dfrac{2}{x} = 0$

31. $\dfrac{1}{1 - x} = 1 - \dfrac{x}{x - 1}$

32. $\dfrac{4t}{3t - 2} + \dfrac{2t}{3t + 2} = 2$

33. $\dfrac{2p}{2p + 3} - \dfrac{2p}{2p - 3} = 1$

34. $\dfrac{12}{x^2 - 16} - \dfrac{24}{x - 4} = 3$

35. $\dfrac{4}{x - 2} - \dfrac{x + 6}{x + 1} = 1$

36. $\dfrac{9}{x - 3} = \dfrac{x - 4}{x - 3} + \dfrac{1}{4}$

37. $\dfrac{x - 4}{x - 2} = \dfrac{x - 2}{x + 2} + \dfrac{1}{x - 2}$

38. $\dfrac{3x - 3}{4x} = \dfrac{6x - 9}{6x} + \dfrac{1}{3x}$

39. $\dfrac{x - 3}{2x} = \dfrac{x - 2}{2x + 1} - \dfrac{1}{2}$

40. $\dfrac{5}{x + 2} = \dfrac{5}{x} + \dfrac{2}{3x}$

41. $\dfrac{6}{a - 7} = \dfrac{a - 49}{a^2 - 7a} + \dfrac{1}{a}$

42. $\dfrac{-2}{x - 1} = \dfrac{2}{x + 2} - \dfrac{4}{x - 3}$

43. $\dfrac{2}{y + 2} - \dfrac{y}{2 - y} = \dfrac{y^2 + 4}{y^2 - 4}$

44. $\dfrac{4x^2}{x^2 - 9} - \dfrac{2x}{x + 3} = \dfrac{3}{x - 3}$

45. $\dfrac{t + 4}{t} + \dfrac{3}{t - 4} = \dfrac{-16}{t^2 - 4t}$

46. $\dfrac{y}{y - 5} + \dfrac{17}{25 - y^2} = \dfrac{1}{y + 5}$

47. $\dfrac{x + 3}{x + 2} = 2 - \dfrac{3}{x^2 + 5x + 6}$

48. $\dfrac{1}{n - 2} = \dfrac{2n + 1}{n^2 + 2n - 8} + \dfrac{2}{n + 4}$

49. $\dfrac{x}{x^2 - 1} + \dfrac{2}{x + 1} = \dfrac{1}{2x - 2}$

Challenge Exercise

50. Find the values of A and B, if $\dfrac{A}{x + 2} + \dfrac{B}{2x - 3} = \dfrac{5x - 11}{2x^2 + x - 6}$.

Use the distance formula to find the distance between each pair of points.

1. $(3, 6)$, $(7, -8)$
2. $(-3, -7)$, $(-6, -9)$

Find $f(-2)$ for each function f.

3. $f(x) = 2x^2 + 3x - 5$
4. $f(x) = x^3 - 3x + 4$
5. $f(x) = x^4 - 2x^3 + 8$
6. $f(x) = 2x^5 - 3$

Find the solutions of each system of equations.

7. $x^2 + y^2 = 25$
   $x = 2$
8. $x - 2 = -2y$
   $y - 1 = 2x + x^2$

Multiply. Write each answer in simplest form.

9. $\dfrac{-7ab}{24c^2} \cdot \dfrac{8cd}{28a}$

10. $\dfrac{3a + 3}{4} \cdot \dfrac{8}{12a}$

Find the center and radius of each circle whose equation is given.

11. $x^2 + y^2 = 25$
12. $(x - 4)^2 + (y - 5)^2 = 100$
13. $x^2 + 4x + y^2 - 10y = -24$

Solve each equation.

14. $\dfrac{1}{y} + \dfrac{3}{y} = 10$

15. $1 + \dfrac{6}{y - 1} = \dfrac{3}{4}$

Find the midpoints of each line segment with endpoints having the following coordinates.

16. $(5, 5)$, $(\sqrt{5}, \sqrt{5})$
17. $(0.4, 0.7)$, $(-0.2, 0.9)$

Solve each inequality.

18. $x^2 - 8x + 7 \geq 0$
19. $2y^2 - 3y - 2 < 0$
20. $4a^2 + 12a + 9 \leq 0$
21. $3m > m^2$

Add. Write each answer in simplest form.

22. $\dfrac{9}{4a} + \dfrac{-7}{5b}$

23. $\dfrac{3}{x} + 4$

Divide. Write each answer in simplest form.

24. $\dfrac{a + b}{x} \div \dfrac{a + b}{x^3}$

25. $\dfrac{x^2 - x - 12}{3a} \div \dfrac{x^2 + 7x + 12}{6a^2}$

Given a function and one of its zeros, find all the zeros of the function.

26. $f(x) = x^3 + 2x^2 + 21x - 58;\ -2 + 5i$
27. $g(x) = x^3 + 5x^2 - 26x - 120;\ -4$

Find all zeros of each function.

28. $f(x) = x^3 + x^2 - 5x + 3$
29. $g(x) = x^3 - 3x^2 - 25x + 42$

30. Find the inverse of $f(x) = \dfrac{(5x + 2)}{2}$.

31. Graph $f(x) = -2x^2 + 4x - 1$. Then name the vertex, axis of symmetry, and direction of opening.

32. Write the simplest polynomial function with integral coefficients whose zeros are -2 and $4 + 3i$.

33. Approximate to the nearest tenth the positive real zero of
    $p(x) = x^4 - 3x^3 - 7x^2 - 5x$.

11-4 Problem Solving: Using Rational Equations

The city swimming pool can be filled from two sources, a well or city water. The pipe for the city water fills the pool in 6 hours. The pipe from the well fills the pool in 10 hours. How long will it take the pool to fill if both sources are piped in at the same time? This question can be answered by solving a rational equation.

explore

Let h stand for the number of hours it will take to fill the pool.

What fraction of the pool can be filled with city water in 1 hour?	$\dfrac{1}{6}$
What fraction of the pool can be filled with well water in 1 hour?	$\dfrac{1}{10}$
What fraction of the pool can be filled with city water in h hours?	$h \cdot \dfrac{1}{6}$ or $\dfrac{h}{6}$
What fraction of the pool can be filled with well water in h hours?	$h \cdot \dfrac{1}{10}$ or $\dfrac{h}{10}$

plan

Now write an equation.

$$\underset{\substack{\textit{fraction filled with}\\ \textit{city water in h hours}}}{\dfrac{h}{6}} + \underset{\substack{\textit{fraction filled with}\\ \textit{well water in h hours}}}{\dfrac{h}{10}} = \underset{\substack{\textit{whole pool filled from}\\ \textit{both sources in h hours}}}{1}$$

solve

Solve the equation.

$$\frac{h}{6} + \frac{h}{10} = 1$$

$$30\left(\frac{h}{6} + \frac{h}{10}\right) = 30(1) \qquad \textit{The LCD is } 2 \cdot 3 \cdot 5 \textit{ or } 30.$$

$$30 \cdot \frac{h}{6} + 30 \cdot \frac{h}{10} = 30 \qquad \textit{Use the distributive property.}$$

$$5h + 3h = 30$$
$$8h = 30$$
$$h = \frac{15}{4} \quad \text{or} \quad 3\frac{3}{4} \qquad \text{It will take } 3\frac{3}{4} \text{ hours.}$$

examine

$$\left(\frac{15}{4}\right)\left(\frac{1}{6}\right) + \left(\frac{15}{4}\right)\left(\frac{1}{10}\right) = \frac{5}{8} + \frac{3}{8} = 1 \quad ✔$$

1 **A car travels 300 kilometers in the same time that a train travels 200 kilometers. The speed of the car is 20 kilometers per hour more than the speed of the train. Find the speed of the car and the speed of the train.**

explore Let r represent the speed of the car.

plan

$$time\ car\ travels = time\ train\ travels$$

$$\frac{distance\ car\ travels}{rate\ car\ travels} = \frac{distance\ train\ travels}{rate\ train\ travels}$$

Since $d = rt$, we know that $t = \frac{d}{r}$.

$$\frac{300}{r} = \frac{200}{r - 20}$$

solve

$$r(r - 20)\frac{300}{r} = r(r - 20)\frac{200}{r - 20} \qquad \textit{The LCD is } r(r - 20).$$

$$300(r - 20) = 200\,r$$
$$300r - 6000 = 200r$$
$$-6000 = -100r$$
$$60 = r$$

The car's speed is 60 kilometers per hour, and the train's speed is $60 - 20$ or 40 kilometers per hour.

examine The car travels 300 kilometers at 60 kilometers per hour. It travels 5 hours. The train travels 200 kilometers at 40 kilometers per hour. It also travels 5 hours. The speeds are correct.

Written Exercises

Solve each problem.

1. One bricklayer can build a wall of a certain size in 5 hours. Another bricklayer can do the same job in 4 hours. If the bricklayers work together, how long will it take to do the job?

2. One hose can fill the Sunshine's small swimming pool in 6 hours. A second, newer hose can fill the pool in 4 hours. If both hoses are used, how long will it take to fill the pool?

3. Jan Zeiss can tile a floor in 14 hours. Together Jan and her helper Bill can tile the same size floor in 9 hours. How long would it take Bill to do the job alone?

4. A tank can be filled by a hose in 10 hours. The tank can be emptied by a drain pipe in 20 hours. If the drain pipe is open while the tank is filling, how long will it take to fill the tank?

5. A painter works on a job for 10 days and is then joined by her helper. Together they finish the job in 6 more days. Her helper could have done the job alone in 30 days. How long would it have taken the painter to do the job alone?

6. Elena Dias can paint a 9 by 12 room in $1\frac{1}{2}$ hours. If Luisa Alicea helps, they can paint the same size room in 1 hour. How long would it take Luisa to paint such a room by herself?

7. The ratio of 4 less than a number to 26 more than that number is 1 to 3. What is the number?

8. Five times the multiplicative inverse of a number is added to the number and the result is $10\frac{1}{2}$. What is the number?

9. The denominator of a fraction is 1 less than twice the numerator. If 7 is added to both numerator and denominator, the resulting fraction has a value of $\frac{7}{10}$. Find the original fraction.

10. Two numbers are in a ratio of 6 to 7. If the first is increased by 2 and the second is increased by 1, the resulting numbers are in the ratio of 4 to 5. Find the original numbers.

11. A plane flies from Chicago to Los Angeles, a distance of 2000 miles, in 4 hours. It flies against a 50 mph wind. It returns in $3\frac{1}{3}$ hours. Find the speed of the plane in still air.

12. The speed of the current in the Mississippi River is 5 miles per hour. A boat travels downstream 26 miles and returns in $10\frac{2}{3}$ hours. What is its speed in still water?

13. A boat travels at a rate of 15 kilometers per hour in still water. It travels 60 kilometers upstream in the same time that it travels 90 kilometers downstream. What is the rate of the current?

14. Increasing the average speed of the Pickerington Express Bus by 13 km/h resulted in the 260 km trip taking an hour less than before. What was the original average speed of the bus?

15. Conrad can jog to Chris's house in 10 minutes. Chris can ride her bike to Conrad's house in 6 minutes. If they start from their houses at the same time, in how many minutes do they meet?

16. The load capacities of two trucks are in the ratio of 5 to 2. The smaller truck has a capacity 3 tons less than that of the larger truck. What is the capacity of the larger truck?

17. The simple interest for one year on a sum of money is $108. Suppose the interest rate is increased by 2%. Then $450 less than the original sum could be invested and yield the same annual interest. How much is the original sum of money, and what is the original rate of interest?

18. A sum of money invested for one year at 4% yields a certain simple interest. If the sum of money is increased by $200 and the interest rate is lowered to 3.5%, then the annual interest is increased by $1. How much is the original sum of money, and how much is the original interest?

19. A chemist needs to make 1000 milliliters of a 30% alcohol solution by mixing 25% and 55% solutions. How much of each should he use?

20. How much of a 45% salt solution must be added to a 20% salt solution to get 750 milliliters of a 30% solution?

21. Pipe A can fill a tank in 4 hours and pipe B can fill the tank in 3 hours. With the tank empty, pipe A is turned on, and one hour later, pipe B is turned on. How long will pipe B run before the tank is full?

Challenge Exercise

22. At what time between 5 o'clock and 6 o'clock do the hands of a clock coincide?

Connie Hardesty is a photographer. To take sharp, clear pictures she must focus the camera.

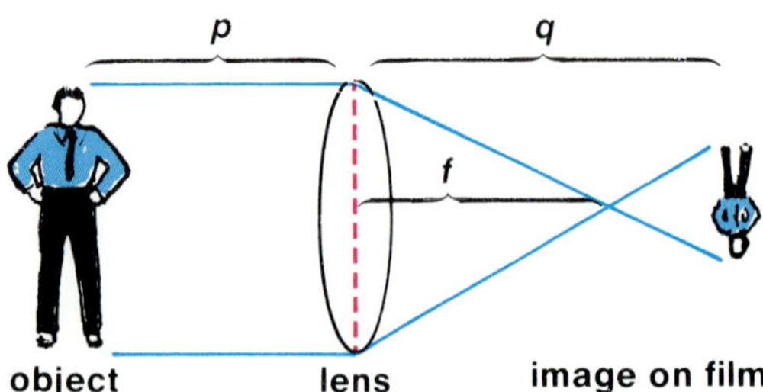

The following formula can be used to determine the distance the lens must be from the film for the camera to be focused.

$$\frac{1}{p} + \frac{1}{q} = \frac{1}{f}$$

p is the distance from the lens to the object.
q is the distance from the lens to the image on the film.
f is the focal length of the lens.

Example: A camera has a lens with a focal length of 10 cm. Connie wants to photograph a flower that is 80 cm away. What must the distance be from the lens to the film for the camera to be in focus?

$$\frac{1}{p} + \frac{1}{q} = \frac{1}{f}$$

$$\frac{1}{80} + \frac{1}{q} = \frac{1}{10} \qquad p = 80,\ f = 10$$

$$q + 80 = 8q \qquad \text{Multiply each side by } 80q.$$

$$80 = 7q \quad \text{or} \quad q \approx 11.4$$

The lens must be about 11.4 cm from the film.

Exercises

For the following values of p and f, find q.

1. $p = 45$ cm, $f = 5$ cm

2. $p = 600$ mm, $f = 60$ mm

3. $p = 28$ in., $f = 6$ in.

4. $p = 50$ cm, $f = 0.5$ cm

Solve each problem.

5. A camera has a lens with a focal length of 10 in. When Usha has the lens 12 in. from the film, the camera is focused, to take a picture of her dog. How far from the lens is the dog?

6. Carl wants to take a picture of his house which is 10 m away. When the camera is focused, the lens is 5 cm from the film. What is the focal length of the lens?

11-5 Graphing Rational Functions

A **rational function** is an equation of the form $f(x) = \dfrac{p(x)}{q(x)}$ where $p(x)$ and $q(x)$ are polynomial functions and $q(x) \neq 0$. Functions such as $f(x) = \dfrac{x}{x-1}$, $f(x) = \dfrac{2}{(x-2)^2}$, and $f(x) = \dfrac{4}{(x+2)(x-3)}$ are examples of rational functions.

The graph of a rational function has two or more branches. The branches of the graph approach lines called **asymptotes**. The lines with equation $x = a$ is a *vertical asymptote* if the rational function is undefined when x is a. The line with equation $y = b$ is a *horizontal asymptote* if the value of the rational function approaches b as the value of $|x|$ increases.

Before graphing a rational function, it is helpful to graph the asymptotes of the function.

Example

1 **Graph** $f(x) = \dfrac{x}{x-1}$.

The function is undefined when x is 1. Thus, the line with equation $x = 1$ is a vertical asymptote. Now, observe the value of the function as the value of $|x|$ increases.

$$f(101) = \frac{101}{101-1} \quad \text{or} \quad 1.01$$

$$f(1001) = \frac{1001}{1001-1} \quad \text{or} \quad 1.001$$

$$f(10{,}001) = \frac{10{,}001}{10{,}001-1} \quad \text{or} \quad 1.0001$$

$$f(-99) = \frac{-99}{-99-1} \quad \text{or} \quad 0.99$$

$$f(-999) = \frac{-999}{-999-1} \quad \text{or} \quad 0.999$$

$$f(-9999) = \frac{-9999}{-9999-1} \quad \text{or} \quad 0.9999$$

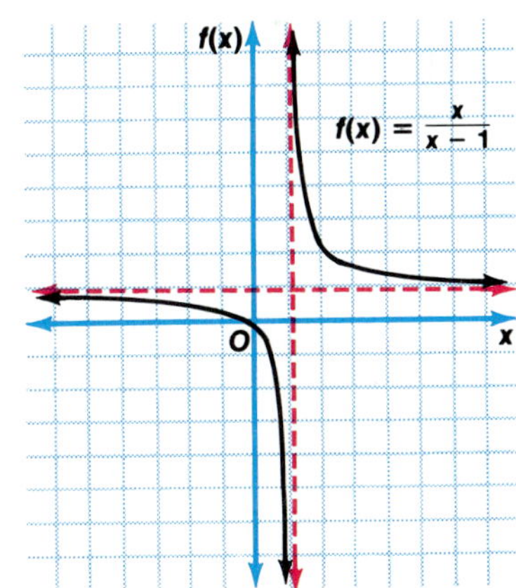

As the value of $|x|$ increases, it appears that the value of the function is approaching 1. The line with the equation $y = 1$ is a horizontal asymptote.

Plot points on either side of each asymptote. Then sketch the graph. Be sure to test values that are close to the asymptotes.

2   **Graph** $f(x) = \dfrac{2}{(x - 2)^2}$.

The line with the equation $x = 2$ is a vertical asymptote.   *Why?*
Now, check for a horizontal asymptote.

$$f(102) = \frac{2}{(102 - 2)^2} = 0.0002 \qquad\qquad f(-98) = \frac{2}{(-98 - 2)^2} = 0.0002$$

$$f(1002) = \frac{2}{(1002 - 2)^2} = 0.000002 \qquad f(-998) = \frac{2}{(-998 - 2)} = 0.000002$$

As the value of $|x|$ increases, it appears that the value of the function is approaching 0. The line with equation $y = 0$ is a horizontal asymptote.

Also note that the value of the function could never be negative.

Plot points above the horizontal asymptote and on either side of the vertical asymptote. Then sketch the graph.

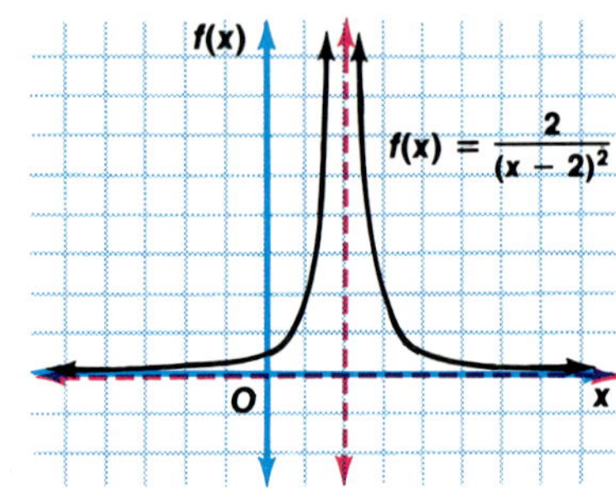

3   **Using Calculators**

Graph $f(x) = \dfrac{4}{(x + 2)(x - 3)}$.

The lines with equations $x = -2$ and $x = 3$ are vertical asymptotes. The procedure for finding horizontal asymptotes can be done quickly using a calculator. The process for finding $f(10)$ is shown below. The other values are computed in a similar manner.

ENTER: 4 ÷ ((10 + 2) × (10 − 3)) =

DISPLAY: 4   10   2   12   10   7   84   0.047619

$$f(10) \approx 0.05 \qquad f(100) \approx 0.0004$$
$$f(1000) \approx 0.000004$$
$$f(-10) \approx 0.04 \qquad f(-100) \approx 0.0004$$
$$f(-1000) \approx 0.000004$$

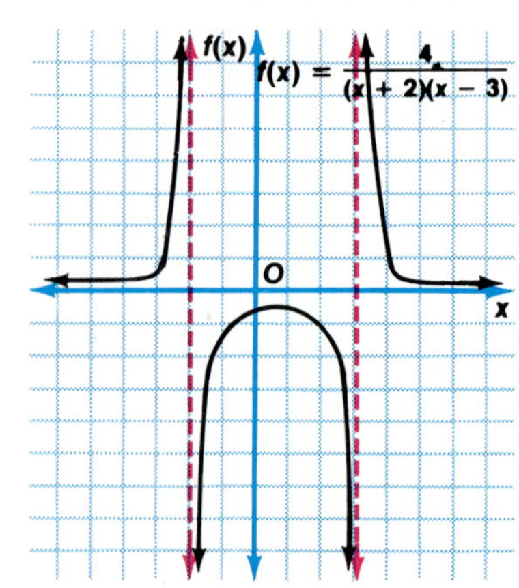

As the value of $|x|$ increases, it appears that the value of the function is approaching 0. The line with equation $y = 0$ is a horizontal asymptote.

Plot points on either side of each asymptote. Then sketch the graph.

State the equations of the vertical and horizontal asymptotes for each rational function.

1. $f(x) = \dfrac{3}{x - 1}$

2. $f(x) = \dfrac{1}{x}$

3. $f(x) = \dfrac{1}{x - 3}$

4. $f(x) = \dfrac{4}{x + 2}$

5. $f(x) = \dfrac{6}{(x - 6)^2}$

6. $f(x) = \dfrac{3}{(x + 1)^3}$

7. $f(x) = \dfrac{4}{(x - 1)(x + 5)}$

8. $f(x) = \dfrac{-6}{x - 3}$

9. $f(x) = \dfrac{-2}{(x - 1)(x - 4)}$

Written Exercises

Graph each rational function. Be sure to sketch the asymptotes on each graph.

10. $f(x) = \dfrac{x - 1}{x - 4}$

11. $f(x) = \dfrac{3}{x + 2}$

12. $f(x) = \dfrac{x - 5}{x + 1}$

13. $f(x) = \dfrac{1}{x}$

14. $f(x) = \dfrac{x}{x - 2}$

15. $f(x) = \dfrac{x}{x - 5}$

16. $f(x) = \dfrac{-4}{x - 1}$

17. $f(x) = \dfrac{-1}{x - 6}$

18. $f(x) = \dfrac{x}{x + 1}$

19. $f(x) = \dfrac{-2}{(x - 3)^2}$

20. $f(x) = \dfrac{4x}{x - 1}$

21. $f(x) = \dfrac{2}{(x - 2)(x + 1)}$

22. $f(x) = \dfrac{-5}{(x - 3)(x + 1)}$

23. $f(x) = \dfrac{3}{(x - 4)^2}$

24. $f(x) = \dfrac{1}{(x + 2)^2}$

25. $f(x) = \dfrac{8}{(x - 1)(x + 3)}$

Challenge Exercises

Graph each rational function. Be sure to sketch the asymptotes on each graph.

26. $f(x) = \dfrac{x}{1 - x^2}$

27. $f(x) = \dfrac{x}{x^2 - 4}$

28. $f(x) = \dfrac{x - 1}{x^2 - 9}$

mini-review

Find all rational zeros for each function.

1. $f(a) = a^3 + 2a^2 - 11a - 12$

2. $f(z) = z^3 + 8z^2 + 4z - 48$

Find the inverse of each function.

3. $g(x) = \dfrac{x + 7}{3}$

4. $f(x) = x^2 - 3$

Perform the indicated operation. Write answer in simplest form.

5. $\dfrac{x^2 - y^2}{x + y} \cdot \dfrac{1}{x - y}$

6. $\dfrac{4}{3a} - \dfrac{7}{5a} + \dfrac{1}{2a}$

7. $\dfrac{13x^2}{40y} \div \dfrac{26x^2}{70y^3}$

Graph each inequality.

8. $y < x^2 + x - 1$

9. $y \geq x^2 - 6x + 4$

10. Find the solutions for the system $2x + 3y = 7$ and $(y - 2)^2 = x + 4$.

Graphing Calculator Application: Graphing Rational Functions

A graphing calculator can generate graphs of rational functions on its graphics screen. When graphing rational functions, it is important to choose range parameter values carefully.

Example

1 **Graph $y = \dfrac{x-1}{x-2}$. State the equations of the vertical and horizontal asymptotes of the graph.**

Since the equation is undefined when x is 2, the graph of $x = 2$ is a vertical asymptote. Choose a set of range parameter values that will produce a complete graph of the equation and make it easy to use the tracing function to determine the horizontal asymptote. One possible set of values is given below. Set the range parameters to these values and graph the equation.

Xmin: -10 Xmax: 10   Xscl: 2
Ymin: -10 Ymax: 10   Yscl: 2

ENTER: GRAPH (X −

1) ÷ (X

− 2) EXE

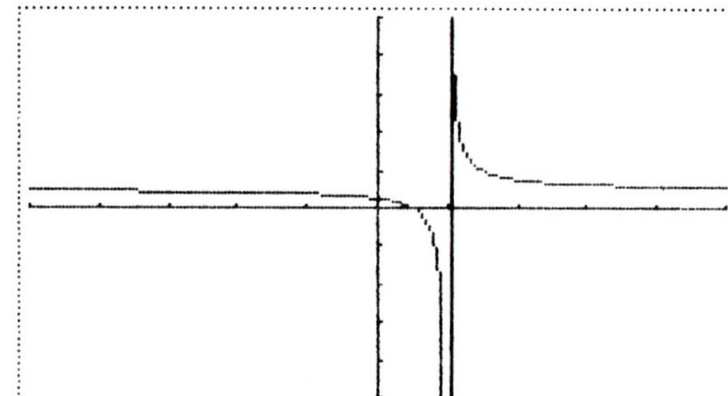

Notice what happened on the graphics screen. The calculator drew a line between the branches of the curve where the vertical asymptote should occur. This line is not part of the graph of $y = \dfrac{x-1}{x-2}$.

Now, try graphing the equation with a different set of range parameter values.

Xmin: -10 Xmax: 10   Xscl: 2
Ymin: $\ \ -8$ Ymax: $\ \ 8$ Yscl: 2

Set the range parameters to these values. Then graph the equation by pressing the EXE (or DRAW) key.

Notice the calculator did not draw a line between the branches of the curve.

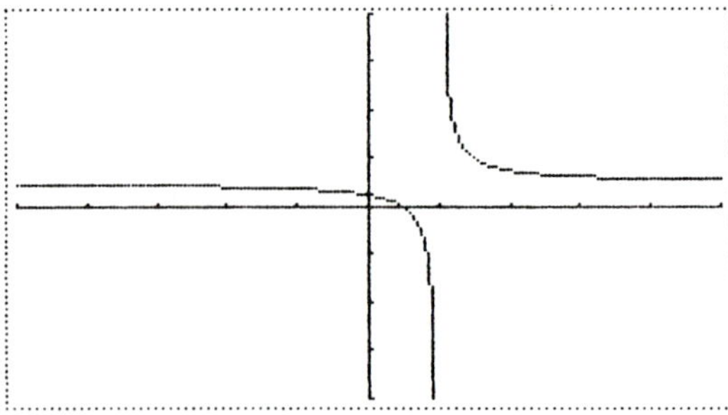

The calculator will draw a line between the branches of the curve only under certain conditions. You will need to experiment with the range parameter values if you want to find a set of values that will not produce the line as part of the graph.

From the graph, it appears that the horizontal asymptote is close to 1. To check this, graph the equation using another set of range parameter values.

Xmin: -100 Xmax: 100   Xscl: 20   Ymin: 0.5   Ymax: 1.5   Yscl: 0.1

Activate the tracing function and change the screen to the y-coordinate value display. Then move the pointer along each branch of the curve so that the value of $|x|$ increases.

It appears that the y-coordinate values are approaching 1 as the value of $|x|$ increases.

The graph of $y = 1$ is a horizontal asymptote.

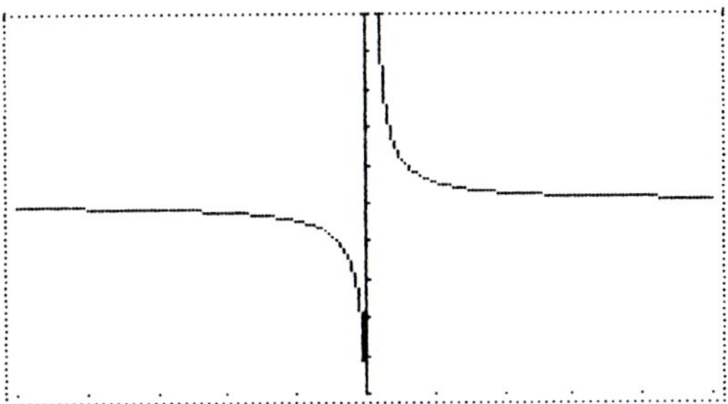

Exploratory Exercises

Graph $y = \dfrac{x^2}{x^2 - 1}$ for each set of range parameter values. Then sketch the graph shown on the graphics screen, indicating the x-axis scale and y-axis scale.

1. Xmin: -10, Xmax: 10, Xscl: 2, Ymin: -10, Ymax: 10, Yscl: 2
2. Xmin: -11.75, Xmax: 11.75, Xscl: 2, Ymin: -5, Ymax: 5, Yscl: 1
3. Xmin: -10, Xmax: 12, Xscl: 2, Ymin: -4, Ymax: 8, Yscl: 1
4. Xmin: -5, Xmax: 5, Xscl: 1, Ymin: -8, Ymax: 8, Yscl: 2

Written Exercises

Graph each equation on a graphing calculator. Then sketch the graph shown on the graphics screen, indicating the x-axis scale and y-axis scale.

5. $y = \dfrac{1}{x}$

6. $y = \dfrac{1}{x^2}$

7. $y = \dfrac{1}{x^3}$

8. $y = \dfrac{-3x}{x - 1}$

9. $y = \dfrac{-3x}{(x - 1)^2}$

10. $y = \dfrac{-3x}{(x - 1)^3}$

11. $y = \dfrac{x^2 - 9}{x + 2}$

12. $y = \dfrac{x^2 - 9}{(x + 2)^2}$

13. $y = \dfrac{x^2 - 9}{(x + 2)^3}$

14. $y = \dfrac{2}{x^2 + 4}$

15. $y = \dfrac{2x}{x^2 + 4}$

16. $y = \dfrac{2x^2}{x^2 + 4}$

17. $y = \dfrac{x + 5}{x^2 - 16}$

18. $y = \dfrac{x + 5}{(x^2 - 16)(x + 1)}$

19. $y = \dfrac{x + 5}{(x^2 - 16)(x^2 - 9)}$

20. $y = \dfrac{x - 4}{x^2 + x - 6}$

21. $y = \dfrac{x - 4}{x^3 + x^2 - 6x}$

22. $y = \dfrac{x - 4}{x^4 + 7x^3 - 36x}$

Use the tracing function on a graphing calculator to determine the value of each function as the value of $|x|$ increases.

23. $f(x) = \dfrac{22}{x}$

24. $f(x) = \dfrac{x - 1}{x + 3}$

25. $f(x) = \dfrac{x - 8}{4x}$

26. $f(x) = \dfrac{-3x}{2x + 5}$

27. $f(x) = \dfrac{-6}{x^2}$

28. $f(x) = \dfrac{x}{x^2 - 4}$

29. $f(x) = \dfrac{x^2 + 9}{-3x^2}$

30. $f(x) = \dfrac{5x^2 - 8}{4x^2 + x}$

11-6 Direct and Inverse Variation

Recall that direct variation is a linear function. Direct variation can be described by an equation of the form $y = kx$ where k is a constant *not* equal to zero. For example, gravity on Earth is about six times as great as the gravity on the moon. Thus, weight on Earth varies directly with weight on the moon. This relationship can be described by the following equation.

$$y = kx \qquad \text{y stands for weight on earth}$$
$$y = 6x \qquad \text{x stands for weight on moon}$$
$$\text{6 is the constant of variation}$$

Direct variation can be related to proportions or products. Suppose that x_1 and y_1 satisfy the equation $y = kx$. Also, suppose that x_2 and y_2 satisfy the equation $y = kx$.

$$y_1 = kx_1 \quad \text{and} \quad y_2 = kx_2$$

$$\frac{y_1}{y_2} = \frac{kx_1}{kx_2} \qquad \textit{Division Property of Equality}$$

$$\frac{x_1}{x_2} = \frac{y_1}{y_2} \qquad \textit{Simplify.} \qquad \textit{What are other forms of this proportion?}$$

$$\text{or} \quad x_1y_2 = x_2y_1 \qquad \textit{Multiply each side by x_2y_2 and simplify.}$$

Example

1 **If y varies directly as x, and $y = 12$ when $x = 15$, find x when $y = 21$.**

Use the proportion $\dfrac{x_1}{x_2} = \dfrac{y_1}{y_2}$.

$$\frac{15}{x_2} = \frac{12}{21} \qquad \textit{Substitute. $x_1 = 15$, $y_1 = 12$, $y_2 = 21$}$$

$$x_2 = 26.25 \qquad \textit{Solve for x_2.}$$

Many quantities are said to be **inversely proportional** or to **vary inversely** with each other. For example, the amount of current in a circuit is inversely proportional to the amount of resistance in the circuit. That is, as the amount of resistance decreases, the amount of current increases proportionally. The following chart shows several corresponding values.

Current (amps)	0.5	1.0	1.5	2.0	2.5	3.0	4.0	5.0
Resistance (ohms)	12	6.0	4.0	3.0	2.4	2.0	1.5	1.2

Let c stand for the amount of current.   *What do you notice about the product of*
Let r stand for the amount of resistance.   *each pair of corresponding values?*

The relationship between these quantities can be described by the equation $c = \dfrac{6}{r}$.

A rational equation in two variables of the form $y = \dfrac{k}{x}$, where k is a constant, is called an **inverse variation**. The constant k is called the **constant of variation**, and y is said to **vary inversely as** x.

Definition of Inverse Variation

Inverse variation can also be related to proportions or products. Suppose x_1 and y_1 satisfy the equation $y = \dfrac{k}{x}$. Also, suppose that x_2 and y_2 satisfy the equation $y = \dfrac{k}{x}$.

$$y_1 = \frac{k}{x_1} \quad \text{and} \quad y_2 = \frac{k}{x_2}$$
$$x_1 y_1 = k \qquad\qquad x_2 y_2 = k$$

$$x_1 y_1 = x_2 y_2 \qquad \text{\textit{Substitution Property of Equality}}$$
$$\frac{x_1}{x_2} = \frac{y_2}{y_1} \qquad \text{\textit{Divide each side by } } y_1 x_2 \text{ \textit{and simplify.}}$$

Note that inverse variation yields a different proportion than direct variation.

Examples

2   **If y varies inversely as x, and $y = 3$ when $x = 4$, find y when $x = 18$.**

Use the proportion $\dfrac{x_1}{x_2} = \dfrac{y_2}{y_1}$.   *The equation $x_1 y_1 = x_2 y_2$ could also be used to solve the problem.*

$$\frac{4}{18} = \frac{y_2}{3} \qquad \text{\textit{Substitute 4 for } } x_1, \text{ \textit{18 for } } x_2, \text{ \textit{and 3 for } } y_1.$$

$$y_2 = \frac{2}{3}$$

3   **The volume of any gas varies inversely with its pressure as long as the temperature remains constant. The volume of a particular gas is 1600 milliliters when the pressure is 25 centimeters of mercury. What is the volume of the gas at the same temperature when the pressure is 40 centimeters of mercury?**

Let V_1 stand for the first volume of gas, 1600 ml.
Let P_1 stand for the first gas pressure, 25 cm of mercury.
Let P_2 stand for the second gas pressure, 40 cm of mercury.

Since volume and pressure are inversely proportional, $V_1 P_1 = V_2 P_2$.

$$1600 \cdot 25 = V_2 \cdot 40$$
$$\frac{40{,}000}{40} = V_2$$
$$1000 = V_2$$

The volume of the gas is 1000 milliliters.

State whether each equation represents direct variation or inverse variation. Then name the constant of variation.

1. $x = 4y$

2. $xy = -3$

3. $y = -4x$

4. $y = \dfrac{7}{x}$

5. $5 = \dfrac{y}{x}$

6. $\dfrac{x}{y} = -6$

7. $\dfrac{3}{4}y = x$

8. $\dfrac{x}{2} = y$

9. $x = \dfrac{9}{y}$

10. $y = \dfrac{3}{x}$

11. $a = 4b$

12. $\dfrac{3}{5}a = -\dfrac{5}{4}b$

Written Exercises

In each of the following, y varies directly as x.

13. If $y = 8$, then $x = 2$.
    Find y when $x = 9$.

14. If $y = 10$, then $x = -3$.
    Find x when $y = 4$.

15. If $x = 4$, then $y = 0.5$.
    Find y when $x = 9$.

16. If $y = -5$, then $x = 0.25$.
    Find x when $y = -7$.

17. If $y = \dfrac{3}{4}$, then $x = \dfrac{2}{5}$.
    Find y when $x = 8$.

18. If $y = 11$, then $x = \dfrac{1}{5}$.
    Find y when $x = \dfrac{2}{5}$.

In each of the following, y varies inversely as x.

19. If $x = 14$, then $y = -6$.
    Find x when $y = -11$.

20. If $y = \dfrac{1}{5}$, then $x = 9$.
    Find y when $x = -3$.

21. If $y = 11$, then $x = 44$.
    Find x when $y = 40$.

22. If $x = 20$, then $y = 10$.
    Find x when $y = 14$.

23. If $y = -2$, then $x = -8$.
    Find x when $y = \dfrac{2}{3}$.

24. If $y = \dfrac{4}{9}$, then $x = \dfrac{3}{8}$.
    Find y when $x = \dfrac{2}{3}$.

25. If $x = 1.5$, then $y = -8$.
    Find x when $y = -3$.

26. If $y = 7$, then $x = -3$.
    Find y when $x = 4$.

Solve each problem.

27. A map is scaled so that 1 cm represents 15 km. How far apart are two towns if they are 7.9 cm apart on the map?

28. A 75-foot tree casts a 40-foot shadow. How tall is a tree that casts a 10-foot shadow at the same time of day?

29. Six feet of steel wire weighs 0.7 kilograms. How much does 100 feet of steel wire weigh?

30. Alan Tokashira invested \$5000 at 7% interest. How much must he invest at $6\frac{1}{2}\%$ interest to obtain the same income?

31. The time to drive a certain distance varies inversely according to the rate of speed. Mary Bronson drives 47 mph for 4 hours. How long would it take her to make the same trip at 55 mph?

32. A volume of gas is 120 cubic feet under 6 pounds of pressure. What is the volume at the same temperature when the pressure is 8 pounds?

33. When air is pumped into an automobile tire, the pressure required varies inversely as the volume. If the pressure is 30 pounds when the volume is 140 cubic inches, find the pressure when the volume is 100 cubic inches.

34. In a closed room, the number of hours of safe oxygen level varies inversely as the number of people in the room. If there are 2 hours of safe oxygen level for 100 people, how many hours of safe oxygen level are there for 600 people?

35. A realtor made a commission of $4800 on a sale of a $90,000 house. At that rate how much would she make on a $129,000 house?

36. At Mel's Diner a 10 kg ham serves 44 people. At Joe's Diner a 16 kg ham serves 68 people. Which diner serves larger portions?

37. The time required to travel a given distance is inversely proportional to speed of travel. If it takes 4 hours to make a trip at 60 kmh, how long will it take to make the same trip at 90 kmh?

38. The intensity of illumination on a surface varies inversely as the square of the distance from the light source. A surface is 12 meters from a light source. How far must the surface be from the source to receive twice as much illumination?

39. If y varies directly as x^2, and $y = 7$ when $x = 9$, then find y when $x = 7$.

40. If y^2 varies inversely as x, and $y = 4$ when $x = 2$, find y when $x = 11$.

mini-review

Define a variable and write a quadratic function to describe each situation.

1. the product of two numbers whose difference is 55

2. the area of a rectangle whose perimeter is 44 inches.

Find the equation for each conic section described below.

3. the parabola with vertex $(7, -5)$ and focus $(5, -5)$.

4. the circle that has a diameter with endpoints at $(5, -4)$ and $(-1, 2)$

For each pair of functions, f and g, find $[f \circ g](2)$ and $[g \circ f](x)$.

5. $f(x) = 3x + 3$
$g(x) = 2x^2 - 5$

6. $f(x) = 4x^2 - 1$
$g(x) = x^2 + 3x - 6$

7. Brittany can type 35 pages of a term paper in 6 hours. Nathan can type the same number of pages in 8 hours. If Brittany and Nathan work together, how long will it take them to type 35 pages?

Find the solutions of each system of equations.

8. $3x^2 - y^2 = 26$
$2 = y - x$

9. $x^2 - y^2 = 144$
$x^2 = 9 + y^2$

10. Simplify $\dfrac{\dfrac{3x + 5}{3x + 1} - 2}{3 + \dfrac{3x}{1 - 2x}}$.

Euclidean Algorithm

The Greek mathematician Euclid (about 300 B.C.) is credited with developing a method for finding the greatest common factor (GCF) of two integers. This method is called the Euclidean Algorithm.

Example: **Find the GCF of 232 and 136.**

First, divide the greater by the lesser and express the division in the form

$$dividend = quotient \cdot divisor + remainder$$
$$232 \quad = \quad 1 \quad \cdot \quad 136 \quad + \quad 96$$

Divide the divisor by the remainder until the remainder is 0. The last nonzero remainder is the GCF.

$$136 = 1 \cdot 96 + 40$$
$$96 = 2 \cdot 40 + 16$$
$$40 = 2 \cdot 16 + 8$$
$$16 = 2 \cdot 8 + 0 \qquad \textit{The GCF of 232 and 136 is 8.}$$

The computer program below performs the Euclidean Algorithm. If the lesser number is the GCF, it will appear without calculations.

```
10    PRINT "ENTER THE GREATER NUMBER"
20    INPUT X
30    PRINT "ENTER THE LESSER NUMBER"
40    INPUT Y
50    IF INT (X / Y) = X / Y THEN 150
60    PRINT "DIVIDEND = QUOTIENT * DIVISOR + REMAINDER"
70    LET Q = INT (X / Y)
80    LET R = X − (Q * Y)
90    PRINT X;" = ";Q;" * ";Y;" + ";R
100   IF R = 0 THEN 150
110   LET X = Y
120   LET Y = R
140   GOTO 70
150   PRINT Y;" IS THE GREATEST COMMON FACTOR."
180   END
```

Exercises

Use the computer program to find the GCF for each pair of integers.

1. 187, 221   **2.** 182, 1690   **3.** 4807, 5083   **4.** 1078, 1547

5. 41, 3   **6.** 199, 24   **7.** 766, 424   **8.** 197, 37

9. If the GCF of two numbers is 1, the numbers are **relatively prime**. Which of the above pairs are relatively prime? Are both numbers prime in any of the pairs?

rational algebraic expression (387)
complex fraction (389)
rational equation (397)
rational function (405)

asymptote (405)
inversely proportional (410)
inverse variation (411)
constant of variation (411)

Chapter Summary

1. To simplify a rational algebraic expression, divide both numerator and denominator by their greatest common factor (GCF). (387)

2. Multiplying Rational Expressions: For all rational expressions $\frac{a}{b}$ and $\frac{c}{d}$, $b \neq 0$ and $d \neq 0$, $\frac{a}{b} \cdot \frac{c}{d} = \frac{ac}{bd}$. (387)

3. Dividing Rational Expressions: For all rational expressions $\frac{a}{b}$ and $\frac{c}{d}$, $b \neq 0$, $c \neq 0$, and $d \neq 0$, $\frac{a}{b} \div \frac{c}{d} = \frac{a}{b} \cdot \frac{d}{c}$. (388)

4. The sum of rational expressions with common denominators is the sum of the numerators over the common denominator. (393)

5. To add or subtract two rational expressions with different denominators, first find two equivalent rational expressions with common denominators. Then add or subtract the equivalent fractions. (393)

6. Before graphing a rational function, it is helpful to graph the asymptotes of the function. (405)

7. A graphing calculator can generate the graphs of rational functions on its graphics screen. Range parameter values must be chosen carefully to produce a good graphical representation of the function. (408–409)

8. Direct variation is a linear function described by $y = kx$ or $f(x) = kx$ where k is a constant *not* equal to zero. (410)

9. A rational equation in two variables of the form $y = \frac{k}{x}$, where k is a constant *not* equal to zero, is called an inverse variation. The constant k is called the constant of variation, and y is said to vary inversely as x. (411)

Chapter Review

11–1 Multiply or divide. Write each answer in simplest form.

1. $\dfrac{-4ab}{21c} \cdot \dfrac{14c^2}{22a^2}$

2. $\dfrac{y-2}{a-x}(a-3)$

3. $\dfrac{3a+b}{4c} \cdot \dfrac{16c^2 d}{3a^2 + ab}$

4. $\dfrac{x+y}{a} \div \dfrac{x+y}{a^3}$

5. $\dfrac{a^2 - b^2}{6b} \div \dfrac{a+b}{36b^2}$

6. $\dfrac{y^2 - y - 12}{y+2} \div \dfrac{y-4}{y^2 - 4y - 12}$

Simplify each expression.

7. $\dfrac{\dfrac{1}{n^2 - 6n + 9}}{\dfrac{n + 3}{2n^2 - 18}}$

8. $\dfrac{\dfrac{x^2 - y^2}{10x^2}}{\dfrac{x - y}{25xy}}$

9. $\dfrac{\dfrac{x^2 + 7x + 10}{x + 2}}{\dfrac{x^2 + 2x - 15}{x + 2}}$

11–2 Add or subtract. Write each answer in simplest form.

10. $-\dfrac{9}{4a} + \dfrac{7}{3b}$

11. $\dfrac{x + 2}{x - 5} + 6$

12. $\dfrac{x - 1}{x^2 - 1} + \dfrac{2}{5x + 5}$

13. $\dfrac{7}{y} - \dfrac{2}{3y}$

14. $\dfrac{7}{y - 2} - \dfrac{11}{2 - y}$

15. $\dfrac{14}{x + y} - \dfrac{9}{y^2 - x^2}$

Simplify each expression.

16. $\dfrac{\dfrac{5x}{4} + \dfrac{2x}{ab}}{\dfrac{6x}{5} \quad \dfrac{3x}{a}}$

17. $\dfrac{x - \dfrac{1}{x}}{\dfrac{1}{x} + 1}$

18. $\dfrac{\dfrac{2a + 4}{a}}{6 + \dfrac{2}{a^2}}$

11–3 Solve each equation.

19. $\dfrac{3}{y} + \dfrac{7}{y} = 9$

20. $1 + \dfrac{5}{y - 1} = \dfrac{7}{6}$

21. $\dfrac{3x + 2}{4} = \dfrac{9}{4} - \dfrac{3 - 2x}{6}$

22. $\dfrac{1}{r^2 - 1} = \dfrac{2}{r^2 + r - 2}$

23. $\dfrac{x}{x^2 - 1} + \dfrac{2}{x + 1} = 1 + \dfrac{1}{2x - 2}$

11–4 Solve each problem.

24. Bob Lopatka can paint his house in 15 hours. His friend, Jack, can paint the house in 20 hours. If they work together, how long will it take them to paint the house?

25. One integer is 2 less than another integer. Three times the reciprocal of the lesser integer plus five times the reciprocal of the greater integer is $\dfrac{7}{8}$.

What are the two integers?

11–5 Graph each rational function.

26. $f(x) = \dfrac{4}{x - 2}$

27. $f(x) = \dfrac{x}{x + 3}$

28. $f(x) = \dfrac{5}{(x + 1)(x - 3)}$

11–6 In each of the following, *y* varies directly as *x*.

29. If $x = 7$, then $y = 21$. Find x when $y = -5$.

30. If $y = 18$, then $x = 28$. Find x when $y = 63$.

In each of the following, *y* varies inversely as *x*.

31. If $y = 9$, then $x = \dfrac{5}{2}$.

Find y when x is $-\dfrac{3}{5}$.

32. If $y = 18$, then $x = 28$. Find x when $y = 63$.

Perform the indicated operation. Write each answer in simplest form.

1. $\dfrac{7ab}{9c} \cdot \dfrac{81c^2}{91a^2b}$

2. $\dfrac{x^2 - y^2}{a^2 - b^2} \cdot \dfrac{a + b}{x - y}$

3. $\dfrac{a^2 - ab}{3a} \div \dfrac{a - b}{15b^2}$

4. $\dfrac{x^2 - 2x + 1}{y - 5} \div \dfrac{x - 1}{y^2 - 25}$

5. $\dfrac{7}{5a} - \dfrac{10}{3ab}$

6. $\dfrac{6}{x - 5} + 7a$

7. $\dfrac{x - y}{a - b} - \dfrac{x + y}{a + b}$

8. $\dfrac{x + 2}{x - 1} + \dfrac{6}{7x - 7}$

Simplify.

9. $\dfrac{\dfrac{2}{x - 4} + \dfrac{5}{x + 1}}{\dfrac{3x}{x^2 - 3x - 4}}$

10. $\dfrac{\dfrac{1}{x} - \dfrac{1}{2x}}{\dfrac{2}{x} + \dfrac{4}{3x}}$

Solve each equation.

11. $\dfrac{3}{x} - \dfrac{7}{x} = 9$

12. $a - \dfrac{5}{a} = 4$

13. $\dfrac{y}{y - 3} + \dfrac{6}{y + 3} = 1$

14. $\dfrac{3}{x} + \dfrac{x}{x + 2} = \dfrac{-2}{x + 2}$

Graph each rational function.

15. $y = \dfrac{2}{x - 1}$

16. $y = \dfrac{3x}{x + 2}$

Solve each problem.

17. Joni Mills can type 75 pages of manuscript in 8 hours. Ted Szatro can type the same number of pages in 13 hours. If Joni and Ted work together, how long will it take them to type 75 pages?

18. The capacities of two barrels are in the ratio of 7 to 4. The larger barrel has a capacity that is 12 gallons less than twice that of the smaller barrel. Find the capacity of each barrel.

19. Suppose y varies directly as x. If $y = 10$, then $x = -3$. Find y when x is 20.

20. Suppose y varies inversely as x. If $y = 9$, then $x = -\dfrac{2}{3}$. Find x when y is -7.

21. On a blueprint a square 12 meters on a side was shown as a square 3 cm on a side. A beam is actually 20 meters long. How long does it appear on the blueprint?

22. If two boxes have the same capacity and depth, the length is inversely proportional to the width. One box is 60 cm long and 40 cm wide. A second box is 5 cm long. What is its width if it has the same capacity and depth as the first box?

CHAPTER 12

Exponential and Logarithmic Functions

Jean Garson needs to determine how long it takes two of a certain strain of bacteria to increase to 1000 bacteria. To find an approximate answer, she uses the general formula for growth and decay in nature, $y = ne^{kt}$. This formula is an example of an exponential equation.

12-1 Real Exponents

Consider the graph of $y = 2^x$ where x is a rational exponent. Since each value of x has a corresponding unique value of y, the equation $y = 2^x$ represents a function.

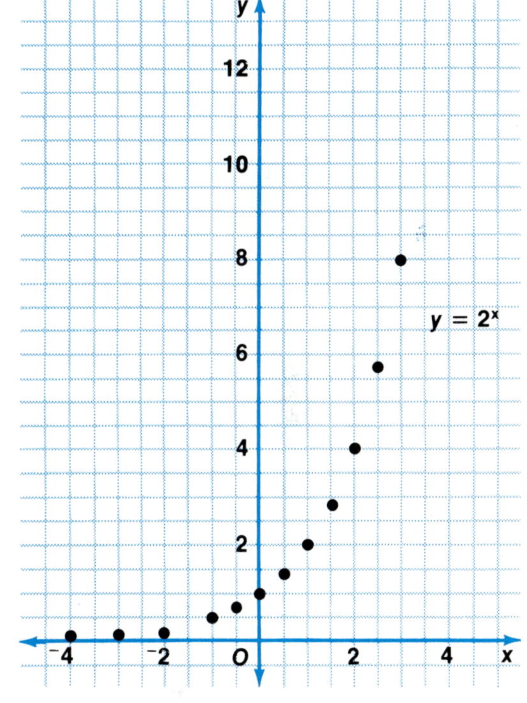

x	2^x or y	y (approximate)
-4	$2^{-4} = \dfrac{1}{16}$	0.06
-3	$2^{-3} = \dfrac{1}{8}$	0.12
-2	$2^{-2} = \dfrac{1}{4}$	0.25
-1	$2^{-1} = \dfrac{1}{2}$	0.5
$-\dfrac{1}{2}$	$2^{-\frac{1}{2}} = \dfrac{1}{2}\sqrt{2}$	0.7
0	$2^0 = 1$	1
$\dfrac{1}{2}$	$2^{\frac{1}{2}} = \sqrt{2}$	1.4
1	$2^1 = 2$	2
$\dfrac{3}{2}$	$2^{\frac{3}{2}} = 2\sqrt{2}$	2.8
2	$2^2 = 4$	4
$\dfrac{5}{2}$	$2^{\frac{5}{2}} = 4\sqrt{2}$	5.7
3	$2^3 = 8$	8

As greater values are selected for x the value of y increases. Thus, the function is increasing.

Since 2^x has not been defined when x is irrational, "holes" still remain in the graph of $y = 2^x$. How could you expand the domain of $y = 2^x$ to include both rational and irrational numbers?

Consider an expression such as $2^{\sqrt{3}}$. Since $1.7 < \sqrt{3} < 1.8$ it follows that $2^{1.7} < 2^{\sqrt{3}} < 2^{1.8}$. By selecting closer approximations for $\sqrt{3}$, closer approximations for $2^{\sqrt{3}}$ are possible.

$$2^{1.7} < 2^{\sqrt{3}} < 2^{1.8}$$
$$2^{1.73} < 2^{\sqrt{3}} < 2^{1.74}$$
$$2^{1.732} < 2^{\sqrt{3}} < 2^{1.733}$$
$$2^{1.7320} < 2^{\sqrt{3}} < 2^{1.7321}$$
$$2^{1.73205} < 2^{\sqrt{3}} < 2^{1.73206}$$

Therefore, it is possible to determine an approximate value for 2^x when x represents an irrational number by using rational approximations for x.

Since 2^x is now defined when x represents an irrational number, the domain of $y = 2^x$ has been expanded to the set of real numbers.

The "holes" have been filled and the graph is now a smooth curve.

By using a large, accurate graph of $y = 2^x$, you could estimate the value of 2^x when x represents any real number.

Examples

1   **Use the graph of $y = 2^x$ to evaluate y to the nearest tenth.**

a. $y = 2^{\sqrt{3}}$

The value of x is $\sqrt{3}$ and $1.7 < \sqrt{3} < 1.8$.

From the graph, the value of y is approximately 3.3.

Verify the value of $2^{\sqrt{3}}$ using a calculator.

$2^{\sqrt{3}} \approx 3.3219971$

b. $y = 2^{2.1}$

From the graph, the value of y is approximately 4.3.

Verify the value of $2^{2.1}$ using a calculator.

$2^{2.1} = 4.2870939$

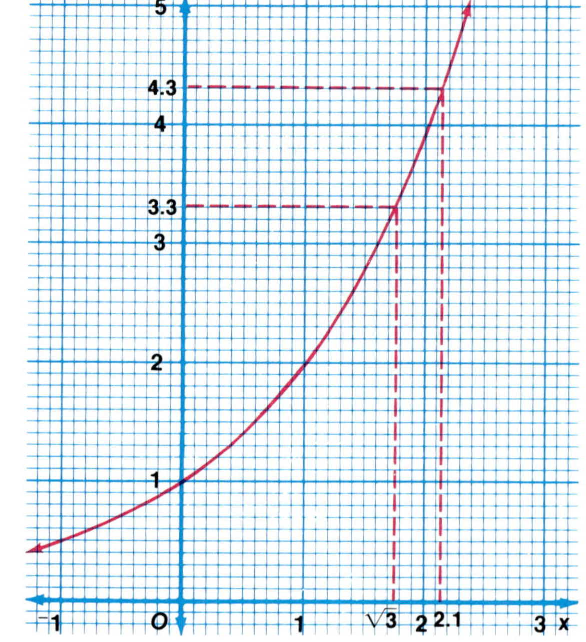

2   **Using Calculators**

Evaluate y to the nearest tenth if $y = 2^{\sqrt{6}}$.

ENTER: 2 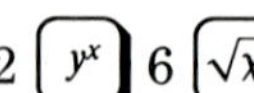 6  $\boxed{=}$

DISPLAY: *2   6 2.4494897 5.4622288*

The value of y is approximately 5.5.

The properties of rational exponents apply to all real exponents.

3 Simplify $3^{\sqrt{2}} \cdot 3^{\sqrt{5}}$.

$3^{\sqrt{2}} \cdot 3^{\sqrt{5}} = 3^{\sqrt{2}+\sqrt{5}}$ *Recall that* $a^m \cdot a^n = a^{m+n}$.

4 Simplify $(4^{\sqrt{3}})^{\sqrt{2}}$.

$(4^{\sqrt{3}})^{\sqrt{2}} = 4^{\sqrt{3}\cdot\sqrt{2}}$ *Recall that* $(a^m)^n = a^{mn}$.
$= 4^{\sqrt{6}}$

An equation of the form $y = a^x$, where $a > 0$ and $a \neq 1$, is called an exponential function.	*Definition of Exponential Function*

The figure at the right shows graphs of several exponential functions. Compare the graphs of functions where $a > 1$ and those where $a < 1$. Notice that when $a > 1$, the value of y increases as the value of x increases. When $a < 1$, the value of y decreases as the value of x increases.

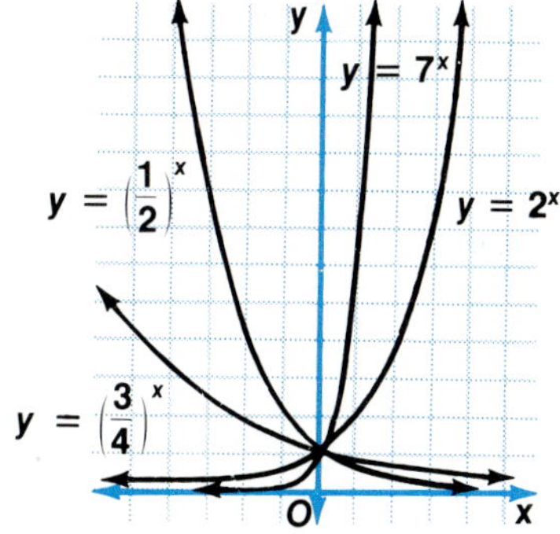

An important property of exponential functions, that is useful for solving equations, is the Property of Equality.

Suppose a is a positive number other than 1. Then $a^{x_1} = a^{x_2}$ if and only if $x_1 = x_2$.	*Property of Equality for Exponential Functions*

Examples

5 Solve $3^5 = 3^{2n-1}$ for n.

$3^5 = 3^{2n-1}$
$5 = 2n - 1$ *Property of Equality for Exponential Functions*
$6 = 2n$
$n = 3$

Check: $3^5 = 3^{2n-1}$
$3^5 \stackrel{?}{=} 3^{2(3)-1}$
$3^5 = 3^5$ ✔

6 Solve $9^{3r} = 27^{r-2}$ for r.

$9^{3r} = 27^{r-2}$ *The bases must be the same.*
$(3^2)^{3r} = (3^3)^{r-2}$ *Replace 9 with 3^2 and 27 with 3^3.*
$3^{6r} = 3^{3r-6}$
$6r = 3r - 6$ *Property of Equality for Exponential Functions*
$3r = -6$
$r = -2$

Check: $9^{3r} = 27^{r-2}$
$9^{3(-2)} \stackrel{?}{=} 27^{-2-2}$
$9^{-6} \stackrel{?}{=} 27^{-4}$
$\dfrac{1}{531441} = \dfrac{1}{531441}$ ✔

Exploratory Exercises

Use the graph of $y = 2^x$ on page 420 to evaluate each expression to the nearest tenth.

1. $2^{0.7}$ **2.** $2^{1.1}$ **3.** $2^{-0.3}$ **4.** 2^{-1} **5.** $2^{\sqrt{2}}$ **6.** $2^{\sqrt{5}}$

Use a calculator to evaluate each expression to the nearest tenth.

7. $2^{\sqrt{5}}$ **8.** $8^{\sqrt{2}}$ **9.** $17^{\sqrt{3}}$ **10.** $64^{\sqrt{5}}$ **11.** $2^{\sqrt[3]{5}}$ **12.** $5^{\sqrt[3]{2}}$

Use the rules for exponents to simplify each expression.

13. $2^{\sqrt{5}} \cdot 2^{3\sqrt{5}}$ **14.** $7^{\sqrt{3}} \cdot 7^{2\sqrt{3}}$ **15.** $(2^{\sqrt{3}})^{\sqrt{3}}$

16. $(9^{\sqrt{5}})^{\sqrt{5}}$ **17.** $3(2^{\sqrt{2}})(2^{-\sqrt{2}})$ **18.** $8^{2\sqrt{3}} \div 8^{\sqrt{12}}$

Solve each equation.

19. $4^x = 4^{-5}$ **20.** $5^x = 125$ **21.** $10^x = 0.001$

22. $7^t = \dfrac{1}{49}$ **23.** $2^{2r} = \dfrac{1}{8}$ **24.** $\left(\dfrac{1}{6}\right)^{a-3} = 216$

Written Exercises

Use the graph of $y = 2^x$ on page 420 or a calculator to evaluate each expression to the nearest tenth.

25. $2^{1.5}$ **26.** $2^{-1.1}$ **27.** $4^{-0.3}$ **28.** $8^{0.6}$

29. $8^{-0.3}$ **30.** $16^{0.4}$ **31.** $2^{\sqrt[3]{2}}$ **32.** $2^{\sqrt[3]{7}}$

Simplify each expression.

33. $(2^{\sqrt{3}})^{\sqrt{12}}$ **34.** $(3^{\sqrt{8}})^{\sqrt{2}}$ **35.** $5^{\sqrt{3}} \cdot 5^{\sqrt{27}}$

36. $11^{\sqrt{5}} \cdot 11^{\sqrt{45}}$ **37.** $16^{\sqrt{7}} \div 2^{\sqrt{7}}$ **38.** $9^{\sqrt{3}} \div 3^{\sqrt{3}}$

39. $8^{\sqrt{3}} \cdot 16^{\sqrt{5}}$ **40.** $64^{\sqrt{2}} \cdot 16^{\sqrt{3}}$ **41.** $y^{\sqrt{5}} \cdot y^{\sqrt{45}}$

42. $(x^{\sqrt{2}})^{\sqrt{8}}$ **43.** $(y^{\sqrt{3}})^{\sqrt{27}}$ **44.** $b^{\sqrt{2}} \cdot b^{\sqrt{32}}$

45. $(m^{\sqrt{2}} \cdot p^{\sqrt{2}})^{\sqrt{2}}$ **46.** $(m^{\sqrt{2}} + n^{\sqrt{2}})^2$

47. $(x^{\sqrt{3}} + y^{\sqrt{2}})^2$ **48.** $(x^{\sqrt{2}} - y^{\sqrt{2}})(x^{\sqrt{2}} + y^{\sqrt{2}})$

Solve each equation.

49. $2^5 = 2^{2x-1}$ **50.** $3^y = 3^{3y+1}$ **51.** $5^{3s+4} = 5^s$

52. $3^x = 9^{x+1}$ **53.** $9^{3y} = 27^{y+2}$ **54.** $8^{r-1} = 16^{3r}$

55. $2^{z+3} = \dfrac{1}{16}$ **56.** $\dfrac{1}{27} = 3^{x-5}$ **57.** $\left(\dfrac{1}{3}\right)^p = 3^{p-6}$

58. $2^{2m-1} = 8^{m+7}$ **59.** $25^{2n} = 125^{n-3}$ **60.** $4^{y-1} = 8^y$

61. $2^{x^2+1} = 32$ **62.** $36^x = 6^{x^2-3}$ **63.** $9^{x^2-2x} = 27^{x^2+1}$

Graph each equation.

64. $y = 3^x$ **65.** $y = \left(\dfrac{1}{3}\right)^x$ **66.** $y = \left(\dfrac{1}{4}\right)^x$ **67.** $y = 4^x$

68. Compare the graphs for Exercises 64 and 65. What do you notice?

69. Compare the graphs for Exercises 66 and 67. What do you notice?

12-2 Logarithms

In the table shown below on the left, x is an exponent. Compute the value of the power of 2 to find y. In the table on the right, the roles of x and y are reversed. You are given x as the value of the power of 2. Now compute the exponent, y, that gives this power of 2.

x	$2^x = y$	y
-1	$2^{-1} = y$	?
2	$2^2 = y$	?
3	$2^3 = y$	?
6	$2^6 = y$	?

Exponent to Power of 2 →

y	$2^y = x$	x
?	$2^y = \frac{1}{2}$	$\frac{1}{2}$
?	$2^y = 4$	4
?	$2^y = 8$	8
?	$2^y = 64$	64

← Exponent from Power of 2

In the relation $2^y = x$, the exponent y is called the logarithm, base 2, of x. This relation is more conveniently written as $\log_2 x = y$. The equation $\log_2 x = y$ is read *the log, base 2, of x is equal to y*. The logarithm corresponds to an exponent.

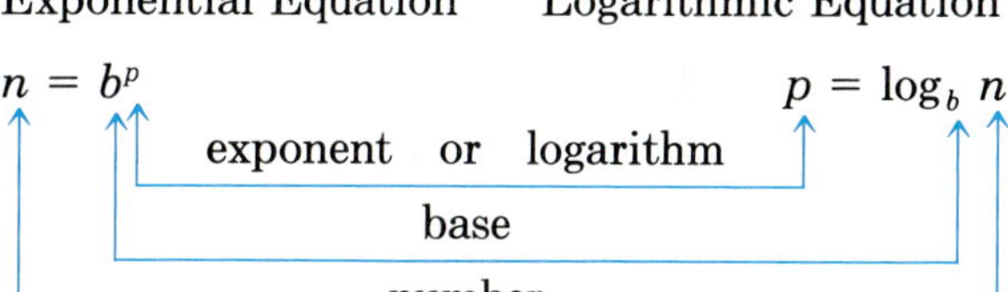

Suppose $b > 0$ and $b \neq 1$. Then for $n > 0$, there is a number p such that $\log_b n = p$ if and only if $b^p = n$.

Definition of Logarithm

Exponential Equation	Logarithmic Equation
$6^2 = 36$	$\log_6 36 = 2$
$10^4 = 10{,}000$	$\log_{10} 10{,}000 = 4$
$3^0 = 1$	$\log_3 1 = 0$
$2^{-3} = \frac{1}{8}$	$\log_2 \frac{1}{8} = -3$
$4^{\frac{1}{2}} = 2$	$\log_4 2 = \frac{1}{2}$

Example

1 Solve $\log_2 64 = y$ for y.

$$\log_2 64 = y$$
$$2^y = 64 \qquad \text{\textit{Definition of Logarithm}}$$
$$2^y = 2^6$$
$$y = 6 \qquad \text{\textit{Property of Equality for Exponential Functions}}$$

2 Solve $\log_9 x = -3$ for x.

$\log_9 x = -3$

$9^{-3} = x$ *Definition of Logarithm*

$x = \dfrac{1}{729}$

3 Solve $\log_b 16 = 2$ for b.

$\log_b 16 = 2$

$b^2 = 16$ *Definition of Logarithm*

$b = 4$ *Since $b > 0$, -4 is not a solution.*

Exploratory Exercises

Write each equation in logarithmic form.

1. $3^3 = 27$ **2.** $4^2 = 16$ **3.** $10^3 = 1000$

4. $10^{-2} = 0.01$ **5.** $2^{-3} = \dfrac{1}{8}$ **6.** $5^{-2} = \dfrac{1}{25}$

Write each equation in exponential form.

7. $\log_4 64 = 3$ **8.** $\log_3 9 = 2$ **9.** $\log_{10} 0.1 = -1$

10. $\log_{10} 0.0001 = -4$ **11.** $\log_9 27 = \dfrac{3}{2}$ **12.** $\log_3 \dfrac{1}{81} = -4$

Written Exercises

Write each equation in logarithmic form.

13. $3^4 = 81$ **14.** $2^6 = 64$ **15.** $5^3 = 125$ **16.** $8^0 = 1$

17. $4^{-2} = \dfrac{1}{16}$ **18.** $3^{-1} = \dfrac{1}{3}$ **19.** $2^{-4} = \dfrac{1}{16}$ **20.** $\left(\dfrac{1}{9}\right)^{-2} = 81$

21. $3^{\frac{1}{2}} = \sqrt{3}$ **22.** $9^{\frac{3}{2}} = 27$ **23.** $36^{\frac{3}{2}} = 216$ **24.** $16^{\frac{3}{4}} = 8$

Write each equation in exponential form.

25. $\log_2 32 = 5$ **26.** $\log_8 64 = 2$ **27.** $\log_{11} 121 = 2$ **28.** $\log_{13} 13 = 1$

29. $\log_5 1 = 0$ **30.** $\log_3 243 = 5$ **31.** $\log_{\frac{1}{2}} 16 = -4$ **32.** $\log_{\frac{1}{3}} 81 = -4$

33. $\log_{10} \dfrac{1}{10} = -1$ **34.** $\log_5 \dfrac{1}{25} = -2$ **35.** $\log_{27} 3 = \dfrac{1}{3}$ **36.** $\log_8 4 = \dfrac{2}{3}$

Evaluate each expression.

37. $\log_{10} 1000$ **38.** $\log_3 81$ **39.** $\log_{12} 144$ **40.** $\log_{10} 0.01$ **41.** $\log_{\frac{1}{4}} 64$

42. $\log_7 \dfrac{1}{343}$ **43.** $\log_2 \dfrac{1}{16}$ **44.** $\log_4 2$ **45.** $\log_9 27$ **46.** $\log_8 16$

Solve each equation.

47. $\log_{\frac{1}{2}} 8 = x$ **48.** $\log_{10} 0.001 = x$ **49.** $\log_b 49 = 2$ **50.** $\log_b 64 = 3$

51. $\log_6 x = 2$ **52.** $\log_9 x = -1$ **53.** $\log_{\frac{1}{2}} 16 = x$ **54.** $\log_3 27 = x$

55. $\log_b 81 = 4$ **56.** $\log_b 18 = 1$ **57.** $\log_5 x = -2$ **58.** $\log_3 x = -3$

59. $\log_{10} \sqrt{10} = x$ **60.** $\log_5 \sqrt{5} = x$ **61.** $\log_{10} \sqrt[3]{10} = x$ **62.** $\log_b 36 = -2$

63. $\log_a \dfrac{1}{27} = -3$ **64.** $\log_x \sqrt{5} = \dfrac{1}{4}$ **65.** $\log_x \sqrt[3]{7} = \dfrac{1}{3}$ **66.** $\log_4 x = -\dfrac{1}{2}$

67. $\log_{\frac{1}{2}} x = -6$ **68.** $\log_2 x = -4$ **69.** $\log_{\sqrt{3}} x = 6$ **70.** $\log_{\sqrt{3}} 27 = x$

Applications in Seismology Earthquakes

Seismology is a science that deals with earthquakes and artificially produced vibrations of the earth. A *logarithmic* scale called the **Richter scale** is used to measure the strength of an earthquake. Each increase of one on the Richter scale corresponds to a ten-times increase in intensity. In other words, an earthquake that registers 8 on the Richter scale is ten times as intense as an earthquake that registers 7. An earthquake that registers 9 is ten times as intense as the one registering 8, and one hundred times as intense as the one registering 7.

The table below gives the effects of earthquakes of various intensities.

Richter Number	Intensity	Effect
1	10^1	only detectable by seismograph
2	10^2	hanging lamps sway
3	10^3	can be felt
4	10^4	glass breaks, buildings shake
5	10^5	furniture collapses
6	10^6	wooden houses damaged
7	10^7	buildings collapse
8	10^8	catastrophic damage

The photo below shows the earth's shift along the San Andreas fault. Scientists claim that shifts in the earth's surface along this fault could cause major earthquakes in the near future. Small earthquakes are a common occurrence in California.

On April 18, 1906, one of the worst California earthquakes in recent history hit San Francisco. Fires following the earthquake burned more than 4 square miles. Hundreds of people died. There was from 250 to 300 million dollars worth of property damage. It is believed that this earthquake would have measured 8.3 on the Richter scale.

Exercises

1. An earthquake with a rating of 7 is how much stronger than one with a rating of 6?

2. An earthquake with a rating of 7 is how much stronger than one with a rating of 4?

3. Which was stronger, the San Francisco earthquake or the Alaska earthquake that rated 8.4?

4. Which was stronger, the Ecuador earthquake that rated 8.9 or the Alaska earthquake?

5. Compare earthquakes rated 6.5 and 7.1. How many times stronger is the latter?
Hint: Divide $10^{7.1}$ by $10^{6.5}$.

12-3 Logarithmic Functions

Study the graphs of $y = 2^x$ and $2^y = x$.

$$y = 2^x \qquad\qquad 2^y = x \quad \text{or} \quad y = \log_2 x$$

x	y
-4	$\frac{1}{16}$
-3	$\frac{1}{8}$
-2	$\frac{1}{4}$
-1	$\frac{1}{2}$
0	1
1	2
2	4
3	8

x	y
$\frac{1}{16}$	-4
$\frac{1}{8}$	-3
$\frac{1}{4}$	-2
$\frac{1}{2}$	-1
1	0
2	1
4	2
8	3

The x and y values are reversed.

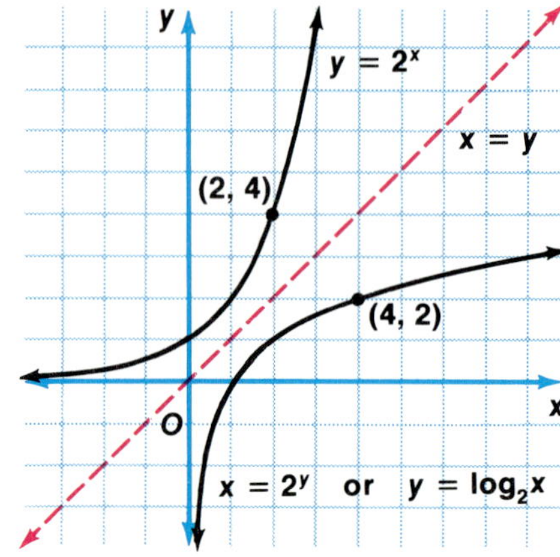

The domain of the relation $y = \log_2 x$ is the set of all positive numbers. The range is the set of all real numbers.

The graphs of $y = 2^x$ and $y = \log_2 x$ are reflections of each other with respect to the graph of $y = x$. Thus, $y = \log_2 x$ is the inverse of $y = 2^x$. From the graph, $y = \log_2 x$ appears to be a function. No vertical line will intersect the graph in more than one point. The relation $y = \log_2 x$ is an example of a **logarithmic function**.

An equation of the form $y = \log_b x$, where $b > 0$ and $b \neq 1$, is called a logarithmic function.	*Definition of Logarithmic Function*

Study the following examples.

$$3^{\log_3 9} = 3^2 \text{ or } 9 \qquad \text{and} \qquad \log_3 3^2 = \log_3 9 \text{ or } 2$$

$$4^{\log_4 16} = 4^2 \text{ or } 16 \qquad \text{and} \qquad \log_4 4^2 = \log_4 16 \text{ or } 2$$

The logarithmic function $f(x) = \log_b x$ and exponential function $g(x) = b^x$ are inverse functions. For functions f and g to be inverse functions, it must be true that $f(g(x)) = x$ and $g(f(x)) = x$.

$$f(g(x)) = x \qquad\qquad g(f(x)) = x$$
$$f(b^x) = x \qquad\qquad g(\log_b x) = x$$
$$\log_b b^x = x \qquad\qquad b^{\log_b x} = x$$

1 Evaluate $\log_7 7^4$.

$\log_7 7^4 = 4$ $\log_b b^x = x$

2 Evaluate $3^{\log_3 (2n-3)}$.

$3^{\log_3 (2n-3)} = 2n - 3$ $b^{\log_b x} = x$

The Property of Equality for Exponential Functions states that $b^{x_1} = b^{x_2}$ if and only if $x_1 = x_2$. A similar property of equality also holds for logarithmic functions.

Suppose $b > 0$ and $b \neq 1$. Then $\log_b x_1 = \log_b x_2$ if and only if $x_1 = x_2$.	*Property of Equality for Logarithmic Functions*

3 Solve $\log_2 (3x - 1) = \log_2 (2x + 4)$.

$$\log_2 (3x - 1) = \log_2 (2x + 4)$$
$$3x - 1 = 2x + 4 \quad \textit{Property of Equality for Logarithmic Functions}$$
$$x = 5$$

The solution is 5.   *Check this result.*

Exploratory Exercises

Evaluate each expression.

1. $\log_5 5^2$ **2.** $\log_9 9^4$ **3.** $9^{\log_9 2}$ **4.** $7^{\log_7 3}$ **5.** $\log_m m^x$ **6.** $b^{\log_b y}$

Solve each equation.

7. $\log_2 x = \log_2 4$ **8.** $\log_4 10 = \log_4 2x$ **9.** $\log_3 (x + 1) = \log_3 (2x)$

10. $\log_9 (3x) = \log_9 (x - 2)$ **11.** $\log_7 (x^2 - 1) = \log_7 3$ **12.** $\log_5 x^2 = \log_5 (6x)$

Written Exercises

Evaluate each expression.

13. $\log_4 4^3$ **14.** $\log_r r^4$ **15.** $6^{\log_6 7}$ **16.** $9^{\log_9 5}$ **17.** $\log_n n^5$ **18.** $3^{\log_3 21}$

Solve each equation.

19. $\log_3 (2x + 1) = \log_3 (3x - 6)$ **20.** $\log_{10} (4 + y) = \log_{10} (2y)$

21. $\log_{10} (3n) = \log_{10} (n + 2)$ **22.** $\log_4 (2x - 3) = \log_4 (x + 2)$

23. $\log_3 (3y - 1) = \log_3 (y + 4)$ **24.** $\log_7 (5x - 1) = \log_7 (3x + 7)$

25. $\log_{10} (x^2 + 36) = \log_{10} 100$ **26.** $\log_{10} (x - 1)^2 = \log_{10} 0.01$

27. $\log_9 (x^2 + 9x) = \log_9 10$ **28.** $\log_5 (4x) = \log_5 (x^2 - 5)$

29. $\log_{12} (7x - 3) = \log_{12} (5 - x^2)$ **30.** $\log_2 (x^2 + 6x) = \log_2 (x - 4)$

Graph each pair of equations on the same set of axes.

31. $y = 3^x$ and $y = \log_3 x$

32. $y = \left(\dfrac{1}{2}\right)^x$ and $y = \log_{\frac{1}{2}} x$

33. $y = 4^x$ and $y = \log_4 x$

34. $y = 10^x$ and $y = \log_{10} x$

Show that each statement is true.

35. $\log_4 4 + \log_4 16 = \log_4 64$

36. $\log_3 27 + \log_3 3 = \log_3 81$

37. $\log_2 32 - \log_2 4 = \log_2 8$

38. $\log_6 36 - \log_6 6 = \log_6 6$

39. $\log_3 27 = 3 \log_3 3$

40. $\log_4 16 = 2 \log_4 4$

41. $\dfrac{1}{2} \log_3 81 = \log_3 9$

42. $\dfrac{1}{3} \log_5 25 = 2 \log_5 \sqrt[3]{5}$

43. $\log_2 8 \cdot \log_8 2 = 1$

44. $\log_5 25 \cdot \log_{25} 5 = 1$

45. $\log_{10} [\log_3 (\log_4 64)] = 0$

46. $\log_2 64 = 3 \log_8 64$

47. $\log_3 81 = \dfrac{4}{3} \log_2 8$

48. $\log_4 [\log_2 (\log_3 81)] = \dfrac{1}{2}$

Challenge Exercises

Solve each equation.

49. $6^{\log_6 x^2} = x + 30$

50. $3^{\log_3 x^3} = 0.125$

51. $\log_2 [\log_4 (\log_3 x)] = -1$

52. $\log_{10} [\log_2 (\log_7 x)] = 0$

mini-review

State whether the graph of each equation is a parabola, a circle, an ellipse, or a hyperbola. Then draw the graph.

1. $16(x - 4)^2 + 9(y - 2)^2 = 144$

2. $x^2 - 4y^2 + 6x + 16y - 11 = 0$

3. Find the solutions for the system $x + 3y = 6$ and $x^2 + 3y^2 = 36$.

If $f(x) = 3x^2 + 2x - 1$, $g(x) = x - 5$, and $h(x) = 2x + 7$, find each value.

4. $h(g(3))$

5. $g(f(2))$

6. $f(g(x))$

Find all zeros of each function.

7. $f(x) = x^3 - 2x^2 - 5x + 6$

8. $g(x) = 2x^4 - 13x^3 + 29x^2 - 18x - 18$

9. State the number of positive real zeros, negative real zeros, and imaginary zeros for $h(x) = -2x^4 + 6x^3 - 7x^2 + 7x - 5$.

10. Approximate to the nearest tenth the real zeros of $f(x) = 3x^3 - 4x^2 + 7x + 28$.

11. Find $\left(\dfrac{2g^2}{h}\right)^3 \cdot \dfrac{5}{6hg^4}$.

12. Find $\dfrac{x^3 - y^3}{y^2 - x^2} \div \dfrac{x^2 + xy}{x^2 + 2xy + y^2}$.

13. Find $\dfrac{k}{k^2 - 10k + 25} + \dfrac{4}{k^2 - 25}$.

14. Solve $\dfrac{2x + 1}{2x - 1} - \dfrac{16}{4x^2 - 1} = \dfrac{2x - 1}{2x + 1}$.

Solve each equation.

15. $2^{2x-3} = 8^{x-4}$

16. $16^{3y-2} = 64^{4y+2}$

17. $\log_b 256 = 4$

18. $\log_9 3 = x$

Logarithms are exponents. Thus, the properties of logarithms can be derived from the properties of exponents.

For example, the product of powers is found by adding exponents. This suggests that the logarithm of a product is found by adding logarithms.

$$\log_2 (8 \cdot 32)$$
$$= \log_2 (2^3 \cdot 2^5)$$
$$= \log_2 (2^{3+5})$$
$$= 3 + 5$$

← Notice that these →
two expressions
are shown to
be equivalent.

$$\log_2 8 + \log_2 32$$
$$= \log_2 2^3 + \log_2 2^5$$
$$= 3 + 5$$

This example illustrates the Product Property of Logarithms.

For all positive numbers m, n, and b, where $b \neq 1$, $$\log_b mn = \log_b m + \log_b n.$$	*Product Property of Logarithms*

To prove this property, let $b^x = m$ and $b^y = n$.
Then $\log_b m = x$ and $\log_b n = y$.

$$b^x b^y = mn$$
$$b^{x+y} = mn \qquad \textit{Property of Exponents}$$
$$\log_b b^{x+y} = \log_b mn \qquad \textit{Property of Equality for Logarithmic Functions}$$
$$x + y = \log_b mn \qquad \textit{Definition of Inverse Functions}$$
$$\log_b m + \log_b n = \log_b mn \qquad \textit{Substitution Property}$$

Example

1 **Given $\log_3 5 = 1.465$, find $\log_3 45$ and $\log_3 25$.**

$$\log_3 45 = \log_3 (3^2 \cdot 5)$$
$$= \log_3 3^2 + \log_3 5$$
$$= 2 + 1.465 \quad \text{or} \quad 3.465$$

$$\log_3 25 = \log_3 (5 \cdot 5)$$
$$= \log_3 5 + \log_3 5$$
$$= 1.465 + 1.465 \quad \text{or} \quad 2.930$$

The quotient of powers is found by subtracting exponents. This suggests that the logarithm of a quotient is found by subtracting logarithms.

$$\log_2 \left(\frac{32}{8} \right)$$
$$= \log_2 \left(\frac{2^5}{2^3} \right)$$
$$= \log_2 (2^{5-3})$$
$$= 5 - 3$$

← Notice that these →
two expressions
are shown to
be equivalent.

$$\log_2 32 - \log_2 8$$
$$= \log_2 2^5 - \log_2 2^3$$
$$= 5 - 3$$

This example illustrates the Quotient Property of Logarithms.

For all positive numbers m, n, and b, where $b \neq 1$,

$$\log_b \frac{m}{n} = \log_b m - \log_b n.$$

Quotient Property of Logarithms

To prove this property, let $b^x = m$ and $b^y = n$.
Then $\log_b m = x$ and $\log_b n = y$.

$$\frac{b^x}{b^y} = \frac{m}{n}$$

$$b^{x-y} = \frac{m}{n} \qquad \textit{Property of Exponents}$$

$$\log_b b^{x-y} = \log_b \frac{m}{n} \qquad \textit{Property of Equality for Logarithmic Functions}$$

$$x - y = \log_b \frac{m}{n} \qquad \textit{Definition of Inverse Functions}$$

$$\log_b m - \log_b n = \log_b \frac{m}{n} \qquad \textit{Substitution Property}$$

Examples

2 **Given $\log_6 2 = 0.3869$ and $\log_6 20 = 1.6720$, find $\log_6 0.1$ and $\log_6 18$.**

$$\log_6 0.1 = \log_6 \left(\frac{2}{20}\right) \qquad\qquad \log_6 18 = \log_6 \left(\frac{6^2}{2}\right)$$

$$= \log_6 2 - \log_6 20 \qquad\qquad\qquad = \log_6 6^2 - \log_6 2$$

$$= 0.3869 - 1.6720 \quad \text{or} \quad -1.2851 \qquad = 2 - 0.3869 \quad \text{or} \quad 1.6131$$

3 **Solve $\log_{12} 72 - \log_{12} 9 = \log_{12} 4m$.**

$$\log_{12} 72 - \log_{12} 9 = \log_{12} 4m$$

$$\log_{12} \frac{72}{9} = \log_{12} 4m \qquad \textit{Quotient Property of Logarithms}$$

$$\frac{72}{9} = 4m \qquad \textit{Property of Equality for Logarithmic Functions}$$

$$2 = m$$

Check:
$$\log_{12} 72 - \log_{12} 9 = \log_{12} 4m$$
$$\log_{12} 72 - \log_{12} 9 \overset{?}{=} \log_{12} (4 \cdot 2)$$
$$\log_{12} 8 = \log_{12} 8 \quad ✔$$

The solution is 2.

The power of a power is found by multiplying the two exponents. This suggests that the logarithm of a power is found by multiplying the logarithm and the exponent.

$\log_2 8^4 \qquad \leftarrow$ *Notice that these* $\rightarrow \quad 4 \log_2 8$

two expressions are shown to be equivalent.

$= \log_2 (2^3)^4 \qquad\qquad\qquad\qquad = (\log_2 8) \cdot 4$

$= \log_2 2^{3 \cdot 4} \qquad\qquad\qquad\qquad = (\log_2 2^3) \cdot 4$

$= 3 \cdot 4 \qquad\qquad\qquad\qquad\qquad = 3 \cdot 4$

This example illustrates the Power Property of Logarithms.

<table>
<tr><td>

For any real number p and positive numbers m and b, where $b \neq 1$,

$$\log_b m^p = p \cdot \log_b m.$$

</td><td>

Power Property of Logarithms

</td></tr>
</table>

To prove this property, let $b^x = m$. Then $\log_b m = x$.

$$(b^x)^p = m^p$$
$$b^{xp} = m^p \qquad \textit{Power Property of Exponents}$$
$$\log_b b^{xp} = \log_b m^p \qquad \textit{Property of Equality for Logarithmic Functions}$$
$$xp = \log_b m^p \qquad \textit{Definition of Inverse Functions}$$
$$p \log_b m = \log_b m^p \qquad \textit{Substitution Property}$$

Examples

4 **Solve $4 \log_5 x - \log_5 4 = \log_5 4$.**

$$4 \log_5 x - \log_5 4 = \log_5 4$$
$$4 \log_5 x = \log_5 4 + \log_5 4$$
$$4 \log_5 x = \log_5 16$$
$$\log_5 x^4 = \log_5 16 \qquad \textit{Power Property of Logarithms}$$
$$x^4 = 16 \qquad \textit{Property of Equality for Logarithmic Functions}$$
$$x = \pm 2$$

Since $\log_5 (-2)$ is not defined, the only solution is 2.   *Check this result.*

5 **Solve $\log_3 (y + 4) + \log_3 (y - 2) = 3$.**

$$\log_3 (y + 4) + \log_3 (y - 2) = 3$$
$$\log_3 (y + 4)(y - 2) = 3 \qquad \textit{Product Property of Logarithms}$$
$$(y + 4)(y - 2) = 3^3 \qquad \textit{Definition of Logarithm}$$
$$y^2 + 2y - 8 = 27$$
$$y^2 + 2y - 35 = 0$$
$$(y - 5)(y + 7) = 0$$
$$y - 5 = 0 \quad \text{or} \quad y + 7 = 0$$
$$y = 5 \quad \text{or} \qquad y = -7$$

Check:

$$\log_3 (y + 4) + \log_3 (y - 2) = 3 \qquad\qquad \log_3 (y + 4) + \log_3 (y - 2) = 3$$
$$\log_3 (5 + 4) + \log_3 (5 - 2) \stackrel{?}{=} 3 \qquad\qquad \log_3 (-7 + 4) + \log_3 (-7 - 2) \stackrel{?}{=} 3$$
$$\log_3 9 + \log_3 3 \stackrel{?}{=} 3 \qquad\qquad \log_3 (-3) + \log_3 (-9) \stackrel{?}{=} 3$$
$$2 + 1 \stackrel{?}{=} 3$$
$$3 = 3 \quad ✔$$

Both $\log_3 (-3)$ and $\log_3 (-9)$ are not defined, so -7 is not a solution.

The only solution is 5.

Solve each equation.

1. $\log_2 3 + \log_2 7 = \log_2 x$
2. $\log_5 4 + \log_5 x = \log_5 36$
3. $\log_4 18 - \log_4 x = \log_4 6$
4. $\log_3 56 - \log_3 8 = \log_3 x$
5. $2 \log_7 3 + 3 \log_7 2 = \log_7 x$
6. $2 \log_6 4 - \frac{1}{3} \log_6 8 = \log_6 x$

Express each logarithm as the sum or difference of simpler logarithmic expressions.

7. $\log_3 xy$
8. $\log_4 rst$
9. $\log_2 m^4 y$
10. $\log_2 \frac{y}{r}$
11. $\log_b \frac{\sqrt{x}}{p}$
12. $\log_4 \frac{xy}{z}$
13. $\log_3 5\sqrt[3]{a}$
14. $\log_{10} (ac)^2$
15. $\log_2 ax^{\frac{1}{2}}$

Evaluate each expression.

16. $5^{\log_5 3 + \log_5 2}$
17. $7^{\log_7 8 - \log_7 4}$
18. $6^{3 \log_6 2}$

Written Exercises

Use $\log_{12} 3 = 0.4421$ and $\log_{12} 7 = 0.7831$ to evaluate each expression.

19. $\log_{12} 21$
20. $\log_{12} \frac{7}{3}$
21. $\log_{12} 49$
22. $\log_{12} 27$
23. $\log_{12} 36$
24. $\log_{12} 0.25$
25. $\log_{12} 252$
26. $\log_{12} 108$
27. $\log_{12} 48$
28. $\log_{12} \frac{36}{49}$
29. $\log_{12} 1.75$
30. $\log_{12} \frac{7}{16}$

Solve each equation.

31. $\log_3 7 + \log_3 x = \log_3 14$
32. $\log_2 10 - \log_2 t = \log_2 2$
33. $\log_3 y - \log_3 2 = \log_3 12$
34. $\log_3 14 + \log_3 m = \log_3 42$
35. $\log_5 x = 3 \log_5 7$
36. $\log_2 p = \frac{1}{2} \log_2 81$
37. $\log_9 x = \frac{1}{2} \log_9 144 - \frac{1}{3} \log_9 8$
38. $\log_7 m = \frac{1}{3} \log_7 64 + \frac{1}{2} \log_7 121$
39. $\log_{10} 7 + \log_{10} (n - 2) = \log_{10} 6n$
40. $\log_{10} (m + 3) - \log_{10} m = \log_{10} 4$
41. $\log_{10} x + \log_{10} x + \log_{10} x = \log_{10} 27$
42. $3 \log_5 x - \log_5 4 = \log_5 16$
43. $\log_2 15 + \log_2 14 - \log_2 105 = \log_2 x$
44. $2 \log_3 x + \log_3 0.1 = \log_3 5 + \log_3 2$
45. $\log_4 (x + 2) + \log_4 (x - 4) = 2$
46. $\log_4 (y - 1) + \log_4 (y - 1) = 2$
47. $\log_{10} (y - 1) + \log_{10} (y + 2) = \log_7 7$
48. $\log_{10} y + \log_{10} (y + 21) = 2$
49. $\log_4 (x + 3) + \log_4 (x - 3) = 2$
50. $\log_2 (9x + 5) - \log_2 (x^2 - 1) = 2$
51. $\log_8 (m + 1) - \log_8 m = \log_8 4$
52. $\log_2 (y + 2) - 1 = \log_2 (y - 2)$

Challenge Exercises

Solve for a.

53. $\log_m 3 + \log_m a = \log_m (y + 3)$
54. $\log_r a = 5 \log_r x + \log_r a^2$
55. $\log_b 2a - \log_b x = \log_b x^3$
56. $2 \log_x a = 2 \log_x (k + 1) - \log_x 4$

1. Find the distance between $(3\sqrt{2}, 5\sqrt{5})$ and $(-\sqrt{2}, -3\sqrt{5})$.

2. Find the radius and the center of the circle whose equation is $x^2 - 6x + y^2 + 8y - 24 = 0$.

3. Find $p(-3)$ if $p(x) = x^3 + 7x^2 - 2x + 5$.

4. Find $f(m - 1)$ if $f(x) = 2x^2 - 3x + 4$.

5. Name the vertex, axis of symmetry, focus, directrix, and direction of opening for the parabola whose equation is $(y + 2)^2 = 8(x - 1)$. Then find the length of the latus rectum and draw the graph.

6. Find the solutions for the system $4x^2 + 25y^2 = 500$ and $x^2 - y^2 = 9$.

7. Find the center, foci, and lengths of the major axis and minor axis for the ellipse whose equation is $x^2 + 4x + 4y^2 - 24y = -24$. Then draw the graph.

For each function, state the number of positive real zeros, negative real zeros, and imaginary zeros.

8. $f(x) = x^3 - 7x^2 + 17x - 15$

9. $g(x) = 2x^4 - x^3 + 6x^2 + 7x - 1$

10. Find the vertices, foci, and slopes of the asymptotes for the hyperbola whose equation is $9y^2 - 36y - 16x^2 - 96x = 252$. Then draw the graph.

11. Determine if $x - 3$ is a factor of $x^3 + 4x^2 + 3x - 3$.

12. Find all rational zeros of $h(x) = x^3 - 4x^2 - 7x + 10$.

Find all zeros of each function.

13. $f(x) = x^3 + 5x^2 + 6x + 2$

14. $g(x) = 2x^4 + 3x^3 - 37x + 18$

15. Approximate to the nearest tenth the real zeros of $f(x) = x^3 + 3x^2 - 8x - 9$.

If $f(x) = 2x^2 - 7$ and $g(x) = 3x + 5$, find each value.

16. $[g \circ f](3)$

17. $[f \circ g](x)$

18. If $h(x) = 3x + 5$, find $h^{-1}(x)$.

Simplify.

19. $(3^{\sqrt{3}})^{\sqrt{12}}$.

20. $\dfrac{y^2 - 4y - 21}{y + 4} \cdot \dfrac{y^2 + y - 12}{y^2 - 10y + 21}$

21. $\dfrac{x^3 + 4x^2 - 9x - 36}{x - 3} \div \dfrac{x + 3}{x + 5}$

22. $\dfrac{1}{x^2 + x - 30} + \dfrac{2x + 1}{x - 5}$

23. $\dfrac{4y - 1}{y - 3} - \dfrac{1}{y^2 + 4y - 21}$

24. $\dfrac{\dfrac{a - 3}{a - 2} + \dfrac{a - 1}{a + 3}}{\dfrac{a - 8}{a^2 + a - 6}}$

Solve each equation.

25. $3^{2x} = 9^{3x-2}$

26. $16^{2x-1} = 32^{x-2}$

27. $\dfrac{3x + 10}{x + 2} = \dfrac{x}{2x - 3} + \dfrac{4x}{x + 2}$

28. $\dfrac{5r - 2}{2r + 1} - \dfrac{2r - 5}{r - 1} = \dfrac{3r - 3}{2r + 1}$

29. $\log_b 36 = 2$

30. $\log_{27} x = \dfrac{2}{3}$

31. $\log_{12}(2x - 5) = \log_{12}(4x + 1)$

32. $\log_4(x - 13) + \log_4(x - 1) = 3$

33. Write $16^{\frac{3}{2}} = 64$ in logarithmic form.

34. Graph $y = \dfrac{2}{x - 4}$.

35. Kris can stuff a batch of envelopes for Meier's Congressional campaign in 120 minutes. Kelly can do the same job in 105 minutes. How long will it take them to do the job together?

12-5   Common Logarithms

Logarithms can make certain computations easier. When logarithms are used, multiplication changes to addition and division changes to subtraction. Logarithms to base 10 are most useful because our number system has base 10. These logarithms are called **common logarithms**. $\text{Log}_{10}\ x$ is written as $\log x$.

A table of common logarithms for numbers between 1 and 10 may be found on pages 673 and 674. To find $\log 1.23$, read across the row labeled 12 and down the column labeled 3.

Common Logarithms of Numbers

n	0	1	2	3	4
10	0000	0043	0086	0128	0170
11	0414	0453	0492	0531	0569
12	0792	0828	0864	0899	0934

$$\log 1.23 = 0.0899$$

The values in the table are rounded to the nearest ten-thousandth.

For numbers greater than 10 or less than 1, scientific notation and the properties of logarithms are used to find the logarithm.

Example

1   Find log 745,000 to the nearest ten-thousandth.

$$745,000 = 7.45 \times 10^5 \quad \text{\textit{Express 745,000 using scientific notation.}}$$

$$\log 745,000 = \log (7.45 \times 10^5) \quad \text{\textit{Property of Equality for Logarithmic Functions}}$$

$$= \log 7.45 + \log 10^5 \quad \text{\textit{Product Property of Logarithms}}$$

$$= \log 7.45 + 5$$

$$= 0.8722 + 5 \quad \text{\textit{From the table, }} \log 7.45 = 0.8722.$$

$$= 5.8722$$

The log of 745,000 is 5.8722.

Every logarithm has two parts, the **characteristic** and the **mantissa**. The mantissa is the logarithm of a number between 1 and 10. When the number is expressed in scientific notation, the characteristic is the power of 10.

$$\log 745,000 = 5.8722$$

$$\underset{\text{characteristic}}{\uparrow} \quad \underset{\text{mantissa}}{\uparrow}$$

The sum of the mantissa and the characteristic is the logarithm.

Mantissas range from 0 to 1.

Since $745,000 = 7.45 \times 10^5$, the characteristic is 5.

The table of logarithms is really a table of mantissas. You must supply the characteristic.

Example

2   **Find log 0.000524.**

$$0.000524 = 5.24 \times 10^{-4}$$
    Express 0.000524 in scientific notation.

$$\log 0.000524 = \log (5.24 \times 10^{-4})$$
    Property of Equality for Logarithmic Functions

$$= \log 5.24 + \log 10^{-4}$$
    Product Property of Logarithms

$$= 0.7193 + (-4)$$
    From the table, log 5.24 = 0.7193.

The logarithm of 0.000524 is approximately $0.7193 - 4$.

Logarithm tables are tables of positive mantissas. To avoid a negative mantissa, do *not* add the -4 and 0.7193. The negative characteristic may be written in many ways.

$$\log 0.000524 = 0.7193 - 4$$

$$\log 0.000524 = 6.7193 - 10$$ *Note $6 - 10 = -4$.*

We usually use $6 - 10$ for -4. But, in some cases it may be more convenient to use another difference such as $26 - 30$.

Sometimes a logarithm is given and you must find the number. To find the number, use the table of mantissas in reverse. The number is called the **antilogarithm.**   *If log x = a, then x = antilog a.*

Example

3   **If log x = 3.5821, find x.**

$\log x = 3.5821$ *3 is the characteristic and 0.5821 is the mantissa.*

$\quad x = \text{antilog } 3.5821$

$\quad\quad = (\text{antilog } 0.5821) \times 10^3$ *The characteristic is the power of 10.*

Find antilog 0.5821 in the table of mantissas. It is in the row labeled 38 and the column labeled 2.

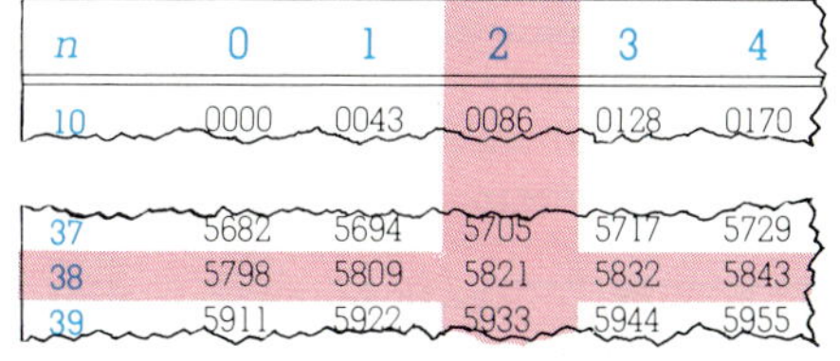

$\quad x = (\text{antilog } 0.5821) \times 10^3$

$\quad\quad = 3.82 \times 10^3$

$\quad\quad = 3820$

      log 3.82 = 0.5821

The value of x is 3820.   *The antilogarithm of 0.5821 is 3820.*

The table of logarithms in this text includes mantissas of numbers with three significant digits. You can approximate logarithms of numbers with four significant digits by a method known as **interpolation**.

4 **Approximate the value of log 1.327.**

n	0	1	2	3	4
10	0000	0043	0086	0128	0170
11	0414	0453	0492	0531	0569
12	0792	0828	0864	0899	0934
13	1139	1173	1206	1239	1271

The logarithm of 1.327 must be between log 1.32 and log 1.33.

Form a porportion of differences.

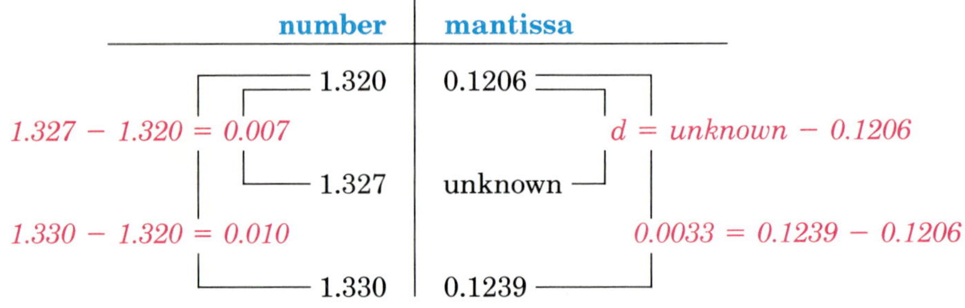

$$\frac{0.007}{0.010} = \frac{d}{0.0033}$$

$$0.00231 = d$$

$$0.0023 \approx d$$

Round d to the nearest ten thousandth since the values in the table are rounded to the nearest ten-thousandth.

Since the values in the table are increasing, add d to the mantissa of 1.320.

$$\begin{aligned} \log 1.327 &= \log 1.320 + d \\ &= 0.1206 + 0.0023 \\ &= 0.1229 \end{aligned}$$

The logarithm of 1.327 is approximately 0.1229.

Interpolation can also be used to find an antilogarithm that cannot be obtained directly from the table.

5 **Find antilog 2.4356.**

antilog 2.4356 = (antilog 0.4356) $\times 10^2$

In the table of mantissas, 4356 lies between 4346 and 4362. So, antilog 0.4356 lies between 2.72 and 2.73.

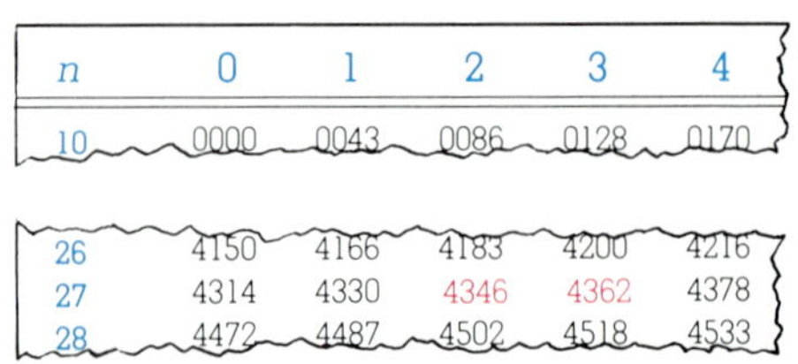

n	0	1	2	3	4
10	0000	0043	0086	0128	0170
26	4150	4166	4183	4200	4216
27	4314	4330	4346	4362	4378
28	4472	4487	4502	4518	4533

Again, form a proportion of differences.

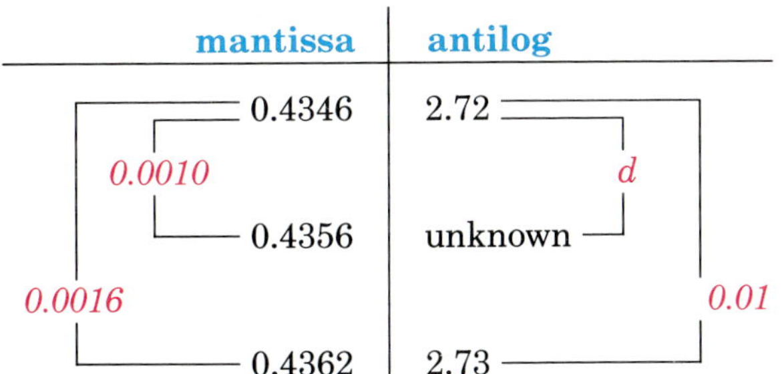

$$\frac{0.0010}{0.0016} = \frac{d}{0.01}$$

$$0.00625 = d$$

$$0.006 \approx d$$

In finding antilogs, interpolation is accurate for one more digit than given in the table. So round d to the nearest thousandth.

Since the values in the table are increasing, add d to the antilog of 0.4346.

$$
\begin{aligned}
\text{antilog } 2.4356 &= (\text{antilog } 0.4356) \times 10^2 \\
&= (\text{antilog } 0.4346 + d) \times 10^2 \\
&= (2.72 + 0.006) \times 10^2 \\
&= 272.6
\end{aligned}
$$

The antilog of 2.4356 is about 272.6.

Exploratory Exercises

If log 483 = 2.6839, find each number.

1. characteristic of log 483
2. mantissa of log 483
3. log 48.3
4. log 4830
5. log 0.004830
6. antilog 0.6839
7. antilog 5.6839
8. antilog $(0.6839 - 4)$

State the characteristic of the logarithm of each number. Then use the table on pages 673 and 674 to find the logarithm.

9. 47.5
10. 370
11. 4.61
12. 0.076
13. 0.209
14. 6870
15. 55
16. 0.00213

State the characteristic of each logarithm and use the table to find the antilogarithm.

17. 1.5527
18. 3.8096
19. $0.8376 - 2$
20. $7.6263 - 10$
21. 4.5955
22. $5.9513 - 10$
23. $0.7910 - 1$
24. 2.1106

State the two numbers whose logarithms would be used to find the logarithm of each number.

25. 7.413
26. 32,520
27. 0.0007463
28. 0.01234

State two numbers between which the antilog of each logarithm lies.

29. 0.6209
30. 2.7295
31. 3.9788
32. $0.7885 - 2$

Written Exercises

Use the table of mantissas on pages 673 and 674 to find the logarithm of each number.

33. 58.2   **34.** 715   **35.** 0.3   **36.** 0.00211

37. 0.0385   **38.** 0.671   **39.** 62,700   **40.** 7420

Use the table to find the antilogarithm of each logarithm.

41. 0.6304   **42.** 1.0899   **43.** $0.7846 - 1$ **44.** $8.1673 - 10$

45. $7.6656 - 10$ **46.** $0.9542 - 2$ **47.** 5.7451   **48.** 3.9581

Interpolate to find the logarithm of each number.

49. 5.273   **50.** 7.003   **51.** 604.7   **52.** 4167

53. 0.7214   **54.** 0.003148   **55.** 0.005364   **56.** 0.04729

57. 8.871   **58.** 60.06   **59.** 19.79   **60.** 80.08

61. 229.4   **62.** 2148×10^3 **63.** 16.57×10^{-2} **64.** 905.4×10^{-4}

Interpolate to find the antilog of each logarithm.

65. 0.4861   **66.** 0.8503   **67.** 3.7248   **68.** 4.2173

69. $0.3353 - 2$ **70.** $0.1177 - 1$ **71.** $9.9709 - 10$ **72.** $8.8787 - 10$

73. 2.7792   **74.** 1.5082   **75.** 4.6279   **76.** $0.4399 - 3$

77. $6.9172 - 10$ **78.** $7.6778 - 10$ **79.** 5.5958   **80.** 3.4170

mini-review

1. Graph the solutions to the system $\dfrac{(x-3)^2}{16} + \dfrac{(y+2)^2}{25} \le 1$ and $\dfrac{5}{16}(x-3)^2 \ge y + 7$.

2. Find all rational zeros of $f(x) = 2x^3 - 3x^2 - 11x + 6$.

3. State the number of positive real zeros, negative real zeros, and imaginary zeros of $g(x) = x^3 - x^2 + 2x + 4$. Then find all the zeros of $g(x)$.

4. Find the equation of the hyperbola with center $(1, -2)$, vertices $(1, -5)$ and $(1, 1)$, and foci $(1, -7)$ and $(1, 3)$.

5. Approximate to the nearest tenth the real zeros of $h(x) = x^3 + 4x^2 - 5x + 2$.

6. If $f(x) = 3x^2 - 1$ and $g(x) = 2x - 7$, find $[f \circ g](x)$ and $[g \circ f](x)$.

7. Solve $9^{y+7} = 27^{3y}$.

8. Solve $\log_5 (x^2 - 4) = \log_5 3x$.

9. Solve $2 \log_3 5 + \log_3 x = \log_3 30$.

10. Solve $\log_6 y + \log_6 (y - 5) = 2$.

11. Simplify $\dfrac{x^2 + 4x + 4}{2x^2 + x - 6} \cdot \dfrac{4x^2 - 9}{3x^2 + x - 10}$.

12. Simplify $\dfrac{g^2 + 5g + 4}{5g + 5} \div \dfrac{g^2 + 8g + 16}{g^2 + g - 12}$.

13. Find $\dfrac{3}{m + 2} - \dfrac{2m - 8}{m^2 - 2m - 8}$.

14. Solve $\dfrac{x - 1}{x + 1} + \dfrac{1}{x} = 1$.

15. The speed of the current in the Penobscot River is 5 mph. A boat travels downstream 28 miles and returns in 9 hours. What is the speed of the boat in still water?

Many problems involving estimation can be solved by using interpolation and extrapolation.

Interpolation is used to estimate a value between two known values. Extrapolation is used to estimate a value that is either greater than or less than two known values.

For example, suppose the deer population of a certain area is known for 1983 and 1988. The deer population in 1986 can be estimated using interpolation while the deer population in 1994 can be estimated using extrapolation.

Year	Number of Deer
1983	4320
1986	unknown
1988	6780

$5\left[\,3\left[\begin{matrix}1983\\1986\\1988\end{matrix}\right.\right.$ ... x ... 2460

Year	Number of Deer
1983	4320
1988	6780
1994	unknown

The increase in the number of deer in 5 years is 2460, or about 2500, but the increase in 3 years or 11 years is unknown. Use the following proportions to estimate the increase for 3 years and for 11 years.

$$\frac{3}{5} = \frac{x}{2500}$$

$$x = 1500$$

$$\frac{11}{5} = \frac{y}{2500}$$

$$y = 5500$$

Increase the original number, 4320, by about 1500 to find the population in 3 years and by about 5500 to find the population in 11 years. Thus, the 1986 population was approximately 5800 and the 1994 population will be approximately 9800.

Exercises

Estimate using interpolation or extrapolation.

1. Ann rented a car for seven days for $210. Another time she paid $75 to rent a car for two days from the same agency. Estimate the cost of renting a car for four days from this agency.

2. A salesman earned $1700 in the month that he sold 35 computers. Another month he received $1300 when he sold 15 computers. Estimate how much he would earn if he sold 42 computers in a month.

3. The phone company charges Richard $17.50 for making 25 calls one month and $22.90 for making 55 calls the next month. Estimate the charges for 42 and for 76 calls in one month.

4. In 1976, the population of a city was 540,025. In 1986, it was 564,871. Estimate the population for 1980 and for 1995.

12-6 Common Logarithms on the Calculator

The calculator is a useful tool for computing logarithms and antilogarithms. Using a calculator instead of a table of logarithms makes interpolation unnecessary.

Examples

Using Calculators

1 **Find the value of log 32.57.**

The $\boxed{\text{LOG}}$ key is used to find common logarithms.

The $\boxed{\text{LOG}}$ key represents $\log_{10}$.

ENTER: 32.57 $\boxed{\text{LOG}}$

DISPLAY: $32.57\ 1.5128178$

The value of log 32.57 is approximately 1.5128.

The characteristic is 1 and the mantissa is 0.5128.

2 **Find the value of the antilogarithm of 3.4557.**

The $\boxed{\text{INV}}$ and $\boxed{\text{LOG}}$ keys are used to find antilogarithms.

ENTER: 3.4557 $\boxed{\text{INV}}$ $\boxed{\text{LOG}}$

DISPLAY: $3.4557\ \ 2855.6173$

The value of the antilogarithm of 3.4557 is approximately 2856.

When a calculator is used to find the logarithm of a number between 0 and 1, some extra steps are needed to determine the characteristic and mantissa of the logarithm.

Remember that the logarithm of a number between 0 and 1 is negative.

Example

3 **Using Calculators**

Find the characteristic and mantissa of log 0.001693.

ENTER: 0.001693 $\boxed{\text{LOG}}$

DISPLAY: $0.001693\ \ -2.77134304$ *The value of log 0.001693 is -2.77134304.*

All the previous work with logarithms has been with logarithms that have positive mantissas. The logarithm -2.77134304 has a negative mantissa. A simple method for rewriting -2.77134304 as a logarithm with a positive mantissa is to add 0 to -2.77134304 by adding, then subtracting, the same number. While any number can be used, the number most often used is 10.

$$-2.77134304 = -2.77134304 + (10 - 10)$$
$$= (-2.77134304 + 10) - 10$$

ENTER: 2.77134304 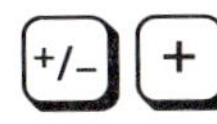 10 =

DISPLAY: *2.77134304*   *−2.77134304*   *10*   *7.22865696*

Thus, the characteristic of log 0.001693 is $7 - 10$ or -3 and the mantissa of log 0.001693 is 0.22865696 or approximately 0.2287.

The calculator can also provide a quick method for verifying that logarithms are exponents. For example, to verify that $10^{\log 72} = 72$, compute log 72 and store it. Then use the y^x key to raise 10 to the power of the logarithm.

ENTER: 72  10 y^x 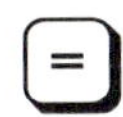 =

DISPLAY: *72 1.8573325 10*   *1.8573325 72*

The result is the original number, 72.

Exploratory Exercises

Use a calculator to find the logarithm of each number, accurate to four decimal places. Then state the characteristic and mantissa of the logarithm.

1. 32.5	**2.** 813.7	**3.** 612,900	**4.** 0.9874	**5.** 0.0472	**6.** 0.00157

Use a calculator to find the antilogarithm of each logarithm, accurate to four significant digits.

7. 2.8837	**8.** 3.6385	**9.** 5.0132	**10.** −0.6478	**11.** −1.2866	**12.** −3.4101

Written Exercises

Use a calculator to find the logarithm of each number, accurate to four decimal places. Then state the characteristic of the logarithm.

13. 93.76	**14.** 398.4	**15.** 46.75	**16.** 76,592
17. 0.8261	**18.** 0.03748	**19.** 0.001125	**20.** 0.97347
21. 213.47	**22.** 716,045	**23.** 0.090317	**24.** 0.164125

25. Compare each logarithm in Exercises 13–24 to its characteristic. What do you notice?

Use a calculator to find the antilogarithm of each logarithm, accurate to four significant digits.

26. 2.4270	**27.** 1.8914	**28.** 3.9921	**29.** 5.5847
30. −1.8684	**31.** −0.1931	**32.** 0.9841 − 3	**33.** 8.7654 − 10
34. 7.13712 − 10	**35.** −4.48553	**36.** 4.61116	**37.** 2.02341

12-7 Exponential Equations

Equations in which the variables appear as exponents are called **exponential equations**. Such equations can be solved by using the Property of Equality for Logarithmic Functions.

> # Examples

1 **Solve $2^x = 27$.**

$$2^x = 27$$

$\log 2^x = \log 27 \qquad$ *Property of Equality for Logarithmic Functions*

$x \log 2 = \log 27 \qquad$ *Power Property of Logarithms*

$$x = \frac{\log 27}{\log 2}$$

$$x = \frac{1.4314}{0.3010}$$

$$x \approx 4.7555$$

The solution is approximately 4.7555.   *Check this result.*

2 **Solve $2^{3y} = 3^{y+1}$.**

$$2^{3y} = 3^{y+1}$$

$\log 2^{3y} = \log 3^{y+1} \qquad$ *Property of Equality for Logarithmic Functions*

$3y \log 2 = (y + 1) \log 3 \qquad$ *Power Property of Logarithms*

$3y \log 2 = y \log 3 + \log 3 \qquad$ *Distributive Property*

$3y \log 2 - y \log 3 = \log 3$

$y(3 \log 2 - \log 3) = \log 3 \qquad$ *Distributive Property*

$$y = \frac{\log 3}{3 \log 2 - \log 3}$$

$$y = \frac{0.4771}{3(0.3010) - 0.4771}$$

$$y \approx 1.1202$$

The solution is approximately 1.1202.   *Check this result.*

3 **Express $\log_3 35$ in terms of common logarithms. Then find its value.**

Let $x = \log_3 35$. Then, $3^x = 35$.

$$3^x = 35$$

$\log 3^x = \log 35 \qquad$ *Property of Equality for Logarithmic Functions*

$x \log 3 = \log 35 \qquad$ *Power Property of Logarithms*

$$x = \frac{\log 35}{\log 3}$$

Thus, $\log_3 35$ may be expressed as $\dfrac{\log 35}{\log 3}$.

$$\log_3 35 = \frac{\log 35}{\log 3}$$

$$= \frac{1.5441}{0.4771}$$

$$\approx 3.2364$$

The value of $\log_3 35$ is approximately 3.2364.

In Example 3 notice that $\log_3 35 = \dfrac{\log_{10} 35}{\log_{10} 3}$.

Example 3 illustrates the Change of Base Formula.

For all positive numbers a, b, and n, where $a \neq 1$ and $b \neq 1$, $$\log_a n = \frac{\log_b n}{\log_b a}.$$	*Change of Base Formula*

The Change of Base Formula can be used to rewrite logarithms with bases other than 10 so that the value of the logarithm can be computed using a calculator.

Example

4 Using Calculators

Find the value of $\log_{12} 51$.

$$\log_{12} 51 = \frac{\log 51}{\log 12} \qquad \textit{Change of Base Formula}$$

ENTER: 51  ÷ 12  =

DISPLAY: 51 1.70757018 12 1.07918125 1.58228303

The value of $\log_{12} 51$ is approximately 1.5823.

Exponential equations occur in compound interest problems. In the compound interest formula below, P is the investment, r is the interest rate per year, n is the number of times the interest is compounded yearly, t is the number of years of the investment, and A is the amount of money accumulated.

$$A = P\left(1 + \frac{r}{n}\right)^{nt}$$	*Compound Interest Formula*

5 **How long would it take to double an investment of \$1000 at 9% interest compounded quarterly?**

$$A = P\left(1 + \frac{r}{n}\right)^{nt}$$

P is \$1000, r is 0.09, and n is 4.
A is 2P or \$2000.

$$2000 = 1000\left(1 + \frac{0.09}{4}\right)^{4t}$$

$$2 = (1.0225)^{4t}$$ *Simplify.*

$$\log 2 = \log (1.0225)^{4t}$$ *Property of Equality for Logarithmic Functions*

$$\log 2 = 4t \log (1.0225)$$ *Power Property of Logarithms*

$$\frac{\log 2}{4 \log (1.0225)} = t$$ *Solve for t.*

$$t \approx 7.7879$$ *Use tables or a calculator to evaluate.*

It would take a little more than $7\frac{3}{4}$ years for the money to double. Since interest is compounded quarterly, the money would double in 8 years.

Exploratory Exercises

State x in terms of common logarithms.

1. $3^x = 55$ **2.** $5^x = 61$ **3.** $7^{2x} = 74$ **4.** $10^{3x} = 191$

5. $x = \log_6 144$ **6.** $x = \log_5 81$ **7.** $x = \log_3 12$ **8.** $x = \log_5 30$

9. $2^{-x} = 10$ **10.** $3^{-x} = 15$ **11.** $3^x = \sqrt{13}$ **12.** $2^x = 3\sqrt{2}$

Written Exercises

13–24. Solve each equation in Exercises 1–12.

Approximate each logarithm to three decimals.

25. $\log_3 7$ **26.** $\log_7 12$ **27.** $\log_4 22$ **28.** $\log_{3.21} 10$

29. $\log_6 11$ **30.** $\log_4 24$ **31.** $\log_6 72$ **32.** $\log_5 104$

33. $\log_{11} 105$ **34.** $\log_{16} 232$ **35.** $\log_{20} 1000$ **36.** $\log_{100} 9872$

Solve each equation using logarithms.

37. $2.7^x = 52.3$ **38.** $4.3^x = 78.5$ **39.** $7.6^{n-2} = 41.7$

40. $2.1^{x-5} = 9.32$ **41.** $9^{x-4} = 6.28$ **42.** $5^{y+2} = 15.3$

43. $x = \log_4 51.6$ **44.** $x = \log_3 19.8$ **45.** $25^{x^2} = 50$

46. $6^{x^2-2} = 48$ **47.** $5^{x-1} = 3^x$ **48.** $7^{x-2} = 5^x$

49. $5^{2x} = 9^{x-1}$ **50.** $12^{x-4} = 4^{2-x}$ **51.** $7^{x-2} = 5^{3-x}$

52. $3^{3x} = 2^{2x+3}$ **53.** $32^{2y} = 5^{4y+1}$ **54.** $2^{5x-1} = 3^{2x+1}$

55. $4^{5y-6} = 3^{2y+5}$ **56.** $\sqrt{3^{b-2}} = 2^b$ **57.** $6^{x-2} = \sqrt[3]{4^{x-1}}$

58. Let x be any real number and n, a, and b be positive real numbers where $a \neq 1$ and $b \neq 1$. Show that if $x = \log_a n$, then $x = \dfrac{\log_b n}{\log_b a}$.

59. Suppose \$2500 is invested at 6% interest compounded quarterly. How long will it take for the amount to triple?

60. Ayako deposited \$500 in an account that earned 8.75% interest compounded monthly. After how long will the account contain \$800?

61. Luis Orta's grandfather invested \$150 at 5% interest compounded quarterly. When the account was recently given to Luis, it contained \$6230. How long ago did Luis' grandfather invest the \$150?

62. Sara deposited \$650 in an account that earned 12% interest compounded monthly. The account contained \$2500 when she withdrew her money. How long was the money in the account?

Challenge Exercise

63. Jan invested some money at 12% interest compounded monthly. At the same time, Joe invested some money at 12.5% interest compounded yearly. Whose money will double more quickly?

mini-review

1. Name the vertex, axis of symmetry, focus, and directrix for the parabola whose equation is $y = \dfrac{1}{12}(x + 4)^2 - 7$. Then find the length of the latus rectum.

2. Write the simplest polynomial function with integral coefficients whose zeros are $3 - i$ and $1 + 3i$.

3. State whether the graph of $4x^2 + 40x + y^2 - 8y + 100 = 0$ is a parabola, a circle, an ellipse, or a hyperbola. Then draw the graph.

4. Find all rational zeros of $f(x) = 4x^3 - 12x^2 - 19x + 12$.

5. Find all zeros of $g(x) = 3x^4 + 10x^3 - 33x^2 - 66x - 18$.

6. Find the solutions for the system $x^2 - 3y^2 = 9$ and $y^2 - x = 3$.

7. Simplify $\dfrac{3z^2 - 15z + 18}{z^2 - 4} \div \dfrac{z^2 - 2z - 3}{z^2 + 4z + 4}$.

8. Graph $\dfrac{(x - 3)^2}{4} - \dfrac{(y + 4)^2}{25} = 1$.

9. If $h(x) = 7x - 5$, find $h^{-1}(x)$.

10. Graph $f(x) = 2x^3 - x^2 - 4x + 3$.

11. Find $\dfrac{y}{2 - y} - \dfrac{3y}{y^2 - 4}$.

12. Solve $\dfrac{x - 1}{x + 2} + \dfrac{x + 1}{4} = 1$.

13. A swimming pool can be filled by a hose in 8 hours. It can be emptied by a drain pipe in 12 hours. If the drain pipe is accidentally left open while the pool is filling, how long will it take to fill the pool?

Solve each equation.

14. $4^{2y+3} = 32^{y+1}$

15. $\log_9 \sqrt{3} = x$

16. $\log_5 (b - 5) = \log_5 3b$

17. $2 \log_{13} 12 - \log_{13} 9 = \log_{13} 2x$

18. $\log_2 (x - 1) + \log_2 (x + 2) = 2$

Graphing Calculator Applications: Graphing Exponential and Logarithmic Functions

The graphs of exponential and logarithmic functions can be drawn on the graphics screen of a graphing calculator.

Examples

1 Graph $y = 2^x$ and $y = \left(\frac{1}{2}\right)^x$ on the same set of axes.

One possible set of range parameter values for graphing the two equations is given below. Set the range parameters to these values and graph the equations.

Xmin: −3.5   Xmax: 3.5   Xscl: 1
Ymin: 0   Ymax: 10   Yscl: 1

ENTER: GRAPH 2 x^y X EXE

GRAPH (1 ÷ 2

) x^y X EXE

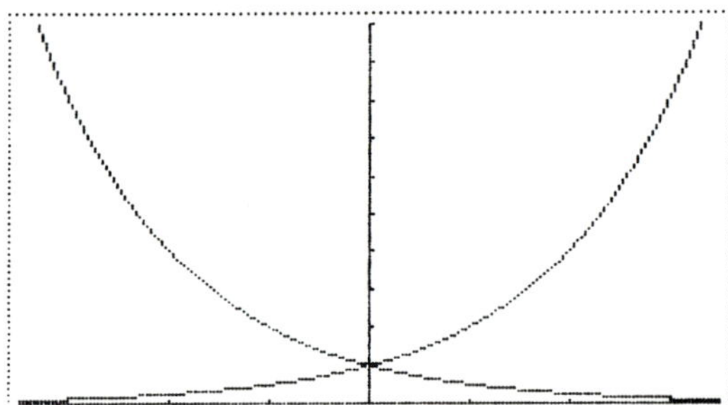

2 Graph $y = 3^x$ and $y = \log_3 x$ on the same set of axes.

The LOG key on the calculator represents the built-in function $y = \log_{10} x$. To graph a logarithmic function with base other than 10, use the Change of Base Formula.

$$\log_3 x = \frac{\log_{10} x}{\log_{10} 3} \qquad \log_a n = \frac{\log_b n}{\log_b a}$$

One possible set of range parameter values for graphing the two equations is given below. Set the range parameters to these values and graph the equations.

Xmin: −2   Xmax: 10   Xscl: 1
Ymin: −2   Ymax: 10   Yscl: 1

ENTER: GRAPH 3 x^y X EXE

GRAPH LOG X ÷

LOG 3 EXE

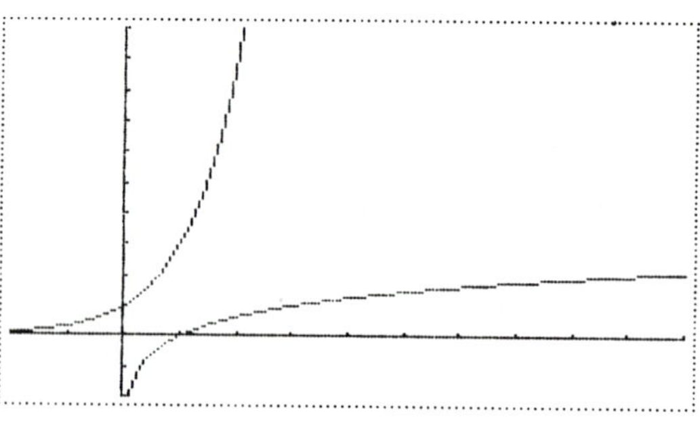

The graphing calculator can also be used to find approximate solutions to exponential equations.

3 **Solve $4^{2x-5} = 3^x$.**

A solution to the equation $4^{2x-5} = 3^x$ will also be a solution to the equation $4^{2x-5} - 3^x = 0$. Thus, the solutions to $4^{2x-5} = 3^x$ are the zeros of the function $f(x) = 4^{2x-5} - 3^x$.

Choose a set of range parameter values that will produce a graph that allows you to roughly approximate the location of the zeros of $f(x)$. One possible set of range parameter values is given below. Set the range parameters to these values and graph $y = 4^{2x-5} - 3^x$.

Xmin: -1 Xmax: 5   Xscl: 0.5   Ymin: -35 Ymax: 10   Yscl: 5

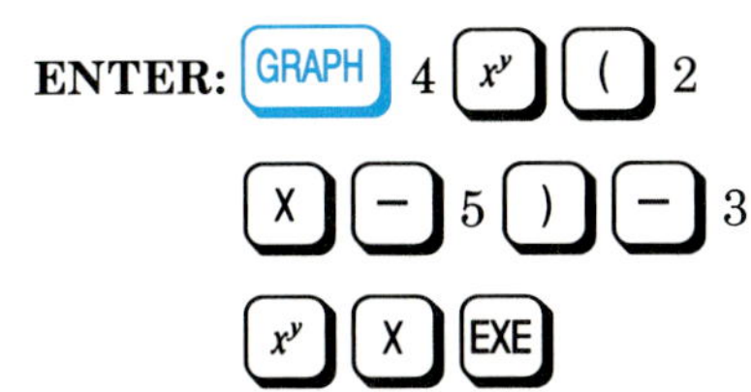

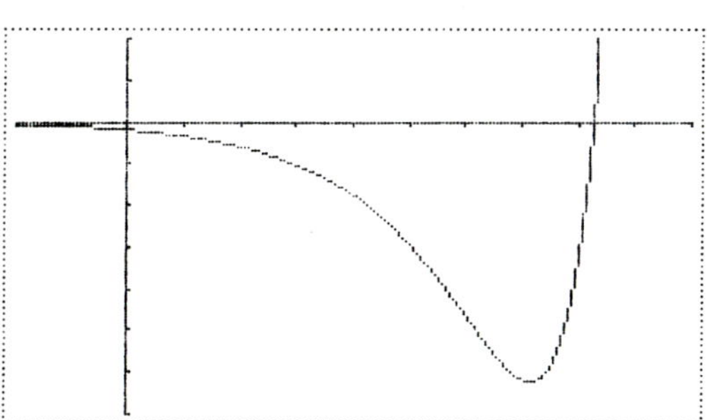

Based on this graph, there is a zero between 4 and 4.5. Now change the range parameters so that Xmin is 4, Xmax is 4.5, and Xscl is 0.05. Then graph $y = 4^{2x-5} - 3^x$ by pressing the 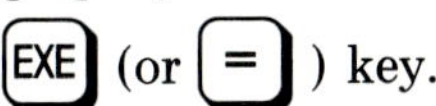 (or =) key.

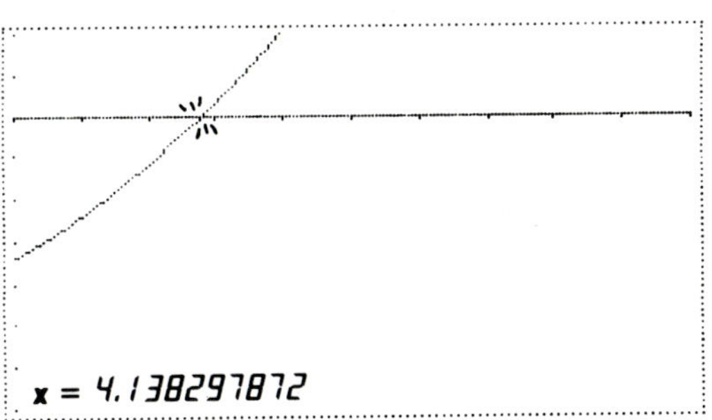

Activate the tracing function and move the pointer to the x-intercept. The x-coordinate is 4.138297872.

Therefore, the solution to $4^{2x-5} = 3^x$ is approximately 4.14.

Written Exercises

Graph each equation on a graphing calculator. Sketch each graph shown on the graphics screen, indicating the x-axis scale and the y-axis scale.

1. $y = 5^x$ **2.** $y = 10^x$ **3.** $y = 20^x$ **4.** $y = 12.5^x$ **5.** $y = 0.2^x$

6. $y = 0.1^x$ **7.** $y = 0.05^x$ **8.** $y = 0.08^x$ **9.** $y = \log_5 x$ **10.** $y = \log_{10} x$

11. $y = \log_{20} x$ **12.** $y = \log_{0.2} x$ **13.** $y = \log_{0.1} x$ **14.** $y = \log_{0.05} x$

Solve each equation using a graphing calculator.

15. $4^x = 100$ **16.** $0.5^x = 75$ **17.** $0.35^{6-x} = 28$ **18.** $25^{2x-1} = 512$

19. $3.32^{3x-2} = 0.5$ **20.** $5^x = 3^{x+2}$ **21.** $0.35^{5x+1} = 0.7^{x-1}$ **22.** $10^{2x} = 17^{1-x}$

23. $0.75^{x+3} = 4^{2x-1}$ **24.** $6^{x-1} = 10^x$ **25.** $2^x = x^2$ **26.** $3^x = x^3$

Before the invention of calculators, logarithms were often used to do computations.

Example 1: **Evaluate** $\dfrac{(673)(549)(13.82)}{147{,}900}$.

Let $A = \dfrac{(673)(549)(13.82)}{147{,}900}$. Then, $\log A = \log\left[\dfrac{(673)(549)(13.82)}{147{,}900}\right]$.

$$\log A = \log\left[\dfrac{(673)(549)(13.82)}{147{,}900}\right]$$

$$\log A = \log 673 + \log 549 + \log 13.82 - \log 147{,}900$$

$$\log A = 2.8280 + 2.7396 + 1.1405 - 5.1700$$

$$\log A = 1.5381$$

$$A = \text{antilog } (1.5381)$$
$$A = 34.52$$

The value is about 34.52.

Example 2: **Use logarithms to estimate the value of $(47.9)^5$.**

Let $A = (47.9)^5$. Then, $\log A = \log (47.9)^5$.

$$\log A = \log (47.9)^5$$
$$\log A = 5 \log (47.9)$$
$$\log A = 5(1.6803)$$

$$\log A = 8.4015$$
$$A = \text{antilog } 8.4015$$
$$A = [\text{antilog } (0.4015)] \times 10^8$$
$$A = 2.520 \times 10^8$$
$$= 252{,}000{,}000$$

The value of $(47.9)^5$ is about 252,000,000.

To evaluate roots, first express the root using rational exponents. For example, $\sqrt[4]{0.0815}$ would be expressed as $(0.0815)^{\frac{1}{4}}$. Then use the procedure demonstrated in Example 2.

Exercises

Use logarithms to estimate each value.

1. $\dfrac{(812)(41.5)}{431}$

2. $\dfrac{(71.63)(313.4)}{(489.2)}$

3. $\dfrac{(665)(899)(0.000172)}{(0.00035)(491)}$

4. $(1790)^5$

5. $(384)^3$

6. $(0.0905)^4$

7. $\sqrt[4]{0.0815}$

8. $\sqrt[4]{594}$

9. $\sqrt[3]{0.0017}$

12-8 Problem Solving: Logarithms

Exponential equations are used to describe growth and decay problems in which the rate of change is not constant. For problems involving biological growth, radioactive decay, compound interest, or the value of equipment or assets, the amount of growth or decay depends on the amount of material present.

The irrational number e is often used in growth and decay problems. This number, which was named after Leonard Euler (1707–1783) and is sometimes called Euler's number, has been found to have many uses in mathematics and science. The value of e is approximately 2.718 and $\log e$ is approximately 0.4343.

Euler is pronounced (ói-ler).

The general formula for growth and decay is

$$y = ne^{kt}.$$

In the formula, y represents the final amount, n represents the initial amount, k represents a constant, and t represents time. The constant k is positive for growth and negative for decay.

Example

1 **For a certain strain of bacteria, k is 0.775 when t is measured in hours. How long will it take 2 bacteria to increase to 1000 bacteria?**

explore The final amount is 1000 bacteria. The initial amount is 2.

plan Use the growth formula $y = ne^{kt}$. Replace y with 1000, k with 0.775, and n with 2. Then solve for t to find the time in hours.

solve

$$y = ne^{kt}$$
$$1000 = 2e^{0.775t}$$
$$500 = e^{0.775t} \qquad \textit{Simplify.}$$
$$\log 500 = \log e^{0.775t} \qquad \textit{Property of Equality for Logarithmic Functions}$$
$$\log 500 = 0.775t \log e \qquad \textit{Power Property of Logarithms}$$
$$\frac{\log 500}{0.775 \log e} = t$$
$$\frac{2.6990}{(0.775)(0.4343)} = t$$
$$t \approx 8.0188$$

It will take 2 bacteria approximately 8.02 hours to increase to 1000 bacteria.

examine Substitute 8.02 for t in the equation $1000 = 2e^{0.775t}$. The value of $2e^{(0.755 \times 8.02)}$ is approximately 1000.

2 Radioactive isotopes decay with time. In 10 years, the mass of a 200-gram sample of an isotope is reduced to 100 grams. This period of time is called the half-life of the isotope. Find the constant k for this isotope.

$$y = ne^{kt}$$ *Since this problem involves decay, k will be negative.*

$$100 = 200e^{k \cdot 10}$$ *Substitute 100 for y, 200 for n, and 10 for t.*

$$0.5 = e^{10k}$$

$$\log 0.5 = \log e^{10k}$$

$$\log 0.5 = 10k \log e$$

$$\frac{\log 0.5}{10 \log e} = k$$

$$k \approx -0.06931$$ *Use tables or a calculator to evaluate.*

The constant k for this element is approximately -0.0693.

When interest is compounded *continuously*, the formula $A = P\left(1 + \dfrac{r}{n}\right)^{nt}$ becomes

$$A = Pe^{rt}.$$

In the formula, A represents the final amount, P represents the beginning investment, r represents the annual interest rate, and t represents the time in years.

Example

3 Assume $100 is deposited in a savings account. The interest rate is 6% compounded continuously. When will the money be double the original amount?

$$A = Pe^{rt}$$ *The final amount will be $200.*

$$200 = 100e^{0.06t}$$ *Substitute 200 for A, 100 for P, and 0.06 for r.*

$$2 = e^{0.06t}$$

$$\log 2 = \log e^{0.06t}$$

$$\log 2 = 0.06t \log e$$

$$\frac{\log 2}{0.06 \log e} = t$$

$$t \approx 11.55$$ *Use tables or a calculator to evaluate.*

The original amount will double in a little over $11\frac{1}{2}$ years.

In business, the formula $V_n = P(1 + r)^n$ can be used to find the value of equipment or assets. In the formula, r represents the fixed rate of appreciation or depreciation, P represents the initial value, and V_n represents the new value at the end of n years.

Appreciation is an increase in value. Depreciation is a decrease in value.

Example

4   **A piece of machinery valued at \$25,000 depreciates at a steady rate of 10% yearly. When will the value be \$5000?**

$$V_n = P(1 + r)^n$$
$$5000 = 25{,}000(1 - 0.10)^n \qquad \textit{Substitute 5000 for } V_n, \textit{ 25,000 for } P, \textit{ and } -0.10 \textit{ for } r.$$
$$0.2 = 0.9^n \qquad \textit{The value of r is negative since it represents depreciation.}$$
$$\log 0.2 = \log 0.9^n$$
$$\log 0.2 = n \log 0.9$$
$$\frac{\log 0.2}{\log 0.9} = n$$
$$n \approx 15.26 \qquad \textit{Use tables or a calculator to evaluate.}$$

The value of the machinery will be \$5000 in a little over $15\frac{1}{4}$ years.

Written Exercises

Solve each problem.

1. A certain culture of bacteria will grow from 500 to 4000 bacteria in 1.5 hours. Find the constant k for the growth formula.

2. For a radioactive substance, the constant k is -0.08042 when t is in years. How long will it take 250 grams of the substance to reduce to 50 grams?

3. If \$500 is invested at 8% annual interest compounded continuously, when will the investment be tripled?

4. In 3 years a deposit of \$1000 grew to \$1276. What was the rate of interest, assuming continuous compounding?

5. After 9 years, half of a 20 milligram sample of a radioactive element is left. Find k for this element.

6. Bacteria of a certain type can grow from 80 to 164 bacteria in 3 hours. Find k for the growth formula.

7. For a certain strain of bacteria, k is 0.783 when t is measured in hours. How long will it take 10 bacteria to increase to 100 bacteria?

8. For a certain strain of bacteria, k is 0.782 when t is measured in hours. How long will it take 10 bacteria to increase to 500 bacteria?

9. Rachel has saved \$2000 to buy a piano that will cost about \$2500. If the money is in a savings account paying 7.25% interest compounded continuously, when will she be able to buy the piano?

10. Citizen's Bank promises to double your money in $8\frac{1}{2}$ years. Assuming continuous compounding of interest, what rate of interest must the bank pay?

11. Suppose \$10 is invested at 8% interest compounded continuously. When will the investment be worth \$100? Worth \$1000?

12. Mr. Winters has \$2500 to invest. If he hopes to have \$5000 after 8 years, what interest rate is needed, assuming continuous compounding?

13. A piece of machinery valued at \$50,000 depreciates 10% per year by the fixed rate method. After how many years will the value have depreciated to \$25,000?

14. Electronic equipment valued at \$150,000 depreciates 20% per year by the fixed rate method. After how many years will the value have depreciated to \$15,000?

15. The output in watts of a power supply is given by $w = 50e^{-0.004t}$ where t is the time in days. In how many days will the power output be reduced to 20 watts?

16. A radioactive substance decays according to the equation $A = A_0 \times 10^{-0.024t}$, where t is in hours. Find the half life of the substance, when $A = 0.5A_0$.

17. Radium-226 decomposes radioactively. Its half-life (the time half the sample takes to decompose) is 1800 years. Find the constant k for the decay formula. Use 100 grams as the original amount.

18. Ed has $500 in a savings account. If he spends 10% of the balance each week, after how many weeks will the balance be under $1? Use $V_n = P(1 + r)^n$, where n is the number of weeks.

19. A new house purchased 10 years ago for $49,000 is worth $120,000 today. Assuming a steady growth rate, by the fixed rate method, what was the yearly rate of appreciation?

20. Ingrid bought a new car 10 years ago for $6000. The car is now worth $600. Assuming a steady depreciation, by the fixed rate method, what was the yearly rate of depreciation?

Excursions in Algebra

Natural Logarithms

Logarithms, base e, are called *natural logarithms*. The natural logarithm of x is sometimes denoted $\log_e x$, but more often $\ln x$. There are tables of natural logarithms, similar to the table of common logarithms, that can be used when doing computations involving natural logarithms. Also, scientific calculators have a natural logarithm key, usually labeled $\boxed{\text{LN}}$.

Exponential equations of the form $y = ne^{kt}$ or $A = Pe^{rt}$ are easier to solve using natural logarithms since $\ln e = 1$. $\ln e = \log_e e$ and $\log_e e = 1$.

Example: **Solve $800 = 400e^{0.0875t}$.**

$$800 = 400\,e^{0.0875t}$$
$$2 = e^{0.0875t} \qquad \textit{Simplify.}$$
$$\ln 2 = \ln e^{0.0875t} \qquad \textit{Property of Equality for Logarithmic Functions}$$
$$\ln 2 = 0.0875t \ln e \qquad \textit{Power Property of Logarithms}$$
$$\frac{\ln 2}{0.0875} = t \qquad \textit{ln e = 1}$$
$$t \approx 7.922 \qquad \textit{Use tables or a calculator to evaluate.}$$

The solution is approximately 7.922.

Exercises

Use natural logarithms to solve each problem.

1. If $500 is invested at 9.9% interest compounded continuously, when will the investment be tripled?

2. In 9 years, a deposit of $875 grew to $2800. What was the interest rate, assuming continuous compounding?

3. For a certain strain of bacteria, k is 0.872 when t is measured in days. How long will it take 9 bacteria to increase to 738 bacteria?

4. After 22 years, 4 grams of a 25 gram sample of a radioactive element are left. Find the value of k for this element.

Accumulated Savings

The compound interest formula $A = P\left(1 + \dfrac{r}{n}\right)^{nt}$ can be used to find the amount of money accumulated when there is a single investment or payment. The compound interest formula below can be used to find the amount of money accumulated when payments are made at regular intervals and interest is compounded just prior to the next payment.

$$S = R\left(\frac{(1 + I)^N - 1}{I}\right)$$

In the formula, R represents the amount of each payment, I represents the interest rate per payment period, N represents the number of payments, and S represents the amount of money accumulated immediately after the final payment. No interest is added to the accumulated savings after the final payment since interest is compounded at the end of each payment period.

The computer program below can be used to determine the number of payments needed to accumulate a specific amount of money given the amount of each payment, the annual interest rate, and the number of payments per year.

Remember that the number of payments per year must be the same as the number of times the interest is compounded per year.

```
 10   INPUT "ENTER THE AMOUNT TO BE ACCUMULATED: $"; S
 20   INPUT "ENTER THE AMOUNT OF EACH PAYMENT: $"; R
 30   INPUT "ENTER THE ANNUAL INTEREST RATE AS A PERCENT: "; IR
 40   INPUT "ENTER THE NUMBER OF PAYMENTS PER YEAR: "; PAY
 50   LET I = IR / (PAY * 100): PRINT          I is the interest rate per payment period.
 60   LET N = LOG (S * I / R + 1) / LOG (1 + I)
 70   IF N = INT (N) THEN 90          The number of payments (N) is rounded to the
 80   LET N = INT (N + 1)            least integer greater than N.
 90   PRINT N;" PAYMENTS ARE NEEDED TO ACCUMULATE $"; S
100   END
```

Run the program for an accumulated savings of $1250, a monthly payment of $50, and an annual interest rate of 9.15% compounded monthly.

```
]RUN
ENTER THE AMOUNT TO BE ACCUMULATED: $1250
ENTER THE AMOUNT OF EACH PAYMENT: $50
ENTER THE ANNUAL INTEREST RATE AS A PERCENT: 9.15
ENTER THE NUMBER OF PAYMENTS PER YEAR: 12

23 PAYMENTS ARE NEEDED TO ACCUMULATE $1250
```

Use the computer program to determine the number of payments needed to accumulate the indicated amount of money, given the amount of each payment, annual interest rate, and number of payments per year.

1. $4000; $300, 9%, 2

2. $7500; $400, 8.5%, 4

3. $8995; $156, 9.25%, 12

4. $14,600; $195, 8.75%, 12

5. $90,000; $850, 9.65%, 12

6. $90,000; $425, 9.65%, 26

7. Michael deposits $100 each month in an account that earns 8.75% interest compounded monthly. How many months will it take him to save $3000?

8. John invests $300 every 3 months at 8.3% interest compounded quarterly. Sue invests $98 each month at 9.3% interest compounded monthly. Who will accumulate $5000 more quickly?

9. Use logarithms to solve $S = R\left(\dfrac{(1 + I)^N - 1}{I}\right)$ for N.

Vocabulary

exponential function (421)	common logarithm (434)	antilogarithm (435)
logarithm (423)	characteristic (434)	interpolation (435)
logarithmic function (426)	mantissa (434)	exponential equation (442)

Chapter Summary

1. If x is an irrational number and $a > 0$, then a^x is the real number between a^{x_1} and a^{x_2}, for all possible choices of rational numbers x_1 and x_2 such that $x_1 < x < x_2$.   (420)

2. An equation of the form $y = a^x$ where $a > 0$ and $a \neq 1$ is called an exponential function.   (421)

3. Property of Equality for Exponential Functions: Suppose a is a positive number other than 1. Then $a^{x_1} = a^{x_2}$ if and only if $x_1 = x_2$.   (421)

4. Suppose $b > 0$ and $b \neq 1$. Then for $n > 0$ there is a number p such that $\log_b n = p$ if and only if $b^p = n$.   (423)

5. An equation of the form $y = \log_b x$ where $b > 0$ and $b \neq 1$ is called a logarithmic function.   (426)

6. Since the logarithmic function $y = \log_b x$ and exponential function $y = b^x$ are inverse functions, $b^{\log_b x} = x$ and $\log_b b^x = x$.   (426)

7. Property of Equality of Logarithmic Functions: Suppose $b > 0$ and $b \neq 1$. Then $\log_b x_1 = \log_b x_2$ if and only if $x_1 = x_2$.   (427)

8. Product Property of Logarithms: For all positive numbers m, n, and b, where $b \neq 1$, $\log_b mn = \log_b m + \log_b n$.   (429)

9. Quotient Property of Logarithms: For all positive numbers m, n, and b, where $b \neq 1$, $\log_b \dfrac{m}{n} = \log_b m - \log_b n$.   (430)

10. Power Property of Logarithms: For any real number p and positive numbers m and b, where $b \neq 1$, $\log_b m^p = p \log_b m$. (431)

11. Common logarithms are logarithms to base 10. To find the common logarithm of a positive number r, first write the number in scientific notation, $r = a \times 10^n$, where $1 \leq a < 10$ and n is an integer. Then, $\log r = \log a + n$. Log a is found in the table of logarithms. It is called the mantissa of $\log r$. The integer n is the characteristic of $\log r$. (434)

12. To find the antilogarithm of any logarithm, use the table of logarithms to find the antilogarithm of the mantissa. Then, multiply this number by the power of ten corresponding to the characteristic. (435)

13. Interpolation is a procedure whereby logarithms and antilogarithms *not* directly found in the table may be calculated. (436)

14. Exponential equations can be solved by using the Property of Equality for Logarithmic Functions. (442)

15. Change of Base Formula: For all positive numbers a, b, and n, where $a \neq 1$ and $b \neq 1$, $\log_a n = \dfrac{\log_b n}{\log_b a}$. (443)

16. The compound interest formula is $A = P\left(1 + \dfrac{r}{n}\right)^{nt}$. In the formula, P is the investment, r is the interest rate per year, n is the number of times the interest is compounded yearly, t is the number of years of the investment, and A is the amount of money accumulated. (443)

17. The general formula for growth and decay is $y = ne^{kt}$. In the formula, y represents the final amount, n represents the initial amount, k represents a constant, and t represents time. The constant k is positive for growth and negative for decay. (449)

Chapter Review

12–1 **Solve each equation.**

  1. $2^{6x} = 4^{5x+2}$ **2.** $(\sqrt{3})^{n+1} = 9^{n-1}$ **3.** $49^{3p+1} = 7^{2p-5}$

Simplify.

  4. $3^{\sqrt{2}} \cdot 3^{\sqrt{2}}$ **5.** $(9^{\sqrt{2}})^{\sqrt{2}}$ **6.** $\dfrac{49^{\sqrt{2}}}{7^{\sqrt{12}}}$

12–2 **Write each equation in logarithmic form.**

  7. $7^3 = 343$ **8.** $5^{-2} = \dfrac{1}{25}$ **9.** $4^0 = 1$ **10.** $4^{\frac{3}{2}} = 8$

Write each equation in exponential form.

  11. $\log_4 64 = 3$ **12.** $\log_8 2 = \dfrac{1}{3}$ **13.** $\log_6 \dfrac{1}{36} = -2$ **14.** $\log_6 1 = 0$

Solve each equation.

  15. $\log_b 9 = 2$ **16.** $\log_b 9 = \dfrac{1}{2}$ **17.** $\log_{16} 2 = x$ **18.** $\log_4 x = -\dfrac{1}{2}$

12–3 Evaluate each expression.

  19. $\log_5 5^7$ **20.** $6^{\log_6 7}$ **21.** $\log_n n^3$ **22.** $n^{\log_n 3}$

Solve each equation.

  23. $\log_6 12 = \log_6 (5x - 3)$ **24.** $\log_3 3y = \log_3 (2y + 5)$
  25. $\log_5 y = \log_5 (14 - y)$ **26.** $\log_3 3x = \log_3 (x + 7)$
  27. $\log_2 (3x - 2) = \log_2 (2x + 6)$ **28.** $\log_4 (1 - 2x) = \log_4 (x + 10)$
  29. $\log_7 (x^2 + x) = \log_7 12$ **30.** $\log_2 (x - 1)^2 = \log_2 7$

12–4 Use $\log_9 7 = 0.8856$ and $\log_9 4 = 0.6309$ to evaluate each expression.

  31. $\log_9 28$ **32.** $\log_9 49$ **33.** $\log_9 144$ **34.** $\log_9 15.75$

Solve each equation.

  35. $\log_3 x - \log_3 4 = \log_3 12$ **36.** $\log_5 z + \log_5 3 = \log_5 15$

  37. $\log_2 y = \frac{1}{3} \log_2 27$ **38.** $\log_5 7 + \frac{1}{2} \log_5 4 = \log_5 x$

  39. $\log_9 8 + \log_9 (n + 4) = \log_9 2n$ **40.** $\log_7 (m + 1) + \log_7 (m - 5) = 1$
  41. $\log_6 (r - 3) + \log_6 (r + 2) = 1$ **42.** $2 \log_2 x - \log_2 (x + 3) = 2$

12–5 Use the table of mantissas on pages 673 and 674 to find the logarithm of each number. Use interpolation when necessary.

  43. 612   **44.** 0.0877   **45.** 0.4111   **46.** 2004

Use the table to find the antilogarithm of each logarithm. Use interpolation when necessary.

  47. 3.7889   **48.** $0.4409 - 2$ **49.** $0.8736 - 3$ **50.** 1.0725

12–6 Use a calculator to find the logarithm of each number.

  51. 0.003141   **52.** 50,030   **53.** 13.27   **54.** 0.0007439

Use a calculator to find the antilogarithm of each logarithm.

  55. 4.2826   **56.** -0.3359 **57.** $0.5574 - 4$ **58.** 2.9876

12–7 Solve each equation using logarithms.

  59. $2^x = 53$ **60.** $4.5^x = 36.2$ **61.** $3.4^{x-2} = 15.6$
  62. $\log_4 11.2 = x$ **63.** $2.3^{x^2} = 66.6$ **64.** $8^{x-2} = 5^x$
  65. $6^{3y} = 8^{y-3}$ **66.** $3^{4x-7} = 4^{2x+3}$ **67.** $\sqrt{3^b} = 2^{b+1}$

12–8 Solve each problem.

  68. If \$200 is invested at 6% annual interest compounded continuously, when will the investment be worth \$300? Use $A = Pe^{rt}$.

  69. A bacteria culture grows from 400 to 5000 bacteria in 2 hours. Find the constant k for the growth formula $y = ne^{kt}$ where t is in hours.

  70. A car valued at \$14,000 depreciates 18% per year by the fixed rate method. After how many years will the value have depreciated to \$1000? Use $V_n = P(1 + r)^n$.

Write each equation in logarithmic form.

1. $6^4 = 1296$

2. $3^7 = 2187$

Write each equation in exponential form.

3. $\log_5 625 = 4$

4. $\log_8 16 = \dfrac{4}{3}$

Evaluate each expression.

5. $\log_{12} 12^2$

6. $4^{\log_4 3}$

7. $\log_b b^{1.6}$

Solve each equation.

8. $4^{3x} = 16^{x-\frac{1}{2}}$

9. $9^x = 3^{3x-2}$

10. $27^{2p+1} = 3^{4p-1}$

11. $\log_2 128 = y$

12. $\log_m 144 = -2$

13. $\log_{\sqrt{7}} x = 4$

14. $\log_3 x - 2\log_3 2 = 3\log_3 3$

15. $\log_5 (8r - 7) = \log_5 (r^2 + 5)$

16. $\log_9 (x + 4) + \log_9 (x - 4) = 1$

Use $\log_4 7 = 1.4037$ and $\log_4 3 = 0.7945$ to evaluate each expression.

17. $\log_4 21$

18. $\log_4 9$

19. $\log_4 36$

20. $\log_4 \dfrac{7}{12}$

Use a table of mantissas or calculator to find the logarithm of each number, accurate to four decimal places. Use interpolation when necessary.

21. 769,000

22. 0.01473

23. 96,930

24. 0.00535

Use a table of mantissas or calculator to find the antilogarithm of each logarithm, accurate to four significant digits. Use interpolation when necessary.

25. $0.8351 - 4$

26. 3.4819

27. 6.3337

28. $0.6128 - 1$

Solve each equation using logarithms.

29. $3^x = 35$

30. $7.6^{x-1} = 431$

31. $\log_4 37 = x$

32. $3^x = 5^{x-1}$

33. $\sqrt{2^{b-4}} = 6^b$

34. $4^{2x-3} = 9^{x+2}$

35. A pilgrim ancestor of Carrie Seltzer deposited \$10 in a savings account at Provident Savings Bank. The annual interest rate was 4% compounded continuously. The account is now worth \$75,000. How long ago was the account started? (Use $A = Pe^{rt}$ and $\log e = 0.4343$.)

CHAPTER 13

Sequences and Series

During a free-fall, a skydiver falls 16 feet in the first second, 48 feet in the second second, 80 feet in the third second, and so on. The number of feet the skydiver falls during each second can be written as the following sequence.

$$16, 48, 80, \ldots$$

Can you tell what the next number will be? What is the pattern?

In this chapter you will learn more about sequences and series.

13-1 Arithmetic Sequences

John Dalton is a race car driver. He enters the straightaway at 91 miles per hour. While on the straightaway he increases his speed by 29.7 mph. After nine seconds, his speed is 120.7 mph.

The speed at each second is shown below.

number of seconds	0	1	2	3	4	5	6	7	8	9
speed in mph	91	94.3	97.6	100.9	104.2	107.5	110.8	114.1	117.4	120.7

Suppose that you graph the speed at each second on a graph as shown at the right. If the points on the graph were connected with a line, what kind of graph would you have?

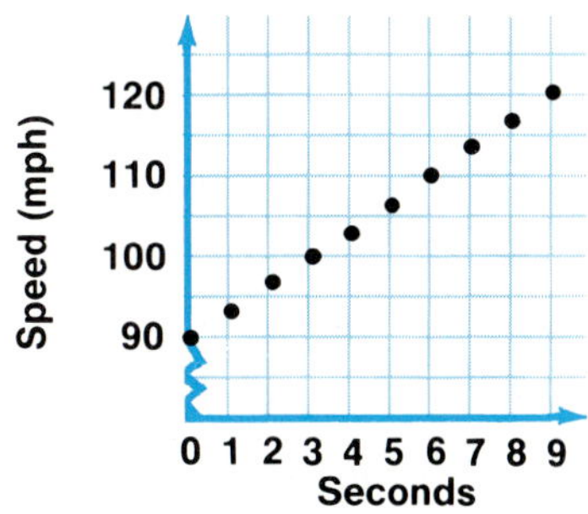

The points in the graph represent speeds at successive seconds. They are an example of a **sequence**. A set of numbers in a specific order is called a sequence. Each number in a sequence is called a **term**. The first term is denoted a_1, the second term a_2, and so on to a_n, the nth term.

A sequence is sometimes called a progression.

symbol	a_1	a_2	a_3	a_4	a_5	a_6	a_7	a_8	a_9	a_{10}
term	91	94.3	97.6	100.9	104.2	107.5	110.8	114.1	117.4	120.7

> **An arithmetic sequence is a sequence in which each term, after the first, is found by adding a constant, called the *common difference*, to the previous term.**

Definition of an Arithmetic Sequence

The common difference, d, is found by subtracting any term from its succeeding term.

Find d for each of the following arithmetic sequences.

$$3, 7, 11, 15, \ldots \qquad d = 4$$

$$2, \tfrac{3}{2}, 1, \tfrac{1}{2}, 0, -\tfrac{1}{2}, -1, \ldots \qquad d = -\tfrac{1}{2}$$

What is d for the arithmetic sequence of speeds listed above?

The three dots mean that the sequence continues indefinitely in the same pattern.

1 **Find the next three terms of the sequence 21, 27, 33, ...**

Find the common difference.

$33 - 27 = 6$ $27 - 21 = 6$ $d = 6$

Add 6 to the third term to get the fourth, and so on.

$33 + 6 = 39$ $39 + 6 = 45$ $45 + 6 = 51$

The next three terms are 39, 45, and 51.

There is a pattern in the way terms of an arithmetic sequence are formed. Consider the terms of the sequence in Example 1.

a_1	a_2	a_3	a_4	a_5
21	27	33	39	45
$21 + 0 \cdot 6$	$21 + 1 \cdot 6$	$21 + 2 \cdot 6$	$21 + 3 \cdot 6$	$21 + 4 \cdot 6$
$a_1 + 0 \cdot d$	$a_1 + 1 \cdot d$	$a_1 + 2 \cdot d$	$a_1 + 3 \cdot d$	$a_1 + 4 \cdot d$

How would you write an addition expression for the term a_n?

The nth term, a_n, of an arithmetic sequence with first term a_1 and common difference d is given by the following formula.

$$a_n = a_1 + (n - 1)d$$

Formula of the nth Term of an Arithmetic Sequence

2 **Suppose the race car driver continues to increase his speed at the same rate. What will be his speed after 15 seconds?**

$a_1 = 91$ $d = 3.3$

Find a_{16} using $a_n = a_1 + (n - 1)d$.   *a_{16} is the speed after 15 seconds. Why?*

$$\begin{aligned} a_{16} &= 91 + (16 - 1)3.3 \\ &= 91 + 15(3.3) \\ &= 91 + 49.5 \\ &= 140.5 \end{aligned}$$

His speed will be 140.5 mph.

The terms between any two nonconsecutive terms of an arithmetic sequence are called **arithmetic means**.

$$12, 21, 30, 39, 48, 57, 66, 75, \ldots$$

Thus 30, 39, and 48 are the three arithmetic means between 21 and 57.

Examples

3 **Find the four arithmetic means between 12 and 47. Use $a_n = a_1 + (n - 1)d$.**

$a_1 = 12$ *12, ____, ____, ____, ____, 47* $a_6 = 47$
$a_6 = a_1 + (5)d$
$47 = 12 + 5d$
$35 = 5d$
$7 = d$ *The common difference is 7.*

$12 + 7 = 19$ $19 + 7 = 26$ $26 + 7 = 33$ $33 + 7 = 40$ *40 + 7 = 47*

The arithmetic means are 19, 26, 33, and 40.

4 **Find the number of multiples of 13 between 29 and 258.**

The least and greatest multiples of 13 between 29 and 258 are 39 and 247. A sequence where $a_1 = 39$, $d = 13$, and $a_n = 247$ will have the required number of terms.
Use $a_n = a_1 + (n - 1)d$.

$247 = 39 + (n - 1)13$ *$a_n = 247$, $a_1 = 39$, $d = 13$*
$208 = (n - 1)13$
$16 = n - 1$ or $n = 17$

There are 17 multiples of 13 between 29 and 258.

Exploratory Exercises

Name the first five terms of each arithmetic sequence described below.

1. $a_1 = 4, d = 3$ **2.** $a_1 = 7, d = 5$ **3.** $a_1 = 16, d = -2$

4. $a_1 = 38, d = -4$ **5.** $a_1 = \frac{3}{4}, d = -\frac{1}{4}$ **6.** $a_1 = \frac{3}{8}, d = \frac{5}{8}$

7. $a_1 = 2.3, d = 1.6$ **8.** $a_1 = 0.88, d = 0$ **9.** $a_1 = -\frac{1}{3}, d = -\frac{2}{3}$

10. $a_1 = -\frac{4}{5}, d = 1$ **11.** $a_1 = -4.2, d = -1.3$ **12.** $a_1 = 2, d = -2.5$

Name the next four terms of each arithmetic sequence.

13. $5, 9, 13, \ldots$ **14.** $11, 14, 17, \ldots$ **15.** $2, -3, -8, \ldots$

16. $21, 15, 9, \ldots$ **17.** $\frac{1}{2}, \frac{3}{2}, \frac{5}{2}, \ldots$ **18.** $-5.4, -1.4, 2.6, \ldots$

19. $-\frac{5}{4}, -\frac{7}{4}, -\frac{9}{4}, \ldots$ **20.** $9.9, 13.7, 17.5, \ldots$ **21.** $-0.06, 2.24, 4.54, \ldots$

Written Exercises

**Find the nth term of each arithmetic sequence described below.
Use $a_n = a_1 + (n - 1)d$.**

22. $a_1 = 7, d = 3, n = 14$

23. $a_1 = -3, d = -9, n = 11$

24. $a_1 = -1, d = -10, n = 25$

25. $a_1 = -7, d = 3, n = 17$

26. $a_1 = 2, d = \frac{1}{2}, n = 8$

27. $a_1 = \frac{3}{4}, d = -\frac{5}{4}, n = 13$

28. $a_1 = 20, d = 4, n = 100$

29. $a_1 = 13, d = 3, n = 101$

30. $a_1 = 27, d = 16, n = 23$

31. $a_1 = 15, d = 80, n = 10$

32. $a_1 = \sqrt{3}, d = -\sqrt{2}, n = 11$

33. $a_1 = 2i, d = -5i, n = 12$

Find the indicated term in each arithmetic sequence.

34. a_{12} for $-17, -13, -9, \ldots$

35. a_{21} for $10, 7, 4, \ldots$

36. a_{32} for $4, 7, 10, 13, \ldots$

37. a_{10} for $8, 3, -2, \ldots$

38. a_{12} for $\frac{3}{4}, \frac{3}{2}, \frac{9}{4}, \ldots$

39. a_{10} for $\frac{5}{6}, \frac{7}{6}, \frac{3}{2}, \ldots$

Answer each question by finding n, the number of the term.

40. Which term of $-2, 5, 12, \ldots$ is 124?

41. Which term of $-3, 2, 7, \ldots$ is 142?

42. Which term of $7, 2, -3, \ldots$ is -28?

43. Which term of $2\frac{1}{4}, 2, 1\frac{3}{4}, \ldots$ is $-\frac{17}{4}$?

Find the missing terms of each sequence. Then show all the terms on a graph.

44. $55, \underline{\quad}, \underline{\quad}, \underline{\quad}, 115$

45. $-8, \underline{\quad}, \underline{\quad}, 3$

46. $-10, \underline{\quad}, \underline{\quad}, \underline{\quad}, \underline{\quad}, 2$

47. $2, \underline{\quad}, \underline{\quad}, \underline{\quad}, \underline{\quad}, \underline{\quad}, 20$

48. $\underline{\quad}, -6, \underline{\quad}, \underline{\quad}, 15, \underline{\quad}$

49. $\underline{\quad}, 49, \underline{\quad}, \underline{\quad}, 28$

Solve each problem.

50. The last term of an arithmetic sequence is 207, the common difference is 3, and the number of terms is 14. What is the first term of the sequence?

51. The third term of an arithmetic sequence is 14 and the ninth term is -1. Find the first four terms.

52. How many multiples of 7 are there between 11 and 391?

53. How many multiples of 12 are there between 16 and 415?

54. During a free-fall, a skydiver falls 16 feet in the first second, 48 feet in the second second, and 80 feet in the third second. If she continues to fall at this rate, how many feet will she fall in the 10th second?

55. A rocket rises 20 feet in the first second, 60 feet in the second second, and 100 feet in the third second. If it continues to rise at this rate, how many feet will it rise in the 20th second?

Challenge Exercises

56. The 5th term of an arithmetic sequence is 19 and the 11th term is 43. Find the first term and the 87th term.

57. Find three numbers that have a sum of 27, a product of 288, and form an arithmetic sequence.

Show the first twelve terms of each arithmetic sequence on a graph.

58. $2, -4, -10, -16, \ldots$

59. $15, 24, 33, 42, \ldots$

13-2 Arithmetic Series

Dana Thompson is starting a savings program. She plans to save five cents the first day, ten cents the second day, fifteen cents the third day, and so on. How much will she save in the first week using this plan?

If the amount Dana saves each day is listed, the list is a sequence.

$$5, 10, 15, 20, 25, 30, 35$$

The sum of this sequence is indicated as follows.

$$5 + 10 + 15 + 20 + 25 + 30 + 35$$

The sum is 140.
Dana will save $1.40.

The indicated sum of the terms of a sequence is called a series.	*Definition of Series*

The chart below contains examples of arithmetic sequences and their corresponding series.

Arithmetic Sequence	Arithmetic Series
3, 6, 9, 12, 15	$3 + 6 + 9 + 12 + 15$
$-4, -1, 2$	$-4 + (-1) + 2$
$\frac{5}{3}, \frac{8}{3}, \frac{11}{3}, \frac{14}{3}$	$\frac{5}{3} + \frac{8}{3} + \frac{11}{3} + \frac{14}{3}$
$a_1, a_2, a_3, a_4, \ldots, a_n$	$a_1 + a_2 + a_3 + a_4 + \ldots + a_n$

The symbol S_n is used to represent the *sum of the first n terms of a series*. For example, S_3 represents the sum of the first three terms of the series $3 + 6 + 9 + 12 + 15$, which is $3 + 6 + 9$ or 18.

If a series has a large number of terms, it is not convenient to find the sum by adding its terms. To write a general formula for the sum of a series of n terms, consider S_7 for the series

$$3 + 7 + 11 + 15 + 19 + 23 + 27.$$

Express the terms of S_7 in ascending order, then descending order, as shown below. Then add each column.

$$S_7 = 3 + 7 + 11 + 15 + 19 + 23 + 27$$
$$+\ S_7 = 27 + 23 + 19 + 15 + 11 + 7 + 3$$
$$2 \cdot S_7 = 30 + 30 + 30 + 30 + 30 + 30 + 30$$

Notice that the sum of each column is 30.

7 sums

$$2 \cdot S_7 = 7 \cdot 30$$

Since $30 = 3 + 27$, each column sum of 30 could be written as the sum of the first and last terms of S_7.

$$2 \cdot S_7 = 7(a_1 + a_7)$$
$$S_7 = \frac{7}{2}(a_1 + a_7) \qquad \textit{Divide each side by 2.}$$

The same method can be used to write a formula for an arithmetic series with n terms. Let $S_n = a_1 + a_2 + a_3 + a_4 + \ldots + a_n$. To write an expression for $2 \cdot S_n$, write each column sum as the sum of the first and last terms of S_n.

$$2 \cdot S_n = (a_1 + a_n) + (a_1 + a_n) + (a_1 + a_n) + \ldots + (a_1 + a_n)$$

n sums

$$2 \cdot S_n = n(a_1 + a_n)$$
$$S_n = \frac{n}{2}(a_1 + a_n) \qquad \textit{Divide each side by 2.}$$

> **The sum, S_n, of the first n terms of an arithmetic series is given by the following formula.**
>
> $$S_n = \frac{n}{2}(a_1 + a_n)$$

Sum of an Arithmetic Series

Example

1 **Find the sum of the first 50 positive integers.**

$$a_1 = 1,\ a_n = 50,\ n = 50$$
$$S_n = \frac{n}{2}(a_1 + a_n)$$
$$S_{50} = \frac{50}{2}(1 + 50) \qquad \textit{Substitute 1 for } a_1, \textit{ 50 for } a_n, \textit{ and 50 for n.}$$
$$= 25(51)$$
$$= 1275$$

The sum of the first 50 positive integers is 1275.

You know that $a_n = a_1 + (n - 1)d$. Using substitution gives another formula for S_n.

$$S_n = \frac{n}{2}(a_1 + a_n)$$

$$= \frac{n}{2}\{a_1 + [a_1 + (n - 1)d]\} \quad \textit{Substitute } a_1 + (n - 1)d \textit{ for } a_n.$$

$$= \frac{n}{2}[2a_1 + (n - 1)d]$$

Examples

2 **Find the sum of the first 60 terms of an arithmetic series where $a_1 = 15$ and $d = 80$.**

$$S_n = \frac{n}{2}[2a_1 + (n - 1)d]$$

$$S_{60} = \frac{60}{2}[2 \cdot 15 + (59)80] \quad \textit{Substitute 15 for } a_1, \textit{ 80 for d, and 60 for n.}$$

$$= 142{,}500$$

3 **A supermarket display consists of cans stacked as shown at the right. The bottom row has 27 cans. Each row above has one less can than the row below it. The display has 15 rows. How many cans are in the display?**

Find the sum of an arithmetic series where $a_1 = 27$, $d = -1$, and $n = 15$.

$$S_n = \frac{n}{2}[2a_1 + (n - 1)d]$$

$$= \frac{15}{2}[2 \cdot 27 + (14)(-1)]$$

$$= 300$$

There are 300 cans in the display.

4 **Find the first three terms of an arithmetic series where $a_1 = 17$, $a_n = 101$, and $S_n = 472$.**

$$S_n = \frac{n}{2}(a_1 + a_n) \qquad\qquad a_n = a_1 + (n - 1)d$$
$$\qquad\qquad\qquad\qquad\qquad 101 = 17 + (8 - 1)d$$
$$472 = \frac{n}{2}(17 + 101) \qquad\qquad 84 = 7d$$
$$\qquad\qquad\qquad\qquad\qquad 12 = d$$
$$472 = 59n$$
$$n = 8$$

$$a_2 = 17 + 12 \text{ or } 29 \qquad a_3 = 29 + 12 \text{ or } 41$$

The first three terms are 17, 29, and 41.

Exploratory Exercises

Evaluate each series.

1. $4 + 7 + 10 + 13 + 16 + 19 + 22 + 25$

2. $1 + 5 + 9 + 13 + 17 + 21 + 25 + 29$

Find S_n for each series.

3. $a_1 = 2, a_n = 200, n = 100$

4. $a_1 = 5, a_n = 100, n = 200$

5. $a_1 = 4, n = 15, d = 3$

6. $a_1 = 50, n = 20, d = -4$

7. $9 + 11 + 13 + 15 + \ldots$ for $n = 12$

8. $-3 + (-7) + (-11) + \ldots$ for $n = 10$

9. the first 100 positive integers

10. the first 100 positive even integers

Written Exercises

Find S_n for each series described below.

11. $a_1 = 11, a_n = 44, n = 23$

12. $a_1 = 3, a_n = -38, n = 8$

13. $a_1 = 5, n = 18, a_n = 73$

14. $a_1 = 85, n = 21, a_n = 25$

15. $a_1 = 3, n = 9, a_n = 27$

16. $a_1 = 34, n = 9, a_n = 2$

17. $a_1 = 9, n = 22, a_n = 101$

18. $a_1 = 76, n = 16, a_n = 31$

19. $a_1 = 5, d = 12, n = 7$

20. $a_1 = 4, d = -1, n = 7$

21. $a_1 = 9, d = -6, n = 14$

22. $a_1 = 5, d = \frac{1}{2}, n = 13$

Find the sum of each series.

23. $7 + 14 + 21 + 28 + \ldots + 98$

24. $6 + 12 + 18 + \ldots + 96$

25. $10 + 4 + (-2) + (-8) + \ldots + (-50)$

26. $34 + 30 + 26 + \ldots + 2$

Find S_n for each series described below.

27. $d = -4, n = 9, a_n = 27$

28. $a_1 = 91, d = -4, a_n = 15$

29. $a_1 = -2, d = \frac{1}{2}, a_n = 5$

30. $d = 5, n = 16, a_n = 72$

Find the first three terms for each series described below.

31. $a_1 = 6, a_n = 306, S_n = 1716$

32. $a_1 = 7, a_n = 139, S_n = 876$

33. $n = 14, a_n = 53, S_n = 378$

34. $n = 21, a_n = 78, S_n = 1008$

Solve each problem.

35. Find the sum of the odd integers from 1 to 100.

36. Find the sum of the positive integers less than 100 that are divisible by 6.

37. A pile of fireplace logs has 1 log in the top layer, 2 logs in the next layer, and so on. How many logs are in the pile if it contains 21 layers?

38. An auditorium has 21 seats in the first row. Each of the other rows has one more seat than the row in front of it. If there are 30 rows of seats, what is the seating capacity of the auditorium?

39. The cost of repairs for a certain automobile increases $60 each year. If the cost of repairs for the first year is $125, what is the total amount spent on repairs for the automobile after 7 years?

40. A person retires after 30 years from a job in which she earned $4,500 the first year and received a raise of $820 at the end of each year she worked. What was her total salary in her 30 years?

Suppose you borrow some money from the local bank. As a general rule, the amount of interest is not the same for each month. You owe more interest the first month than you do the last month. Do you see why? For a one-year loan, the following series may be used in figuring the amount of interest for each month.

$$12 + 11 + 10 + 9 + 8 + 7 + 6 + 5 + 4 + 3 + 2 + 1 \text{ or } 78$$

How many months are there in a year? How many addends are there in the series shown above?

The example below shows how this series is used to figure amounts of interest.

Example: What part of the interest is paid during the first month of a one-year loan? What part of the interest is paid during the first three months of a one-year loan? What part of the interest is paid during the first six months of a one-year loan?

$\frac{12}{78}$ of the total interest is paid in the first month.

$\frac{12 + 11 + 10}{78}$ or $\frac{33}{78}$ is paid in the first three months.

$\frac{12 + 11 + 10 + 9 + 8 + 7}{78}$ or $\frac{57}{78}$ is paid in the first six months.

The above method is called the **rule of 78**. Do you see why?

For a two-year loan, find the number of months in two years. The interest series has how many addends? What are the addends?

Exercises

Find the part of the interest owed for each of the following using the rule of 78.

1. first two months of a one-year loan
2. first four months of a one-year loan
3. last month of a one-year loan
4. first month of a two-year loan
5. first three months of a two-year loan
6. first six months of a two-year loan
7. first month of a three-year loan
8. last month of a three-year loan

13-3 Geometric Sequences

Ann read about a *foolproof way to save a million dollars*. The plan is simple. Save a penny the first day. Then each day save double the amount saved on the previous day. How much should Ann save on the tenth day?

Ann makes a graph as shown at the right to show the amount of savings for each successive day. If the points on the graph were connected with a smooth curve, what kind of graph would Ann have?

The amounts in the savings program form a **geometric sequence**.

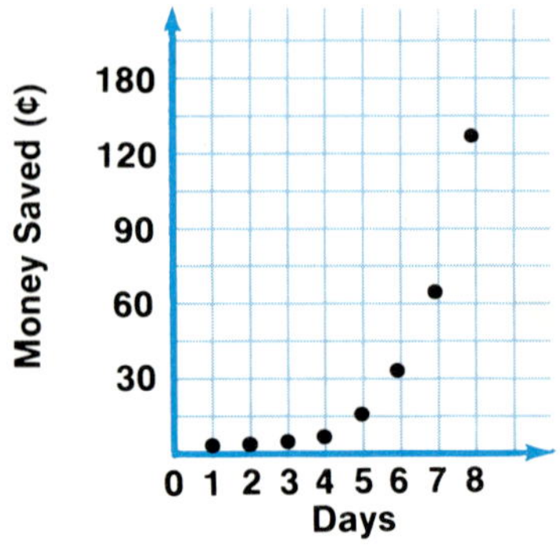

1, 2, 4, 8, 16, 32, 64, . . .   *What is the next term?*   *How is it found?*

A geometric sequence is a sequence in which each term, after the first, is found by multiplying the previous term by a constant called the *common ratio*.	*Definition of Geometric Sequence*

In any geometric sequence, the constant or common ratio, r, is found by dividing any term by the previous term.

Example

1 **Find the common ratio and the next two terms of the geometric sequence 4, 12, 36, ...**

$12 \div 4 = 3$ $36 \div 12 = 3$

The common ratio is 3. So, multiply the third term by 3 to get the fourth, and so on.

$36 \cdot 3 = 108$ $108 \cdot 3 = 324$

The next two terms are 108 and 324.

A geometric sequence containing n terms and having common ratio r can be written as follows. The second term, a_2, is found by multiplying the first term, a_1, by the common ratio, r.

a_1	a_2	a_3	a_4	$\cdots$	a_n	the sequence
a_1	$a_1 r$	$a_2 r$	$a_3 r$	$\cdots$	$a_{n-1} r$	the sequence in terms of the previous term and the common ratio
a_1	$a_1 r$	$a_1 r^2$	$a_1 r^3$	$\cdots$	$a_1 r^{n-1}$	the sequence in terms of the first term and the common ratio

> The nth term, a_n, of a geometric sequence with first term, a_1, and common ratio, r, is given by either of the following formulas.
>
> $$a_n = a_{n-1}r \qquad a_n = a_1 r^{n-1}$$

Formulas for the nth Term of a Geometric Sequence

Examples

2 **Write the first six terms of a geometric sequence in which $a_1 = 4$ and $r = 3$.**

Write each term using the formula $a_n = a_1 r^{n-1}$.

a_1	a_2	a_3	a_4	a_5	a_6
4	$4 \cdot 3$	$4 \cdot 3^2$	$4 \cdot 3^3$	$4 \cdot 3^4$	$4 \cdot 3^5$
4	12	36	108	324	972

The six terms are 4, 12, 36, 108, 324, and 972.

3 **Find the seventh term of a geometric sequence in which $a_3 = 7$ and $r = 2$.**

The seventh term of a sequence beginning with a_3 is the same as the fifth term of a sequence beginning with a_1. *Why?*

Use $a_n = a_1 r^{n-1}$.

$$a_5 = a_1 \cdot 2^{5-1} \qquad \textit{Substitute 2 for r and 5 for n.}$$
$$= 7 \cdot 2^4 \qquad \textit{Substitute 7 for } a_1.$$
$$= 112$$

The seventh term is 112.

The terms between any two nonconsecutive terms of a geometric sequence are called **geometric means**. In the sequence 3, 12, 48, 192, 768, 3072, . . . , the three geometric means between 3 and 768 are 12, 48, and 192.

Example

4 **Find the missing geometric means in the sequence 81, ____, ____, 3.**

$$a_4 = a_1 \cdot r^3 \qquad \textit{Use } a_n = a_1 r^{n-1}. \textit{ Substitute 4 for n.}$$
$$3 = 81 \cdot r^3 \qquad \textit{Substitute 81 for } a_1 \textit{ and 3 for } a_4.$$
$$\frac{1}{3} = r$$

$$a_2 = 81\left(\frac{1}{3}\right) \qquad a_3 = 81\left(\frac{1}{3}\right)^2$$
$$= 27 \qquad\qquad = 9$$

The missing terms are 27 and 9.

Sometimes there is more than one way to supply the missing geometric means for a particular sequence.

5 **Find the missing geometric means in the sequence 6, ____, ____, ____, 96.**

$a_5 = a_1 r^4$ $a_n = a_1 r^{n-1}$
$96 = 6r^4$
$16 = r^4$
$\pm 2 = r$

$$a_2 = 6(2) \qquad\qquad a_3 = 6(2)^2 \qquad\qquad a_4 = 6(2)^3$$
$$= 12 \qquad\qquad\qquad = 24 \qquad\qquad\qquad = 48$$

or   or   or

$$a_2 = 6(-2) \qquad\qquad a_3 = 6(-2)^2 \qquad\qquad a_4 = 6(-2)^2$$
$$= -12 \qquad\qquad\qquad = 24 \qquad\qquad\qquad = -48$$

The missing terms are 12, 24, and 48 or -12, 24, and -48.

6 **A vacuum pump removes $\frac{1}{5}$ of the air from a sealed container on each stroke of its piston. How much of the air remains after five strokes of the piston?**

Let 1 represent the original amount of air. After the first stroke, $1 - \frac{1}{5}$ or $\frac{4}{5}$ of the air remains. The second stroke removes $\frac{1}{5}$ of the remaining air. Thus, the amount that remains after two strokes is $\frac{4}{5}\left(1 - \frac{1}{5}\right) = \frac{4}{5} \cdot \frac{4}{5} = \frac{16}{25}$.

The sequence can be indicated as follows.

$$0 \quad 1 \quad 2 \quad\quad 3 \quad\quad 4 \quad\quad 5 \qquad \textit{number of each stroke}$$
$$1, \quad \frac{4}{5}, \quad \frac{16}{25}, \quad \text{---}, \quad \text{---}, \quad \text{---},$$
$$a_1 \quad a_2 \quad a_3 \quad a_4 \quad a_5 \quad a_6 \qquad \textit{number of each term}$$

Use $a_n = a_1 \cdot r^{n-1}$.

$$a_6 = 1 \cdot \left(\frac{4}{5}\right)^5 \text{ or } \frac{4^5}{5^5} \qquad \textit{Substitute 1 for } a_1, \text{ 6 for } n, \text{ and } \frac{4}{5} \text{ for } r.$$
$$= \frac{1024}{3125} \text{ or about } 0.328$$

Thus, about 32.8% of the air remains.

Exploratory Exercises

Tell whether each sequence is geometric. If so, find the common ratio.

1. 4, 20, 100, 500

2. $9, 6, 4, \frac{8}{3}$

3. $\frac{3}{2}, \frac{9}{4}, \frac{27}{8}, \frac{81}{16}$

4. 2, 4, 6, 8

5. 7, 14, 21, 28

6. 1, 4, 9, 16, 25

Find the missing terms of each geometric sequence.

7. 5, 15, 45, ____, ____

8. 2, 10, 50, ____, ____

9. ____, ____, 3, 9, 27

Written Exercises

Find the next two terms of each geometric sequence.

10. 2, 6, 18, . . .

11. 729, 243, 81, . . .

12. 20, 30, 45, . . .

13. 90, 30, 10, . . .

14. $\frac{1}{27}, \frac{1}{9}, \frac{1}{3}, \ldots$

15. $-\frac{1}{4}, \frac{1}{2}, -1, \ldots$

Find the first four terms of each geometric sequence described below.

16. $a_1 = \frac{3}{2}, r = 2$

17. $a_1 = 3, r = -2$

18. $a_1 = 12, r = \frac{1}{2}$

19. $a_1 = 27, r = -\frac{1}{3}$

Find the *n*th term of each geometric sequence described below. Use $a_n = a_1 \cdot r^{n-1}$.

20. $a_1 = 7, n = 4, r = 2$

21. $a_1 = 4, n = 3, r = 5$

22. $a_1 = 2, n = 5, r = 2$

23. $a_1 = 243, n = 5, r = -\frac{1}{3}$

24. $a_3 = 32, n = 6, r = -\frac{1}{2}$

25. $a_4 = 16, n = 8, r = \frac{1}{2}$

Find the missing terms of each sequence. Then show all the terms on a graph.

26. 3, ____, ____, ____, 48

27. 1, ____, ____, 8

28. 8, ____, ____, ____, ____, $\frac{1}{4}$

29. 3, ____, 75

30. 5, ____, ____, ____, 80

31. 7, ____, ____, ____, 112

32. ____, ____, −12, ____, ____, 96

33. ____, ____, ____, 24, ____, ____, ____, 384

Solve each problem.

34. A vacuum pump removes $\frac{1}{10}$ of the air from a space capsule on each stroke of its piston. What percent of the air remains after ten strokes of the piston?

35. A laboratory vacuum pump removes $\frac{1}{100}$ of the air from a chamber on each stroke of its piston. What percent of the air remains after 50 strokes of the piston?

36. The population of Sunville increases by 10% each year. It is now 20,000. What will be the expected population after five years (to the nearest 100 people)?

37. The population of Fairfield increases by 3% per year. At the last census, its population was 37,521. What is the expected population at the next census in ten years?

38. Mr. Culligan invested in business equipment worth $40,000. The equipment depreciates at the rate of 20% per year. What will be the value of his equipment at the end of the sixth year (to the nearest dollar)?

39. Mr. Olsen invested in computer equipment worth $90,000. The equipment depreciates at the rate of 25% per year. What will the value of his equipment be at the end of four years?

Solve each problem.

40. A vacuum pump removes $\frac{1}{20}$ of the air in a sealed jar on each stroke of its piston. How many strokes of the piston are required to remove 99% of the air from the jar?

41. A piece of paper is cut in half. Then each half is cut in half again. If this process is repeated 8 more times, what is the area of each piece compared to the original?

Show the first six terms of each geometric sequence on a graph.

42. 1.2, 1.8, 2.7, 4.05, . . .

43. $-1, 2, -4, 8, . . .$

mini-review

1. Find $f[g(2)]$ if $f(x) = x^2 + 2x - 5$ and $g(x) = 4x^2 + 12x - 17$.

Find all rational zeros of each function.

2. $f(x) = x^3 - 2x^2 - 23x + 60$

3. $f(x) = 2x^3 + x^2 - 13x + 6$

4. For $f(x) = 4x^3 - 7x^2 - 7x - 1$, state the number of positive real zeros, negative real zeros, and imaginary zeros.

Simplify.

5. $\dfrac{\dfrac{3x^2 + 10x - 8}{x + 3}}{\dfrac{3x^2 + x - 2}{x^2 + x - 6}}$

6. $\dfrac{\dfrac{3}{2x} + \dfrac{1}{x}}{4 - \dfrac{1}{2x}}$

7. Find $\dfrac{6x}{x - 5} + \dfrac{2x + 1}{3x + 4}$.

Solve each equation.

8. $\log_7 x = 3$

9. $\log_2 x = \frac{1}{2} \log_2 81$

Evaluate each expression.

10. $4^{\log_4 4}$

11. $n^{\log_n 5}$

Solve each problem.

12. Suppose y varies directly as x, and $y = 4$ when $x = 6$. Find y when $x = 9$.

13. Anne Marie earned $750 delivering papers last year. She deposited the money in a savings account paying 8.5% annual interest compounded monthly. How long will it take for the deposit to grow to $1,000? Use $A = P\left(1 + \dfrac{r}{n}\right)^{nt}$.

Problem Solving

Geometric Sequences

Several steps are needed to explore problems involving sequences.

(1) Look for a description of the pattern.
(2) Determine the type of sequence.
(3) Draw a "picture" by writing the numbers in a sequence with blanks for the missing numbers.

Example: If the Wilsons pay $70,000 for their house in 1990 and its value increases at a yearly rate of 10%, how much will their house be worth in five years?

explore

1. The pattern is 10% increase each year for five years.
2. It is a geometric sequence where $r = 110\%$ or 1.10.
3. The sequence looks like the following.

If the original price is a_1, the value in 5 years will be a_6.

1990	1991	1992	1993	1994	1995
a_1	$a_1 r$	$a_1 r^2$	$a_1 r^3$	$a_1 r^4$	$a_1 r^5$

$70,000 . . .

plan

Use the definition of the nth term of a geometric sequence.

$$a_n = a_1 r^{n-1}$$
$$a_1 = 70,000 \qquad r = 1.1$$

solve

$$a_n = a_1 r^{n-1}$$
$$a_6 = 70,000(1.1)^5$$
$$= 112,735.70$$

The house will be worth $112,735.70 after 5 years.

examine

One way to examine the solution is to estimate. Ten percent of $70,000 is $7000. $7000 each year for 5 years would be $35,000. Add the original $70,000 to $35,000. The value should be greater than the sum, $105,000, since the value increase is compounded. An exact check would be to find each term.

Exercises

1. A tank holds 4000 gallons. Each day one half of the fluid is removed. How much fluid remains after 8 days?

2. Suppose a $20,000 sports car depreciates 20% per year. What will its value be after six years from date of purchase?

3. The population of Niota increases 8% each year. If it is now 5000, what will it be after 4 years?

4. The number of bacteria in a culture increases at a rate of 100% per three hours. If there are 300 bacteria now, how many will there be after 24 hours?

5. In a game of chance, there are 25 squares. The first square is worth 1¢ and each successive square is worth twice as much as the previous square. How much is the 24th square worth?

13-4 Geometric Series

Anne began her savings plan by saving a penny the first day, two cents the second day, and so on. On the seventh day, she saved 64 cents. How much had she saved altogether? The amounts saved on each day form a geometric sequence in which $a_1 = 1$ and $r = 2$.

The geometric sequence 1, 2, 4, 8, 16, 32, 64 is determined by Anne's plan. The corresponding **geometric series** is shown below.

For the sequence and series, the first term is a_1, the second term is a_2, and so on.

$$1 + 2 + 4 + 8 + 16 + 32 + 64$$

Let S_7 be the sum of the first seven terms indicated above. The following method can be used to calculate the sum.

$$
\begin{aligned}
S_7 &= 1 + 2 + 4 + 8 + 16 + 32 + 64 \\
2 \cdot S_7 &= 2 + 4 + 8 + 16 + 32 + 64 + 128 \\
\hline
S_7 - 2S_7 &= 1 + 0 + 0 + 0 + 0 + 0 + 0 + (-128) \\
(1 - 2)S_7 &= 1 - 128 \\
S_7 &= 127
\end{aligned}
$$

Subtract $2 \cdot S_7$ from S_7.

Factor $S_7 - 2S_7$ to get $(1 - 2)S_7$.

Anne saved 127 cents, or \$1.27.

In this case, $S_7 = \dfrac{1 - 128}{1 - 2}$.

Substitute names of terms of sequence, and r for 2.

$$S_7 = \frac{a_1 - a_8}{1 - r}$$

Note that $a_8 = a_1 r^7$

$$= \frac{a_1 - a_1 r^7}{1 - r}$$

Can r = 1? Why or why not?

The same method can be used to write a formula for a geometric series with n terms.

Let $S_n = a_1 + a_1 r + a_1 r^2 + \ldots + a_1 r^{n-1}$.

$$
\begin{aligned}
S_n &= a_1 + a_1 r + a_1 r^2 + \ldots + a_1 r^{n-1} \\
r \cdot S_n &= a_1 r + a_1 r^2 + \ldots + a_1 r^{n-1} + a_1 r^n \\
\hline
S_n - rS_n &= a_1 + 0 + 0 + \ldots + \phantom{a_1r^{n-1}}0 - a_1 r^n \\
(1 - r)S_n &= a_1 - a_1 r^n \\
S_n &= \frac{a_1 - a_1 r^n}{1 - r}
\end{aligned}
$$

Multiply each term of the series by r. Write the second row so that like terms are in columns. Subtract.

The sum, S_n, of the first n terms of a geometric series is given by the following formula.

$$S_n = \frac{a_1 - a_1 r^n}{1 - r} \quad \text{or} \quad S_n = \frac{a_1(1 - r^n)}{1 - r} \quad \text{where } r \neq 1$$

Sum of a Geometric Series

Study the following examples.

Example

1 **Find the sum of the first six terms of a geometric series for which $a_1 = 3$ and $r = -2$. Use the formula $S_n = \dfrac{a_1 - a_1 r^n}{1 - r}$.**

$$S_n = \frac{a_1 - a_1 r^n}{1 - r}$$

$$S_6 = \frac{3 - 3(-2)^6}{1 - (-2)} \qquad \textit{Substitute 3 for } a_1, -2 \textit{ for } r, \textit{ and 6 for } n.$$

$$= -63$$

The sum of the first six terms is -63.

You know that $a_n = a_1 r^{n-1}$. Then $a_n \cdot r = a_1 r^{n-1} \cdot r$ or $a_1 r^n$. Replacing $a_1 r^n$ by $a_n r$ gives another formula for finding the value of S_n.

$$S_n = \frac{a_1 - a_1 r^n}{1 - r}$$

$$= \frac{a_1 - a_n r}{1 - r} \qquad \begin{array}{l} \textit{Substitute } a_n r \textit{ for } a_1 r^n. \\ r \neq 1 \end{array}$$

Example

2 **Find the sum of a geometric series for which $a_1 = 48$, $a_n = 3$, and $r = -\dfrac{1}{2}$. Use the formula $S_n = \dfrac{a_1 - a_n r}{1 - r}$.**

$$S_n = \frac{a_1 - a_n r}{1 - r}$$

$$= \frac{48 - 3\left(-\dfrac{1}{2}\right)}{1 - \left(-\dfrac{1}{2}\right)} \qquad \textit{Substitute 48 for } a_1, \textit{ 3 for } a_n, \textit{ and } -\dfrac{1}{2} \textit{ for } r.$$

$$= \frac{\dfrac{99}{2}}{\dfrac{3}{2}}$$

$$= 33$$

The sum S_n is 33.

When $r = 1$, S_n is found by the formula $S_n = n \cdot a_1$. For example, the sum of the series $3 + 3 + 3 + 3 + 3 + 3$ is given by $S_n = 6 \cdot 3$ or 18.

Exploratory Exercises

For each geometric series described below, state the first term, the common ratio, the last term, and the number of terms.

1. $9 - 18 + 36 - 72$

2. $3 + 1.5 + 0.75 + 0.375$

3. $a_1 + 8 + 32 + 128 + a_5$

4. $a_1 + 6 - 3 + \frac{3}{2} + a_5$

5. $a_1 = 20, a_4 = -\frac{5}{16}, n = 5$

6. $a_2 = 6, a_4 = 24, n = 5$

Find the sum of each geometric series described below.

7. $2 + (-6) + 18 + \ldots$ to 6 terms.

8. $3 + 6 + 12 + \ldots$ to 6 terms.

9. $8 + 4 + 2 + \ldots$ to 6 terms.

10. $\frac{1}{9} - \frac{1}{3} + 1 - \ldots$ to 5 terms.

11. $1296 - 216 + 36 - \ldots$ to 5 terms.

12. $7 + 7 + 7 + \ldots$ to 9 terms.

Written Exercises

Find the sum of each geometric series described below. Use a calculator as necessary.

13. $75 + 15 + 3 + \ldots$ to 10 terms.

14. $16 + 16 + 16 + \ldots$ to 11 terms.

15. $a_1 = 7, r = 2, n = 14$

16. $a_1 = 5, r = 3, n = 12$

17. $a_1 = 12, a_5 = 972, r = -3$

18. $a_1 = 256, r = 0.75, n = 9$

19. $a_1 = 243, r = -\frac{2}{3}, n = 5$

20. $a_1 = 16, r = -\frac{1}{2}, n = 10$

21. $a_1 = 625, a_5 = 81, r = \frac{3}{5}$

22. $a_1 = 625, r = \frac{2}{5}, n = 8$

23. $a_1 = 4, a_6 = \frac{1}{8}, r = \frac{1}{2}$

24. $a_1 = 1, a_5 = \frac{1}{16}, r = -\frac{1}{2}$

25. $a_1 = 125, a_5 = \frac{1}{5}, r = \frac{1}{5}$

26. $a_1 = 343, a_4 = -1, r = -\frac{1}{7}$

27. $a_3 = \frac{3}{4}, a_6 = \frac{3}{32}, n = 6$

28. $a_2 = 1.5, a_5 = 0.1875, n = 9$

29. $a_3 = \frac{5}{4}, a_4 = -\frac{5}{16}, n = 6$

30. $a_2 = -12, a_5 = -324, n = 10$

Find a_1 for each geometric series described below.

31. $S_n = 32, r = 2, n = 6$

32. $S_n = 244, r = -3, n = 5$

33. $a_n = 324, r = 3, S_n = 484$

34. $S_n = 635, a_n = 320, r = 2$

35. $S_n = 1022, r = 2, n = 9$

36. $S_n = 15.75, r = 0.5, a_n = 0.25$

Solve each problem.

37. Find the sum of the first nine terms of the geometric series whose 4th term is 20 and whose 8th term is 1620.

38. One minute after it is released, a hot air balloon rises 80 feet. In each succeeding minute the balloon rises only 60% as far as it rose in the previous minute. How far will the balloon rise in 6 minutes?

39. The teaching staff of Fairmeadow High School informs its members of school cancellation by telephone. The principal calls 2 teachers, each of whom in turn calls 2 other members, and so on. This process must be repeated 6 times counting the principal's calls as the first time. How many teachers, including the principal, work at Fairmeadow High?

13-5 Infinite Geometric Series

The first swing of a pendulum measures 25 centimeters. The lengths of the successive swings of the pendulum form the geometric sequence 25, 20, 16, 12.8, ...

Suppose this pendulum continues to swing back and forth indefinitely. Then the sequence shown above is called an **infinite geometric sequence**.

The distances traveled by the pendulum can be added.

$$25 + 20 + 16 + 12.8 + \cdots$$

The sum of these distances is an infinite geometric series with $a_1 = 25$ and $r = \frac{20}{25}$ or 0.8.

The series can be written as follows.

$$25 + 25(0.8)^1 + 25(0.8)^2 + 25(0.8)^3 + 25(0.8)^4 + 25(0.8)^5 + 25(0.8)^6 + \cdots$$

Notice what happens as 0.8 is raised to various powers.

$$(0.8)^1 = 0.8 \qquad (0.8)^{10} \approx 0.1074 \qquad (0.8)^{50} \approx 0.00001$$

As the values of n become greater, what happens to $(0.8)^n$?

What happens to the terms of the series as the exponents increase?

$$a_1 = 25 \qquad a_{10} \approx 3.3554 \qquad a_{50} \approx 0.0004$$

What happens to the sums S_n as n becomes greater?

$$S_1 = 25 \qquad S_{10} \approx \frac{25 - 25(0.1074)}{0.2} \qquad S_{50} = \frac{25 - 25(0.00001)}{0.2}$$

$$\approx \frac{25 - 2.685}{0.2} \qquad \approx \frac{25 - 0.00025}{0.2}$$

$$\approx 111.575 \qquad \approx 124.9988$$

Find S_{100}. Do you think that S_{500} is greater than 125?

In any geometric series, the sum of the first n terms can be found by using either of the following formulas.

$$S_n = \frac{a_1 - a_1 r^n}{1 - r} \quad \text{or} \quad S_n = \frac{a_1(1 - r^n)}{1 - r}, r \neq 1$$

Suppose the value of r^n is very close to zero. Then the value of S_n is very close to the following expression, called the **sum of an infinite series**.

$$\frac{a_1(1 - 0)}{1 - r} \quad \text{or} \quad \frac{a_1}{1 - r}$$

> **The sum, S, of an infinite geometric series where $-1 < r < 1$ is given by the following formula.**
>
> $$S = \frac{a_1}{1 - r}$$

Sum of an Infinite Geometric Series

An infinite geometric series where r is not between -1 and 1 does not have a sum.

1 **Find the sum of the series $25 + 20 + 16 + 12.8 + \ldots$.**

$r = \dfrac{20}{25}$ or 0.8 *Since $-1 < 0.8 < 1$, you can use the formula $S = \dfrac{a_1}{1-r}$.*

$S = \dfrac{25}{1 - 0.8}$ *Substitute 25 for a_1 and 0.8 for r.*

$ = 125$

The sum of the series is 125.

2 **Find the sum of the infinite geometric series $\dfrac{4}{3} - \dfrac{2}{3} + \dfrac{1}{3} - \dfrac{1}{6} + \cdots$.**

$a_2 = a_1 r$ *First, find r.*

$-\dfrac{2}{3} = \left(\dfrac{4}{3}\right)r$ *Substitute $-\dfrac{2}{3}$ for a_2 and $\dfrac{4}{3}$ for a_1.*

$-\dfrac{1}{2} = r$

$S = \dfrac{\dfrac{4}{3}}{1 - \left(-\dfrac{1}{2}\right)} = \dfrac{\dfrac{4}{3}}{\dfrac{3}{2}} = \dfrac{8}{9}$ *Next, find the sum using $S = \dfrac{a_1}{1-r}$.*

Substitute $\dfrac{4}{3}$ for a_1 and $\left(-\dfrac{1}{2}\right)$ for r.

The sum is $\dfrac{8}{9}$.

3 **Express the repeating decimal $0.11111\ldots$ or $0.\overline{1}$ as a fraction.**

Write $0.\overline{1}$ as an infinite geometric series.
$0.\overline{1} = 0.1 + 0.01 + 0.001 + \ldots$
Then $a_1 = 0.1$ and $r = 0.1$ since $0.01 = 0.1(r)$.

$S = \dfrac{0.1}{1 - 0.1}$ *Use $S = \dfrac{a_1}{1-r}$.*

$ = \dfrac{1}{9}$ *Substitute 0.1 for a_1 and 0.1 for r.*

The repeating decimal $0.\overline{1}$ is equal to the common fraction $\dfrac{1}{9}$.

4 **A rubber ball dropped 30 feet bounces $\dfrac{2}{5}$ of the height from which it fell on each bounce. How far will it travel before coming to rest?**

downward distance

$S = \dfrac{30}{1 - 0.4}$ *Note that a_1 is 30 and r is 0.4.*

$ = \dfrac{30}{0.6}$

$ = 50$

upward distance

$S = \dfrac{12}{1 - 0.4}$ *Note that a_1 is $\dfrac{2}{5}$ of 30, or 12, and r is 0.4.*

$ = \dfrac{12}{0.6}$

$ = 20$

The total distance is $50 + 20$ or 70 feet.

Find a_1 and r for each series. Then find the sum, if it exists.

1. $\frac{1}{2} + \frac{1}{3} + \frac{2}{9} + \frac{4}{27} + \cdots$

2. $12 + 3 + \frac{3}{4} + \frac{3}{16} + \cdots$

3. $1 - \frac{1}{3} + \frac{1}{9} - \frac{1}{27} + \cdots$

4. $1 - 3 + 9 - 27 + \cdots$

5. $1 + \frac{3}{2} + \frac{9}{4} + \frac{27}{8} + \cdots$

6. $48 + 16 + \frac{16}{3} + \frac{16}{9} + \cdots$

Express each repeating decimal as an infinite geometric series. State the ratio for each.

7. $0.\overline{7}$　　**8.** $0.\overline{3}$　　**9.** $0.\overline{73}$　　**10.** $0.\overline{8}$　　**11.** $0.\overline{152}$　　**12.** $0.\overline{746}$　　**13.** $0.\overline{93}$　　**14.** $0.\overline{75}$

Written Exercises

Find the sum of each infinite geometric series described below.

15. $a_1 = 6, r = \frac{11}{12}$

16. $a_1 = 18, r = -\frac{2}{7}$

17. $a_1 = 7, r = -\frac{3}{4}$

18. $a_1 = 27, r = -\frac{4}{5}$

19. $9 + 6 + 4 + \ldots$

20. $\frac{1}{3} + \frac{1}{9} + \frac{1}{27} + \ldots$

21. $3 - 2 + \frac{4}{3} - \ldots$

22. $\frac{3}{4} + \frac{1}{2} + \frac{1}{3} + \ldots$

23. $12 - 4 + \frac{4}{3} - \frac{4}{9} + \ldots$

24. $1 - \frac{1}{4} + \frac{1}{16} - \ldots$

25. $10 - \frac{5}{2} + \frac{5}{8} - \ldots$

26. $2 + 6 + 18 + 54 + \ldots$

27. $3 - 9 + 27 - \ldots$

28. $12 + 6 + 3 + \ldots$

29. $10 - 1 + 0.1 - \ldots$

Find a common fraction equivalent to each repeating decimal below.

30. $0.\overline{3}$　　**31.** $0.\overline{9}$　　**32.** $0.\overline{15}$　　**33.** $0.\overline{31}$　　**34.** $0.\overline{075}$　　**35.** $0.\overline{410}$　　**36.** $0.3\overline{7}$　　**37.** $0.4\overline{5}$

Find the first three terms of each infinite geometric series described below.

38. $S = 9, r = \frac{1}{3}$

39. $S = 16, r = \frac{3}{4}$

40. $S = 28, r = -\frac{2}{7}$

41. $S = \frac{27}{4}, r = -\frac{1}{3}$

Solve each problem.

42. The end of a swinging pendulum 90 cm long moves through 50 cm on its first swing. Each succeeding swing is $\frac{9}{10}$ of the preceding one. How far will the pendulum travel before coming to rest?

43. The end of a swinging pendulum 30 cm long moves through 20 cm on its first swing. Each succeeding swing is $\frac{10}{11}$ of the preceding one. How far will it travel before coming to rest?

44. A silicon ball dropped 12 feet rebounds $\frac{7}{10}$ of the height from which it fell on each bounce. How far will it travel before coming to rest?

45. A hot-air balloon rises 80 feet in the first minute of flight. If in each succeeding minute the balloon rises only 90 percent as far as in the previous minute, what will be its maximum altitude?

Challenge Exercises

46. A side of an equilateral triangle is 20 inches. The midpoints of its sides are joined to form an inscribed equilateral triangle. If this process is continued without end, find the sum of the perimeters of the triangles.

47. Find the sum of the areas of the series of triangles in the previous exercise.

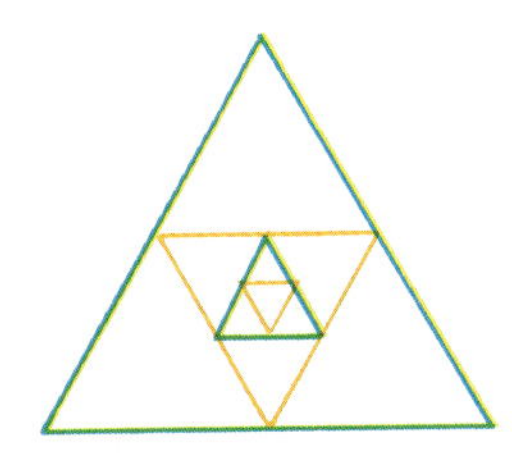

Cumulative Review

1. Find $p(-4)$ if $p(x) = 3x^2 - 7x + 1$.

2. Find $f(m + 2)$ if $f(x) = 2x^3 - 5x^2$.

3. State the possible number of positive real zeros, negative real zeros, and imaginary zeros for
$f(x) = x^3 - 7x^2 + 17x - 15$.

4. Find all rational zeros of
$f(x) = x^3 - 4x^2 - 11x + 30$.

5. Approximate to the nearest tenth the positive real zero of
$f(x) = x^3 + x^2 - 17x - 33$.

If $f(x) = 2x + 3$, $g(x) = x - 1$, and $h(x) = x^2 + 4$, find each value.

6. $[g \circ h](2)$

7. $[f \circ g](3)$

8. $h[f(x)]$

9. $g[f(-1)]$

Simplify.

10. $\dfrac{\dfrac{x^2 - 4x - 21}{x + 2}}{\dfrac{x^2 - 2x - 15}{x + 2}}$

11. $\dfrac{\dfrac{1}{a} - b}{\dfrac{1}{a} + b}$

12. $\dfrac{6}{x^2 + 4x + 4} + \dfrac{5}{x + 2}$

13. $\dfrac{13}{a + b} - \dfrac{7}{b^2 - a^2}$

14. Write $7^{-2} = \dfrac{1}{49}$ in logarithmic form.

15. Write $\log_{16} 2 = \dfrac{1}{4}$ in exponential form.

Solve each equation using logarithms.

16. $3^x = 48$

17. $7^{x-1} = 3^{2x}$

Solve each equation.

18. $\log_7 2x + \log_7 18 = \log_7 54$

19. $\log_{13} (4m + 1) = 1$

20. Find the twelfth term in the arithmetic sequence $-17, -13, -9, \ldots$.

21. Find S_n for $d = 5$, $n = 16$, and $a_n = 72$.

22. Find the missing geometric means in the sequence $3, \underline{\quad}, \underline{\quad}, \underline{\quad}, 48$.

23. Which term of the sequence $-7, -3, 1, 5, \ldots$ is 69?

Solve each problem.

24. The number of a certain type of bacteria can increase from 80 to 164 in 3 hours. Find the value of k in the growth formula.

25. A tank can be filled by a hose in 12 hours. The tank can be emptied by a drain pipe in 24 hours. If the drain pipe is left open while the tank is filling, how long will it take to fill the tank?

26. Danny and Sonia run a lawn service during the summer. Sonia can cut the lawns of all customers in 15 hours. Danny could do the same in 12 hours. Sonia works alone for 8 hours, after which Danny joins her. How long will it take both of them to finish the job?

27. A salesman receives \$25 for every vacuum cleaner he sells. If he sells more than 10 vacuum cleaners he will receive an additional \$1.75 for each successive sale until he is paid a maximum of \$46 per vacuum cleaner. How many must he sell to reach this maximum?

13-6 Sigma Notation

Consider the sum of the first ten positive even integers.

$$2 + 4 + 6 + 8 + \ldots + 20$$

This series can be written in a more concise way. First, notice the following pattern.

$$2(1) + 2(2) + 2(3) + 2(4) + \ldots + 2(10)$$

The Greek letter sigma, Σ, can be used to indicate this sum.

stop at 10 $\longrightarrow$ $\displaystyle\sum_{n=1}^{10} 2n$ This is read *the summation from 1 to 10 of 2n.*

start n at 1 $\longrightarrow$

Using this technique to indicate the sum of a series is called **sigma** or **summation notation**. When using this notation, the variable defined at the bottom of the sigma is called the **index of summation**. The terms of the series above are generated by successively replacing the index of summation with 1, 2, 3, . . ., 10.

Any letter can be used for the index of summation.

Examples

1   Write $\displaystyle\sum_{k=1}^{5} 4k$ in expanded form.

Replace k with 1, 2, 3, 4, 5 successively.

$$\sum_{k=1}^{5} 4k = \overset{k=1}{4 \cdot 1} + \overset{k=2}{4 \cdot 2} + \overset{k=3}{4 \cdot 3} + \overset{k=4}{4 \cdot 4} + \overset{k=5}{4 \cdot 5}$$
$$= 4 + 8 + 12 + 16 + 20$$

2   Write $\displaystyle\sum_{n=3}^{7} (3n - 2)$ in expanded form and find the sum.

$$\sum_{n=3}^{7} (3n - 2) = \overset{n=3}{(3 \cdot 3 - 2)} + \overset{n=4}{(3 \cdot 4 - 2)} + \overset{n=5}{(3 \cdot 5 - 2)} + \overset{n=6}{(3 \cdot 6 - 2)} + \overset{n=7}{(3 \cdot 7 - 2)}$$
$$= (9 - 2) + (12 - 2) + (15 - 2) + (18 - 2) + (21 - 2)$$
$$= (7 + 10 + 13 + 16 + 19) \text{ or } 65$$

The sum is 65.

3   Write $\displaystyle\sum_{j=1}^{5} 2(4)^{j-1}$ in expanded form and find the sum.

$$\sum_{j=1}^{5} 2(4)^{j-1} = \overset{j=1}{2 \cdot 4^{1-1}} + \overset{j=2}{2 \cdot 4^{2-1}} + \overset{j=3}{2 \cdot 4^{3-1}} + \overset{j=4}{2 \cdot 4^{4-1}} + \overset{j=5}{2 \cdot 4^{5-1}}$$
$$= 2 \cdot 4^0 + 2 \cdot 4^1 + 2 \cdot 4^2 + 2 \cdot 4^3 + 2 \cdot 4^4$$
$$= 2 \cdot 1 + 2 \cdot 4 + 2 \cdot 16 + 2 \cdot 64 + 2 \cdot 256$$
$$= (2 + 8 + 32 + 128 + 512) \text{ or } 682$$

The sum is 682.

4 **Find the sum of** $\sum\limits_{x=1}^{9} [3 + (x - 1)5]$.

$$\sum_{x=1}^{9} [3 + (x - 1)5] = (3 + 0 \cdot 5) + (3 + 1 \cdot 5) + (3 + 2 \cdot 5) + \ldots + (3 + 8 \cdot 5)$$
$$= \quad 3 \quad + \quad 8 \quad + \quad 13 \quad + \ldots + \quad 43$$

Since this is an arithmetic series, use $S_n = \frac{n}{2}(a_1 + a_n)$.

$S_9 = \frac{9}{2}(3 + 43)$ $a_1 = 3, a_n = 43, n = 9.$

$\quad = \frac{9}{2}(46)$

$\quad = 207$

The sum is 207.

5 **Use sigma notation to express 5 + 9 + 13 + 17 + 21.**

First, search for a pattern.

$$\quad 5 \quad + \quad 9 \quad + \quad 13 \quad + \quad 17 \quad + \quad 21$$
$$= (5 + 0 \cdot 4) + (5 + 1 \cdot 4) + (5 + 2 \cdot 4) + (5 + 3 \cdot 4) + (5 + 4 \cdot 4)$$

The pattern is $5 + (n - 1)4$, if n is replaced successively with 1, 2, 3, 4, 5. Therefore, the series can be expressed as:

$\sum\limits_{n=1}^{5} [5 + (n - 1)4]$ or $\sum\limits_{n=1}^{5} (4n + 1)$ *Why are parentheses necessary around 4n + 1?*

6 **Use sigma notation to express 1 − 3 + 9 − 27 + 81 − 243.**

Since $1 - 3 + 9 - 27 + 81 - 243$ is a geometric series, use $a_n = a_1 r^{n-1}$.
$a_n = 1 \cdot (-3)^{n-1}$ *$a_1 = 1$ and $r = -3$*

Therefore, the series can be expressed as:

$$\sum_{n=1}^{6} (-3)^{n-1}$$

Exploratory Exercises

State the index and the number of terms in each series. Then, state each series in expanded form.

1. $\sum\limits_{j=1}^{4} (j + 2)$ **2.** $\sum\limits_{k=3}^{5} 4k$ **3.** $\sum\limits_{r=1}^{3} (r - 1)$ **4.** $\sum\limits_{k=2}^{6} (3 - k)$

5. $\sum\limits_{i=0}^{4} 2i$ **6.** $\sum\limits_{m=6}^{8} (-m)$ **7.** $\sum\limits_{p=4}^{7} (p + 2)$ **8.** $\sum\limits_{i=2}^{5} (i + 9)$

Written Exercises

Write each expression in expanded form and find the sum.

9. $\displaystyle\sum_{t=0}^{4} (13 + 7t)$

10. $\displaystyle\sum_{i=1}^{5} (1 + 7i)$

11. $\displaystyle\sum_{p=3}^{7} (wp - 1)$

12. $\displaystyle\sum_{j=0}^{6} (24 - 9j)$

13. $\displaystyle\sum_{b=2}^{6} (2b + 1)$

14. $\displaystyle\sum_{y=5}^{11} (3y - 5)$

15. $\displaystyle\sum_{i=1}^{7} 2i$

16. $\displaystyle\sum_{r=3}^{6} (r + 2)$

17. $\displaystyle\sum_{n=4}^{8} 4^n$

18. $\displaystyle\sum_{k=1}^{7} 2^{k-2}$

19. $\displaystyle\sum_{s=1}^{4} 24\left(-\frac{1}{2}\right)^s$

20. $\displaystyle\sum_{b=1}^{4} 2\left(\frac{3}{8}\right)^b$

Find the sum of each series.

21. $\displaystyle\sum_{n=1}^{25} 2n$

22. $\displaystyle\sum_{n=1}^{30} (2n - 1)$

23. $\displaystyle\sum_{n=1}^{40} (3n + 2)$

24. $\displaystyle\sum_{n=10}^{50} (3n - 1)$

25. $\displaystyle\sum_{n=21}^{75} (2n + 5)$

26. $\displaystyle\sum_{j=1}^{20} (-2)^j$

27. $\displaystyle\sum_{k=1}^{19} 2^{k-4}$

28. $\displaystyle\sum_{s=1}^{81} \left(\frac{3}{10}\right)^s$

Use sigma notation to express each series.

29. $7 + 10 + 13 + 16 + 19$

30. $3 + 10 + 17 + 24 + 31$

31. $15 + 11 + 7 + 3 + (-1)$

32. $7 + 6\frac{1}{2} + 6 + 5\frac{1}{2} + 5$

33. $1^2 + 2^2 + 3^2 + 4^2 + 5^2$

34. $\frac{1}{2} + \frac{1}{3} + \frac{1}{4} + \frac{1}{5} + \frac{1}{6}$

35. $8 + 4 + 2 + 1 + \frac{1}{2} + \frac{1}{4}$

36. $2 - 6 + 18 - 54 + 162 - 486$

37. $243 - 162 + 108 - 72 + 48 - 32$

38. $625 + 375 + 225 + 135 + 81$

Challenge Exercises

State whether the sums are equivalent.

39. $\displaystyle\sum_{a=1}^{4} a^2 \overset{?}{=} \sum_{a=3}^{6} (a - 2)^2$

40. $\displaystyle 2\sum_{k=3}^{7} k^2 \overset{?}{=} \sum_{k=3}^{7} 2k^2$

mini-review

1. For $f(x) = 7x^5 + 4x^4 - 3x^3 - 2x^2 + 7x + 1$, state the number of positive real zeros, negative real zeros, and imaginary zeros.

Simplify.

2. $\dfrac{3x^2 - 13x - 10}{x - 6} \div \dfrac{x^2 - 2x - 15}{x^2 - 3x - 18}$

3. $\dfrac{x - 5}{6x + 7} + \dfrac{3}{2x - 3}$

Solve each equation.

4. $\log_3 4x = \frac{1}{2} \log_3 121 + \frac{1}{3} \log_3 64$

5. $13^{y+1} = 47^y$

6. For a certain single-celled organism, $k = 0.732$ when time (t) is measured in days. How long will it take 15 organisms to increase to 1,000? Use $y = ne^{kt}$.

13-7 The General Term

Sequences and series often are described by giving a formula for their general term. This formula may show how to find the nth term.

Sequence type	Formula for the nth term
Arithmetic	$a_n = 7 + (n - 1) \cdot 3$
Geometric	$a_n = 5 \cdot 4^{n-1}$

Sometimes a sequence or series may be described recursively. A **recursive formula** depends on knowing one or more previous terms.

In a recursive formula, the first term of the sequence is given, and the relationship between any term and its successor is defined.

Sequence type	Recursive formula
Arithmetic	$a_{n+1} = a_n + 3,\ a_1 = 7$
Geometric	$a_{n+1} = a_n \cdot 4,\ a_1 = 5$

When a sequence or series is described by a formula for its nth term, any term can be computed directly.

Example

1 **Find the 99th term of the sequence in which $a_n = 7 + (n - 1)3$.**

$a_n = 7 + (n - 1)3$
$a_{99} = 7 + (99 - 1)3$ *Substitute 99 for n.*
$\quad = 7 + 98 \cdot 3$
$\quad = 7 + 294$ or 301

The 99th term of the sequence is 301.

When a sequence or series is described by a recursive formula, you may need to compute several terms to find the terms desired.

Example

2 **Find the first six terms of the sequence in which $a_1 = 1$, $a_2 = 1$, and $a_{n+2} = a_n + 2 \cdot a_{n+1}$.**

$a_3 = a_{1+2}$ $a_4 = a_{2+2}$
$\quad = a_1 + 2 \cdot a_2$ $\quad = a_2 + 2 \cdot a_3$
$\quad = 1 + 2(1)$ or 3 $\quad = 1 + 2(3)$ or 7

$a_5 = a_{3+2}$ $a_6 = a_{4+2}$
$\quad = a_3 + 2 \cdot a_4$ $\quad = a_4 + 2 \cdot a_5$
$\quad = 3 + 2(7)$ or 17 $\quad = 7 + 2(17)$ or 41

The first six terms are 1, 1, 3, 7, 17, and 41.

3 **Find the 51st through the 54th terms of the sequence for which $a_n = 2n + 1$.**

$a_{51} = 2(51) + 1$ or 103 $a_{52} = 2(52) + 1$ or 105
$a_{53} = 2(53) + 1$ or 107 $a_{54} = 2(54) + 1$ or 109

The 51st through the 54th terms are 103, 105, 107, and 109.

4 **Find both a recursive formula and a formula for the nth term of the sequence 3, -6, 12, -24, 48, ...**

3	-6	12	-24	48
$3 \cdot (-2)$	$-6 \cdot (-2)$	$12 \cdot (-2)$	$-24 \cdot (-2)$	

Consider the ratio of two consecutive terms.

A recursive formula is $a_{n+1} = a_n(-2)$ and $a_1 = 3$.
A formula for the nth term is $a_n = 3(-2)^{n-1}$.   *Recall $a_n = a_1 r^{n-1}$.*

5 **Using Calculators**

Find the 100th term of the sequence 3, -6, 12, -24, 48, . . . using a calculator.

ENTER: 3 [×] 2 [+/−] [y^x] [(] 100 [−] 1 [)] [=]

DISPLAY: 3 $2 -2$ 100 $1\ 99$ $-1.9014759\ 00^{30}$

The 100th term of the sequence is approximately -1.90×10^{30}.

6 **Find a formula for the nth term of the series $7 + 9 + 11 + 13 + 15 + 17$ and express the series using sigma notation.**

Since $7 + 9 + 11 + 13 + 15 + 17$ is an arithmetic series, use
$a_n = a_1 + (n - 1)d$.
$a_n = 7 + (n - 1)2$ *$a_1 = 7$ and $d = 2$*
$a_n = 5 + 2n$ *$a_n = 7 + 2n - 2$*

Therefore, the series can be expressed as: $\displaystyle\sum_{n=1}^{6} (5 + 2n)$.

Exploratory Exercises

Find the ninth and tenth terms of each sequence.

1. $a_n = n(n + 2)$

2. $a_n = 3n - 4$

3. $a_n = n^2 - 1$

4. $a_n = (-1)^n$

Find the first four terms of each sequence.

5. $a_1 = 8$, $a_{n+1} = a_n - 1$

6. $a_1 = 13$, $a_{n+1} = a_n + 2$

7. $a_1 = -2$, $a_{n+1} = 3a_n$

8. $a_1 = 7$, $a_{n+1} = 2a_n$

9. $a_1 = -4$, $a_{n+1} = (-1)^{n+1} a_n$

10. $a_1 = 3$, $a_{n+1} = (-1)^n a_n$

11. $a_1 = 3$, $a_2 = 1$, $a_{n+2} = a_n + a_{n+1}$

12. $a_1 = 0$, $a_2 = 1$, $a_{n+2} = a_n + a_{n+1}$

Find a formula for the *n*th term of each sequence.

13. 2, 4, 6, 8, 10, 12, 14, . . .

14. 3, 5, 7, 9, 11, 13, . . .

15. $\dfrac{2}{1}, \dfrac{3}{2}, \dfrac{4}{3}, \dfrac{5}{4}, \dfrac{6}{5}, \ldots$

16. $\dfrac{1}{3}, \dfrac{1}{5}, \dfrac{1}{7}, \dfrac{1}{9}, \dfrac{1}{11}, \dfrac{1}{13}, \ldots$

Find a recursive formula for each sequence.

17. $1, \dfrac{1}{3}, \dfrac{1}{9}, \dfrac{1}{27}, \ldots$

18. 1, −1, 1, −1, . . .

19. $1, -\dfrac{1}{2}, \dfrac{1}{4}, -\dfrac{1}{8}, \ldots$

Written Exercises

Find the eighth, ninth, and tenth terms of each sequence.

20. $a_n = 4n - 3$

21. $a_n = \dfrac{n}{n + 1}$

22. $a_n = \dfrac{2n + 1}{n + 2}$

23. $a_n = \dfrac{n(n - 1)}{3}$

24. $a_n = \dfrac{2}{n}$

25. $a_n = 4n^2$

26. $a_n = (-1)^{n+1}2n$

27. $a_n = \dfrac{1}{2}(n^2 + n + 4)$

28. $a_n = n^2 + 2n + 1$

Find the first six terms of each sequence.

29. $a_1 = 2,\ a_{n+1} = 3a_n$

30. $a_1 = 7,\ a_{n+1} = a_n + 5$

31. $a_1 = 3,\ a_2 = 5,\ a_{n+2} = a_n + a_{n+1}$

32. $a_1 = 1,\ a_2 = 2,\ a_{n+2} = a_n + a_{n+1}$

33. $a_1 = 2,\ a_2 = 3,\ a_{n+2} = 2a_n + a_{n+1}$

34. $a_1 = 5,\ a_2 = 11,\ a_{n+2} = a_{n+1} - a_n$

Find both a recursive formula and a formula for the *n*th term of each of the following sequences.

35. 3, 7, 11, 15, 19, . . .

36. 4, 9, 14, 19, 24, . . .

37. 3, 15, 75, 375, 1875, . . .

38. $\dfrac{3}{2}, \dfrac{3}{4}, \dfrac{3}{8}, \dfrac{3}{16}, \dfrac{3}{32}, \ldots$

39. 5, 10, 15, 20, 25, . . .

40. $\dfrac{7}{2}, \dfrac{7}{10}, \dfrac{7}{50}, \dfrac{7}{250}, \ldots$

41–46. Find the 100th term for each sequence of Exercises 35–40 using a calculator. Round your answers to the nearest tenth.

Express each series using sigma notation.

47. $3 + 10 + 17 + 24 + 31$

48. $\dfrac{3}{3} + \dfrac{6}{4} + \dfrac{9}{5} + 2 + \dfrac{15}{7}$

49. $\dfrac{3}{4} + \dfrac{3}{2} + \dfrac{9}{4} + 3 + \dfrac{15}{4} + \dfrac{9}{2}$

50. $2 \cdot 5 + 4 \cdot 7 + 6 \cdot 9 + 8 \cdot 11 + 10 \cdot 13$

51. $\dfrac{4}{5} + \dfrac{7}{5} + 2 + \dfrac{13}{5} + \dfrac{16}{5}$

52. $6 - 2 + \dfrac{2}{3} - \dfrac{2}{9} + \dfrac{1}{27}$

53. $2 + 2\dfrac{1}{2} + 3\dfrac{1}{3} + 4\dfrac{1}{4} + 5\dfrac{1}{5} + 6\dfrac{1}{6}$

54. $1 + \left(-\dfrac{1}{3}\right) + \dfrac{1}{5} + \left(-\dfrac{1}{7}\right) + \dfrac{1}{9} + \left(-\dfrac{1}{27}\right)$

55. $1 - 1 + 1 - 1 + 1 - 1 + 1$

56. $\dfrac{1}{2} - \dfrac{1}{4} + \dfrac{1}{6} - \dfrac{1}{8} + \dfrac{1}{10}$

13-8 Special Sequences and Series

The base of this pine cone shows an example of a pattern that is often found in nature.

Count the number of strips that spiral to the left.

Count the number of strips that spiral to the right.

These two numbers shown in color below belong to a very special sequence.

This sequence is named after its discoverer, Leonardo Fibonacci.

$$1, 1, 2, 3, 5, 8, 13, 21, 34, 55, 89, 144, \ldots$$

Can you see what the next term will be? What do you think is the pattern used in the **Fibonacci sequence**?

Let F_n be a term of the sequence.

$F_1 = 1 \qquad F_2 = 1 \qquad F_3 = 2$ or $1 + 1$

$F_4 = 3$ or $2 + 1 \qquad F_5 = 5$ or $3 + 2$

How do you find the next term from the two previous terms?

Then $F_5 = 3 + 2$. Thus, $F_5 = F_4 + F_3$.

In general, if F_n is the nth term of the Fibonacci sequence, then

$$F_n = F_{n-1} + F_{n-2}.$$

The Fibonacci sequence is the basis of other sequences as well. One of these is the sequence of ratios found by dividing each term of the Fibonacci sequence by the preceding term.

$$\frac{1}{1}, \frac{2}{1}, \frac{3}{2}, \frac{5}{3}, \frac{8}{5}, \frac{13}{8}, \frac{21}{13}, \frac{34}{21}, \frac{55}{34}, \frac{89}{55}, \frac{144}{89}$$

Notice that the spirals of the pine cone have the ratio $\frac{13}{8}$. Some other pine cones have the ratio $\frac{8}{5}$. Some daisy centers have the ratio $\frac{34}{21}$.

Sunflower heads have spirals of seeds which may have ratios of $\frac{21}{13}, \frac{34}{21}$, or $\frac{55}{34}$.

The following examples illustrate methods for finding patterns.

Examples

1 **Find the missing terms of the sequence**
6, 10, 15, 21, 28, ____, ____, ____, ...

6,   10,   15,   21,   28,   ____,   ____,   ____

$+4$ $+5$ $+6$ $+7$ $+8$ $+9$ $+10$

Find the difference of consecutive terms.

The difference increases by one for each term.

The missing terms are 36, 45, and 55.

2 **Find the missing terms of the sequence 1, 2, 6, 24, 120, ____, ____, ...**

1,   2,   6,   24,   120,   ____,   ____

$\times 2$ $\times 3$ $\times 4$ $\times 5$ $\times ?$ $120 \times 6 = 720$ $720 \times 7 = 5040$

Note that each term is multiplied by an integer.

The integers increase by one each time.

The next two terms are 720 and 5040.

3 **Complete the sequence 4, 9, 16, 25, ____, ____, ____, ...**

4,   9,   16,   25,   ____,   ____,   ____

↓   ↓   ↓   ↓

2^2 3^2 4^2 5^2

Notice that each term is a perfect square.

The next terms are $6^2 = 36$, $7^2 = 49$, and $8^2 = 64$.

4 **Complete the sequence 1, 1, 4, 10, 28, 76, ____, ____, ...**

1,   1,   4,   10,   28,   76,   ____,   ____

$2(1 + 1), 2(4 + 1), 2(10 + 4), 2(28 + 10)$

Notice that each term seems to be double the sum of the two previous terms.

The next two terms are 208 and 568.

Some special series often are used in more advanced mathematics. One of these is the Leibniz series for calculating π.

$$\frac{\pi}{4} = 1 + \left(-\frac{1}{3}\right) + \frac{1}{5} + \left(-\frac{1}{7}\right) + \frac{1}{9} + \cdots + \frac{(-1)^{n-1}}{2n - 1} + \cdots$$

An approximation for π can be found by taking a finite number of terms in the sum.

Another special series can be used to find an approximation for the exponential function.

$$e^x = 1 + x + \frac{x^2}{2 \cdot 1} + \frac{x^3}{3 \cdot 2 \cdot 1} + \cdots + \frac{x^n}{n(n - 1)(n - 2) \cdots 3 \cdot 2 \cdot 1} + \cdots$$

Find the missing terms of each sequence.

1. 52, 156, 468, ____, ____
2. 1, 2, 4, 7, 11, ____, ____
3. 2, 2.5, 2.75, ____, 2.9375
4. 2, 6, 30, 210, ____
5. 1, 3, 7, 13, 21, ____, ____
6. 64, 32, 8, 1, ____, ____
7. 1, 5, 14, 30, 55, ____, ____
8. 1, 8, 27, 64, 125, ____
9. 1, 3, 4, 7, 11, ____, ____, ____
10. 1, 1, 3, 7, 17, 41, ____, ____

Solve each problem.

11. Find the first twenty terms of the Fibonacci sequence.

12. Find the first fifteen terms of the ratios of the Fibonacci sequence $\frac{1}{1}, \frac{2}{1}, \frac{3}{2}, \frac{5}{3}, \ldots$.

    Express each as a decimal rounded to the nearest hundredth.

13. Find the sum of the first eight terms of the series e^2.

14. Find the sum of the first thirteen terms of the series e^5.

15. The Lucas sequence is 1, 3, 4, 7, 11, 18, 29, 47. Let L_n be a term of the Lucas sequence. Describe L_n in terms of the Fibonacci sequence.

16. Use the first ten terms of Leibniz's series to find an approximation for π.

mini-review

1. Find $g[f(-2)]$ if $f(x) = 3x^2 + 12x - 5$ and $g(x) = x^3 + 5x^2 - 4x + 12$.

Find all rational zeros for each function.

2. $f(x) = 6x^3 + 7x^2 - 9x + 2$

3. $f(x) = x^3 + 4x^2 - 7x - 10$

Simplify.

4. $\dfrac{1}{2x - 3} - \dfrac{1}{3x - 2}$

5. $\dfrac{3x^2 - 5x + 2}{2x^2 - 5x - 3} \div \dfrac{x^2 + x - 2}{2x^2 - x - 3}$

Evaluate each expression.

6. $\log_8 8^5$

7. $4^{\log_4 5}$

8. $n^{\log_n 2}$

9. Solve $\log_5 (3x + 7) = \log_5 (x^2 - 4x - 1)$.

10. Find the sum of the infinite series $\dfrac{2}{3} + \dfrac{1}{3} + \dfrac{1}{6} + \ldots$.

11. Find the 57th term of the sequence 6, 15, 24, 33, ...

12. A Pilgrim ancestor of Kevin White left \$150 in a savings account. Interest was compounded continuously at 4%. The account is now worth \$24,000,000. How long ago was the account started? Use $A = Pe^{rt}$.

What is the sum of the series formed by the first n positive odd integers?

$$1 + 3 + 5 + 7 + \ldots + (2n - 1)$$

One approach is to make a table.

n	1	2	3	4	5 ...
Positive Odd Integers	1	3	5	7	9 ...
Cumulative Sums	1	4	9	16	25 ...

Notice that each cumulative sum is the square of the number of terms, n.

$n = 1$	$n = 2$	$n = 3$	$n = 4$
$1^2 = 1$	$2^2 = 4$	$3^2 = 9$	$4^2 = 16$

This pattern suggests that $1 + 3 + 5 + 7 + \ldots + (2n - 1) = n^2$. However, this is only a hypothesis based on observation. A hypothesis such as this becomes a theorem only when it is proved.

A common method of proof is called **induction**. A proof by mathematical induction is much like climbing a ladder. You must get on the first step. Then you must show that you can always advance to the next step and so on up the ladder for all steps.

The sum of the series formed by the first n positive odd integers can be proved by mathematical induction.

$$1 + 3 + 5 + \ldots + (2n - 1) = n^2$$

Step 1   **First verify that the formula is valid for the first possible case, usually $n = 1$.**

For $n = 1$, it is true that $2 \cdot 1 - 1 = 1^2$. Therefore, $S_n = n^2$ is valid for the first case.

Step 2   **Assume that the formula is valid for $n = k$. Using this information, prove that it is also valid for $n = k + 1$.**

Assume that $1 + 3 + 5 + 7 + \ldots (2k - 1) = k^2$ is true. Then if the $(k + 1)st$ integer is added to both sides the result is an equivalent equation.

$$\underbrace{1 + 3 + 5 + 7 + \ldots + 2k - 1}_{\substack{k \text{ integers} \\ \text{by assumption}}} + \underbrace{[2(k + 1) - 1]}_{\substack{(k + 1)st \\ \text{integer}}} = k^2 + [2(k + 1) - 1]$$

Add $[2(k + 1) - 1]$ to each side.

$$= k^2 + 2k + 1$$
$$= (k + 1)^2$$

The formula is valid for the $(k + 1)st$ integer, since the sum of $(k + 1)$ odd integers is $(k + 1)^2$.

Conclusion   The formula is valid for $n = 1$. Step 2 illustrates that the formula is valid for the next positive integer $n + 1$ or 2. Since the formula is valid when $n = 2$, it is also valid for $n + 1$ or 3, and so on, indefinitely.

What is the sum of the first n positive integers?

n	1 2 3 4 5...
Positive Integers	1 2 3 4 5...
Cumulative Sums	1 3 6 10 15...

This pattern suggests $1 + 2 + 3 + \ldots + n = \dfrac{n(n + 1)}{2}$. Prove this by mathematical induction.

Step 1 $\dfrac{1 \cdot (1 + 1)}{2} = 1$ So the formula is valid for $n = 1$.

Step 2   Assume that the formula is then valid for $n = k$.

$$1 + 2 + 3 + \ldots + k = \frac{k(k + 1)}{2}$$

Prove that it is also valid for $n = k + 1$.

$$1 + 2 + 3 + \ldots + k = \frac{k(k + 1)}{2}$$

$$1 + 2 + 3 + \ldots + k + (k + 1) = \frac{k(k + 1)}{2} + (k + 1) \qquad \textit{Add k + 1 to each side.}$$

$$= \frac{k(k + 1) + 2(k + 1)}{2}$$

$$= \frac{(k + 1)(k + 2)}{2}$$

If $k + 1$ is substituted for n in the original formula, the same result is obtained.

$$1 + 2 + \ldots + (k + 1) = \frac{(k + 1)[(k + 1) + 1]}{2}$$

$$= \frac{(k + 1)(k + 2)}{2}$$

Thus, the formula is valid for $n = k + 1$.

So $1 + 2 + 3 + \ldots + n = \dfrac{n(n + 1)}{2}$.

Exercises

Prove that each statement is true for all positive integers n.

1. $2 + 4 + 6 + \ldots + 2n = n(n + 1)$

2. $1 + 2 + 3 + \ldots + n = \dfrac{n(n + 1)}{2}$

3. $-\dfrac{1}{2} - \dfrac{1}{4} - \dfrac{1}{8} - \ldots - \dfrac{1}{2^n} = \dfrac{1}{2^n} - 1$

4. $3 + 6 + 9 + \ldots + 3n = \dfrac{3n(n + 1)}{2}$

5. $1^3 + 2^3 + 3^3 + \ldots + n^3 = \dfrac{n^2(n + 1)^2}{4}$

6. $\dfrac{1}{2} + \dfrac{1}{2^2} + \dfrac{1}{2^3} + \ldots + \dfrac{1}{2^n} = 1 - \dfrac{1}{2^n}$

7. $2 + 2^2 + 2^3 + \ldots + 2^n = 2^{n+1} - 2$

8. $1 + 2 + 4 + 8 + \ldots + 2^{n-1} = 2^n - 1$

9. $1^2 + 2^2 + 3^2 + \ldots + n^2 = \dfrac{n(n + 1)(2n + 1)}{6}$

10. $1^2 + 3^2 + 5^2 + \ldots + (2n - 1)^2 = \dfrac{n(2n - 1)(2n + 1)}{3}$

13-9 The Binomial Theorem

The binomial expression $(a + b)$ can be raised to various powers.
There are patterns to be found in the powers of $(a + b)$ listed below.

$$(a + b)^0 = 1a^0b^0$$
$$(a + b)^1 = 1a^1b^0 + 1a^0b^1$$
$$(a + b)^2 = 1a^2b^0 + 2a^1b^1 + 1a^0b^2$$
$$(a + b)^3 = 1a^3b^0 + 3a^2b^1 + 3a^1b^2 + 1a^0b^3$$
$$(a + b)^4 = 1a^4b^0 + 4a^3b^1 + 6a^2b^2 + 4a^1b^3 + 1a^0b^4$$

$a + b \neq 0$

Note that the coefficients are one.

Why can b^0 and a^0 be written here?

What happened to the powers of a?

What about powers of b?

Note the sum of the exponents in any term of $(a + b)^4$.
How many terms are in the expansion of $(a + b)^4$?

The following patterns are seen in the expansion of $(a + b)^n$.

1. The exponent of $(a + b)^n$ is the exponent of a in the first term and the exponent of b in the last term.

2. In successive terms, the exponent of a decreases by one. It is n in the first term and zero in the last term.

3. In successive terms, the exponent of b increases by one. It is zero in the first term and n in the last term.

4. The sum of the exponents of each term is n.

5. The coefficients are symmetric. They increase at the beginning and decrease at the end of the expansion.

If the coefficients are displayed alone, a definite pattern appears.

$(a + b)^0$	1
$(a + b)^1$	1 1
$(a + b)^2$	1 2 1
$(a + b)^3$	1 3 3 1
$(a + b)^4$	1 4 6 4 1
$(a + b)^5$	1 5 10 10 5 1
$(a + b)^6$	1 6 15 20 15 6 1

This is known as Pascal's Triangle. Each new row is formed by adding elements of the previous row in pairs as marked. Each row begins and ends with 1. The triangle can go on indefinitely.

Notice that the expansion of $(a + b)^n$ has $n + 1$ terms.

Example

1 Use the pattern to write $(a + b)^7$ in expanded form.   *($(a + b)^7$ will have 8 terms.*

The next line of Pascal's Triangle is

$$1 \quad 7 \quad 21 \quad 35 \quad 35 \quad 21 \quad 7 \quad 1.$$

$$(a + b)^7 = 1a^7b^0 + 7a^6b^1 + 21a^5b^2 + 35a^4b^3 + 35a^3b^4 + 21a^2b^5 + 7a^1b^6 + 1a^0b^7$$
$$(a + b)^7 = a^7 + 7a^6b + 21a^5b^2 + 35a^4b^3 + 35a^3b^4 + 21a^2b^5 + 7ab^6 + b^7$$

Here is another way to show the coefficients.

$(a + b)^0$ 1

$(a + b)^1$ 1 $\dfrac{1}{1}$

$(a + b)^2$ 1 $\dfrac{2}{1}$ $\dfrac{2 \cdot 1}{1 \cdot 2}$

$(a + b)^3$ 1 $\dfrac{3}{1}$ $\dfrac{3 \cdot 2}{1 \cdot 2}$ $\dfrac{3 \cdot 2 \cdot 1}{1 \cdot 2 \cdot 3}$

$(a + b)^4$ 1 $\dfrac{4}{1}$ $\dfrac{4 \cdot 3}{1 \cdot 2}$ $\dfrac{4 \cdot 3 \cdot 2}{1 \cdot 2 \cdot 3}$ $\dfrac{4 \cdot 3 \cdot 2 \cdot 1}{1 \cdot 2 \cdot 3 \cdot 4}$

This pattern can provide the coefficients of a binomial expansion without writing the previous rows of coefficients. The Binomial Theorem summarizes these patterns.

If n is a positive integer, then

$$(a + b)^n = 1a^n b^0 + \frac{n}{1}a^{n-1}b^1 + \frac{n(n-1)}{1 \cdot 2}a^{n-2}b^2 + \ldots + 1a^0 b^n.$$

The Binomial Theorem

Example

2   **Use the Binomial Theorem to find the terms in the expansion of $(x + y)^8$.**

Find the first five terms. Then, use symmetry to find the remaining terms.

$$(x + y)^8 = 1 \cdot x^8 y^0 + \frac{8}{1}x^7 y^1 + \frac{8 \cdot 7}{1 \cdot 2}x^6 y^2 + \frac{8 \cdot 7 \cdot 6}{1 \cdot 2 \cdot 3}x^5 y^3 + \frac{8 \cdot 7 \cdot 6 \cdot 5}{1 \cdot 2 \cdot 3 \cdot 4}x^4 y^4 + \cdots$$

$$= x^8 + 8x^7 y + 28x^6 y^2 + 56x^5 y^3 + 70x^4 y^4 + \cdots$$

$$= x^8 + 8x^7 y + 28x^6 y^2 + 56x^5 y^3 + 70x^4 y^4 + 56x^3 y^5 + 28x^2 y^6 + 8xy^7 + y^8$$

Note that in terms having the same coefficients the exponents are reversed, as in $28x^6 y^2$ and $28x^2 y^6$.

In Example 2, some of the denominators are written as shown below.

$$1 \cdot 2 \cdot 3 \cdot 4 = 4 \cdot 3 \cdot 2 \cdot 1$$
$$1 \cdot 2 \cdot 3 = 3 \cdot 2 \cdot 1$$

The product $4 \cdot 3 \cdot 2 \cdot 1$ is called 4 factorial and is expressed as 4!.

If n is a positive integer, the expression $n!$ (n factorial) is defined as follows.

$$n! = n(n - 1)(n - 2) \ldots (1)$$

Definition of n Factorial

The expression $0!$ is equal to 1, by definition.

3 Evaluate $\dfrac{8!}{2!6!}$.

$$\frac{8!}{2!6!} = \frac{8 \cdot 7 \cdot 6 \cdot 5 \cdot 4 \cdot 3 \cdot 2 \cdot 1}{2 \cdot 1 \cdot 6 \cdot 5 \cdot 4 \cdot 3 \cdot 2 \cdot 1} = 28$$

Notice that the coefficients in Example 2 are equivalent to the factorial expressions below.

$$\frac{8}{1} = \frac{8!}{1!7!} \qquad \frac{8 \cdot 7}{1 \cdot 2} = \frac{8!}{2!6!} \qquad \frac{8 \cdot 7 \cdot 6}{1 \cdot 2 \cdot 3} = \frac{8!}{3!5!}$$

Thus, another way to write the expansion is:

$$(x + y)^8 = \frac{8!}{0!8!}x^8 + \frac{8!}{1!7!}x^7y^1 + \frac{8!}{2!6!}x^6y^2 + \frac{8!}{3!5!}x^5y^3 + \frac{8!}{4!4!}x^4y^4 + \frac{8!}{5!3!}x^3y^5 +$$

$$\frac{8!}{6!2!}x^2y^6 + \frac{8!}{7!1!}x^1y^7 + \frac{8!}{8!0!}y^8$$

An equivalent form of the Binomial Theorem uses both sigma and factorial notation.

$$(a + b)^n = \frac{n!}{0!(n - 0)!}a^n + \frac{n!}{1!(n - 1)!}a^{n-1}b^1 + \frac{n!}{2!(n - 2)!}a^{n-2}b^2 + \ldots$$

$$= \sum_{k=0}^{n} \frac{n!}{k!(n - k)!}a^{n-k}b^k \qquad \textit{Here n is a positive integer.}$$
$$\textit{k is a positive integer or zero.}$$

4 **Express $(2s + t)^4$ using sigma notation. Then find the terms in the expansion.**

$$(2s + t)^4 = \sum_{k=0}^{4} \frac{4!}{k!(4 - k)!}(2s)^{4-k}t^k \qquad \textit{Now construct each term.}$$

$$= \frac{4!}{0!(4 - 0)!}(2s)^{4-0}t^0 + \frac{4!}{1!(4 - 1)!}(2s)^{4-1}t^1 + \frac{4!}{2!(4 - 2)!}(2s)^{4-2}t^2 +$$

$$\frac{4!}{3!(4 - 3)!}(2s)^{4-3}t^3 + \frac{4!}{4!(4 - 4)!}(2s)^{4-4}t^4$$

$$= \frac{4 \cdot 3 \cdot 2 \cdot 1}{1 \cdot 4 \cdot 3 \cdot 2 \cdot 1}(2s)^4 + \frac{4 \cdot 3 \cdot 2 \cdot 1}{1 \cdot 3 \cdot 2 \cdot 1}(2s)^3t + \frac{4 \cdot 3 \cdot 2 \cdot 1}{2 \cdot 1 \cdot 2 \cdot 1}(2s)^2t^2 +$$

$$\frac{4 \cdot 3 \cdot 2 \cdot 1}{3 \cdot 2 \cdot 1 \cdot 1}(2s)t^3 + \frac{4 \cdot 3 \cdot 2 \cdot 1}{4 \cdot 3 \cdot 2 \cdot 1 \cdot 1}t^4$$

$$= 16s^4 + 32s^3t + 24s^2t^2 + 8st^3 + t^4$$

Sometimes a particular term in the binomial expansion is needed. In the example above, k is 0 for the first term, 1 for the second term, and so on. In general, the value of k is one less than the number of the term.

Example

5 **Find the fifth term of $(p + q)^9$.**

$$(p + q)^9 = \sum_{k=0}^{9} \frac{9!}{k!(9 - k)!}p^{9-k}q^k \qquad \textit{In the fifth term, k will be 4 since k starts at zero.}$$

The fifth term, $\dfrac{9!}{4!(9 - 4)!}p^{9-4}q^4$, is $\dfrac{9 \cdot 8 \cdot 7 \cdot 6}{1 \cdot 2 \cdot 3 \cdot 4}p^5q^4$ or $126p^5q^4$.

Exploratory Exercises

Evaluate each expression.

1. $7!$ **2.** $9!$ **3.** $10!$ **4.** $12!$

5. $\dfrac{10!}{8!}$ **6.** $\dfrac{31!}{28!}$ **7.** $\dfrac{6!}{3!}$ **8.** $\dfrac{10!}{4!6!}$

State the number of terms for the expanded form of each expression. Then, find the fourth term for each.

9. $(r + s)^4$ **10.** $(a + b)^5$

11. $(k - m)^7$ **12.** $(a - 3)^4$

13. $(x - 2)^5$ **14.** $(b - z)^5$

Written Exercises

Expand each binomial.

15. $(x + m)^4$ **16.** $(r + s)^6$

17. $(y + p)^7$ **18.** $(x - y)^3$

19. $(b - z)^5$ **20.** $(r - m)^6$

21. $(2m + y)^5$ **22.** $(3r + y)^4$

23. $(2b + x)^6$ **24.** $(2x + 3y)^4$

25. $(3x - 2y)^5$ **26.** $(2m - 3)^6$

27. $\left(2 + \dfrac{x}{2}\right)^6$ **28.** $\left(\dfrac{y}{3} + 3\right)^6$

Find the requested term of each of the following.

29. fifth term of $(x + y)^7$

30. fourth term of $(2x + 3y)^9$

31. seventh term of $(x - y)^{15}$

32. fifth term of $(x - 2)^{10}$

33. sixth term of $(2m + 3n)^{12}$

34. eighth term of $(3a + 5b)^{11}$

35. fifth term of $(4x - 5)^{10}$

36. sixth term of $(m + k)^{20}$

37. fourth term of $(0.25h + 0.75n)^6$

38. fifth term of $(0.32h + 0.68n)^7$

Solve each problem.

39. Jorma Johnson invested \$5000 at 8% annual interest for 3 years. The interest is compounded semiannually. Find the value of Jorma's investment after 3 years. Use $A = 5000(1 + 0.04)^6$.

40. Joanne Mauch owns a tree plantation. The value of her trees increases about 10% each year. The trees are now worth \$10,000. What will be their value 6 years from now? Use $V = 10,000(1 + 0.10)^6$.

Challenge Exercises

Simplify.

41. $\dfrac{k!}{(k - 1)!}$

42. $\dfrac{(k + 3)!}{(k + 2)!}$

43. $(k + 1)!(k + 2)$

44. $\dfrac{3!4(k - 3)!}{(k - 2)!}$

Excursions in Algebra

Limits of Sequences

Certain types of sequences approach a specific number as more and more terms are found. This number is called a **limit**. You can use your calculator to find the limit of the geometric sequence $1, \dfrac{1}{2}, \dfrac{1}{4}, \dfrac{1}{8}, \dfrac{1}{16}, \dfrac{1}{32}, \cdots \left(\dfrac{1}{2}\right)^n$. Try several values for n to determine what happens as n increases. Use $a_n = a_1 r^{n-1}$ where $a_1 = 1$ and $r = \dfrac{1}{2}$.

$$a_5 = 1 \cdot \left(\dfrac{1}{2}\right)^{5-1} = (0.5)^4 \qquad a_{10} = 1 \cdot \left(\dfrac{1}{2}\right)^{10-1} = (0.5)^9 \qquad a_{20} = 1 \cdot \left(\dfrac{1}{2}\right)^{20-1} = (0.5)^{19}$$

ENTER: .5 $\boxed{y^x}$ 4 $\boxed{=}$ .5 $\boxed{y^x}$ 9 $\boxed{=}$ .5 $\boxed{y^x}$ 19    $\boxed{=}$

DISPLAY: $.5 \qquad 4\ 0.0625 \qquad .5 \qquad 9\ 0.00195 \qquad .5 \qquad 19\ 0.00000191$

Notice that as n increases, the sequence approaches 0.   *The limit is 0.*

Exercises

Use a calculator to find the limit of each sequence.

1. $\dfrac{1}{2}, \dfrac{2}{3}, \dfrac{3}{4}, \dfrac{4}{5}, \dfrac{5}{6}, \cdots \dfrac{n}{n + 1}$

2. $\dfrac{2}{3}, \dfrac{4}{9}, \dfrac{8}{27}, \dfrac{16}{81}, \cdots \left(\dfrac{2}{3}\right)^n$

3. $\dfrac{1}{5}, \dfrac{2}{7}, \dfrac{3}{9}, \cdots \dfrac{n}{2n + 3}$

4. $2, \dfrac{2}{3}, \dfrac{2}{9}, \dfrac{2}{27}, \dfrac{2}{81}, \cdots 2\left(\dfrac{1}{3}\right)^{n-1}$

Pascal's Triangle

Blaise Pascal (1623–1662) was a French mathematician. At age nineteen he invented a computing machine, a forerunner of today's computers and calculators. He also devised a quick method for finding the coefficients of the expansion for $(a + b)^n$.

You can make the following observations in the triangle of coefficients.
1. *Each row begins and ends with 1.*
2. *Each coefficient is the sum of the two coefficients to the left and right in the row directly above.*

Power	Coefficients of the Expansion						
$(a + b)^0$				1			
$(a + b)^1$			1		1		
$(a + b)^2$		1		2		1		$1 + 1 = 2$
$(a + b)^3$	1		3		3		1
$(a + b)^4$	1	4		6		4		1
$(a + b)^5$	1	5	10	10	5	1	$6 + 4 = 10$

The program at the right uses Pascal's triangle. It provides the coefficients for the expansion of $(a + b)^n$ when $n = 0$ to 10.

Enter and run this program. Your output will be in slightly different form. The first six lines are shown below.

```
] RUN
1
1  1
1  2  1
1  3  3  1
1  4  6  4  1
1  5  10  10  5  1
```

```
10   FOR N = 0 TO 10
20   FOR R = 0 TO N
30   LET C = 1
40   IF N < N − R + 1 THEN 80
50   FOR X = N TO N − R + 1 STEP − 1
60   LET C = C * X / (N − X + 1)
70   NEXT X
80   PRINT C;"    ";
90   NEXT R
100   PRINT
110   NEXT N
120   END
```

Example: **Expand $(x + y)^5$.** *Notice the patterns of coefficients and exponents.*

$$(x + y)^5 = 1 \cdot x^5y^0 + 5x^4y^1 + 10x^3y^2 + 10x^2y^3 + 5x^1y^4 + 1x^0y^5$$
$$= x^5 + 5x^4y + 10x^3y^2 + 10x^2y^3 + 5xy^4 + y^5$$

Exercises

Use the computer program for Pascal's triangle to find the coefficients for the expansion of the following expressions.

1. $(a + b)^4$ **2.** $(a + b)^6$ **3.** $(a + b)^7$

4. $(a + b)^8$ **5.** $(a + b)^9$ **6.** $(x − y)^6$

Make the necessary changes in the computer program to find the coefficients for the expansion of the following expressions.

7. $(a + b)^{12}$ **8.** $(x − y)^{12}$

sequence (459)
term (459)
arithmetic sequence (459)
common difference (459)
arithmetic means (461)
series (463)
arithmetic series (463)
sum of an arithmetic series (464)
geometric sequence (468)
common ratio (468)
geometric means (469)

geometric series (474)
sum of a geometric series (474)
infinite geometric series (477)
sum of an infinite geometric
  series (477)
sigma notation (481)
index of summation (481)
recursive formula (484)
Fibonacci sequence (487)
Binomial Theorem (493)
n factorial (493)

Chapter Summary

1. An arithmetic sequence is a sequence in which each term, after the first, is found by adding a constant, called the *common difference*, to the previous term. (459)

2. The nth term, a_n, of an arithmetic sequence with first term, a_1, and common difference, d, is given by $a_n = a_1 + (n - 1)d$. (460)

3. The indicated sum of the terms of a sequence is called a series. (463)

4. The sum, S_n, of the first n terms of an arithmetic series is given by $S_n = \frac{n}{2}(a_1 + a_n)$. (464)

5. A geometric sequence is a sequence in which each term, after the first, is found by multiplying the previous term by a constant called the *common ratio*. (468)

6. The nth term, a_n, of a geometric sequence with first term, a_1, and common ratio, r, is given by either $a_n = a_1 r^{n-1}$ or $a_n = a_{n-1}r$. (469)

7. The sum, S_n, of the first n terms of a geometric series is given by $S_n = \frac{a_1 - a_1 r^n}{1 - r}$ where $r \neq 1$. (474)

8. The sum, S, of an infinite geometric series where $-1 < r < 1$ is given by $S = \frac{a_1}{1 - r}$. (477)

9. The Binomial Theorem: If n is a positive integer, then $(a + b)^n = 1a^n b^0 + \frac{n}{1}a^{n-1}b^1 + \frac{n(n - 1)}{1 \cdot 2}a^{n-2}b^2 + \ldots + 1a^0 b^n$. (493)

10. Definition of n Factorial: $n! = n(n - 1)(n - 2) \ldots (1)$. (493)

11. The following equivalent form of the Binomial Theorem uses both sigma and factorial notation.

$$(a + b)^n = \sum_{k=0}^{n} \frac{n!}{k!(n - k)!}a^{n-k}b^k \quad (494)$$

13–1 **1.** Find the first five terms of the arithmetic sequence when $a_1 = 6$, $d = 8$.

2. Find the next four terms of the arithmetic sequence 9, 12, 15, ...

3. Find the nth term of the arithmetic sequence when $a_1 = 3$, $d = 7$, and $n = 34$.

4. Which term of -5, 2, 9, . . . is 142?

5. Find the missing terms for the arithmetic sequence -7, ____, ____, ____, 9.

6. A stack of boxes in a warehouse is arranged so that there are 5 boxes in the top row, 7 boxes in the second row, 9 boxes in the third row, and so on. How many boxes are in the twentieth row?

13–2 **Find S_n for each arithmetic series described below.**

7. $a_1 = 12$, $a_n = 117$, $n = 36$ **8.** $a_1 = 4$, $d = 6$, $n = 18$

9. Find the sum of the series $7 + 10 + 13 + \ldots + 97$.

13–3 **10.** Find the common ratio of the sequence $\dfrac{2}{3}, \dfrac{4}{3}, \dfrac{8}{3}, \dfrac{16}{3}, \ldots$.

11. Find the next two terms of the geometric sequence 7.5, 15, 30, ...

12. Find the 5th term of the geometric sequence in which $a_1 = 7$ and $r = 3$.

13. Find the geometric means of 4, ____, ____, ____, 324.

13–4 **Find the sum of each geometric series described below.**

14. $a_1 = 6$, $r = 3$, $n = 5$ **15.** $a_1 = 625$, $a_n = 16$, $r = \dfrac{2}{5}$

16. A ball dropped from a height of 21 feet rebounds $\dfrac{2}{3}$ of the distance from which it was dropped on each bounce. How far has the ball traveled after 6 bounces (rebounds)?

17. For a geometric series, find a_1 given that $S_n = 1441$, $r = \dfrac{3}{5}$, and $n = 5$.

13–5 **18.** Find the sum of the infinite geometric series when $a_1 = -2$ and $r = -\dfrac{5}{8}$.

19. Find the sum of the series $\dfrac{1}{2} + \dfrac{1}{3} + \dfrac{2}{9} + \dfrac{4}{27} + \cdots$.

20. Find a common fraction equivalent to $0.1\overline{7}$.

13–6 **21.** Write the sum $\displaystyle\sum_{k=8}^{11} (3k - 4)$ in expanded form.   **22.** Evaluate $\displaystyle\sum_{r=0}^{10} (5 + 8r)$.

13–7 **23.** Find the first five terms of this sequence:
$a_1 = 1$, $a_2 = 3$, $a_{n+2} = a_{n+1} + 2 \cdot a_n$.

24. Express the following series using sigma notation:
$2 + 6 + 12 + 20 + 30 + 42$.

13–8 **25.** Find a pattern and complete the sequence for 3, 7, 12, 18, 25, ____, ____, ____, ____.

13–9 **26.** Find the fourth term of $(x + 2y)^6$.   **27.** Expand $(3a + b)^5$.

Find S_n for each arithmetic series.

1. $a_1 = 7, n = 31, a_n = 127$

2. $a_1 = 13, d = -2, n = 17$

Find the sum of each geometric series.

3. $a_1 = 125, r = \frac{2}{5}, n = 4$

4. $a_1 = 16, a_n = -\frac{1}{2}, r = -\frac{1}{2}$

Solve each problem.

5. Find the next four terms of the arithmetic sequence 42, 37, 32, ...

6. Find the 27th term of an arithmetic sequence when $a_1 = 2, d = 6$.

7. Which term of the arithmetic sequence 7, 13, 19, . . . is 193?

8. How many integers between 26 and 415 are multiples of 9?

9. Find the first three terms of this arithmetic series: $a_1 = 7, n = 13, S_n = 1027$.

10. Find the sum of the series $91 + 85 + 79 + \ldots + (-29)$.

11. Find the next two terms of the geometric sequence $\frac{1}{81}, \frac{1}{27}, \frac{1}{9}, \ldots$.

12. Find the sixth term of a geometric sequence if $a_1 = 5$ and $r = -2$.

13. Find the geometric means of 7, ____, ____, 189.

14. A vacuum pump removes $\frac{1}{7}$ of the air from a jar on each stroke of its piston. What percent of the air remains after 4 strokes of the piston?

15. Find the sum of the series $12 - 6 + 3 - \frac{3}{2} + \ldots$.

16. Find common fractions equivalent to the repeating decimals $0.\overline{7}$ and $0.3\overline{2}$.

17. Describe the sequence 2, 6, 18, 54, 162, . . . in terms of n.

18. Write $\sum\limits_{k=2}^{6} (3k^2 - 1)$ in expanded form.

19. Find the sum of $\sum\limits_{k=3}^{15} (14 - 2k)$.

20. Describe this sequence recursively: 2, 6, 18, 54, 162, ...

21. Find the first five terms of this sequence: $a_1 = 3, a_2 = 1, a_{n+2} = a_{n+1} + 2 \cdot a_n$.

22. Find a pattern and complete the sequence 3, 3, 6, 18, 72, ____, ____, ____.

23. Find the third term of $(x + y)^8$.

24. Expand $(2s + 3t)^5$.

25. A grocery stock boy makes a display of cans of corn for a sale. He puts 20 cans in the bottom row and each row above it contains 3 fewer cans than the previous row. He continues until there are only 2 cans in the top row. How many rows are there and what is the total number of cans in the display?

26. A tank contains 9000 gallons of water. Each day $\frac{2}{3}$ of the water is removed. How much water has been removed by the end of the fifth day?

The test questions on this page deal with rational expressions and radicals. Keep in mind that many questions can be solved by writing the expression in a different form.

Directions: Choose the one best answer. Write A, B, C, or D.

1. If $\frac{x}{y} = z$ and $y = z$, find y in terms of x.

   (A) y **(B)** $\pm\sqrt{y}$

   (C) $\pm\sqrt{x}$ **(D)** $\pm\sqrt{xz}$

2. If the average of x and y equals the average of x, y, and z, then express z in terms of x and y.

   (A) $x + y$ **(B)** $2(x + y)$

   (C) $\frac{x + y}{2}$ **(D)** $\frac{x + y}{3}$

3. If the product of a number and $2b$ is increased by y, the result is p. Find the number in terms of b, y, and p.

   (A) $2by - p$ **(B)** $\frac{2b}{y - p}$

   (C) $\frac{p - y}{2b}$ **(D)** $\frac{y - b}{2b}$

4. If $\frac{1}{p} = \sqrt{0.25}$, then $p^2 =$

   (A) 0.25   **(B)** 4   **(C)** 25   **(D)** 400

5. Of the following numbers, which is the greatest?

   (A) $\frac{1}{3\sqrt{3}}$ **(B)** $\frac{1}{3}$ **(C)** $\frac{\sqrt{3}}{3}$ **(D)** $\sqrt{3}$

6. If $\frac{a + b}{a} = \frac{5}{4}$, then $\frac{b}{a} =$

   (A) $\frac{1}{4}$ **(B)** $\frac{5}{4}$ **(C)** $\frac{7}{4}$ **(D)** $\frac{9}{4}$

7. Of the following, which is closest to the value of $\frac{65.9 \times 0.49}{3.3}$?

   (A) 10   **(B)** 80   **(C)** 100   **(D)** 450

Most standardized tests have a time limit, so you must budget your time carefully. Some questions will be much easier than others. If you cannot answer a question within a few minutes, go on to the next one. If there is still time left when you get to the end of the test, go back to the ones that you skipped.

8. The reciprocal of $\frac{5}{b - 1} + \frac{3}{b}$ is

   (A) $\frac{b^2 - b}{15}$ **(B)** $\frac{b - 1}{2}$

   (C) $\frac{b^2 - b}{8b - 3}$ **(D)** $\frac{2b - 1}{8}$

9. Simplify $\dfrac{1 \div \frac{1}{b}}{\frac{1}{b}}$.

   (A) 1   **(B)** $\frac{1}{b^2}$ **(C)** b **(D)** b^2

10. If $4b - 3a = 0$, then what is the value of $\frac{16b^2}{a^2}$?

   (A) $\frac{1}{9}$ **(B)** 9   **(C)** 16   **(D)** $\frac{256}{9}$

11. If $xyz = 8$ and $y = z$, then $x =$

   (A) y^2 **(B)** $\frac{8}{y^2}$ **(C)** $8y^2$ **(D)** $\frac{1}{y^2}$

12. If $\frac{2b}{5a} = 12$, then $\frac{2b - 10a}{5a} =$

   (A) 5   **(B)** 10   **(C)** 14   **(D)** 24

13. If $\frac{x}{6} + 4 = 1$, the value of $\frac{x}{3}$ is

   (A) -36 **(B)** -18 **(C)** -6 **(D)** 6

14. If $3 + \frac{d}{4} = 8\frac{1}{2}$, then $d =$

   (A) -2 **(B)** 5   **(C)** 16   **(D)** 22

CHAPTER 14

Probability

The plants produced by these milkweed seeds will not all be exactly alike. In studying generations of plants, botanists often gather extensive data. Sometimes they can discover genetic patterns and use probability to predict characteristics of future generations.

14-1 Counting

Jana Lee is ordering a new automobile. She still has three choices to make.

    1. 4-cylinder or 6-cylinder engine?

    2. Standard or automatic transmission?

    3. Maroon, white, or tan?

These three choices are called **independent events**. That is, the choice of one of them does *not* affect the others. Jana's possible choices can be shown in a diagram.

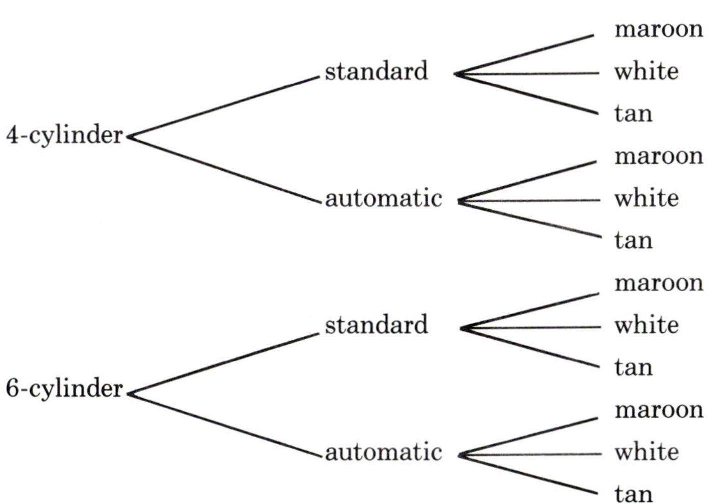

This diagram is called a tree diagram.

One of the choices is a 6-cylinder car with standard transmission and maroon paint.

How many other choices are there?

You can find the total number of choices that Jana has without drawing a diagram.

	Choices:	4-cyl. or 6-cyl.	standard or automatic	maroon or white or tan
Number of choices:		2	2	3

Multiplication can be used to find the total number of choices that Jana has.

$$2 \cdot 2 \cdot 3 = 12$$

> **Suppose an event can occur in p different ways. Another event can occur in q different ways. There are $p \cdot q$ ways both events can occur.**
>
> *Basic Counting Principle*

This principle can be extended to any number of events. Some events are **dependent**. In dependent events the number of choices is altered by the choice in the previous event.

The choice of one event does affect the others.

Examples

1 **How many different 3-letter patterns can be formed using the letters a, b, and c, if a letter can be used more than once?**

Letters:	1st	2nd	3rd
Number of choices:	3	3	3

There are $3 \cdot 3 \cdot 3$ or 27 possible patterns.

A pattern is a selection of three letters. Each one must be an a, b, or c. Since each choice is not affected by a previous choice, these events are called independent events.

2 **How many different 3-letter patterns can be formed using the letters a, b, and c if each letter is used exactly once?**

Letters:	1st	2nd	3rd
Number of choices:	3	2	1

There are $3 \cdot 2 \cdot 1 = 3!$ or 6 patterns.

Note that after the first letter is chosen, it may not be chosen again. These events are called dependent events.

3 **How many 7-digit phone numbers can begin with the prefix 457?**

Digit in phone number:	4th	5th	6th	7th
Number of choices:	10	10	10	10

There are $10 \cdot 10 \cdot 10 \cdot 10 = 10^4$ or 10,000 numbers.

The choices are 0, 1, 2, 3, 4, 5, 6, 7, 8, and 9.

Exploratory Exercises

Tell whether each choice is independent or dependent.

1. Choose color and size to order an item of clothing.

2. Choose a president, secretary, and treasurer for a club.

3. Choose five numbers in a bingo game.

4. Choose the winner and loser of a chess game.

5. Each of five people guess the total number of runs in a baseball game. They write down the guess, without telling what it is.

6. The numerals 0 through 9 are written on pieces of paper and placed in a jar. Three of them are selected one after the other, without replacement.

Written Exercises

Solve each problem.

7. The letters g, h, j, k, and l are to be used to form 5-letter patterns. How many patterns can be formed if repetitions are allowed?

8. A license plate must have two letters (not I or O) followed by three digits. The last digit cannot be zero. How many possible plates are there?

9. There are five roads from Albany to Briscoe, six from Briscoe to Chadwick, and three from Chadwick to Dover. How many different routes are there from Albany to Dover?

10. A store has 15 sofas, 12 lamps, and 10 tables at half price. How many different combinations of a sofa, a lamp, and a table can be bought at the sale?

11. A restaurant serves 5 main dishes, 3 salads, and 4 desserts. How many different meals could be ordered if each has a main dish, a salad, and a dessert?

12. A car dealer offers a choice of 6 vinyl top colors, 18 body colors, and 7 upholstery colors. How many color combinations are there?

13. Four ferry boats make the crossing between Harrod and Lafayette. How many different ways can a traveler make a round trip?

14. Using the ferry boats in Exercise 13, how many different ways can a traveler make a round trip, but return on a different ferry boat from the one she went on?

15. How many ways can six different books be placed on a shelf?

16. How many ways can six books be placed on a shelf if the only dictionary must be on an end?

17. How many different 4-letter patterns can be formed from the letters, a, e, i, o, r, s, and t if no letter occurs more than once?

18. How many of the patterns in Exercise 17 begin with a vowel and end with a consonant?

19. How many 4-digit patterns are there in which all the digits are different?

20. Three different colored dice are tossed. How many distinct outcomes can occur?

21. Using the letters from the word *equation*, how many 5-letter patterns can be formed in which q is followed immediately by u?

22. How many 5-digit numbers exist between and including 65,000 and 69,999 if no digit is to be repeated?

23. Draw a tree diagram to show the possibilities for boys and girls in a family with 2 children.

24. Draw a tree diagram to show the possibilities for boys and girls in a family with 3 children.

14-2 Linear Permutations

Suppose a group of objects are placed in an arrangement. The arrangement of things in a certain order is called a **permutation**. In a permutation, the *order* of the objects is very important. The arrangement of n objects in a line is called a **linear permutation**.

If 5 students enter a room in which there are 8 chairs in a row, how many different ways are there for the students to be seated? The first student can choose from the 8 chairs. The second student has 7 chairs to choose from, and so on.

Students:	1st	2nd	3rd	4th	5th
Number of Choices:	8	7	6	5	4

There are $8 \cdot 7 \cdot 6 \cdot 5 \cdot 4$ or 6720 possible seating arrangements.

The number of ways to arrange 8 things taken 5 at a time is written $P(8, 5)$. Thus, we speak of taking n objects, r at a time.

Note that $8 \cdot 7 \cdot 6 \cdot 5 \cdot 4 = \dfrac{8!}{(8-5)!}$ *or* $\dfrac{8!}{3!}$

> **The number of permutations of n objects taken r at a time is defined as follows.**
> $$P(n, r) = \frac{n!}{(n-r)!}$$

Definition of $P(n, r)$

If all the objects are taken at once, you write $P(n, n)$.

$$P(n, n) = \frac{n!}{(n-n)!} \text{ or } \frac{n!}{1} \qquad \textit{Recall that } 0! = 1.$$

Example

1 **The seven high school cheerleaders are to have their pictures taken. How many different pictures can be taken if there are to be three cheerleaders in each picture?**

Find the number of permutations of 7 people, taken 3 at a time.

$$P(n, r) = \frac{n!}{(n-r)!}$$

$$P(7, 3) = \frac{7!}{(7-3)!} \qquad \textit{Substitute 7 for n and 3 for r.}$$

$$= \frac{7 \cdot 6 \cdot 5 \cdot 4 \cdot 3 \cdot 2 \cdot 1}{4 \cdot 3 \cdot 2 \cdot 1} \text{ or } 210$$

There are 210 ways to arrange the cheerleaders for the pictures.

2   **How many ways can 4 algebra, 3 chemistry, and 5 history books be placed on a shelf if the books are arranged according to subject?**

Think of the books as being in 3 groups. There are $P(3, 3)$ or 3! ways of arranging these groups.   *$P(n, n) = n!$*

Also, there are $P(4, 4)$ or 4! ways of arranging the books in the algebra group, $P(3, 3)$ or 3! ways of arranging the books in the chemistry group, and $P(5, 5)$ or 5! ways of arranging the books in the history group.   *Basic Counting Principle*

There are $3! \cdot 4! \cdot 3! \cdot 5!$ or 103,680 ways of arranging the books according to subject.

How many different arrangements can be made from the letters of the word *free*? The four letters can be arranged in $P(4, 4)$ ways. $P(4, 4) = 4!$ or $4 \cdot 3 \cdot 2 \cdot 1$ or 24. There are 24 ways. However, some of these arrangements look the same. The two e's are not distinguishable. If we call them e_1 and e_2, then $e_1 fre_2$ and $e_2 fre_1$ are different. Drop the subscripts and the two appear the same: *efre*. The two e's can be arranged in $P(2, 2)$ ways. $P(2, 2) = 2!$ or 2.

To find the number of different arrangements, divide by 2!.

$$\frac{P(4, 4)}{P(2, 2)} = \frac{4!}{2!}$$

$$= \frac{4 \cdot 3 \cdot 2 \cdot 1}{2 \cdot 1} \text{ or } 12$$

There are 12 ways to arrange the letters.

When some objects are alike, use the following rule to find the number of permutations of those objects.

> **The number of permutations of n objects of which p are alike and q are alike is**
>
> $$\frac{n!}{p!q!}.$$
>
> *Permutations with Repetitions*

Example

3   **How many 7-letter patterns can be formed from the letters of *benzene*?**

Find the number of permutations of 7 objects of which 3 are e's and 2 are n's.

$$\frac{7!}{3!2!} = \frac{7 \cdot 6 \cdot 5 \cdot 4 \cdot 3 \cdot 2 \cdot 1}{3 \cdot 2 \cdot 1 \cdot 2 \cdot 1} \text{ or } 420$$

There are 420 7-letter patterns.

Exploratory Exercises

Determine whether each statement is true or false.

1. $5! - 3! = 2!$ **2.** $6 \cdot 5! = 6!$ **3.** $\dfrac{6!}{3!} = 2!$ **4.** $(6 - 3)! = 6! - 3!$

5. $\dfrac{6!}{30} = 4!$ **6.** $\dfrac{6!}{8!} \cdot \dfrac{8!}{6!} = 1$ **7.** $3! + 4! = 5 \cdot 3!$

8. $1!2!3!2! = 4!$ **9.** $\dfrac{P(9,\ 9)}{9!} = 1$ **10.** $\dfrac{3!}{3} = \dfrac{2!}{2}$

Written Exercises

How many different ways can the letters of the following words be arranged?

11. FLOWER   **12.** STUDY   **13.** POP

14. SEE   **15.** PEGGY   **16.** LEVEL

17. MISSISSIPPI   **18.** ALASKA   **19.** ALGEBRA

20. PARALLEL   **21.** ESSENTIAL   **22.** PERPENDICULAR

Find each value.

23. $\dfrac{P(6,\ 4)}{P(5,\ 3)}$ **24.** $\dfrac{P(10,\ 3)}{P(5,\ 3)}$ **25.** $\dfrac{P(6,\ 3) \cdot P(4,\ 2)}{P(5,\ 2)}$ **26.** $\dfrac{P(5,\ 3)}{P(8,\ 5)P(5,\ 5)}$

Solve each problem.

27. Don has 5 pennies, 3 nickels, and 4 dimes. The coins of each denomination are indistinguishable. How many ways can he arrange the coins in a row?

28. Estelle has 8 quarters, 5 dimes, 3 nickels, and a penny. The coins of each denomination are indistinguishable. How many ways can she place the coins in a straight line?

29. Ten scores received on a test were 82, 91, 75, 83, 91, 64, 83, 77, 91, and 75. In how many different orders might they be recorded?

30. How many 6-digit numbers can be made using the digits from 833,284?

31. There are 3 identical red flags and 5 identical white flags that are used to send signals. All 8 flags must be used. How many signals can be given?

32. Five algebra and four geometry books are to be placed on a shelf. How many ways can they be arranged if all the algebra books are together?

33. How many ways can 4 nickels and 5 dimes be distributed among 9 children if each is to receive one coin?

34. There are 4 green, 1 red, and 1 blue books on a shelf. How many ways can they be arranged if the red book and the blue book are separated?

Challenge Exercises

Find n in each equation.

35. $n[P(5,\ 3)] = P(7,\ 5)$ **36.** $P(n,\ 4) = 3[P(n,\ 3)]$

37. $P(n,\ 4) = 40[P(n - 1,\ 2)]$ **38.** $7[P(n,\ 5)] = P(n,\ 3) \cdot P(9,\ 3)$

39. $9P(n,\ 5) = P(n,\ 3) \cdot P(9,\ 3)$ **40.** $208P(n,\ 2) = P(16,\ 4)$

14-3 Circular Permutations

A food vending machine has 6 items on each of the revolving trays. One such tray has an orange, an apple, a can of juice, a salad, a cup of yogurt, and a boiled egg. How many ways can these items be arranged on the tray?

Think of each tray as a circle. Three possible arrangements of the various items on the tray are shown below.

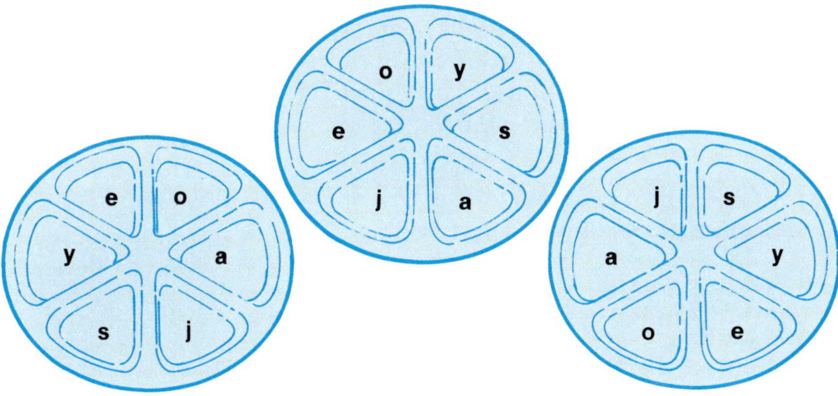

How does the first arrangement change as the tray is turned? Which arrangement is *really* different from the other two?

When 6 objects are placed in a line, there are 6! or 720 arrangements of the 6 objects taken 6 at a time. However, when they are arranged in a circle, some of the arrangements are alike. These arrangements fall into groups of six. Once an arrangement is determined, the rest of the group is formed by turning the circle. Then the objects can be placed into a new arrangement and the same process of turning the circle continues. Thus, the number of really different arrangements around a circle is $\frac{1}{6}$ of the total number of arrangements in a line.

$P(6, 6) = \frac{6!}{1} = 720$

$$\frac{1}{6} \cdot 6! = \frac{6 \cdot 5 \cdot 4 \cdot 3 \cdot 2 \cdot 1}{6}$$
$$= 5 \cdot 4 \cdot 3 \cdot 2 \cdot 1$$
$$= 5! \text{ or } (6 - 1)!$$

$\frac{1}{6}$ of $P(6, 6)$

There are $(6 - 1)!$ arrangements of 6 objects in a circle.

If _n_ distinct objects are arranged in a circle, then there are $\frac{n!}{n}$ or $(n - 1)!$ permutations of the objects around the circle.

Circular Permutations

Example

1 **Five people are to be seated at a round table. How many seating arrangements are possible?**

$(5 - 1)! = 4!$ *Use $(n - 1)!$*

$\qquad = 4 \cdot 3 \cdot 2 \cdot 1$ or 24

There are 24 seating arrangements possible.

Suppose the people are seated around the table. Everyone moves one chair to the left. Each person is still sitting next to the same two people as before.

Suppose n objects are in a circular arrangement, but the position of the objects is related to a fixed point. Rotating the circle will relate a different object to the fixed point and will make a new arrangement of the objects. Because of the fixed point, the permutations are now considered linear and there are $n!$ permutations. Consider the following example.

Suppose now that five people are to be seated at a round table. One of them is seated close to the door as shown. How many arrangements are possible?

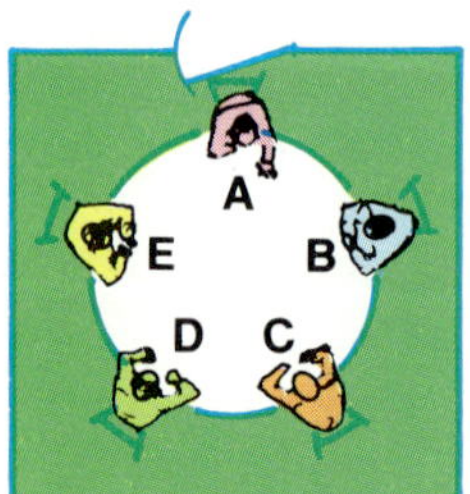

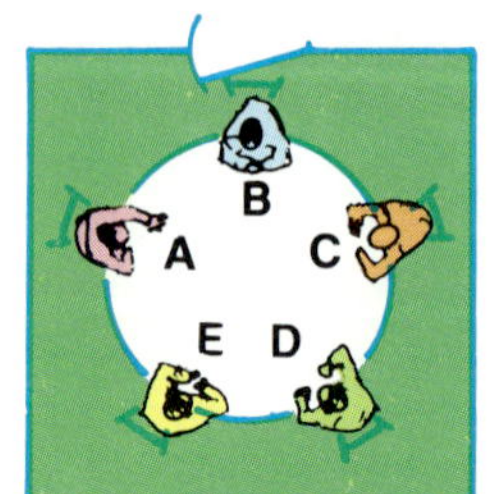

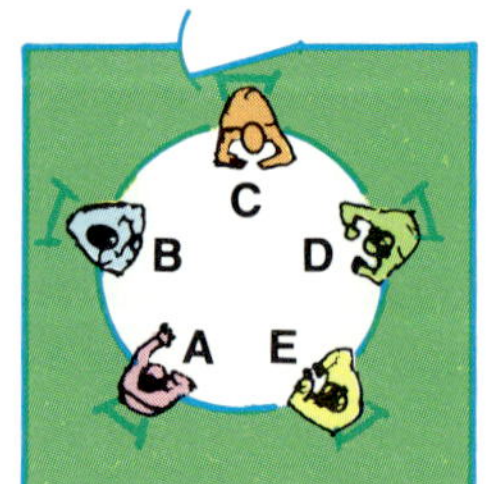

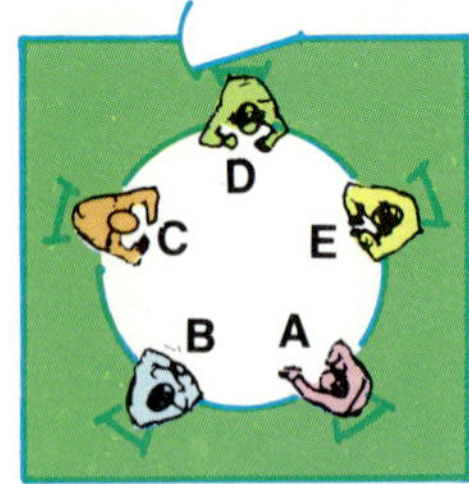

These arrangements can be considered different. In each one, a different person sits closest to the door. Thus, there are $P(5, 5)$ or $5!$ arrangements relative to a fixed point which is the door.

$$5! = 5 \cdot 4 \cdot 3 \cdot 2 \cdot 1 \text{ or } 120$$

Suppose three keys are placed on a key ring. Then it appears that there are at most $(3 - 1)!$ or 2 different arrangements of keys on the ring.

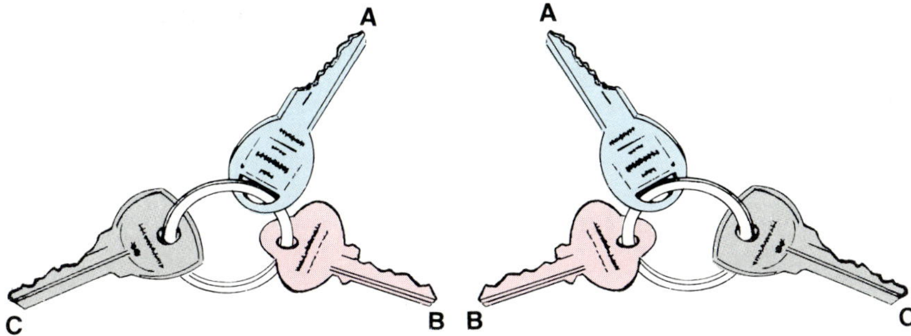

But what happens if the key ring with the first arrangement is turned over? The second arrangement appears. Then there is really only one arrangement of the three keys. These two arrangements are **reflections** of one another. There are only half as many arrangements when reflections are possible.

$$\frac{(3 - 1)!}{2} = \frac{2}{2} \text{ or } 1$$

Examples

2 **How many different ways can 5 charms be placed on a bracelet that has no clasp?**

This is a circular permutation. Because the bracelet can be turned over, it is also reflective.

$$\frac{(5 - 1)!}{2} = \frac{4!}{2} \qquad \frac{(n - 1)!}{2}$$

$$= \frac{4 \cdot 3 \cdot 2 \cdot 1}{2} \text{ or } 12$$

There are 12 different ways to arrange the charms.

3 **How many different ways can 5 charms be placed on a bracelet that has a clasp?**

This is no longer a circular permutation since objects are arranged with respect to a fixed point, the clasp. However, it is still reflective.

$$\frac{5!}{2} = \frac{5 \cdot 4 \cdot 3 \cdot 2 \cdot 1}{2} \qquad \frac{n!}{2}$$

$$= \frac{5 \cdot 4 \cdot 3 \cdot 2 \cdot 1}{2} \text{ or } 60$$

There are 60 different ways to arrange the charms.

Exploratory Exercises

State whether arrangements of the following objects are reflective or not reflective. Then, state whether their permutations are linear or circular.

1. charms on a bracelet, having no clasp
2. a football huddle of 11 players
3. placing 6 coins in a circle on a table
4. beads on a necklace with no clasp
5. chairs in a row
6. a pearl necklace that is open
7. four people seated around a square table relative to each other
8. people seated around a square table relative to one chair
9. a baseball team's batting order
10. a list of students in a given class

Written Exercises

Find each value.

11. $8! - 6!$
12. $6! + 5!$
13. $5!3!$
14. $\dfrac{6!}{4!}$
15. $\dfrac{8! + 6!}{8! - 6!}$
16. $P(8, 5)$

17. $P(10, 4)$
18. $3!P(6, 5) \cdot P(9, 2)$
19. $4!3!P(7, 3)$
20. $\dfrac{P(6, 4) \cdot P(5, 2)}{5!}$

21. $\dfrac{P(8, 3) \cdot P(5, 4)}{P(6, 6)}$
22. $\dfrac{P(12, 6)}{P(12, 3) \cdot P(8, 2)}$
23. $\dfrac{P(10, 8)}{5!P(8, 5)}$
24. $\dfrac{P(7, 6)}{7!P(9, 3)}$

Solve each problem.

25. How many ways can 6 keys be arranged on a key ring?
26. How many ways can 6 people be around a campfire?
27. How many ways can 8 charms be arranged on a bracelet that has no clasp?
28. How many ways can 4 men and 4 women be seated alternately at a round table?
29. How many ways can 5 people be seated at a round table relative to each other?
30. How many ways can 6 people be seated at a round table relative to a door?
31. How many ways can 5 people be seated around a circular table if 2 of the people must be seated next to each other?
32. Twenty beads are strung in a circle. Fourteen are brown and 6 are red. How many ways can the beads be strung in the circle if a clasp is used?
33. There are 8 chairs around a table. One chair is reserved for the President. How many seating arrangements are possible for the President and his 7 advisors?
34. A charm bracelet has 20 links and no clasp. How many ways can 6 charms be arranged on the bracelet if each charm requires one link?

mini-review

Solve each equation.

1. $x^{\frac{3}{2}} = 0.008$
2. $4^{3x} = 8^{5x-1}$
3. $\dfrac{6}{x + 8} = \dfrac{9}{x + 3}$
4. $\dfrac{4x^2}{x^2 - 25} - \dfrac{3x}{x + 5} = \dfrac{10}{x - 5}$
5. $\log_6 (x + 5) + \log_6 (x - 4) = 2$

6. Find the missing geometric means in the sequence 6, ___, ___, 93.75.
7. Find the missing terms in the arithmetic sequence 87, ___, ___, 198.

14-4 Combinations

Suppose that from a group of nine girls, five are chosen to start the basketball game. In this case, the order in which the girls are chosen is not important. Such a selection is called a **combination**.

The combination of nine things taken five at a time is written $C(9, 5)$.

You know that five things can be arranged in 5! ways. These arrangements are eliminated when finding the number of combinations.

$$C(9, 5) = \frac{P(9, 5)}{5!}$$

$$= \frac{9!}{(9 - 5)! \cdot 5!} \qquad P(9, 5) = \frac{9!}{(9 - 5)!}$$

$$= \frac{9!}{4! \cdot 5!} \text{ or } 126$$

The number of combinations of **n** distinct objects taken **r** at a time is defined as follows.

$$C(n, r) = \frac{n!}{(n - r)!r!}$$

Definition of C(n, r)

The main difference between a permutation and a combination is whether order is considered (permutation) or not (combination).

Example

1 From a group of 6 men and 4 women, how many committees of 2 men and 3 women can be formed?

Order is not considered. The questions are: How many ways can 2 men be chosen from 6? How many ways can 3 women be chosen from 4?

$$C(6, 2) \cdot C(4, 3) = \frac{6!}{(6 - 2)!2!} \cdot \frac{4!}{(4 - 3)!3!}$$

$$= \frac{6!}{4!2!} \cdot \frac{4!}{1!3!}$$

$$= \frac{6 \cdot 5 \cdot 4 \cdot 3 \cdot 2 \cdot 1}{4 \cdot 3 \cdot 2 \cdot 1 \cdot 2 \cdot 1} \cdot \frac{4 \cdot 3 \cdot 2 \cdot 1}{1 \cdot 3 \cdot 2 \cdot 1}$$

$$= \frac{6 \cdot 5}{2 \cdot 1} \cdot \frac{4}{1}$$

$$= 15 \cdot 4 \text{ or } 60$$

There are 60 possible committees.

2 **A box contains 8 red, 5 white, and 4 blue discs. How many ways can 5 discs be chosen so that 2 are red, 1 is white, and 2 are blue?**

$C(8, 2)$ Select 2 of the 8 red discs.

$C(5, 1)$ Select 1 of the 5 white discs.

$C(4, 2)$ Select 2 of the 4 blue discs.

$$C(8, 2) \cdot C(5, 1) \cdot C(4, 2) = \frac{8!}{(8-2)!2!} \cdot \frac{5!}{(5-1)!1!} \cdot \frac{4!}{(4-2)!2!}$$

$$= \frac{8 \cdot 7}{2 \cdot 1} \cdot 5 \cdot \frac{4 \cdot 3}{2 \cdot 1}$$

Note $\dfrac{8!}{6!2!} \cdot \dfrac{5!}{4!1!} \cdot \dfrac{4!}{2!2!} =$

$$= 28 \cdot 5 \cdot 6 \text{ or } 840$$

$\dfrac{8 \cdot 7 \cdot 6!}{6!2!} \cdot \dfrac{5 \cdot 4!}{4!1!} \cdot \dfrac{4 \cdot 3 \cdot 2!}{2!2!}$

There are 840 ways to choose the discs.

3 **Find the total number of diagonals that can be drawn in a decagon.**

Each diagonal has two endpoints. Suppose one has endpoints A and B. Then segments AB and BA are the same. Thus, order is not considered, and the combination of 10 points, taken two at a time, is desired. This gives the total number of line segments. But 10 of them are sides, so the number of diagonals is as follows.

$$C(10, 2) - 10 = \frac{10!}{(10-2)!2!} - 10$$

$$= \frac{10!}{8!2!} - 10$$

Note $10! = 10 \cdot 9 \cdot 8 \cdot 7 \cdot 6 \cdot 5 \cdot 4 \cdot 3 \cdot 2 \cdot 1$
$= 10 \cdot 9 \cdot 8!$

$$= \frac{10 \cdot 9 \cdot 8!}{8!2!} - 10$$

$$= \frac{10 \cdot 9}{2 \cdot 1} - 10 \text{ or } 35$$

There are 35 diagonals.

4 **From a deck of 52 cards, how many ways can 5 cards be drawn so that 3 are of 1 suit and 2 are of another?**

$P(4, 2)$ First, select 2 suits from the 4 suits.

$C(13, 3)$ Then, select 3 cards from 1 suit of 13 cards.

$C(13, 2)$ Next, select 2 cards from the other suit.

This is a permutation since the 3 cards come from either of the two suits selected.

$$P(4, 2) \cdot C(13, 3) \cdot C(13, 2) = \frac{4!}{(4-2)!} \cdot \frac{13!}{(13-3)!3!} \cdot \frac{13!}{(13-2)!2!}$$

$$= \frac{4!}{2!} \cdot \frac{13!}{10!3!} \cdot \frac{13!}{11!2!}$$

$$= \frac{4 \cdot 3}{1} \cdot \frac{13 \cdot 12 \cdot 11}{3 \cdot 2 \cdot 1} \cdot \frac{13 \cdot 12}{2 \cdot 1}$$

$$= 12 \cdot 286 \cdot 78 \text{ or } 267,696$$

There are 267,696 ways to draw the cards.

State whether arrangements of the following objects represent a combination or a permutation.

1. a team of 5 people, chosen from a group of 12 people
2. three-letter patterns, chosen from the letters of the word *algebra*
3. a hand of 5 cards
4. a batting order in baseball
5. seating students in a row
6. the answers on a true-false test
7. a committee of 4 men and 5 women, chosen from 8 men and 7 women
8. people seated around a table

Written Exercises

Find each value.

9. $C(8, 3)$
10. $C(8, 5) \cdot C(7, 3)$
11. $C(7, 2)$
12. $C(24, 21)$

Find the value of n.

13. $C(n, 3) = C(n, 8)$
14. $C(n, 5) = C(n, 7)$
15. $C(n, 12) = C(30, 18)$
16. $C(14, 3) = C(n, 11)$

Solve each problem.

17. From a list of 12 books, how many groups of 5 books can be selected?
18. How many baseball teams of 9 members can be formed from 14 players?
19. Suppose there are 9 points on a circle. How many different 4-sided polygons can be formed by joining any 4 of these points?
20. There are 85 telephones at Kennedy High School. How many 2-way connections can be made among the school telephones?
21. How many different groups of 25 people can be formed from 27 people?
22. Suppose there are 8 points in a plane, no 3 of which are collinear. How many distinct triangles could be formed with these points as vertices?
23. From a deck of 52 playing cards, how many different 5-card hands can have 5 cards of the same suit?
24. From a deck of 52 playing cards, how many different 4-card hands can have each card from a different suit?

A bag contains 4 red, 6 white, and 9 blue marbles. How many ways can 5 marbles be selected to meet the following conditions?

25. All the marbles are white.
26. All the marbles are blue.
27. All the marbles are red.
28. Two are red, 2 are white, and 1 is blue.
29. Two must be blue.
30. Two are 1 color and 3 are another color.

From a group of 8 men and 10 women, a committee of 5 is to be formed. How many committees can be formed if the committee is to be comprised as follows?

31. All are men.
32. There are 3 men and 2 women.
33. There is 1 man and 4 women.
34. All are women.

Solve each equation.

1. $\dfrac{2x + 5}{x + 3} = \dfrac{x + 4}{x} + \dfrac{2x - 1}{2x}$

2. $\dfrac{x + 4}{x - 8} = 3 - \dfrac{4x - 20}{x^2 - 3x - 40}$

3. $\dfrac{2x + 16}{x + 6} = \dfrac{x + 3}{x}$

4. $\log_{27} x = \dfrac{1}{3}$

5. $\log_4 (x^2 + 2x) = \log_4 35$

6. $\log_2 x + \log_2 (x - 4) = 5$

Simplify.

7. $\dfrac{\dfrac{4a}{a^2 - 4a - 12}}{\dfrac{5}{a^2 - 9a + 18}}$

8. $\dfrac{\dfrac{12b}{3b^2 + 5b - 12}}{\dfrac{8}{3b^2 - 4b}}$

9. $(c^{\sqrt{5}} + d^{\sqrt{6}})^2$

10. $y^{\sqrt{7}} \cdot y^{\sqrt{28}}$

Find each value.

11. $5^{\log_5 10}$

12. $\dfrac{P(7, 5)}{P(5, 2)}$

13. $C(7, 5) \cdot C(5, 2)$

14. $\displaystyle\sum_{j=3}^{8} 2^{j-5}$

Perform the indicated operations. Write each answer in simplest form.

15. $\dfrac{x + 4}{x^2} \cdot \dfrac{3x}{x^2 - 2x - 24}$

16. $\dfrac{2z^2}{z^2 - 2z - 35} - \dfrac{z + 7}{z + 5}$

17. $\dfrac{2r + 5}{r + 4} + \dfrac{r - 8}{3r - 2}$

18. $\dfrac{a^2 - b^2}{b - a} \div \dfrac{b^2 - a^2}{b - a}$

19. Find the sum of the first 30 terms of $17 + 24 + 31 + 38 + \ldots$.

20. Graph $y = \dfrac{3x}{x + 2}$.

21. Suppose y varies directly as x, and $x = 11$ when $y = 4$. Find x when y is 2.

22. Interpolate to find the antilog of 4.6422.

23. Find the 27th term of $3 + 6 + 12 + 24 \ldots$.

24. Expand $(4r + s)^5$.

25. If y varies inversely as x, and $y = -9$ when $x = -5$, find y when $x = \dfrac{1}{2}$.

26. Write $\dfrac{2}{3} + \dfrac{3}{5} + \dfrac{4}{7} + \dfrac{5}{9} + \cdots + \dfrac{21}{41}$ using sigma notation.

27. Show that $\log_{36} (\log_2 64) = \dfrac{1}{2}$.

Solve each problem.

28. A plane flies from Baltimore to Omaha with a tailwind of 120 kilometers per hour in $2\dfrac{1}{2}$ hours. On the return trip flying into the same wind takes $3\dfrac{3}{4}$ hours. In still air, the plane would have the same speed both ways. Find the distance from Baltimore to Omaha.

29. A piece of farm equipment valued at $60,000 depreciates 10% per year by the fixed rate method. After how many years will the value have depreciated to $45,000?

30. How many ways can 5 books be placed on a shelf?

31. From a group of 8 men and 7 women, how many different committees of 3 men and 4 women can be formed?

14-5 Probability

When a coin is tossed, only two outcomes are possible. Either the coin will show a *head* or a *tail*. The desired outcome is called a **success**. Any other outcome is called a **failure**.

An event is a set of outcomes.

> **If an event can succeed in *s* ways and fail in *f* ways, then the probabilities of success, *P(s)*, and of failure, *P(f)*, are as follows.**
>
> $$P(s) = \frac{s}{s + f} \qquad P(f) = \frac{f}{s + f}$$

Probability of Success and of Failure

If the event cannot succeed, $P(s) = 0$. If the event cannot fail, $P(s) = 1$. Thus, $P(s)$ is always between 0 and 1, inclusive.

$s + f$ = total number of outcomes.

$$P(s) + P(f) = \frac{s}{s + f} + \frac{f}{s + f}$$
$$= \frac{s + f}{s + f} \text{ or } 1$$

This is an important property of probabilities.

Because their sum is 1, $P(s)$ and $P(f)$ are called *complements*. For example, if $P(s)$ is $\frac{1}{3}$, then $P(f)$ is $1 - \frac{1}{3}$ or $\frac{2}{3}$.

Examples

1 **A bag contains 5 blue marbles and 4 white marbles. If one marble is chosen at random, what is the probability that it is blue?**

$$P(\text{blue marble}) = \frac{s}{s + f}$$
$$= \frac{5}{5 + 4} \text{ or } \frac{5}{9}$$

P(blue marble) is read the probability of selecting a blue marble.
A blue marble is a success.
A white marble is a failure.

The probability of selecting a blue marble is $\frac{5}{9}$ or approximately 0.556.

2 **A committee of 2 is to be selected from a group of 6 men and 3 women. What is the probability that the 2 selected are women?**

$$P(\text{two women}) = \frac{C(3, 2)}{C(9, 2)}$$

There are C(3, 2) ways to select 2 of 3 women.
There are C(9, 2) ways to select 2 of 9 people.

$$= \frac{\frac{3!}{1!2!}}{\frac{9!}{7!2!}}$$
$$= \frac{3}{36} \text{ or } \frac{1}{12}$$

The probability that the 2 selected are women is $\frac{1}{12}$ or approximately 0.083.

The odds of the successful outcome of an event is expressed as the ratio of the number of ways it can succeed to the number of ways it can fail.

Odds = the ratio of s to f or $\frac{s}{f}$

Definition of Odds

Examples

3 **What are the odds of tossing a die and getting a 3?**

The number 3 is on only one face of the die. The other five faces have a number other than 3.

Odds $= \frac{1}{5}$ *A 3 can appear only 1 way.*   *Remember that*
Other numbers can appear 5 ways.   *$s + f$ = total number of outcomes.*

The odds of getting a 3 are 1 to 5.

4 **Suppose Michael draws 5 cards from a deck of 52 cards. What are the odds that the first 4 cards will be of one suit and the fifth of another suit?**

$P(4, 2)$ Select 2 suits among 4.   *Since a different number of cards are to be selected from each suit, order is important.*
$C(13, 4)$ Select 4 cards from a suit containing 13 cards.
$C(13, 1)$ Select 1 card from the other suit.

The number of ways to select the first 4 cards from one suit and the fifth card from another is found as follows.

$$P(4, 2) \cdot C(13, 4) \cdot C(13, 1) = \frac{4!}{2!} \cdot \frac{13!}{4!9!} \cdot \frac{13!}{1!12!}$$

First, calculate the number of successful outcomes.

$$= \frac{4 \cdot 3}{1} \cdot \frac{13 \cdot 12 \cdot 11 \cdot 10}{4 \cdot 3 \cdot 2 \cdot 1} \cdot \frac{13}{1}$$

$$= 12 \cdot 715 \cdot 13 \text{ or } 111{,}540$$

Thus the number of outcomes that can be considered successful is 111,540. The total number of outcomes is the combination of 52 cards taken 5 at a time.

$$C(52, 5) = \frac{52!}{47!5!}$$

Then, calculate the total number of outcomes.

$$= \frac{52 \cdot 51 \cdot 50 \cdot 49 \cdot 48}{5 \cdot 4 \cdot 3 \cdot 2 \cdot 1}$$

$$= \frac{311{,}875{,}200}{120} \text{ or } 2{,}598{,}960$$

The total number of outcomes, including successful outcomes and failures, is 2,598,960. Thus, the number of failures is $2{,}598{,}960 - 111{,}540$ or 2,487,420. The odds of selecting the first 4 cards from one suit and the fifth card from another suit are $\frac{111{,}540}{2{,}487{,}420}$ or $\frac{143}{3189}$. This is approximately $\frac{1}{22}$ or 0.045.

Exploratory Exercises

State the odds of an event occurring given the probability of the event occurring.

1. $\frac{1}{2}$
2. $\frac{3}{4}$
3. $\frac{1}{7}$
4. $\frac{5}{8}$
5. $\frac{7}{15}$
6. $\frac{8}{9}$

State the probability of an event occurring given the odds of the event occurring.

7. $\frac{3}{4}$
8. $\frac{5}{1}$
9. $\frac{6}{5}$
10. $\frac{3}{7}$
11. $\frac{5}{11}$
12. $\frac{1}{1}$

Solve each problem.

13. The odds are 6-to-1 *against* an event occurring. What is the probability that it will occur?

14. The probability of an event occurring is $\frac{3}{4}$. What are the odds that it will not occur?

Written Exercises

A bag contains 7 pennies, 4 nickels, and 5 dimes. Three coins are selected at random. Find the probability of each selection.

15. all 3 pennies
16. all 3 nickels
17. all 3 dimes
18. 2 pennies, 1 dime
19. 1 penny, 1 dime, 1 nickel
20. 1 dime, 2 nickels

A bag contains 5 red, 9 blue, and 6 white marbles. Two are selected at random. Find the probability of each selection.

21. 2 red
22. 2 blue
23. 1 red and 1 white

There are 5 fudgesicles and 8 popsicles in the freezer. If 2 are selected at random, find the probability of each selection.

24. 2 fudgesicles
25. 2 popsicles
26. 1 fudgesicle and 1 popsicle

Suppose you select 2 letters from the word *algebra*. What is the probability of making each selection.

27. 1 vowel and 1 consonant
28. 2 vowels
29. 2 consonants

Sharon has 8 mystery books and 9 science fiction books. Four are selected. Find the probability of each selection.

30. 4 mystery books
31. 4 science fiction books
32. 2 mysteries and 2 science fiction
33. 3 mysteries and 1 science fiction

From a deck of 52 cards, 5 cards are dealt. What are the odds of the following?

34. 5 aces
35. 5 face cards
36. 5 from one suit

From a deck of 52 cards, 5 cards are dealt. What are the odds against the following?

37. 5 hearts
38. 5 face cards
39. 5 from one suit
40. 5 queens
41. the first 3 cards from one suit and the last 2 from another suit

Applications in Probability Games

Companies often use games of chance to increase sales. By law, the company must explain how many game tickets will be distributed and how many prizes will be awarded. From this information, you can compute the odds of winning a certain prize.

Here is an easy way to compute the odds of winning:

  1. Find the probability and express it as a unit fraction.
  2. Subtract 1 from the denominator to obtain the odds.

This method works because of the relationship between probability $\frac{s}{s+f}$ and odds $\frac{s}{f}$. When s is 1, the denominator for probability is one greater than the denominator for odds.

Example: **Suppose a company distributes 2400 game tickets, including 50 for wristwatches and 30 for radios.**

    a. What are the odds of winning a wristwatch?

$$\text{Probability} = \frac{50}{2400} = \frac{1}{48} \qquad \text{Odds} = \frac{1}{48 - 1} = \frac{1}{47}$$

The odds are $\frac{1}{47}$ or 1 to 47.

    b. What are the odds of winning a wristwatch or a radio?

$$\text{Probability} = \frac{50 + 30}{2400} = \frac{1}{30} \qquad \text{Odds} = \frac{1}{30 - 1} = \frac{1}{29}$$

The odds are $\frac{1}{29}$ or 1 to 29.

Exercises

A company plans to give out 24,864,840 game tickets. Winning tickets can be exchanged for cash prizes of \$25, \$10, or \$5. The following chart shows the number of winning tickets that will be distributed. Complete the chart by finding the values of the winning tickets, the probability of winning, and the odds.

Prize Category	Quantity of Winning Tickets	Value of Winning Tickets	Probability of Winning	Odds of Winning
\$25	1656	\$25 × 1656 = \$41,400	1 to 15,015	1 to 15, 014
\$10	4620	**1.**	**2.**	**3.**
\$ 5	6210	**4.**	**5.**	**6.**

7. If all tickets are exchanged for cash, how much money will the company pay in prizes?

8. What are the odds of winning anything?

9. If the company's sales increase by \$500,000 and 30% of this amount is profit, how much money will the company earn as a result of the game? Assume that all prizes will be awarded.

14-6 Multiplying Probabilities

The Basic Counting Principle can be used to help find probabilities. Suppose you toss a red and a blue die. The probability that the red die shows a 2 is $\frac{1}{6}$. The probability that the blue die shows a 2 is $\frac{1}{6}$. The probability that both dice show 2 is $\frac{1}{6} \cdot \frac{1}{6}$ or $\frac{1}{36}$. Since the outcome of tossing the red die does *not* affect the outcome of tossing the blue die, the events are *independent*.

If two events, **A** and **B**, are independent, then the probability of both events occurring is found as follows. $$P(A \text{ and } B) = P(A) \cdot P(B)$$	*Probability of Two Independent Events*

Example

1 A bag contains 5 red marbles and 4 white marbles. A marble is to be selected, and replaced in the bag. A second selection is then made. What is the probability of selecting 2 red marbles?

These events are independent because the first marble selected is replaced. The outcome of the second selection is not affected by the results of the first selection.

$$P(both\ red) = P(red) \cdot P(red)$$

The probability is a little less than $\frac{1}{3}$.

$$= \frac{5}{9} \cdot \frac{5}{9} \text{ or } \frac{25}{81}$$

$\frac{25}{81} \approx 0.309$

What is the probability of selecting 2 red marbles from 5 red ones and 4 white ones if the first selection is *not* replaced? These events are *dependent* because the outcome of the first selection affects the outcome of the second selection. Suppose the first selection is red.

first selection $\qquad$ second selection

Notice that there is one less red marble so there is one less marble in the bag.

$$P(red) = \frac{5}{9} \qquad\qquad P(red) = \frac{4}{8}$$

$$P(both\ red) = P(red) \cdot P(red\ following\ red)$$

$$= \frac{5}{9} \cdot \frac{4}{8} \text{ or } \frac{5}{18}$$

$\frac{5}{18} \approx 0.278$

If two events, **A** and **B**, are dependent, then the probability of both events occurring is found as follows. $$P(A \text{ and } B) = P(A) \cdot P(B \text{ following } A)$$	*Probability of Two Dependent Events*

2 There are 5 nickels, 7 dimes, and 9 pennies in a coin purse. Suppose two coins are to be selected, without replacing the first one. Find the probability of each event.

a. penny and then a dime

Because the coins are not replaced, the events are dependent. Thus, P(A and B) = P(A) · P(B following A).

$P(penny\ and\ dime) = P(penny) \cdot P(dime\ following\ penny)$

$$= \frac{9}{21} \cdot \frac{7}{20}$$

$$= \frac{63}{420} \text{ or } \frac{3}{20}$$

The probability is $\frac{3}{20}$ or 0.150.

b. two nickels

$P(2\ nickels) = P(nickel) \cdot P(nickel\ following\ nickel)$

$$= \frac{5}{21} \cdot \frac{4}{20}$$

$$= \frac{20}{420} \text{ or } \frac{1}{21}$$

The probability is $\frac{1}{21}$ or approximately 0.048.

3 From a deck of 52 cards, an eight, a nine, and then another eight are to be selected in that order. Find the probability of this event occurring under each given condition.

a. Replacement occurs each time.   *The events are independent.*

$P(eight,\ nine,\ eight) = P(eight) \cdot P(nine) \cdot P(eight)$ $P(eight) = \frac{4}{52}$ or $\frac{1}{13}$

$$= \frac{1}{13} \cdot \frac{1}{13} \cdot \frac{1}{13}$$ $P(nine) = \frac{4}{52}$ or $\frac{1}{13}$

$$= \frac{1}{2197}$$

The probability is $\frac{1}{2197}$ or approximately 0.0005.

b. No replacement occurs.   *The events are dependent.*

$P(eight,\ nine,\ different\ eight)$

$= P(eight) \cdot P(nine\ following\ eight) \cdot P(eight\ after\ nine\ following\ eight)$

$$= \frac{4}{52} \cdot \frac{4}{51} \cdot \frac{3}{50}$$

$$= \frac{48}{132,600} \text{ or } \frac{1}{2763}$$

The probability is $\frac{1}{2763}$ or approximately 0.0004.

Identify the events in each problem as *independent* or *dependent*.

1. In a bag are 5 red, 3 green, and 8 blue marbles. Three are selected in sequence without replacement. What is the probability of selecting a red, green, and blue, in that order?

2. There are 4 glasses of root beer and 3 glasses of ice tea on the counter. Bill drinks 2 of them. What is the probability that he drank 2 root beers?

3. In a bag are 5 apricots and 4 plums. Marie selects one, replaces it, and selects another. What is the probability that both selections were apricots?

4. When James plays Ted in cribbage, the odds are 3 to 2 that he will win. What is the probability that he will win the next 4 games?

Written Exercises

5–8. Solve each problem in Exploratory Exercises 1–4.

A bag contains 5 red, 3 white, and 7 blue marbles. If 3 marbles are selected in succession, what is the probability that they are red, white, and blue, in that order?

9. Suppose no marbles are replaced.

10. Suppose each marble is replaced.

A bag contains 5 red, 3 blue, and 7 black marbles. Three marbles are chosen, one after the other. What is the probability that there is one of each color under the following conditions?

11. No replacement occurs.

12. Replacement occurs each time.

One hundred tickets, numbered consecutively 1 to 100, are placed in a box. What is the probability that in 5 separate drawings, the following selections occur?

13. 5 odd numbers, if replacement occurs

14. 5 odd numbers, if no replacement occurs

15. 5 consecutive numbers if no replacement occurs

The letters A, B, E, I, J, K, and M are written on cards that are placed in a box. Two letters are selected. What is the probability that the following occurs?

16. two vowels, if no replacement occurs

17. two vowels, if replacement occurs

18. two the same letter, if no replacement occurs

There are 6 plates, 5 saucers, and 5 cups on the counter. Charlie accidentally knocks off two and breaks them. What is the probability that he broke the following?

19. 2 plates

20. 2 cups

21. a cup and a saucer, in that order

22. a cup and a saucer, in any order

A red and a green die are tossed. What is the probability that the following occurs?

23. both show 3

24. neither show 3

25. the red shows a 3 and the green shows a 4

26. the red shows a 3 and the green shows any other number

27. both show the same number

28. both show different numbers

mini-review

Perform the indicated operations. Write each answer in simplest form.

1. $10g + \dfrac{g + 1}{g - 8}$

2. $\dfrac{9}{c + 6} - \dfrac{3}{c^2 + 12c + 36}$

3. $\dfrac{a^2 - a - 20}{a + 7} \div \dfrac{a - 5}{a^3 - 49a}$

4. $\dfrac{n^2 + 3n - 28}{n^2 + 12n + 35} \cdot \dfrac{n^2 + 10n + 25}{n^2 - 8n + 16}$

Solve each equation.

5. $\log_7 y = 4$

6. $\log_3 (x + 6) = 2 \log_3 x$

7. $9^{3y} = 27^{y-1}$

8. $\dfrac{x}{x + 1} + \dfrac{1}{x - 2} = \dfrac{3x + 1}{3x - 2}$

9. $3 + \dfrac{y + 2}{y} = \dfrac{y - 10}{2y}$

10. Find the 19th term of an arithmetic sequence if the 7th term is 11 and the 16th term is 56.

11. How many ways can five people be seated around a circular table relative to each other?

12. A ball dropped 120 feet bounces $\frac{2}{3}$ of the height from which it fell on each bounce. How far will it travel before coming to rest?

13. A fraction has a value of $\frac{6}{7}$. If 1 is added to its numerator, its value is $\frac{7}{8}$. Find the original fraction.

14. If $200 is invested at 8% interest compounded continuously, when will the investment double? Use $A = Pe^{rt}$.

15. A piece of machinery valued at $75,000 depreciates at a steady rate of 8% yearly. When will the value be $15,000? Use $V_n = P(1 + r)^n$.

Excursions in Algebra
History

Maria Agnesi (ähn yā′ zē) was an Italian mathematician who lived from 1718 to 1799. At one time, she was a professor of mathematics at Bologna, Italy. In 1748, she wrote about a special set of curves that she called *versiera*. The general equation for those curves is $yx^2 = a^2(a - y)$.

The popular name for this type of curve is the Witch of Agnesi, because the curve resembles the outline of a witch's hat.

Exercises

Graph each versiera. State the value of a.

1. $yx^2 = 4(2 - y)$

2. $yx^2 = -64 - 16y$

14-7 Adding Probabilities

Suppose a card is drawn from a standard deck of 52 cards. What is the probability of drawing an ace or a king? Since no card is both an ace and a king, the events are said to be **mutually exclusive**. That is, the two events cannot occur simultaneously.

probability of drawing an ace   **probability of drawing a king**

There are 4 aces in a deck. $\dfrac{4}{52}$ or $\dfrac{1}{13}$ $\dfrac{4}{52}$ or $\dfrac{1}{13}$ *There are 4 kings in a deck.*

probability of drawing an ace or a king

ace   king   ace or king

$$\dfrac{1}{13} + \dfrac{1}{13} = \dfrac{2}{13} \qquad \dfrac{2}{13} \approx 0.154$$

> **The probability of one of two mutually exclusive events, A and B, occurring is the sum of their probabilities.**
>
> $$P(A \text{ or } B) = P(A) + P(B)$$

Probability of Mutually Exclusive Events

This rule can be extended to any number of mutually exclusive events. **Inclusive events** are *not* mutually exclusive. Therefore, the two events can occur simultaneously.

What is the probability of drawing an ace or a red card? Since there are two red aces, the events are inclusive.

probability of drawing an ace   **probability of drawing a red card**

There is an ace in each suit, hearts, diamonds, spades, and clubs. $\dfrac{4}{52}$ $\dfrac{26}{52}$ *Hearts and diamonds are red.*

probability of drawing a red ace

$\dfrac{2}{52}$ *There are two red aces*

The probability of drawing a red ace is counted twice, once for an ace and once for a red card.

probability of drawing an ace or a red card

ace   red   red ace   ace or red

$$\dfrac{4}{52} + \dfrac{26}{52} - \dfrac{2}{52} = \dfrac{28}{52}$$
$$= \dfrac{7}{13} \qquad \dfrac{7}{13} \approx 0.538$$

> The probability of one of two inclusive events, *A* and *B*, occurring is the sum of the individual probabilities decreased by the probability of both occurring.
>
> $$P(A \text{ or } B) = P(A) + P(B) - P(A \text{ and } B)$$

Probability of Inclusive Events

Examples

1 Vivian has 6 nickels, 4 pennies, and 3 dimes in her purse. She selects one. What is the probability it is a penny or a nickel?

$$P(penny \text{ or } nickel) = P(penny) + P(nickel)$$
$$= \frac{4}{13} + \frac{6}{13}$$
$$= \frac{10}{13}$$

There is no coin that is both a penny and a nickel. These events are mutually exclusive.

The probability of selecting a penny or a nickel is $\frac{10}{13}$ or about 0.769.

2 A card is selected from a deck of 52 cards. What is the probability that it is a red card or a face card?

$$P(red \text{ or } face\ card) = P(red) + P(face\ card) - P(red\ face\ card)$$
$$= \frac{26}{52} + \frac{12}{52} - \frac{6}{52}$$
$$= \frac{32}{52} \text{ or } \frac{8}{13}$$

There are 6 face cards that are red. Thus, these events are inclusive.

The probability of selecting a red card or a face card is $\frac{8}{13}$ or about 0.615.

3 A committee of 5 people is to be formed from a group of 7 men and 6 women. What is the probability that the committee will have at least 3 women?

At least 3 women means that the committee may have 3, or 4, or 5 women. It is not possible to select a group of 3, a group of 4, and a group of 5 women all to be on the same 5-member committee. The events are mutually exclusive.

$$P(at\ least\ 3\ women) = P(3\ women) + P(4\ women) + P(5\ women)$$

3 women, 2 men   4 women, 1 man   5 women, 0 men
$$= \frac{C(6, 3) \cdot C(7, 2)}{C(13, 5)} + \frac{C(6, 4) \cdot C(7, 1)}{C(13, 5)} + \frac{C(6, 5) \cdot C(7, 0)}{C(13, 5)}$$
$$= \frac{140}{429} + \frac{35}{429} + \frac{2}{429}$$
$$= \frac{177}{429} \text{ or } \frac{59}{143} \qquad \frac{59}{143} \approx 0.413$$

The probability of at least 3 women on the committee is $\frac{59}{143}$ or about 0.413.

Exploratory Exercises

Identify each event as *inclusive* or *exclusive*.

1. A box contains slips of paper numbered from 1 to 10. A slip of paper is drawn and a die is tossed. What is the probability of getting a 2 on only one of them?

2. Two cards are drawn from a standard deck of playing cards. What is the probability that the 2 cards are both kings or both queens?

3. In her pocket, Linda has 5 nickels, 3 dimes, and 7 pennies. She selects 3 coins. What is the probability that she has selected 3 nickels or 3 pennies?

4. The Dodger pitching staff has 4 left-handers and 7 right-handers. If 2 are selected, what is the probability that at least one of them is a left-hander?

5. From a standard deck of playing cards, 2 cards are drawn. What is the probability of having drawn a black card or an ace?

6. Five coins are dropped. What is the probability of having at least 3 heads?

7. In one class, 3 of the 12 girls are redheads and 2 of the 15 boys are redheads. What is the probability of selecting a boy or a redhead?

8. There are 8 red, 3 blue, and 12 black marbles in a bag. If 3 are selected, what is the probability that all are red or all are blue?

Written Exercises

A bag contains 6 red and 5 white marbles. Three are selected. What is the probability that the following occurs?

9. all 3 red or all 3 white

10. at least 2 red

11. at least 2 white

12. exactly 2 white

Two cards are drawn from a standard deck of cards. What is the probability that the following occurs?

13. both aces or both face cards

14. both black or both face cards

15. both aces or both red

16. both either red or an ace

Seven coins are tossed. What is the probability that the following occurs?

17. 3 heads or 2 tails

18. at least 5 heads

19. 3 heads or 3 tails

20. all tails or all heads

From a group of 6 men and 8 women, a committee of 6 is to be selected. What is the probability of the following?

21. all men or all women

22. 5 men or 5 women

23. 3 men and 3 women

24. 4 men or 4 women

The numerals 1 through 25 are written on slips of paper and placed in a bag. The numerals 20 through 40 are written on slips of paper and placed in a different bag. One slip of paper is selected at random from each bag. What is the probability that the following occurs?

25. Both numerals are 20.

26. Neither numeral is 20.

27. Both numerals are greater than 10.

28. At least one of the numerals is 22.

29–36. Solve each problem in Exploratory Exercises 1–8.

Recall that the binomial expansion of $(a + b)^n$ can be written using sigma and factorial notation.

$$(a + b)^n = \frac{n!}{0!(n - 0)!}a^n + \frac{n!}{1!(n - 1)!}a^{n-1}b^1 + \frac{n!}{2!(n - 2)!}a^{n-2}b^2 + \ldots$$

$$= \sum_{k=0}^{n} \frac{n!}{k!(n - k)!}a^{n-k}b^k$$

Here n is a positive integer. k is a positive integer or zero.

Study the expansion of $(x + y)^5$.

$$(x + y)^5 = \sum_{k=0}^{5} \frac{5!}{k!(5 - k)!}x^{5-k}y^k$$

$$= \frac{5!}{0!5!}x^5 + \frac{5!}{1!4!}x^4y + \frac{5!}{2!3!}x^3y^2 + \frac{5!}{3!2!}x^2y^3 + \frac{5!}{4!1!}xy^4 + \frac{5!}{5!0!}y^5$$

$$x^5 + 5x^4y + 10x^3y^2 + 10x^2y^3 + 5xy^4 + y^5$$

The coefficient of each term may be expressed using combination notation.

Recall that $C(n, r)$ means $\dfrac{n!}{(n - r)!r!}$.

$$(x + y)^5 = C(5, 5)x^5 + C(5, 4)x^4y + C(5, 3)x^3y^2 + C(5, 2)x^2y^3 + C(5, 1)xy^4 + C(5, 0)y^5$$

$$1 \qquad 5 \qquad 10 \qquad 10 \qquad 5 \qquad 1$$

The coefficients of a binomial expansion are symmetric. Study the values of the coefficients of $(x + y)^5$.

$$C(5, 0) = C(5, 5) \qquad C(5, 1) = C(5, 4) \qquad C(5, 2) = C(5, 3)$$

In general, $C(n, k)$ is equivalent to $C(n, n - k)$. Thus, the following two formulas for binomial expansions are equivalent. Both formulas use sigma notation and combination notation.

$$(a + b)^n = \sum_{k=0}^{n} C(n, n - k)a^{n-k}b^k \qquad \text{or} \qquad (a + b)^n = \sum_{k=0}^{n} C(n, k)a^{n-k}b^k$$

Exercises

Write each expansion using combination notation for the coefficients. Then simplify by evaluating each coefficient.

1. $(r + s)^3$ **2.** $(m + n)^6$ **3.** $(v + w)^4$

4. $(c + d)^7$ **5.** $(f + g)^9$ **6.** $(p + q)^8$

Are the two expressions equivalent? Write *yes* or *no*.

7. $C(6, 4)$ and $C(6, 2)$ **8.** $C(4, 3)$ and $C(3, 1)$

9. $C(5, 5)$ and $C(6, 6)$ **10.** $C(7, 1)$ and $C(7, 7)$

11. $C(n, 0)$ and $C(n, n)$ **12.** $C(a, b)$ and $C(a, a - b)$

14-8　Binomial Trials

Arthur normally wins 1 out of every 3 backgammon games he plays. In other words, the probability that Arthur wins when he plays backgammon is $\frac{1}{3}$.

Suppose Arthur plays 4 games. What is the probability that he will win 3 and lose only one?

The possible ways of winning 3 games and losing 1 are shown at the right. The illustration shows the combinations of four things (games) taken three at a time (wins). That is, $C(4, 3)$.

$$\begin{array}{cccc} L & W & W & W \\ W & L & W & W \\ W & W & L & W \\ W & W & W & L \end{array}$$

The terms of the binomial expansion of $(W + L)^4$ can be used to find probabilities.

$$(W + L)^4 = W^4 + 4W^3L + 6W^2L^2 + 4WL^3 + L^4$$

term	meaning	coefficient
W^4	1 way to win all 4 games	$C(4, 4) = 1$
$4W^3L$	4 ways to win 3 games and lose 1 game	$C(4, 3) = 4$
$6W^2L^2$	6 ways to win 2 games and lose 2 games	$C(4, 2) = 6$
$4WL^3$	4 ways to win 1 game and lose 3 games	$C(4, 1) = 4$
L^4	1 way to lose all 4 games	$C(4, 0) = 1$

Notice that the coefficient of each term can also be found using $C(n, r)$.

The probability that Arthur wins when he plays is $\frac{1}{3}$. And, thus, the probability that he loses is $\frac{2}{3}$. Substitute $\frac{1}{3}$ for W and $\frac{2}{3}$ for L in the term $4W^3L$. For example, the probability of winning 3 out of 4 games is $4\left(\frac{1}{3}\right)^3\left(\frac{2}{3}\right)$ or $\frac{8}{81}$.　　*$\frac{8}{81} \approx 0.099$*

What is the probability of winning 2 games and losing 2 games?

Problems that can be solved using a binomial expansion are called **binomial trials**.

A binomial trial exists if and only if the following conditions occur.

1. **There are only two possible outcomes.**
2. **The events are independent.**

Conditions of Binomial Trials

1 **What is the probability that 3 coins show heads and 2 show tails when 5 coins are tossed?**

There are only 2 possible outcomes: heads (H) or tails (T). The tosses of 5 coins are independent events. When $(H + T)^5$ is expanded, the term containing H^3T^2, which represents 3 heads and 2 tails, is used to get the desired probabilities.

The coefficient of H^3T^2 is $C(5, 3)$.

$$P(3\ heads,\ 2\ tails) = C(5, 3)H^3T^2 \qquad \textit{Replace H with P(H) which is } \tfrac{1}{2}.$$

$$= \frac{5 \cdot 4}{2 \cdot 1}\left(\frac{1}{2}\right)^3\left(\frac{1}{2}\right)^2 \qquad \textit{Replace T with P(T) which is } \tfrac{1}{2}.$$

$$= \frac{5}{16} \qquad\qquad \tfrac{5}{16} \approx \textit{0.313}$$

The probability of 3 heads and 2 tails is $\frac{5}{16}$ or approximately 0.313.

2 **Suppose Amy and Marla play 7 games. The probability that Amy wins a game is $\frac{1}{5}$, and that Marla wins is $\frac{4}{5}$. What is the probability that Amy will win at least 3 of the games?**

There are only two outcomes of each game: Amy wins (A) or Marla wins (M). The binomial expansion of $(A + M)^7$ follows.

$$(A + M)^7 = A^7 + 7A^6M + 21A^5M^2 + 35A^4M^3 + 35A^3M^4 + 21A^2M^5 + 7AM^6 + M^7$$

Amy must win 7, 6, 5, 4, or 3 games, so use these terms from the expansion.

$P(Amy\ wins\ at\ least\ 3\ games)$

$$= A^7 + 7A^6M + 21A^5M^2 + 35A^4M^3 + 35A^3M^4 \qquad \textit{Substitute } \tfrac{1}{5} \textit{ for A.}$$

$$= \left(\frac{1}{5}\right)^7 + 7 \cdot \left(\frac{1}{5}\right)^6\left(\frac{4}{5}\right) + 21\left(\frac{1}{5}\right)^5\left(\frac{4}{5}\right)^2 + 35\left(\frac{1}{5}\right)^4\left(\frac{4}{5}\right)^3 + 35\left(\frac{1}{5}\right)^3\left(\frac{4}{5}\right)^4 \qquad \textit{Substitute } \tfrac{4}{5} \textit{ for M.}$$

$$= \frac{1}{78125} + 7\left(\frac{1}{15625}\right)\left(\frac{4}{5}\right) + 21\left(\frac{1}{3125}\right)\left(\frac{16}{25}\right) + 35\left(\frac{1}{625}\right)\left(\frac{64}{125}\right) + 35\left(\frac{1}{125}\right)\left(\frac{256}{625}\right)$$

$$= \frac{1 + 28 + 336 + 2240 + 8960}{78,125}$$

$$= \frac{11,565}{78,125} \text{ or } \frac{2313}{15,625} \qquad \tfrac{2313}{15,625} \approx \textit{0.148 which is a little greater than } \tfrac{1}{7}.$$

The probability that Amy will win at least 3 games is $\frac{2313}{15,625}$ or approximately 0.148.

Exploratory Exercises

Tell whether each situation in Exercises 1–9 represents a binomial trial or not. Solve those that represent a binomial trial.

1. Ann tosses a coin 3 times. What is the probability of 2 heads and 1 tail?

Jess draws 4 cards from a deck of 52 playing cards. What is the probability of drawing 4 aces if the following occurs?

2. He replaces the card.

3. He does not replace the card.

There are 8 algebra books, 4 geometry books, and 6 trigonometry books on a shelf. If 2 are selected, with replacement after the first selection, what is the probability of the following?

4. both algebra

5. both geometry

6. both trigonometry

7. one algebra, one geometry

8. one algebra, one trigonometry

9. one geometry, one trigonometry

Written Exercises

A coin is tossed 4 times. What is the probability of the following?

10. no heads

11. 2 heads and 2 tails

12. 3 or more tails

A die is tossed 5 times. What is the probability of the following?

13. only one 4

14. at least three 4's

15. no more than two 4's

Cathy Black has a bent coin. The probability of heads is $\frac{2}{3}$ with this coin. She flips the coin 4 times. What is the probability of the following?

16. no heads

17. at least 3 heads

18. no more than 2 heads

Joey Diller guesses on all 10 questions on a true-false test. What is the probability of the following?

19. 7 correct

20. at least 6 correct

21. all incorrect

A batter is now batting 0.200 (meaning 200 hits in 1000 times at bat). In the next 5 at-bats, what is the probability of having the following?

22. exactly 3 hits

23. at least 4 hits

24. at least 2 hits

Three coins are tossed. What is the probability of the following?

25. 3 heads

26. 3 tails

27. at least 2 heads

28. exactly 2 tails

If a tack is dropped, the probability that it will land point up is 0.4. Ten tacks are dropped. What is the probability of the following?

29. all point up

30. exactly 3 point up

31. at least 6 point up

Harold is a skeet shooter. He hits the clay pigeon 9 of 10 times. If he shoots 12 times, what is the probability of the following?

32. all misses

33. exactly 7 hits

34. all hits

35. at least 10 hits

Coin Tossing Simulation

Probability is used to predict the outcome in games of chance involving tossing coins, rolling dice, selecting cards, or winning sweepstakes. A computer program can be used to do the actual counting in simulated situations where large samples are needed.

```
10    LET T2 = 0: LET T3 = 0
20    LET H2 = 0: LET H3 = 0
30    FOR I = 1 TO 100
40    LET T = 0: LET H = 0
50    FOR N = 1 TO 3
60    LET R = RND (1)
70    IF R < 0.5 THEN 110
80    PRINT "TAIL";" ";
90    LET T = T + 1
100   GOTO 100
110   PRINT "HEAD";" ";
120   LET H = H + 1
130   NEXT N
134   PRINT
140   IF H = 3 THEN 180
150   IF T = 3 THEN 200
160   IF H = 2 AND T = 1 THEN 220
170   IF H = 1 AND T = 2 THEN 240
180   LET H3 = H3 + 1
190   GOTO 250
200   LET T3 = T3 + 1
210   GOTO 250
220   LET H2 = H2 + 1
230   GOTO 250
240   LET T2 = T2 + 1
250   NEXT I
254   PRINT
260   PRINT "HHH: ";H3
270   PRINT "TTT: ";T3
280   PRINT "HHT: ";H2
290   PRINT "HTT: ";T2
300   END
```

The program at the left simulates tossing three coins, prints the outcomes, and keeps totals of the possible outcomes of the 100 samples. Each time the program is run, the totals may be different due to the random selection of values in line 60.

The random numbers selected by the computer lie between 0 and 1. The program defines heads as numbers less than 0.5. Tails are defined as numbers greater than or equal to 0.5.

The output shows each toss of the coin and lists the totals for the four possible combinations; 3 heads; 3 tails; 2 heads and 1 tail; or 1 head and 2 tails.

Enter and run the program several times. Notice the change in the totals.

In the exercises below it will take the computer several minutes to complete the large samples. Be prepared to wait for the totals.

A possible outcome is given below.

```
HHH: 11
TTT: 19
HHT: 38
HTT: 32
```

Exercises

1. Run the above program once. Write the actual percentage of HHH, TTT, HHT, and HTT.

2. Calculate the expected probabilities of HHH, TTT, HHT, and HTT by the methods in this chapter.

3. Change the program to increase the sample size to 1000. Delete lines 80 and 110. Then run the program and find the percentage of each outcome.

independent events   (503)
dependent events   (504)
linear permutation   (506)
permutation   (506)
circular permutation   (510)

reflection   (511)
combination   (513)
probability   (517)
success   (517)
failure   (517)

odds   (518)
mutually exclusive
   events   (525)
inclusive events   (525)
binomial trials   (529)

Chapter Summary

1. Two events are independent if the result of the first event has no effect on the second.   (503)

2. Basic Counting Principle: Suppose an event can occur in p different ways. Another event can occur in q different ways. There are $p \cdot q$ ways both events can occur.   (504)

3. Two events are dependent if the result of the first event has an effect on the second.   (504)

4. The number of permutations of n objects, taken r at a time, is defined as follows.

$$P(n, r) = \frac{n!}{(n - r)!}   \quad (506)$$

5. Permutations with Repetition: The number of permutations of n objects of which p are alike and q are alike is $\frac{n!}{p!q!}$ (507)

6. Circular Permutations: If n distinct objects are arranged in a circle, then there are $\frac{n!}{n}$ or $(n - 1)!$ permutations of the objects around the circle.   (510)

7. The number of combinations of n distinct objects taken r at a time is defined as follows.

$$C(n, r) = \frac{n!}{(n - r)!r!}   \quad (513)$$

8. Probability of Success and of Failure: If an event can succeed in s ways and fail in f ways, then the probabilities of success, $P(s)$, and of failure, $P(f)$, are as follows.

$$P(s) = \frac{s}{s + f}   \qquad P(f) = \frac{f}{s + f}   \quad (517)$$

9. The odds of the successful outcome of an event is expressed as the ratio of the number of ways it can succeed to the number of ways it can fail.

$$\text{Odds} = \text{the ratio of } s \text{ to } f \text{ or } \frac{s}{f}   \quad (518)$$

10. Probability of Two Independent Events: If two events, A and B, are independent, then the probability of both events occurring is found as follows.

$$P(A \text{ and } B) = P(A) \cdot P(B) \quad (521)$$

11. Probability of Two Dependent Events: If two events, A and B, are dependent, then the probability of both events occurring is found as follows.

$$P(A \text{ and } B) = P(A) \cdot P(B \text{ following } A) \quad (521)$$

12. Probability of Mutually Exclusive Events: The probability of one of two mutually exclusive events, A and B, occurring is the sum of their probabilities.

$$P(A \text{ or } B) = P(A) + P(B) \quad (525)$$

13. Probability of Inclusive Events: The probability of one of two inclusive events, A and B, is the sum of the individual probabilities decreased by the probability of both occurring.

$$P(A \text{ or } B) = P(A) + P(B) - P(A \text{ and } B) \quad (526)$$

14. Conditions of Binomial Trials: A binomial trial problem exists if and only if the following conditions hold.

1. There are only two possible outcomes.
2. The events are independent. (529)

Chapter Review

14–1 **Using only the digits 0, 1, 2, 3, and 4, how many 3-digit patterns can be formed under the following conditions?**

1. Repetitions are allowed.

2. No repetitions are allowed.

14–2 **On a shelf are 8 mystery and 7 romance novels. How many ways can they be arranged as follows?**

3. all mysteries together

4. all mysteries together, romances together.

5. Evaluate $\dfrac{P(8,\ 5)}{P(5,\ 3)}$.

6. Evaluate $\dfrac{P(7,\ 3)}{P(5,\ 2)}$.

Solve each problem.

14–3 **7.** How many ways can 8 people be seated at a round table?

8. How many ways can 10 charms be placed on a bracelet that has a clasp?

14–4 **9.** How many baseball teams can be formed from 15 players if only 3 pitch while the others play the remaining 8 positions?

10. From a deck of 52 cards, how many different 4-card hands exist?

14–5 **11.** A card is selected from a deck of 52 cards. What is the probability that it is a queen?

12. In a bag are 6 red and 2 white marbles. If two marbles are selected, what is the probability that one is red and the other is white?

14–6 **13.** In his pocket, Jose has 5 dimes, 7 nickels, and 4 pennies. He selects 4 coins. What is the probability that he has 2 dimes and 2 pennies?

14. Ben has 6 blue socks and 4 black socks in a drawer. One dark morning he pulls out 2 socks. What is the probability that he has 2 black socks?

14–7 **15.** From a deck of 52 cards, one card is selected. What is the probability that it is an ace or a face card?

16. If a letter is selected at random from the alphabet, what is the probability that it is a letter from the words CAT or SKATE?

14–8 **17.** Four coins are tossed. What is the probability that they show 3 heads and 1 tail?

18. A die is tossed 5 times. What is the probability of at least two 3's?

Chapter Test

Solve each problem

1. From 8 shirts, 6 pairs of slacks, and 4 jackets, how many different outfits can be made?

2. In a row are 8 chairs. How many ways can 5 people be seated?

3. How many ways can 11 books be arranged on a shelf?

4. How many ways can the letters from the word *television* be arranged?

5. How many ways can 6 keys be placed on a key ring?

6. How many different basketball teams could be formed from a group of 12 girls?

7. Nine points are placed on a circle. How many triangles can be formed using these points, three at a time, as vertices?

8. From a group of 4 men and 5 women, a committee of 3 is to be formed. What is the probability that it will have 2 men and 1 woman?

9. A red die and a green die are tossed. What is the probability that the red will show even and the green will show a number greater than four?

10. From a deck of cards, what is the probability of selecting a 4 followed by a 7 if no replacement occurs?

11. While shooting arrows, William Tell can hit an apple 9 out of 10 times. What is the probability that he will hit it exactly 4 out of the next 7 times?

12. Five bent coins are tossed. The probability of heads is $\frac{2}{3}$ for each of them. What is the probability that no more than 2 will show heads?

Find each value.

13. $P(8, 3)$

14. $P(6, 4)$

15. $C(8, 3)$

16. $C(6, 4)$

CHAPTER 15

Statistics

Insurance companies use statistics extensively in calculating rates. For example, an automobile insurance company compiles data concerning number of accidents, amount of damages, condition of roads, age of drivers, and so on.

15-1 Data, Tables, and Graphs

Statistics provide techniques for collecting, organizing, analyzing, and interpreting numerical information called **data**. Organized data is easier to read and interpret. One way to organize data is by using tables. The following table shows the normal monthly precipitation for selected cities in the United States.

Normal Monthly Precipitation in Inches

City	Jan.	Feb.	Mar.	Apr.	May	Jun.	July	Aug.	Sep.	Oct.	Nov.	Dec.	Total
Albuquerque, NM	0.30	0.39	0.47	0.48	0.53	0.50	1.39	1.34	0.77	0.79	0.29	0.52	7.77
Boston, MA	3.69	3.54	4.01	3.49	3.47	3.19	2.74	3.46	3.16	3.02	4.51	4.24	42.52
Chicago, IL	1.85	1.59	2.73	3.75	3.41	3.95	4.09	3.14	3.00	2.62	2.20	2.11	34.44
Houston, TX	3.57	3.54	2.68	3.54	5.10	4.52	4.12	4.35	4.65	4.05	4.03	4.04	48.19
Mobile, AL	4.71	4.76	7.07	5.59	4.52	6.09	8.86	6.93	6.59	2.55	3.39	5.92	66.98
San Francisco, CA	4.37	3.04	2.54	1.59	0.41	0.13	0.01	0.03	0.16	0.98	2.29	3.98	19.53

In this table, the data is organized in such a manner that you can quickly answer questions like the following.

What city has the most precipitation in November?   *Boston*

What is the driest month in Houston?   *March*

What is the greatest monthly precipitation for any of the cities?   *8.86 inches (Mobile)*

Which city has the most precipitation in one year?   *Mobile*

How many of the cities have more than 30 inches of precipitation in one year?   *4 cities*

The data provided in a table or chart may be the result of a survey. For example, a pollster surveyed a **sample** of the city's population of 550,000 at various months to obtain the data in the chart below.

Would a sample of 50 people have been large enough to represent the city's population? Would a local Republican headquarters have been a good place to conduct such a survey?

A good sample is taken randomly and is representative of the larger group. It must be large enough to provide results that reflect the larger group.

Preference for Mayor			
Month			
April	June	Aug.	Oct.
Candidate			
Alvarez 223	286	294	347
Lewis 167	202	387	399
Rinehart 287	268	168	157
Others 209	152	64	31
Undecided 114	92	87	66

Graphs are often used to present data and show relationships. The following types of graphs are frequently used in newspapers and magazines because of their descriptive appearance. These graphs can present the same data in different ways.

Average Motor Fuel Consumption in U.S.
gallons per automobile

Year	1960	1965	1970	1975	1980	1985
Consumption	661	667	738	712	603	549

This graph is called a **bar graph**. It shows how specific quantities compare.

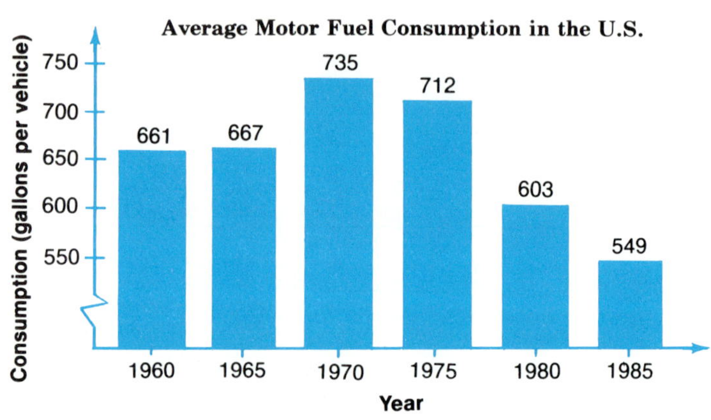

This graph is called a **line graph**. It is helpful for showing trends or changes.

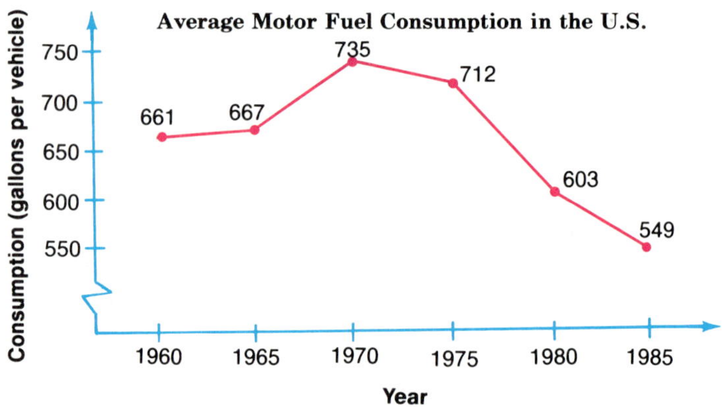

Example

1 **Draw a bar graph and a line graph to present the following data.**

Home Runs Hit by Home Run Champion in Each League

Year	1960	1965	1970	1975	1980	1985
National League	41	52	45	38	48	37
American League	40	32	44	36	41	40

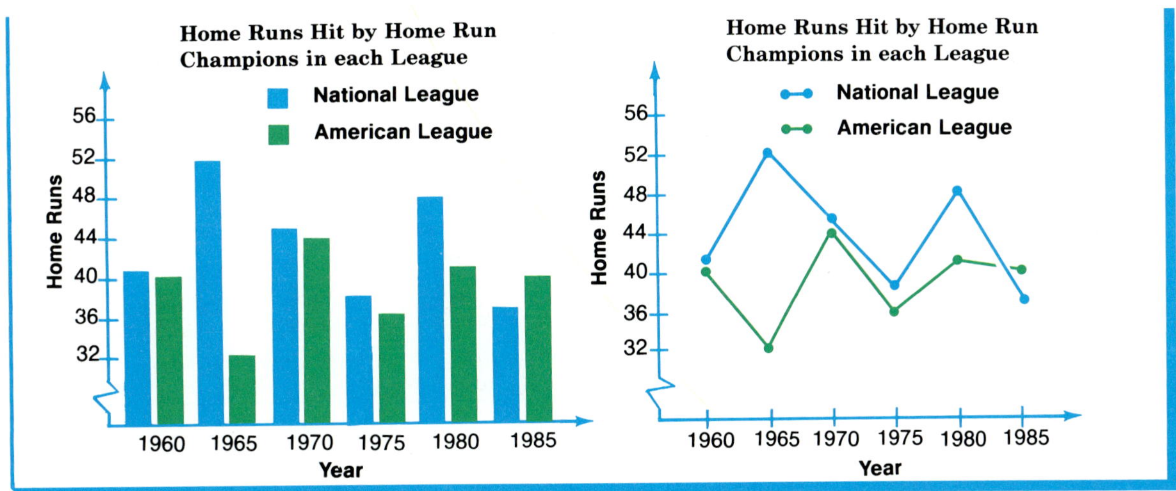

Circle graphs show how parts are related to the whole. For example, the circle graph at the right shows how the cost of a $4 paperback book is broken down.

The circle is separated into proportional parts. For example, 25% of the book's cost goes to royalties and profit. Thus, 25% of the circle, 90°, is used to show this part of the book's cost.

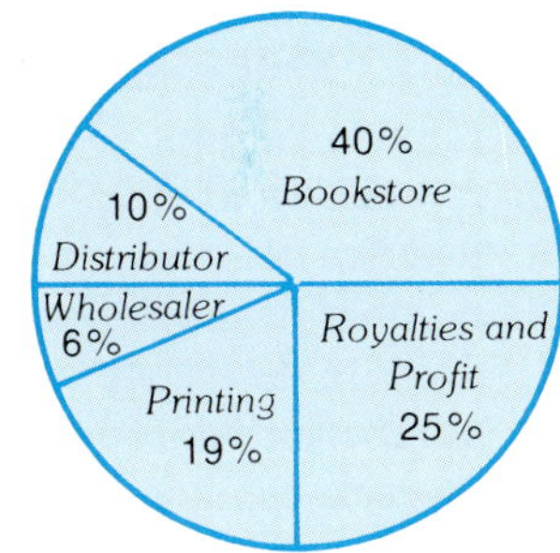

Example

2 **Draw a circle graph to show how the cost of a $9.98 record is broken down. The manufacturer's cost is $3.09. The manufacturer's income is $1.10. The distributor's income is $0.70. The retailer's income is $5.09.**

First, find the percent of a circle represented by each cost. Then, figure the number of degrees represented by each cost and draw the graph.

Cost	Percent of Circle	Approximate Number of Degrees
Manufacturer's cost	$\frac{3.09}{9.98}$ or 31%	360 × 31% or 111.6
Manufacturer's income	$\frac{1.10}{9.98}$ or 11%	360 × 11% or 39.6
Distributor's income	$\frac{0.70}{9.98}$ or 7%	360 × 7% or 25.2
Retailer's income	$\frac{5.09}{9.98}$ or 51%	360 × 51% or 183.6

Exploratory Exercises

Use the table to solve each problem.

Cars in Use, by Age, 1986

Age of Autos	Number (millions)	Percent
Under 3 years	27.9	23.8
3–5 years	22.7	19.4
6–8 years	26.5	22.6
9–11 years	18.7	15.9
12 years and over	21.5	18.3

1. What percent of cars in use were between 9 and 11 years old in 1986?
2. How many cars in use were 12 years old or older in 1986?
3. How many cars in use were between 3 and 11 years old in 1986?
4. What percent of cars in use were over 8 years old in 1986?
5. How many cars in use were less than 6 years old in 1986?
6. Draw a bar graph to show the number of cars in use, by age, in 1986.

State whether the given location would be a good place to take the indicated poll.

	Poll	**Location**
7.	number of pets	apartment building
8.	favorite sandwich	shopping center
9.	favorite color of car	state fair
10.	number of movies seen	movie theater
11.	favorite detergent	laundromat
12.	number of bedrooms	suburban development
13.	favorite candidate	retirement community

14. Predict the favorite soft drink in your school by surveying a sample of the students.

Written Exercises

The total annual precipitation in inches for 100 cities was recorded in 1988.

31	26	35	20	38	30	41	21	23	25	24	27	30	19	27
38	30	31	33	20	22	30	33	27	25	33	25	27	31	27
17	38	46	33	22	27	22	19	25	33	36	30	45	31	45
35	23	25	40	36	20	30	22	26	41	35	25	30	30	27
33	25	28	27	24	45	26	21	41	26	22	31	37	38	26
20	22	26	25	20	27	25	23	27	31	35	27	25	40	24
41	30	17	22	26	19	33	36	30	28					

15. Organize the data into a table with the headings **Precipitation** and **Number of Cities.** Under **Precipitation**, group the data by threes beginning with 17-19.
16. How many cities had from 32-34 inches of precipitation?
17. How many cities had from 23-25 inches of precipitation?
18. How many cities had from 23-31 inches of precipitation?
19. How many cities had from 29-37 inches of precipitation?
20. How many cities had at least 37 inches of precipitation?
21. How many cities had at most 34 inches of precipitation?
22. Draw a bar graph to present the data from the table made for Exercise 15.
23. Draw a line graph to present the data from the table made for Exercise 15.

Draw bar graphs and line graphs for the data in each table.

24. High School Graduates (in thousands)

Year	1960	1965	1970	1975	1980	1985
Number	1864	2665	2896	3140	3058	2683

25. Indianapolis 500-Winner's Speed (in miles per hour)

Year	1960	1964	1968	1972	1976	1980	1984	1988
Speed	138.7	147.4	152.9	163.5	148.7	142.9	163.6	144.8

26. U.S. Imports and Exports (in millions of dollars), 1986

Region	Africa	Asia	Oceania	Europe	N. America	S. America
Imports	10,348	153,869	3,717	91,826	91,724	18,477
Exports	5,978	64,822	6,659	63,631	64,461	11,050

27. U.S. Population Distribution, by Age

Year	Under 5	5–19	20–44	45–64	65 and over
1900	12.1%	32.3%	37.7%	13.7%	4.2%
1950	10.7%	23.2%	37.6%	20.3%	8.2%
1985	7.5%	22.3%	39.5%	18.8%	11.9%

28. U.S. Energy Consumption

Year	Coal	Refined Petroleum	Natural Gas	Other
1970	19%	44%	33%	4%
1975	18%	46%	28%	8%
1980	20%	45%	27%	8%
1985	24%	43%	22%	11%

29. Draw a circle graph to show the population distribution by age in 1985. Use the data from the table in Exercise 27.

30. Draw a circle graph to show the U.S. energy consumption in 1985. Use the data from the table in Exercise 28.

The data in the table below gives a breakdown of the civilian labor force in the United States for various years, in millions of persons.

Employment Status	1965	1975	1985
Employed in nonagricultural industries	66.7	82.4	104.0
Employed in agriculture	4.4	3.4	3.2
Unemployed	3.4	7.9	8.3
Total civilian labor force	74.5	93.7	115.5

31–33. For each year, draw a circle graph to show the breakdown of the civilian labor force.

Sometimes graphs contain information on two or more related topics and are given to show the comparisons between them. These **comparative graphs** can be quite helpful for showing trends in certain areas. Analysis of these comparative data requires that you read the graphs carefully.

Exercises

Use the graphs at the right to answer each question.

1. Did the percentage of people under 18 increase or decrease from 1970 to 1986?

2. Which age group had the least number of people in both years?

3. If the population in 1970 was 203 million and in 1986 it was 241 million, how many more people were in the 18–64 age group in 1986 than in 1970?

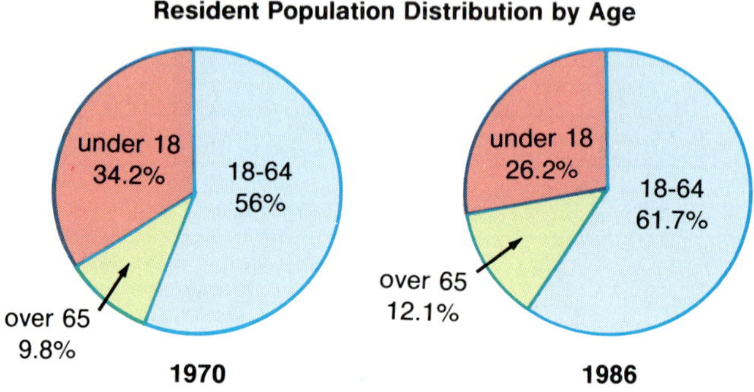

Use the graphs below to answer each question.

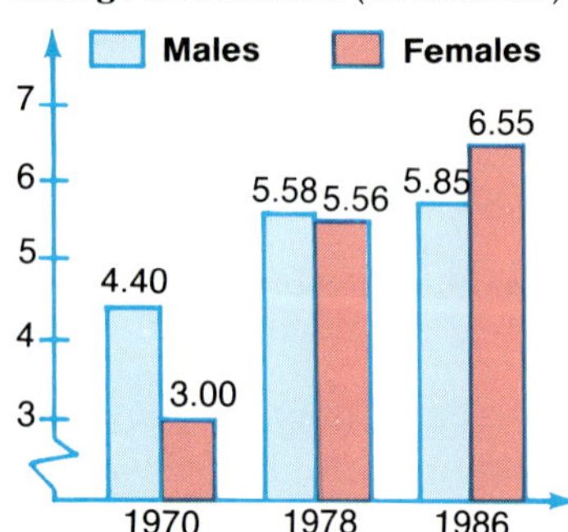

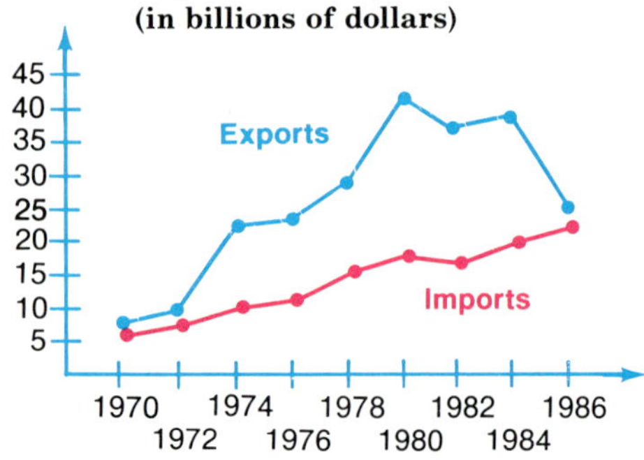

4. How many more men were enrolled in college than women in 1970?

5. Find the percent of increase in female enrollment from 1970 to 1986.

6. Find the percent of increase in male enrollment from 1978 to 1986.

7. How many more women were enrolled in college than men in 1986?

8. What were the values of U.S. agricultural exports and imports in 1986?

9. In what year was the difference in value between exports and imports greatest?

10. Between what two years was the increase in value of exports greatest?

11. Between what two years did the value of agricultural imports decrease?

12. In general, has the value of U.S. agricultural exports increased or decreased between 1970 and 1986? Has the value of agricultural imports increased or decreased?

15-2 Line Plots and Stem and Leaf Plots

Numerical data is often organized and displayed using either **line plots** or **stem and leaf plots**. In a line plot, data is recorded and displayed using a number line.

Example

1 **The pass receiving leaders in professional football from 1960 to 1987 are listed in the following table. Make a line plot of this data.**

Year	Player	Number of Catches	Year	Player	Number of Catches
1960	Raymond Berry	74	1974	Lydell Mitchell	72
1961	Lionel Taylor	100	1975	Chuck Foreman	73
1962	Bobby Mitchell	72	1976	MacArthur Lane	66
1963	Lionel Taylor	78	1977	Lydell Mitchell	71
1964	Charley Hennigan	101	1978	Rickey Young	88
1965	David Parks	80	1979	Joe Washington	82
1966	Lance Alworth	73	1980	Kellen Winslow	89
1967	George Sauer	75	1981	Kellen Winslow	88
1968	Clifton McNeil	71	1982	Dwight Clark	60
1969	Dan Abramowicz	73	1983	Todd Christensen	92
1970	Dick Gordon	71	1984	Art Monk	106
1971	Fred Biletnikoff	61	1985	Roger Craig	92
1972	Harold Jackson	62	1986	Todd Christensen	95
1973	Harold Carmichael	67	1987	J. T. Smith	91

The data range from 60 to 106. The number line must be drawn large enough to contain all of these values. An "x" is used to indicate the number of catches for each year.

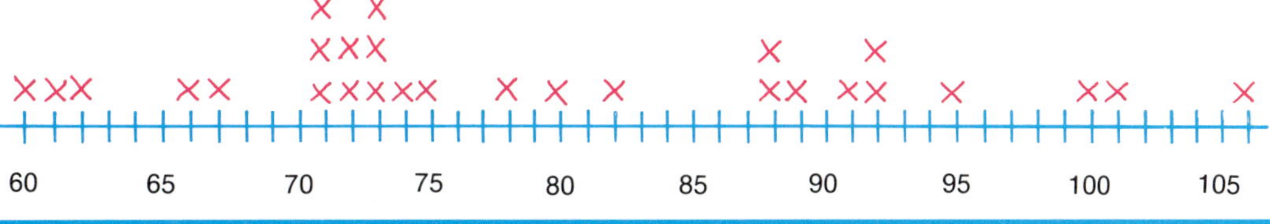

In a stem and leaf plot, each piece of data is separated into two numbers that are used to form the stem and leaf for the piece of data. The data is organized into two columns. The column on the left is the stem. It usually consists of the digits in the greatest common place value of all the data. For example, if the greatest common place value is hundreds, then the stem of 980 is 9 and the stem of 1130 is 11. The column on the right is the leaf. It usually consists of the other part of each piece of data.

Leaves are always one-digit numbers.

The leaf of 980 is 8.
The leaf of 1130 is 3.

2 **Make a stem and leaf plot of the number of catches made by the pass receiving leaders from 1960 to 1987. Use the data from the table in Example 1.**

The data ranges from 60 to 106 and the greatest common place value of all the data is tens.
Thus, the stems are the numbers 6 to 10.
To plot the number 73, use 7 as the stem and 3 as the leaf.

Two stem and leaf plots can be done for this data. In the plot at the left, the leaves are in the order of the given data. In the plot at the right, the leaves are arranged in numerical order.

Stem	Leaf
6	1 2 7 6 0
7	4 2 8 3 5 1 3 1 2 3 1
8	0 8 2 9 8
9	2 2 5 1
10	0 1 6

9 | 2 represents 92 catches.

Stem	Leaf
6	0 1 2 6 7
7	1 1 1 2 2 3 3 3 4 5 8
8	0 2 8 8 9
9	1 2 2 5
10	0 1 6

3 **The costs per cup of twenty-nine liquid laundry detergents are 28¢, 17¢, 16¢, 18¢, 19¢, 21¢, 26¢, 15¢, 19¢, 19¢, 16¢, 14¢, 21¢, 12¢, 26¢, 17¢, 17¢, 28¢, 13¢, 18¢, 22¢, 14¢, 10¢, 19¢, 12¢, 9¢, 15¢, 12¢, and 10¢. Make a stem and leaf plot of the cost per cup.**

Since the data range from 9 to 28, the stems can be 0, 1, and 2. However, the stems 1 and 2 will each have a great number of leaves. In order to better evaluate the data from the stem and leaf plot, each stem can be broken into two parts to represent the data.

Let 2 | represent the costs 20¢ to 24¢, and
let · | represent the costs 25¢ to 29¢.

Stem	Leaf
0	
·	9
1	0 0 2 2 2 3 4 4
·	5 5 6 6 7 7 7 8 8 9 9 9 9
2	1 1 2
·	6 6 8 8

The leaves have been ordered from least to greatest.

2 | 1 represents 21¢.

For some data, the numbers will have more than two digits. Each piece of data must then be rounded or truncated before plotting so that the numbers used for leaves will have exactly one digit. For example, if the greatest common place value is thousands, then the leaf for 1489 is formed using 489. If 1489 is rounded to 1500, the stem is 1 and the leaf is 5. If 1489 is truncated to 1400, the stem is 1 and the leaf is 4.

Remember that leaves are always one-digit numbers.

Back-to-back stem and leaf plots are sometimes used to compare two
sets of data.

Example

4 **The populations of selected capital cities in the U.S. are listed in the following table. Make a back-to-back stem and leaf plot of the rounded and truncated values of the data.**

Capital City	Pop. (thousands)	Capital City	Pop. (thousands)
Albany, NY	841	Honolulu, HI	814
Austin, TX	695	Jackson, MS	385
Baton Rouge, LA	544	Little Rock, AR	499
Boise, ID	192	Madison, WI	342
Charleston, WV	269	Montgomery, AL	289
Des Moines, IA	372	Oklahoma City, OK	976
Harrisburg, PA	573	Salem, OR	259
Hartford, CT	732	Tallahassee, FL	214

Rounded Leaf	Stem	Truncated Leaf
9	1	9
9 7 6 1	2	1 5 6 8
9 7 4	3	4 7 8
	4	9
7 4 0	5	4 7
	6	9
3 0	7	3
4 1	8	1 4
8	9	7

*Using rounded data,
4 | 3 | represents
335,000 to
344,000 people.*

*Using truncated data,
| 3 | 4 represents
340,000 to
349,000 people.*

Exploratory Exercises

State the stems that would be used to plot each set of data.

1. 61, 47, 82, 27, 31, 64, 52, 49, 57, 63   **2.** 37, 41, 31, 38, 45, 42, 40, 39, 38, 46

3. 865, 421, 582, 693, 529, 267, 841, 629, 441, 763, 1036

4. 129, 137, 124, 147, 126, 148, 136, 142, 137, 151, 131, 151, 129, 149

Use the line plot at the right to answer each question.

5. What was the highest score on the test?

6. What was the lowest score on the test?

7. How many students took the test?

8. How many students scored in the 40's?

9. What test score was received by the
most students?

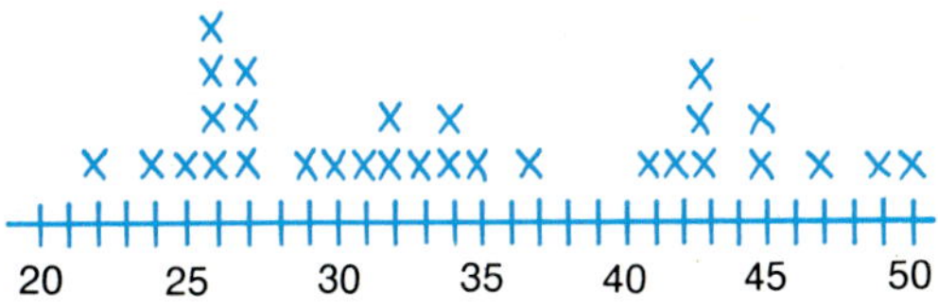

Use the stem and leaf plot at the right to answer each question.

10. What was the highest temperature recorded?
11. What was the lowest temperature recorded?
12. How many temperatures were recorded?
13. On how many days was the high temperature in the 70's?
14. Which temperature occurred most frequently?

Daily High Temperature in April

Stem	Leaf
8	0 1 2 2 5
7	0 3 3 4 5 7 7 7 8 9 9
6	1 2 4 5 5 6 7 8 8 9
5	6 7 8 8

Written Exercises

Each number below represents the age of a U.S. president on his first inauguration.

57	61	57	57	58	57	61	54	68	51	49	64	50	48
65	52	56	46	54	49	50	47	55	55	54	42	51	56
55	51	54	51	60	62	43	55	56	61	52	69	64	

15. Make a line plot of the ages of U.S. presidents on their first inauguration.
16. What was the age of the youngest president on his first inauguration?
17. What was the age of the oldest president on his first inauguration?
18. What was the difference in age between the youngest and oldest presidents?
19. How many ages are given?
20. Which age(s) occurred most frequently?
21. How many presidents were in their 60's on their first inauguration?

The following table gives the cost, number of calories, and milligrams of sodium per 5-tablespoon serving of certain pancake syrups.

Pancake Syrup	Cost	Calories	mg of Sodium
Camp Pure Maple	$0.90	277	37
Reese Pure Maple	$1.24	269	29
Vermont Maple Orchards Pure Maple	$0.92	271	21
Vermont Maid	$0.24	273	34
Golden Griddle	$0.26	278	47
Kroger	$0.24	276	52
Log Cabin (2% Maple Syrup)	$0.26	270	99
Log Cabin Country Kitchen	$0.19	280	63
Staley	$0.26	280	134
Cost Cutter (Kroger)	$0.13	281	58
Mrs. Butterworth's (Thick n' Rich)	$0.25	280	91
Kroger Buttered Syrup	$0.24	277	84
Karo Syrup	$0.19	319	123
Log Cabin Buttered Syrup	$0.27	284	195

22. Make a stem and leaf plot of the cost per serving of pancake syrup.
23. Which syrup is the most expensive?
24. Which syrup is the least expensive?
25. What is the difference in cost between the most and the least expensive syrups?
26. Which cost(s) occurred most frequently?
27. How many syrups cost from $0.20 to $0.29 per serving?

28. Make a stem and leaf plot of the number of calories per serving of pancake syrup.

29. Which syrup contains the least number of calories per serving?

30. Which syrup contains the most number of calories per serving?

31. What is the difference in the number of calories between the syrups with the most and the least number of calories?

32. Which number(s) of calories occurred most frequently?

33. How many syrups contain 270 to 279 calories per serving?

34. Make a stem and leaf plot of the milligrams of sodium per serving of pancake syrup.

35. Which syrup contains the least amount of sodium per serving?

36. Which syrup contains the most amount of sodium per serving?

37. What is the difference in the amount of sodium between the syrups with the most and the least amount of sodium?

38. Which amount of sodium occurred most frequently?

39. How many syrups contain less than 80 milligrams of sodium per serving?

mini-review

Solve each equation.

1. $\log_8 x = \log_8 (x^2 - 2)$

2. $\log_3 (x + 4) + \log_3 (x - 4) = 2$

3. Find a pattern and complete the sequence 1, 3, 8, 16, 27, ____, ____, ____.

4. Write $\log_{12} 144 = 2$ in exponential form.

5. What are the odds of rolling a pair of dice and getting a sum of 5?

6. Find the 37th term of the arithmetic sequence if $a_1 = -3$ and $d = 7$.

7. How many ways can the letters of the word STATISTICS be arranged?

8. Find the missing geometric means of 5, ____, ____, -1080.

9. Bacteria of a certain type can grow from 1200 to 2550 in 9 days. Find k for the growth formula.

10. Find the first five terms of this sequence: $a_1 = -2$, $a_2 = 1$, $a_{n+2} = 3 \cdot a_{n+1} - a_n$

11. From a deck of 52 cards, what is the probability of selecting a 9, a king, then another 9, then another king, in that order, if no replacement occurs?

Find each value.

12. $6!3!$

13. $P(7, 4)$

14. $C(7, 4)$

15. How many 3-digit patterns can be formed if all the digits are different?

16. Find the sum of the series $13.5 - 9 + 6 - 4 + \ldots$.

17. Find the fifth term of $(x - 3z)^6$.

18. A red die and a green die are tossed. What is the probability that the red die shows an odd number or the green die shows a number less than 3?

19. Jerry can make 8 out of every 10 free throws. What is the probability that he will make exactly 4 of his next 6 free throws?

20. From a group of 7 men and 11 women, a coed softball team of 10 is to be formed. How many teams can be formed if the team must have the same number of men and women?

15-3 Central Tendency: Median, Mode, and Mean

During a cold spell lasting 43 days, the following high temperatures (in degrees Fahrenheit) were recorded in Chicago. What temperature is most representative of the high temperatures for that period?

26	17	12	5	4	25	17	23	13
6	25	19	27	22	26	20	31	24
12	27	16	27	16	30	7	31	16
5	29	18	16	22	29	8	31	
13	24	5	−7	20	29	18	12	

The most representative temperature, the *average* temperature, is neither the greatest nor the least temperature. It is a value somewhere in the middle of the group.

The most commonly used averages are the **median**, **mode**, and **mean**.

> **The median of a set of data is the middle value. If there are two middle values, it is the value halfway between.**
>
> **The mode of a set of data is the most frequent value. Some sets of data have multiple modes.**
>
> **The mean of a set of data is the sum of all the values divided by the number of values.**

Definition of Median, Mode, and Mean

To find the median of the Chicago temperatures, arrange the values in descending order, as shown in the margin. Then, find the middle value. In this case, the median temperature is 19.

To find the mode, determine how many times each particular high temperature occurred. Then find the most frequently occurring value. In this case, the mode is 16.

To find the mean, add all the values. Then divide by 43, the number of values. In this case, the mean temperature to the nearest tenth is 18.5. *This example shows that median, mode, and mean are not always the same value.*

temperatures		
31	24	16
31	24	16
31	23	13
30	22	13
29	22	12
29	20	12
29	20	12
27	19	8
27	18	7
27	18	6
26	17	5
26	17	5
25	16	5
25	16	4
		−7

The value of every item in a set of data affects the value of the mean. Thus, when extreme values are included, the mean may become less representative of the set. The values of the median and the mode are *not* affected by extreme values.

1 **Find the mean of {1, 2, 4, 93} and {24, 25, 25, 26}.**

$$\{1, 2, 4, 93\}$$

$$\text{mean} = \frac{1 + 2 + 4 + 93}{4}$$

$$= \frac{100}{4} \text{ or } 25$$

$$\{24, 25, 25, 26\}$$

$$\text{mean} = \frac{24 + 25 + 25 + 26}{4}$$

$$= \frac{100}{4} \text{ or } 25$$

The mean is not close to any one of the four values in this set. In this case, it *is not* a particularly representative value.

There are *no* extreme values in this set. In this case, the mean is a representative value.

The mean for both sets is 25.

2 **Find the median, mode, and mean of the hourly wages of 80 workers. Five workers make \$4.60 per hour, fifteen make \$4.40 per hour, thirty make \$5.70 per hour, ten make \$6.60 per hour, and twenty make \$4.50 per hour.**

Arrange the wages in descending order.

Since this set of data has 80 values, there are two middle values, the 40th value, \$5.70, and the 41st value, \$4.60. The median is halfway between these values. Thus, the median is $\frac{\$5.70 + \$4.60}{2}$ or \$5.15.

\$6.60	10 workers
\$5.70	30 workers
\$4.60	5 workers
\$4.50	20 workers
\$4.40	15 workers

More workers make \$5.70 per hour than any other wage. So it is the most frequently occurring value. The mode is \$5.70.

$$\text{mean} = \frac{10(6.60) + 30(5.70) + 5(4.60) + 20(4.50) + 15(4.40)}{80}$$

$$= \frac{66.00 + 171.00 + 23.00 + 90.00 + 66.00}{80}$$

$$= \frac{416.00}{80} \text{ or } 5.20$$

The median is \$5.15, the mode is \$5.70, and the mean is \$5.20.

3 **The stem and leaf plot below represents the high temperature for selected cities on March 29. Find the median, mode, and mean for the temperatures.**

There are 23 leaves in this plot.

The median is the 12th temperature, 79°.

Both 79 and 80 occur most frequently. Thus, there are two modes, 79° and 80°.

The sum of all the values is 1812.

Thus the mean is $\frac{1812°}{23}$ or about 78.8°

Stem	Leaf
6	6 7 8 9
7	0 3 4 6 8 8 9 9 9
8	0 0 0 2 4 4 5
9	0 3 8

9 | 3 represents 93°

Exploratory Exercises

Find the median, mode, and mean for each set of data.

1. {1, 2, 3, 4, 5}
2. {2, 4, 6, 8, 10}
3. {7, 7, 7, 7, 7, 7, 7}
4. {1, 1, 2, 4, 1}
5. {8, 43, 2, 56, 44}
6. {399, 299, 399, 239, 318}

Written Exercises

Find the median, mode, and mean for each set of data.

7. {399, 219, 216, 177, 179, 399, 189}
8. {7.1, 5.0, 2.7, 9.1, 8.1, 6.3, 8.5}
9. {2.1, 4.8, 2.1, 5.7, 2.1, 4.8, 2.1}
10. {1, 7, 7, 0, 2, 0, 4, 1, 3, 7, 7, 5, 4, 1, 8}
11. {11, 10, 13, 12, 12, 13, 15}
12. {50, 75, 65, 70, 55, 65, 50, 80}

13. A die was tossed 25 times with the results shown at the right. Find the median, mode, and mean for the tosses.

5	3	1	6	5	2	1	5	4
6	5	6	3	6	4	4	4	1
1	6	6	4	1	2	2		

14. Two dice were tossed 64 times with the following sums. Find the median, mode, and mean for the tosses.

8	11	10	8	8	7	10	3	9	10	8	2	9
5	2	3	5	7	11	7	11	10	11	10	6	7
5	6	4	4	5	10	8	6	7	4	8	5	10
5	11	9	12	4	7	2	7	4	3	9	2	11
5	11	9	8	12	3	7	8	5	5	7	6	

15. The heights in feet of 21 of the highest mountains in the world are given below. Find the median, mode, and mean for the heights.

20,270	29,002	14,255	18,700	28,146	22,835	21,201
18,481	13,653	14,431	28,250	13,202	14,408	14,110
15,781	25,263	19,344	19,565	19,887	14,701	14,495

16. Find the median, mode, and mean of the hourly wages of 200 workers. One hundred workers make $4.00 per hour, ten make $5.50 per hour, ten make $6.75 per hour, twenty make $3.80 per hour, and sixty make $5.25 per hour.

17. Find the median, mode, and mean of the hourly wages of 500 workers. Two hundred workers make $3.75 per hour, two hundred make $4.25 per hour, sixty make $6.75 per hour, and forty make $10.50 per hour.

18. The stem and leaf plot below represents the points scored by the winning teams in the first 22 Super Bowls. Find the median, mode, and mean for the scores.

Stem	Leaf	2	1 represents 21 points.
1	4 6 6 6		
2	1 3 4 4 6 7 7 7		
3	1 2 3 5 5 8 8 9		
4	2 6		

19. The stem and leaf plot below represents the points scored by the losing teams in the first 22 Super Bowls. Find the median, mode, and mean for the scores.

Stem	Leaf	2	1 represents 21 points.
0	3 6 7 7 7 7 9		
1	0 0 0 0 3 4 4 6 7 7 9		
2	0 1		
3	1		

 Variation: Range, Interquartile Range, and Outliers

If 10,000 family incomes in a city were all the same, you would know all there is to know about the incomes. However, values in a set of data usually vary. The **variation** is called *dispersion*.

There are several kinds of measures of variation. The simplest measure is called the **range**.

> **The range of a set of data is the difference between the greatest and least values in the set.**

Definition of Range

Example

1 **The heights of some young trees in a reforestation plot are 58 cm, 56 cm, 51 cm, 54 cm, 49 cm, 61 cm, 54 cm, and 49 cm. Find the range.**

The greatest value is 61 centimeters.
The least value is 49 centimeters.

range = 61 − 49   *The range is the difference between the greatest and least values.*
      = 12

The range is 12 centimeters.

Because the range is the difference between the greatest and least values in a set of data, it is affected by unusually extreme values. In such cases, it is not a good measure of variation.

Another commonly used measure of variation is called the **interquartile range**. In a set of data, **quartiles** are the values that separate the data into four equal parts. The median of a set of data separates the data into two equal parts. These two sets of data each have a median that separates each set into two equal parts. These two medians are called the *upper quartile* and the *lower quartile* for the entire set of data. One-fourth of the data is less than the lower quartile and three-fourths of the data is less than the upper quartile.

The median, upper quartile, and lower quartile are the quartiles of a set of data.

The difference between the upper quartile and the lower quartile is the interquartile range. It represents the middle 50%, or half, of the data.

> **The interquartile range of a set of data is the difference between the upper quartile and lower quartile of the set.**

Definition of Interquartile Range

2 The points scored by the winning team in each National Basketball Association game over a two-day period were 114, 106, 108, 131, 109, 118, 132, 123, 107, 114, 99, 131, 115, 108, and 102. Find the median, lower quartile, upper quartile, and interquartile range for the scores.

Arrange the 15 scores in ascending order. The median is the 8th score.

99   102   106   107   108   109   109   114   114   115   118   123   131   131   132
median

There are 7 scores above and below the median. The median for each half of the data is the 4th score for that set of data.

99   102   106   107   108   109   109   114   114   115   118   123   131   131   132
lower quartile   median   upper quartile

The median is 114 points, the lower quartile is 107 points, and the upper quartile is 123 points. Thus, the interquartile range is $123 - 107$ or 16 points. The middle 50% of the points scored by the winning NBA teams vary by 16 points.

3 The stem and leaf plot below represents the winning times at the Kentucky Derby from 1965 to 1988. Find the interquartile range for the times.

There are 24 leaves in this plot.

The median is halfway between the 12th and 13th values. Thus, the lower quartile is between the 6th and 7th values and the upper quartile is between the 18th and 19th values.

Stem	Leaf
119	4
120	2 6
121	2 2 6|8 8
122	0 0 0 0|2 2 2 4 4 4|4 8
123	2 4 4
124	0

121 | 2 represents 121.2 seconds.

$$\text{lower quartile} = \frac{121.6 + 121.8}{2}$$

$$= 121.7$$

$$\text{upper quartile} = 122.4$$

The interquartile range is $122.4 - 121.7$ or 0.7 seconds. Thus, the middle 50% of the winning times at the Kentucky Derby vary by 0.7 seconds.

In Example 3, the interquartile range is quite small. This means that the middle 50% of the data is closely clustered around the median. Since most of the times are close to the median time of 122.1 seconds, times such as 119.4 seconds and 124.0 seconds seem inconsistent with the other times. When such extreme values exist in a set of data, the values are called **outliers**.

> An outlier is any value in a set of data that is at least 1.5 interquartile ranges beyond the upper or lower quartiles.

Definition of Outlier

4 Find any outliers in the winning times at the Kentucky Derby from 1965 to 1988. Use the data from Example 3.

The interquartile range is 0.7 seconds. Find the times that are 1.5 interquartile ranges beyond the upper and lower quartiles. The upper quartile is 122.4 seconds and the lower quartile is 121.7 seconds.

$$122.4 + 1.5(0.7) = 123.45 \qquad\qquad 121.7 - 1.5(0.7) = 120.65$$

Any time above 123.45 seconds or below 120.65 seconds is an outlier. Therefore, 119.4, 120.2, 120.6, and 124.0 are all outliers.

5 The stem and leaf plot below represents the scores on a 100-point test in Mrs. Garrick's geometry class. Find the interquartile range for the scores. Then determine if any score is an outlier.

There are 27 leaves in this plot. The median is the 14th score. Thus, the lower quartile is the 7th score, 60, and the upper quartile is the 21st score, 80.

The interquartile range is $80 - 60$ or 20 points.

$$80 + 1.5(20) = 110 \qquad 60 - 1.5(20) = 30$$

Any score above 110 or below 30 is an outlier. Therefore, none of the scores is an outlier. *No student can score above 100.*

Stem	Leaf
4	1 5 9
5	2 7 8
6	0 1 2 2 6 8 9
7	1 1 3 4 7 7 8
8	0 2 3 3 8
9	6
10	0

8 | 3 represents a test score of 83.

Exploratory Exercises

Find the range for each set of data.

1. {6, 12, 10, 4, 9}

2. {250, 275, 325, 300, 200, 225, 175}

3. {39, 47, 51, 38, 45, 29, 37, 40, 36, 48}

4. {9, 12, 31, 19, 22, 3, 15, 13, 19, 7, 82, 8}

5.

Stem	Leaf
4	1 3 9
5	2 3 6 9
6	4 4 5
7	2 4 7

5 | 2 represents 52.

6.

Stem	Leaf
3	0 0 1 2 4
·	5 6 6 6 8 9
4	1 1 3 4 4
·	5 5 6

3 | 4 represents 34.
· | 5 represents 35.

7–12. Find the quartiles for each set of data in Exercises 1–6.

Written Exercises

Find the range, quartiles, and interquartile range for each set of data.

13. {1, 7, 7, 11, 2, 20, 4, 1, 3, 7, 7, 5, 4, 1, 8}

14. {23, 47, 49, 57, 51, 82, 49, 47, 54, 58}

15. {1050, 1175, 835, 1075, 1025, 1145, 1100, 1125, 975, 1005, 1125, 1095, 1075, 1055}

16. {50, 92, 79, 61, 76, 83, 65, 98, 82, 64, 76, 63, 57, 96, 75, 53, 66, 88, 59, 85, 95, 65, 81, 71}

17. Stem | **Leaf**
0	1 1 8 8 9 *2	4 represents 240.*
1	4 5 5 7 7 7 9
2	1 4 4 5 8 8 8 8 9
3	0 1 3 6 6 7 8
4	4 5 6 9

18. Stem | **Leaf**
4	0 *5	9 represents 5.9.*
5	7 9
6	0 4 5 6 6 7 8 8
7	0 1 1 2 2 3
8	0 5 9

19–24. Find any outliers in each set of data in Exercises 13–18.

25. The record high temperatures in degrees for each state in the U.S. are given below. Find the range, quartiles, and interquartile range for the temperatures. Then determine if any temperature is an outlier.

112	100	127	120	134	118	105	110	109	113	100	118	117
116	118	121	114	105	109	107	112	114	115	118	117	118
114	122	106	110	116	108	109	121	113	120	119	111	104
111	120	116	105	110	118	112	114	114	113	120		

26. The stem and leaf plot at the right represents the number of passengers served by certain airports in 1988. Find the range, quartiles, and interquartile range for the data. Then determine if there are any outliers.

Stem | **Leaf**
1	6 6 7 7 8 8 8 9
2	0 0 2 2 2 7 7 9 9
3	1 5
4	0 1 5 *3	1 = 31,000,000*
5	3

The following table gives the cost and the milligrams of caffeine per cup for certain brands of coffee.

Brand of Coffee	Cost	Caffeine	Brand of Coffee	Cost	Caffeine
Zabar's Colombian	11¢	99	S & W Hawaiian Kona	16¢	106
Kroger's Colombian	9¢	82	Brown Gold Colombian	12¢	109
Yuban	6¢	52	Zabar's Special Blend	11¢	102
A & P Colombian	9¢	83	Brown & Jenkins Colombian Supremo	19¢	115
S & W Colombian	12¢	93	MJB Premium Colombian	7¢	62
Pathmark Premium	4¢	71	Brown & Jenkins Special Blend	19¢	110
Hills Bros Gold	4¢	48	Martinson Mr. Automatic	4¢	123
Edwards	8¢	103	Melitta Gourmet Filter	6¢	80
MJB Premium	6¢	62	Maryland Club Filter Blend	5¢	63
Folgers	4¢	63	A & P Eight O'Clock	6¢	79

27. Make a stem and leaf plot of the cost per cup of coffee.

28. Find the range and quartiles for the cost per cup of coffee.

29. Find the interquartile range for the cost per cup of coffee.

30. Determine if any of the costs per cup is an outlier.

31. Make a stem and leaf plot of the amount of caffeine per cup of coffee.

32. Find the range and quartiles for the amount of caffeine per cup of coffee.

33. Find the interquartile range for the amount of caffeine per cup of coffee.

34. Determine if any of the amounts of caffeine is an outlier.

15-5 Box and Whisker Plots

Numerical data can also be displayed in a **box and whisker plot**. In a box and whisker plot, the quartiles and extreme values of a set of data are displayed using a number line.

Box and whisker plots are sometimes called box plots.

Example

1 **The winning average speeds, in miles per hour, of the Daytona 500 from 1975 to 1988 are 154, 152, 153, 160, 144, 178, 170, 154, 156, 151, 172, 148, 177, and 138. Make a box and whisker plot of the speeds.**

Arrange the 14 speeds in ascending order.

138 144 148 151 152 153 154 | 154 156 160 170 172 177 178

The median is halfway between the 7th and 8th values. Since each of these values is 154, the median (m) is 154. The lower quartile (LQ) is the 4th value, 151, and the upper quartile (UQ) is the 12th value, 170. The extreme values are the least value (LV), 138, and the greatest value (GV), 178.

Draw a number line. Then plot the quartiles and extreme values.

The number line should be drawn large enough to easily plot the quartiles and extreme values.

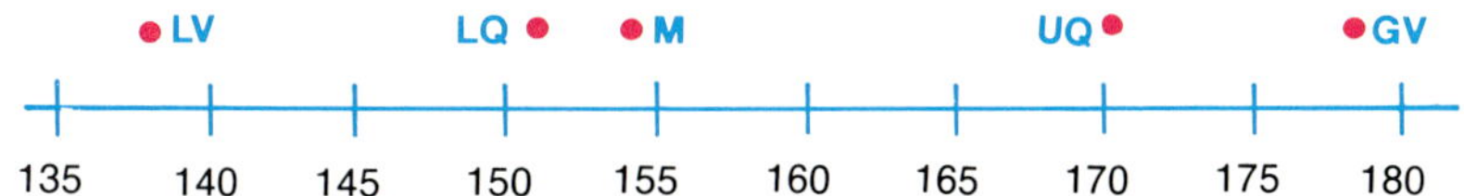

Draw a box to designate the interquartile range. Mark the median by drawing a segment containing its point in the box.

Draw segments (whiskers) connecting the lower quartile to the least value and the upper quartile to the greatest value.

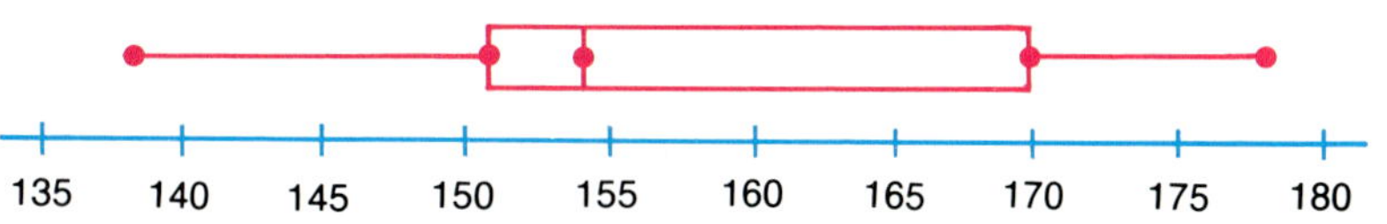

Notice that the whiskers are different sizes and the median line did not divide the box into equal parts.

A box and whisker plot can be drawn horizontally, as in Example 1, or vertically, as shown at the right. Notice that the box contains 50% of the data, the middle 50%. Also, each whisker contains 25% of the data.

In Example 1, the interquartile range of the speeds was $170 - 151$ or 19. Since each speed was either below $170 + 1.5(19)$ or 198.5, or above $170 - 1.5(19)$ or 122.5, there were no outliers.

Outliers can be displayed in a box and whisker plot. Each outlier is represented by a point only, and the whisker is extended only to the last value of the data that is not an outlier.

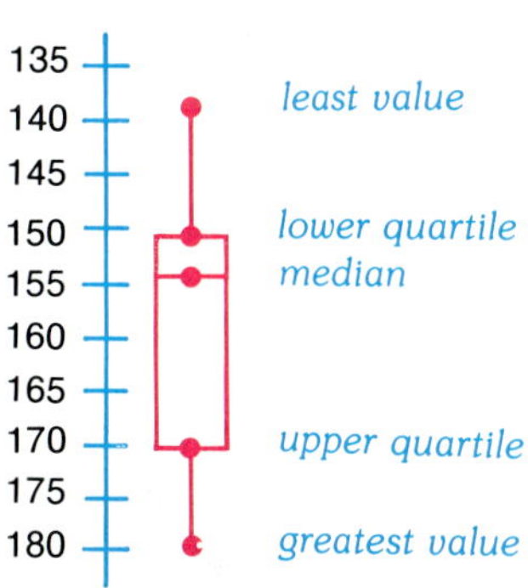

2 **The stem and leaf plot below represents the 1987 batting averages of certain players for the World Series champion Minnesota Twins. Find any outliers in the averages. Then draw a box and whisker plot of the data.**

There are 14 leaves in this plot. The median is halfway between the 7th and 8th values. Thus, the median is $\frac{0.257 + 0.259}{2}$ or 0.258. The lower quartile is the 4th value, 0.245, and the upper quartile is the 11th value, 0.267.

The interquartile range is $0.267 - 0.245$ or 0.022.

$$0.267 + 1.5(0.022) = 0.300$$
$$0.245 - 1.5(0.022) = 0.212$$

The values 0.332 and 0.191 are outliers.

Stem	Leaf
19	1
22	1
23	8
24	5 9
25	3 7 9
26	5 6 7
27	5
28	5
33	2

23 | 8 represents 0.238.

Draw a number line and plot the quartiles and outliers. Also plot 0.285 and 0.221, the last values of the data that are not outliers. Extend the whiskers to these two values and leave the outliers as single points.

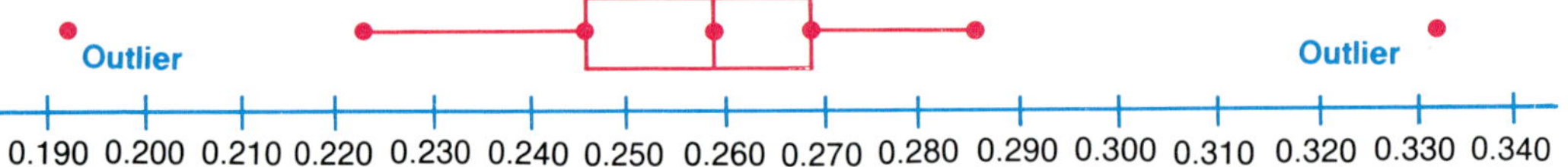

Exploratory Exercises

Use the following box and whisker plots to answer each question.

1. What is the range of the data?

2. What is the median of the data?

3. What percent of the data is greater than 28?

4. Between what two values of the data is the middle 50% of the data?

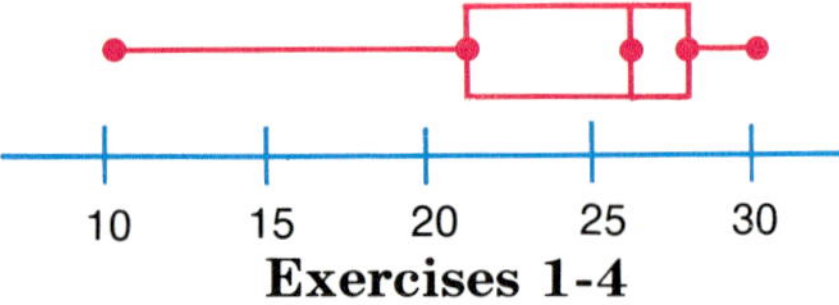

5. What percent of the data is less than 76?

6. What percent of the data is less than 92?

7. What percent of the data is greater than 64 and less than 92?

8. Under what conditions would a set of data have this type of box and whisker plot?

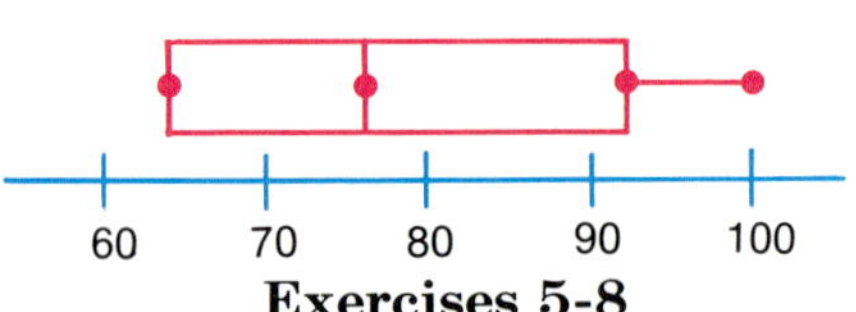

9. What is the range of the data?

10. What values of the data are outliers?

11. What percent of the data is greater than 560?

12. What percent of the data is less than 580?

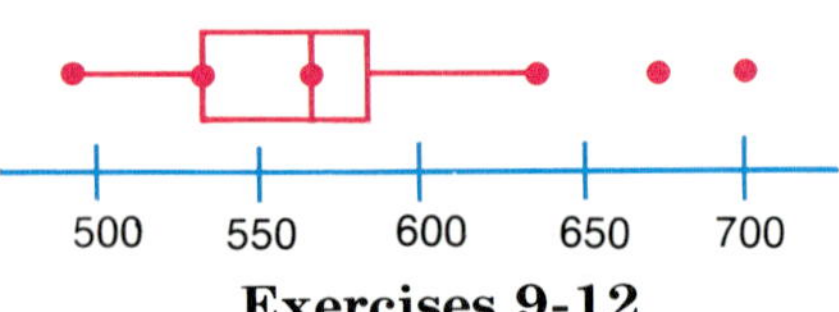

Written Exercises

13. The number of medals won by the top 16 countries at the 1984 Summer Olympics is given below. Make a box and whisker plot of the data, labeling any outliers.

174	59	53	44	37	32	32	32
27	24	19	19	18	13	13	11

14. The number of shares, in millions, of the 20 most active stocks bought and sold on the New York Stock Exchange for 1986 is given below. Make a box and whisker plot of the data, labeling any outliers.

418	350	329	212	190	187	185	184	179	179
173	168	167	165	163	161	159	157	156	148

15. The number of passes caught by the top 20 all-time leading pass receivers through the 1986 National Football League season is given below. Make a box and whisker plot of the data, labeling any outliers.

750	694	649	633	631	590	589	579	567	542
541	530	521	516	510	503	501	496	495	495

16. The number of years of life expected at birth for women in certain countries is given below. Make a box and whisker plot of the data, labeling any outliers.

77.2	76.8	76.0	74.3	77.5	78.8	78.4	75.4	77.5
76.0	73.7	75.6	77.2	79.5	79.5	75.0	72.9	76.2
79.9	79.6	74.0	77.6	75.6	75.9	73.2		

17. The stem and leaf plot below represents the number of home runs hit by the American League home run champion from 1962 to 1987. Make a box and whisker plot of the data, labeling any outliers.

Stem	Leaf
2	2
·	
3	2 2 2 2 3
·	6 7 9 9 9
4	0 0 1 3 4 4 4
·	5 5 6 8 9 9 9 9

3 | 2 represents 32.
· | 6 represents 36.

18. The stem and leaf plot below represents the number of home runs hit by the National League home run champion from 1962 to 1987. Make a box and whisker plot of the data, labeling any outliers.

Stem	Leaf
3	1
·	6 6 6 7 7 7 8 8 9
4	0 0 0 4 4 4
·	5 5 7 8 8 8 9 9
5	2 2

3 | 1 represents 31.
· | 6 represents 36.

19. The amount of sodium, in milligrams, per serving of certain brands of peanut butter is given below. Make a box and whisker plot of the data, labeling any outliers.

195	225	195	225	188	233	225	180	240	178	225	210	188
210	255	225	195	191	225	210	225	195	255	225	225	205
180	225	203	195	210	248	240	240	189	225	194	195	

20. Janine's bowling scores for each week of her summer bowling league are given below. Make a box and whisker plot of the data, labeling any outliers.

153	167	154	172	167	166	201	158	166	134	163	167	188
187	144	154	176	170	129	139	190	221	165	160	171	170
149	168	197	161	166	165	169	200	204	151	178	161	162

1. Graph $3x + 2y \leq 7$.

2. Graph the system. Name the vertices of the region formed.
$$y \leq 0, \quad x \geq 0, \quad y \geq 2x - 4$$

3. Write the equation of a plane with x-intercept 5, y-intercept 8, and z-intercept 4.

4. Solve using Cramer's Rule
$$x - 4y + 2z = 8$$
$$-2x + 5y = -14$$
$$3x + 2y - z = 3$$

5. Multiply $4y^2 - 7y + 3$ by $y - 2$.

6. Divide $a^4 - 2a^3 - 5a^2 - 9a - 12$ by $a - 4$.

7. Solve $\sqrt{2y + 12} = 10$.

8. Simplify $\dfrac{\sqrt{2} + i\sqrt{3}}{\sqrt{2} - i\sqrt{3}}$.

9. Solve $a^3 = 81a$.

10. Solve $x^2 + 8x - 9 > 0$.

11. Graph $f(x) = 2x^2 - 12x + 13$ and $f(x) = -2x^2 + 12x - 13$ on the same set of axes.

12. Graph $y^2 + 2 + x^2 + 6x - 12y = 4$. Identify the conic section.

13. Graph $16(y + 2)^2 - 9(x - 1)^2 = 144$. Identify the conic section.

14. Given $f(x) = 3x^2 - 2x$, find the values of $f(-2)$ and $f(x + 2)$.

15. Graph $g(x) = 2x^3 + 5x^2 - 7$. Use a calculator to help find points on the graph.

16. Simplify $\dfrac{(a - 1)^2}{a^2 + 3a + 2} \div \dfrac{a^2 - 1}{3a^2 + 12a + 12}$.

17. Suppose y varies directly as x. If $y = 9$, then $x = -5$. Find y when x is 42.

18. Solve $\log_{\sqrt{2}} 16 = x$.

19. Solve $\log_4 (3x - 2) = \log_4 (6x + 1)$.

20. Find the first four terms of this arithmetic series: $a_1 = 7$, $n = 21$, $S_n = 2331$.

21. Find the sum of the series $27 - 21.6 + 17.28 - 13.824 + \cdots$.

22. Find the third term of $(2m - 5n)^4$.

The number of points scored by players on the Mavricks in a recent game was 34, 15, 11, 2, 4, 14, 6, 12, 24, and 2.

23. Make a stem and leaf plot of the data.

24. Find the median, mode, and mean for the data.

25. Find the range and interquartile range for the data.

26. Determine if any value is an outlier.

Solve each problem.

27. Paul wants to invest \$5000, some in stocks at 9% interest annually, and the rest in bonds at 7% interest annually. If he wishes to earn at least \$420 in interest this year, what is the minimum he should invest in stocks?

28. A rectangular lawn is 48 feet by 52 feet. A sidewalk of uniform width will be built along the inside edge of two adjacent sides. If the remaining lawn has an area of 2112 square feet, find the width of the sidewalk.

29. A car dealer offers a choice of 8 vinyl top colors, 16 body colors, and 10 upholstery colors. How many color combinations are there?

30. How many softball teams of 10 players can be formed from 15 people?

31. From a group of 7 men and 11 women, a committee of 5 is to be formed. What is the probability that the committee will have at least 3 women?

32. Approximate to the nearest tenth the real zeros of $f(x) = x^3 - 4x + 4$.

15-6 Variation: Standard Deviation

The most commonly used measure of variation is called the **standard deviation**. The standard deviation for a set of data is an average measure of how much each value differs from the mean.

The Greek letter sigma, σ, is often used to represent the standard deviation.

From a set of data, the standard deviation is calculated by following these steps.

1. Find the mean.
2. Find the difference between each value in the set of data and the mean.
3. Square each difference.
4. Find the mean of the squares.
5. Take the positive square root of this mean.

From a set of data with n values, if x_i represents a value such that $1 \le i \le n$, and $\bar{x}$ represents the mean, then the standard deviation can be found as follows.

$$\text{standard deviation} = \sqrt{\frac{\sum_{i=1}^{n} (x_i - \bar{x})^2}{n}}$$

Definition of Standard Deviation

Example

1 **The heights of some young trees in a reforestation plot are 58 cm, 56 cm, 51 cm, 54 cm, 49 cm, 61 cm, 54 cm, and 49 cm. Find the standard deviation.**

First find the mean height of the trees.

$$\text{mean height} = \frac{58 + 56 + 51 + 54 + 49 + 61 + 54 + 49}{8} \text{ or } 54 \qquad \bar{x} = 54$$

Then find the standard deviation.

$$\text{standard deviation} = \sqrt{\frac{\sum_{i=1}^{n} (x_i - \bar{x})^2}{n}}$$

n is 8
$x_1 - \bar{x}$ is $58 - 54$ or 4
$x_2 - \bar{x}$ is $56 - 54$ or 2 and so on.

$$= \sqrt{\frac{(4)^2 + (2)^2 + (-3)^2 + (0)^2 + (-5)^2 + (7)^2 + (0)^2 + (-5)^2}{8}}$$

$$= \sqrt{\frac{16 + 4 + 9 + 0 + 25 + 49 + 0 + 25}{8}}$$

$$= \sqrt{\frac{128}{8}}$$

$$= \sqrt{16} \text{ or } 4$$

The standard deviation is 4.

2 Using Calculators

The points scored by the winning team in each National Football League game over a three-day period were 17, 24, 23, 30, 21, 30, 24, 20, 34, 24, 23, 30, 40, and 31. Find the standard deviation of the scores.

First calculate the mean score for these 14 scores and store it.

Remember, $\sigma = \sqrt{\dfrac{\Sigma\,(x_i - \bar{x})^2}{n}}$

ENTER: (17 + 24 + $\cdots$ + 40 + 31) $\div$ 14 = STO

DISPLAY: 17 17 24 41   300 40 340 31 371 371 14 26.5 26.5

Then calculate the standard deviation.

ENTER: ((17 − RCL) x^2 + $\cdots$ + (31 −

DISPLAY: 17 17 26.5 − 9.5 90.25 90.25   481.25   31 31

ENTER: RCL) x^2) $\div$ 14 = $\sqrt{x}$

DISPLAY: 26.5 4.5 20.25 501.5 501.5 14 35.82142857 5.985100548

The standard deviation is approximately 6 points.

When studying standard deviation of a set of data, it is important to keep the mean in mind. For example, suppose a firm manufactures televisions and the standard deviation of the average monthly prices for televisions sold in the last two years is $50.

If the mean price over the last two years was $200, the standard deviation indicates a great deal of variation. If the mean price over the last two years was $600, the standard deviation indicates very little variation.

Written Exercises

Find the standard deviation for each set of data.

1. {1, 4, 11, 7, 2}   **2.** {2, 2, 8, 14, 6, 4}

3. {39, 47, 51, 38, 45, 29, 37, 40, 36, 48}   **4.** {250, 275, 325, 300, 200, 225, 175}

5. {81, 80, 87, 97, 82, 86, 85, 82, 72, 80, 85, 84, 84, 63, 90, 82, 85, 79, 95, 81}

6. {1050, 1175, 835, 1075, 1025, 1145, 1100, 1125, 975, 1005, 1125, 1095, 1075, 1055}

7.

Stem	Leaf
4	1 3 9
5	2 3 6 9
6	4 4 5 7 8
7	2 4 7

5 | 2 represents 52.

8.

Stem	Leaf
3	0 0 1 2 4
·	5 6 6 6 8 9
4	1 1 3 4 4
·	5 5 6 7

3 | 4 represents 3.4.
· | 5 represents 3.5.

The weights in pounds of 11 players on three college football teams are given below.

LaSalle College: 160, 180, 190, 200, 210, 170, 250, 220, 180, 200, 240

LaRoche College: 160, 190, 210, 230, 240, 220, 150, 190, 210, 160, 240

LaRouge College: 170, 185, 205, 210, 205, 215, 185, 220, 205, 170, 250

9. Find the standard deviation of the weights for the LaSalle College team.

10. Find the standard deviation of the weights for the LaRoche College team.

11. Find the standard deviation of the weights for the LaRouge College team.

The mileage in miles per gallon obtained by three utility companies are given below.

Electric Company: 25, 13, 24, 18, 29, 12, 30, 16, 25, 21, 28, 25, 33, 11, 22, 12, 30, 16, 28, 23

Gas Company: 32, 16, 22, 24, 23, 13, 23, 31, 15, 21, 24, 27, 30, 21, 12, 24

Water Company: 22, 14, 33, 11, 25, 11, 22, 14, 36, 35, 28, 20, 36, 15, 21, 12, 22, 10

12. Find the standard deviation of the mileage for the Electric Company cars.

13. Find the standard deviation of the mileage for the Gas Company cars.

14. Find the standard deviation of the mileage for the Water Company cars.

15. The average points scored per game by the leading scorer in the NBA from 1959 to 1988 are given below. Make a stem and leaf plot of the data. Then find the mode, mean, quartiles, interquartile range, standard deviation, and any outliers in the data.

29.2	37.9	38.4	50.4	44.8	36.5	34.7	33.5	35.6	27.1
28.4	31.2	31.7	34.8	34.0	30.6	34.5	31.1	31.1	27.2
29.6	33.1	30.7	32.3	28.4	30.6	32.9	30.3	37.1	35.0

mini-review

Solve each equation.

1. $8^{2c-1} = 32^{c+2}$

2. $\log_8 \sqrt{2} = x$

3. $\log_b 27 = -3$

4. Find the sum of the series $33 + 29 + 25 + \cdots + (-3)$.

5. Find the sum of the series $2 + 6 + 18 + \cdots + 4374$.

6. Use sigma notation to express $1 \cdot 3 + 2 \cdot 9 + 3 \cdot 27 + 4 \cdot 81$.

7. Describe this sequence recursively: 2, 2, 4, 12, 48, ...

8. From a bag of 4 blue, 5 red, and 2 green marbles, what is the probability of selecting a blue, a red, and a blue marble, in that order, if no replacement occurs?

Three cards are drawn from a standard deck of 52 cards. What is the probability that the following occurs?

9. all face cards

10. all aces or all red

11. at least 2 from one suit

Use the stem and leaf plot at the right to solve each problem.

12. Find the median, mode, and mean of the data.

13. Find the range and interquartile range of the data.

14. Determine if any value in the data is an outlier.

15. Find the standard deviation of the data.

Monthly Precipitation

Stem	Leaf
2	6 8
3	0 0 1 3 3 4 6 8
4	2 2

2 | 8 represents 2.8 inches.

Statistics can be misleading. Graphs for a set of data can look very different from one another. Compare the following graphs.

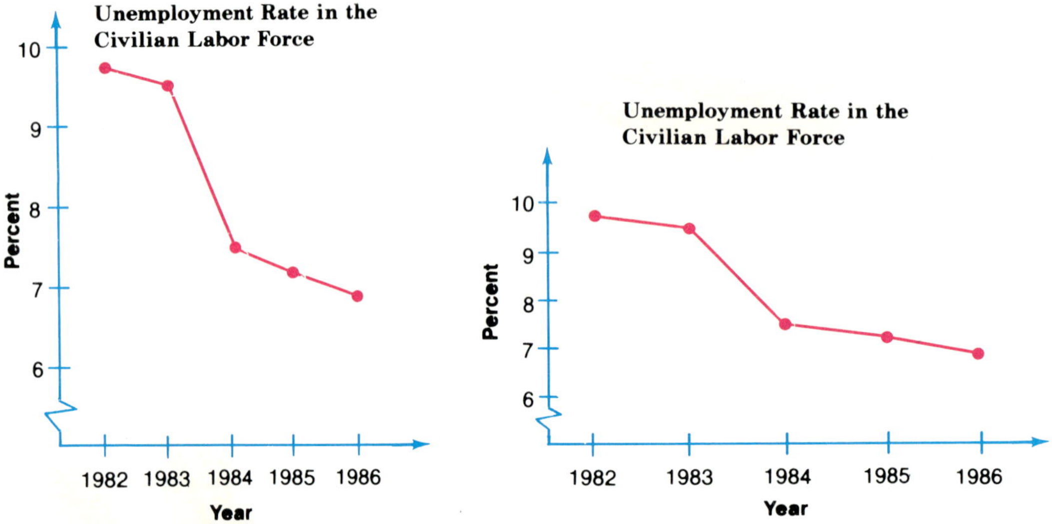

Notice that the two graphs show the same data, but the spacing in the vertical and horizontal scales differ. Scales can be cramped or spread out to make a graph that gives a certain impression. Which graph would you use to give the impression that the unemployment rate dropped dramatically from 1983 to 1984?

Companies use advertisements to make consumers buy more of their products. Often, these advertisements use graphs that have no scales, or only one scale. These graphs seem to show statistics, but actually they give no information.

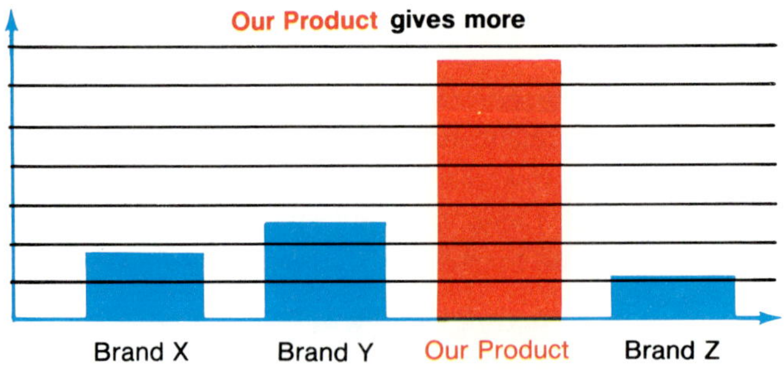

How many people should be surveyed before making the following statement?

> The average height of men in the United States is 6 feet 4 inches.

Suppose the study surveyed only male college basketball players. Do you think the outcome is reliable? The answer is no. On the average, college basketball players are taller than the rest of the male population.

Sampling is a very important part of a statistical study. A sample must be representative of all members of the population if the statistical study is to be reliable.

How many cars should be tested before an advertiser can make the following claim?

Suppose four people were asked which car used less oil. If 75% agreed, how many people thought that *Our Car* used less oil?

The advertisement above is misleading in other ways. For example, what was the condition of the cars tested? Suppose *Car Z* was old and out of tune, and used more oil than *Our Car*, which was brand new. Is it reasonable to assume that a brand new *Car Z* would use more oil than a brand new *Our Car*?

Exercises

Solve each problem.

1–3. Draw two line graphs for each set of data in problems 24–26 on page 541. Use different scales.

4. List additional ways the advertisement about *Our Car* is misleading.

Suppose an advertiser claims that 90% of all of one brand of cars sold in the last 10 years are still on the road.

5. If 10,000 cars were sold, how many are still on the road?

6. If 1000 cars were sold, how many are still on the road?

7. Find an example to show how you think averages could be used in a misleading way.

15-7 **The Normal Distribution**

One way of analyzing data is to consider the frequency with which each value occurs. The table on the right gives the frequencies of certain scores on a mechanical aptitude test taken by 175 people.

The following bar graph shows the frequencies of the scores in the table.

Score	Number of People
40–49	5
50–59	0
60–69	0
70–79	2
80–89	8
90–99	35
100–109	50
110–119	40
120–129	15
130–139	5
140–149	9
150–159	1
160–169	0
170–179	3
180–189	0
190–199	2

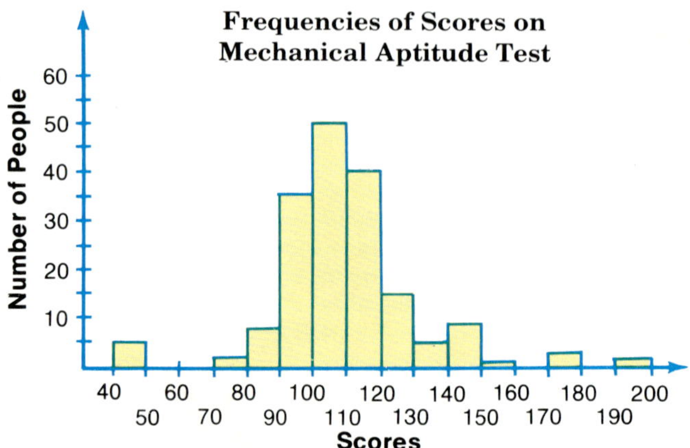

The bar graph shows the **frequency distribution** of the scores. In other words, it shows how the scores are spread out. A bar graph that shows a frequency distribution is called a **histogram**.

Frequency distributions are often shown by curves rather than histograms, especially when the distribution contains a large number of values. These curves may be of many different shapes. Many distributions have graphs like the one at the right. Distributions with such a graph are called **normal distributions**.

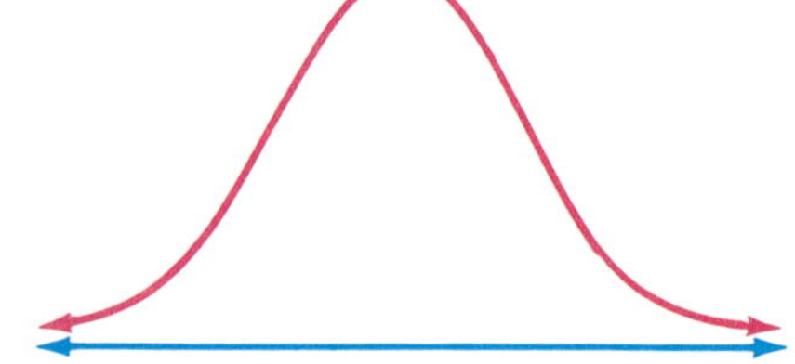

The curve is bell-shaped and symmetric. The shape of the curve indicates that frequencies in a normal distribution are concentrated in the center portion of the distribution. What does this tell you about the mean?

Normal distributions have these properties.

1. The graph is maximized at the mean.

2. About 68% of the items are within one standard deviation from the mean.

Of the 68%, by symmetry, 34% are greater than the mean and 34% are less.

3. About 95% of the items are within two standard deviations from the mean.

Of the 95%, by symmetry, 47.5% are greater than the mean, and 47.5% are less.

4. About 99% of the items are within three standard deviations from the mean.

Of the 99%, by symmetry, 49.5% are greater than the mean, and 49.5% are less.

Suppose the weights of 600 people are recorded, and the frequency of these weights is normally distributed. If the mean weight is 100 pounds and the standard deviation is 20 pounds, then the graph at the right approximates the curve for the frequency distribution of the weights.

As this graph shows, the mean, 100 pounds, is the most frequent weight. Out of the 600 people, about 408 (68%) have weights between 80 pounds and 120 pounds. About 570 (95%) have weights between 60 pounds and 140 pounds. About 594 (99%) have weights between 40 pounds and 160 pounds.

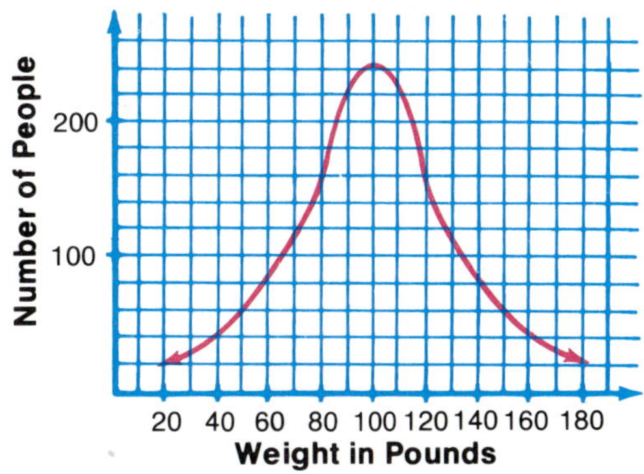

Normal distributions occur very frequently. For example, the diameter of a hole made by a drill press, the number of errors made by a typist, the tosses in a dart game if the player aims at the bull's-eye, the scores on tests, the grain yield on a farm, and the length of a newborn child can all be approximated by a normal distribution provided the number of data is sufficiently great.

Examples

1 **The approximate number of hours worked per week for 100 people is normally distributed. The mean is 40 hours per week and the standard deviation is 2 hours per week. About how many people work more than 42 hours per week?**

This frequency distribution is shown by the following curve. The percentages represent the percentage of 100 people working the number of hours within the given interval.

The percentage of people working more than 42 hours per week is
13.5% + 2% + 0.5% or 16%.

$$100 \times 16\% = 16$$

Thus, 16 people work more than 42 hours per week.

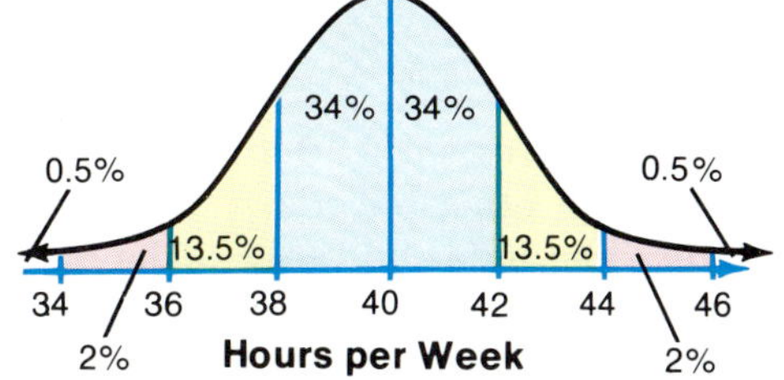

2 **The lengths of babies born at Valle Hospital in the last year were normally distributed. The mean length was 50 cm and the standard deviation was 3.5 cm. What is the probability that the length of a baby born at the hospital is between 43 cm and 53.5 cm?**

Of the babies born at the hospital last year, 68% had lengths between 46.5 cm and 53.5 cm and another 13.5% had lengths between 43 cm and 46.5 cm.
Therefore, the probability that the length of a baby born at the hospital is between 43 cm and 53.5 cm is 0.68 + 0.135 or 0.815.

Exploratory Exercises

Suppose 500 items are normally distributed. Solve each problem.

1. How many items are within one standard deviation from the mean?

2. How many items are within two standard deviations from the mean?

3. How many items are within three standard deviations from the mean?

4. How many items are within one standard deviation less than the mean?

5. How many items are within one standard deviation greater than the mean?

6. How many items are within two standard deviations greater than the mean?

Written Exercises

The lifetimes of 10,000 light bulbs are normally distributed. The mean lifetime is 300 days, and the standard deviation is 40 days.

7. How many light bulbs will last between 260 and 340 days?

8. How many light bulbs will last between 220 and 380 days?

9. How many light bulbs will last less than 300 days?

10. How many light bulbs will last more than 300 days?

11. How many light bulbs will last more than 380 days?

12. How many light bulbs will last less than 180 days?

The diameters of metal fittings produced by a machine are normally distributed. The mean diameter is 7.5 centimeters and the standard deviation is 0.5 centimeters.

13. What is the probability that a fitting has a diameter between 7.0 centimeters and 8.0 centimeters?

14. What is the probability that a fitting has a diameter between 7.5 centimeters and 8.0 centimeters?

15. What is the probability that a fitting has a diameter between 6.0 centimeters and 8.5 centimeters?

16. What is the probability that a fitting has a diameter greater than 6.5 centimeters?

A fair coin is tossed 100 times and the number of heads obtained is recorded. If this experiment is repeated many times, the number of heads obtained is distributed almost normally. The mean number of heads is 50 and the standard deviation is 5.

17. What percentage of the experiments will show less than 50 heads?

18. What percentage of the experiments will show more than 50 heads?

19. What percentage of the experiments will show between 35 and 45 heads?

20. What percentage of the experiments will show between 40 and 60 heads?

The number of hours of TV watched weekly by 3000 families in Gahanna is normally distributed. The mean number of hours is 22 and the standard deviation is 7.5 hours.

21. How many families watch TV at least 22 hours per week?

22. How many families watch TV between 7 and 29.5 hours per week?

23. What is the probability that a family watches TV more than 37 hours per week?

24. What is the probability that a family watches TV at most 14.5 hours per week?

15-8 Scatter Plots and Predictions

The first step in determining how quantities are related often is making a **scatter plot**. Such a diagram shows visually the nature of a relationship, both its shape and variation.

Suppose, for example, you wish to predict the quantity of a food product sold based on its weekly selling price. The following table shows the quantity sold for each of the last ten weeks and its selling price.

Quantity Sold (dozens)	30	47	38	28	49	23	47	46	39	42
Price (cents per dozen)	28	22	29	32	20	35	21	20	24	29

The graph on the left is a scatter plot for the data. The scatter of dots suggests a straight line that slopes downward from the upper left corner to the lower right corner.

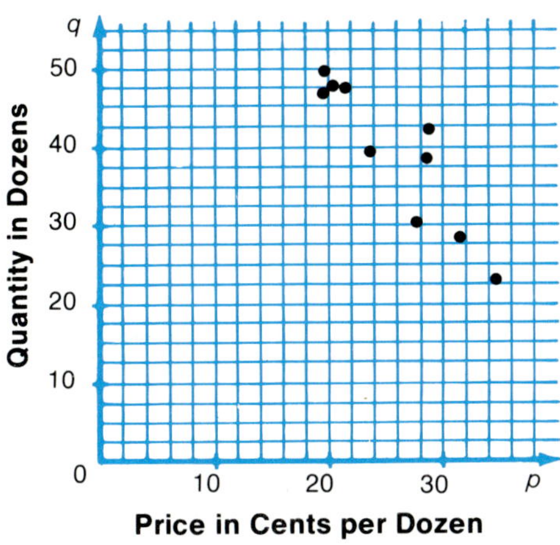

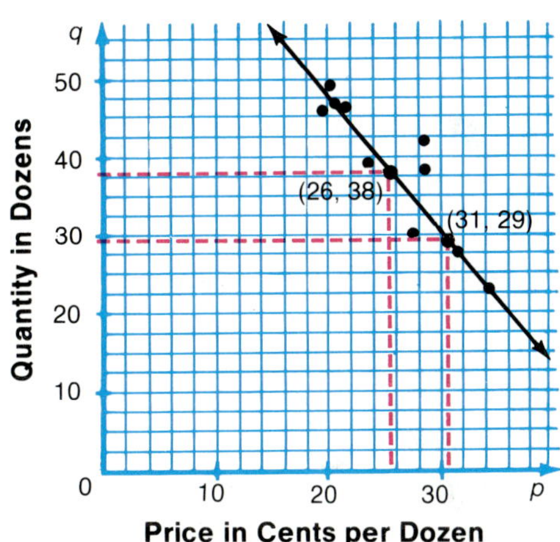

You can draw the line that is suggested by the dots. It represents the relationship between quantity and price. By choosing several points on the line, you can find the equation of the line. This equation is called the **prediction equation** for the relationship.

$$\text{slope} = \frac{38 - 29}{26 - 31} \text{ or } -1.8$$

$q = -1.8p + b$ *q stands for quantity and p stands for price*

$38 = -1.8(26) + b$ *y = mx + b*

$84.8 = b$ The equation is $q = -1.8p + 84.8$.

Now, suppose that next week, the price of the food product is 30 cents. Using the *prediction equation*, you can estimate that 30.8 dozen items will be sold.

$$q = -1.8p + 84.8$$
$$= -1.8(30) + 84.8 \text{ or } 30.8$$

1 Draw a scatter plot and find a prediction equation to show how typing speed and experience are related. Use the data in the following table.

Typing speed (wpm)	33	45	46	20	40	30	38	22	52	44	42	55
Experience (weeks)	4	7	8	1	6	3	5	2	9	6	7	10

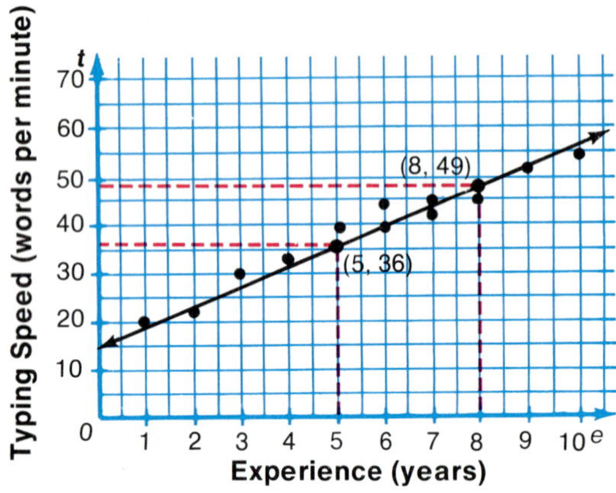

The black line is suggested by the pattern of dots. This line passes through (5, 36) and (8, 49).

$$\text{slope} = \frac{49 - 36}{8 - 5}$$

$$= \frac{13}{3} \text{ or about } 4.3$$

Let e stand for experience.
Let t stand for typing speed.

$$t = 4.3e + b \qquad y = mx + b$$
$$36 = 4.3(5) + b$$
$$14.5 = b$$

A prediction equation is $t = 4.3e + 14.5$.

2 Draw a scatter plot and find a prediction equation to show how the number of years of college education and annual income are related. Use the data in the following table.

Income (thousands of dollars)	15	20	22	47	19	18	35	10
College Education (years)	3	2	4	6	2.5	7.5	6.5	1

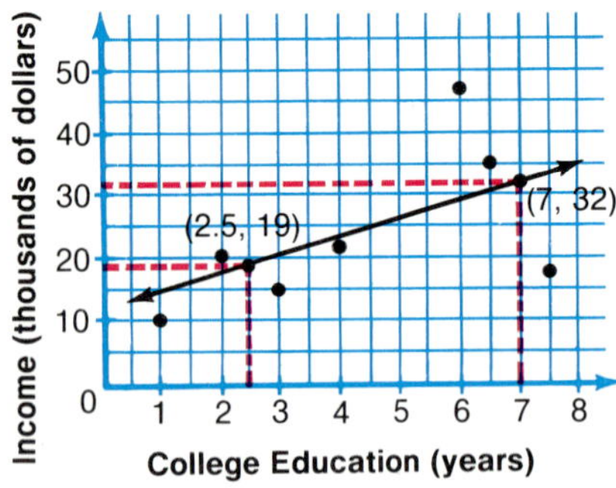

The black line is suggested by the pattern of dots. This line passes through (2.5, 19) and (7, 32).

$$\text{slope} = \frac{32 - 19}{7 - 2.5}$$

$$= \frac{13}{4.5} \text{ or about } 2.9$$

Let i stand for annual income.
Let c stand for years of college education.

$$i = 2.9c + b$$
$$32 = 2.9(7) + b$$
$$11.7 = b$$

A prediction equation is $i = 2.9c + 11.7$.

The procedure for determining a prediction equation is dependent on your judgment. Such an equation is satisfactory only when rough prediction is desired. Statisticians normally use other, more precise procedures to determine prediction equations.

Exploratory Exercises

The prediction equation in a study of the relationship between plant height in centimeters, h, and number of times watered per month, t, is $h = 0.5t + 0.5$. Predict the plant height for each of the following numbers of waterings.

1. 1
2. 3
3. 5
4. 8
5. 9
6. 10
7. 12
8. 15

According to a certain prediction equation, if Acme Soap spends \$20,000 on advertising, sales will be \$10,000,000. If Acme Soap spends \$50,000 on advertising, sales will be \$22,000,000. Let x stand for advertising expenditure and y stand for sales revenue.

9. Find the slope of the prediction equation.

10. Find the y-intercept of the prediction equation.

11. Find the prediction equation.

12. Predict sales revenue if \$10,000 is spent on advertising.

13. Predict sales revenue if \$15,000 is spent on advertising.

14. Predict sales revenue if \$35,000 is spent on advertising.

Written Exercises

A certain study claims that the number of yearly visits to a public health clinic is related to a family's weekly income. According to the study's prediction equation, a family that earns \$170 a week will visit the clinic 11 times a year. A family that earns \$220 a week will visit the clinic 6 times a year. Let x stand for family income and y stand for number of visits.

15. Find the prediction equation.

16. Predict the number of yearly visits if a family earns \$140 a week.

17. Predict the number of yearly visits if a family earns \$250 a week.

18. Predict the number of yearly visits if a family earns \$200 a week.

The following table shows the amount of sales for each of eight sales representatives during a given period and the years of sales experience for each representative.

Amount of Sales	$9000	$6000	$4000	$3000	$3000	$5000	$8000	$2000
Years of Experience	6	5	3	1	4	3	6	2

19. Draw a scatter plot to show how amount of sales and years of experience are related.

20. Find a prediction equation to show how amount of sales and years of experience are related.

21. Predict the amount of sales for a representative with 8 years of experience.

22. Predict the amount of sales for a representative with no experience.

23. Predict the years of experience for a representative that sells \$7300.

The following table shows the statistics grades and the economics grades for a group of college students at the end of a given semester.

Statistics Grades	95	51	49	27	42	52	67	48	46
Economics Grades	88	70	65	50	60	80	68	49	40

24. Draw a scatter plot to show how statistics grades and economics grades are related.

25. Find a prediction equation to show how statistics grades and economics grades are related.

26. Predict the economics grade of a student who receives a 75 in statistics.

27. Predict the statistics grade of a student who receives an 85 in economics.

The following table shows the heights and weights of each of twelve players on a professional basketball team.

Height (inches)	75	82	80	75	74	78	81	76	80	79	75	80
Weight (pounds)	180	235	205	184	185	195	215	205	221	230	185	230

28. Draw a scatter plot to show how the heights and weights are related.

29. Find a prediction equation to show how the heights and weights are related.

30. Predict the weight of a player that is 77 inches tall.

31. Predict the weight of a player that is 7 feet tall.

32. Predict the height of a player that weighs 210 pounds.

mini-review

Find the nth term of each arithmetic sequence described below.

1. $a_1 = 2$, $d = 5$, $n = 20$

2. $a_1 = -3$, $d = -7$, $n = 15$

Find the sum of each geometric series described below.

3. $a_1 = 10$, $a_n = 781{,}250$, $n = 8$

4. $54 + 18 + 6 + \ldots$ to 6 terms.

5. A car valued at \$18,000 depreciates 14% per year by the fixed rate method. After how many years will the value have depreciated to \$2000? Use $V_n = P(1 + r)^n$.

6. A safe deposit box contains 4 diamonds, 3 emeralds, and 6 rubies. How many ways can 1 diamond, 2 emeralds, and 3 rubies be chosen?

7. Suppose Steve and Rocio play five games of chess. The probability that Steve wins a game is $\frac{1}{4}$ and that Rocio wins is $\frac{3}{4}$. What is the probability that Steve will win at least 2 of the games?

The lengths of 4400 babies born at Grant Hospital in the past year were normally distributed. The mean length was 49 cm and the standard deviation was 4 cm.

8. What is the probability that a baby's length was greater than 53 cm?

9. What is the probability that a baby's length was less than 57 cm?

10. How many babies' lengths were between 37 cm and 45 cm?

Graphing Calculator Application: **Regression Lines**

A graphing calculator can be used to draw scatter plots and to draw a line that best fits the points in the scatter plot. This line is called a regression line. Once the regression line is generated, the tracing function can be used to make predictions about the data.

Think of the regression line as the graph of the best prediction equation for a given scatter plot.

Examples

1 Draw a scatter plot and regression line for the data in the following table.

Typing speed (wpm)	33	45	46	20	40	30	38	22	52	44	42	55
Experience (weeks)	4	7	8	1	6	3	5	2	9	6	7	10

Set the calculator to the proper statistical mode. Then clear the statistical memories and the graphics screen. The values of the data suggest that the range parameters be set to the following values. *Can you see why these values were chosen?*

Xmin: 0   Xmax: 10   Xscl: 1   Ymin: 0   Ymax: 60   Yscl: 10

Enter the data and draw the scatter plot and regression line.
DT stands for DATA.

ENTER: 4 [,] 33 [DT] 7 [,] 45 [DT] 8 [,] 46 [DT] 1 [,] 20 [DT]

6 [,] 40 [DT] 3 [,] 30 [DT] 5 [,] 38 [DT] 2 [,] 22 [DT]

9 [,] 52 [DT] 6 [,]

44 [DT] 7 [,] 42 [DT]

10 [,] 55 [DT]

[GRAPH] [LINE] 1 [EXE]

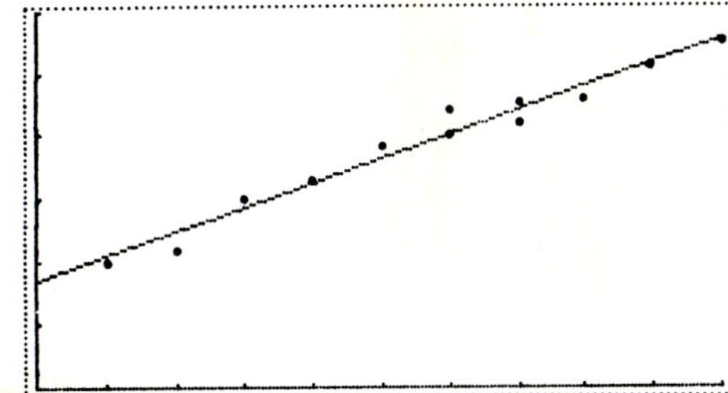

2 Use the tracing function on the regression line found in Example 1 to predict the typing speed of a person with 2.5 weeks of experience.

Activate the tracing function. Then use the arrow keys to find the *x*-coordinate value that best approximates 2.5. Press the [X↔Y] key to display the corresponding typing speed.

The value displayed is approximately 26.8. Based on this regression line, a person with 2.5 weeks of experience types about 27 words per minute.

Exploratory Exercises

Use a graphing calculator to draw a scatter plot and a regression line for the data in the following tables.

1.

x	0.1	2	3	4	5
y	1	0.2	−1	−1.5	−2

2.

x	−2	−1	0.5	1	2.5
y	−2	1	−1	2	0.5

3. Compare the graphs for Exercises 1 and 2. Which scatter plot appears to "lie" closer to its regression line?

Written Exercises

Use a graphing calculator to draw a scatter plot and a regression line for the data in the following tables.

4.

x	−4	−2	1	3	8	5
y	−1	1	3	−2	4	2

5.

x	1.5	0.5	−2	−4	4.5	5
y	−1	0.5	0.6	1.7	−1	1.5

6.

x	25	48	53	69	88	95
y	53	66	81	70	90	80

7.

x	900	600	400	500	300	850
y	6	5	3	1	2	6

8.

x	68	96	64	52	79	46	93
y	69	89	50	28	52	40	93

9.

x	1	3	6	8	7	2	5	4
y	6	5	8	2	4	9	3	7

The following table shows the average monthly income for each of eight families, and the percentage of that income spent on food for each family.

Avg. Monthly Income	$870	$1430	$1920	$2460	$2850	$3240	$3790	$4510
Percentage spent on food	44	39	40	35	43	38	37	33

10. Use a graphing calculator to draw a scatter plot and regression line to show how average monthly income and percentage of income spent on food are related.

11. Use the tracing function to predict the percentage of income spent on food for a family with an average monthly income of $3000.

12. Use the tracing function to predict the percentage of income spent on food for a family with an average monthly income of $1725.

The following table shows the number of miles driven per week by each of twelve people, and the amount of fuel used by each of their vehicles.

Miles Driven	120	322	250	300	350	135	50	150	180	70	315	175
Fuel Used (gallons)	7.5	14	11	10	10	3.5	2.3	5	6.2	3.8	11	6.5

13. Use a graphing calculator to draw a scatter plot and regression line to show how number of miles driven and amount of fuel used are related.

14. Use the tracing function to predict the amount of fuel used by a person who drives 280 miles per week.

15. Use the tracing function to predict the number of miles that can be driven per week using 8.5 gallons of fuel.

Hypothesis Testing

A claim appears in a consumers magazine that 30% of American households have at least one personal computer. A student believes that this percentage is incorrect in her neighborhood. To prove the validity of her statement, she conducted a survey in her neighborhood. She went to twenty households and asked if there was a personal computer at home. Eight of them answered YES and the rest answered NO.

The computer program below can be used to analyze the data to see if it provides enough evidence to accept or reject the claim that 30% of American households have at least one personal computer. This claim is what statisticians call the *null hypothesis*.

The program will work if the hypothetical percentages of YES and NO answers are greater than five.

```
10   READ P,Q,Z
12   DATA  0.3,0.7,1.96
13   INPUT ''WHAT IS THE NUMBER OF
        SURVEYED PEOPLE? '';N
14   INPUT ''WHAT IS THE NUMBER OF
        YES ANSWERS? '';C
15   LET X1 = N * P
20   LET X2 = N * Q
25   IF X1 < = 5 THEN 85
26   IF X2 < = 5 THEN 85
30   LET SD = SQR (X1 * Q)
35   LET C1 = X1 + Z * SD
45   LET C2 = X1 − Z * SD
50   IF ABS (C − X1) > Z * SD THEN 65
55   PRINT ''ACCEPT THE NULL HYPOTHESIS
        SINCE '';C2;'' ≤ '';C; '' ≤ '';C1;''.''
60   GOTO 85
65   PRINT ''REJECT THE NULL HYPOTHESIS
        SINCE '';C;'' > '';C1;'' OR
        '';C;'' < '';C2;''.''
85   END
```

In this program, P represents the percentage of American households with a personal computer. Q is the percentage of American households without a personal computer. Z is the number of standard deviations from the mean to a YES answer.

Line 25 of the program checks if the hypothetical percentages of YES and NO answers are less or equal to 5.

Run the program for $N = 20$ and $C = 8$.

```
WHAT IS THE NUMBER OF SURVEYED
PEOPLE? 20
WHAT IS THE NUMBER OF YES
ANSWERS? 8
ACCEPT THE NULL HYPOTHESIS SINCE
1.9831953 ≤ 8 ≤ 10.0168047.
```

Exercises

Use the computer program above to answer the following.

1. If $N = 30$ and $C = 17$, is the null hypothesis rejected or accepted?

2. What would happen if you try to run the program for $N = 10$?

3. A claim appears in a newspaper that 45% of American households have at least one VCR. To prove this percentage is incorrect, fifty-five people from different households were surveyed. Thirty of them had a VCR and the rest did not have one. Is the claim correct or incorrect? (Use $P = 0.45$, $Q = 0.55$, and $Z = 1.96$)

data (537)
bar graph (538)
line graph (538)
circle graph (539)
line plot (543)
stem and leaf plot (543)
median (548)

mode (548)
mean (548)
variation (551)
range (551)
interquartile range (551)
quartiles (551)
outlier (552)

box and whisker plot (555)
standard deviation (559)
frequency distribution (564)
histogram (564)
normal distribution (564)
scatter plot (567)
prediction equation (567)

Chapter Summary

1. Bar graphs, line graphs, and circle graphs are often used to present data and show relationships. (538, 539)
2. In a line plot, data is displayed using a number line. (543)
3. In a stem and leaf plot, data are organized into a two-column table by separating each piece of data into two numbers. (543)
4. The median of a set of data is the middle value. If there are two middle values, it is the value halfway between. (548)
5. The mode of a set of data is the most frequent value. (548)
6. The mean of a set of data is the sum of all the values divided by the number of values. (548)
7. The range of a set of data is the difference between the greatest and least values in the set. (551)
8. Quartiles are the values in a set of data that separate the data into four equal parts. (551)
9. The interquartile range of a set of data is the difference between the upper quartile and the lower quartile in the set. (551)
10. An outlier is any value in a set of data that is at least 1.5 interquartile ranges beyond the upper or lower quartiles. (552)
11. In a box and whisker plot, the quartiles and the extreme values of a set of data are displayed using a number line. (555)
12. From a set of data with n values, if x_1 represents a value such that $1 \leq i \leq n$, and $\bar{x}$ represents the mean, then the standard deviation is

$$\sqrt{\frac{\sum_{i=1}^{n} (x_i - \bar{x})^2}{n}}.\ \ (559)$$

13. In a normal distribution, the graph of the distribution is maximized at the mean. About 68% of the items are within one standard deviation from the mean. About 95% of the items are within two standard deviations from the mean. About 99% of the items are within three standard deviations from the mean. (559)
14. Scatter plots picture how quantities are related. Prediction equations give an approximate description of the relationship. (567)

15–1

1. How many times did the AL champion have more RBIs than the NL champion?

2. Find each entry for a new column on the table with the heading, **Difference**.

3. Draw a line graph to present the data in the table.

15–2

4. Make a line plot for the RBIs by the NL champion.

5. Make a line plot for the RBIs by the AL champion.

6. Make a stem and leaf plot for the RBIs by the NL champion.

7. Make a stem and leaf plot for the RBIs by the AL champion.

15–3

8. Find the median, mode, and mean for the RBIs by the NL champion.

9. Find the median, mode, and mean for the RBIs by the AL champion.

15–4

10. Find the range for the RBIs by the NL champion.

11. Find the range for the RBIs by the AL champion.

12. Find the interquartile range for the RBIs by the NL champion.

13. Find the interquartile range for the RBIs by the AL champion.

14. Find any outliers for the RBIs by the NL champion.

15. Find any outliers for the RBIs by the AL champion.

Runs Batted in (RBIs) by the NL and AL RBI Champions

Year	NL	AL
1970	148	126
1971	137	119
1972	125	113
1973	119	117
1974	129	118
1975	120	109
1976	121	109
1977	149	119
1978	120	139
1979	118	139
1980	121	122
1981	91	78
1982	109	133
1983	121	126
1984	106	123
1985	125	145
1986	119	121
1987	137	134

15–5

16. Make a box and whisker plot for the RBIs by the NL champion.

17. Make a box and whisker plot for the RBIs by the AL champion.

15–6

18. Find the standard deviation for the RBIs by the NL champion.

19. Find the standard deviation for the RBIs by the AL champion.

15–7 **The monthly incomes of 10,000 workers in King City are distributed normally. Suppose the mean monthly income is \$1250 and the standard deviation is \$250.**

20. How many workers earn more than \$1500 a month?

21. How many workers earn less than \$750 a month?

22. What is the probability that a worker earns between \$500 and \$1750 a month?

15–8 **According to a certain prediction equation, a person 180 centimeters tall weighs about 76 kilograms. A person 160 centimeters tall weighs about 57 kilograms. Let x stand for height in centimeters, and y stand for weight in kilograms.**

23. Find the prediction equation.

24. Predict the weight of a person who is 174 centimeters tall.

25. Predict the height of a person who weighs 88 kilograms.

The following high tempeatures in degrees Fahrenheit were recorded during a cold spell in Cleveland lasting 40 days.

29	26	17	12	5	4	25	17	23	18
13	6	25	20	27	22	26	30	31	20
2	12	27	16	27	16	30	6	16	5
0	5	29	18	16	22	29	8	23	24

1. Organize the data into a table with the headings Temperature in Degrees Fahrenheit and Number of Days.
2. How many days was the high temperature 29 degrees?
3. How many days was the high temperature less than 20 degrees?
4. Draw a line graph to present the data from the table made in Exercise 1.
5. Make a line plot for the daily high temperatures.
6. Make a stem and leaf plot for the daily high temperatures.
7. How many days was the high temperature in the 10's?
8. Find the median for the daily high temperatures.
9. Find the mode for the daily high temperatures.
10. Find the mean for the daily high temperatures.
11. Find the range for the daily high temperatures.
12. Find the upper quartile and the lower quartile for the daily high temperatures.
13. Find the interquartile range for the daily high temperatures.
14. Find any outliers for the daily high temperatures.
15. Make a box and whisker plot for the daily high temperatures.
16. Find the standard deviation for the daily high temperatures.

The frequencies of the scores on a college entrance examination are normally distributed. Suppose the mean score is 510 and the standard deviation is 80. And suppose 50,000 people took the examination.

17. What is the probability that a person's score is above 750?

18. What is the probability that a person's score is less than 670?

19. How many people scored between 430 and 590?

The following table shows the age in years and the systolic blood pressure in millimeters for a group of ten people tested at a hospital.

Age	50	35	24	34	55	48	26	30	41	37
Blood Pressure	135	128	108	119	146	140	104	122	132	121

20. Draw a scatter plot to show how age and systolic blood pressure are related.

21. Find a prediction equation to show how age and systolic blood pressure are related.

22. Predict the systolic blood pressure of a person who is 45 years old.

23. Predict the age of a person who has a systolic blood pressure of 120 mm.

The questions on this page involve comparing two quantities, one in Column A and one in Column B. In certain questions, information related to one or both quantities is centered above them. All variables used stand for real numbers.

Directions:
Write A if the quantity in Column A is greater.
Write B if the quantity in Column B is greater.
Write C if the quantities are equal.
Write D if there is not enough information to determine the relationship.

	Column A	Column B
1.	0.4	$\sqrt{0.4}$
2.	$0 < x < 7$	
	$0 < y < 9$	
	y	x
3.	$n < 0$	
	$b < 0$	
	$n + b$	$n - b$
4.	One tenth of the product of the first ten integers.	The product of the first nine integers.
5.	$a > b > c > d > 0$	
	$a - d$	$b - c$
6.	$\dfrac{4}{k} < 0$	
	$-\dfrac{1}{k}$	$4k$
7.	$y \neq 0$	
	$\dfrac{y}{3}$	$\dfrac{3}{y}$
8.	$\dfrac{\frac{2}{3}}{\frac{5}{7}}$	$\dfrac{6}{17}$

Examples

	Column A	Column B
I.	$\dfrac{1}{k} < 0$	
	$-\dfrac{1}{k}$	k

The answer is A because $-\dfrac{1}{k}$ is positive, while k is negative.

	Column A	Column B
II.	$\dfrac{3^4 + 3^5}{3^4}$	$\dfrac{3^2 + 3^3}{3^2}$

The answer is C because the simplest form of each quantity is $1 + 3$ or 4.

$$Hint: \frac{3^4 + 3^5}{3^4} = \frac{3^4}{3^4} + \frac{3^5}{3^4}$$

	Column A	Column B
9.	Let $\star y$ denote the least integer equal to or greater than y.	
	$\star 0.4$	$\star 1.0$
10.	$b = d + 1$	
	the average of a, b, and c.	The average of a, c, and d.
11.	c and d are integers greater than 1.	
	$[1 + (-c)]^d$	$(-c)^d$
12.	$b < 0$	
	$\dfrac{b^9}{b^4}$	$\dfrac{b^{10}}{b^5}$
13.	$\dfrac{5}{b} = 2;\ 5 = \dfrac{2}{c}$	
	$c + \dfrac{1}{3}$	$b - \dfrac{11}{6}$
14.	12% of 1600	16.5% of 1200

CHAPTER 16

Trigonometric Functions

It has been found that musical sounds are made of precise patterns of waves. These vibrational waves can be described by trigonometric functions.

16-1 Angles and the Unit Circle

Consider a circle, centered at the origin, with two rays extending from the center as shown. These rays form an angle. One ray is fixed along the positive x-axis and is called the **initial side** of the angle. The other ray can rotate about the center and is called the **terminal side** of the angle.

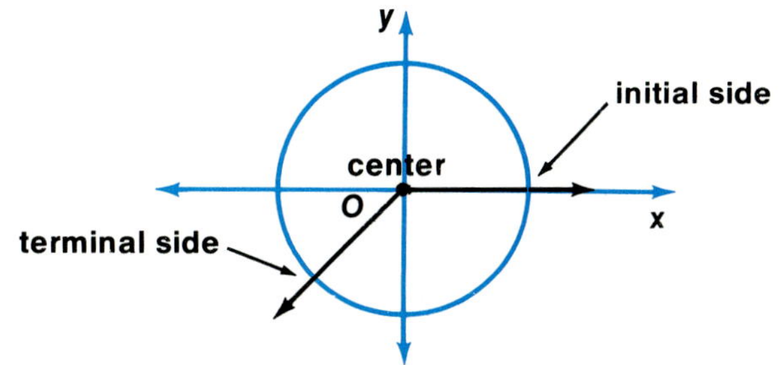

An angle with its vertex at the origin and its initial side along the positive x-axis is said to be in **standard position**.

If the initial side of the angle lies along the positive x-axis and the terminal side rotates counterclockwise, the measure of the angle formed is positive.

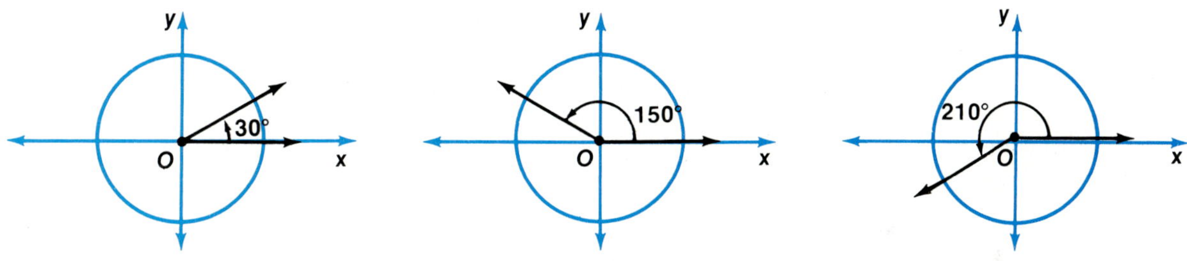

If the initial side of the angle lies along the positive x-axis and the terminal side rotates clockwise, the measure of the angle formed is negative.

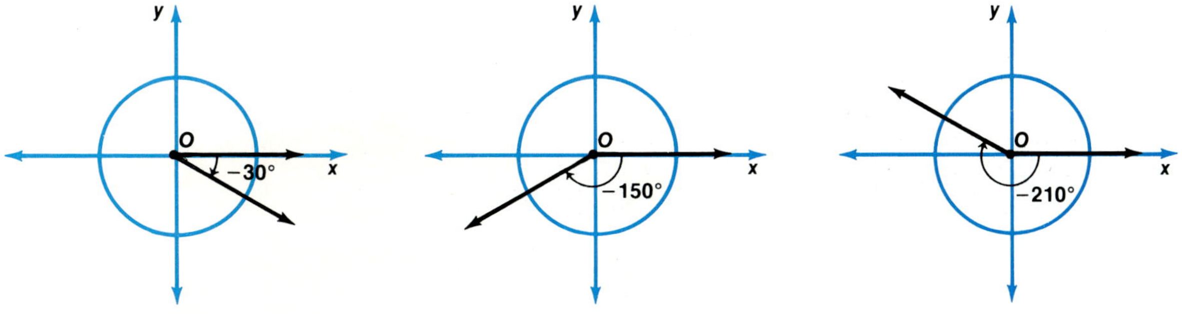

The rotation of the terminal side of the angle may include one or more complete revolutions about the center. The measurement of an angle representing one complete revolution of the circle is 360°.

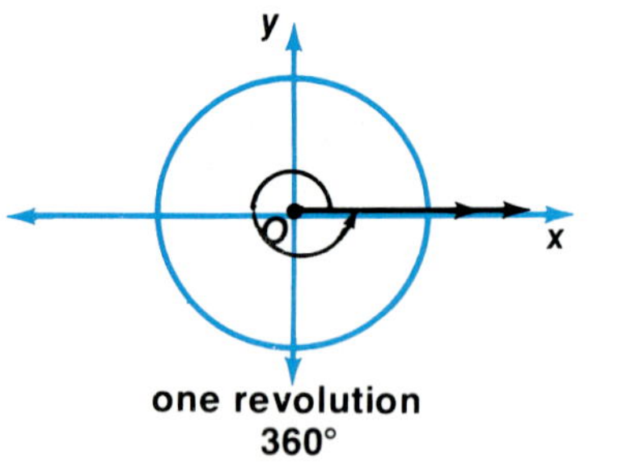

one revolution
360°

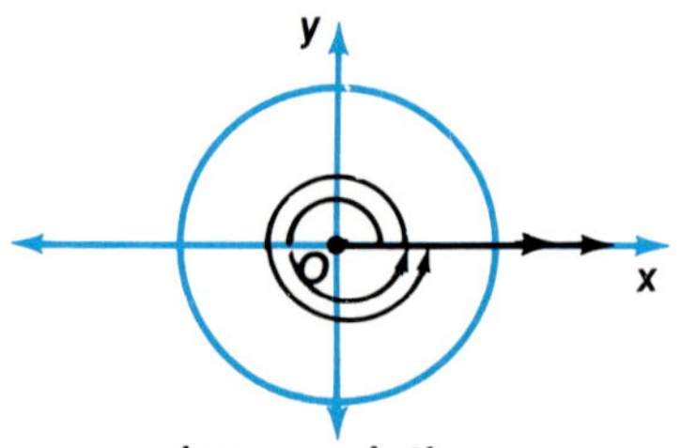

two revolutions
360° • 2 or 720°

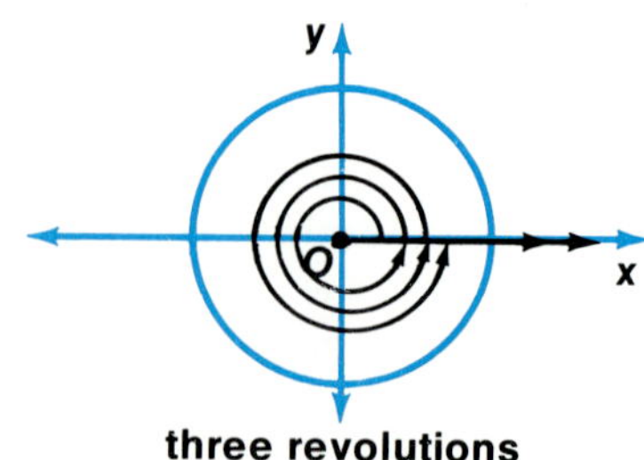

three revolutions
360° • 3 or 1,080°

Angles in standard position that have the same terminal side are called **coterminal angles**. For example, 60°, 420°, and 780° are coterminal angles.

A unit of measure other than the degree may be used in angle measurements.

Suppose a circle with a radius of 1 unit is centered at the origin. This circle is called a **unit circle**. Form an angle in standard position so that it intercepts an arc whose length is 1 unit. This angle is given the measurement 1 **radian**.

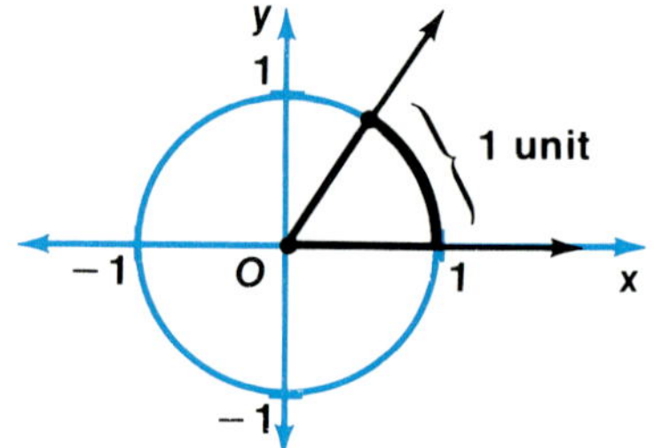

The circumference of a circle with a radius of 1 unit is $2\pi(1)$ or 2π units. Thus, an angle representing one complete revolution of the circle is 2π radians or 360°.

1 radian is $\dfrac{360}{2\pi}$ or $\dfrac{180}{\pi}$ degrees.   1 degree is $\dfrac{2\pi}{360}$ or $\dfrac{\pi}{180}$ radians.

To change radian measure to degree measure, multiply the number of radians by $\dfrac{180}{\pi}$.

To change degree measure to radian measure, multiply the number of degrees by $\dfrac{\pi}{180}$.

Example

1 Change the degree measures 45°, 240°, and −150° to radians.

$$45 \cdot \frac{\pi}{180} = \frac{45\pi}{180} \text{ or } \frac{\pi}{4} \qquad 240 \cdot \frac{\pi}{180} = \frac{240\pi}{180} \text{ or } \frac{4\pi}{3} \qquad -150 \cdot \frac{\pi}{180} = -\frac{150\pi}{180} \text{ or } -\frac{5\pi}{6}$$

Example

2 **Change the radian measures $\frac{5\pi}{3}$, $-\frac{4\pi}{3}$, and $\frac{3}{4}$ to degrees.**

$$\frac{5\pi}{3} \cdot \frac{180}{\pi} = \left(\frac{900\pi}{3\pi}\right)^{\circ} \qquad\qquad -\frac{4\pi}{3} \cdot \frac{180}{\pi} = \left(-\frac{720\pi}{3\pi}\right)^{\circ} \qquad \frac{3}{4} \cdot \frac{180}{\pi} = \left(\frac{540}{4\pi}\right)^{\circ}$$

$$= 300^{\circ} \qquad\qquad\qquad\qquad = -240^{\circ} \qquad\qquad = \frac{135^{\circ}}{\pi}$$

$$\approx 42.97^{\circ}$$

Exploratory Exercises

Suppose angles with the following measurements are in standard position. For each angle, name the quadrant that contains the terminal side.

1. 245°	**2.** 397°	**3.** 800°	**4.** 275°
5. $\frac{\pi}{3}$	**6.** $\frac{3}{5}\pi$	**7.** $\frac{11}{3}\pi$	**8.** $2\frac{1}{3}\pi$
9. -240°	**10.** -32°	**11.** 440°	**12.** 300°
13. $\frac{5}{3}\pi$	**14.** $-\frac{12}{5}\pi$	**15.** $-\frac{4}{7}\pi$	**16.** $\frac{5}{9}\pi$
17. 945°	**18.** -210°	**19.** 198°	**20.** -94°

Written Exercises

Determine whether each pair of angles is coterminal.

21. $30^{\circ}, 750^{\circ}$	**22.** $434^{\circ}, 794^{\circ}$	**23.** $\frac{\pi}{2}, \frac{3\pi}{2}$	**24.** $3\pi, 5\pi$

Change each degree measure to radian measure.

25. 90°	**26.** 120°	**27.** -45°	**28.** 60°
29. 450°	**30.** -300°	**31.** 150°	**32.** -600°
33. 45°	**34.** -120°	**35.** 330°	**36.** -240°
37. 270°	**38.** -135°	**39.** 180°	**40.** -210°
41. 405°	**42.** 810°	**43.** -315°	**44.** -270°

Change each radian measure to degree measure.

45. π	**46.** $-\frac{\pi}{2}$	**47.** $\frac{\pi}{4}$	**48.** $-\frac{\pi}{6}$
49. 3π	**50.** $-\frac{5}{4}\pi$	**51.** $-\frac{8}{3}\pi$	**52.** $-\frac{7}{4}\pi$
53. $\frac{\pi}{6}$	**54.** $\frac{5}{6}\pi$	**55.** $-\frac{\pi}{4}$	**56.** $\frac{3}{4}\pi$
57. $\frac{11\pi}{6}$	**58.** $\frac{7\pi}{4}$	**59.** 5	**60.** 2
61. $5\frac{1}{2}\pi$	**62.** $-2\frac{1}{3}$	**63.** $6\frac{1}{2}$	**64.** $3\frac{1}{3}$

16-2 Sine and Cosine

The display below is called an oscillogram. The curves represent musical tones.

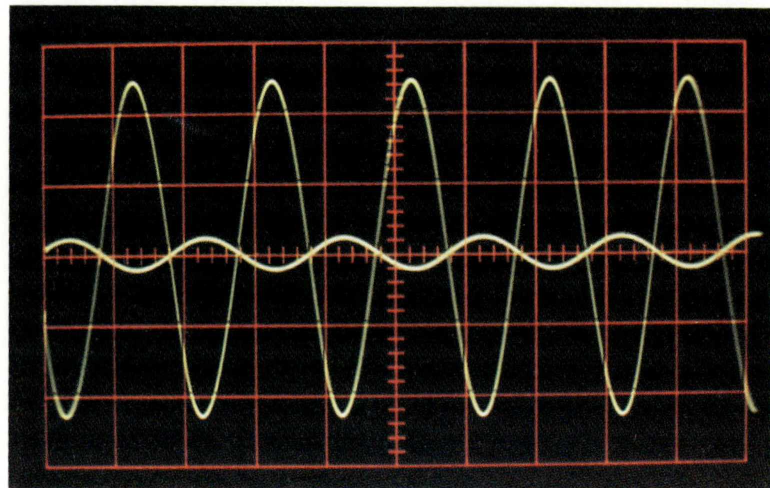

Both the **sine** function and the **cosine** function are used to describe phenomena like musical tones. These functions can be defined in terms of the unit circle.

Consider an angle in standard position. Let the Greek letter θ (theta) stand for the measurement of the angle. The terminal side of this angle intersects the unit circle at a particular point. The x-coordinate of the point is called cosine θ. The y-coordinate of the point is called sine θ.

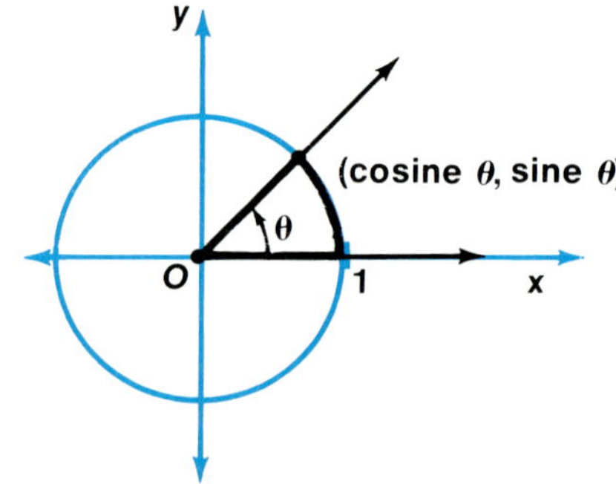

> **Let θ stand for the measurement of an angle in standard position. Let (x, y) represent the coordinates of the point where the terminal side intersects the unit circle. Then the following equations hold.**
>
> **cosine $\theta = x$ and sine $\theta = y$**

Definition of Sine and Cosine

Sine is abbreviated *sin*. Cosine is abbreviated *cos*.

You can use geometry to find values of the sine and cosine functions for certain angles.

1 Find sin 45°.

Consider the right triangle formed by two sides and a diagonal of a square. One side of this right triangle is part of the initial side of a 45° angle, and the hypotenuse is along the terminal side.

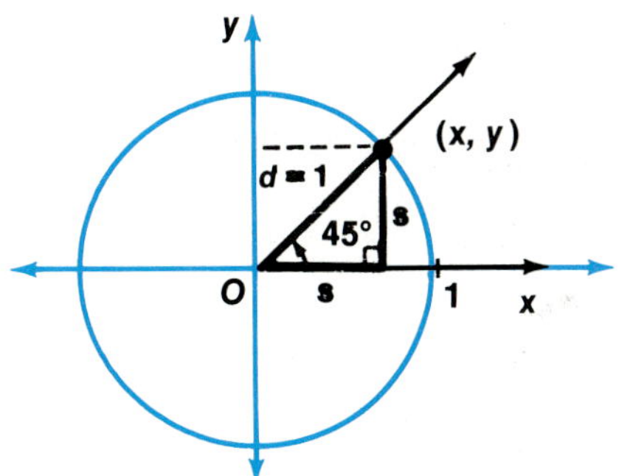

$d = s\sqrt{2}$ *Property of 45°–45° right triangles.*

$1 = s\sqrt{2}$ *d = 1*

$s = \dfrac{1}{\sqrt{2}}$ or $\dfrac{\sqrt{2}}{2}$ *The length of each side is $\dfrac{\sqrt{2}}{2}$ units.*

The triangle formed is an isosceles triangle

The coordinates of the point labeled (x, y) are $\left(\dfrac{\sqrt{2}}{2}, \dfrac{\sqrt{2}}{2}\right)$. Therefore, $\sin 45° = \dfrac{\sqrt{2}}{2}$.

2 Find cos 60°.

Look at the graph on the right. The dashed line segment cuts the x-axis and the terminal side of the angle to form a 30°–60° right triangle.

The length of the radius of the circle is 1 unit. Thus, $s = 1$ unit and $\dfrac{s}{2} = \dfrac{1}{2}$ unit.

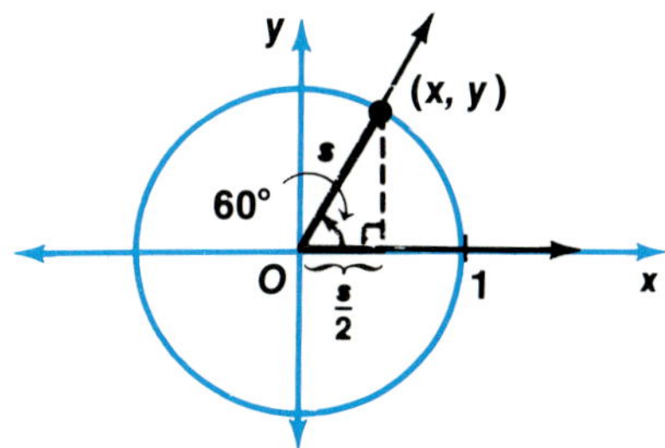

The x-coordinate of the point (x, y) is $\dfrac{1}{2}$. Therefore, $\cos 60° = \dfrac{1}{2}$.

3 Find cos 210°.

Look at the graph on the right. The dashed line segment cuts the x-axis and the terminal side of the angle to form a 30°–60° right triangle.

The length of the radius is 1 unit. Thus, $s = 1$ unit.

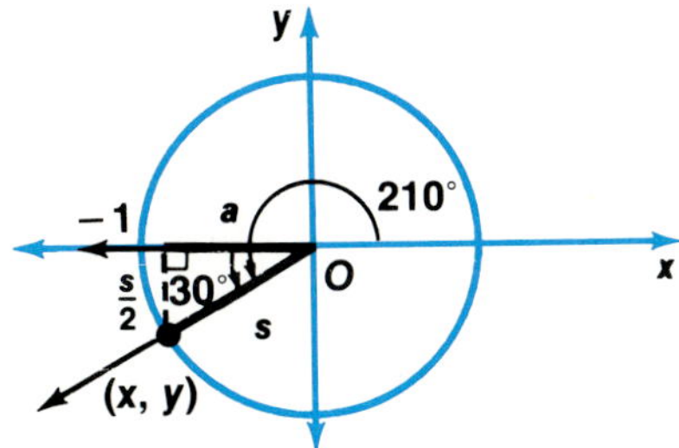

$a = \dfrac{s\sqrt{3}}{2}$ *Property of 30°–60° right triangles.*

$= \dfrac{\sqrt{3}}{2}$ *The length of side a is $\dfrac{\sqrt{3}}{2}$ units.*

Since a lies on the negative side of the x-axis, the x-coordinate of the point (x, y) is $-\dfrac{\sqrt{3}}{2}$. Therefore, $\cos 210° = -\dfrac{\sqrt{3}}{2}$.

The table at the right lists the sign of the sine and cosine functions in each of the four quadrants.

	Quadrant II	Quadrant I	
$\cos \theta$	$-$	$\cos \theta$	$+$
$\sin \theta$	$+$	$\sin \theta$	$+$
	Quadrant III	Quadrant IV	
$\cos \theta$	$-$	$\cos \theta$	$+$
$\sin \theta$	$-$	$\sin \theta$	$-$

Remember that $\cos \theta = x$ and $\sin \theta = y$.

Using the technique shown in the examples, you should be able to complete a chart like the following.

degrees	0	30	45	60	90	120	135	150	180	210	225	240	270	300	315	330	360
radians	0	$\dfrac{\pi}{6}$	$\dfrac{\pi}{4}$	$\dfrac{\pi}{3}$	$\dfrac{\pi}{2}$	$\dfrac{2\pi}{3}$	$\dfrac{3\pi}{4}$	$\dfrac{5\pi}{6}$	π	$\dfrac{7\pi}{6}$	$\dfrac{5\pi}{4}$	$\dfrac{4\pi}{3}$	$\dfrac{3\pi}{2}$	$\dfrac{5\pi}{3}$	$\dfrac{7\pi}{4}$	$\dfrac{11\pi}{6}$	2π
$\sin \theta$	0	$\dfrac{1}{2}$	$\dfrac{\sqrt{2}}{2}$	$\dfrac{\sqrt{3}}{2}$	1	$\dfrac{\sqrt{3}}{2}$	$\dfrac{\sqrt{2}}{2}$	$\dfrac{1}{2}$	0	$-\dfrac{1}{2}$	$-\dfrac{\sqrt{2}}{2}$	$-\dfrac{\sqrt{3}}{2}$	-1	$-\dfrac{\sqrt{3}}{2}$	$-\dfrac{\sqrt{2}}{2}$	$-\dfrac{1}{2}$	0
$\cos \theta$	1	$\dfrac{\sqrt{3}}{2}$	$\dfrac{\sqrt{2}}{2}$	$\dfrac{1}{2}$	0	$-\dfrac{1}{2}$	$-\dfrac{\sqrt{2}}{2}$	$-\dfrac{\sqrt{3}}{2}$	-1	$-\dfrac{\sqrt{3}}{2}$	$-\dfrac{\sqrt{2}}{2}$	$-\dfrac{1}{2}$	0	$\dfrac{1}{2}$	$\dfrac{\sqrt{2}}{2}$	$\dfrac{\sqrt{3}}{2}$	1

The chart below lists the same information for angles from 360° to 720°. What can be concluded about the values of $\sin \theta$ and $\cos \theta$ when compared to the first chart?

degrees	360	390	405	420	450	480	495	510	540	570	585	600	630	660	675	690	720
radians	2π	$\dfrac{13\pi}{6}$	$\dfrac{9\pi}{4}$	$\dfrac{7\pi}{3}$	$\dfrac{5\pi}{2}$	$\dfrac{8\pi}{3}$	$\dfrac{11\pi}{4}$	$\dfrac{17\pi}{6}$	3π	$\dfrac{19\pi}{6}$	$\dfrac{13\pi}{4}$	$\dfrac{10\pi}{3}$	$\dfrac{7\pi}{2}$	$\dfrac{11\pi}{3}$	$\dfrac{15\pi}{4}$	$\dfrac{23\pi}{6}$	4π
$\sin \theta$	0	$\dfrac{1}{2}$	$\dfrac{\sqrt{2}}{2}$	$\dfrac{\sqrt{3}}{2}$	1	$\dfrac{\sqrt{3}}{2}$	$\dfrac{\sqrt{2}}{2}$	$\dfrac{1}{2}$	0	$-\dfrac{1}{2}$	$-\dfrac{\sqrt{2}}{2}$	$-\dfrac{\sqrt{3}}{2}$	-1	$-\dfrac{\sqrt{3}}{2}$	$-\dfrac{\sqrt{2}}{2}$	$-\dfrac{1}{2}$	0
$\cos \theta$	1	$\dfrac{\sqrt{3}}{2}$	$\dfrac{\sqrt{2}}{2}$	$\dfrac{1}{2}$	0	$-\dfrac{1}{2}$	$-\dfrac{\sqrt{2}}{2}$	$-\dfrac{\sqrt{3}}{2}$	-1	$-\dfrac{\sqrt{3}}{2}$	$-\dfrac{\sqrt{2}}{2}$	$-\dfrac{1}{2}$	0	$\dfrac{1}{2}$	$\dfrac{\sqrt{2}}{2}$	$\dfrac{\sqrt{3}}{2}$	1

Every 360°, or 2π radians, represents one complete revolution of a circle. Every 360°, or 2π radians, the sine and cosine functions repeat their values. Because of this, we say that the sine and cosine functions are periodic. Each has a **period** of 360°, or 2π radians.

> **A function f is called periodic if there is a number a such that $f(x) = f(x + a)$ for all x in the domain of the function. The least positive value of a for which $f(x) = f(x + a)$ is the period of the function.**
>
> *Definition of Periodic Function*

4 **Find sin 570°.**

$$\sin 570° = \sin (210 + 360)°$$ *570° = 210° + 360°*
$$= \sin 210°$$ *The sine function has a period of 360°.*
$$= -\frac{1}{2}$$ *sin θ is negative in the third quadrant.*

5 **Find $\cos\left(-\frac{5\pi}{6}\right)$.**

$$\cos\left(-\frac{5\pi}{6}\right) = \cos\left(-\frac{5\pi}{6} + 2\pi\right)$$ *Adding 2π determines the least positive angle that is coterminal.*
$$= \cos\left(\frac{7\pi}{6}\right) \text{ or } \cos 210°$$ *The cosine function has a period of 2π.*
$$= -\frac{\sqrt{3}}{2}$$ *cos θ is negative in the third quadrant.*

Exploratory Exercises

State whether the value of each function is positive or negative.

1. $\sin 300°$ **2.** $\sin 240°$ **3.** $\cos (-210°)$ **4.** $\cos (-45°)$

5. $\sin 225°$ **6.** $\cos (-135°)$ **7.** $\sin (-270°)$ **8.** $\sin 315°$

9. $\sin \frac{\pi}{3}$ **10.** $\cos \frac{7\pi}{3}$ **11.** $\cos \frac{5}{3}\pi$ **12.** $\sin \left(-\frac{3}{4}\pi\right)$

Written Exercises

For each of the following, find the least positive angle measurement that is coterminal.

13. $420°$ **14.** $-40°$ **15.** $1020°$ **16.** $-450°$ **17.** 3π

18. $\frac{11}{4}\pi$ **19.** $\frac{9}{2}\pi$ **20.** $\frac{11}{5}\pi$ **21.** $-\frac{\pi}{4}$ **22.** $\frac{13}{3}\pi$

23. $1200°$ **24.** $1400°$ **25.** $600°$ **26.** $960°$ **27.** $-120°$

28. $680°$ **29.** $-600°$ **30.** $1240°$ **31.** $-300°$ **32.** $-240°$

33. $-\frac{7}{4}\pi$ **34.** $\frac{31}{6}\pi$ **35.** $\frac{27}{4}\pi$ **36.** $-\frac{8}{9}\pi$ **37.** $-760°$

Use the properties of 30°–60° and 45°–45° right triangles to find each value.

38. $\cos 150°$ **39.** $\cos -150°$ **40.** $\cos \frac{11}{3}\pi$ **41.** $\sin \frac{17}{4}\pi$

42. $\cos \left(-\frac{3}{4}\pi\right)$ **43.** $\sin \left(-\frac{5}{3}\pi\right)$ **44.** $\sin \frac{3\pi}{2}$ **45.** $\cos \frac{7}{4}\pi$

46. $\cos 390°$ **47.** $\sin (-240°)$ **48.** $\cos 1560°$ **49.** $\sin 660°$

50. $\sin 300°$ **51.** $\cos 900°$ **52.** $\cos 330°$ **53.** $\sin (-180°)$

54. $\cos (-60°)$ **55.** $\sin \left(-\frac{\pi}{6}\right)$ **56.** $\sin \frac{4}{3}\pi$ **57.** $\cos \left(-\frac{7}{4}\pi\right)$

58. $\frac{\sin 30° + \cos 60°}{2}$ **59.** $\frac{4 \sin 300° + 2 \cos 30°}{3}$ **60.** $4(\sin 30°)(\cos 60°)$

61. $\sin 30° + \sin 60°$ **62.** $(\sin 60°)^2 + (\cos 60°)^2$ **63.** $8(\sin 120°)(\cos 120°)$

16-3 Graphing the Sine and Cosine Functions

To graph the sine or cosine function, use the horizontal axis for the values of θ in either degrees or radians. Use the vertical axis for values of $\sin \theta$ or $\cos \theta$.

degrees	0	30	45	60	90	120	135	150	180	210	225	240	270	300	315	330	360
$\sin \theta$	0	$\frac{1}{2}$	$\frac{\sqrt{2}}{2}$	$\frac{\sqrt{3}}{2}$	1	$\frac{\sqrt{3}}{2}$	$\frac{\sqrt{2}}{2}$	$\frac{1}{2}$	0	$-\frac{1}{2}$	$-\frac{\sqrt{2}}{2}$	$-\frac{\sqrt{3}}{2}$	-1	$-\frac{\sqrt{3}}{2}$	$-\frac{\sqrt{2}}{2}$	$-\frac{1}{2}$	0
nearest tenth	0	0.5	0.7	0.9	1	0.9	0.7	0.5	0	-0.5	-0.7	-0.9	-1	-0.9	-0.7	-0.5	0
$\cos \theta$	1	$\frac{\sqrt{3}}{2}$	$\frac{\sqrt{2}}{2}$	$\frac{1}{2}$	0	$-\frac{1}{2}$	$-\frac{\sqrt{2}}{2}$	$-\frac{\sqrt{3}}{2}$	-1	$-\frac{\sqrt{3}}{2}$	$-\frac{\sqrt{2}}{2}$	$-\frac{1}{2}$	0	$\frac{1}{2}$	$\frac{\sqrt{2}}{2}$	$\frac{\sqrt{3}}{2}$	1
nearest tenth	1	0.9	0.7	0.5	0	-0.5	-0.7	-0.9	-1	-0.9	-0.7	-0.5	0	0.5	0.7	0.9	1

After plotting several points, complete the graphs by connecting the points with a smooth, continuous curve.

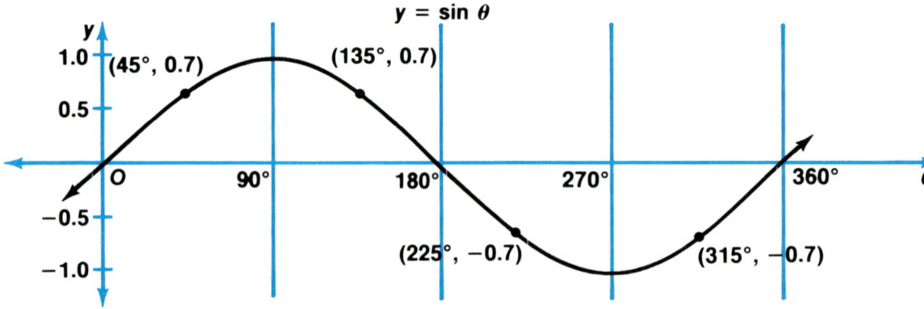

Negative values of θ would be represented to the left of zero.

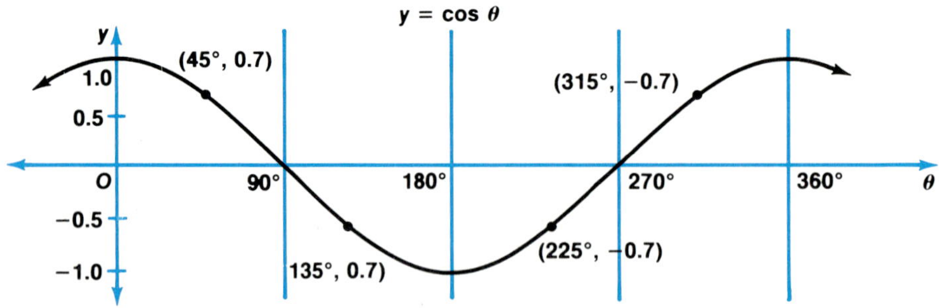

Recall that both the sine and cosine functions have a period of 360° or 2π radians. The graph of either function repeats itself every 360°. The following examples are variations of the sine function that have periods other than 360°.

Examples

1 **Graph $y = \sin 2\theta$. State the period.**

First, complete a table of values.

θ	0°	15°	30°	45°	60°	75°	90°	105°	120°	135°	150°	165°	180°
2θ	0°	30°	60°	90°	120°	150°	180°	210°	240°	270°	300°	330°	360°
$\sin 2\theta$	0	$\frac{1}{2}$	$\frac{\sqrt{3}}{2}$	1	$\frac{\sqrt{3}}{2}$	$\frac{1}{2}$	0	$-\frac{1}{2}$	$-\frac{\sqrt{3}}{2}$	-1	$-\frac{\sqrt{3}}{2}$	$-\frac{1}{2}$	0

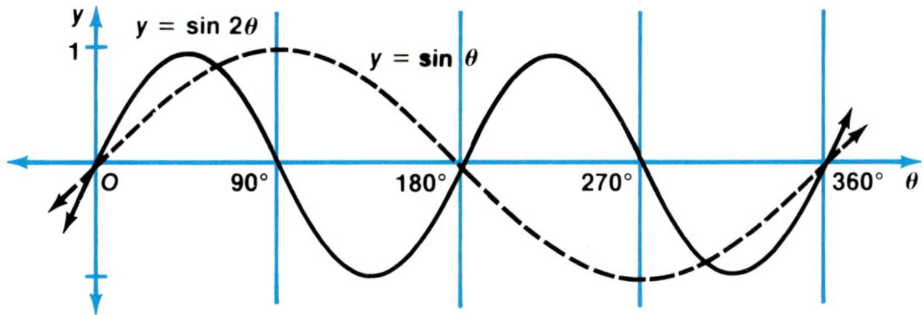

The period of $y = \sin 2\theta$ is 180° or π radians.

2 **Graph $y = \sin \frac{1}{2}\theta$. State the period.**

θ	0°	60°	90°	120°	180°	240°	270°	300°	360°
$\frac{1}{2}\theta$	0°	30°	45°	60°	90°	120°	135°	150°	180°
$\sin \frac{1}{2}\theta$	0	$\frac{1}{2}$	$\frac{\sqrt{2}}{2}$	$\frac{\sqrt{3}}{2}$	1	$\frac{\sqrt{3}}{2}$	$\frac{\sqrt{2}}{2}$	$\frac{1}{2}$	0

For the graph below, note that the θ-axis is drawn to a different scale than on the above graph.

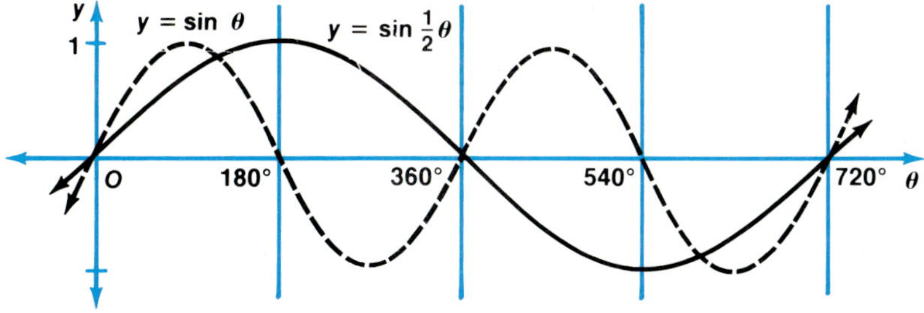

The period of $y = \sin \frac{1}{2}\theta$ is 720° or 4π radians.

All the trignometric functions you have graphed so far have a maximum value of 1 and a minimum value of -1. The **amplitude** of a periodic function is half the difference between its maximum and minimum values. So the amplitude of these graphs is 1. The graphs in the following examples have amplitudes other than 1.

$$amplitude = \frac{1 - (-1)}{2} = 1$$

Examples

3 **Graph $y = 2 \cos \theta$. State the amplitude.**

θ	0°	30°	60°	90°	120°	150°	180°	210°	240°	270°	300°	330°	360°
$\cos \theta$	1	$\frac{\sqrt{3}}{2}$	$\frac{1}{2}$	0	$-\frac{1}{2}$	$-\frac{\sqrt{3}}{2}$	-1	$-\frac{\sqrt{3}}{2}$	$-\frac{1}{2}$	0	$\frac{1}{2}$	$\frac{\sqrt{3}}{2}$	1
$2 \cos \theta$	2	$\sqrt{3}$	1	0	-1	$-\sqrt{3}$	-2	$-\sqrt{3}$	-1	0	1	$\sqrt{3}$	2

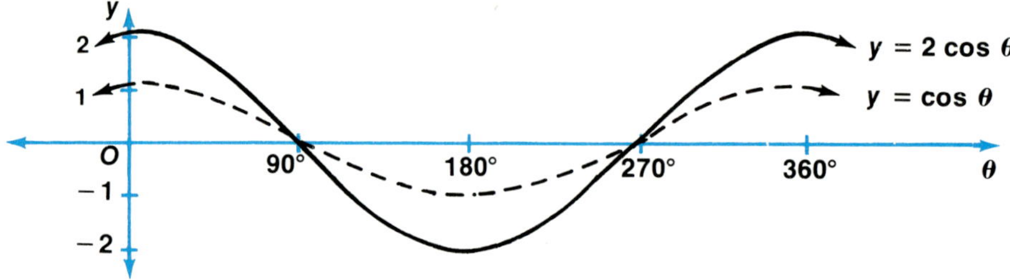

The amplitude of $y = 2 \cos \theta$ is 2.

4 **Graph $y = \frac{1}{2} \cos \theta$. State the amplitude.**

θ	0°	30°	60°	90°	120°	150°	180°	210°	240°	270°	300°	330°	360°
$\cos \theta$	1	$\frac{\sqrt{3}}{2}$	$\frac{1}{2}$	0	$-\frac{1}{2}$	$-\frac{\sqrt{3}}{2}$	-1	$-\frac{\sqrt{3}}{2}$	$-\frac{1}{2}$	0	$\frac{1}{2}$	$\frac{\sqrt{3}}{2}$	1
$\frac{1}{2} \cos \theta$	$\frac{1}{2}$	$\frac{\sqrt{3}}{4}$	$\frac{1}{4}$	0	$-\frac{1}{4}$	$-\frac{\sqrt{3}}{4}$	$-\frac{1}{2}$	$-\frac{\sqrt{3}}{4}$	$-\frac{1}{4}$	0	$\frac{1}{4}$	$\frac{\sqrt{3}}{4}$	$\frac{1}{2}$

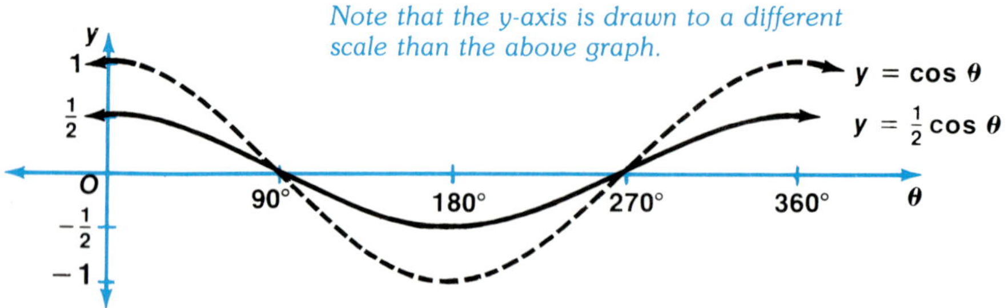

The amplitude of $y = \frac{1}{2} \cos \theta$ is $\frac{1}{2}$.

From these examples the following generalizations can be made.

Example

5 State the amplitude and period of $y = \dfrac{3}{2} \cos \dfrac{1}{2}\theta$. Then draw the graph.

The amplitude is $\left|\dfrac{3}{2}\right|$ or $\dfrac{3}{2}$. The period is $\dfrac{360°}{\left|\frac{1}{2}\right|}$ or 720°.   *720° = 4π radians*

The graph has a shape like $y = \cos \theta$.

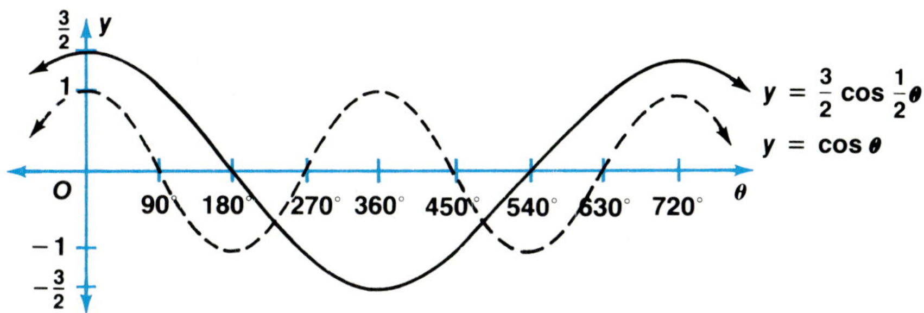

Exploratory Exercises

State the amplitude and period of each function.

1. $y = \sin \theta$

2. $y = \dfrac{1}{2} \cos \theta$

3. $y = \dfrac{2}{3} \cos \theta$

4. $y = 3 \sin \theta$

5. $y = 6 \sin \dfrac{2}{3}\theta$

6. $y = 3 \cos \dfrac{1}{2}\theta$

7. $y = 4 \cos \dfrac{3}{4}\theta$

8. $y = 2 \sin \dfrac{1}{5}\theta$

9. $y = 5 \sin \theta$

10. $y = \sin 4\theta$

11. $y = \cos 3\theta$

12. $y = \cos 2\theta$

13. $y = 4 \sin \dfrac{1}{2}\theta$

14. $y = -2 \sin \theta$

15. $y = -3 \sin \dfrac{2}{3}\theta$

16. $y = -6 \sin 2\theta$

17. $y = -\dfrac{1}{2} \cos \dfrac{3}{4}\theta$

18. $3y = 2 \sin \dfrac{1}{2}\theta$

19. $\dfrac{3}{4}y = \dfrac{2}{3} \sin \dfrac{3}{5}\theta$

20. $\dfrac{5}{3}y = \dfrac{1}{2} \cos \dfrac{1}{5}\theta$

Written Exercises

Graph each function.

21. $y = \sin \theta$

22. $y = \frac{1}{2} \cos \theta$

23. $y = \frac{2}{3} \cos \theta$

24. $y = 3 \sin \theta$

25. $y = 6 \sin \frac{2}{3}\theta$

26. $y = 3 \cos \frac{1}{2}\theta$

27. $y = 4 \cos \frac{3}{4}\theta$

28. $y = 2 \sin \frac{1}{5}\theta$

29. $y = 5 \sin \theta$

30. $y = \sin 4\theta$

31. $y = \cos 3\theta$

32. $y = \cos 2\theta$

33. $y = 4 \sin \frac{1}{2}\theta$

34. $y = -2 \sin \theta$

35. $y = -3 \sin \frac{2}{3}\theta$

36. $y = -6 \sin 2\theta$

37. $y = -\frac{1}{2} \cos \frac{3}{4}\theta$

38. $3y = 2 \sin \frac{1}{2}\theta$

39. $\frac{1}{2}y = 3 \sin 2\theta$

40. $\frac{3}{4}y = \frac{2}{3} \sin \frac{3}{5}\theta$

mini-review

Expand each binomial.

1. $(f + g)^3$

2. $(m - 2x)^4$

3. $\left(\frac{a}{3} + 3\right)^5$

Find the indicated term.

4. fourth term of $(x - y)^7$

5. sixth term of $(3y - 4)^{10}$

Evaluate.

6. $\frac{3!4!}{5!}$

7. $\frac{10!}{8!2!}$

8. $\frac{21!}{16!}$

Find each value.

9. $P(8, 4)$

10. $C(9, 4)$

11. $P(5, 3)$

Change each degree measure to radians.

12. $500°$

13. $750°$

14. $-280°$

Find each value.

15. $\cos 120°$

16. $\sin \frac{7\pi}{4}$

17. $\cos \frac{17\pi}{4}$

Solve.

18. The probability of rain is 65%. What are the odds it will rain?

19. A card is selected from a deck of 52 cards. What is the probability that it is not a 7?

The approximate number of hours worked per week for 200 people is normally distributed. The mean is 37.5 hours per week and the standard deviation is 1 hour per week.

20. How many people worked less than 36.5 hours per week?

21. What is the probability that a person works at most 36.5 hours per week?

Applications in Music
Simple Tone

We have defined periodic functions as those functions whose graphs are **periodic**, or repetitive. For example, the period of $y = \sin \theta$ and of $y = \cos \theta$ is 360°. Music is also periodic in nature. In fact, musical tones measured on an oscilloscope closely resemble the sine curve. Hence, the sine curve is called pure or simple tone by musicians.

Graphs of the form $y = \sin b\theta$ all have similar shapes. However they differ in period and frequency (the number of repetitions during a given interval). In music, frequency is measured in Hertz (cycles/second). A higher frequency produces a higher pitch. The lowest frequency is called the fundamental, and any multiple of it is called a harmonic.

The illustration at the right shows how the vibration of a string produces a musical tone. Doubling the frequency raises the pitch one octave. Therefore, the second harmonic is an octave above the fundamental, and the fourth harmonic is an octave above the second. For example, when the fundamental is 440 Hertz, the second harmonic is 880 Hertz, and so on.

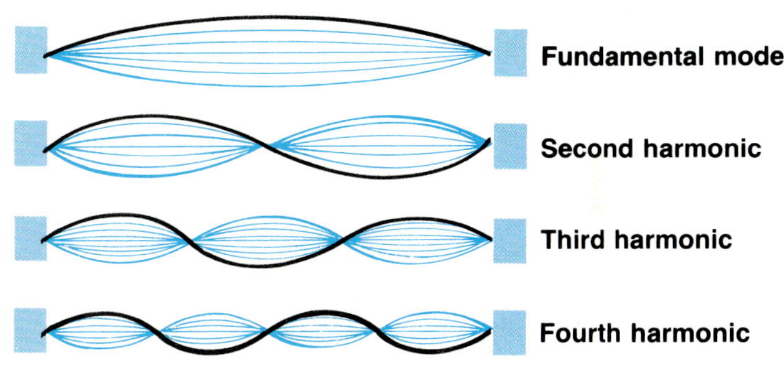

Exercises

Write an equation for each of the oscillograms below of the form $y = \sin b\theta$. State the period.

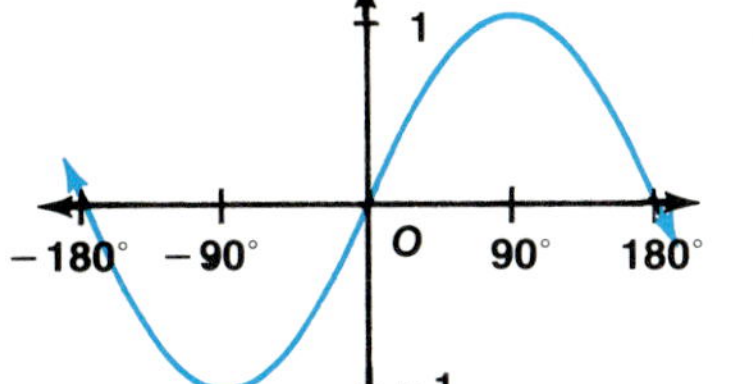
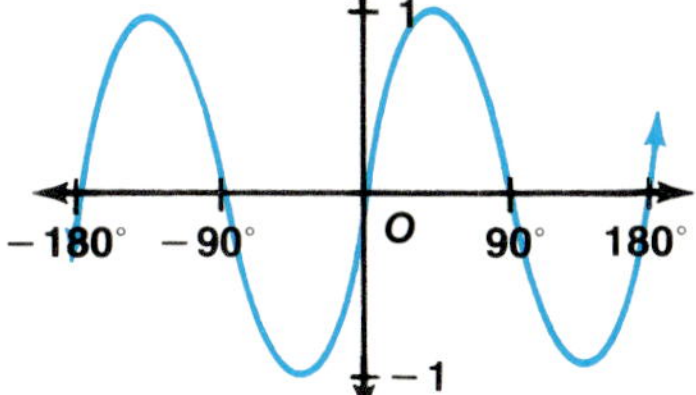
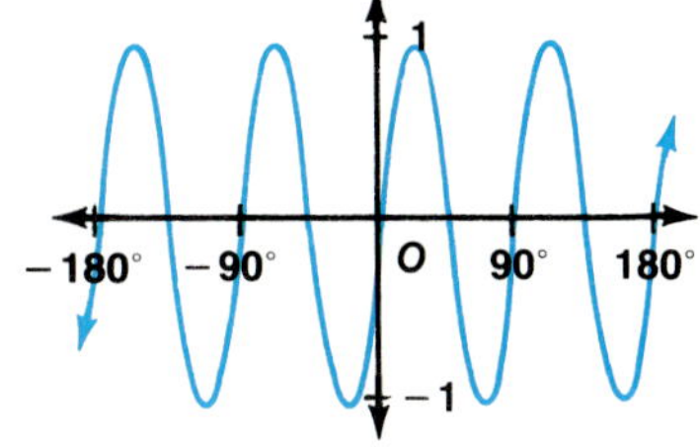

Given the notes below and their frequencies, state the frequencies of the same notes x octaves higher. (Frequencies given in Hertz.)

4. A, 440; $x = 4$

5. C, 66; $x = 3$

6. G, 196; $x = -2$

7. E, 1319; $x = 1$

16-4 Other Trigonometric Functions

Other trigonometric functions are defined using sine and cosine.

Let θ stand for the measurement of an angle in standard position on the unit circle. Then the following equations hold.

If $\cos \theta \neq 0$, $\tan \theta = \dfrac{\sin \theta}{\cos \theta}$ and $\sec \theta = \dfrac{1}{\cos \theta}$.

If $\sin \theta \neq 0$, $\cot \theta = \dfrac{\cos \theta}{\sin \theta}$ and $\csc \theta = \dfrac{1}{\sin \theta}$.

Definition of Tangent (tan), Cotangent (cot), Secant (sec), and Cosecant (csc)

Examples

1 Find tan 150°.

$$\tan 150° = \frac{\sin 150°}{\cos 150°}$$

$$= \frac{\frac{1}{2}}{-\frac{\sqrt{3}}{2}}$$

$$= -\frac{1}{\sqrt{3}} \text{ or } -\frac{\sqrt{3}}{3}$$

2 Find sec $\dfrac{\pi}{4}$.

$$\sec \frac{\pi}{4} = \frac{1}{\cos \frac{\pi}{4}}$$

$$= \frac{1}{\frac{\sqrt{2}}{2}}$$

$$= \frac{2}{\sqrt{2}} \text{ or } \sqrt{2}$$

After completing a table of values, you can graph $y = \tan \theta$. *Why is tan 90° not defined?*

θ	0°	30°	45°	60°	90°	120°	135°	150°	180°	210°	225°	240°	270°	300°	315°	330°	360°
$\tan \theta$	0	$\frac{\sqrt{3}}{3}$	1	$\sqrt{3}$	not defined	$-\sqrt{3}$	-1	$-\frac{\sqrt{3}}{3}$	0	$\frac{\sqrt{3}}{3}$	1	$\sqrt{3}$	not defined	$-\sqrt{3}$	-1	$\frac{\sqrt{3}}{3}$	0

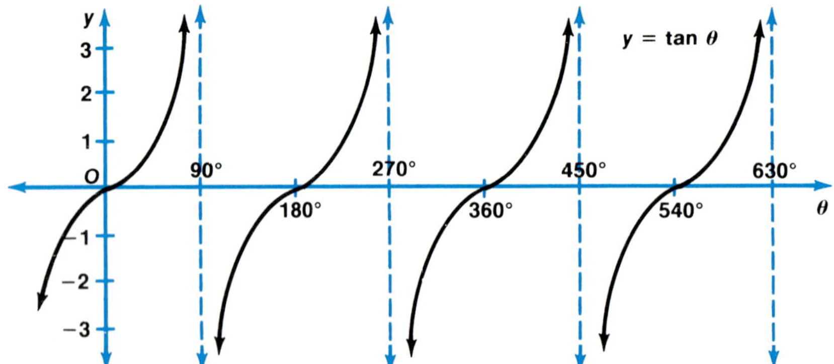

The tangent function is not defined for 90°, 270°, and so on. (We say that it is not defined for 90° + k · 180°, where k is an integer.) The graph is separated by vertical asymptotes, indicated by dashed lines.

The period of the tangent function is 180° or π radians. What is the amplitude?

The following are graphs of the secant, cosecant, and cotangent functions. Compare them to the graphs of the cosine, sine, and tangent functions, shown in red curves.

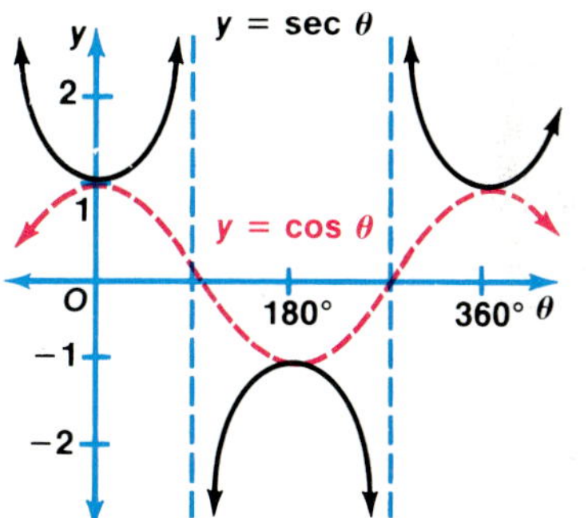

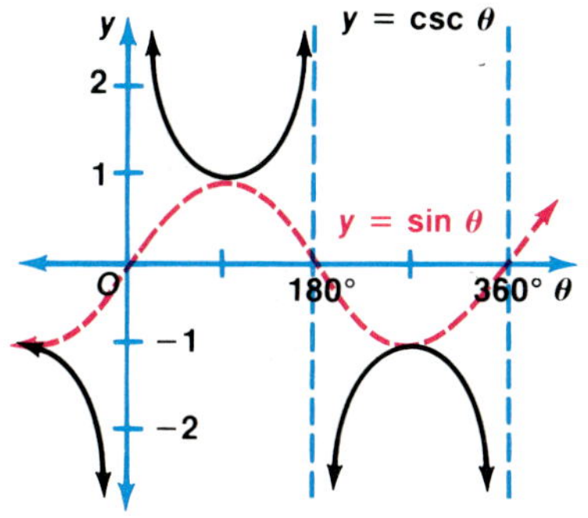

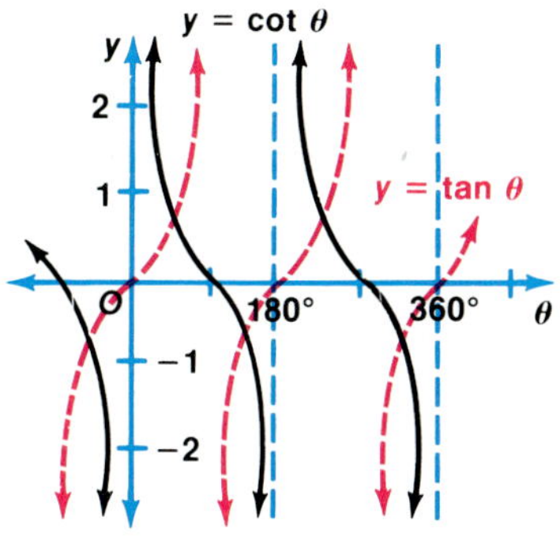

Note that the periods of the secant and cosecant functions are 360° or 2π radians. The period of the cotangent function is 180° or π radians. What are the amplitudes of the secant, cosecant, and cotangent functions?

Example

3 Graph $y = -\frac{1}{2} \csc 2\theta$.

θ	0°	15°	30°	45°	60°	75°	90°	105°	120°	135°	150°	165°	180°
2θ	0°	30°	60°	90°	120°	150°	180°	210°	240°	270°	300°	330°	360°
$-\frac{1}{2}\csc 2\theta$	not defined	-1	$-\frac{\sqrt{3}}{3}$	$-\frac{1}{2}$	$-\frac{\sqrt{3}}{3}$	-1	not defined	1	$\frac{\sqrt{3}}{3}$	$\frac{1}{2}$	$\frac{\sqrt{3}}{3}$	1	not defined

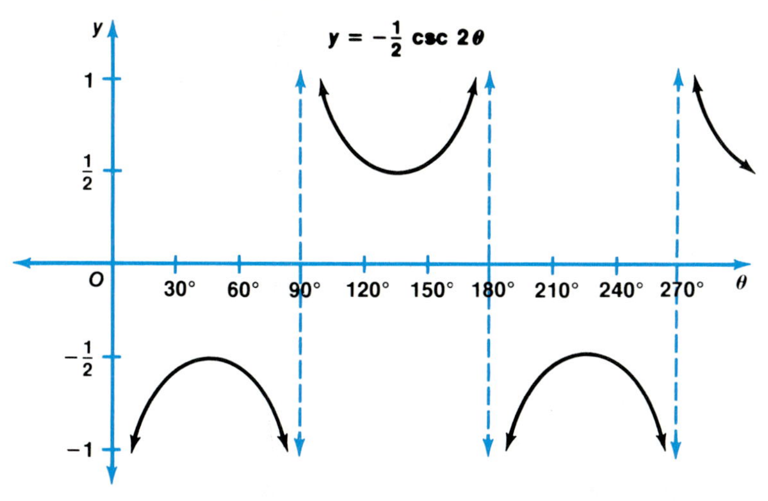

The period is 180° or π radians. This function is not defined for 0°, 90°, 180°, and so on. Thus, it has no amplitude.

Exploratory Exercises

State the values of θ for $0° \leq \theta \leq 360°$ for which each expression is not defined.

1. $\sin \theta$
2. $\cos \theta$
3. $\tan \theta$
4. $\cot \theta$
5. $\sec \theta$
6. $\csc \theta$

State the period of each function.

7. $y = \tan \frac{1}{2}\theta$
8. $y = \sec 3\theta$
9. $y = \csc 2\theta$
10. $y = \cot 5\theta$

11. $y = \tan 3\theta$
12. $y = 4 \sec \theta$
13. $y = \csc \frac{3}{4}\theta$
14. $y = \cot \frac{1}{3}\theta$

15. $y = 3 \tan \theta$
16. $y = \frac{1}{2} \sec \frac{1}{2}\theta$
17. $y = \frac{3}{4} \csc \frac{2}{3}\theta$
18. $y = 6 \cot 2\theta$

Written Exercises

State whether values of each function are increasing or decreasing in each of the four quadrants.

19. $y = \sin \theta$
20. $y = \cos \theta$
21. $y = \tan \theta$
22. $y = \cot \theta$
23. $y = \sec \theta$
24. $y = \csc \theta$

Find the value of each function.

25. $\sec 60°$
26. $\tan 120°$
27. $\cot 135°$
28. $\csc 45°$
29. $\tan \left(-\frac{\pi}{3}\right)$
30. $\csc (-210°)$
31. $\sec 300°$
32. $\cot (-60°)$

33. $\sec (-120°)$
34. $\cot \left(-\frac{\pi}{6}\right)$
35. $\csc \left(-\frac{\pi}{6}\right)$
36. $\tan \frac{7}{6}\pi$

37. $\cot \frac{7}{4}\pi$
38. $\tan (-300°)$
39. $\sec 240°$
40. $\cot 210°$

41. $\cot 540°$
42. $\csc 180°$
43. $\tan \frac{9}{4}\pi$
44. $\csc \frac{\pi}{2}$

45. $\tan \left(-\frac{5}{6}\pi\right)$
46. $\csc \frac{4}{3}\pi$
47. $\cot 270°$
48. $\sec 390°$

49. $\cot (-600°)$
50. $\sec (-30°)$
51. $\tan 405°$
52. $\csc \left(-\frac{7}{6}\pi\right)$

Evaluate each expression.

53. $\tan \frac{13}{4}\pi - \sec \pi$
54. $\dfrac{\tan 300°}{\csc 540°}$
55. $\left(\csc\left(-\frac{5\pi}{2}\right)\right)\left(\sec\left(-\frac{11}{6}\pi\right)\right)$

Graph each function.

56. $y = 3 \sec \theta$
57. $y = \csc \frac{1}{3}\theta$
58. $y = \frac{1}{3} \sec \theta$
59. $y = \cot \theta$

60. $y = \sec 3\theta$
61. $y = \csc 2\theta$
62. $y = 2 \sec \theta$
63. $y = 2 \tan \theta$

64. $y = \frac{1}{2} \tan \theta$
65. $y = -\frac{1}{2} \cot 2\theta$
66. $y = 3 \csc \frac{1}{2}\theta$
67. $y = -\cot \theta$

1. Find the first 4 terms of the arithmetic sequence when $a_1 = -2$ and $d = 7$.

2. Find the common difference of the sequence $-2x, 0, 2x, \ldots$.

3. Find the missing geometric means of 3, ——, ——, ——, 1875.

4. Find the sum of the geometric series $\frac{2}{3} + \frac{1}{2} + \frac{3}{8} + \frac{9}{32} + \cdots$.

5. Find the sum of this series:
$1 + \frac{1}{2} + \frac{1}{4} + \frac{1}{8} + \cdots$.

6. Express the repeating decimal $0.0\overline{3}$ as a fraction.

7. Express the following series using sigma notation:
$-2 + 1 + 6 + 13 + 22$.

Evaluate.

8. $\dfrac{P(9, 5)}{P(5, 4)}$

9. $\dfrac{P(7, 5) \cdot P(6, 4)}{P(5, 5)}$

Find each value of n.

10. $C(n, 4) = C(n, 6)$

11. $C(n, 12) = C(n, 16)$

Use the frequency distribution for Exercises 12–15.

Item	1	2	3	4	5	6	7	8	9
Frequency	4	2	2	3	4	1	2	4	3

12. Draw a bar graph to show the data.
13. Find the range.
14. Find the mean.
15. Find the standard deviation.

16. What are the odds of tossing a die and getting a 3 or a 5?

17. How many Social Security numbers (nine digits) can begin with 127?

18. Find the range and standard deviation for $\{85, 60, 93, 57, 82, 97, 100\}$.

19. Change $-720°$ to radians.

20. Change $3\frac{1}{2}\pi$ to degrees.

Evaluate each expression.

21. $\dfrac{\sin 45° + 2 \cos 60°}{2}$

22. $3(\tan 30°)(\sin 30°)$

Find each value.

23. $\sec\left(-\dfrac{\pi}{6}\right)$

24. $\csc\left(\dfrac{9}{4}\pi\right)$

Solve each problem.

25. How many ways can 6 books be placed on a shelf from a selection of 15 different books?

26. A red die and a green die are tossed. What is the probability that the red shows a 4 and the green shows an odd number?

27. The number of baskets made by the five-man basketball team is normally distributed. The mean is 9 baskets and the standard deviation is 2 baskets. What percentage of the players made between 5 and 13 baskets?

Graphing Calculator Application: Graphing Trigonometric Functions

A graphing calculator can generate graphs of trigonometric functions on its graphics screen. On most graphing calculators, $y = \sin x$, $y = \cos x$, and $y = \tan x$ are built-in functions.

Examples

1 **Graph $y = \sin x$.**

ENTER: GRAPH   SIN   EXE

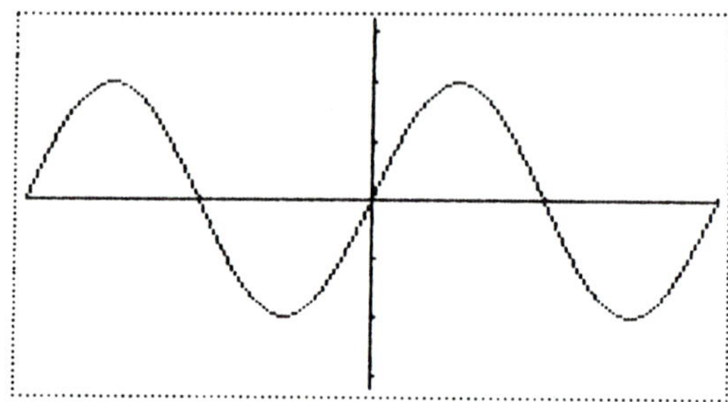

A graphing calculator will produce the same graph for y = sin x in degree mode as it will in radian mode. This graph was done in degree mode. The range parameter values are as follows:

Xmin: −360   Xmax: 360   Xscl: 180
Ymin: −1.6   Ymax: 1.6   Yscl: 0.5

2 **Graph $y = 3 \sin 2x$.**

When choosing range parameter values for graphing trigonometric functions, it is important to consider the period and amplitude (if it exists) of the function.

To see this, graph $y = 3 \sin 2x$ using the range parameter values for $y = \sin x$ set in Example 1.

ENTER: GRAPH   3   SIN   2

X   EXE

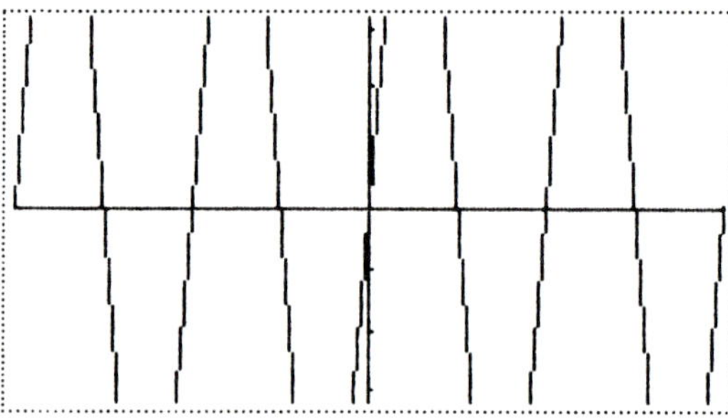

The amplitude of y = 3 sin 2x is 3. The y-axis for the built-in function y = sin x only ranges from −1.6 to 1.6.

Now clear the graphics screen and change the calculator to radian mode. Then graph $y = 3 \sin 2x$ again.

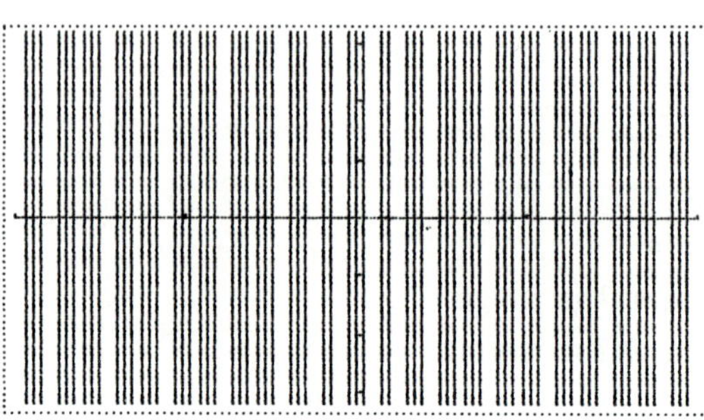

The period of y = 3 sin 2x is π radians, or about 3.14 radians. The x-axis for the built-in function y = sin 2x ranges from −360 to 360.

One possible set of range parameter values for producing a complete graph of $y = 3 \sin 2x$ is given below. Change back to degree mode and set the range parameters to these values.

Xmin: -360 Xmax: 360 Xscl: 90
Ymin: -3.6 Ymax: 3.6 Yscl: 1

Now graph $y = 3 \sin 2x$.

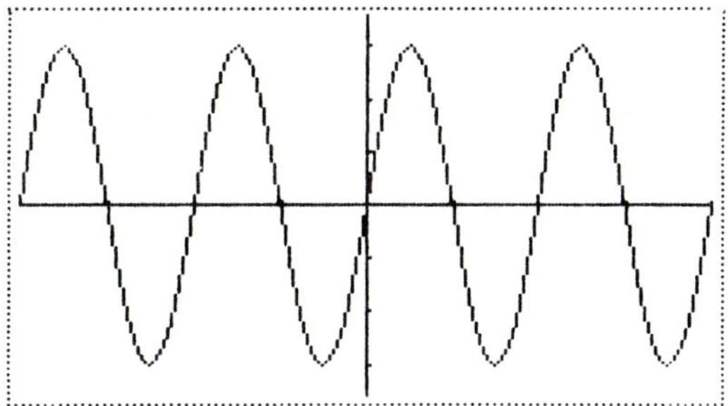

3 **Graph $y = 2 \cot (-x)$.**

Two possible sets of range parameter values for graphing $y = 2 \cot (-x)$ are given below.

Xmin: 0 Xmax: 720 Xscl: 180
Ymin: -4 Ymax: 4 Yscl: 1

Xmin: -9 Xmax: 9 Xscl: 1.5
Ymin: -9 Ymax: 9 Yscl: 1.5

Graph the equation using the first set of range parameter values.

ENTER: 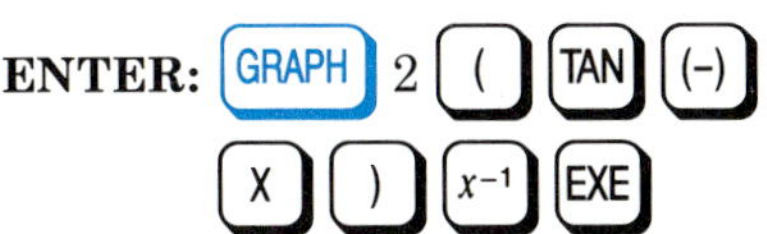

Now change to radian mode. Then graph the equation using the other set of range parameter values.

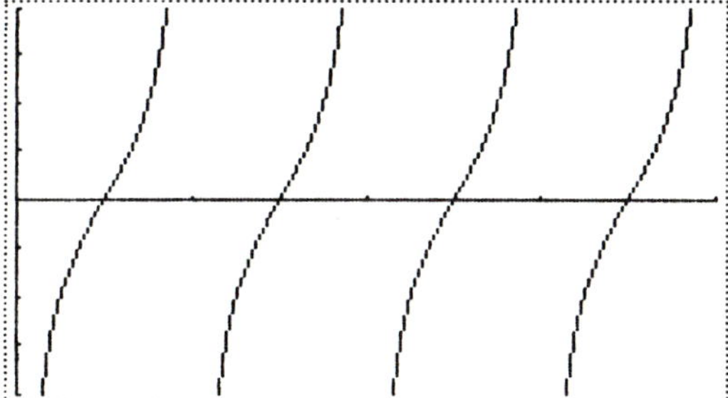

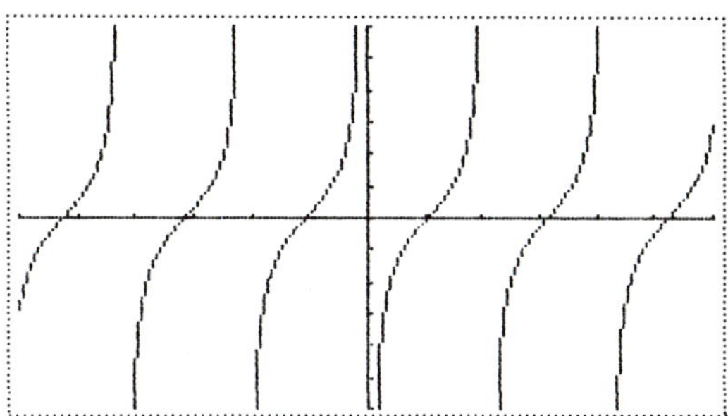

Written Exercises

Graph each equation on a graphing calculator. Then sketch the graph as shown on the graphics screen, indicating the x-axis scale and the y-axis scale.

1. $y = \cos (-x)$ **2.** $y = -2 \sin 3x$ **3.** $y = -\tan (-x)$ **4.** $y = \cot x$

5. $y = \sec x$ **6.** $y = \csc x$ **7.** $y = 5 \csc (-2x)$ **8.** $y = 6 \cot 2x$

9. $y = 3 \sin \frac{2}{3}x$ **10.** $y = 2 \cos \frac{1}{2}x$ **11.** $y = \frac{8}{3} \sec \left(-\frac{3}{2}x\right)$ **12.** $y = 62 \tan \frac{5}{4}x$

13. $y = \frac{3}{4} \cos \frac{2}{3}x$ **14.** $y = \frac{8}{9} \sin \frac{3}{5}x$ **15.** $y = \frac{81}{2} \csc \frac{4}{3}x$ **16.** $y = \frac{81}{2} \sec \frac{3}{4}x$

17. $y = -9 \cot 3.5x$ **18.** $y = 3 \tan (-30x)$ **19.** $y = 10 \sin 10x$ **20.** $y = 15 \cos 15x$

16-5 Inverse Trigonometric Functions

The concept of inverse functions can be applied to find the inverse trigonometric functions.

Consider the graph of $y = \cos x$ and the corresponding graph where all the values of x and y are reversed.

y = cos x

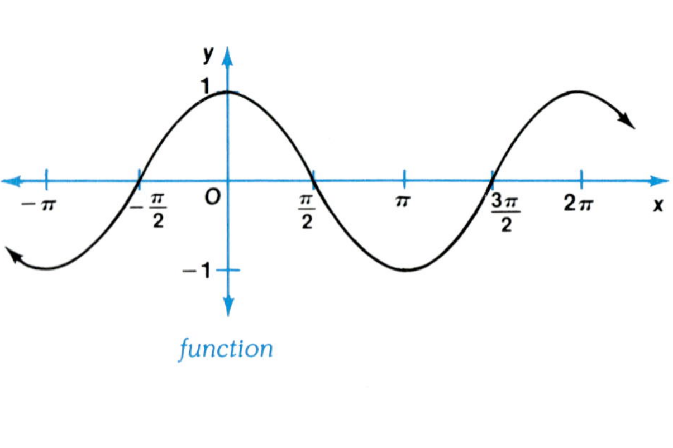

x = cos y

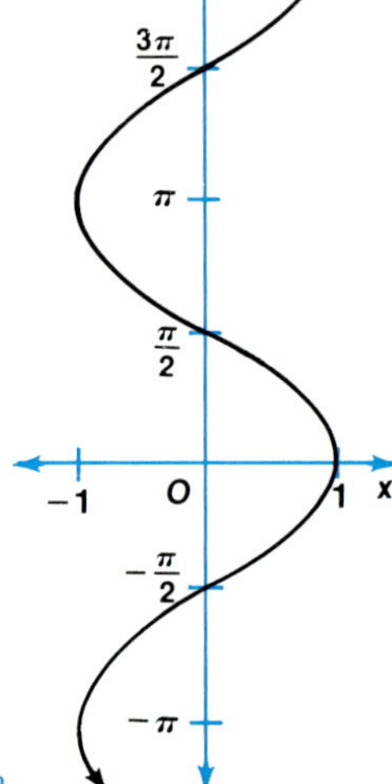

Consider only a part of the domain of the cosine function, namely any x so that $0 \le x \le \pi$. It is possible to define a new function, called Cosine, whose inverse is a function.

$$y = \text{Cos } x \text{ if and only if } y = \cos x \text{ and } 0 \le x \le \pi$$

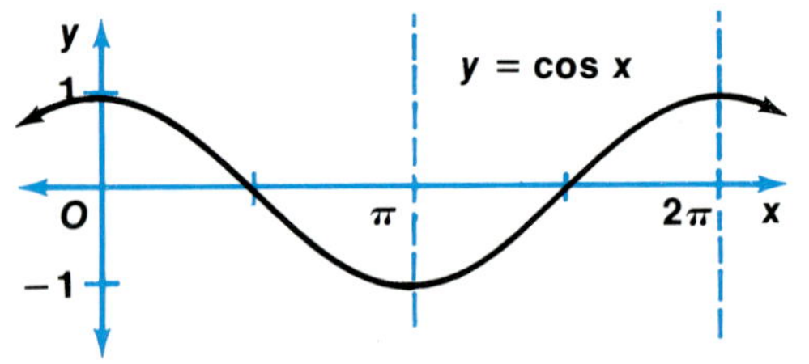

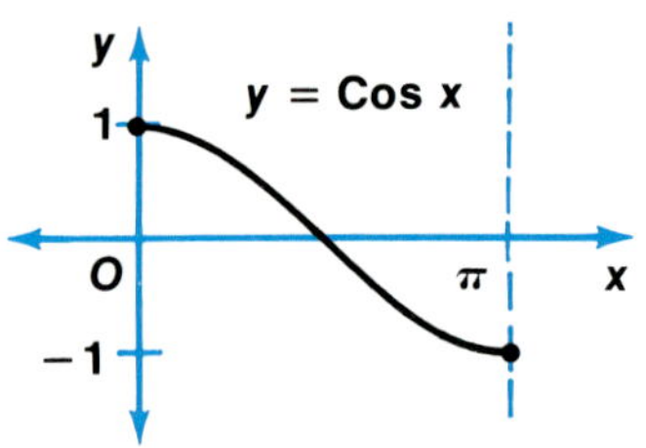

The domain is $\{x | x \text{ is a real number}\}$.
The range is $\{y | -1 \le y \le 1\}$.
The inverse is *not* a function.

The domain is $\{x | 0 \le x \le \pi\}$.
The range is $\{y | -1 \le y \le 1\}$.
The inverse is a function.

The values in the domain of Cosine are called **principal values**. Other new functions that have inverses can be defined in a similar way.

$$y = \text{Sin } x \text{ if and only if } y = \sin x \text{ and } -\frac{\pi}{2} \le x \le \frac{\pi}{2}.$$

$$y = \text{Tan } x \text{ if and only if } y = \tan x \text{ and } -\frac{\pi}{2} < x < \frac{\pi}{2}.$$

The principal values of x in $y = \text{Sin } x$ are $\left\{ x \mid -\frac{\pi}{2} \le x \le \frac{\pi}{2} \right\}$.

The principal values of x in $y = \text{Tan } x$ are $\left\{ x \mid -\frac{\pi}{2} < x < \frac{\pi}{2} \right\}$.

The inverse cosine function is also called the Arccosine function and is symbolized by **Cos**$^{-1}$ or **Arccos**.

Given $y = $ Cos x, the inverse cosine function is defined by the following equation. $$y = \text{Cos}^{-1} x \text{ or } y = \text{Arccos } x$$	*Definition of Inverse Cosine*

The Arccosine function has the following characteristics.

1. Its domain is the set of real numbers from -1 to 1.
2. Its range is the set of angle measurements from 0 to π.
3. Cos $x = y$ if and only if Cos^{-1} $y = x$.
4. $(\text{Cos}^{-1} \circ \text{Cos})(x) = (\text{Cos} \circ \text{Cos}^{-1})(x) = x$

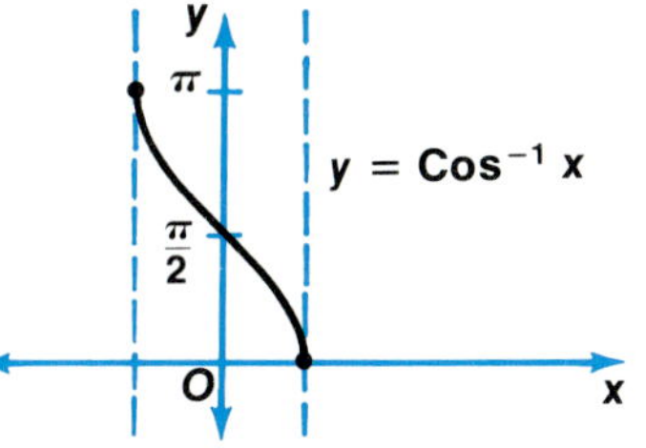

The Arcsine and Arctangent functions are defined similarly.

Given $y = $ Sin x, the inverse sine function is defined by the following equation. $$y = \text{Sin}^{-1} x \text{ or } y = \text{Arcsin } x$$	*Definition of Inverse Sine*

Given $y = $ Tan x, the inverse tangent function is defined by the following equation. $$y = \text{Tan}^{-1} x \text{ or } y = \text{Arctan } x$$	*Definition of Inverse Tangent*

The graphs of the inverse sine and inverse tangent functions are shown at the right.

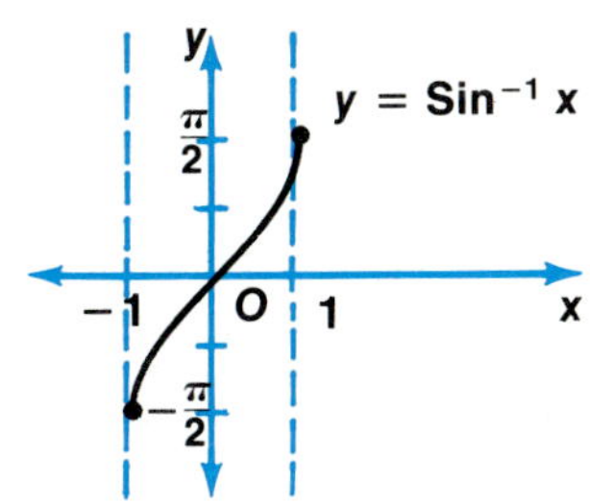 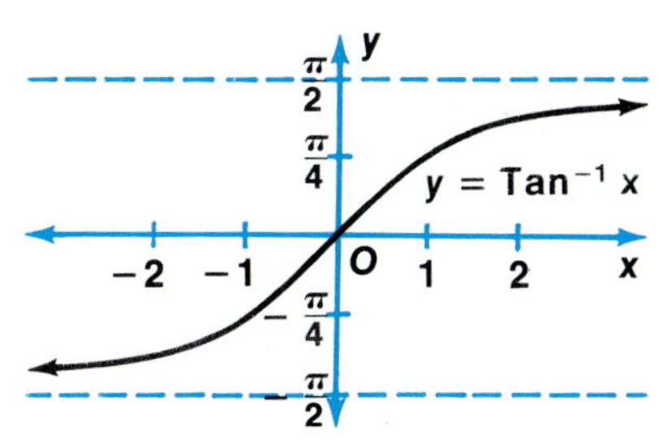

1 **Find $\mathrm{Cos}^{-1}\left(-\dfrac{\sqrt{3}}{2}\right)$.**

$$\theta = \mathrm{Cos}^{-1}\left(-\frac{\sqrt{3}}{2}\right)$$

$$\mathrm{Cos}\,\theta = -\frac{\sqrt{3}}{2}$$

$$\theta = \frac{5\pi}{6} \qquad \textit{Why is } \theta \textit{ not } \frac{7\pi}{6}?$$

2 **Find $\cos\left(\mathrm{Arcsin}\,\dfrac{1}{2}\right)$.**

Let $\theta = \mathrm{Arcsin}\,\dfrac{1}{2}$. Then $\mathrm{Sin}\,\theta = \dfrac{1}{2}$ and $\theta = \dfrac{\pi}{6}$. $\qquad -\dfrac{\pi}{2} \le \theta \le \dfrac{\pi}{2}$

$$\cos\left(\mathrm{Arcsin}\,\frac{1}{2}\right) = \cos\frac{\pi}{6}$$

$$= \frac{\sqrt{3}}{2}$$

In the examples thus far, the principal value of each inverse trigonometric function was known to be the value of a trigonometric function for some angle. Sometimes it is *not* known which angle the principal value corresponds to. In such a case, a diagram may be helpful.

3 **Find $\sin\left(\mathrm{Cos}^{-1}\left(\dfrac{2}{3}\right)\right)$.**

Let $\theta = \mathrm{Cos}^{-1}\left(\dfrac{2}{3}\right)$

From the diagram, we see that $\left(\dfrac{2}{3}\right)^2 + \sin^2\theta = 1$.

$$\sin^2\theta = 1 - \frac{4}{9} \qquad \textit{Solve for } \sin\theta.$$

$$\sin^2\theta = \frac{5}{9}$$

$$\sin\theta = \frac{\sqrt{5}}{3}$$

Therefore, $\sin\left(\mathrm{Cos}^{-1}\left(\dfrac{2}{3}\right)\right) = \dfrac{\sqrt{5}}{3}$.

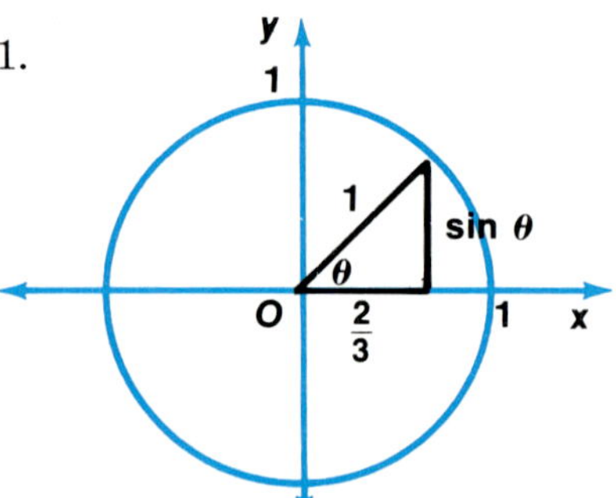

Write each of the following in the form of an inverse function.

1. $\sin y = x$

2. $a = \cos b$

3. $\tan \alpha = \beta$

4. $-\frac{4}{3} = \tan z$

5. $\sin 30° = \frac{1}{2}$

6. $2 \cos 45° = y$

Find each value.

7. $\operatorname{Sin} \frac{\pi}{6}$

8. $\operatorname{Cos}^{-1} \frac{1}{2}$

9. $\operatorname{Sin}^{-1} \left(-\frac{\sqrt{3}}{2} \right)$

10. $\operatorname{Cos} 300°$

11. $\operatorname{Tan} \pi$

12. $\operatorname{Arctan} 1$

13. $\operatorname{Tan} \frac{\pi}{4}$

14. $\operatorname{Sin}^{-1} \left(-\frac{1}{2} \right)$

15. $\operatorname{Cos}^{-1} \left(-\frac{\sqrt{3}}{2} \right)$

16. $\operatorname{Sin} 270°$

17. $\operatorname{Tan}^{-1} 1$

18. $\operatorname{Sin}^{-1} 0$

19. $\operatorname{Cos} \left(-\frac{3}{4}\pi \right)$

20. $\operatorname{Arcsin} \frac{\sqrt{3}}{2}$

21. $\operatorname{Sin}^{-1} \frac{\sqrt{3}}{2}$

22. $\operatorname{Tan}^{-1} \frac{\sqrt{3}}{3}$

23. $\operatorname{Cos} 45°$

24. $\operatorname{Sin}^{-1} 1$

25. $\operatorname{Sin}^{-1} (-1)$

26. $\operatorname{Sin} \frac{5}{6}\pi$

27. $\operatorname{Cos}^{-1} 0$

28. $\operatorname{Sin}^{-1} \frac{1}{2}$

29. $\operatorname{Sin} 0°$

30. $\operatorname{Tan} \left(-\frac{\pi}{4} \right)$

Written Exercises

Find each value.

31. $\operatorname{Sin} \frac{\pi}{6}$

32. $\operatorname{Tan}^{-1} (-1)$

33. $\operatorname{Sin}^{-1} 1$

34. $\operatorname{Cos}^{-1} \left(-\frac{1}{2} \right)$

35. $\sin \left(\operatorname{Sin}^{-1} \frac{1}{2} \right)$

36. $\operatorname{Sin}^{-1} \left(\cos \frac{\pi}{2} \right)$

37. $\cos \left(\operatorname{Cos}^{-1} \frac{1}{2} \right)$

38. $\cos \left(\operatorname{Cos}^{-1} \frac{4}{5} \right)$

39. $\tan \left(\operatorname{Sin}^{-1} \frac{5}{13} \right)$

40. $\tan \left(\operatorname{Cos}^{-1} \frac{6}{7} \right)$

41. $\cot \left(\operatorname{Sin}^{-1} \frac{5}{6} \right)$

42. $\cot \left(\operatorname{Sin}^{-1} \frac{7}{9} \right)$

43. $\operatorname{Arccos} \frac{\sqrt{3}}{2}$

44. $\sin \left(2 \operatorname{Cos}^{-1} \frac{3}{5} \right)$

45. $\operatorname{Arctan} \sqrt{3}$

46. $\sin \left(\operatorname{Arctan} \frac{\sqrt{3}}{3} \right)$

47. $\cos \left(\operatorname{Arcsin} \frac{3}{5} \right)$

48. $\tan (\operatorname{Arctan} 3)$

49. $\sin \left(2 \operatorname{Sin}^{-1} \frac{1}{2} \right)$

50. $\operatorname{Sin}^{-1} \left(\tan \frac{\pi}{4} \right)$

51. $\cos \left[\operatorname{Cos}^{-1} \left(-\frac{\sqrt{2}}{2} \right) - \frac{\pi}{2} \right]$

52. $\sin \left(\operatorname{Sin}^{-1} \frac{\sqrt{3}}{2} \right)$

53. $\tan \left[\operatorname{Cos}^{-1} \left(-\frac{3}{5} \right) \right]$

54. $\sin [\operatorname{Arctan} (-\sqrt{3})]$

55. $\cos (\operatorname{Tan}^{-1} \sqrt{3})$

56. $\cos \left[\operatorname{Arcsin} \left(-\frac{1}{2} \right) \right]$

57. $\cos (\operatorname{Tan}^{-1} 1)$

58. $\operatorname{Tan}^{-1} \left(\cos \frac{3\pi}{2} \right)$

59. $\operatorname{Cos}^{-1} \left(\sin \frac{3\pi}{2} \right)$

60. $\operatorname{Sin}^{-1} \left(\cos \frac{\pi}{6} \right)$

In physics the concept of *velocity*, a rate of change of position, is used frequently. Velocity is defined by its speed, or *magnitude*, and its *direction*. This is an example of a *vector*.

In general, a vector quantity is one that has both magnitude and direction. Graphically, it is represented by an arrow. The length of the arrow, related to a given scale, represents the magnitude of the vector. The direction of the arrow indicates the direction of the vector.

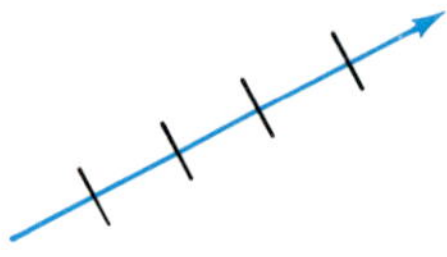

This vector has a magnitude of 5 units and a direction corresponding to the direction of the arrow.

Trigonometry is used to solve problems involving the addition of vectors. A boat is heading due west at 7 meters per second (m/s) across a river flowing due south at 3 m/s. Since each velocity acts on the boat at the same time, the resulting velocity is different from these two. In other words, in one second this resultant velocity will carry the boat 7 m due west and 3 m due south.

A figure is drawn to add these two velocity vectors. In this case, a scale of 1 unit = 1 m/s is used for the vectors. The two vectors are placed head-to-tail. The resultant vector is drawn from the tail of the first vector to the head of the second vector.

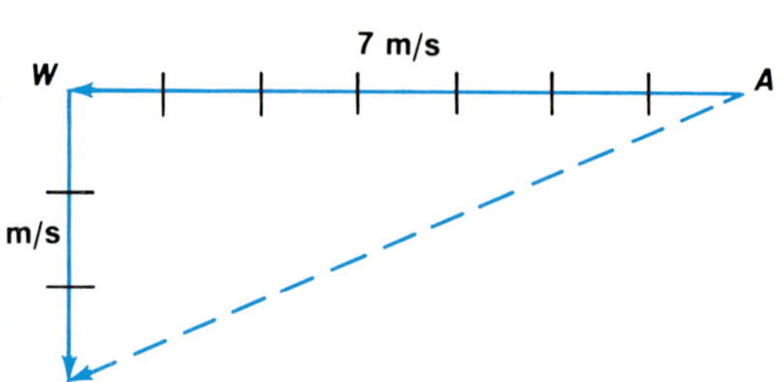

Trigonometry can now be used to solve for the resultant vector.

$$\tan A = \frac{3}{7}$$

$$\tan A = 0.4286$$
$$\tan 23° = 0.4286$$
$$m\angle A = 23$$

$$\sin A = \sin 23°$$

$$\sin 23° = \frac{3}{AS}$$

$$(\sin 23°)(AS) = 3$$
$$(0.3907)(AS) = 3$$

$$AS = \frac{3}{0.3907} \text{ or } 7.6785$$

Therefore, the boat is traveling at about 7.7 m/s, 23° south of west.

Exercise

Use trigonometry and vectors to solve.

A plane flying due north at 120 m/s is blown due east at 50 m/s by a strong wind. Find the plane's resultant velocity (speed and direction).

16-6 Trigonometric Functions and the Calculator

The examples provided in this lesson show how the calculator may be used to find trigonometric values of trigonometric functions and their inverses.

Examples

Using Calculators

1 **If sin x = 0.5346, find x.**   *Set your calculator in degree mode.*

   ENTER: 0.5346 [INV] [SIN]

   DISPLAY: 0.5346 32.31678847

   Therefore, x is approximately 32.3168°.   *Round to four decimal places.*

2 **Find csc 24°.**

   $$\csc x = \frac{1}{\sin x}$$

   ENTER: 24 [SIN]   

   DISPLAY: 24 0.406736643 2.458593336

   csc 24° is approximately 2.4586.   *Round to four decimal places.*

3 **If sec x = 1.7239, find x.**

   $$\sec x = \frac{1}{\cos x}, \cos x = \frac{1}{1.7239}.$$

   ENTER: 1.7239 [$^1/_x$]   [INV] [COS]

   DISPLAY: 1.7239 0.580080051 54.54382680

   Therefore, x is approximately 54.5438°.   *Round to four decimal places.*

4 **Find $\cot\left(\dfrac{4\pi}{3}\right)$.**   *Set your calculator in radian mode.*

   ENTER: 4 [÷] 3 [×] [π]      [TAN]   [$^1/_x$]

   DISPLAY: 4 3 3.1415927 4.1887902 0.0732387 13.6539817

   $\cot\left(\dfrac{4\pi}{3}\right)$ is approximately 13.6540.   *Round to four decimal places.*

5 Using Calculators

Find $\sin\left[\text{Cos}^{-1}\left(-\frac{\sqrt{3}}{2}\right)\right]$. *Make sure your calculator is in degree mode.*

ENTER: 3 $\sqrt{}$ $\div$ 2 $=$ $+/-$ INV COS SIN

DISPLAY: *3 1.732050808   2 0.866025404 − 0.866025404   150 0.5*

Therefore, $\sin\left[\text{Cos}^{-1}\left(-\frac{\sqrt{3}}{2}\right)\right]$ is 0.5.

Written Exercises

Use a calculator to find the value of x, in degrees, for each trigonometric function. Round your answers to four decimal places.

1. $\sin x = 0.6712$ **2.** $\cos x = 0.1389$ **3.** $\tan x = 0.8999$

4. $\sec x = 1.777$ **5.** $\csc x = 1.2955$ **6.** $\cot x = 1.9876$

Use a calculator to find each value to four decimal places.

7. $\csc 85°$ **8.** $\tan -90°$ **9.** $\sin 530°$ **10.** $\cos 5\pi$

11. $\tan \frac{-11\pi}{8}$ **12.** $\sin \frac{7\pi}{3}$ **13.** $\cot \frac{-3\pi}{5}$ **14.** $\sec \frac{-8\pi}{5}$

15. $\sec 600°$ **16.** $\cot 250°$ **17.** $\csc 7\pi$ **18.** $\csc (-890°)$

19. $(\sin 95°)(\tan 37°)$ **20.** $\left(\sin \frac{\pi}{3}\right)\left(\cos \frac{\pi}{8}\right)$ **21.** $\dfrac{3 (\sin 50°) + 9 (\cos 10°)}{\tan 290°}$

22. $\dfrac{3 \tan \frac{\pi}{6}}{2 \sin \frac{\pi}{3} - \sec 3\pi}$ **23.** $\text{Sin}^{-1} \frac{15}{16}$ **24.** $\text{Cos}^{-1} 0.89$

25. $\text{Tan}^{-1} \frac{3}{2}$ **26.** $\tan \left(\text{Cos}^{-1} \frac{\sqrt{3}}{3}\right)$ **27.** $\sin [\text{Tan}^{-1} (-\sqrt{6})]$

mini-review

1. Find a pattern and complete the sequence 3, 6, 18, 72, 360, _______, _______, _______, _______.

2. Two lab experiments are to be chosen from a group of 3 biology, 2 chemistry, and 4 physics experiments. What is the probability that the two chosen are 1 physics experiment and 1 biology experiment?

3. According to a certain prediction equation, a person with 12.5 years of education makes $14,500 a year. A person has 11 years of education and makes $13,200 a year. Predict how much money a person with 8 years of education will make a year.

State whether the two angles are coterminal.

4. $-59°, 661°$ **5.** $29°, 9361°$ **6.** $75°, -275°$

Coterminal Angles

We have defined coterminal angles as those angles whose measures differ by a multiple of 360°. If the degree measure of an angle is entered into the program at the right, the coterminal angle between 0° and 360° will be printed.

The program uses loops to cause it to keep adding or subtracting 360 until the angle measure is between 0° and 360°.

```
10   PRINT "ENTER A DEGREE MEASURE:"
20   INPUT A
30   IF A < 360 THEN 60
40   LET A = A − 360
50   GOTO 30
60   IF A > = 0 THEN 90
70   LET A = A + 360
80   GOTO 60
90   PRINT "ITS COTERMINAL ANGLE
         MEASURES ";A;" DEGREES."
100  END
```

When the program is run, the results should resemble those shown below.

```
] RUN
ENTER A DEGREE MEASURE:
?-1276
ITS COTERMINAL ANGLE MEASURES 164 DEGREES.
```

Exercises

Find the coterminal angle between 0° and 360°.

1. 720°   **2.** 1373.56°   **3.** −981°   **4.** −13,472°

State whether the two angles are coterminal.

5. −59°, 661°   **6.** 29°, 9361°   **7.** 49°, 1129°   **8.** 75°, −275°

9. How would you change lines 30, 40, and 70 if the angle measure given was in radians?

Vocabulary

initial side   (579)

terminal side   (579)

standard position   (579)

coterminal angles   (580)

unit circle   (580)

radian   (580)

sine   (582)

cosine   (582)

period   (584, 589)

periodic function   (584)

amplitude   (588, 589)

tangent   (592)

cotangent   (592)

secant   (592)

cosecant   (592)

principal values   (598)

Arccosine   (599)

Arcsine   (599)

Arctangent   (599)

1. Let θ stand for the measurement of an angle in standard position. Let (x, y) represent the coordinates of the point where the terminal side of the angle intersects the unit circle. Then sine $\theta = y$ and cosine $\theta = x$.   (582)

2. A function f is called periodic if there is a number a such that $f(x) = f(x + a)$ for all x in the domain of the function. The least positive value of a for which $f(x) = f(x + a)$ is the period of the function.   (584)

3. The amplitude of a periodic function is half the difference between its maximum and minimum values.   (588)

4. For functions of the form $y = a \sin b\theta$ and $y = a \cos b\theta$, the amplitude is $|a|$ and the period is $\dfrac{2\pi}{|b|}$.   (589)

5. Let θ stand for the measurement of an angle in standard position on a unit circle. Then the following equations hold.

    If $\cos \theta \neq 0$, $\tan \theta = \dfrac{\sin \theta}{\cos \theta}$ and $\sec \theta = \dfrac{1}{\cos \theta}$.

    If $\sin \theta \neq 0$, $\cot \theta = \dfrac{\cos \theta}{\sin \theta}$ and $\csc \theta = \dfrac{1}{\sin \theta}$.   (592)

6. Given $y = \text{Cos } x$, the inverse cosine function is defined by $y = \text{Cos}^{-1} x$ or $y = \text{Arccos } x$.   (599)

7. Given $y = \text{Sin } x$, the inverse sine function is defined by $y = \text{Sin}^{-1} x$ or $y = \text{Arcsin } x$.   (599)

8. Given $y = \text{Tan } x$, the inverse tangent function is defined by $y = \text{Tan}^{-1} x$ or $y = \text{Arctan } x$.   (599)

9. A calculator can be used to find the values of trigonometric functions and their inverses.   (603)

10. Trigonometric functions can be drawn on the graphics screen of a graphing calculator.   (596-597)

16-1 **Change each degree measure to radians.**

1. $120°$ **2.** $-315°$ **3.** $270°$ **4.** $255°$

Change each radian measure to degrees.

5. $\dfrac{\pi}{3}$ **6.** $-\dfrac{5}{12}\pi$ **7.** $\dfrac{4}{3}$ **8.** $\dfrac{7}{4}\pi$

16-2 **For each of the following, find the least positive angle that is coterminal.**

9. $-155°$ **10.** $830°$ **11.** $540°$ **12.** $945°$

13. $\dfrac{20}{3}\pi$ **14.** $-\dfrac{4}{3}\pi$ **15.** $-\dfrac{2}{9}\pi$ **16.** $-\dfrac{11}{6}\pi$

Find each value.

17. $\sin 120°$ **18.** $\cos 210°$ **19.** $\cos 3\pi$ **20.** $\sin (-150°)$

21. $\sin (-30°)$ **22.** $\sin \dfrac{5}{4}\pi$ **23.** $\cos (-135°)$

24. $\cos (300°)$ **25.** $(\sin 30°)^2 + (\cos 30°)^2$ **26.** $(\sin 45°)(\sin 225°)$

16-3 **Graph each function. State the amplitude and period for each graph.**

27. $y = \sin x$ **28.** $y = -\dfrac{1}{2}\cos \theta$ **29.** $y = 4 \sin 2\theta$

16-4 **Find each of the following.**

30. $\csc \pi$ **31.** $\sec (-30°)$ **32.** $\csc 135°$ **33.** $\cos 600°$

34. $\sin \dfrac{4}{3}\pi$ **35.** $\cot \dfrac{7}{6}\pi$ **36.** $\tan 120°$ **37.** $\sec (-60°)$

16-5 **Find each value.**

38. $\mathrm{Cos}^{-1} \left(\dfrac{\sqrt{3}}{2}\right)$ **39.** $\mathrm{Sin}^{-1} (-1)$ **40.** $\mathrm{Tan}^{-1} \sqrt{3}$

41. $\mathrm{Sin}^{-1} \left(\tan \dfrac{\pi}{4}\right)$ **42.** $\cos (\mathrm{Sin}^{-1} 1)$ **43.** $\sin \left(2\, \mathrm{Sin}^{-1} \dfrac{1}{2}\right)$

16-6 **Use a calculator to find the value of x for each trigonometric function.**

44. $\cos x = 0.4630$ **45.** $\csc x = 1.3729$

Use a calculator to find each value.

46. $\left(\cos \dfrac{2\pi}{3}\right)\left(\sin \dfrac{\pi}{4}\right)$ **47.** $\tan [\mathrm{Cos}^{-1} 0.8 + \mathrm{Sin}^{-1} (-0.4)]$

Chapter Test

Change each degree measure to radians.

1. $135°$ **2.** $275°$ **3.** $-150°$ **4.** $-4°$

Change each radian measure to degrees.

5. $\frac{4}{5}\pi$ **6.** $\frac{12}{5}\pi$ **7.** $-\frac{7}{4}\pi$ **8.** 7

For each of the following, find the least positive angle that is coterminal.

9. $620°$ **10.** $-260°$ **11.** $595°$ **12.** $-1270°$

Find each value.

13. $\sin 225°$ **14.** $\cos(-120°)$ **15.** $\cos\frac{3}{4}\pi$ **16.** $\sin\frac{7}{4}\pi$

Graph each function. Then state the amplitude and period.

17. $y = 2\sin 2x$ **18.** $y = \frac{3}{4}\cos\frac{2}{3}x$

Find each value.

19. $\tan 225°$ **20.** $\csc(-120°)$ **21.** $\cos\frac{2}{3}\pi$ **22.** $\sec 150°$

Find each value.

23. $\operatorname{Tan}^{-1}\frac{\sqrt{3}}{3}$ **24.** $\operatorname{Sin}^{-1}\left(-\frac{1}{2}\right)$ **25.** $\operatorname{Cos}^{-1}(\sin 30°)$ **26.** $\sin 2\left(\operatorname{Cos}^{-1}\frac{1}{2}\right)$

Use a calculator to find the value of x for each trigonometric function.

27. $\sin x = 0.7169$ **28.** $\cot x = 1.0099$

Use a calculator to find each value.

29. $\sin\left(\operatorname{Tan}^{-1} 1 - \operatorname{Cos}^{-1} 0.15\right)$ **30.** $\cot\left(\operatorname{Sin}^{-1}\frac{2}{5} + \operatorname{Cos}^{-1}\frac{2}{3}\right)$

The test questions on this page deal with a variety of concepts from arithmetic, algebra, and geometry.

Most standardized tests have a time limit, so you must budget your time carefully. Some questions will be much easier than others. If you cannot answer a question within a few minutes, go on to the next one. If there is still time left when you get to the end of the test, go back to the ones that you skipped.

Directions: Choose the one best answer. Write A, B, C, D, or E.

1. The length of a rectangle is 1 unit and the width is w. If the width is increased by 3 units, by how many units will the perimeter be increased?

(A) 2   (B) 4   (C) 6

(D) $2w + 3$ (E) $2w + 6$

2. O is the center of the circle below. $\overline{MO}$ is perpendicular to $\overline{NO}$ and the area of triangle MON is 40. What is the area of circle O?

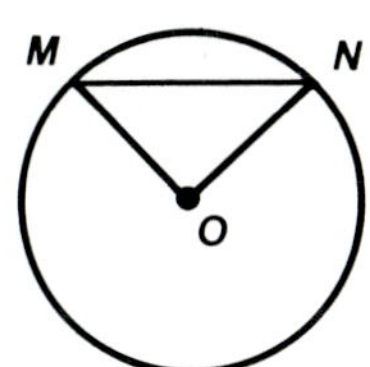

(A) 80π (B) 40π (C) 20π

(D) 160π (E) 240π

3. The distance between point $X(4, 0)$ and Y is 8. The coordinates of point Y could be any of the following except

(A) $(-4, 0)$ (B) $(0, 4\sqrt{3})$

(C) $(0, -4\sqrt{3})$ (D) $(8, 0)$

(E) $(4, 8)$

4. $x^2 + y^2 = 16$
$$xy = 8$$
$$(x + y)^2 = ?$$

(A) 32   (B) 22   (C) 16

(D) 24   (E) 40

5. If $x = -6$ and $\dfrac{1}{2y} = -12$, what is the value of y in terms of x?

(A) $2x$ (B) $x - 6$ (C) $2x - 6$

(D) $\dfrac{1}{2x}$ (E) $\dfrac{1}{4x}$

6. 5% of what number is 81?

(A) 405   (B) 16   (C) 1620

(D) 16,200   (E) 0.405

7. The area of triangle ABC is 8 square units. Angle C is 45°. What is AC?

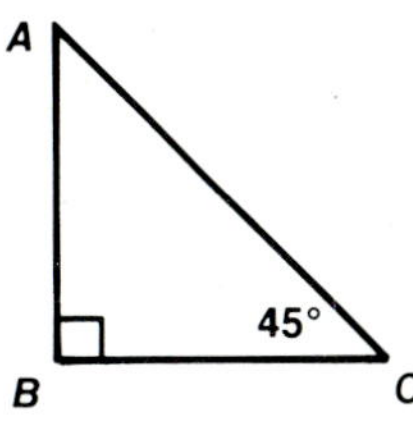

(A) $16\sqrt{2}$ (B) $4\sqrt{2}$ (C) $2\sqrt{2}$

(D) $8\sqrt{2}$ (E) $24\sqrt{2}$

8. A person travels five miles north, fifteen miles west, and fifteen miles north. How far (to the nearest mile) is he from the starting point?

(A) 35   (B) 20   (C) 15

(D) 30   (E) 25

CHAPTER 17

Trigonometric Identities and Equations

Rays of light change direction, or refract, as they move from one medium to another. For example, when white light moves through air onto a prism, the waves of each frequency of light refract different amounts and the colors are separated.

Angles of refraction can be computed by solving certain trigonometric equations. In this chapter, you will study methods for solving trigonometric equations.

17-1 Trigonometric Identities

Let θ be the measurement of an angle in standard position. Let (x, y) be the coordinates of the point of intersection of the terminal side of the angle and the unit circle. Then the following equations hold.

$$\cos \theta = x \text{ and } \sin \theta = y$$

The equation for the unit circle is $x^2 + y^2 = 1$. By substituting $\cos \theta$ for x and $\sin \theta$ for y, you obtain the following equation.

$$(\cos \theta)^2 + (\sin \theta)^2 = 1$$

Usually, this equation is written in the following way.

$$\sin^2 \theta + \cos^2 \theta = 1$$

An equation like $\sin^2 \theta + \cos^2 \theta = 1$ is called an **identity** because it is true for all values of θ. Some other trigonometric identities are given below.

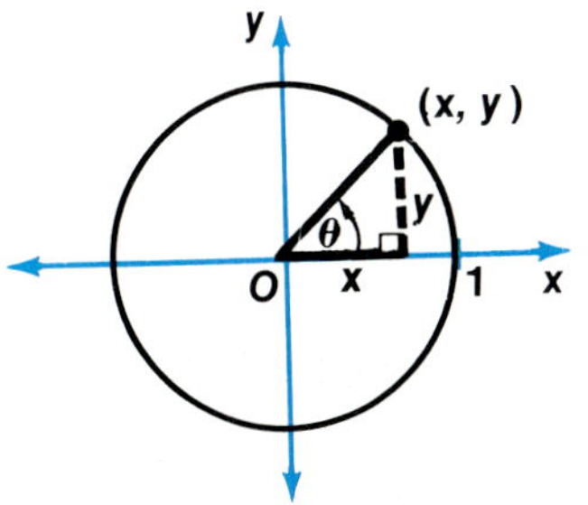

Equations that are true for all values of the variables for which they are defined are called identities.

The following trigonometric identities hold for all values of θ except those for which either side of the equation is undefined.

$$\sin^2 \theta + \cos^2 \theta = 1$$
$$1 + \tan^2 \theta = \sec^2 \theta$$
$$1 + \cot^2 \theta = \csc^2 \theta$$

Basic Trigonometric Identities

Example

1 Show that $1 + \tan^2 \theta = \sec^2 \theta$.

$$1 + \tan^2 \theta = 1 + \left(\frac{\sin \theta}{\cos \theta}\right)^2 \qquad \text{Definition of } \tan \theta.$$

$$= 1 + \frac{\sin^2 \theta}{\cos^2 \theta}$$

$$= \frac{\cos^2 \theta}{\cos^2 \theta} + \frac{\sin^2 \theta}{\cos^2 \theta} \qquad 1 = \frac{\cos^2 \theta}{\cos^2 \theta}$$

$$= \frac{\cos^2 \theta + \sin^2 \theta}{\cos^2 \theta}$$

$$= \frac{1}{\cos^2 \theta} \qquad \sin^2 \theta + \cos^2 \theta = 1$$

$$= \sec^2 \theta$$

These steps show the development of a basic trigonometric identity. The expression on one side of the equation is transformed into the form of the expression on the other side.

Trigonometric identities can be used to solve problems involving trigonometric functions. They can also be used to evaluate or simplify expressions containing trigonometric functions.

Examples

2 **If $\cot \theta = \frac{3}{5}$, find $\csc \theta$ for values of θ between $0°$ and $90°$.**

$$\csc^2 \theta = 1 + \cot^2 \theta \quad \textit{Trigonometric Identity}$$
$$= 1 + \left(\frac{3}{5}\right)^2$$
$$= \frac{34}{25}$$
$$\csc\theta = \sqrt{\frac{34}{25}} \quad \textit{csc } \theta \textit{ is positive for values of } \theta \textit{ between 0° and 90°.}$$
$$= \frac{\sqrt{34}}{5}$$

3 **If $\tan a = -\frac{3}{5}$, find $\cos a$ for values of a between $90°$ and $180°$.**

$$\sec^2 a = 1 + \tan^2 a \quad \textit{Trigonometric Identity}$$
$$= 1 + \left(-\frac{3}{5}\right)^2$$
$$\left(\frac{1}{\cos a}\right)^2 = \frac{34}{25} \quad \sec a = \frac{1}{\cos a}$$
$$\cos^2 a = \frac{25}{34}$$
$$\cos a = -\sqrt{\frac{25}{34}} \quad \textit{cos } a \textit{ is negative for values of } a \textit{ between 90° and 180°.}$$
$$= -\frac{5\sqrt{34}}{34}$$

4 **Simplify $\dfrac{1}{1 + \sin x} + \dfrac{1}{1 - \sin x}$.**

$$\frac{1}{1 + \sin x} + \frac{1}{1 - \sin x} = \frac{(1 - \sin x) + (1 + \sin x)}{(1 + \sin x)(1 - \sin x)}$$
$$= \frac{2}{1 - \sin^2 x}$$
$$= \frac{2}{\cos^2 x} \quad \sin^2 x + \cos^2 x = 1$$
$$= 2 \sec^2 x \quad \sec x = \frac{1}{\cos x}$$

Simplifying an expression that contains trigonometric functions means that the expression is written as a numerical value or in terms of a single trigonometric function, if possible.

Exploratory Exercises

Simplify each expression.

1. $\csc^2 \theta - \cot^2 \theta$

2. $\tan \theta \cos^2 \theta$

3. $\dfrac{\sin^2 \theta + \cos^2 \theta}{\sin^2 \theta}$

4. $\dfrac{\tan x}{\sin x}$

5. $\csc^2 \gamma - \cot^2 \gamma$

6. $\cos \alpha \csc \alpha$

7. $\sin \theta \cot \theta$

8. $\tan x \csc x$

9. $\dfrac{\cos x \csc x}{\tan x}$

10. $\dfrac{\tan \beta}{\cot \beta}$

11. $\dfrac{1 - \sin^2 \alpha}{\sin^2 \alpha}$

12. $\dfrac{1 + \tan^2 x}{1 + \cot^2 x}$

Written Exercises

Solve for values of θ between 0° and 90°.

13. If $\sin \theta = \frac{1}{2}$, find $\cos \theta$.

14. If $\cos \theta = \frac{2}{3}$, find $\sin \theta$.

15. If $\sin \theta = \frac{4}{5}$, find $\cos \theta$.

16. If $\sin \theta = \frac{3}{4}$, find $\sec \theta$.

17. If $\cos \theta = \frac{2}{3}$, find $\csc \theta$.

18. If $\cos \theta = \frac{4}{5}$, find $\tan \theta$.

19. If $\tan \theta = 4$, find $\sin \theta$.

20. If $\cot \theta = 2$, find $\tan \theta$.

Solve for values of θ between 90° and 180°.

21. If $\sin \theta = \frac{3}{5}$, find $\cos \theta$.

22. If $\sin \theta = \frac{1}{2}$, find $\tan \theta$.

23. If $\cos \theta = -\frac{3}{5}$, find $\csc \theta$.

24. If $\tan \theta = -2$, find $\sec \theta$.

Solve for values of θ between 180° and 270°.

25. If $\cot \theta = \frac{1}{4}$, find $\csc \theta$.

26. If $\sec \theta = -3$, find $\tan \theta$.

27. If $\sin \theta = -\frac{1}{2}$, find $\cos \theta$.

28. If $\cos \theta = -\frac{3}{5}$, find $\csc \theta$.

Solve for values of θ between 270° and 360°.

29. If $\cos \theta = \frac{5}{13}$, find $\sin \theta$.

30. If $\tan \theta = -1$, find $\sec \theta$.

31. If $\sec \theta = \frac{5}{3}$, find $\cos \theta$.

32. If $\csc \theta = -\frac{5}{3}$, find $\cos \theta$.

Simplify each expression.

33. $\tan \beta \cot \beta$

34. $\sec^2 \theta - 1$

35. $\sin x + \cos x \tan x$

36. $\csc a \cos a \tan a$

37. $\dfrac{1}{\sin^2 \theta} - \dfrac{\cos^2 \theta}{\sin^2 \theta}$

38. $\sin \beta (1 + \cot^2 \beta)$

39. $2(\csc^2 \theta - \cot^2 \theta)$

40. $\dfrac{\tan^2 \theta - \sin^2 \theta}{\tan^2 \theta \sin^2 \theta}$

Show that each equation is an identity.

41. $1 + \cot^2 \theta = \csc^2 \theta$

42. $\dfrac{\sec \theta}{\csc \theta} = \tan \theta$

43. $\sin x \sec x = \tan x$

44. $\sec a - \cos a = \sin a \tan a$

Challenge Exercises

45. If $\cos x = \frac{1}{4}$, find $\sin x$.

46. If $\sin \theta = \frac{1}{3}$, find $\tan \theta$.

47. If $\sin \theta = \frac{1}{3}$, find $\dfrac{\cos \theta \tan \theta}{\csc \theta}$.

48. If $\tan \beta = \frac{3}{4}$, find $\dfrac{\sin \beta \sec \beta}{\cot \beta}$.

17-2 Verifying Trigonometric Identities

You can use the basic trigonometric identities, along with the definitions of the trigonometric functions, to verify other identities. For example, suppose you wish to know if $\sin\theta \sec\theta \cot\theta = 1$ is an identity. To find out, simplify the expression on the left side of the equation by using the identities and definitions.

$$\sin\theta \sec\theta \cot\theta \overset{?}{=} 1$$

$$\sin\theta \cdot \frac{1}{\cos\theta} \cdot \frac{1}{\tan\theta} \overset{?}{=} 1 \qquad \sec\theta = \frac{1}{\cos\theta} \text{ and } \cot\theta = \frac{1}{\tan\theta}$$

$$\frac{\sin\theta}{\cos\theta} \cdot \frac{1}{\tan\theta} \overset{?}{=} 1 \qquad \text{Multiply } \sin\theta \text{ and } \frac{1}{\cos\theta}.$$

$$\tan\theta \cdot \frac{1}{\tan\theta} \overset{?}{=} 1 \qquad \tan\theta = \frac{\sin\theta}{\cos\theta}$$

$$1 = 1$$

Thus, $\sin\theta \sec\theta \cot\theta = 1$ is an identity.

In a way, verifying an identity is like checking the solution to an equation. You do not know if the expressions on each side are equal. That is what you are trying to verify. So, you must simplify one or both sides of the sentence *separately* until they are the same.

Often it is easier to work with only one side of the sentence. You may choose either side.

Examples

1 **Verify $\tan^2 x - \sin^2 x = \tan^2 x \sin^2 x$.**

$$\tan^2 x - \sin^2 x \overset{?}{=} \tan^2 x \sin^2 x$$

$$\left(\frac{\sin x}{\cos x}\right)^2 - \sin^2 x \overset{?}{=} \tan^2 x \sin^2 x \qquad \tan x = \frac{\sin x}{\cos x}$$

$$\left(\frac{1}{\cos^2 x} - 1\right)\sin^2 x \overset{?}{=} \tan^2 x \sin^2 x \qquad \textit{Distributive Property}$$

$$(\sec^2 x - 1)\sin^2 x \overset{?}{=} \tan^2 x \sin^2 x \qquad \sec x = \frac{1}{\cos x}$$

$$\tan^2 x \sin^2 x = \tan^2 x \sin^2 x \qquad 1 + \tan^2 x = \sec^2 x$$

Thus, the identity has been verified.

2 **Verify $1 - \cot^4 x = 2\csc^2 x - \csc^4 x$.**

$$1 - \cot^4 x \overset{?}{=} 2\csc^2 x - \csc^4 x$$

$$(1 - \cot^2 x)(1 + \cot^2 x) \overset{?}{=} 2\csc^2 x - \csc^4 x \qquad \textit{Factor.}$$

$$[1 - (\csc^2 x - 1)][\csc^2 x] \overset{?}{=} 2\csc^2 x - \csc^4 x \qquad 1 + \cot^2 x = \csc^2 x$$

$$(2 - \csc^2 x)(\csc^2 x) \overset{?}{=} 2\csc^2 x - \csc^4 x \qquad \textit{Simplify.}$$

$$2\csc^2 x - \csc^4 x = 2\csc^2 x - \csc^4 x$$

Thus, the identity has been verified.

The following suggestions are helpful in verifying trigonometric identities. Study the example to see how these suggestions can be used to verify an identity.

Suggestions for
Verifying Identities

- **Start with the more complicated side of the equation. Transform the expression into the form of the simpler side.**

- **Work with each side of the equation at the same time. Transform each expression separately into the same form.**

- **Substitute one or more basic trigonometric identities to simplify the expression.**

- **Try factoring or multiplying to simplify the expression.**

- **Multiply both numerator and denominator by the same trigonometric expression.**

There is often more than one way to verify an identity. Remember that verifying an identity is not the same as solving an equation. An identity is true for all values of the variable except those values for which either side of the equation is undefined.

Example

3 Verify that $\dfrac{1 - \cos x}{\sin x} = \dfrac{\sin x}{1 + \cos x}$.

$$\dfrac{1 - \cos x}{\sin x} \stackrel{?}{=} \dfrac{\sin x}{1 + \cos x}$$

$$\dfrac{1 - \cos x}{\sin x} \stackrel{?}{=} \dfrac{\sin x(1 - \cos x)}{(1 + \cos x)(1 - \cos x)}$$ *Multiply both numerator and denominator by $1 - \cos x$.*

$$\dfrac{1 - \cos x}{\sin x} \stackrel{?}{=} \dfrac{\sin x(1 - \cos x)}{1 - \cos^2 x}$$ *Simplify the denominator.*

$$\dfrac{1 - \cos x}{\sin x} \stackrel{?}{=} \dfrac{\sin x(1 - \cos x)}{\sin^2 x}$$ *Substitute $\sin^2 x$ for $1 - \cos^2 x$.*

$$\dfrac{1 - \cos x}{\sin x} = \dfrac{1 - \cos x}{\sin x}$$ *Simplify.*

Thus, the identity has been verified.

Verify each identity.

1. $\tan \beta(\cot \beta + \tan \beta) = \sec^2 \beta$

2. $\tan^2 \theta \cos^2 \theta = 1 - \cos^2 \theta$

3. $\csc x \sec x = \cot x + \tan x$

4. $\sec^2 x - \tan^2 x = \tan x \cot x$

5. $\dfrac{\sec \theta}{\sin \theta} - \dfrac{\sin \theta}{\cos \theta} = \cot \theta$

6. $\dfrac{1}{\sec^2 \theta} + \dfrac{1}{\csc^2 \theta} = 1$

7. $\dfrac{\sin \alpha}{1 - \cos \alpha} + \dfrac{1 - \cos \alpha}{\sin \alpha} = 2 \csc \alpha$

8. $\dfrac{\sec \alpha + \csc \alpha}{1 + \tan \alpha} = \csc \alpha$

9. $\dfrac{\cos^2 x}{1 - \sin x} = 1 + \sin x$

10. $\dfrac{1 - \cos \theta}{1 + \cos \theta} = (\csc \theta - \cot \theta)^2$

11. $\dfrac{\sin \theta}{\sec \theta} = \dfrac{1}{\tan \theta + \cot \theta}$

12. $\dfrac{\sec \theta + 1}{\tan \theta} = \dfrac{\tan \theta}{\sec \theta - 1}$

13. $\dfrac{\cot x + \csc x}{\sin x + \tan x} = \cot x \csc x$

14. $\dfrac{1 - 2 \cos^2 \theta}{\sin \theta \cos \theta} = \tan \theta - \cot \theta$

15. $\cos^2 x + \tan^2 x \cos^2 x = 1$

16. $\dfrac{\cos x}{1 + \sin x} + \dfrac{\cos x}{1 - \sin x} = 2 \sec x$

17. $\dfrac{1 + \tan^2 \theta}{\csc^2 \theta} = \tan^2 \theta$

18. $\cot x(\cot x + \tan x) = \csc^2 x$

19. $\dfrac{1 + \sin x}{\sin x} = \dfrac{\cot^2 x}{\csc x - 1}$

20. $\dfrac{1 + \tan \gamma}{1 + \cot \gamma} = \dfrac{\sin \gamma}{\cos \gamma}$

21. $\cos^4 x - \sin^4 x = \cos^2 x - \sin^2 x$

22. $1 + \sec^2 x \sin^2 x = \sec^2 x$

23. $\dfrac{\tan^2 x}{\sec x - 1} = 1 + \dfrac{1}{\cos x}$

24. $\sin \theta + \cos \theta = \dfrac{1 + \tan \theta}{\sec \theta}$

mini-review

A committee of 5 is to be selected from a group of 7 women and 6 men. What is the probability of the following?

1. all men

2. at least 4 women

3. How many ways can the letters of the word NATHAN be arranged?

4. How many ways can 7 people be arranged around a circular table?

5. Change $\dfrac{11\pi}{6}$ to degrees.

6. Change $-30°$ to radians.

7. Graph $y = 2 \sin \theta$. State the amplitude and period of the graph.

8. Graph $y = \cot x$. State the amplitude and period of the graph.

Use the stem and leaf plot at the right to solve each problem.

9. Find the median, mode, and mean of the scores.

10. Find the range and interquartile range of the scores.

11. Find any outliers in the scores.

12. Find the standard deviation of the scores.

English Test Scores

Stem	Leaf
2	4 6 9 9 9
3	1 3 6 8 9 9
4	1 2 3 5 7 8 9
5	0 0

$3 \mid 8 = 38$

In most problems, a set of conditions or facts is given and you must arrive at a solution. However, in some cases, you are given the final solution or goal and then asked for an intermediate condition. In other cases, it may be faster to determine how the problem ends and then work backwards rather than to start from the beginning. Consider the following example.

Example: **Find the sum of the reciprocals of two numbers whose sum is 2 and whose product is 3.**

A first reaction might be to set up the following system of equations.

$$x + y = 2$$
$$xy = 3$$

But solving the system is complicated. Rather than using this approach, work backwards. The desired outcome is $\frac{1}{x} + \frac{1}{y}$.

$$\frac{1}{x} + \frac{1}{y} = \frac{y}{xy} + \frac{x}{xy} \qquad \textit{The LCD is xy.}$$
$$= \frac{x + y}{xy}$$

The two original equations immediately reveal the numerator and denominator of this fraction.

$$\frac{x + y}{xy} = \frac{2}{3} \qquad \textit{Substitute.}$$

The sum of the reciprocals is $\frac{2}{3}$.

Exercises

1. A pirate found a treasure chest containing silver coins. He buried half of them and gave half of the remaining coins to his mother. If he was left with 4550 coins, how many were in the treasure chest that he found?

2. If the sum of two numbers is 2 and the product of the same two numbers is 3, find the sum of the squares of the reciprocals of these numbers.

3. Paul, Eric, and Garnet are playing a card game. They have a rule that when a player loses a hand, he must subtract enough points from his score to double each of the other players' scores. First Paul loses a hand, then Eric, and then Garnet. Each player now has 8 points. Who lost the most points?

4. Tim collects model cars. He decides to give them away. First he gives half of them plus half a car more to Amy. Then he gives half of what is left plus half a car more to Tina. Then he has one car left which he gives to Aaron. How many cars did Tim start with? (Assume that no car is cut in half.)

Graphing Calculator Application: Verifying Trigonometric Identities

A graphing calculator can be used to help determine whether an equation is an identity.

The expression on each side of an equation can be used to form two functions. For example, the equation $\dfrac{1 + \sin x}{\cos x} = \dfrac{\cos x}{1 - \sin x}$ yields the functions $y = \dfrac{1 + \sin x}{\cos x}$ and $y = \dfrac{\cos x}{1 - \sin x}$. The two functions can be graphed on the same set of axes on a graphing calculator.

If the two graphs do not match, then the equation *is not* an identity.

If the two graphs match, then the equation *may* be an identity. In that case, it is necessary to verify algebraically that the equation is an identity.

Examples

1   **Determine whether $\dfrac{1 + \sin x}{\cos x} = \dfrac{\cos x}{1 - \sin x}$ may be an identity.**

Choose a set of range parameter values that will produce complete graphs of $y = \dfrac{1 + \sin x}{\cos x}$ and $y = \dfrac{\cos x}{1 - \sin x}$. One possible set of values is given below. Set the range parameters to these values and graph the equations.

Xmin: -270 Xmax: 810   Xscl: 90   Ymin: -4 Ymax: 4   Yscl: 1

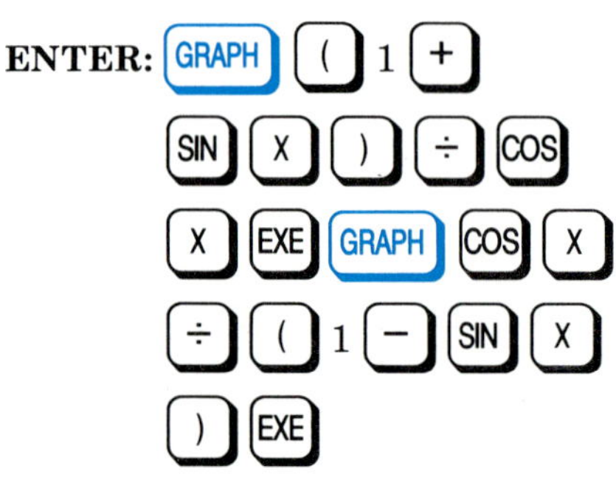

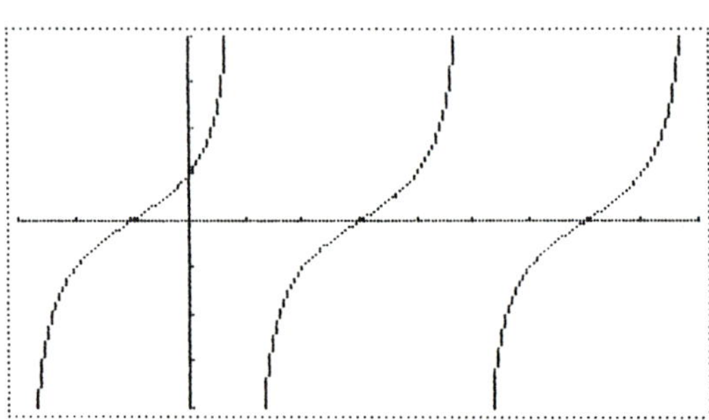

The graphs of $y = \dfrac{1 + \sin x}{\cos x}$ and $y = \dfrac{\cos x}{1 - \sin x}$ match. Thus, the equation may be an identity.

2   **Determine whether $\tan \dfrac{x}{2} = \dfrac{2 \sin x}{1 + 2 \cos x}$ may be an identity.**

Choose a set of range parameter values that will produce complete graphs of $y = \tan \dfrac{x}{2}$ and $y = \dfrac{2 \sin x}{1 + 2 \cos x}$. One possible set of values is given on the next page. Set the range parameters to these values and graph the equations.

$$\text{Xmin: } -180° \quad \text{Xmax: } 540 \quad \text{Xscl: } 180 \quad \text{Ymin: } -6 \quad \text{Ymax: } 6 \quad \text{Yscl: } 2$$

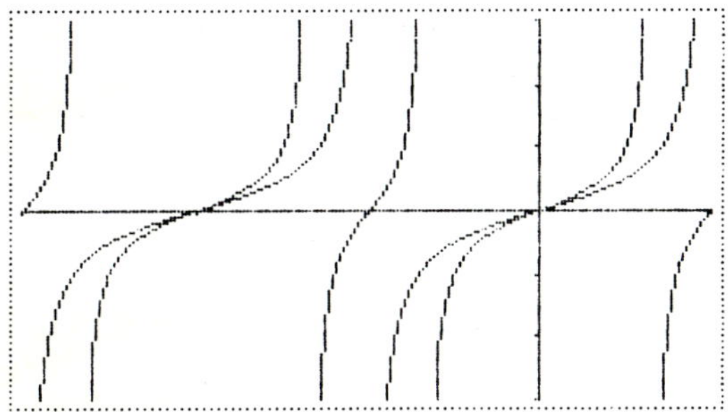

ENTER: GRAPH TAN (X ÷ 2) EXE GRAPH 2 SIN X ÷ (1 + 2 COS X) EXE

The graphs of $y = \tan \dfrac{x}{2}$ and $y = \dfrac{2 \sin x}{1 + 2 \cos x}$ *do not* match.

Thus, the equation *is not* an identity.

Exploratory Exercises

Graph $y = 1 - \sin^2 x$ and $y = \cos^2 x$ for each set of range parameter values. Then sketch the graphs shown on the graphics screen, indicating the x-axis scale and y-axis scale. Use degree mode for Exercises 1–3 and radian mode for Exercise 4.

1. Xmin: -360, Xmax: 360, Xscl: 90, Ymin: -1.6, Ymax: 1.6, Yscl: 0.5
2. Xmin: 0, Xmax: 1080, Xscl: 90, Ymin: 0, Ymax: 1, Yscl: 0.2
3. Xmin: -360, Xmax: 0, Xscl: 45, Ymin: -0.2, Ymax: 1, Yscl: 0.2
4. Xmin: -9.5, Xmax: 9.5, Xscl: 2, Ymin: 0, Ymax: 1.5, Yscl: 0.1

5. Based on the results of Exercises 1–4, does the choice of range parameter values affect whether two graphs match?

Written Exercises

Use a graphing calculator to determine whether each equation *may* be an identity. Write *yes* or *no*.

6. $\cot x + \tan x = \csc x \cot x$
7. $\sec^2 x + \csc^2 x = \sec^2 x \csc^2 x$
8. $\cos^2 x + \tan^2 x \cos^2 x = 1$
9. $\tan x(\cot x + \tan x) = \sec^2 x$
10. $\cot^2 x \sec^2 x = 1 + \cot^2 x$
11. $\sin x + \cos x \tan x = 2 \sin x$
12. $\cos x \sin x \tan x = 1$
13. $\csc x - \cos x = \sin x \tan x$
14. $\sin (90° - x) = \cos x$
15. $\sin (x - 90°) = \cos x$
16. $\dfrac{1}{\sec x} + \dfrac{1}{\csc x} = 1$
17. $\dfrac{1}{\sin^2 x} + \dfrac{1}{\cos^2 x} = 1$
18. $\dfrac{\sec x}{\sin x} - \dfrac{\sin x}{\cos x} = \cot x$
19. $\dfrac{1 + \tan x}{1 + \cot x} = \dfrac{\sin x}{\cos x}$
20. $\dfrac{1 - \cos x}{\sin x} = \dfrac{\sin x}{1 + \cos x}$
21. $\dfrac{\cos x}{\sec x + 1} + \dfrac{\cos x}{\sec x - 1} = 2 \cot^2 x$
22. $\cos 2x + 2 \sin^2 x = 1$
23. $\cos 2x + 1 = 2 \cos^2 x$
24. $\dfrac{1 + \sin x}{\sin x} = \dfrac{\csc^2 x}{\csc x - 1}$
25. $\dfrac{\sin x}{\sin x + \cos x} = \dfrac{\tan x}{1 + \tan x}$

17-3 Differences and Sums

It is often helpful to use formulas for the trigonometric values of the difference or sum of two angles. For example, you could find sin 15° by evaluating sin (45 − 30)°.

The figure at the right shows two angles, α and β, in standard position on the unit circle.

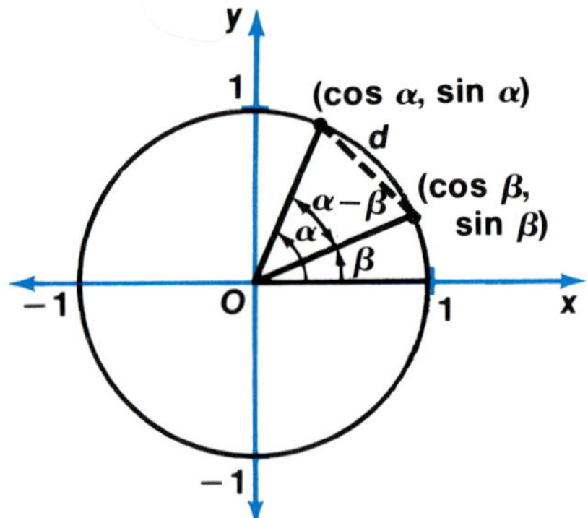

Use the distance formula to find d.

$$d = \sqrt{(\cos \alpha - \cos\beta)^2 + (\sin \alpha - \sin \beta)^2}$$

$$\begin{aligned}
d^2 &= (\cos \alpha - \cos \beta)^2 + (\sin \alpha - \sin \beta)^2 \\
&= (\cos^2 \alpha - 2 \cos \alpha \cos \beta + \cos^2 \beta) + (\sin^2 \alpha - 2 \sin \alpha \sin \beta + \sin^2 \beta) \\
&= \cos^2 \alpha + \sin^2 \alpha + \cos^2 \beta + \sin^2 \beta - 2 \cos \alpha \cos \beta - 2 \sin \alpha \sin \beta \\
&= 1 + 1 - 2 \cos \alpha \cos \beta - 2 \sin \alpha \sin \beta \qquad \sin^2 \alpha + \cos^2 \alpha = 1 \text{ and } \sin^2 \beta + \cos^2 \beta = 1 \\
&= 2 - 2 \cos \alpha \cos \beta - 2 \sin \alpha \sin \beta
\end{aligned}$$

The figure below shows the angle having measure $\alpha - \beta$ in standard position on the unit circle.

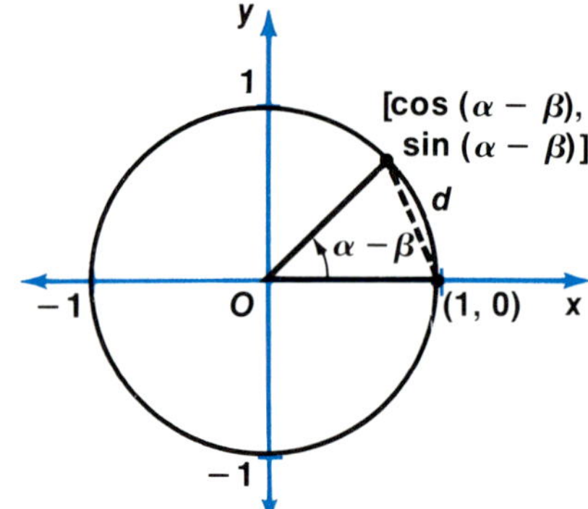

$$d = \sqrt{[\cos (\alpha - \beta) - 1]^2 + [\sin (\alpha - \beta) - 0]^2}$$

$$\begin{aligned}
d^2 &= [\cos (\alpha - \beta) - 1]^2 + [\sin (\alpha - \beta) - 0]^2 \\
&= [\cos^2(\alpha - \beta) - 2 \cos (\alpha - \beta) + 1] + [\sin^2(\alpha - \beta)] \\
&= \cos^2(\alpha - \beta) + \sin^2(\alpha - \beta) - 2 \cos (\alpha - \beta) + 1 \\
&= \qquad\qquad 1 \qquad\qquad - 2 \cos (\alpha - \beta) + 1 \\
&= 2 - 2 \cos (\alpha - \beta)
\end{aligned}$$

By equating the two expressions obtained for d^2, it is possible to find a formula for cos $(\alpha - \beta)$.

$$2 - 2 \cos (\alpha - \beta) = 2 - 2 \cos \alpha \cos \beta - 2 \sin \alpha \sin \beta$$
$$-1 + \cos (\alpha - \beta) = -1 + \cos \alpha \cos \beta + \sin \alpha \sin \beta \qquad \textit{Divide each side by } -2.$$
$$\cos (\alpha - \beta) = \cos \alpha \cos \beta + \sin \alpha \sin \beta \qquad \textit{Add 1 to each side.}$$

Use the formula for cos $(\alpha - \beta)$ to find a formula for cos $(\alpha + \beta)$.

$$\begin{aligned}
\cos (\alpha + \beta) &= \cos[\alpha - (-\beta)] \\
&= \cos \alpha \cos (-\beta) + \sin \alpha \sin (-\beta) \\
&= \cos \alpha \cos \beta - \sin \alpha \sin \beta \qquad \cos (-\beta) = \cos \beta \text{ and } \sin (-\beta) = -\sin \beta
\end{aligned}$$

1 Use the formula for $\cos(\alpha - \beta)$ to find $\cos(90° - \theta)$.

$$\cos(90° - \theta) = \cos 90° \cos \theta + \sin 90° \sin \theta$$
$$= 0 \cdot \cos \theta + 1 \cdot \sin \theta$$
$$= \sin \theta$$

2 Use the formula for $\cos(90° - \theta)$ (Example 1) to find $\sin(90° - \gamma)$.

$$\sin(90° - \gamma) = \cos[90° - (90° - \gamma)] \qquad \textit{Substitute } (90° - \gamma) \textit{ for } \theta \textit{ in}$$
$$= \cos(90° - 90° + \gamma) \qquad \sin \theta = \cos(90° - \theta).$$
$$= \cos \gamma$$

3 Use the formulas for $\cos(90° - \theta)$ and $\sin(90° - \theta)$ (Examples 1 and 2) and for $\cos(\alpha - \beta)$ to find $\sin(\alpha - \beta)$.

$$\sin(\alpha - \beta) = \cos[90° - (\alpha - \beta)] \quad \textit{Substitute } (\alpha - \beta) \textit{ for } \theta \textit{ in } \sin \theta = \cos(90° - \theta).$$
$$= \cos[(90° - \alpha) + \beta] \quad \textit{Distributive and Associative Properties}$$
$$= \cos(90° - \alpha)\cos \beta - \sin(90° - \alpha)\sin \beta$$
$$= \sin \alpha \cos \beta - \cos \alpha \sin \beta$$

The following identities hold for all values of α and β.

$$\cos(\alpha \pm \beta) = \cos \alpha \cos \beta \mp \sin \alpha \sin \beta$$
$$\sin(\alpha \pm \beta) = \sin \alpha \cos \beta \pm \cos \alpha \sin \beta$$

Difference and Sum Formulas

The examples below show how to evaluate expressions using the sum and difference formulas.

Examples

4 Evaluate $\sin 15°$.

$$\sin 15° = \sin(45° - 30°)$$
$$= \sin 45° \cos 30° - \cos 45° \sin 30° \quad \sin(\alpha - \beta) = \sin \alpha \cos \beta - \cos \alpha \sin \beta$$
$$= \frac{\sqrt{2}}{2} \cdot \frac{\sqrt{3}}{2} - \frac{\sqrt{2}}{2} \cdot \frac{1}{2} \text{ or } \frac{\sqrt{6} - \sqrt{2}}{4} \quad \frac{\sqrt{6} - \sqrt{2}}{4} \approx 0.2588$$

5 Evaluate $\cos 175° \cos 50° - \sin 175° \sin 50°$.

$$\cos 175° \cos 50° - \sin 175° \sin 50° = \cos(175° + 50°) \quad \cos \alpha \cos \beta - \sin \alpha \sin \beta$$
$$= \cos 225° \qquad\qquad\qquad\quad = \cos(\alpha + \beta)$$
$$= -\frac{\sqrt{2}}{2} \qquad -\frac{\sqrt{2}}{2} \approx -0.7071$$

Exploratory Exercises

Write each angle measure in terms of sums or differences of 30°, 45°, 60°, and 90° or their multiples.

1. 105° **2.** −15° **3.** −165° **4.** 165°

5. 75° **6.** −75° **7.** 285° **8.** 255°

Written Exercises

Evaluate each expression.

9. $\sin 75°$ **10.** $\sin 165°$ **11.** $\sin 285°$ **12.** $\cos 105°$

13. $\cos 195°$ **14.** $\cos 255°$ **15.** $\cos 15°$ **16.** $\sin 105°$

17. $\cos 165°$ **18.** $\sin 195°$ **19.** $\cos 345°$ **20.** $\cos 285°$

21. $\cos 25° \cos 5° - \sin 25° \sin 5°$

22. $\sin 40° \cos 20° + \cos 40° \sin 20°$

23. $\cos 80° \cos 20° + \sin 80° \sin 20°$

24. $\sin 65° \cos 35° - \cos 65° \sin 35°$

Verify each identity.

25. $\sin (270° - \theta) = -\cos \theta$

26. $\cos (270° - \theta) = -\sin \theta$

27. $\sin (180° + \theta) = -\sin \theta$

28. $\cos (180° + \theta) = -\cos \theta$

29. $\sin (90° + \theta) = \cos \theta$

30. $\cos (90° + \theta) = -\sin \theta$

31. $\sin (x + y) \sin (x - y) = \sin^2 x - \sin^2 y$

32. $\sin \left(\theta + \dfrac{\pi}{3} \right) - \cos \left(\theta + \dfrac{\pi}{6} \right) = \sin \theta$

33. $\sin (60° + \theta) + \sin (60° - \theta) = \sqrt{3} \cos \theta$

34. $\cos (x + y) + \cos (x - y) = 2 \cos x \cos y$

35. $\cos (x + y) \cos (x - y) = \cos^2 y - \sin^2 x$

36. $\sin (x + 30°) + \cos (x + 60°) = \cos x$

Use the identity $\tan (\alpha - \beta) = \dfrac{\tan \alpha - \tan \beta}{1 + \tan \alpha \tan \beta}$ **to evaluate each expression.**

37. $\tan (225° - 120°)$ **38.** $\tan (315° - 120°)$ **39.** $\tan (225° - 240°)$

40. $\tan (315° + 60°)$ **41.** $\tan (30° + 30°)$ **42.** $\tan (210° + 120°)$

43. $\tan 285°$ **44.** $\tan 195°$ **45.** $\tan 165°$

46. $\tan 75°$ **47.** $\tan (180° - \theta)$ **48.** $\tan (45° + \beta)$

Challenge Exercises

49. Use the formulas for $\sin (\alpha + \beta)$ and $\cos (\alpha + \beta)$ to derive the formula for $\tan (\alpha + \beta)$.

(*Hint:* Divide all terms of the expression by $\cos \alpha \cos \beta$.)

50. Use the sum and difference formulas for sin and cos to verify the following identities.

$$\cos (-\alpha) = \cos \alpha$$
$$\sin (-\alpha) = -\sin \alpha$$

Perform the indicated operations. Write each answer in simplest form.

1. $\dfrac{2y^2 + 13y - 7}{3y^2 + 5y + 2} \cdot \dfrac{3y^2 - 7y - 6}{y^2 + 4y - 21}$

2. $\dfrac{8x + 12}{3x^2 + 17x + 10} \div \dfrac{4x^2 - 6x - 18}{x^2 + 2x - 15}$

3. $\dfrac{n - 3}{n^2 - 1} + \dfrac{6}{4n + 4}$

4. $4d + 5 - \dfrac{3d}{4d + 5}$

5. Find the first three terms of this arithmetic series: $a_1 = 9$, $n = 12$, $S_n = 438$.

6. Find the missing geometric means of the sequence 8, _____, _____, _____, 128.

7. Evaluate $\displaystyle\sum_{r=1}^{6} (4 + 3r)$.

8. Find the sum of the series
$\dfrac{1}{3} + \dfrac{1}{4} + \dfrac{3}{16} + \dfrac{9}{64} + \dots$

State whether the graph of each equation is a parabola, a circle, an ellipse, or a hyperbola.

9. $2x^2 + y^2 - 4x + 4y - 10 = 0$
10. $x^2 - 6x + y^2 - 10y + 34 = 16$
11. $5x^2 - 30x - y^2 - 10y = 15$
12. $y^2 = 7x + 21$

13. Find the solutions to the system $x = 2y^2 - 3$ and $x^2 + 9y^2 = 36$.

14. How many ways can nine different books be arranged on a shelf?

15. How many basketball teams of 5 members can be formed from 12 players?

16. A blue and a red die are tossed. What is the probability that neither show 4?

17. Suppose two letters are selected from the word *mathematics*. What is the probability of selecting two vowels?

Solve each equation.

18. $\dfrac{5}{x + 2} = \dfrac{4}{x} + \dfrac{5}{3x}$

19. $\log_{32} 2 = x$

20. $\log_7 35 = \log_7 (x^2 + 2x)$

21. $\log_3 (x + 4) + \log_3 (x - 4) = 2$

22. Mr. Block has \$6000 to invest. If he wants to have \$10,000 after 8 years, what interest rate is needed, assuming continuous compounding?

23. Is $a - 3$ a factor of $a^4 - a^3 - 6a^2 + 8a - 6$?

24. Find all rational zeros of $f(x) = 2x^3 + x^2 - 16x - 15$.

25. Graph $f(x) = x^3 - 4x^2 - 1$.

26. If $f(x) = x^2 + 3$ and $g(x) = 2x - 1$, find $[g \circ f](x)$.

27. Find the median, mode, and mean of $\{15, 7, 1, 7, 6, 16, 22, 27, 6, 30, 1, 6\}$.

28. Find the range for the set of data in Exercise 27.

29. Find the upper and lower quartiles for the set of data in Exercise 27.

30. Find any outliers for the set of data in Exercise 27.

31. Graph $y = -\cos \frac{1}{2}x$.

32. State the amplitude and period of $y = -3 \sin \frac{2}{3}x$.

33. Change $\dfrac{11\pi}{12}$ to degrees.

34. Find $\operatorname{Sin}^{-1}(\cos 60°)$.

35. If $\sin x = \frac{2}{3}$, find $\sec x$ for values of x between 90° and 180°.

36. Verify $\dfrac{\cos x}{\sec x + 1} + \dfrac{\cos x}{\sec x - 1} = 2 \cot^2 x$.

37. Find the exact value of $\sin 285°$.

17-4 Double Angles and Half Angles

You can use the formula for $\sin(\alpha + \beta)$ to find $\sin 2\theta$ and the formula for $\cos(\alpha + \beta)$ to find $\cos 2\theta$.

$$\begin{aligned}
\sin 2\theta &= \sin(\theta + \theta) \\
&= \sin\theta\cos\theta + \cos\theta\sin\theta \\
&= 2\sin\theta\cos\theta
\end{aligned} \qquad \begin{aligned}
\cos 2\theta &= \cos(\theta + \theta) \\
&= \cos\theta\cos\theta - \sin\theta\sin\theta \\
&= \cos^2\theta - \sin^2\theta
\end{aligned}$$

Alternate forms for $\cos 2\theta$ can also be found by making substitutions into the expression $\cos^2\theta - \sin^2\theta$.

$\cos^2\theta - \sin^2\theta = (1 - \sin^2\theta) - \sin^2\theta$ or $1 - 2\sin^2\theta$ *Substitute $1 - \sin^2\theta$ for $\cos^2\theta$.*

$\cos^2\theta - \sin^2\theta = \cos^2\theta - (1 - \cos^2\theta)$ or $2\cos^2\theta - 1$ *Substitute $1 - \cos^2\theta$ for $\sin^2\theta$.*

These formulas are known as the **double-angle formulas**.

The following identities hold for all values of θ.

$$\sin 2\theta = 2\sin\theta\cos\theta \qquad \begin{aligned}
\cos 2\theta &= \cos^2\theta - \sin^2\theta \\
&= 1 - 2\sin^2\theta \\
&= 2\cos^2\theta - 1
\end{aligned}$$

Double-Angle Formulas

Example

1 **Suppose x is between $90°$ and $180°$ and $\sin x = \dfrac{3}{5}$. Find $\sin 2x$.**

Since $\sin 2x = 2\sin x\cos x$, find $\cos x$ first. Use $\cos^2 x + \sin^2 x = 1$.

$$\cos^2 x + \sin^2 x = 1$$

$$\cos^2 x + \left(\frac{3}{5}\right)^2 = 1 \qquad \textit{Substitute } \tfrac{3}{5} \textit{ for sin x.}$$

$$\cos^2 x = 1 - \left(\frac{3}{5}\right)^2$$

$$= \frac{16}{25}$$

$$\cos x = \pm\sqrt{\frac{16}{25}} \text{ or } \pm\frac{4}{5}$$

Since x is between $90°$ and $180°$, $\cos x$ is negative.

$$\sin 2x = 2\sin x\cos x$$

$$= 2\left(\frac{3}{5}\right)\left(-\frac{4}{5}\right)$$

$$= -\frac{24}{25}$$

There are also formulas for $\cos \frac{\alpha}{2}$ and $\sin \frac{\alpha}{2}$.

$$2 \cos^2 \theta - 1 = \cos 2\theta$$ *Use double-angle formulas.* $$1 - 2 \sin^2 \theta = \cos 2\theta$$

$$2 \cos^2 \frac{\alpha}{2} - 1 = \cos \alpha$$ *Substitute α for 2θ and $\frac{\alpha}{2}$ for θ.* $$1 - 2 \sin^2 \frac{\alpha}{2} = \cos \alpha$$

$$\cos^2 \frac{\alpha}{2} = \frac{1 + \cos \alpha}{2}$$ *Solve for the squared term.* $$\sin^2 \frac{\alpha}{2} = \frac{1 - \cos \alpha}{2}$$

$$\cos \frac{\alpha}{2} = \pm \sqrt{\frac{1 + \cos \alpha}{2}}$$ *Take the square root of both sides.* $$\sin \frac{\alpha}{2} = \pm \sqrt{\frac{1 - \cos \alpha}{2}}$$

These formulas are known as the **half-angle formulas**.

The following identities hold for all values of α.

$$\cos \frac{\alpha}{2} = \pm \sqrt{\frac{1 + \cos \alpha}{2}} \quad \text{and} \quad \sin \frac{\alpha}{2} = \pm \sqrt{\frac{1 - \cos \alpha}{2}}$$

Half-Angle Formulas

Example

2   **Suppose $\sin x = -\frac{1}{2}$ and x is in the fourth quadrant. Find $\cos \frac{x}{2}$.**

Since $\cos \frac{x}{2} = \pm \sqrt{\frac{1 + \cos x}{2}}$, find $\cos x$ first. Use $\cos^2 x + \sin^2 x = 1$.

$$\cos^2 x + \sin^2 x = 1$$

$$\cos^2 x + \left(-\frac{1}{2}\right)^2 = 1 \qquad \text{\textit{Substitute} } -\tfrac{1}{2} \text{ \textit{for} } \sin x.$$

$$\cos^2 x = \frac{3}{4}$$

$$\cos x = \pm \sqrt{\frac{3}{4}} \text{ or } \pm \frac{\sqrt{3}}{2}$$

Since x is in the fourth quadrant, $\cos x$ is positive.

$$\cos \frac{x}{2} = \pm \sqrt{\frac{1 + \cos x}{2}}$$

$$= \pm \sqrt{\frac{1 + \frac{\sqrt{3}}{2}}{2}}$$

$$= \pm \sqrt{\frac{2 + \sqrt{3}}{4}} \text{ or } \pm \frac{\sqrt{2 + \sqrt{3}}}{2}$$

Since x is in the fourth quadrant, x is between $270°$ and $360°$.

Thus, $\frac{x}{2}$ is between $135°$ and $180°$, and $\cos \frac{x}{2}$ is negative.

The solution is $-\dfrac{\sqrt{2 + \sqrt{3}}}{2}$.

Exploratory Exercises

1. x is a first quadrant angle. In which quadrant does the terminal side for $2x$ lie?

2. x is a second quadrant angle. In which quadrant does the terminal side for $2x$ lie?

3. x is a third quadrant angle. In which quadrant does the terminal side for $2x$ lie?

4. x is a fourth quadrant angle. In which quadrant does the terminal side for $2x$ lie?

5. x is a first quadrant angle. In which quadrant does the terminal side for $\frac{x}{2}$ lie?

6. x is a second quadrant angle. In which quadrant does the terminal side for $\frac{x}{2}$ lie?

7. x is a third quadrant angle. In which quadrant does the terminal side for $\frac{x}{2}$ lie?

8. x is a fourth quadrant angle. In which quadrant does the terminal side for $\frac{x}{2}$ lie?

Written Exercises

Find $\sin 2x$, $\cos 2x$, $\sin \frac{x}{2}$, and $\cos \frac{x}{2}$ for each of the following.

9. $\sin x = \frac{1}{2}$, x is in the first quadrant

10. $\cos x = \frac{3}{5}$, x is in the first quadrant

11. $\cos x = -\frac{2}{3}$, x is in the third quadrant

12. $\sin x = \frac{4}{5}$, x is in the second quadrant

13. $\sin x = \frac{5}{13}$, x is in the second quadrant

14. $\cos x = \frac{1}{5}$, x is in the fourth quadrant

15. $\sin x = -\frac{3}{4}$, x is in the fourth quadrant

16. $\cos x = -\frac{1}{3}$, x is in the third quadrant

17. $\cos x = -\frac{1}{4}$, x is in the second quadrant

18. $\sin x = -\frac{3}{5}$, x is in the third quadrant

19. $\sin x = -\frac{3}{8}$, x is in the fourth quadrant

20. $\cos x = \frac{1}{6}$, x is in the first quadrant

21. $\sin x = -\frac{1}{4}$, x is in the third quadrant

22. $\cos x = -\frac{1}{3}$, x is in the second quadrant

Evaluate each expression using the Half-Angle Formulas.

23. $\sin 105°$

24. $\cos 165°$

25. $\cos \frac{\pi}{8}$

26. $\sin \frac{7\pi}{12}$

27. $\sin 22\frac{1}{2}°$

28. $\cos 157\frac{1}{2}°$

29. $\cos \frac{19\pi}{12}$

30. $\sin \frac{7\pi}{8}$

31. $\sin 195°$

32. $\cos 345°$

Verify each identity.

33. $\cos^2 2x + 4 \sin^2 x \cos^2 x = 1$

34. $(\sin x + \cos x)^2 = 1 + \sin 2x$

35. $\sin^4 x - \cos^4 x = 2 \sin^2 x - 1$

36. $\sin 2x = 2 \cot x \sin^2 x$

37. $\sin^2 \theta = \frac{1}{2}(1 - \cos 2\theta)$

38. $\dfrac{1}{\sin x \cos x} - \dfrac{\cos x}{\sin x} = \tan x$

39. $\tan^2 \frac{x}{2} = \dfrac{1 - \cos x}{1 + \cos x}$

40. $2 \cos^2 \frac{x}{2} = 1 + \cos x$

Challenge Exercise

41. If $\sin \theta = x$ and θ is in the third quadrant, find $\sin 2\theta$.

17-5 Solving Trigonometric Equations

Trigonometric identities are true for *all* values of the variable for which both sides are defined. Most **trigonometric equations**, like most algebraic equations, are true for *some* but *not all* values of the variable.

Example

1 Solve $\sin 2\theta + \sin \theta = 0$ if $0° \le \theta < 360°$.

$$\sin 2\theta + \sin \theta = 0$$
$$2 \sin \theta \cos \theta + \sin \theta = 0 \qquad \textit{sin } 2\theta = 2 \sin \theta \cos \theta$$
$$\sin \theta (2 \cos \theta + 1) = 0 \qquad \textit{Factor.}$$

$\sin \theta = 0$ or $2 \cos \theta + 1 = 0$

$\theta = 0°$ or $\theta = 180°$ $\qquad\qquad \cos \theta = -\dfrac{1}{2}$

$$\theta = 120° \text{ or } \theta = 240°$$

The solutions are $0°$, $120°$, $180°$, and $240°$.

Usually trigonometric equations are solved for values of the variable between $0°$ and $360°$ or 0 radians and 2π radians. There are solutions outside that interval. These other solutions differ by integral multiples of the period of the function.

Example

2 Solve $\cos \theta + 1 = 0$ for all values of θ if θ is measured in radians.

$$\cos \theta + 1 = 0$$
$$\cos \theta = -1$$

Look at the graph of $y = \cos \theta$ to find solutions to $\cos \theta = -1$.

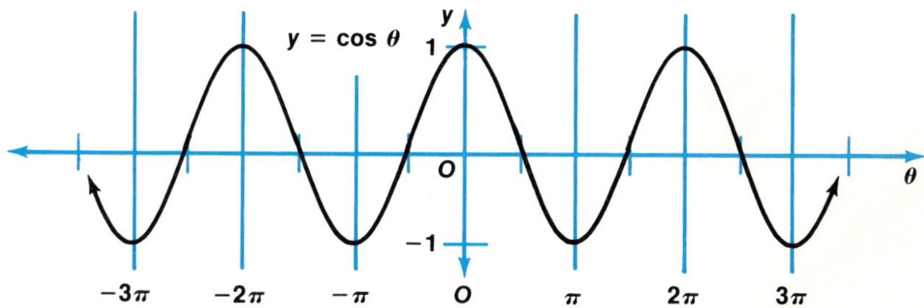

The solutions are π, 3π, 5π, and so on, and $-\pi$, -3π, -5π, and so on. The only solution in the interval 0 radians to 2π radians is π. The period of cosine is 2π radians. So the solutions can be written as $\pi + 2n\pi$ where n is any integer.

If an equation cannot be easily solved by factoring, try writing the expression in terms of only one trigonometric function.

Example

3 **Solve $4 \sin x \cos x = -\sqrt{3}$.**

$$4 \sin x \cos x = -\sqrt{3}$$

$$2(2 \sin x \cos x) = -\sqrt{3}$$

$$2 \sin x \cos x = -\frac{\sqrt{3}}{2}$$

$$\sin 2x = -\frac{\sqrt{3}}{2} \qquad \textit{2 sin x cos x = sin 2x}$$

$2x = 240° + n \cdot 360°$ or $2x = 300° + n \cdot 360°$ where n is any integer.

$x = 120° + n \cdot 180°$ or $x = 150° + n \cdot 180°$ where n is any integer.

The solutions are $120° + n \cdot 180°$ and $150° + n \cdot 180°$ where n is any integer.

Some trigonometric equations have *no solution*. In other words, there is no replacement for the variable that will make the sentence true. For example, the equation $\sin x = -3$ has no solution since all values of $\sin x$ are between -1 and 1, inclusive. Thus, the solution set for $\sin x = -3$ is $\varnothing$.

Example

4 **Solve $2 \cos^2 \theta - 3 \cos \theta - 2 = 0$ if $0 \le \theta < 2\pi$.**

$$2 \cos^2 \theta - 3 \cos \theta - 2 = 0$$

$$(\cos \theta - 2)(2 \cos \theta + 1) = 0 \qquad \textit{Factor.}$$

$\cos \theta - 2 = 0$ or $2 \cos \theta + 1 = 0$

$\qquad \cos \theta = 2$ $2 \cos \theta = -1$

There is no solution to $\cos \theta = 2$ since all values of $\cos \theta$ are between -1 and 1, inclusive.

$$\cos \theta = -\frac{1}{2}$$

$$\theta = \frac{2\pi}{3} \text{ or } \theta = \frac{4\pi}{3}$$

The solutions are $\dfrac{2\pi}{3}$ and $\dfrac{4\pi}{3}$.

It is important to check your solutions. Some algebraic operations may result in answers that are *not* solutions of the original equation.

5 Solve $\sin x = 1 - \cos x$ if $0° \leq x < 360°$.

$$\sin x = 1 - \cos x$$
$$\sin^2 x = (1 - \cos x)^2 \qquad \textit{Square each side.}$$
$$\sin^2 x = 1 - 2 \cos x + \cos^2 x$$
$$1 - \cos^2 x = 1 - 2 \cos x + \cos^2 x \qquad \textit{sin}^2 \textit{ x} = 1 - \textit{cos}^2 \textit{ x}$$
$$0 = 2 \cos^2 x - 2 \cos x$$
$$0 = 2 \cos x(\cos x - 1)$$

$$\begin{array}{ccc} 2 \cos x = 0 & \text{or} & \cos x - 1 = 0 \\ \cos x = 0 & & \cos x = 1 \\ x = 90° \text{ or } x = 270° & & x = 0° \end{array}$$

Check:

$$\begin{array}{ccc}
\sin x = 1 - \cos x & \sin x = 1 - \cos x & \sin x = 1 - \cos x \\
\sin 90° \stackrel{?}{=} 1 - \cos 90° & \sin 270° \stackrel{?}{=} 1 - \cos 270° & \sin 0° \stackrel{?}{=} 1 - \cos 0° \\
1 \stackrel{?}{=} 1 - 0 & -1 \stackrel{?}{=} 1 - 0 & 0 \stackrel{?}{=} 1 - 1 \\
1 = 1 \ \checkmark & -1 \neq 1 & 0 = 0 \ \checkmark
\end{array}$$

The solutions are $0°$ and $90°$.

Exploratory Exercises

How many solutions does each equation have if $0° \leq \theta < 360°$?

1. $\sin \theta = 1$ **2.** $\sin \theta = \frac{1}{2}$ **3.** $\cos \theta = -\frac{\sqrt{3}}{2}$ **4.** $\tan \theta = 1$

5. $\tan \theta = -3$ **6.** $\tan^2 \theta = 1$ **7.** $\sin^2 \theta = 3$ **8.** $\cos^2 \theta = 1$

9. $\sin 2\theta = \frac{1}{2}$ **10.** $\cos 2\theta = \frac{3}{2}$ **11.** $\sin 2\theta = -\frac{\sqrt{3}}{2}$ **12.** $\sin 3\theta = -2$

13. $\cos 8\theta = 1$ **14.** $\sin \frac{1}{2}\theta = -\frac{\sqrt{3}}{2}$ **15.** $\sin^2 2\theta = \frac{1}{2}$ **16.** $\tan^2 2\theta = 3$

Written Exercises

Find all solutions if $0° \leq x < 360°$.

17. $2 \sin^2 x - 1 = 0$ **18.** $4 \cos^2 x = 1$

19. $2 \sin^2 x + \sin x = 0$ **20.** $2 \cos^2 x = \sin x + 1$

21. $\sin 2x = \cos x$ **22.** $\sin 2x = 2 \cos x$

23. $\sin^2 x + \cos 2x - \cos x = 0$ **24.** $4 \sin^2 x - 4 \sin x + 1 = 0$

Find all solutions if $0 \leq \theta < 2\pi$.

25. $2 \cos \theta - 1 = 0$ **26.** $2 \sin \theta = -1$

27. $2 \sin \theta + \sqrt{3} = 0$ **28.** $4 \sin^2 \theta = 1$

29. $\sin^2 \theta = \cos^2 \theta - 1$ **30.** $2 \sin^2 \theta - \sin \theta = 1$

31. $2 \cos^2 \theta - 5 \cos \theta = -2$ **32.** $\sin^2 \theta - 3 \sin \theta = -2$

Solve each equation for all values of x if x is measured in degrees.

33. $\cos 2x = \cos x$

34. $\sin^2 x - 2 \sin x - 3 = 0$

35. $\sin x = \cos x$

36. $\tan x = \sin x$

37. $3 \cos 2x - 5 \cos x = 1$

38. $\tan^2 x - \sqrt{3} \tan x = 0$

39. $\sin x = 1 + \cos x$

40. $\cos 2x + \cos x + 1 = 0$

41. $\sin \frac{x}{2} + \cos x = 1$

42. $\sin \frac{x}{2} + \cos \frac{x}{2} = \sqrt{2}$

43. $\sin^2 x - \sin x = 0$

44. $\cos x \tan x - \sin^2 x = 0$

Solve each equation for all values of θ if θ is measured in radians.

45. $3 \sin^2 \theta - \cos^2 \theta = 0$

46. $\cos 2\theta + 3 \cos \theta - 1 = 0$

47. $2 \sin^2 \theta - 3 \sin \theta - 2 = 0$

48. $2 \sin^2 \theta - \cos \theta - 1 = 0$

49. $\cos^2 \theta - \frac{7}{2} \cos \theta - 2 = 0$

50. $\cos^2 \theta - \frac{5}{2} \cos \theta - 3 = 0$

51. $2 \cos^2 \theta + 3 \sin \theta - 3 = 0$

52. $\cos \theta = 3 \cos \theta - 2$

53. $4 \cos^2 \theta - 4 \cos \theta + 1 = 0$

54. $\cos 2\theta = 1 - \sin \theta$

mini-review

Evaluate.

1. $P(5, 3) \cdot P(8, 4)$

2. $7! + 4!$

3. $C(6, 3)$

Two cards are selected from a deck of 52 cards. What is the probability that the following occurs?

4. both 5s or both black

5. both red or both face cards

Find each of the following.

6. $\tan \left(\text{Arcsin } \frac{4}{5} \right)$

7. $\csc 3\pi$

8. $\sec^2 45° + \tan^2 45°$

9. Graph $y = 3 \cos 2\theta$.

10. Graph $y = \csc x$.

The frequency of the scores on a college entrance examination are normally distributed. Suppose 100,000 people took the exam, the mean score is 475, and the standard deviation is 65.

11. What percentage of the scores is between 410 and 605?

12. How many people scored less than 540?

According to a certain prediction equation, a person 70 inches tall weighs about 165 pounds. A person 75 inches tall weighs about 184 pounds. Let x stand for height in inches and y stand for weight in pounds.

13. Find the prediction equation.

14. Predict the weight of a person who is 72 inches tall.

15. Predict the height of a person who weighs 180 pounds.

A refractometer measures the change in direction, or bending, of a beam of light as it moves from one medium to another. The measurement is called the *index of refraction*. This instrument is used in a variety of fields. In gemology, a refractometer helps in identification of gems. In chemistry, it is used to determine the composition of unknown solutions or substances.

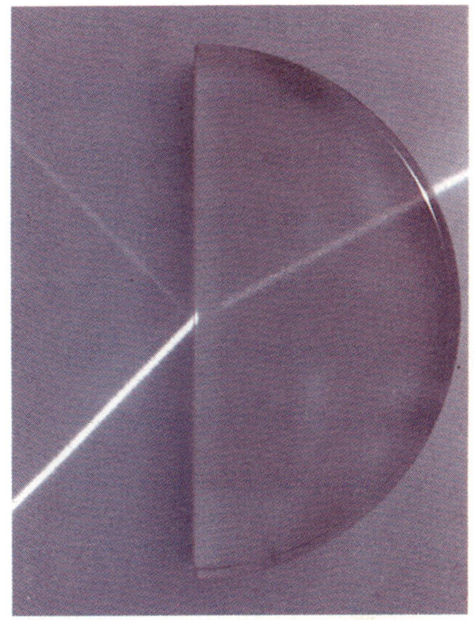

A refractometer uses the basic principles of **Snell's Law**. If n represents the index of refraction, i represents the angle of incidence, and r represents the angle of refraction, then Snell's Law states their relationship.

$$n = \frac{\sin i}{\sin r} \qquad 0° < i \le 90° \\ \qquad\qquad 0° < r \le 90°$$

Example: A beam of light moves from a vacuum to glass. The angle of incidence is 45° and the index of refraction is $\sqrt{2}$. What is the angle of refraction?

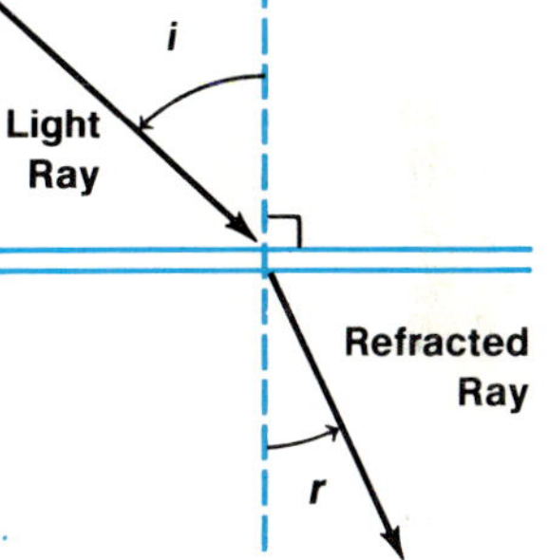

$$n = \frac{\sin i}{\sin r}$$

$$\sqrt{2} = \frac{\sin 45°}{\sin r} \qquad \textit{Substitute 45° for i and } \sqrt{2} \textit{ for n.}$$

$$\sqrt{2} = \frac{\frac{\sqrt{2}}{2}}{\sin r}$$

$$\sin r = \frac{1}{2}$$

$$r = 30°$$

The angle of refraction is 30°.

Exercises

1. A beam of light moves from a vacuum to water. If the angle of incidence is 60° and the angle of refraction is 30°, what is the index of refraction?

2. A beam of light moves from a vacuum to diamond. If the angle of refraction is 15° and the index of refraction 3.346, what is the angle of incidence?

Graphing Calculator Application: Solving Trigonometric Equations

The method for solving exponential equations presented in Chapter 12 can be used for solving trigonometric equations.

Example

1 **Solve $2 \sin^2 x + 3 \sin x - 1 = 0$ if $0° \leq x < 360°$.**

The solutions to the equation $2 \sin^2 x + 3 \sin x - 1 = 0$ are also zeros of the function $f(x) = 2 \sin^2 x + 3 \sin x - 1$.

Choose a set of range parameter values to roughly approximate the location of the zeros of $f(x)$ for $0° \leq x < 360°$. One possible set of values is given below. Set the range parameters to these values and graph $y = 2 \sin^2 x + 3 \sin x - 1$.

Use degree mode for these range parameters.

Xmin: -10 Xmax: 370   Xscl: 30   Ymin: -0.5 Ymax: 3.5   Yscl: 0.5

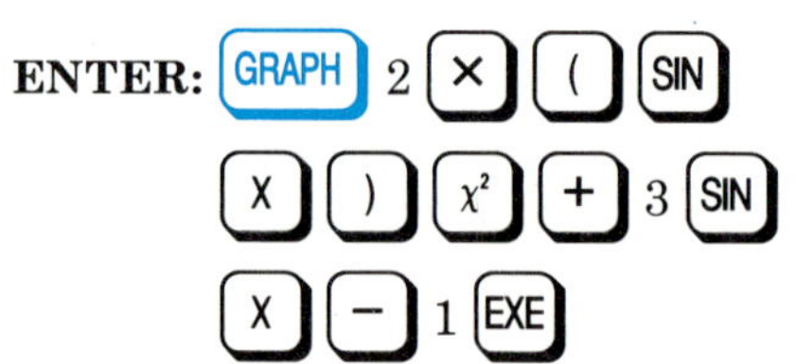

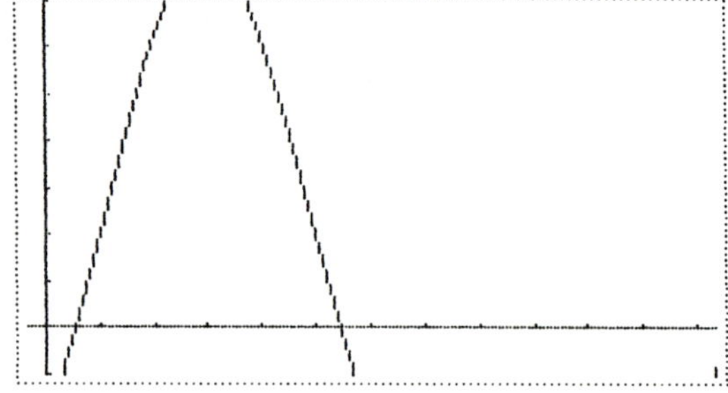

Based on this graph, there is a zero between $0°$ and $30°$ and a zero between $150°$ and $180°$. Now, "zoom-in" on each zero.

To do this, graph the function using the following sets of range parameter values.

For zero between $0°$ and $30°$:

Xmin: 10   Xmax: 20   Xscl: 1
Ymin: -0.2 Ymax: 0.2   Yscl: 0.4

For zero between $150°$ and $180°$:

Xmin: 160   Xmax: 170   Xscl: 1
Ymin: -0.2 Ymax: 0.2   Yscl: 0.4

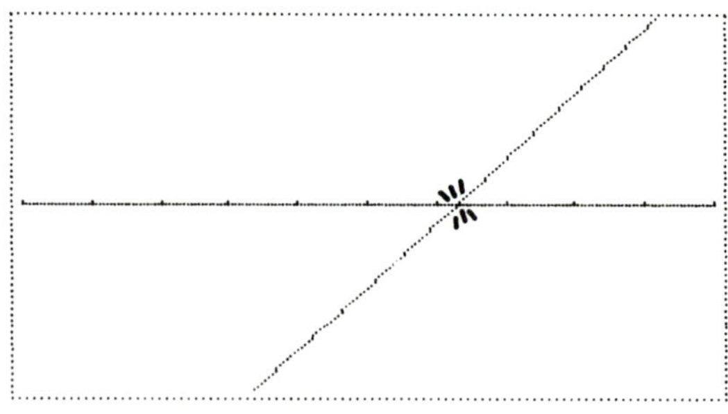

The graph for the first set of values is shown at the right.

Activate the tracing function and move the pointer to the x-intercept. The x-coordinate is 16.27659574. Now, repeat the process for the other set of range parameter values. The x-coordinate of that x-intercept is 163.7234043.

Therefore, the solutions to $2 \sin^2 x + 3 \sin x - 1 = 0$ are approximately $16°$ and $164°$.

2   **Solve $\sin \frac{x}{2} = \cos \frac{x}{2}$ if $-360° \le x < 360°$.**

The solutions to $\sin \frac{x}{2} = \cos \frac{x}{2}$ are also the x-coordinates of the point(s) of intersection of the graphs of $y = \sin \frac{x}{2}$ and $y = \cos \frac{x}{2}$. Choose a set of range parameter values that will produce complete graphs of both equations for $-360° \le x < 360°$. One possible set of values is given below. Set the range parameters to these values and graph the equations.   *Use degree mode.*

Xmin: -360 Xmax: 360 Xscl: 90
Ymin: -1.6 Ymax: 1.6 Yscl: 0.5

ENTER: GRAPH COS (X ÷ 2) EXE GRAPH SIN (X ÷ 2) EXE

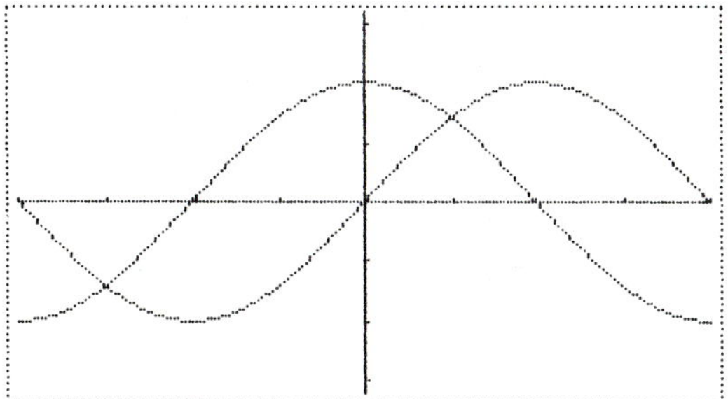

Based on this graph, the points of intersection are near $x = -270°$ and $x = 90°$.

Now, "zoom-in" on each point of intersection using the procedure from Example 1. Graph the equations using the following sets of range parameter values.

For point of intersection near $x = -270°$:
Xmin: -275 Xmax: -265 Xscl: 1 Ymin: -0.75 Ymax: -0.65 Yscl: 0.01

For point of intersection near $x = 90°$:
Xmin: 85 Xmax: 95 Xscl: 1 Ymin: 0.65 Ymax: 0.75 Yscl: 0.01

Now, use the tracing function to find the x-coordinate of the point of intersection for each graph. For these sets of range parameter values, the x-coordinates of the points of intersection are $-270°$ and $90°$.

Therefore, the solutions to $\sin \frac{x}{2} = \cos \frac{x}{2}$ are approximately $-270°$ and $90°$.

Written Exercises

Solve each equation for the indicated values of x.

1. $\sin x = \tan x,\ 0° \le x < 360°$

2. $\cos 2x = \cos x,\ 0° \le x < 360°$

3. $\sin 2x = \sin x,\ -360° \le x < 360°$

4. $\cos 3x = \tan 2x,\ -360° \le x < 360°$

5. $2 \cos^2 x - \sin x = 1,\ 0° \le x < 360°$

6. $2 \sin^2 x - 5 \sin x = 2,\ 0° \le x < 360°$

7. $3 \sin 2x - 5 \sin x = 1,\ -360° \le x < 360°$

8. $2 \cos 2x = 3 \cos x,\ -360° \le x < 360°$

9. $\cos 2x + \sin x = -1,\ 0° \le x < 360°$

10. $\sin 2x + \cos x = -1,\ 0° \le x < 360°$

Solving Trigonometric Equations

A computer program can be used to find solutions to trigonometric equations.

```
10   PRINT "FIND THE SOLUTIONS OF "
15   PRINT "2 * COS(X) ^ 2 + 3 * COS(X) −
        2 = 0.": PRINT
20   INPUT "ENTER THE LOWEST VALUE OF
        X, IN DEGREES, FOR THE INTERVAL
        TO BE TESTED: "; X1
30   INPUT " ENTER THE HIGHEST VALUE
        OF X, IN DEGREES, FOR THE
        INTERVAL TO BE TESTED: ";X2
35   PRINT
40   FOR J = X1 TO X2
50   LET Y = J / 57.2957795: GOSUB 100
60   NEXT J
70   IF X = 1 THEN 160
80   PRINT "NO SOLUTIONS FOUND
        BETWEEN "; X1;" DEGREES AND "
        ;X2;" DEGREES."
90   GOTO 160
100   LET F = 2 * COS (Y) ^ 2 +
        3 * COS (Y) − 2
110   IF ABS (F) > 0.0000001 THEN 140
120   LET X = 1
130   PRINT "ONE SOLUTION IS ";J;
        " DEGREES."
140   RETURN
150   PRINT "THESE ARE THE ONLY
        SOLUTIONS BETWEEN ";X1;
        " DEGREES AND ";X2: "DEGREES."
160   END
```

The program at the left finds the solutions to $2 \cos^2 x + 3 \cos x - 2 = 0$ within a given interval that you specify. This interval must be given in degrees.

Enter and run this program. Use the interval $-360° \le x \le 360°$.

```
] RUN
FIND THE SOLUTIONS OF
2 * COS(X) ^ 2 + 3 * COS(X) − 2 = 0.

ENTER THE LOWEST VALUE OF X, IN DEGREES,
FOR THE INTERVAL TO BE TESTED: −360
ENTER THE HIGHEST VALUE OF X, IN DEGREES,
FOR THE INTERVAL TO BE TESTED: 360

ONE SOLUTION IS −300 DEGREES.
ONE SOLUTION IS −60 DEGREES.
ONE SOLUTION IS 60 DEGREES.
ONE SOLUTION IS 300 DEGREES.
THESE ARE THE ONLY SOLUTIONS BETWEEN
−360 DEGREES AND 360 DEGREES.
```

Any trigonometric equation can be solved using this program by changing statements 15 and 100.

Exercises

Make the necessary changes in the program to solve each equation for all values of x in the indicated interval.

1. $2 \cos x - \sin 2x = 0, \ 0° \le x \le 360°$

2. $\cos x + 1 - \sin x = 0, \ -180° \le x \le 180°$

3. $2 \sin^2 x + \sin x - 1 = 0, \ 180° \le x \le 240°$

4. $\sin x + \cos x \tan^2 x = 0, \ -360° \le x \le 0°$

5. $\tan^2 x - \sqrt{3} \tan x = 0, \ 90° \le x \le 450°$

6. $\sin \frac{x}{2} + \cos x = 1, \ -360° \le x \le 360°$

7. $\cos 3x = \tan 2x, \ -180° \le x \le 0°$

8. $\cos 2x + \sin x = 1, \ 0° \le x \le 720°$

identity   (611)   trigonometric equations   (627)

Chapter Summary

1. Suggestions for Verifying Identities: (615)

   - Start with the more complicated side of the equation. Transform the expression into the form of the simpler side.
   - Work with each side of the equation at the same time. Transform each expression separately into the same form.
   - Substitute one or more basic trigonometric identities to simplify the expression.
   - Try factoring or multiplying to simplify the expression.
   - Multiply both numerator and denominator by the same trigonometric expression.

2. Difference and Sum Formulas: The following identities hold for all values of α and β.   (621)

$$\cos(\alpha \pm \beta) = \cos\alpha\cos\beta \mp \sin\alpha\sin\beta$$
$$\sin(\alpha \pm \beta) = \sin\alpha\cos\beta \pm \cos\alpha\sin\beta$$

3. Double-Angle Formulas: The following identities hold for all values of θ. (624)

$$\sin 2\theta = 2\sin\theta\cos\theta \qquad \cos 2\theta = \cos^2\theta - \sin^2\theta$$
$$= 1 - 2\sin^2\theta$$
$$= 2\cos^2\theta - 1$$

4. Half-Angle Formulas: The following identities hold for all values of α. (625)

$$\cos\frac{\alpha}{2} = \pm\sqrt{\frac{1 + \cos\alpha}{2}} \text{ and } \sin\frac{\alpha}{2} = \pm\sqrt{\frac{1 - \cos\alpha}{2}}$$

17–1 **Solve for values of θ between 90° and 180°.**

1. If $\sin \theta = \frac{1}{2}$, find $\cos \theta$.

2. If $\csc \theta = \frac{5}{3}$, find $\cos \theta$.

3. If $\sec \theta = -3$, find $\tan \theta$.

4. If $\cot \theta = -\frac{1}{4}$, find $\csc \theta$.

Solve for values of θ between 270° and 360°.

5. If $\sin \theta = -\frac{4}{5}$, find $\cos \theta$.

6. If $\sec \theta = 1$, find $\tan \theta$.

7. If $\csc \theta = -\frac{5}{3}$, find $\cot \theta$.

8. If $\sin \theta = -\frac{1}{2}$, find $\sec \theta$.

17–2 **Verify each identity.**

9. $\sin^4 x - \cos^4 x = \sin^2 x - \cos^2 x$

10. $\dfrac{\sin \theta}{\tan \theta} + \dfrac{\cos \theta}{\cot \theta} = \cos \theta + \sin \theta$

11. $\dfrac{\sin \theta}{1 - \cos \theta} = \csc \theta + \cot \theta$

12. $\tan x + \cot x = \sec x \csc x$

13. $\sec x(\sec x - \cos x) = \tan^2 x$

14. $\cot^2 \theta \sec^2 \theta = 1 + \cot^2 \theta$

15. $\dfrac{\csc \theta + 1}{\cot \theta} = \dfrac{\cot \theta}{\csc \theta - 1}$

16. $\dfrac{\cos x}{\csc x} - \dfrac{\sin x}{\cos x} = -\sin^2 x \tan x$

17–3 **Evaluate each expression.**

17. $\sin 105°$ **18.** $\cos 285°$ **19.** $\cos 15°$ **20.** $\sin (-255°)$

Verify each identity.

21. $\cos (90° - \theta) = \sin \theta$

22. $\cos (60° + \theta) + \cos (60° - \theta) = \cos \theta$

17–4 **23.** If $\sin x = -\frac{3}{5}$ and x is in the third quadrant, find $\sin 2x$.

24. If $\sin x = \frac{1}{4}$ and x is in the first quadrant, find $\cos 2x$.

25. If $\cos 2x = -\frac{17}{25}$ and $\cos x = \frac{2}{5}$, find $\sin x$.

26. Verify $\cos^2 \theta = \frac{1}{2}(\cos 2\theta + 1)$.

17–5 **Find all solutions if $0° \leq x < 360°$.**

27. $2 \cos^2 x + \sin^2 x = 2 \cos x$

28. $\cos 2x = \cos x$

29. $\cos x = 1 - \sin x$

30. $2 \sin 2x = 1$

31. $\cos 2x \sin x = 1$

32. $\tan^2 x + \tan x = 0$

Solve each equation for all values of θ if θ is measured in radians.

33. $6 \sin^2 \theta - 5 \sin \theta - 4 = 0$

34. $\sqrt{3} \sin \theta = \sin 2\theta$

Solve each of the following for values of θ between 180° and 270°.

1. If $\sin \theta = -\frac{1}{2}$, find $\tan \theta$.

2. If $\cot \theta = \frac{3}{4}$, find $\sec \theta$.

Evaluate each expression.

3. $\sin 255°$

4. $\cos 165°$

5. $\cos 225° \cos 15° + \sin 225° \sin 15°$

6. $2 \sin 75° \cos 75°$

7. If x is in the first quadrant and $\cos x = \frac{3}{4}$, find $\sin \frac{1}{2}x$.

8. If $\cos 2x = \frac{2}{9}$ and $\sin x = \frac{1}{3}$, find $\cos x$.

9. If x is in the third quadrant and $\sin x = -\frac{3}{5}$, find $\sin 2x$.

Find all solutions if $0° \leq x < 360°$.

10. $2 \sin x \cos x - \sin x = 0$

11. $\cos 2x + \sin x = 1$

12. $\sec x = 1 + \tan x$

13. $2 \cos^2 2x + \cos 2x - 1 = 0$

Solve each equation for all values of θ if θ is measured in radians.

14. $3 \tan^2 \theta - \sqrt{3} \tan \theta = 0$

15. $\sin \theta = 1 + \cos \theta$

Verify each identity.

16. $\dfrac{\cos x}{1 - \sin^2 x} = \sec x$

17. $\dfrac{\sec x}{\sin x} - \dfrac{\sin x}{\cos x} = \cot x$

18. $\dfrac{1 + \tan^2 \theta}{\cos^2 \theta} = \sec^4 \theta$

19. $(\sin x - \cos x)^2 = 1 - \sin 2x$

20. $\sin (90° + \theta) + \sin (90 - \theta) = 2 \cos \theta$

Triangle Trigonometry

For maximum strength, structures must be constructed as designed. Civil engineers often use surveying equipment and trigonometry to check building construction.

18-1 Right Triangles

Trigonometry can be used to find the missing measures of triangles. Consider the right triangle below.

The **hypotenuse** of the triangle is side AB. Its length is c units.

The side **opposite** angle A is side BC. Its length is a units.

The side **adjacent** to angle A is side AC. Its length is b units.

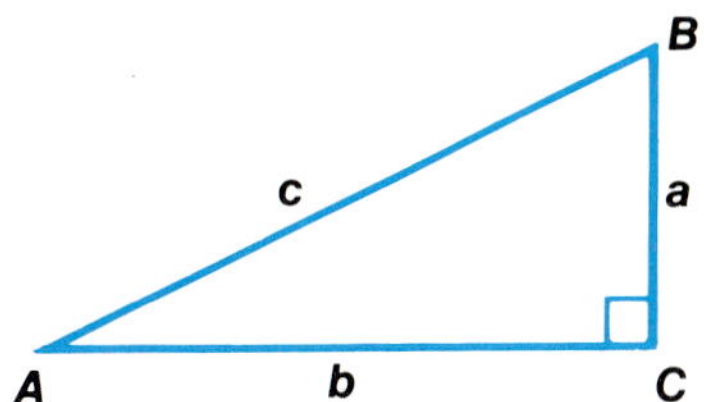

For convenience of notation, we refer to the angle with vertex at A as angle A and use A to stand for its measurement. Similarly, we refer to angle B and its measurement, B, and angle C and its measurement, C.

Suppose a unit circle is drawn with its center at vertex A as shown below.

The triangle in the interior of the unit circle is similar to triangle ABC. Hence, the measures of the corresponding sides of the triangles are proportional.

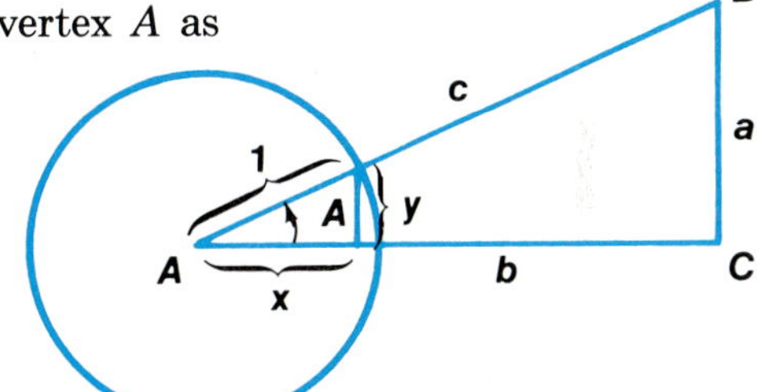

In this figure, $\sin A = y$. Since the two triangles are similar, $\dfrac{y}{a} = \dfrac{1}{c}$. Hence, $y = \dfrac{a}{c}$ and $\sin A = \dfrac{a}{c}$. Using the figure, trigonometric values can be defined in the following way.

$$\sin A = \frac{a}{c} \qquad \cos A = \frac{b}{c} \qquad \tan A = \frac{a}{b}$$

$$\csc A = \frac{c}{a} \qquad \sec A = \frac{c}{b} \qquad \cot A = \frac{b}{a}$$

SOH-CAH-TOA is a helpful mnemonic device for remembering the first three equations.

$$sin = \frac{opposite}{hypotenuse}$$
$$cos = \frac{adjacent}{hypotenuse}$$
$$tan = \frac{opposite}{adjacent}$$

Example

1 **Find the sine, cosine, tangent, cosecant, secant, and cotangent of angle A rounded to four decimal places.**

$\sin A = \dfrac{6}{10}$ or 0.6000 $\qquad\qquad$ $\csc A = \dfrac{10}{6}$ or 1.6667

$\cos A = \dfrac{8}{10}$ or 0.8000 $\qquad\qquad$ $\sec A = \dfrac{10}{8}$ or 1.2500

$\tan A = \dfrac{6}{8}$ or 0.7500 $\qquad\qquad$ $\cot A = \dfrac{8}{6}$ or 1.3333

Example

2 **Find sin A rounded to four decimal places.**

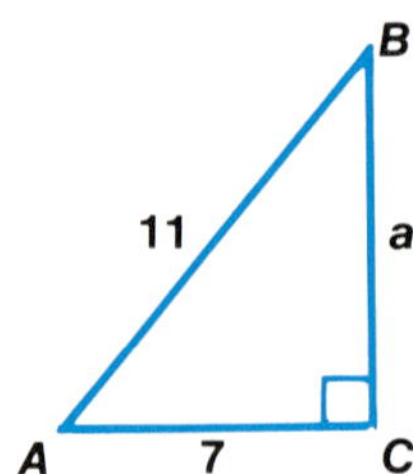

$$a^2 + 7^2 = 11^2 \qquad \text{Use the Pythagorean Theorem}$$
$$a^2 = 72 \qquad \text{to find the value of } a.$$
$$a = \sqrt{72}$$

$$\sin A = \frac{a}{c}$$

$$= \frac{\sqrt{72}}{11} \text{ or } 0.7714$$

The sine of A is about 0.7714.

Consider two special right triangles. Triangle ABC is an isosceles right triangle. Assume the congruent sides are each 1 unit long. Find the length of the hypotenuse.

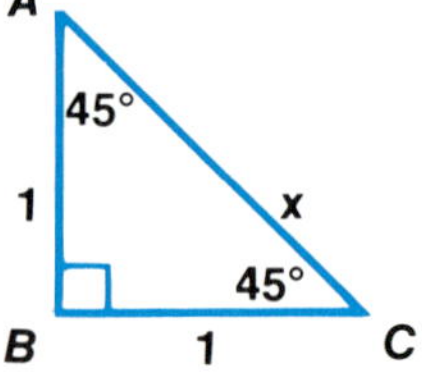

$$1^2 + 1^2 = x^2 \qquad \text{Use the Pythagorean Theorem.}$$
$$2 = x^2$$
$$\sqrt{2} = x \qquad \text{The hypotenuse is } \sqrt{2} \text{ units long.}$$

Now, write the trigonometric values.

$$\sin 45° = \frac{1}{\sqrt{2}} \text{ or } \frac{\sqrt{2}}{2} \qquad \cos 45° = \frac{1}{\sqrt{2}} \text{ or } \frac{\sqrt{2}}{2} \qquad \tan 45° = \frac{1}{1} \text{ or } 1$$

$$\csc 45° = \frac{\sqrt{2}}{1} \text{ or } \sqrt{2} \qquad \sec 45° = \frac{\sqrt{2}}{1} \text{ or } \sqrt{2} \qquad \cot 45° = \frac{1}{1} \text{ or } 1$$

Triangle DEG is an equilateral triangle. Assume each side is 2 units long. The altitude EF forms a triangle whose angle measurements are 30°, 60°, and 90°. Since altitude EF is the perpendicular bisector of side DG, the length of side DF is 1. Find the length of side EF.

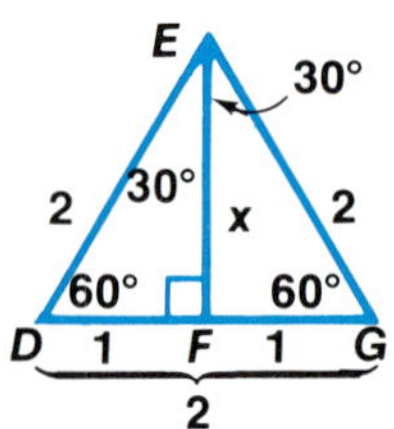

$$x^2 + 1^2 = 2^2$$
$$x^2 = 3$$
$$x = \sqrt{3} \qquad \text{Side } EF \text{ is } \sqrt{3} \text{ units long.}$$

Now, write the trigonometric values.

$$\sin 30° = \frac{1}{2} \qquad \cos 30° = \frac{\sqrt{3}}{2} \qquad \tan 30° = \frac{1}{\sqrt{3}} \text{ or } \frac{\sqrt{3}}{3}$$

$$\csc 30° = \frac{2}{1} \text{ or } 2 \qquad \sec 30° = \frac{2}{\sqrt{3}} \text{ or } \frac{2\sqrt{3}}{3} \qquad \cot 30° = \frac{\sqrt{3}}{1} \text{ or } \sqrt{3}$$

$$\sin 60° = \frac{\sqrt{3}}{2} \qquad \cos 60° = \frac{1}{2} \qquad \tan 60° = \frac{\sqrt{3}}{1} \text{ or } \sqrt{3}$$

$$\csc 60° = \frac{2}{\sqrt{3}} \text{ or } \frac{2\sqrt{3}}{3} \qquad \sec 60° = \frac{2}{1} \text{ or } 2 \qquad \cot 60° = \frac{1}{\sqrt{3}} \text{ or } \frac{\sqrt{3}}{3}$$

Exploratory Exercises

For each triangle, give the sine of each acute angle. State each answer in fraction form.

1.

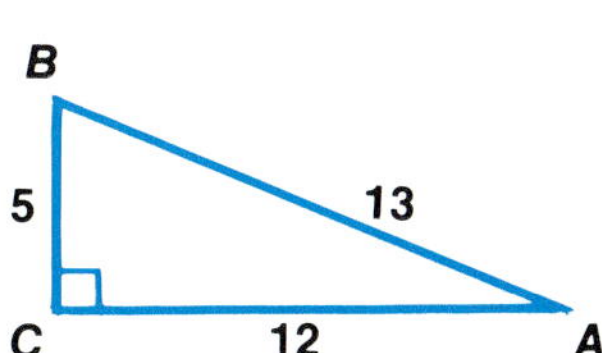

2.

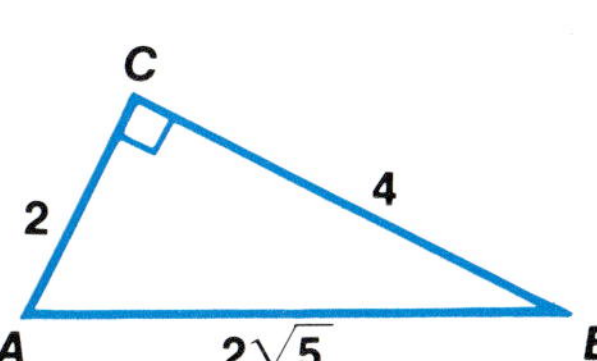

3.

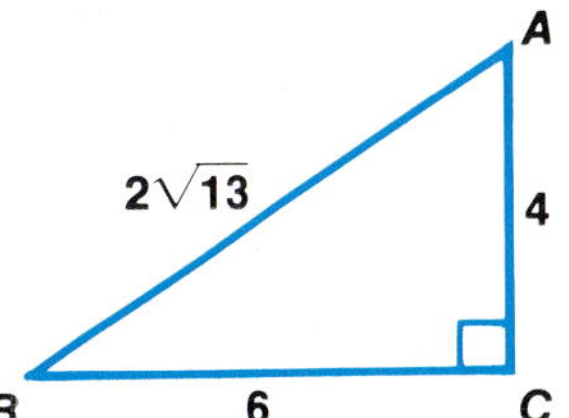

4–6. Give the cosine of each acute angle in problems **1–3**.

7–9. Give the tangent of each acute angle in problems **1–3**.

Written Exercises

Find each value rounded to four decimal places.

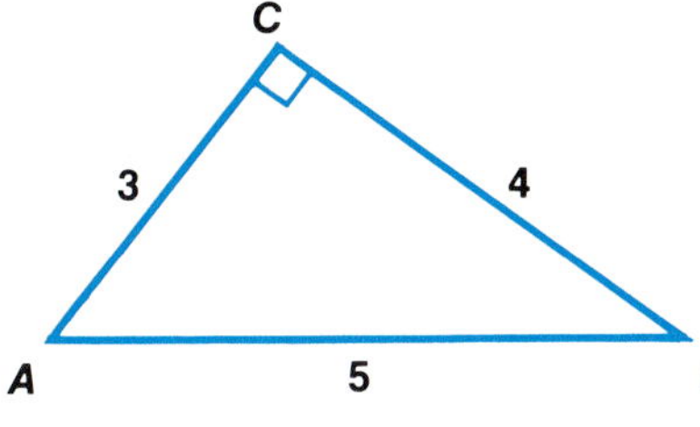

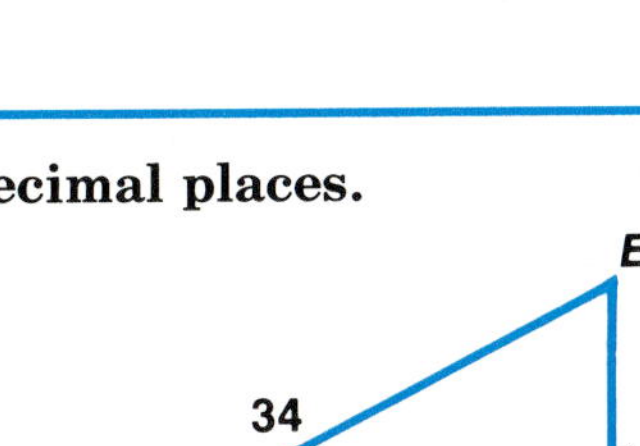

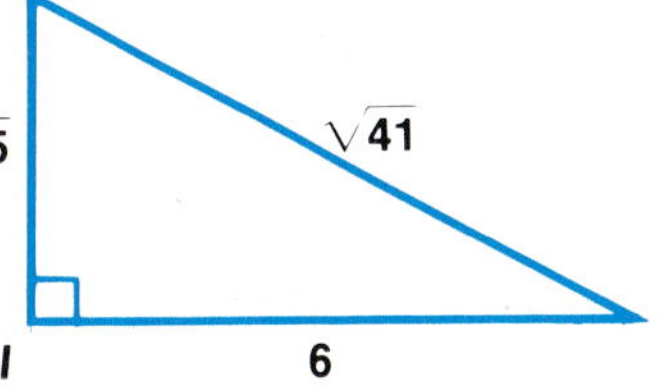

10. $\sin A$	**11.** $\cos A$	**12.** $\tan A$	**13.** $\sin B$
14. $\cos B$	**15.** $\tan B$	**16.** $\sin D$	**17.** $\sin E$
18. $\cos D$	**19.** $\cos E$	**20.** $\tan D$	**21.** $\tan E$
22. $\cos G$	**23.** $\tan G$	**24.** $\sin G$	**25.** $\cot G$
26. $\tan H$	**27.** $\sin H$	**28.** $\cot B$	**29.** $\sec D$
30. $\csc H$	**31.** $\csc A$	**32.** $\sec E$	**33.** $\cot A$

For each triangle, find sin A, cos A, and tan A rounded to four decimal places.

25.

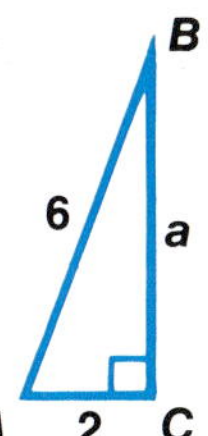

26.

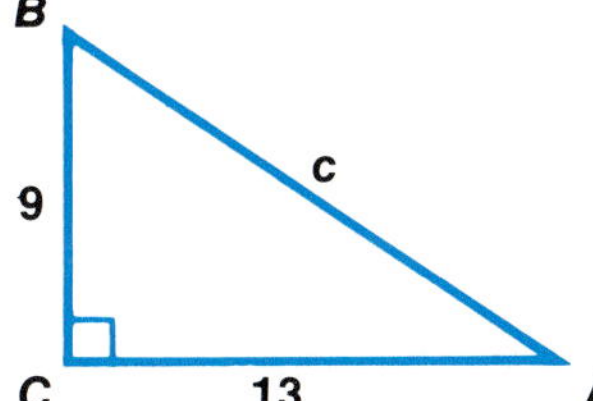

27.

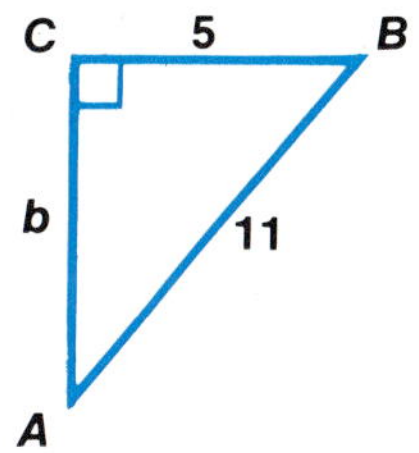

Find the value of each expression. Write each answer in fraction form.

37. $2 \cos 30°$

38. $-\sin 60°$

39. $\sin^2 45° + \cos^2 45°$

40. $2 \sin 60° \cos 60°$

41. $\cos^2 30° - \sin^2 30°$

42. $\sin 30° \cos 60° - \sin 60° \cos 30°$

43. $\csc 45° - \sec 30°$

44. $\sec 60° + \cot 30°$

18-2 Trigonometric Tables

Decimal approximations for values of trigonometric functions can be found by using the trigonometric tables in the back of this book. The tables provide values for angle measurements between 0° and 90° in intervals of ten minutes.

Angle measurements from 0°00′ to 45°00′ are listed on the left-hand side of the tables, and from 45°00′ to 90°00′ on the right-hand side. Part of the table of values is shown below.

Values of Trigonometric Functions

Angle	Sin	Cos	Tan	Cot	Sec	Csc	
27°00′	0.4540	0.8910	0.5095	1.963	1.122	2.203	63°00′
10′	0.4566	0.8897	0.5132	1.949	1.124	2.190	50′
20′	0.4592	0.8884	0.5169	1.935	1.126	2.178	40′
30′	0.4617	0.8870	0.5206	1.921	1.127	2.166	30′
40′	0.4643	0.8857	0.5243	1.907	1.129	2.154	20′
50′	0.4669	0.8843	0.5280	1.894	1.131	2.142	10′
28°00′	0.4695	0.8829					
						1.736	50′
20′	0.5783	0.8158	0.7089	1.411	1.226	1.729	40′
30′	0.5807	0.8141	0.7133	1.402	1.228	1.722	30′
40′	0.5831	0.8124	0.7177	1.393	1.231	1.715	20′
50′	0.5854	0.8107	0.7221	1.385	1.233	1.708	10′
36°00′	0.5878	0.8090	0.7265	1.376	1.236	1.701	54°00′
	Cos	Sin	Cot	Tan	Csc	Sec	Angle

Degrees are separated into minutes. Sixty minutes are equivalent to one degree. For example, $1\frac{1}{2}$ degrees can be expressed as 1 degree 30 minutes, and is abbreviated 1°30′.

Examples

1 **Find tan 27°40′.**

Read across the row labeled 27°40′ and down the column labeled Tan.

The value of tan 27°40′ is 0.5243.

For angle measurements between 0°00′ and 45°00′, use the column titles at the top.

2 **Find sin 54°10′.**

Read across the row labeled 54°10′ and up the column labeled Sin.

The value of sin 54°10′ is 0.8107.

For angle measurements between 45°00′ and 90°00′, use the column titles at the bottom.

A more accurate approximation of trigonometric values can be found by using **interpolation.**

3 **Find sin 28°23′.**

Use the table to find sin 28°20′ and sin 28°30′.

$$
10' \left\{ \begin{array}{l} 3' \left[\begin{array}{l} \sin 28°20' \; = \; 0.4746 \\ \sin 28°23' \; = \; \text{unknown} \end{array} \right] d \\ \sin 28°30' \; = \; 0.4772 \end{array} \right\} 0.0026
$$

d stands for the difference between 0.4746 and the unknown value.

Then set up a proportion and solve for *d*.

$$\frac{3}{10} = \frac{d}{0.0026}$$
$$10d = 3(0.0026)$$
$$d = 0.00078$$
$$\text{or } 0.0008$$

The table gives the values rounded to four decimal places. Therefore, round the value for d to four decimal places.

sin 28°23′ = 0.4746 + 0.0008 or 0.4754   *Add 0.0008 to the value of 28°20′.*

The value of sin 28°23′ is about 0.4754.

4 **Suppose sin x = 0.7820. Find the value of x to the nearest minute.**

$$
10' \left\{ \begin{array}{l} d \left[\begin{array}{l} \sin 51°20' \; = \; 0.7808 \\ \sin x \qquad = \; 0.7820 \end{array} \right] 0.0012 \\ \sin 51°30' \; = \; 0.7826 \end{array} \right\} 0.0018
$$

Use the table to find those values which the given value is between.

$$\frac{d}{10} = \frac{0.0012}{0.0018}$$
$$0.0018d = 0.0120$$
$$d = 6.66$$
$$\approx 7 \qquad \text{Round to the nearest whole number.}$$
$$x \approx 51°20' + 7' \text{ or } 51°27'$$

The value of x is about 51°27′.

The sum and difference formulas for sine and cosine can be used to find the trigonometric values of angles whose measures are not listed in the table.

$$\cos (A \pm B) = \cos A \cos B \mp \sin A \sin B$$
$$\sin (A \pm B) = \sin A \cos B \pm \cos A \sin B$$

5 **Find sin 153°.**

$$\begin{aligned}
\sin 153° &= \sin (180° - 27°) \\
&= \sin 180° \cos 27° - \cos 180° \sin 27° \\
&= 0 \cdot \cos 27° - (-1) \sin 27° \\
&= \sin 27° \\
&= 0.4540
\end{aligned}$$

153° is not listed in the table.

sin (A − B) = sin A cos B − cos A sin B

The value of sin 153° is about 0.4540.

A calculator can also be used to find decimal approximations of trigonometric functions.

Using Calculators

6 **Find tan 58°21′.**

Minutes must be changed to decimal form. This is done by dividing the minutes portion by 60.

ENTER: 58 [+] 21 [÷] 60 [=] [TAN]

DISPLAY: *58   21   60 58.35 1.6223029*

Therefore, tan 58°21′ is approximately 1.6223.

7 **Suppose cos x = 0.5030. Find the value of x to the nearest minute.**

To find the angle measure, use the [INV] key.

The decimal portion must be changed to minutes. This is done by multiplying the decimal portion by 60.

ENTER: 0.503 [INV] [COS]   [−]   59 [=]   [×] 60 [=]

DISPLAY: *0.503   59.801322   59 0.801322   60 48.079337*

Therefore, x is approximately 59°48′.

Exploratory Exercises

Use the table to find each trigonometric value.

1. sin 42°   **2.** cos 81°   **3.** sin 68°   **4.** tan 5°

5. tan 89°50′   **6.** cos 42°20′   **7.** tan 49°30′   **8.** sin 3°10′

Round each angle measurement to the nearest ten minutes. Then approximate each value.

9. cos 63°18′   10. tan 77°14′   11. sin 73°46′   12. cos 73°58′

13. cos 18°2′   14. tan 43°51′   15. sin 27°18′   16. sin 53°43′

Written Exercises

Approximate each trigonometric value. Use interpolation when necessary.

17. cos 38°   18. tan 12°   19. sin 68°   20. tan 88°

21. sin 16°20′   22. tan 85°16′   23. tan 77°30′   24. sin 38°15′

25. cos 59°10′   26. sin 127°40′   27. tan 88°52′   28. sec 47°10′

29. csc 33°33′   30. cot 44°44′   31. tan 110°55′   32. sec 11°11′

33. sin 194°35′   34. cot 47°18′   35. tan 42°42′   36. csc 273°18′

Find the value of x to the nearest minute.

37. cos x = 0.5132   38. tan x = 1.705   39. sin x = 0.3291   40. tan x = 0.3147

41. cos x = 0.7193   42. sin x = 0.1111   43. tan x = 0.2222   44. cos x = 0.3333

45. sin x = 0.8081   46. tan x = 42.71   47. csc x = 1.412   48. cot x = 0.1234

49. sec x = 1.319   50. csc x = 1.319   51. cot x = 1.384   52. sec x = 2.8672

Use a calculator to find each trigonometric value.

53. sin 136.25°   54. tan 65°14′   55. csc 98°36′   56. cot 10°56′

Use a calculator to find the value of x to the nearest minute.

57. cos x = 0.2489   58. sin x = 0.4683   59. sec x = 1.1573   60. tan x = 9.876

mini-review

Find the median, mode, and mean for each set of data.

1. {549, 467, 366, 210, 467, 140}   2. {21, 23, 26, 28, 29, 30, 26, 32}

Graph each function. Then state its amplitude and period.

3. $y = \frac{1}{2} \cos 2\theta$ 4. $y = -2 \sin \frac{1}{2}\theta$

Find all solutions if $0° \le x < 360°$.

5. $\sin^2 2x - \cos^2 x = 0$ 6. $\tan^2 x + \sec^2 x = 1$

Find the value of each expression.

7. csc 30° + cot 45°   8. 2 sin 60° − sec 60°

9. $\cos^2 60° - \sin^2 30°$ 10. sec 60° cos 60° − tan 45° sin 45°

18-3 Solving Right Triangles

Trigonometric functions can be used to solve problems involving right triangles. To solve a right triangle means to find all the measures of the sides and angles.

Examples

1 **Solve the right triangle.**

$$\sin A = \frac{8}{10}$$

$$\sin A = 0.800$$

$$A = 53°08'$$ *Round to the nearest minute.*

$$53°08' + B = 90°$$

$$B = 36°52'$$

Therefore, $A = 53°08'$ and $B = 36°52'$.

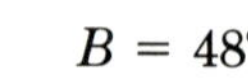

2 **Solve the right triangle. Round measures of sides to the nearest tenth.**

$$\frac{a}{14} = \sin 42° \qquad \frac{b}{14} = \cos 42°$$

$$\frac{a}{14} = 0.6691 \qquad \frac{b}{14} = 0.7431$$

$$a = 9.4 \qquad b = 10.4$$

$$42° + B = 90° \qquad$$ *Angles A and B are*
$$B = 48° \qquad$$ *complementary.*

Therefore, $a = 9.4$, $b = 10.4$, and $B = 48°$.

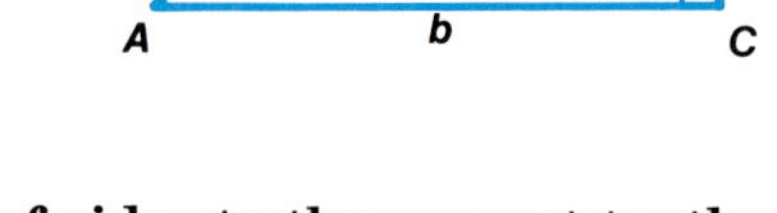

3 **Solve the right triangle. Round measures of sides to the nearest tenth.**

$$\frac{7}{c} = \sin 49° \qquad \frac{7}{b} = \tan 49°$$

$$7 = 0.7547c \qquad 7 = 1.150b$$

$$c = 9.3 \qquad b = 6.1$$

$$49° + B = 90°$$

$$B = 41°$$

Therefore, $c = 9.3$, $b = 6.1$, and $B = 41°$.

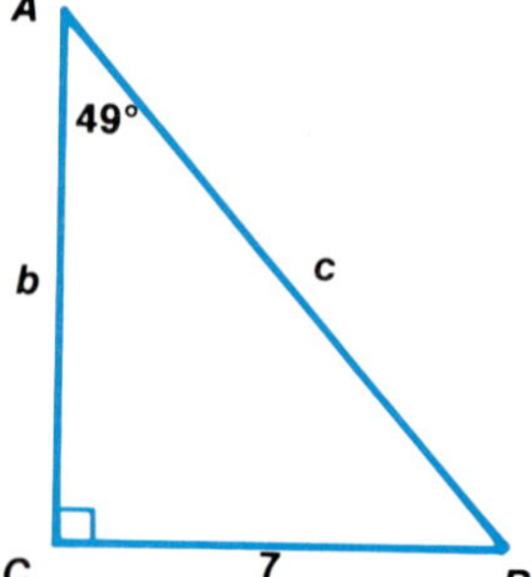

Example

4 **Solve the right triangle.**

$$7^2 + b^2 = 16^2 \qquad \textit{Use the Pythagorean Theorem.}$$
$$49 + b^2 = 256$$
$$b^2 = 207$$
$$b = 14.4 \qquad \textit{Round to the nearest tenth.}$$

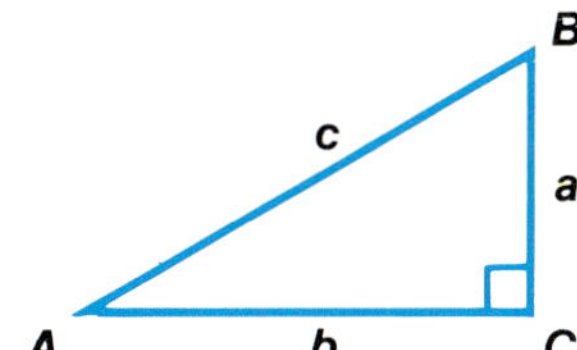

$$\sin A = \frac{7}{16} \qquad\qquad \cos B = \frac{7}{16}$$
$$\sin A = 0.4375 \qquad\qquad \cos B = 0.4375$$
$$A = 25°57' \qquad\qquad B = 64°3' \qquad \textit{Round angle measurements to the nearest minute.}$$

Therefore, $b = 14.4$, $A = 25°57'$, and $B = 64°3'$.

Exploratory Exercises

State equations that would enable you to solve each problem. Use the triangle below.

1. If $A = 15°$ and $c = 37$, find a.
2. If $A = 76°$ and $a = 13$, find b.
3. If $A = 49°13'$ and $a = 10$, find c.
4. If $a = 21.2$ and $A = 71°13'$, find b.
5. If $a = 13$ and $B = 16°$, find c.
6. If $A = 19°07'$ and $b = 11$, find c.
7. If $c = 16$ and $a = 7$, find b.
8. If $b = 10$ and $c = 20$, find a.
9. If $a = 7$ and $b = 12$, find A.
10. If $a = b$ and $c = 12$, find B.

Written Exercises

Solve each triangle described. Round measures of sides to the nearest hundredth and measures of angles to the nearest minute.

11. $a = 2$, $b = 7$
12. $c = 10$, $a = 8$
13. $c = 13$, $a = 12$
14. $a = 11$, $b = 21$
15. $b = 6$, $c = 13$
16. $c = 21$, $b = 18$
17. $A = 16°$, $c = 14$
18. $A = 63°$, $a = 9.7$
19. $A = 37°15'$, $b = 11$
20. $B = 64°$, $c = 19.2$
21. $B = 42°10'$, $a = 9$
22. $B = 83°$, $b = \sqrt{31}$
23. $c = 6$, $B = 13°$
24. $a = 9$, $B = 49°$
25. $b = 42$, $A = 77°$
26. $b = 22$, $A = 22°22'$
27. $a = 33$, $B = 33°$
28. $a = 44$, $B = 44°44'$
29. $A = 55°55'$, $c = 16$
30. $B = 18°$, $a = \sqrt{15}$
31. $A = 45°$, $c = 7\sqrt{2}$
32. $B = 30°$, $b = 11$
33. $c = 25$, $A = 15°$
34. $a = 7$, $A = 27°$
35. $\tan A = \frac{7}{8}$, $a = 7$
36. $\tan B = \frac{8}{6}$, $b = 8$
37. $\sin A = \frac{1}{3}$, $a = 5$

38. Solve the isosceles right triangle shown below. $CA = CB$.

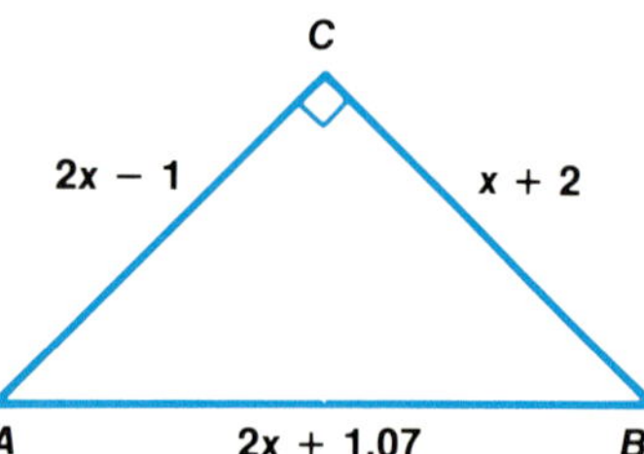

39. Solve the equilateral triangle shown below.

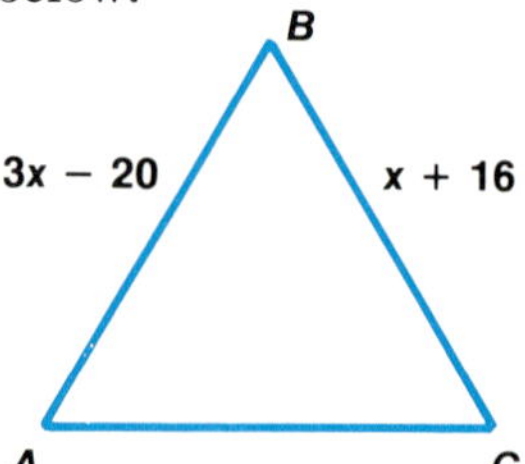

40. Solve the isosceles triangle shown at the right. The measure of the altitude from C is 12.26.

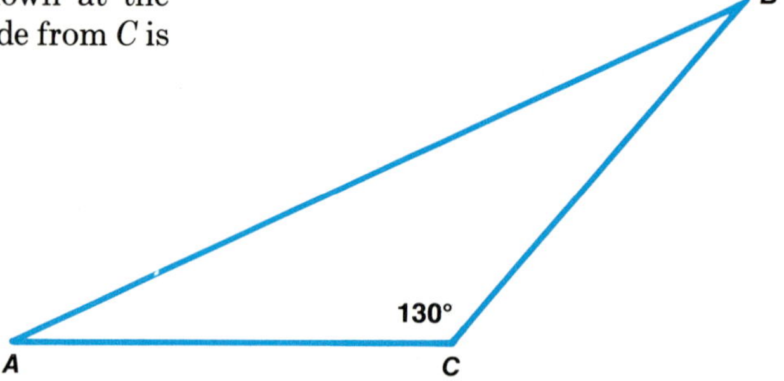

Excursions in Algebra   History

Eratosthenes, an astronomer who lived in Greece in the 3rd century B.C., is credited with providing the first accurate measure of the earth's circumference. To do this, Eratosthenes reasoned as follows.

At noon on the day of the summer solstice, the sun is directly over the city of Syene. At the same time, in Alexandria, which is north of Syene, the sun is 7°12′ south of being directly overhead. Since the distance between the two cities is 5,000 stadia, the following proportion can be written.

$$\frac{360°}{7°12'} = \frac{c}{5,000}$$

Solving the proportion gives the value of c, the earth's circumference, as 250,000 stadia, or 24,661 miles. This is only 158 miles less than the currently accepted value.

The stadium (singular form of stadia) is an ancient unit of measurement.

One of the keys to solving problems using trigonometry is drawing a figure, or **model**, of the problem. Read the problem given below.

> Sandra is watching Johnny from a window 100 feet above the ground while Johnny plays 20 feet from the base of the building. What is the measurement of the angle Sandra's line of sight forms with the ground where Johnny is playing?

You can draw a figure to represent the problem above. The problem states that the window is 100 feet above the ground, so label that distance on your figure. The problem also states that Johnny is playing 20 feet from the base of the building, so label that distance. Finally, find the measurement of the angle Sandra's line of sight forms with the ground where Johnny is playing.

Write an equation for the problem. Since you are given the opposite and adjacent sides to the angle, use the tangent function to find x.

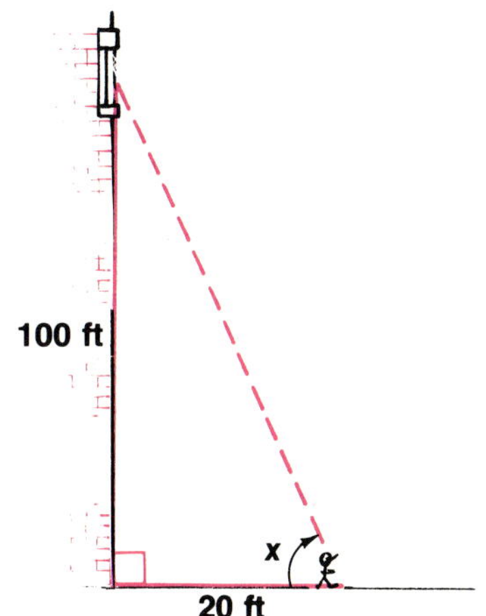

$$\tan x = \frac{100}{20}$$
$$\tan x = 5$$
$$x = 78°41'$$

Her line of sight forms an angle measuring $78°41'$ with the ground.

Exercises

Draw a figure and solve each problem.

1. The tortoise and the hare are 60 meters apart. A boy sitting in a tree at the finish line can see the hare as the tortoise crosses the line, winning the race. The measurement of the angle his line of sight forms with the tree is $75°29'$. How far above the ground is the boy?

2. Patty, a basketball player, knows that the rim of the basket is 10 feet from the floor. From where she is standing, the angle of elevation to the rim is $33°33'$. Find the distance from Patty's feet to the rim.

3. Two cars leave the same point at the same time. Car A goes east at 50 miles per hour, and car B goes south at 30 miles per hour. After one hour, what is the measurement of the angle the path of car A forms with the line between cars A and B?

4. A tent has a center pole that is 6 feet high, and a floor 8 feet wide. Find the measurement of the angle the tent side forms with the ground.

18-4 Problem Solving: Using Right Triangles

Right triangles and the trigonometric functions can be used to solve many problems.

Example

1 **A ladder 14 meters long rests against the wall of a house. The foot of the ladder rests on level ground 2 meters from the wall. What angle does the ladder form with the ground?**

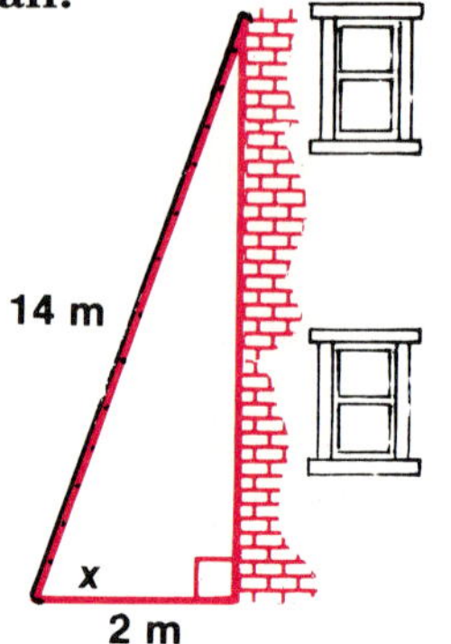

explore Draw a diagram.
Let x represent the angle measurement.

plan $\cos x = \dfrac{2}{14}$

solve $\cos x = 0.1429$
$x = 81°47'$

examine According to the drawing, the measurement of angle x is less than 90°. The answer 81°47′ seems reasonable.

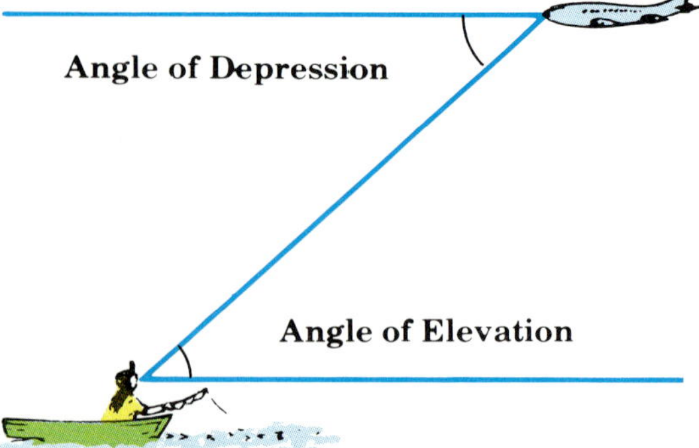

A person fishing on a lake sees an airplane flying overhead. The angle formed by her line of sight to the plane and a horizontal is called the **angle of elevation**. The angle formed by the line of sight from the pilot to the boat and a horizontal is called the **angle of depression**. The angles of elevation and depression are alternate interior angles and have equal measures.

Example

2 **Two hikers are 300 meters from the base of a radio tower. The measurement of the angle of elevation to the top of the tower is 40°. How high is the tower to the nearest meter?**

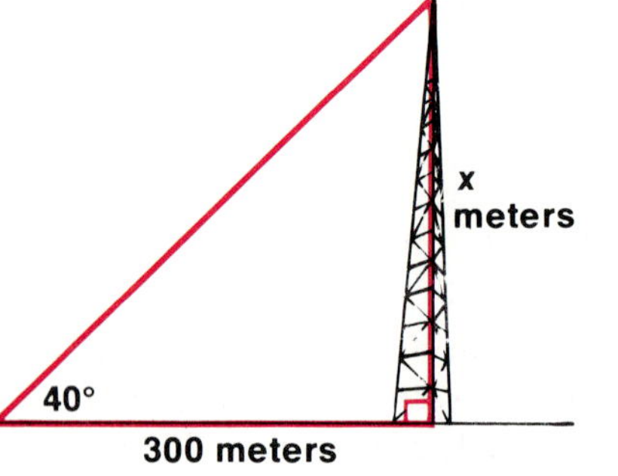

$$\frac{x}{300} = \tan 40°$$

$$x = 300(0.8391) \quad \textit{tan } 40° = 0.8391$$
$$= 251.73$$

The height of the tower is about 252 meters.

3   Robert is standing on top of a cliff 200 feet above a lake. The measurement of the angle of depression to a boat on the lake is 21°. How far is the boat from the base of the cliff to the nearest foot?

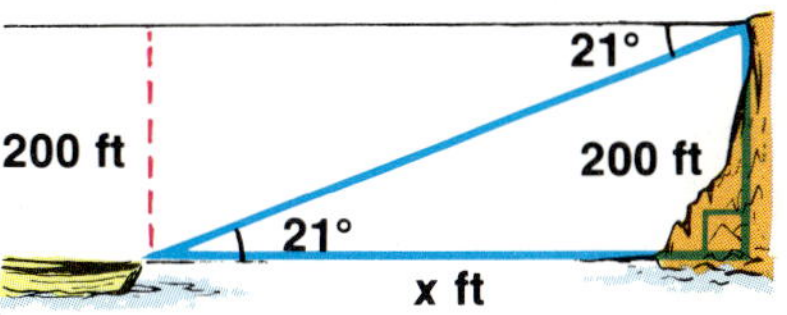

$$\frac{200}{x} = \tan 21°$$

$200 = x(0.3839)$ *tan 21° = 0.3839*

$521 \approx x$

The boat is about 521 feet from the base of the cliff.

4   The base of a television antenna and two points on the ground are in a straight line. The two points are 100 feet apart. From the two points, the measurements of the angles of elevation to the top of the antenna are 30° and 20°. Find the height of the antenna to the nearest foot.

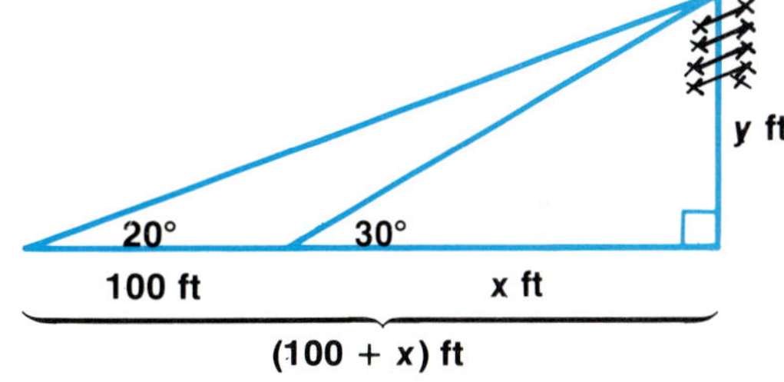

$\tan 30° = \dfrac{y}{x}$ *Solve for y.* $\tan 20° = \dfrac{y}{x + 100}$ *Solve for y.*

$x \tan 30° = y$ $0.3640(x + 100) = y$ *tan 20° = 0.3640*

$x(0.5774) = y$ *tan 30° = 0.5774* $0.3640x + 36.40 = y$

$$0.5774x = 0.3640x + 36.40$$
$$0.2134x = 36.40$$
$$x \approx 170.6 \qquad \textit{Round to the nearest tenth.}$$

$$\frac{y}{170.6} = \tan 30°$$
$$y \approx 98.5 \qquad \textit{Round to the nearest tenth.}$$

The height of the antenna is about 99 feet.

Written Exercises

Solve each problem. Round measures of lengths to the nearest hundredth and measures of angles to the nearest minute.

1. At a point on the ground that is 30 meters from the base of a tree, the measurement of the angle of elevation to the top of the tree is 65°. How tall is the tree?

2. A flagpole casts a shadow 40 feet long when the measurement of the angle of elevation to the sun is 31°20′. How tall is the flagpole?

3. The measurement of the angle of depression of an aircraft carrier from a plane 1000 feet above the water is 63°18′. How far is the plane from the carrier?

4. At the point from which it is being flown, the measurement of the angle of elevation of a kite is 70°. It is held by a string 65 meters long. How far is the kite above the ground?

5. A 24-foot ladder leans against a building. It forms an angle with the building measuring 18°. How far is the foot of the ladder from the base of the building?

6. The top of a lighthouse is 120 meters above sea level. From the top of the lighthouse, the measurement of the angle of depression of a boat at sea is 43°. Find the distance of the boat from the foot of the lighthouse.

7. A tree is broken by the wind. The top touches the ground 13 meters from the base. It makes an angle with the ground measuring 29°. How tall was the tree before it was broken?

8. The pilot of a plane flying 5000 feet above sea level observes two ships in line due east. The measurements of the angles of depression are 30° and 39°. How far apart are the ships?

9. In a parking garage, each floor is 20 feet apart. The ramp to each floor is 120 feet long. What is the measurement of the angle of elevation of the ramp?

10. A railroad track rises 10 feet for every 400 feet along the track. What is the measurement of the angle the track forms with the horizontal?

11. The Washington Monument is 555 feet high. What is the measurement of the angle of elevation of the top when observed from a point $\frac{1}{4}$ mile from the base? (1 mile = 5280 feet)

12. A train travels 5000 meters along a track whose angle of elevation has a measurement of 3°. How much did it rise during this distance?

13. The diagram shows square $ABCD$. The midpoint of side AD is E. Find the values of x, y, and z to the nearest minute.

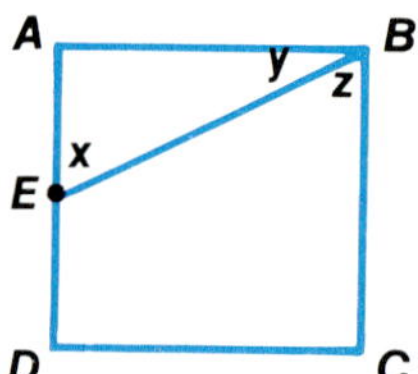

14. The measurement of the angle of elevation to the top of a building from a point on the ground is 38°20′. From a point 50 feet closer to the building, the measurement of the angle of elevation is 45°. What is the height of the building?

15. Two buildings are separated by an alley. Joe is looking out of a window 60 feet above the ground in one building. He observes the measurement of the angle of depression of the base of the second building to be 50°, and that of the angle of elevation of the top to be 40°. How high is the second building?

16. A television antenna sits atop a building. From a point 200 feet from the base of the building, the measurement of the angle of elevation of the top of the antenna is 80°. That of the angle of elevation of the bottom of the antenna from the same point is 75°. How tall is the antenna?

17. Two observers 200 feet apart are in line with the base of a flagpole. The measurement of the angle of elevation of the top from one observer is 30° and from the other 60°. How far is the flagpole from each observer?

18. To find the height of a mountain peak two points, A and B, were located on a plain in line with the peak. The angles of elevation were measured from each point. The angle at A was 36°40′ and the angle at B was 21°10′. The distance from A to B was 720 feet. How high is the peak above the level of the plain?

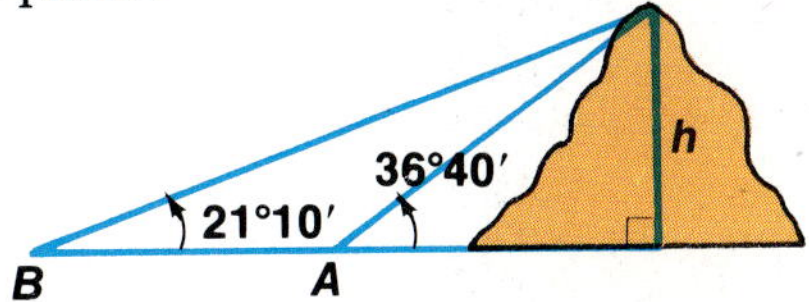

19. The isosceles triangle RST below has base TS measuring 10 centimeters and base angles each measuring 39°. Find the length of the altitude QR.

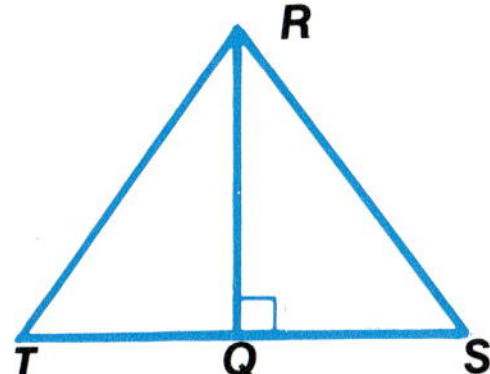

20. A pendulum 50 centimeters long is moved 40° from the vertical. How far did the tip of the pendulum rise?

21. A ship sails due north from port for 90 kilometers, then 40 kilometers east, and then 70 kilometers north. How far is the ship from port?

mini-review

1. Draw a circle graph to show how the cost of a $4.46 plant is broken down. The manufacturer's cost is $1.16. The manufacturer's income is $0.50. The distributor's income is $0.30. The retailer's income is $2.50.

Verify each identity.

2. $\dfrac{1}{\sec^2 \theta} + \dfrac{1}{\csc^2 \theta} = 1$

3. $\sin (m + 30°) + \cos (m + 60°) = \cos m$

Find the value of each function.

4. $\sin 600°$

5. $\cot (-120°)$

6. $\cos 105°$

Solve each right triangle.

7. $A = 50°,\ b = 6$

8. $a = 5,\ b = 10$

Find the value of x to the nearest minute.

9. $\cos x = 0.1030$

10. $\cot x = 1.4445$

11. In a parking garage, each floor is 22 feet apart. The ramp to each floor is 140 feet long. What is the angle of elevation of a ramp?

1. Solve the equation $3(x - 2) = 5x + 4$.

2. Graph the function $g(x) = -[x]$.

3. Find an equation of the line that passes through $(6, 4)$ and is parallel to the graph of $y = 2x - 1$.

4. Solve the system. Use augmented matrices.
$$5x - 2y + 3z = -7$$
$$-2x + 4y + z = 4$$
$$3x - 3y + 4z = 2$$

5. Divide using synthetic division.
$(x^3 + 2x^2 - 5x - 6) \div (x - 2)$

6. Simplify $\sqrt{\dfrac{3}{8}} + \sqrt{72} - \sqrt{\dfrac{8}{3}}$.

7. Simplify $\sqrt{\dfrac{4}{7}}$.

8. Solve $6x^2 - x - 1 = 0$.

9. Find the value of the discriminant for $6a^2 - a + 2 = 0$. Describe the nature of the roots.

10. Graph $f(x) = -3x^2$.

11. Graph $x^2 - 2x + 4y + 5 = 0$. Then state the axis of symmetry, vertex, and direction of opening.

12. Use the distance formula to find the distance between $A(3\sqrt{5}, \ 2\sqrt{5})$ and $B(3\sqrt{5}, -3\sqrt{5})$.

13. Graph the following system of inequalities.
$$(y - 3)^2 \leq x + 2$$
$$y \geq x^2 - 4$$

14. Use synthetic division to find the factors of $x^4 - x^3 - 17x^2 + 21x + 36$.

15. Find the rational zeros of $f(x) = 2x^3 + 7x^2 - 42x - 72$.

16. Graph the equation $y = \dfrac{x}{x - 2}$. Show the asymptotes.

Find the logarithm of each number.

17. 25,300

18. 0.0001582

Find the antilogarithm of each logarithm.

19. 4.4232

20. $(0.4600 - 3)$

21. Find the first 4 terms of this sequence: $a_1 = 2$, $a_2 = 3$, $a_{n+2} = a_n + 2 \cdot a_{n+1}$

22. For a geometric series, find a_1 given that $S_n = 203$, $r = \dfrac{2}{5}$, and $n = 4$.

23. Write $\displaystyle\sum_{k=1}^{3} (2k - 1)^k$ in expanded form and find the sum.

How many different ways can the letters in the following words be arranged?

24. LETTER

25. GEOMETRY

Find all solutions if $0° \leq x < 360°$.

26. $\sin 2x - \cos x = 0$

27. $\sin^2 2x + \cos 2x = 1$

Solve each problem.

28. A 24-foot ladder leans against a building. It forms an angle of $22°$ with the building. How far is the foot of the ladder from the base of the building?

29. Eric can rake the leaves in his yard in 10 hours. Eric and Mark together can rake the leaves in 4 hours. How long will it take Mark to rake the leaves alone?

18-5 Law of Sines

The trigonometric functions also can be used to solve problems involving triangles that are *not* right triangles.

Consider $\triangle ABC$ with height h units and sides with lengths a units, b units, and c units. The area of this triangle is given by the equation area $= \frac{1}{2}bh$. Also, $\sin A = \frac{h}{c}$. By combining these equations, you can find a new formula for the area of the triangle.

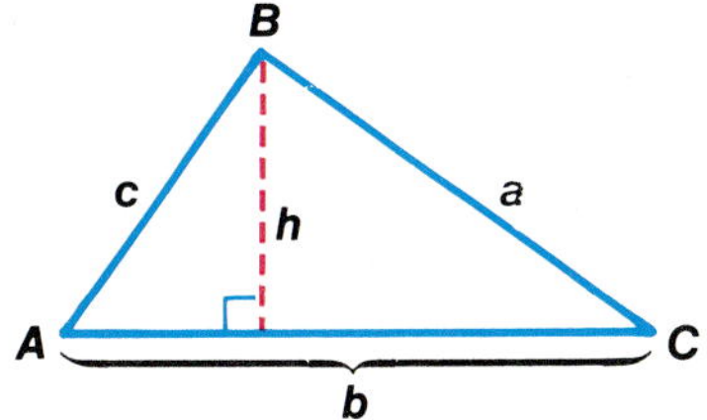

$$\text{area} = \frac{1}{2}bh$$

$$= \frac{1}{2}b(c \sin A) \qquad \sin A = \frac{h}{c}, \text{ so } h = c \sin A$$

In a similar way, you can find two other formulas for the area of a triangle.

$$\text{area} = \frac{1}{2}ac \sin B \qquad\qquad \text{area} = \frac{1}{2}ab \sin C$$

All of these formulas represent the area of the same triangle. Thus, the following must be true.

$$\frac{1}{2}bc \sin A = \frac{1}{2}ac \sin B = \frac{1}{2}ab \sin C$$

The **Law of Sines** is obtained by dividing each of the above expressions by $\frac{1}{2}abc$.

$$\frac{\sin A}{a} = \frac{\sin B}{b} = \frac{\sin C}{c}$$

> **Let $\triangle ABC$ be any triangle with a, b, and c representing the measures of sides opposite angles with measurements A, B, and C respectively. Then,**
> $$\frac{\sin A}{a} = \frac{\sin B}{b} = \frac{\sin C}{c}.$$
>
> *Law of Sines*

Example

1 Find the area of $\triangle ABC$ if $a = 6$, $b = 10$, and $C = 40°$.

$$\text{area} = \frac{1}{2}(6)(10) \sin 40° \qquad \sin 40° = 0.6428$$

$$= 19.284$$

To the nearest whole unit, the area is 19 square units.

2 **Solve the triangle.**

$$\frac{\sin B}{14} = \frac{\sin 105°}{18}$$

$$\sin B = \frac{14 \sin 105°}{18}$$

$$= 0.7513$$

$$B = 48°42' \quad \textit{Round to the nearest minute.}$$

$$48°42' + 105° + C = 180°$$

$$C = 26°18'$$

$$\frac{c}{\sin 26°18'} = \frac{18}{\sin 105°}$$

$$c = \frac{18 \sin 26°18'}{\sin 105°}$$

$$= 8.3 \quad \textit{Round to the nearest tenth.}$$

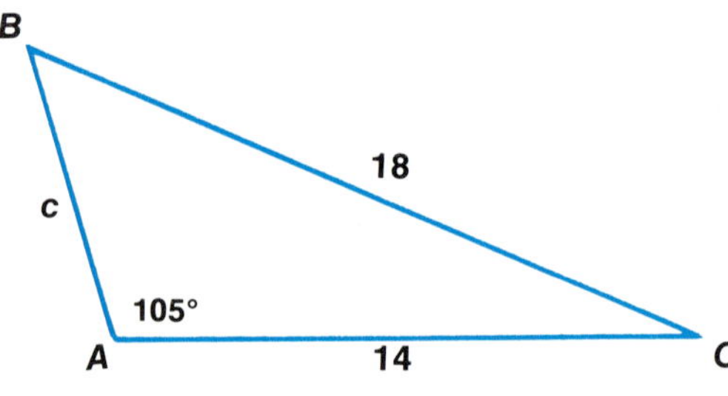

Therefore, $B = 48°42'$, $C = 26°18'$, and $c = 8.3$.

3 **A surveyor measures a fence 440 meters long. She takes bearings of a landmark C from A and B and finds that $A = 48°$ and $B = 75°$. Find the distance from A to C to the nearest meter.**

$$48° + 75° + C = 180°$$

$$C = 57°$$

$$\frac{b}{\sin 75°} = \frac{440}{\sin 57°}$$

$$b = \frac{440 \sin 75°}{\sin 57°}$$

$$= 506.8$$

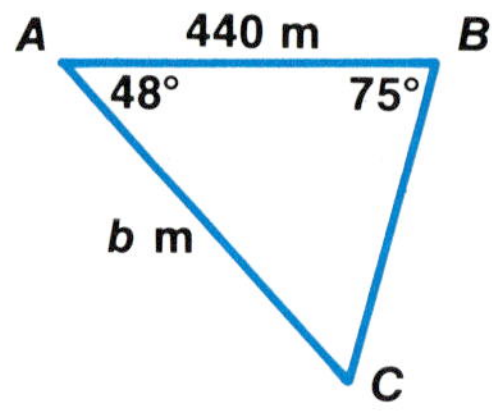

The distance is approximately 507 meters.

Exploratory Exercises

State an equation that would enable you to find the area of each triangle.

1. $a = 10$, $b = 17$, $C = 46°$

2. $b = 15$, $c = 20$, $A = 63°$

3. $a = 15$, $b = 30$, $C = 90°$

4. $a = 6$, $c = 4$, $B = 52°$

State an equation that would enable you to solve each triangle described below.

5. If $b = 10$, $a = 14$, and $A = 50°$, find B.

6. If $A = 40°$, $B = 60°$, and $a = 20$, find b.

7. If $b = 2.8$, $A = 53°$, and $B = 61°$, find a.

8. If $b = 16$, $c = 12$, and $B = 42°$, find C.

Written Exercises

Find the area of each triangle described below.

9. $a = 12$, $b = 12$, $C = 50°$

10. $a = 15$, $b = 22$, $C = 90°$

11. $b = 11.5$, $c = 14$, $A = 20°$

12. $a = 11$, $c = 5$, $B = 50°6'$

13. $a = 11$, $b = 13$, $C = 31°10'$

14. $b = 4$, $c = 19$, $A = 73°24'$

15. $a = 9.4$, $c = 13.5$, $B = 95°$

16. $b = 17.3$, $c = 12.4$, $A = 110°$

Solve each triangle described below.

17. $a = 8$, $A = 49°$, $B = 57°$

18. $A = 45°$, $a = 83$, $b = 79$

19. $A = 83°10'$, $a = 80$, $b = 70$

20. $A = 40°$, $B = 60°$, $c = 20$

21. $B = 70°$, $C = 58°$, $a = 84$

22. $A = 30°$, $C = 70°$, $c = 8$

23. $b = 15$, $c = 17$, $C = 64°40'$

24. $a = 23$, $A = 73°25'$, $C = 24°30'$

25. $B = 36°36'$, $C = 119°$, $b = 8$

26. $a = 14$, $b = 7.5$, $A = 103°$

Solve each problem.

27. An isosceles triangle has a base of 22 centimeters and a vertex angle measuring 36°. Find its perimeter.

28. The longest side of a triangle is 34 yards. Two angles of the triangle are 40° and 65°. Find the length of the other two sides.

29. A triangular lot faces two streets that meet at an angle measuring 85°. The sides of the lot facing the streets are each 160 feet in length. Find the perimeter of the lot.

30. A ship is sighted at sea from two observation points A and B. Points A and B are 30 miles apart. The angle at A between line AB and the ship is 34°. The angle at B is 45°34'. How far is the ship from point B?

31. Two planes leave an airport at the same time. Each flies at a speed of 110 miles per hour. One flies in the direction 60° east of north. The other flies in the direction 40° east of south. How far apart are the planes after 3 hours?

32. Points X and Y are on opposite sides of a valley. Point C is 60 kilometers from point X. Angle YXC is 108°, and angle YCX is 35°. Find the width of the valley.

33. A building 60 feet tall is on top of a hill. A surveyor is at a point on the hill and observes that the angle of elevation to the top of the building has measurement 42° and to the bottom of the building has measurement 18°. How far is the surveyor from the building?

34. A flower bed is in the shape of an obtuse triangle. One angle is 45° and the opposite side is 28 feet long. The longest side is 36 feet long. Find the measures of the remaining angles and side.

35. Find the lengths of the diagonals and the perimeter for the parallelogram shown at the right.

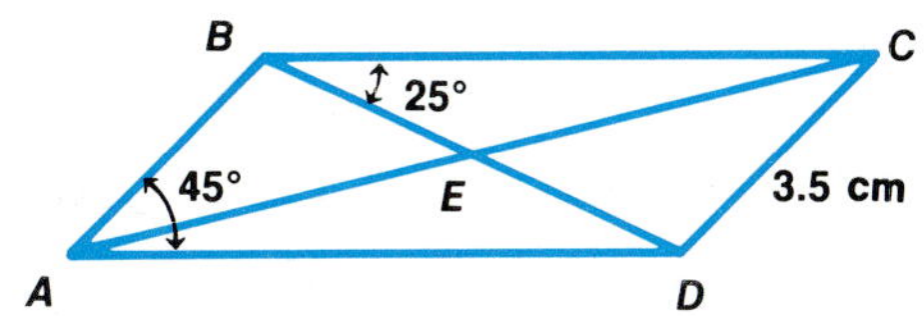

18-6 Examining Solutions

To solve a triangle, you need to analyze the data you have in order to determine whether there is a solution or not.

When the lengths of two sides of a triangle and the measurement of the angle opposite one of them are given, a single solution does not always exist. In such a case, one of the following will be true.

    1. No triangle exists.

    2. Exactly one triangle exists.

    3. Two triangles exist.

In other words, there may be no solution, one solution, or two solutions.

Suppose you are given a, b, and A. First, consider the case where $A < 90°$.

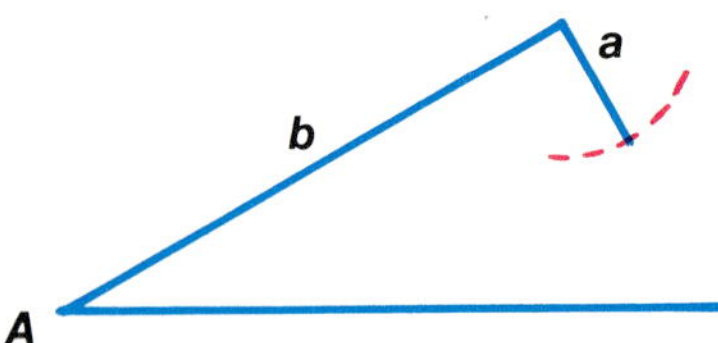

If $a < b \sin A$, no solution exists.

If $a = b \sin A$, one solution exists.

If $a = b \sin A$, the solution is a right triangle.

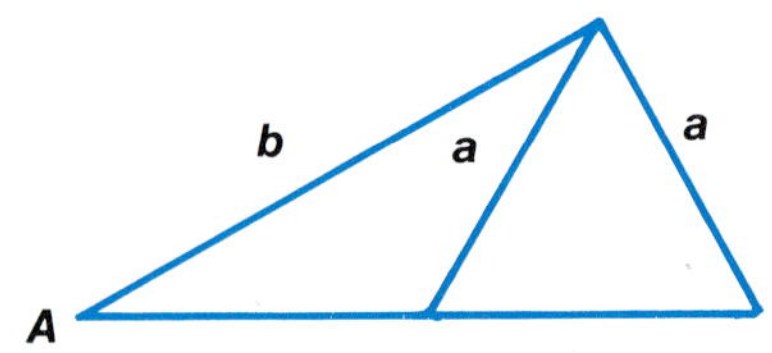

If $b \sin A < a < b$, two solutions exist.

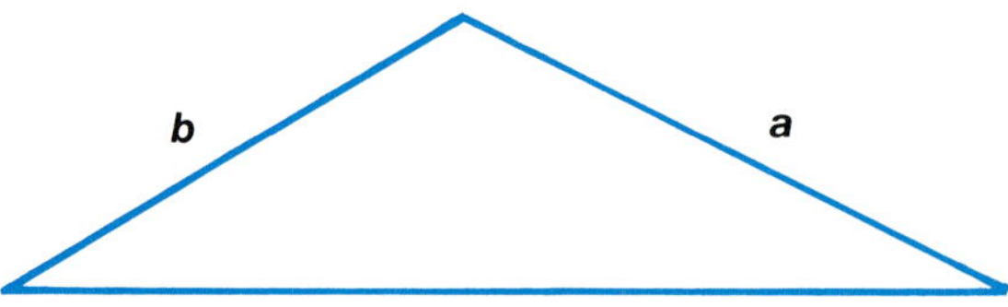

If $a > b$, one solution exists.

Consider the case where $A \geq 90°$.

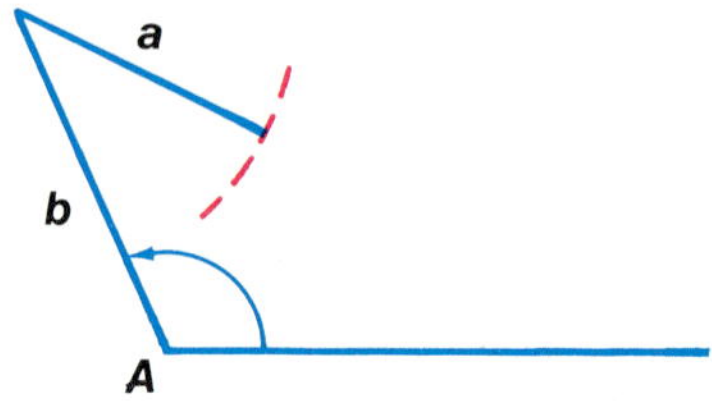

If $a \leq b$, no solution exists.

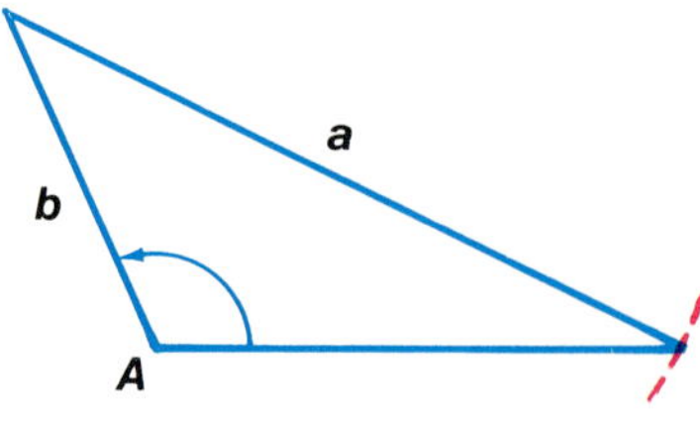

If $a > b$, one solution exists.

1 **Solve the triangle where $A = 50°$, $b = 10$, and $a = 2$.**

$$b \sin A = 10 \sin 50°$$
$$= 10(0.7660)$$
$$= 7.66$$

Since $50° < 90°$ and $2 < 7.66$, no solution exists.

2 **Solve the triangle where $A = 40°$, $b = 10$, and $a = 8$.**

$$b \sin A = 10 \sin 40°$$
$$= 10(0.6428)$$
$$= 6.428$$

Since $40° < 90°$ and $6.428 < 8 < 10$, two solutions exist.

$$\frac{\sin 40°}{8} = \frac{\sin B}{10} \qquad \textit{Use the Law of Sines.}$$

$$\sin B = \frac{10 \sin 40°}{8} \qquad \textit{sin 40° = 0.6428}$$

$$= 0.8035$$

$B = 53°28'$ or $126°32'$ *Round to the nearest minute.*

Since two solutions exist, there must be two values for B. The equation sin (180° − A) = sin A is used to find the second value.

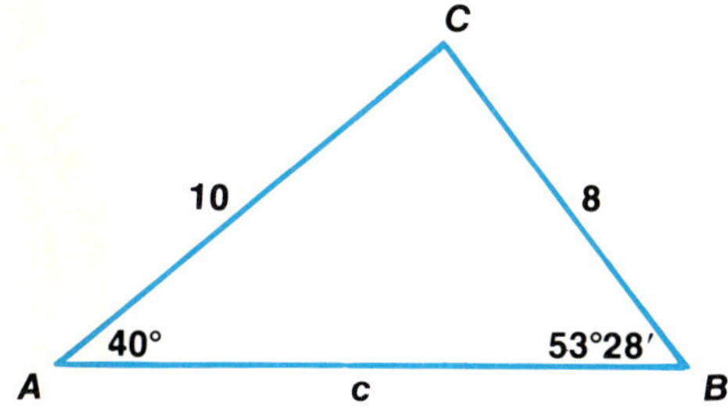

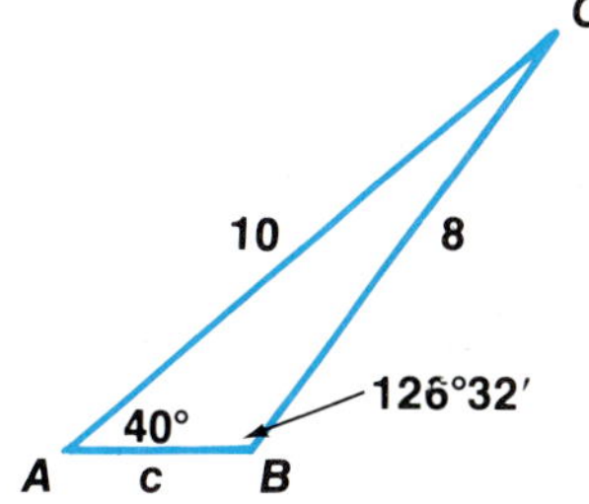

$$40° + 53°28' + C = 180°$$
$$C = 86°32'$$

$$c = \frac{8 \sin 86°32'}{\sin 40°} \qquad \textit{sin 86°32' = 0.9981}$$
$$= 12.4 \quad \textit{Round to the nearest tenth.}$$

One solution is $B = 53°28'$, $C = 86°32'$, and $c = 12.4$.

$$40° + 126°32' + C = 180°$$
$$C = 13°28'$$

$$c = \frac{8 \sin 13°28'}{\sin 40°} \qquad \textit{sin 13°28' = 0.2328}$$
$$= 2.9$$

Another solution is $B = 126°32'$, $C = 13°28'$, and $c = 2.9$.

3 **Solve the triangle where $A = 40°$, $b = 10$, and $a = 14$.**

Since $40° < 90°$ and $14 > 10$, one solution exists.

$$\frac{\sin 40°}{14} = \frac{\sin B}{10} \qquad \textit{Use the Law of Sines.}$$

$$\sin B = \frac{10 \sin 40°}{14} \qquad \textit{sin 40° = 0.6428}$$

$$= 0.4591$$

$$B = 27°20' \qquad \textit{Round to the nearest minute.}$$

$$40° + 27°20' + C = 180°$$

$$C = 112°40'$$

$$\frac{\sin 40°}{14} = \frac{\sin 112°40'}{c}$$

$$c = \frac{14 \sin 112°40'}{\sin 40°} \qquad \textit{sin 112°40' = 0.9228}$$

$$= 20.1 \qquad \textit{Round to the nearest tenth.}$$

Therefore, $B = 27°20'$, $C = 112°40'$, and $c = 20.1$.

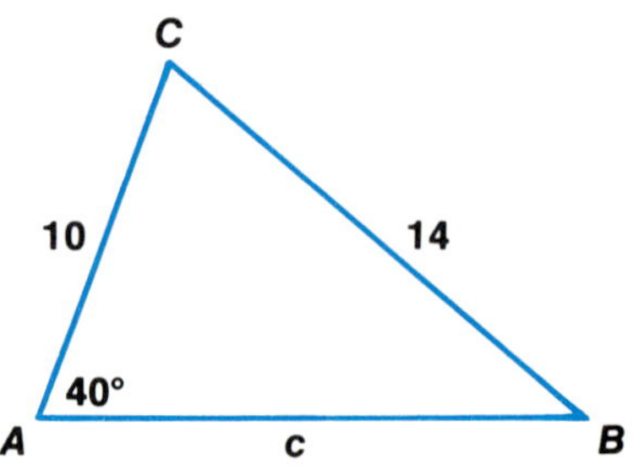

Exploratory Exercises

State if the given information determines one triangle, two triangles, or no triangle.

1. $A = 140°$, $b = 10$, $a = 3$

2. $A = 118°$, $b = 11$, $a = 17$

3. $A = 30°$, $a = 4$, $b = 8$

4. $A = 43°$, $b = 20$, $a = 11$

5. $A = 58°$, $a = 17$, $b = 13$

6. $A = 38°$, $b = 10$, $a = 8$

Written Exercises

Determine the number of possible solutions. If a solution exists, solve the triangle.

7. $a = 6$, $b = 10$, $A = 36°52'$

8. $a = 6$, $b = 8$, $A = 150°$

9. $a = 12$, $b = 19$, $A = 57°$

10. $a = 7$, $b = 6$, $A = 30°$

11. $a = 64$, $c = 90$, $C = 98°$

12. $a = 26$, $b = 29$, $A = 58°$

13. $b = 40$, $a = 32$, $A = 125°20'$

14. $a = 9$, $b = 20$, $A = 31°$

15. $a = 12$, $b = 14$, $A = 90°$

16. $A = 25°$, $a = 125$, $b = 150$

17. $A = 40°$, $b = 16$, $a = 10$

18. $A = 76°$, $a = 5$, $b = 20$

19. $B = 34°20'$, $b = 5$, $a = 11$

20. $A = 120°$, $b = 20$, $a = 18$

Solve each problem.

21. Ana Galo needs to draw a triangle for her geometry class. She makes one side 40 mm long, and another side 32 mm long with opposite angle measuring $48°19'$. What would be the length of the third side?

22. An engineering student was given an assignment to construct a triangular model of three steel girders. Two of the girders measured 20 cm and 15 cm, and the angle opposite the 15 cm girder had to be 61°. Could he construct the triangle? If so, how long did the third girder have to be?

mini-review

1. If $\sin x = -\frac{2}{3}$ and x is in the third quadrant, find $\cos 2x$.

Find each value.

2. $\text{Tan}^{-1} \frac{\sqrt{2}}{2}$

3. $\text{Cos}^{-1} (\sin 60°)$

4. $\sin \left(2 \, \text{Cos}^{-1} \frac{\sqrt{2}}{2} \right)$

Solve each right triangle. Round measures of sides to the nearest hundredth and measures of angles to the nearest minute.

5. $a = 4.4$, $b = 6.6$

6. $A = 20°$, $c = 15$

7. $B = 31°16'$, $a = 8$

8. Two buildings are separated by an alley. Carla is looking out of a window 50 feet above the ground in one building. She observes the measurement of the angle of depression of the base of the second building to be 40° and that of the angle of elevation of the top to be 30°. How tall is the second building?

9. The average points scored per event in the Women's Gymnastics Competition of the 1988 Olympic Games by four teams are given below. Make a stem and leaf plot of the data. Then find the mode, mean, quartiles, interquartile range, standard deviation, and any outliers in the data.

USSR	9.925	9.975	10.000	9.925	9.850	10.000
ROMANIA	9.900	9.975	9.925	10.000	10.000	9.875
USA	9.600	9.375	9.800	9.825	9.300	9.575
GDR	9.375	9.825	9.800	9.575	9.600	9.310

Jon Thomas is a civil engineer. Part of his job is to determine if machinery is being overloaded. Overloading can ruin equipment or make it wear out sooner. Sometimes, it is a potential danger to human safety.

The boom hoist shown below carries a load of 2000 pounds. Jon needs to find the tension, t, in the cable and the compression, c, in the boom. To do this, he uses the following method.

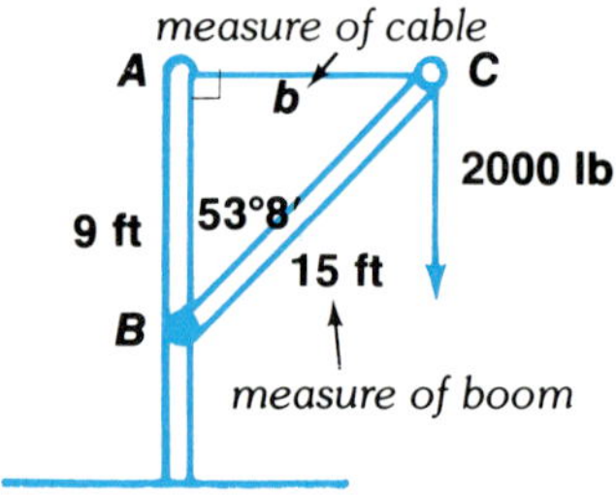

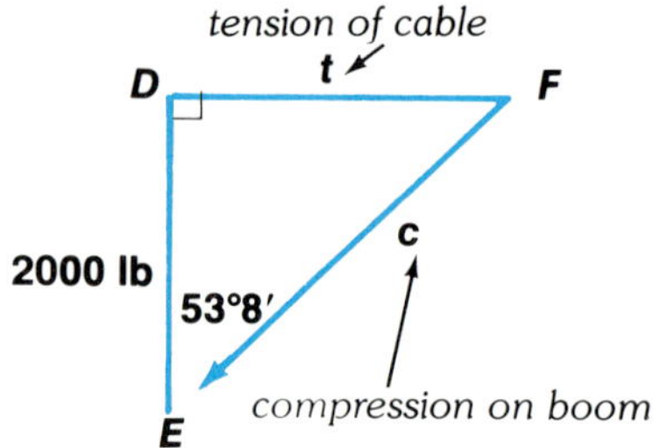

First, assume triangle ABC is a right triangle, and find b. Use the Pythagorean Theorem.

$$b^2 = 15^2 - 9^2$$
$$= 225 - 81$$
$$= 144$$
$$b = 12 \qquad \text{\textit{the measure of the cable}}$$

Since triangle ABC is similar to triangle DEF, the following proportions can be used.

$$\frac{t}{2000} = \frac{12}{9} \qquad \frac{DF}{DE} = \frac{AC}{AB}$$
$$9t = 24{,}000$$
$$t \approx 2667$$

$$\frac{c}{2000} = \frac{15}{9} \qquad \frac{EF}{DE} = \frac{BC}{AB}$$
$$9c = 30{,}000$$
$$c \approx 3333$$

The tension is about 2667 pounds.

The compression is about 3333 pounds.

Exercises

Find the tension and compression for each load on the hoist described above.

1. 500 pounds

2. 1000 pounds

3. 1200 pounds

4. 1500 pounds

18-7 Law of Cosines

If two sides and the included angle, or three sides of a triangle, are given, the Law of Sines cannot be used to solve the triangle. Another formula is needed.

Consider $\triangle ABC$ with height h units and sides with lengths a units, b units, and c units. Suppose segment AD is x units long. Then segment DC is $(b - x)$ units long.

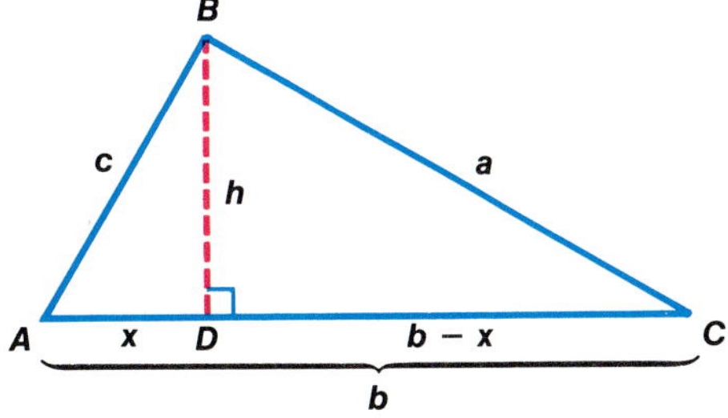

How are A, a, b, and c related?

$$
\begin{aligned}
a^2 &= (b - x)^2 + h^2 && \textit{Use the Pythagorean Theorem for } \triangle BDC. \\
&= b^2 - 2bx + x^2 + h^2 && \textit{Expand } (b - x)^2. \\
&= b^2 - 2bx + c^2 && \textit{In } \triangle ADB,\ c^2 = x^2 + h^2. \\
&= b^2 - 2b(c \cos A) + c^2 && \cos A = \tfrac{x}{c} \textit{ so } x = c \cos A. \\
&= b^2 + c^2 - 2bc \cos A
\end{aligned}
$$

In a similar way, two other formulas can be found relating the lengths of sides to the cosine of B and C. All three formulas, the **Law of Cosines**, can be summarized as follows.

> **Let $\triangle ABC$ be any triangle with a, b, and c representing the measures of sides opposite angles with measurements A, B, and C respectively. Then, the following equations are true.**
>
> $$a^2 = b^2 + c^2 - 2bc \cos A$$
> $$b^2 = a^2 + c^2 - 2ac \cos B$$
> $$c^2 = a^2 + b^2 - 2ab \cos C$$

Law of Cosines

Use the Law of Cosines to solve a triangle in the following cases.

1. To find the length of the third side of any triangle if the lengths of two sides and the measurement of the included angle are given.
2. To find the measurement of an angle of a triangle if the lengths of three sides are given.

1 **Solve the triangle where $A = 35°$, $b = 16$, and $c = 19$.**

$a^2 = 16^2 + 19^2 - 2(16)(19) \cos 35°$ *Use the Law of Cosines.*
$\quad = 118.93$
$a = 10.9$

$\dfrac{\sin 35°}{10.9} = \dfrac{\sin B}{16}$ *Use the Law of Sines.*

$\sin B = \dfrac{16 \sin 35°}{10.9}$ *$\sin 35° = 0.5736$*

$\quad = 0.8420$
$B = 57°21'$ *Round to the nearest minute.*

$35° + 57°21' + C = 180°$
$\qquad\qquad\quad C = 87°39'$

Therefore, $a = 10.9$, $B = 57°21'$, and $C = 87°39'$.

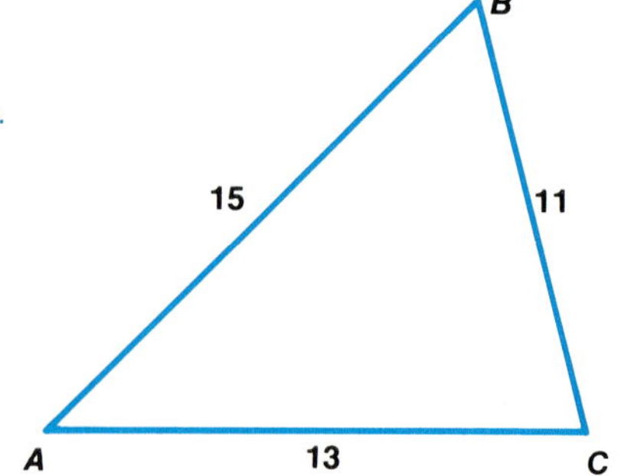

2 **Solve the triangle where $a = 11$, $b = 13$, and $c = 15$.**

$11^2 = 13^2 + 15^2 - 2(13)(15) \cos A$ *Use the Law of Cosines.*
$2(13)(15) \cos A = 13^2 + 15^2 - 11^2$

$\cos A = \dfrac{13^2 + 15^2 - 11^2}{2(13)(15)}$

$\quad = 0.7000$
$A = 45°34'$ *Round to the nearest minute.*

$\dfrac{\sin 45°34'}{11} = \dfrac{\sin B}{13}$ *Use the Law of Sines.*

$\sin B = \dfrac{13 \sin 45°34'}{11}$ *$\sin 45°34' = 0.7141$*

$\quad = 0.8439$
$B = 57°33'$ *Round to the nearest minute.*

$45°34' + 57°33' + C = 180°$
$\qquad\qquad\qquad C = 76°53'$

Therefore, $A = 45°34'$, $B = 57°33'$, and $C = 76°53'$.

Exploratory Exercises

Determine whether the Law of Sines or the Law of Cosines should be used first to solve each triangle described below.

1. $A = 40°, b = 6, c = 7$

2. $a = 10, A = 40°, c = 8$

3. $a = 14, b = 15, c = 16$

4. $A = 40°, C = 70°, c = 14$

5. $C = 35°, a = 11, b = 10.5$

6. $c = 21, a = 14, B = 60°$

7. $c = 10.3, a = 21\frac{1}{2}, b = 16.71$

8. $b = 17, B = 42°58', a = 11$

9. $c = 14.1, A = 29°, b = 7.6$

10. $A = 28°50', b = 5, c = 4.9$

Written Exercises

Solve each triangle described below.

11. $a = 140, b = 185, c = 166$

12. $A = 51°, b = 40, c = 45$

13. $a = 5, b = 6, c = 7$

14. $a = 5, b = 12, c = 13$

15. $a = 20, c = 24, B = 47°$

16. $b = 13, a = 21.5, C = 39°20'$

17. $A = 40°, B = 59°, c = 14$

18. $B = 19°, a = 51, c = 61$

19. $a = 345, b = 648, c = 442$

20. $A = 25°26', a = 13.7, B = 78°$

Solve each problem.

21. A triangular plot of land has two sides which have lengths 400 feet and 600 feet. The measurement of the angle between those sides is $46°20'$. Find its perimeter and area.

22. The sides of a triangular city lot have lengths 50 meters, 70 meters, and 85 meters. Find the measurement of the angle opposite the short side.

23. A pilot is flying from Chicago to Columbus, a distance of 300 miles. He starts his flight 15° off course and flies on this course for 75 miles. How far is he from Columbus and by how much must he correct his error?

24. Two ships leave San Francisco at the same time. One travels 40° west of north at a speed of 20 knots. The other travels 10° west of south at a speed of 15 knots. How far apart are they after 11 hours? (1 knot = 1 nautical mile per hour)

25. A ship at sea is 70 miles from one radio transmitter and 130 miles from another. The measurement of the angle between the signals is 130°. How far apart are the transmitters?

26. A 40-foot television antenna stands on top of a building. From a point on the ground, the angles of elevation of the top and bottom of the antenna, respectively, have measurements of 56° and 42°. How tall is the building?

27. The sides of a triangle are 6.8 cm, 8.4 cm, and 4.9 cm. Find the measure of the smallest angle.

28. The sides of a parallelogram are 55 cm and 71 cm. Find the length of each diagonal if the larger angle is 106°.

29. A plane flew 1200 kilometers north. It then changed direction by turning 15 degrees clockwise and flew for another 850 kilometers. How far was the plane from its starting point?

30. Circle Q has a radius of 15 cm. Two radii, $\overline{QA}$ and $\overline{QB}$, form an angle of 123°. Find the length of chord AB.

Examining Solutions

The computer program below determines the number of solutions of a triangle given the lengths of two sides and the measurement of the angle opposite one of them.

Lesson 18-6 was used as the basis for the program. If you study the lesson carefully, you can understand how the program works.

Be sure to enter the degree measure of angle A and the lengths of sides a and b (called A1 and B1 in the program) in the order given.

```
10   INPUT A,A1,B1
20   LET B = B1 * SIN (A * 3.1415927 / 180)
30   PRINT "B SIN A= ";B
40   IF A > = 90 THEN 110
50   IF A1 < B THEN 120
60   IF A1 = B THEN 130
70   IF A1 < B1 THEN 100
80   IF A1 > B1 THEN 130
100   PRINT "TWO SOLUTIONS EXIST." : GOTO 140
110   IF A1 > B1 THEN 130
120   PRINT "NO SOLUTION EXISTS.": GOTO 140
130   PRINT "ONE SOLUTION EXISTS."
140   END
```

```
] RUN
?26, 3, 7
B SIN A=3.06859807
NO SOLUTION EXISTS.
```

Exercises

Determine the number of possible solutions. State the value of $b \sin A$ to four decimal places.

1. $A = 103°, a = 5, b = 2$

2. $A = 175°, a = 19, b = 2$

3. $A = 77°, a = 1, b = 5$

4. $A = 130°, a = 8, b = 10$

5. $A = 38.62°, a = 15, b = 16$

6. $A = 10.524°, a = 1, b = 2$

7. $A = 53°20', a = 9, b = 6$

8. $A = 9°55', a = 3, b = 6$

9. Why is A multiplied by 3.1415927/180 in line 20?

10. What happens when the values for A, A1, and B1 are entered out of order? Why?

hypotenuse   (639)
opposite side   (639)
adjacent side   (639)
interpolation   (643)

angle of elevation   (650)
angle of depression   (650)
Law of Sines   (655)
Law of Cosines   (663)

Chapter Summary

1. The trigonometric functions relate the sides and acute angles of a right triangle as follows.   (639)

$$\sin A = \frac{a}{c}$$

$$\cos A = \frac{b}{c}$$

$$\tan A = \frac{a}{b}$$

$$\csc A = \frac{c}{a}$$

$$\sec A = \frac{c}{b}$$

$$\cot A = \frac{b}{a}$$

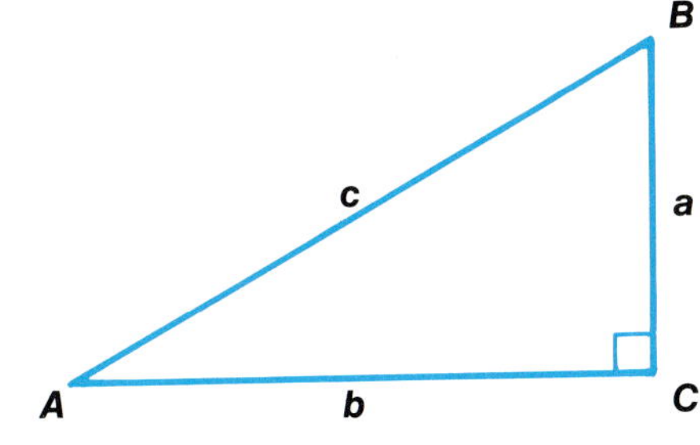

2. The special triangle with angle measurements of 45°, 45°, and 90° has sides whose lengths are in the ratio of 1 to 1 to $\sqrt{2}$.   (640)

3. The special triangle with angle measurements of 30°, 60°, and 90° has sides whose lengths are in the ratio of 1 to $\sqrt{3}$ to 2.   (640)

4. The values of trigonometric functions may be found in a table. Sometimes interpolation is needed to find a value which is between consecutive entries in the table.   (642)

5. Trigonometric functions can be used to solve right triangles.   (646)

6. Trigonometric functions may be used to solve many problems including those involving angles of elevation and depression.   (650)

7. Law of Sines: Let triangle ABC be any triangle with a, b, and c representing the measures of sides opposite angles with measurements A, B, and C respectively. Then the following equations are true.

$$\frac{\sin A}{a} = \frac{\sin B}{b} = \frac{\sin C}{c} \quad (655)$$

8. When the lengths of two sides of a triangle and the measurement of the angle opposite one of them are given, one solution does not always exist. No triangle may exist, one triangle may exist, or two triangles may exist.   (658)

9. Law of Cosines: Let triangle ABC be any triangle with a, b, and c representing the measures of sides opposite angles with measurements A, B, and C respectively. Then, the following equations are true.

$$a^2 = b^2 + c^2 - 2bc \cos A$$
$$b^2 = a^2 + c^2 - 2ac \cos B$$
$$c^2 = a^2 + b^2 - 2ab \cos C \quad (663)$$

Chapter Review

18–1 Find each value to the nearest four decimal places.

1. sin A
2. sin B
3. cos A
4. cos B
5. tan A
6. tan B
7. csc A
8. sec A
9. cot B

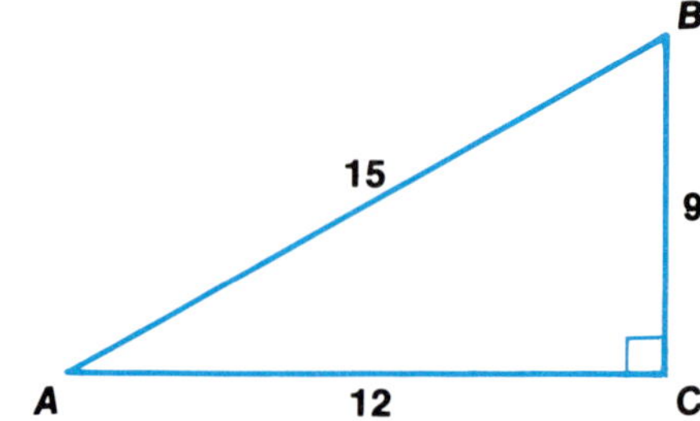

Find the value of each expression. Write each answer in fraction form.

10. cos 30°   11. tan 60°   12. sin 45°

18–2 Use a table to approximate each trigonometric value. Use interpolation when necessary.

13. sin 70°   14. sin 18°20′   15. cos 35°
16. cos 81°40′   17. tan 47°   18. tan 16°35′

Find the value of x to the nearest minute.

19. sin $x = 0.9272$ 20. sin $x = 0.2164$ 21. cos $x = 0.9171$
22. cos $x = 0.5150$ 23. tan $x = 0.3476$ 24. tan $x = 1.664$

Use a calculator to find each trigonometric value.

25. sin 82°14′   **26.** cos 43°   **27.** tan 67°33′

28. sec 91°54′   **29.** csc 39°12′   **30.** cot 19°26′

Use a calculator to find the value of x to the nearest minute.

31. sin x = 0.7432   **32.** cos x = 0.2659   **33.** tan x = 1.1499

34. csc x = 1.6004   **35.** cot x = 1.5599   **36.** sec x = 1.7842

18–3 Solve each right triangle.

37. $A = 25°$, $c = 6$ **38.** $A = 50°$, $a = 11$

39. $B = 85°$, $a = 6.21$ **40.** $B = 31°$, $c = 12$

41. $a = 1$, $b = 3$ **42.** $a = 15$, $c = 20$

43. $b = 7$, $c = 10$ **44.** $a = 10$, $b = 24$

18–4 Solve each problem.

45. From a point on the ground 50 meters from the base of a flagpole, the measurement of the angle of elevation of the top is 48°. How tall is the flagpole?

46. A pilot 3000 feet above the ocean notes the measurement of the angle of depression of a ship is 42°. How far is the plane from the ship?

47. A building is 80 feet tall. Find the measurement of the angle of elevation to the top of the building from a point on the ground 100 feet from the base of the building.

48. The base of a monument and two points on the ground are in a straight line. The two points are 50 meters apart. The measurements of the angles of elevation to the top of the monument are 45° and 25°. Find the height of the monument.

18–5 Use the Law of Sines to solve each triangle.

49. $A = 50°$, $b = 12$, $a = 10$ **50.** $A = 83°10′$, $a = 80$, $b = 70$

51. $B = 46°$, $C = 83°$, $b = 65$ **52.** $A = 45°$, $B = 30°$, $b = 20$

18–6 Determine the number of possible solutions. If a solution exists, solve the triangle.

53. $A = 36°$, $a = 2$, $b = 14$ **54.** $A = 40°$, $a = 8$, $b = 10$

55. $A = 46°$, $a = 10$, $b = 8$ **56.** $a = 130°$, $a = 25$, $b = 16$

18–7 Use the Law of Cosines to solve each triangle.

57. $A = 60°$, $b = 2$, $c = 5$ **58.** $C = 65°$, $a = 4$, $b = 7$

59. $C = 40°$, $a = 6$, $b = 7$ **60.** $B = 24°$, $a = 42$, $c = 6.5$

Chapter Test

Find each value rounded to four decimal places.

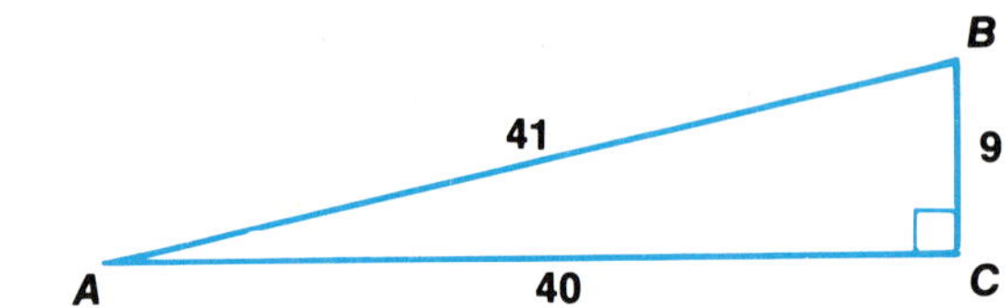

1. $\sin A$	**2.** $\cos A$
3. $\tan A$	**4.** $\cos B$

Find the value of each expression. Write each answer in fraction form.

5. $\cos 60°$	**6.** $\tan 45°$

Use a table or a calculator to approximate each trigonometric value. Use interpolation when necessary.

7. $\sin 67°$	**8.** $\cos 38°10'$
9. $\tan 59°38'$	**10.** $\sin 13°3'$

Find the value of x to the nearest minute.

11. $\cos x = 0.4384$	**12.** $\tan x = 0.4734$
13. $\sin x = 0.4423$	**14.** $\tan x = 1.635$

Solve each right triangle.

15. $A = 36°$, $b = 14$	**16.** $B = 75°$, $b = 6$
17. $A = 22°$, $c = 8$	**18.** $a = 7$, $c = 12$

Solve each problem.

19. A 32-foot ladder leans against a building. The top touches the building 26 feet above the ground. What is the measurement of the angle formed by the ladder with the ground?

20. From the top of a cliff, a camper sees a deer. The measurement of the angle of depression to the deer is 70°. The cliff is 50 meters high. How far is the deer from the base of the cliff?

21. A plane flew 1000 kilometers north. It then changed direction by turning 20 degrees clockwise and flew for another 700 kilometers. How far was the plane from its starting point?

22. The longest side of a triangle is 23 yards. Two of the angles are 23° and 49°. Find the measures of the other two sides and the other angle.

Determine the number of possible solutions. If a solution exists, solve each triangle.

23. $a = 13$, $b = 11$, $c = 17$	**24.** $A = 46°$, $B = 77°$, $a = 6$
25. $A = 75°$, $b = 21$, $a = 30$	**26.** $A = 65°$, $b = 21$, $a = 6$
27. $A = 70°$, $B = 31°$, $c = 17$	**28.** $A = 44°$, $a = 12$, $b = 14$
29. $A = 140°$, $b = 10$, $a = 7$	**30.** $C = 48°$, $a = 7$, $b = 9$

The test questions on this page deal with expressions and equations. The information at the right may help you with some of the questions.

1. Many problems can be solved without much calculating if the basic mathematical concepts are understood. Always look carefully at what is asked, and think of possible shortcuts for solving the problem.

2. Check your solutions by substituting values for the variables.

Directions: Choose the one best answer. Write A, B, C, or D.

1. If $\frac{x}{6} > x$, which could be a value for x?

   (A) -6 **(B)** 0 **(C)** 5 **(D)** 6

2. If $n^2 - 4 = -3n$, what is the value of $\left(n + \frac{3}{2}\right)^2$?

   (A) 5 **(B)** $6\frac{1}{4}$ **(C)** 11

   (D) cannot be determined

3. If b is an odd integer greater than one, which of the following must be an odd integer?

   (A) $b^3 - 1$ **(B)** $b^3 - b^2$

   (C) $1 + b^3$ **(D)** $\frac{3b - 3}{b - 1}$

4. If $0 < n < 1$, which of the following increases as n increases?

   I. $1 - n^2$
   II. $n - 1$
   III. $\frac{1}{n^2}$

   (A) II only   **(B)** III only
   (C) I and III only   **(D)** II and III only

5. If $-a < 0 < -b$, which of the following is true?

   (A) $0 < b < 0$ **(B)** $a < 0 < b$
   (C) $b < 0 < a$ **(D)** $0 < b < a$

6. If $x^2 = 25$, then 2^{x-1} could equal

   (A) 2 **(B)** 4 **(C)** 8 **(D)** 16

7. If k is any odd integer and $x = 6k$, then $\frac{x}{2}$ will always be

   (A) odd   **(B)** even
   (C) positive   **(D)** negative

8. If $a + b = 8$ and $3b - 4 = -13a$, then what is the value of a?

   (A) $-\frac{13}{3}$ **(B)** -3 **(C)** -2 **(D)** 10

9. $q * t$ is defined as $q^2 + t^2 + qt$. What is the value of $4 * (-2)$?

   (A) 4 **(B)** 12 **(C)** 26 **(D)** 28

10. If $x > y$ and $z < 0$, which of the following are true?

   I. $xz < yz$
   II. $x + z > y + z$
   III. $x - z < y - z$

   (A) I only   **(B)** II only
   (C) I and II only   **(D)** I, II and III

11. If $5y + 1$ is an odd integer, what is the next consecutive odd integer?

   (A) $3y + 1$ **(B)** $5y + 3$
   (C) $7y + 1$ **(D)** $7y + 3$

12. A man owns $\frac{1}{4}$ of a business. He sells half of his share for $12,000. What is the total value of the business?

   (A) $1500   **(B)** $48,000
   (C) $96,000   **(D)** $108,000

Square Roots and Cube Roots

n	$\sqrt{n}$	$\sqrt{10n}$	$\sqrt[3]{n}$	$\sqrt[3]{10n}$	$\sqrt[3]{100n}$	n	$\sqrt{n}$	$\sqrt{10n}$	$\sqrt[3]{n}$	$\sqrt[3]{10n}$	$\sqrt[3]{100n}$
1.0	1.000	3.162	1.000	2.154	4.642	5.5	2.345	7.416	1.765	3.803	8.193
1.1	1.049	3.317	1.032	2.224	4.791	5.6	2.366	7.483	1.776	3.826	8.243
1.2	1.095	3.464	1.063	2.289	4.932	5.7	2.387	7.550	1.786	3.849	8.291
1.3	1.140	3.606	1.091	2.351	5.066	5.8	2.408	7.616	1.797	3.871	8.340
1.4	1.183	3.742	1.119	2.410	5.192	5.9	2.429	7.681	1.807	3.893	8.387
1.5	1.225	3.873	1.145	2.466	5.313	6.0	2.449	7.746	1.817	3.915	8.434
1.6	1.265	4.000	1.170	2.520	5.429	6.1	2.470	7.810	1.827	3.936	8.481
1.7	1.304	4.123	1.193	2.571	5.540	6.2	2.490	7.874	1.837	3.958	8.527
1.8	1.342	4.243	1.216	2.621	5.646	6.3	2.510	7.937	1.847	3.979	8.573
1.9	1.378	4.359	1.239	2.668	5.749	6.4	2.530	8.000	1.857	4.000	8.618
2.0	1.414	4.472	1.260	2.714	5.848	6.5	2.550	8.062	1.866	4.021	8.662
2.1	1.449	4.583	1.281	2.759	5.944	6.6	2.569	8.124	1.876	4.041	8.707
2.2	1.483	4.690	1.301	2.802	6.037	6.7	2.588	8.185	1.885	4.062	8.750
2.3	1.517	4.796	1.320	2.844	6.127	6.8	2.608	8.246	1.895	4.082	8.794
2.4	1.549	4.899	1.339	2.884	6.214	6.9	2.627	8.307	1.904	4.102	8.837
2.5	1.581	5.000	1.357	2.924	6.300	7.0	2.646	8.367	1.913	4.121	8.879
2.6	1.612	5.099	1.375	2.962	6.383	7.1	2.665	8.426	1.922	4.141	8.921
2.7	1.643	5.196	1.392	3.000	6.463	7.2	2.683	8.485	1.931	4.160	8.963
2.8	1.673	5.292	1.409	3.037	6.542	7.3	2.702	8.544	1.940	4.179	9.004
2.9	1.703	5.385	1.426	3.072	6.619	7.4	2.720	8.602	1.949	4.198	9.045
3.0	1.732	5.477	1.442	3.107	6.694	7.5	2.739	8.660	1.957	4.217	9.086
3.1	1.761	5.568	1.458	3.141	6.768	7.6	2.757	8.718	1.966	4.236	9.126
3.2	1.789	5.657	1.474	3.175	6.840	7.7	2.775	8.775	1.975	4.254	9.166
3.3	1.817	5.745	1.489	3.208	6.910	7.8	2.793	8.832	1.983	4.273	9.205
3.4	1.844	5.831	1.504	3.240	6.980	7.9	2.811	8.888	1.992	4.291	9.244
3.5	1.871	5.916	1.518	3.271	7.047	8.0	2.828	8.944	2.000	4.309	9.283
3.6	1.897	6.000	1.533	3.302	7.114	8.1	2.846	9.000	2.008	4.327	9.322
3.7	1.924	6.083	1.547	3.332	7.179	8.2	2.864	9.055	2.017	4.344	9.360
3.8	1.949	6.164	1.560	3.362	7.243	8.3	2.881	9.110	2.025	4.362	9.398
3.9	1.975	6.245	1.574	3.391	7.306	8.4	2.898	9.165	2.033	4.380	9.435
4.0	2.000	6.325	1.587	3.420	7.368	8.5	2.915	9.220	2.041	4.397	9.473
4.1	2.025	6.403	1.601	3.448	7.429	8.6	2.933	9.274	2.049	4.414	9.510
4.2	2.049	6.481	1.613	3.476	7.489	8.7	2.950	9.327	2.057	4.431	9.546
4.3	2.074	6.557	1.626	3.503	7.548	8.8	2.966	9.381	2.065	4.448	9.583
4.4	2.098	6.633	1.639	3.530	7.606	8.9	2.983	9.434	2.072	4.465	9.619
4.5	2.121	6.708	1.651	3.557	7.663	9.0	3.000	9.487	2.080	4.481	9.655
4.6	2.145	6.782	1.663	3.583	7.719	9.1	3.017	9.539	2.088	4.498	9.691
4.7	2.168	6.856	1.675	3.609	7.775	9.2	3.033	9.592	2.095	4.514	9.726
4.8	2.191	6.928	1.687	3.634	7.830	9.3	3.050	9.644	2.103	4.531	9.761
4.9	2.214	7.000	1.698	3.659	7.884	9.4	3.066	9.695	2.110	4.547	9.796
5.0	2.236	7.071	1.710	3.684	7.937	9.5	3.082	9.747	2.118	4.563	9.830
5.1	2.258	7.141	1.721	3.708	7.990	9.6	3.098	9.798	2.125	4.579	9.865
5.2	2.280	7.211	1.732	3.733	8.041	9.7	3.114	9.849	2.133	4.595	9.899
5.3	2.302	7.280	1.744	3.756	8.093	9.8	3.130	9.899	2.140	4.610	9.933
5.4	2.324	7.348	1.754	3.780	8.143	9.9	3.146	9.950	2.147	4.626	9.967

Common Logarithms of Numbers

n	0	1	2	3	4	5	6	7	8	9
10	0000	0043	0086	0128	0170	0212	0253	0294	0334	0374
11	0414	0453	0492	0531	0569	0607	0645	0682	0719	0755
12	0792	0828	0864	0899	0934	0969	1004	1038	1072	1106
13	1139	1173	1206	1239	1271	1303	1335	1367	1399	1430
14	1461	1492	1523	1553	1584	1614	1644	1673	1703	1732
15	1761	1790	1818	1847	1875	1903	1931	1959	1987	2014
16	2041	2068	2095	2122	2148	2175	2201	2227	2253	2279
17	2304	2330	2355	2380	2405	2430	2455	2480	2504	2529
18	2553	2577	2601	2625	2648	2672	2695	2718	2742	2765
19	2788	2810	2833	2856	2878	2900	2923	2945	2967	2989
20	3010	3032	3054	3075	3096	3118	3139	3160	3181	3201
21	3222	3243	3263	3284	3304	3324	3345	3365	3385	3404
22	3424	3444	3464	3483	3502	3522	3541	3560	3579	3598
23	3617	3636	3655	3674	3692	3711	3729	3747	3766	3784
24	3802	3820	3838	3856	3874	3892	3909	3927	3945	3962
25	3979	3997	4014	4031	4048	4065	4082	4099	4116	4133
26	4150	4166	4183	4200	4216	4232	4249	4265	4281	4298
27	4314	4330	4346	4362	4378	4393	4409	4425	4440	4456
28	4472	4487	4502	4518	4533	4548	4564	4579	4594	4609
29	4624	4639	4654	4669	4683	4698	4713	4728	4742	4757
30	4771	4786	4800	4814	4829	4843	4857	4871	4886	4900
31	4914	4928	4942	4955	4969	4983	4997	5011	5024	5038
32	5051	5065	5079	5092	5105	5119	5132	5145	5159	5172
33	5185	5198	5211	5224	5237	5250	5263	5276	5289	5302
34	5315	5328	5340	5353	5366	5378	5391	5403	5416	5428
35	5441	5453	5465	5478	5490	5502	5514	5527	5539	5551
36	5563	5575	5587	5599	5611	5623	5635	5647	5658	5670
37	5682	5694	5705	5717	5729	5740	5752	5763	5775	5786
38	5798	5809	5821	5832	5843	5855	5866	5877	5888	5899
39	5911	5922	5933	5944	5955	5966	5977	5988	5999	6010
40	6021	6031	6042	6053	6064	6075	6085	6096	6107	6117
41	6128	6138	6149	6160	6170	6180	6191	6201	6212	6222
42	6232	6243	6253	6263	6274	6284	6294	6304	6314	6325
43	6335	6345	6355	6365	6375	6385	6395	6405	6415	6425
44	6435	6444	6454	6464	6474	6484	6493	6503	6513	6522
45	6532	6542	6551	6561	6571	6580	6590	6599	6609	6618
46	6628	6637	6646	6656	6665	6675	6684	6693	6702	6712
47	6721	6730	6739	6749	6758	6767	6776	6785	6794	6803
48	6812	6821	6830	6839	6848	6857	6866	6875	6884	6893
49	6902	6911	6920	6928	6937	6946	6955	6964	6972	6981
50	6990	6998	7007	7016	7024	7033	7042	7050	7059	7067
51	7076	7084	7093	7101	7110	7118	7126	7135	7143	7152
52	7160	7168	7177	7185	7193	7202	7210	7218	7226	7235
53	7243	7251	7259	7267	7275	7284	7292	7300	7308	7316
54	7324	7332	7340	7348	7356	7364	7372	7380	7388	7396

The values given are mantissas correct to four decimal places. For example, log 5.42 = 0.7340.

Common Logarithms of Numbers

n	0	1	2	3	4	5	6	7	8	9
55	7404	7412	7419	7427	7435	7443	7451	7459	7466	7474
56	7482	7490	7497	7505	7513	7520	7528	7536	7543	7551
57	7559	7566	7574	7582	7589	7597	7604	7612	7619	7627
58	7634	7642	7649	7657	7664	7672	7679	7686	7694	7701
59	7709	7716	7723	7731	7738	7745	7752	7760	7767	7774
60	7782	7789	7796	7803	7810	7818	7825	7832	7839	7846
61	7853	7860	7868	7875	7882	7889	7896	7903	7910	7917
62	7924	7931	7938	7945	7952	7959	7966	7973	7980	7987
63	7993	8000	8007	8014	8021	8028	8035	8041	8048	8055
64	8062	8069	8075	8082	8089	8096	8102	8109	8116	8122
65	8129	8136	8142	8149	8156	8162	8169	8176	8182	8189
66	8195	8202	8209	8215	8222	8228	8235	8241	8248	8254
67	8261	8267	8274	8280	8287	8293	8299	8306	8312	8319
68	8325	8331	8338	8344	8351	8357	8363	8370	8376	8382
69	8388	8395	8401	8407	8414	8420	8426	8432	8439	8445
70	8451	8457	8463	8470	8476	8482	8488	8494	8500	8506
71	8513	8519	8525	8531	8537	8543	8549	8555	8561	8567
72	8573	8579	8585	8591	8597	8603	8609	8615	8621	8627
73	8633	8639	8645	8651	8657	8663	8669	8675	8681	8686
74	8692	8698	8704	8710	8716	8722	8727	8733	8739	8745
75	8751	8756	8762	8768	8774	8779	8785	8791	8797	8802
76	8808	8814	8820	8825	8831	8837	8842	8848	8854	8859
77	8865	8871	8876	8882	8887	8893	8899	8904	8910	8915
78	8921	8927	8932	8938	8943	8949	8954	8960	8965	8971
79	8976	8982	8987	8993	8998	9004	9009	9015	9020	9025
80	9031	9036	9042	9047	9053	9058	9063	9069	9074	9079
81	9085	9090	9096	9101	9106	9112	9117	9122	9128	9133
82	9138	9143	9149	9154	9159	9165	9170	9175	9180	9186
83	9191	9196	9201	9206	9212	9217	9222	9227	9232	9238
84	9243	9248	9253	9258	9263	9269	9274	9279	9284	9289
85	9294	9299	9304	9309	9315	9320	9325	9330	9335	9340
86	9345	9350	9355	9360	9365	9370	9375	9380	9385	9390
87	9395	9400	9405	9410	9415	9420	9425	9430	9435	9440
88	9445	9450	9455	9460	9465	9469	9474	9479	9484	9489
89	9494	9499	9504	9509	9513	9518	9523	9528	9533	9538
90	9542	9547	9552	9557	9562	9566	9571	9576	9581	9586
91	9590	9595	9600	9605	9609	9614	9619	9624	9628	9633
92	9638	9643	9647	9652	9657	9661	9666	9671	9675	9680
93	9685	9689	9694	9699	9703	9708	9713	9717	9722	9727
94	9731	9736	9741	9745	9750	9754	9759	9763	9768	9773
95	9777	9782	9786	9791	9795	9800	9805	9809	9814	9818
96	9823	9827	9832	9836	9841	9845	9850	9854	9859	9863
97	9868	9872	9877	9881	9886	9890	9894	9899	9903	9908
98	9912	9917	9921	9926	9930	9934	9939	9943	9948	9952
99	9956	9961	9965	9969	9974	9978	9983	9987	9991	9996

Values of Trigonometric Functions

Angle	Sin	Cos	Tan	Cot	Sec	Csc	
0°00′	0.0000	1.0000	0.0000	—	1.000	—	90°00′
10′	0.0029	1.0000	0.0029	343.8	1.000	343.8	50′
20′	0.0058	1.0000	0.0058	171.9	1.000	171.9	40′
30′	0.0087	1.0000	0.0087	114.6	1.000	114.6	30′
40′	0.0116	0.9999	0.0116	85.94	1.000	85.95	20′
50′	0.0145	0.9999	0.0145	68.75	1.000	68.76	10′
1°00′	0.0175	0.9998	0.0175	57.29	1.000	57.30	89°00′
10′	0.0204	0.9998	0.0204	49.10	1.000	49.11	50′
20′	0.0233	0.9997	0.0233	42.96	1.000	42.98	40′
30′	0.0262	0.9997	0.0262	38.19	1.000	38.20	30′
40′	0.0291	0.9996	0.0291	34.37	1.000	34.38	20′
50′	0.0320	0.9995	0.0320	31.24	1.001	31.26	10′
2°00′	0.0349	0.9994	0.0349	28.64	1.001	28.65	88°00′
10′	0.0378	0.9993	0.0378	26.43	1.001	26.45	50′
20′	0.0407	0.9992	0.0407	24.54	1.001	24.56	40′
30′	0.0436	0.9990	0.0437	22.90	1.001	22.93	30′
40′	0.0465	0.9989	0.0466	21.47	1.001	21.49	20′
50′	0.0494	0.9988	0.0495	20.21	1.001	20.23	10′
3°00′	0.0523	0.9986	0.0524	19.08	1.001	19.11	87°00′
10′	0.0552	0.9985	0.0553	18.07	1.002	18.10	50′
20′	0.0581	0.9983	0.0582	17.17	1.002	17.20	40′
30′	0.0610	0.9981	0.0612	16.35	1.002	16.38	30′
40′	0.0640	0.9980	0.0641	15.60	1.002	15.64	20′
50′	0.0669	0.9978	0.0670	14.92	1.002	14.96	10′
4°00′	0.0698	0.9976	0.0699	14.30	1.002	14.34	86°00′
10′	0.0727	0.9974	0.0729	13.73	1.003	13.76	50′
20′	0.0756	0.9971	0.0758	13.20	1.003	13.23	40′
30′	0.0785	0.9969	0.0787	12.71	1.003	12.75	30′
40′	0.0814	0.9967	0.0816	12.25	1.003	12.29	20′
50′	0.0843	0.9964	0.0846	11.83	1.004	11.87	10′
5°00′	0.0872	0.9962	0.0875	11.43	1.004	11.47	85°00′
10′	0.0901	0.9959	0.0904	11.06	1.004	11.10	50′
20′	0.0929	0.9957	0.0934	10.71	1.004	10.76	40′
30′	0.0958	0.9954	0.0963	10.39	1.005	10.43	30′
40′	0.0987	0.9951	0.0992	10.08	1.005	10.13	20′
50′	0.1016	0.9948	0.1022	9.788	1.005	9.839	10′
6°00′	0.1045	0.9945	0.1051	9.514	1.006	9.567	84°00′
10′	0.1074	0.9942	0.1080	9.255	1.006	9.309	50′
20′	0.1103	0.9939	0.1110	9.010	1.006	9.065	40′
30′	0.1132	0.9936	0.1139	8.777	1.006	8.834	30′
40′	0.1161	0.9932	0.1169	8.556	1.007	8.614	20′
50′	0.1190	0.9929	0.1198	8.345	1.007	8.405	10′
7°00′	0.1219	0.9925	0.1228	8.144	1.008	8.206	83°00′
10′	0.1248	0.9922	0.1257	7.953	1.008	8.016	50′
20′	0.1276	0.9918	0.1287	7.770	1.008	7.834	40′
30′	0.1305	0.9914	0.1317	7.596	1.009	7.661	30′
40′	0.1334	0.9911	0.1346	7.429	1.009	7.496	20′
50′	0.1363	0.9907	0.1376	7.269	1.009	7.337	10′
8°00′	0.1392	0.9903	0.1405	7.115	1.010	7.185	82°00′
10′	0.1421	0.9899	0.1435	6.968	1.010	7.040	50′
20′	0.1449	0.9894	0.1465	6.827	1.011	6.900	40′
30′	0.1478	0.9890	0.1495	6.691	1.011	6.765	30′
40′	0.1507	0.9886	0.1524	6.561	1.012	6.636	20′
50′	0.1536	0.9881	0.1554	6.435	1.012	6.512	10′
9°00′	0.1564	0.9877	0.1584	6.314	1.012	6.392	81°00′
	Cos	Sin	Cot	Tan	Csc	Sec	Angle

For the values of cos, sin, tan, and so on for angles greater than 45°, use the angle measures listed on the right and the functions on the bottom. For example, cos 81° = 0.1564.

Values of Trigonometric Functions

Angle	Sin	Cos	Tan	Cot	Sec	Csc	
9°00′	0.1564	0.9877	0.1584	6.314	1.012	6.392	81°00′
10′	0.1593	0.9872	0.1614	6.197	1.013	6.277	50′
20′	0.1622	0.9868	0.1644	6.084	1.013	6.166	40′
30′	0.1650	0.9863	0.1673	5.976	1.014	6.059	30′
40′	0.1679	0.9858	0.1703	5.871	1.014	5.955	20′
50′	0.1708	0.9853	0.1733	5.769	1.015	5.855	10′
10°00′	0.1736	0.9848	0.1763	5.671	1.015	5.759	80°00′
10′	0.1765	0.9843	0.1793	5.576	1.016	5.665	50′
20′	0.1794	0.9838	0.1823	5.485	1.016	5.575	40′
30′	0.1822	0.9833	0.1853	5.396	1.017	5.487	30′
40′	0.1851	0.9827	0.1883	5.309	1.018	5.403	20′
50′	0.1880	0.9822	0.1914	5.226	1.018	5.320	10′
11°00′	0.1908	0.9816	0.1944	5.145	1.019	5.241	79°00′
10′	0.1937	0.9811	0.1974	5.066	1.019	5.164	50′
20′	0.1965	0.9805	0.2004	4.989	1.020	5.089	40′
30′	0.1994	0.9799	0.2035	4.915	1.020	5.016	30′
40′	0.2022	0.9793	0.2065	4.843	1.021	4.945	20′
50′	0.2051	0.9787	0.2095	4.773	1.022	4.876	10′
12°00′	0.2079	0.9781	0.2126	4.705	1.022	4.810	78°00′
10′	0.2108	0.9775	0.2156	4.638	1.023	4.745	50′
20′	0.2136	0.9769	0.2186	4.574	1.024	4.682	40′
30′	0.2164	0.9763	0.2217	4.511	1.024	4.620	30′
40′	0.2193	0.9757	0.2247	4.449	1.025	4.560	20′
50′	0.2221	0.9750	0.2278	4.390	1.026	4.502	10′
13°00′	0.2250	0.9744	0.2309	4.331	1.026	4.445	77°00′
10′	0.2278	0.9737	0.2339	4.275	1.027	4.390	50′
20′	0.2306	0.9730	0.2370	4.219	1.028	4.336	40′
30′	0.2334	0.9724	0.2401	4.165	1.028	4.284	30′
40′	0.2363	0.9717	0.2432	4.113	1.029	4.232	20′
50′	0.2391	0.9710	0.2462	4.061	1.030	4.182	10′
14°00′	0.2419	0.9703	0.2493	4.011	1.031	4.134	76°00′
10′	0.2447	0.9696	0.2524	3.962	1.031	4.086	50′
20′	0.2476	0.9689	0.2555	3.914	1.032	4.039	40′
30′	0.2504	0.9681	0.2586	3.867	1.033	3.994	30′
40′	0.2532	0.9674	0.2617	3.821	1.034	3.950	20′
50′	0.2560	0.9667	0.2648	3.776	1.034	3.906	10′
15°00′	0.2588	0.9659	0.2679	3.732	1.035	3.864	75°00′
10′	0.2616	0.9652	0.2711	3.689	1.036	3.822	50′
20′	0.2644	0.9644	0.2742	3.647	1.037	3.782	40′
30′	0.2672	0.9636	0.2773	3.606	1.038	3.742	30′
40′	0.2700	0.9628	0.2805	3.566	1.039	3.703	20′
50′	0.2728	0.9621	0.2836	3.526	1.039	3.665	10′
16°00′	0.2756	0.9613	0.2867	3.487	1.040	3.628	74°00′
10′	0.2784	0.9605	0.2899	3.450	1.041	3.592	50′
20′	0.2812	0.9596	0.2931	3.412	1.042	3.556	40′
30′	0.2840	0.9588	0.2962	3.376	1.043	3.521	30′
40′	0.2868	0.9580	0.2994	3.340	1.044	3.487	20′
50′	0.2896	0.9572	0.3026	3.305	1.045	3.453	10′
17°00′	0.2924	0.9563	0.3057	3.271	1.046	3.420	73°00′
10′	0.2952	0.9555	0.3089	3.237	1.047	3.388	50′
20′	0.2979	0.9546	0.3121	3.204	1.048	3.356	40′
30′	0.3007	0.9537	0.3153	3.172	1.049	3.326	30′
40′	0.3035	0.9528	0.3185	3.140	1.049	3.295	20′
50′	0.3062	0.9520	0.3217	3.108	1.050	3.265	10′
18°00′	0.3090	0.9511	0.3249	3.078	1.051	3.236	72°00′
	Cos	Sin	Cot	Tan	Csc	Sec	Angle

Values of Trigonometric Functions

Angle	Sin	Cos	Tan	Cot	Sec	Csc	
18°00′	0.3090	0.9511	0.3249	3.078	1.051	3.236	72°00′
10′	0.3118	0.9502	0.3281	3.047	1.052	3.207	50′
20′	0.3145	0.9492	0.3314	3.018	1.053	3.179	40′
30′	0.3173	0.9483	0.3346	2.989	1.054	3.152	30′
40′	0.3201	0.9474	0.3378	2.960	1.056	3.124	20′
50′	0.3228	0.9465	0.3411	2.932	1.057	3.098	10′
19°00′	0.3256	0.9455	0.3443	2.904	1.058	3.072	71°00′
10′	0.3283	0.9446	0.3476	2.877	1.059	3.046	50′
20′	0.3311	0.9436	0.3508	2.850	1.060	3.021	40′
30′	0.3338	0.9426	0.3541	2.824	1.061	2.996	30′
40′	0.3365	0.9417	0.3574	2.798	1.062	2.971	20′
50′	0.3393	0.9407	0.3607	2.773	1.063	2.947	10′
20°00′	0.3420	0.9397	0.3640	2.747	1.064	2.924	70°00′
10′	0.3448	0.9387	0.3673	2.723	1.065	2.901	50′
20′	0.3475	0.9377	0.3706	2.699	1.066	2.878	40′
30′	0.3502	0.9367	0.3739	2.675	1.068	2.855	30′
40′	0.3529	0.9356	0.3772	2.651	1.069	2.833	20′
50′	0.3557	0.9346	0.3805	2.628	1.070	2.812	10′
21°00′	0.3584	0.9336	0.3839	2.605	1.071	2.790	69°00′
10′	0.3611	0.9325	0.3872	2.583	1.072	2.769	50′
20′	0.3638	0.9315	0.3906	2.560	1.074	2.749	40′
30′	0.3665	0.9304	0.3939	2.539	1.075	2.729	30′
40′	0.3692	0.9293	0.3973	2.517	1.076	2.709	20′
50′	0.3719	0.9283	0.4006	2.496	1.077	2.689	10′
22°00′	0.3746	0.9272	0.4040	2.475	1.079	2.669	68°00′
10′	0.3773	0.9261	0.4074	2.455	1.080	2.650	50′
20′	0.3800	0.9250	0.4108	2.434	1.081	2.632	40′
30′	0.3827	0.9239	0.4142	2.414	1.082	2.613	30′
40′	0.3854	0.9228	0.4176	2.394	1.084	2.595	20′
50′	0.3881	0.9216	0.4210	2.375	1.085	2.577	10′
23°00′	0.3907	0.9205	0.4245	2.356	1.086	2.559	67°00′
10′	0.3934	0.9194	0.4279	2.337	1.088	2.542	50′
20′	0.3961	0.9182	0.4314	2.318	1.089	2.525	40′
30′	0.3987	0.9171	0.4348	2.300	1.090	2.508	30′
40′	0.4014	0.9159	0.4383	2.282	1.092	2.491	20′
50′	0.4041	0.9147	0.4417	2.264	1.093	2.475	10′
24°00′	0.4067	0.9135	0.4452	2.246	1.095	2.459	66°00′
10′	0.4094	0.9124	0.4487	2.229	1.096	2.443	50′
20′	0.4120	0.9112	0.4522	2.211	1.097	2.427	40′
30′	0.4147	0.9100	0.4557	2.194	1.099	2.411	30′
40′	0.4173	0.9088	0.4592	2.177	1.100	2.396	20′
50′	0.4200	0.9075	0.4628	2.161	1.102	2.381	10′
25°00′	0.4226	0.9063	0.4663	2.145	1.103	2.366	65°00′
10′	0.4253	0.9051	0.4699	2.128	1.105	2.352	50′
20′	0.4279	0.9038	0.4734	2.112	1.106	2.337	40′
30′	0.4305	0.9026	0.4770	2.097	1.108	2.323	30′
40′	0.4331	0.9013	0.4806	2.081	1.109	2.309	20′
50′	0.4358	0.9001	0.4841	2.066	1.111	2.295	10′
26°00′	0.4384	0.8988	0.4877	2.050	1.113	2.281	64°00′
10′	0.4410	0.8975	0.4913	2.035	1.114	2.268	50′
20′	0.4436	0.8962	0.4950	2.020	1.116	2.254	40′
30′	0.4462	0.8949	0.4986	2.006	1.117	2.241	30′
40′	0.4488	0.8936	0.5022	1.991	1.119	2.228	20′
50′	0.4514	0.8923	0.5059	1.977	1.121	2.215	10′
27°00′	0.4540	0.8910	0.5095	1.963	1.122	2.203	63°00′
	Cos	Sin	Cot	Tan	Csc	Sec	Angle

Values of Trigonometric Functions

Angle	Sin	Cos	Tan	Cot	Sec	Csc	
27°00′	0.4540	0.8910	0.5095	1.963	1.122	2.203	63°00′
10′	0.4566	0.8897	0.5132	1.949	1.124	2.190	50′
20′	0.4592	0.8884	0.5169	1.935	1.126	2.178	40′
30′	0.4617	0.8870	0.5206	1.921	1.127	2.166	30′
40′	0.4643	0.8857	0.5243	1.907	1.129	2.154	20′
50′	0.4669	0.8843	0.5280	1.894	1.131	2.142	10′
28°00′	0.4695	0.8829	0.5317	1.881	1.133	2.130	62°00′
10′	0.4720	0.8816	0.5354	1.868	1.134	2.118	50′
20′	0.4746	0.8802	0.5392	1.855	1.136	2.107	40′
30′	0.4772	0.8788	0.5430	1.842	1.138	2.096	30′
40′	0.4797	0.8774	0.5467	1.829	1.140	2.085	20′
50′	0.4823	0.8760	0.5505	1.816	1.142	2.074	10′
29°00′	0.4848	0.8746	0.5543	1.804	1.143	2.063	61°00′
10′	0.4874	0.8732	0.5581	1.792	1.145	2.052	50′
20′	0.4899	0.8718	0.5619	1.780	1.147	2.041	40′
30′	0.4924	0.8704	0.5658	1.767	1.149	2.031	30′
40′	0.4950	0.8689	0.5696	1.756	1.151	2.020	20′
50′	0.4975	0.8675	0.5735	1.744	1.153	2.010	10′
30°00′	0.5000	0.8660	0.5774	1.732	1.155	2.000	60°00′
10′	0.5025	0.8646	0.5812	1.720	1.157	1.990	50′
20′	0.5050	0.8631	0.5851	1.709	1.159	1.980	40′
30′	0.5075	0.8616	0.5890	1.698	1.161	1.970	30′
40′	0.5100	0.8601	0.5930	1.686	1.163	1.961	20′
50′	0.5125	0.8587	0.5969	1.675	1.165	1.951	10′
31°00′	0.5150	0.8572	0.6009	1.664	1.167	1.942	59°00′
10′	0.5175	0.8557	0.6048	1.653	1.169	1.932	50′
20′	0.5200	0.8542	0.6088	1.643	1.171	1.923	40′
30′	0.5225	0.8526	0.6128	1.632	1.173	1.914	30′
40′	0.5250	0.8511	0.6168	1.621	1.175	1.905	20′
50′	0.5275	0.8496	0.6208	1.611	1.177	1.896	10′
32°00′	0.5299	0.8480	0.6249	1.600	1.179	1.887	58°00′
10′	0.5324	0.8465	0.6289	1.590	1.181	1.878	50′
20′	0.5348	0.8450	0.6330	1.580	1.184	1.870	40′
30′	0.5373	0.8434	0.6371	1.570	1.186	1.861	30′
40′	0.5398	0.8418	0.6412	1.560	1.188	1.853	20′
50′	0.5422	0.8403	0.6453	1.550	1.190	1.844	10′
33°00′	0.5446	0.8387	0.6494	1.540	1.192	1.836	57°00′
10′	0.5471	0.8371	0.6536	1.530	1.195	1.828	50′
20′	0.5495	0.8355	0.6577	1.520	1.197	1.820	40′
30′	0.5519	0.8339	0.6619	1.511	1.199	1.812	30′
40′	0.5544	0.8323	0.6661	1.501	1.202	1.804	20′
50′	0.5568	0.8307	0.6703	1.492	1.204	1.796	10′
34°00′	0.5592	0.8290	0.6745	1.483	1.206	1.788	56°00′
10′	0.5616	0.8274	0.6787	1.473	1.209	1.781	50′
20′	0.5640	0.8258	0.6830	1.464	1.211	1.773	40′
30′	0.5664	0.8241	0.6873	1.455	1.213	1.766	30′
40′	0.5688	0.8225	0.6916	1.446	1.216	1.758	20′
50′	0.5712	0.8208	0.6959	1.437	1.218	1.751	10′
35°00′	0.5736	0.8192	0.7002	1.428	1.221	1.743	55°00′
10′	0.5760	0.8175	0.7046	1.419	1.223	1.736	50′
20′	0.5783	0.8158	0.7089	1.411	1.226	1.729	40′
30′	0.5807	0.8141	0.7133	1.402	1.228	1.722	30′
40′	0.5831	0.8124	0.7177	1.393	1.231	1.715	20′
50′	0.5854	0.8107	0.7221	1.385	1.233	1.708	10′
36°00′	0.5878	0.8090	0.7265	1.376	1.236	1.701	54°00′
	Cos	Sin	Cot	Tan	Csc	Sec	Angle

Values of Trigonometric Functions

Angle	Sin	Cos	Tan	Cot	Sec	Csc	
36°00′	0.5878	0.8090	0.7265	1.376	1.236	1.701	54°00′
10′	0.5901	0.8073	0.7310	1.368	1.239	1.695	50′
20′	0.5925	0.8056	0.7355	1.360	1.241	1.688	40′
30′	0.5948	0.8039	0.7400	1.351	1.244	1.681	30′
40′	0.5972	0.8021	0.7445	1.343	1.247	1.675	20′
50′	0.5995	0.8004	0.7490	1.335	1.249	1.668	10′
37°00′	0.6018	0.7986	0.7536	1.327	1.252	1.662	53°00′
10′	0.6041	0.7969	0.7581	1.319	1.255	1.655	50′
20′	0.6065	0.7951	0.7627	1.311	1.258	1.649	40′
30′	0.6088	0.7934	0.7673	1.303	1.260	1.643	30′
40′	0.6111	0.7916	0.7720	1.295	1.263	1.636	20′
50′	0.6134	0.7898	0.7766	1.288	1.266	1.630	10′
38°00′	0.6157	0.7880	0.7813	1.280	1.269	1.624	52°00′
10′	0.6180	0.7862	0.7860	1.272	1.272	1.618	50′
20′	0.6202	0.7844	0.7907	1.265	1.275	1.612	40′
30′	0.6225	0.7826	0.7954	1.257	1.278	1.606	30′
40′	0.6248	0.7808	0.8002	1.250	1.281	1.601	20′
50′	0.6271	0.7790	0.8050	1.242	1.284	1.595	10′
39°00′	0.6293	0.7771	0.8098	1.235	1.287	1.589	51°00′
10′	0.6316	0.7753	0.8146	1.228	1.290	1.583	50′
20′	0.6338	0.7735	0.8195	1.220	1.293	1.578	40′
30′	0.6361	0.7716	0.8243	1.213	1.296	1.572	30′
40′	0.6383	0.7698	0.8292	1.206	1.299	1.567	20′
50′	0.6406	0.7679	0.8342	1.199	1.302	1.561	10′
40°00′	0.6428	0.7660	0.8391	1.192	1.305	1.556	50°00′
10′	0.6450	0.7642	0.8441	1.185	1.309	1.550	50′
20′	0.6472	0.7623	0.8491	1.178	1.312	1.545	40′
30′	0.6494	0.7604	0.8541	1.171	1.315	1.540	30′
40′	0.6517	0.7585	0.8591	1.164	1.318	1.535	20′
50′	0.6539	0.7566	0.8642	1.157	1.322	1.529	10′
41°00′	0.6561	0.7547	0.8693	1.150	1.325	1.524	49°00′
10′	0.6583	0.7528	0.8744	1.144	1.328	1.519	50′
20′	0.6604	0.7509	0.8796	1.137	1.332	1.514	40′
30′	0.6626	0.7490	0.8847	1.130	1.335	1.509	30′
40′	0.6648	0.7470	0.8899	1.124	1.339	1.504	20′
50′	0.6670	0.7451	0.8952	1.117	1.342	1.499	10′
42°00′	0.6691	0.7431	0.9004	1.111	1.346	1.494	48°00′
10′	0.6713	0.7412	0.9057	1.104	1.349	1.490	50′
20′	0.6734	0.7392	0.9110	1.098	1.353	1.485	40′
30′	0.6756	0.7373	0.9163	1.091	1.356	1.480	30′
40′	0.6777	0.7353	0.9217	1.085	1.360	1.476	20′
50′	0.6799	0.7333	0.9271	1.079	1.364	1.471	10′
43°00′	0.6820	0.7314	0.9325	1.072	1.367	1.466	47°00′
10′	0.6841	0.7294	0.9380	1.066	1.371	1.462	50′
20′	0.6862	0.7274	0.9435	1.060	1.375	1.457	40′
30′	0.6884	0.7254	0.9490	1.054	1.379	1.453	30′
40′	0.6905	0.7234	0.9545	1.048	1.382	1.448	20′
50′	0.6926	0.7214	0.9601	1.042	1.386	1.444	10′
44°00′	0.6947	0.7193	0.9657	1.036	1.390	1.440	46°00′
10′	0.6967	0.7173	0.9713	1.030	1.394	1.435	50′
20′	0.6988	0.7153	0.9770	1.024	1.398	1.431	40′
30′	0.7009	0.7133	0.9827	1.018	1.402	1.427	30′
40′	0.7030	0.7112	0.9884	1.012	1.406	1.423	20′
50′	0.7050	0.7092	0.9942	1.006	1.410	1.418	10′
45°00′	0.7071	0.7071	1.000	1.000	1.414	1.414	45°00′
	Cos	Sin	Cot	Tan	Csc	Sec	Angle

Glossary

absolute value   The absolute value of a number is the number of units that it is from zero on the number line. For any number a:
   If $a \geq 0$, then $|a| = a$.
   If $a < 0$, then $|a| = -a$.   (23)

additive identity   Zero is the additive identity. The sum of any number and zero is identical to the original number.   (11)

additive inverse   If the sum of two numbers is 0, they are called additive inverses of each other. (11)

algebraic expressions   Algebraic expressions are mathematical expressions having at least one variable.   (4)

amplitude   For functions of the form $y = a \sin b\theta$ and $y = a \cos b\theta$, the amplitude is $|a|$.   (589)

angle of depression   An angle of depression is the angle formed by a horizontal line and the line of sight to an object at a lower level.   (650)

angle of elevation   The angle of elevation is the angle formed by a horizontal line, and the line of sight to an object at a higher level.   (650)

antilogarithm   If $\log x = a$, then $x = \text{antilog } a$. (435)

Arccosine   Given $y = \text{Cos } x$, the inverse cosine function is defined by $y = \text{Cos}^{-1} x$.   (599)

Arcsine   Given $y = \text{Sin } x$, the inverse sine function is defined by $y = \text{Sin}^{-1} x$.   (599)

Arctangent   Given $y = \text{Tan } x$, the inverse tangent function is defined by $y = \text{Tan}^{-1} x$.   (599)

arithmetic means   The terms between any two nonconsecutive terms of an arithmetic sequence are called arithmetic means.   (461)

arithmetic sequence   An arithmetic sequence is a sequence in which the difference between any two consecutive terms is the same.   (459)

arithmetic series   The indicated sum of the terms of an arithmetic sequence is called an arithmetic series.   (463)

associativity   The way you group, or associate, three or more numbers does not change their sum or their product. That is, for all numbers a, b, and c, $(a + b) + c = a + (b + c)$ and $(a \cdot b) \cdot c = a \cdot (b \cdot c)$.   (11)

asymptote   Asymptotes are lines that a curve approaches.   (318, 405)

augmented matrix   An augmented matrix is a matrix representation of a system of equations. Each row of the matrix corresponds to an equation in the system. Each column corresponds to the coefficients of a given variable or the constant term.   (139)

axis of symmetry   An axis of symmetry is the line about which a figure is symmetric.   (270)

bar graph   A bar graph shows how specific quantities compare to one another.   (538)

binomial   A polynomial with two unlike terms is a binomial.   (166)

Binomial Theorem   If n is a positive integer, then the following is true.

$$(a + b)^n = 1a^n b^0 + \frac{n}{1}a^{n-1}b^1$$
$$+ \frac{n(n-1)}{1 \cdot 2}a^{n-2}b^2 + \cdots$$
$$+ \frac{n}{1}a^1 b^{n-1} + 1a^0 b^n \quad (493)$$

binomial trial   A binomial trial exists if and only if the following conditions occur.   **1.** There are only two possible outcomes.   **2.** The events are independent.   (529)

box and whisker plot   In a box and whisker plot, the quartiles and extreme values of a set of data are displayed using a number line.   (555)

center of ellipse The center of an ellipse is the point where the two axes of symmetry intersect. (312)

center of hyperbola The center of hyperbola is the midpoint of the segment connecting the foci of a hyperbola. (318)

characteristic The characteristic is the power of 10 by which that number is multiplied when the number is expressed in scientific notation. (434)

circle A circle is a set of points in a plane each of which is the same distance from a given point. The given distance is the radius of the circle and the given point is the center of the circle. (308)

circle graph A circle graph shows how parts are related to the whole. (539)

circular permutations If n objects are arranged in a circle, then there are $\frac{n!}{n}$ or $(n-1)!$ permutations of the n objects around the circle. (510)

coefficient The numerical factor of a monomial is the coefficient. (157)

coefficient matrix A coefficient matrix is a matrix representation of the coefficients of the variables in a system of equations. Each row of the matrix corresponds to an equation in the system. Each column corresponds to the coefficients of a given variable. (139)

combination The number of combinations of n objects, taken r at a time, is defined as follows.

$$C(n, r) = \frac{n!}{(n-r)!r!} \quad (513)$$

common difference The common difference of an arithmetic sequence is the constant that is the difference between successive terms. (459)

common logarithms Common logarithms are logarithms to base 10. (434)

common ratio The common ratio of a geometric sequence is the constant that is the ratio of successive terms. (468)

commutativity The order in which two numbers are added or multiplied does not change their sum or product. That is, for all numbers a and b, $a + b = b + a$ and $a \cdot b = b \cdot a$. (11)

complex conjugates Complex conjugates are complex numbers of the form $a + bi$ and $a - bi$. (224)

complex fraction A complex rational expression, also called a complex fraction, is an expression whose numerator or denominator, or both, contain rational expressions. (389)

complex number A complex number is any number that can be written in the form $a + bi$ where a and b are real numbers and i is the imaginary unit. a is the real part and bi is the imaginary part. (220)

composition of functions Given functions f and g, the composite function $f \circ g$ can be described by the following equation.

$$[f \circ g](x) = f[g(x)] \quad (372)$$

compound interest The compound interest formula is $A = P\left(1 + \frac{r}{n}\right)^{nt}$ where P is the investment, r is the interest rate, n is the number of times the interest is compounded yearly, t is the number of years of the investment, and A is the amount of money accumulated. (443)

conic section A conic section is a curve formed by slicing a hollow double cone with a plane. The equation of a conic section can be written in the form $Ax^2 + Bxy + Cy^2 + Dx + Ey + F = 0$ where A, B, and C are not all zero. (324)

conjugate axis The conjugate axis of a hyperbola is the segment perpendicular to the transverse axis at its center. (318)

conjugates Binomials of the form $a + b\sqrt{c}$ and $a - b\sqrt{c}$ are conjugates of each other. (205)

consistent and dependent system A system of equations where the graphs of the equations are the same line is called a consistent and dependent system. There is an infinite number of solutions to this system of equations. (85)

consistent and independent system A system of equations that has one ordered pair as its solution is a consistent and independent system. (85)

constant A monomial that contains no variable is a constant. (157)

constant function A constant function is a function of the form $f(x) = b$ where the slope is zero. (63)

constant of variation The constant k in either of the equations $y = kx$ or $y = \frac{k}{x}$ is called the constant of variation. (411)

coordinate plane The plane determined by the perpendicular axes is called the coordinate plane. (43)

coordinates Each point in the coordinate plane corresponds to an ordered pair of numbers called its coordinates. (43)

cosecant Let θ stand for the measurement of an angle in standard position on the unit circle. Then the following equation holds whenever it is defined.

$$\csc \theta = \frac{1}{\sin \theta} \quad (592)$$

cosine Let θ stand for the measurement of an angle in standard position on the unit circle. Let (x, y) represent the point where the terminal side intersects the unit circle. Then the following equation holds.

$$\cos \theta = x \quad (582)$$

cotangent Let θ stand for the measurement of an angle in standard position on the unit circle. Then the following equation holds whenever it is defined.

$$\cot \theta = \frac{\cos \theta}{\sin \theta} \quad (592)$$

coterminal angles Angles in standard position that have the same terminal side are called coterminal angles. (580)

Cramer's Rule The solution to the system of equations $\begin{aligned} ax + by &= c \\ dx + ey &= f \end{aligned}$ is (x, y) where

$$x = \frac{\begin{vmatrix} c & b \\ f & e \end{vmatrix}}{\begin{vmatrix} a & b \\ d & e \end{vmatrix}} \text{ and } y = \frac{\begin{vmatrix} a & c \\ d & f \end{vmatrix}}{\begin{vmatrix} a & b \\ d & e \end{vmatrix}} \text{ and } \begin{vmatrix} a & b \\ d & e \end{vmatrix} \neq 0.$$

$$(93)$$

data Numerical observations are called data. (537)

degree of monomial The degree of a monomial is the sum of the exponents of its variables. (157)

degree of polynomial The degree of a polynomial is the degree of the monomial of greatest degree. (166)

dependent events Two events are dependent when the outcome of the first event affects the outcome of the second event. If two events, A and B, are dependent, then the probability of both occurring is found as follows.

$$P(A \text{ and } B) = P(A) \cdot P(B \text{ following } A) \quad (504)$$

depressed polynomial A polynomial whose degree is less than the original polynomial is called a depressed polynomial. It is the result of factoring out a factor of the original polynomial. (350)

determinant A determinant is a square array of numbers having a numerical value. (92)

dimension In a matrix consisting of n rows and m columns, the matrix is said to have dimension $n \times m$ (read "n by m"). (121)

direct variation A direct variation is a linear function described by $y = mx$ or $f(x) = mx$ where $m \neq 0$. (63, 410)

discriminant In the quadratic formula, the expression under the radical sign, $b^2 - 4ac$, is called the discriminant. (243)

distance The distance between two points with coordinates (x_1, y_1) and (x_2, y_2) is given by the following formula.

$$d = \sqrt{(x_2 - x_1)^2 + (y_2 - y_2)^2} \quad (302)$$

distributive property For all numbers a, b, and c, $a(b + c) = ab + ac$ and $(b + c)a = ba + ca$. (12)

domain The domain is the set of all first coordinates of the ordered pairs of a relation. (45)

e e is the base of the natural system of logarithms. Its numerical value is $2.7182818284\ldots$, which is an irrational number. (449)

ellipse An ellipse is the set of all points in a plane such that the sum of the distances from two given points in the plane, called the foci, is constant. The standard equation of an ellipse that has center (h, k) and major axis of length $2a$ is $\dfrac{(x - h)^2}{a^2} + \dfrac{(y - k)^2}{b^2} = 1$ when the major axis is parallel to the x-axis. The equation is $\dfrac{(x - h)^2}{b^2} + \dfrac{(y - k)^2}{a^2} = 1$ when the major axis is parallel to the y-axis. For an ellipse, $b^2 = a^2 - c^2$. (311)

equation A statement of equality between two mathematical expressions is called an equation. (14)

expansion by minors Expansion by minors is a method that can be used to find the value of any third or higher order determinant. (121)

exponent An exponent is a numeral written to the right and above a number indicating how many times the number is used as a factor. (3)

exponential equation An equation in which the variables appear as exponents is called an exponential equation. (442)

exponential function An equation in the form $y = a^x$, where $a > 0$ and $a \neq 1$, is called an exponential function. (421)

Fibonacci sequence A fibonacci sequence is a special sequence often found in nature. It is named after its discoverer, Leonardo Fibonacci. (487)

FOIL FOIL is a method used to multiply binomials. The product of 2 binomials is the sum of the products of

    F the first terms
    O the outer terms
    I the inner terms
    L the last terms. (167)

formula A mathematical sentence about the relationships among certain quantities is called a formula. (4)

frequency distribution A frequency distribution shows how data are spread out. (564)

function A function is a relation in which each element of the domain is paired with exactly one element of the range. (46)

geometric means The terms between any two nonconsecutive terms of a geometric sequence are called geometric means. (469)

geometric sequence A geometric sequence is a sequence in which each term after the first is the product of the preceding term and the common ratio. (468)

geometric series The indicated sum of the terms of a geometric sequence is called a geometric series. (474)

greatest integer The greatest integer of x is written $|x|$ and means the greatest integer *not* greater than x. (64)

histogram A histogram is a bar graph that shows a frequency distribution. (564)

hyperbola A hyperbola is the set of all points in a plane such that the absolute value of the difference of the distances from any point on the hyperbola to two given points in the plane, called the foci, is constant. The standard equation of a hyperbola that has center (h, k) and a horizontal transverse axis of length $2a$ is $\dfrac{(x - h)^2}{a^2} - \dfrac{(y - k)^2}{b^2} = 1$. The equation is $\dfrac{(y - k)^2}{a^2} - \dfrac{(x - h)^2}{b^2} = 1$ when the transverse axis is vertical. For the hyperbola, $b^2 = c^2 - a^2$. (317)

hypotenuse The hypotenuse is the side opposite the right angle in a right triangle. (639)

identity An identity is an equation that is true for all values of the variable for which both sides of the equation are defined. (611)

identity function An identity function is a linear function described by $y = x$ or $f(x) = x$. (63)

identity matrix for multiplication The identity matrix, I, for multiplication is a square matrix with a 1 for every element of the principal diagonal and a 0 in all other positions. (129)

imaginary number An imaginary number is a complex number of the form $a + bi$, where $b \neq 0$. (220)

imaginary unit The imaginary unit i is defined by $i^2 = -1$. (218)

inclusive events Two events are inclusive if the outcomes of the events may be the same. The probability of two inclusive events, A and B, occurring is found as follows.

$$P(A \text{ or } B) = P(A) + P(B) - P(A \text{ and } B) \quad (525)$$

inconsistent system An inconsistent system is a system of equations where the graph of the equations is parallel lines. There is no solution to this system of equations. (85)

independent events Two events are independent if the outcome of one event does not affect the outcome of the other event. The probability of two independent events, A and B, occurring is found as follows.

$$P(A \text{ and } B) = P(A) \cdot P(B) \quad (503)$$

index of summation An index of summation is a variable used with the summation symbol (Σ). (481)

infinite geometric series An infinite geometric series is the indicated sum of the terms of an infinite geometric sequence. (477)

integers (Z) The set of numbers $\{\ldots, -3, -2, -1, 0, 1, 2, 3, \ldots\}$. (8)

interpolation Interpolation is a method for approximating values that are between given consecutive entries in a table, such as a table of logarithms or a table of trigonometric values. (435, 643)

interquartile range The interquartile range of a set of data is the difference between the upper and lower quartiles of the set. (551)

inverse functions Two polynomial functions f and g are inverse functions if and only if both their compositions are the identity function. That is,

$$[f \circ g](x) = [g \circ f](x) = x. \quad (375)$$

inverse matrix For a matrix A, with a nonzero determinant, the inverse matrix, A^{-1}, of A is that matrix with the property

$$A \cdot A^{-1} = A^{-1} \cdot A = I \quad (129)$$

inverse relations Two relations are inverse relations if and only if whenever one relation contains the element (a, b), the other relation contains the element (b, a). (376)

inverse variation A rational equation in two variables of the form $y = \dfrac{k}{x}$, where k is a constant, is called an inverse variation. The constant k is called the constant of variation, and y is said to vary inversely as x. (411)

irrational numbers (I) Irrational numbers are real numbers that cannot be written as terminating or repeating decimals. (8)

latus rectum A latus rectum is the line segment through the focus of a parabola perpendicular to its axis of symmetry with endpoints on the parabola. (304)

Laws of Cosines Let triangle ABC be any triangle with a, b, and c representing the measures of sides opposite angles with measurements A, B, and C, respectively. Then the following equations are true.

$$a^2 = b^2 + c^2 - 2bc \cos A$$
$$b^2 = a^2 + c^2 - 2ac \cos B$$
$$c^2 = a^2 + b^2 - 2ab \cos C \quad (663)$$

Law of Sines Let triangle ABC be any triangle with a, b, and c representing the measures of sides opposite angles with measurements A, B, and C, respectively. Then the following equations are true.

$$\frac{\sin A}{a} = \frac{\sin B}{b} = \frac{\sin C}{c} \quad (655)$$

like terms Two monomials that are the same or differ only by their coefficients are called like terms. (12, 157)

linear equation A linear equation is an equation whose graph is a straight line. (50)

linear function A linear function can be defined by $f(x) = mx + b$ where m and b are real numbers. Any function whose ordered pairs satisfy a linear equation in two variables is a linear function. (52)

linear permutation The arrangement of n objects in a certain linear order is called a linear permutation. The number of linear permutations of n objects, taken r at a time, is defined as follows.

$$P(n, r) = \frac{n!}{(n - r)!} \quad (506)$$

linear programming Linear programming is a method for finding the maximum or the minimum value of a function in two variables subject to given constraints on the variables. (99)

line graph A line graph shows trends or changes. (538)

line plot In a line plot, data are recorded and displayed using a number line. (543)

logarithm Suppose $b > 0$ and $b \neq 1$. Then for $n > 0$, there is a number p such that $\log_b n = p$ if and only if $b^p = n$. (423)

logarithm function An equation of the form $y = \log_b x$ where $b > 0$ and $b \neq 1$ is called a logarithm function. (426)

major axis The major axis of an ellipse is the segment with endpoints at the vertices of the ellipse. (312)

mantissa The mantissa is the logarithm of a number between 1 and 10. (434)

mapping A mapping illustrates how each element in the domain of a relation is paired with an element in the range. (45)

matrix A matrix is a rectangular arrangement of terms in rows and columns enclosed in brackets or large parentheses. (121)

matrix equation The matrix equation form of a system of equations is an equation of the form $AX = C$. A is the coefficient matrix for the system. X is the column matrix consisting of the variables of the system. C is the column matrix consisting of the constant terms of the system. (142)

mean The mean of a set of data is the sum of all the values divided by the number of values. (548)

median The median of a set of data is the middle value. If there are two middle values, it is the value halfway between. (548)

midpoint The midpoint of a line segment with endpoints (x_1, y_1) and (x_2, y_2) has coordinates $\left(\dfrac{x_1 + x_2}{2}, \dfrac{y_1 + y_2}{2} \right)$. (302)

minor A minor is the determinant formed when the row and column containing the element are deleted. (122)

minor axis The minor axis of an ellipse is the shorter axis. (312)

mode The mode of a set of data is the most frequent value. (548)

monomial A monomial is an expression that is a number, a variable, or the product of a number and one or more variables. (157)

multiplicative identity One is the multiplicative identity. The product of any number and 1 is identical to the original number. (11)

multiplicative inverses If the product of two numbers is 1, they are called multiplicative inverses or reciprocals of each other. (11)

mutually exclusive events Two events are mutually exclusive if their outcomes can never be the same. The probability of two mutually exclusive events, A and B, occurring is found as follows.
$$P(A \text{ or } B) = P(A) + P(B) \quad (525)$$

natural numbers (N) The set of numbers $\{1, 2, 3, 4 \ldots\}$. (8)

negative integer exponents For any number a, except $a = 0$, and for any positive integer n, $a^{-n} = \dfrac{1}{a^n}$ and $\dfrac{1}{a^{-n}} = a^n$. (161)

***n* factorial** If n is a positive integer, the expression $n!$ (n factorial) is defined as follows.
$$n! = n(n - 1)(n - 2) \cdots (1) \quad (493)$$

normal distribution Normal distributions have bell-shaped, symmetric graphs. About 68% of the items are within one standard deviation from the mean. About 95% of the items are within two standard deviations from the mean. About 99% of the items are within three standard deviations from the mean. (564)

***n*th root** For any numbers a and b, and any positive integer n, if $a^n = b$, then a is an nth root of b. (195)

***n*th term** The nth term of an arithmetic sequence with the first term a_1 and common difference d is given by the following equation.
$$a_n = a_1 + (n - 1)d \quad (460)$$

octant Three mutually perpendicular planes separate space into eight regions, each called an octant. (107)

odds The odds of the successful outcome of an event are expressed as the ratio of the number of ways it can succeed to the number of ways it can fail.

$$\text{Odds} = \text{the ratio of } s \text{ to } f \text{ or } \frac{s}{f} \quad (518)$$

ordered pair Points in a plane can be located by using ordered pairs of real numbers. The ordered pair are the coordinates of the point. (43)

ordered triple **1.** The solution to an equation in three variables is called an ordered triple. (107) **2.** Each point in space corresponds to three numbers called an ordered triple. (107)

origin The origin is the point on the coordinate plane whose coordinates are $(0, 0)$. (43)

outlier An outlier is any value in a set of data that is at least 1.5 interquartile ranges beyond the upper or lower quartiles. (552)

parabola **1.** The general shape of the graph of a quadratic function is called a parabola. (270) **2.** A parabola is the set of all points that are the same distance from a given point and a given line. The point is called the focus. The line is called the directrix. The standard equation of a parabola with vertex at (h, k) is $y = a(x - h)^2 + k$ when the directrix is horizontal. The equation is $x = a(y - k)^2 + h$ when the directrix is vertical. (304)

parallel lines In a plane, lines with the same slope are called parallel lines. Also, vertical lines are parallel. (81)

period For a function f, the least positive value of a for which $f(x) = f(x + a)$ is the period of the function. For functions of the form $y = a \sin b\theta$ and $y = a \cos b\theta$, the period is $\dfrac{2\pi}{|b|}$. (589)

periodic function A function f is called periodic if there is a number a such that $f(x) = f(x + a)$. (584)

permutation A permutation is the arrangement of things in a certain order. (506)

perpendicular lines Two nonvertical lines are perpendicular if and only if the product of their slopes is -1. Any vertical line is perpendicular to any horizontal line. (82)

polynomial A polynomial is a monomial or the sum or difference of monomials. (162)

polynomial function A polynomial equation in the form $p(x) = a_n x^n + a_{n-1} x^{n-1} + \cdots + a_1 x + a_0$ is a polynomial function. The coefficients $a_0, a_1, a_2, \ldots, a_{n-1}, a_n$ are real numbers, a_0 is not zero, and n is a nonnegative integer. (346)

polynomial in one variable A polynomial in one variable, x, is an expression of the form $a_0 x^n + a_1 x^{n-1} + \cdots + a_{n-2} x^2 + a_{n-1} x + a_n$. The coefficients a_0, a_1, a_2 are real numbers, a_0 is not zero, and n represents a nonnegative integer. (345)

prediction equation The equation of the line suggested by the dots on a scatter plot is a prediction equation. (567)

principal values The values in the domain of the functions like Cosine, Sine, and Tangent are the principal values. (598)

probability If an event can succeed in s ways and fail in f ways, then the probabilities of success $P(s)$ and of failure $P(f)$ are as follows.

$$P(s) = \frac{s}{s + f} \qquad P(f) = \frac{f}{s + f} \quad (517)$$

pure imaginary number For any positive real number b, $\sqrt{-(b^2)} = \sqrt{b^2}\sqrt{-1}$ or bi where i is a number whose square is -1. bi is called a pure imaginary number. (218)

quadrants Two perpendicular number lines separate the plane into four parts called quadrants. (43)

quadratic equation Any equation that can be written in the form $ax^2 + bx + c = 0$, where a, b, and c are complex numbers and $a \neq 0$, is a quadratic equation. (233)

quadratic form For any numbers a, b, and c, except $a = 0$, an equation that may be written as $a[f(x)]^2 + b[f(x)] + c = 0$, where $f(x)$ is some expression in x, is in quadratic form. (253)

quadratic formula The solutions of a quadratic equation of the form $ax^2 + bx + c = 0$ with $a \neq 0$ are given by the quadratic formula.

$$x = \frac{-b \pm \sqrt{b^2 - 4ac}}{2a} \quad (239)$$

quadratic function A quadratic function is a function described by an equation of the form $f(x) = ax^2 + bx + c$ where $a \neq 0$. (261)

quartiles In a set of data, quartiles are the values that separate the data into four equal parts. (551)

radian A radian is an angle that intercepts an arc whose length is 1 unit. (580)

radical equations Radical equations are equations containing variables in the radicands. (214)

radical sign A radical sign, $\sqrt{}$, indicates a square root. (195)

range 1. The range is a set of all second coordinates of the ordered pairs of a relation. (45)
2. The range of a set of data is the difference between the greatest and least values in the set. (551)

rational algebraic expression A rational algebraic expression can be expressed as the quotient of two polynomials. (387)

rational equation A rational equation is an equation that contains one or more rational expressions. (397)

rational exponents For any nonzero number b, and any integers m and n, $n > 1$

$$b^{\frac{m}{n}} = \sqrt[n]{b^m} = (\sqrt[n]{b})^m$$

except when $\sqrt[n]{b}$ does not represent a real number. (209)

rational function A rational function is an equation of the form $f(x) = \dfrac{p(x)}{q(x)}$ where $p(x)$ and $q(x)$ are polynomial functions and $q(x) \neq 0$. (405)

rational numbers (Q) Rational numbers are numbers that can be expressed in the form $\dfrac{a}{b}$, where a and b are integers and b is not zero. (8)

rationalizing the denominator Rationalizing the denominator is changing the form of a rational expression to one without radicals in the denominator. (201)

real numbers (R) Irrational numbers together with rational numbers form the set of real numbers. (8)

reciprocal *See* multiplicative inverse.

recursive formula A recursive formula depends on knowing one or more previous terms. (484)

relation A relation is a set of ordered pairs. (45)

relative maximum or minimum A relative maximum or minimum is a point that represents the maximum or minimum respectively for a certain interval. (370)

root A root is a solution of an equation. (243)

row operations on matrices The row operations on matrices are as follows.
1. Interchange any two rows.
2. Replace any row with a nonzero multiple of that row.
3. Replace any row with the sum of that row and a multiple of another row. (139)

scatter plot A scatter plot shows visually the nature of a relationship, both shape and closeness. (567)

scientific notation A number is expressed in scientific notation when it is in the form $a \times 10^n$ where $1 \leq a < 10$ and n is an integer. (164)

secant Let θ stand for the measurement of an angle in standard position on the unit circle. Then the following equation holds whenever it is defined.

$$\sec \theta = \frac{1}{\cos \theta} \quad (592)$$

sequence A sequence is a set of numbers in a specific order. (459)

series The indicated sum of the terms of a sequence is called a series. (463)

sigma notation The Σ symbol is used to indicate a sum of a series. (481)

signed minor The signed minor of an element of a matrix is the minor for that element multiplied by its sign. (131)

simplified expression An expression is simplified when:
1. It has no negative exponents.
2. It has no fractional exponents in the denominator.
3. It is not a complex fraction.
4. The index of any remaining radical is as small as possible. (212)

sine Let θ stand for the measurement of an angle in standard position on the unit circle. Let (x, y) represent the point where the terminal side intersects the unit circle. Then the following equation holds.

$$\sin \theta = y \quad (582)$$

slope The slope of a line described by $f(x) = mx + b$ is m. Slope is also given by the following expression.

$$m = \frac{y_2 - y_1}{x_2 - x_1} \quad (54)$$

slope-intercept form The equation $y = mx + b$ is in slope-intercept form. The slope is m and the y-intercept is b. (59)

solution The replacements for the variable that make a sentence true are each a solution of the open sentence. (14)

solution set A solution set is the set of all replacements for variables that make an open sentence true. (14)

solving a right triangle Solving a right triangle means finding all the measures of the sides and angles. (646)

square matrix A matrix that has the same number of rows as columns is called a square matrix. (121)

square root For any numbers a and b, if $a^2 = b$, then a is a square root of b. (195)

standard deviation From a set of data with n values, if x_i represents a value such that $1 \leq i \leq n$, and $\bar{x}$ represents the mean, then the standard deviation is

$$\sqrt{\frac{\sum_{i=1}^{n} (x_i - \bar{x})^2}{n}}. \quad (559)$$

standard form The standard form of a linear equation is $ax + by = c$ where a, b, and c are real numbers and a and b are not both zero. (50)

standard position An angle with its vertex at the origin and its initial side along the positive x-axis is said to be in standard position. (579)

stem and leaf plot In a stem and leaf plot, each piece of data is separated into two numbers that are used to form the stem and leaf. The data are organized into two columns. The column on the left is the stem and the column on the right is the leaf. (543)

sum of a geometric series The sum, S_n, of the first n terms of a geometric series is given by the following formula.

$$S_n = \frac{a_1 - a_1 r^n}{1 - r} \text{ where } r \neq 1 \quad (464)$$

sum of an arithmetic series The sum, S_n, of the first n terms of an arithmetic series is given by the following formula.

$$S_n = \frac{n}{2}(a_1 + a_n) \quad (464)$$

sum of an infinite geometric series The sum of an infinite geometric series is given by the formula

$$S = \frac{a_1}{1 - r} \text{ where } -1 < r < 1. \quad (477)$$

synthetic division A shortcut method used to divide polynomials by binomials is called synthetic division. (186)

synthetic substitution Synthetic substitution is the process of using the Remainder Theorem and synthetic division to find the value of a function. (349)

tangent Let θ stand for the measurement of an angle in standard position on the unit circle. Then the following equation holds whenever it is defined.

$$\tan \theta = \frac{\sin \theta}{\cos \theta} \quad (592)$$

term **1.** Each monomial in a polynomial is called a term. (166) **2.** A term is each number in a sequence. (459)

trace A trace is the line formed by the intersection of a plane with one of the three coordinate planes. (109)

transpose The transpose of any matrix is the matrix formed by interchanging the rows and the columns of the original matrix. (131)

transverse axis The line segment of a hyperbola of length $2a$ that has its endpoints at the vertices is called the transverse axis. (318)

trichotomy property For any two numbers a and b, one of the following statements is true.

$$a < b, \, a = b, \, a > b \quad (27)$$

trinomial A polynomial with three unlike terms is called a trinomial. (166)

unit circle   A unit circle has a radius of 1 unit. (580)

variable   A variable is a symbol that represents an unknown quantity.   (3)

vertex of hyperbola   The point on each branch of the hyperbola nearest the center is called a vertex of the hyperbola.   (318)

vertex of parabola   The point of intersection of the parabola and its axis of symmetry is called the vertex of the parabola.   (270)

vertical line test   If any vertical line drawn on the graph of a relation passes through no more than one point of that graph, then the relation is a function.   (47)

whole numbers (W)   The set of numbers $\{0, 1, 2, 3, \ldots\}$.   (8)

x-axis   The horizontal number line in a plane is called the x-axis.   (43)

x-intercept   An x-intercept is the value of x when the value of the function or y is zero.   (56)

y-axis   The vertical number line in a plane is called the y-axis.   (43)

y-intercept   A y-intercept is the value of a function when x is 0.   (56)

zero exponent   For any number a, except $a = 0$, $a^0 = 1$.   (160)

zero of function   For any polynomial function $f(x)$, if $f(a) = 0$, then a is a zero of the function. (353)

zero product property   For any real numbers a and b, if $ab = 0$ then $a = 0$ or $b = 0$.   (233)

Selected Answers

CHAPTER 1   EQUATIONS AND INEQUALITIES

Pages 5-6   Lesson 1-1
1. 7 **3.** -44 **5.** 13 **7.** 21 **9.** 21 **11.** 0 **13.** 41
15. 14 **17.** 6 **19.** 60 **21.** 10 **23.** 22 **25.** 4
27. $-3\frac{2}{3}$ **29.** 148 **31.** $-\frac{19}{2}$ **33.** -7.9 **35.** $\frac{5}{4}$
37. 94 **39.** -72 **41.** -7.2 **43.** $\frac{3}{7}$ **45.** 272.16
47. \$180 **49.** \$300 **51.** \$17,400 **53.** 610.42
55. 2578.82 **57.** 6.45 **59.** 128 **61.** 93.6 **63.** $53\frac{1}{4}$
65. $\frac{88}{8} + \frac{8}{8} + \frac{8}{8} + \frac{8}{8}$

Pages 9-10   Lesson 1-2
1. Z, Q, R **3.** Q, R **5.** Q, R **7.** I, R **9.** W, Z, Q, R
11. Z, Q, R **13.** Z, Q, R **15.** I, R
17. 1; N, W, Z, Q, R **19.** -9; Z, Q, R
21. -24; Z, Q, R **23.** 1; N, W, Z, Q, R
25. 5; N, W, Z, Q, R **27.** -90; Z, Q, R
29. 0; W, Z, Q, R **31.** 7; N, W, Z, Q, R **33.** $\frac{3}{2}$; Q, R
35. 12; N, W, Z, Q, R **37.** $\sqrt{14}$; I, R **39.** true
41. false **43.** true **45.** true **47.** $\sqrt{5}, \sqrt{6}, \sqrt{7}, \sqrt{8}$
49. $\sqrt{22}, \sqrt{23}$

Pages 12-13   Lesson 1-3
1. associative, $+$ **3.** distributive **5.** additive inverse
7. commutative, $+$ **9.** multiplicative inverse
11. commutative, $\times$ **13.** distributive **15.** commutative, $+$ **17.** commutative, $+$ **19.** distributive **21.** additive identity **23.** multiplicative inverse **25.** additive identity **27.** commutative, $+$ **29.** multiplicative inverse **31.** 20 **33.** $2x + 17y$ **35.** $31a + 10b$
37. $12 + 20a$ **39.** $\frac{2}{3}a + 2\frac{1}{2}b$ **41.** $4.4m - 2.9n$ **43.** 0, additive identity **45.** 1, multiplicative identity
47. yes **49.** no **51.** yes

Pages 16-17   Lesson 1-4
1. reflexive **3.** subtraction **5.** transitive
7. symmetric **9.** addition **11.** symmetric
13. symmetric **15.** transitive **17.** reflexive
19. substitution **21.** distributive **23.** division
25. a. distributive **b.** substitution **c.** division
27. a. subtraction, addition **b.** distributive, substitution **c.** substitution **d.** division **29.** 7.5
31. 14 **33.** $\frac{20}{21}$ **35.** $\frac{35}{8}$ **37.** 7 **39.** -12 **41.** 12
43. $-\frac{1}{6}$ **45.** $\frac{3}{10}$ **47.** 8 **49.** 2.1 **51.** 3 **53.** $\frac{1}{2}$
55. 11 **57.** -4 **59.** 10 **61.** -3 **63.** -4 **65.** -46
67. -11 **69.** 18.5

Pages 21-22   Lesson 1-5
1. $4x$ **3.** $2x + 11$ **5.** $2(x + 7)$ **7.** $2x + 7$ **9.** $12 - x^2$
11. $\frac{1}{5}(4 + x)$ **13.** $8(x + x^2)$ **15.** $(x + 11)^2$ **17.** 118
19. 26 **21.** 85% **23.** 18 years **25.** 19 years, 43 years
27. 88% **29.** 5 **31.** 8 adults, 12 students **33.** 3 hours **35.** 43 mph **37.** 24 m, 36 m **39.** 625 ft

Pages 25-26   Lesson 1-6
1. 5 **3.** 10 **5.** 36 **7.** 5 **9.** 0 **11.** 0 **13.** 2
15. $-2, 2$ **17.** 0, 1, 2 **19.** 2 **21.** 31, -53
23. 16, -6 **25.** 5, -19 **27.** 14, -8 **29.** $-20, 12$
31. $-\frac{13}{3}, \frac{8}{3}$ **33.** $-\frac{39}{2}, \frac{21}{2}$ **35.** no solution **37.** $-\frac{19}{5}, 3$
39. $-\frac{10}{21}, -\frac{20}{7}$ **41.** $\frac{7}{2}, \frac{21}{2}$ **43.** 1, -9 **45.** 1 **47.** $\frac{3}{2}, \frac{19}{16}$
49. no solution **51.** 31 **53.** 0.6 **55.** 2 **57.** $-4, -\frac{5}{2}$
59. $7, \frac{5}{4}$ **61.** $-\frac{1}{7}, -\frac{1}{3}$

Page 29   Lesson 1-7
7. $\{x | x > 12\}$ **9.** $\{n | n \geq 7\}$ **11.** $\{r | r > 3.2\}$
13. $\{x | x > -12\}$ **15.** $\{y | y > 17.6\}$ **17.** $\left\{y \middle| y > \frac{5}{7}\right\}$
19. $\{x | x < 6\}$ **21.** $\{x | x > -4\}$ **23.** $\{x | x < 0\}$
25. $\{p | p \geq 9\}$ **27.** $\left\{x \middle| x < -\frac{9}{2}\right\}$ **29.** $\left\{x \middle| x > \frac{9}{4}\right\}$
31. $\left\{x \middle| x \geq \frac{3}{7}\right\}$ **33.** $\{r | r \geq 5\}$ **35.** $\left\{z \middle| z > -\frac{9}{2}\right\}$
37. $\left\{m \middle| m > \frac{4}{9}\right\}$ **39.** $\{y | y < -1\}$ **41.** $\left\{x \middle| x < \frac{2}{3}\right\}$
43. $\{x | x \leq -1.425\}$ **45.** $\left\{w \middle| w \leq \frac{4}{3}\right\}$ **47.** $\left\{x \middle| x > \frac{40}{7}\right\}$
49. $\{x | -1 < x < 3\}$ **51.** $\emptyset$

Pages 32-33   Lesson 1-8
1. true **3.** false **5.** true **7.** $c \geq 30$ **9.** $s \leq 48$
11. $v \geq \$0.99$ or $v \geq 99¢$ **13.** $-3 < x$ and $x < 2$
15. $1 < 3y$ and $3y \leq 13$
17. $5 \leq 3 - 2g$ and $3 - 2g < 1$ **19.** $-2 < x < 10$
21. $-2 \leq x \leq 10$ **23.** $-5 \leq m \leq 5$ **25.** \$60.37
27. 21.25 **29.** \$2,400 **31.** $513\frac{2}{3}$ miles **33.** 81 **35.** 4 more **37.** at least 4 gallons **39.** $-2 < y < 7$
41. $x < 5$ or $x > -1$ (all reals) **43.** $3 < x < 6$
45. $x < -2$ or $x \geq 1$

Page 36   Lesson 1-9
1. $|x| < 3$ **3.** $|x| > 6$ **5.** $|x| > 3$ **7.** $|x| \leq 4$
9. $|x| < 6$ **11.** $|x| < 2$ **13.** $\{x | -9 < x < 9\}$
15. $\{x | x > 2$ or $x < -4\}$ **17.** $\{x | -4 \leq x \leq 12\}$
19. $\{x | x \leq -3$ or $x \geq 3\}$ **21.** $\{x | x > 7$ or $x < -7\}$
23. $\{x | -13 \leq x \leq 13\}$ **25.** $\emptyset$
27. $\{x | x < -20$ or $x > 14\}$ **29.** $\{x | -30 < x < 54\}$
31. $\{x | -9 \leq x \leq 18\}$ **33.** $\left\{x \middle| x \geq \frac{15}{4}$ or $x \leq -\frac{9}{4}\right\}$

35. $\{x \mid -17.6 < x < 14.8\}$ **37.** $\{x \mid -1 \le x \le 6\}$ **39.** all reals **41.** $\{x \mid x \ge 0\}$ **43.** all reals **45.** $\{x \mid -1 \le x \le 1\}$ **47.** $\{x \mid -2 \le x \le 2\}$

1. 92 **3.** $\frac{211}{3}$ **5.** $\frac{13}{12}$ **7.** Q, R **9.** N, W, Z, Q, R **11.** commutative, $+$ **13.** additive identity **15.** 402 **17.** $-3p + 13q$ **19.** substitution **21.** symmetric **23.** $-\frac{33}{13}$ **25.** $\frac{8}{3}$ **27.** $\frac{56}{5}$ **29.** length 21 m, width 11 m **31.** 4 blouses, 2 skirts **33.** 11, 26 **35.** $-2.5, 5.5$ **37.** $\left\{x \mid x \ge -\frac{23}{5}\right\}$ **39.** 15 gallons **41.** $\frac{5}{3} < y \le 5$ **43.** $\left\{x \mid -\frac{9}{2} \le x \le -\frac{1}{2}\right\}$ **45.** all reals

CHAPTER 2   LINEAR RELATIONS AND FUNCTIONS

1. I **3.** III **5.** none **27.** $(-5, 1)$ **29.** $(1, 3)$ **31.** $(-3, -3)$

1. domain = $\{1, 3, 4\}$, range = $\{1, 3, 4\}$; yes **3.** domain = $\{-17, 4, 8\}$, range = $\{-2, 3, 4, 8\}$; no **5.** domain = $\{-3, -2, 2, 4\}$, range = $\{-3, -2, 2, 4\}$; yes **7.** domain = $\{-3, 5\}$, range = $\{-3, 5\}$; yes **9.** $\{(8, 3), (-1, 3), (2, 3)\}$; yes **11.** $\{(15, 15), (15, 30), (15, 45), (15, 60)\}$; no **13.** $\{(-3, 3), (-2, 2), (-1, 1), (0, 0), (1, 1), (2, 2), (3, 3)\}$; domain = $\{-3, -2, -1, 0, 1, 2, 3\}$, range = $\{3, 2, 1, 0\}$ **15.** $\{(2, y)$ such that y is any real number$\}$; domain = $\{2\}$, range = $\{$all reals$\}$ **17.** yes **19.** no **21.** no **23.** no **25.** $\frac{7}{10}$ **27.** $-\frac{7}{3}$ **29.** $-\frac{7}{2}$ **31.** $-\frac{14}{3}$ **33.** $\frac{7}{a-2}$ **35.** $\frac{7}{3a-2}$ **37.** 0 **39.** $-\frac{15}{2}$ **41.** $4t^3 + 2t^2 + t - 7$ **43.** $\frac{20}{3}$ **45.** $-\frac{19}{15}$ **47.** $\frac{a(a+7)}{a+4}$ **49.** 55 **51.** 0.8164 **53.** -0.4489 **55.** $\{$all reals except 5$\}$ **57.** $\{$all reals except 0$\}$ **59.** $\{$all non-negative reals$\}$ **61.** $\{$all non-negative reals$\}$

1. no **3.** yes **5.** yes **7.** yes **9.** no **11.** yes **13.** no **15.** yes **17.** $2x - y = 6$ **19.** $x = 5$ **21.** $5x - 8y = -8$ **23.** $3x - y = 0$

25. 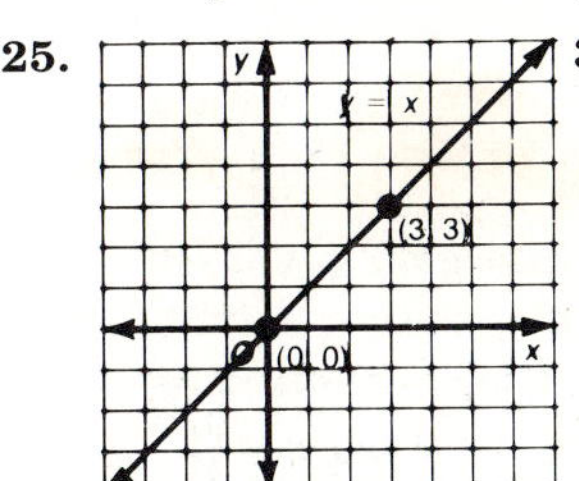 **31.** 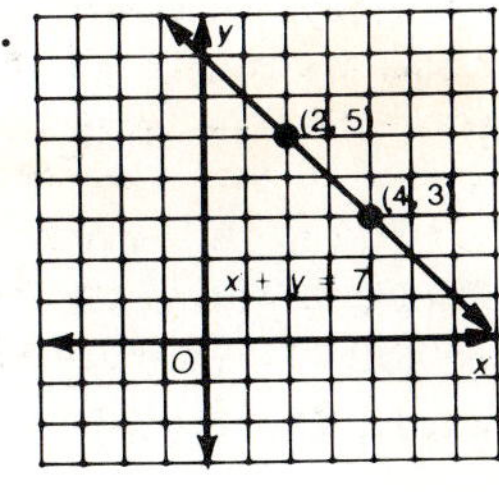

1. $1, 2, -2$ **3.** $-4, 4, 1$ **5.** $0, 2$, none **7.** $2, -4, 2$ **9.** $-\frac{5}{2}$ **11.** $\frac{2}{11}$ **13.** $\frac{5}{2}$ **15.** 8 **17.** 0 **19.** undefined **21.** $-9, \frac{9}{5}$ **23.** $1, -\frac{1}{7}$ **25.** $-5, -\frac{15}{2}$ **27.** -2, none **29.** $-2, 2$ **31.** $\frac{5}{2}, 5$ **33.** $-6, 4$ **35.** $\frac{7}{2}, 7$ **37.** $-4, 6$ **51.** 1 **53.** -8 **55.** 9 **57.** 23

1. $y = 5x - 3$ **3.** $y = -x + 4$ **5.** $y = \frac{2}{3}x - 7$ **7.** $y = 2.5x$ **9.** $y = 0$ **11.** $-4, 6$ **13.** $-\frac{3}{4}, -3$ **15.** $14, -2$ **17.** $-1, 0$ **19.** $y = -\frac{2}{5}x + 2$ **21.** $y = -\frac{2}{3}x + \frac{4}{3}$ **23.** $y = x - 2$ **25.** $2, 15$ **27.** $\frac{5}{3}, -0.2$ **29.** undefined, none **31.** $6, 15$ **33.** $y = \frac{1}{2}x + 1$ **35.** $y = \frac{2}{3}x - 6$ **37.** $y = 4x - 11$ **39.** $y = -\frac{5}{2}x + 16$ **41.** $y = \frac{2}{11}x + \frac{6}{11}$ **43.** $y = \frac{5}{2}x + 18$ **45.** $y = 2x + 6$ **47.** $y = -10x + 6$ **49.** $x = 0$; no slope-intercept form **51.** $x - y = -1$ **53.** $2x + y = 2$ **55.** $4x - 3y = 7$ **57.** $2x + y = 7$ **59.** $9x - 2y = 13$ **61.** $2x + 3y = 13$ **63.** $3x + 4y = 12$ **65.** $5x - 6y = -30$ **67.** $x = 0$ **69.** 7 **71.** 10 **73.** $\frac{5}{7}$

1. A **3.** D **5.** G **7.** A **9.** D **11.** C **13.** 5 **15.** 3 **17.** 2 **19.** D **21.** C **23.** G **25.** G **27.** A

35. 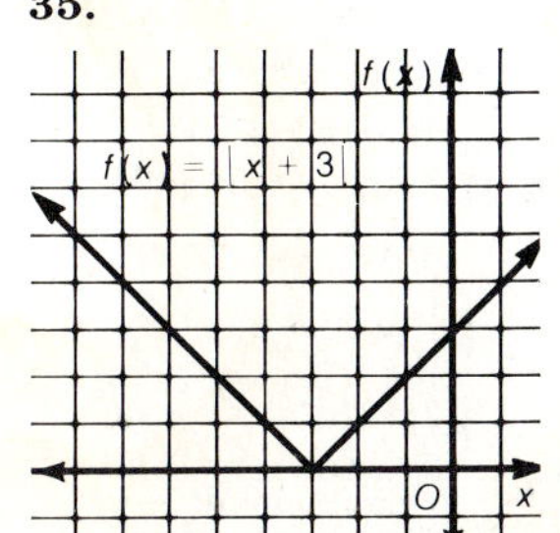

37. 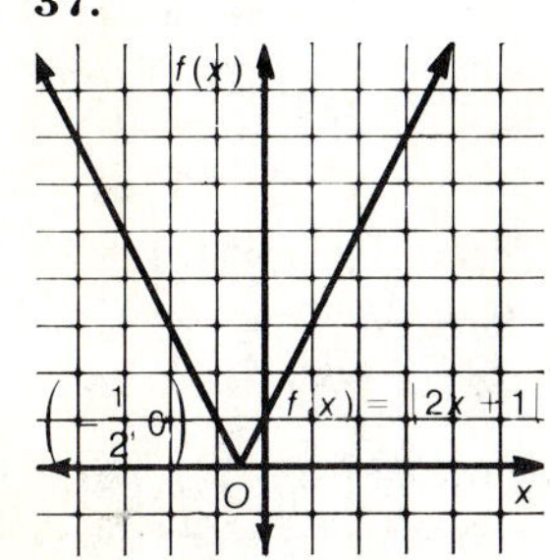

39. 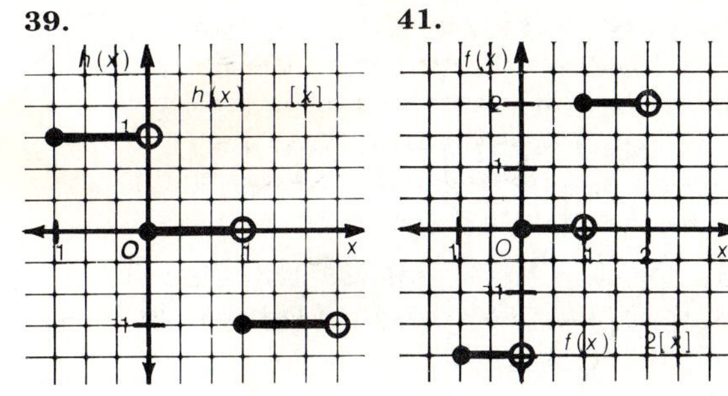

 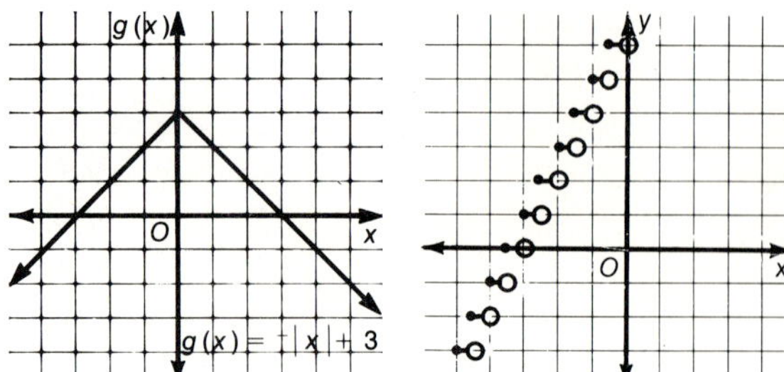

53. The graphs are identical. **55.** The graphs are identical. **57.** The graphs are identical. **59.** The graph of $y = |x + b|$ is like the graph of $y = |x| + b$ moved b units right or left and b units up or down. **61.** The graph of $y = -2|x|$ is like the graph of $y = |-2x|$ but opens down.

Pages 69-70 Lesson 2-7

1. $s = 10r$ **3.** 4.8 inches **5.** 45 inches **7.** \$92 **9.** 1.75 hours **11.** $\frac{2}{3}$, 1, $1\frac{1}{3}$, $1\frac{2}{3}$, $3\frac{1}{3}$, 5, $6\frac{2}{3}$, $8\frac{1}{3}$, 10, $13\frac{1}{3}$, $16\frac{2}{3}$, 20 **13.** $p = 550y + 47{,}000$ **15.** $d = 6000 - 75x$ **17.** $c = 0.18t + 161$ **19.** $c = 1.25n$; \$5.00, \$6.25, \$7.50 **21.** $c = 0.36 + 0.34m$; \$2.40, \$3.08, \$3.76, \$5.46, \$7.16, \$10.56, \$15.66, \$20.76

Pages 72-73 Lesson 2-8

1. all three **3.** (0, 0), (2, −3) **5.** none **7.** (2, −3)

11. 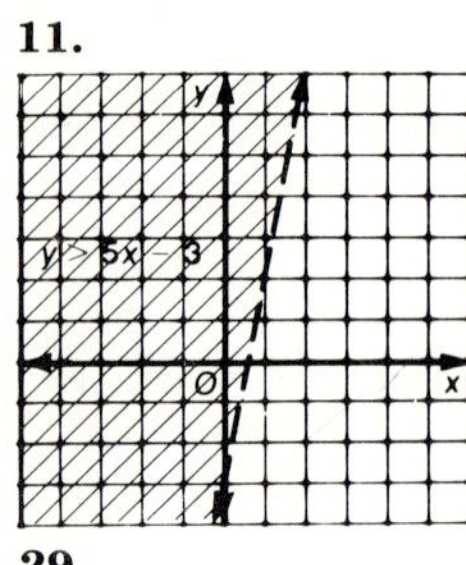 **19.** 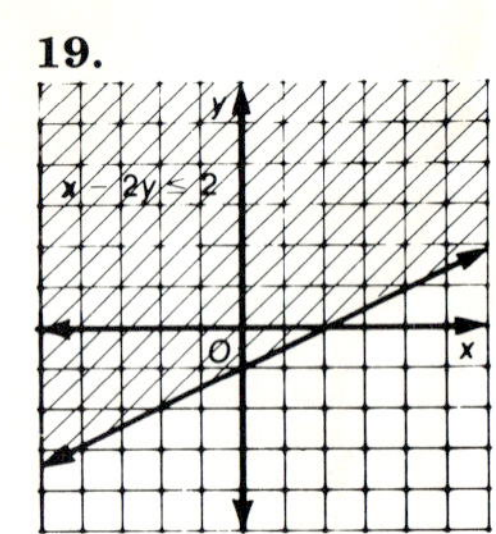

29. 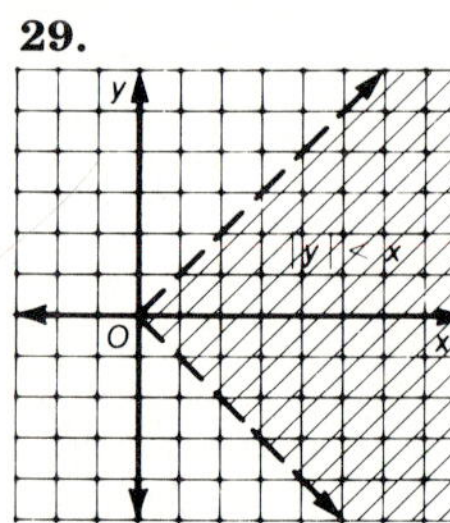 **31.** 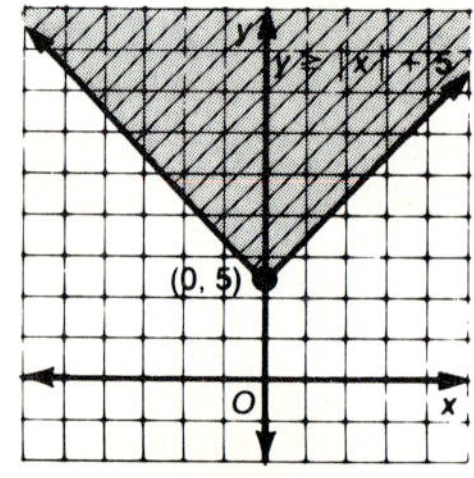

Pages 76-77 Chapter Review

5. domain = {1, 2, 3, 4, 5}; range = {−4.5, −3.5, 4.5}; yes **7.** no **9.** no **11.** −29 **13.** no **15.** yes **21.** $-\frac{3}{7}, \frac{2}{7}, \frac{2}{3}$ **23.** $\frac{1}{2}, -\frac{15}{2}, 15$ **25.** −3 **27.** 2 **29.** 6 **33.** $y = 5x - 7$ **35.** $y = -\frac{6}{7}x - \frac{18}{7}$ **37.** $4x - y = 22$ **39.** $3x - y = 6$ **47.** $c = 4.50 + p$

CHAPTER 3 SYSTEMS OF EQUATIONS AND INEQUALITIES

Page 83 Lesson 3-1

1. 4 **3.** $\frac{3}{2}$ **5.** $\frac{8}{9}$ **7.** $-\frac{1}{2}$ **9.** $-\frac{3}{8}$ **11.** $-\frac{8}{3}$ **13.** parallel **15.** parallel **17.** neither **19.** parallel **21.** neither **23.** $y = \frac{1}{3}x$ **25.** $y = 3x$ **27.** $y = \frac{3}{2}x - \frac{23}{2}$ **29.** $y = -x + \frac{5}{6}$ **31.** $y = \frac{1}{3}x + \frac{2}{3}$ **33.** $y = \frac{5}{3}x + \frac{5}{3}$ **35.** $y = -\frac{1}{2}x - 4$ **37.** $y = \frac{2}{3}x - 2$ **39.** $y = -\frac{1}{15}x - \frac{23}{5}$ **41.** $-\frac{3}{4}$ **43.** 4 **45.** A line connecting (−6, 5) and (−2, 7) has slope $\frac{1}{2}$. A line connecting (5, 3) and (1, 1) has slope $\frac{1}{2}$. These two lines are parallel since their slopes are the same. A line connecting (−6, 5) and (1, 1) has slope $-\frac{4}{7}$. A line connecting (−2, 7) and (5, 3) has slope $-\frac{4}{7}$. These two lines are parallel since their slopes are the same. **47.** (−1, 11) and (7, 15), (7, −5) and (15, −1), or (5, 9) and (9, 1)

Page 86 Lesson 3-2

1. (−3, 6) **3.** (5, 2) **5.** (2, −4) **7.** (−3, 6) **9.** (3, 1); consistent and independent **11.** no solutions; inconsistent **13.** {(x, y)|x + 2y = 5}; consistent and dependent **15.** (2, −6); consistent and independent **17.** (8, 6); consistent and independent **19.** no solutions; inconsistent **21.** $\left(-\frac{4}{3}, -\frac{14}{3}\right)$; consistent and independent **23.** {(x, y)|2x + 3y = 5}; consistent and dependent **25.** no solutions; inconsistent **27.** (1, −2); consistent and independent **29.** {(x, y)|9x − 5 = 7y}; consistent and dependent **31.** $a = 3, b = 8$ **33.** $r = 3, s = 9$

Pages 89-90 Lesson 3-3

11. $\left(\frac{1}{3}, 2\right)$ **13.** (3, 5) **15.** (−1, −1) **17.** (3, −1) **19.** (5, −3) **21.** $\left(\frac{1}{4}, \frac{2}{3}\right)$ **23.** $\left(\frac{14}{5}, \frac{4}{5}\right)$ **25.** (4, −28) **27.** (3, 3) **29.** $\left(\frac{21}{4}, -\frac{27}{4}\right)$ **31.** $\left(\frac{1}{3}, 2\right)$ **33.** (4, 6) **35.** $\left(\frac{10}{3}, \frac{8}{3}\right)$ **37.** (−6, −8) **39.** (6, 9) **41.** (4, 6) **43.** (41, 39) **45.** (−10.5, 48.2) **47.** (−7.6, 5.13)

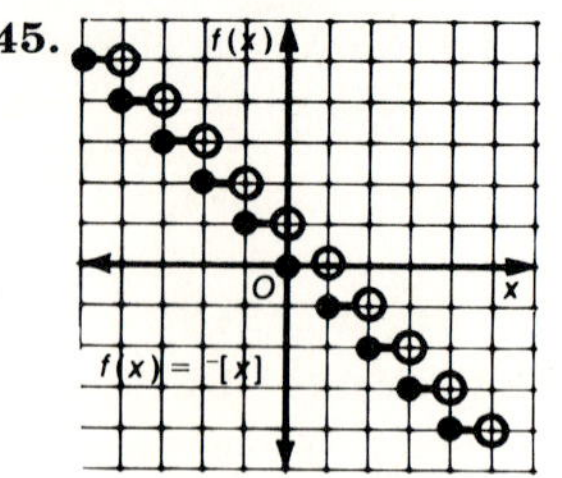 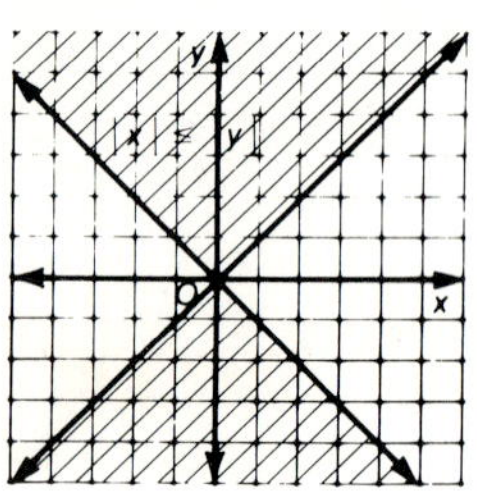

49. $(4.27, -5.11)$ **51.** 27, 15 **53.** width $= 15$ cm; length $= 28$ cm **55.** $(11, 4), (3, -4), \left(-\frac{11}{7}, \frac{4}{7}\right)$
57. $(-5, -2), (1, 10), (4, 4), (-2, -8)$

1. 74 **3.** 0 **5.** 1 **7.** 14

9. $\dfrac{\begin{vmatrix} 5 & 2 \\ 3 & -1 \end{vmatrix}}{\begin{vmatrix} 3 & 2 \\ 4 & -1 \end{vmatrix}}, \dfrac{\begin{vmatrix} 3 & 5 \\ 4 & 3 \end{vmatrix}}{\begin{vmatrix} 3 & 2 \\ 4 & -1 \end{vmatrix}}$ **11.** $\dfrac{\begin{vmatrix} 16 & -4 \\ 21 & -5 \end{vmatrix}}{\begin{vmatrix} 2 & -4 \\ 3 & -5 \end{vmatrix}}, \dfrac{\begin{vmatrix} 2 & 16 \\ 3 & 21 \end{vmatrix}}{\begin{vmatrix} 2 & -4 \\ 3 & -5 \end{vmatrix}}$

13. $\dfrac{\begin{vmatrix} -8 & 1 \\ -14 & -2 \end{vmatrix}}{\begin{vmatrix} 3 & 1 \\ 4 & -2 \end{vmatrix}}, \dfrac{\begin{vmatrix} 3 & -8 \\ 4 & -14 \end{vmatrix}}{\begin{vmatrix} 3 & 1 \\ 4 & -2 \end{vmatrix}}$ **15.** -18 **17.** 343

19. $(3, -4)$ **21.** $(2, 3)$ **23.** $\left(\frac{13}{7}, -\frac{2}{7}\right)$ **25.** $\left(\frac{2}{3}, 0\right)$
27. $(-3, 1)$ **29.** $\left(\frac{9}{64}, \frac{49}{16}\right)$ **31.** $\left(\frac{7}{2}, -4\right)$ **33.** Both the numerator and denominator are zero, therefore no unique solution exists.

1. yes **3.** yes

5. 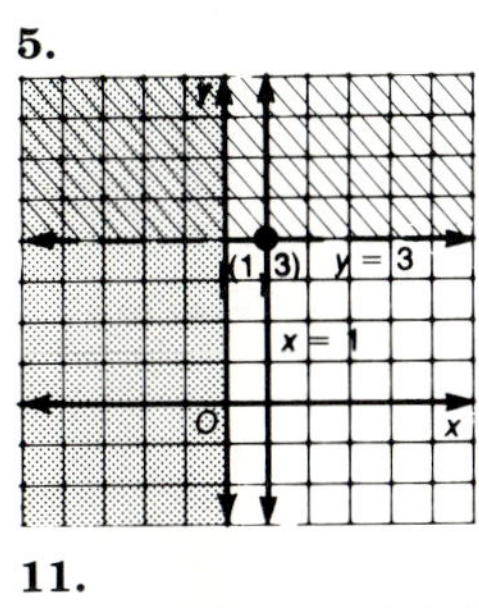   **7.** 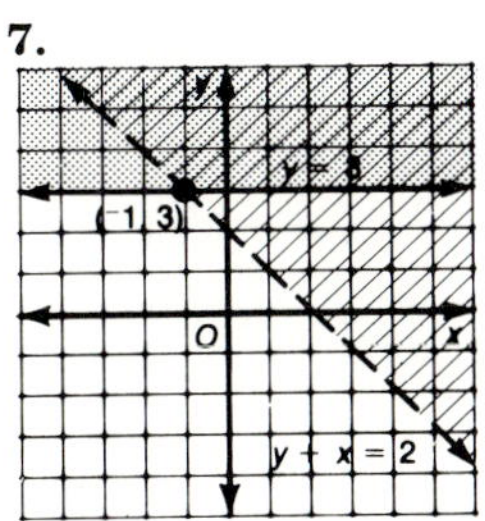

11. 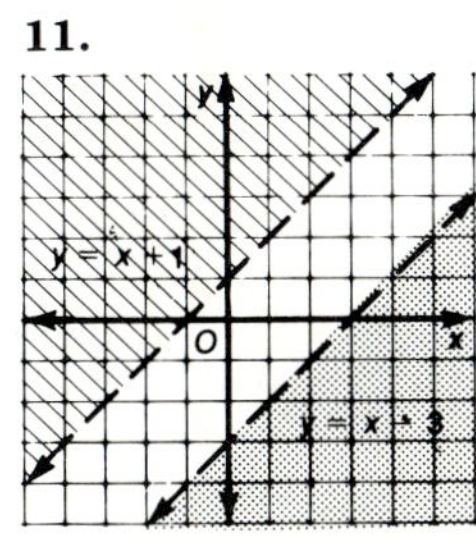   **15.** 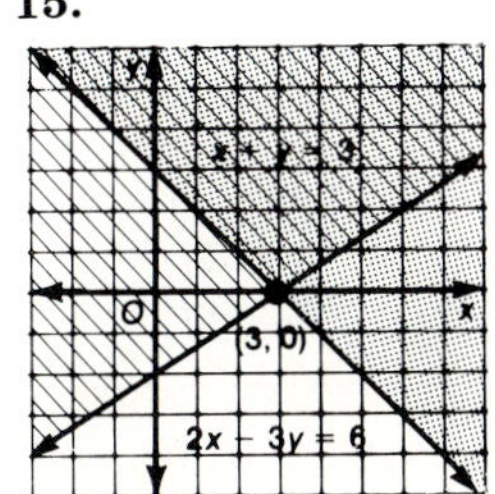

1. 13 **3.** -4 **5.** -10 **7.** 3.9 **9.** 18 **11.** -33
13. 31 **15.** 0.6 **17.** max.: $f(4, 6) = 10$; min.: $f(0, 0) = 0$ **19.** max.: $f(5, 0) = 2\frac{1}{2}$; min.: $f(0, 6) = -6$ **21.** max.: $f(0, 0) = 0$; min.: $f(4, 6) = -22$ **23.** $(0, 1), (1, 3), (6, 3), (10, 1)$
25. $(0, 1), (6, 13), (6, 1)$ **27.** vertices: $(1, 2), (1, 4), (5, 8), (5, 2)$; max.: 11, min.: -5 **29.** vertices: $(2, 2), (2, 8), (6, 12), (6, -6)$; max.: 30, min.: 8 **31.** vertices: $(1, 5), (3, 7), (5, 7), (5, 1)$; max.: 7, min.: -15 **33.** vertices: $(0, 0), (0, 2), (2, 4), (5, 1), (5, 0)$; max.: 25, min.: -6

35. vertices: $(-4.08, -2), (-1.58, 1), (-0.4, 1), (-0.4, -2)$; max.: 0.4; min.: -20.32

1. $2.50c + 7.50b$ **3.** $6c + 30b \le 600$
5. $(0, 0), (0, 20), (50, 10), (60, 0)$ **7.** 50 cars, 10 buses
9a. $x \ge 0, y \ge 0, 4x + y \le 32, x + 6y \le 54$
9c. 14 gallons (6A, 8B) **11a.** 30 dresses, 10 pantsuits
13. 30 footballs, 0 basketballs

1. 1 **3.** none **5.** 5 **7.** 6 **9.** 7 **11.** 8
13. $x = 1, y = -5, z = 2$ **15.** $x = 1, y = -1, z = -1$
17. $x = 3, y = \frac{6}{5}, z = 6$ **19.** $x = 8, y = 4, z = -3$
21. $x = \frac{5}{2}, y = \frac{5}{4}, z = -5$ **23.** no x-intercept, $y = 4, z = 6$ **25.** 4 **27.** 8 **29.** 5 **31.** 3 **33.** 1
35. $x = 6, y = -4, z = 12; 2x - 3y = 12; 2x + z = 12; -3y + z = 12$ **37.** $x = -8, y = 4, z = -6; x - 2y = -8; -3x - 4z = 24; -3y - 2z = 12$
39. $x = \frac{5}{2}, y = -10, z = 5; 4x - y = 10; 2x + z = 5; -y + 2z = 10$ **41.** no x-intercept, $y = 4, z = 10; y = 4; z = 10; 5y + 2z = 20$ **43.** $x = 4, y = \frac{3}{2}, z = -\frac{12}{5}; 3x + 8y = 12; 3x - 5z = 12; 8y - 5z = 12$
45. $x = -\frac{12}{5}, y = \frac{3}{2}$, no z-intercept; $5x - 8y = -125x = -12; 2y = 3$

35. 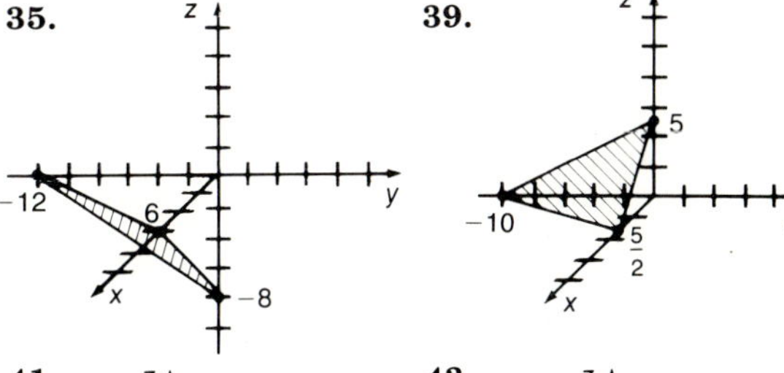   **39.**

41. 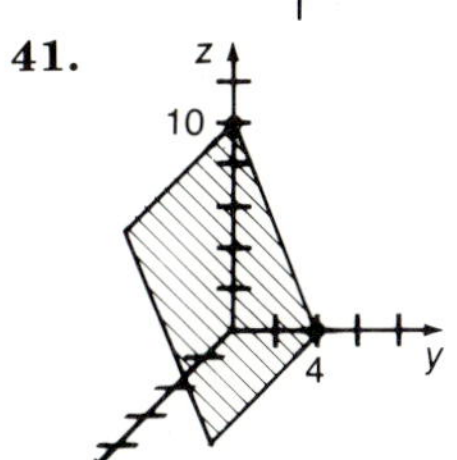   **43.** 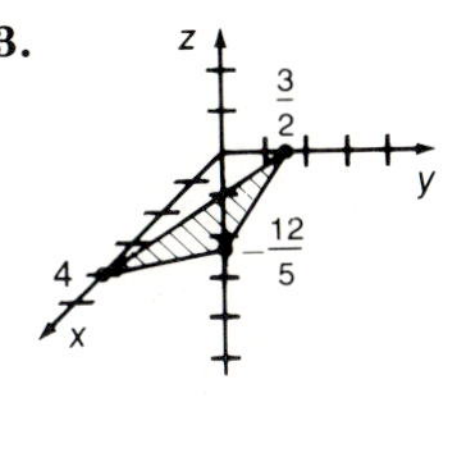

47. $5x - 5y + 2z = 10$ **49.** $6x + 2y + 3z = -12$
51. $4x + 28y + 7z = 28$ **53.** $-x + 6y + 4z = 4$
55. $8x - 3y + 18z = 6$ **57.** $40x + 15y + 32z = -20$
59. $4x + 3y - z = 12$ **61.** $5x + 3y + 3z = 15$
63. $2x + 3y - 5z = 6$ **65.** $2x - 8y - z = 2$
67. $3y - 4z = 6$ **69.** $5x - 2z = -10$

1. no **3.** yes **5.** yes **7.** $(-9, 2, 4)$ **9.** $(3, -1, 5)$
11. $(-8, 13, -5)$ **13.** $(10, 1, -5)$ **15.** $(0, 2, -1)$
17. $(4, -8, 3)$ **19.** $\left(\frac{43}{14}, -\frac{1}{14}, -\frac{33}{14}\right)$ **21.** $(1, 4, 9)$

23. $\left(\frac{3}{14}, -\frac{59}{14}, -\frac{41}{7}\right)$ **25.** 2, 1, 3 **27.** $1.05, $1.75, $0.50
29. 8 ft, 6 ft, 4 ft

1. $y = 3x - 6$ **3.** $y = 6x + 5$ **5.** $y = -\frac{1}{2}x + \frac{13}{2}$
7. $y = \frac{2}{3}x - \frac{1}{3}$ **9.** $(-2, -6)$, consistent and independent
11. $(6, 5)$ consistent and independent **13.** $\left(5\frac{1}{4}, \frac{3}{4}\right)$
15. $\left(\frac{14}{5}, \frac{4}{5}\right)$ **17.** -2 **19.** $\left(\frac{1}{29}, -\frac{39}{29}\right)$ **25.** vertices:
$(0, 0), (0, 3), \left(\frac{3}{2}, \frac{3}{2}\right), (2, 0)$; max.: 12, min.: 0
27. 30 acres of peanuts, 60 acres of corn **29.** 4 **31.** 8
33. $x = 4, y = 4, z = -2$;
$x + y = 4, x - 2z = 4, y - 2z = 4$
35. $x = \frac{11}{3}, y = \frac{11}{4}, z = -\frac{11}{7}$;
$3x + 4y = 11, 3x - 7z = 11, 4y - 7z = 11$ **37.** $(1, -6, 4)$

CHAPTER 4   MATRICES

1. 4×3 **3.** 2×2 **5.** 31 **7.** 6 **9.** 2×2; 3
11. 2×1; no determinant **13.** 3×3; -33
15. 3×3; 24 **17.** 3×3; 29 **19.** 3×2; no
determinant **21.** 3×3; -21 **23.** 3×3; 5,125

1. $\begin{bmatrix} 12 & 3 \\ -6 & 9 \end{bmatrix}$ **3.** $\begin{bmatrix} \frac{5}{2} & -2 & \frac{\sqrt{2}}{2} \end{bmatrix}$ **5.** $\begin{bmatrix} 6 \\ -8 \\ 2 \end{bmatrix}$ **7.** $\begin{bmatrix} 6 & -2 \\ 0 & 7 \end{bmatrix}$
9. $\begin{bmatrix} 2 & -5 & 6 & -13 \end{bmatrix}$ **11.** $\begin{bmatrix} 12 \\ 2 \\ 1 \end{bmatrix}$ **13.** $M_{3\times3}$ **15.** $M_{4\times5}$
17. $M_{3\times2}$ **19.** $M_{5\times7}$ **21.** $M_{4\times4}$ **23.** $M_{2\times3}$
25. $\begin{bmatrix} 6 & -2 \\ 4 & 8 \end{bmatrix}$ **27.** $\begin{bmatrix} -2 & \frac{4}{3} \\ -\frac{2}{3} & \frac{8}{3} \\ 1 & 0 \end{bmatrix}$ **29.** $\begin{bmatrix} 2 & -1 \\ 5 & 11 \end{bmatrix}$ **31.** not
defined **33.** $\begin{bmatrix} -13 & 3 \\ 6 & 16 \end{bmatrix}$ **35.** $\begin{bmatrix} 5 & 5 & -18 \\ 36 & -20 & 30 \end{bmatrix}$
37. $\begin{bmatrix} 7 & -7 \\ 14 & 14 \end{bmatrix}$ **39.** $\begin{bmatrix} -6 & -7 \\ 10 & 28 \end{bmatrix}$ **41.** $\begin{bmatrix} -33 & 16 \\ -5 & -12 \end{bmatrix}$
43. not defined **45.** $\begin{bmatrix} 9 & 5 & -21 \\ 43 & -25 & 39 \end{bmatrix}$ **47.** not defined
49. $x = 1, y = 5, z = 10$ **51.** $x = 5, y = 3, z = 2$
53. $x = 4, y = -3$ **55.** $x = 2, y = 7, z = -2$
57. $A = \begin{bmatrix} a & b \\ c & d \end{bmatrix}, B = \begin{bmatrix} e & f \\ g & h \end{bmatrix}$,
$A + B = \begin{bmatrix} a + e & b + f \\ c + g & d + h \end{bmatrix}$,
$B + A = \begin{bmatrix} e + a & f + b \\ g + c & h + d \end{bmatrix}$
$= \begin{bmatrix} a + e & b + f \\ c + g & d + h \end{bmatrix}$
So, $A + B = B + A$.

1. $+$ **3.** $+$ **5.** $+$ **7.** $\begin{bmatrix} 1 & 0 \\ 0 & 1 \end{bmatrix}$ **9.** yes
11. No. The determinant is not defined.
13. $\begin{bmatrix} 0 & 1 \\ 1 & 0 \end{bmatrix}$; $-1\begin{bmatrix} 0 & -1 \\ -1 & 0 \end{bmatrix} = \begin{bmatrix} 0 & 1 \\ 1 & 0 \end{bmatrix}$
15. $\begin{bmatrix} 2 & 6 \\ -5 & 1 \end{bmatrix}$; $\frac{1}{32}\begin{bmatrix} 1 & 5 \\ -6 & 2 \end{bmatrix}$ **17.** $\begin{vmatrix} -2 & 6 \\ 0 & -6 \end{vmatrix}$
19. $-\begin{vmatrix} 3 & 6 \\ 8 & -6 \end{vmatrix}$ **21.** $\begin{vmatrix} 1 & -2 \\ 8 & -6 \end{vmatrix}$
23. $e_{12}:$ $-\begin{vmatrix} 4 & -2 \\ 4 & -3 \end{vmatrix}$; $e_{22}:$ $\begin{vmatrix} 2 & 1 \\ 4 & -3 \end{vmatrix}$; $e_{32}:$ $-\begin{vmatrix} 2 & 1 \\ 4 & -2 \end{vmatrix}$ **25.** -2
27. -2 **29.** $\begin{bmatrix} 4 & -2 \\ 3 & 8 \end{bmatrix}$; $\frac{1}{38}\begin{bmatrix} 8 & -3 \\ 2 & 4 \end{bmatrix}$ **31.** $\begin{bmatrix} 2 & -3 \\ 1 & 4 \\ 5 & 7 \end{bmatrix}$;
no inverse **33.** $\begin{bmatrix} 9 & 1 \\ 3 & 4 \end{bmatrix}$; $\frac{1}{33}\begin{bmatrix} 4 & -3 \\ -1 & 9 \end{bmatrix}$
35. $\begin{bmatrix} 12 & -7 & 9 \end{bmatrix}$; no inverse
37. $\begin{bmatrix} 3 & -2 & 3 \\ 1 & 0 & 5 \\ 2 & 4 & 2 \end{bmatrix}$; $-\frac{1}{64}\begin{bmatrix} -20 & 8 & 4 \\ 16 & 0 & -16 \\ -10 & -12 & 2 \end{bmatrix}$
39. $\begin{bmatrix} 1 & 0 & 3 \\ 0 & 4 & 5 \\ 2 & 2 & 0 \end{bmatrix}$; $-\frac{1}{34}\begin{bmatrix} -10 & 10 & -8 \\ 6 & -6 & -2 \\ -12 & -5 & 4 \end{bmatrix}$
41. $\begin{bmatrix} 4 & 0 & 5 \\ -2 & 6 & -2 \\ 5 & 1 & 3 \end{bmatrix}$; $-\frac{1}{80}\begin{bmatrix} 20 & -4 & -32 \\ 5 & -13 & -4 \\ -30 & -2 & 24 \end{bmatrix}$

1. yes **3.** no **5.** yes **7.** $\dfrac{\begin{vmatrix} -4 & -2 & 3 \\ -1 & -1 & 4 \\ 1 & 3 & 5 \end{vmatrix}}{\begin{vmatrix} 1 & -2 & 3 \\ 2 & -1 & 4 \\ 2 & 3 & 5 \end{vmatrix}}$,
$\dfrac{\begin{vmatrix} 1 & -4 & 3 \\ 2 & -1 & 4 \\ 2 & 1 & 5 \end{vmatrix}}{\begin{vmatrix} 1 & -2 & 3 \\ 2 & -1 & 4 \\ 2 & 3 & 5 \end{vmatrix}}$, $\dfrac{\begin{vmatrix} 1 & -2 & -4 \\ 2 & -1 & -1 \\ 2 & 3 & 1 \end{vmatrix}}{\begin{vmatrix} 1 & -2 & 3 \\ 2 & -1 & 4 \\ 2 & 3 & 5 \end{vmatrix}}$ **9.** $(3, 6, -2)$
11. $(3, 1, 6)$ **13.** $\emptyset$ **15.** $\left(6, -\frac{1}{2}, 2\right)$ **17.** $\left(\frac{4}{3}, 0, -\frac{1}{2}\right)$
19. $\left(\frac{1}{2}, \frac{2}{3}, \frac{1}{4}\right)$ **21.** 3 pennies, 5 nickels, 8 dimes
23. cheeseburger: 79¢; fries: 69¢; milkshake: 84¢

7. $(2, -3)$ **9.** $(1, 1)$ **11.** $\left(\frac{1}{4}, -\frac{1}{2}\right)$ **13.** $(-1, 2, -3)$
15. $(5, 6, 7)$ **17.** $(-4, 1, 3)$ **19.** $\left(-3, \frac{1}{3}, 1\right)$
21. $\left(\frac{3}{4}, -\frac{2}{3}, \frac{1}{2}\right)$ **23.** $\left(\frac{43}{14}, -\frac{1}{14}, -\frac{33}{14}\right)$

1. $3x + y = 13$ **3.** $2x - 11y = 3$
$\quad\; 4x - 2y = 24$ $\qquad\; x + 2y = 9$

5. $x - 2y \qquad\;\; = -8$ **7.** $x + y + 2z = \quad\; 5$
$\;\;\; 3x + y + 2z = \quad 9$ $\qquad 3x - 6y + 4z = -17$
$\;\;\; 4x - 3y + 3z = \quad 1$ $\qquad 4x - 5y - 2z = \quad\; 4$

9. $\begin{bmatrix} 2 & -1 \\ 3 & 2 \end{bmatrix}\begin{bmatrix} x \\ y \end{bmatrix} = \begin{bmatrix} 11 \\ -1 \end{bmatrix}$ **11.** $\begin{bmatrix} 2 & 5 \\ 3 & 4 \end{bmatrix}\begin{bmatrix} x \\ y \end{bmatrix} = \begin{bmatrix} 1 \\ 12 \end{bmatrix}$

13. $\begin{bmatrix} 6 & 9 \\ 4 & 6 \end{bmatrix}\begin{bmatrix} a \\ b \end{bmatrix} = \begin{bmatrix} 6 \\ 8 \end{bmatrix}$ **15.** $\begin{bmatrix} 1 & 1 & -1 \\ 2 & 3 & -4 \\ 1 & -6 & 1 \end{bmatrix}\begin{bmatrix} x \\ y \\ z \end{bmatrix} = \begin{bmatrix} 2 \\ -1 \\ 1 \end{bmatrix}$

17. $\begin{bmatrix} 2 & -3 & 3 \\ 3 & 1 & -1 \\ 1 & -7 & 7 \end{bmatrix}\begin{bmatrix} a \\ b \\ c \end{bmatrix} = \begin{bmatrix} 8 \\ -3 \\ 19 \end{bmatrix}$ **19.** $(-3, 3)$

21. $(-8, 0)$ **23.** $(8, 0, -3)$ **25.** $(7, -9, 4)$ **27.** $(4, 7)$
29. $(-6, 1)$ **31.** no unique solution   **33.** $(5, 0, -2)$
35. $\left(\frac{3}{4}, \frac{1}{2}\right)$ **37.** $\left(\frac{1}{3}, 3, -5\right)$ **39.** $\left(\frac{1}{3}, -\frac{1}{2}, \frac{2}{3}\right)$ **41.** $\left(\frac{3}{4}, \frac{2}{3}\right)$
43. $\left(\frac{3}{2}, 4, -1\right)$ **45.** $\left(-\frac{5}{3}, -\frac{5}{6}, \frac{5}{2}\right)$

1. $r + b - 14 = 2m$ **3.** $a + 2b - 80 = c$
5. $a + b - 62 = c$ **7.** 269   **9.** single: 17, double: 24,
triple: 11   **11.** Model A: \$20, Model B: \$32, Model C:
\$24   **13.** chicken: \$0.95, salad: \$1.05, roll: \$0.35
15. Arturo 14, Benny 17, Carlos 10   **17.** $24°, 48°, 108°$
19. 13 in., 31 in., 39 in.

1. 2×2; 207   **3.** 3×3; 15   **5.** $\begin{bmatrix} 12 & 3 & -6 & -3 \\ 6 & -9 & 12 & 0 \end{bmatrix}$

7. $\begin{bmatrix} 1 & -10 \\ 5 & -4 \end{bmatrix}$ **9.** $x = -3; y = -4$ **11.** $\begin{bmatrix} 4 & 7 \\ -3 & 0 \\ 5 & -2 \end{bmatrix}$; no

inverse   **13.** $\begin{bmatrix} 1 & 1 & 2 \\ 3 & 4 & 1 \\ 0 & -2 & 2 \end{bmatrix}$; $-\frac{1}{8}\begin{bmatrix} 10 & -6 & -6 \\ -6 & 2 & 2 \\ -7 & 5 & 1 \end{bmatrix}$

15. $(1, -6, 4)$ **17.** $(4, -3)$ **19.** $\left(-3, 0, \frac{1}{2}\right)$
21. $(-4, 2, 1)$ **23.** 747

CHAPTER 5   POLYNOMIALS

1. yes, 7, 1   **3.** no   **5.** yes, $\frac{11}{7}$, 2   **7.** no   **9.** yes, 5, 3
11. yes, 5, 9   **13.** yes, 17, 0   **15.** yes, -24, 5   **17.** no
19. $8m$ **21.** $5d^3$ **23.** $39x^2 - 3y^2$ **25.** y^{12} **27.** 2^7
29. 8^{14} **31.** y^{10} **33.** $81a^4$ **35.** $20m^3k^5$ **37.** $\frac{1}{2}x^5y^{10}$
39. $\frac{16}{15}c^4d^2f$ **41.** $-10{,}368a^{11}b^6$ **43.** $2k^2r^3t^2$ **45.** $-a^7b^3$
47. $9m^4n^3p^2$

1. r^3 **3.** $\frac{1}{x^2}$ **5.** 3   **7.** m^2 **9.** $\frac{1}{3}$ **11.** $\frac{1}{125}$ **13.** 10,000
15. $\frac{729}{b^6}$ **17.** $\frac{64}{k^3}$ **19.** $\frac{1}{m^5}$ **21.** $3n^5$ **23.** $2n^2$ **25.** $\frac{y^7}{x^3}$
27. $\frac{5}{b^3}$ **29.** $\frac{1}{5b^2}$ **31.** $\frac{1}{3x^7}$ **33.** $4b^2c^3$ **35.** 2   **37.** $\frac{9}{t^5}$
39. $\frac{d^2}{125c^3}$ **41.** n^4m^2 **43.** 243   **45.** x^2 **47.** $\frac{1}{x^3y^2}$
49. $\frac{b^3x^4}{5^3y}$ **51.** $3rs^3$ **53.** $\frac{x}{1296z^4}$ **55.** $\frac{1}{xk}$ **57.** $\frac{1}{x^6y^9}$
59. $\frac{z^2}{xy}$ **61.** 29   **63.** $-\frac{r^4s^9}{3} \cdot$ **65.** $4r^{14}$ **67.** $\frac{1}{m} + m - \frac{1}{m^4}$
69. $\frac{1}{y^6} - \frac{1}{y^7} - \frac{1}{y^4}$

1. 2.1×10^{-3} **3.** 8.104×10^2 **5.** 9×10^9
7. 7.21×10^7 **9.** 6,000   **11.** 0.00057   **13.** 3,210,000
15. 0.0427   **17.** 5,832,000,000; 5.832×10^9
19. 5; 5×10^0 **21.** 6; 6×10^0 **23.** 4.93; 4.93×10^0
25. 190.4; 1.904×10^2 **27.** $93{,}333.\overline{3}$; $9.\overline{3} \times 10^4$
29. 2,100,000; 2.1×10^6 **31.** 5.865696×10^{12} m
33. 4.4×10^{49} (approx.)   **35.** 1.32×10^9 km
37. 5.524×10^{27} grams (approx.)

1. 2   **3.** 8   **5.** 5   **7.** 3   **9.** 9   **11.** $(a + b)^2 =$
$(a + b)(a + b) = a^2 + ab + ab + b^2 = a^2 + 2ab + b^2$
13. $(a + b)(a - b) = a^2 - ab + ab - b^2 = a^2 - b^2$
15. $-19y$ **17.** $5m - 2a$ **19.** $4a^2 - 10d + 20$
21. $5y^2 - 2y + 3$ **23.** $7a^2 - b^2$ **25.** $k^2 + 3k - 9$
27. $7n^2 - 8nt + 4t^2$ **29.** $11b^5 + 11b^3 + 7b^2 - b - 1$
31. $-6x^3 - 4x^2y + 13xy^2$ **33.** $4gf^3 - 4bhf$
35. $15m^3n^3 - 30m^4n^3 + 15m^5n^6$
37. $-68b^5d^4 - 187b^6d^5 - 85b^4d^6$ **39.** $r^2 - 2 + \frac{1}{r^4}$
41. $xy^3 + y + \frac{1}{x}$ **43.** $m^2 - 2m - 35$
45. $y^4 + y^3 + 5y^2 + 5y$ **47.** $6x^2 + 31x + 35$
49. $6x^2 - xy - 15y^2$ **51.** $m^2 + 8m + 16$
53. $y^2 - 4y + 4$ **55.** $y^2 - 25$ **57.** $x^2 - 6xy + 9y^2$
59. $16m^2 - 24mn + 9n^2$ **61.** $1 + 8r + 16r^2$
63. $16a^2 - 4b^2$ **65.** $x^6 - y^2$ **67.** $2x^3 - 9x^2 - 7x + 24$
69. $a^3 + b^3$ **71.** $2t^3 + 9t^2 - 19t - 40$
73. $p^3 + 4p^2 - 5p$ **75.** $6x^3 - 7x^2 - 7x + 6$
77. $2a^3 - 5a^2b - 4ab^2 + 12b^3$
79. $2k^3 - 11k^2 + 21k + 63$
81. $z^3 + z^3r + z^2r^2 + zr + zr^2 + r^3$
83. $y^3 - 3y^2 - 16y + 48$
85. $15r^4 - 6r^3 - 13r^2 - 2r + 10r^2d - 4rd - 12d - 6$
87. $4m^5 + 7m^4 + 18m^3 + 29m^2 + 20m + 32$

1. $6(a + b)$ **3.** $a(b + c)$ **5.** $(r + 3)(r - 3)$
7. $(y + 4k)(3y - 2)$ **9.** $(9r - w)(t + v)$
11. $(y - 9z)(y + 9z)$ **13.** $x(x + y + 3)$
15. $r^2(r^2 + rs + s^2)$ **17.** $(7s - 10)(7s + 10)$
19. $(2x^2 + 1)(x - 3)$ **21.** $(b - 1)(3b - 2y)$
23. $(b - 3)(b^2 + 3b + 9)$ **25.** $(b - 12)(b + 12)$

27. $(1 + r)(1 - r + r^2)$ **29.** $(a - 2b)(3x - 4)$

31. $(3m + k)(m - 7)$ **33.** $(r - p)(r - 4)$

35. $(2 - x)(4 + 2x + x^2)$ **37.** $2y(y - 7)(y + 7)$

39. $(2 + x)(4 - 2x + x^2)$ **41.** $(b - 1)(3b - 2y)$

43. $(2b - 3x)(4b^2 + 6bx + 9x^2)$

45. $ab(1 - a)(1 + a + a^2)$ **47.** $s^3(r - 2)(r^2 + 2r + 4)$

49. $(k - 3)(k + 3)(k + 4)$ **51.** $(x - y)(x + y - 4)$

53. $(4y - 1)(16y^2 + 4y + 1)$

55. $(4y^2 + k^2)(2y - k)(2y + k)$

57. $4(2x^2 - 7y^2)(2x^2 + 7y^2)$

59. $3r(1 - 3r)(1 + 3r + 9r^2)$

61. $(a + b - m)(a + b + m)$

63. $((c + d) + f)((c + d)^2 - (c + d)f + f^2)$

1. $10d(2d + 1)$ **3.** $(ab + bd)(c - d)$

5. $(a + 2)(a + 3)$ **7.** $(y + 3)^2$ **9.** $(3x - 4)(a - 2b)$

11. $(b + 1)(b + 6)$ **13.** $p(7m + 2p - 14x)$

15. $(k - 4)^2$ **17.** $(f - 9)^2$ **19.** $(a + 5)(a + 7)$

21. $(k + 6)^2$ **23.** $(3y + 2)(y + 1)$

25. $(2x + 3)(2x - 3)$ **27.** $(2z - 7)(2z - 3)$

29. $(a + 2b)^2$ **31.** $3(d + 4)(d - 4)$ **33.** $(p - 2b)^2$

35. $(2r - 5s)^2$ **37.** $x(x + 7)(x - 5)$

39. $3(2d - 3)(d + 7)$ **41.** $(x - 3)(x + 3)(x - 2)(x + 2)$

43. $2y(y + 3)(y - 7)$ **45.** $(9d + 4)(2d - 3)$

47. $2(r - 2s)(r^2 + 2rs + 4s^2)$

49. $\left(x + y - \dfrac{1}{2}\right)\left(x + y + \dfrac{1}{2}\right)$

51. $(2a + b - y)(2a + b + y)$

53. $(m - k + 3)(m + k - 3)$ **55.** $(a + b)(1 + 3a - 3b)$

57. $(2a - 1)(b + m)$ **59.** $\left(a - \dfrac{1}{2} - y\right)\left(a - \dfrac{1}{2} + y\right)$

61. $(b + m)(2a + b - m)$ **63.** $(2x + 7)(2a - 5b)$

65. $(x + y)(x - y)^2$ **67.** $(r - p)(r + p)^2$

69. $(5x - 7y)(2x - 3)$ **71.** $(a^n - 8)(a^n + 8)$

73. $a(a - 2b)(a^2 - 10ab + 4b^2)$

75. $a(a + 3)(a - 4)(a + 4)$ **77.** $y(x + 2)(x - 4)(x - 1)$

1. $6y - 3 + 2x$ **3.** $2rs + s - 3r$ **5.** $\dfrac{(x + 3)^2}{5}$ **7.** $c + 5$

9. $r + 4$ **11.** $-\dfrac{c^2}{c + 1}$ **13.** $6p^3q + 4p + 5q^2$

15. $2k^2 - 3py + 4p^2y$ **17.** $5r + \dfrac{23}{3}s + \dfrac{2s^2}{r}$ **19.** $a - 12$

21. $a - 2b$ **23.** $x - 15$ **25.** $3y - 1$

27. $4a + 3 - \dfrac{2}{2a + 7}$ **29.** $4y + 5 + \dfrac{3}{7y - 3}$

31. $-a - 10 + \dfrac{44}{6 - a}$ **33.** $8x - 44 + \dfrac{231}{x + 5}$

35. $7m - 8 + \dfrac{3}{8m - 7}$ **37.** $2y^2 + 5y + 2$

39. $3a^2 - 2a + 3$ **41.** $m^2 + m + 1$

43. $y^2 - 6y + 9 - \dfrac{1}{y - 3}$ **45.** $2a^2 - 3a - 2$

47. $m + 3$ **49.** $x^2 - 2x + 8 - \dfrac{20}{x + 2}$ **51.** $x^2 + 2x + 2$

53. $a^2 - a - 1$ **55.** yes   **57.** $a + 1, a - 1$

59. Both are 15.   **61.** -48 **63.** 9   **65.** $\dfrac{y - 7}{y + 3}$

1. no   **3.** yes   **5.** yes   **7.** no   **9.** $2x^2 + x + 5 + \dfrac{6}{x - 2}$

11. $2a^2 - a - 1 + \dfrac{4}{a + 1}$ **13.** $x^3 + x - 1$

15. $6k^2 - k - 2$ **17.** $2b^2 - 5b - 3$

19. $y^3 - 11y^2 + 31y - 21$

21. $2x^3 + x^2 + 3x - 1 + \dfrac{5}{x - 3}$

23. $y^3 + 3y^2 - 16y + 55 - \dfrac{166}{y + 3}$

25. $2x^3 + x^2 - 2x + \dfrac{3}{2x - 1}$ **27.** $2x^2 - 8x + 1 + \dfrac{5}{3x - 2}$

29. $x^4 - 2x^3 + 4x^2 - 8x + 16$ **31.** Both are 12.

33. $4x^2 + 7.5x - 5.8 + \dfrac{0.29}{x - 0.2}$

35. $x^2 - 0.4x + 0.16 - \dfrac{2.864}{x + 0.4}$ **37.** 21

1. y^{11} **3.** x^6 **5.** $114a^4b^4$ **7.** a^4 **9.** $\dfrac{n}{2}$ **11.** 1   **13.** m^6

15. 3.176×10^9 **17.** 2.4×10^8 **19.** $p^4 - p^3 + 4p^2 + 2$

21. $9m^2 - 42m + 49$ **23.** $2x^3 + 11x^2 - 54x + 35$

25. $4z^3 - 24z^2 - 25z + 150$ **27.** $35(a - 5)^2$

29. $(x^2 + y^2)(x + y)(x - y)$ **31.** $q^3(p - 3)(p^2 + 3p + q)$

33. $(a + 4b)(3a - 2)$ **35.** $(2x + y)(x + 3y)$

37. $r(r + 2s)(r + 4s)$ **39.** $(3x + 4y)(2x + y)$

41. $(3a + b)(a + b)$ **43.** $r^2 + 4r - \dfrac{21}{2} + \dfrac{27}{2(2r + 3)}$

45. Since the remainder upon division is 0, $x - 3$ is a factor of $2x^3 - 11x^2 + 12x + 9$ **47.** $r^2 + r - 6$

CHAPTER 6   ROOTS

1. 11   **3.** 2   **5.** y **7.** y^2 **9.** $2n$ **11.** $|x - 2|$

13. 6.856   **15.** 3.107   **17.** -8.185 **19.** -9 **21.** 15

23. 3   **25.** -1 **27.** 0.7   **29.** $11|n|$ **31.** $9s^2$ **33.** 24

35. $8|a|b^2$ **37.** $-2bm$ **39.** $4a^2b$ **41.** $|3p + q|$

43. $z + a$ **45.** $2m - 3$ **47.** $|x + 3|$ **49.** $|3x + 1|$

51. $|2x + 3y|$ **53.** 9.110   **55.** 3.448   **57.** 2.008

59. -3.915 **61.** 2.893   **63.** 2.141   **65.** 3.001   **67.** No, if x is a negative number, $\sqrt[4]{(-x)^4} = -x$.   **69.** When n is even and $x \geq 0$ or when n is odd and $x = 0$.

1. $2\sqrt{2}$ **3.** $5|x|\sqrt{2}$ **5.** $2\sqrt[3]{2}$ **7.** $2\sqrt[4]{3}$ **9.** $|b|\sqrt{b}$

11. $|a|\sqrt[4]{a}$ **13.** $r\sqrt[5]{r^2}$ **15.** $3\sqrt{5}$ **17.** $3\sqrt[4]{2}$

19. $5 - \sqrt{15}$ **21.** $\sqrt{2}$ **23.** $\sqrt[3]{3y}$ **25.** $\dfrac{\sqrt{5}}{2}$ **27.** $\dfrac{\sqrt[3]{5}}{2}$

29. $\dfrac{\sqrt{3}}{\sqrt{3}}$ **31.** $\dfrac{\sqrt{a}}{\sqrt{a}}$ **33.** $\dfrac{\sqrt[3]{2}}{\sqrt[3]{2}}$ **35.** $\dfrac{\sqrt[3]{3}}{\sqrt[3]{3}}$

37. $15\sqrt{6}$ **39.** $2\sqrt[3]{3}$ **41.** $9\sqrt{2}$ **43.** $-4\sqrt[3]{3}$ **45.** $36\sqrt{6}$

47. $48\sqrt{7}$ **49.** 22   **51.** $3\sqrt{2} - 2\sqrt{3}$ **53.** $7\sqrt{2} + 7\sqrt{3}$

55. 14   **57.** $2ab^2\sqrt[3]{ab}$ **59.** $3|m|p\sqrt[4]{p}$ **61.** $3x^2z^2\sqrt{5}$

63. $5mb^2\sqrt[4]{m}$ **65.** $20|m|^3n^3\sqrt{10n}$ **67.** $r + r\sqrt{rs}$

69. $\sqrt{mp} + m\sqrt{q}$ **71.** 2   **73.** $\sqrt{3}$ **75.** $\sqrt[3]{9}$ **77.** $\dfrac{\sqrt{7}}{4}$

79. $\dfrac{\sqrt{7}}{2}$ **81.** $\dfrac{\sqrt[3]{2}}{3}$ **83.** $\dfrac{2\sqrt[3]{2}}{3}$ **85.** $\dfrac{\sqrt[4]{7}}{3}$ **87.** $\dfrac{\sqrt{5}}{5}$ **89.** $\dfrac{\sqrt{3r}}{r}$

91. $\dfrac{\sqrt{10b}}{8b}$ **93.** $\dfrac{\sqrt[3]{18m}}{2m}$ **95.** 2.72 seconds **97.** 7 inches

Page 206   Lesson 6-3
1. $1 - \sqrt{3}$ **3.** $1 + \sqrt{2}$ **5.** $5 - 2\sqrt{5}$ **7.** $2\sqrt{7} + 5$
9. $\sqrt{7} - \sqrt{2}$ **11.** $11\sqrt[3]{6}$ **13.** $11\sqrt{y}$ **15.** $4\sqrt[3]{2}$
17. $3\sqrt[3]{x}$ **19.** 0 **21.** $12 - 2\sqrt{3}$ **23.** 34
25. $m^2 + 2m\sqrt{y} + y$ **27.** $5\sqrt{5} - 10\sqrt{2}$ **29.** $-7\sqrt{7}$
31. $11\sqrt[3]{5b}$ **33.** $3\sqrt[3]{2a}$ **35.** $7\sqrt[3]{2} + 6\sqrt[3]{150}$
37. $14\sqrt{6} + 2\sqrt[3]{3}$ **39.** $5\sqrt{2}$ **41.** $15\sqrt[4]{2}$
43. $(1 + |x|)\sqrt[4]{x^2}$ **45.** $3|yz|\sqrt[4]{z^2}$ **47.** $|z| + z^2 + z^4$
49. $17 + 8\sqrt{2}$ **51.** $25 - 5\sqrt{2} + 5\sqrt{6} - 2\sqrt{3}$
53. $49 - 11p$ **55.** $1 + \sqrt{15}$ **57.** $4 - 2\sqrt{3}$
59. $3a\sqrt{5} - 3\sqrt{ab} + 30ab - b\sqrt{6}$ **61.** $16\sqrt[3]{3} - 12$
63. $x^3 - 3$ **65.** $8 + k$ **67.** $\dfrac{15 + 3\sqrt{2}}{23}$ **69.** $\dfrac{28 + 7\sqrt{3}}{13}$
71. $-5 - 2\sqrt{6}$ **73.** $\dfrac{-25 + 14\sqrt{5}}{5}$
75. $\dfrac{(2m + p)(\sqrt{a}) + mp + 2a}{4a - p^2}$
77. $\dfrac{4\sqrt{3} + 4n\sqrt{6} + n\sqrt{6n} + \sqrt{3n}}{16 - n}$ **79.** $\dfrac{3\sqrt[3]{2}}{2}$
81. $(x + \sqrt{5})(x - \sqrt{5})$ **83.** $(b - 5\sqrt{2})^2$

Page 210   Lesson 6-4
1. 8 **3.** $\dfrac{1}{2}$ **5.** $\dfrac{1}{2}$ **7.** 32 **9.** 4 **11.** 3 **13.** 16 **15.** 6
17. $\sqrt{6}$ **19.** $\sqrt[3]{9}$ **21.** $21^{\frac{1}{2}}$ **23.** $32^{\frac{1}{6}}$ **25.** $y^{\frac{1}{3}}$
27. $2mr^2$ **29.** $27^{\frac{1}{4}}$ **31.** $n^{\frac{2}{3}}$ **33.** $\sqrt{6}$ **35.** $\sqrt[3]{6}$
37. $ab^2\sqrt{ab}$ **39.** $2x^2\sqrt[3]{4x}$ **41.** $\sqrt[3]{5p^2q}$ **43.** $r^2\sqrt[4]{r^2q^3}$
45. $\sqrt[6]{x^2y^3}$ **47.** $25\sqrt[4]{b^2c}$ **49.** $\sqrt{3}$ **51.** $\sqrt{2}$ **53.** 11
55. $\dfrac{7}{4}$ **57.** 12 **59.** 36 **61.** 0.25 **63.** 0.3 **65.** $2\sqrt[12]{2^{11}}$
67. $3\sqrt[6]{3}$ **69.** 2 **71.** $\sqrt{3}$

Page 213   Lesson 6-5
1. $\dfrac{3^{\frac{1}{2}}}{3^{\frac{1}{2}}}$ **3.** $\dfrac{4^{\frac{1}{2}}}{4^{\frac{1}{2}}}$ **5.** $\dfrac{y^{\frac{1}{3}}}{y^{\frac{1}{3}}}$ **7.** $\dfrac{p^{\frac{1}{2}}}{p^{\frac{1}{2}}}$ **9.** $\dfrac{m^{\frac{1}{2}} - p}{m^{\frac{1}{2}} - p}$ **11.** $\dfrac{t^{\frac{3}{2}} - s^{\frac{1}{2}}}{t^{\frac{3}{2}} - s^{\frac{1}{2}}}$

13. $2 \cdot 3^{\frac{1}{2}}$ **15.** 2 **17.** $\dfrac{y^{\frac{1}{3}}}{y}$ **19.** $\dfrac{p^{\frac{1}{2}}}{p^2}$

21. $\dfrac{m^{\frac{3}{2}} - mp + m^{\frac{1}{2}}p - p^2}{m - p^2}$ **23.** $\dfrac{2(t^{\frac{3}{2}} - s^{\frac{1}{2}})}{t^3 - s}$ **25.** $\dfrac{y^{\frac{3}{5}}}{y}$ **27.** $\dfrac{b^{\frac{3}{4}}}{b}$

29. $3 \cdot 5^{\frac{1}{3}}$ **31.** $\dfrac{rm^{\frac{1}{2}}b^{\frac{1}{2}}}{b^2}$ **33.** $\dfrac{b^2 + 3}{b}$ **35.** $3x^{\frac{5}{3}} + 4x^{\frac{8}{3}}$

37. $r^{\frac{1}{9}}$ **39.** $\dfrac{r^2 - 2r^{\frac{3}{2}}}{r - 4}$ **41.** $\dfrac{m^{\frac{5}{6}} - m^{\frac{1}{3}}p^{\frac{3}{2}} - m^{\frac{1}{2}}p^{\frac{1}{3}} + p^{\frac{5}{6}}}{m - p^3}$

43. $\dfrac{r^{\frac{1}{2}}s}{1 + r}$ **45.** $\dfrac{1}{b - 1}$ **47.** $3xy^3$ **49.** $\dfrac{25x^{21}}{y^8}$

51. $\dfrac{3x^{\frac{1}{3}}y - 2x}{xy}$ **53.** $-\dfrac{16}{9} \cdot 2^{\frac{1}{6}}$ **55.** $\dfrac{x^{\frac{2}{3}} + x^{\frac{1}{3}}y^{\frac{1}{3}} + y^{\frac{2}{3}}}{x - y}$

57. $\dfrac{a^{\frac{4}{3}} + a^{\frac{2}{3}}b^{\frac{2}{3}} + b^{\frac{4}{3}}}{a^2 - b^2}$

Pages 216-217   Lesson 6-6
1. 4 **3.** 64 **5.** 1 **7.** 29 **9.** $\dfrac{7}{2}(\sqrt{3} + 1)$ **11.** $1 + \sqrt{3}$

13. $\dfrac{-\sqrt{2}}{2}$ **15.** $\dfrac{-7\sqrt{5}}{30}$ **17.** $\dfrac{-\sqrt{10}}{25}$ **19.** $\dfrac{(1 + \sqrt{5})}{-2}$
21. $\dfrac{(7\sqrt{2} - 14)}{2}$ **23.** $\dfrac{39 - 13\sqrt{5}}{4}$ **25.** 13 **27.** 28
29. $5\dfrac{1}{3}$ **31.** 7 **33.** 4 **35.** -1 **37.** no solution
39. no solution **41.** 4 **43.** 2 **45.** 7.41 **47.** no
solution **49.** $\dfrac{6(11 + 2\sqrt{3})}{109}$ **51.** 4
53. $\pm\sqrt{y^2 - s^2}$ if $y \geq s$ **55.** $\dfrac{2mM}{r^3}$ if $r \neq 0$
57. $\dfrac{T}{4v^2 - 1}$ if $4v^2 - 1 \neq 0$ **59.** $r = 14$ cm

Page 219   Lesson 6-7
1. $6i$ **3.** $4i\sqrt{2}$ **5.** -3 **7.** $5i$ **9.** $6i$ **11.** -1 **13.** $9i$
15. $5i\sqrt{2}$ **17.** $\dfrac{2}{3}i$ **19.** $\dfrac{i\sqrt{3}}{3}$ **21.** i **23.** $-i$ **25.** $-i$
27. -1 **29.** -4 **31.** $-7\sqrt{2}$ **33.** -3 **35.** $-3i\sqrt{3}$
37. 24 **39.** $-216i$ **41.** $-18i$ **43.** $\pm 4i$ **45.** $\pm 13i$
47. $\pm i\sqrt{3}$ **49.** $\pm 2i$ **51.** $\pm 5i$ **53.** $\pm i\dfrac{\sqrt{5}}{2}$

Page 221   Lesson 6-8
1. $8 + 11i$ **3.** 3 **5.** 10 **7.** $20 + 12i$ **9.** $-10 + 10i$
11. $14 + 5i$ **13.** $x = 5, y = -6$ **15.** $x = 7, y = 2$
17. $x = 3, y = 0$ **19.** $7 + 7i$ **21.** $6 + 4i$
23. $2 + i\sqrt{7} + i\sqrt{2}$ **25.** $8 - 15i$ **27.** $5 - 3i\sqrt{3}$
29. $-21 - 2i$ **31.** $13 - 13i$ **33.** $32 - 24i$
35. $37 + 2i\sqrt{2}$ **37.** $5 + 12i$ **39.** 3 **41.** 7
43. $20 + 15i$ **45.** $148 - 222i$ **47.** $109 - 37i$
49. $x = 2, y = 3$ **51.** $x = -1, y = -3$
53. $x = 3, y = 1$ **55.** $-a - bi$
57. $(a + bi) \cdot 1 = (a \cdot 1) + (bi \cdot 1) = a + bi$

Page 225   Lesson 6-9
1. $2 - i$ **3.** $5 + 4i$ **5.** $-4i$ **7.** $5i$ **9.** 6 **11.** $5 + 6i$
13. $(3 + 2i)\left(\dfrac{3 - 2i}{13}\right) = \dfrac{9 - 4i^2}{13} = \dfrac{9 + 4}{13} = \dfrac{13}{13} = 1$
15. $(6 + 8i)\left(\dfrac{3 - 4i}{50}\right) = \dfrac{18 - 24i + 24i - 32i^2}{50} = \dfrac{18 + 32}{50} =$
$\dfrac{50}{50} = 1$ **17.** 61 **19.** 290 **21.** 98 **23.** 100 **25.** $\dfrac{9 + i}{2}$
27. $\dfrac{9 + i}{41}$ **29.** $\dfrac{7 + 4i}{3}$ **31.** $\dfrac{1 - 2i}{3}$ **33.** $\dfrac{12 - 10i}{61}$
35. $\dfrac{7\sqrt{2} + 21i}{11}$ **37.** $\dfrac{-1 + 2i\sqrt{2}}{3}$ **39.** $\dfrac{-3 - 4i\sqrt{7}}{11}$ **41.** 9
43. $\dfrac{-44 + 117i}{50}$ **45.** $\dfrac{2 + 5i}{29}$ **47.** $\dfrac{3 - 7i}{58}$ **49.** $\dfrac{-1 - 5i}{2}$
51. $\dfrac{-4 + 3i}{3}$

Pages 229-230   Chapter Review
1. $7|a|$ **3.** $2xy^2$ **5.** 3.391 **7.** 941.192 **9.** $2\sqrt[3]{6a^2}$
11. $45\sqrt{30}$ **13.** $5\sqrt{2} + 2\sqrt{5}$ **15.** $2\sqrt[3]{6} + 3$
17. $2xy^2\sqrt[3]{9}$ **19.** $2\sqrt[3]{5}$ **21.** $\dfrac{4\sqrt{3} + 4\sqrt{2} - \sqrt{6} - 3}{13}$
23. $60 + 10\sqrt{2} + 6\sqrt{5} + \sqrt{10}$ **25.** $r^2\sqrt[6]{s^2y^3}$
27. $\sqrt[4]{9x^3}$ **29.** $3mn^{\frac{2}{3}}$ **31.** 36 **33.** 0.04 **35.** $\dfrac{z^{\frac{2}{5}}}{z}$
37. $\dfrac{z^{\frac{2}{3}}}{z - 1}$ **39.** 1 **41.** 4 **43.** $-i$ **45.** $-30i$
47. $6, 16i, 73$ **49.** $\dfrac{-3 - 7i}{2}$

CHAPTER 7   QUADRATIC EQUATIONS

Page 234   Lesson 7-1

1. yes **3.** yes **5.** no **7.** no **9.** yes **11.** 3, −7
13. 1, 8 **15.** 0, −4 **17.** $-\frac{5}{2}$, 1 **19.** −2, −4 **21.** 4, 5
23. −5, 2 **25.** −2 **27.** 0, −3 **29.** 4, −1 **31.** −6, 5
33. $-\frac{3}{2}$, −1 **35.** $-\frac{3}{2}$, 3 **37.** 0, $\frac{5}{3}$ **39.** $-\frac{3}{2}$, $-\frac{2}{3}$
41. $-\frac{1}{4}$, 3 **43.** −8, 5 **45.** $\frac{2}{3}$, 4 **47.** $-\frac{1}{4}$, $\frac{5}{3}$ **49.** $\frac{5}{6}$, $-\frac{3}{2}$
51. $-\frac{3}{4}$, 4 **53.** $\frac{3}{2}$ **55.** $\frac{11}{4}$, $-\frac{11}{4}$ **57.** 0, 3, −3
59. 0, $-\frac{6}{7}$, $\frac{2}{5}$ **61.** 0, $\frac{3}{5}$ **63.** no real solutions

Page 238   Lesson 7-2

1. yes **3.** no **5.** yes **7.** 9 **9.** 100 **11.** 36 **13.** $\frac{9}{4}$
15. $\frac{121}{4}$ **17.** $\frac{1}{4}$ **19.** 9 **21.** $\frac{1}{16}$ **23.** $\frac{1}{4}$ **25.** $\frac{9}{4}$ **27.** $\frac{49}{4}$
29. 625 **31.** 6, −4 **33.** −11, 8 **35.** 3, 5 **37.** −10, 2
39. 3, 4 **41.** 6, −14 **43.** −15, 12 **45.** $4 \pm \sqrt{2}$
47. $\frac{7 \pm \sqrt{29}}{2}$ **49.** $\frac{5 \pm \sqrt{65}}{2}$ **51.** $-\frac{1}{3}$, −2 **53.** $-\frac{3}{2}$, $\frac{1}{3}$
55. $\frac{3}{2}$, −7 **57.** $\frac{5}{3}$, −3 **59.** $2 \pm \frac{2}{3}\sqrt{6}$
61. $\frac{-a \pm \sqrt{a^2 - 4b}}{2}$ **63.** $\frac{\pm\sqrt{-4ab}}{2a} = \pm\frac{i\sqrt{ab}}{a}$

Pages 241-242   Lesson 7-3

1. 5, −3, 7 **3.** 1, 2, −1 **5.** 3, −2, 7 **7.** 6, −5
9. −5, 3 **11.** 6, 4 **13.** 4, 1 **15.** 4, $-\frac{5}{3}$ **17.** $\frac{5}{3}$, $-\frac{3}{2}$
19. $\frac{1}{7}$, $-\frac{5}{2}$ **21.** $\frac{1}{4}$, $-\frac{2}{5}$ **23.** $\frac{-4 \pm i\sqrt{14}}{6}$ **25.** $\frac{5 \pm i\sqrt{7}}{4}$
27. $\frac{-1 \pm i\sqrt{5}}{2}$ **29.** $\frac{1 \pm \sqrt{57}}{14}$ **31.** 0, $\frac{24}{7}$ **33.** $\frac{3 \pm i\sqrt{15}}{3}$
35. $1 \pm i\sqrt{3}$, −2 **37.** $-2 \pm 2i\sqrt{3}$, 4 **39.** $\frac{-1 \pm i\sqrt{3}}{2}$, 1
41. −0.23, −8.77 **43.** 1.64, −0.55 **45.** 15.75, 0.25
47. 3.415 meters

Page 244   Lesson 7-4

1. 33 **3.** 1 **5.** 0 **7.** 81 **9.** 1 **11.** −16 **13.** 144; 2
real; rational; 7, −5 **15.** 0; 1 real; rational; 2 **17.** 12; 2
real; irrational; $2 \pm \sqrt{3}$ **19.** 16; 2 real; rational; $-\frac{3}{2}$,
$-\frac{1}{2}$ **21.** 73; 2 real; irrational; $\frac{-11 \pm \sqrt{73}}{6}$ **23.** −16; 2
imaginary; $1 \pm 2i$ **25.** 89; 2 real; irrational; $\frac{-9 \pm \sqrt{89}}{2}$
27. 36; 2 real; rational; 0, 6 **29.** −144; 2 imaginary;
$\frac{2 \pm 3i}{2}$ **31.** 289; 2 real; rational; 6, $\frac{1}{3}$ **33.** −3; 2
imaginary; $\frac{1 \pm i\sqrt{3}}{2}$ **35.** 21; 2 real; irrational; $\frac{-1 \pm \sqrt{21}}{2}$

Page 248   Lesson 7-5

1. −7, −4 **3.** 3, 5 **5.** $-\frac{7}{3}$, −3 **7.** $\frac{3}{5}$, 0 **9.** 0, $-\frac{3}{5}$
11. $\frac{2}{3}$, $\frac{11}{3}$ **13.** $\frac{1}{4}$, $\frac{1}{3}$ **15.** $\frac{1}{15}$, $-\frac{4}{15}$ **17.** −5; 6; −3, −2
19. $-\frac{14}{3}$; $-\frac{5}{3}$; −5, $\frac{1}{3}$ **21.** $\frac{5}{2}$; $\frac{1}{2}$; $\frac{5 \pm \sqrt{17}}{2}$
23. $\frac{8}{3}$; $-\frac{35}{3}$; 5, $-\frac{7}{3}$ **25.** −9; 25; $\frac{-9 \pm i\sqrt{19}}{2}$
27. 0; −16; ±4 **29.** $-\frac{3}{4}$, $\frac{1}{8}$; $-\frac{1}{2}$, $-\frac{1}{4}$

31. $-\frac{27}{4}$; $-\frac{7}{4}$, $\frac{1}{4}$, −7 **33.** −4; −77; 7, −11
35. $-\frac{5}{7}$; $-\frac{1}{7}$; $\frac{-5 \pm \sqrt{53}}{14}$ **37.** $x^2 - 6x - 16 = 0$
39. $x^2 - 10x + 24 = 0$ **41.** $x^2 - 36 = 0$
43. $2x^2 - 7x + 3 = 0$ **45.** $4x^2 - 45x - 36 = 0$
47. $4x^2 - 1 = 0$ **49.** $32x^2 - 28x + 5 = 0$
51. $x^2 - 3\sqrt{3}x + 6 = 0$ **53.** $x^2 - 4x + 1 = 0$
55. $x^2 + 9 = 0$ **57.** $x^2 - 6x + 58 = 0$
59. $2x^2 - 2x - 3 = 0$ **61.** 4 **63.** 27 **65.** $-\frac{49}{3}$

Pages 251-252   Lesson 7-6

1. 26, 27 or −27, −26 **3.** 37, 39 or −39, −37 **5.** 15,
17 **7.** 11, 12 or −12, −11 **9.** 10 meters **11.** 18 feet by
24 feet **13.** 0 or 7 **15.** 8 or −9 **17.** $\frac{2}{3}$ or $\frac{1}{3}$ **19.** 3 or
$\frac{1}{3}$ **21.** 2 meters **23.** $\frac{6}{5}$ or $-\frac{5}{6}$ **25.** 6 cm by 4 cm
27. 0 or 7 **29.** 4 meters **31.** 0.618

Page 255   Lesson 7-7

1. yes **3.** no **5.** yes **7.** 8 **9.** 25 **11.** $\frac{1}{5}$ **13.** 4
15. $\frac{1}{13}$ **17.** $1(x^{-3})^2 - 8(x^{-3}) + 16 = 0$
19. $1(x^{\frac{1}{4}})^2 + 7(x^{\frac{1}{4}}) + 12 = 0$
21. $1(y^{\frac{1}{4}})^2 - 10(y^{\frac{1}{4}}) + 16 = 0$
23. $1(s^{\frac{1}{3}})^2 - 9(s^{\frac{1}{3}}) + 20 = 0$
25. $1(k^{-\frac{2}{3}})^2 - 10(k^{-\frac{2}{3}}) + 21 = 0$ **27.** ±1, $\pm\sqrt{2}$
29. ±6, ±2 **31.** ±2, ±2i **33.** $\pm\sqrt{6}$, $\pm i\sqrt{6}$
35. 0, ±3 **37.** ±2, ±$i\sqrt{2}$ **39.** $\pm2\sqrt{2}$, $\pm\sqrt{3}$
41. 81, 16 **43.** 64 **45.** 2, −2, $\pm1 \pm i\sqrt{3}$
47. −3, 1, $\frac{3 \pm 3i\sqrt{3}}{2}$, $\frac{-1 \pm i\sqrt{3}}{2}$ **49.** 1, $\frac{-1 \pm i\sqrt{3}}{2}$
51. 27, 125 **53.** 125, 8 **55.** 9, 4 **57.** 8, 27
59. $\frac{1}{32}$ **61.** 4 **63.** $\frac{625}{16}$, 256 **65.** $\frac{4}{9}$
67. $-\frac{27}{8}$, 64 **69.** 5, −1 **71.** 10, 5, 3, −2

Page 258   Chapter Review

1. $-\frac{3}{2}$, $\frac{1}{3}$ **3.** $-\frac{3}{2}$, −1 **5.** $-\frac{6}{5}$, $\frac{1}{3}$ **7.** 49 **9.** 15, 5
11. $-\frac{7}{2}$, 3 **13.** 2, $\frac{5}{3}$ **15.** $\frac{\pm3\sqrt{2}}{2}$ **17.** 1200; 2 real;
irrational; $\frac{10 \pm 5\sqrt{3}}{2}$ **19.** 0; 1 real; rational; 4 **21.** 12;
−45; 15, −3 **23.** 0; $-\frac{11}{3}$; $\frac{\pm\sqrt{33}}{3}$ **25.** $x^2 + 2x - 24 = 0$
27. $x^2 - 10x + 34 = 0$ **29.** 32 or −12 **31.** 3.5 feet
(24.5 is not a reasonable answer.) **33.** ±3, $\pm\sqrt{3}$
35. no solutions **37.** 64, 125

CHAPTER 8   QUADRATIC RELATIONS AND FUNCTIONS

Pages 262-263   Lesson 8-1

1. yes **3.** no **5.** yes **7.** yes **9.** no **11.** x^2; $3x$; $-\frac{1}{4}$
13. x^2; $-3x$; $-\frac{1}{4}$ **15.** $3a^2$; 0; −2 **17.** z^2; $3z$; 0

19. x^2; $6x$; 9　**21.** r^2; $-4r$; 9　**23.** $f(x) = x^2 - 4x + 4$
25. $f(x) = 9x^2 + 12x + 4$　**27.** $f(x) = 32x^2 + 16x + 2$
29. $f(x) = 3x^2 - 24x + 42$　**31.** $f(x) = 45x^2 - 60x + 24$
33. $f(x) = 20x^2 - 20x + 13$　**35.** $A = \pi r^2$　**37.** x = one
of the numbers; product = $40x - x^2$　**39.** x = the lesser
number; product = $64x + x^2$　**41.** x = length of rectangle;
area = $10x - x^2$　**43.** x = one of the numbers; sum =
$2x^2 - 20x + 100$　**45.** x = length of adjacent to shed;
area = $120x - 2x^2$　**47.** p = number of \$1.00 price
increases; $I(p) = 2400 + 140p - 20p^2$

Page 268　Graphing Calculator Application
5. 5 units to the right　**7.** 3 units above　**9.** 2 units to
the right　**11.** 8 units above　**13.** 10 units to the right, 9
units above　**15.** 9 units to the left, 4 units above
17. narrower　**19.** opens downward not upward
21. narrower　**23.** 14 units above, wider, opens
downward not upward　**25.** narrower, opens downward
not upward　**27.** 5 units to the left, wider
29. narrower　**31.** 12 units to the left, 12 units below,
wider, opens downward not upward　**33.** $|h|$ units to the
left or right　**35.** $|k|$ units above or below　**37.** $|h|$ units
to the left or right and $|k|$ units above or below
39. narrower if $|a| > 1$, wider if $|a| < 1$, opens downward
not upward if $a < 0$　**41.** narrower if $|a| > 1$, wider if
$|a| < 1$, opens downward not upward if $a < 0$

Page 273　Lesson 8-2
1. $(0, 0)$; $x = 0$　**3.** $(8, 0)$; $x = 8$　**5.** $(0, 11)$; $x = 0$
7. $(-1.5, -3.2)$; $x = -1.5$　**9.** $\left(\frac{1}{5}, 1\right)$; $x = \frac{1}{5}$
11. $y = (x + 2)^2$　**13.** $y = (x + 2)^2 - 2$
15. $f(x) = (x + 3)^2$; $(-3, 0)$; $x = -3$
17. $f(x) = x^2 - 7$; $(0, 7)$; $x = 0$
19. $f(x) = (x - 6)^2 - 36$; $(6, -36)$; $x = 6$
21. $f(x) = \left(x - \frac{3}{2}\right)^2 + \frac{3}{4}$; $\left(\frac{3}{2}, \frac{3}{4}\right)$; $x = \frac{3}{2}$

25.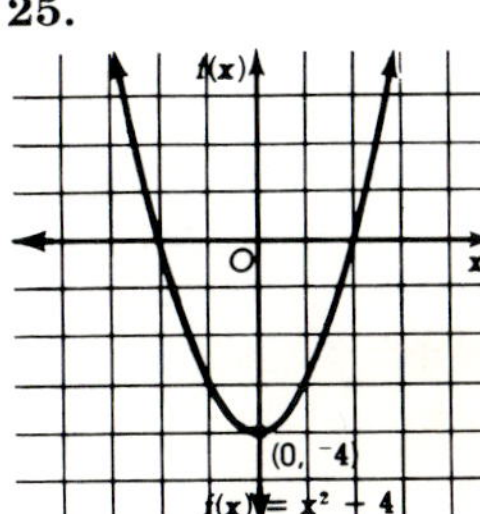　**31.**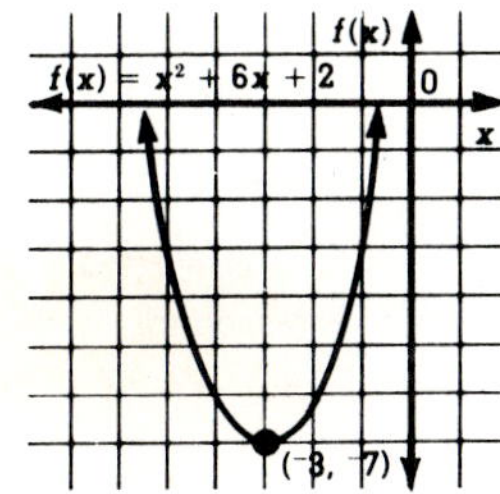

39. $(0, -5)$; $y = x^2 - 5$　**41.** $(-3, 2)$; $y = (x + 3)^2 + 2$

Pages 277-278　Lesson 8-3
1. $(0, 0)$; $x = 0$; up　**3.** $(-2, 0)$; $x = -2$; down
5. $(0, -6)$; $x = 0$; up　**7.** $(2, -2)$; $x = 2$; down
9. $\left(\frac{1}{2}, \frac{1}{4}\right)$; $x = \frac{1}{2}$; up　**11.** $f(x) = -2x^2 + 4$
13. $f(x) = 5(x - 1)^2 - 8$

15. $f(x) = -(x + 1)^2 + 3$; $(-1, 3)$; $x = -1$; down
17. $f(x) = 4(x + 3)^2 - 36$; $(-3, -36)$; $x = -3$; up
19. $f(x) = 3(x - 3)^2 - 16$; $(3, -16)$; $x = 3$; up
21. $f(x) = -\frac{1}{2}(x - 5)^2 - 1$; $(5, -1)$; $x = 5$; down
23. $y = -2x^2$　**25.** $y = \frac{1}{3}x^2 + 5$
27. $y = -\frac{3}{4}(x - 4)^2 + 1$　**45.** $y = 2(x - 8)^2 - 5$
33.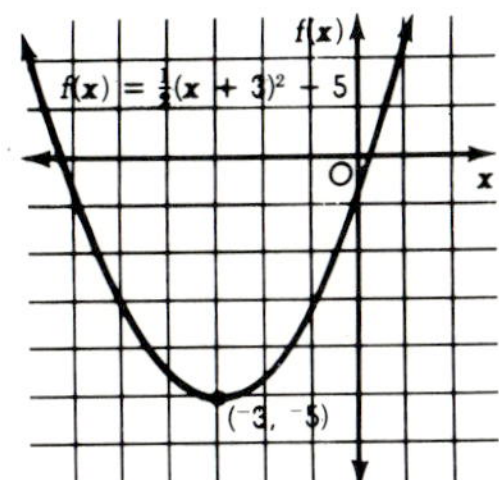　**37.**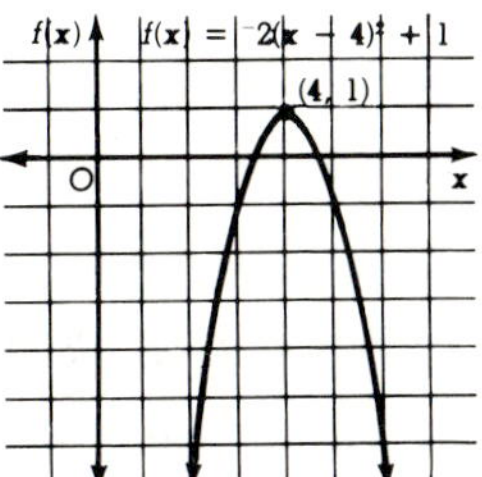

Pages 280-282　Lesson 8-4
1. $40 + 10x$　**3.** income = $(40 + 10x)(50,000 - 5000x)$
5. 70¢
7. $x(300 - 2x) - (20x + 1000)$ or $-2x^2 + 280x - 1000$
11. $-20, 20$　**13.** $\frac{37}{2}, \frac{37}{2}$　**15.** $-10, 10$　**17.** 10 cm by
10 cm; 100 cm^2　**19.** 60 m by 30 m　**21.** \$11.50
23. 300 ft; 2.5 sec　**25.** 300 m by 200 m
27. 5 cm by 4 cm　**29.** 16 in.　**31.** \$3.50

Pages 287-288　Graphing Calculator Application
1. $(23.82978723, -548.9710276)$
3. $(23.93617021, -548.9959258)$　*For Exercises 5-30,*
answers may vary. Answers given are exact values.
5. $(-8, -5)$　**7.** $(7, 21)$　**9.** $(12, -9)$
11. $(43.5, -1915.25)$　**13.** $(-118, -13,822)$
15. $(-33.875, 86,330.8125)$　**17.** $(0.5625, 5.7890625)$
19. $(-74.4, -128,129)$　**21.** 143.5 m, 71.75 m
23. about \$14.27
25. 393.75 cm by 393.75 cm; 155,039.0625 cm^2
27. 1909.5 m, 1273 m　**29.** 48.75 in.

Page 291　Lesson 8-7
1. yes　**3.** yes　**5.** yes　**7.** no　**9.** no

17.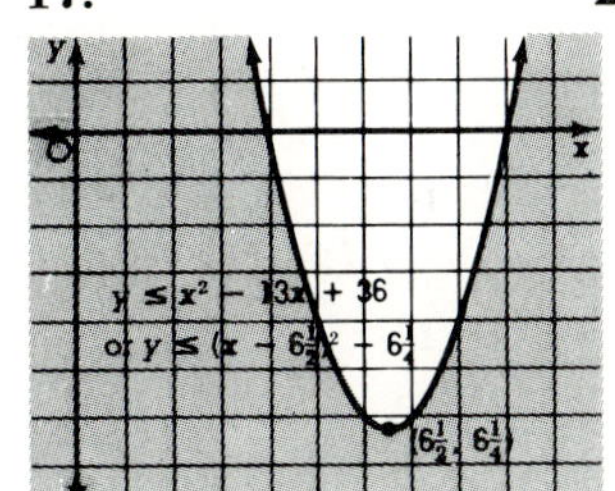　**23.**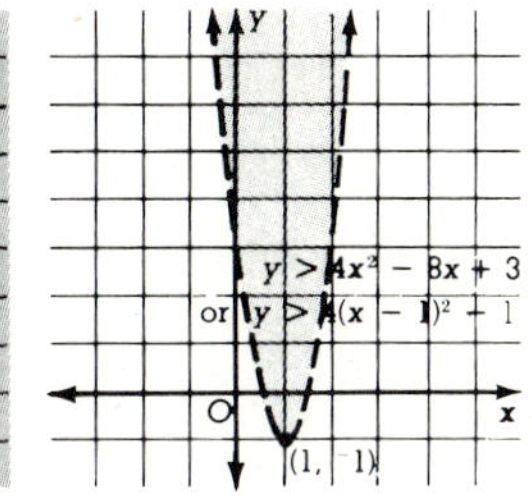

1. one > 0 and one < 0 **3.** both > 0 or both < 0
5. both ≥ 0 or both ≤ 0 **7.** one ≤ 0 and one ≥ 0
9. both ≥ 0 or both ≤ 0 **11.** $\{y|5 \leq y \leq 12\}$
13. $\{x|x > 2 \text{ or } x < -3\}$ **15.** $\{p|p \geq 4 \text{ or } p \leq -6\}$
17. $\left\{b\left|-\frac{3}{2} < b < 2\right.\right\}$ **19.** $\{x|0 \leq x \leq 4\}$
21. $\{t|-6 \leq t \leq 6\}$ **23.** $\{r|-9 \leq r \leq -3\}$
25. $\left\{x\left|x > \frac{5\sqrt{2}}{2} \text{ or } x < \frac{-5\sqrt{2}}{2}\right.\right\}$ **27.** $\left\{n\left|n = \frac{1}{3}\right.\right\}$
29. $\left\{y\left|y > \frac{1}{3} \text{ or } y < -2\right.\right\}$ **31.** $\{m|-\sqrt{3} \leq m \leq \sqrt{3}\}$
33. $\left\{s\left|s > \frac{-1 + \sqrt{3}}{3} \text{ or } s < \frac{-1 - \sqrt{3}}{3}\right.\right\}$
35. $\{x|x < -6 \text{ or } -2 < x < 3\}$
37. $\{x|-7 \leq x \leq -6 \text{ or } x \geq -5\}$
39. $\{x|x < -5 \text{ or } -1 < x < 4 \text{ or } x > 6\}$

Page 297   Chapter Review

1. $f(x) = 3x^2 + 12x + 5$ **3.** $f(x) = -2x^2 + 12x - 9$
5. $x =$ the lesser number, product $= 110x + x^2$
7. $(3, 0); x = 3$ **9.** $(-3, -7); x = -3$
11. $\left(\frac{5}{2}, -\frac{1}{4}\right); x = \frac{5}{2}$ **13.** $(0, -4); x = 0;$ down
15. $(-2, -1); x = -2;$ up   **17.** $(1, 7); x = 1;$ up
19. $\frac{5}{2}, -\frac{5}{2}$ **21.** \$13.75   **27.** $\{x|-2 < x < 4\}$
29. $\{y|y \leq 0 \text{ or } y \geq 12\}$ **31.** all reals

CHAPTER 9   CONICS

Page 303   Lesson 9-1

1. 2 units   **3.** 9 units   **5.** 11 units   **7.** 16 units   **9.** 31.1
units   **11.** $22\frac{7}{10}$ units   **13.** $2\sqrt{53}$ units   **15.** $\sqrt{58}$ units
17. $\sqrt{53}$ units   **19.** $\frac{\sqrt{4594}}{15}$ units   **21.** $\frac{\sqrt{5}}{10}$ units
23. $\sqrt{16.85}$ units   **25.** 1 unit   **27.** $5\sqrt{3}$ units
29. 3 or 11   **31.** 7.1 or 13.1   **33.** $\left(\frac{3}{2}, -\frac{5}{2}\right)$ **35.** $\left(\frac{2}{3}, -\frac{3}{4}\right)$
37. $\left(\frac{3 + \sqrt{2}}{2}, \frac{3 - \sqrt{2}}{2}\right)$ **39.** $2\sqrt{106}$ and $2\sqrt{146}$ units
41. $AB = \sqrt{4 + 16}$ or $\sqrt{20}; AC = \sqrt{16 + 4}$ or $\sqrt{20}$;
Thus, $\triangle ABC$ is isosceles.
43. midpoint of hypotenuse $= \left(\frac{4 + 0}{2}, \frac{1 + 7}{2}\right) = (2, 4)$
distance from $(2, 4)$ to $D = \sqrt{4 + 9} = \sqrt{13}$ units
distance from $(2, 4)$ to $E = \sqrt{4 + 9} = \sqrt{13}$ units
distance from $(2, 4)$ to $F = \sqrt{4 + 9} = \sqrt{13}$ units
47. $(-4, 14)$

Page 306   Lesson 9-2

1. 4   **3.** 16   **5.** $\frac{9}{4}$ **7.** $\frac{49}{4}$ **9.** $y = \frac{1}{10}x^2$
11. $y = (x - 3)^2 + 24$ **13.** $y = 3(x - 4)^2 + 2$
15. $x = \frac{1}{6}y^2$ **17.** $x = (y + 4)^2 + 4$
19. $x = \frac{1}{4}(y - 1)^2 - \frac{13}{4}$
21. $(0, 0); x = 0; \left(0, \frac{3}{2}\right); y = -\frac{3}{2};$ up; 6
23. $(-2, 3); x = -2; \left(-2, 3\frac{1}{4}\right); y = 2\frac{3}{4};$ up; 1

25. $(8, -1); x = 8; \left(8, -\frac{7}{8}\right); y = -\frac{9}{8};$ up; $\frac{1}{2}$
27. $(0, 1); x = 0; \left(0, \frac{5}{4}\right); y = \frac{3}{4};$ up; 1
29. $(2, -3); y = -3; (3, -3); x = 1;$ right; 4
31. $(3, 24); x = 3; \left(3, 24\frac{1}{4}\right); y = 23\frac{3}{4};$ up; 1
33. $(-24, 7); y = 7; \left(-23\frac{3}{4}, 7\right); x = -24\frac{1}{4};$ right; 1
35. $\left(-\frac{13}{4}, 1\right); y = 1; \left(-\frac{9}{4}, 1\right); x = -\frac{17}{4};$ right; 4
37. $(4, 2); x = 4; \left(4, 2\frac{1}{12}\right); y = 1\frac{11}{12};$ up; $\frac{1}{3}$
39. $y = -\frac{1}{4}(x - 2)^2 + 5$ **41.** $y = -\frac{1}{8}(x - 8)^2 + 2$
43. $y = \frac{1}{16}(x - 5)^2 + 1$ **45.** $x = \frac{1}{10}(y + 1)^2 + \frac{1}{2}$
47. $x = -\frac{1}{2}(y - 4)^2 + \frac{1}{2}$ **49.** $x = -\frac{1}{8}(y + 1)^2 + 5$
51. $y = -\frac{1}{6}(x + 7)^2 + 4$ **53.** $x = -\frac{1}{4}(y - 6)^2 + 4$
55. $x = \frac{1}{12}(y + 5)^2 - 1$

Pages 309-310   Lesson 9-3

1. circle   **3.** parabola   **5.** circle   **7.** $(0, 0); 4$
9. $(2, 0); 3$ **11.** $(10, -10); 10$ **13.** $\left(-4, \frac{1}{2}\right); \sqrt{6}$
15. $(-5, 2); \frac{\sqrt{3}}{2}$ **17.** $(2, 0); 3$ **19.** $(0, 8); 8$
21. $(0, 0); 3$ **23.** $(2, 5); 4$ **25.** $(-8, 3); 5$
27. $(-1, -9); 6$ **29.** $(6, 8); 4$ **31.** $(-4, 3); 5$
33. $(2, 0); \sqrt{13}$ **35.** $\left(-\frac{3}{2}, -1\right); \frac{\sqrt{141}}{6}$ **37.** $\left(-\frac{9}{2}, 5\right); 7$
39. $(-1, -2); \sqrt{14}$ **41.** $(-1, 0); \sqrt{11}$
43. $(x - 6)^2 + (y - 2)^2 = 25$ **45.** $x^2 + (y - 3)^2 = 4$
47. $(x + 6)^2 + (y - 2)^2 = \frac{1}{16}$
49. $(x - 1)^2 + (y - 5)^2 = 26$
51. $(x - 2)^2 + (y - 2)^2 = 9$
53. $(x + 3)^2 + (y - 8)^2 = 64$

Pages 315-316   Lesson 9-4

1. $(0, 0);$ H   **3.** $(0, 0);$ V   **5.** $(0, 5);$ H   **7.** $(2, -5);$ H
9. $(-2, 3);$ V   **11.** $(\pm\sqrt{15}, 0)$ **13.** $(0, \pm\sqrt{26})$
15. $(1, 0), (-7, 0)$ **17.** $\frac{x^2}{45} + \frac{y^2}{15} = 1$
19. $\frac{(x - 2)^2}{49} + \frac{(y + 3)^2}{24.5} = 1$ **21.** $\frac{(x + 1)^2}{24} + \frac{(y + 1)^2}{8} = 1$
23. $\frac{x^2}{36} + \frac{y^2}{20} = 1$ **25.** $\frac{(x + 3)^2}{25} + \frac{(y - 3)^2}{9} = 1$
27. $\frac{(x - 4.5)^2}{12.25} + \frac{(y - 6)^2}{3.25} = 1$
29. $(0, 0); (0, \pm\sqrt{21}); 10, 4$ **31.** $(0, 0); (\pm 4, 0); 10, 6$
33. $(0, 0); (\pm\sqrt{7}, 0); 8, 6$ **35.** $(0, 0); (\pm\sqrt{5}, 0); 6, 4$
37. $(0, 0); (\pm 3\sqrt{5}, 0); 18, 12$
39. $(0, 0); (0, \pm\sqrt{6}); 6, 2\sqrt{3}$
41. $(-2, -3); (-2, -3 \pm 2\sqrt{5}); 4\sqrt{10}, 4\sqrt{5}$
43. $(-2, 3); (-2 \pm \sqrt{3}, 3); 2\sqrt{5}, 2\sqrt{2}$
45. $(2, 3); (2 \pm \sqrt{7}, 3); 8, 6$
47. $(2, 2); (4, 2), (0, 2); 2\sqrt{7}, 2\sqrt{3}$
49. $(-1, 3); (2, 3), (-4, 3); 10, 8$ **51.** $\frac{x^2}{169} + \frac{y^2}{25} = 1$
53. $\frac{(x + 2)^2}{16} + \frac{(y - 3)^2}{36} = 1$ **55.** $\frac{(x + 2)^2}{49} + \frac{(y - 2)^2}{9} = 1$
57. $\frac{(x - 3)^2}{32} + \frac{(y - 1)^2}{81} = 1$ **59.** $\frac{(x + 1)^2}{9} + \frac{(y + 1)^2}{34} = 1$

1. ellipse   **3.** hyperbola   **5.** hyperbola   **7.** ellipse

9. $\dfrac{x^2}{6} - \dfrac{y^2}{34} = 1$ **11.** $\dfrac{x^2}{12} - \dfrac{y^2}{18} = 1$

13. $\dfrac{(x-2)^2}{42} - \dfrac{(y-1)^2}{168} = 1$ **15.** $\dfrac{(y+5)^2}{44} - \dfrac{(x-1)^2}{8.8} = 1$

17. $\dfrac{x^2}{9} - \dfrac{y^2}{16} = 1$ **19.** $\dfrac{(x-1)^2}{4} - \dfrac{(y+2)^2}{5} = 1$

21. $\dfrac{(y+5.5)^2}{6.25} - \dfrac{x^2}{6} = 1$ **23.** $(\pm 3, 0)$; $(\pm\sqrt{34}, 0)$; $\pm\dfrac{5}{3}$

25. $(\pm 6, 0)$; $(\pm\sqrt{37}, 0)$; $\pm\dfrac{1}{6}$

27. $(0, \pm 9)$; $(0, \pm\sqrt{106})$; $\pm\dfrac{9}{5}$

29. $(\pm 2, 0)$; $(\pm\sqrt{13}, 0)$; $\pm\dfrac{3}{2}$

31. $(\pm 9, 0)$; $(\pm\sqrt{117}, 0)$; $\pm\dfrac{2}{3}$ **33.** $(\pm 3, 0)$; $(\pm 5, 0)$; $\pm\dfrac{4}{3}$

35. $(\pm 2, 0)$; $(\pm\sqrt{29}, 0)$; $\pm\dfrac{5}{2}$

37. $(\pm\sqrt{2}, 0)$; $(\pm\sqrt{3}, 0)$; $\pm\dfrac{\sqrt{2}}{2}$

39. $(0, \pm 6)$; $(0, \pm 3\sqrt{5})$; ± 2

41. $(0, -3), (-12, -3)$; $(-6 \pm 3\sqrt{5}, -3)$; $\pm\dfrac{1}{2}$

43. $(-2, 0), (-2, 8)$; $(-2, -1), (-2, 9)$; $\pm\dfrac{4}{3}$

45. $(4 \pm 2\sqrt{5}, -2)$; $(4 \pm 3\sqrt{5}, -2)$; $\pm\dfrac{\sqrt{5}}{2}$

47. $(1, -3 \pm 2\sqrt{6})$; $(1, -3 \pm 4\sqrt{2})$; $\pm\sqrt{3}$

49. $\dfrac{x^2}{1} - \dfrac{y^2}{16} = 1$ **51.** $\dfrac{(y-2)^2}{36} - \dfrac{(x+2)^2}{64} = 1$

53. $\{x \mid x \text{ is real}, x \neq 0\}$ **57.** $\dfrac{(x-1)^2}{9} - \dfrac{(y+2)^2}{25} = 1$

1. circle   **3.** parabola   **5.** ellipse   **7.** hyperbola

9. $y = \dfrac{1}{8}x^2$; parabola   **11.** $x^2 + y^2 = 27$; circle

13. $\dfrac{x^2}{4} + \dfrac{(y+1)^2}{3} = 1$; ellipse   **15.** $\dfrac{y^2}{16} - \dfrac{x^2}{8} = 1$;

hyperbola   **17.** $x^2 + (y-4)^2 = 5$; circle

19. $y = -1\left(x - \dfrac{1}{2}\right)^2 + \dfrac{9}{4}$; parabola

21. $\dfrac{(x-3)^2}{25} + \dfrac{(y-1)^2}{9} = 1$; ellipse

23. $\dfrac{(y+4)^2}{2} - \dfrac{(x+1)^2}{6} = 1$; hyperbola

25. $y = -3x$, $y = 3x$, intersecting lines   **27.** $y = 0$, $y = 1$, parallel lines

7. $y = \pm\sqrt{100 - x^2}$; circle

9. $y = \pm 5\sqrt{\dfrac{x^2}{49} - 1}$; hyperbola

11. $y = \pm\sqrt{9 - \dfrac{x^2}{3}}$; ellipse

13. $y = \pm 2\sqrt{3x^2 - 6}$; hyperbola

15. $y = -6 \pm\sqrt{16 - 0.4(x - 3)^2}$; ellipse

17. $y = -2.5 \pm\sqrt{x + 8}$; parabola

19. $y = -11 \pm\sqrt{1.25x^2 - 5}$; hyperbola

21. $y = 7 \pm\sqrt{20 - 2.5(x + 15)^2}$; ellipse

23. $y = -9 \pm\sqrt{144 - 2x}$; parabola

25. $y = -\dfrac{1}{2} \pm\sqrt{\dfrac{11}{3} + \dfrac{2}{3}\left(x - \dfrac{5}{2}\right)^2}$; hyperbola

1. $(\pm 3.5, 2)$ **3.** $(4, 12), (-1, -3)$ **5.** $(3, 0), (-5, -4)$

7. $(0, -1), (-3, 2)$ **9.** $(\pm 5.2, 6)$ **11.** no solution

13. no solution   **15.** $(-2, 0), (2, 4)$

17. $(3, 3), (-1, -1)$ **19.** $(3, 7), (8, 4)$

21. $(\pm 1, 5), (\pm 1, -5)$ **23.** $(\pm 2, 1.7), (\pm 2, -1.7)$ **25.** no solution   **27.** no solution   **29.** $(\pm 8, 0)$

31. $(\pm 4, 3), (\pm 4, -3)$ **33.** no solution   **35.** no solution

1. $(\pm 3.5, 2)$ **3.** $(4, 12), (-1, -3)$ **5.** $(3, 0), (-5, -4)$

7. $(0, -1), (-3, 2)$ **9.** $(\pm 5.2, 6)$ **11.** no solution

13. no solution   **15.** $(-2, 0), (2, 4)$

17. $(3, 3), (-1, -1)$ **19.** $(3, 7), (8, 4)$

21. $(\pm 1, 5), (\pm 1, -5)$ **23.** $(\pm 2, 1.7), (\pm 2, -1.7)$ **25.** no solution   **27.** no solution   **29.** $(\pm 8, 0)$

31. $(\pm 4, 3), (\pm 4, -3)$ **33.** no solution   **35.** no solution

1. $(2, \pm 2\sqrt{3})$ **3.** $(2, 2), (-2, -2)$ **5.** $(3, 0), (-5, -4)$

7. $(-3, 2), (0, -1)$ **9.** $(\pm 3\sqrt{3}, 6)$ **11.** no solutions

13. no solutions   **15.** $(-2, 0), (2, 4)$ **17.** $(6, 9), (2, 1)$

19. $(3, 7), (8, 4)$ **21.** $(1, \pm 5), (-1, \pm 5)$

23. $(2, \pm\sqrt{3}), (-2, \pm\sqrt{3})$ **25.** no solutions   **27.** no solutions   **47.** $(0, \sqrt{2}), (-2, -2)$

1. $2\sqrt{53}$ **3.** $\sqrt{16.85}$ **5.** $\left(1, \dfrac{3}{2}\right)$ **7.** $(0.25, 0.5)$

9. $(0, 0)$; $x = 0$; $(0, 1)$; $y = -1$; up; 4

11. $(4, 8)$; $y = 8$; $(3, 8)$; $x = 5$; left; 4   **13.** $(3, -7)$; 9

15. $(0, 0)$; $(0, \pm 2\sqrt{2})$; 8, $4\sqrt{2}$ **17.** $(0, 0)$; $(\pm\sqrt{7}, 0)$; 8, 6

19. $(1, -6 \pm 2\sqrt{5})$; $(1, -6 \pm 3\sqrt{5})$; $\pm\dfrac{2\sqrt{5}}{5}$

21. parabola   **23.** ellipse   **25.** $(-1.6, 2.6), (2.6, -1.6)$

27. $(-2, 0), (2, 4)$

CHAPTER 10   POLYNOMIAL FUNCTIONS

1. no   **3.** no   **5.** no   **7.** yes   **9.** no   **11.** yes   **13.** 4

15. 1   **17.** -4 **19.** $-\dfrac{11}{2}$ **21.** $\dfrac{19}{3}$ **23.** -5 **25.** -6

27. 28   **29.** 2   **31.** 13   **33.** 144   **35.** 35   **37.** -12

39. 74   **41.** $5x + 5h - 10$

43. $x^2 + 2hx + h^2 - 7x - 7h + 4$

45. $\dfrac{4}{3}x^3 + 4hx^2 + 4h^2x + \dfrac{4}{3}h^3 - 1$

47. $x^3 + 3hx^2 + 3h^2x + h^3 - 4x^2 - 8hx - 4h^2$

49. $9x + 6$ **51.** $3x^3 - 9x^2 + 9x - 3$

53. $5x^2 - 10x + \dfrac{5}{2}$ **55.** $-5x - 12$ **57.** $-x^2 + x + 2$

59. $-\dfrac{4}{5}x^3 - \dfrac{77}{10}x^2 - \dfrac{51}{5}x - \dfrac{19}{10}$ **61.** $\dfrac{(x+2)^2}{6}$

63. $\dfrac{2(x+2)^2}{3(x-2)^2}$ **65.** $\dfrac{(x-1)^2(x^2 + x + 1)^2(x+3)}{(x+1)^4}$

1. $x^2 - 3x + 1 = (x - 1)(x - 2) - 1$ **3.** $x^3 - 8x^2 + 2x - 1 = (x^2 - 9x + 11)(x + 1) - 12$ **5.** $x^5 + x^4 + 2x - 1 = (x^4 + 3x^3 + 6x^2 + 12x + 26)(x - 2) + 51$ **7.** $x^5 + 32 = (x^4 - 2x^3 + 4x^2 - 8x + 16)(x + 2) + 0$ **9.** $2x^3 + 8x^2 - 3x - 1 = (2x^2 + 12x + 21)(x - 2) + 41$ **11.** $x^4 - 16 = (x^3 + 2x^2 + 4x + 8)(x - 2) + 0$ **13.** $4x^4 + 3x^3 - 2x^2 + x + 1 = (4x^3 + 7x^2 + 5x + 6)(x - 1) + 7$
15. $3x^3 + 2x^2 - 4x - 1 = \left(3x^2 + \frac{1}{2}x - \frac{17}{4}\right)\left(x + \frac{1}{2}\right) + \frac{9}{8}$
17. 37, 7 **19.** $-46, -31$ **21.** 314, 79 **23.** 461, -94
25. $x + 2, x - 2$ **27.** $x + 1, x + 2$ **29.** $x - 1, x + 2$
31. $x - 2, x^2 + 2x + 4$ **33.** $x - 1, x + 1, x^2 + 1$
35. -17 **37.** $\frac{25}{2}$ **39.** 5, -8

Page 356   Lesson 10-3

1. 3 or 1; 1 **3.** 4, 2, or 0; 0 **5.** 2 or 0; 2 or 0 **7.** 5, 3, or 1; 1 **9.** 1; 3 or 1 **11.** 3, -5, $-\frac{5}{2}$ **13.** 3, 3, -2, $\frac{1}{2}$, $\frac{2}{3}$
15. $\frac{2}{3}$, 5, -1 **17.** 3 or 1; 0; 0 or 2 **19.** 2 or 0; 2 or 0; 0, 2, or 4 **21.** 2 or 0; 1; 2 or 4 **23.** 2 or 0; 0; 2 or 4; 0 is a zero. **25.** 1; 0; 2 **27.** 5, 3, or 1; 5, 3, or 1; 0, 2, 4, 6, or 8
29. $3 + i, 3 - i, 4$ **31.** $1 + 2i, 1 - 2i, -4$
33. $-2, -2 + 3i, -2 - 3i$ **35.** $-\frac{3}{2}, 1 + 4i, 1 - 4i$
37. $3 - 2i, 3 + 2i, -1, 1$ **39.** $f(x) = x^3 - 4x^2 + 6x - 4$
41. $f(x) = x^4 - 4x^3 + 4x^2 + 4x - 5$
43. $f(x) = x^4 + 2x^3 + 7x^2 + 30x + 50$
45. $f(x) = x^6 - 2x^5 + 15x^4 - 26x^3 + 62x^2 - 72x + 72$
47. $-4, -1, 1$ **49.** $-6, 3, 7$ **51.** $-4, -3, 1, 2$

Page 360   Lesson 10-4

1. $\pm 1, \pm 2$ **3.** $\pm 1, \pm 2, \pm 3, \pm 6$ **5.** $\pm 1, \pm 2, \pm 4, \pm 8$
7. ± 1 **9.** $\pm 1, \pm 2, \pm 4, \pm 5, \pm 10, \pm 20$
11. $\pm 1, \pm \frac{1}{2}, \pm \frac{1}{3}, \pm \frac{1}{6}$ **13.** $\pm 1, \pm 2, \pm 4, \pm \frac{1}{3}, \pm \frac{2}{3}, \pm \frac{4}{3}$
15. $-2, -4, 7$ **17.** 3, 3, $-\frac{1}{2}$ **19.** 0, 3 **21.** $-2, -4$
23. 1, -1 **25.** $-1, -2, 5$ **27.** -6 **29.** 0, $-\frac{1}{3}, \frac{1}{2}, -\frac{1}{2}$
31. 0, 2, -2 **33.** $\frac{2}{3}, \frac{-3 \pm \sqrt{17}}{4}$ **35.** 0, $\frac{4}{5}, \frac{5 \pm i\sqrt{3}}{2}$
37. 3, $\frac{2}{3}, \frac{3}{4}, -\frac{1}{2}$ **39.** 3, $\frac{2}{3}, -\frac{2}{3}, \frac{-3 \pm \sqrt{13}}{2}$
41. height = 2 m, width = 4 m, length = 9 m

Page 364   Lesson 10-5

1. -1.3 **3.** 0.6 **5.** 1.7, -1.3, -2.4 **7.** 1.4 **9.** -0.8
11. 0.1, 2.5 **13.** 1, 0.8, -1.4 **15.** -1 **17.** -2.4
19. 0.4, 1.3, 2.6, 4.7

Page 367   Graphing Calculator Application

21. 1.71 **23.** -1.34 **25.** -0.75 **27.** 1.43
29. $-1, 0.72$ **31.** $-4.08, 1.54$ **33.** $-2.38, 0.27$
35. $-1.38, 0.82, 1$ **37.** $-1.73, -1, 1, 1.73$
39. $-3.60, -1.62, -0.66, 0.62, 1.31$

Page 370   Lesson 10-6

3.
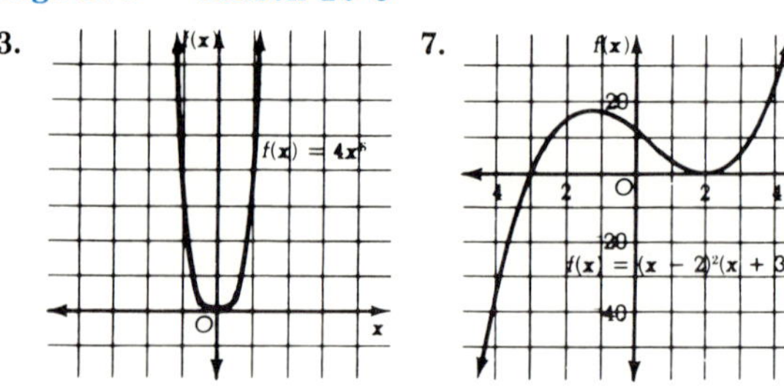

7.

11.
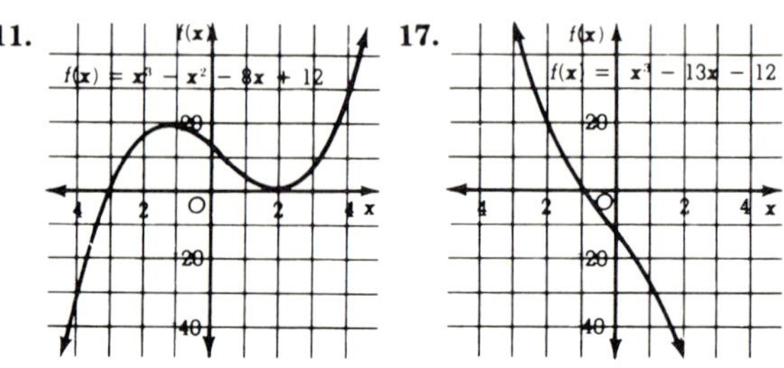

17.

Page 374   Lesson 10-7

1. $-1, -5, -3$ **3.** 1, 9, 1 **5.** 4, 4, 2 **7.** 10, 10, -6
9. 30, -34, 2 **11.** 4, 9, 3 **13.** 4, 4 **15.** 63, 27
17. 129, 360 **19.** $2x - 5, 2x - 2$
21. $4x^2 - 4x + 4, 2x^2 + 5$
23. $-x^4 + 2x^2 - 9, x^4 + 16x^2 + 63$ **25.** $-1, -1$
27. 50, 2 **29.** $-7, 9$ **31.** 9 **33.** 8 **35.** 12 **37.** -9
39. $\frac{9}{4}$ **41.** $6 + 4\sqrt{2}$ **43.** $9x^2$ **45.** $9x^2 - 18x + 9$
47. $\{(3, 6), (4, 4), (6, 6), (7, -8)\}$; does not exist
49. $\{(1, -8), (-1, -2), (5, -8), (9, -3)\}$; does not exist

Pages 378-379   Lesson 10-8

1. $\{(1, 3), (4, 2), (5, 1)\}$; yes
3. $\{(8, 3), (-2, 4), (-3, 5)\}$; yes
5. $\{(1, -3), (4, 2), (8, 7)\}$; yes
7. $\{(3, -5), (7, -2), (3, 0), (11, 6)\}$; no **9.** $y = \frac{1}{2}x$
11. $g^{-1}(x) = -\frac{1}{6}x - \frac{5}{6}$ **13.** $x = 0$ **15.** $g^{-1}(x) = 2x - 8$
17. $f^{-1}(x) = -\frac{1}{3}x - \frac{1}{3}$ **19.** $y = \pm\sqrt{x + 4}$

21.
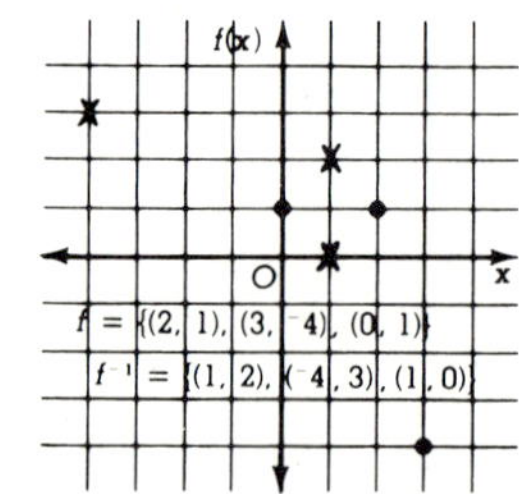

25.
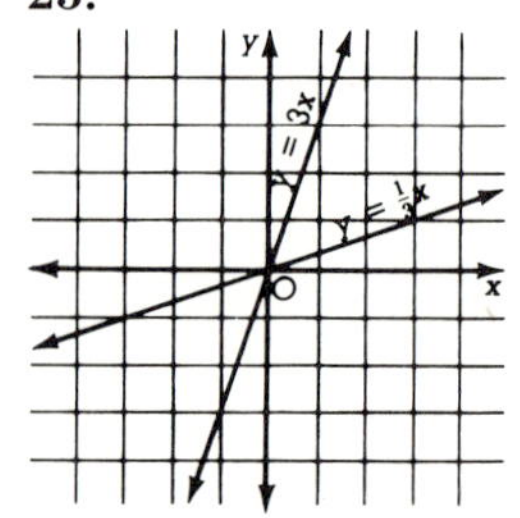

33. no **35.** yes **37.** yes **39.** yes **41.** no **43.** yes
45. no **47.** 3 **49.** a **51.** $a + 1$

1. -13 **3.** $m^2 + 7m + 3$ **5.** $3h$

7. $x^2 + 5x + 6 = (x + 4)(x + 1) + 2$ **9.** $x^5 + 8x^3 + 2 = (x^4 - 2x^3 + 12x^2 - 24x + 48)(x + 2) - 94$

11. $(x + 2)(x - 1)(x + 1)$ **13.** $2x + 1, x - 4$

15. 3 or 1; 1; 2 or 0   **17.** 1; 0; 2   **19.** $2 + i, 2 - i, 3$

21. $2, \dfrac{-2 \pm i\sqrt{5}}{3}$ **23.** $-\dfrac{1}{2}, 3, 4$ **25.** $-3, 2, 4$

27. $5, \dfrac{1}{3}, \dfrac{3}{2}$ **29.** $1, 1, -\dfrac{2}{3}, \dfrac{1}{4}$ **31.** -0.8 **33.** $-1.3, -0.2$

39. $25; 25; 6x + 7$ **41.** $2; 15; x^2 - 6x + 11$

43. $\{(2, 1), (6, 6), (1, 2)\}; \{(2, 3), (3, 2), (-1, -1)\}$

49. yes

CHAPTER 11   RATIONAL POLYNOMIAL EXPRESSIONS

1. $6y; \dfrac{7}{3x}$ **3.** $2a; \dfrac{19a}{21b}$ **5.** $4x^5y; -2xy^2$ **7.** $x^4y^2; 16y^2$

9. $m + 5; \dfrac{1}{2}$ **11.** $\dfrac{9y}{7x}$ **13.** $\dfrac{2}{x + y}$ **15.** $\dfrac{(x - 3)^2}{(x + 4)^2}$

17. $x - y$ **19.** $\dfrac{a^2}{b^2} \cdot \dfrac{a^2}{b^2}$ **21.** $\dfrac{3h}{h - 1} \cdot \dfrac{1}{h - 2}$ **23.** $\dfrac{a^2}{14}$

25. $\dfrac{x + 1}{1 - x^2}; 1 - x$ **27.** $\dfrac{a^2 - b^2}{a^2 + 2ab + b^2}; \dfrac{a - b}{a + b}$

29. $\dfrac{x^2 + x - 12}{2x^2 + 8x}; \dfrac{x - 3}{2x}$ **31.** $\dfrac{x^2 + 7x + 12}{x^2 - x - 20}; \dfrac{x + 3}{x - 5}$

33. $\dfrac{2a^2 + 3a + 1}{a^2 + 2a + 1}; \dfrac{2a + 1}{a + 1}$ **35.** $\dfrac{3a^2}{2cb}$ **37.** b **39.** $\dfrac{-ad^2}{2}$

41. cd^2x **43.** $\dfrac{5x^3}{3y^2}$ **45.** $\dfrac{1}{2}$ **47.** 4 **49.** 1

51. $-y(x + y)$ **53.** $\dfrac{2(a + 5)}{(a - 2)(a + 2)}$ **55.** $\dfrac{-y}{(w^2 - y^2)}$

57. 11 **59.** $\dfrac{x - 2}{x + 2}$ **61.** $(a - b)^2$ **63.** $\dfrac{w - 3}{w - 4}$

65. $2(x + y)$ **67.** $\dfrac{a(a + 2)}{a + 1}$ **69.** $\dfrac{2y(y - 2)}{3(y + 2)}$ **71.** -1

73. 1 **75.** $\dfrac{x(x + 2)}{3(x + 3)(x - 1)}$ **77.** $\dfrac{1}{a - 2}$

1. 756 **3.** 2000 **5.** $14a^2b$ **7.** $x(x - 2)(x + 2)$

9. $(x + 1)^2(x - 3)(x + 3)$ **11.** $xy(x - 8)(y - 8)$

13. $\dfrac{31}{12a}$ **15.** $\dfrac{5 + 7a}{a}$ **17.** $\dfrac{x}{y - x}$ **19.** $\dfrac{-12x + 21xy - 4y}{6x^2y}$

21. $\dfrac{3(x - 1)}{2(x - 3)(x + 3)}$ **23.** $\dfrac{5a - 13}{(a - 2)(a - 3)}$ **25.** $\dfrac{3(2x + 11)}{(x - 5)(x + 5)}$

27. $\dfrac{7x + 38}{2(x - 7)(x + 4)}$ **29.** $\dfrac{7y + 11}{(y - 5)(y + 1)(y + 3)}$ **31.** 0

33. $\dfrac{2x^2 + x - 4}{(x - 1)(x - 2)}$ **35.** $\dfrac{110a - 423}{90a}$ **37.** $\dfrac{13}{y - 8}$ **39.** $\dfrac{y + 4}{y - 8}$

41. $\dfrac{x(x - 9)}{(x + 3)(x - 3)}$ **43.** $\dfrac{6}{y - 4}$ **45.** $\dfrac{-4x^2 - 5x - 2}{(x + 1)^2}$

47. $\dfrac{3(3m^2 - 14m + 27)}{(m + 3)(m - 3)^2}$ **49.** $\dfrac{3}{10}$ **51.** y **53.** $\dfrac{2(x + 5)}{x^2(2x + 1)}$

55. $\dfrac{a + 7}{a + 2}$ **57.** $\dfrac{m^2 + 6m + 20}{4m - 1}$ **59.** $\dfrac{2x^2 - x + 2}{4}$

61. $\dfrac{y^2 - 3y - 7}{2y - 6}$ **63.** $\dfrac{n - 1}{n + 2}$ **65.** $\dfrac{2}{2x - 5}$

67. $\dfrac{3x^2 + 7x + 2}{(x + 5)(x - 5)(x - 2)}$ **69.** $\dfrac{7x^3 + 16x^2 + 24x + 10}{x(2x - 1)(2x + 1)(x + 2)}$

1. $2x; x \neq 0$ **3.** $8s; s \neq 0$

5. $(x + 5)(x - 3); x \neq 3, x \neq -5$ **7.** $7(2 + m); m \neq 2$

9. $0; 3$ **11.** none; $-3, \dfrac{1}{6}$ **13.** none; 2 **15.** none; 4

17. none; 5 **19.** none; 54 **21.** $0; -6, 1$

23. $\pm 1, -2; 0$ **25.** $0; 2, 6$ **27.** $-\dfrac{3}{2}; -\dfrac{21}{20}$ **29.** $0; -6, \dfrac{3}{2}$

31. 1; all reals except 1 **33.** $\pm\dfrac{3}{2}; \dfrac{-3 \pm 3\sqrt{2}}{2}$

35. $2, -1; \dfrac{1 \pm \sqrt{145}}{4}$ **37.** $\pm 2; 14$ **39.** $0, -\dfrac{1}{2}; \pm\dfrac{\sqrt{6}}{2}$

41. $0, 7; -14$ **43.** ± 2; no solution   **45.** $4, 0; -3$

47. $-2, -3; 0, -4$ **49.** ± 1; no solution

1. $2\dfrac{2}{9}$ hours   **3.** $25\dfrac{1}{5}$ hours   **5.** 20 days   **7.** 19   **9.** $\dfrac{7}{13}$

11. 550 mph   **13.** 3 km/h   **15.** $3\dfrac{3}{4}$ minutes   **17.** \$1800 at 6%   **19.** $166\dfrac{2}{3}$ ml of 55% solution, $833\dfrac{1}{3}$ ml of 25% solution   **21.** $1\dfrac{2}{7}$ hours

1. $x = 1, y = 0$ **3.** $x = 3, y = 0$ **5.** $x = 6, y = 0$

7. $x = 1, x = -5, y = 0$ **9.** $x = 1, x = 4, y = 0$

For Exercises 11-27, equations for asymptotes are given.

11. $x = -2, y = 0$ **13.** $x = 0, y = 0$ **15.** $x = 5, y = 1$

17. $x = 6, y = 0$ **19.** $x = 3, y = 0$

21. $x = 2, x = -1, y = 0$ **23.** $x = 4, y = 0$

25. $x = 1, x = -3, y = 0$ **27.** $x = 2, x = -2, y = 1$

23. 0 **25.** $\dfrac{1}{4}$ **27.** 0 **29.** $-\dfrac{1}{3}$

1. direct; $\dfrac{1}{4}$ **3.** direct; -4 **5.** direct; 5 **7.** direct; $\dfrac{4}{3}$

9. inverse; 9 **11.** direct; 4 **13.** 36 **15.** 1.125 **17.** 15

19. $\dfrac{84}{11}$ **21.** $\dfrac{121}{10}$ **23.** 24 **25.** 4 **27.** 118.5 km

29. $11\dfrac{2}{3}$ kg   **31.** $3\dfrac{23}{55}$ hours   **33.** 42 pounds   **35.** \$6880

37. $2\dfrac{2}{3}$ hours   **39.** $\dfrac{343}{81}$

1. $-\dfrac{4bc}{33a}$ **3.** $\dfrac{4cd}{a}$ **5.** $6b(a - b)$ **7.** $\dfrac{2}{n - 3}$ **9.** $\dfrac{x + 2}{x - 3}$

11. $\dfrac{7(x - 4)}{x - 5}$ **13.** $\dfrac{19}{3y}$ **15.** $\dfrac{14y - 14x - 9}{y^2 - x^2}$ **17.** $x - 1$

19. $\dfrac{10}{9}$ **21.** 3 **23.** $\dfrac{3}{2}$ **25.** 8 and 10   **27.** The asymptotes are $x = -3$ and $y = 1$.   **29.** $-\dfrac{5}{3}$ **31.** $-\dfrac{75}{2}$

CHAPTER 12   EXPONENTIAL AND LOGARITHMIC FUNCTIONS

1. 1.6 **3.** 0.8 **5.** 2.7 **7.** 4.7 **9.** 135.3 **11.** 3.3

13. $2^{4\sqrt{5}}$ **15.** 2^3 **17.** 3 **19.** -5 **21.** -3 **23.** $-\dfrac{3}{2}$

25. 2.8 **27.** 0.7 **29.** 0.5 **31.** 2.4 **33.** 64 **35.** $5^{4\sqrt{3}}$
37. $2^{3\sqrt{7}}$ **39.** $2^{3\sqrt{3}+4\sqrt{5}}$ **41.** $y^{4\sqrt{5}}$ **43.** y^9 **45.** m^2p^2
47. $x^{2\sqrt{3}} + 2x^{\sqrt{3}}y^{\sqrt{2}} + y^{2\sqrt{2}}$ **49.** 3 **51.** -2 **53.** 2
55. -7 **57.** 3 **59.** -9 **61.** 2, -2 **63.** $-1, -3$
69. The graphs are reflections over the y-axis.

Page 424   Lesson 12-2

1. $\log_3 27 = 3$ **3.** $\log_{10} 1000 = 3$ **5.** $\log_2 \frac{1}{8} = -3$
7. $4^3 = 64$ **9.** $10^{-1} = 0.1$ **11.** $9^{\frac{3}{2}} = 27$
13. $\log_3 81 = 4$ **15.** $\log_5 125 = 3$ **17.** $\log_4 \frac{1}{16} = -2$
19. $\log_2 \frac{1}{16} = -4$ **21.** $\log_3 \sqrt{3} = \frac{1}{2}$ **23.** $\log_{36} 216 = \frac{3}{2}$
25. $2^5 = 32$ **27.** $11^2 = 121$ **29.** $5^0 = 1$
31. $\left(\frac{1}{2}\right)^{-4} = 16$ **33.** $10^{-1} = \frac{1}{10}$ **35.** $27^{\frac{1}{3}} = 3$ **37.** 3
39. 2 **41.** -3 **43.** -4 **45.** $\frac{3}{2}$ **47.** -3 **49.** 7
51. 36 **53.** -4 **55.** 3 **57.** $\frac{1}{25}$ **59.** $\frac{1}{2}$ **61.** $\frac{1}{3}$ **63.** 3
65. 7 **67.** 64 **69.** 27

Pages 427-428   Lesson 12-3

1. 2 **3.** 2 **5.** x **7.** 4 **9.** 1 **11.** ± 2 **13.** 3 **15.** 7
17. 5 **19.** 7 **21.** 1 **23.** 2.5 **25.** ± 8 **27.** 1, -10 **29.** 1
35. $\log_4 4 + \log_4 16 \overset{?}{=} \log_4 64$
$\qquad 1 + 2 = 3$ ✔
37. $\log_2 32 - \log_2 4 \overset{?}{=} \log_2 8$
$\qquad 5 - 2 = 3$ ✔
39. $\log_3 27 \overset{?}{=} 3 \log_3 3$
$\qquad 3 \overset{?}{=} 3(1)$
$\qquad 3 = 3$ ✔
41. $\frac{1}{2} \log_3 81 \overset{?}{=} \log_3 9$
$\qquad \frac{1}{2}(4) \overset{?}{=} 2$
$\qquad\qquad 2 = 2$ ✔
43. $\log_2 8 \cdot \log_8 2 \overset{?}{=} 1$
$\qquad 3 \cdot \frac{1}{3} \overset{?}{=} 1$
$\qquad\qquad 1 = 1$ ✔
45. $\log_{10} [\log_3 (\log_4 64)] \overset{?}{=} 0$
$\qquad \log_{10} (\log_3 3) \overset{?}{=} 0$
$\qquad\qquad \log_{10} 1 \overset{?}{=} 0$
$\qquad\qquad\qquad 0 = 0$ ✔
47. $\log_3 81 \overset{?}{=} \frac{4}{3} \log_2 8$
$\qquad 4 \overset{?}{=} \frac{4}{3}(3)$
$\qquad 4 = 4$ ✔
49. 6, -5 **51.** 9

Page 432   Lesson 12-4

1. 21 **3.** 3 **5.** 72 **7.** $\log_3 x + \log_3 y$
9. $4 \log_2 m + \log_2 y$ **11.** $\frac{1}{2} \log_b x - \log_b p$
13. $\log_3 5 + \frac{1}{3} \log_3 a$ **15.** $\log_2 a + \frac{1}{2} \log_2 x$ **17.** 2
19. 1.2252 **21.** 1.5662 **23.** 1.4421 **25.** 2.2252
27. 1.5579 **29.** 0.2252 **31.** 2 **33.** 24 **35.** 343
37. 6 **39.** 14 **41.** 3 **43.** 2 **45.** 6 **47.** 3 **49.** 5
51. $\frac{1}{3}$ **53.** $\frac{y}{3} + 1$ **55.** $\frac{x^4}{2}$

Pages 437-438   Lesson 12-5

1. 2 **3.** 1.6839 **5.** $0.6839 - 3$ **7.** 483,000
9. 1; 1.6767 **11.** 0; 0.6637 **13.** -1; $0.3201 - 1$
15. 1; 1.7404 **17.** 1; 35.70 **19.** -2; 0.0688
21. 4; 39,400 **23.** -1; 0.618 **25.** 7.41, 7.42

27. 0.000746, 0.000747 **29.** 4.17, 4.18 **31.** 9520, 9530
33. 1.7649 **35.** $0.4771 - 1$ **37.** $0.5855 - 2$ **39.** 4.7973
41. 4.27 **43.** 0.609 **45.** 0.00463 **47.** 556,000
49. 0.7220 **51.** 2.7816 **53.** $0.8581 - 1$ **55.** $0.7295 - 3$
57. 0.9480 **59.** 1.2965 **61.** 2.3606 **63.** $0.2193 - 1$
65. 3.063 **67.** 5306 **69.** 0.02164 **71.** 0.9352
73. 601.4 **75.** 42,450 **77.** 0.0008264 **79.** 394,300

Page 441   Lesson 12-6

1. 1.5119; 1; 0.5119 **3.** 5.7874; 5; 0.7874 **5.** -1.3261;
-2; 0.6739 **7.** 765.1 **9.** 103,100 **11.** 0.05169
13. 1.9720; 1 **15.** 1.6698; 1 **17.** -0.0830; -1
19. -2.9488; -3 **21.** 2.3293; 2 **23.** -1.0442; -2
25. The characteristic is the greatest integer less than the
logarithm. **27.** 77.88 **29.** 384,300 **31.** 0.6416
33. 0.05826 **35.** 0.00003269 **37.** 105.5

Pages 444-445   Lesson 12-7

1. $\frac{\log 55}{\log 3}$ **3.** $\frac{\log 74}{2 \log 7}$ **5.** $\frac{\log 144}{\log 6}$ **7.** $\frac{\log 12}{\log 3}$ **9.** $\frac{-1}{\log 2}$
11. $\frac{\frac{1}{2} \log 13}{\log 3}$ or $\frac{\log 13}{2 \log 3}$ **13.** 3.6479 **15.** 1.1059
17. 2.7736 **19.** 2.2619 **21.** -3.3222 **23.** 1.1674
25. 1.771 **27.** 2.230 **29.** 1.338 **31.** 2.387 **33.** 1.941
35. 2.306 **37.** 3.9839 **39.** 3.8394 **41.** 4.8363
43. 2.8446 **45.** ± 1.1024 **47.** 3.1501 **49.** -2.1507
51. 2.4527 **53.** 3.2618 **55.** 2.9169 **57.** 2.3475
59. $18\frac{1}{2}$ years **61.** 75 years ago **63.** Jan's

Page 447   Graphing Calculator Application

15. 3.32 **17.** 9.17 **19.** 0.47 **21.** -0.28 **23.** 0.17
25. 2, 4, -0.77

Pages 451-452   Lesson 12-8

1. 1.3863 **3.** 13.73 years **5.** -0.0770 **7.** 2.94 hours
9. 3.08 years **11.** 28.78 years; 57.56 years **13.** 6.57
years **15.** 229.07 days **17.** -0.000385 **19.** 9.37%

Pages 455-456   Chapter Review

1. -1 **3.** $-\frac{7}{4}$ **5.** 81 **7.** $\log_7 343 = 3$ **9.** $\log_4 1 = 0$
11. $4^3 = 64$ **13.** $6^{-2} = \frac{1}{36}$ **15.** 3 **17.** $\frac{1}{4}$ **19.** 7
21. 3 **23.** 3 **25.** 7 **27.** 8 **29.** $-4, 3$ **31.** 1.5165
33. 2.2618 **35.** 48 **37.** 3 **39.** no solution **41.** 4
43. 2.7868 **45.** $0.6139 - 1$ **47.** 6150 **49.** 0.007475
51. -2.5029 **53.** 1.1229 **55.** 19,170 **57.** 0.0003609
59. 5.7286 **61.** 4.2448 **63.** ± 2.2452 **65.** -1.8928
67. -4.8188 **69.** 1.2628

CHAPTER 13   SEQUENCES AND SERIES

Pages 461-462   Lesson 13-1

1. 4, 7, 10, 13, 16 **3.** 16, 14, 12, 10, 8 **5.** $\frac{3}{4}, \frac{1}{2}, \frac{1}{4}, 0, -\frac{1}{4}$

7. 2.3, 3.9, 5.5, 7.1, 8.7   **9.** $-\frac{1}{3}, -1, -\frac{5}{3}, -\frac{7}{3}, -3$

11. $-4.2, -5.5, -6.8, -8.1, -9.4$ **13.** 17, 21, 25, 29

15. $-13, -18, -23, -28$ **17.** $\frac{7}{2}, \frac{9}{2}, \frac{11}{2}, \frac{13}{2}$

19. $-\frac{11}{4}, -\frac{13}{4}, -\frac{15}{4}, -\frac{17}{4}$ **21.** 6.84, 9.14, 11.44, 13.74

23. -93 **25.** 41   **27.** $-14\frac{1}{4}$ **29.** 313   **31.** 735

33. $-53i$ **35.** -50 **37.** -37 **39.** $\frac{23}{6}$ **41.** 30

43. 27   **45.** $-\frac{13}{3}, -\frac{2}{3}$ **47.** 5, 8, 11, 14, 17

49. 56, 42, 35   **51.** 19, $16\frac{1}{2}$, 14, $11\frac{1}{2}$ **53.** 33

55. 780 feet   **57.** 2, 9, 16

Page 466   Lesson 13-2

1. 116   **3.** 10,100   **5.** 375   **7.** 240   **9.** 5050   **11.** 632.5

13. 702   **15.** 135   **17.** 1210   **19.** 287   **21.** -420

23. 735   **25.** -220 **27.** 387   **29.** $\frac{45}{2}$ **31.** 6, 36, 66

33. 1, 5, 9   **35.** 2500   **37.** 231   **39.** $2,135

Pages 470-472   Lesson 13-3

1. yes, 5   **3.** yes, $\frac{3}{2}$ **5.** no   **7.** 135, 405   **9.** $\frac{1}{3}$, 1

11. 27, 9   **13.** $\frac{10}{3}, \frac{10}{9}$ **15.** 2, -4 **17.** 3, -6, 12, -24

19. 27, -9, 3, -1 **21.** 100   **23.** 3   **25.** 1   **27.** 2, 4

29. 15 or -15 **31.** 14, 28, 56, or -14, 28, -56

33. 3, 6, 12, 48, 96, 192 or -3, 6, -12, -48, 96, -192

35. 60.5%   **37.** 50,425   **39.** $28,477

41. $\frac{1}{1024}$ of original

Page 476   Lesson 13-4

1. 9; -2; 144; 5   **3.** 2; 4; 512; 5   **5.** 20; $-\frac{1}{4}$; $\frac{5}{64}$; 5

7. -364 **9.** $15\frac{3}{4}$ **11.** 1111   **13.** 93.75   **15.** 114,681

17. 732   **19.** 165   **21.** 1441   **23.** $\frac{63}{8}$ **25.** $\frac{781}{5}$

27. $\frac{189}{32}$ **29.** $\frac{4095}{256}$ **31.** $\frac{32}{63}$ **33.** 4   **35.** 2   **37.** $\frac{196820}{27}$

39. 127

Page 479   Lesson 13-5

1. $\frac{1}{2}; \frac{2}{3}; \frac{3}{2}$ **3.** 1; $-\frac{1}{3}; \frac{3}{4}$ **5.** 1; $\frac{3}{2}$; no sum

7. $\frac{7}{10} + \frac{7}{100} + \frac{7}{1000} + \cdots = \frac{7}{9}$

9. $\frac{73}{100} + \frac{73}{10,000} + \frac{73}{1,000,000} + \cdots = \frac{73}{99}$

11. $\frac{152}{1000} + \frac{152}{1,000,000} + \frac{152}{1,000,000,000} + \cdots = \frac{152}{999}$

13. $\frac{93}{100} + \frac{93}{10,000} + \frac{93}{1,000,000} + \cdots = \frac{31}{33}$ **15.** 72   **17.** 4

19. 27   **21.** $\frac{9}{5}$ **23.** 9   **25.** 8   **27.** no sum   **29.** $\frac{100}{11}$

31. 1   **33.** $\frac{31}{99}$ **35.** $\frac{410}{999}$ **37.** $\frac{41}{90}$ **39.** $4 + 3 + \frac{9}{4}$

41. $9 + (-3) + 1$ **43.** 220 cm   **45.** 800 feet

47. $\frac{400\sqrt{3}}{3}$ square inches

Pages 482-483   Lesson 13-6

1. j; 4; $3 + 4 + 5 + 6$ **3.** r; 3; $0 + 1 + 2$

5. i; 5; $0 + 2 + 4 + 6 + 8$ **7.** p; 4; $6 + 7 + 8 + 9$

9. $13 + 20 + 27 + 34 + 41 = 135$ **11.** $5 + 7 + 9 + 11 + 13 = 45$

13. $5 + 7 + 9 + 11 + 13 = 45$

15. $2 + 4 + 6 + 8 + 10 + 12 + 14 = 56$

17. $256 + 1024 + 4096 + 16,384 + 65,536 = 87,296$

19. $-12 + 6 + (-3) + \frac{3}{2} = -7\frac{1}{2}$ **21.** 650

23. 2540   **25.** 5555   **27.** 65,535.875   **29.** $\sum_{n=1}^{5} (3n + 4)$

31. $\sum_{n=1}^{5} (-4n + 19)$ **33.** $\sum_{n=1}^{5} n^2$ **35.** $\sum_{n=1}^{6} 2^{4-n}$

37. $\sum_{n=1}^{6} 243\left(-\frac{2}{3}\right)^{n-1}$ **39.** yes

Pages 485-486   Lesson 13-7

1. 99, 120   **3.** 80, 99   **5.** 8, 7, 6, 5

7. $-2, -6, -18, -54$ **9.** $-4, -4, 4, 4$ **11.** 3, 1, 4, 5

13. $a_n = 2n$ **15.** $a_n = \frac{n + 1}{n}$ **17.** $a_{n+1} = \frac{1}{3}a_n, a_1 = 1$

19. $a_{n+1} = \left(-\frac{1}{2}\right)a_n, a_1 = 1$ **21.** $\frac{8}{9}, \frac{9}{10}, \frac{10}{11}$

23. $\frac{56}{3}$, 24, 30   **25.** 256, 324, 400   **27.** 38, 47, 57

29. 2, 6, 18, 54, 162, 486   **31.** 3, 5, 8, 13, 21, 34

33. 2, 3, 7, 12, 27, 53

35. $a_{n+1} = a_n + 4, a_1 = 3; a_n = 4n - 1$

37. $a_{n+1} = 5a_n; a_1 = 3; a_n = 3(5)^{n-1}$

39. $a_{n+1} = a_n + 5, a_1 = 5; a_n = 5n$ **41.** 399

43. 4.7×10^{69} **45.** 500   **47.** $\sum_{n=1}^{5} (7n - 4)$ **49.** $\sum_{n=1}^{6} \frac{3}{4}n$

51. $\sum_{n=1}^{5} \frac{3n + 1}{5}$ **53.** $\sum_{n=1}^{6} \frac{n^2 + 1}{n}$ **55.** $\sum_{n=1}^{7} (-1)^{n-1}$

Page 489   Lesson 13-8

1. 1404, 4212   **3.** 2.875   **5.** 31, 43   **7.** 91, 140

9. 18, 29, 47   **11.** 1, 1, 2, 3, 5, 8, 13, 21, 34, 55, 89, 144, 233, 377, 610, 987, 1597, 2584, 4181, 6765

13. 7.3810   **15.** $L_n = F_{n+1} + F_{n-1}$ for $n \geq 2$

Pages 495-496   Lesson 13-9

1. 5040   **3.** 3,628,800   **5.** 90   **7.** 120   **9.** 5; $4rs^3$

11. 8; $-35k^4m^3$ **13.** 6; $-80x^2$

15. $x^4 + 4x^3m + 6x^2m^2 + 4xm^3 + m^4$ **17.** $y^7 + 7y^6p + 21y^5p^2 + 35y^4p^3 + 35y^3p^4 + 21y^2p^5 + 7yp^6 + p^7$

19. $b^5 - 5b^4z + 10b^3z^2 - 10b^2z^3 + 5bz^4 - z^5$

21. $32m^5 + 80m^4y + 80m^3y^2 + 40m^2y^3 + 10my^4 + y^5$

23. $64b^6 + 192b^5x + 240b^4x^2 + 160b^3x^3 + 60b^2x^4 + 12bx^5 + x^6$ **25.** $243x^5 - 810x^4y + 1080x^3y^2 - 720x^2y^3 + 240xy^4 - 32y^5$ **27.** $64 + 96x + 60x^2 + 20x^3 + \frac{15}{4}x^4 + \frac{3}{8}x^5 + \frac{1}{64}x^6$ **29.** $35x^3y^4$ **31.** $5005x^9y^6$

33. $24,634,368m^7n^5$ **35.** $537,600,000x^6$ **37.** $0.1318h^3n^3$

39. $6326.60   **41.** k **43.** $(k + 2)!$

Page 499   Chapter Review

1. 6, 14, 22, 30, 38   **3.** 234   **5.** -3, 1, 5   **7.** 2322

9. 1612   **11.** 60, 120   **13.** 12, 36, 108 or -12, 36, -108

15. 1031   **17.** 625   **19.** $\frac{3}{2}$ **21.** $20 + 23 + 26 + 29$

23. 1, 3, 5, 11, 21   **25.** 33, 42, 52, 63

27. $243a^5 + 405a^4b + 270a^3b^2 + 90a^2b^3 + 15ab^4 + b^5$

Page 505   Lesson 14-1

1. independent **3.** dependent **5.** independent **7.** 3125 patterns **9.** 90 routes **11.** 60 meals **13.** 16 ways **15.** 720 ways **17.** 840 patterns **19.** 5040 patterns **21.** 480 patterns

Page 508   Lesson 14-2

1. false **3.** false **5.** true **7.** true **9.** true **11.** 720 **13.** 3 **15.** 60 **17.** 34.650 **19.** 2520 **21.** 90,720 **23.** 6 **25.** 72 **27.** 27,720 **29.** 151,200 **31.** 56 **33.** 126 **35.** 42 **37.** 8 **39.** 11

Page 512   Lesson 14-3

1. reflective, circular **3.** not reflective, circular **5.** not reflective, linear **7.** not reflective, circular **9.** not reflective, linear **11.** 39,600 **13.** 720 **15.** 1.0364 **17.** 5040 **19.** 30,240 **21.** 56 **23.** 2.25 **25.** 60 **27.** 2520 **29.** 24 **31.** 12 **33.** 5040

Page 515   Lesson 14-4

1. combination **3.** combination **5.** permutation **7.** combination **9.** 56 **11.** 21 **13.** 11 **15.** 30 **17.** 792 **19.** 126 **21.** 351 **23.** 5148 **25.** 6 **27.** 0 **29.** 4320 **31.** 56 **33.** 1680

Page 519   Lesson 14-5

1. 1 to 1 **3.** 3 to 1 **5.** 7 to 8 **7.** $\frac{3}{7}$ **9.** $\frac{6}{11}$ **11.** $\frac{5}{16}$ **13.** $\frac{1}{7}$ **15.** $\frac{1}{16} \approx 0.063$ **17.** $\frac{1}{56} \approx 0.018$ **19.** $\frac{1}{4} = 0.250$ **21.** $\frac{1}{19} \approx 0.035$ **23.** $\frac{3}{19} \approx 0.158$ **25.** $\frac{14}{39} \approx 0.359$ **27.** $\frac{4}{7} \approx 0.571$ **29.** $\frac{2}{7} \approx 0.286$ **31.** $\frac{9}{170} \approx 0.053$ **33.** $\frac{18}{85} \approx 0.212$ **35.** 33 to 108,257 **37.** 66,607 to 33 **39.** 16,627 to 33 **41.** 3736 to 429

Pages 523-524   Lesson 14-6

1. dependent **3.** independent **5.** $\frac{1}{28} \approx 0.036$ **7.** $\frac{25}{81} \approx 0.309$ **9.** $\frac{1}{26} \approx 0.038$ **11.** $\frac{3}{13} \approx 0.231$ **13.** $\frac{1}{32} \approx 0.031$ **15.** $\frac{1}{94,109,400} \approx 1.06 \times 10^{-8}$ **17.** $\frac{9}{49} \approx 0.184$ **19.** $\frac{1}{8} = 0.125$ **21.** $\frac{5}{48} \approx 0.104$ **23.** $\frac{1}{36} \approx 0.028$ **25.** $\frac{1}{36} \approx 0.028$ **27.** $\frac{1}{6} \approx 0.167$ **29.** $\frac{1}{635,013,559,600} \approx 1.57 \times 10^{-12}$ **31.** $\frac{19}{1,160,054} \approx 1.64 \times 10^{-5}$

Page 527   Lesson 14-7

1. inclusive **3.** exclusive **5.** inclusive **7.** inclusive **9.** $\frac{2}{11} \approx 0.181$ **11.** $\frac{14}{33} \approx 0.424$ **13.** $\frac{12}{221} \approx 0.054$ **15.** $\frac{55}{221} \approx 0.249$ **17.** $\frac{7}{16} \approx 0.438$ **19.** $\frac{35}{64} \approx 0.567$ **21.** $\frac{29}{3003} \approx 0.010$ **23.** $\frac{160}{429} \approx 0.373$ **25.** $\frac{1}{525} \approx 0.002$

27. $\frac{3}{5} = 0.60$ **29.** $\frac{1}{4} = 0.250$ **31.** $\frac{9}{91} \approx 0.099$ **33.** $\frac{183}{221} \approx 0.828$ **35.** $\frac{2}{3} \approx 0.667$

Page 531   Lesson 14-8

1. binomial; $\frac{3}{8}$ **3.** not binomial **5.** binomial; $\frac{4}{81}$ **7.** not binomial **9.** not binomial **11.** $\frac{3}{8}; = 0.375$ **13.** $\frac{3125}{7776} \approx 0.063$ **15.** $\frac{625}{648} \approx 0.965$ **17.** $\frac{16}{27} \approx 0.593$ **19.** $\frac{15}{128} \approx 0.117$ **21.** $\frac{1}{1024} \approx 0.001$ **23.** $\frac{21}{3125} \approx 0.007$ **25.** $\frac{1}{8} = 0.125$ **27.** $\frac{1}{2} = 0.5$ **29.** 0.0001049 **31.** 0.1662386 **33.** 0.0037881 **35.** 0.8891293

Pages 534-535   Chapter Review

1. 125 **3.** 1,625,702,400 **5.** 112 **7.** 5040 **9.** 1485 **11.** $\frac{1}{13} \approx 0.077$ **13.** $\frac{3}{91} \approx 0.033$ **15.** $\frac{4}{13} \approx 0.308$ **17.** $\frac{1}{4} = 0.250$

CHAPTER 15   STATISTICS

Pages 540-541   Lesson 15-1

1. 15.9% **3.** 67.9 million **5.** 50.6 million **7.** no **9.** yes **11.** yes **13.** no **17.** 16 cities **21.** 22 cities

Pages 545-546   Lesson 15-2

1. 2, 3, 4, 5, 6, 7, 8 **3.** 2, 3, 4, 5, 6, 7, 8, 9, 10 **5.** 50 **7.** 30 **9.** 26 **11.** 56° **13.** 11 **17.** 69 **19.** 40 **21.** 9 **23.** Reese Pure Maple **25.** $1.11 **27.** 8 **29.** Reese Pure Maple **31.** 50 **33.** 7 **35.** Vermont Maple Orchards Pure Maple **37.** 174 **39.** 8

Page 550   Lesson 15-3

1. 3; none; 3 **3.** 7; 7; 7 **5.** 43; none; 30.6 **7.** 216; 399; 254 **9.** 2.1; 2.1, about 3.4 **11.** 12; 12, 13; about 12.3 **13.** 4; 6; about 3.7 **15.** 18,700; none; about 19,047 **17.** $4.25; $3.75 and $4.25; $4.85 **19.** 10; 10; about 12.6

Pages 553-554   Lesson 15-4

1. 8 **3.** 22 **5.** 36 **7.** 5, 9, 11 **9.** 37, 39.5, 47 **11.** 50.5, 59, 68.5 **13.** 19; 2, 5, 7; 5 **15.** 340; 1025, 1075, 1125; 100 **17.** 480; 160, 265, 345; 185 **19.** 20 **21.** 835 **23.** none **25.** 34°; 10°, 114°, 118°; 8°; 134° **29.** 6¢ **33.** 41.5 mg

Pages 556-557   Lesson 15-5

1. 20 **3.** 25% **5.** 50% **7.** 50% **9.** 210 **11.** 50% *For Exercises 13-19, outliers are given.* **13.** 174 **15.** no outliers **17.** 22 **19.** no outliers

1. 3.6 **3.** 6.3 **5.** 7.0 **7.** 10.6 **9.** 27 **11.** 22
13. 5.9 **15.** mode: 28.4, 30.6, 31.1; mean: about 33.4;
quartiles: 30.6, 32.6, 35.0; interquartile range: 4.4;
standard deviation: about 4.9; outliers: 44.8, 50.4

1. 340 **3.** 495 **5.** 170 **7.** 6800 **9.** 5000 **11.** 250
13. 0.68 **15.** 0.97 **17.** 50% **19.** 15.5% **21.** 1500
23. 0.025

1. 1.0 **3.** 3.0 **5.** 5.0 **7.** 6.5 **9.** 400
11. $y = 400x + 2,000,000$ **13.** $8,000,000
15. $y = -\frac{1}{10}x + 28$ **17.** 3 **21.** $10,666 **23.** about 5.5
years **25.** $y = \frac{4}{7}x + 32\frac{1}{7}$ **27.** 92.5 **29.** $y = 7x - 339$
31. 249 pounds

11. about 37.8% **15.** about 231 miles

1. 8 **9.** 121.5; 109, 119, 126, 139; about 121.7 **11.** 67
13. 16 **15.** 78 **19.** about 14.6 **21.** 250
23. $y = 0.95x - 95$ **25.** 192.6 cm

CHAPTER 16   TRIGONOMETRIC FUNCTIONS

1. III **3.** I **5.** I **7.** IV **9.** II **11.** I **13.** IV
15. III **17.** III **19.** III **21.** yes **23.** no **25.** $\frac{\pi}{2}$
27. $-\frac{\pi}{4}$ **29.** $\frac{5\pi}{2}$ **31.** $\frac{5\pi}{6}$ **33.** $\frac{\pi}{4}$ **35.** $\frac{11\pi}{6}$ **37.** $\frac{3\pi}{2}$
39. π **41.** $\frac{9\pi}{4}$ **43.** $-\frac{7\pi}{4}$ **45.** 180° **47.** 45°
49. 540° **51.** −480° **53.** 30° **55.** −45° **57.** 330°
59. $\frac{900°}{\pi} \approx 286.48°$ **61.** 990° **63.** $\frac{1170°}{\pi} \approx 372.42°$

1. − **3.** − **5.** − **7.** + **9.** + **11.** + **13.** 60°
15. 300° **17.** π **19.** $\frac{\pi}{2}$ **21.** $\frac{7\pi}{4}$ **23.** 120° **25.** 240°
27. 240° **29.** 120° **31.** 60° **33.** $\frac{\pi}{4}$ **35.** $\frac{3\pi}{4}$
37. 320° **39.** $-\frac{\sqrt{3}}{2}$ **41.** $\frac{\sqrt{2}}{2}$ **43.** $\frac{\sqrt{3}}{2}$ **45.** $\frac{\sqrt{2}}{2}$
47. $\frac{\sqrt{3}}{2}$ **49.** $-\frac{\sqrt{3}}{2}$ **51.** −1 **53.** 0 **55.** $-\frac{1}{2}$ **57.** $\frac{\sqrt{2}}{2}$
59. $\frac{\sqrt{3}}{3}$ **61.** $\frac{(1 + \sqrt{3})}{2}$ **63.** $-2\sqrt{3}$

1. 1, 2π **3.** $\frac{2}{3}$, 2π **5.** 6, 3π **7.** 4, $\frac{8\pi}{3}$ **9.** 5, 2π
11. 1, $\frac{2\pi}{3}$ **13.** 4, 4π **15.** 3, 3π **17.** $\frac{1}{2}$, $\frac{8\pi}{3}$ **19.** $\frac{8}{9}$, $\frac{10\pi}{3}$

27. 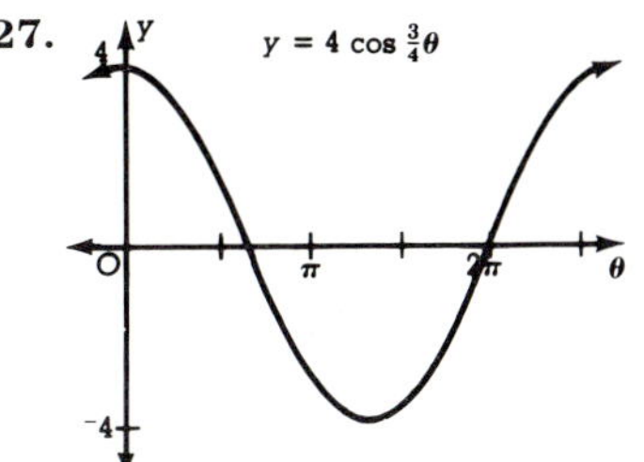

31. 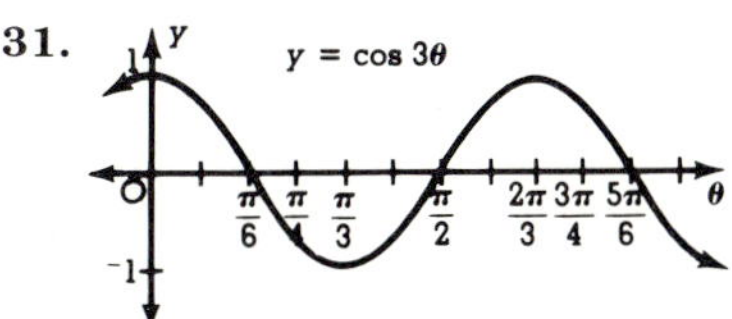

35. 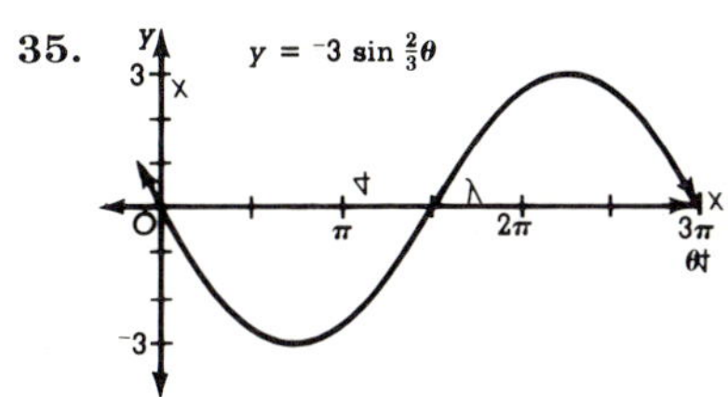

37. 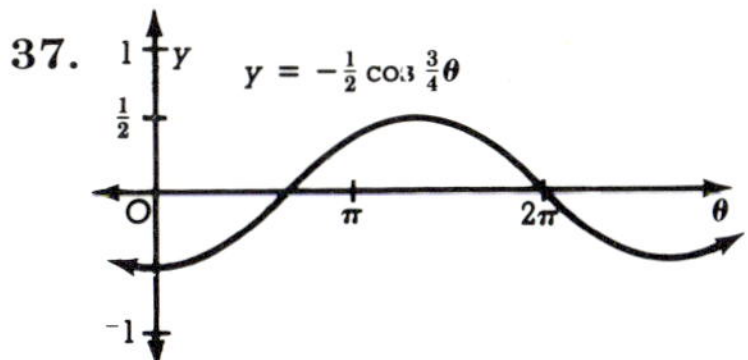

1. none **3.** 90°, 270° **5.** 90°, 270° **7.** 2π **9.** π
11. $\frac{\pi}{3}$ **13.** $\frac{8\pi}{3}$ **15.** π **17.** 3π **19.** I, increasing; II,
decreasing; III, decreasing; IV, increasing **21** I,
increasing; II, increasing; III, increasing; IV, increasing
23. I, increasing; II, increasing; III, decreasing; IV,
decreasing **25.** 2 **27.** −1 **29.** $-\sqrt{3}$ **31.** 2 **33.** −2
35. −2 **37.** −1 **39.** −2 **41.** undefined **43.** 1
45. $\frac{\sqrt{3}}{3}$ **47.** 0 **49.** $-\frac{\sqrt{3}}{3}$ **51.** 1 **53.** 2 **55.** $\frac{-2\sqrt{3}}{3}$

61. 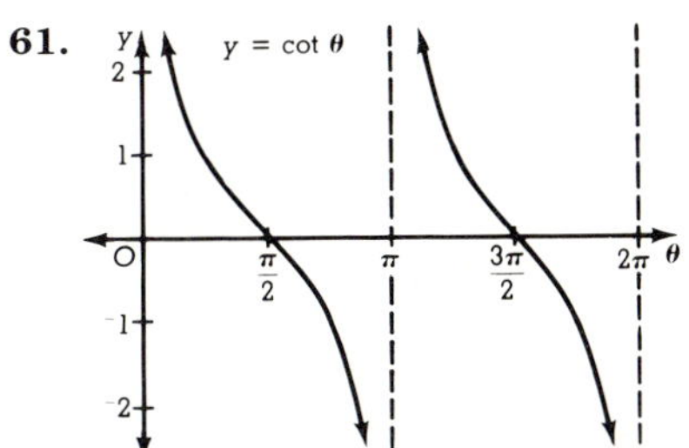

63. 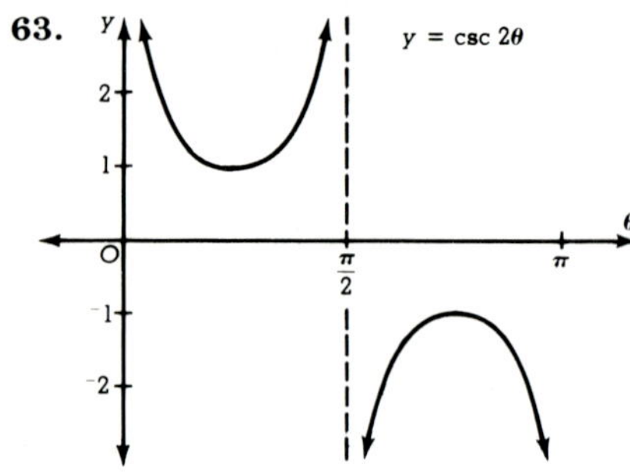

For Exercises 1-20, graphs may vary. Sample graphs are given.

9. Xmin: -360, Xmax: 360, Xscl: 90
Ymin: -5, Ymax: 5, Yscl: 1

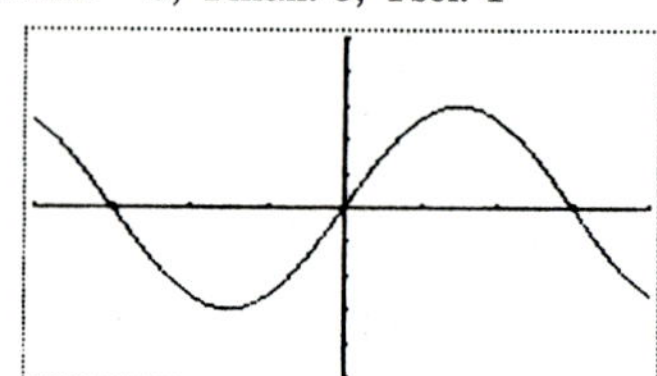

17. Xmin: -90, Xmax: 270, Xscl: 90
Ymin: -25, Ymax: 25, Yscl: 5

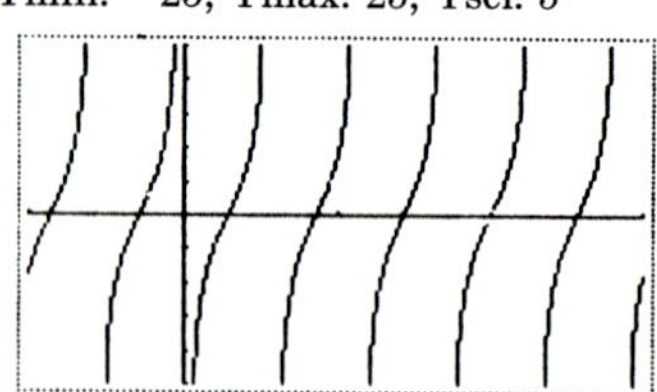

1. $y = \mathrm{Sin}^{-1} x$　**3.** $\alpha = \mathrm{Tan}^{-1} \beta$　**5.** $30° = \mathrm{Sin}^{-1} \frac{1}{2}$

7. $\frac{1}{2}$　**9.** $-60°$　**11.** undefined　**13.** 1　**15.** $150°$

17. $45°$　**19.** undefined　**21.** $60°$　**23.** $\frac{\sqrt{2}}{2}$　**25.** $-90°$

27. $90°$　**29.** 0　**31.** $\frac{1}{2}$　**33.** $90°$　**35.** $\frac{1}{2}$　**37.** $\frac{1}{2}$　**39.** $\frac{5}{12}$

41. $\frac{\sqrt{11}}{5}$　**43.** $30°$　**45.** $60°$　**47.** $\frac{4}{5}$　**49.** $\frac{\sqrt{3}}{2}$　**51.** $\frac{\sqrt{2}}{2}$

53. $-\frac{4}{3}$　**55.** $\frac{1}{2}$　**57.** $\frac{\sqrt{2}}{2}$　**59.** π or $180°$

1. $42.1597°$　**3.** $41.9840°$　**5.** $50.5251°$　**7.** 1.0038
9. 0.1736　**11.** -2.4142　**13.** 0.3249　**15.** -2
17. undefined　**19.** 0.7507　**21.** -4.0625　**23.** $69.6359°$
25. $56.3099°$　**27.** -0.8258

1. $\frac{2\pi}{3}$　**3.** $\frac{3\pi}{2}$　**5.** $60°$　**7.** $\left(\frac{240}{\pi}\right)^{\circ}$　**9.** $205°$　**11.** $180°$

13. $\frac{2\pi}{3}$　**15.** $\frac{16\pi}{9}$　**17.** $\frac{\sqrt{3}}{2}$　**19.** -1　**21.** $-\frac{1}{2}$
23. $-\frac{\sqrt{2}}{2}$　**25.** 1　**27.** $1; 2\pi$　**29.** $4; \pi$　**31.** $\frac{2\sqrt{3}}{3}$
33. $-\frac{1}{2}$　**35.** $\sqrt{3}$　**37.** 2　**39.** $-90°$　**41.** $90°$　**43.** $\frac{\sqrt{3}}{2}$
45. $46.7512°$　**47.** 0.2362

CHAPTER 17　　TRIGONOMETRIC IDENTITIES AND EQUATIONS

1. 1　**3.** $\csc^2 \theta$　**5.** 1　**7.** $\cos \theta$　**9.** $\cot^2 x$　**11.** $\cot^2 \alpha$

13. $\frac{\sqrt{3}}{2}$　**15.** $\frac{3}{5}$　**17.** $\frac{3\sqrt{5}}{5}$　**19.** $\frac{4\sqrt{17}}{17}$　**21.** $-\frac{4}{5}$　**23.** $\frac{5}{4}$

25. $-\frac{\sqrt{17}}{4}$　**27.** $-\frac{\sqrt{3}}{2}$　**29.** $-\frac{12}{13}$　**31.** $\frac{3}{5}$　**33.** 1

35. $2 \sin x$　**37.** 1　**39.** 2

41.
$$1 + \cot^2 \theta \overset{?}{=} \csc^2 \theta$$
$$\frac{\sin^2 \theta + \cos^2 \theta}{\sin^2 \theta} \overset{?}{=} \csc^2 \theta$$
$$\frac{1}{\sin^2 \theta} \overset{?}{=} \csc^2 \theta$$
$$\csc^2 \theta = \csc^2 \theta$$

43.
$$\sin x \sec x \overset{?}{=} \tan x$$
$$\sin x \cdot \frac{1}{\cos x} \overset{?}{=} \tan x$$
$$\frac{\sin x}{\cos x} \overset{?}{=} \tan x$$
$$\tan x = \tan x$$

45. $\pm \frac{\sqrt{15}}{4}$　**47.** $\frac{1}{9}$

1.
$$\tan \beta (\cot \beta + \tan \beta) \overset{?}{=} \sec^2 \beta$$
$$\tan \beta \left(\frac{1}{\tan \beta} + \tan \beta\right) \overset{?}{=} \sec^2 \beta$$
$$1 + \tan^2 \beta \overset{?}{=} \sec^2 \beta$$
$$\sec^2 \beta = \sec^2 \beta$$

3.
$$\csc x \sec x \overset{?}{=} \cot x + \tan x$$
$$\frac{1}{\sin x} \cdot \frac{1}{\cos x} \overset{?}{=} \frac{\cos x}{\sin x} + \frac{\sin x}{\cos x}$$
$$\frac{1}{\sin x} \cdot \frac{1}{\cos x} \overset{?}{=} \frac{\cos^2 x + \sin^2 x}{\sin x \cos x}$$
$$\frac{1}{\sin x \cos x} = \frac{1}{\sin x \cos x}$$

5.
$$\frac{\sec \theta}{\sin \theta} - \frac{\sin \theta}{\cos \theta} \overset{?}{=} \cot \theta$$
$$\frac{1}{\cos \theta \sin \theta} - \frac{\sin \theta}{\cos \theta} \overset{?}{=} \cot \theta$$
$$\frac{1 - \sin^2 \theta}{\cos \theta \sin \theta} \overset{?}{=} \cot \theta$$
$$\frac{\cos^2 \theta}{\cos \theta \sin \theta} \overset{?}{=} \cot \theta$$
$$\frac{\cos \theta}{\sin \theta} \overset{?}{=} \cot \theta$$
$$\cot \theta = \cot \theta$$

7.
$$\frac{\sin \alpha}{1 - \cos \alpha} + \frac{1 - \cos \alpha}{\sin \alpha} \overset{?}{=} 2 \csc \alpha$$
$$\frac{\sin^2 \alpha + (1 - \cos \alpha)}{(1 - \cos \alpha)(\sin \alpha)} \overset{?}{=} \frac{2}{\sin \alpha}$$
$$\frac{\sin^2 \alpha + 1 - 2 \cos \alpha + \cos^2 \alpha}{(1 - \cos \alpha) \sin \alpha} \overset{?}{=} \frac{2}{\sin \alpha}$$
$$\frac{2 - 2 \cos \alpha}{(1 - \cos \alpha) \sin \alpha} \overset{?}{=} \frac{2}{\sin \alpha}$$
$$\frac{2(1 - \cos \alpha)}{(1 - \cos \alpha) \sin \alpha} \overset{?}{=} \frac{2}{\sin \alpha}$$
$$\frac{2}{\sin \alpha} = \frac{2}{\sin \alpha}$$

9.
$$\frac{\cos^2 x}{1 - \sin x} \overset{?}{=} 1 + \sin x$$
$$\frac{\cos^2 x}{1 - \sin x} \cdot \frac{1 + \sin x}{1 + \sin x} \overset{?}{=} 1 + \sin x$$
$$\frac{\cos^2 x \,(1 + \sin x)}{1 - \sin^2 x} \overset{?}{=} 1 + \sin x$$
$$\frac{\cos^2 x \,(1 + \sin x)}{\cos^2 x} \overset{?}{=} 1 + \sin x$$
$$1 + \sin x = 1 + \sin x$$

11.
$$\frac{\sin \theta}{\sec \theta} \overset{?}{=} \frac{1}{\tan \theta + \cot \theta}$$
$$\frac{\sin \theta}{\dfrac{1}{\cos \theta}} \overset{?}{=} \frac{1}{\dfrac{\sin \theta}{\cos \theta} + \dfrac{\cos \theta}{\sin \theta}}$$
$$\sin \theta \cdot \cos \theta \overset{?}{=} \frac{1}{\dfrac{\sin^2 \theta + \cos^2 \theta}{\cos \theta \cdot \sin \theta}}$$
$$\sin \theta \cdot \cos \theta \overset{?}{=} \frac{1}{\dfrac{1}{\cos \theta \cdot \sin \theta}}$$
$$\sin \theta \cdot \cos \theta = \sin \theta \cdot \cos \theta$$

13.
$$\frac{\cot x + \csc x}{\sin x + \tan x} \overset{?}{=} \cot x \cdot \csc x$$
$$\frac{\dfrac{\cos x}{\sin x} + \dfrac{1}{\sin x}}{\sin x + \dfrac{\sin x}{\cos x}} \overset{?}{=} \frac{\cos x}{\sin x} \cdot \frac{1}{\sin x}$$
$$\frac{\dfrac{\cos x + 1}{\sin x}}{\dfrac{\sin x \cos x + \sin x}{\cos x}} \overset{?}{=} \frac{\cos x}{\sin^2 x}$$
$$\frac{\dfrac{\cos x + 1}{\sin x}}{\dfrac{\sin x (\cos x + 1)}{\cos x}} \overset{?}{=} \frac{\cos x}{\sin^2 x}$$
$$\frac{\cos x + 1}{\sin x} \cdot \frac{\cos x}{\sin x (\cos x + 1)} \overset{?}{=} \frac{\cos x}{\sin^2 x}$$
$$\frac{\cos x}{\sin^2 x} = \frac{\cos x}{\sin^2 x}$$

15.
$$\cos^2 x + \tan^2 x \cos^2 x \overset{?}{=} 1$$
$$\cos^2 x + \frac{\sin^2 x}{\cos^2 x} \cos^2 x \overset{?}{=} 1$$
$$\cos^2 x + \sin^2 x \overset{?}{=} 1$$
$$1 = 1$$

17.
$$\frac{1 + \tan^2 \theta}{\csc^2 \theta} \overset{?}{=} \tan^2 \theta$$
$$\frac{\sec^2 \theta}{\csc^2 \theta} \overset{?}{=} \tan^2 \theta$$
$$\frac{\dfrac{1}{\cos^2 \theta}}{\dfrac{1}{\sin^2 \theta}} \overset{?}{=} \tan^2 \theta$$
$$\frac{1}{\cos^2 \theta} \cdot \frac{\sin^2 \theta}{1} \overset{?}{=} \tan^2 \theta$$
$$\frac{\sin^2 \theta}{\cos^2 \theta} \overset{?}{=} \tan^2 \theta$$
$$\tan^2 \theta = \tan^2 \theta$$

19.
$$\frac{1 + \sin x}{\sin x} \overset{?}{=} \frac{\cot^2 x}{\csc x - 1}$$
$$\frac{1}{\sin x} + \frac{\sin x}{\sin x} \overset{?}{=} \frac{\csc^2 x - 1}{\csc x - 1}$$
$$\csc x + 1 \overset{?}{=} \frac{(\csc x - 1)(\csc x + 1)}{\csc x - 1}$$
$$\csc x + 1 = \csc x + 1$$

21.
$$\cos^4 x - \sin^4 x \overset{?}{=} \cos^2 x - \sin^2 x$$
$$(\cos^2 x + \sin^2 x)(\cos^2 x - \sin^2 x) \overset{?}{=} \cos^2 x - \sin^2 x$$
$$1(\cos^2 x - \sin^2 x) \overset{?}{=} \cos^2 x - \sin^2 x$$
$$\cos^2 x - \sin^2 x = \cos^2 x - \sin^2 x$$

23.
$$\frac{\tan^2 x}{\sec x - 1} \overset{?}{=} 1 + \frac{1}{\cos x}$$
$$\frac{\tan^2 x(\sec x + 1)}{(\sec x - 1)(\sec x + 1)} \overset{?}{=} 1 + \frac{1}{\cos x}$$
$$\frac{\tan^2 x(\sec x + 1)}{\tan^2 x} \overset{?}{=} 1 + \frac{1}{\cos x}$$
$$\frac{1}{\cos x} + 1 = 1 + \frac{1}{\cos x}$$

Page 619 **Graphing Calculator Application**

5. no **7.** yes **9.** yes **11.** yes **13.** no **15.** no
17. no **19.** no **21.** yes **23.** yes **25.** yes

Page 622 **Lesson 17-3**

1. $45° + 60°$ **3.** $-135° - 30°$ **5.** $30° + 45°$
7. $225° + 60°$ **9.** $\dfrac{\sqrt{6} + \sqrt{2}}{4}$ **11.** $\dfrac{-\sqrt{2} - \sqrt{6}}{4}$
13. $\dfrac{-\sqrt{6} - \sqrt{2}}{4}$ **15.** $\dfrac{\sqrt{6} + \sqrt{2}}{4}$ **17.** $\dfrac{-\sqrt{6} - \sqrt{2}}{4}$
19. $\dfrac{\sqrt{2} + \sqrt{6}}{4}$ **21.** $\dfrac{\sqrt{3}}{2}$ **23.** $\dfrac{1}{2}$
25. $\sin (270° - \theta) = \sin 270° \cos \theta - \cos 270° \sin \theta$
$$= -1 \cdot \cos \theta - 0 \cdot \sin \theta$$
$$= -\cos \theta$$
27. $\sin (180° + \theta) = \sin 180° \cos \theta + \cos 180° \sin \theta$
$$= 0 \cdot \cos \theta - 1 \cdot \sin \theta$$
$$= -\sin \theta$$
29. $\sin (90° + \theta) = \sin 90° \cos \theta + \cos 90° \sin \theta$
$$= 1 \cdot \cos \theta + 0 \cdot \sin \theta$$
$$= \cos \theta$$
37. $-2 - \sqrt{3}$ **39.** $-2 + \sqrt{3}$ **41.** $\sqrt{3}$ **43.** $-2 - \sqrt{3}$
45. $-2 + \sqrt{3}$ **47.** $-\tan \theta$ **49.** $\dfrac{\tan \alpha - \tan \beta}{1 + \tan \alpha \tan \beta}$

Page 626 **Lesson 17-4**

1. I or II **3.** I or II **5.** I **7.** II **9.** $\dfrac{\sqrt{3}}{2}; \dfrac{1}{2}; \dfrac{\sqrt{2 - \sqrt{3}}}{2};$
$\dfrac{\sqrt{2 + \sqrt{3}}}{2}$ **11.** $\dfrac{4}{9}\sqrt{5}; -\dfrac{1}{9}; \dfrac{\sqrt{30}}{6}; -\dfrac{\sqrt{6}}{6}$ **13.** $-\dfrac{120}{169}; \dfrac{119}{169}; \dfrac{5\sqrt{26}}{26};$
$\dfrac{\sqrt{26}}{26}$ **15.** $-\dfrac{3\sqrt{7}}{8}; -\dfrac{1}{8}; \dfrac{\sqrt{8 - 2\sqrt{7}}}{4}; -\dfrac{\sqrt{8 + 2\sqrt{7}}}{4}$ **17.** $-\dfrac{\sqrt{15}}{8};$
$-\dfrac{7}{8}; \dfrac{\sqrt{10}}{4}; \dfrac{\sqrt{6}}{4}$ **19.** $-\dfrac{3\sqrt{55}}{32}; \dfrac{23}{32}; \dfrac{\sqrt{8 - \sqrt{55}}}{4}; -\dfrac{\sqrt{8 + \sqrt{55}}}{4}$
21. $\dfrac{\sqrt{15}}{8}; \dfrac{7}{8}; \dfrac{\sqrt{8 + 2\sqrt{15}}}{4}; -\dfrac{\sqrt{8 - 2\sqrt{15}}}{4}$ **23.** $\sqrt{2 + \sqrt{3}}$
25. $\sqrt{2 + \sqrt{2}}$ **27.** $\sqrt{2 - \sqrt{2}}$ **29.** $-\sqrt{2 - \sqrt{3}}$
31. $\sqrt{2 - \sqrt{3}}$
33. $\cos^2 2x + 4 \sin^2 x \cos^2 x \overset{?}{=} 1$
$$\cos^2 2x + \sin^2 2x \overset{?}{=} 1$$
$$1 = 1$$
35.
$$\sin^4 x - \cos^4 x \overset{?}{=} 2 \sin^2 x - 1$$
$$(\sin^2 x - \cos^2 x)(\sin^2 x + \cos^2 x) \overset{?}{=} 2 \sin^2 x - 1$$
$$(\sin^2 x - \cos^2 x) \cdot 1 \overset{?}{=} 2 \sin^2 x - 1$$
$$[\sin^2 x - (1 - \sin^2 x)] \cdot 1 \overset{?}{=} 2 \sin^2 x - 1$$
$$2 \sin^2 x - 1 = 2 \sin^2 x - 1$$

37. $\sin^2 \theta \stackrel{?}{=} \dfrac{1}{2}(1 - \cos 2\theta)$

$\quad \sin^2 \theta \stackrel{?}{=} \dfrac{1}{2}[1 - (1 - 2\sin^2 \theta)]$

$\quad \sin^2 \theta \stackrel{?}{=} \dfrac{1}{2}[2\sin^2 \theta]$

$\quad \sin^2 \theta = \sin^2 \theta$

41. $-2x\sqrt{1 - x^2}$

Pages 629-630   Lesson 17-5

1. 1 **3.** 2 **5.** 2 **7.** 0 **9.** 4 **11.** 4 **13.** 8 **15.** 8
17. 45°, 135°, 225°, 315° **19.** 0°, 180°, 210°, 330°
21. 30°, 90°, 150°, 270° **23.** 0°, 90°, 270° **25.** $\dfrac{\pi}{3}, \dfrac{5\pi}{3}$
27. $\dfrac{4\pi}{3}, \dfrac{5\pi}{3}$ **29.** 0, π **31.** $\dfrac{\pi}{3}, \dfrac{5\pi}{3}$ **33.** $0° + n \cdot 120°$
where n is any integer **35.** $45° + n \cdot 180°$ where n is any
integer **37.** $120° + n \cdot 360°$ where n is any integer,
$240° + n \cdot 360°$ where n is any integer **39.** $90° + n \cdot 360°$
where n is any integer, $180° + n \cdot 360°$ where n is any
integer **41.** $0° + n \cdot 360°$ where n is any integer,
$60° + n \cdot 360°$ where n is any integer, $300° + n \cdot 360°$
where n is any integer **43.** $0° + n \cdot 180°$ where n is any
integer, $90° + n \cdot 180°$ where n is any integer
45. $\dfrac{\pi}{6} + n\pi$ where n is any integer, $\dfrac{5\pi}{6} + n\pi$ where n is
any integer **47.** $\dfrac{7\pi}{6} + 2n\pi$ where n is any integer,
$\dfrac{11\pi}{6} + 2n\pi$ where n is any integer **49.** $\dfrac{2\pi}{3} + 2n\pi$ where
n is any integer, $\dfrac{4\pi}{3} + 2n\pi$ where n is any integer
51. $\dfrac{\pi}{6} + 2n\pi$ where n is any integer, $\dfrac{5\pi}{6} + 2n\pi$ where n
is any integer, $\dfrac{\pi}{2} + 2n\pi$ where n is any integer
53. $\dfrac{\pi}{3} + 2n\pi$ where n is any integer, $\dfrac{5\pi}{3} + 2n\pi$ where n
is any integer

Page 633   Graphing Calculator Application

1. 0°, 180°, 360° **3.** −360°, −300°, 180°, −60°, 0°, 60°,
180°, 300°, 360° **5.** 30°, 150°, 270° **7.** −174.8°, −51.6°,
185.2°, 308.4° **9.** 231.3°, 308.7°

Page 636   Chapter Review

1. $-\dfrac{\sqrt{3}}{2}$ **3.** $-2\sqrt{2}$ **5.** $\dfrac{3}{5}$ **7.** $-\dfrac{4}{3}$
9.

$\quad (\sin^4 x - \cos^4 x) \stackrel{?}{=} \sin^2 x - \cos^2 x$

$(\sin^2 x - \cos^2 x)(\sin^2 x + \cos^2 x) \stackrel{?}{=} \sin^2 x - \cos^2 x$

$\quad\quad (\sin^2 x - \cos^2 x) \cdot 1 \stackrel{?}{=} \sin^2 x - \cos^2 x$

$\quad\quad\quad \sin^2 x - \cos^2 x = \sin^2 x - \cos^2 x$

11.

$\quad\quad \dfrac{\sin \theta}{1 - \cos \theta} \stackrel{?}{=} \csc \theta + \cot \theta$

$\quad \dfrac{\sin \theta (1 + \cos \theta)}{(1 - \cos \theta)(1 + \cos \theta)} \stackrel{?}{=} \csc \theta + \cot \theta$

$\quad \dfrac{\sin \theta + \sin \theta \cos \theta}{1 - \cos^2 \theta} \stackrel{?}{=} \csc \theta + \cot \theta$

$\quad \dfrac{\sin \theta + \sin \theta \cos \theta}{\sin^2 \theta} \stackrel{?}{=} \csc \theta + \cot \theta$

$\dfrac{\sin \theta}{\sin^2 \theta} + \dfrac{\sin \theta \cos \theta}{\sin^2 \theta} \stackrel{?}{=} \dfrac{1}{\sin \theta} + \dfrac{\cos \theta}{\sin \theta}$

$\quad \dfrac{1}{\sin \theta} + \dfrac{\cos \theta}{\sin \theta} = \dfrac{1}{\sin \theta} + \dfrac{\cos \theta}{\sin \theta}$

13. $\sec x (\sec x - \cos x) \stackrel{?}{=} \tan^2 x$

$\quad\quad\quad \sec^2 x - 1 \stackrel{?}{=} \tan^2 x$

$\quad\quad\quad\quad\quad \tan^2 x = \tan^2 x$

15. $\dfrac{\csc \theta + 1}{\cot \theta} \stackrel{?}{=} \dfrac{\cot \theta}{\csc \theta - 1}$

$\quad \dfrac{\csc \theta + 1}{\cot \theta} \stackrel{?}{=} \dfrac{\cot \theta (\csc \theta + 1)}{(\csc \theta - 1)(\csc \theta + 1)}$

$\quad \dfrac{\csc \theta + 1}{\cot \theta} \stackrel{?}{=} \dfrac{\cot \theta (\csc \theta + 1)}{\csc^2 \theta - 1}$

$\quad \dfrac{\csc \theta + 1}{\cot \theta} \stackrel{?}{=} \dfrac{\cot \theta (\csc \theta + 1)}{\cot^2 \theta}$

$\quad \dfrac{\csc \theta + 1}{\cot \theta} = \dfrac{\csc \theta + 1}{\cot \theta}$

17. $\dfrac{\sqrt{2} + \sqrt{6}}{4}$ **19.** $\dfrac{\sqrt{6} + \sqrt{2}}{4}$ **23.** $\dfrac{24}{25}$ **25.** $\dfrac{\sqrt{21}}{5}$
27. 0° **29.** 0°, 90° **31.** 270° **33.** $\dfrac{7\pi}{6} + 2n\pi$ where n is
any integer, $\dfrac{11\pi}{6} + 2n\pi$ where n is an integer

CHAPTER 18   TRIANGLE TRIGONOMETRY

Page 641   Lesson 18-1

1. $\sin A = \dfrac{5}{13}$; $\sin B = \dfrac{12}{13}$
3. $\sin A = \dfrac{3\sqrt{13}}{13}$; $\sin B = \dfrac{2\sqrt{13}}{13}$
5. $\cos A = \dfrac{\sqrt{5}}{5}$; $\cos B = \dfrac{2\sqrt{5}}{5}$
7. $\tan A = \dfrac{5}{12}$; $\tan B = \dfrac{12}{5}$ **9.** $\tan A = \dfrac{3}{2}$; $\tan B = \dfrac{2}{3}$
11. 0.6000 **13.** 0.6000 **15.** 0.7500 **17.** 0.8824
19. 0.4706 **21.** 1.8750 **23.** 2.6833 **25.** 0.3727
27. 0.3492 **29.** 1.1333 **31.** 1.2500 **33.** 0.7500
35. $\sin A = 0.5692$, $\cos A = 0.8222$, $\tan A = 0.6923$
37. $\sqrt{3}$ **39.** 1 **41.** $\dfrac{1}{2}$ **43.** $\dfrac{(3\sqrt{2} - 2\sqrt{3})}{3}$

Pages 644-645   Lesson 18-2

1. 0.6691 **3.** 0.9272 **5.** 343.8 **7.** 1.171 **9.** 0.4488
11. 0.9605 **13.** 0.9511 **15.** 0.4592 **17.** 0.7880
19. 0.9272 **21.** 0.2812 **23.** 4.511 **25.** 0.5125
27. 50.54 **29.** 1.809 **31.** −2.616 **33.** −0.2518
35. 0.9228 **37.** 59°7′ **39.** 19°13′ **41.** 44°00′
43. 12°32′ **45.** 53°55′ **47.** 45°5′ **49.** 40°42′
51. 35°51′ **53.** 0.6915 **55.** 1.0114 **57.** 75°35′
59. 30°13′

Pages 647-648   Lesson 18-3

1. $\sin 15° = \dfrac{a}{37}$ **3.** $\sin 49°13′ = \dfrac{10}{c}$ **5.** $\cos 16° = \dfrac{13}{c}$
7. $7^2 + b^2 = 16^2$ **9.** $\tan A = \dfrac{7}{12}$
11. $c = \sqrt{53}$, $A = 15°57′$, $B = 74°04′$
13. $b = 5$, $A = 67°23′$, $B = 22°37′$
15. $a = \sqrt{133}$, $A = 62°31′$, $B = 27°29′$
17. $a = 3.86$, $b = 13.46$, $B = 74°$
19. $B = 52°45′$, $c = 13.82$, $a = 8.36$
21. $A = 47°50′$, $c = 12.14$, $b = 8.15$

23. $A = 77°$, $a = 5.85$, $b = 1.35$
25. $B = 13°$, $a = 181.92$, $c = 186.71$
27. $A = 57°$, $c = 39.35$, $b = 21.43$
29. $B = 34°5'$, $a = 13.25$, $b = 8.97$
31. $B = 45°$, $b = 7$, $a = 7$
33. $B = 75°$, $a = 6.47$, $b = 24.15$
35. $A = 41°11'$, $B = 48°49'$, $b = 8$, $c = 10.63$
37. $A = 19°28'$, $B = 70°32'$, $c = 15$, $b = 14.14$

1. 64.35 m   **3.** 1119.36 ft   **5.** 7.42 ft   **7.** 22.07 m
9. $9°36'$ **11.** $22°48'$
13. $x = 63°26'$, $y = 26°34'$, $z = 63°26'$ **15.** 102.25 ft
17. 100 ft, 300 ft   **19.** 4.05 cm   **21.** 164.92 km

1. area $= \left(\dfrac{1}{2}\right)(10)(17)(\sin 46°)$
3. area $= \left(\dfrac{1}{2}\right)(15)(30)(\sin 90°)$ **5.** $\dfrac{\sin 50°}{14} = \dfrac{\sin B}{10}$
7. $\dfrac{\sin 53°}{a} = \dfrac{\sin 61°}{2.8}$ **9.** 55.152   **11.** 27.531   **13.** 37.001
15. 63.2089   **17.** $C = 74°$, $b = 8.8904$, $c = 10.19$
19. $B = 60°19'$, $C = 36°31'$, $c = 47.95$
21. $A = 52°$, $b = 100.17$, $c = 90.40$
23. $B = 52°53'$, $A = 62°27'$, $a = 16.68$
25. $A = 24°24'$, $a = 5.54$, $c = 11.74$ **27.** 93.13 cm
29. 536.19 ft   **31.** 424.26 mi   **33.** 109.63 ft
35. $BD = 5.9$ cm, $CA = 11$ cm; perimeter $= 22.6$ cm

1. no triangle   **3.** 1 triangle   **5.** 1 triangle
7. 1; $B = 90°$, $C = 53°08'$, $c = 8$ **9.** none
11. 1; $A = 44°46'$, $B = 37°14'$, $b = 54.99$ **13.** none
15. none   **17.** none   **19.** none   **21.** 38.3 mm or 15 mm

1. Law of Cosines   **3.** Law of Cosines   **5.** Law of
Cosines   **7.** Law of Cosines   **9.** Law of Cosines
11. $A = 46°37'$, $B = 73°50'$, $C = 59°33'$
13. $A = 44°25'$, $B = 57°07'$, $C = 78°28'$
15. $b = 17.92$, $A = 54°42'$, $C = 78°18'$
17. $C = 81°$, $a = 9.11$, $b = 12.15$
19. $A = 29°58'$, $B = 110°15'$, $C = 39°47'$
21. $P = 1434.26$ ft.; $A = 86{,}804.28$ ft^2 **23.** 228.38 mi;
$19°53'$ **25.** 183.03 mi.   **27.** $35°41'$ **29.** 2033 km

1. 0.6000   **3.** 0.8000   **5.** 0.7500   **7.** 1.6667   **9.** 0.7500
11. $\sqrt{3}$ **13.** 0.9397   **15.** 0.8192   **17.** 1.0724
19. $68°00'$ **21.** $23°30'$ **23.** $19°10'$ **25.** 0.9908
27. 2.4202   **29.** 1.5822   **31.** $48°00'$ **33.** $48°59'$
35. $32°40'$ **37.** $B = 65°$, $a = 2.54$, $b = 5.44$
39. $A = 5°$, $b = 70.98$, $c = 71.25$
41. $c = 3.16$, $A = 18°26'$, $B = 71°34'$
43. $a = 7.14$, $A = 45°34'$, $B = 44°26'$ **45.** 55.53 m
47. $38°39'$ **49.** $c = 11.65$, $C = 63°11'$, $B = 66°49'$
51. $A = 51°$, $a = 70.22$, $c = 89.69$
57. $a = 4.36$, $B = 23°24'$, $C = 96°35'$ **59.** $c = 4.54$,
$A = 58°05'$, $B = 81°55'$ **53.** no solution
55. $B = 35°08'$, $C = 98°52'$, $c = 13.74$

Index